Lecture Notes in Artificial Intelligence 13385

Subseries of Lecture Notes in Computer Science

More information about this subseries at https://link.springer.com/bookseries/1244

Jasmin Blanchette · Laura Kovács · Dirk Pattinson (Eds.)

Automated Reasoning

11th International Joint Conference, IJCAR 2022
Haifa, Israel, August 8–10, 2022
Proceedings

Springer

Editors
Jasmin Blanchette
Vrije Universiteit Amsterdam
Amsterdam, The Netherlands

Laura Kovács
Vienna University of Technology
Wien, Austria

Dirk Pattinson
Australian National University
Canberra, ACT, Australia

ISSN 0302-9743 ISSN 1611-3349 (electronic)
Lecture Notes in Artificial Intelligence
ISBN 978-3-031-10768-9 ISBN 978-3-031-10769-6 (eBook)
https://doi.org/10.1007/978-3-031-10769-6

LNCS Sublibrary: SL7 – Artificial Intelligence

This Springer imprint is published by the registered company Springer Nature Switzerland AG
The registered company address is: Gewerbestrasse 11, 6330 Cham, Switzerland

Preface

This volume contains the papers presented at the 11th International Joint Conference on Automated Reasoning (IJCAR 2022) held during August 8–10, 2022, in Haifa, Israel. IJCAR was part of the Federated Logic Conference (FLoC 2022), which took place from July 31 to August 12, 2022, in Haifa.

IJCAR is the premier international joint conference on all aspects of automated reasoning, including foundations, implementations, and applications, comprising several leading conferences and workshops. IJCAR 2022 united the Conference on Automated Deduction (CADE), the International Symposium on Frontiers of Combining Systems (FroCoS), and the International Conference on Automated Reasoning with Analytic Tableaux and Related Methods (TABLEAUX). Previous IJCAR conferences were held in Siena, Italy, in 2001, Cork, Ireland, in 2004, Seattle, USA, in 2006, Sydney, Australia, in 2008, Edinburgh, UK, in 2010, Manchester, UK, in 2012, Vienna, Austria, in 2014, Coimbra, Portugal, in 2016, Oxford, UK, in 2018, and Paris, France, in 2020 (virtual).

There were 85 submissions. Each submission was assigned to at least three Program Committee members and was reviewed in single-blind mode. The committee decided to accept 41 papers: 32 regular papers and nine system descriptions.

The program also included two invited talks, by Elvira Albert and Gilles Dowek, as well as a plenary FLoC talk by Aarti Gupta.

We acknowledge the FLoC sponsors:

- Diamond sponsors: Amazon Web Services, Meta, Intel
- Gold sponsors: Google, Nvidia, Synopsys
- Silver sponsor: Cadence
- Bronze sponsors: DLVSystem, Veridise
- Other sponsors: Technion, The Henry and Marilyn Taub Faculty of Computer Science

We also acknowledge the generous sponsorship of Springer and the Trakhtenbrot family, as well as the invaluable support provided by the EasyChair developers. We finally thank the FLoC 2022 organization team for assisting us with local organization and general conference management.

May 2022

Jasmin Blanchette
Laura Kovács
Dirk Pattinson

Preface

Organization

Program Committee

Erika Abraham	RWTH Aachen University, Germany
Carlos Areces	Universidad Nacional de Córdoba, Spain
Bernhard Beckert	Karlsruhe Institute of Technology, Germany
Alexander Bentkamp	Chinese Academy of Sciences, China
Armin Biere	University of Freiburg, Germany
Nikolaj Bjørner	Microsoft, USA
Jasmin Blanchette (Co-chair)	Vrije Universiteit Amsterdam, The Netherlands
Frédéric Blanqui	Inria, France
Maria Paola Bonacina	Università degli Studi di Verona, Italy
Kaustuv Chaudhuri	Inria, France
Agata Ciabattoni	Vienna University of Technology, Austria
Stéphane Demri	CNRS, LMF, ENS Paris-Saclay, France
Clare Dixon	University of Manchester, UK
Huimin Dong	Sun Yat-sen University, China
Katalin Fazekas	Vienna University of Technology, Austria
Mathias Fleury	University of Freiburg, Austria
Pascal Fontaine	Université de Liège, Belgium
Nathan Fulton	IBM, USA
Silvio Ghilardi	Università degli Studi di Milano, Italy
Jürgen Giesl	RWTH Aachen University, Germany
Rajeev Gore	Australian National University, Australia
Marijn Heule	Carnegie Mellon University, USA
Radu Iosif	Verimag, CNRS, Université Grenoble Alpes, France
Mikolas Janota	Czech Technical University in Prague, Czech Republic
Moa Johansson	Chalmers University of Technology, Sweden
Cezary Kaliszyk	University of Innsbruck, Austria
Laura Kovacs (Co-chair)	Vienna University of Technology, Austria
Orna Kupferman	Hebrew University, Israel
Cláudia Nalon	University of Brasília, Brazil
Vivek Nigam	Huawei ERC, Germany
Tobias Nipkow	Technical University of Munich, Germany
Jens Otten	University of Oslo, Norway
Dirk Pattinson (Co-chair)	Australian National University, Australia
Nicolas Peltier	CNRS, LIG, France

Brigitte Pientka	McGill University, Canada
Elaine Pimentel	University College London, UK
André Platzer	Carnegie Mellon University, USA
Giles Reger	Amazon Web Services, USA, and University of Manchester, UK
Andrew Reynolds	University of Iowa, USA
Simon Robillard	Université de Montpellier, France
Albert Rubio	Universidad Complutense de Madrid, Spain
Philipp Ruemmer	Uppsala University, Sweden
Renate A. Schmidt	University of Manchester, UK
Stephan Schulz	DHBW Stuttgart, Germany
Roberto Sebastiani	University of Trento, Italy
Martina Seidl	Johannes Kepler University Linz, Austria
Viorica Sofronie-Stokkermans	University of Koblenz-Landau, Germany
Lutz Straßburger	Inria, France
Martin Suda	Czech Technical University in Prague, Czech Republic
Tanel Tammet	Tallinn University of Technology, Estonia
Sophie Tourret	Inria, France, and Max Planck Institute for Informatics, Germany
Uwe Waldmann	Max Planck Institute for Informatics, Germany
Christoph Weidenbach	Max Planck Institute for Informatics, Germany
Sarah Winkler	Free University of Bozen-Bolzano, Italy
Yoni Zohar	Bar-Ilan University, Israel

Additional Reviewers

László Antal
Paolo Baldi
Lionel Blatter
Brandon Bohrer
Marius Bozga
Chad Brown
Lucas Bueri
Guillaume Burel
Marcelo Coniglio
Riccardo De Masellis
Warren Del-Pinto
Zafer Esen
Michael Färber
Sicun Gao
Jacques Garrigue
Thibault Gauthier
Samir Genaim
Alessandro Gianola
Raúl Gutiérrez
Fajar Haifani
Alejandro Hernández-Cerezo
Ullrich Hustadt
Jan Jakubuv
Martin Jonas
Michael Kirsten
Gereon Kremer
Roman Kuznets
Jonathan Laurent
Chencheng Liang
Enrico Lipparini
Florin Manea
Marco Maratea

Sonia Marin
Enrique Martin-Martin
Andrea Mazzullo
Stephan Merz
Antoine Miné
Sibylle Möhle
Cristian Molinaro
Markus Müller-Olm
Jasper Nalbach
Joel Ouaknine
Tobias Paxian
Wolfram Pfeifer
Andrew Pitts
Amaury Pouly
Stanisław Purgał
Michael Rawson
Giselle Reis
Clara Rodríguez-Núñez
Daniel Skurt
Giuseppe Spallitta
Sorin Stratulat
Petar Vukmirović
Alexander Weigl
Richard Zach
Anna Zamansky
Michal Zawidzki

Contents

Invited Talks

Using Automated Reasoning Techniques for Enhancing the Efficiency and Security of (Ethereum) Smart Contracts 3
Elvira Albert, Pablo Gordillo, Alejandro Hernández-Cerezo, Clara Rodríguez-Núñez, and Albert Rubio

From the Universality of Mathematical Truth to the Interoperability of Proof Systems 8
Gilles Dowek

Satisfiability, SMT Solving, and Arithmetic

Flexible Proof Production in an Industrial-Strength SMT Solver 15
Haniel Barbosa, Andrew Reynolds, Gereon Kremer, Hanna Lachnitt, Aina Niemetz, Andres Nötzli, Alex Ozdemir, Mathias Preiner, Arjun Viswanathan, Scott Viteri, Yoni Zohar, Cesare Tinelli, and Clark Barrett

CTL* Model Checking for Data-Aware Dynamic Systems with Arithmetic 36
Paolo Felli, Marco Montali, and Sarah Winkler

SAT-Based Proof Search in Intermediate Propositional Logics 57
Camillo Fiorentini and Mauro Ferrari

Clause Redundancy and Preprocessing in Maximum Satisfiability 75
Hannes Ihalainen, Jeremias Berg, and Matti Järvisalo

Cooperating Techniques for Solving Nonlinear Real Arithmetic in the cvc5 SMT Solver (System Description) 95
Gereon Kremer, Andrew Reynolds, Clark Barrett, and Cesare Tinelli

Preprocessing of Propagation Redundant Clauses 106
Joseph E. Reeves, Marijn J. H. Heule, and Randal E. Bryant

Reasoning About Vectors Using an SMT Theory of Sequences 125
Ying Sheng, Andres Nötzli, Andrew Reynolds, Yoni Zohar, David Dill, Wolfgang Grieskamp, Junkil Park, Shaz Qadeer, Clark Barrett, and Cesare Tinelli

Calculi and Orderings

An Efficient Subsumption Test Pipeline for BS(LRA) Clauses 147
Martin Bromberger, Lorenz Leutgeb, and Christoph Weidenbach

Ground Joinability and Connectedness in the Superposition Calculus 169
André Duarte and Konstantin Korovin

Connection-Minimal Abduction in $\mathcal{EL}$ via Translation to FOL 188
Fajar Haifani, Patrick Koopmann, Sophie Tourret, and Christoph Weidenbach

Semantic Relevance 208
Fajar Haifani and Christoph Weidenbach

SCL(EQ): SCL for First-Order Logic with Equality 228
Hendrik Leidinger and Christoph Weidenbach

Term Orderings for Non-reachability of (Conditional) Rewriting 248
Akihisa Yamada

Knowledge Representation and Justification

EVONNE: Interactive Proof Visualization for Description Logics (System Description) 271
Christian Alrabbaa, Franz Baader, Stefan Borgwardt, Raimund Dachselt, Patrick Koopmann, and Julián Méndez

Actions over Core-Closed Knowledge Bases 281
Claudia Cauli, Magdalena Ortiz, and Nir Piterman

GK: Implementing Full First Order Default Logic for Commonsense Reasoning (System Description) 300
Tanel Tammet, Dirk Draheim, and Priit Järv

Hypergraph-Based Inference Rules for Computing $\mathcal{EL}^+$-Ontology Justifications 310
Hui Yang, Yue Ma, and Nicole Bidoit

Choices, Invariance, Substitutions, and Formalizations

Sequent Calculi for Choice Logics 331
Michael Bernreiter, Anela Lolic, Jan Maly, and Stefan Woltran

Lash 1.0 (System Description) 350
Chad E. Brown and Cezary Kaliszyk

Goéland: A Concurrent Tableau-Based Theorem Prover (System Description) 359
Julie Cailler, Johann Rosain, David Delahaye, Simon Robillard, and Hinde Lilia Bouziane

Binary Codes that Do Not Preserve Primitivity 369
Štěpán Holub, Martin Raška, and Štěpán Starosta

Formula Simplification via Invariance Detection by Algebraically Indexed Types 388
Takuya Matsuzaki and Tomohiro Fujita

Synthetic Tableaux: Minimal Tableau Search Heuristics 407
Michał Sochański, Dorota Leszczyńska-Jasion, Szymon Chlebowski, Agata Tomczyk, and Marcin Jukiewicz

Modal Logics

Paraconsistent Gödel Modal Logic 429
Marta Bílková, Sabine Frittella, and Daniil Kozhemiachenko

Non-associative, Non-commutative Multi-modal Linear Logic 449
Eben Blaisdell, Max Kanovich, Stepan L. Kuznetsov, Elaine Pimentel, and Andre Scedrov

Effective Semantics for the Modal Logics K and KT via Non-deterministic Matrices 468
Ori Lahav and Yoni Zohar

Local Reductions for the Modal Cube 486
Cláudia Nalon, Ullrich Hustadt, Fabio Papacchini, and Clare Dixon

Proof Systems and Proof Search

Cyclic Proofs, Hypersequents, and Transitive Closure Logic 509
Anupam Das and Marianna Girlando

Equational Unification and Matching, and Symbolic Reachability Analysis in Maude 3.2 (System Description) 529
Francisco Durán, Steven Eker, Santiago Escobar, Narciso Martí-Oliet, José Meseguer, Rubén Rubio, and Carolyn Talcott

Leśniewski's Ontology – Proof-Theoretic Characterization 541
Andrzej Indrzejczak

Bayesian Ranking for Strategy Scheduling in Automated Theorem Provers 559
Chaitanya Mangla, Sean B. Holden, and Lawrence C. Paulson

A Framework for Approximate Generalization in Quantitative Theories 578
Temur Kutsia and Cleo Pau

Guiding an Automated Theorem Prover with Neural Rewriting 597
Jelle Piepenbrock, Tom Heskes, Mikoláš Janota, and Josef Urban

Rensets and Renaming-Based Recursion for Syntax with Bindings 618
Andrei Popescu

Finite Two-Dimensional Proof Systems for Non-finitely Axiomatizable Logics 640
Vitor Greati and João Marcos

Vampire Getting Noisy: Will Random Bits Help Conquer Chaos? (System Description) 659
Martin Suda

Evolution, Termination, and Decision Problems

On Eventual Non-negativity and Positivity for the Weighted Sum of Powers of Matrices 671
S. Akshay, Supratik Chakraborty, and Debtanu Pal

Decision Problems in a Logic for Reasoning About Reconfigurable Distributed Systems 691
Marius Bozga, Lucas Bueri, and Radu Iosif

Proving Non-Termination and Lower Runtime Bounds with LoAT (System Description) 712
Florian Frohn and Jürgen Giesl

Implicit Definitions with Differential Equations for KeYmaera X: (System Description) 723
James Gallicchio, Yong Kiam Tan, Stefan Mitsch, and André Platzer

Automatic Complexity Analysis of Integer Programs via Triangular Weakly Non-Linear Loops 734
Nils Lommen, Fabian Meyer, and Jürgen Giesl

Author Index 755

Invited Talks

Using Automated Reasoning Techniques for Enhancing the Efficiency and Security of (Ethereum) Smart Contracts

Elvira Albert[1,2](✉), Pablo Gordillo[1], Alejandro Hernández-Cerezo[1], Clara Rodríguez-Núñez[1], and Albert Rubio[1,2]

[1] Complutense University of Madrid, Madrid, Spain
[2] Instituto de Tecnología del Conocimiento, Madrid, Spain
elvira@fdi.ucm.es

The use of the Ethereum blockchain platform [17] has experienced an enormous growth since its very first transaction back in 2015 and, along with it, the verification and optimization of the programs executed in the blockchain (known as Ethereum *smart contracts*) have raised considerable interest within the research community. As for any other kind of programs, the main properties of smart contracts are their *efficiency* and *security*. However, in the context of the blockchain, these properties acquire even more relevance. As regards efficiency, due to the huge volume of transactions, the cost and response time of the Ethereum blockchain platform have increased notably: the processing capacity of the transactions is limited and it is providing low transaction ratios per minute together with increased costs per transaction. Ethereum is aware of such limitations and it is currently working on solutions to improve scalability with the goal of increasing its capacity. As regards security, due to the public nature and immutability of smart contracts and the fact that their public functions can be executed by any user at any time, programming errors can be exploited by attackers and have a high economic impact [7,13]. Verification is key to ensure the security of smart contract's execution and provide safety guarantees. This talk will present our work on the use of automated reasoning techniques and tools to enhance the security and efficiency [2–4,6] of Ethereum smart contracts along the two directions described below.

Security. Our main focus on security will be to detect and avoid potential *reentrancy* attacks, one of the best known and exploited vulnerabilities that have caused infamous attacks in the Ethereum ecosystem due to they economic impact [9,11,15]. Reentrancy attacks might occur on programs with callbacks, a mechanism that allows making calls among contracts. Callbacks occur when a method of a contract invokes a method of another contract and the latter, either directly or indirectly, invokes one or more methods of the former before the original method invocation returns. While this mechanism is useful and powerful

This work was funded partially by the Ethereum Foundation (Grant FY21-0372), the Spanish MCIU, AEI and FEDER (EU) project RTI2018-094403-B-C31 and by the CM project S2018/TCS-4314 co-funded by EIE Funds of the European Union.

J. Blanchette et al. (Eds.): IJCAR 2022, LNAI 13385, pp. 3–7, 2022.
https://doi.org/10.1007/978-3-031-10769-6_1

in event-driven programming, it has been used to exploit vulnerabilities. Our approach to detect potential reentrancy problems is to ensure that the program meets the Effectively Callback Freeness (ECF) property [10]. ECF guarantees the modularity of a contract in the sense that executions with callbacks cannot result in new states that are not reachable by callback free executions. This implies that the use of callbacks will not lead to unpredicted, potentially dangerous, states. In order to ensure the ECF property, we use commutation and projection of fragments of code [6]. Intuitively, given a function fragment A followed by B (denoted $A.B$), in case we can receive a callback to some function f between these fragments (that is, $A.f.B$), we ensure safety by proving that this execution that contains callbacks is equivalent to a callback free execution: either to $A.B$ (projection), $f.A.B$ (left-commutation) or $A.B.f$ (right-commutation). The use of automated reasoning techniques enables proving this kind of properties. Inspired by the use of SMT solvers to prove redundancy of concurrent executions [1,8,16], we have implemented such checks using state-of-the-art SMT solvers.

The ECF property can be generalized to allow callbacks to introduce new behaviors as long as they are benign, as [5] does by defining the notion of R-ECF. The main difference between ECF and R-ECF is that while ECF checks that the states reached by executions with callbacks are exactly the same as the ones reached by executions that do not contain callbacks, R-ECF checks that they satisfy a relation with respect to the states reached without callbacks. This way, R-ECF is able to recognize and distinguish the benign behaviors introduced by callbacks from the ones that are potentially dangerous, while ECF cannot. The main application of R-ECF is that, from a particular invariant of the program, it allows reducing the problem of verifying the invariant in the presence of callbacks, to the callback-free setting. For example, if we consider the invariant $\texttt{balance} \geq 0$ and prove that the contract is R-ECF with respect to the relation $\texttt{balance}_{cb} \geq \texttt{balance}_{cbfree}$ (i.e., the balance reached by executions with callbacks is greater than the one reached without callbacks), then we only need to consider callback free executions in order to prove the preservation of the invariant.

We considered as benchmarks the top-150 contracts based on volume of usage, and studied the modularity of their functions in terms of ECF and R-ECF. A total of 386 of their functions were susceptible to have callbacks, from which 62.7% were verified to be ECF. The R-ECF approach was able to increase the accuracy of the analysis, being able to prove the correctness of an extra 2% of functions [5,6].

Efficiency. The main focus on efficiency will be on optimizing the resource consumption of smart contract executions. On the Ethereum blockchain, the resource consumption is measured in terms of *gas*, a unit introduced in the system to quantify the computational effort and charge a fee accordingly in order to have a transaction executed. To understand how we can optimize gas, we need to discuss it (and do it) at the level of the Ethereum bytecode. Smart contracts in Ethereum are executed using the Ethereum Virtual Machine (EVM). The EVM is a simple stack-based architecture which uses 256-bit words and has its own repertory of instructions (EVM opcodes). In the EVM, the memory model is split into two different structures: the *storage*, which is persistent between transactions and expensive to use; and the *memory*, which does not persist between transactions and is cheaper. Each opcode has a gas cost associated to its execution. Besides, an additional fee must be paid for each byte when the smart contract is deployed. Thus, the resource to be optimized can be either the total amount of gas in a program or its size. Even though both criteria are usually related, there are some situations in which they do not correlate. For instance, pushing a big number in the stack consumes a small amount of gas and increases significantly the bytecode size, whereas obtaining the same value using arithmetic operations is more expensive but involves fewer bytes.

Among all possible techniques to optimize code, we have used the technique known as superoptimization [12]. The main idea of superoptimization is automatically finding an equivalent optimal sequence of instructions to another given loop-free sequence. In order to achieve this goal, we enumerate all possible candidates and determine the best option among them *wrt.*the optimization criteria. In the context of EVM, there exists several superoptimizers: EBSO [14], SYRUP [3,4] and GASOL [2]. The techniques presented in this work correspond to the ones implemented in GASOL, which are an improvement and extension of the ones in SYRUP. We apply two kinds of automated reasoning techniques to superoptimize Ethereum smart contracts, symbolic execution and Max-SMT as described next.

- Symbolic execution is used to obtain a a representation on how the stack and memory evolves *wrt.* to an initial stack. We determine the lowest size of the stack needed to perform all the operations in a block and apply symbolic execution to an initial stack containing that number of unknown stack variables. Opcodes representing operations that don't manage the stack are left as uninterpreted functions. Then, we apply as many simplification rules as possible from a fixed set of rules. Depending on the chosen criteria, some rules are disabled if they lead to worse candidates. Moreover, we apply static analysis regarding memory opcodes to determine whether there are some redundant store or load operations inside a block that can be safely removed or replaced. This leads to a simplified specification of the optimal block.
- The second technique involves synthesizing the optimal block from a given symbolic representation using a Max-SMT solver. The synthesis problem is expressed as a first-order formula in which every model corresponds to a valid equivalent block. Our encoding is expressed in the simple logic *QF_IDL*, so that the Max-SMT solver can reason effectively on EVM blocks. In this encoding, the length of the sequence of instructions is fixed by an upper bound so that quantifiers are avoided. NOP operations are considered in the encoding to allow shorter sequences. The state of the stack is represented explicitly for each position in the sequence. Every instruction in the block and every basic stack operation have a constraint that reflects the impact they have on the stack for each possible position. Memory accesses are encoded as a partial order relation that synthesizes the dependencies among them. Regarding the optimization process, we express the cost (gas or bytes-size) of

each instruction using soft constraints. For both criteria, the corresponding set of soft constraints satisfies that an optimal model returned by the solver corresponds to an optimal block for that criteria.

Combining both approaches, we obtain significant savings for both criteria. For a subset of 30 smart contracts, selected among the latest published in Etherscan as of June 21, 2021 and optimized using the compiler solc v0.8.9, GASOL still manages to reduce 0.72% the amount of gas with the gas criteria enabled, and decreases the overall size by 3.28% with the size criteria enabled.

Future work. The current directions for future work include enhancing the performance of the smart contract optimizer in both accuracy and scalability of the process while keeping the efficiency. For the accuracy we are currently working on adding further reasoning on non-stack operations while staying in a quite simple logic. This will allow us to consider a wider set of equivalent blocks and hence increase the savings. Scalability can be threatened when we consider blocks of code of large size. We are investigating different approaches to scale better, including heuristics to partition the blocks in smaller sub-blocks, more efficient SMT encodings, among others. Finally, another direction for future work is to formally prove the correctness of the optimizer, *i.e.*developing a checker that can formally prove the equivalence of the optimized and the original (Ethereum) bytecode. For this, we are planning to use the Coq proof assistant in which we will develop a checker that, given an original bytecode –that corresponds a block of the control flow graph– and its optimization, it can formally prove their equivalence for any possible execution, and optionally it can generate a soundness proof that can be used as certificate.

References

1. Albert, E., Gómez-Zamalloa, M., Isabel, M., Rubio, A.: Constrained dynamic partial order reduction. In: Chockler, H., Weissenbacher, G. (eds.) CAV 2018. LNCS, vol. 10982, pp. 392–410. Springer, Cham (2018). https://doi.org/10.1007/978-3-319-96142-2_24
2. Albert, E., Gordillo, P., Hernández-Cerezo, A., Rubio, A.: A Max-SMT superoptimizer for EVM handling memory and storage. In: Fisman, D., Rosu, G. (eds) Tools and Algorithms for the Construction and Analysis of Systems. TACAS 2022. LNCS, vol. 13243. Springer, Cham (2022). https://doi.org/10.1007/978-3-030-99524-9_11
3. Albert, E., Gordillo, P., Hernández-Cerezo, A., Rubio, A., Schett, M.A.: Superoptimization of smart contracts. ACM Trans. Softw. Eng. Methodol. (2022)
4. Albert, E., Gordillo, P., Rubio, A., Schett, M.A.: Synthesis of super-optimized smart contracts using Max-SMT. In: Lahiri, S.K., Wang, C. (eds.) CAV 2020. LNCS, vol. 12224, pp. 177–200. Springer, Cham (2020). https://doi.org/10.1007/978-3-030-53288-8_10
5. Albert, E., Grossman, S., Rinetzky, N., Nunez, C.R., Rubio, A., Sagiv, M.: Relaxed effective callback freedom: a parametric correctness condition for sequential modules with callbacks. IEEE Trans. Dependable Secure Comput. (2022)

6. Albert, E., Grossman, S., Rinetzky, N., Rodríguez-Núñez, C., Rubio, A., Sagiv, M.: Taming callbacks for smart contract modularity. In: Proceedings of the ACM SIGPLAN Conference on Object-Oriented Programming Systems, Languages and Applications, OOPSLA 2020, vol. 4, pp. 209:1–209:30 (2020)
7. Atzei, N., Bartoletti, M., Cimoli, T.: A survey of attacks on ethereum smart contracts (SoK). In: Maffei, M., Ryan, M. (eds.) POST 2017. LNCS, vol. 10204, pp. 164–186. Springer, Heidelberg (2017). https://doi.org/10.1007/978-3-662-54455-6_8
8. Bansal, K., Koskinen, E., Tripp, O.: Automatic generation of precise and useful commutativity conditions. In: Beyer, D., Huisman, M. (eds.) TACAS 2018. LNCS, vol. 10805, pp. 115–132. Springer, Cham (2018). https://doi.org/10.1007/978-3-319-89960-2_7
9. Daian, P.: Analysis of the DAO exploit (2016). http://hackingdistributed.com/2016/06/18/analysis-of-the-dao-exploit/
10. Grossman, S., et al.: Online detection of effectively callback free objects with applications to smart contracts. PACMPL, 2(POPL) (2018)
11. Liu, M.: Urgent: OUSD was hacked and there has been a loss of funds (2020). https://medium.com/originprotocol/urgent-ousd-has-hacked-and-there-has-been-a-loss-of-funds-7b8c4a7d534c. Accessed 29 Jan 2021
12. Massalin, H.: Superoptimizer - a look at the smallest program. In: Proceedings of the Second International Conference on Architectural Support for Programming Languages and Operating Systems (ASPLOS II), pp. 122–126 (1987)
13. Mehar, M.I., et al.: Understanding a revolutionary and flawed grand experiment in blockchain: the DAO attack. J. Cases Inf. Technol. **21**(1), 19–32 (2019)
14. Nagele, J., Schett, M.A.: Blockchain superoptimizer. In: Proceedings of 29th International Symposium on Logic-Based Program Synthesis and Transformation (LOPSTR) (2019). https://arxiv.org/abs/2005.05912
15. Tarasov, A.: Millions lost: the top 19 DeFi cryptocurrency hacks of 2020 (2020). https://cryptobriefing.com/50-million-lost-the-top-19-defi-cryptocurrency-hacks-2020/2. Accessed 29 Jan 2021
16. Wang, C., Yang, Z., Kahlon, V., Gupta, A.: Peephole partial order reduction. In: Ramakrishnan, C.R., Rehof, J. (eds.) TACAS 2008. LNCS, vol. 4963, pp. 382–396. Springer, Heidelberg (2008). https://doi.org/10.1007/978-3-540-78800-3_29
17. Wood, G.: Ethereum: a secure decentralised generalised transaction ledger (2019)

From the Universality of Mathematical Truth to the Interoperability of Proof Systems

Gilles Dowek(✉)

Inria and ENS Paris-Saclay, Paris, France
gilles.dowek@ens-paris-saclay.fr

1 Yet Another Crisis of the Universality of Mathematical Truth

The development of computerized proof systems, such as Coq, Matita, Agda, Lean, HOL 4, HOL Light, Isabelle/HOL, Mizar, etc. is a major step forward in the never ending quest of mathematical rigor. But it jeopardizes the universality of mathematical truth [5]: we used to have proofs of Fermat's little theorem, we now have Coq proofs of Fermat's little theorem, Isabelle/HOL proofs of Fermat's little theorem, PVS proofs of Fermat's little theorem, etc. Each proof system: Coq, Isabelle/HOL, PVS, etc. defining its own language for mathematical statements and its own truth conditions for these statements.

This crisis can be compared to previous ones, when mathematicians have disagreed on the truth of some mathematical statements: the discovery of the incommensurability of the diagonal and side of a square, the introduction of infinite series, the non-Euclidean geometries, the discovery of the independence of the axiom of choice, and the emergence of constructivity. All these past crises have been resolved.

2 Predicate Logic and Other Logical Frameworks

One way to resolve a crisis, such as that of non-Euclidean geometries, or that of the axiom of choice, is to view geometry, or set theory, as an axiomatic theory. The judgement that the statement *the sum of the angles in a triangle equals the straight angle* is true evolves to that that it is a consequence of the parallel axiom and of the other axioms of geometry. Thus, the truth conditions must be defined, not for the statements of geometry, but for arbitrary sequents: pairs $\Gamma \vdash A$ formed with a theory, a set of axioms, Γ and a statement A.

This induces a separation between the definition of the truth conditions of a sequent: the logical framework and the definition of the various geometries as theories in this logical framework. This logical framework, Predicate logic, was made precise by Hilbert and Ackermann [13], in 1928, more than a century after the beginning of the crisis of non-Euclidean geometries. The invention of

J. Blanchette et al. (Eds.): IJCAR 2022, LNAI 13385, pp. 8–11, 2022.
https://doi.org/10.1007/978-3-031-10769-6_2

Predicate Logic was a huge step forward. But Predicate Logic also has some limitations.

To overcome these limitation, it has been modernized in various ways in the last decades. First, λ-PROLOG [15] and ISABELLE [17] have extended Predicate logic with variable binding function symbols, such as the symbol λ in the term $\lambda x\ x$. Then, the $\lambda\Pi$-calculus [12] has permitted to explicitly represent proof-trees, using the so-called Brouwer-Heyting-Kolmogorov algorithmic interpretation of proofs and Curry-de Bruijn-Howard correspondence. In a second stream of research, Deduction modulo theory [4,6] has introduced a distinction between computation and deduction, in such a way that the statement $27 \times 37 = 999$ computes to $999 = 999$, with the algorithm of multiplication, and then to $\top$, with the algorithm of natural number comparison. It thus has a trivial proof. A third stream of research has extended classical Predicate logic to an Ecumenical predicate logic [3,9–11,14,18,19] with both constructive and classical logical constants.

These streams of research have merged, to provide a logical framework, the $\lambda\Pi$-calculus modulo theory [2], also called Martin-Löf's logical framework [16]. This framework permits function symbols to bind variables, it includes an explicit representation for proof-trees, it distinguishes computation from deduction, and it permits to define both constructive and classical logical constants. It is the basis of the language DEDUKTI, where Simple type theory, Martin-Löf's type theory, the Calculus of constructions, etc. can easily be expressed.

3 The Theory $\mathcal{U}$

The expression in DEDUKTI of Simple type theory, Simple type theory with polymorphism, Simple type theory with predicate subtyping, the Calculus of constructions, etc. use symbol declarations and computation rules that play the *rôle* of axioms in Predicate logic. But, just like the various geometries or the various set theories share a lot of axioms and distinguish by a few, these theories share a lot of symbols and rules. This remark leads to defining a large theory, the theory $\mathcal{U}$ [1], that contains Simple type theory, Simple type theory with polymorphism, Simple type theory with predicate subtyping, and the Calculus of constructions, etc. as sub-theories.

Many proofs developed in proof processing systems can be expressed in the theory $\mathcal{U}$ and depending on the symbols and rules they use they can be translated to more common formulations of the theories implemented in these systems.

For instance, F. Thiré has expressed a large library of arithmetic, originally developed in MATITA, in an sub-theory of the theory $\mathcal{U}$, corresponding to Simple type theory with polymorphism and translated these proofs to the language of seven proof systems [20], Y. Géran has expressed the first book of Euclid's elements originally developed in COQ, in a sub-theory of the theory $\mathcal{U}$, corresponding to Predicate logic, and translated these proofs to the language of many proof systems, including predicate logic ones [8], and T. Felicissimo has shown that a large library of proofs originally developed in MATITA, including a proof of Bertrand's postulate, could be expressed in predicative type theory and expressed in Agda [7].

References

1. Blanqui, F., Dowek, G., Grienenberger, É., Hondet, G., Thiré, F.: Some axioms for mathematics. In: Kobayashi, N. (ed.) Formal Structures for Computation and Deduction, vol. 195, pp. 20:1–20:19. LIPIcs. Schloss Dagstuhl - Leibniz-Zentrum für Informatik (2021)
2. Cousineau, D., Dowek, G.: Embedding pure type systems in the lambda-pi-calculus modulo. In: Della Rocca, S.R. (ed.) TLCA 2007. LNCS, vol. 4583, pp. 102–117. Springer, Heidelberg (2007). https://doi.org/10.1007/978-3-540-73228-0_9
3. Dowek, G.: On the definition of the classical connectives and quantifiers. In: Haeusler, E.H., de Campos Sanz, W., Lopes, B. (eds.) Why is this a Proof? Festschrift for Luiz Carlos Pereira. College Publications (2015)
4. Dowek, G., Hardin, T., Kirchner, C.: Theorem proving modulo. J. Autom. Reason. **31**, 33–72 (2003). https://doi.org/10.1023/A:1027357912519
5. Dowek, G., Thiré, F.: The universality of mathematical truth jeopardized by the development of computerized proof systems. In: Arana, A., Pataut, F. (eds.) Proofs, To be published
6. Dowek, G., Werner, B.: Proof normalization modulo. J. Symb. Log. **68**(4), 1289–1316 (2003)
7. Felicissimo, T., Blanqui, F., Kumar Barnawal, A.: Predicativize: sharing proofs with predicative systems. Manuscript (2022)
8. Géran, Y.: Mathématiques inversées de Coq. l'exemple de GeoCoq. Master thesis (2021)
9. Gilbert, F.: Extending higher-order logic with predicate subtyping: application to PVS. (Extension de la logique d'ordre supérieur avec le sous-typage par prédicats). PhD thesis, Sorbonne Paris Cité, France (2018)
10. Girard, J.-Y.: On the unity of logic. Ann. Pure Appl. Logic **59**(3), 201–217 (1993)
11. Grienenberger, É.: A logical system for an ecumenical formalization of mathematics. Manuscript (2020)
12. Harper, R., Honsell, F., Plotkin, G.: A framework for defining logics. J. ACM **40**(1), 143–184 (1993)
13. Hilbert, D., Ackermann, W.: Grundzüge der theoretischen Logik. Springer-Verlag (1928)
14. Liang, C., Miller, D.: Unifying classical and intuitionistic logics for computational control. In: 28th Symposium on Logic in Computer Science, pp. 283–292 (2013)
15. Miller, D., Nadathur, G.: Programming with Higher-Order Logic. Cambridge University Press (2012)
16. Nordström, B., Petersson, K., Smith, J.M.: Programming in Martin-Löf's type theory. Oxford University Press (1990)
17. Paulson, L.C.: Isabelle: the next 700 theorem provers. In: Odifreddi, P. (ed.) Logic and Computer Science, pp. 361–386. Academic Press (1990)
18. Pereira, L.C., Rodriguez, R.O.: Normalization, soundness and completeness for the propositional fragment of Prawitz'ecumenical system. Rev. Port. Filos. **73**(3–4), 1153–1168 (2017)
19. Prawitz, D.: Classical versus intuitionistic logic. In: Haeusler, E.H., de Campos Sanz, W., Lopes, B. (eds.) Why is this a Proof? Festschrift for Luiz Carlos Pereira. College Publications (2015)

20. Thiré, F.: Sharing a library between proof assistants: reaching out to the HOL family. In: Blanqui, F., Reis, G. (eds.) Proceedings of the 13th International Workshop on Logical Frameworks and Meta-Languages, vol. 274, pp. 57–71. EPTCS (2018)

Satisfiability, SMT Solving, and Arithmetic

Flexible Proof Production in an Industrial-Strength SMT Solver

Haniel Barbosa[1], Andrew Reynolds[2], Gereon Kremer[3], Hanna Lachnitt[3], Aina Niemetz[3], Andres Nötzli[3], Alex Ozdemir[3], Mathias Preiner[3], Arjun Viswanathan[2], Scott Viteri[3], Yoni Zohar[4(✉)], Cesare Tinelli[2], and Clark Barrett[3]

[1] Universidade Federal de Minas Gerais, Belo Horizonte, Brazil
[2] The University of Iowa, Iowa City, USA
[3] Stanford University, Stanford, USA
[4] Bar-Ilan University, Ramat Gan, Israel
yoni206@gmail.com

Abstract. Proof production for SMT solvers is paramount to ensure their correctness independently from implementations, which are often prohibitively difficult to verify. Historically, however, SMT proof production has struggled with performance and coverage issues, resulting in the disabling of many crucial solving techniques and in coarse-grained (and thus hard to check) proofs. We present a flexible proof-production architecture designed to handle the complexity of versatile, industrial-strength SMT solvers and show how we leverage it to produce detailed proofs, including for components previously unsupported by any solver. The architecture allows proofs to be produced modularly, lazily, and with numerous safeguards for correctness. This architecture has been implemented in the state-of-the-art SMT solver cvc5. We evaluate its proofs for SMT-LIB benchmarks and show that the new architecture produces better coverage than previous approaches, has acceptable performance overhead, and supports detailed proofs for most solving components.

1 Introduction

SMT solvers [9] are widely used as backbones of formal methods tools in a variety of applications, often safety-critical ones. These tools rely on the solver's correctness to guarantee the validity of their results such as, for instance, that an access policy does not inadvertently give access to sensitive data [4]. However, SMT solvers, particularly industrial-strength ones, are often extremely complex pieces of engineering. This makes it hard to ensure that implementation issues do not affect results. As the industrial use of SMT solvers increases, it is paramount to be able to convince non-experts of the trustworthiness of their results.

A solution is to decouple confidence from the implementation by coupling results with machine-checkable certificates of their correctness. For SMT solvers,

This work was partially supported by the Office of Naval Research (Contract No. 68335-17-C-0558), a gift from Amazon Web Services, and by NSF-BSF grant numbers 2110397 (NSF) and 2020704 (BSF).

J. Blanchette et al. (Eds.): IJCAR 2022, LNAI 13385, pp. 15–35, 2022.
https://doi.org/10.1007/978-3-031-10769-6_3

this amounts to providing proofs of unsatisfiability. The main challenges are justifying a combination of theory-specific algorithms while keeping the solver performant and providing enough details to allow *scalable* proof checking, i.e., checking that is fundamentally simpler than solving. Moreover, while proof production is well understood for propositional reasoning and common theories, that is not the case for more expressive theories, such as the theory of strings, or for more advanced solver operations such as formula preprocessing.

We present a new, flexible proof-production architecture for versatile, industrial-strength SMT solvers and discuss its integration into the cvc5 solver [5]. The architecture (Sect. 2) aims to facilitate the implementation effort via modular proof production and internal proof checking, so that more critical components can be enabled when generating proofs. We provide some details on the core proof calculus and how proofs are produced (Sect. 3), in particular how we support eager and lazy proof production with built-in proof reconstruction (Sect. 3.2). This feature is particularly important for substitution and rewriting techniques, facilitating the instrumentation of notoriously challenging functionalities, such as simplification under global assumptions [6, Section 6.1] and string solving [40, 46, 48], to produce detailed proofs. Finally, we describe (Sect. 5) how the architecture is leveraged to produce detailed proofs for most of the theory reasoning, critical preprocessing, and underlying SAT solving of cvc5. We evaluate proof production in cvc5 (Sect. 6) by measuring the proof overhead and the proof quality over an extensive set of benchmarks from SMT-LIB [8].

In summary, *our contributions* are a flexible proof-producing architecture for state-of-the-art SMT solvers, its implementation in cvc5, the production of detailed proofs for simplification under global assumptions and the full theory of strings, and initial experimental evidence that proof-production overhead is acceptable and detailed proofs can be generated for a majority of the problems.

Preliminaries. We assume the usual notions and terminology of many-sorted first-order logic with equality ($\approx$) [29]. We consider signatures Σ all containing the distinguished Boolean sort Bool. We adopt the usual definitions of well-sorted Σ-terms, with literals and formulas as terms of sort Bool, and Σ-interpretations. A *Σ-theory* is a pair $T = (\Sigma, \mathbf{I})$ where $\mathbf{I}$, the *models* of T, is a class of Σ-interpretations closed under variable reassignment. A Σ-formula φ is *T-valid* (resp., *T-unsatisfiable*) if it is satisfied by all (resp., no) interpretations in $\mathbf{I}$. Two Σ-terms s and t of the same sort are *T-equivalent* if $s \approx t$ is T-valid. We write $\vec{a}$ to denote a tuple $(a_1, \ldots, a_n)$ of elements, with $n \geq 0$. Depending on context, we will abuse this notation and also denote the set of the tuple's elements or, in case of formulas, their conjunction. Similarly, for term tuples $\vec{s}, \vec{t}$ of the same length and sort, we will write $\vec{s} \approx \vec{t}$ to denote the conjunction of equalities between their respective elements.

2 Proof-Production Architecture

Our proof-production architecture is intertwined with the CDCL($\mathcal{T}$) architecture [43], as shown in Fig. 1. Proofs are produced and stored modularly by each solving component, which also checks they meet the expected proof structure

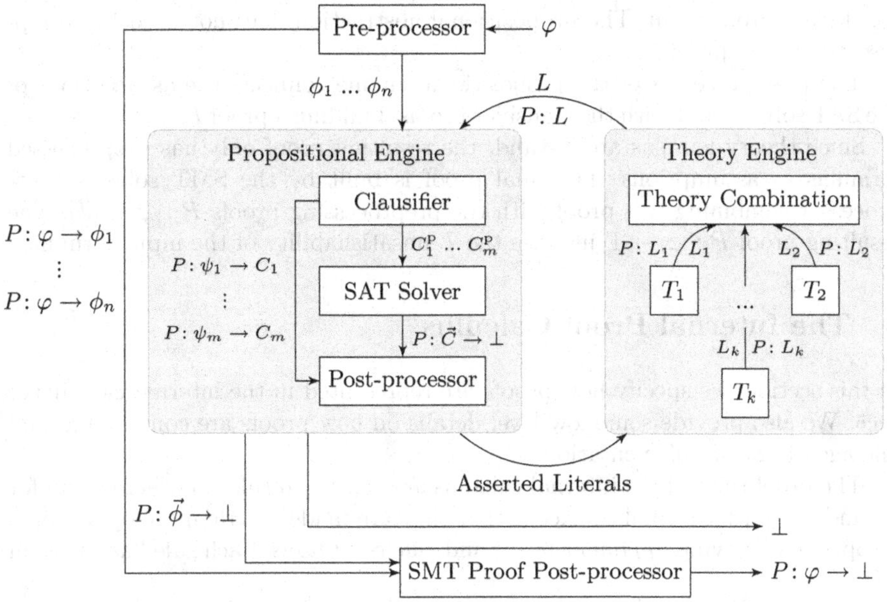

Fig. 1. Flexible proof-production architecture for CDCL($\mathcal{T}$)-based SMT solvers. In the above, $\psi_i \in \{\vec{\phi}, \vec{L}\}$ for each i, with ψ_i not necessarily distinct from ψ_{i+1}.

for that component, as described below. Proofs are combined only when needed, via post-processing. The *pre-processor* receives an input formula φ and simplifies it in a variety of ways into formulas $\phi_1, \ldots, \phi_n$. For each ϕ_i, the pre-processor stores a proof $P : \varphi \to \phi_i$ justifying its derivation from φ.

The *propositional engine* receives the preprocessed formulas, and its *clausifier* converts them into a conjunctive normal form $C_1 \wedge \cdots \wedge C_l$. A proof $P : \psi \to C_i$ is stored for each clause C_i, where ψ is a preprocessed formula. Note that several clauses may derive from each formula. Corresponding propositional clauses $C_1^{\mathrm{p}}, \ldots, C_l^{\mathrm{p}}$, where first-order atoms are abstracted as Boolean variables, are sent to the SAT solver, which checks their joint satisfiability. The propositional engine enters a loop with the *theory engine*, which considers a set of literals asserted by the SAT solver (corresponding to a model of the propositional clauses) and verifies its satisfiability modulo a *combination of theories* T. If the set is T-unsatisfiable, a lemma L is sent to the propositional engine together with its proof $P : L$. Note that since lemmas are T-valid, their proofs have no assumptions. The propositional engine stores these proofs and clausifies the lemmas, keeping the respective clausification proofs in the clausifier. The clausified and abstracted lemmas are sent to the SAT solver to block the current model and cause the assertion of a different set of literals, if possible. If no new set is asserted, then all the clauses $C_1, \ldots, C_m$ generated until then are jointly unsatisfiable, and the SAT solver yields a proof $P : C_1 \wedge \cdots \wedge C_m \to \bot$. Note that the proof is in terms of the first-order clauses, as are the derivation rules that conclude $\bot$ from them. The propositional abstraction does not need to be represented in the proof.

The post-processor of the propositional engine connects the assumptions of the SAT solver proof with the clausifier proofs, building a proof $P : \phi_1 \wedge \cdots \wedge \phi_n \rightarrow \bot$. Since theory lemmas are T-valid, the resulting proof only has preprocessed formulas as assumptions. The final proof is built by the SMT solver's post-processor combining this proof with the preprocessing proofs $P : \varphi \rightarrow \phi_i$. The resulting proof $P : \varphi \rightarrow \bot$ justifies the T-unsatisfiability of the input formula.

3 The Internal Proof Calculus

In this section, we specify how proofs are represented in the internal calculus of cvc5. We also provide some low-level details on how proofs are constructed and managed in our implementation.

The proof rules of the internal calculus are similar to rules in other calculi for ground first-order formulas, except that they are made a little more operational by optionally having *argument* terms and *side conditions*. Each rule has the form

$$r \; \frac{\varphi_1 \;\cdots\; \varphi_n}{\psi} \qquad \text{or} \qquad r \; \frac{\varphi_1 \;\cdots\; \varphi_n \;\mid\; t_1, \ldots, t_m}{\psi} \;\text{if}\; C$$

with *identifier* r, *premises* $\varphi_1, \ldots, \varphi_n$, *arguments* $t_1, \ldots, t_m$, *conclusion* ψ, and *side condition* C. The argument terms are used to construct the conclusion from the premises and can be used in the side condition together with the premises.

3.1 Proof Checkers and Proofs

The semantics of each proof rule r is provided operationally in terms of a *proof-rule checker* for r. This is a procedure that takes as input a list of argument terms $\vec{t}$ and a list of premises $\vec{\varphi}$ for r. It returns fail if the input is malformed, i.e., it does not match the rule's arguments and premises or does not satisfy the side condition. Otherwise, it returns a conclusion formula ψ expressing the result of applying the rule. All proof rules of the internal calculus have an associated proof-rule checker. We say that a proof rule *proves* a formula ψ, from given arguments and premises, if its checker returns ψ.

cvc5 has an internal proof checker built modularly out of the individual proof-rule checkers. This checker is meant mostly for internal debugging during development, to help guarantee that the constructed proofs are correct. The expectation is that users will rely instead on third-party tools to check the proof certificates emitted by the solver.

A proof object is constructed internally using a data structure that we will describe abstractly here and call a *proof node*. This is a triple $(r, \vec{N}, \vec{t})$ consisting of a rule identifier r; a sequence $\vec{N}$ of proof nodes, its *children*; and a sequence $\vec{t}$ of terms, its *arguments*. The relationships between proof nodes and their children induces a directed graph over proof nodes, with edges from proofs nodes to their children. We call a single-root graph rooted at node N a *proof*. A proof P is

$$\textsf{refl}\ \frac{-\mid t}{t \approx t} \qquad \textsf{trans}\ \frac{r \approx s \quad s \approx t}{r \approx t} \qquad \textsf{cong}\ \frac{\vec{s} \approx \vec{t} \mid f}{f(\vec{s}) \approx f(\vec{t})}\ \text{if } f(\vec{s}) \text{ is well sorted}$$

$$\textsf{symm}\ \frac{s \approx t}{t \approx s} \qquad \textsf{sr}\ \frac{\varphi \quad \vec{\varphi} \mid \mathcal{S}, \mathcal{R}, \mathcal{D}, \psi}{\psi}\ \text{if } \mathcal{S}(\varphi, \mathcal{D}(\vec{\varphi}))\uparrow\downarrow_{\mathcal{R}} = \mathcal{S}(\psi, \mathcal{D}(\vec{\varphi}))\uparrow\downarrow_{\mathcal{R}}$$

$$\textsf{eq\_res}\ \frac{\varphi \quad \varphi \approx \psi}{\psi} \qquad \textsf{atom\_rewrite}\ \frac{-\mid \mathcal{R}, s}{s \approx t}\ \text{if } s\downarrow_{\mathcal{R}} = t \qquad \textsf{witness}\ \frac{-\mid k}{k \approx k\uparrow}$$

$$\textsf{assume}\ \frac{-\mid \varphi}{\varphi} \qquad \textsf{scope}\ \frac{\varphi \mid \varphi_1, \ldots, \varphi_n}{\varphi_1 \wedge \cdots \wedge \varphi_n \Rightarrow \varphi}$$

Fig. 2. Core proof rules of the internal calculus.

well-formed if it is finite, acyclic, and there is a total mapping Ψ from the nodes of P to formulas such that, for each node $N = (r, (N_1, \ldots, N_m), \vec{t})$, $\Psi(N)$ is the formula returned by the proof checker for rule r when given premises $\Psi(N_1), \ldots, \Psi(N_n)$ and arguments $\vec{t}$. For a well-formed proof P with root N and mapping Ψ, the *conclusion* of P is the formula $\Psi(N)$; a *subproof* of P is any proof rooted at a descendant of N in P. For convenience, we will identify a well-formed proof with its root node from now on.

3.2 Core Proof Rules

In total, the internal calculus of cvc5 consists of 155 proof rules,[1] which cover all reasoning performed by the SMT solver, including theory-specific rules, rules for Boolean reasoning, and others. In the remainder of this section, we describe the *core* rules of the internal calculus, which are used throughout the system, and are illustrated in Fig. 2.

Proof Rules for Equality. Many theory solvers in cvc5 perform theory-specific reasoning on top of basic equational reasoning. The latter is captured by the proof rules eq_res, refl, symm, trans, and cong. The first rule is used to prove a formula ψ from a formula φ that was proved equivalent to ψ. The rest are the standard rules for computing the congruence closure of a set of term equalities.

Proof Rules for Rewriting, Substitution and Witness Forms. A single *coarse-grained* rule, sr, is used for tracking justifications for core utilities in the SMT solver such as *rewriting* and *substitution*. This rule, together with other non-core rules with side conditions (omitted for brevity), allows the generation of coarse-grained proofs that trust the correctness of complex side conditions. Those conditions involve rewriting and substitution operations performed by cvc5 during solving. More fine-grained proofs can be constructed from coarse-grained ones by justifying the various rewriting and substitution steps in terms of simpler proof rules. This is done with the aid of the equality rules mentioned above and the additional core rules atom_rewrite and witness. To describe atom_rewrite, witness, and sr, we first need to introduce some definitions and notations.

[1] See https://cvc5.github.io/docs/cvc5-1.0.0/proofs/proof_rules.html.

A *rewriter* $\mathcal{R}$ is a function over terms that preserves equivalence in the background theory T, i.e., returns a term $t{\downarrow_\mathcal{R}}$ T-equivalent to its input t. We call $t{\downarrow_\mathcal{R}}$ the *rewritten* form of t with respect to $\mathcal{R}$. Currently, cvc5 uses a handful of specialized rewriters for various purposes, such as evaluating constant terms, preprocessing input formulas, and normalizing terms during solving. Each individual rewrite step executed by a rewriter $\mathcal{R}$ is justified in fine-grained proofs by an application of the rule atom_rewrite, which takes as argument both (an identifier for) $\mathcal{R}$ and the term s the rewrite was applied to. Note that the rule's soundness requires that the rewrite step be equivalence preserving.

A *(term) substitution* σ is a finite sequence $(t_1 \mapsto s_1, \ldots, t_n \mapsto s_n)$ of oriented pairs of terms of the same sort. A *substitution method* $\mathcal{S}$ is a function that takes a term r and a substitution σ and returns a new term that is the result of *applying* σ to r, according to some strategy. We write $\mathcal{S}(r, \sigma)$ to denote the resulting term. We distinguish three kinds of substitution methods for σ: *simultaneous*, which returns the term obtained by simultaneously replacing every occurrence of term t_i in r with s_i, for $i = 1, \ldots, n$; *sequential*, which splits σ into n substitutions $(t_1 \mapsto s_1), \ldots, (t_n \mapsto s_n)$ and applies them in sequence to r using the simultaneous strategy above; and *fixed-point*, which, starting with r, repeatedly applies σ with the simultaneous strategy until no further subterm replacements are possible. For example, consider the application $\mathcal{S}(y, (x \mapsto u, y \mapsto f(z), z \mapsto g(x)))$. The steps the substitution method takes in computing its result are the following: $y \rightsquigarrow f(z)$ if $\mathcal{S}$ is simultaneous; $y \rightsquigarrow f(z) \rightsquigarrow f(g(x))$ if $\mathcal{S}$ is sequential; $y \rightsquigarrow f(z) \rightsquigarrow f(g(x)) \rightsquigarrow f(g(u))$ if $\mathcal{S}$ is fixed-point.

In cvc5, we use a *substitution derivation method* $\mathcal{D}$ to derive a *contextual* substitution $(t_1 \mapsto s_1, \ldots, t_n \mapsto s_n)$ from a collection $\vec{\varphi}$ of derived formulas. The substitution essentially orients a selection of term equalities $t_i \approx s_i$ entailed by $\vec{\varphi}$ and, as such, can be applied soundly to formulas derived from $\vec{\varphi}$.[2] We write $\mathcal{D}(\vec{\varphi})$ to denote the substitution computed by $\mathcal{D}$ from $\vec{\varphi}$.

Finally, cvc5 often introduces fresh variables, or *Skolem* variables, which are implicitly globally existentially quantified. This happens as a consequence of Skolemization of existential variables, lifting of if-then-else terms, and some kinds of flattening. Each Skolem variable k is associated with a term $k{\uparrow}$ of the same sort containing no Skolem variables, called its *witness term*. This global map from Skolem variables to their witness term allows cvc5 to detect when two Skolem variables can be equated, as a consequence of their respective witness terms becoming equivalent in the current context [47]. Witness terms can also be used to eliminate Skolem variables at proof output time. We write $t{\uparrow}$ to denote the *witness form* of term t, which is obtained by replacing every Skolem variable in t by its witness term. For example, if k_1 and k_2 are Skolem variables with associated witness terms $\mathsf{ite}(x \approx z, y, z)$ and $y - z$, respectively, and φ is the formula $\mathsf{ite}(x \approx k_2, k_1 \approx y, k_1 \approx z)$, the witness form $\varphi{\uparrow}$ of φ is the formula $\mathsf{ite}(x \approx y - z, \mathsf{ite}(x \approx z, y, z) \approx y, \mathsf{ite}(x \approx z, y, z) \approx z)$. When a Skolem variable k

[2] Observe that substitutions are generated dynamically from the formulas being processed, whereas rewrite rules are hard-coded in cvc5's rewriters.

appears in a proof, the witness proof rule is used to explicitly constrain its value to be the same as that of the term $k\uparrow$ it abstracts.[3]

We can now explain the sr proof rule, which is parameterized by a substitution method $\mathcal{S}$, a rewriter $\mathcal{R}$, and substitution derivation method $\mathcal{D}$. The rule is used to transform the proof of a formula φ into one of a formula ψ provided that the two formulas are equal up to rewriting under a substitution derived from the premises $\vec{\varphi}$. Note that this rule is quite general because its conclusion ψ, which is provided as an argument, can be any formula that satisfies the side condition.

Proof Rules for Scoped Reasoning. Two of the core proof rules, assume and scope, enable local reasoning. Together they achieve the effect of the $\Rightarrow$-introduction rule of Natural Deduction. However, separating the local assumption functionality in assume provides more flexibility. That rule has no premises and introduces a local assumption φ provided as an argument. The scope rule is used to *close the scope* of the local assumptions $\varphi_1, \ldots, \varphi_n$ made to prove a formula φ, inferring the formula $\varphi_1 \wedge \cdots \wedge \varphi_n \Rightarrow \varphi$.

We say that φ is a *free assumption* in proof P if P has a node (assume, (), φ) that is not a subproof of a scope node with φ as one of its arguments. A proof is *closed* if it has no free assumptions, and *open* otherwise.

Soundness. All proof rules other than assume are *sound* with respect to the background theory T in the following sense: if a rule proves a formula ψ from premises $\vec{\varphi}$, every model of T that satisfies $\vec{\varphi}$, and assigns the same values to Skolem variables and their respective witness term, satisfies ψ as well. Based on this and a simple structural induction argument, one can show that well-formed closed proofs have T-valid conclusions. In contrast, open proofs have conclusions that are T-valid only under assumptions. More precisely, in general, if $\vec{\varphi}$ are all the free assumptions of a well-formed proof P with conclusion ψ and $\vec{k}$ are all the Skolem variables introduced in P, then $\vec{k} \approx \vec{k}\uparrow \wedge \vec{\varphi} \Rightarrow \psi$ is T-valid.

3.3 Constructing Proof Nodes

We have implemented a library of *proof generators* that encapsulates common patterns for constructing proof nodes. We assume a method getProof that takes the proof generator g and a formula φ as input and returns a proof node with conclusion φ based on the information in g. During solving, cvc5 uses a combination of *eager* and *lazy* proof generation. In general terms, eager proof generation involves constructing proof nodes for inference steps at the time those steps are taken during solving. Eager proof generation may be required if the computation state pertinent to that inference cannot be easily recovered later. In contrast, lazy proof generation occurs for inferred formulas associated with proof generators that can do internal bookkeeping to be able to construct proof nodes for the formula *after* solving is completed. Depending on the formula, different kinds of proof generators are used. For brevity, we only describe in detail (see Sect. 3.2)

[3] The proof rules that account for the introduction of Skolem variables in the first place are not part of the core set and so are not discussed here.

Algorithm 1. Proof generation for term-conversion generators, rewrite-once policy. B is a lazy proof builder, R a map from terms to their converted form, and $c_{\mathsf{pre}}, c_{\mathsf{post}}$ are sets of pairs of equalities and the proof generators justifying them.

$\mathsf{getProof}(g, \varphi)$ where g contains $c_{\mathsf{pre}}, c_{\mathsf{post}}$ and φ is $t_1 \approx t_2$

$B := \emptyset,\ R := \emptyset$
$\mathsf{getTermConv}(t_1, c_{\mathsf{pre}}, c_{\mathsf{post}}, B, R)$
if $R[t_1] \neq t_2$ **then** fail **else return** $\mathsf{getProof}(B, t_1 \approx R[t_1])$

$\mathsf{getTermConv}(s, c_{\mathsf{pre}}, c_{\mathsf{post}}, B, R)$, where $s = f(s_1, \ldots, s_n)$

if s in $\mathsf{dom}(R)$ **then return**
if $(s \approx s', g') \in c_{\mathsf{pre}}$ for some s', g' **then**
    $R[s] := s'$, $\mathsf{addLazyStep}(B, s \approx s', g')$
    return
for $1 \leq i \leq n$ **do** $\mathsf{getTermConv}(s_i, c_{\mathsf{pre}}, c_{\mathsf{post}}, B, R)$
$R[s] := r$, where $r = f(R[s_1], \ldots, R[s_n])$
if $s \neq r$ **then** $\mathsf{addStep}(B, \mathsf{cong}, (s_1 \approx R[s_1], \ldots, s_n \approx R[s_n]), f)$
else $\mathsf{addStep}(B, \mathsf{rfl}, (), s \approx s)$
if $(r \approx r', g') \in c_{\mathsf{post}}$ for some r', g' **then**
    $R[s] := r'$, $\mathsf{addLazyStep}(B, r \approx r', g')$, $\mathsf{addStep}(B, \mathsf{trans}, (s \approx r, r \approx r'), ())$

the proof generator most relevant to the core calculus, the *term-conversion proof generator*, targeted for substitution and rewriting proofs.

4 Proof Reconstruction for Substitution and Rewriting

Once it determines that the input formulas $\varphi_1, \ldots, \varphi_n$ are jointly unsatisfiable, the SMT solver has a reference to a proof node P that concludes $\bot$ from the free assumptions $\varphi_1, \ldots, \varphi_n$. After the post-processor is run, the (closed) proof $(\mathsf{scope}, P', (\varphi_1, \ldots, \varphi_n))$ is then generated as the final proof for the user, where P' is the result of optionally expanding coarse-grained steps (in particular, applications of the rule sr) in P into fine-grained ones. To do so, we require the following algorithm for generating *term-conversion* proofs.

In particular, we focus on equalities $t \approx s$ whose proof can be justified by a set of steps that replace subterms of t until it is syntactically equal to s. We assume these steps are provided to a *term-conversion proof generator*. Formally, a term-conversion proof generator g is a pair of sets c_{pre} and c_{post}. The set c_{pre} (resp., c_{post}) contains pairs of the form $(t \approx s, g_{t,s})$ indicating that t should be replaced by s in a preorder (resp., postorder) traversal of the terms that g processes, where $g_{t,s}$ is a proof generator that can prove the equality $t \approx s$. We require that neither c_{pre} nor c_{post} contain multiple entries of the form $(t \approx s_1, g_1)$ and $(t \approx s_2, g_2)$ for distinct (s_1, g_1) and (s_2, g_2).

The procedure for generating proofs from a term-conversion proof generator g is given in Algorithm 1. When asked to prove an equality $t_1 \approx t_2$, $\mathsf{getProof}$ traverses the structure of t_1 and applies steps from the sets c_{pre} and c_{post} from g.

The traversal is performed by the auxiliary procedure getTermConv which relies on two data structures. The first is a *lazy proof builder* B that stores the intermediate steps in the overall proof of $t_1 \approx t_2$. The proof builder is given these steps either via addStep, as a concrete triple with the proof rule, a list of premise formulas, and a list of argument terms, or as a *lazy* step via addLazyStep, with a formula and a reference to another generator that can prove that formula. The second data structure is a mapping R from terms to terms that is updated (using array syntax in the pseudo-code) as the converted form of terms is computed by getTermConv. For any term s, executing getTermConv$(s, c_{\mathsf{pre}}, c_{\mathsf{post}}, B, R)$ will result in $R[s]$ containing the converted form of s according to the rewrites in c_{pre} and c_{post}, and B storing a proof step for $s \approx R[s]$. Thus, the procedure getProof succeeds when, after invoking getTermConv$(t_1, c_{\mathsf{pre}}, c_{\mathsf{post}}, B, R)$ with B and R initially empty, the mapping R contains t_2 as the converted form of t_1. The proof for the equality $t_1 \approx R[t_1]$ can then be constructed by calling getProof on the lazy proof builder B, based on the (lazy) steps stored in it.

Each subterm s of t_1 is traversed only once by getTermConv by checking whether R already contains the converted form of s. When that is not the case, s is first preorder processed. If c_{pre} contains an entry indicating that s rewrites to s', this rewrite step is added to the lazy proof builder and the converted form $R[s]$ of s is set to s'. Otherwise, the immediate subterms of s, if any, are traversed and then s is postorder processed. The converted form of s is set to some term r of the form $f(R[s_1], \ldots, R[s_n])$, considering how its immediate subterms were converted. Note that B will contain steps for $\vec{s} \approx R[\vec{s}]$. Thus, the equality $s \approx r$ can be proven by congruence for function f with these premises if $s \neq r$, and by reflexivity otherwise. Furthermore, if c_{post} indicates that r rewrites to r', then this step is added to the lazy proof builder; a transitivity step is added to prove $s \approx r'$ from $t \approx r$ and $r \approx r'$; and the converted form $R[s]$ is set to r'.

Example 1. Consider the equality $t \approx \bot$, where $t = f(b)+f(a) < f(a-0)+f(b)$, and suppose the conversion of t is justified by a term-conversion proof generator g containing the sets $c_{\mathsf{pre}} = \{(f(b) + f(a) \approx f(a) + f(b), g^{\mathsf{AC}}), (a - 0 \approx a, g_0^{\mathsf{Arith}})\}$ and $c_{\mathsf{post}} = \{(f(a)+f(b) < f(a)+f(b) \approx \bot, g_1^{\mathsf{Arith}})\}$. The generator g^{AC} provides a proof based on associative and commutative reasoning, whereas g_0^{Arith} and g_1^{Arith} provide proofs based on arithmetic reasoning. Invoking getProof$(g, t \approx \bot)$ initiates the traversal with getTermConv$(t, c_{\mathsf{pre}}, c_{\mathsf{post}}, \emptyset, \emptyset)$. Since t is not in the conversion map, it is preorder processed. However, as it does not occur in c_{pre}, nothing is done and its subterms are traversed. The subterm $f(b) + f(a)$ is equated to $f(a) + f(b)$ in c_{pre}, justified by g^{AC}. Therefore R is updated with $R[f(b) + f(a)] = f(a) + f(b)$ and the respective lazy step is added to B. The subterms of $f(b)+f(a)$ are not traversed, therefore the next term to be traversed is $f(a-0)+f(b)$. Since it does not occur in c_{pre}, its subterm $f(a-0)$ is traversed, which analogously leads to the traversal of $a-0$. As $a-0$ does occur in c_{pre}, both R and B are updated accordingly and the processing of its parent $f(a-0)$ resumes. A congruence step added to B justifies its conversion to $f(a)$ being added to R.

No more additions happen since $f(a)$ does not occur in c_{post}. Analogously, R and B are updated with $f(b)$ not changing and $f(a-0)+f(b)$ being converted into $f(a)+f(b)$. Finally, the processing returns to the initial term t, which has been converted to $R[f(b)+f(a)] < R[f(a+0)+f(b)]$, i.e., $f(a)+f(b) < f(a)+f(b)$. Since this term is equated to $\bot$ in c_{post}, justified by g_1^{Arith}, the respective lazy step is added to B, as well as a transitivity step to connect $f(b)+f(a) < f(a-0)+f(b) \approx f(a)+f(b) < f(a)+f(b)$ and $f(a)+f(b) < f(a)+f(b) \approx \bot$. At this point, the execution terminates with $R[f(b)+f(a) < f(a+0)+f(b)] = \bot$, as expected. A proof for $t \approx \bot$ with the following structure can then be extracted from B:

$$P_0 : \mathsf{cong}\,\frac{\mathsf{Lazy}\,\dfrac{g^{\mathsf{AC}}}{f(b)+f(a)\approx f(a)+f(b)} \quad P_1 \mid <}{f(b)+f(a) < f(a-0)+f(b) \approx f(a)+f(b) < f(a)+f(b)} \qquad P_2 : \mathsf{refl}\,\frac{- \mid f(b)\approx f(b)}{f(b)\approx f(b)}$$

$$\mathsf{trans}\,\frac{P_0 \quad \mathsf{Lazy}\,\dfrac{g_1^{\mathsf{Arith}}}{f(a)+f(b) < f(a)+f(b) \approx \bot}}{f(b)+f(a) < f(a-0)+f(b) \approx \bot} \qquad P_1 : \mathsf{cong}\,\frac{\mathsf{cong}\,\dfrac{\mathsf{Lazy}\,\dfrac{g_0^{\mathsf{Arith}}}{a-0\approx a} \mid f}{f(a-0)\approx f(a)} \quad P_2 \mid +}{f(a-0)+f(b)\approx f(a)+f(b)}$$

We use several extensions to the procedures in Algorithm 1. Notice that this procedure follows the policy that terms on the right-hand side of conversion steps (equalities from c_{pre} and c_{post}) are not traversed further. The procedure getTermConv is used by term-conversion proof generators that have the *rewrite-once* policy. A similar procedure which additionally traverses those terms is used by term-conversion proof generators that have a *rewrite-to-fixpoint* policy.

We now show how the term-conversion proof generator can be used for reconstructing fine-grained proofs from coarse-grained ones. In particular we focus on proofs P_{ψ_1} of the form $(\mathsf{sr}, (Q_{\psi_0}, \vec{Q}), (\mathcal{S}, \mathcal{R}, \mathcal{D}, \psi))$. Recall from Fig. 2 that the proof rule sr concludes a formula ψ that can be shown equivalent to the formula ψ_0 proven by Q_{ψ_0} based on a substitution derived from the conclusions of the nodes $\vec{Q}$. A proof like P_{ψ_1} above can be transformed to one that involves (atomic) theory rewrites and equality rules only. We show this transformation in two phases. In the first phase, the proof is expanded to:

$$(\mathsf{eq\_res}, (Q_{\psi_0}, (\mathsf{trans}, (R_0, (\mathsf{symm}, R_1)))))$$

with $R_i = (\mathsf{trans}, ((\mathsf{subs}, \vec{Q}_{\vec{\varphi}}, (\mathcal{S}, \mathcal{D}, \psi_i)), (\mathsf{rewrite}, (), (\mathcal{R}, \mathcal{S}(\psi_i, \mathcal{D}(\vec{\varphi}))))))$ for $i \in \{0,1\}$ where $\vec{\varphi}$ are the conclusions of $\vec{Q}_{\vec{\varphi}}$, and subs and rewrite are auxiliary proof rules used for further expansion in the second phase. We describe them next.

Substitution Steps. Let $P_{t\approx s}$ be the subproof $(\mathsf{subs}, \vec{Q}_{\vec{\varphi}}, (\mathcal{S}, \mathcal{D}, t))$ of R_i above proving $t \approx s$ with $s = \mathcal{S}(\psi_i, \mathcal{D}(\vec{\varphi}))$ and $\mathcal{D}(\vec{\varphi}) = (t_1 \mapsto s_1, \ldots, t_n \mapsto s_n)$. Substitution steps can be expanded to fine-grained proofs using a term-conversion proof generator. First, for each $j = 1, \ldots, n$, we construct a proof of $t_j \approx s_j$, which involves simple transformations on the proofs of $\vec{\varphi}$. Suppose we store all of these in an eager proof generator g. If $\mathcal{S}$ is a simultaneous or fixed-point substitution, we then build a single term-conversion proof generator C, which recall is modeled as a pair of mappings $(c_{\mathsf{pre}}, c_{\mathsf{post}})$. We add $(t_j \approx s_j, g)$ to c_{pre} for all j. We use the rewrite-once policy for C if $\mathcal{S}$ is a simultaneous substitution, and the rewrite-fixed-point policy for C otherwise. We then replace the proof $P_{t\approx s}$ by $\mathsf{getProof}(C, t \approx s)$, which runs the procedure in Algorithm 1. Otherwise, if $\mathcal{S}$ is a sequential substitution, we construct a term-conversion generator C_j for *each* j, initializing it so that its c_{pre} set contains the single rewrite step $(t_j \approx s_j, g)$ and uses a rewrite-once policy. We then replace the proof $P_{t\approx s}$ by $(\mathsf{trans}, (P_1, \ldots, P_n))$ where, for $j = 1, \ldots, n$: P_j is generated by $\mathsf{getProof}(C_j, s_{j-1} \approx s_j)$; $s_0 = t$; s_i is the result of the substitution $\mathcal{D}(\vec{\varphi})$ after the first i steps; and $s_n = s$.

Rewrite Steps. Let P be the proof node $(\mathsf{rewrite}, (), (\mathcal{R}, t))$, which proves the equality $t \approx t{\uparrow}{\downarrow}_{\mathcal{R}}$. During reconstruction, we replace P with a proof involving only fine-grained rules, depending on the rewrite method $\mathcal{R}$. For example, if $\mathcal{R}$ is the core rewriter, we run the rewriter again on t in proof tracking mode. Normally, the core rewriter performs a term traversal and applies atomic rewrites to completion. In proof tracking mode, it also return two lists, for pre- and post-rewrites, of steps $(t_1 \approx s_1, g), \ldots, (t_n \approx s_n, g)$ where g is a proof generator that returns $(\mathsf{atom\_rewrite}, (), (\mathcal{R}, t_i))$ for all equalities $t_i \approx s_i$. Furthermore, for each Skolem k that is a subterm of t, we construct the rewrite steps $(k \approx k{\uparrow}, g')$ where g' is a proof generator that returns $(\mathsf{witness}, (), (k))$ for equalities $k \approx k{\uparrow}$. We add these rewrite proof steps to a term-conversion generator C with rewrite-fixed-point policy, and replace P by $\mathsf{getProof}(C, t \approx t{\uparrow}{\downarrow}_{\mathcal{R}})$.

5 SMT Proofs

Here we briefly describe each component shown in Sect. 2 and how it produces proofs with the infrastructure from Sects. 3 and 3.2.

5.1 Preprocessing Proofs

The *pre-processor* transforms an input formula φ into a list of formulas to be given to the core solver. It applies a sequence of *preprocessing passes*. A pass may *replace* a formula φ_i with another one ϕ_i, in which case it is responsible for providing a proof of $\varphi_i \approx \phi_i$. It may also append a new formula ϕ to the list, in which case it is responsible for providing a proof for it. We use a (lazy) proof generator that tracks these proofs, maintaining the invariant that a proof can be provided for all (preprocessed) formulas when requested. We have instrumented proof production for the most common preprocessing passes, relying heavily on the sr rule to model transformations such as expansion of function definitions and, with witness forms, Skolemization and if-then-else elimination [6].

Simplification Under Global Assumptions. cvc5 aggressively learns literals that hold globally by performing Boolean constraint propagation over the input formula. When a learned literal corresponds to a variable elimination (e.g., $x \approx 5$ corresponds to $x \mapsto 5$) or a constant propagation (e.g., $P(x)$ corresponds to

$P(x) \mapsto \top$), we apply the corresponding (term) substitution to the input. This application is justified via sr, while the derivation of the globally learned literals is justified via clausification and resolution proofs, as explained in Sect. 5.3.

The key features of our architecture that make it feasible to produce proofs for this simplification are the automatic reconstruction of sr steps and the ability to customize the strategy for substitution application during reconstruction, as detailed in Sect. 3.2. When a new variable elimination $x \mapsto t$ is learned, old ones need to be normalized to eliminate any occurrences of x in their right-hand sides. Computing the appropriate simultaneous substitution for all eliminations requires quadratically many traversals over those terms. We have observed that the size of substitutions generated by this preprocessing pass can be very large (with thousands of entries), which makes this computation prohibitively expensive. Using the fixed-point strategy, however, the reconstruction for the sr steps can apply the substitution efficiently and its complexity depends on how many applications are necessary to reach a fix-point, which is often low in practice.

5.2 Theory Proofs

The theory engine produces lemmas, as disjunctions of literals, from an individual theory or a combination of them. In the first case, the lemma's proof is provided directly by the corresponding theory solver. In the second case, a theory solver may produce a lemma ψ containing a literal ℓ derived by some other theory solver from literals $\vec{\ell}$. A lemma over the combined theory is generated by replacing ℓ in ψ by $\vec{\ell}$. This regression process, which is similar to the computation of *explanations* during solving, is repeated until the lemma contains only input literals. The proof of the final lemma then uses rules like sr to combine the proofs of the intermediate literals derived locally in various theories and their replacement by input literals in the final lemma.

Equality and Uninterpreted Function (EUF) Proofs. The EUF solver can be easily instrumented to produce proofs [31,42] with equality rules (see Fig. 2). In cvc5, term equivalences are also derived via rewriting in some other theory T: when a function from T has all of its arguments inferred to be congruent to T-values, it may be rewritten into a T-value itself, and this equivalence asserted. Such equivalences are justified via sr steps. Since generating equality proofs incurs minimal overhead [42] and rewriting proofs are reconstructed lazily, EUF proofs are generated during solving and stored in an eager proof generator.

Extensional Arrays and Datatypes Proofs. While these two theories differ significantly, they both combine equality reasoning with rules for handling their particular operators. For arrays, these are rules for select, store, and array extensionality (see [36, Sec. 5]). For datatypes, they are rules reflecting the properties of *constructors* and *selectors*, as well as acyclicity. The justifications for lemmas are also generated eagerly and stored in an eager proof generator.

Bit-Vector Proofs. The bit-vector solver applies bit-blasting to reduce bit-vector problems to equisatisfiable propositional problems. Thus, its lemmas amount to the rewriting of the bit-vector literals into Boolean formulas, which will be solved and proved by the propositional engine. The bit-vector lemmas are proven lazily, analogous to sr steps, with the difference that the reconstruction uses the bit-blaster in the bit-vector solver instead of the rewriter.

Arithmetic Proofs. The *linear* arithmetic solver is based on the simplex algorithm [24], and each of its lemmas is the negation of an unsatisfiable conjunction of inequalities. Farkas' lemma [30,49] guarantees that there exists a linear combination of these inequalities equivalent to $\bot$. The coefficients of the combination are computed during solving with minimal overhead [38], and the equivalence is proven with an sr step. To allow the rewriter to prove this equivalence, the bounds of the inequalities are scaled by constants and summed during reconstruction. Integer reasoning is proved through rules for branching and integer bound tightening, recorded eagerly.

Non-linear arithmetic lemmas are generated from incremental linearization [16] or cylindrical algebraic coverings [1]. The former can be proven via propositional and basic arithmetic rules, with only a few, such as the tangent plane lemma, needing a dedicated proof rule. The latter requires two complex rules that are not inherently simpler than solving, albeit not as complex as those for regular CAD-based theory solvers [2]. We point out that checking these rules would require a significant portion of CAD-related theory, whose proper formalization is still an open, if actively researched, problem [18,25,34,41,53].

Quantifier Proofs. Quantified formulas not Skolemized during pre-processing are handled via instantiation, which produces theory lemmas of the form $(\forall \vec{x}\, \varphi) \Rightarrow \varphi\sigma$, where σ is a grounding substitution. An instantiation rule proves them independently of how the substitution was actually derived, since any well-typed one suffices for soundness.

String Proofs. The strings solver applies a layered approach, distinguishing between core [40] and extended operators [48]. The core operators consist of (dis)equalities between string concatenations and length constraints. Reasoning over them is proved by a combination of equality and linear integer arithmetic proofs, as well as specific string rules. The extended operators are reduced to core ones via formulas with bounded quantifiers. The reductions are proven with rules defining each extended function's semantics, and sr steps justifying the reductions. Finally, regular membership constraints are handled by string rules that unfold occurrences of the Kleene star operator and split up regular expression concatenations into different parts. Overall, the proofs for the strings theory solver encompass not only string-specific reasoning but also equality, linear integer arithmetic, and quantifier reasoning, as well as substitution and rewriting.

Unsupported. The theory solvers for the theories of floating-point arithmetic, sequences, sets and relations, and separation logic are currently not proof-producing in cvc5. These are relatively new or non-standard theories in SMT and have not been our focus, but we intend to produce proofs for them in the future.

Table 1. Cumulative solving times (s) on benchmarks solved by all configurations, with the slowdown versus CVC+S in parentheses.

Logics	#	CVC+OS	CVC+S	CVC+SP	CVC+SPR
NON-BVs	116,321	164k	166k	284k (1.7×)	299k (1.8×)
BVs	29,192	45k	57k	150k (2.6×)	224k (3.9×)

5.3 Propositional Proofs

Propositional proofs justify both the conversion of preprocessed input formulas and theory lemmas into conjunctive normal form (CNF) and the derivation of $\bot$ from the resulting clauses. CNF proofs are a combination of Boolean transformations and introductions of Boolean formulas representing the definition of Tseytin variables, used to ensure that the CNF conversion is polynomial. The clausifier uses a lazy proof builder which stores the clausification steps eagerly, with the preprocessed input formulas as assumptions, and the theory lemmas as lazy steps, with associated proof generators.

For Boolean reasoning, cvc5 uses a version of MiniSat [27] instrumented to produce resolution proofs. It uses a lazy proof builder to record resolution steps for learned clauses as they are derived (see [7, Chap 1] for more details) and to lazily build a refutation with only the resolution steps necessary for deriving $\bot$. The resolution rule, however, is ground first-order resolution, since the proofs are in terms of the first-order clauses rather than their propositional abstractions.

6 Evaluation

In this section, we discuss an initial evaluation of our implementation in cvc5 of the proof-production architecture presented in this paper. In the following, we denote different configurations of cvc5 by CVC plus some suffixes. A configuration using variable and clause elimination in the SAT solver [26], symmetry breaking [23] in the EUF solver, and black-box SAT solving in the bit-vector (BV) solver, is denoted by the suffix O. These techniques are currently incompatible with the proof production architecture. Other cvc5 techniques for which we do not yet support fine-grained proofs, however, are active and have their inferences registered in the proofs as trusted steps. A configuration that includes simplification under global assumptions is denoted by S; one that includes producing proofs by P; and one that additionally reconstructs proofs by R. The default configuration of cvc5 is CVC+OS.

We split our evaluation into measuring the proof-production cost as well as the performance impact of making key techniques proof-producing; the proof reconstruction overhead; and the coverage of the proof production. We also comment on how cvc5's proofs compare with CVC4's proofs. Note that the internal proof checking described in Sect. 3, which was invaluable for a correct implementation, is disabled for evaluating performance. Experiments ran on a cluster with

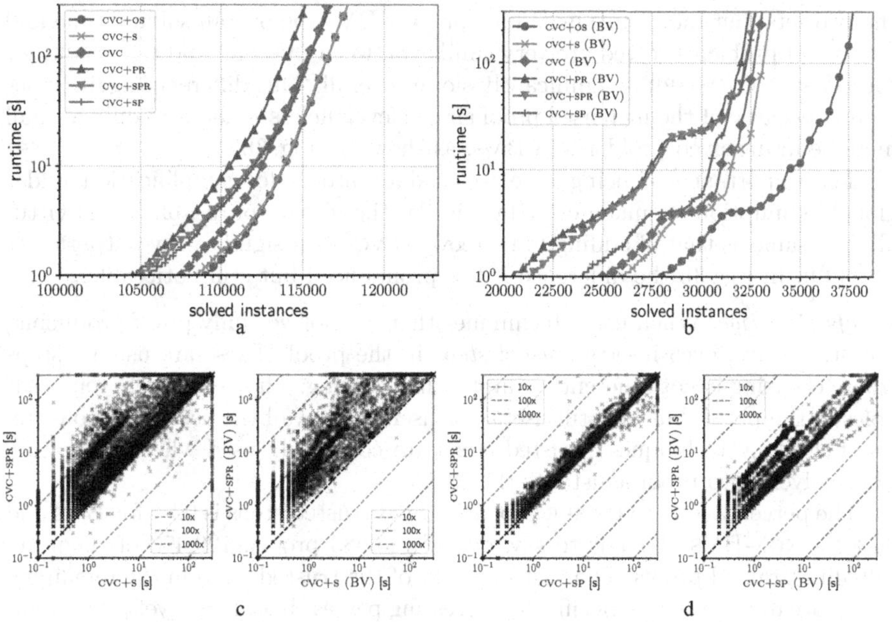

Fig. 3. (a) Cactus plot for NON-BVs (b) Cactus plot for BVs (c) Scatter plot of overall proof cost (d) Reconstruction cost

Intel Xeon E5-2620 v4 CPUs, with 300s and 8GB of RAM for each solver and benchmark pair. We consider 162,060 unsatisfiable problems from SMT-LIB [8], across all logics except those with floating point arithmetic, as determined by cvc5 [5, Sec. 4]. We split them into 38,732 problems with the BV theory (the BVs set) and 123,328 problems without (the NON-BVs set).

Proof Production Cost. The cost of proof production is summarized in Table 1 and Figs. 3a to 3d. The impact of running without o is negligible overall in NON-BVs, but steep for BVs, both in terms of solving time and number of problems solved, as evidenced by the table and Fig. 3b respectively. This is expected given the effectiveness of combining bit-blasting with black-box SAT solvers. The overhead of P is similar for both sets, although more pronounced in BVs. While the total time is around double that of cvc+s, Fig. 3c shows a finer distribution, with most problems having a less significant overhead. Moreover, the total number of problems solved is quite similar, as shown in Figs. 3a and 3b, particularly for NON-BVs. The difference in overhead due to P between the BVs and NON-BVs sets can be attributed to the cost of managing large proofs, which are more common in BVs. This stems from the well-known blow-up in problem size incurred by bit-blasting, which is reflected in the proofs.

The cost of generating fine-grained steps for the sr rule and for the similarly reconstructed theory-specific steps mentioned in Sect. 5, varies again between the two sets, but more starkly. While for NON-BVs the overall solving time and number of problems solved are very similar between CVC+SP and CVC+SPR, for the BVs set CVC+SPR is significantly slower overall. This difference again arises mainly because of the increased proof sizes. Nevertheless, R leads to only a small increase in unsolved problems in BVs, as shown in Fig. 3b.

The importance of being able to produce proofs for simplification under global assumptions is made clear by Fig. 3a: the impact of disabling S is virtually the same as that of adding P; moreover, CVC+SPR significantly outperforms CVC+PR. In Fig. 3b the difference is less pronounced but still noticeable.

Proofs Coverage. When using techniques that are not yet fully proof-producing, but still active, cvc5 inserts *trusted steps* in the proof. These are usually steps whose checking is not inherently simpler than solving. They effectively represent holes in the proof, but are still useful for users who avail themselves of powerful proof-checking techniques. Trusted steps are commonly used when integrating SMT solvers into proof assistants [11,28,51].

The percentage of CVC+SPR proofs *without* trusted steps is 92% for BVs and 80% for NON-BVs. That is to say, out of 145,683 proofs, 120,473 of them are fully fine-grained proofs. The vast majority of the trusted steps in the remaining proofs are due to theory-specific preprocessing passes that are not yet fully proof-producing. In NON-BVs, the occurrence of trusted steps is heavily dependent on the specific SMT-LIB logic, as expected. Common offenders are logics with datatypes, with trusted steps for acyclicity checks, and quantified logics, with trusted steps for certain α-equivalence eliminations. In non-linear real arithmetic logics, all cylindrical algebraic coverings proofs are built with trusted steps (see Sect. 5.2), but we note this is the state of the art for CAD-based proofs. As for non-linear integer arithmetic logics, our proof support is still in its early stages, so a significant portion of their theory lemmas are trusted steps.

We stress the extent of our coverage for string proofs, which were previously unsupported by any SMT solver. In the string logics without length constraints, 100% of the proofs are fully fine-grained. This rate goes down to 80% in the logics with length. For the remaining 20%, the overwhelming majority of the trusted steps are for theory-specific preprocessing or some particular string or linear arithmetic inference within the proof of a theory lemma.

Comparison with CVC4 Proofs. We compare the proof coverage of cvc5 versus CVC4. The cvc5 proof production replaces CVC4's [32,36], which was incomplete and monolithic. CVC4 did not produce proofs at all for strings, substitutions, rewriting, preprocessing, quantifiers, datatypes, or non-linear arithmetic. In particular, simplification over global assumptions had to be disabled when producing proofs. In fragments supported by both systems, CVC4's proofs are at most as detailed as cvc5's. The only superior aspect of CVC4's proof production was to support proofs from external SAT solvers [45] used in the BV solver, which are very significant for solving performance, as shown above. Integrating this feature into cvc5 is left as future work, but we note that there is no limitation in the proof architecture that would prevent it. We also point out that cvc5 produces resolution proofs for the bit-blasted BV constraints, which can be checked in polynomial time, whereas external SAT solvers produce DRAT proofs [33] (or reconstructions of them via other tools [19,20,37,39]), which can take exponential time to check. So there is a significant trade-off to be considered.

7 Related Work

Two significant proof-producing state-of-the-art SMT solvers are z3 [22] and veriT [14]. Both can have their proofs successfully reconstructed in proof assistants [3,12,13,51]. They can produce detailed proofs for the propositional and theory reasoning in EUF and linear arithmetic, as well as for quantifiers. However, z3's proofs are coarse-grained for preprocessing and rewriting, and for bit-vector reasoning, which complicates proof checking. Moreover, to the best of our knowledge, z3 does not produce proofs for its other theories. In contrast, veriT can produce fine-grained proofs for preprocessing and rewriting [6], which has led to a better integration with Isabelle/HOL [51]. However, it does so eagerly, which requires a tight integration between the preprocessing and the proof-production code. In addition, it does not support simplification under global assumptions when producing proofs, which significantly impacts its performance. Other proof-producing SMT solvers are MathSAT5 [17] and SMTInterpol [15]. They produce resolution proofs and theory proofs for EUF, linear arithmetic, and, in SMTInterpol's case, array theories. Their proofs are tailored towards unsatisfiable core and interpolant generation, rather than external certification. Moreover, they do not seem to provide proofs for preprocessing, clausification or rewriting.

While cvc5 is possibly the only proof-producing solver for the full theory of strings, CertiStr [35] is a certified solver for the fragment with concatenation and regular expressions. It is automatically generated from Isabelle/HOL [44] but is significantly less performant than cvc5, although a proper comparison would need to account for proof-checking time in cvc5's case.

8 Conclusion and Future Work

We presented and evaluated a flexible proof production architecture, showing it is capable of producing proofs with varying levels of granularity in a scalable manner for a state-of-the-art and industrial-strength SMT solver like cvc5.

Since currently, there is no standard proof format for SMT solvers, our architecture is designed to support multiple proof formats via a final post-processing transformation to convert internal proofs accordingly. We are developing backends for the LFSC [52] proof checker and the proof assistants Lean 4 [21], Isabelle/HOL [44], and Coq [10], the latter two via the Alethe proof format [50]. Since using these tools requires mechanizing the respective target proof calculi in their languages, besides external checking, another benefit is to decouple confidence on the soundness of the proof calculi from the internal cvc5 proof calculus.

A considerable challenge for SMT proofs is the plethora of rewrite rules used by the solvers, which are specific for each theory and vary in complexity. In particular, string rewrites can be very involved [46] and hard to check. We are also developing an SMT-LIB-based DSL for specifying rewrite rules, to be used during proof reconstruction to decompose rewrite steps in terms of them, thus providing more fine-grained proofs for rewriting.

Finally, we plan to incorporate into the proof-production architecture the unsupported theories and features mentioned in Sects. 5.2 and 6, particularly those relevant for solving performance that currently either leave holes in proofs, such as theory pre-processing or non-linear arithmetic reasoning, or that have to be disabled, such as the use of external SAT solvers in the BV theory.

References

1. Ábrahám, E., Davenport, J.H., England, M., Kremer, G.: Deciding the consistency of non-linear real arithmetic constraints with a conflict driven search using cylindrical algebraic coverings. J. Log. Algebr. Methods Program. **119**, 100633 (2021)
2. Abrahám, E., Davenport, J.H., England, M., Kremer, G.: Proving UNSAT in SMT: the case of quantifier free non-linear real arithmetic. arXiv preprint arXiv:2108.05320 (2021)
3. Armand, M., Faure, G., Grégoire, B., Keller, C., Théry, L., Werner, B.: A modular integration of SAT/SMT solvers to coq through proof witnesses. In: Jouannaud, J.-P., Shao, Z. (eds.) CPP 2011. LNCS, vol. 7086, pp. 135–150. Springer, Heidelberg (2011). https://doi.org/10.1007/978-3-642-25379-9_12
4. Backes, J., et al.: Semantic-based automated reasoning for AWS access policies using SMT. In: Bjørner, N., Gurfinkel, A. (eds.) Formal Methods in Computer-Aided Design (FMCAD), pp. 1–9. IEEE (2018)
5. Barbosa, H., et al.: cvc5: a versatile and industrial-strength SMT solver. In: Fisman, D., Rosu, G. (eds.) Tools and Algorithms for Construction and Analysis of Systems (TACAS). LNCS, Springer, Cham (2022). https://doi.org/10.1007/978-3-030-99524-9_24
6. Barbosa, H., Blanchette, J.C., Fleury, M., Fontaine, P.: Scalable fine-grained proofs for formula processing. J. Autom. Reason. **64**(3), 485–510 (2020)
7. Barrett, C., de Moura, L., Fontaine, P.: Proofs in satisfiability modulo theories. All About Proofs Proofs All (APPA) **55**(1), 23–44 (2014)
8. Barrett, C., Fontaine, P., Tinelli, C.: The SMT-LIB standard: version 2.6. Technical report, Department of Computer Science, The University of Iowa (2017). www.SMT-LIB.org
9. Barrett, C., Tinelli, C.: Satisfiability Modulo Theories. In: Clarke, E., Henzinger, T., Veith, H., Bloem, R. (eds.) Handbook of Model Checking, pp. 305–343. Springer, Cham (2018). https://doi.org/10.1007/978-3-319-10575-8_11
10. Bertot, Y., Castéran, P.: Interactive Theorem Proving and Program Development - Coq'Art: The Calculus of Inductive Constructions. Texts in Theoretical Computer Science. An EATCS Series, Springer, Heidelberg (2004). https://doi.org/10.1007/978-3-662-07964-5
11. Blanchette, J.C., Böhme, S., Paulson, L.C.: Extending sledgehammer with SMT solvers. J. Autom. Reason. **51**(1), 109–128 (2013)
12. Böhme, S., Fox, A.C.J., Sewell, T., Weber, T.: Reconstruction of Z3's bit-vector proofs in HOL4 and Isabelle/HOL. In: Jouannaud, J.-P., Shao, Z. (eds.) CPP 2011. LNCS, vol. 7086, pp. 183–198. Springer, Heidelberg (2011). https://doi.org/10.1007/978-3-642-25379-9_15

13. Böhme, S., Weber, T.: Fast LCF-style proof reconstruction for Z3. In: Kaufmann, M., Paulson, L.C. (eds.) ITP 2010. LNCS, vol. 6172, pp. 179–194. Springer, Heidelberg (2010). https://doi.org/10.1007/978-3-642-14052-5_14
14. Bouton, T., Caminha B. de Oliveira, D., Déharbe, D., Fontaine, P.: veriT: an open, trustable and efficient SMT-solver. In: Schmidt, R.A. (ed.) CADE 2009. LNCS (LNAI), vol. 5663, pp. 151–156. Springer, Heidelberg (2009). https://doi.org/10.1007/978-3-642-02959-2_12
15. Christ, J., Hoenicke, J., Nutz, A.: SMTInterpol: an interpolating SMT solver. In: Donaldson, A., Parker, D. (eds.) SPIN 2012. LNCS, vol. 7385, pp. 248–254. Springer, Heidelberg (2012). https://doi.org/10.1007/978-3-642-31759-0_19
16. Cimatti, A., Griggio, A., Irfan, A., Roveri, M., Sebastiani, R.: Satisfiability modulo transcendental functions via incremental linearization. In: de Moura, L. (ed.) CADE 2017. LNCS (LNAI), vol. 10395, pp. 95–113. Springer, Cham (2017). https://doi.org/10.1007/978-3-319-63046-5_7
17. Cimatti, A., Griggio, A., Schaafsma, B.J., Sebastiani, R.: The MathSAT5 SMT solver. In: Piterman, N., Smolka, S.A. (eds.) TACAS 2013. LNCS, vol. 7795, pp. 93–107. Springer, Heidelberg (2013). https://doi.org/10.1007/978-3-642-36742-7_7
18. Cohen, C.: Construction of real algebraic numbers in Coq. In: Beringer, L., Felty, A. (eds.) ITP 2012. LNCS, vol. 7406, pp. 67–82. Springer, Heidelberg (2012). https://doi.org/10.1007/978-3-642-32347-8_6
19. Cruz-Filipe, L., Heule, M.J.H., Hunt, W.A., Kaufmann, M., Schneider-Kamp, P.: Efficient certified RAT verification. In: de Moura, L. (ed.) CADE 2017. LNCS (LNAI), vol. 10395, pp. 220–236. Springer, Cham (2017). https://doi.org/10.1007/978-3-319-63046-5_14
20. Cruz-Filipe, L., Marques-Silva, J., Schneider-Kamp, P.: Efficient certified resolution proof checking. In: Legay, A., Margaria, T. (eds.) TACAS 2017. LNCS, vol. 10205, pp. 118–135. Springer, Heidelberg (2017). https://doi.org/10.1007/978-3-662-54577-5_7
21. de Moura, L., Ullrich, S.: The lean 4 theorem prover and programming language. In: Platzer, A., Sutcliffe, G. (eds.) CADE 2021. LNCS (LNAI), vol. 12699, pp. 625–635. Springer, Cham (2021). https://doi.org/10.1007/978-3-030-79876-5_37
22. de Moura, L.M., Bjørner, N.: Proofs and refutations, and Z3. In: Rudnicki, P., Sutcliffe, G., Konev, B., Schmidt, R.A., Schulz, S. (eds.) Logic for Programming, Artificial Intelligence, and Reasoning (LPAR) Workshops. CEUR Workshop Proceedings, vol. 418. CEUR-WS.org (2008)
23. Déharbe, D., Fontaine, P., Merz, S., Woltzenlogel Paleo, B.: Exploiting symmetry in SMT problems. In: Bjørner, N., Sofronie-Stokkermans, V. (eds.) CADE 2011. LNCS (LNAI), vol. 6803, pp. 222–236. Springer, Heidelberg (2011). https://doi.org/10.1007/978-3-642-22438-6_18
24. Dutertre, B., de Moura, L.: A fast linear-arithmetic solver for DPLL(T). In: Ball, T., Jones, R.B. (eds.) CAV 2006. LNCS, vol. 4144, pp. 81–94. Springer, Heidelberg (2006). https://doi.org/10.1007/11817963_11
25. Eberl, M.: A decision procedure for univariate real polynomials in Isabelle/HOL. In: Proceedings of the 2015 Conference on Certified Programs and Proofs, CPP 2015, pp. 75–83. Association for Computing Machinery, New York (2015)
26. Eén, N., Biere, A.: Effective preprocessing in SAT through variable and clause elimination. In: Bacchus, F., Walsh, T. (eds.) SAT 2005. LNCS, vol. 3569, pp. 61–75. Springer, Heidelberg (2005). https://doi.org/10.1007/11499107_5
27. Eén, N., Sörensson, N.: An extensible SAT-solver. In: Giunchiglia, E., Tacchella, A. (eds.) SAT 2003. LNCS, vol. 2919, pp. 502–518. Springer, Heidelberg (2004). https://doi.org/10.1007/978-3-540-24605-3_37

28. Ekici, B., et al.: SMTCoq: a plug-in for integrating SMT solvers into Coq. In: Majumdar, R., Kunčak, V. (eds.) CAV 2017. LNCS, vol. 10427, pp. 126–133. Springer, Cham (2017). https://doi.org/10.1007/978-3-319-63390-9_7
29. Enderton, H.B.: A Mathematical Introduction to Logic, 2nd edn. Academic Press, Cambridge (2001)
30. Farkas, G.: A Fourier-féle mechanikai elv alkamazásai. Mathematikaiés Természettudományi Értesítö **12**, 457–472 (1894). Reference from Schrijver's Combinatorial Optimization textbook (Hungarian)
31. Fontaine, P., Marion, J.-Y., Merz, S., Nieto, L.P., Tiu, A.: Expressiveness + automation + soundness: towards combining SMT solvers and interactive proof assistants. In: Hermanns, H., Palsberg, J. (eds.) TACAS 2006. LNCS, vol. 3920, pp. 167–181. Springer, Heidelberg (2006). https://doi.org/10.1007/11691372_11
32. Hadarean, L., Barrett, C., Reynolds, A., Tinelli, C., Deters, M.: Fine grained SMT proofs for the theory of fixed-width bit-vectors. In: Davis, M., Fehnker, A., McIver, A., Voronkov, A. (eds.) LPAR 2015. LNCS, vol. 9450, pp. 340–355. Springer, Heidelberg (2015). https://doi.org/10.1007/978-3-662-48899-7_24
33. Heule, M.J.H.: The DRAT format and drat-trim checker. CoRR, abs/1610.06229 (2016)
34. Joosten, S.J.C., Thiemann, R., Yamada, A.: A verified implementation of algebraic numbers in Isabelle/HOL. J. Autom. Reason. **64**, 363–389 (2020)
35. Kan, S., Lin, A.W., Rümmer, P., Schrader, M.: Certistr: a certified string solver. In: Popescu, A., Zdancewic, S. (eds.) Certified Programs and Proofs (CPP), pp. 210–224. ACM (2022)
36. Katz, G., Barrett, C., Tinelli, C., Reynolds, A., Hadarean, L.: Lazy proofs for DPLL(T)-based SMT solvers. In: Piskac, R., Talupur, M. (eds.) Formal Methods in Computer-Aided Design (FMCAD), pp. 93–100. IEEE (2016)
37. Kiesl, B., Rebola-Pardo, A., Heule, M.J.H.: Extended resolution simulates DRAT. In: Galmiche, D., Schulz, S., Sebastiani, R. (eds.) IJCAR 2018. LNCS (LNAI), vol. 10900, pp. 516–531. Springer, Cham (2018). https://doi.org/10.1007/978-3-319-94205-6_34
38. King, T.: Effective algorithms for the satisfiability of quantifier-free formulas over linear real and integer arithmetic (2014)
39. Lammich, P.: Efficient verified (UN)SAT certificate checking. In: de Moura, L. (ed.) CADE 2017. LNCS (LNAI), vol. 10395, pp. 237–254. Springer, Cham (2017). https://doi.org/10.1007/978-3-319-63046-5_15
40. Liang, T., Reynolds, A., Tinelli, C., Barrett, C., Deters, M.: A DPLL(T) theory solver for a theory of strings and regular expressions. In: Biere, A., Bloem, R. (eds.) CAV 2014. LNCS, vol. 8559, pp. 646–662. Springer, Cham (2014). https://doi.org/10.1007/978-3-319-08867-9_43
41. Mahboubi, A.: Implementing the cylindrical algebraic decomposition within the coq system. Math. Struct. Comput. Sci. **17**(1), 99–127 (2007)
42. Nieuwenhuis, R., Oliveras, A.: Proof-producing congruence closure. In: Giesl, J. (ed.) RTA 2005. LNCS, vol. 3467, pp. 453–468. Springer, Heidelberg (2005). https://doi.org/10.1007/978-3-540-32033-3_33
43. Nieuwenhuis, R., Oliveras, A., Tinelli, C.: Solving SAT and SAT modulo theories: from an abstract Davis-Putnam-Logemann-Loveland procedure to DPLL(T). J. ACM **53**(6), 937–977 (2006)
44. Nipkow, T., Wenzel, M., Paulson, L.C. (eds.): Isabelle/HOL. LNCS, vol. 2283. Springer, Heidelberg (2002). https://doi.org/10.1007/3-540-45949-9

45. Ozdemir, A., Niemetz, A., Preiner, M., Zohar, Y., Barrett, C.: DRAT-based bit-vector proofs in CVC4. In: Janota, M., Lynce, I. (eds.) SAT 2019. LNCS, vol. 11628, pp. 298–305. Springer, Cham (2019). https://doi.org/10.1007/978-3-030-24258-9_21
46. Reynolds, A., Nötzli, A., Barrett, C., Tinelli, C.: High-level abstractions for simplifying extended string constraints in SMT. In: Dillig, I., Tasiran, S. (eds.) CAV 2019. LNCS, vol. 11562, pp. 23–42. Springer, Cham (2019). https://doi.org/10.1007/978-3-030-25543-5_2
47. Reynolds, A., Nötzli, A., Barrett, C.W., Tinelli, C.: Reductions for strings and regular expressions revisited. In: Formal Methods in Computer-Aided Design (FMCAD), pp. 225–235. IEEE (2020)
48. Reynolds, A., Woo, M., Barrett, C., Brumley, D., Liang, T., Tinelli, C.: Scaling up DPLL(T) string solvers using context-dependent simplification. In: Majumdar, R., Kunčak, V. (eds.) CAV 2017. LNCS, vol. 10427, pp. 453–474. Springer, Cham (2017). https://doi.org/10.1007/978-3-319-63390-9_24
49. Schrijver, A.: Theory of Linear and Integer Programming. Wiley, Hoboken (1998)
50. Schurr, H.-J., Fleury, M., Barbosa, H., Fontaine, P.: Alethe: towards a generic SMT proof format (extended abstract). CoRR, abs/2107.02354 (2021)
51. Schurr, H.-J., Fleury, M., Desharnais, M.: Reliable reconstruction of fine-grained proofs in a proof assistant. In: Platzer, A., Sutcliffe, G. (eds.) CADE 2021. LNCS (LNAI), vol. 12699, pp. 450–467. Springer, Cham (2021). https://doi.org/10.1007/978-3-030-79876-5_26
52. Stump, A., Oe, D., Reynolds, A., Hadarean, L., Tinelli, C.: SMT proof checking using a logical framework. Formal Methods Syst. Des. **42**(1), 91–118 (2013)
53. Thiemann, R., Yamada, A.: Algebraic numbers in Isabelle/HOL. In: Blanchette, J.C., Merz, S. (eds.) ITP 2016. LNCS, vol. 9807, pp. 391–408. Springer, Cham (2016). https://doi.org/10.1007/978-3-319-43144-4_24

CTL* Model Checking for Data-Aware Dynamic Systems with Arithmetic

Paolo Felli, Marco Montali, and Sarah Winkler(✉)

Free University of Bolzano-Bozen, Bolzano, Italy
{pfelli,montali,winkler}@inf.unibz.it

Abstract. The analysis of complex dynamic systems is a core research topic in formal methods and AI, and combined modelling of systems with data has gained increasing importance in applications such as business process management. In addition, process mining techniques are nowadays used to automatically mine process models from event data, often without correctness guarantees. Thus verification techniques for linear and branching time properties are needed to ensure desired behavior.

Here we consider data-aware dynamic systems with arithmetic (DDSAs), which constitute a concise but expressive formalism of transition systems with linear arithmetic guards. We present a CTL* model checking procedure for DDSAs that addresses a generalization of the classical verification problem, namely to compute conditions on the initial state, called *witness maps*, under which the desired property holds. Linear-time verification was shown to be decidable for specific classes of DDSAs where the constraint language or the control flow are suitably confined. We investigate several of these restrictions for the case of CTL*, with both positive and negative results: witness maps can always be found for monotonicity and integer periodicity constraint systems, but verification of bounded lookback systems is undecidable. To demonstrate the feasibility of our approach, we implemented it in an SMT-based prototype, showing that many practical business process models can be effectively analyzed.

Keywords: Verification · CTL* · Counter systems · Constraints · SMT

1 Introduction

The study of complex dynamic systems is a core research topic in AI, with a long tradition in formal methods. It finds application in a variety of domains, such as notably business process management (BPM), where studying the interplay between control-flow and data has gained momentum [9,10,24,46]. Processes are increasingly mined by automatic techniques [1,3] that lack any correctness guarantees, making verification even more important to ensure the desired behavior.

This work is partially supported by the UNIBZ projects DaCoMan, QUEST, SMART-APP, VERBA, and WineId.

J. Blanchette et al. (Eds.): IJCAR 2022, LNAI 13385, pp. 36–56, 2022.
https://doi.org/10.1007/978-3-031-10769-6_4

However, the presence of data pushes verification to the verge of undecidability due to an infinite state space. This is aggravated by the use of arithmetic, in spite of its importance for practical applications [24]. Indeed, model checking of transition systems operating on numeric data variables with arithmetic constraints is known to be undecidable, as it is easy to model a two-counter machine.

In this work, we focus on the concise but expressive framework of data-aware dynamic systems with arithmetic (DDSAs) [28,38], also known as counter systems [13,20,34]. Several classes of DDSAs have been isolated where specific verification tasks are decidable, notably reachability [6,13,29,34] and linear-time model checking [14,20,22,28,38]. Fewer results are known about the case of branching time, except for flat counter systems [21], gap-order systems where constraints are restricted to the form $x - y \geq 2$ [8,42], and systems with a *nice symbolic valuation abstraction* [31]. However, many processes in BPM and beyond fall into neither of these classes, as illustrated by the next example.

Example 1. The following DDSA $\mathcal{B}$ models a management process for road fines by the Italian police [41]. It maintains seven so-called *case data* variables (i.e., variables local to each process instance, called "case" in the BPM literature): a (amount), t (total amount), d (dismissal code), p (points deducted), e (expenses), and time durations ds, dp, dj. The process starts by creating a case, upon which the offender is notified within 90 days, i.e., 2160h (send fine). If the offender pays a sufficient amount t, the process terminates via silent actions τ_1, τ_2, or τ_3. For the less happy paths, the credit collection action is triggered if the payment was insufficient; while appeal to judge and appeal to prefecture reflect filed protests by the offender, which again need to respect certain time constraints.

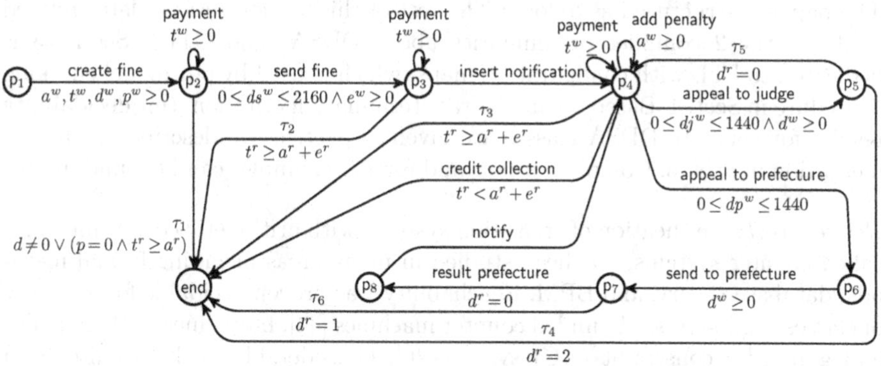

This model was generated from real-life logs by automatic process mining techniques paired with domain knowledge [41], but without any correctness guarantee. For instance, *data-aware soundness* [4,25] requires that the process can always reach a final state from any reachable configuration, expressed by the branching-time property $\mathsf{A\,G\,E\,F}\,\mathsf{end}$. This property is false here, as $\mathcal{B}$ can get stuck in state p_7 if $d > 1$. In addition, process-specific linear-time properties are needed, e.g., that a send fine event is always followed by a sufficient payment (i.e., $\langle\mathsf{send\ fine}\rangle\top \rightarrow \mathsf{F}\,\langle\mathsf{payment}\rangle(t \geq a)$, where $\langle\alpha\rangle$ is the next operator via action α).

This example highlights how both linear-time and branching-time verification are needed. In this paper, we present a CTL* model checking algorithm for DDSAs, adopting a finite-trace semantics (CTL_f^*) [44] to reflect the nature of processes as in Example 1. More precisely, our approach can synthesize conditions on the initial variable assignment such that a given property χ holds, called *witness maps*. If such a witness map can be found, it is in particular decidable what is more commonly called the *verification problem*, namely whether χ is satisfied in a designated initial configuration. We derive an abstract criterion on the computability of witness maps, which is satisfied by two practical DDSA classes that restrict the constraint language to (a) monotonicity constraints [20,25], i.e., variable-to-variable or variable-to-constant comparisons over $\mathbb{Q}$ or $\mathbb{R}$, and (b) integer periodicity constraints [18,22], i.e., variable-to-constant and restricted variable-to-variable comparisons with modulo operators. On the other hand, we show that the verification problem is undecidable for *bounded lookback* systems [28], a control flow restriction that generalizes *feedback freedom* [14].

In summary, we make the following contributions:

1. We present a model checking algorithm to generate a witness map for a given DDSA and CTL_f^* property;
2. We prove an abstract termination criterion for this algorithm (Corollary 1);
3. This result is used to show that witness maps can be effectively computed for monotonicity constraint and integer periodicity constraint systems;
4. CTL_f^* verification is shown undecidable for bounded-lookback systems;
5. We implemented our approach in the prototype ada using SMT solvers as backends and tested it on a range of business processes from the literature.

The paper is structured as follows: The rest of this section recapitulates related work. Section 2 compiles preliminaries about DDSAs and CTL_f^*. Section 3 is dedicated to LTL with configuration maps, which is used by our model checking procedure in Sect. 4. Based on an abstract termination criterion, (un)decidability results for concrete DDSA classes are given in Sect. 5. We describe our implementation in Sect. 6. Complete proofs and further examples can be found in [27].

Related work. Verification of transition systems with arithmetic constraints, also called counter systems, has been studied in many areas including formal methods, database theory, and BPM. Reachability was proven decidable for a variety of classes, e.g., reversal-bounded counter machines [34], finite linear [29], flat [13], and gap-order constraint (GC) systems [6]. Considerable work has also been dedicated to linear-time verification: LTL model checking is decidable for monotonicity constraint (MC) systems [20]. LTL verification is also decidable for integer periodicity constraint (IPC) systems, even with past-time operators [18,22]; and feedback-free systems, for an enriched constraint language referring to a read-only database [14]. DDSAs with MCs are also considered in [25] from the perspective of LTL with a finite-run semantics (LTL_f), giving a procedure to compute finite, faithful abstractions. LTL_f is moreover decidable for systems with the abstract *finite summary* property [28], which includes MC, GC, and systems with bounded lookback, where the latter generalizes feedback freedom.

Branching-time verification was less studied: Decidability of CTL* was proven for flat counter systems with Presburger-definable loop iteration [21], even in NP [19]. Moreover, it was shown that CTL* verification is decidable for pushdown systems, which can model counter systems with a single integer variable [30]. For integer relational automata (IRA), i.e., systems with constraints $x \geq y$ or $x > y$ and domain $\mathbb{Z}$, CTL model checking is undecidable while the existential and universal fragments of CTL* remain decidable [12]. For GC systems, which extend IRAs to constraints of the form $x - y \geq k$, the existential fragment of CTL* is decidable while the universal one is not [8]. A similar dichotomy holds for the EF and EG fragments of CTL [42]. A subclass of IRAs was considered in [7,11], allowing only periodicity and monotonicity constraints. While satisfiability of CTL* was proven decidable, model checking is not (as already shown in [12]), though it is decidable for CEF$^+$ properties, an extension of the EF fragment [7]. In contrast, rather than restricting temporal operators, we show decidability of model checking under an abstract property of the DDSA and the verified property, which can be guaranteed by suitably constraining the constraint class or the control flow. More closely related is work by Gascon [31], who shows decidability of CTL* model checking for DDSAs that admit a *nice symbolic valuation abstraction*, an abstract property which includes MC and IPC systems. The relationship between our decidability criterion and the property defined by Gascon will need further investigation. Another difference is that we here adopt a finite-path semantics for CTL* as e.g. considered in [47], since for the analysis of real-world processes such as business processes it is sufficient to consider finite traces. On a high level, our method follows a common approach to CTL*: the verification property is processed bottom-up, computing solutions for each subproperty. These are then used to solve an equivalent linear-time problem [2, p. 429]. For the latter, we partially rely on earlier work [28].

2 Background

We start by defining the set of constraints over expressions of sort *int*, *rat*, or *real*, with associated domains $dom(int) = \mathbb{Z}$, $dom(rat) = \mathbb{Q}$, and $dom(real) = \mathbb{R}$.

Definition 1. *For a given set of sorted variables V, expressions e_s of sort s and atoms a are defined as follows:*

$$e_s := v_s \mid k_s \mid e_s + e_s \mid e_s - e_s \qquad a := e_s = e_s \mid e_s < e_s \mid e_s \leq e_s \mid e_{int} \equiv_n e_{int}$$

where $k_s \in dom(s)$, $v_s \in V$ has sort s, and $\equiv_n$ denotes equality modulo some $n \in \mathbb{N}$. A constraint *is then a quantifier-free boolean expression over atoms a.*

The set of all constraints built from atoms over variables V is denoted by $\mathcal{C}(V)$. For instance, $x \neq 1$, $x < y - z$, and $x - y = 2 \wedge y \neq 1$ are valid constraints independent of the sort of $\{x, y, z\}$, while $u \equiv_3 v + 1$ is a constraint for integer variables u and v. We write $\mathcal{V}ar(\varphi)$ for the set of variables in a formula φ. For an assignment α with domain V that maps variables to values in their domain, and a formula φ we write $\alpha \models \varphi$ if α satisfies φ.

We are thus in the realm of SMT with linear arithmetic, which is decidable and admits *quantifier elimination* [45]: if φ is a formula in $\mathcal{C}(X \cup \{y\})$, thus having free variables $X \cup \{y\}$, there is a quantifier-free φ' with free variables X that is equivalent to $\exists y.\varphi$, i.e., $\varphi' \equiv \exists y.\varphi$, where $\equiv$ denotes logical equivalence.

2.1 Data-Aware Dynamic Systems with Arithmetic

From now on, V is a fixed, finite set of variables. We consider two disjoint, marked copies of V, denoted $V^r = \{v^r \mid v \in V\}$ and $V^w = \{v^w \mid v \in V\}$, called the *read* and *write* variables. They will refer to variable values before and after a transition, respectively. We also write $\overline{V}$ for a vector that orders V in an arbitrary but fixed way, and $\overline{V}^r$ and $\overline{V}^w$ for vectors ordering V^r and V^w in the same way.

Definition 2. *A DDSA $\mathcal{B} = \langle B, b_I, \mathcal{A}, T, B_F, V, \alpha_I, guard\rangle$ is a labeled transition system where (i) B is a finite set of* control states, *with $b_I \in B$ the initial one; (ii) $\mathcal{A}$ is a set of* actions; *(iii) $T \subseteq B \times \mathcal{A} \times B$ is a* transition relation; *(iv) $B_F \subseteq B$ are* final states; *(v) V is the set of* process variables; *(vi) α_I the* initial variable assignment; *(vii) $guard\colon \mathcal{A} \mapsto \mathcal{C}(V^r \cup V^w)$ specifies* executability constraints *for actions over variables $V^r \cup V^w$.*

Example 2. We consider the following DDSAs $\mathcal{B}$, $\mathcal{B}_{bl}$, and $\mathcal{B}_{ipc}$, where x, y have domain $\mathbb{Q}$ and u, v, s have domain $\mathbb{Z}$. Initial and final states have incoming arrows and double borders, respectively; α_I is not fixed for now.

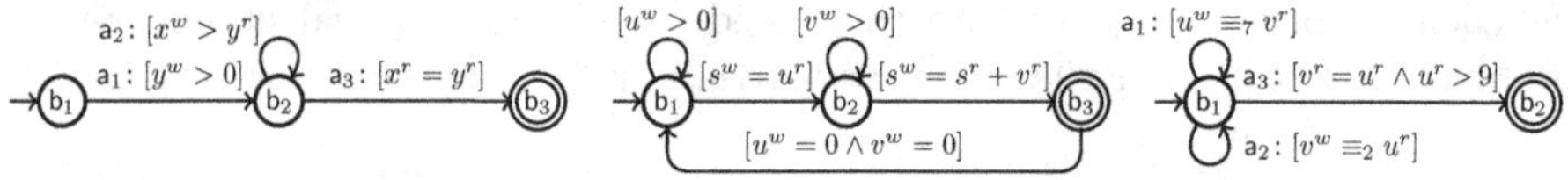

Also the system in Example 1 represents a DDSA. If state b admits a transition to b' via action a, namely $(b, a, b') \in \Delta$, this is denoted by $b \xrightarrow{a} b'$. A *configuration* of $\mathcal{B}$ is a pair (b, α) where $b \in B$ and α is an assignment with domain V. A *guard assignment* is an assignment β with domain $V^r \cup V^w$. For an action a, let $write(a) = \mathcal{V}ar(guard(a)) \cap V^w$. As defined next, an action a transforms a configuration (b, α) into a new configuration (b', α') by updating the assignment α according to the action guard, which can at the same time evaluate conditions on the current values of variables and write new values:

Definition 3. *A DDSA $\mathcal{B} = \langle B, b_I, \mathcal{A}, T, B_F, V, \alpha_I, guard\rangle$ admits a* step *from configuration (b, α) to (b', α') via action a, denoted $(b, \alpha) \xrightarrow{a} (b', \alpha')$, if $b \xrightarrow{a} b'$, $\alpha'(v) = \alpha(v)$ for all $v \in V \setminus write(a)$, and the guard assignment β given by $\beta(v^r) = \alpha(v)$ and $\beta(v^w) = \alpha'(v)$ for all $v \in V$, satisfies $\beta \models guard(a)$.*

For instance, for $\mathcal{B}$ in Example 2 and initial assignment $\alpha_I(x) = \alpha_I(y) = 0$, the initial configuration admits a step $(\mathsf{b}_1, \left[\begin{smallmatrix} x=0 \\ y=0 \end{smallmatrix}\right]) \xrightarrow{\mathsf{a}_1} (\mathsf{b}_2, \left[\begin{smallmatrix} x=0 \\ y=3 \end{smallmatrix}\right])$ with $\beta(x^r) = \beta(x^w) = \beta(y^r) = 0$ and $\beta(y^w) = 3$.

A *run* ρ of a DDSA $\mathcal{B}$ of length n from configuration (b, α) is a sequence of steps $\rho\colon (b, \alpha) = (b_0, \alpha_0) \xrightarrow{a_1} (b_1, \alpha_1) \xrightarrow{a_2} \dots \xrightarrow{a_n} (b_n, \alpha_n)$. We also associate with ρ the *symbolic run* $\sigma\colon b_0 \xrightarrow{a_1} b_1 \xrightarrow{a_2} \dots \xrightarrow{a_n} b_n$ where state and action sequences are recorded without assignments, and say that σ is the *abstraction* of ρ (or, σ *abstracts* ρ). For some $m < n$, $\sigma|_m$ denotes the prefix of σ that has m steps.

2.2 History Constraints

In this section, we fix a DDSA $\mathcal{B} = \langle B, b_I, \mathcal{A}, T, B_F, V, \alpha_I, \mathit{guard}\rangle$. We aim to build an abstraction of $\mathcal{B}$ that covers the (potentially infinite) set of configurations by finitely many nodes of the form (b, φ), where $b \in B$ is a control state and φ a formula that expresses conditions on the variables V. A state (b, φ) thus represents all configurations (b, α) s.t. $\alpha \models \varphi$. To express how such a formula φ is modified by executing an action, let the *transition formula* of action a be $\Delta_a(\overline{V}^r, \overline{V}^w) = \mathit{guard}(a) \wedge \bigwedge_{v \in V \setminus \mathit{write}(a)} v^w = v^r$. This states conditions on variables before and after executing a: $\mathit{guard}(a)$ must hold and the values of all variables that are not written are propagated by inertia. We write $\Delta_a(\overline{X}, \overline{Y})$ for the formula obtained from Δ_a by replacing $\overline{V}^r$ by $\overline{X}$ and $\overline{V}^w$ by $\overline{Y}$. Let a variable vector $\overline{U}$ be a *fresh copy* of $\overline{V}$ if it has the same length as $|\overline{V}|$ and $U \cap V = \emptyset$. To mimic steps on the abstract level, we define the following *update* function:

Definition 4. *For a formula φ with free variables V and action a, $\mathit{update}(\varphi, a) = \exists \overline{U}. \varphi(\overline{U}) \wedge \Delta_a(\overline{U}, \overline{V})$, where $\overline{U}$ is a fresh copy of $\overline{V}$.*

Our approach will generate formulas of a special shape called *history constraints* [28], obtained by iterated *update* operations in combination with a sequence of *verification constraints* $\overline{\vartheta}$. Intuitively, the latter depends on the verification property. For now it suffices to consider $\overline{\vartheta}$ an arbitrary sequence of constraints with free variables V. Its prefix of length k is denoted by $\overline{\vartheta}|_k$. We need a fixed set of placeholder variables V_0 disjoint from V, and assume an injective variable renaming $\nu\colon V \mapsto V_0$. Let φ_ν be the formula $\varphi_\nu = \bigwedge_{v \in V} v = \nu(v)$.

Definition 5. *For a symbolic run $\sigma\colon b_0 \xrightarrow{a_1} b_1 \xrightarrow{a_2} \dots \xrightarrow{a_n} b_n$, and verification constraint sequence $\overline{\vartheta} = \langle \vartheta_0, \dots, \vartheta_n\rangle$, the* history constraint *$h(\sigma, \overline{\vartheta})$ is given by $h(\sigma, \overline{\vartheta}) = \varphi_\nu \wedge \vartheta_0$ if $n = 0$, and $h(\sigma, \overline{\vartheta}) = \mathit{update}(h(\sigma|_{n-1}, \overline{\vartheta}|_{n-1}), a_n) \wedge \vartheta_n$ if $n > 0$.*

Thus, history constraints are formulas with free variables $V \cup V_0$. Satisfying assignments for history constraints are closely related to assignments in runs:[1]

Lemma 1. *For a symbolic run $\sigma\colon b_0 \xrightarrow{a_1} b_1 \xrightarrow{a_2} \dots \xrightarrow{a_n} b_n$ and $\overline{\vartheta} = \langle \vartheta_0, \dots, \vartheta_n\rangle$, $h(\sigma, \overline{\vartheta})$ is satisfied by assignment α with domain $V \cup V_0$ iff σ abstracts a run $\rho\colon (b_0, \alpha_0) \xrightarrow{a_1} \dots \xrightarrow{a_n} (b_n, \alpha_n)$ such that (i) $\alpha_0(v) = \alpha(\nu(v))$, and (ii) $\alpha_n(v) = \alpha(v)$ for all $v \in V$, and (iii) $\alpha_i \models \vartheta_i$ for all i, $0 \leq i \leq n$.*

[1] Lemma 1 is a slight variation of [28, Lemma 3.5]: Definition 5 differs from history constraints in [28] in that the initial assignment is not fixed. A proof can be found in [27].

2.3 CTL$^*_f$

For a DDSA $\mathcal{B}$ as above, we consider the following verification properties:

Definition 6. *CTL$^*_f$ state formulas χ and path formulas ψ are defined by the following grammar, for constraints $c \in \mathcal{C}(V)$ and control states $b \in B$:*

$$\chi := \top \mid c \mid b \mid \chi \wedge \chi \mid \neg\chi \mid \mathsf{E}\,\psi \qquad \psi := \chi \mid \psi \wedge \psi \mid \neg\psi \mid \mathsf{X}\,\psi \mid \mathsf{G}\,\psi \mid \psi\,\mathsf{U}\,\psi$$

We use the usual abbreviations $\mathsf{F}\,\psi = \top\;\mathsf{U}\;\psi$, $\chi_1 \vee \chi_2 = \neg(\neg\chi_1 \wedge \neg\chi_2)$, and $\mathsf{A}\,\psi = \neg\mathsf{E}\,\neg\psi$. To simplify the presentation, we do not explicitly treat next state operators $\langle a\rangle$ via a specific action a, as used in Example 1, though this would be possible (cf. [28]). However, such an operator can be encoded by adding a fresh data variable x to V, the conjunct $x^w = 1$ to $guard(a)$, and $x^w = 0$ to all other guards, and replacing $\langle a\rangle\psi$ in the verification property by $\mathsf{X}\,(\psi \wedge x = 1)$.

The maximal number of nested path quantifiers in a formula ψ is called the *quantifier depth* of ψ, denoted by $qd(\psi)$. We adopt a finite path semantics for CTL* [44]: For a control state $b \in B$ and a state assignment α, let $FRuns(b, \alpha)$ be the set of *final runs* ρ: $(b, \alpha) = (b_0, \alpha_0) \xrightarrow{a_1} \dots \xrightarrow{a_n} (b_n, \alpha_n)$ such that $b_n \in F$ is a final state. The i-th configuration (b_i, α_i) in ρ is denoted by ρ_i.

Definition 7. *The semantics of CTL$^*_f$ is inductively defined as follows. For a DDSA $\mathcal{B}$ with configuration (b, α), state formulas χ, χ', and path formulas ψ, ψ':*

$$\begin{array}{ll}
(b,\alpha) \models \top & \\
(b,\alpha) \models c & \textit{iff } \alpha \models c \\
(b,\alpha) \models b' & \textit{iff } b = b' \\
(b,\alpha) \models \chi \wedge \chi' & \textit{iff } (b,\alpha) \models \chi \textit{ and } (b,\alpha) \models \chi' \\
(b,\alpha) \models \neg\chi & \textit{iff } (b,\alpha) \not\models \chi \\
(b,\alpha) \models \mathsf{E}\,\psi & \textit{iff } \exists\rho \in FRuns(b,\alpha) \textit{ such that } \rho \models \psi
\end{array}$$

where $\rho \models \psi$ iff $\rho, 0 \models \psi$ holds, and for a run ρ of length n and all i, $0 \le i \le n$:

$$\begin{array}{ll}
\rho, i \models \chi & \textit{iff } \rho_i \models \chi \\
\rho, i \models \neg\psi & \textit{iff } \rho,\, i \not\models \psi \\
\rho, i \models \psi \wedge \psi' & \textit{iff } \rho, i \models \psi \textit{ and } \rho, i \models \psi' \\
\rho, i \models \mathsf{X}\,\psi & \textit{iff } i < n \textit{ and } \rho, i+1 \models \psi \\
\rho, i \models \mathsf{G}\,\psi & \textit{iff for all } j,\ i \le j \le n, \textit{ it holds that } \rho, j \models \psi \\
\rho, i \models \psi\,\mathsf{U}\,\psi' & \textit{iff } \exists k \textit{ with } i + k \le n \textit{ such that } \rho, i+k \models \psi' \\
& \textit{and for all } j, 0 \le j < k, \textit{ it holds that } \rho, i+j \models \psi.
\end{array}$$

Instead of simply checking whether the initial configuration of a DDSA $\mathcal{B}$ satisfies a CTL$^*_f$ property χ, we try to determine, for every state $b \in B$, which constraints on variables need to hold in order to satisfy χ. As the number of configurations (b, α) of a DDSA $\mathcal{B}$ is usually infinite, configuration sets cannot be enumerated explicitly. Instead, we represent a set of configurations as a *configuration map* $K\colon B \mapsto \mathcal{C}(V)$ that associates with every control state $b \in B$ a formula $K(b) \in \mathcal{C}(V)$, representing all configurations (b, α) such that $\alpha \models K(b)$.

We now define when a configuration captures the maximal set of configurations in which a formula χ holds. We call these witness maps.

Definition 8. *For a DDSA $\mathcal{B}$ and state formula χ, a configuration map K is a* witness map *if it holds that $(b, \alpha) \models \chi$ iff $\alpha \models K(b)$, for all $b \in B$ and all α.*

For instance, for $\mathcal{B}$ from Example 2 and $\chi_1 = \mathsf{A}\,\mathsf{G}\,(x \geq 2)$, a witness map is given by $K = \{\mathsf{b}_1 \mapsto \bot,\ \mathsf{b}_2 \mapsto x \geq 2 \wedge y \geq 2,\ \mathsf{b}_3 \mapsto x \geq 2\}$. For $\chi_2 \models \mathsf{E}\,\mathsf{X}\,(\mathsf{A}\,\mathsf{G}\,(x \geq 2))$, a solution is $K' = \{\mathsf{b}_1 \mapsto x \geq 2,\ \mathsf{b}_2 \mapsto y \geq 2,\ \mathsf{b}_3 \mapsto \bot\}$. As b_1 is the initial state, $\mathcal{B}$ satisfies χ_2 with every initial assignment that sets $\alpha_I(x) \geq 2$.

In this paper we address the problem of finding a witness map for $\mathcal{B}$ and χ. Note that a witness map in particular allows to decide what is commonly called the *verification problem*, namely to check whether $(b_I, \alpha_I) \models \chi$ holds, by testing $\alpha_I \models K(b_I)$. It remains to investigate whether there exist a DDSA $\mathcal{B}$ and χ for which no witness map exists, as the configuration set satisfying χ is not finitely representable. Even if it exists, finding it is in general undecidable. However, in this paper we identify DDSA classes where a witness map can always be found.

3 LTL with Configuration Maps

Following a common approach to CTL* verification, our technique processes the property χ bottom-up, computing solutions for each subformula $\mathsf{E}\,\psi$, before solving a linear-time model checking problem χ' in which the solutions to subformulas appear as atoms. Given our representation of sets of configurations, we use LTL formulas where atoms are configuration maps, and denote this specification language by $\mathrm{LTL}_f^{\mathcal{B}}$. For a given DDSA $\mathcal{B}$, it is formally defined as follows:

$$\psi \ := \ K \mid \psi \wedge \psi \mid \neg\psi \mid \mathsf{X}\,\psi \mid \mathsf{G}\,\psi \mid \psi \,\mathsf{U}\, \psi$$

where $K \in \mathcal{K}_{\mathcal{B}}$, for $\mathcal{K}_{\mathcal{B}}$ is the set of configuration maps for $\mathcal{B}$.

Definition 9. *A run ρ of length n* satisfies *an* $\mathrm{LTL}_f^{\mathcal{B}}$ *formula ψ, denoted $\rho \models_{\mathcal{K}} \psi$, iff $\rho, 0 \models_{\mathcal{K}} \psi$ holds, where for all i, $0 \leq i \leq n$:*

$\rho, i \models_{\mathcal{K}} K$ iff $\rho_i = (b, \alpha)$ and $\alpha \models K(b)$;
$\rho, i \models_{\mathcal{K}} \psi \wedge \psi'$ iff $\rho, i \models_{\mathcal{K}} \psi$ and $\rho, i \models_{\mathcal{K}} \psi'$;
$\rho, i \models_{\mathcal{K}} \neg\psi$ iff $\rho, i \not\models_{\mathcal{K}} \psi$;
$\rho, i \models_{\mathcal{K}} \mathsf{X}\,\psi$ iff $i < n$ and $\rho, i{+}1 \models_{\mathcal{K}} \psi$;
$\rho, i \models_{\mathcal{K}} \mathsf{G}\,\psi$ iff $\rho, i \models_{\mathcal{K}} \psi$ and ($i = n$ or $\rho, i{+}1 \models_{\mathcal{K}} \mathsf{G}\,\psi$);
$\rho, i \models_{\mathcal{K}} \psi \,\mathsf{U}\, \psi'$ iff $\rho, i \models_{\mathcal{K}} \psi'$ or ($i < n$ and $\rho, i \models_{\mathcal{K}} \psi$ and $\rho, i{+}1 \models_{\mathcal{K}} \psi \,\mathsf{U}\, \psi'$).

Our approach to $\mathrm{LTL}_f^{\mathcal{B}}$ verification proceeds along the lines of the LTL_f procedure from [28], with the difference that simple constraint atoms are replaced by configuration maps. In order to express the requirements on a run of a DDSA $\mathcal{B}$ to satisfy an $\mathrm{LTL}_f^{\mathcal{B}}$ formula χ, we use a nondeterministic automaton (NFA) $\mathcal{N}_\psi = (Q, \Sigma, \varrho, q_0, Q_F)$, where the states Q are a set of subformulas of ψ, $\Sigma = 2^{\mathcal{K}_{\mathcal{B}}}$ is the alphabet, ϱ is the transition relation, $q_0 \in Q$ is the initial state, and $Q_F \subseteq Q$ is the set of final states. The construction of $\mathcal{N}_\psi$ is standard [15,28], treating configuration maps for the time being as propositions; but for completeness it is described in [27, Appendix C]. For instance, for a configuration map K, $\psi = \mathsf{F}\,K$

corresponds to the NFA 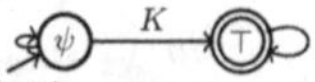 and $\psi' = \mathsf{X} K$ to 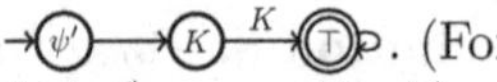. (For simplicity, edges labels $\{K\}$ are shown as K, and edge labels $\emptyset$ are omitted.)

For $w_i \in \Sigma$, i.e., w_i is a set of configuration maps, $w_i(b)$ denotes the formula $\bigwedge_{K \in w} K(b)$. Moreover, for $w = w_0, \ldots, w_n \in \Sigma^*$ and a symbolic run $\sigma\colon b_0 \xrightarrow{a_1} b_1 \xrightarrow{a_2} \ldots \xrightarrow{a_n} b_n$, let $w \otimes \sigma$ denote the sequence of formulas $\langle w_0(b_0), \ldots, w_n(b_n)\rangle$, i.e., the component-wise application of w to the control states of σ. A word $w_0, \ldots, w_n \in \Sigma^*$ is *consistent* with a run $(b_0, \alpha_0) \xrightarrow{a_1} (b_1, \alpha_1) \xrightarrow{a_2} \ldots \xrightarrow{a_n} (b_n, \alpha_n)$ if $\alpha_i \models w_i(b_i)$ for all i, $0 \leq i \leq n$. The key correctness property of $\mathcal{N}_\psi$ is the following (cf. [28, Lemma 4.4], and see [27] for the proof adapted to $\mathrm{LTL}_f^{\mathcal{B}}$):

Lemma 2. *$\mathcal{N}_\psi$ accepts a word that is consistent with a run ρ iff $\rho \models_{\mathcal{K}} \psi$.*

Product Construction. As a next step in our verification procedure, given a control state b of $\mathcal{B}$, we aim to find (a symbolic representation of) all configurations (b, α) that satisfy an $\mathrm{LTL}_f^{\mathcal{B}}$ formula ψ. To that end, we combine $\mathcal{N}_\psi$ with $\mathcal{B}$ to a cross-product automaton $\mathcal{N}_{\mathcal{B},b}^{\psi}$. For technical reasons, when performing the product construction, the steps in $\mathcal{B}$ need to be shifted by one with respect to the steps in $\mathcal{N}_\psi$. Hence, given $b \in B$, let $\mathcal{B}_b$ be the DDSA obtained from $\mathcal{B}$ by adding a dummy initial state $\underline{b}$, so that $\mathcal{B}_b$ has state set $B' = B \cup \{\underline{b}\}$ and transition relation $T' = T \cup \{(\underline{b}, a_0, b)\}$ for a fresh action a_0 with $guard(a_0) = \top$.

Definition 10. *The* product automaton *$\mathcal{N}_{\mathcal{B},b}^{\psi}$ is defined for an $\mathrm{LTL}_f^{\mathcal{B}}$ formula ψ, a DDSA $\mathcal{B}$, and a control state $b \in B$. Let $\mathcal{B}_b = \langle B', \underline{b}, \mathcal{A}, T', B_F, V, \alpha_I, guard\rangle$ and $\mathcal{N}_\psi$ as above. Then $\mathcal{N}_{\mathcal{B},b}^{\psi} = (P, R, p_0, P_F)$ is as follows:*

- *$P \subseteq B' \times Q \times \mathcal{C}(V \cup V_0)$, i.e., states in P are triples (b, q, φ) such that*
- *the initial state is $p_0 = (\underline{b}, q_0, \varphi_\nu)$;*
- *if $b \xrightarrow{a} b'$ in T', $q \xrightarrow{w} q'$ in $\mathcal{N}_\psi$, and $update(\varphi, a) \wedge w(b')$ is satisfiable, there is a transition $(b, q, \varphi) \xrightarrow{a,w} (b', q', \varphi')$ in R such that $\varphi' \equiv update(\varphi, a) \wedge w(b')$;*
- *(b', q', φ') is in the set of final states $P_F \subseteq P$ iff $b' \in B_F$, and $q' \in Q_F$.*

Example 3. Consider the DDSA $\mathcal{B}$ from Example 2, and let $K = \{\mathsf{b}_1 \mapsto \bot,\ \mathsf{b}_2 \mapsto x \geq 2 \wedge y \geq 2,\ \mathsf{b}_3 \mapsto x \geq 2\}$. The property $\psi = \mathsf{X} K$ is captured by the NFA 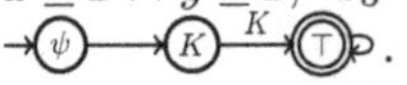. The product automata $\mathcal{N}_{\mathcal{B},\mathsf{b}_1}^{\psi}$ and $\mathcal{N}_{\mathcal{B},\mathsf{b}_2}^{\psi}$ are as follows:

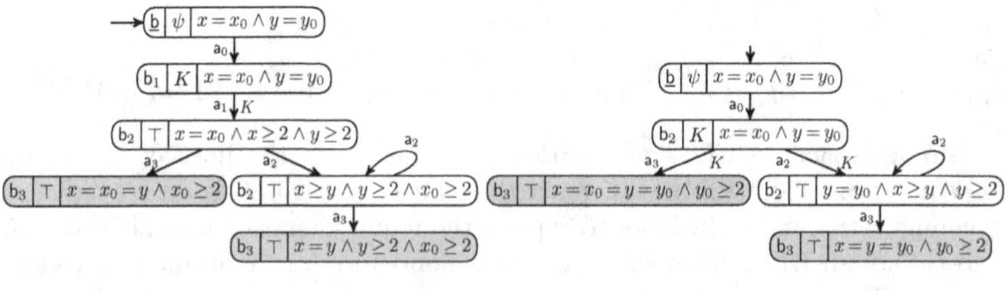

where the shaded nodes are final. The formulas in nodes were obtained by applying quantifier elimination to the formulas built using *update* according to Definition 10. $\mathcal{N}_{\mathcal{B},\mathsf{b}_3}^{\psi}$ consists only of the dummy transition and has no final states.

Definition 10 need not terminate if infinitely many non-equivalent formulas occur in the construction. In Sect. 4 we will identify a criterion that guarantees termination. First, we state the key correctness property, which lifts [28, Theorem 4.7] to LTL with configuration maps. Its proof is similar to the respective result in [28], and can be found in [27].

Theorem 1. *Let $\psi \in \mathrm{LTL}_f^{\mathcal{B}}$ and $b \in B$ such that there is a finite product automaton $\mathcal{N}_{\mathcal{B},b}^{\psi}$. Then there is a final run $\rho\colon (b, \alpha_0) \to^* (b_F, \alpha_F)$ of $\mathcal{B}$ such that $\rho \models_{\mathcal{K}} \psi$, iff $\mathcal{N}_{\mathcal{B},b}^{\psi}$ has a final state (b_F, q_F, φ) for some q_F and φ such that φ is satisfied by assignment γ with $\gamma(\overline{V_0}) = \alpha_0(\overline{V})$ and $\gamma(\overline{V}) = \alpha_F(\overline{V})$.*

Thus, witnesses for ψ correspond to paths to final states in the product automaton: e.g., in $\mathcal{N}_{\mathcal{B},\mathsf{b}_1}^{\psi}$ in Example 3 the formula in the left final node is satisfied by $\gamma(x_0) = \gamma(x) = \gamma(y) = 3$ and $\gamma(y_0) = 0$. For α_0 and α_2 such that $\alpha_0(\overline{V}) = \gamma(\overline{V}_0) = \{x \mapsto 3, y \mapsto 0\}$ and $\alpha_2(\overline{V}) = \gamma(\overline{V}) = \{x \mapsto 3, y \mapsto 3\}$ there is a witness run for ψ from (b_1, α_0) to (b_1, α_2), e.g., $(\mathsf{b}_1, \left[\begin{smallmatrix} x=3 \\ y=0 \end{smallmatrix}\right]) \xrightarrow{\mathsf{a}_1} (\mathsf{b}_2, \left[\begin{smallmatrix} x=3 \\ y=3 \end{smallmatrix}\right]) \xrightarrow{\mathsf{a}_3} (\mathsf{b}_3, \left[\begin{smallmatrix} x=3 \\ y=3 \end{smallmatrix}\right])$.

4 Model Checking Procedure

Using the results of Sect. 3, we define a model checking procedure, shown in Fig. 1. First, we explain the tasks achieved by the three mutually recursive functions:

• *checkState*(χ) returns a configuration map representing the set of configurations that satisfy a state formula χ. In the base cases, it returns a function that checks the respective condition, for boolean operators we recurse on the arguments, and for a formula $\mathsf{E}\,\psi$ we proceed to the *checkPath* procedure.

• *checkPath*(ψ) returns a configuration map K that represents all configurations from which a path satisfying ψ exists. First, *toLTL*$_{\mathcal{K}}$ is used to obtain an equivalent $\mathrm{LTL}_f^{\mathcal{B}}$ formula ψ' (which entails the computation of solutions for all subproperties $\mathsf{E}\,\eta$). Then solution K is constructed as follows: For every control state b, we build the product automaton $\mathcal{N}_{\mathcal{B},b}^{\psi'}$, and collect the set Φ_F of formulas in final states. Every $\varphi \in \Phi_F$ encodes runs from b to a final state of $\mathcal{B}$ that satisfy ψ'. The variables $\overline{V_0}$ and $\overline{V}$ in φ act as placeholders for the initial and the final values of the runs, respectively. We rename variables in φ to use $\overline{V}$ at the start and $\overline{U}$ at the end, we quantify existentially over $\overline{U}$ (as the final valuation is irrelevant), and take the disjunction over all $\varphi \in \Phi_F$. The resulting formula φ' encodes all final runs from b that satisfy ψ', so we set $K(b) := \varphi'$.

• *toLTL*$_{\mathcal{K}}(\psi)$ computes an $\mathrm{LTL}_f^{\mathcal{B}}$ formula equivalent to a path formula ψ. To this end, it performs two kinds of replacements in ψ: (a) $\top$, $b \in B$, and constraints c are represented as configuration maps; and (b) subformulas $\mathsf{E}\,\eta$ are replaced by their solutions $K_{\mathsf{E}\eta}$, which are computed by a recursive call to *checkPath*.

To represent the base cases of formulas as configuration maps in Fig. 1, we define $K_\top := (\lambda\_.\top)$, $K_b := (\lambda b'. b = b' ? \top : \bot)$ for all $b \in B$, and $K_c := (\lambda\_.c)$ for constraints c. We also write $\neg K$ for $(\lambda b.\neg K(b))$ and $K \wedge K'$ for $(\lambda b.K(b) \wedge K'(b))$. The next example illustrates the approach.

```
procedure checkState(χ)
    switch χ do
        case ⊤, b ∈ B, or c ∈ C: return K_χ
        case χ_1 ∧ χ_2:      return checkState(χ_1) ∧ checkState(χ_2)
        case ¬χ:             return ¬checkState(χ)
        case E ψ:            return checkPath(ψ)

procedure checkPath(ψ)
    ψ' := toLTL_K(ψ)
    for b ∈ B do
        (P, R, p_0, P_F) := N^{ψ'}_{B,b}        ▷ product automaton for ψ', B, and b
        Φ := {φ | (b_F, q_F, φ) ∈ P_F}          ▷ collect formulas in final states
        K(b) := ⋁_{φ∈Φ} ∃Ū.φ(V̄, Ū)
    return K

procedure toLTL_K(ψ)
    switch ψ do
        case ⊤, b ∈ B, or c ∈ C: return K_ψ
        case ψ_1 ∧ ψ_2:      return toLTL_K(ψ_1) ∧ toLTL_K(ψ_2)
        case ¬ψ:             return ¬toLTL_K(ψ)
        case E ψ:            return checkPath(ψ)
        case X ψ:            return X toLTL_K(ψ)
        case G ψ:            return G toLTL_K(ψ)
        case ψ_1 U ψ_2:      return toLTL_K(ψ_1) U toLTL_K(ψ_2)
```

Fig. 1. Model checking procedure.

Example 4. Consider $\chi = \mathsf{E}\,\mathsf{X}\,(\mathsf{A}\,\mathsf{G}\,(x \geq 2))$ and the DDSA $\mathcal{B}$ in Example 2. To get a solution K_1 to *checkState*(χ) = *checkPath*(ψ_1) for $\psi_1 = \mathsf{X}\,(\mathsf{A}\,\mathsf{G}\,(x \geq 2))$, we first compute an equivalent $\text{LTL}_f^{\mathcal{B}}$ formula $\psi_1' = \mathsf{X}\,K_2$, where K_2 is a solution to $\mathsf{A}\,\mathsf{G}\,(x \geq 2) \equiv \neg\mathsf{E}\,\mathsf{F}\,(x < 2)$. To this end, we run *checkPath*(ψ_2) for $\psi_2 = \mathsf{F}\,(x < 2)$, which is represented in $\text{LTL}_f^{\mathcal{B}}$ as $\psi_2' = \mathsf{F}\,(K_{x<2})$ with NFA 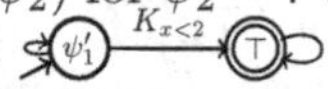. Next, *checkPath* builds $\mathcal{N}_{\mathcal{B},b}^{\psi_2'}$ for all states b. For instance, for b_2 we get:

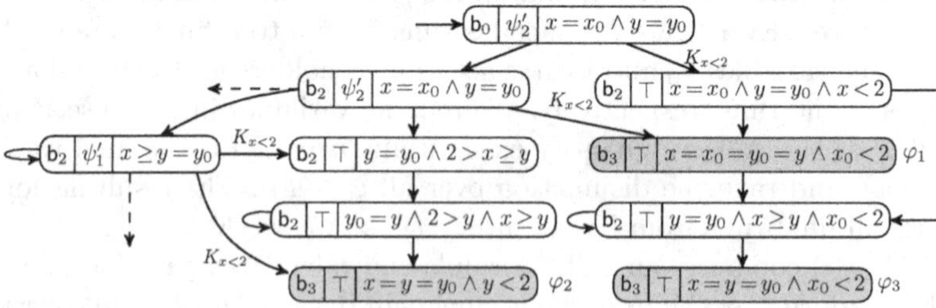

where dashed arrows indicate transitions to non-final sink states. For $\overline{U} = \langle \hat{x}, \hat{y} \rangle$, and the formulas φ_1, φ_2, and φ_3 in final nodes, we compute

$$
\begin{aligned}
\exists \overline{U}.\, \varphi_1(\overline{V}, \overline{U}) &= \exists \hat{x}\, \hat{y}.\ \hat{x} = x = \hat{y} = y \wedge x < 2 \equiv x < 2 \\
\exists \overline{U}.\, \varphi_2(\overline{V}, \overline{U}) &= \exists \hat{x}\, \hat{y}.\ \hat{x} = \hat{y} = y \wedge \hat{y} < 2 \qquad \equiv y < 2 \\
\exists \overline{U}.\, \varphi_3(\overline{V}, \overline{U}) &= \exists \hat{x}\, \hat{y}.\ \hat{x} = \hat{y} = y \wedge x < 2 \qquad \equiv x < 2
\end{aligned}
$$

so that $K_3 := checkPath(\psi_2)$ sets $K_3(\mathsf{b}_2) = \bigvee_{i=1}^{3} \exists \overline{U}.\ \varphi_i(\overline{V}, \overline{U}) \equiv x<2 \vee y<2$. For reasons of space, the constructions for b_1 and b_3 are shown in [27, Appendix B]; we obtain $K_3(\mathsf{b}_1) = \top$ and $K_3(\mathsf{b}_3) = x < 2$. By negation, the solution K_2 to $\mathsf{A}\,\mathsf{G}\,(x \geq 2)$ is $K_2 = \neg K_3 = \{\mathsf{b}_1 \mapsto \bot,\ \mathsf{b}_2 \mapsto x \geq 2 \wedge y \geq 2,\ \mathsf{b}_3 \mapsto x \geq 2\}$. Now we can proceed with $checkPath(\psi_1)$. The NFA and product automata for $\psi_1' = \mathsf{X}\,K_2$ are as shown in Example 3 and in a similar way as above we obtain the solution K_1 for $\mathsf{E}\,\mathsf{X}\,\mathsf{A}\,\mathsf{G}\,(x \geq 2)$ as $K_1 = \{\mathsf{b}_1 \mapsto x\geq 2,\ \mathsf{b}_2 \mapsto y \geq 2,\ \mathsf{b}_3 \mapsto \bot\}$. Thus, $\mathcal{B}$ satisfies the property for any initial assignment α_I with $\alpha_I(x) \geq 2$.

Next we prove correctness of $checkState(\chi)$ under the condition that it is defined, i.e., all required product automata are finite. First we state our main result, but before giving its proof we show helpful properties of $toLTL_{\mathcal{K}}$ and $checkPath$.

Theorem 2. *For every configuration (b, α) of the DDSA $\mathcal{B}$ and every state property χ, if $checkState(\chi)$ is defined then $(b, \alpha) \models \chi$ iff $\alpha \models checkState(\chi)(b)$.*

Lemma 3. *Let ψ be a path formula with $qd(\psi) = k$. Suppose that for all configurations (b, α) and path formulas ψ' with $qd(\psi') < k$, there is a $\rho' \in FRuns(b, \alpha)$ with $\rho' \models \psi'$ iff $\alpha \models checkPath(\psi')(b)$. Then $\rho \models \psi$ iff $\rho \models_{\mathcal{K}} toLTL_{\mathcal{K}}(\psi)$.*

Proof (sketch). By induction on ψ. The base cases are by the definitions of $K_\top$, K_b, and K_c. In the induction step, if $\psi = \mathsf{E}\,\psi'$ then $\rho \models \psi$ iff $\exists \rho' \in FRuns(b_0, \alpha_0)$ with $\rho' \models \psi'$, for $\rho_0 = (b_0, \alpha_0)$. As $qd(\psi') < qd(\psi)$, this holds by assumption iff $\alpha_0 \models checkPath(\psi')(b_0)$. This is equivalent to $\rho \models_{\mathcal{K}} toLTL_{\mathcal{K}}(\psi) = checkPath(\psi')$. All other cases are by the induction hypothesis and Definitions 7 and 9.

Lemma 4. *If $\psi' = toLTL_{\mathcal{K}}(\psi)$ such that for all runs ρ it is $\rho \models \psi$ iff $\rho \models_{\mathcal{K}} \psi'$, there is a run $\rho \in FRuns(b, \alpha)$ with $\rho \models \psi$ iff $\alpha \models checkPath(\psi)(b)$.*

Proof. ($\Longrightarrow$) Suppose there is a run $\rho \in FRuns(b, \alpha)$ with $\rho \models \psi$, so ρ is of the form $(b, \alpha) \to^* (b_F, \alpha_F)$ for some $b_F \in B_F$. By assumption, this implies $\rho \models_{\mathcal{K}} \psi'$, so that by Theorem 1, $\mathcal{N}_{\mathcal{B},b}^{\psi'}$ has a final state (b_F, q_F, φ) where φ is satisfied by an assignment γ with domain $V \cup V_0$ such that $\gamma(\overline{V_0}) = \alpha(\overline{V})$ and $\gamma(\overline{V}) = \alpha_F(\overline{V})$. By definition, $checkPath(\psi)(b)$ contains a disjunct $\exists \overline{U}.\ \varphi(\overline{V}, \overline{U})$. As γ satisfies φ and $\gamma(\overline{V_0}) = \alpha(\overline{V})$, $\alpha \models checkPath(\psi)(b)$. ($\Longleftarrow$) If $\alpha \models checkPath(\psi)(b)$, by definition of $checkPath$ there is a formula φ such that $\alpha \models \exists \overline{U}.\ \varphi(\overline{V}, \overline{U})$ and φ occurs in a final state (b_F, q_F, φ) of $\mathcal{N}_{\mathcal{B},b}^{\psi'}$. Hence there is an assignment γ with domain $V \cup V_0$ and $\gamma(\overline{V_0}) = \alpha(\overline{V})$ such that $\gamma \models \varphi$. By Theorem 1, there is a run $\rho \colon (b, \alpha) \to^* (b_F, \alpha_F)$ such that $\rho \models_{\mathcal{K}} \psi'$. By the assumption, we have $\rho \models \psi$. □

At this point the main theorem can be proven:

Proof (of Theorem 2). We first show ($\star$): for any path formula ψ, there is a run $\rho \in FRuns(b, \alpha)$ with $\rho \models \psi$ iff $\alpha \models checkPath(\psi)(b)$. The proof is by induction on $qd(\psi)$. If ψ contains no path quantifiers, Lemma 3 implies that $\rho \models \psi$ iff $\rho \models_{\mathcal{K}} toLTL_{\mathcal{K}}(\psi)$ for all runs ρ, so ($\star$) follows from Lemma 4. In the induction step, we conclude from Lemma 3, using the induction hypothesis of

($\star$) as assumption, that $\rho \models \psi$ iff $\rho \models_\mathcal{K} toLTL_\mathcal{K}(\psi)$ for all runs ρ. Again, ($\star$) follows from Lemma 4.

The theorem is then shown by induction on χ: The base cases $\top$, $b' \in B$, $c \in \mathcal{C}$ are easy to check, and for properties of the form $\neg\chi'$ and $\chi_1 \wedge \chi_2$ the claim follows from the induction hypothesis and the definitions. Finally, for $\chi = \mathsf{E}\,\psi$, $(b, \alpha) \models \chi$ iff there is a run $\rho \in FRuns(b, \alpha)$ such that $\rho \models \psi$. By ($\star$) this is the case iff $\alpha \models checkPath(\psi)(b) = checkState(\chi)(b)$. □

Termination. We next show that the formulas generated in our procedure all have a particular shape, to obtain an abstract termination result. For a set of formulas $\Phi \subseteq \mathcal{C}(V)$ and a symbolic run σ, let a history constraint $h(\sigma, \overline{\vartheta})$ be *over basis* Φ if $\overline{\vartheta} = \langle \vartheta_0, \dots, \vartheta_n \rangle$ and for all i, $1 \le i \le n$, there is a subset $T_i \subseteq \Phi$ s.t. $\vartheta_i = \bigwedge T_i$. Moreover, for a set of formulas Φ, let $\Phi^\pm = \Phi \cup \{\neg\varphi \mid \varphi \in \Phi\}$.

Definition 11. *For a DDSA $\mathcal{B}$, a constraint set $\mathcal{C}$ over free variables V, and $k \ge 0$, the formula sets Φ_k are inductively defined by $\Phi_0 = \mathcal{C} \cup \{\top, \bot\}$ and*

$$\Phi_{k+1} = \{\bigvee\nolimits_{\varphi \in H} \exists \overline{U}.\ \varphi(\overline{V}, \overline{U}) \mid H \subseteq \mathcal{H}_k\}$$

where $\mathcal{H}_k$ is the set of all history constraints of $\mathcal{B}$ with basis $\bigcup_{i \le k} \Phi_i^\pm$.

Note that formulas in Φ_k have free variables V, while those in $\mathcal{H}_k$ have free variables $V_0 \cup V$. We next show that these sets correspond to the formulas generated by our procedure, if all constraints in the verification property are in $\mathcal{C}$.

Lemma 5. *Let $\mathsf{E}\,\psi$ have quantifier depth k, $\psi' = toLTL_\mathcal{K}(\psi)$, and $\mathcal{N}_{\mathcal{B},b}^{\psi'}$ be a constraint graph constructed in checkPath(ψ) for some $b \in B$. Then,*

(1) for all nodes (b', q, φ) in $\mathcal{N}_{\mathcal{B},b}^{\psi'}$ there is some $\varphi' \in \mathcal{H}_k$ such that $\varphi \equiv \varphi'$,
(2) checkPath(ψ)(b) is equivalent to a formula in Φ_{k+1}.

The statements are proven by induction on k, using the results on the product construction ([27, Lemma 6]). From part (1) of this lemma and Theorem 2 we thus obtain an abstract criterion for decidability that will be useful in the next section:

Corollary 1. *For a DDSA $\mathcal{B}$ as above and a state formula χ, if $\mathcal{H}_j(b)$ is finite up to equivalence for all $j < qd(\chi)$ and $b \in B$, a witness map can always be computed.*

Proof. By the assumption about the sets $\mathcal{H}_j(b)$ for $j < qd(\chi)$, all product automata constructions in recursive calls *checkPath*(ψ) of *checkState*(χ) terminate if logical equivalence of formulas is checked eagerly. Thus *checkState*(χ) is defined, and by Theorem 2 the result is a witness map. □

The property that all sets $\mathcal{H}_j(b)$, $j < qd(\chi)$, are finite might not be decidable itself. However, in the next section we will show means to guarantee this property. Moreover, we remark that finiteness of all $\mathcal{H}_j(b)$ implies a *finite history set*, a decidability criterion identified for the linear-time case [28, Definition 3.6]; but Example 5 below illustrates that the requirement on the $\mathcal{H}_j(b)$'s is strictly stronger.

5 Decidability of DDSA Classes

We here illustrate restrictions on DDSAs, either on the control flow or on the constraint language, that render our approach a decision procedure for CTL^*_f.

Monotonicity constraints (MCs) restrict constraints (Definition 1) as follows: MCs over variables V and domain D have the form $p \odot q$ where $p, q \in D \cup V$ and $\odot$ is one of $=, \neq, \leq, <, \geq$, or $>$. The domain D may be $\mathbb{R}$ or $\mathbb{Q}$. We call a boolean formula whose atoms are MCs an *MC formula*, a DDSA where all atoms in guards are MCs an *MC-DDSA*, and a CTL^*_f property whose constraint atoms are MCs an *MC property*. For instance, $\mathcal{B}$ in Example 2 is an MC-DDSA.

We exploit a useful quantifier elimination property: If φ is an MC formula over a set of constants L and variables $V \cup \{x\}$, there is some $\varphi' \equiv \exists x.\, \varphi$ such that φ' is a quantifier-free MC formula over V and L. Such a φ' can be obtained by writing φ in disjunctive normal form and applying a Fourier-Motzkin procedure [36, Sect. 5.4] to each disjunct, which guarantees that all constants in φ' also occur in φ.

Theorem 3. *For any DDSA $\mathcal{B}$ and property χ over monotonicity constraints, a witness map is computable.*

Proof. Let χ be an MC property, and L the finite set of constants in constraints in χ, α_0, and guards of $\mathcal{B}$. Let moreover MC_L be the set of quantifier-free formulas whose atoms are MCs over $V \cup V_0$ and L, so MC_L is finite up to equivalence.

We show the following property ($\star$): all history constraints $h(\sigma, \overline{\vartheta})$ over basis MC_L are equivalent to a formula in MC_L. For a symbolic run $\sigma\colon b_0 \to^* b_{n-1} \xrightarrow{a} b_n$ and a sequence $\overline{\vartheta} = \langle \vartheta_0, \dots, \vartheta_n \rangle$ over MC_L, the proof is by induction on n. In the base case, $h(\sigma, \overline{\vartheta}) = \varphi_\nu \wedge \vartheta_0$ is in MC_L because φ_ν is a conjunction of equalities between $V \cup V_0$, and $\vartheta_0 \in \text{MC}_L$ by assumption. In the induction step, $h(\sigma, \overline{\vartheta}) = \mathit{update}(h(\sigma|_{n-1}, \overline{\vartheta}|_{n-1}), a_n) \wedge \vartheta_n$. By induction hypothesis, $h(\sigma|_{n-1}, \overline{\vartheta}|_{n-1}) \equiv \varphi$ for some φ in MC_L. Thus $h(\sigma, \overline{\vartheta}) \equiv \exists \overline{U}.\varphi(\overline{U}) \wedge \Delta_a(\overline{U}, \overline{V}) \wedge \vartheta_n$. As $\mathcal{B}$ is an MC-DDSA, $\Delta_a(\overline{U}, \overline{V})$ is a conjunction of MCs over $V \cup U$ and constants L, and $\vartheta_n \in \text{MC}_L$ by assumption. By the quantifier elimination property, there exists a quantifier-free MC-formula φ' over variables $V_0 \cup V$ that is equivalent to $\exists \overline{U}.\varphi(\overline{U}) \wedge \Delta_a(\overline{U}, \overline{V}) \wedge \vartheta_n$, and mentions only constants in L, so $\varphi' \in \text{MC}_L$.

For $\mathcal{C}$ the set of constraints in χ, we now show that $\mathcal{H}_j \subseteq \text{MC}_L$ for all $j \geq 0$, by induction on j. In the base case ($j = 0$), the claim follows from ($\star$), as all constraints in Φ_0, i.e., in χ, are in MC_L. For $j > 0$, consider first a formula $\widehat{\varphi} \in \Phi_j$ for some $b \in B$. Then $\widehat{\varphi}$ is of the form $\widehat{\varphi} = \bigvee_{\varphi \in H} \exists \overline{U}.\ \varphi(\overline{V}, \overline{U})$ for some $H \subseteq \mathcal{H}_{j-1}$. By the induction hypothesis, $H \subseteq \text{MC}_L$, so by the quantifier elimination property of MC formulas, $\widehat{\varphi}$ is equivalent to an MC-formula over V and L in MC_L. As $\mathcal{H}_j$ s built over basis Φ_j, the claim follows from ($\star$). □

Notably, the above quantifier elimination property fails for MCs over integer variables; indeed, CTL model checking is undecidable in this case [42, Theorem 4.1].

Integer periodicity constraint systems confine the constraint language to variable-to-constant comparisons and restricted forms of variable-to-variable comparisons, and are for instance used in calendar formalisms [18,22]. More precisely, *integer periodicity constraint* (IPC) atoms have the form $x = y$, $x \odot d$ for $\odot \in \{=, \neq, <, >\}$, $x \equiv_k y + d$, or $x \equiv_k d$, for variables x, y with domain $\mathbb{Z}$ and $k, d \in \mathbb{N}$. A boolean formula whose atoms are IPCs is an *IPC formula*, a DDSA whose guards are conjunctions of IPCs an *IPC-DDSA*, and a CTL_f^* formula whose constraint atoms are IPCs an *IPC property.* For instance, $\mathcal{B}_{ipc}$ in Example 2 is an IPC-DDSA.

Using Corollary 1 and a known quantifier elimination property for IPCs [18, Theorem 2], one can show that witness maps are also computable for IPC-DDSAs, in a proof that resembles the one of Theorem 3 (see [27, Theorem 4]).

Theorem 4. *For any DDSA $\mathcal{B}$ and property χ over integer periodicity constraints, a witness map is computable.*

The proofs of both Theorems 3 and 4 rely on the fact that all transition guards and constraints in the verification property are in a *finite* set of constraints C that is closed under quantifier elimination, so that for all $\varphi \in C$ and actions a, $update(\varphi, a)$ is again equivalent to a formula in C. However, this is not the only way to ensure the requirements of Corollary 1: For a simple example, these requirements are satisfied by a loop-free DDSA, where the number of runs is finite. Interestingly, while the cases of MC and IPC systems are also captured by the abstract decidability criterion by Gascon [31], this need not apply to loop-free DDSAs. A clarification of the relationship between the criteria in Corollary 1 and [31, Thm 4.5] requires further investigation.

Bounded lookback [28] restricts the control flow of a DDSA rather than the constraint language, and is a generalization of the earlier *feedback-freedom* property [14]. Intuitively, k-bounded lookback demands that the behavior of a DDSA at any point in time depends only on k events from the past. We refer to [28, Definition 5.9] for the formal definition. Systems that enjoy bounded lookback allow for decidable linear-time verification [28, Theorem 5.10]. However, we next show that this result does not extend to branching time.

Example 5. We reduce control state reachability of two-counter machines (2CM) to the verification problem of CTL_f^* formulas in bounded lookback systems, inspired by [42, Theorem 4.1]. 2CMs have a finite control structure and two counters x_1, and x_2 that can be incremented, decremented, and tested for 0. It is undecidable whether a 2CM will ever reach a designated control state f [43]. For a 2CM $\mathcal{M}$, we build a feedback-free DDSA $\mathcal{B} = \langle B, b_I, \mathcal{A}, T, B_F, V, \alpha_I, guard\rangle$ and a CTL_f^* property χ such that $\mathcal{B}$ satisfies χ iff f is reachable in $\mathcal{M}$. The set B consists of the control states of $\mathcal{M}$, together with an error state e and auxiliary states b_t for transitions t of $\mathcal{M}$, and $B_F = \{f, e\}$. The set V consists of x_1, x_2 and auxiliary variables p_1, p_2, m_1, m_2. Zero-test transitions of $\mathcal{M}$ are directly modeled in $\mathcal{B}$, whereas a step $q \to q'$ that increments x_i by one is modeled as:

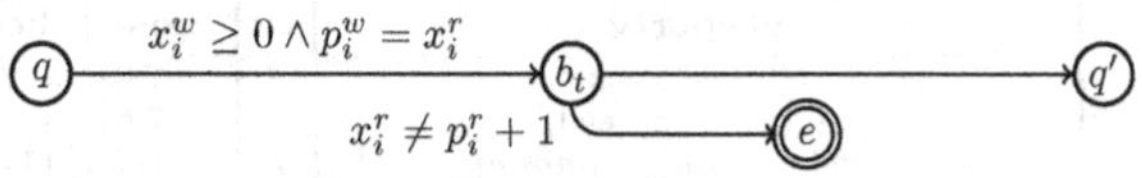

The step $q \rightarrow b_t$ writes x_i, storing its previous value in p_i, but if the write was not an increment by exactly 1, a step to state e is enabled. Decrements are modeled similarly. Intuitively, bounded lookback holds because variable dependencies are limited: in a run of $\mathcal{M}$, a variable dependency that is not an equality extends over at most two time points. (More formally, non-equality paths in the computation graph have at most length 1.) As increments are not exact, $\mathcal{B}$ overapproximates $\mathcal{M}$. However, $\chi = \mathsf{E}\,\mathsf{G}\,(\neg \mathsf{E}\,\mathsf{X}\,e)$ asserts existence of a path that never allows for a step to e (i.e., it properly simulates $\mathcal{M}$) but reaches the final state f. Thus, $\mathcal{B}$ satisfies χ iff f is reachable in $\mathcal{M}$.

6 Implementation

We implemented our approach in the prototype **ada** (arithmetic DDS analyzer) in Python; source code, benchmarks, and a web interface are available (https://ctlstar.adatool.dev). As input, the tool takes a CTL* property χ together with a DDSA in JSON format; alternatively, a given (bounded) Petri net with data (DPN) in PNML format [5] can be transformed into a DDSA. The tool then applies the algorithm in Fig. 1. If successful, it outputs the configuration map returned by *checkState*(χ), and it can visualize the product constructions. For SMT checks and quantifier elimination, **ada** interfaces CVC5 [23] and Z3 [17]. Besides numeric variables, **ada** also supports variables of type boolean and string; for the latter, only equality comparison is supported, so different constants can be represented by distinct integers. In addition to the operations in Definition 6, **ada** allows next operators $\langle a \rangle$ via an action a, which are useful for verification.

We tested **ada** on a set of business process models presented as Data Petri nets (DPNs) in the literature. As these nets are bounded, they can be transformed into DDSAs. The results are reported in the table below. We indicate whether the system belongs to a decidable class, the verified property and whether it is satisfied by the initial configuration, the verification time, the number of SMT checks, and the number of nodes in the DDSA $\mathcal{B}$ and the sum of all product constructions, respectively. We used CVC5 as SMT solver; times are without visualization, which tends to be time-consuming for large graphs. All tests were run on an Intel Core i7 with 4×2.60 GHz and 19 GB RAM.

process	property	sat	time	checks	$\lvert\mathcal{B}\rvert$	$\lvert\mathcal{N}^{\psi}_{\mathcal{B},b}\rvert$
(a) road fines	No deadlock	✗	7.0s	8161	9	2052
	$\mathsf{AG}\,(\mathsf{p}_7 \to \mathsf{E\,F\,end})$	✓	7.6s	7655		1987
	$\mathsf{AG}\,(\mathsf{end} \to \mathit{total} \leq \mathit{amount})$	✗	1m12s	111139		3622
(b) road fines	No deadlock	✓	15m27s	247563	9	4927
	$\mathsf{AG}\,(\mathsf{p}_7 \to \mathsf{E\,F\,end})$	✓	16m7s	246813		4927
(c) road fines	No deadlock	✗	9s	9179	9	1985
	$\mathsf{AG}\,(\mathsf{p}_7 \to \mathsf{E\,F\,end})$	✓	6.6s	6382		1597
	$\psi_{c1} = \mathsf{EF}\,(dS \geq 2160)$	✗	11.5s	17680		1280
	$\psi_{c2} = \mathsf{EF}\,(dP \geq 1440)$	✗	10.0s	15187		1280
	$\psi_{c3} = \mathsf{EF}\,(dJ \geq 1440)$	✗	10.5	16000		1280
(d) hospital billing	No deadlock	✓	20m59s	1234928	17	23147
	$\psi_{d1} = \mathsf{EF}\,(\mathsf{p16} \wedge \neg \mathit{closed})$	✓	10m20s	669379		10654
(e) sepsis	No deadlock	✓	1m36s	139	301	44939
	$\psi_{e1} = \mathsf{AG}\,(\mathsf{sink} \to t_{tr} < t_{ab})$	✗	30.1s	170		22724
	$\psi_{e2} = \mathsf{AG}\,(\mathsf{sink} \to t_{tr}{+}60 \geq t_{ab})$	✓	32s	153		22538
(f) sepsis	No deadlock	✓	7m24	4524	301	161242
	$\psi_{f1} = \mathsf{A}\,(\neg \mathit{lacticAcid}\mathsf{G}\,\langle \mathsf{diagnostic} \rangle \top)$	✓	3m53s	5734		74984
(g) board: register	No deadlock	✓	1.4s	12	7	27
(h) board: transfer	No deadlock	✓	1.4s	27	7	51
(i) board: discharge	No deadlock	✓	1.5s	25	6	67
	$\psi_{i1} = \mathsf{AG}\,(\mathsf{p}_2 \wedge o_1{=}207 \to \mathsf{AG}\; o_1{=}207)$	✓	1.5s	94		91
	$\psi_{i2} = \mathsf{A}\,(\mathsf{EF}\,\langle \mathsf{tra} \rangle \top \wedge \mathsf{EF}\,\langle \mathsf{his} \rangle \top)$	✓	1.5s	27		98
	$\psi_{i3} = \neg \mathsf{E}\,(\mathsf{F}\,\langle \mathsf{tra} \rangle \top \wedge \mathsf{F}\,\langle \mathsf{his} \rangle \top)$	✓	1.4s	56		43
(j) credit approval	No deadlock	✓	1.7s	470	6	230
	$\psi_{j1} = \mathsf{AG}\,(\langle \mathsf{openLoan} \rangle \top \to \mathit{ver} \wedge \mathit{dec})$	✓	13.2s	14156		645
	$\psi_{j2} = \mathsf{A}\,(\mathsf{F}\,(\mathit{ver} \wedge \mathit{dec}) \to \mathsf{F}\,\langle \mathsf{openLoan} \rangle \top)$	✗	3.7s	3128		316
	$\psi_{j3} = \mathsf{A}\,(\mathsf{F}\,(\mathit{ver} \wedge \mathit{dec}) \to \mathsf{EF}\,\langle \mathsf{openLoan} \rangle \top)$	✓	5.6s	4748		548
(k) package handling	No deadlock	✓	2.7ss	1025	16	693
	No deadlock (τ_1)	✓	2.5s	1079		398
	$\psi_{k1} = \mathsf{EF}\,\langle \mathsf{fetch} \rangle \top$	✗	2.6s	850		343
	$\psi_{k2} = \mathsf{EF}\,\langle \tau_6 \rangle \top$	✗	2.4s	875		336
(l) auction	No deadlock	✗	10.8s	1683	5	186
	$\mathsf{EF}\,(\mathsf{sold} \wedge d > 0 \wedge o \leq t)$	✗	6.4s	1180		79
	$\mathsf{EF}\,(b = 1 \wedge o > t \wedge \mathsf{F}\,(\mathsf{sold} \wedge b > 1))$	✓	26.5s	4000		263

We briefly comment on the benchmarks and some properties: For all examples we checked *no deadlock*, which abbreviates $\mathsf{AG\,EF}\,\chi_f$ where χ_f is a disjunction of all final states. This is one of the two requirements of the crucial *soundness* property (cf. Example 1). Weak soundness [4] relaxes this requirement to demand only that if a transition is reachable, it does not lead to deadlocks; this is called here *no deadlock*(a), expressed by $\mathsf{EF}\,(\langle a \rangle \top) \to \mathsf{AG}\,(\langle a \rangle \top \to \mathsf{F}\,\chi_f)$. One can also check whether a specific state p is deadlock-free, via $\mathsf{AG}\,(p \to \mathsf{EF}\,\chi_f)$.

(a)-(c) are versions of the road fine management process (cf. Example 1); (a) [40, Fig. 12.7] and (b) [37, Fig. 13] were mined automatically from logs, while (c) is the normative version [41, Fig. 7] shown in Example 1. While in (a) and (c) *no deadlock* is violated, this issue was fixed in version (b). The fact that ψ_{c1}, ψ_{c2}, and ψ_{c3} hold confirm that the time constraints are never violated.

(d) models a billing process in a hospital [40, Fig. 15.3], which is deadlock-free.

(e) is a normative model for a sepsis triage process in a hospital [40, Fig. 13.3], and (f) is a variation that was mined purely automatically from logs [40, Fig. 13.6]. According to [40, Sect. 13], triage should happen before antibiotics are administered, expressed by ψ_{e1}, which is actually not satisfied. However, the desired time constraint expressed by ψ_{e2} holds.

(g)–(i) reflect activities in patient logistics of a hospital, based on logs of real-life processes [40, Fig. 14.3]. While the *no deadlock* property is satisfied by all initial configurations, the output of ada reveals that for (h) this need not hold for other initial assignments.

(j) is a credit approval process [16, Fig. 3]. It is deadlock-free; ψ_{j1} and ψ_{j2} verify desirable conditions under which a loan is granted to a client.

(k) is a package handling routine [26, Fig. 5]. The fact that the properties ψ_{k1} and ψ_{k2} are not satisfied shows that the transitions τ_6 and fetch are dead.

(l) models an auction process [28, Example 1.1], for which ada reveals a deadlock. Results for two further properties from [28, Example 1.1] are listed as well.

Seven systems are in a decidable class wrt. the listed properties: (a), (b), (d), (f), (h), (i), (k) are MC, while (d), (h), (i), (k) are IPC. This is due to the fact that automatic mining techniques often produce monotonicity constraints [39].

7 Conclusion

This paper presents a technique to compute witness maps for a given DDSA and CTL_f^* property, where a witness map specifies conditions on the initial variable assignment such that the property holds. The addressed problem is thus a slight generalization of the common verification problem. While our model checking procedure need not terminate in general, we show that it does if an abstract property on history constraints holds. Moreover, witness maps always exist for monotonicity and integer periodicity constraint systems. However, this result does not extend to bounded lookback systems. We implemented our approach in the tool ada and showed its usefulness on a range of business process models.

We see various opportunities to extend this work. A richer verification language could support past time operators [18] and future variable values [20]. Further decidable fragments could be sought using covers [33], or aiming for compatibility with locally finite theories [32]. Moreover, a restricted version of the bounded lookback property could guarantee decidability of CTL_f^*, similarly to the way feedback freedom was strengthened in [35]. The implementation could be improved to avoid the computation of many similar formulas, thus gaining efficiency. Finally, the complexity class that our approach implies for CTL_f^* in the decidable classes is yet to be clarified.

References

1. van der Aalst, W.M.P.: Process Mining: Data Science in Action. Springer (2016). https://doi.org/10.1007/978-3-662-49851-4
2. Baier, C., Katoen, J.: Principles of Model Checking. MIT Press (2008)
3. Baral, C., De Giacomo, G.: Knowledge representation and reasoning: what's hot. In: Proceedings of the 29th AAAI, pp. 4316–4317 (2015)
4. Batoulis, K., Haarmann, S., Weske, M.: Various notions of soundness for decision-aware business processes. In: Mayr, H.C., Guizzardi, G., Ma, H., Pastor, O. (eds.) ER 2017. LNCS, vol. 10650, pp. 403–418. Springer, Cham (2017). https://doi.org/10.1007/978-3-319-69904-2_31

5. Billington, J., et al.: The petri net markup language: concepts, technology, and tools. In: van der Aalst, W.M.P., Best, E. (eds.) ICATPN 2003. LNCS, vol. 2679, pp. 483–505. Springer, Heidelberg (2003). https://doi.org/10.1007/3-540-44919-1_31
6. Bozga, M., Gîrlea, C., Iosif, R.: Iterating octagons. In: Kowalewski, S., Philippou, A. (eds.) TACAS 2009. LNCS, vol. 5505, pp. 337–351. Springer, Heidelberg (2009). https://doi.org/10.1007/978-3-642-00768-2_29
7. Bozzelli, L., Gascon, R.: Branching-time temporal logic extended with qualitative Presburger constraints. In: Hermann, M., Voronkov, A. (eds.) LPAR 2006. LNCS (LNAI), vol. 4246, pp. 197–211. Springer, Heidelberg (2006). https://doi.org/10.1007/11916277_14
8. Bozzelli, L., Pinchinat, S.: Verification of gap-order constraint abstractions of counter systems. Theor. Comput. Sci. **523**, 1–36 (2014). https://doi.org/10.1016/j.tcs.2013.12.002
9. Calvanese, D., De Giacomo, G., Montali, M.: Foundations of data-aware process analysis: a database theory perspective. In: Proceedings of the 32nd PODS, pp. 1–12 (2013). https://doi.org/10.1145/2463664.2467796
10. Calvanese, D., De Giacomo, G., Montali, M., Patrizi, F.: First-order μ-calculus over generic transition systems and applications to the situation calculus. Inf. Comput. **259**(3), 328–347 (2018). https://doi.org/10.1016/j.ic.2017.08.007
11. Carapelle, C., Kartzow, A., Lohrey, M.: Satisfiability of ECTL* with constraints. J. Comput. Syst. Sci. **82**(5), 826–855 (2016). https://doi.org/10.1016/j.jcss.2016.02.002
12. Čerāns, K.: Deciding properties of integral relational automata. In: Abiteboul, S., Shamir, E. (eds.) ICALP 1994. LNCS, vol. 820, pp. 35–46. Springer, Heidelberg (1994). https://doi.org/10.1007/3-540-58201-0_56
13. Comon, H., Jurski, Y.: Multiple counters automata, safety analysis and Presburger arithmetic. In: Hu, A.J., Vardi, M.Y. (eds.) CAV 1998. LNCS, vol. 1427, pp. 268–279. Springer, Heidelberg (1998). https://doi.org/10.1007/BFb0028751
14. Damaggio, E., Deutsch, A., Vianu, V.: Artifact systems with data dependencies and arithmetic. ACM Trans. Database Syst. **37**(3), 22:1–22:36 (2012). https://doi.org/10.1145/2338626.2338628
15. de Giacomo, G., De Masellis, R., Montali, M.: Reasoning on LTL on finite traces: insensitivity to infiniteness. In: Proceedings of the 28th AAAI, pp. 1027–1033 (2014)
16. de Leoni, M., Mannhardt, F.: Decision discovery in business processes. In: Encyclopedia of Big Data Technologies, pp. 1–12. Springer (2018). https://doi.org/10.1007/978-3-319-63962-8_96-1
17. de Moura, L., Bjørner, N.: Z3: an efficient SMT solver. In: Ramakrishnan, C.R., Rehof, J. (eds.) TACAS 2008. LNCS, vol. 4963, pp. 337–340. Springer, Heidelberg (2008). https://doi.org/10.1007/978-3-540-78800-3_24
18. Demri, S.: LTL over integer periodicity constraints. Theor. Comput. Sci. **360**(1–3), 96–123 (2006). https://doi.org/10.1016/j.tcs.2006.02.019
19. Demri, S., Dhar, A.K., Sangnier, A.: Equivalence between model-checking flat counter systems and Presburger arithmetic. Theor. Comput. Sci. **735**, 2–23 (2018). https://doi.org/10.1016/j.tcs.2017.07.007
20. Demri, S., D'Souza, D.: An automata-theoretic approach to constraint LTL. Inform. Comput. **205**(3), 380–415 (2007). https://doi.org/10.1016/j.ic.2006.09.006
21. Demri, S., Finkel, A., Goranko, V., van Drimmelen, G.: Model-checking CTL* over flat Presburger counter systems. J. Appl. Non Class. Logics **20**(4), 313–344 (2010). https://doi.org/10.3166/jancl.20.313-344

22. Demri, S., Gascon, R.: Verification of qualitative Z constraints. Theor. Comput. Sci. **409**(1), 24–40 (2008). https://doi.org/10.1016/j.tcs.2008.07.023
23. Deters, M., Reynolds, A., King, T., Barrett, C.W., Tinelli, C.: A tour of CVC4: how it works, and how to use it. In: Proceedings of the 14th FMCAD, p. 7 (2014). https://doi.org/10.1109/FMCAD.2014.6987586
24. Deutsch, A., Hull, R., Li, Y., Vianu, V.: Automatic verification of database-centric systems. ACM SIGLOG News **5**(2), 37–56 (2018). https://doi.org/10.1145/3212019.3212025
25. Felli, P., de Leoni, M., Montali, M.: Soundness verification of decision-aware process models with variable-to-variable conditions. In: Proceedings of the 19th ACSD, pp. 82–91. IEEE (2019). https://doi.org/10.1109/ACSD.2019.00013
26. Felli, P., de Leoni, M., Montali, M.: Soundness verification of data-aware process models with variable-to-variable conditions. Fund. Inform. **182**(1), 1–29 (2021). https://doi.org/10.3233/FI-2021-2064
27. Felli, P., Montali, M., Winkler, S.: CTL* model checking for data-aware dynamic systems with arithmetic (extended version) (2022). https://doi.org/10.48550/arXiv.2205.08976
28. Felli, P., Montali, M., Winkler, S.: Linear-time verification of data-aware dynamic systems with arithmetic. In: Proceedings of the 36th AAAI (2022). https://doi.org/10.48550/arXiv.2203.07982
29. Finkel, A., Leroux, J.: How to compose Presburger-accelerations: applications to broadcast protocols. In: Agrawal, M., Seth, A. (eds.) FSTTCS 2002. LNCS, vol. 2556, pp. 145–156. Springer, Heidelberg (2002). https://doi.org/10.1007/3-540-36206-1_14
30. Finkel, A., Willems, B., Wolper, P.: A direct symbolic approach to model checking pushdown systems. In: Proc. 2nd INFINITY. ENTCS, vol. 9, pp. 27–37 (1997). https://doi.org/10.1016/S1571-0661(05)80426-8
31. Gascon, R.: An automata-based approach for CTL* with constraints. In: Proceedings of the INFINITY 2006, 2007 and 2008. ENTCS, vol. 239, pp. 193–211 (2009). https://doi.org/10.1016/j.entcs.2009.05.040
32. Ghilardi, S., Nicolini, E., Ranise, S., Zucchelli, D.: Combination methods for satisfiability and model-checking of infinite-state systems. In: Pfenning, F. (ed.) CADE 2007. LNCS (LNAI), vol. 4603, pp. 362–378. Springer, Heidelberg (2007). https://doi.org/10.1007/978-3-540-73595-3_25
33. Gulwani, S., Musuvathi, M.: Cover algorithms and their combination. In: Drossopoulou, S. (ed.) ESOP 2008. LNCS, vol. 4960, pp. 193–207. Springer, Heidelberg (2008). https://doi.org/10.1007/978-3-540-78739-6_16
34. Ibarra, O.H., Su, J.: Counter machines: decision problems and applications. In: Jewels are Forever: Contributions on Theoretical Computer Science in Honor of Arto Salomaa, pp. 84–96 (1999)
35. Koutsos, A., Vianu, V.: Process-centric views of data-driven business artifacts. J. Comput. Syst. Sci. **86**, 82–107 (2017). https://doi.org/10.1016/j.jcss.2016.11.012
36. Kroening, D., Strichman, O.: Decision Procedures - An Algorithmic Point of View. Second Edition. Springer (2016). https://doi.org/10.1007/978-3-662-50497-0
37. de Leoni, M., Felli, P., Montali, M.: A holistic approach for soundness verification of decision-aware process models. In: Trujillo, J.C., Davis, K.C., Du, X., Li, Z., Ling, T.W., Li, G., Lee, M.L. (eds.) ER 2018. LNCS, vol. 11157, pp. 219–235. Springer, Cham (2018). https://doi.org/10.1007/978-3-030-00847-5_17
38. de Leoni, M., Felli, P., Montali, M.: Strategy synthesis for data-aware dynamic systems with multiple actors. In: Proceedings of the 17th KR, pp. 315–325 (2020). https://doi.org/10.24963/kr.2020/32

39. de Leoni, M., Felli, P., Montali, M.: Integrating BPMN and DMN: modeling and analysis. J. Data Semant. **10**(1), 165–188 (2021). https://doi.org/10.1007/s13740-021-00132-z
40. Mannhardt, F.: Multi-perspective process mining. Ph.D. thesis, Technical University of Eindhoven (2018)
41. Mannhardt, F., de Leoni, M., Reijers, H.A., van der Aalst, W.M.P.: Balanced multi-perspective checking of process conformance. Computing **98**(4), 407–437 (2015). https://doi.org/10.1007/s00607-015-0441-1
42. Mayr, R., Totzke, P.: Branching-time model checking gap-order constraint systems. Fundam. Informaticae **143**(3–4), 339–353 (2016). https://doi.org/10.3233/FI-2016-1317
43. Minsky, M.: Computation: Finite and Infinite Machines. Prentice-Hall (1967)
44. Murano, A., Parente, M., Rubin, S., Sorrentino, L.: Model-checking graded computation-tree logic with finite path semantics. Theor. Comput. Sci. **806**, 577–586 (2020). https://doi.org/10.1016/j.tcs.2019.09.021
45. Presburger, M.: Über die Vollständigkeit eines gewissen Systems der Arithmetik ganzer Zahlen, in welchem die Addition als einzige Operation hervortritt. In: Comptes Rendus du I congres de Mathem. des Pays Slaves, pp. 92–101 (1929)
46. Reichert, M.: Process and data: two sides of the same coin? In: Meersman, R., Panetto, H., Dillon, T., Rinderle-Ma, S., Dadam, P., Zhou, X., Pearson, S., Ferscha, A., Bergamaschi, S., Cruz, I.F. (eds.) OTM 2012. LNCS, vol. 7565, pp. 2–19. Springer, Heidelberg (2012). https://doi.org/10.1007/978-3-642-33606-5_2
47. Sorrentino, L., Rubin, S., Murano, A.: Graded CTL* over finite paths. In: Proceedings of the 19th ICTCS. CEUR Workshop Proceedings, vol. 2243, pp. 152–161. CEUR-WS.org (2018)

SAT-Based Proof Search in Intermediate Propositional Logics

Camillo Fiorentini[1] and Mauro Ferrari[2(✉)]

[1] Department of Computer Science, Università degli Studi di Milano, Milan, Italy
[2] Department of Theoretical and Applied Sciences, Università degli Studi dell'Insubria, Varese, Italy
mauro.ferrari@uninsubria.it

Abstract. We present a decision procedure for intermediate logics relying on a modular extension of the SAT-based prover intuitR for IPL (Intuitionistic Propositional Logic). Given an intermediate logic L and a formula α, the procedure outputs either a Kripke countermodel for α or the instances of the characteristic axioms of L that must be added to IPL in order to prove α. The procedure exploits an incremental SAT-solver; during the computation, new clauses are learned and added to the solver.

1 Introduction

Recently, Claessen and Rosén have introduced `intuit` [4], an efficient decision procedure for Intuitionistic Propositional Logic (IPL) based on the Satisfiability Modulo Theories (SMT) approach. The prover language consists of (flat) clauses of the form $\bigwedge A_1 \rightarrow \bigvee A_2$ (with A_i a set of atoms), which are fed to the SAT-solver, and implication clauses of the form $(a \rightarrow b) \rightarrow c$ (a, b, c atoms); thus, we need an auxiliary clausification procedure to preprocess the input formula. The search is performed via a proper variant of the DPLL($\mathcal{T}$) procedure [16], by exploiting an incremental SAT-solver; during the computation, whenever a semantic conflict is thrown, a new clause is learned and added to the SAT-solver. As discussed in [9], there is a close connection between the `intuit` approach and the known proof-theoretic methods. Actually, the decision procedure mimics the standard root-first proof search strategy for a sequent calculus strongly connected with Dyckhoff's calculus LJT [5] (alias G4ip). To improve performances, we have re-designed the prover by adding a restart operation, thus obtaining `intuitR` [8] (`intuit` with Restart). Differently from `intuit`, the `intuitR` procedure has a simple structure, consisting of two nested loops. Given a formula α, if α is provable in IPL the call `intuitR(`α`)` yields a derivation of α in the sequent calculus introduced in [8], a plain calculus where derivations have a single branch. If α is not provable in IPL, the outcome of `intuitR(`α`)` is a (typically small) countermodel for α, namely a Kripke model falsifying α. We stress that `intuitR` is highly performant: on the basis of a standard benchmarks suite, it outperforms `intuit` and other state-of-the-art provers (in particular, `fCube` [6] and `intHistGC` [12]).

J. Blanchette et al. (Eds.): IJCAR 2022, LNAI 13385, pp. 57–74, 2022.
https://doi.org/10.1007/978-3-031-10769-6_5

In this paper we present `intuitRIL`, an extension of `intuitR` to Intermediate Logics, namely propositional logics extending IPL and contained in CPL (Classical Propositional Logic). Specifically, let α be a formula and L an axiomatizable intermediate logic having Kripke semantics; the call intuitRIL(α,L) tries to prove the validity of α in L. To this aim, the prover searches for a set Ψ containing instances of $\mathrm{Ax}(L)$, the characteristic axioms of L, such that α can be proved in IPL from Ψ. Note that this is different from other approaches, where the focus is on the synthesis of specific inference rules for the logic at hand (see, e.g., [17]). Basically, intuitRIL(α,L) searches for a countermodel $\mathcal{K}$ for α, exploiting the search engine of `intuitR`: whenever we get $\mathcal{K}$, we check whether $\mathcal{K}$ is a model of L. If this is the case, we conclude that α is not valid in L (and $\mathcal{K}$ is a witness to this). Otherwise, the prover selects an instance ψ of $\mathrm{Ax}(L)$ falsified in $\mathcal{K}$ (there exists at least one); ψ is acknowledged as learned axiom and, after clausification, it is fed to the SAT-solver. We stress that a naive implementation of the procedure, where at each iteration of the main loop the computation restarts from scratch, would be highly inefficient: each time the SAT-solver should be initialized by inserting all the clauses encoding the input problem and all the clauses learned so far. Instead, we exploit an incremental SAT-solver, where clauses can be added but never deleted (hence, all the simplifications and optimisations performed by the solver are preserved); note that this prevents us from exploiting strategies based on standard sequent/tableaux calculi, where backtracking is required.

If the call intuitRIL(α,L) succeeds, by tracking the computation we get a derivation $\mathcal{D}$ of α in the sequent calculus C_L (see Fig. 1); from $\mathcal{D}$ we can extract all the axioms learned during the computation. We stress that the procedure is quite modular: to handle a logic L, one has only to implement a specific learning mechanism for L (namely: if $\mathcal{K}$ is not a model of L, pick an instance of $\mathrm{Ax}(L)$ falsified in $\mathcal{K}$). The main drawback is that there is no general way to bound the learned axioms, thus termination must be investigated on a case-by-case basis. We guarantee termination for some relevant intermediate logics, such as Gödel-Dummett Logic GL, the family GL_n ($n \geq 1$) of Gödel-Dummett Logics with depth bounded by n (GL_1 coincides with Here and There Logic, well known for its applications in Answer Set Programming [15]) and Jankov Logic (for a presentation of such logics see [2]). As a corollary, for each of the mentioned logic L we get a bounding function [3], namely: given α, we compute a bounded set Ψ_α of instances of $\mathrm{Ax}(L)$ such that α is valid in L iff α is provable in IPL from assumptions Ψ_α; in general we improve the bounds in [1,3]. The `intuitRIL` Haskell implementation and other additional material (e.g., the omitted proofs) can be downloaded at https://github.com/cfiorentini/intuitRIL.

2 Basic Definitions

Formulas, denoted by lowercase Greek letters, are built from an enumerable set of propositional variables $\mathcal{V}$, the constant $\bot$ and the connectives $\wedge$, $\vee$, $\rightarrow$; moreover, $\neg\alpha$ stands for $\alpha \rightarrow \bot$ and $\alpha \leftrightarrow \beta$ stands for $(\alpha \rightarrow \beta) \wedge (\beta \rightarrow \alpha)$. Elements of the set $\mathcal{V} \cup \{\bot\}$ are called *atoms* and are denoted by lowercase Roman letters, uppercase Greek letters denote sets of formulas. By $\mathcal{V}_\alpha$ we denote the set of propositional variables occurring in α. The notation is extended to sets: $\mathcal{V}_\Gamma$ is the union of $\mathcal{V}_\alpha$ such that $\alpha \in \Gamma$; $\mathcal{V}_{\Gamma,\Gamma'}$ and $\mathcal{V}_{\Gamma,\alpha}$ stand for $\mathcal{V}_{\Gamma\cup\Gamma'}$ and $\mathcal{V}_{\Gamma\cup\{\alpha\}}$ respectively. A *substitution* is a map from propositional variables to formulas. By $[p_1 \mapsto \alpha_1, \ldots, p_n \mapsto \alpha_n]$ we denote the substitution χ such that $\chi(p) = \alpha_i$ if $p = p_i$ and $\chi(p) = p$ otherwise; the set $\{p_1, \ldots, p_n\}$ is the *domain* of χ, denoted by $\mathrm{Dom}(\chi)$; ϵ is the substitution having empty domain. The application of χ to a formula α, denoted by $\chi(\alpha)$, is defined as usual; $\chi(\Gamma)$ is the set of $\chi(\alpha)$ such that $\alpha \in \Gamma$. The *composition* $\chi_1 \cdot \chi_2$ is the substitution mapping p to $\chi_1(\chi_2(p))$.

A *(classical) interpretation* M is a subset of $\mathcal{V}$, identifying the propositional variables assigned to true. By $M \models \alpha$ we mean that α is true in M; $M \models \Gamma$ iff $M \models \alpha$ for every $\alpha \in \Gamma$. Classical Propositional Logic (CPL) is the set of formulas true in every interpretation. We write $\Gamma \vdash_c \alpha$ iff $M \models \Gamma$ implies $M \models \alpha$, for every M. Note that α is CPL-valid (namely, $\alpha \in \mathrm{CPL}$) iff $\emptyset \vdash_c \alpha$.

A (rooted) Kripke model is a quadruple $\langle W, \leq, r, \vartheta\rangle$ where W is a finite and non-empty set (the set of *worlds*), $\leq$ is a reflexive and transitive binary relation over W, the world r (the *root* of $\mathcal{K}$) is the minimum of W w.r.t. $\leq$, and $\vartheta : W \mapsto 2^{\mathcal{V}}$ (the *valuation* function) is a map obeying the persistence condition: for every pair of worlds w_1 and w_2 of $\mathcal{K}$, $w_1 \leq w_2$ implies $\vartheta(w_1) \subseteq \vartheta(w_2)$; the triple $\langle W, \leq, r\rangle$ is called *(Kripke) frame*. The valuation ϑ is extended to a *forcing* relation between worlds and formulas as follows:

$$w \Vdash p \text{ iff } p \in \vartheta(w), \forall p \in \mathcal{V} \qquad w \nVdash \bot \qquad w \Vdash \alpha \wedge \beta \text{ iff } w \Vdash \alpha \text{ and } w \Vdash \beta$$

$$w \Vdash \alpha \vee \beta \text{ iff } w \Vdash \alpha \text{ or } w \Vdash \beta \qquad w \Vdash \alpha \to \beta \text{ iff } \forall w' \geq w, w' \Vdash \alpha \text{ implies } w' \Vdash \beta.$$

By $w \Vdash \Gamma$ we mean that $w \Vdash \alpha$ for every $\alpha \in \Gamma$. A formula α is *valid* in the frame $\langle W, \leq, r\rangle$ iff for every valuation ϑ, $r \Vdash \alpha$ in the model $\langle W, \leq, r, \vartheta\rangle$. Propositional Intuitionistic Logic (IPL) is the set of formulas valid in all frames. Accordingly, if there is a model $\mathcal{K}$ such that $r \nVdash \alpha$ (here and below r designates the root of $\mathcal{K}$), then α is not IPL-valid; we call $\mathcal{K}$ a *countermodel* for α. We write $\Gamma \vdash_i \delta$ iff, for every model $\mathcal{K}$, $r \Vdash \Gamma$ implies $r \Vdash \delta$; thus, α is IPL-valid iff $\emptyset \vdash_i \alpha$.

Let L be one of the logics IPL and CPL; then, L is closed under modus ponens ($\{\alpha, \alpha \to \beta\} \subseteq L$ implies $\beta \in L$) and under substitution (for every χ, $\alpha \in L$ implies $\chi(\alpha) \in L$). An *intermediate logic* is any set of formulas L such that $\mathrm{IPL} \subseteq L \subseteq \mathrm{CPL}$, L is closed under modus ponens and under substitution. A model $\mathcal{K}$ is an L-model iff $r \Vdash L$; if $r \nVdash \alpha$, we say that $\mathcal{K}$ is an *L-countermodel* for α. An intermediate logic L can be characterized by a set of CPL-valid formulas, called the *L-axioms* and denoted by $\mathrm{Ax}(L)$. An L-axiom ψ of $\mathrm{Ax}(L)$ must be understood as a schematic formula, representing all the formulas of the kind $\chi(\psi)$; we call $\chi(\psi)$ an *instance* of ψ. Formally, $\mathrm{IPL} + \mathrm{Ax}(L)$ is the intermediate logic collecting the formulas α such that $\Psi \vdash_i \alpha$, where Ψ is a finite set of instances of L-axioms from $\mathrm{Ax}(L)$. A *bounding function* for L is a map that, given α, yields a finite set Ψ_α of instances of L-axioms such that $\Psi_\alpha \vdash_i \alpha$. If L admits a computable bounding function, we can reduce L-validity to IPL-validity (see [3] for an in-depth discussion). Let $\mathcal{F}$ be a class of frames and let $\mathrm{Log}(\mathcal{F})$ be the set of formulas valid in all frames of $\mathcal{F}$; then, $\mathrm{Log}(\mathcal{F})$ is an intermediate logic. A logic L has *Kripke semantics* iff there exists a class of frames $\mathcal{F}$ such that $L = \mathrm{Log}(\mathcal{F})$; we also say that L is characterized by $\mathcal{F}$. Henceforth, when we mention a logic L, we leave understood that L is an axiomatizable intermediate logic having Kripke semantics.

Example 1 (GL). A well-known intermediate logic is Gödel-Dummett logic GL [2], characterized by the class of linear frames. An axiomatization of GL is obtained by adding the linearity axiom $\mathbf{lin} = (a \to b) \vee (b \to a)$ to IPL. Using the terminology of [3], GL is formula-axiomatizable: a bounding function for GL is obtained by mapping α to the set Ψ_α of instances of **lin** where a and b are replaced with subformulas of α. In [1] it is proved that it is sufficient to consider the subformulas of α of the kind $p \in \mathcal{V}_\alpha$, $\neg\beta$, $\beta_1 \to \beta_2$. In Lemma 4 we further improve this bound tacking as bounding function the following map:

$$\begin{aligned}\mathrm{Ax_{GL}}(\alpha) &= \{\, (a \to b) \vee (b \to a) \mid a, b \in \mathcal{V}_\alpha \,\} \cup \{\, (a \to \neg a) \vee (\neg a \to a) \mid a \in \mathcal{V}_\alpha \,\} \\ &\quad \cup \{\, (a \to (a \to b)) \vee ((a \to b) \to a)) \mid a, b \in \mathcal{V}_\alpha \,\}\end{aligned}$$

Thus, if $\mathcal{V}_\alpha = \{a\}$, the only instance of **lin** to consider is $(a \to \neg a) \vee (\neg a \to a)$, independently of the size of α (the other instances are IPL-valid and can be omitted). As pointed out in [3], GL is not variable-axiomatizable, namely: it is not sufficient to consider instances of **lin** obtained by replacing a and b with variables from $\mathcal{V}_\alpha$. As an example, let $\alpha = \neg a \vee \neg\neg a$; α is GL-valid, the only variable-replacement instance of **lin** is $\psi_\alpha = (a \to a) \vee (a \to a)$ and $\psi_\alpha \nvdash_{\mathrm{i}} \alpha$. ◊

We review the main concepts about the clausification procedure described in [4]. *Clauses* φ and *implication clauses* λ are defined as

$$\begin{aligned}&\varphi := \bigwedge A_1 \to \bigvee A_2 \mid \bigvee A_2 && \emptyset \subset A_k \subseteq \mathcal{V} \cup \{\bot\}, \mathit{for} k \in \{1,2\} \\ &\lambda := (a \to b) \to c && a \in \mathcal{V},\ \{b, c\} \subseteq \mathcal{V} \cup \{\bot\}\end{aligned}$$

where $\bigwedge A_1$ and $\bigvee A_2$ denote the conjunction and the disjunction of the atoms in A_1 and A_2 respectively ($\bigwedge\{a\} = \bigvee\{a\} = a$). Henceforth, $\bigwedge\emptyset \to \bigvee A_2$ must be read as $\bigvee A_2$; R, R_1, ... denote sets of clauses, X, X_1, ... sets of implication clauses. Given a set of implication clauses X, the *closure* of X, denoted by $(X)^\star$, is the set of clauses $b \to c$ such that $(a \to b) \to c \in X$.

The following lemma states some properties of clauses and closures.

Lemma 1. *(i)* $R \vdash_{\mathrm{i}} g$ *iff* $R \vdash_{\mathrm{c}} g$*, for every set of clauses* R *and every atom* g*.*
(ii) $X \vdash_{\mathrm{i}} b \to c$*, for every* $b \to c \in (X)^\star$*.*
(iii) $\Gamma \vdash_{\mathrm{i}} \alpha$ *iff* $\alpha \leftrightarrow g, \Gamma \vdash_{\mathrm{i}} g$*, where* $g \notin \mathcal{V}_{\Gamma,\alpha}$*.*

Clausification. We assume a procedure `Clausify` that, given a formula α, computes sets of clauses R and X equivalent to α w.r.t. IPL. Formally, let α be a formula and let V be a set of propositional variables such that $\mathcal{V}_\alpha \subseteq V$. The procedure `Clausify(`$\alpha$`,`$V$`)` computes a triple (R, X, χ) satisfying:

(C1) $\Gamma, \alpha \vdash_{\mathrm{i}} \delta$ iff $\Gamma, R, X \vdash_{\mathrm{i}} \delta$, for every Γ and δ such that $\mathcal{V}_{\Gamma,\delta} \subseteq V$.
(C2) $\mathrm{Dom}(\chi) = \mathcal{V}_{R,X} \setminus V$ and $\mathcal{V}_{\chi(p)} \subseteq V$ for every $p \in \mathrm{Dom}(\chi)$.
(C3) $R, X \vdash_{\mathrm{i}} p \leftrightarrow \chi(p)$ for every $p \in \mathrm{Dom}(\chi)$.

$$\frac{R \vdash_c g}{R, X \Rightarrow g}\ \mathrm{cpl}_0 \qquad \frac{R, A \vdash_c b \qquad R, \varphi, X \Rightarrow g}{R, X \Rightarrow g}\ \mathrm{cpl}_1(\lambda) \quad \begin{array}{l} \lambda = (a \to b) \to c \in X \\ A \subseteq \mathcal{V}_{R,X,g} \\ \varphi \;=\; \bigwedge(A \setminus \{a\}) \to c \end{array}$$

$$\frac{R, (X)^{\star}, X \Rightarrow g}{\Rightarrow \alpha}\ \mathrm{Claus}_0(g, \chi) \qquad \begin{array}{l} g \notin \mathcal{V}_\alpha \\ (R, X, \chi) = \mathtt{Clausify}(\alpha \leftrightarrow g,\ \mathcal{V}_{\alpha,g}) \end{array}$$

$$\frac{R, R', (X')^{\star}, X, X' \Rightarrow g}{R, X \Rightarrow g}\ \mathrm{Claus}_1(\psi, \chi) \qquad \begin{array}{l} \psi \in \mathrm{Ax}(L,\ \mathcal{V}_{R,X,g}) \\ (R', X', \chi) = \mathtt{Clausify}(\psi,\ \mathcal{V}_{R,X,g}) \end{array}$$

R is a set of clauses
X is a set of implication clauses
g is an atom

$$\pi(\rho) \;=\; \begin{cases} \langle \emptyset,\ [g \mapsto \alpha] \cdot \chi \rangle & \text{if } \rho = \mathrm{Claus}_0(g, \chi) \\ \langle \{\psi\},\ \chi \rangle & \text{if } \rho = \mathrm{Claus}_1(\psi, \chi) \\ \langle \emptyset,\ \epsilon \rangle & \text{otherwise} \end{cases}$$

Fig. 1. The sequent calculus C_L.

Basically, clausification introduces new propositional variables to represent subformulas of α; as a result we obtain a substitution χ which tracks the mapping on the new variables. Condition (C1) states that α can be replaced by $R \cup X$ in IPL reasoning. By (C2) the domain of χ consists of the new variables introduced in the clausification process. The following properties easily follow by (C1)–(C3):

(P1) $R, X \vdash_i \alpha$. (P2) $R, X \vdash_i \beta \leftrightarrow \chi(\beta)$ for every formula β.

We exploit a `Clausify` procedure essentially similar to the one described in [4], with slight modifications in order to match (C3). As discussed in [4], in IPL we can use a weaker condition (either $R, X \vdash_i p \to \chi(p)$ or $R, X \vdash_i \chi(p) \to p$ according to the case). It is not obvious whether the weaker condition should be more efficient; in many cases strong equivalences are more performant, maybe because they trigger more simplifications in the SAT-solver.

Example 2. Let $\alpha = (a \to b) \vee (b \to a)$ and $V = \{a, b\}$. The call `Clausify(`α,V`)` introduces the new variables $\tilde{p}_0$ and $\tilde{p}_1$ associated with the subformulas $a \to b$ and $b \to a$ respectively. Accordingly, the obtained sets R and X must satisfy $R, X \vdash_i \tilde{p}_0 \leftrightarrow (a \to b)$ and $R, X \vdash_i \tilde{p}_1 \leftrightarrow (b \to a)$. We get:

$$R = \{\, \tilde{p}_0 \vee \tilde{p}_1,\ \tilde{p}_0 \wedge a \to b,\ \tilde{p}_1 \wedge b \to a \,\} \qquad \chi = [\, \tilde{p}_0 \mapsto a \to b,\ \tilde{p}_1 \mapsto b \to a \,]$$
$$X = \{\, (a \to b) \to \tilde{p}_0,\ (b \to a) \to \tilde{p}_1 \,\}$$

◇

3 The Calculus C_L

Let L be an intermediate logic; we introduce the sequent calculus C_L to prove L-validity. We assume that L is axiomatized by a set $\mathrm{Ax}(L)$ of L-axioms; by

$$
\dfrac{\cdots \quad \dfrac{\dfrac{R_{n-1} \vdash_{\mathrm{c}} g}{R_{n-1}, X_{n-1} \Rightarrow g}\ \rho_n = \mathrm{cpl}_0}{R_{n-2}, X_{n-2} \Rightarrow g}\ \rho_{n-1}}{\vdots}
$$

$$
\dfrac{\dfrac{\cdots \quad R_1, X_1 \Rightarrow g}{R_0, X_0 \Rightarrow g}\ \rho_1}{\Rightarrow \alpha}\ \rho_0 = \mathrm{Claus}_0
$$

$\forall i \in \{1, \ldots, n-1\}$, $\rho_i = \mathrm{cpl}_1$ or $\rho_i = \mathrm{Claus}_1$

$\pi(\mathcal{D}) = \langle\, \Psi_0 \cup \cdots \cup \Psi_n\,,\ \chi_0 \cdot \ldots \cdot \chi_n \,\rangle$

where $\langle \Psi_j, \chi_j \rangle = \pi(\rho_j)$

Fig. 2. A C_L-derivation of $\Rightarrow \alpha$.

$\mathrm{Ax}(L, V)$ we denote the set of instances ψ of L-axioms such that $\mathcal{V}_\psi \subseteq V$. The calculus relies on a clausification procedure `Clausify` satisfying conditions (C1)–(C3) and acts on sequents $\Gamma \Rightarrow \delta$ such that:

- either $\Gamma = \emptyset$ or $\Gamma = R \cup X$ and $(X)^\star \subseteq R$ and δ is an atom.

Rules of C_L are displayed in Fig. 1. Rule cpl_0 (initial rule) can only be applied if the condition $R \vdash_{\mathrm{c}} g$ holds; if this is the case, the conclusion $R, X \Rightarrow g$ is an initial sequent, namely a top sequent of a derivation. The other rules depend on parameters that are made explicit in the rule name. A bottom-up application of cpl_1 requires the choice of an implication clause $\lambda = (a \to b) \to c$ from X, we call the *main formula*, and the selection of a set of atoms $A \subseteq \mathcal{V}_{R,X,g}$ such that $R, A \vdash_{\mathrm{c}} b$, where b is the middle variable in λ. As discussed in [8,9], cpl_1 is a sort of generalization of the rule $L \to\to$ of the sequent calculus LJT/G4ip for IPL [5,18]. Rules Claus_0 and Claus_1 exploit the clausification procedure. Rule Claus_0 requires the clausification of the formula $\alpha \leftrightarrow g$, with g a new atom ($g \notin \mathcal{V}_\alpha$); in rule Claus_1, the clausified formula ψ is selected from $\mathrm{Ax}(L, \mathcal{V}_{R,X,g})$. In both cases, the clauses returned by `Clausify` are stored in the premise of the applied rule and the computed substitution χ is displayed in the rule name; moreover, Claus_0 is annotated with the new atom g and Claus_1 with the chosen L-axiom ψ. To recover the relevant information associated with the application of a rule ρ, in Fig. 1 we define the pair $\pi(\rho) = \langle \Psi, \chi \rangle$, where Ψ is a set of instances of L-axioms and χ is a substitution. C_L-trees and C_L-derivations are defined as usual (see e.g. [18]); a sequent σ is provable in C_L iff there exists a C_L-derivation having root sequent σ. Let us consider a C_L-derivation $\mathcal{D}$ of $\Rightarrow \alpha$ (see Fig. 2). Reading the derivation bottom-up, the first applied rule is Claus_0. After such an application, the obtained sequents have the form $\sigma_k = R_k, X_k \Rightarrow g$, where $R_k \cup X_k$ is non-empty, thus rule Claus_0 cannot be applied any more; the rule applied at the top is cpl_0. Note that $\mathcal{D}$ contains a unique branch, consisting of the sequents $\Rightarrow \alpha$, $\sigma_0, \ldots, \sigma_{n-1}$. In Fig. 2 we also define the pair $\pi(\mathcal{D}) = \langle \Psi, \chi \rangle$: Ψ collects the (instances of) L-axioms selected by rule Claus_1, χ is obtained by composing the substitutions associated with the applied rules. The definition of $\pi(\mathcal{T})$, with $\mathcal{T}$ a C_L-tree, is similar. By $\mathcal{T}(\alpha; R, X \Rightarrow g)$ we denote a C_L-tree having root $\Rightarrow \alpha$ and leaf $R, X \Rightarrow g$. Given a C_L-tree $\mathcal{T}$, $\mathcal{V}_\mathcal{T}$ is the set of variables occurring in $\mathcal{T}$. We state some properties about C_L-trees:

Lemma 2. *Let $\mathcal{T} = \mathcal{T}(\alpha; R, X \Rightarrow g)$ and let $\pi(\mathcal{T}) = \langle \Psi, \chi \rangle$.*

(i) $\mathcal{V}_{\chi(p)} \subseteq \mathcal{V}_\alpha$, *for every* $p \in \mathcal{V}_\mathcal{T}$.
(ii) $R, X \vdash_\mathrm{i} \beta \leftrightarrow \chi(\beta)$, *for every formula* β.
(iii) *If* $R, X, \Gamma \vdash_\mathrm{i} g$ *and* $\mathcal{V}_\Gamma \subseteq \mathcal{V}_\alpha$, *then* $\Gamma, \chi(\Psi) \vdash_\mathrm{i} \alpha$.

Proposition 1. *Let $\mathcal{D}$ be a C_L-derivation of $\Rightarrow \alpha$ and let $\pi(\mathcal{D}) = \langle \Psi, \chi \rangle$. Then, $\mathcal{V}_{\chi(\Psi)} \subseteq \mathcal{V}_\alpha$ and $\chi(\Psi) \vdash_\mathrm{i} \alpha$.*

$$\mathcal{D} = \begin{array}{c} \dfrac{R \vdash_\mathrm{c} g}{R, X \Rightarrow g}\,\mathrm{cpl}_0 \\ \vdots\, \mathcal{T} \\ \Rightarrow \alpha \end{array}$$

Proof. Since $\mathcal{D}$ is a C_L-derivation, $\mathcal{D}$ has the form depicted on the right where $\mathcal{T} = \mathcal{T}(\alpha; R, X \Rightarrow g)$; note that $\pi(\mathcal{T}) = \pi(\mathcal{D}) = \langle \Psi, \chi \rangle$. Since $R \vdash_\mathrm{c} g$, by Lemma 1(i) we get $R \vdash_\mathrm{i} g$, hence $R, X \vdash_\mathrm{i} g$. We can apply Lemma 2 and claim that $\mathcal{V}_{\chi(\Psi)} \subseteq \mathcal{V}_\alpha$ and $\chi(\Psi) \vdash_\mathrm{i} \alpha$. □

Given a C_L-derivation $\mathcal{D}$ of $\Rightarrow \alpha$, Prop. 1 exhibits how to extract a set of instances Ψ_α of the L-axioms such that $\Psi_\alpha \vdash_\mathrm{i} \alpha$. If $\mathcal{D}$ does not contain applications of rule Claus_1, Ψ_α is empty, and this ascertains that α is IPL-valid; actually, $\mathcal{D}$ can be immediately embedded into the calculus for IPL introduced in [8]. As an immediate consequence of Prop. 1, we get the soundness of C_L: if $\Rightarrow \alpha$ is provable in C_L, then α is L-valid.

Even though C_L-derivations have a simple structure, the design of a root-first proof search strategy for C_L is far from being trivial. After having applied rule Claus_0 to the root sequent $\Rightarrow \alpha$, we enter a loop where at each iteration k we search for a derivation of $\sigma_k = R_k, X_k \Rightarrow g$. It is convenient to firstly check whether $R_k \vdash_\mathrm{c} g$ so that, by applying rule cpl_0, we immediately close the derivation at hand. To check classical provability, we exploit a SAT-solver; each time the solver is invoked, the set R_k has increased, thus it is advantageous to use an incremental SAT-solver. If $R_k \nvdash_\mathrm{c} g$, we have to apply either rule cpl_1 or rule Claus_1, but it is not obvious which strategy should be followed. First, we have to select one between the two rules. If rule cpl_1 is chosen, we have to guess proper λ and A; otherwise, we have to apply Claus_1, and this requires the selection of an instance ψ of an L-axiom. In any case, if we followed a blind choice, the procedure would be highly inefficient. To guide proof search, we follow a different approach based on countermodel construction; to this aim, we introduce a representation of Kripke models where worlds are classical interpretations ordered by inclusion.

Countermodels. Let W be a finite set of interpretations with minimum M_0, namely: $M_0 \subseteq M$ for every $M \in W$. By $\mathcal{K}(W)$ we denote the Kripke model $\langle W, \leq, M_0, \vartheta \rangle$ where $\leq$ coincides with the subset relation $\subseteq$ and ϑ is the identity map, thus $M \Vdash p$ (in $\mathcal{K}(W)$) iff $p \in M$. We introduce the following *realizability relation* $\triangleright_W$ between elements of W and implication clauses:

$$M \triangleright_W (a \to b) \to c \text{ iff } (a \in M) \text{ or } (b \in M) \text{ or } (c \in M) \text{ or}$$
$$(\ \exists M' \in W \text{ s.t. } M \subset M' \text{ and } a \in M' \text{ and } b \notin M'\ ).$$

By $M \rhd_W X$ we mean that $M \rhd_W \lambda$ for every $\lambda \in X$. We state the crucial properties of the model $\mathcal{K}(W)$:

Proposition 2. *Let $\mathcal{K}(W)$ be the model generated by W and let $w \in W$. Let φ be a clause and $\lambda = (a \to b) \to c$ an implication clause.*

(i) If $w' \models \varphi$, for every $w' \in W$ such that $w \leq w'$, then $w \Vdash \varphi$.
(ii) If $w' \models b \to c$ and $w' \rhd_W \lambda$, for every $w' \in W$ such that $w \leq w'$, then $w \Vdash \lambda$.

Let $\mathcal{K}(W)$ be a model with root r, and assume that every interpretation w in W is a model of R; our goal is to get $r \Vdash R \cup X$ (where $(X)^* \subseteq R$), possibly by filling W with new worlds. To this aim, we exploit Prop. 2. By our assumption and point (i), we claim that $r \Vdash R$. Suppose that there is $w \in W$ and $\lambda = (a \to b) \to c \in X$ such that $w \not\rhd_W \lambda$; is it possible to amend $\mathcal{K}(W)$ in order to match (ii) and conclude $r \Vdash X$? By definition of $\rhd_W$, none of the atoms a, b, c belongs to w; moreover $\mathcal{K}(W)$ lacks a world w' such that $w \subset w'$ and $a \in w'$ and $b \notin w'$. We can try to fix $\mathcal{K}(W)$ by inserting the missing world w'; to preserve (i), we also need $w' \models R$. Accordingly, such a w' exists if and only if $R, w, a \nvdash_{\mathrm{c}} b$. This can be checked by querying a SAT-solver; moreover, if $R, w, a \nvdash_{\mathrm{c}} b$, the solver also computes the required w'. This completion process must be iterated until $\mathcal{K}(W)$ has been saturated with all the missing worlds or we get stuck. It is easy to check that the process eventually terminates. This is one of the key ideas beyond the procedure `intuitRIL` we present in next section.

4 The Procedure `intuitRIL`

We present the procedure `intuitRIL` (`intuit` with Restart for Intermediate Logics) that, given a formula α and a logic $L = \mathrm{IPL} + \mathrm{Ax}(L)$, returns either a set of L-axioms Ψ_α or a model $\mathcal{K}(W)$ with the following properties:

(Q1) If `intuitRIL(`α,L`)` returns Ψ_α, then $\Psi_\alpha \subseteq \mathrm{Ax}(L, \mathcal{V}_\alpha)$ and $\Psi_\alpha \vdash_{\mathrm{i}} \alpha$.
(Q2) If `intuitRIL(`α,L`)` returns $\mathcal{K}(W)$, then $\mathcal{K}(W)$ is an L-countermodel for α.

Thus, α is L-valid in the former case, not L-valid in the latter. If `intuitRIL(`α,L`)` returns Ψ_α, by tracing the computation we can build a C_L-derivation $\mathcal{D}$ of $\Rightarrow \alpha$ such that $\Psi_\alpha = \chi(\Psi)$, where $\langle \Psi, \chi \rangle = \pi(\mathcal{D})$; this certificates that $\Psi_\alpha \vdash_{\mathrm{i}} \alpha$.

The procedure is described by the flowchart in Fig. 3 and exploits a single incremental SAT-solver s: clauses can be added to s but not removed; by $\mathrm{R}(s)$ we denote the set of clauses stored in s. The SAT-solver is required to support the following operations:

- `newSolver(`R`)` creates a new SAT-solver initialized with the clauses in R.
- `addClauses(`s, R`)` adds the clauses in R to the SAT-solver s.
- `satProve(`s, A, g`)` calls s to decide whether $\mathrm{R}(s), A \vdash_{\mathrm{c}} g$ (A is a set of propositional variables). The solver outputs one of the following answers:
  - Yes(A'): thus, $A' \subseteq A$ and $\mathrm{R}(s), A' \vdash_{\mathrm{c}} g$;
  - No(M): thus, $A \subseteq M \subseteq \mathcal{V}_{\mathrm{R}(s)} \cup A$ and $M \models \mathrm{R}(s)$ and $g \notin M$.

  In the former case it follows that $\mathrm{R}(s), A \vdash_{\mathrm{c}} g$, in the latter $\mathrm{R}(s), A \nvdash_{\mathrm{c}} g$.

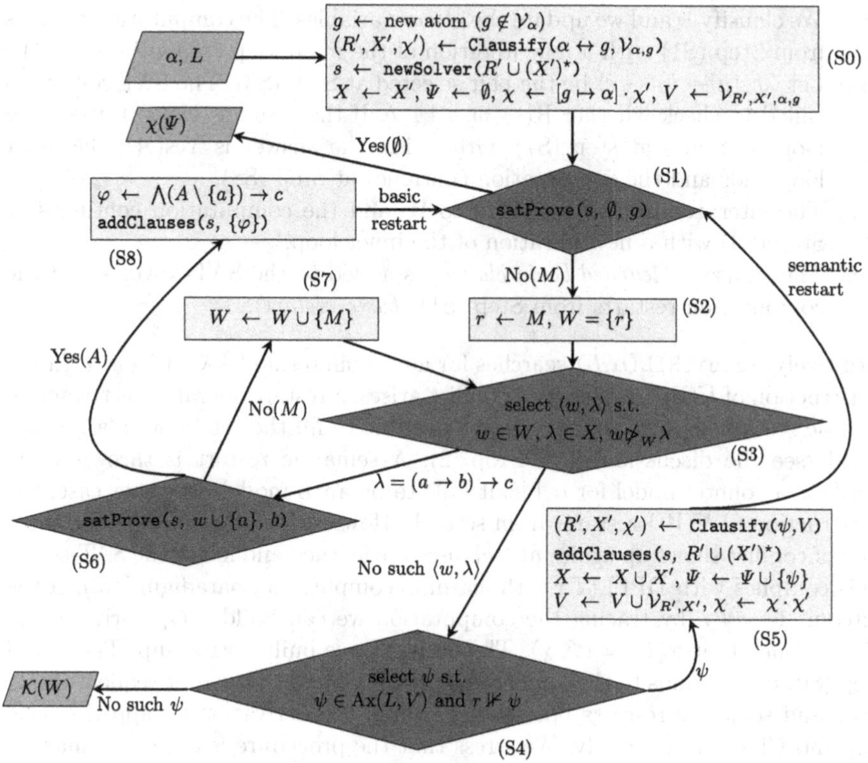

Fig. 3. Computation of `intuitRIL(`α, L`)`.

The computation of `intuitRIL(`α,L`)` consists of the following steps:

(S0) The formula $\alpha \leftrightarrow g$, with g new propositional variable, is clausified. The outcome (R', X', χ') is used to create a new SAT-solver s and to properly initialize the global variables X (set of implication clauses), Ψ (set of L-axiom instances), V (set of propositional variables) and χ (substitution).

(S1) A loop starts *(main loop)*. The SAT-solver s is called to check whether $\mathrm{R}(s) \vdash_{\mathrm{c}} g$. If the answer is Yes($\emptyset$), the computation stops yielding $\chi(\Psi)$. Otherwise, the output is No(M) and the computation continues at Step (S2).

(S2) We set $r = M$ (the root of $\mathcal{K}(W)$) and $W = \{r\}$.

(S3) A loop starts *(inner loop)*. We have to select a pair $\langle w, \lambda \rangle$ such that $w \in W$, $\lambda \in X$ and $w \not\triangleright_W \lambda$. If such a pair does not exist, the inner loop ends and next step is (S4), otherwise the inner loop continues at Step (S6).

(S4) As we show in Lemma 3, at this point $\mathcal{K}(W)$ is a countermodel for α. If all the axioms in $\mathrm{Ax}(L, V)$ are forced at the root r of $\mathcal{K}(W)$, then $\mathcal{K}(W)$ is an L-countermodel for α and the computation ends returning $\mathcal{K}(W)$. Otherwise, we select ψ from $\mathrm{Ax}(L, V)$ such that $r \nVdash \psi$ and the computation continues at Step (S5); we call ψ the *learned axiom*.

(S5) We clausify ψ and we update the global variables. The computation restarts from Step (S1) with a new iteration of the main loop (*semantic restart*).

(S6) Let $\langle w, (a \to b) \to c\rangle$ be the pair selected at Step (S3). The SAT-solver s is called to check whether $\mathrm{R}(s), w, a \vdash_{\mathrm{c}} b$. If the result is $\mathrm{No}(M)$, the inner loop continues at step (S7). Otherwise, the answer is $\mathrm{Yes}(A)$; the inner loop ends and the computation continues at Step (S8).

(S7) The interpretation M is added to W and the computation continues at Step (S3) with a new iteration of the inner loop.

(S8) The clause φ (*learned basic clause*) is added to the SAT-solver s and the computation restarts from Step (S1) (*basic restart*).

Intuitively, `intuitRIL(`α,L`)` searches for an L-countermodel $\mathcal{K}(W)$ for α. In the construction of $\mathcal{K}(W)$, whenever a conflict arises, a restart operation is triggered. A basic restart happens when it is not possible to fill the set W with a missing world (see the discussion after Prop. 2). A semantic restart is thrown when $\mathcal{K}(W)$ is a countermodel for α but it fails to be an L-model. In either case, the construction of $\mathcal{K}(W)$ restarts from scratch. However, to prevent that the same kind of conflict shows up again, new clauses are learned and fed to the SAT-solver (this complies with DPLL($\mathcal{T}$) with learning computation paradigm [16]). If the outcome is $\chi(\Psi)$, by tracing the computation we can build a C_L-derivation $\mathcal{D}$ of $\Rightarrow \alpha$ such that $\pi(\mathcal{D}) = \langle \Psi, \chi\rangle$. The derivation is built bottom-up. The initial Step (S0) corresponds to the application of rule Claus_0 to the root sequent $\Rightarrow \alpha$; basic and semantic restarts bottom-up expand the derivation by applying rule cpl_1 and Claus_1 respectively. We stress that the procedure is quite modular; to treat a specific logic L one has only to provide a concrete implementation of Step (S4). For $L = \mathrm{IPL}$, Step (S4) is trivial, since the set $\mathrm{Ax}(\mathrm{IPL}, V)$ is empty. Actually, `intuitRIL` applied to IPL has the same behaviour as the procedure `intuitR` introduced in [8].

Example 3. Let us consider *Jankov axiom* **wem** $= \neg a \lor \neg\neg a$ [2,13] (aka *weak excluded middle*), which holds in all frames having a single maximal world (thus, **wem** is GL-valid). The trace of the execution of `intuitRIL(`**wem**,GL`)` is shown in Fig. 4. The initial clausification yields $(R_0, X_0, \tilde{g})$, where X_0 consists of the implication clauses λ_0, λ_1 in Fig. 4 and R_0 contains the 7 clauses below:

$$\tilde{g} \to \tilde{p}_2,\quad \tilde{p}_0 \to \tilde{p}_2,\quad a \land \tilde{p}_0 \to \bot,\quad \tilde{p}_1 \to \tilde{p}_2,\quad \tilde{p}_0 \land \tilde{p}_1 \to \bot,\quad \tilde{p}_2 \to \tilde{g},\quad \tilde{p}_2 \to \tilde{p}_0 \lor \tilde{p}_1.$$

Each row in Fig. 4 displays the validity tests performed by the SAT-solver and the computed answers. If the result is $\mathrm{No}(M)$, the last two columns show the worlds w_k in the current set W and, for each w_k, the list of λ such that $w \not\triangleright_W \lambda$; the pair selected for the next step is underlined. For instance, after call (1) we have $W = \{w_0\}$, $w_0 \not\triangleright_W \lambda_0$ and $w_0 \not\triangleright_W \lambda_1$; the selected pair is $\langle w_0, \lambda_0\rangle$. After call (2), the set W is updated by adding the world w_1; we have $w_1 \triangleright_W \lambda_0$, $w_1 \triangleright_W \lambda_1$, $w_0 \triangleright_W \lambda_0$ and $w_0 \not\triangleright_W \lambda_1$. Whenever the SAT-solver outputs $\mathrm{Yes}(A)$, we display the learned clause ψ_k. The SAT-solver is invoked 18 times and there are 6 restarts (1 semantic, 5 basic). After (3), we get $W = \{w_0, w_1, w_2\}$ and no pair $\langle w, \lambda\rangle$ can be selected, hence the model $\mathcal{K}(W)$ (displayed in the figure) is a countermodel for **wem**. However, $\mathcal{K}(W)$ is not a GL-model (indeed, it is not linear), hence we choose an instance of the linearity axiom not forced at w_0, namely ψ_0, and we force a semantic restart. The clausification of ψ_0 produces 6 new clauses and the new implication clauses λ_2, λ_3, λ_4. After each restart, the sets R_j are:

$$\begin{aligned} R_1 &= R_0 \cup \{\, \tilde{p}_3 \to \tilde{p}_4,\ a \to \tilde{p}_5,\ \tilde{p}_3 \wedge \tilde{p}_5 \to a,\ a \wedge \tilde{p}_4 \to \tilde{p}_3,\ a \wedge \tilde{p}_3 \to \bot,\ \tilde{p}_4 \vee \tilde{p}_5 \,\} \\ R_j &= R_{j-1} \cup \{\psi_{j-1}\} \quad \text{for } 2 \le j \le 6 \text{ (the } \psi_j'\text{s are defined in Fig. 4).} \end{aligned}$$

The C_{GL}-derivation of $\Rightarrow \neg a \vee \neg\neg a$ extracted from the computation is:

$$\dfrac{\dfrac{\dfrac{R_1, a, \tilde{p}_0 \vdash_{\mathrm{c}} \bot \qquad \dfrac{R_2, a, \tilde{p}_0 \vdash_{\mathrm{c}} \bot \qquad \dfrac{R_3, a, \tilde{p}_3 \vdash_{\mathrm{c}} \bot \qquad \dfrac{R_4, \tilde{p}_0, \tilde{p}_5 \vdash_{\mathrm{c}} \bot \qquad \dfrac{R_5, a, \tilde{p}_4 \vdash_{\mathrm{c}} \bot \qquad \dfrac{R_6 \vdash_{\mathrm{c}} \tilde{g}}{R_6, X_1 \Rightarrow \tilde{g}}\,\mathrm{cpl}_0}{R_5, X_1 \Rightarrow \tilde{g}}\,\mathrm{cpl}_1(\lambda_1)}{R_4, X_1 \Rightarrow \tilde{g}}\,\mathrm{cpl}_1(\lambda_0)}{R_3, X_1 \Rightarrow \tilde{g}}\,\mathrm{cpl}_1(\lambda_1)}{R_2, X_1 \Rightarrow \tilde{g}}\,\mathrm{cpl}_1(\lambda_0)}{R_1, X_1 \Rightarrow \tilde{g}}\,\mathrm{cpl}_1(\lambda_3)}{R_0, X_0 \Rightarrow \tilde{g}}\,\mathrm{Claus}_1(\psi_0, \chi_1)}{\Rightarrow \neg a \vee \neg\neg a}\,\mathrm{Claus}_0(\tilde{g}, \chi_0)$$

◊

Now, we discuss partial correctness and termination of `intuitRIL`. Let us denote with $\sim_{\mathrm{c}}$ classical equivalence ($\alpha \sim_{\mathrm{c}} \beta$ iff $\vdash_{\mathrm{c}} \alpha \leftrightarrow \beta$) and with $\sim_{\mathrm{i}}$ intuitionistic equivalence ($\alpha \sim_{\mathrm{i}} \beta$ iff $\vdash_{\mathrm{i}} \alpha \leftrightarrow \beta$). We introduce some notation.

(†) The following terms refer to the configuration at the beginning of iteration k ($k \ge 0$), just after the execution of Step (S2):
- Φ_k is the set collecting all the learned basic clauses;
- R_k is the set of clauses stored in the SAT-solver s;
- X_k, Ψ_k, V_k, χ_k, r_k are the values of the corresponding global variables.

In Fig. 5 we inductively define the C_L-tree $\mathcal{T}_k$, having the form $\mathcal{T}(\alpha; R_k, X_k \Rightarrow g)$. In the application of rule Claus_0, g and χ' are defined as in Step (S0). In rule cpl_1, λ is the implication clause selected at iteration $k-1$ (of the main loop) in the last execution of Step (S3); A is the value computed at Step (S6) of iteration $k-1$. In the application of rule Claus_1, ψ and χ' are defined as in the execution of Step (S4) and (S5) of iteration $k-1$. One can easily check that the applications of the rules are sound. If Step (S1) yields Yes($\emptyset$), we can turn $\mathcal{T}_k$ into a C_L-derivation by applying rule cpl_0.

Next lemma states some relevant properties of the computations of `intuitRIL`.

Lemma 3. *Let us consider the execution of iteration k of the main loop ($k \ge 0$).*

(i) $(X_k)^\star \cup \Phi_k \subseteq R_k$.
(ii) $V_k = \mathcal{V}_{\mathcal{T}_k}$ *and* $\Psi_k \subseteq \mathrm{Ax}(L, V_k)$ *and* $\pi(\mathcal{T}_k) = \langle \Psi_k, \chi_k \rangle$.
(iii) $\mathcal{V}_{\chi_k(p)} \subseteq \mathcal{V}_\alpha$, *for every* $p \in V_k$, *and* $R_k, X_k \vdash_{\mathrm{i}} \beta \leftrightarrow \chi_k(\beta)$, *for every* β.

$\lambda_0 = (\tilde{p}_0 \to \bot) \to \tilde{p}_1 \qquad \lambda_1 = (a \to \bot) \to \tilde{p}_0$
$\lambda_2 = (a \to \tilde{p}_3) \to \tilde{p}_4 \qquad \lambda_3 = (a \to \bot) \to \tilde{p}_3 \qquad \lambda_4 = (\tilde{p}_3 \to a) \to \tilde{p}_5$

$w_0 = \emptyset \quad w_1 = \{\tilde{g}, \tilde{p}_0, \tilde{p}_2\} \quad w_2 = \{a, \tilde{g}, \tilde{p}_1, \tilde{p}_2\} \quad w_3 = \{\tilde{p}_4\} \quad w_4 = \{\tilde{g}, \tilde{p}_0, \tilde{p}_2, \tilde{p}_4\}$
$w_5 = \{\tilde{g}, \tilde{p}_0, \tilde{p}_2, \tilde{p}_3, \tilde{p}_4\} \quad w_6 = \{a, \tilde{p}_5\} \quad w_7 = \{\tilde{p}_3, \tilde{p}_4\} \quad w_8 = \{\tilde{g}, \tilde{p}_0, \tilde{p}_2, \tilde{p}_3, \tilde{p}_4\}$
$w_9 = \{\tilde{p}_5\} \quad w_{10} = \{\tilde{p}_4\} \quad w_{11} = \{\tilde{g}, \tilde{p}_0, \tilde{p}_2, \tilde{p}_3, \tilde{p}_4\}$

$\chi_0 = [\tilde{g} \mapsto \neg a \vee \neg\neg a,\ \tilde{p}_0 \mapsto \neg a,\ \tilde{p}_1 \mapsto \neg\neg a,\ \tilde{p}_2 \mapsto \neg a \vee \neg\neg a]$
$\chi_1 = [\tilde{p}_3 \mapsto \neg a,\ \tilde{p}_4 \mapsto a \to \neg a,\ \tilde{p}_5 \mapsto \neg a \to a]$

	@SAT	Answer	W	λ s.t. $w \not\triangleright_W \lambda$
Start	(1) $R_0 \vdash_c \tilde{g}$?	No(w_0)	$\underline{w_0}$	$\underline{\lambda_0}, \lambda_1$
	(2) $R_0,\ w_0,\ \tilde{p}_0 \vdash_c \bot$?	No(w_1)	w_1	$\emptyset$
			$\underline{w_0}$	$\underline{\lambda_1}$
	(3) $R_0,\ w_0,\ a \vdash_c \bot$?	No(w_2)	w_2	$\emptyset$
			w_1	$\emptyset$
			w_0	$\emptyset$

Semantic failure

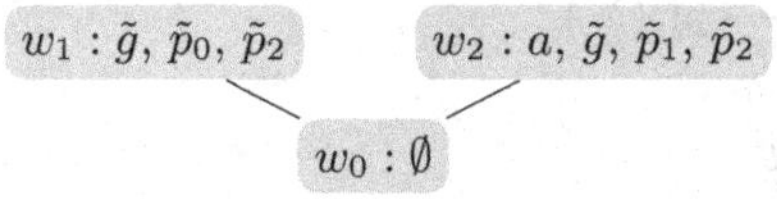

Learned axiom:
$\psi_0 = (a \to \neg a) \vee (\neg a \to a)$

	@SAT	Answer	W	λ s.t. $w \not\triangleright_W \lambda$
SRest 1	(4) $R_1 \vdash_c \tilde{g}$?	No(w_3)	$\underline{w_3}$	$\underline{\lambda_0}, \lambda_1, \lambda_3, \lambda_4$
	(5) $R_1,\ w_3,\ \tilde{p}_0 \vdash_c \bot$?	No(w_4)	$\underline{w_4}$	$\lambda_3, \underline{\lambda_4}$
			w_3	$\lambda_1, \lambda_3, \lambda_4$
	(6) $R_1,\ w_4,\ \tilde{p}_3 \vdash_c a$?	No(w_5)	w_5	$\emptyset$
			$\underline{w_4}$	$\underline{\lambda_3}$
			w_3	λ_1, λ_3
	(7) $R_1,\ w_4,\ a \vdash_c \bot$?	Yes($\{a, \tilde{p}_0\}$)	$\psi_1 = \tilde{p}_0 \to \tilde{p}_3$	
BRest 2	(8) $R_2 \vdash_c \tilde{g}$?	No(w_6)	$\underline{w_6}$	$\underline{\lambda_0}$
	(9) $R_2,\ w_6,\ \tilde{p}_0 \vdash_c \bot$?	Yes($\{a, \tilde{p}_0\}$)	$\psi_2 = a \to \tilde{p}_1$	
BRest 3	(10) $R_3 \vdash_c \tilde{g}$?	No(w_7)	$\underline{w_7}$	$\underline{\lambda_0}, \lambda_1$
	(11) $R_3,\ w_7,\ \tilde{p}_0 \vdash_c \bot$?	No(w_8)	w_8	$\emptyset$
			$\underline{w_7}$	$\underline{\lambda_1}$
	(12) $R_3,\ w_7,\ a \vdash_c \bot$?	Yes($\{a, \tilde{p}_3\}$)	$\psi_3 = \tilde{p}_3 \to \tilde{p}_0$	
BRest 4	(13) $R_4 \vdash_c \tilde{g}$?	No(w_9)	$\underline{w_9}$	$\underline{\lambda_0}, \lambda_1, \lambda_2, \lambda_3$
	(14) $R_4,\ w_9,\ \tilde{p}_0 \vdash_c \bot$?	Yes($\{\tilde{p}_0, \tilde{p}_5\}$)	$\psi_4 = \tilde{p}_5 \to \tilde{p}_1$	
BRest 5	(15) $R_5 \vdash_c \tilde{g}$?	No(w_{10})	$\underline{w_{10}}$	$\underline{\lambda_0}, \lambda_1, \lambda_3, \lambda_4$
	(16) $R_5,\ w_{10},\ \tilde{p}_0 \vdash_c \bot$?	No(w_{11})	w_{11}	$\emptyset$
			$\underline{w_{10}}$	$\underline{\lambda_1}, \lambda_3$
	(17) $R_5,\ w_{10},\ a \vdash_c \bot$?	Yes($\{a, \tilde{p}_4\}$)	$\psi_5 = \tilde{p}_4 \to \tilde{p}_0$	
BRest 6	(18) $R_6 \vdash_c \tilde{g}$?	Yes($\emptyset$)	**Proved**	

Fig. 4. Computation of `intuitRIL`($\neg a \vee \neg\neg a$, GL).

$$\mathcal{T}_0 = \quad \frac{R_0, X_0 \Rightarrow g}{\Rightarrow \alpha}\ \mathrm{Claus}_0(g, \chi')$$

$$\mathcal{T}_k = \begin{array}{c} \dfrac{R_{k-1}, A \vdash_c b \quad R_k, X_k \Rightarrow g}{R_{k-1}, X_{k-1} \Rightarrow g}\ \mathrm{cpl}_1(\lambda) \\ \vdots\ \mathcal{T}_{k-1} \\ \Rightarrow \alpha \end{array} \qquad \mathcal{T}_k = \begin{array}{c} \dfrac{R_k, X_k \Rightarrow g}{R_{k-1}, X_{k-1} \Rightarrow g}\ \mathrm{Claus}_1(\psi, \chi') \\ \vdots\ \mathcal{T}_{k-1} \\ \Rightarrow \alpha \end{array}$$

if $k > 0$ and iteration $k-1$ ends with a basic restart (thus $X_k = X_{k-1}$)

if $k > 0$ and iteration $k-1$ ends with a semantic restart

Fig. 5. Definition of $\mathcal{T}_k$ ($k \geq 0$).

(iv) *At every step after (S2), $w \models R_k$, for every $w \in W$.*
(v) *At every step after (S2), r_k is the root of $\mathcal{K}(W)$ and $r_k \Vdash R_k$ and $r_k \nVdash g$.*
(vi) *At Step (S4), $r_k \Vdash R_k \cup X_k \cup \Psi_k$ and $r_k \nVdash g$ (in $\mathcal{K}(W)$).*
(vii) *Assume that iteration k ends with a basic restart and let φ be the learned basic clause. For every $\varphi' \in \Phi_k$, $\varphi \not\sim_c \varphi'$.*
(viii) *Assume that iteration k ends with a semantic restart and let ψ be the learned axiom. For every $\psi' \in \Psi_k$, $\chi_k(\psi) \not\sim_i \chi_k(\psi')$.*

Proof. We only sketch the proof of the non-trivial points.

(iii). By Lemma 2 applied to $\mathcal{T}_k$.

(v). Every interpretation M generated at Step (S6) is a superset of r_k, thus after Step (S2) r_k is the minimum element of W and the root of $\mathcal{K}(W)$. By (iv) and Prop. 2(i), $r_k \Vdash R_k$. Since $g \notin r_k$, we get $r_k \nVdash g$.

(vi). At Step (S4), $w \triangleright_W \lambda$ for every $w \in W$ and $\lambda \in X_k$. Since $(X_k)^\star \subseteq R_k$, by Prop. 2(ii) we get $r_k \Vdash X_k$. Let $\psi \in \Psi_k$; then, ψ has been learned at some iteration $k' < k$. Let (R', X', χ') be the output of `Clausify(`ψ,V`)` at Step (S5) of iteration k' . Since $R' \subseteq R_k$ and $X' \subseteq X_k$, it holds that $r_k \Vdash R' \cup X'$. By (P1) $R', X' \vdash_i \psi$, hence $r_k \Vdash \psi$, which proves $r_k \Vdash \Psi_k$.

(vii). Let $\varphi' \in \Phi_k$; we show that $\varphi \not\sim_c \varphi'$. Let $\varphi = \bigwedge(A \setminus \{a\}) \to c$; then, there are $w \in W$ and $\lambda = (a \to b) \to c \in X_k$ such that $\langle w, \lambda \rangle$ has been selected at Step (S3) and the outcome of `satProve(`s,$w \cup \{a\}$,b`)` at Step (S6) is Yes(A). Note that $w \not\triangleright_W \lambda$, hence $c \notin w$; since $A \subseteq w \cup \{a\}$, we get $w \not\models \varphi$. On the other hand, $w \models \varphi'$, since $\varphi' \in \Phi_k$ and $\Phi_k \subseteq R_k$. We conclude $\varphi \not\sim_c \varphi'$.

(viii). Let $\psi' \in \Psi_k$ and let $\mathcal{K}(W)$ be the model obtained at Step (S4) of iteration k. By (iii) $R_k, X_k \vdash_i \psi \leftrightarrow \chi_k(\psi)$ and $R_k, X_k \vdash_i \psi' \leftrightarrow \chi_k(\psi')$. Since $r_k \nVdash \psi$ and $r_k \Vdash \psi'$ (indeed, $\psi' \in \Psi_k$ and $r_k \Vdash \Psi_k$) and $r_k \Vdash R_k \cup X_k$, we get $r_k \nVdash \chi_k(\psi)$ and $r_k \Vdash \chi_k(\psi')$. We conclude $\chi_k(\psi) \not\sim_i \chi_k(\psi')$. □

The following proposition proves the partial correctness of `intuitRIL`:

Proposition 3. `intuitRIL(`α,L`)` *satisfies properties (Q1) and (Q2).*

Proof. Let us assume that the computation ends at iteration k with output Ψ_α. Then, the call to the SAT-solver at Step (S0) yields Yes($\emptyset$), meaning that $R_k \vdash_c g$. We can build the following C_L-derivation $\mathcal{D}$ of $\Rightarrow \alpha$:

$$\mathcal{D} = \begin{array}{c} \dfrac{R_k \vdash_{\rm c} g}{R_k, X_k \Rightarrow g}\,{\rm cpl}_0 \\ \vdots\; \mathcal{T}_k \\ \Rightarrow \alpha \end{array} \qquad \pi(\mathcal{D}) = \pi(\mathcal{T}_k) = \langle \Psi_k, \chi_k \rangle$$

Note that $\Psi_\alpha = \chi_k(\Psi_k)$. Accordingly, by Prop. 1 we get (Q1).

Let us assume that the output is the model $\mathcal{K}(W)$, having root r. Then, $\mathcal{K}(W)$ is an L-model (otherwise, Step (S4) should have forced a semantic restart). By Lemma 3(vi) we get $r \Vdash R_0 \cup X_0$ and $r \nVdash g$. Since at Step (S0) we have clausified the formula $\alpha \leftrightarrow g$, by (P1) we get $R_0, X_0 \vdash_{\rm i} \alpha \leftrightarrow g$, which implies $r \Vdash \alpha \leftrightarrow g$. We conclude that $r \nVdash \alpha$, hence (Q2) holds. □

It seems challenging to provide a general proof of termination, and each logic must be treated apart. We can only state some general properties about the termination of the inner loop and of consecutive basic restarts.

Proposition 4. *(i) The inner loop is terminating.*
(ii) The number of consecutive basic restarts is finite.

Proof. Let us assume, by absurd, that the inner loop is not terminating. For every $j \geq 0$, by W_j we denote the value of W at Step (S3) of iteration j of the inner loop; note that the value of the variable V does not change during the iterations. We show that $W_j \subset W_{j+1}$, for every $j \geq 0$. At iteration j, the outcome of Step (S6) is No(M). Thus, there are $w \in W_j$ and $\lambda = (a \to b) \to c \in X$ such that the pair $\langle w, \lambda \rangle$ has been selected at Step (S3); accordingly, $w \not\triangleright_{W_j} \lambda$ and $w \cup \{a\} \subseteq M$ and $b \notin M$. We have $M \notin W_j$, otherwise we would get $w \triangleright_{W_j} \lambda$, a contradiction. Since $W_{j+1} = W_j \cup \{M\}$, this proves that $W_j \subset W_{j+1}$. We have shown that $W_0 \subset W_1 \subset W_2 \ldots$. This leads to a contradiction since, for every $j \geq 0$ and every $w \in W_j$, w is a subset of V and V is finite. We conclude that the inner loop is terminating, and this proves (i).

Let us assume, by contradiction, that there is an infinite sequence of consecutive basic restarts. Then, there is $n \geq 0$ such that, for every $k \geq n$, the iteration k of the main loop ends with a basic restart. Let φ_k be the clause learned at iteration k. Note that an iteration ending with a basic restart does not introduce new atoms, thus $\mathcal{V}_{\varphi_k} \subseteq V_n$ for every $k \geq n$ (where V_n is defined as in (†)). We get a contradiction, since V_n is finite and, by Lemma 3(vi), the clauses φ_k are pairwise non $\sim_{\rm c}$-equivalent; this proves (ii). □

Lemma 3(vii) guarantees that the learned axioms are pairwise distinct, but this is not sufficient to prove termination since in general we cannot set a bound on the size and on the number of learned axioms. In next section we present some relevant logics where the procedure is terminating.

5 Termination

Let GL = IPL + **lin** be the Gödel-Dummett logic presented in Ex. 1; we show that every call `intuitRIL(`α`,GL)` is terminating. To this aim, we exploit the bounding function $\mathrm{Ax}_{\mathrm{GL}}(\alpha)$ presented in the mentioned example.

Lemma 4. *Let us consider the computation of* `intuitRIL(`α`,GL)` *and assume that at iteration k of the main loop Step (S4) is executed and that the obtained model $\mathcal{K}(W)$ is not linear. Then, there exists $\psi \in \mathrm{Ax}_{\mathrm{GL}}(\alpha)$ such that $r_k \nVdash \psi$.*

Proof. Let us assume that $\mathcal{K}(W)$ has two distinct maximal worlds w_1 and w_2; note that $w_1 \subseteq V_k$ and $w_2 \subseteq V_k$ (with V_k defined as in (†)). We show that:

(a) $w_1 \cap \mathcal{V}_\alpha \neq w_2 \cap \mathcal{V}_\alpha$.

Suppose by contradiction $w_1 \cap \mathcal{V}_\alpha = w_2 \cap \mathcal{V}_\alpha$; let $p \in V_k$ and $\beta = \chi_k(p)$ (with χ_k defined as in (†)). By Lemma 3(iii), $R_k, X_k \vdash_{\mathrm{i}} p \leftrightarrow \beta$; by Lemma 3(vi) we get $w_1 \Vdash p \leftrightarrow \beta$ and $w_2 \Vdash p \leftrightarrow \beta$. Since $\mathcal{V}_\beta \subseteq \mathcal{V}_\alpha$ (see Lemma 3(iii)) and we are assuming $w_1 \cap \mathcal{V}_\alpha = w_2 \cap \mathcal{V}_\alpha$, it holds that $w_1 \Vdash \beta$ iff $w_2 \Vdash \beta$, thus $w_1 \Vdash p$ iff $w_2 \Vdash p$, namely $p \in w_1$ iff $p \in w_2$. Since p is any element of V_k, we get $w_1 = w_2$, a contradiction; this proves (a). By (a) there is $a \in \mathcal{V}_\alpha$ such that either $a \in w_1 \setminus w_2$ or $a \in w_2 \setminus w_1$. We consider the former case (the latter one is symmetric), corresponding to Case 1 in Fig. 6. We have $w_1 \Vdash a$ and $w_2 \Vdash \neg a$; setting $\psi = (a \to \neg a) \vee (\neg a \to a)$, we conclude $r_k \nVdash \psi$.

Assume that $\mathcal{K}(W)$ has only one maximal world; since it is not linear, there are three distinct worlds w_1, w_2, w_3 as in Case 2 in Fig. 6, namely: w_1 is an immediate successor of w_2 and w_3 (i.e., for $j \in \{2,3\}$, $w_j < w_1$ and, if $w_j < w$, then $w_1 \leq w$), $w_2 \nleq w_3$, $w_3 \nleq w_2$. Reasoning as in (a), we get:

(b) $w_2 \cap \mathcal{V}_\alpha \neq w_3 \cap \mathcal{V}_\alpha$. (c) $w_2 \cap \mathcal{V}_\alpha \subset w_1 \cap \mathcal{V}_\alpha$ and $w_3 \cap \mathcal{V}_\alpha \subset w_1 \cap \mathcal{V}_\alpha$.

By (b) there is $a \in \mathcal{V}_\alpha$ such that either $a \in w_2 \setminus w_3$ or $a \in w_3 \setminus w_2$. Let us consider the former case (the latter one is symmetric). By (c), there is $b \in \mathcal{V}_\alpha$ such that $b \in w_1 \setminus w_2$. If $b \in w_3$ (Case 2.1 in Fig. 6), we get $a \in w_2$, $b \notin w_2$, $a \notin w_3$, $b \in w_3$. Setting $\psi = (a \to b) \vee (b \to a)$, we conclude $r_k \nVdash \psi$. Finally, let us assume $b \notin w_3$ (Case 2.2). We have $\{a, b\} \subseteq w_1$, $a \in w_2$, $b \notin w_2$, $a \notin w_3$ and $b \notin w_3$. It is easy to check that $w_3 \Vdash a \to b$ (recall that $w_3 < w$ implies $w_1 \leq w$), thus $w_3 \nVdash (a \to b) \to a$. On the other hand $w_2 \nVdash a \to (a \to b)$. Setting $\psi = (a \to (a \to b)) \vee ((a \to b) \to a)$, we get $r_k \nVdash \psi$. □

We exploit Lemma 4 to implement Step (S4). If $\mathcal{K}(W)$ is linear, then $\mathcal{K}(W)$ is a GL-model and we are done. Otherwise, the proof of Lemma 4 hints an effective method to select an instance ψ of **lin** from $\mathrm{Ax}_{\mathrm{GL}}(\alpha)$.

Proposition 5. *The computation of* `intuitRIL(`α`,GL)` *is terminating.*

Proof. Assume that `intuitRIL(`α`,GL)` is not terminating. Since the number of iterations of the inner loop and of the consecutive basic restarts is finite (see Prop. 4), Step (S4) must be executed infinitely many times. This leads to a contradiction, since the axioms selected at Step (S4) are pairwise distinct (see Lemma 3(vii)) and such axioms are chosen from the finite set $\mathrm{Ax}_{\mathrm{GL}}(\alpha)$. □

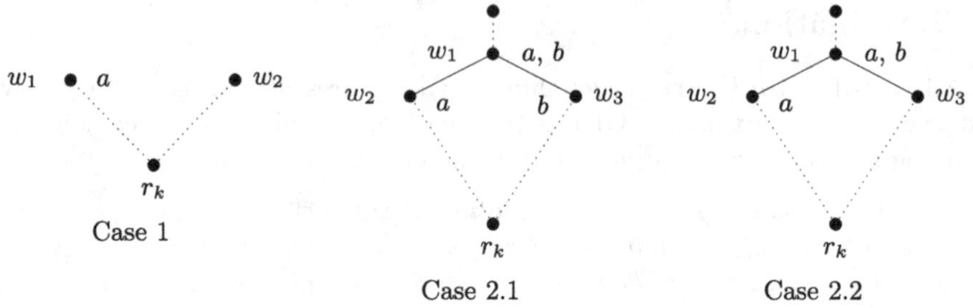

Fig. 6. Proof of Lemma 4, case analysis.

As a corollary, we get that $\mathrm{Ax}_{\mathrm{GL}}(\alpha)$ is a bounding function for GL:

Proposition 6. *If α is* GL*-valid, there is $\Psi_\alpha \subseteq \mathrm{Ax}_{\mathrm{GL}}(\alpha)$ such that $\Psi_\alpha \vdash_{\mathrm{i}} \alpha$.*

Other proof-search strategies for GL are discussed in [10,14]. This technique can be extended to other notable intermediate logics. Among these, we recall the logics GL_n (Gödel Logic of depth n), obtained by adding to GL the axioms $\mathbf{bd}_n$ (bounded depth) where: $\mathbf{bd}_0 = a_0 \vee \neg a_0$, $\mathbf{bd}_{n+1} = a_{n+1} \vee (a_{n+1} \to \mathbf{bd}_n)$. Semantically, GL_n is the logic characterized by linear frames having depth at most n. We are not able to prove termination for the logics $\mathrm{IPL} + \mathbf{bd}_n$, but we can implement the following terminating strategy for GL_n. Let $\mathcal{K}(W)$ be the model obtained at Step (S4) of the computation of intuitRIL(α,GL_n):

- If $\mathcal{K}(W)$ is not linear, we select the axiom ψ from $\mathrm{Ax}_{\mathrm{GL}}(\alpha)$.
- Otherwise, assume that $\mathcal{K}(W)$ is linear but not a GL_n-model. Then, $\mathcal{K}(W)$ contains a chain of worlds $w_0 \subset w_1 \subset \cdots \subset w_{n+1}$. The crucial point is that $w_{j+1} \setminus w_j$ contains at least a propositional variable from $\mathcal{V}_\alpha$, for every $0 \leq j \leq n$. Thus, we can choose a proper renaming of $\mathbf{bd}_n$ as ψ.

Another terminating logic is the Jankov Logic (see Ex. 3); actually, also in this case the learned axiom can be chosen by renaming the **wem** axiom. In general, all the logics BTW_n (Bounded Top Width, at most n maximal worlds, see [2]) are terminating. An intriguing case is Scott Logic ST [2]: even though the class of ST-frames is not first-order definable, we can implement a learning procedure for ST-axioms arguing as in [7] (see Sec. 2.5.2). Some of the mentioned logics have been implemented in intuitRIL[1].

One may wonder whether this method can be applied to other non-classical logics or to fragments of predicate logics (these issues have been already raised in the seminal paper [4]). A significant work in this direction is [11], where the procedure has been applied to some modal logics. However, the main difference with the original approach is that it is not possible to use a single SAT-solver, but one needs a supply of SAT-solvers. This is primarily due to the fact that forcing relation of modal Kripke models is not persistent; thus worlds are loosely related and must be handled by independent solvers.

[1] Available at https://github.com/cfiorentini/intuitRIL.

References

1. Avellone, A., Moscato, U., Miglioli, P., Ornaghi, M.: Generalized tableau systems for intermediate propositional logics. In: Galmiche, D. (ed.) TABLEAUX 1997. LNCS, vol. 1227, pp. 43–61. Springer, Heidelberg (1997). https://doi.org/10.1007/BFb0027404
2. Chagrov, A.V., Zakharyaschev, M.: Modal Logic, Oxford Logic Guides, vol. 35. Oxford University Press (1997)
3. Ciabattoni, A., Lang, T., Ramanayake, R.: Bounded-analytic sequent calculi and embeddings for hypersequent logics. J. Symb. Log. **86**(2), 635–668 (2021)
4. Claessen, K., Rosén, D.: SAT modulo intuitionistic implications. In: Davis, M., Fehnker, A., McIver, A., Voronkov, A. (eds.) LPAR 2015. LNCS, vol. 9450, pp. 622–637. Springer, Heidelberg (2015). https://doi.org/10.1007/978-3-662-48899-7_43
5. Dyckhoff, R.: Contraction-free sequent calculi for intuitionistic logic. J. Symb. Log. **57**(3), 795–807 (1992)
6. Ferrari, M., Fiorentini, C., Fiorino, G.: fCube: an efficient prover for intuitionistic propositional logic. In: Fermüller, C.G., Voronkov, A. (eds.) LPAR 2010. LNCS, vol. 6397, pp. 294–301. Springer, Heidelberg (2010). https://doi.org/10.1007/978-3-642-16242-8_21
7. Fiorentini, C.: Kripke completeness for intermediate logics. Ph.D. thesis, Università degli Studi di Milano (2000)
8. Fiorentini, C.: Efficient SAT-based proof search in intuitionistic propositional logic. In: Platzer, A., Sutcliffe, G. (eds.) CADE 2021. LNCS (LNAI), vol. 12699, pp. 217–233. Springer, Cham (2021). https://doi.org/10.1007/978-3-030-79876-5_13
9. Fiorentini, C., Goré, R., Graham-Lengrand, S.: A proof-theoretic perspective on SMT-solving for intuitionistic propositional logic. In: Cerrito, S., Popescu, A. (eds.) TABLEAUX 2019. LNCS (LNAI), vol. 11714, pp. 111–129. Springer, Cham (2019). https://doi.org/10.1007/978-3-030-29026-9_7
10. Fiorino, G.: Terminating calculi for propositional dummett logic with subformula property. J. Autom. Reason. **52**(1), 67–97 (2013). https://doi.org/10.1007/s10817-013-9276-7
11. Goré, R., Kikkert, C.: CEGAR-tableaux: improved modal satisfiability via modal clause-learning and SAT. In: Das, A., Negri, S. (eds.) TABLEAUX 2021. LNCS (LNAI), vol. 12842, pp. 74–91. Springer, Cham (2021). https://doi.org/10.1007/978-3-030-86059-2_5
12. Goré, R., Thomson, J., Wu, J.: A history-based theorem prover for intuitionistic propositional logic using global caching: IntHistGC system description. In: Demri, S., Kapur, D., Weidenbach, C. (eds.) IJCAR 2014. LNCS (LNAI), vol. 8562, pp. 262–268. Springer, Cham (2014). https://doi.org/10.1007/978-3-319-08587-6_19
13. Jankov, V.: The calculus of the weak "law of excluded middle.". Math. USSR **8**, 648–650 (1968)
14. Larchey-Wendling, D.: Gödel-dummett counter-models through matrix computation. Electron. Notes Theory Comput. Sci. **125**(3), 137–148 (2005)
15. Lifschitz, V., Pearce, D., Valverde, A.: Strongly equivalent logic programs. ACM Trans. Comput. Log. **2**(4), 526–541 (2001)
16. Nieuwenhuis, R., Oliveras, A., Tinelli, C.: Solving SAT and SAT modulo theories: from an abstract Davis-Putnam-Logemann-Loveland procedure to DPLL(T). J. ACM **53**(6), 937–977 (2006)

17. Schmidt, R.A., Tishkovsky, D.: Automated synthesis of tableau calculi. Log. Methods Comput. Sci. **7**(2) (2011)
18. Troelstra, A.S., Schwichtenberg, H.: Basic Proof Theory, Cambridge Tracts in Theoretical Computer Science, vol. 43, 2nd edn. Cambridge University Press, Cambridge (2000)

Clause Redundancy and Preprocessing in Maximum Satisfiability

Hannes Ihalainen, Jeremias Berg(✉), and Matti Järvisalo

HIIT, Department of Computer Science, University of Helsinki, Helsinki, Finland
hannes.ihalainen@helsinki.fi, jeremias.berg@helsinki.fi, matti.jarvisalo@helsinki.fi

Abstract. The study of clause redundancy in Boolean satisfiability (SAT) has proven significant in various terms, from fundamental insights into preprocessing and inprocessing to the development of practical proof checkers and new types of strong proof systems. We study liftings of the recently-proposed notion of propagation redundancy—based on a semantic implication relationship between formulas—in the context of maximum satisfiability (MaxSAT), where of interest are reasoning techniques that preserve optimal cost (in contrast to preserving satisfiability in the realm of SAT). We establish that the strongest MaxSAT-lifting of propagation redundancy allows for changing in a controlled way the set of minimal correction sets in MaxSAT. This ability is key in succinctly expressing MaxSAT reasoning techniques and allows for obtaining correctness proofs in a uniform way for MaxSAT reasoning techniques very generally. Bridging theory to practice, we also provide a new MaxSAT preprocessor incorporating such extended techniques, and show through experiments its wide applicability in improving the performance of modern MaxSAT solvers.

Keywords: Maximum satisfiability · Clause redundancy · Propagation redundancy · Preprocessing

1 Introduction

Building heavily on the success of Boolean satisfiability (SAT) solving [13], maximum satisfiability (MaxSAT) as the optimization extension of SAT constitutes a viable approach to solving real-world NP-hard optimization problems [6,35]. In the context of SAT, the study of fundamental aspects of clause redundancy [20,21,23,28,29,31,32] has proven central for developing novel types of preprocessing and inprocessing-style solving techniques [24,29] as well as in enabling efficient proof checkers [7,15,16,18,19,41,42] via succinct representation of most practical SAT solving techniques. Furthermore, clause redundancy notions have

Work financially supported by Academy of Finland under grants 322869, 328718 and 342145. The authors wish to thank the Finnish Computing Competence Infrastructure (FCCI) for supporting this project with computational and data storage resources.

J. Blanchette et al. (Eds.): IJCAR 2022, LNAI 13385, pp. 75–94, 2022.
https://doi.org/10.1007/978-3-031-10769-6_6

been shown to give rise to very powerful proof systems, going far beyond resolution [22,23,30]. In contrast to viewing clause redundancy through the lens of logical entailment, the redundancy criteria developed in this line of work are based on a semantic implication relationship between formulas, making them desirably efficient to decide and at the same time are guaranteed to merely preserve satisfiability rather than logical equivalence.

The focus of this work is the study of clause redundancy in the context of MaxSAT through lifting recently-proposed variants of the notion of *propagation redundancy* [23] based on a semantic implication relationship between formulas from the realm of SAT. The study of such liftings is motivated from several perspectives. Firstly, earlier it has been shown that a natural MaxSAT-lifting called SRAT [10] of the redundancy notion of the notion of *resolution asymmetric tautologies* (RAT) [29] allows for establishing the general correctness of MaxSAT-liftings of typical preprocessing techniques in SAT solving [14], alleviating the need for correctness proofs for individual preprocessing techniques [8]. However, the need for preserving the *optimal cost* in MaxSAT—as a natural counterpart for preserving satisfiability in SAT—allows for developing MaxSAT-centric preprocessing and solving techniques which cannot be expressed through SRAT [2,11]. Capturing more generally such cost-aware techniques requires developing more expressive notions of clause redundancy. Secondly, due to the fundamental connections between solutions and so-called minimal corrections sets (MCSes) of MaxSAT instances [8,25], analyzing the effect of clauses that are redundant in terms of expressive notions of redundancy on the MCSes of MaxSAT instances can provide further understanding on the relationship between the different notions and their fundamental impact on the solutions of MaxSAT instances. Furthermore, in analogy with SAT, more expressive redundancy notions may prove fruitful for developing further practical preprocessing and solving techniques for MaxSAT.

Our main contributions are the following. We propose natural liftings of the three recently-proposed variants PR, LPR and SPR of propagation redundancy in the context of SAT to MaxSAT. We provide a complete characterization of the relative expressiveness of the lifted notions CPR, CLPR and CSPR (C standing for cost for short) and of their impact on the set of MCSes in MaxSAT instances. In particular, while removing or adding clauses redundant in terms of CSPR and CLPR (the latter shown to be equivalent with SRAT) do not influence the set of MCSes underlying MaxSAT instances, CPR can in fact have an influence on MCSes. In terms of solutions, this result implies that CSPR or CLPR clauses can not remove minimal (in terms of sum-of-weights of falsified soft clauses) solutions of MaxSAT instances, while CPR clauses can.

The—theoretically greater—effect that CPR clauses have on the solutions of MaxSAT instances is key for succinctly expressing further MaxSAT reasoning techniques via CPR and allows for obtaining correctness proofs in a uniform way for MaxSAT reasoning techniques very generally; we give concrete examples of how CPR captures techniques not in the reach of SRAT. Bridging to practical preprocessing in MaxSAT, we also provide a new MaxSAT preprocessor extended with such techniques. Finally, we provide large-scale empirical evidence on the positive impact of the preprocessor on the runtimes of various modern MaxSAT solvers, covering both complete and incomplete approaches, suggesting that extensive preprocessing going beyond the scope of SRAT appears beneficial to integrate for speeding up modern MaxSAT solvers.

An extended version of this paper, with formal proofs missing from this version, is available via the authors' homepages.

2 Preliminaries

SAT. For a Boolean variable x there are two literals, the positive x and the negative $\neg x$, with $\neg\neg l = l$ for a literal l. A clause C is a set (disjunction) of literals and a CNF formula F a set (conjunction) of clauses. We assume that all clauses are non-tautological, i.e., do not contain both a literal and its negation. The set $\texttt{var}(C) = \{x \mid x \in C \text{ or } \neg x \in C\}$ consists of the variables of the literals in C. The set of variables and literals, respectively, of a formula are $\texttt{var}(F) = \bigcup_{C \in F} \texttt{var}(C)$ and $\texttt{lit}(F) = \bigcup_{C \in F} C$, respectively. For a set L of literals, the set $\neg L = \{\neg l \mid l \in L\}$ consists of the negations of the literals in L.

A *(truth) assignment* τ is a set of literals for which $x \notin \tau$ or $\neg x \notin \tau$ for any variable x. For a literal l we denote $l \in \tau$ by $\tau(l) = 1$ and $\neg l \in \tau$ by $\tau(l) = 0$ or $\tau(\neg l) = 1$ as convenient, and say that τ assigns l the value 1 and 0, respectively. The set $\texttt{var}(\tau) = \{x \mid x \in \tau \text{ or } \neg x \in \tau\}$ is the range of τ, i.e., it consists of the variables τ assigns a value for. For a set L of literals and an assignment τ, the assignment $\tau_L = (\tau \setminus \neg L) \cup L$ is obtained from τ by setting $\tau_L(l) = 1$ for all $l \in L$ and $\tau_L(l) = \tau(l)$ for all $l \notin L$ assigned by τ. For a literal l, τ_l stands for $\tau_{\{l\}}$. An assignment τ satisfies a clause C ($\tau(C) = 1$) if $\tau \cap C \neq \emptyset$ or equivalently if $\tau(l) = 1$ for some $l \in C$, and a CNF formula F ($\tau(F) = 1$) if it satisfies each clause $C \in F$. A CNF formula is satisfiable if there is an assignment that satisfies it, and otherwise unsatisfiable. The empty formula $\top$ is satisfied by any truth assignment and the empty clause $\bot$ is unsatisfiable. The Boolean satisfiability problem (SAT) asks to decide whether a given CNF formula F is satisfiable.

Given two CNF formulas F_1 and F_2, F_1 entails F_2 ($F_1 \models F_2$) if any assignment τ that satisfies F_1 and only assigns variables of F_1 (i.e. for which $\texttt{var}(\tau) \subset \texttt{var}(F_1)$) can be extended into an assignment $\tau^2 \supset \tau$ that satisfies F_2. The formulas are equisatisfiable if F_1 is satisfiable iff F_2 is. An assignment τ is complete for a CNF formula F if $\texttt{var}(F) \subset \texttt{var}(\tau)$, and otherwise partial for F. The restriction $F|_\tau$ of F wrt a partial assignment τ is a CNF formula obtained by (i) removing from F all clauses that are satisfied by τ and (ii) removing from the remaining clauses of F literals l for which $\tau(l) = 0$. Applying unit propagation on F refers to iteratively restricting F by $\tau = \{l\}$ for a unit clause (clause with a single literal) $(l) \in F$ until the resulting (unique) formula, denoted by $\texttt{UP}(F)$, contains no unit clauses or some clause in F becomes empty. We say that unit propagation on F derives a conflict if $\texttt{UP}(F)$ contains the empty clause. The formula F_1 implies F_2 under unit propagation ($F_1 \vdash_1 F_2$) if, for each $C \in F_2$, unit propagation derives a conflict in $F_1 \wedge \{(\neg l) \mid l \in C\}$. Note that $F_1 \vdash_1 F_2$ implies $F_1 \models F_2$, but not vice versa in general.

Maximum Satisfiability. An instance $\mathcal{F} = (\mathcal{F}_H, \mathcal{F}_S, w)$ of (weighted partial) maximum satisfiability (MaxSAT for short) consists of two CNF formulas, the hard clauses $\mathcal{F}_H$ and the soft clauses $\mathcal{F}_S$, and a weight function $w\colon \mathcal{F}_S \to \mathbb{N}$ that assigns a positive weight to each soft clause.

Without loss of generality, we assume that every soft clause $C \in \mathcal{F}_S$ is unit[1]. The set of *blocking* literals $\mathcal{B}(\mathcal{F}) = \{l \mid (\neg l) \in \mathcal{F}_S\}$ consists of the literals l the negation of which occurs in $\mathcal{F}_S$. The weight function w is extended to blocking literals by $w(l) = w((\neg l))$. Without loss of generality, we also assume that $l \in \mathtt{lit}(\mathcal{F}_H)$ for all $l \in \mathcal{B}(\mathcal{F})$[2]. Instead of using the definition of MaxSAT in terms of hard and soft clauses, we will from now on view a MaxSAT instance $\mathcal{F} = (\mathcal{F}_H, \mathcal{B}(\mathcal{F}), w)$ as a set $\mathcal{F}_H$ of hard clauses, a set $\mathcal{B}(\mathcal{F})$ of blocking literals and a weight function $w\colon \mathcal{B}(\mathcal{F}) \to \mathbb{N}$.

Any complete assignment τ over $\mathtt{var}(\mathcal{F}_H)$ that satisfies $\mathcal{F}_H$ is a solution to $\mathcal{F}$. The cost $\mathtt{COST}(\mathcal{F}, \tau) = \sum_{l \in \mathcal{B}(\mathcal{F})} \tau(l) w(l)$ of a solution τ is the sum of weights of blocking literals it assigns to 1[3]. The cost of a complete assignment τ that does not satisfy $\mathcal{F}_H$ is defined as ∞. The cost of a partial assignment τ over $\mathtt{var}(\mathcal{F}_H)$ is defined as the cost of smallest-cost assignments that are extensions of τ. A solution τ^o is optimal if $\mathtt{COST}(\mathcal{F}, \tau^o) \leq \mathtt{COST}(\mathcal{F}, \tau)$ holds for all solutions τ of $\mathcal{F}$. The cost of the optimal solutions of a MaxSAT instance is denoted by $\mathtt{COST}(\mathcal{F})$, with $\mathtt{COST}(\mathcal{F}) = \infty$ iff $\mathcal{F}_H$ is unsatisfiable. In MaxSAT the task is to find an optimal solution to a given MaxSAT instance.

Example 1. Let $\mathcal{F} = (\mathcal{F}_H, \mathcal{B}(\mathcal{F}), w)$ be a MaxSAT instance with $\mathcal{F}_H = \{(x \vee b_1), (\neg x \vee b_2), (y \vee b_3 \vee b_4), (z \vee \neg y \vee b_4), (\neg z)\}$, $\mathcal{B}(\mathcal{F}) = \{b_1, b_2, b_3, b_4\}$ having $w(b_1) = w(b_4) = 1$, $w(b_2) = 2$ and $w(b_3) = 8$. The assignment $\tau = \{b_1, b_4, \neg b_2, \neg b_3, \neg x, \neg z, y\}$ is an example of an optimal solution of $\mathcal{F}$ and has $\mathtt{COST}(\mathcal{F}, \tau) = \mathtt{COST}(\mathcal{F}) = 2$.

With a slight abuse of notation, we denote by $\mathcal{F} \wedge C = (\mathcal{F}_H \cup \{C\}, \mathcal{B}(\mathcal{F} \wedge C), w)$ the MaxSAT instance obtained by adding a clause C to an instance $\mathcal{F} = (\mathcal{F}_H, \mathcal{B}(\mathcal{F}), w)$. Adding clauses may introduce new blocking literals but not change the weights of already existing ones, i.e., $\mathcal{B}(\mathcal{F}) \subset \mathcal{B}(\mathcal{F} \wedge C)$ and $w^{\mathcal{F}}(l) = w^{\mathcal{F} \wedge C}(l)$ for all $l \in \mathcal{B}(\mathcal{F})$.

Correction Sets. For a MaxSAT instance $\mathcal{F}$, a subset $\mathrm{cs} \subset \mathcal{B}(\mathcal{F})$ is a minimal correction set (MCS) of $\mathcal{F}$ if (i) $\mathcal{F}_H \wedge \bigwedge_{l \in \mathcal{B}(\mathcal{F}) \setminus \mathrm{cs}} (\neg l)$ is satisfiable and (ii) $\mathcal{F}_H \wedge \bigwedge_{l \in \mathcal{B}(\mathcal{F}) \setminus \mathrm{cs}_s} (\neg l)$ is unsatisfiable for every $\mathrm{cs}_s \subsetneq \mathrm{cs}$. In words, cs is an MCS if it

[1] A soft clause C can be replaced by the hard clause $C \vee x$ and soft clause $(\neg x)$, where x is a variable not in $\mathtt{var}(\mathcal{F}_H \wedge \mathcal{F}_S)$, without affecting the costs of solutions.

[2] Otherwise the instance can be simplified by unit propagating $\neg l$ without changing the costs of solutions. As a consequence, any complete assignment for $\mathcal{F}_H$ will be complete for $\mathcal{F}_H \wedge \mathcal{F}_S$ as well.

[3] This is equivalent to the sum of weights of soft clauses not satisfied by τ.

is a subset-minimal set of blocking literals that is included in some solution τ of $\mathcal{F}$.[4] We denote the set of MCSes of $\mathcal{F}$ by $\mathtt{mcs}(\mathcal{F})$.

There is a tight connection between the MCSes and solutions of MaxSAT instances. Given an optimal solution τ^o of a MaxSAT instance $\mathcal{F}$, the set $\tau^o \cap \mathcal{B}(\mathcal{F})$ is an MCS of $\mathcal{F}$. In the other direction, for any cs $\in \mathtt{mcs}(\mathcal{F})$, there is a (not necessary optimal) solution τ^{cs} such that cs $= \mathcal{B}(\mathcal{F}) \cap \tau^{\text{cs}}$ and $\mathtt{COST}(\mathcal{F}, \tau^{\text{cs}}) = \sum_{l \in \text{cs}} w(l)$.

Example 2. Consider the instance $\mathcal{F}$ from Example 1. The set $\{b_1, b_4\} \in \mathtt{mcs}(\mathcal{F})$ is an MCS of $\mathcal{F}$ that corresponds to the optimal solution τ described in Example 1. The set $\{b_2, b_3\} \in \mathtt{mcs}(\mathcal{F})$ is another example of an MCS that instead corresponds to the solution $\tau_2 = \{b_2, b_3, \neg b_1, \neg b_4, x, \neg z, \neg y\}$ for which $\mathtt{COST}(\mathcal{F}, \tau) = 10$.

3 Propagation Redundancy in MaxSAT

We extend recent work [23] on characterizing redundant clauses using semantic implication in the context of SAT to MaxSAT. In particular, we provide natural counterparts for several recently-proposed strong notions of redundancy in SAT to the context of MaxSAT and analyze the relationships between them.

In the context of SAT, the most general notion of clause redundancy is seemingly simple: a clause C is redundant for a formula F if it does not affect its satisfiability, i.e., clause C is redundant wrt a CNF formula F if F and $F \wedge \{C\}$ are equisatisfiable [20,29]. This allows for the set of satisfying assignments to change, and does not require preserving logical equivalence; we are only interested in satisfiability.

A natural counterpart for this general view in MaxSAT is that the *cost* of optimal solutions (rather than the set of optimal solutions) should be preserved.

Definition 1. *A clause C is redundant wrt a MaxSAT instance $\mathcal{F}$ if* $\mathtt{COST}(\mathcal{F}) = \mathtt{COST}(\mathcal{F} \wedge C)$.

This coincides with the counterpart in SAT whenever $\mathcal{B}(\mathcal{F}) = \emptyset$, since then the cost of a MaxSAT instance $\mathcal{F}$ is either 0 (if $\mathcal{F}_H$ is satisfiable) or ∞ (if $\mathcal{F}_H$ is unsatisfiable). Unless explicitly specified, we will use the term "redundant" to refer to Definition 1.

Following [23], we say that a clause C *blocks* the assignment $\neg C$ (and all assignments τ for which $\neg C \subset \tau$). As shown in the context of SAT [23], a clause C is redundant (in the equisatisfiability sense) for a CNF formula F if C does not block all of its satisfying assignments. The counterpart that arises in the context of MaxSAT from Definition 1 is that the cost of at least one of the solutions not blocked by C is no greater than the cost of $\neg C$.

Proposition 1. *A clause C is redundant wrt a MaxSAT instance $\mathcal{F}$ if and only if there is an assignment τ for which* $\mathtt{COST}(\mathcal{F} \wedge C, \tau) = \mathtt{COST}(\mathcal{F}, \tau) \leq \mathtt{COST}(\mathcal{F}, \neg C)$.

[4] This is equivalent to a subset-minimal set of soft clauses falsified by τ.

The equality $\texttt{COST}(\mathcal{F} \wedge C, \tau) = \texttt{COST}(\mathcal{F}, \tau)$ of Proposition 1 is necessary, as witnessed by the following example.

Example 3. Consider the MaxSAT instance $\mathcal{F}$ detailed in Example 1, the clause $C = (b_5)$ with $b_5 \in \mathcal{B}(\mathcal{F} \wedge C)$ and the assignment $\tau = \{b_5\}$. Then $2 = \texttt{COST}(\mathcal{F}, \tau) \leq \texttt{COST}(\mathcal{F}, \neg C) = 2$ but C is not redundant since $\texttt{COST}(\mathcal{F} \wedge C) = 2 + w^{\mathcal{F} \wedge C}(b_5) > 2 = \texttt{COST}(\mathcal{F})$.

Proposition 1 provides a sufficient condition for a clause C being redundant. Further requirements on the assignment τ can be imposed without loss of generality.

Theorem 1. *A non-empty clause C is redundant wrt a MaxSAT instance $\mathcal{F} = (\mathcal{F}_H, \mathcal{B}(\mathcal{F}), w)$ if and only if there is an assignment τ such that*
(i) $\tau(C) = 1$, *(ii)* $\mathcal{F}_H|_{\neg C} \models \mathcal{F}_H|_{\tau}$ *and*
(iii) $\texttt{COST}(\mathcal{F} \wedge C, \tau) = \texttt{COST}(\mathcal{F}, \tau) \leq \texttt{COST}(\mathcal{F}, \neg C)$.

As we will see later, a reason for including two additional conditions in Theorem 1 is to allow defining different restrictions of redundancy notions, some of which allow for efficiently identifying redundant clauses.

Example 4. Consider the instance $\mathcal{F} = (\mathcal{F}_H, \mathcal{B}(\mathcal{F}), w)$ detailed in Example 1, a clause $C = (\neg x \vee b_5)$ for a $b_5 \in \mathcal{B}(\mathcal{F} \wedge C)$ and an assignment $\tau = \{\neg x, b_1\}$. Then: $\tau(C) = 1$, $\{(b_2), (y \vee b_3 \vee b_4), (z \vee \neg y \vee b_4), (\neg z)\} = \mathcal{F}_H|_{\neg C} \models \mathcal{F}_H|_{\tau} = \{(y \vee b_3 \vee b_4), (z \vee \neg y \vee b_4), (\neg z)\}$, and $2 = \texttt{COST}(\mathcal{F} \wedge C, \tau) = \texttt{COST}(\mathcal{F}, \tau) \leq \texttt{COST}(\mathcal{F}, \neg C) = 3$. We conclude that C is redundant.

In the context of SAT, imposing restrictions on the entailment operator and the set of assignments has been shown to give rise to several interesting redundancy notions which hold promise of practical applicability. These include three variants (LPR, SPR, and PR) of so-called (literal/set) propagation redundancy [23]. For completeness we restate the definitions of these three notions. A clause C is LPR wrt a CNF formula F if there is a literal $l \in C$ for which $F|_{\neg C} \vdash_1 F|_{(\neg C)_l}$, SPR if the same holds for a subset $L \subset C$, and PR if there exists an assignment τ that satisfies C and for which $F|_{\neg C} \vdash_1 F|_{\tau}$. With the help of Theorem 1, we obtain counterparts for these notions in the context of MaxSAT.

Definition 2. *With respect to an instance $\mathcal{F} = (\mathcal{F}_H, \mathcal{B}(\mathcal{F}), w)$, a clause C is*

- ***cost literal propagation redundant (CLPR)*** *(on l) there is a literal $l \in C$ for which either (i) $\bot \in \texttt{UP}(\mathcal{F}_H|_{\neg C})$ or (ii) $l \notin \mathcal{B}(\mathcal{F} \wedge C)$ and $\mathcal{F}_H|_{\neg C} \vdash_1 \mathcal{F}_H|_{(\neg C)_l}$;*
- ***cost set propagation redundant (CSPR)*** *(on L) if there is a set $L \subset C \setminus \mathcal{B}(\mathcal{F} \wedge C)$ of literals for which $\mathcal{F}_H|_{\neg C} \vdash_1 \mathcal{F}_H|_{(\neg C)_L}$; and*
- ***cost propagation redundant (CPR)*** *if there is an assignment τ such that*
  (i) $\tau(C) = 1$, *(ii)* $\mathcal{F}_H|_{\neg C} \vdash_1 \mathcal{F}_H|_{\tau}$ *and*
  (iii) $\texttt{COST}(\mathcal{F} \wedge C, \tau) = \texttt{COST}(\mathcal{F}, \tau) \leq \texttt{COST}(\mathcal{F}, \neg C)$.

Example 5. Consider again $\mathcal{F} = (\mathcal{F}_H, \mathcal{B}(\mathcal{F}), w)$ from Example 1. The clause $D = (b_1 \vee b_2)$ is CLPR wrt $\mathcal{F}$ since $\bot \in \texttt{UP}(\mathcal{F}_H|_{\neg D})$ as $\{(x), (\neg x)\} \subset \mathcal{F}_H|_{\neg D}$. As for the redundant clause C and assignment τ detailed in Example 3, we have that C is CPR, since $\mathcal{F}_H|_\tau \subset \mathcal{F}_H|_{\neg C}$ which implies $\mathcal{F}_H|_{\neg C} \vdash_1 \mathcal{F}_H|_\tau$.

We begin the analysis of the relationship between these redundancy notions by showing that CSPR (and by extension CLPR) clauses also satisfy the MaxSAT-centric condition (iii) of Theorem 1. Assume that C is CSPR wrt a instance $\mathcal{F} = (\mathcal{F}_H, \mathcal{B}(\mathcal{F}), w)$ on the set L.

Lemma 1. *Let $\tau \supset \neg C$ be a solution of $\mathcal{F}$. Then, $\texttt{COST}(\mathcal{F}, \tau) \geq \texttt{COST}(\mathcal{F}, \tau_L)$.*

The following corollary of Lemma 1 establishes that CSPR and CLPR clauses are redundant according to Definition 1.

Corollary 1. $\texttt{COST}(\mathcal{F} \wedge C, (\neg C)_L) = \texttt{COST}(\mathcal{F}, (\neg C)_L) \leq \texttt{COST}(\mathcal{F}, \neg C)$.

The fact that CPR clauses are redundant follows trivially from the fact that $\mathcal{F}_H|_{\neg C} \vdash_1 \mathcal{F}_H|_\tau$ implies $\mathcal{F}_H|_{\neg C} \models \mathcal{F}_H|_\tau$. However, given a solution ω that does not satisfy a CPR clause C, the next example demonstrates that the assignment ω_τ need not have a cost lower than ω. Stated in another way, the example demonstrates that an observation similar to Lemma 1 does not hold for CPR clauses in general.

Example 6. Consider a MaxSAT instance $\mathcal{F} = (\mathcal{F}_H, \mathcal{B}(\mathcal{F}), w)$ having $\mathcal{F}_H = \{(x \vee b_1), (\neg x, b_2)\}$, $\mathcal{B}(\mathcal{F}) = \{b_1, b_2\}$ and $w(b_1) = w(b_2) = 1$. The clause $C = (x)$ is CPR wrt $\mathcal{F}$, the assignment $\tau = \{x, b_2\}$ satisfies the three conditions of Definition 2. Now $\delta = \{\neg x, b_1\}$ is a solution of $\mathcal{F}$ that does not satisfy C for which $\delta_\tau = \{x, b_1, b_2\}$ and $1 = \texttt{COST}(\mathcal{F}, \delta) < 2 = \texttt{COST}(\mathcal{F}, \delta_\tau)$.

Similarly as in the context of SAT, verifying that a clause is CSPR (and by extension CLPR) can be done efficiently. However, in contrast to SAT, we conjecture that verifying that a clause is CPR can not in the general case be done efficiently, *even if the assignment τ is given.* While we will not go into detail on the complexity of identifying CPR clauses, the following proposition gives some support for our conjecture.

Proposition 2. *Let $\mathcal{F}$ be an instance and $k \in \mathbb{N}$. There is another instance $\mathcal{F}^M$, a clause C, and an assignment τ such that C is CPR wrt $\mathcal{F}^M$ if and only if $\texttt{COST}(\mathcal{F}) \geq k$.*

As deciding if $\texttt{COST}(\mathcal{F}) \geq k$ is NP-complete in the general case, Proposition 2 suggests that it may not be possible to decide in polynomial time if an assignment τ satisfies the three conditions of Definition 2 unless P=NP. This is in contrast to SAT, where verifying propagation redundancy can be done in polynomial time if the assignment τ is given, but is NP-complete if not [24].

The following observations establish a more precise relationship between the redundancy notions. For the following, let RED($\mathcal{F}$) denote the set of clauses that are redundant wrt a MaxSAT instance $\mathcal{F}$ according to Definition 1. Analogously, the sets CPR($\mathcal{F}$), CSPR($\mathcal{F}$) and CLPR($\mathcal{F}$) consist of the clauses that are CPR, CSPR and CLPR wrt $\mathcal{F}$, respectively.

Observation 1 CLPR($\mathcal{F}$) $\subset$ CSPR($\mathcal{F}$) $\subset$ CPR($\mathcal{F}$) $\subset$ RED($\mathcal{F}$) *holds for any MaxSAT instance* $\mathcal{F}$.

Observation 2 *There are MaxSAT instances* $\mathcal{F}_1, \mathcal{F}_2$ *and* $\mathcal{F}_3$ *for which* CLPR($\mathcal{F}_1$) $\subsetneq$ CSPR($\mathcal{F}_1$), CSPR($\mathcal{F}_2$) $\subsetneq$ CPR($\mathcal{F}_2$) *and* CPR($\mathcal{F}_3$) $\subsetneq$ RED($\mathcal{F}_3$).

The proofs of Observations 1 and 2 follow directly from known results in the context of SAT [23] by noting that any CNF formula can be viewed as an instance of MaxSAT without blocking literals.

For a MaxSAT-centric observation on the relationship between the redundancy notions, we note that the concept of redundancy and CPR coincide for any MaxSAT instance that has solutions.

Observation 3 CPR($\mathcal{F}$) = RED($\mathcal{F}$) *holds for any MaxSAT instance* $\mathcal{F}$ *with* $\texttt{COST}(\mathcal{F}) < \infty$.

We note that a result similar to Observation 3 could be formulated in the context of SAT. The SAT-counterpart would state that the concept of redundancy (in the equisatisfiability sense) coincides with the concept of propagation redundancy for SAT solving (defined e.g. in [23]) for *satisfiable* CNF formulas. However, assuming that a CNF formula is satisfiable is very restrictive in the context of SAT. In contrast, it is natural to assume that a MaxSAT instance admits solutions.

We end this section with a simple observation: adding a redundant clause C to a MaxSAT instance $\mathcal{F}$ preserves not only optimal cost, but optimal solutions of $\mathcal{F} \wedge C$ are also optimal solutions of $\mathcal{F}$. However, the converse need not hold; an instance $\mathcal{F}$ might have optimal solutions that do not satisfy C.

Example 7. Consider an instance $\mathcal{F} = (\mathcal{F}_H, \mathcal{B}(\mathcal{F}), w)$ with $\mathcal{F}_H = \{(b_1 \vee b_2)\}$, $\mathcal{B}(\mathcal{F}) = \{b_1, b_2\}$ and $w(b_1) = w(b_2) = 1$. The clause $C = (\neg b_1)$ is CPR wrt $\mathcal{F}$. In order to see this, let $\tau = \{\neg b_1, b_2\}$. Then τ satisfies C (condition (i) of Definition 2). Furthermore, τ satisfies $\mathcal{F}_H$, implying $\mathcal{F}_H|_{\neg C} \vdash_1 \mathcal{F}_H|_{\tau}$ (condition (ii)). Finally, we have that $1 = \texttt{COST}(\mathcal{F}, \tau) = \texttt{COST}(\mathcal{F} \wedge C, \tau) \leq \texttt{COST}(\mathcal{F}, \neg C) = 1$ (condition (iii)). The assignment $\delta = \{b_1, \neg b_2\}$ is an example of an optimal solution of $\mathcal{F}$ that is not a solution of $\mathcal{F} \wedge C$.

4 Propagation Redundancy and MCSes

In this section, we analyze the effect of adding redundant clauses on the MCSes of MaxSAT instances. As the main result, we show that adding CSPR (and by extension CLPR) clauses to a MaxSAT instance $\mathcal{F}$ preserves all MCSes while adding CPR clauses does not in general. Stated in terms of solutions, this means that adding CSPR clauses to $\mathcal{F}$ preserves not only all optimal solutions, but all solutions τ for which $(\tau \cap \mathcal{B}(\mathcal{F})) \in \texttt{mcs}(\mathcal{F})$, while adding CPR clauses only preserves at least one optimal solution.

Effect of CLPR Clauses on MCSes. MaxSAT-liftings of four specific SAT solving techniques (including bounded variable elimination and self-subsuming resolution) were earlier proposed in [8]. Notably, the correctness of the liftings was shown individually for each of the techniques by arguing individually that applying one of the liftings does not change the set of MCSes of any MaxSAT instance. Towards a more generic understanding of optimal cost preserving MaxSAT preprocessing, in [10] the notion of solution resolution asymmetric tautologies (SRAT) was proposed as a MaxSAT-lifting of the concept of resolution asymmetric tautologies (RAT). In short, a clause C is a SRAT clause for a MaxSAT instance $\mathcal{F} = (\mathcal{F}_H, \mathcal{B}(\mathcal{F}), w)$ if there is a literal $l \in C \setminus \mathcal{B}(\mathcal{F} \wedge C)$ such that $\mathcal{F}_H \vdash_1 ((C \vee D) \setminus \{\neg l\})$ for every $D \in \mathcal{F}_H$ for which $\neg l \in D$.

In analogy with RAT [29], SRAT was shown in [10] to allow for a general proof of correctness for natural MaxSAT-liftings of a wide range of SAT preprocessing techniques, covering among other the four techniques for which individual correctness proofs were provided in [8]. The generality follows essentially from the fact that the addition and removal of SRAT clauses preserves MCSes. The same observations apply to CLPR, as CLPR and SRAT are equivalent.

Proposition 3. *A clause C is CLPR wrt $\mathcal{F}$ iff it is SRAT wrt $\mathcal{F}$.*

The proof of Proposition 3 follows directly from corresponding results in the context of SAT [23]. Informally speaking, a clause C is SRAT on a literal l iff it is RAT [29] on l and $l \notin \mathcal{B}(\mathcal{F})$. Similarly, a clause C is CLPR on a literal l iff it is LPR as defined in [23] on l and $l \notin \mathcal{B}(\mathcal{F})$. Proposition 3 together with previous results from [10] implies that the MCSes of MaxSAT instances are preserved under removing and adding CLPR clauses.

Corollary 2. *If C is CLPR wrt $\mathcal{F}$, then* $\texttt{mcs}(\mathcal{F}) = \texttt{mcs}(\mathcal{F} \wedge C)$.

Effect of CPR Clauses on MCSes. We turn our attention to the effect of CPR clauses on the MCSes of MaxSAT instances. Our analysis makes use of the previously-proposed MaxSAT-centric preprocessing rule known as *subsumed label elimination* (SLE) [11,33][5].

Definition 3. *(Subsumed Label Elimination [11,33]) Consider a MaxSAT instance $\mathcal{F} = (\mathcal{F}_H, \mathcal{B}(\mathcal{F}), w)$ and a blocking literal $l \in \mathcal{B}(\mathcal{F})$ for which $\neg l \notin$ $\texttt{lit}(\mathcal{F}_H)$. Assume that there is another blocking literal $l_s \in \mathcal{B}(\mathcal{F})$ for which (1) $\neg l_s \notin \texttt{lit}(\mathcal{F}_H)$, (2) $\{C \in \mathcal{F}_H \mid l \in C\} \subset \{C \in \mathcal{F}_H \mid l_s \in C\}$ and (3) $w(l) \geq w(l_s)$. The subsumed label elimination (SLE) rule allows adding $(\neg l)$ to $\mathcal{F}_H$.*

A specific proof of correctness of SLE was given in [11]. The following proposition provides an alternative proof based on CPR.

Proposition 4 (Proof of correctness for SLE). *Let $\mathcal{F}$ be a MaxSAT instance and assume that the blocking literals $l, l_s \in \mathcal{B}(\mathcal{F})$ satisfy the three conditions of Definition 3. Then, the clause $C = (\neg l)$ is CPR wrt $\mathcal{F}$.*

[5] Rephrased here using our notation.

Proof. We show that $\tau = \{\neg l, l_s\}$ satisfies the three conditions of Definition 2. First τ satisfies C (condition (i)). Conditions (1) and (2) of Definition 3 imply $\mathcal{F}_H|_\tau \subset \mathcal{F}_H|_{\neg C}$ which in turn implies $\mathcal{F}_H|_{\neg C} \vdash_1 \mathcal{F}_H|_\tau$ (condition (ii)).

As for condition (iii), the requirement $\texttt{COST}(\mathcal{F} \wedge C, \tau) = \texttt{COST}(\mathcal{F}, \tau)$ follows from $\mathcal{B}(\mathcal{F} \wedge C) = \mathcal{B}(\mathcal{F})$. Let $\delta \supset \neg C$ be a complete assignment of $\mathcal{F}_H$ for which $\texttt{COST}(\mathcal{F}, \delta) = \texttt{COST}(\mathcal{F}, \neg C)$. If $\texttt{COST}(\mathcal{F}, \delta) = \infty$ then $\texttt{COST}(\mathcal{F}, \tau) \leq \texttt{COST}(\mathcal{F}, \neg C)$ follows trivially. Otherwise $\delta \setminus \neg C$ satisfies $\mathcal{F}_H|_{\neg C}$ so by $\mathcal{F}_H|_{\neg C} \vdash_1 \mathcal{F}_H|_\tau$ it satisfies $\mathcal{F}_H|_\tau$ as well. Thus $\delta^R = ((\delta \setminus \neg C) \setminus \{\neg l \mid l \in \tau\}) \cup \tau = (\delta \setminus \{l, \neg l, \neg l_s\}) \cup \{\neg l, l_s\}$ is an extension of τ that satisfies $\mathcal{F}_H$ and for which $\texttt{COST}(\mathcal{F}, \tau) \leq \texttt{COST}(\mathcal{F}, \delta^R) \leq \texttt{COST}(\mathcal{F}, \delta)$ by condition (3) of Definition 3. Thereby τ satisfies the conditions of Definition 2 so C is CPR wrt $\mathcal{F}$. □

Example 8. The blocking literals $b_3, b_4 \in \mathcal{B}(\mathcal{F})$ of the instance $\mathcal{F}$ detailed in Example 1 satisfy the conditions of Definition 3. By Proposition 4 the clause $(\neg b_3)$ is CPR wrt $\mathcal{F}$.

In [11] it was shown that SLE does not preserve MCSes in general. By Corollary 2, this implies that SLE can not be viewed as the addition of CLPR clauses. Furthermore, by Proposition 4 we obtain the following.

Corollary 3. *There is a MaxSAT instance $\mathcal{F}$ and a clause C that is CPR wrt $\mathcal{F}$ for which* $\texttt{mcs}(\mathcal{F}) \neq \texttt{mcs}(\mathcal{F} \wedge C)$.

Effect of CSPR Clauses on MCSes. Having established that CLPR clauses preserve MCSes while CPR clauses do not, we complete the analysis by demonstrating that CSPR clauses preserve MCSes.

Theorem 2. *Let $\mathcal{F}$ be a MaxSAT instance and C a CSPR clause of $\mathcal{F}$. Then* $\texttt{mcs}(\mathcal{F}) = \texttt{mcs}(\mathcal{F} \wedge C)$.

Theorem 2 follows from the following lemmas and propositions. In the following, let C be a clause that is CSPR wrt a MaxSAT instance $\mathcal{F}$ on a set $L \subset C \setminus \mathcal{B}(\mathcal{F} \wedge C)$.

Lemma 2. *Let $cs \subset \mathcal{B}(\mathcal{F})$. If $\mathcal{F}_H \wedge \bigwedge_{l \in \mathcal{B}(\mathcal{F}) \setminus cs}(\neg l)$ is satisfiable, then $(\mathcal{F}_H \wedge C) \wedge \bigwedge_{l \in \mathcal{B}(\mathcal{F} \wedge C) \setminus cs}(\neg l)$ is satisfiable.*

Lemma 2 helps in establishing one direction of Theorem 2.

Proposition 5. $\texttt{mcs}(\mathcal{F}) \subset \texttt{mcs}(\mathcal{F} \wedge C)$.

Proof. Let $\text{cs} \in \texttt{mcs}(\mathcal{F})$. Then $\mathcal{F}_H \wedge \bigwedge_{l \in \mathcal{B}(\mathcal{F}) \setminus \text{cs}}(\neg l)$ is satisfiable, which by Lemma 2 implies that $(\mathcal{F}_H \wedge C) \wedge \bigwedge_{l \in \mathcal{B}(\mathcal{F} \wedge C) \setminus \text{cs}}(\neg l)$ is satisfiable.

To show that $(\mathcal{F}_H \wedge C) \wedge \bigwedge_{l \in \mathcal{B}(\mathcal{F} \wedge C) \setminus \text{cs}_s}(\neg l)$ is unsatisfiable for any $\text{cs}_s \subsetneq \text{cs} \subset \mathcal{B}(\mathcal{F})$, we note that any assignment satisfying $(\mathcal{F}_H \wedge C) \wedge \bigwedge_{l \in \mathcal{B}(\mathcal{F} \wedge C) \setminus \text{cs}_s}(\neg l)$ would also satisfy $\mathcal{F}_H \wedge \bigwedge_{l \in \mathcal{B}(\mathcal{F}) \setminus \text{cs}_s}(\neg l)$, contradicting $\text{cs} \in \texttt{mcs}(\mathcal{F})$. □

The following lemma is useful for showing inclusion in the other direction.

Lemma 3. *Let* $cs \in \texttt{mcs}(\mathcal{F} \wedge C)$*. Then* $cs \subset \mathcal{B}(\mathcal{F})$*.*

Lemma 3 allows for completing the proof of Theorem 2.

Proposition 6. $\texttt{mcs}(\mathcal{F} \wedge C) \subset \texttt{mcs}(\mathcal{F})$.

Proof. Let $\text{cs} \in \texttt{mcs}(\mathcal{F} \wedge C)$, which by Lemma 3 implies $\text{cs} \subset \mathcal{B}(\mathcal{F})$. Let τ be a solution that satisfies $(\mathcal{F}_H \wedge C) \wedge \bigwedge_{l \in \mathcal{B}(\mathcal{F} \wedge C) \setminus \text{cs}} (\neg l)$. Then τ satisfies $\mathcal{F}_H \wedge \bigwedge_{l \in \mathcal{B}(\mathcal{F}) \setminus \text{cs}} (\neg l)$. For contradiction, assume that $\mathcal{F}_H \wedge \bigwedge_{l \in \mathcal{B}(\mathcal{F}) \setminus \text{cs}_s} (\neg l)$ is satisfiable for some $\text{cs}_s \subsetneq \text{cs}$. Then by Lemma 2, $(\mathcal{F}_H \wedge C) \wedge \bigwedge_{l \in \mathcal{B}(\mathcal{F} \wedge C) \setminus \text{cs}_s} (\neg l)$ is satisfiable as well, contradicting $\text{cs} \in \texttt{mcs}(\mathcal{F} \wedge C)$. Thereby $\text{cs} \in \texttt{mcs}(\mathcal{F})$. □

Theorem 2 implies that SLE can not be viewed as the addition of CSPR clauses. In light of this, an interesting remark is that—in contrast to CPR clauses in general (recall Example 6)—the assignment τ used in the proof of Proposition 4 can be used to convert any assignment that does not satisfy the CPR clause detailed in Definition 3 into one that does, without increasing its cost.

Observation 4 *Let* $\mathcal{F}$ *be a MaxSAT instance and assume that the blocking literals* $l, l_s \in \mathcal{B}(\mathcal{F})$ *satisfy the three conditions of Definition 3. Let* $\tau = \{\neg l, l_s\}$ *and consider any solution* $\delta \supset \neg C$ *of* $\mathcal{F}$ *that does not satisfy the CPR clause* $C = (\neg l)$*. Then* δ_τ *is a solution of* $\mathcal{F} \wedge C$ *for which* $\texttt{COST}(\mathcal{F}, \delta_\tau) \leq \texttt{COST}(\mathcal{F}, \delta)$*.*

5 CPR-Based Preprocessing for MaxSAT

Mapping the theoretical observations into practical preprocessing, in this section we discuss through examples how CPR clauses can be used as a unified theoretical basis for capturing a wide variety of known MaxSAT reasoning rules, and how they could potentially help in the development of novel MaxSAT reasoning techniques.

Our first example is the so-called *hardening rule* [2,8,17,26]. In terms of our notation, given a solution τ to a MaxSAT instance $\mathcal{F} = (\mathcal{F}_H, \mathcal{B}(\mathcal{F}), w)$ and a blocking literal $l \in \mathcal{B}(\mathcal{F})$ for which $w(l) > \texttt{COST}(\mathcal{F}, \tau)$, the hardening rule allows adding the clause $C = (\neg l)$ to $\mathcal{F}_H$.

The correctness of the hardening rule can be established with CPR clauses. More specifically, as $\texttt{COST}(\mathcal{F}, \tau) < w(l)$ it follows that $\tau(C) = 1$ (condition (i) of Definition 2). Since τ satisfies $\mathcal{F}$, we have that $\mathcal{F}_H\big|_\tau = \top$ so $\mathcal{F}_H\big|_{\neg C} \vdash_1 \mathcal{F}_H\big|_\tau$ (condition (ii)). Finally, as $\texttt{COST}(\mathcal{F}, \delta) \geq w(l) > \texttt{COST}(\mathcal{F}, \tau)$ holds for all $\delta \supset \neg C$ it follows that $\texttt{COST}(\mathcal{F}, \neg C) > \texttt{COST}(\mathcal{F}, \tau) = \texttt{COST}(\mathcal{F} \wedge C, \tau)$. As such, $(\neg l)$ is CPR clause wrt $\mathcal{F}$. If fact, instead of assuming $w(l) > \texttt{COST}(\mathcal{F}, \tau)$ it suffices to assume $w(l) \geq \texttt{COST}(\mathcal{F}, \tau)$ and $\tau(l) = 0$.

The hardening rule can not be viewed as the addition of CSPR or CLPR clauses because it does not in general preserve MCSes.

Example 9. Consider the MaxSAT instance $\mathcal{F}$ from Example 1 and a solution $\tau = \{b_1, b_2, b_4, \neg b_3, \neg z, x, y\}$. Since $\texttt{COST}(\mathcal{F}, \tau) = 3 < 8 = w(b_3)$, the clause $(\neg b_3)$ is CPR. However, $\texttt{mcs}(\mathcal{F}) \neq \texttt{mcs}(\mathcal{F} \wedge C)$ since the set $\{b_2, b_3\} \in \texttt{mcs}(\mathcal{F})$ is not an MCS of $\mathcal{F} \wedge C$ as $(\mathcal{F}_H \wedge C) \wedge \bigwedge_{l \in \mathcal{B}(\mathcal{F}) \setminus \text{cs}} (\neg l) = (\mathcal{F}_H \wedge (\neg b_3)) \wedge (\neg b_1) \wedge (\neg b_4)$ is not satisfiable.

Viewing the hardening rule through the lens of CPR clauses demonstrates novel aspects of the MaxSAT-liftings of propagation redundancy. In particular, instantiated in the context of SAT, an argument similar to the one we made for hardening shows that given a CNF formula F, an assignment τ satisfying F, and a literal l for which $\tau(l) = 0$, the clause $(\neg l)$ is redundant (wrt equisatisfiability). While formally correct, such a rule is not very useful for SAT solving. In contrast, in the context of MaxSAT the hardening rule is employed in various modern MaxSAT solvers and leads to non-trivial performance-improvements [4,5].

As another example of capturing MaxSAT-centric reasoning with CPR, consider the so-called TrimMaxSAT rule [39]. Given a MaxSAT instance $\mathcal{F} = (\mathcal{F}_H, \mathcal{B}(\mathcal{F}), w)$ and a literal $l \in \mathcal{B}(\mathcal{F})$ for which $\tau(l) = 1$ for all solutions of $\mathcal{F}$, the TrimMaxSAT rule allows adding the clause $C = (l)$ to $\mathcal{F}_H$. In this case the assumptions imply that all solutions of $\mathcal{F}$ also satisfy C, i.e., that $\mathcal{F}_H\big|_{\neg C}$ is unsatisfiable. As such, any assignment τ that satisfies C and $\mathcal{F}_H$ will also satisfy the three conditions of Definition 2 which demonstrates that C is CPR. It is, however, not CSPR since the only literal in C is blocking.

As a third example of capturing (new) reasoning techniques with CPR, consider an extension of the central variable elimination rule that allows (to some extent) for eliminating blocking literals.

Definition 4. *Consider a MaxSAT instance $\mathcal{F}$ and a blocking literal $l \in \mathcal{B}(\mathcal{F})$. Let* BBVE($\mathcal{F}$) *be the instance obtained by (i) adding the clause $C \vee D$ to $\mathcal{F}$ for every pair $(C \vee l), (D \vee \neg l) \in \mathcal{F}_H$ and (ii) removing all clauses $(D \vee \neg l) \in \mathcal{F}_H$. Then* $\texttt{COST}(\mathcal{F}) = \texttt{COST}(\text{BBVE}(\mathcal{F}))$ *and* $\texttt{mcs}(\mathcal{F}) = \texttt{mcs}(\text{BBVE}(\mathcal{F}))$.

On the Limitations of CPR. Finally, we note that while CPR clauses significantly generalize existing theory on reasoning and preprocessing rules for MaxSAT, there are known reasoning techniques that can not (at least straightforwardly) be viewed through the lens of propagation redundancy. For a concrete example, consider the so-called intrinsic atmost1 technique [26].

Definition 5. *Consider a MaxSAT instance $\mathcal{F}$ and a set $L \subset \mathcal{B}(\mathcal{F})$ of blocking literals. Assume that (i) $|\tau \cap \{\neg l \mid l \in L\}| \leq 1$ holds for any solution τ of $\mathcal{F}$ and (ii) $w(l) = 1$ for each $l \in L$. Now form the instance* AT-MOST-ONE($\mathcal{F}, L$) *by (i) removing each literal $l \in L$ from $\mathcal{B}(\mathcal{F})$, and (ii) adding the clause $\{(\neg l) \mid l \in L\} \cup \{l_L\}$ to $\mathcal{F}$, where l_L is a fresh blocking literal with $w(l_L) = 1$.*

It has been established that any optimal solution of AT-MOST-ONE($\mathcal{F}, L$) is an optimal solution of $\mathcal{F}$ [26]. However, as the next example demonstrates, the preservation of optimal solutions is in general not due to the clauses added being redundant, as applying the technique can affect optimal cost.

Example 10. Consider the MaxSAT instance $\mathcal{F} = (\mathcal{F}_H, \mathcal{B}(\mathcal{F}), w)$ with $\mathcal{F}_H = \{(l_i) \mid i = 1 \ldots n\}$, $\mathcal{B}(\mathcal{F}) = \{l_1 \ldots l_n\}$ and $w(l) = 1$ for all $l \in \mathcal{B}(\mathcal{F})$. Then $|\tau \cap \neg\mathcal{B}(\mathcal{F})| = 0 \leq 1$ holds for all solutions τ of $\mathcal{F}$ so the intrinsic-at-most-one technique can be used to obtain the instance $\mathcal{F}^2 = \text{AT-MOST-ONE}(\mathcal{F}, \mathcal{B}(\mathcal{F})) = (\mathcal{F}^2_H, \mathcal{B}(\mathcal{F}^2), w^2)$ with $\mathcal{F}^2_H = \mathcal{F}_H \cup \{(\neg l_1 \vee \ldots \vee \neg l_n \vee l_L)\}$, $\mathcal{B}(\mathcal{F}^2) = \{l_L\}$ and

$w^2(l_L) = 1$. Now $\delta = \{l \mid l \in \mathcal{B}(\mathcal{F})\} \cup \{l_L\}$ is an optimal solution to both $\mathcal{F}^2$ and $\mathcal{F}$ for which $1 = \texttt{COST}(\mathcal{F}^2, \delta) < \texttt{COST}(\mathcal{F}, \delta) = n$.

Example 10 implies that the intrinsic atmost1 technique can not be viewed as the addition or removal of redundant clauses. Generalizing CPR to cover weight changes could lead to further insights especially due to potential connections with core-guided MaxSAT solving [1,36–38].

6 MaxPre 2: More General Preprocesssing in Practice

Connecting to practice, we extended the MaxSAT preprocessor MaxPre [33] version 1 with support for techniques captured by propagation redundancy. The resulting MaxPre version 2, as outlined in the following, hence includes techniques which have previously only been implemented in specific solver implementations rather than in general-purpose MaxSAT preprocessors.

First, let us mention that the earlier MaxPre [33] version 1 assumes that any blocking literals only appear in a single polarity among the hard clauses. Removing this assumption—supported by theory developed in Sects. 3–4—decreases the number of auxiliary variables that need to be introduced when a MaxSAT instance is rewritten to only include unit soft clauses. For example, consider a MaxSAT instance $\mathcal{F}$ with $\mathcal{F}_H = \{(\neg x \vee y), (\neg y \vee x)\}$ and $\mathcal{F}_S = \{(x), (\neg y)\}$. For preprocessing the instance, MaxPre 1 extends both soft clauses with a new, auxiliary variable and runs preprocessing on the instance $\mathcal{F} = \{(\neg x \vee y), (\neg y \vee x), (x \vee b_1), (\neg y \vee b_2)\}$ with $\mathcal{B}(\mathcal{F}) = \{b_1, b_2\}$. In contrast, MaxPre 2 detects that the clauses in $\mathcal{F}_S$ are unit and reuses them as blocking literals, invoking preprocessing on $\mathcal{F} = \{(\neg x \vee y), (\neg y \vee x)\}$ with $\mathcal{B}(\mathcal{F}) = \{\neg x, y\}$.

In addition to the techniques already implemented in MaxPre 1, MaxPre 2 includes the following additional techniques: hardening [2], a variant TrimMaxSAT [39] that works on all literals of a MaxSAT instance, the intrinsic atmost1 technique [26] and a MaxSAT-lifting of failed literal elimination [12]. In short, failed literal elimination adds the clause $(\neg l)$ to the hard clauses $\mathcal{F}_H$ of an instance in case unit-propagation derives a conflict in $\mathcal{F}_H \wedge \{(l)\}$. Additionally, the implementation of failed literal elimination attempts to identify implied equivalences between literals that can lead to further simplification.

For computing the solutions required by TrimMaxSAT and detecting the cardinality constraints required by intrinsic-at-most-one constraints, MaxPre 2 uses the Glucose 3.0 SAT-solver [3]. For computing solutions required by hardening, MaxPre 2 additionally uses the SatLike incomplete MaxSAT solver [34] within preprocessing. MaxPre 2 is available in open source at https://bitbucket.org/coreo-group/maxpre2/.

We emphasize that, while the additional techniques implemented by MaxPre 2 have been previously implemented as heuristics in specific solver implementations, MaxPre 2 is—to the best of our understanding—the first stand-alone implementation supporting techniques whose correctness cannot be established with previously-proposed MaxSAT redundancy notions (i.e., SRAT). The goal of our empirical evaluation presented in the next section is to demonstrate the potential of viewing expressive reasoning techniques not only as solver heuristics, but as a separate step in the MaxSAT solving process whose correctness can be established via propagation redundancy.

7 Empirical Evaluation

We report on results from an experimental evaluation of the potential of incorporating more general reasoning in MaxSAT preprocessing. In particular, we evaluated both complete solvers (geared towards finding provably-optimal solutions) and incomplete solvers (geared towards finding relatively good solutions fast) on standard heterogenous benchmarks from recent MaxSAT Evaluations. All experiments were run on 2.60-GHz Intel Xeon E5-2670 8-core machines with 64 GB memory and CentOS 7. All reported runtimes include the time used in preprocessing (when applicable).

7.1 Impact of Preprocessing on Complete Solvers

We start by considering recent representative complete solvers covering three central MaxSAT solving paradigms: the core-guided solver CGSS [27] (as a recent improvement to the successful RC2 solver [26]), and the MaxSAT Evaluation 2021 versions of the implicit hitting set based solver MaxHS [17] and the solution-improving solver Pacose [40]. For each solver S we consider the following variants.

- S: S in its default configuration.
- S `no preprocess`: S with the solver's own internal preprocessing turned off (when applicable).
- S+`maxpre1`: S after applying MaxPre 1 using its default configuration.
- S+`maxpre2/none`: S after applying MaxPre 2 using the default configuration of MaxPre 1.
- S + `maxpre2/<TECH>`: S after applying MaxPre 2 using the standard configuration of MaxPre 1 and additional techniques integrated into MaxPre 2 (as detailed in Section 6) as specified by `<TECH>`.

More precisely, `<TECH>` specifies which of the techniques HTVGR are applied: H for hardening, T and V for TrimMaxSAT on blocking and non-blocking literals, respectively, G for intrinsic-at-most-one-constraints and R for failed literal elimination. It should be noted that an exhaustive evaluation of all subsets and application orders of these techniques is infeasible in practice. Based on preliminary experiments, we observed that the following choices were promising: HRT for CGSS and MaxHS, and HTVGR for Pacose; we report results using these individual configurations.

As benchmarks, we used the combined set of weighted instances from the complete tracks of MaxSAT Evaluation 2020 and 2021. After removing duplicates, this gave a total of 1117 instances. We enforced a per-instance time limit of

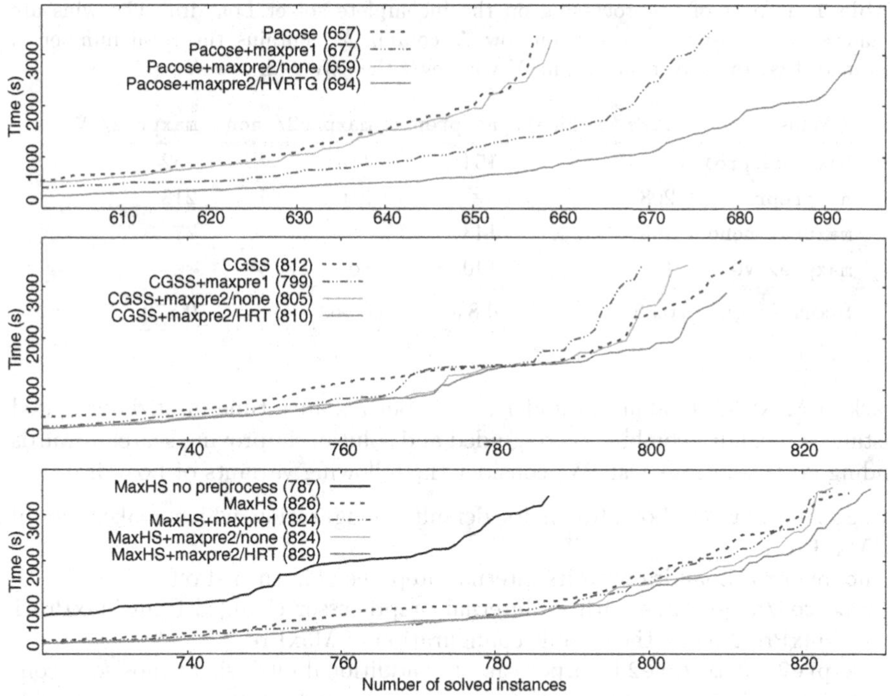

Fig. 1. Impact of preprocessing on complete solvers. For each solver, the number of instances solved within a 60-min per-instance time limit in parentheses.

60 minutes and memory limit of 32 GB. Furthermore, we enforced a per-instance 120-second time limit on preprocessing.

An overview of the results is shown in Fig. 1, illustrating for each solver the number of instances solved (x-axis) under different per-instance time limits (y-axis). We observe that for both CGSS and MaxHS, S+`maxpre1` and S+`maxpre2/none` leads to less instances solved compared to S. In contrast, S+`maxpre2/HRT`, i.e., incorporating the stronger reasoning techniques of MaxPre 2, performs best of all preprocessing variants and improves on MaxHS also in terms of the number of instances solved. For Pacose, we observe that both Pacose+`maxpre1` and Pacose+`maxpre2/new` (without the stronger reasoning techniques) already improve the performance of Pacose, leading to more instances solved. Incorporating the stronger reasoning rules further significantly improves performance, with Pacose+`maxpre2/HVRTG` performing the best among all of the Pacose variants.

7.2 Impact of Preprocessing on Incomplete MaxSAT Solving

As a representative incomplete MaxSAT solver we consider the MaxSAT Evaluation 2021 version of Loandra [9], as the best-performing solver in the incomplete

Table 1. Impact of preprocessing on the incomplete solver Loandra. The wins are organized column-wise, the cell on row X column Y contains the total number of instances that the solver on column Y wins over the solver on row X.

#Wins	`base (maxpre1)`	`no-prepro`	`maxpre2/ none`	`maxpre2/ VG`
`base (maxpre1)`	—	154	135	152
`no-prepro`	208	—	216	218
`maxpre2/none`	105	143	—	77
`maxpre2/VG`	110	140	80	—
Score (avg):	0.852	0.840	0.863	0.870

track of MaxSAT Evaluation under a 300s per-instance time limit on weighted instances. Loandra combines core-guided and solution-improving search towards finding good solutions fast. We consider the following variants of Loandra.

- `base (maxpre1)`: Loandra in its default configuration which makes use of MaxPre 1.
- `no-prepro`: Loandra with its internal preprocessing turned off.
- `maxpre2/none`: `base` with its internal preprocessor changed from MaxPre 1 to MaxPre 2 using the default configuration of MaxPre 1.
- `maxpre2/VG`: `maxpre2` incorporating the additional intrinsic-at-most-one constraints technique and the extension of TrimMaxSAT to non-blocking literals (cf. Sect. 6), found promising in preliminary experimentation.

As benchmarks, we used the combined set of weighted instances from the incomplete tracks of MaxSAT Evaluation 2020 and 2021. After removing duplicates, this gave a total of 451 instances. When reporting results, we consider for each instance and solver the cost of the best solution found by the solver within 300 s (including time spent preprocessing and solution reconstruction).

We compare the relative runtime performance of the solver variants using two metrics: *#wins* and the average *incomplete score*. Assume that τ_x and τ_y are the lowest-cost solutions computed by two solvers X and Y on a MaxSAT instance $\mathcal{F}$ and that $\texttt{best-cost}(\mathcal{F})$ is the lowest cost of a solution of $\mathcal{F}$ found either in our evaluation or in the MaxSAT Evaluations. Then X wins over Y if $\texttt{COST}(\mathcal{F}, \tau_x) < \texttt{COST}(\mathcal{F}, \tau_y)$. The incomplete score, $\texttt{score}(\mathcal{F}, X)$, obtained by solver X on $\mathcal{F}$ is the ratio between the cost of the solution found by X and $\texttt{best-cost}(\mathcal{F})$, i.e., $\texttt{score}(\mathcal{F}, X) = (\texttt{best-cost}(\mathcal{F}) + 1)/(\texttt{COST}(\mathcal{F}, \tau_x) + 1)$. The score of X on $\mathcal{F}$ is 0 if X is unable to find any solutions within 300 s.

An overview of the results is shown in Table 1. The upper part of the table shows a pairwise comparison on the number of wins over all benchmarks. The wins are organized column-wise, i.e., the cell on row X column Y contains the total number of instances that the solver on column Y wins over the solver on row X. The last row contains the average score obtained by each solver over all instances. We observe that any form of preprocessing improves the performance of Loandra, as witnessed by the fact that `no-prepro` is clearly the worst-performing variant. The variants that make use of MaxPre 2 outperform the

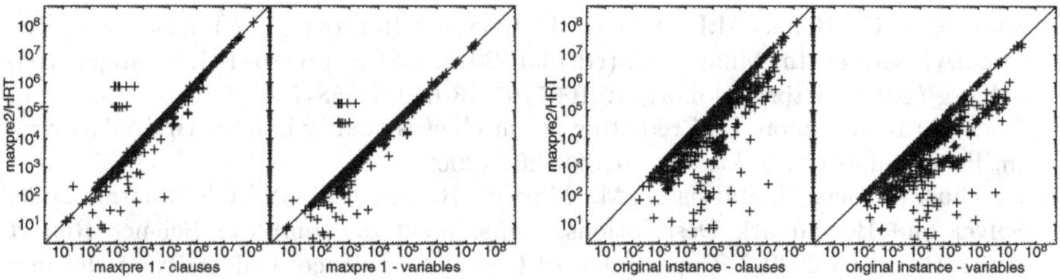

Fig. 2. Impact of preprocessing on instance size.

baseline under both metrics; both `maxpre2 no new` and `maxpre2-w:VG` obtain a higher average score and win on more instances over `base`. The comparison between `maxpre2/none` and `maxpre2/VG` is not as clear. On one hand, the score obtained by `maxpre2/VG` is higher. On the other hand, `maxpre2/none` wins on 80 instances over `maxpre2/VG` and looses on 77. This suggests that the quality of solutions computed by `maxpre2/VG` is on average higher, and that on the instances on which `maxpre2/none` wins the difference is smaller.

7.3 Impact of Preprocessing on Instance Sizes

In addition to improved solver runtimes, we note that MaxPre 2 has a positive effect on the size of instances (both in terms of the number of variables and clauses remaining) when compared to preprocessing with MaxPre 1; see Fig. 2 for a comparison, with `maxpre2/HRT` compared to `maxpre1` (left) and to original instance sizes (right).

8 Conclusions

We studied liftings of variants of propagation redundancy from SAT in the context of maximum satisfiability where—more fine-grained than in SAT—of interest are reasoning techniques that preserve optimal cost. We showed that CPR, the strongest MaxSAT-lifting, allows for changing minimal corrections sets in MaxSAT in a controlled way, thereby succinctly expressing MaxSAT reasoning techniques very generally. We also provided a practical MaxSAT preprocessor extended with techniques captured by CPR and showed empirically that extended preprocessing has a positive overall impact on a range of MaxSAT solvers. Interesting future work includes the development of new CPR-based preprocessing rules for MaxSAT capable of significantly affecting the MaxSAT solving pipeline both in theory and practice, as well as developing an understanding of the relationship between redundancy notions and the transformations performed by MaxSAT solving algorithms.

References

1. Ansótegui, C., Bonet, M., Levy, J.: SAT-based MaxSAT algorithms. Artif. Intell. **196**, 77–105 (2013)

2. Ansótegui, C., Bonet, M.L., Gabàs, J., Levy, J.: Improving SAT-based weighted MaxSAT solvers. In: Milano, M. (ed.) CP 2012. LNCS, pp. 86–101. Springer, Heidelberg (2012). https://doi.org/10.1007/978-3-642-33558-7_9
3. Audemard, G., Simon, L.: Predicting learnt clauses quality in modern SAT solvers. In: Proceedings of the IJCAI, pp. 399–404 (2009)
4. Bacchus, F., Berg, J., Järvisalo, M., Martins, R. (eds.): MaxSAT Evaluation 2020: Solver and Benchmark Descriptions, Department of Computer Science Report Series B, vol. B-2020-2. Department of Computer Science, University of Helsinki (2020)
5. Bacchus, F., Järvisalo, M., Martins, R. (eds.): MaxSAT Evaluation 2019: Solver and Benchmark Descriptions, Department of Computer Science Report Series B, vol. B-2019-2. Department of Computer Science, University of Helsinki (2019)
6. Bacchus, F., Järvisalo, M., Martins, R.: Maximum satisfiability (chap. 24). In: Biere, A., Heule, M., van Maaren, H., Walsh, T. (eds.) Handbook of Satisfiability. Frontiers in Artificial Intelligence and Applications, pp. 929–991. IOS Press (2021)
7. Baek, S., Carneiro, M., Heule, M.J.H.: A flexible proof format for sat solver-elaborator communication. In: TACAS 2021. LNCS, vol. 12651, pp. 59–75. Springer, Cham (2021). https://doi.org/10.1007/978-3-030-72016-2_4
8. Belov, A., Morgado, A., Marques-Silva, J.: SAT-based preprocessing for MaxSAT. In: McMillan, K., Middeldorp, A., Voronkov, A. (eds.) LPAR 2013. LNCS, vol. 8312, pp. 96–111. Springer, Heidelberg (2013). https://doi.org/10.1007/978-3-642-45221-5_7
9. Berg, J., Demirović, E., Stuckey, P.J.: Core-boosted linear search for incomplete MaxSAT. In: Rousseau, L.-M., Stergiou, K. (eds.) CPAIOR 2019. LNCS, vol. 11494, pp. 39–56. Springer, Cham (2019). https://doi.org/10.1007/978-3-030-19212-9_3
10. Berg, J., Järvisalo, M.: Unifying reasoning and core-guided search for maximum satisfiability. In: Calimeri, F., Leone, N., Manna, M. (eds.) JELIA 2019. LNCS (LNAI), vol. 11468, pp. 287–303. Springer, Cham (2019). https://doi.org/10.1007/978-3-030-19570-0_19
11. Berg, J., Saikko, P., Järvisalo, M.: Subsumed label elimination for maximum satisfiability. In: Proceedings of the ECAI. Frontiers in Artificial Intelligence and Applications, vol. 285, pp. 630–638. IOS Press (2016)
12. Bhalla, A., Lynce, I., de Sousa, J.T., Marques-Silva, J.: Heuristic-based backtracking for propositional satisfiability. In: Pires, F.M., Abreu, S. (eds.) EPIA 2003. LNCS (LNAI), vol. 2902, pp. 116–130. Springer, Heidelberg (2003). https://doi.org/10.1007/978-3-540-24580-3_19
13. Biere, A., Heule, M., van Maaren, H., Walsh, T.: Handbook of Satisfiability (Second Edition): Volume 336 Frontiers in Artificial Intelligence and Applications. IOS Press, Amsterdam, The Netherlands (2021)
14. Biere, A., Järvisalo, M., Kiesl, B.: Preprocessing in SAT solving (chap. 9). In: Biere, A., Heule, M., van Maaren, H., Walsh, T. (eds.) Handbook of Satisfiability. Frontiers in Artificial Intelligence and Applications, pp. 391–435. IOS Press (2021)
15. Cruz-Filipe, L., Heule, M.J.H., Hunt, W.A., Kaufmann, M., Schneider-Kamp, P.: Efficient certified RAT verification. In: de Moura, L. (ed.) CADE 2017. LNCS (LNAI), vol. 10395, pp. 220–236. Springer, Cham (2017). https://doi.org/10.1007/978-3-319-63046-5_14
16. Cruz-Filipe, L., Marques-Silva, J., Schneider-Kamp, P.: Efficient certified resolution proof checking. In: Legay, A., Margaria, T. (eds.) TACAS 2017. LNCS, vol. 10205, pp. 118–135. Springer, Heidelberg (2017). https://doi.org/10.1007/978-3-662-54577-5_7

17. Davies, J., Bacchus, F.: Exploiting the power of MIP solvers in MAXSAT. In: Järvisalo, M., Van Gelder, A. (eds.) SAT 2013. LNCS, vol. 7962, pp. 166–181. Springer, Heidelberg (2013). https://doi.org/10.1007/978-3-642-39071-5_13
18. Heule, M., Hunt, W., Kaufmann, M., Wetzler, N.: Efficient, verified checking of propositional proofs. In: Ayala-Rincón, M., Muñoz, C.A. (eds.) ITP 2017. LNCS, vol. 10499, pp. 269–284. Springer, Cham (2017). https://doi.org/10.1007/978-3-319-66107-0_18
19. Heule, M., Hunt, W.A., Jr., Wetzler, N.: Bridging the gap between easy generation and efficient verification of unsatisfiability proofs. Softw. Test. Verif. Reliab. **24**(8), 593–607 (2014)
20. Heule, M., Järvisalo, M., Lonsing, F., Seidl, M., Biere, A.: Clause elimination for SAT and QSAT. J. Artif. Intell. Res. **53**, 127–168 (2015)
21. Heule, M., Kiesl, B.: The potential of interference-based proof systems. In: Proceedings of the ARCADE@CADE. EPiC Series in Computing, vol. 51, pp. 51–54. EasyChair (2017)
22. Heule, M.J.H., Kiesl, B., Biere, A.: Short proofs without new variables. In: de Moura, L. (ed.) CADE 2017. LNCS (LNAI), vol. 10395, pp. 130–147. Springer, Cham (2017). https://doi.org/10.1007/978-3-319-63046-5_9
23. Heule, M.J.H., Kiesl, B., Biere, A.: Strong extension-free proof systems. J. Autom. Reason. **64**(3), 533–554 (2020)
24. Heule, M.J.H., Kiesl, B., Seidl, M., Biere, A.: PRuning through satisfaction. In: HVC 2017. LNCS, vol. 10629, pp. 179–194. Springer, Cham (2017). https://doi.org/10.1007/978-3-319-70389-3_12
25. Hou, A.: A theory of measurement in diagnosis from first principles. Artif. Intell. **65**(2), 281–328 (1994)
26. Ignatiev, A., Morgado, A., Marques-Silva, J.: RC2: an efficient MaxSAT solver. J. Satisf. Boolean Model. Comput. **11**(1), 53–64 (2019)
27. Ihalainen, H., Berg, J., Järvisalo, M.: Refined core relaxation for core-guided MaxSAT solving. In: Proceedings of the CP. LIPIcs, vol. 210, pp. 28:1–28:19. Schloss Dagstuhl - Leibniz-Zentrum für Informatik (2021)
28. Järvisalo, M., Biere, A., Heule, M.: Simulating circuit-level simplifications on CNF. J. Autom. Reason. **49**(4), 583–619 (2012)
29. Järvisalo, M., Heule, M.J.H., Biere, A.: Inprocessing rules. In: Gramlich, B., Miller, D., Sattler, U. (eds.) IJCAR 2012. LNCS (LNAI), vol. 7364, pp. 355–370. Springer, Heidelberg (2012). https://doi.org/10.1007/978-3-642-31365-3_28
30. Kiesl, B., Rebola-Pardo, A., Heule, M.J.H., Biere, A.: Simulating strong practical proof systems with extended resolution. J. Autom. Reason. **64**(7), 1247–1267 (2020)
31. Kiesl, B., Seidl, M., Tompits, H., Biere, A.: Super-blocked clauses. In: Olivetti, N., Tiwari, A. (eds.) IJCAR 2016. LNCS (LNAI), vol. 9706, pp. 45–61. Springer, Cham (2016). https://doi.org/10.1007/978-3-319-40229-1_5
32. Kiesl, B., Seidl, M., Tompits, H., Biere, A.: Local redundancy in SAT: generalizations of blocked clauses. Log. Methods Comput. Sci. **14**(4) (2018)
33. Korhonen, T., Berg, J., Saikko, P., Järvisalo, M.: MaxPre: an extended MaxSAT preprocessor. In: Gaspers, S., Walsh, T. (eds.) SAT 2017. LNCS, vol. 10491, pp. 449–456. Springer, Cham (2017). https://doi.org/10.1007/978-3-319-66263-3_28
34. Lei, Z., Cai, S.: Solving (weighted) partial MaxSAT by dynamic local search for SAT. In: Proceedings of the IJCAI, pp. 1346–1352 (2018). ijcai.org
35. Li, C., Manyà, F.: MaxSAT, hard and soft constraints. In: Biere, A., Heule, M., van Maaren, H., Walsh, T. (eds.) Handbook of Satisfiability, pp. 613–631. IOS Press (2009)

36. Morgado, A., Dodaro, C., Marques-Silva, J.: Core-guided MaxSAT with soft cardinality constraints. In: O'Sullivan, B. (ed.) CP 2014. LNCS, vol. 8656, pp. 564–573. Springer, Cham (2014). https://doi.org/10.1007/978-3-319-10428-7_41
37. Morgado, A., Heras, F., Liffiton, M., Planes, J., Marques-Silva, J.: Iterative and core-guided MaxSAT solving: a survey and assessment. Constraints **18**(4), 478–534 (2013)
38. Narodytska, N., Bacchus, F.: Maximum satisfiability using core-guided MaxSAT resolution. In: Proceedings of the AAAI, pp. 2717–2723. AAAI Press (2014)
39. Paxian, T., Raiola, P., Becker, B.: On preprocessing for weighted MaxSAT. In: Henglein, F., Shoham, S., Vizel, Y. (eds.) VMCAI 2021. LNCS, vol. 12597, pp. 556–577. Springer, Cham (2021). https://doi.org/10.1007/978-3-030-67067-2_25
40. Paxian, T., Reimer, S., Becker, B.: Dynamic polynomial watchdog encoding for solving weighted MaxSAT. In: Beyersdorff, O., Wintersteiger, C.M. (eds.) SAT 2018. LNCS, vol. 10929, pp. 37–53. Springer, Cham (2018). https://doi.org/10.1007/978-3-319-94144-8_3
41. Rebola-Pardo, A., Cruz-Filipe, L.: Complete and efficient DRAT proof checking. In: Proceedings of the FMCAD, pp. 1–9. IEEE (2018)
42. Yolcu, E., Wu, X., Heule, M.J.H.: Mycielski graphs and PR proofs. In: Pulina, L., Seidl, M. (eds.) SAT 2020. LNCS, vol. 12178, pp. 201–217. Springer, Cham (2020). https://doi.org/10.1007/978-3-030-51825-7_15

Cooperating Techniques for Solving Nonlinear Real Arithmetic in the cvc5 SMT Solver (System Description)

Gereon Kremer[1], Andrew Reynolds[2(✉)], Clark Barrett[1], and Cesare Tinelli[2]

[1] Stanford University, Stanford, USA
[2] The University of Iowa, Iowa City, USA
andrew.j.reynolds@gmail.com

Abstract. The cvc5 SMT solver solves quantifier-free nonlinear real arithmetic problems by combining the cylindrical algebraic coverings method with incremental linearization in an abstraction-refinement loop. The result is a complete algebraic decision procedure that leverages efficient heuristics for refining candidate models. Furthermore, it can be used with quantifiers, integer variables, and in combination with other theories. We describe the overall framework, individual solving techniques, and a number of implementation details. We demonstrate its effectiveness with an evaluation on the SMT-LIB benchmarks.

Keywords: Satisfiability modulo theories · Nonlinear real arithmetic · Abstraction refinement · Cylindrical algebraic coverings

1 Introduction

SMT solvers are used as back-end engines for a wide variety of academic and industrial applications [2,19,20]. Efficient reasoning in the theory of real arithmetic is crucial for many such applications [5,8]. While modern SMT solvers have been shown to be quite effective at reasoning about *linear* real arithmetic problems [21,43], *nonlinear* problems are typically much more difficult. This is not surprising, given that the worst-case complexity for deciding the satisfiability of nonlinear real arithmetic formulas is doubly-exponential in the number of variables in the formula [15]. Nevertheless, a variety of techniques have been proposed and implemented, each attempting to target a class of formulas for which reasonable performance can be observed in practice.

Related Work. All complete decision procedures for nonlinear real arithmetic (or the *theory of the reals*) originate in computer algebra, the most prominent being cylindrical algebraic decomposition (CAD) [11]. While alternatives exist [6,25,41], they have not seen much use [27], and CAD-based methods are the only sound and complete methods in practical use today. CAD-based methods used

J. Blanchette et al. (Eds.): IJCAR 2022, LNAI 13385, pp. 95–105, 2022.
https://doi.org/10.1007/978-3-031-10769-6_7

in modern SMT solvers include incremental CAD implementations [34,36] and cylindrical algebraic coverings [3], both of which are integrated in the traditional CDCL(T) framework for SMT [40].

In contrast, the NLSAT [30] calculus and the generalized MCSAT [28,39] framework provide for a much tighter integration of a conflict-driven CAD-based theory solver into a theory-aware core solver. This has been the dominant approach over the last decade due to its strong performance in practice. However, it has the significant disadvantage of being difficult to integrate with CDCL(T)-based frameworks for theory combination.

A number of *incomplete* techniques are also used by various SMT solvers: incremental linearization [9] gradually refines an abstraction of the nonlinear formula obtained via a naive linearization by refuting spurious models of the abstraction; interval constraint propagation [24,36,45] employs interval arithmetic to narrow down the search space; subtropical satisfiability [22] provides sufficient linear conditions for nonlinear solutions in the exponent space of the polynomials; and virtual substitution [12,31,46] makes use of parametric solution formulas for polynomials of bounded degree. Though all of these techniques have limitations, each of them is useful for certain subclasses of nonlinear real arithmetic or in combination with other techniques.

Contributions. We present an integration of cylindrical algebraic coverings and incremental linearization, implemented in the `cvc5` SMT solver. Crucial to the success of the integration is an abstraction-refinement loop used to combine the two techniques cooperatively. The solution is effective in practice, as witnessed by the fact that `cvc5` won the nonlinear real arithmetic category of SMT-COMP 2021 [44], the first time a non-MCSAT-based technique has won since 2013. Our integrated technique also has the advantage of being very flexible: in particular, it fits into the regular CDCL(T) schema for theory solvers and theory combination, it supports (mixed) integer problems, and it can be easily extended using further subsolvers that support additional arithmetic operators beyond the scope of traditional algebraic routines (e.g., transcendental functions).

2 Nonlinear Solving Techniques

The nonlinear arithmetic solver implemented in `cvc5` generally follows the abstraction-refinement framework introduced by Cimatti et al. [9] and depicted in Fig. 1. The input assertions are first checked by the linear arithmetic solver, where they are linearized implicitly by treating every application of multiplication as if it were an arithmetic variable. For example, given input assertions $x \cdot y > 0 \wedge x > 1 \wedge y < 0$, the linear solver treats the expression $x \cdot y$ as a variable. It may then find the (spurious) model: $x \mapsto 2$, $y \mapsto -1$, and $x \cdot y \mapsto 1$. We call the candidate model returned by the linear arithmetic solver, where applications of multiplication are treated as variables, a *linear model*. If a linear model does not exist, i.e., the input is unsatisfiable according to the linear solver, the linear solver generates a conflict that is immediately returned to the CDCL(T) engine.

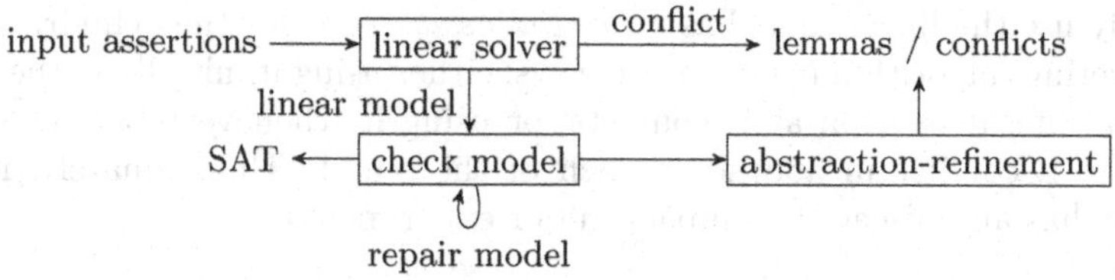

Fig. 1. Structural overview of the nonlinear solver

When a linear model does exist, we check whether it already satisfies the input assertions or try to *repair* it to do so. We only apply a few very simple heuristics for repairs such as updating the value for z in the presence of a constraint like $z = x \cdot y$ based on the values of x and y.

If the model can not be repaired, we refine the abstraction for the linear solver [9]. This step constructs lemmas, or conflicts, based on the input assertions and the linear model, to advance the solving process by blocking either the current linear model or the current Boolean model, that is, the propositional assignment generated by the SMT solver's SAT engine. The Boolean model is usually eliminated only by the coverings approach, while the incremental linearization technique generates lemmas with new literals that target the linear model, e.g., the lemma $x > 0 \wedge y < 0 \Rightarrow x \cdot y < 0$ in the example above. We next describe our implementation of cylindrical algebraic coverings and incremental linearization, and how they are combined in cvc5.

2.1 Cylindrical Algebraic Coverings

Cylindrical algebraic coverings is a technique recently proposed by Ábrahám et al. [3] and is heavily inspired by CAD. While the way the computation proceeds is very different from traditional CAD, and instead somewhat similar to NLSAT [30], their mathematical underpinnings are essentially identical. The cylindrical algebraic coverings subsolver in cvc5 closely follows the presentation in [3]. Below, we discuss some differences and extensions. For this discussion, we must refer the reader to [3] for the relevant background material because of space constraints. We note that cvc5 relies on the libpoly library [29] to provide most of the computational infrastructure for algebraic reasoning.

Square-Free Basis. As with most CAD projection schemas, the set of projection polynomials needs to be a square-free basis when computing the characterization for an interval in [3, Algorithm 4]. However, the resultants computed in this algorithm combine polynomials from different sets, which are not necessarily coprime. The remedy is to either make these sets of polynomials pairwise square-free or to fully factor all projection polynomials. We adopt the former approach.

Starting Model. Although the linear model may not satisfy the nonlinear constraints, we may expect it to be in the vicinity of a proper model. We thus optionally use the linear model as an *initial assignment* for the cylindrical algebraic coverings algorithm in one of two ways: either using it initially in the search and discarding it as soon as it conflicts; or using it whenever possible, even if it leads to a conflict in another branch of the search. Unfortunately, neither technique has any discernible impact in our experiments.

Interval Pruning. As already noted in [3], a covering may contain two kinds of redundant intervals: intervals fully contained in another interval, or intervals contained in the union of other intervals. Removing the former kind of redundancies is not only clearly beneficial, but also required for how the characterizations are computed. It is not clear, however, if it is worthwhile to remove redundancies of the second kind because, while it can simplify the characterization locally, it may also make the resulting interval smaller, slowing down the overall solving process. Note that there may not be a unique redundant interval: e.g., if multiple intervals overlap, it may be possible to remove one of two intervals, but not both of them. We have implemented a simple heuristic to detect redundancies of the second kind, always removing the smallest interval with respect to the interval ordering given in [3]. Even if these redundancies occur in about 7.5% of all `QF_NRA` benchmarks, using this technique has only a very limited impact. It may be that for certain kinds of benchmarks, underrepresented in SMT-LIB, the technique is valuable. Or it may be that some variation of the technique is more broadly helpful. These are interesting directions for future work.

Lifting and Coefficient Selection with Lazard. The original cylindrical algebraic coverings technique is based on McCallum's projection operator [37], which is particularly well-studied, but also (refutationally) unsound: polynomial nullification may occur when computing the real roots, possibly leading to the loss of real roots and thus solution candidates. One then needs to check for these cases and fall back to a more conservative, albeit more costly, projection schema such as those due to Collins [11] or Hong [26].

Lazard's projection schema [35], which has been proven correct only recently [38], provides very small projection sets and is both sound and complete. This comes at the price of a different mathematical background and a modified lifting procedure, which corresponds to a modified procedure for real root isolation. Although the local projections employed in cylindrical algebraic coverings have not been formally verified for Lazard's projection schema yet, we expect no significant issues there. Adopting it seems to be a logical improvement, as already mentioned in [3]. The modified real root isolation procedure is a significant hurdle in practice, as it requires additional nontrivial algorithms [32, Section 5.3.2]. We implemented it using CoCoALib [1] in `cvc5` [33], achieving soundness without any discernible negative performance impact.

Using Lazard's projection schema, for all its benefits, may seem questionable for the following reasons: (*i*) the unsoundness of McCallum's projection operator is virtually never witnessed in practice [32,33, Section 6.5], and (*ii*) the projection sets computed by Lazard's and McCallums's projection operator are identical on more than 99.5% on all of `QF_NRA` [33]. We argue, though, that working in the domain of formal verification warrants the effort of obtaining a (provably) correct result, especially if it does not incur a performance overhead.

Proof Generation. Recently, generating formal proofs to certify the result of SMT solvers has become an area of focus. In particular, there is a large and ongoing effort to produce proofs in cvc5. The incremental linearization approach can be seen as an oracle which produces lemmas that are easy to prove individually, so cvc5 does generate proofs for them; the complex part is finding those lemmas and making sure they actually help the solver make progress.

The situation is very different for cylindrical algebraic coverings: the produced lemma is the infeasible subset, and we usually have no simpler proof than the computations relying on CAD theory. That said, cylindrical algebraic coverings appear to be more amenable to automatic proof generation than traditional CAD-based approaches [4, 14]. In fact, although making these proofs detailed enough for automated verification is still an open problem, they are already broken into smaller parts that closely follow the tree-shaped computation of the algorithm. This allows cvc5 to produce at least a proof skeleton in that case.

2.2 Incremental Linearization

Our theory solver for nonlinear (real) arithmetic optionally uses lemma schemas following the incremental linearization approaches described by Cimatti et al. [9] and Reynolds et al. [42]. These schemas incrementally refine candidate models from the linear arithmetic solver by introducing selected quantifier-free lemmas that express properties of multiplication, such as signedness (e.g., $x > 0 \wedge y > 0 \Rightarrow x \cdot y > 0$) or monotonicity (e.g., $|x| > |y| \Rightarrow x \cdot x > y \cdot y$). They are generated as needed to refute spurious models that violate these properties.

Most lemma schemas built-in in cvc5 are crafted so as to avoid introducing new monomial terms or coefficients, since that could lead to non-termination in the CDCL(T) search. As a notable exception, we rely on a lemma schema for *tangent planes* for multiplication [9], which can be used to refute the candidate model for any application of the multiplication operator $\cdot$ whose value in the linear model is inconsistent with the standard interpretation of $\cdot$. Note that since these lemmas depend upon the current model value chosen for arithmetic variables, tangent plane lemmas may introduce an unbounded number of new literals into the search. The set of lemma schemas used by the solver is user-configurable, as described in the following section.

2.3 Strategy

The overall theory solver for nonlinear arithmetic is built from several subsolvers, implementing the techniques described above, using a rather naive strategy, as summarized in Algorithm 1. After a spurious linear model has been constructed that cannot be repaired, we first apply a subset of the lemma schemas that do not introduce an unbounded number of new terms (with procedure `IncLinearizationLight`); then, we continue with the remaining lemma schemas

```
1 Function NlSolve(assertions)
2     if not LinearSolve(assertions) then return linear conflict
3     M = linear model for assertions
4     if RepairModel(assertions, M) then return repaired model
5     if IncLinearizationLight(assertions, M) then return lemmas
6     if IncLinearizationFull(assertions, M) then return lemmas
7     return Coverings(assertions, M)
```

Algorithm 1: Strategy for nonlinear arithmetic solver

(with procedure `IncLinearizationFull`); finally, we resort to the coverings solver which is guaranteed to find either a conflict or a model. Internally, each procedure sequentially tries its assigned lemma schemas from [9,42] until it constructs a lemma that can block the spurious model.

The approach is dynamically configured based on input options and the logic of the input formula. For example, by default, we disable `IncLinearizationFull` for `QF_NRA` as it tends to diverge in cases where the coverings solver quickly terminates.

2.4 Beyond `QF_NRA`

The presented solver primarily targets quantifier-free nonlinear real arithmetic, but is used also in the presence of quantifiers and with multiple theories.

Quantified Logics. Solving quantified logics for nonlinear arithmetic requires solving quantifier-free subproblems, and thus any improvement to quantifier-free solving also benefits solving with quantifiers. In practice, however, the instantiation heuristics are just as important for overall solver performance.

Multiple Theories. The theory combination framework as implemented in `cvc5` requires evaluating equalities over the combined model. To support this functionality, real algebraic numbers had to be properly integrated into the entire solver; in particular, the ability to compute with these numbers could not be local to the cylindrical algebraic coverings module or even the nonlinear solver.

3 Experimental Results

We evaluate our implementation within `cvc5` (commit id `449dd7e`) in comparison with other SMT solvers on all 11552 benchmarks in the quantifier-free nonlinear real arithmetic (`QF_NRA`) logic of SMT-LIB. We consider three configurations of `cvc5`, each of which runs a subset of steps from Algorithm 1. All the configurations run lines 2–4. In addition, `cvc5.cov` runs line 7, `cvc5.inclin` runs lines 5 and 6, and `cvc5` runs lines 5 and 7. All experiments were conducted on Intel Xeon E5-2637v4 CPUs with a time limit of 20 min and 8 GB memory.

We compare `cvc5` with recent versions of all other SMT solvers that participated in the `QF_NRA` logic of SMT-COMP 2021 [44]: `MathSAT` 5.6.6 [10], `SMT-RAT` 19.10.560 [13], `veriT` [7] (`veriT+raSAT+Redlog`), `Yices2` 2.6.4 [18] (`Yices-QS` for quantified logics), and `z3` 4.8.14 [16]. `MathSAT` employs an abstraction-refinement mechanism very similar to the one described in Sect. 2.2; `veriT` [23] forwards non-linear arithmetic problems to the external tools `raSAT` [45], which uses interval constraint propagation, and `Redlog/Reduce` [17], which focuses on virtual substitution and cylindrical algebraic decomposition; `SMT-RAT`, `Yices2`, and `z3` all implement some variant of MCSAT [30]. Note that `SMT-RAT` also implements the cylindrical algebraic coverings approach, but it is less effective than `SMT-RAT`'s adaptation of MCSAT [3].

QF_NRA	sat	unsat	solved
cvc5	**5137**	**5596**	**10733**
Yices2	4966	5450	10416
z3	5136	5207	10343
cvc5.cov	5001	5077	10078
SMT-RAT	4828	5038	9866
veriT	4522	5034	9556
MathSAT	3645	5357	9002
cvc5.inclin	3421	5376	8797

(a)

Beyond QF_NRA		sat	unsat	solved
NRA	Yices2	231	**3817**	**4048**
	z3	**236**	3812	**4048**
	cvc5.cov	**236**	3809	4045
	cvc5	221	3809	4030
	cvc5.inclin	120	3786	3906
QF_UFNRA	z3	**24**	**11**	**35**
	Yices2	23	**11**	34
	cvc5	20	**11**	31
	cvc5.inclin	12	**11**	23
	cvc5.cov	2	**11**	13

(b)

Fig. 2. (a) Experiments for `QF_NRA` (b) Experiments for `NRA` and `QF_UFNRA`

Figure 2a shows that `cvc5` significantly outperforms all other `QF_NRA` solvers. Both the coverings approach (`cvc5.cov`) and the incremental linearization approach (`cvc5.inclin`) contribute substantially to the overall performance of the unified solver in `cvc5`, with coverings solving many satisfiable instances, and incremental linearization helping on unsatisfiable ones. Even though `cvc5.inclin` closely follows [9], it outperforms `MathSAT` on unsatisfiable benchmarks, those where `cvc5` relies on incremental linearization the most.

Comparing `cvc5` and `Yices2` is particularly interesting, as the coverings approach in `cvc5` and the NLSAT solver in `Yices2` both rely on libpoly [29], thus using the same implementation of algebraic numbers and operations over them. Our integration of incremental linearization and algebraic coverings is compatible with the traditional CDCL(T) framework and outperforms the alternative NLSAT approach, which is specially tailored to nonlinear real arithmetic.

Going beyond `QF_NRA`, we also evaluate the performance of our solver in the context of theory combination (with all 37 benchmarks from `QF_UFNRA`) and quantifiers (with all 4058 benchmarks from `NRA`). There, `cvc5` is a close runner-up to `Yices2` and `z3`, thanks to the coverings subsolver which significantly improves `cvc5`'s performance. We conjecture that the remaining gap is due to components other than the nonlinear arithmetic solver, such as the solver for equality and uninterpreted functions, details of theory combination, or quantifier instantiation heuristics. Interestingly, the sets of unsolved instances in NRA are almost disjoint for cvc5.cov, Yices2 and z3, indicating that each tool could solve the remaining benchmarks with reasonable extra effort.

4 Conclusion

We have presented an approach for solving quantifier-free nonlinear real arithmetic problems that combines previous approaches based on incremental linearization [9] and cylindrical algebraic coverings [3] into one coherent abstraction-refinement loop. The resulting implementation is very effective, outperforming other state-of-the-art solver implementations, and integrates seamlessly in the CDCL(T) framework.

The general approach also applies to integer problems, quantified formulas, and instances with multiple theories, and can additionally be used in combination with transcendental functions [9] and bitwise conjunction for integers [47]. Further evaluations of these combinations are left to future work.

References

1. Abbott, J., Bigatti, A.M., Palezzato, E.: New in CoCoA-5.2.4 and CoCoALib-0.99600 for SC-square. In: Satisfiability Checking and Symbolic Computation. CEUR Workshop Proceedings, vol. 2189, pp. 88–94. CEUR-WS.org (2018). http://ceur-ws.org/Vol-2189/paper4.pdf
2. Ábrahám, E., Corzilius, F., Johnsen, E.B., Kremer, G., Mauro, J.: Zephyrus2: on the fly deployment optimization using SMT and CP technologies. In: Fränzle, M., Kapur, D., Zhan, N. (eds.) SETTA 2016. LNCS, vol. 9984, pp. 229–245. Springer, Cham (2016). https://doi.org/10.1007/978-3-319-47677-3_15
3. Ábrahám, E., Davenport, J.H., England, M., Kremer, G.: Deciding the consistency of non-linear real arithmetic constraints with a conflict driven search using cylindrical algebraic coverings. J. Logic. Algebr. Methods Program. **119**(100633) (2021). https://doi.org/10.1016/j.jlamp.2020.100633
4. Abrahám, E., Davenport, J.H., England, M., Kremer, G.: Proving UNSAT in SMT: the case of quantifier free non-linear real arithmetic. arXiv preprint arXiv:2108.05320 (2021)
5. Arnett, T.J., Cook, B., Clark, M., Rattan, K.: Fuzzy logic controller stability analysis using a satisfiability modulo theories approach. In: 19th AIAA Non-Deterministic Approaches Conference, p. 1773 (2017)
6. Basu, S., Pollack, R., Roy, M.F.: On the combinatorial and algebraic complexity of quantifier elimination. J. ACM **43**, 1002–1045 (1996). https://doi.org/10.1145/235809.235813
7. Bouton, T., Caminha B. de Oliveira, D., Déharbe, D., Fontaine, P.: veriT: an open, trustable and efficient SMT-solver. In: Schmidt, R.A. (ed.) CADE 2009. LNCS (LNAI), vol. 5663, pp. 151–156. Springer, Heidelberg (2009). https://doi.org/10.1007/978-3-642-02959-2_12
8. Cashmore, M., Magazzeni, D., Zehtabi, P.: Planning for hybrid systems via satisfiability modulo theories. J. Artif. Intell. Res. **67**, 235–283 (2020). https://doi.org/10.1613/jair.1.11751

9. Cimatti, A., Griggio, A., Irfan, A., Roveri, M., Sebastiani, R.: Incremental linearization for satisfiability and verification modulo nonlinear arithmetic and transcendental functions. ACM Trans. Comput. Logic **19**, 19:1–19:52 (2018). https://doi.org/10.1145/3230639
10. Cimatti, A., Griggio, A., Schaafsma, B.J., Sebastiani, R.: The MathSAT5 SMT solver. In: Piterman, N., Smolka, S.A. (eds.) TACAS 2013. LNCS, vol. 7795, pp. 93–107. Springer, Heidelberg (2013). https://doi.org/10.1007/978-3-642-36742-7_7
11. Collins, G.E.: Quantifier elimination for real closed fields by cylindrical algebraic decompostion. In: Brakhage, H. (ed.) GI-Fachtagung 1975. LNCS, vol. 33, pp. 134–183. Springer, Heidelberg (1975). https://doi.org/10.1007/3-540-07407-4_17
12. Corzilius, F., Ábrahám, E.: Virtual substitution for SMT-solving. In: Owe, O., Steffen, M., Telle, J.A. (eds.) FCT 2011. LNCS, vol. 6914, pp. 360–371. Springer, Heidelberg (2011). https://doi.org/10.1007/978-3-642-22953-4_31
13. Corzilius, F., Kremer, G., Junges, S., Schupp, S., Ábrahám, E.: SMT-RAT: an open source C++ toolbox for strategic and parallel SMT solving. In: Heule, M., Weaver, S. (eds.) SAT 2015. LNCS, vol. 9340, pp. 360–368. Springer, Cham (2015). https://doi.org/10.1007/978-3-319-24318-4_26
14. Davenport, J., England, M., Kremer, G., Tonks, Z., et al.: New opportunities for the formal proof of computational real geometry? arXiv preprint arXiv:2004.04034 (2020)
15. Davenport, J.H., Heintz, J.: Real quantifier elimination is doubly exponential. J. Symb. Comput. **5**(1), 29–35 (1988). https://doi.org/10.1016/S0747-7171(88)80004-X
16. de Moura, L., Bjørner, N.: Z3: an efficient SMT solver. In: Ramakrishnan, C.R., Rehof, J. (eds.) TACAS 2008. LNCS, vol. 4963, pp. 337–340. Springer, Heidelberg (2008). https://doi.org/10.1007/978-3-540-78800-3_24
17. Dolzmann, A., Sturm, T.: REDLOG: computer algebra meets computer logic. ACM SIGSAM Bull. **31**(2), 2–9 (1997). https://doi.org/10.1145/261320.261324
18. Dutertre, B.: Yices 2.2. In: Biere, A., Bloem, R. (eds.) CAV 2014. LNCS, vol. 8559, pp. 737–744. Springer, Cham (2014). https://doi.org/10.1007/978-3-319-08867-9_49
19. Ermon, S., Le Bras, R., Gomes, C.P., Selman, B., van Dover, R.B.: SMT-aided combinatorial materials discovery. In: Cimatti, A., Sebastiani, R. (eds.) SAT 2012. LNCS, vol. 7317, pp. 172–185. Springer, Heidelberg (2012). https://doi.org/10.1007/978-3-642-31612-8_14
20. Fagerberg, R., Flamm, C., Merkle, D., Peters, P.: Exploring chemistry using SMT. In: Milano, M. (ed.) CP 2012. LNCS, pp. 900–915. Springer, Heidelberg (2012). https://doi.org/10.1007/978-3-642-33558-7_64
21. Faure, G., Nieuwenhuis, R., Oliveras, A., Rodríguez-Carbonell, E.: SAT modulo the theory of linear arithmetic: exact, inexact and commercial solvers. In: Kleine Büning, H., Zhao, X. (eds.) SAT 2008. LNCS, vol. 4996, pp. 77–90. Springer, Heidelberg (2008). https://doi.org/10.1007/978-3-540-79719-7_8
22. Fontaine, P., Ogawa, M., Sturm, T., Vu, X.T.: Subtropical satisfiability. In: Dixon, C., Finger, M. (eds.) FroCoS 2017. LNCS (LNAI), vol. 10483, pp. 189–206. Springer, Cham (2017). https://doi.org/10.1007/978-3-319-66167-4_11
23. Fontaine, P., Ogawa, M., Sturm, T., Vu, X.T., et al.: Wrapping computer algebra is surprisingly successful for non-linear SMT. In: SC-square 2018-Third International Workshop on Satisfiability Checking and Symbolic Computation (2018)
24. Gao, S., Kong, S., Clarke, E.M.: dReal: an SMT solver for nonlinear theories over the reals. In: Bonacina, M.P. (ed.) CADE 2013. LNCS (LNAI), vol. 7898, pp. 208–214. Springer, Heidelberg (2013). https://doi.org/10.1007/978-3-642-38574-2_14

25. Grigor'ev, D.Y., Vorobjov, N.: Solving systems of polynomial inequalities in subexponential time. J. Symb. Comput. **5**, 37–64 (1988). https://doi.org/10.1016/S0747-7171(88)80005-1
26. Hong, H.: An improvement of the projection operator in cylindrical algebraic decomposition. In: International Symposium on Symbolic and Algebraic Computation, pp. 261–264 (1990). https://doi.org/10.1145/96877.96943
27. Hong, H.: Comparison of several decision algorithms for the existential theory of the reals. RES report, Johannes Kepler University (1991)
28. Jovanović, D., Barrett, C., de Moura, L.: The design and implementation of the model constructing satisfiability calculus. In: Formal Methods in Computer-Aided Design, pp. 173–180. IEEE (2013). https://doi.org/10.1109/FMCAD.2013.7027033
29. Jovanovic, D., Dutertre, B.: LibPoly: a library for reasoning about polynomials. In: Satisfiability Modulo Theories. CEUR Workshop Proceedings, vol. 1889. CEUR-WS.org (2017). http://ceur-ws.org/Vol-1889/paper3.pdf
30. Jovanović, D., de Moura, L.: Solving non-linear arithmetic. In: Gramlich, B., Miller, D., Sattler, U. (eds.) IJCAR 2012. LNCS (LNAI), vol. 7364, pp. 339–354. Springer, Heidelberg (2012). https://doi.org/10.1007/978-3-642-31365-3_27
31. Košta, M., Sturm, T.: A generalized framework for virtual substitution. CoRR abs/1501.05826 (2015)
32. Kremer, G.: Cylindrical algebraic decomposition for nonlinear arithmetic problems. Ph.D. thesis, RWTH Aachen University (2020). https://doi.org/10.18154/RWTH-2020-05913
33. Kremer, G., Brandt, J.: Implementing arithmetic over algebraic numbers: a tutorial for Lazard's lifting scheme in CAD. In: Symbolic and Numeric Algorithms for Scientific Computing, pp. 4–10 (2021). https://doi.org/10.1109/SYNASC54541.2021.00013
34. Kremer, G., Ábrahám, E.: Fully incremental cylindrical algebraic decomposition. J. Symb. Comput. **100**, 11–37 (2020). https://doi.org/10.1016/j.jsc.2019.07.018
35. Lazard, D.: An improved projection for cylindrical algebraic decomposition. In: Bajaj, C.L. (ed.) Algebraic Geometry and Its Applications, pp. 467–476. Springer, New York (1994). https://doi.org/10.1007/978-1-4612-2628-4_29
36. Loup, U., Scheibler, K., Corzilius, F., Ábrahám, E., Becker, B.: A symbiosis of interval constraint propagation and cylindrical algebraic decomposition. In: Bonacina, M.P. (ed.) CADE 2013. LNCS (LNAI), vol. 7898, pp. 193–207. Springer, Heidelberg (2013). https://doi.org/10.1007/978-3-642-38574-2_13
37. McCallum, S.: An improved projection operation for cylindrical algebraic decomposition. In: Caviness, B.F. (ed.) EUROCAL 1985. LNCS, vol. 204, pp. 277–278. Springer, Heidelberg (1985). https://doi.org/10.1007/3-540-15984-3_277
38. McCallum, S., Parusiński, A., Paunescu, L.: Validity proof of Lazard's method for cad construction. J. Symb. Comput. **92**, 52–69 (2019). https://doi.org/10.1016/j.jsc.2017.12.002
39. de Moura, L., Jovanović, D.: A model-constructing satisfiability calculus. In: Giacobazzi, R., Berdine, J., Mastroeni, I. (eds.) VMCAI 2013. LNCS, vol. 7737, pp. 1–12. Springer, Heidelberg (2013). https://doi.org/10.1007/978-3-642-35873-9_1
40. Nieuwenhuis, R., Oliveras, A., Tinelli, C.: Solving SAT and SAT modulo theories: from an abstract Davis-Putnam-Logemann-Loveland procedure to DPLL(T). J. ACM **53**(6), 937–977 (2006). https://doi.org/10.1145/1217856.1217859
41. Renegar, J.: A faster PSPACE algorithm for deciding the existential theory of the reals. In: Symposium on Foundations of Computer Science, pp. 291–295 (1988). https://doi.org/10.1109/SFCS.1988.21945

42. Reynolds, A., Tinelli, C., Jovanović, D., Barrett, C.: Designing theory solvers with extensions. In: Dixon, C., Finger, M. (eds.) FroCoS 2017. LNCS (LNAI), vol. 10483, pp. 22–40. Springer, Cham (2017). https://doi.org/10.1007/978-3-319-66167-4_2
43. Roselli, S.F., Bengtsson, K., Åkesson, K.: SMT solvers for job-shop scheduling problems: models comparison and performance evaluation. In: International Conference on Automation Science and Engineering (CASE), pp. 547–552 (2018). https://doi.org/10.1109/COASE.2018.8560344
44. SMT-COMP 2021 (2021). https://smt-comp.github.io/2021/
45. Tung, V.X., Van Khanh, T., Ogawa, M.: raSAT: an SMT solver for polynomial constraints. In: Olivetti, N., Tiwari, A. (eds.) IJCAR 2016. LNCS (LNAI), vol. 9706, pp. 228–237. Springer, Cham (2016). https://doi.org/10.1007/978-3-319-40229-1_16
46. Weispfenning, V.: Quantifier elimination for real algebra-the quadratic case and beyond. Appl. Algebra Eng. Commun. Comput. **8**(2), 85–101 (1997). https://doi.org/10.1007/s002000050055
47. Zohar, Y., et al.: Bit-precise reasoning via Int-blasting. In: Finkbeiner, B., Wies, T. (eds.) VMCAI 2022. LNCS, vol. 13182, pp. 496–518. Springer, Cham (2022). https://doi.org/10.1007/978-3-030-94583-1_24

Preprocessing of Propagation Redundant Clauses

Joseph E. Reeves(✉), Marijn J. H. Heule, and Randal E. Bryant

Carnegie Mellon University, Pittsburgh, PA, USA
{jereeves,mheule,randy.bryant}@cs.cmu.edu

Abstract. The *propagation redundant* (PR) proof system generalizes the *resolution* and *resolution asymmetric tautology* proof systems used by *conflict-driven clause learning* (CDCL) solvers. PR allows short proofs of unsatisfiability for some problems that are difficult for CDCL solvers. Previous attempts to automate PR clause learning used hand-crafted heuristics that work well on some highly-structured problems. For example, the solver SADICAL incorporates PR clause learning into the CDCL loop, but it cannot compete with modern CDCL solvers due to its fragile heuristics. We present PRELEARN, a preprocessing technique that learns short PR clauses. Adding these clauses to a formula reduces the search space that the solver must explore. By performing PR clause learning as a preprocessing stage, PR clauses can be found efficiently without sacrificing the robustness of modern CDCL solvers. On a large portion of SAT competition benchmarks we found that preprocessing with PRELEARN improves solver performance. In addition, there were several satisfiable and unsatisfiable formulas that could only be solved after preprocessing with PRELEARN. PRELEARN supports proof logging, giving a high level of confidence in the results.

1 Introduction

Conflict-driven clause learning (CDCL) [27] is the standard paradigm for solving the satisfiability problem (SAT) in propositional logic. CDCL solvers learn clauses implied through *resolution* inferences. Additionally, all competitive CDCL solvers use pre- and in-processing techniques captured by the *resolution asymmetric tautology* (RAT) proof system [21]. As examples, the well-studied pigeonhole and mutilated chessboard problems are challenging benchmarks with exponentially-sized resolution proofs [1,12]. It is possible to construct small hand-crafted proofs for the pigeonhole problem using *extended resolution* (ER) [8], a proof system that allows the introduction of new variables [32]. ER can be expressed in RAT but has proved difficult to automate due to the large search space. Even with modern inprocessing techniques, many CDCL solvers struggle on these seemingly simple problems. The *propagation redundant* (PR) proof system allows short proofs for these problems [14,15], and unlike in ER, no new variables are required. This makes PR an attractive candidate for automation.

At a high level, CDCL solvers make decisions that typically yield an unsatisfiable branch of a problem. The clause that prunes the unsatisfiable branch from the search space is learned, and the solver continues by searching another branch. PR extends this

The authors are supported by the NSF under grant CCF-2108521.

J. Blanchette et al. (Eds.): IJCAR 2022, LNAI 13385, pp. 106–124, 2022.
https://doi.org/10.1007/978-3-031-10769-6_8

paradigm by allowing more aggressive pruning. In the PR proof system a branch can be pruned as long as there exists another branch that is at least as satisfiable. As an example, consider the mutilated chessboard. The mutilated chessboard problem involves finding a covering of 2×1 dominos on an $n \times n$ chessboard with two opposite corners removed (see Section 5.4). Given two horizontally oriented dominoes covering a 2×2 square, two vertically oriented dominos could cover the same 2×2 square. For any solution that uses the dominos in the horizontal orientation, replacing them with the dominos in the vertical orientation would also be a solution. The second orientation is as satisfiable as the first, and so the first can be pruned from the search space. Even though the number of possible solutions may be reduced, the pruning is satisfiability preserving. This is a powerful form of reasoning that can efficiently remove many symmetries from the mutilated chessboard, making the problem much easier to solve [15].

The *satisfaction-driven clause learning* (SDCL) solver SADICAL [16] incorporates PR clause learning into the CDCL loop. SADICAL implements hand-crafted decision heuristics that exploit the canonical structure of the pigeonhole and mutilated chessboard problems to find short proofs. However, SADICAL's performance deteriorates under slight variations to the problems including different constraint encodings [7]. The heuristics were developed from a few well-understood problems and do not generalize to other problem classes. Further, the heuristics for PR clause learning are likely ill-suited for CDCL, making the solver less robust.

In this paper, we present PRELEARN, a preprocessing technique for learning PR clauses. PRELEARN alternates between finding and learning PR clauses. We develop multiple heuristics for finding PR clauses and multiple configurations for learning some subset of the found PR clauses. As PR clauses are learned we use failed literal probing [11] to find unit clauses implied by the formula. The preprocessing is made efficient by taking advantage of the inner/outer solver framework in SADICAL. The learned PR clauses are added to the original formula, aggressively pruning the search space in an effort to guide CDCL solvers to short proofs. With this method PR clauses can be learned without altering the complex heuristics that make CDCL solvers robust. PRELEARN focuses on finding short PR clauses and failed literals to effectively reduce the search space. This is done with general heuristics that work across a wide range of problems.

Most SAT solvers support logging proofs of unsatisfiability for independent checking [17,20,33]. This has proved valuable for verifying solutions independent of a (potentially buggy) solver. Modern SAT solvers log proofs in the DRAT proof system (RAT [21] with deletions). DRAT captures all widely used pre- and in-processing techniques including bounded variable elimination [10], bounded variable addition [26], and extended learning [4,32]. DRAT can express the common symmetry-breaking techniques, but it is complicated [13]. PR can compactly express some symmetry-breaking techniques [14,15], yielding short proofs that can be checked by the proof checker DPR-TRIM [16]. PR gives a framework for strong symmetry-breaking inferences and also maintains the highly desirable ability to independently verify proofs.

The contributions of this paper include: (1) giving a high-level algorithm for extracting PR clauses, (2) implementing several heuristics for finding and learning PR clauses, (3) evaluating the effectiveness of different heuristic configurations, and (4) assessing the impact of PRELEARN on solver performance. PRELEARN improves the performance of the CDCL solver KISSAT on a quarter of the satisfiable and unsatisfiable competition benchmarks we considered. The improvement is significant for a number of instances that can only be solved by KISSAT after preprocessing. Most of them come from hard combinatorial problems with small formulas. In addition, PRELEARN directly produces refutation proofs for the mutilated chessboard problem containing only unit and binary PR clauses.

2 Preliminaries

We consider propositional formulas in *conjunctive normal form* (CNF). A CNF formula ψ is a conjunction of *clauses* where each clause is a disjunction of *literals*. A literal l is either a variable x (positive literal) or a negated variable $\overline{x}$ (negative literal). For a set of literals L the formula $\psi(L)$ is the clauses $\{C \in \psi \mid C \cap L \neq \emptyset\}$.

An *assignment* is a mapping from variables to truth values 1 (*true*) and 0 (*false*). An assignment is *total* if it assigns every variable to a value, and *partial* if it assigns a subset of variables to values. The set of variables occurring in a formula, assignment, or clause is given by $\mathsf{var}(\psi)$, $\mathsf{var}(\alpha)$, or $\mathsf{var}(C)$. For a literal l, $\mathsf{var}(l)$ is a variable.

An assignment α *satisfies* a positive (negative) literal l if α maps $\mathsf{var}(l)$ to true (α maps $\mathsf{var}(l)$ to false, respectively), and *falsifies* it if α maps $\mathsf{var}(l)$ to false (α maps $\mathsf{var}(l)$ to true, respectively). We write a finite partial assignment as the set of literals it satisfies. An assignment satisfies a clause if the clause contains a literal satisfied by the assignment. An assignment satisfies a formula if every clause in the formula is satisfied by the assignment. A formula is *satisfiable* if there exists a satisfying assignment, and *unsatisfiable* otherwise. Two formula are *logically equivalent* if they share the same set of satisfying assignments. Two formulas are *satisfiability equivalent* if they are either both satisfiable or both unsatisfiable.

If an assignment α satisfies a clause C we define $C|\alpha = \top$, otherwise $C|\alpha$ represents the clause C with the literals falsified by α removed. The empty clause is denoted by $\bot$. The formula ψ reduced by an assignment α is given by $\psi|\alpha = \{C|\alpha \mid C \in \psi \text{ and } C|\alpha \neq \top\}$. Given an assignment $\alpha = l_1 \dots l_n$, $C = (\bar{l}_1 \vee \cdots \vee \bar{l}_n)$ is the clause that *blocks* α. The assignment *blocked* by a clause is the negation of the literals in the clause. The literals touched by an assignment is defined by $\mathsf{touched}_\alpha(C) = \{l \mid l \in C \text{ and } \mathsf{var}(l) \in \mathsf{var}(\alpha)\}$ for a clause. For a formula ψ, $\mathsf{touched}_\alpha(\psi)$ is the union of touched variables for each clause in ψ. A *unit* is a clause containing a single literal. The *unit clause rule* takes the assignment α of all units in a formula ψ and generates $\psi|\alpha$. Iteratively applying the unit clause rule until fixpoint is referred to as *unit propagation*. In cases where unit propagation yields $\bot$ we say it derived a *conflict*. A formula ψ *implies* a formula ψ', denoted $\psi \models \psi'$, if every assignment satisfying ψ satisfies ψ'. By $\psi \vdash_1 \psi'$ we denote that for every clause $C \in \psi'$, applying unit propagation to the assignment blocked by C in ψ derives a conflict. If unit propagation derives a conflict on the formula $\psi \cup \{\{l\}\}$, we say l is a *failed literal* and the unit $\bar{l}$ is logically implied by the formula. Failed literal probing [11] is the process of successively assigning literals to check if units are implied by the formula. In its simplest form, probing involves assigning a literal l and learning the unit $\bar{l}$ if unit propagation derives a conflict, otherwise l is unassigned and the next literal is checked.

To evaluate the satisfiability of a formula, a CDCL solver [27] iteratively performs the following operations: First, the solver performs unit propagation, then tests for a conflict. Unit propagation is made efficient with two-literal watch pointers [28]. If there is no conflict and all variables are assigned, the formula is satisfiable. Otherwise, the solver chooses an unassigned variable through a variable decision heuristic [6,25], assigns a truth value to it, and performs unit propagation. If, however, there is a conflict, the solver performs conflict analysis potentially learning a short clause. In case this clause is the empty clause, the formula is unsatisfiable.

3 The PR Proof System

A clause C is *redundant* w.r.t. a formula ψ if ψ and $\psi \cup \{C\}$ are *satisfiability equivalent*. The clause sequence $\psi, C_1, C_2, \ldots, C_n$ is a clausal proof of C_n if each clause C_i $(1 \leq i \leq n)$ is redundant w.r.t. $\psi \cup \{C_1, C_2, \ldots, C_{i-1}\}$. The proof is a refutation of ψ if C_n is $\bot$. Clausal proof systems may also allow deletion. In a refutation proof clauses can be deleted freely because the deletion cannot make a formula less satisfiable.

Clausal proof systems are distinguished by the kinds of redundant clauses they allow to be added. The standard SAT solving paradigm CDCL learns clauses implied through *resolution*. These clauses are logically implied by the formula, and fall under the *reverse unit propagation* (RUP) proof system. The *Resolution Asymmetric Tautology* (RAT) proof system generalizes RUP. All commonly used inprocessing techniques emit DRAT proofs. The *propagation redundant* (PR) proof system generalizes RAT by allowing the pruning of branches *without loss of satisfaction*.

Let C be a clause in the formula ψ and α the assignment blocked by C. Then C is PR w.r.t. ψ if and only if there exists an assignment ω such that $\psi|_\alpha \vdash_1 \psi|_\omega$ and ω satisfies C. Intuitively, this allows inferences that block a partial assignment α as long as another assignment ω is as satisfiable. This means every assignment containing α that satisfies ψ can be transformed to an assignment containing ω that satisfies ψ.

Clausal proofs systems must be checkable in polynomial time to be useful in practice. RUP and RAT are efficiently checkable due to unit propagation. In general, determining if a clause is PR is an NP-complete problem [18]. However, a PR proof is checkable in polynomial time if the witness assignments ω are included. A clausal proof with witnesses will look like ψ, (C_1, ω_1), (C_2, ω_2), $\ldots$, (C_n, ω_n). The proof checker DPR-TRIM can efficiently check PR proofs that include witnesses. Further, DPR-TRIM can emit proofs in the LPR format. They can be validated by the formally-verified checker CAKE-LPR [31], which was used to validate results in recent SAT competitions.

4 Pruning Predicates and SADICAL

Determining if a clause is PR is NP-complete and can naturally be formulated in SAT. Given a clause C and formula ψ, a *pruning predicate* is a formula such that if it is satisfiable, the clause C is redundant w.r.t. ψ. SADICAL uses two pruning predicates to determine if a clause is PR: *positive reduct* and *filtered positive reduct*. If either predicate is satisfiable, the satisfying assignment serves as the witness showing the clause is PR.

Given a formula ψ and assignment α, the *positive reduct* is the formula $G \wedge C$ where C is the clause that blocks α and $G = \{\mathsf{touched}_\alpha(D) \mid D \in \psi \text{ and } D|\alpha = \top\}$. If the positive reduct is satisfiable, the clause C is PR w.r.t. ψ. The positive reduct is satisfiable iff the clause blocked by α is a *set-blocked* clause [23].

Given a formula ψ and assignment α, the *filtered positive reduct* is the formula $G \wedge C$ where C is the clause that blocks α and $G = \{\mathsf{touched}_\alpha(D) \mid D \in \psi \text{ and } D|\alpha \nvdash_1 \mathsf{touched}_\alpha(D)\}$. If the filtered positive reduct is satisfiable, the clause C is PR w.r.t. ψ. The filtered positive reduct is a subset of the positive reduct and is satisfiable iff the clause blocked by α is a *set-propagation redundant* clause [14]. Example 1 shows a formula for which the positive and filtered positive reducts are different, and only the filtered positive reduct is satisfiable.

Example 1. Given the formula $(x_1 \vee x_2) \wedge (\overline{x}_1 \vee x_2)$, the positive reduct with $\alpha = x_1$ is $(x_1) \wedge (\overline{x}_1)$, which is unsatisfiable. The clause (x_1) can be filtered, giving the filtered positive reduct $(\overline{x}_1)$, which is satisfiable.

SADICAL [16] uses satisfaction-driven clause learning (SDCL) that extends CDCL by learning PR clauses [18] based on (filtered) positive reducts. SADICAL uses an inner/outer solver framework. The outer solver attempts to solve the SAT problem with SDCL. SDCL diverges from the basic CDCL loop when unit propagation after a decision does not derive a conflict. In this case a reduct is generated using the current assignment, and the inner solver attempts to solve the reduct using CDCL. If the reduct is satisfiable, the PR clause blocking the current assignment is learned, and the SDCL loop continues. The PR clause can be simplified by removing all non-decision variables from the assignment. SADICAL emits PR proofs by logging the satisfying assignment of the reduct as the witness, and these proofs are verified with DPR-TRIM. The key to SADICAL finding good PR clauses leading to short proofs is the decision heuristic, because variable selection builds the candidate PR clauses. Hand-crafted decision heuristics enable SADICAL to find short proofs on pigeonhole and mutilated chessboard problems. However, these heuristics differ significantly from the score-based heuristics in most CDCL solvers. Our experiences with SaCiDaL suggest that improving the heuristics for SDCL reduces the performance of CDCL and the other way around. This may explain why SADICAL performs worse than standard CDCL solvers on the majority of the SAT competition benchmarks. While SADICAL integrates finding PR clauses of arbitrary size in the main search loop, our tool focuses on learning short PR clauses as a preprocessing step. This allows us to develop good heuristics for PR learning without compromising the main search loop.

5 Extracting PR Clauses

The goal of PRELEARN is to find useful PR clauses that improve the performance of CDCL solvers on both satisfiable and unsatisfiable instances. Figure 1 shows how a SAT problem is solved using PRELEARN. For some preset time limit, PR clauses are found and then added to the original formula. Interleaved in this process is failed literal probing to check if unit clauses can be learned. When the preprocessing stage ends, the new formula that includes learned PR clauses is solved by a CDCL solver. If the

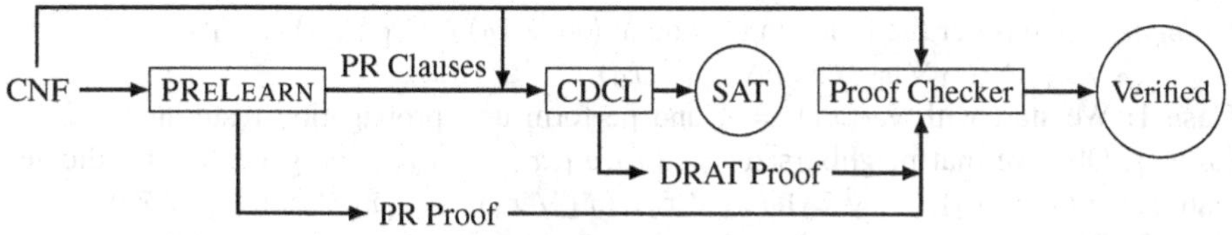

Fig. 1. Solving a formula with PRELEARN and a CDCL solver.

formula is satisfiable, the solver will produce a satisfying assignment. If the formula is unsatisfiable, a refutation proof of the original formula can be computed by combining the satisfaction preserving proof from PRELEARN and the refutation proof emitted by the CDCL solver. The complete proof can be verified with DPR-TRIM.

PRELEARN alternates between finding PR clauses and learning PR clauses. Candidate PR clauses are found by iterating over each variable in the formula, and for each variable constructing clauses that include that variable. To determine if a clause is PR, the positive reduct generated by that clause is solved. It can be costly to generate and solve many positive reducts, so heuristics are used to find candidate clauses that are more likely to be PR. It is possible to find multiple PR clauses that conflict with each other. PR clauses are conflicting if adding one of the PR clauses to the formula makes the other no longer PR. Learning PR clauses involves selecting PR clauses that are non-conflicting. The selection may maximize the number of PR clauses learned or optimize for some other metric. Adding PR clauses and units derived from probing may cause new clauses to become PR, so the entire process is iterated multiple times.

5.1 Finding PR Clauses

PR clauses are found by constructing a set of candidate clauses and solving the positive reduct generated by each clause. In SADICAL the candidates are the clauses blocking the partial assignment of the solver after each decision in the SDCL loop that does not derive a conflict. In effect, candidates are constructed using the solver's variable decision heuristic. We take a more general approach, constructing sets of candidates for each variable based on unit propagation and the partial assignment's neighbors.

For a variable x, $\mathsf{neighbors}(x)$ denotes the set of variables occurring in clauses containing literal x or $\overline{x}$, excluding variable x. For a partial assignment α, $\mathsf{neighbors}(\alpha)$ denotes $\bigcup_{x \in \mathsf{var}(\alpha)} \mathsf{neighbors}(x) \setminus \mathsf{var}(\alpha)$. Candidate clauses for a literal l are generated in the following way:

- Let α be the partial assignment found by unit propagation starting with the assignment that makes l true.
- Generate the candidate PR clauses $\{(\bar{l} \vee y), (\bar{l} \vee \overline{y}) \mid y \in \mathsf{neighbors}(\alpha)\}$.

Example 2 shows how candidate binary clauses are constructed using both polarities of an initial variable x. In Example 3 the depth is expanded to reach more variables and create larger sets of candidate clauses. The depth parameter is used in Section 5.4.

Example 2. Consider the following formula: $(x_1 \vee \overline{x}_2) \wedge (\overline{x}_1 \vee x_3) \wedge (x_1 \vee x_4 \vee x_5) \wedge (x_2 \vee x_6 \vee x_7) \wedge (x_3 \vee x_7 \vee x_8) \wedge (x_8 \vee x_9)$,
Case 1: We start with $\mathsf{var}(x_1) = 1$ and perform unit propagation resulting in $\alpha = \{x_1 x_3\}$. Observe that $\mathsf{neighbors}(\alpha) = \{x_2, x_4, x_5, x_7, x_8\}$. The generated candidate clauses are $(\overline{x}_1 \vee x_2), (\overline{x}_1 \vee \overline{x}_2), (\overline{x}_1 \vee x_4), (\overline{x}_1 \vee \overline{x}_4), \ldots, (\overline{x}_1 \vee x_8), (\overline{x}_1 \vee \overline{x}_8)$.
Case 2: We start with $\mathsf{var}(x_1) = 0$ and perform unit propagation resulting in $\alpha = \{\overline{x}_1 \overline{x}_2\}$. Observe that $\mathsf{neighbors}(\alpha) = \{x_3, x_4, x_5, x_6, x_7\}$. The generated candidate clauses are $(x_1 \vee x_3), (x_1 \vee \overline{x}_3), (x_1 \vee x_4), (x_1 \vee \overline{x}_4), \ldots, (x_1 \vee x_7), (x_1 \vee \overline{x}_7)$.

Example 3. Take the formula from Example 2 and assignment of $\mathsf{var}(x_1) = 1$ as in case 1. The set of candidate clauses can be expanded by also considering the unassigned neighbors of the variables in $\mathsf{neighbors}(\alpha)$. For example, $\mathsf{neighbors}(x_8) = \{x_3, x_7, x_9\}$, of which x_9 is new and unassigned. This adds $(\overline{x}_1 \vee x_9)$ and $(\overline{x}_1 \vee \overline{x}_9)$ to the set of candidate clauses. This can be iterated by including neighbors of new unassigned variables from the prior step.

We consider both polarities when constructing candidates for a variable. After all candidates for a variable are constructed, the positive reduct for each candidate is generated and solved in order. Note that propagated literals appearing in the partial assignment do not appear in the PR clause. The satisfying assignment is stored as the witness and the PR clause may be learned immediately depending on the learning configuration.

This process is naturally extended to ternary clauses. The binary candidates are generated, and for each candidate $(x \vee y)$, x and y are assigned to false in the first step. The variables $z \in \mathsf{neighbors}(\alpha)$ yield clauses $(x \vee y \vee z)$ and $(x \vee y \vee \overline{z})$. This approach can generate many candidate ternary clauses depending on the connectivity of the formula since each candidate binary clause is expanded. A filtering operation would be useful to avoid the blow-up in number of candidates. There are likely diminishing returns when searching for larger PR clauses because (1) there are more possible candidates, (2) the positive reducts are likely larger, and (3) each clause blocks less of the search space. We consider only unit and binary candidate clauses in our main evaluation.

Ideally, we should construct candidate clauses that are likely PR to reduce the number of failed reducts generated. Note, the (filtered) positive reduct can only be satisfiable if given the partial assignment there exists a reduced, satisfied clause. By focusing on neighbors, we guarantee that such a clause exists. The *reduced* heuristic in SADICAL finds variables in all reduced but unsatisfied clauses. The idea behind this heuristic is to direct the assignment towards conditional autarkies that imply a satisfiable positive reduct [18]. The neighbors approach generalizes this to variables in all reduced clauses whether or not they are unsatisfied. A comparison can be found in our repository.

5.2 Learning PR Clauses

Given multiple clauses that are PR w.r.t. the same formula, it is possible that some of the clauses conflict with each other and cannot be learned simultaneously. Example 4 shows how learning one PR clause may invalidate the witness of another PR clause. It may be that a different witness exists, but finding it requires regenerating the positive reduct to include the learned PR clause and solving it. The simplest way to avoid conflicting PR clause is to learn PR clauses as they are found. When a reduct is satisfiable, the PR clauses is added to the formula and logged with its witness in the proof. Then subsequent reducts will be generated from the formula including all added PR clauses. Therefore, a satisfiable reduct ensures a PR clause can be learned.

Alternatively, clauses can be found in batches, then a subset of nonconflicting clauses can be learned. The set of conflicts between PR clauses can be computed in polynomial time. For each pair of PR clauses C and D, if the assignment that generated the pruning predicate for D touches C and C is not satisfied by the witness of D, then C conflicts with D. In some cases reordering the two PR clauses may avoid a conflict. In Example 4 learning the second clause would not affect the validity of the first clauses' witness. Once the conflicts are known, clauses can be learned based on some heuristic ordering. Batch learning configurations are discussed more in the following section.

Example 4. Assume the following clause witness pairs are valid in a formula ψ: $\{(x_1 \vee x_2 \vee x_3), x_1\overline{x}_2\overline{x}_3\}$, and $\{(x_1 \vee \overline{x}_2 \vee x_4), \overline{x}_1\overline{x}_2x_4\}$. The first clause conflicts with the second. If the first clause is added to ψ, the clause $(x_1 \vee x_2)$ would be in the positive reduct for the second clause, but it is not satisfied by the witness of the second clause.

5.3 Additional Configurations

The sections above describe the PRELEARN configuration used in the main evaluation, i.e., finding candidate PR clauses with the neighbors heuristic and learning clauses instantly as the positive reducts are solved. In this section we present several additional configurations. The time-constrained reader may skip ahead to Section 5.4 for the presentation of our main results.

In batch learning a set of PR clauses are found in batches then learned. Learning as many nonconflicting clauses as possible coincides with the maximum independent set problem. This problem is NP-Hard. We approximate the solution by adding the clause causing the fewest conflicts with unblocked clauses. When a clause is added, the clauses it blocks are removed from the batch and conflict counts are recalculated Alternatively, clauses can be added in a random order. Random ordering requires less computation at the cost of potentially fewer learned PR clauses.

The neighbors heuristic for constructing candidate clauses can be modified to include a depth parameter. $\mathsf{neighbors}(i)$ indicates the number of iterations expanding the variables. For example, $\mathsf{neighbors}(2)$ expands on the variables in $\mathsf{neighbors}(1)$, seen in Example 3. We also implement the reduced heuristic, shown in Example 5. Detailed evaluations and comparisons can be found in our repository. In general, we found that the additional configurations did not improve on our main configuration. More work needs to be done to determine when and how to apply these additional configurations.

Example 5. Given the set of clauses $(x_1 \vee x_2 \vee x_3) \wedge (\overline{x}_1 \vee x_3 \vee x_4) \wedge (x_3 \vee x_5)$, and initial assignment $\alpha = x_1$, only the second clause is reduced and not satisfied, giving $\mathsf{reduced}(\alpha) = \{x_3, x_4\}$ and candidate clauses $(\overline{x}_1 \vee x_3)$, $(\overline{x}_1 \vee x_4)$, $(\overline{x}_1 \vee \overline{x}_3)$, $(\overline{x}_1 \vee \overline{x}_4)$.

5.4 Implementation

PRELEARN was implemented using the inner/outer-solver framework in SADICAL. The inner solver acts the same as in SADICAL, solving pruning predicates using CDCL.

The outer solver is not used for SDCL, but the SDCL data-structures are used to find and learn PR clauses. The outer solver is initialized with the original formula and maintains the list of variables, clauses, and watch pointers. By default, the outer solver has no variables assigned other than learned units. When finding candidates, the variables in the partial clause are assigned in the outer solver. Unit propagation makes it possible to find all reduced clauses in the formula with a single pass. This is necessary for constructing the positive reduct. After a candidate clause has been assigned and the positive reduct solved, the variables are unassigned. This returns the outer solver to the top-level before examining the next candidate. When a PR clause is learned, it is added to the formula along with its watch pointers. Additionally, failed literals are found if assigning a variable at the top-level causes a conflict through unit propagation. The negation of a failed literal is a unit that can be added to the formula.

In a single iteration each variable in the formula is processed in a breadth-first search (BFS) starting from the first variable in the numbering. When a variable is encountered it is first checked whether either assignment of the variable is a failed literal or a unit PR clause. If not, binary candidates are generated based on the selected heuristic and PR clauses are learned based on the learning configuration. Variables are added to the frontier of the BFS as they are encountered during candidate clause generation, but they are not repeated. Optionally, after all variables have been encountered the BFS restarts, now constructing ternary candidates. The repetition continues to the desired clause length. Then another iteration begins again with binary clauses. Running PRELEARN multiple times is important because adding PR clauses in one iteration may allow additional clauses to be added in the next.

6 Mutilated Chessboard

The *mutilated chessboard* is an $n \times n$ grid of alternating black and white squares with two opposite corners removed. The problem is whether or not the the board can be covered with 2×1 dominoes. This can be encoded in CNF by using variables to represent

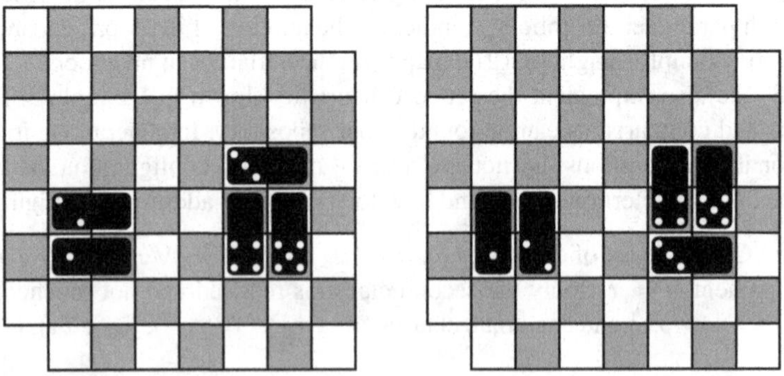

Fig. 2. Occurrences of two horizontal dominoes may be replaced by two vertical dominos in a solution. Similarly, occurrences of a horizontal domino atop two vertical dominos can be replaced by shifting the horizontal domino down.

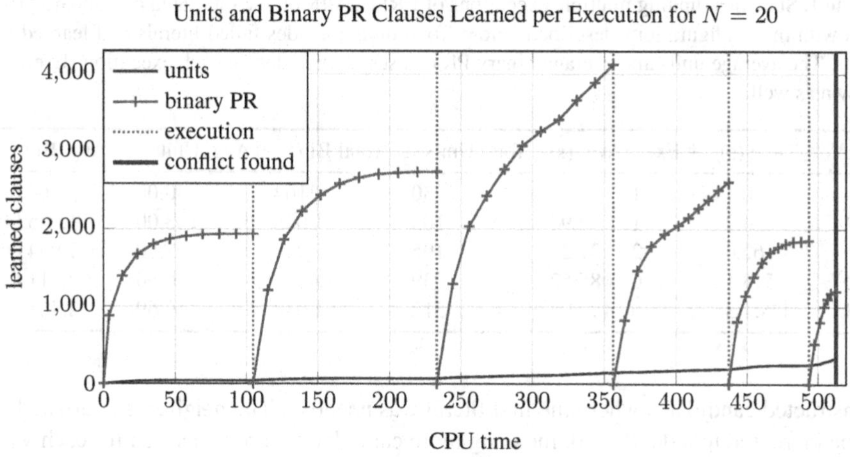

Fig. 3. Unit and binary PR clauses learned each execution (red-dotted line) until a contradiction was found. Markers on binary PR lines represent an iteration within an execution.

domino placements on the board. At-most-one constraints (using the pairwise encodings) say only one domino can cover each square, and at-least-one constraints (using a disjunction) say some domino must cover each square.

In recent SAT competitions, no proof-generating SAT solver could deal with instances larger than $N = 18$. In ongoing work, we found refutation proofs that contain only units and binary PR clauses for some boards of size $N \leq 30$. PRELEARN can be modified to automatically find proofs of this type. Running iterations of PRELEARN until *saturation*, meaning no new binary PR clauses or units can be found, yields some set of units and binary PR clauses. Removing the binary PR clauses from the formula and rerunning PRELEARN will yield additional units and a new set of binary PR clauses. Repeating the process of removing binary PR clauses and keeping units will eventually derive the empty clause for this problem. Figure 3 gives detailed values for $N = 20$. Within each execution (red dotted lines) there are at most 10 iterations (red tick markers), and each iteration learns some set of binary PR clauses (red). Some executions saturate binary PR clauses before the tenth iteration and exit early. At the end of each execution the binary PR clauses are deleted, but the units (blue) are kept for the following execution. A complete DPR proof (PR with deletion) can be constructed by adding deletion information for the binary PR clauses removed between each execution when concatenating the PRELEARN proofs. The approach works for mutilated chess because in each execution there are many binary PR clauses that can be learned and will lead to units, but they are mutually exclusive and cannot be learned simultaneously. Further, adding units allows new binary PR clauses to be learned in following executions.

Table 1 shows the statistics for PRELEARN. Achieving these results required some modifications to the configuration of PRELEARN. First, notice in Figure 2 the PR clauses that can be learned involve blocking one domino orientation that can be replaced by a symmetric orientation. To optimize for these types of PR clauses, we only

Table 1. Statistics running multiple executions of PRELEARN on the mutilated chessboard problem with the configurations described below. Total units includes failed literals and learned PR units. The average units and average binary PR clauses learned during each execution (Exe.) are shown as well.

N	Time (s)	# Exe.	Avg. (s)	Total Units	Total Bin.	Avg. Units	Avg. Bin.
8	0.14	1	0.14	30	164	30.00	164.00
12	4.94	1	4.94	103	1,045	103.00	1,045.00
16	62.47	2	31.23	195	3,988	97.50	1,994.00
20	513.12	6	85.52	339	1,4470	56.50	2,411.67
24	4,941.38	26	190.05	512	64,038	19.69	2,463.00

constructed candidates where the first literal was negative. The neighbors heuristic had to be increased to a depth of 6, meaning more candidates were generated for each variable. Intuitively, the proof is constructed by adding binary PR clauses in order to find negative units (dominos that cannot be placed) around the borders of the board. Following iterations build more units inwards, until a point is reached where units cover almost the entire board. This forces an impossible domino placement leading to a contradiction. Complete proofs using only units and binary PR clauses were found for boards up to size $N = 24$ within 5,000 seconds. We verified all proofs using DPR-TRIM. The mutilated chessboard has a high degree of symmetry and structure, making it suitable for this approach. For most problems it is not expected that multiple executions while keeping learned units will find new PR clauses.

Experiments were done with several configurations (see Section 5.3) to find the best results. We found that increasing the depth of neighbors was necessary for larger boards including $N = 24$. Increasing the depth allows more binary PR clauses to be found, at the cost of generating more reducts. This is necessary to find units. The reduced heuristic (a subset of neighbors) did not yield complete proofs. We also tried incrementing the depth after each execution starting with 1 and reseting at 9. In this approach, the execution times for depth greater than 6 were larger but did not yield more unit clauses on average. We attempted batch learning on every 500 found clauses using either random or the sorted heuristic. In each batch many of the 500 PR clauses blocked each other because many conflicting PR clauses can be found on a small set of variables in mutilated chess. The PR clauses that were blocked would be found again in following iterations, leading to more reducts generated and solved. This caused much longer execution times. Adding PR clauses instantly is a good configuration for reducing execution time when there are many conflicting clauses. However, for some less symmetric problems it may be worth the tradeoff to learn the clauses in batches, because learning a few bad PR clauses may disrupt the subsequent iterations.

7 SAT Competition Benchmarks

We evaluated PRELEARN on previous SAT competition formulas. Formulas from the '13, '15, '16, '19, '20, and '21 SAT competitions' main tracks were grouped by size. **0-10k** contains the 323 formulas with less than 10,000 clauses and **10k-50k** contains

Table 2. Fraction of benchmarks where PR clauses were learned, average runtime of PRELEARN, generated positive reducts and satisfiable positive reducts (PR clauses learned), and number of failed literals found.

Set	Benches	Avg. (s)	Generated Reducts	Sat. Reducts	% Sat.	Failed Lits
0-10k	221/323	22.36	104,850,011	548,417	0.52%	3,416
10k-50k	237/348	71.08	163,014,068	789,281	0.48%	6,290

the 348 formulas with between 10,000 and 50,000 clauses. In general, short PR proofs have been found for hard combinatorial problems typically having few clauses (0-10k). These include the pigeonhole and mutilated chessboard problems, some of which appear in 0-10k benchmarks. The PR clauses that can be derived for these formulas are intuitive and almost always beneficial to solvers. Less is known about the impact of PR clauses on larger formulas, motivating our separation of test sets by size. The repository containing the preprocessing tool, experiment configurations, and experiment data can be found at https://github.com/jreeves3/PReLearn.

We ran our experiments on StarExec [30]. The specs for the compute nodes can be found online.[1] The compute nodes that ran our experiments were Intel Xeon E5 cores with 2.4 GHz, and all experiments ran with 64 GB of memory and a 5,000 second timeout. We run PRELEARN for 50 iterations over 100 seconds, exiting early if no new PR clauses were found in an iteration.

PRELEARN was executed as a stand-alone program, producing a derivation proof and a modified CNF. For experiments, the CDCL solver KISSAT [5] was called once on the original formula and once on the modified CNF. KISSAT was selected because of its high-rankings in previous SAT competitions, but we expect the results to generalize to other CDCL SAT solvers.

Derivation proofs from PRELEARN were verified in all solved instances using the independent proof checker DPR-TRIM using a forward check. This can be extended to complete proofs in the following way. In the unsatisfiable case the proof for the learned PR clauses is concatenated to the proof traced by KISSAT, and the complete proof is verified against the original formula. In the satisfiable case the partial proof for the learned PR clauses is verified using a forward check in DPR-TRIM, and the satisfying assignment found by KISSAT is verified by the StarExec post-processing tool. Due to resource limitations, we verified a subset of complete proofs in DPR-TRIM. This is more costly because it involves running KISSAT with proof logging, then running DPR-TRIM on the complete proof.

Table 2 shows the cumulative statistics for running PRELEARN on the benchmark sets. Note the number of satisfiable reducts is the number of learned PR clauses, because PR clauses are learned immediately after the reduct is solved. These include both unit and binary PR clauses. A very small percentage of generated reducts is satisfiable, and subsequently learned. This is less important for small formulas when reducts can be computed quickly and there are fewer candidates to consider. However, for the 10k-50k formulas the average runtime more than triples but the number of generated reducts

[1] https://starexec.org/starexec/public/about.jsp

Table 3. Number of total solved instances and exclusive solved instances running KISSAT with and without PRELEARN. Number of improved instances running KISSAT with PRELEARN. PRELEARN execution times were included in total execution times.

	0-10k SAT	0-10k UNSAT	10k-50k SAT	10k-50k UNSAT
Total w/ PRELEARN	84	149	143	89
Total w/o PRELEARN	80	141	143	91
Exclusively w/ PRELEARN	4	10	4	1
Exclusively w/o PRELEARN	0	2	4	3
Improved w/ PRELEARN	20	44	25	13

less than doubles. PR clauses are found in about two thirds of the formulas, showing our approach generalizes beyond the canonical problems for which we knew PR clauses existed. Expanding the exploration and increasing the time limit did not help to find PR clauses in the remaining one third.

Table 3 gives a high-level picture of PRELEARN's impact on KISSAT. PRELEARN significantly improves performance on 0-10k SAT and UNSAT benchmarks. These contain the hard combinatorial problems including pigeonhole that PRELEARN was expected to perform well on. There were 4 additional SAT formulas solved with PRELEARN that KISSAT alone could not solve. This shows that PRELEARN impacts not only hard unsatisfiable problems but satisfiable problems as well. On the other hand, the addition of PR clauses makes some problems more difficult. This is clear with the 10k-50k results, where 5 benchmarks are solved exclusively with PRELEARN and 7 are solved exclusively without. Additionally, PRELEARN improved KISSAT's performance on 102 of 671 or approx. 15% of benchmarks. This is a large portion of benchmarks, both SAT and UNSAT, for which PRELEARN is helpful.

Figure 4 gives a more detailed picture on the impact of PRELEARN per benchmark. In the scatter plot the left-hand end of each line indicates the KISSAT execution time, while the length of the line indicates the PRELEARN execution time, and so the right-hand end gives the total time for PRELEARN plus KISSAT. Lines that cross the diagonal indicate that the preprocessing improved KISSAT's performance but ran for longer than the improvement. PRELEARN improved performance for points above the diagonal. Points on the dotted-lines (timeout) are solved by one configuration and not the other.

The top plot gives the results for the 0-10k formulas, with many points on the top timeout line as expected. These are the hard combinatorial problems that can only be solved with PRELEARN. In general, the unsatisfiable formulas benefit more than the satisfiable formulas. PR clauses can reduce the number of solutions in a formula and this may explain the negative impact on many satisfiable formulas. However, there are still some satisfiable formulas that are only solved with PRELEARN.

In the bottom plot, formulas that take a long time to solve (above the diagonal in the upper right-hand corner) are helped more by PRELEARN. In the bottom half of the plot, many lines cross the diagonal meaning the addition of PR clauses provided a negligible benefit. For this set there are more satisfiable formulas for which PRELEARN is helpful.

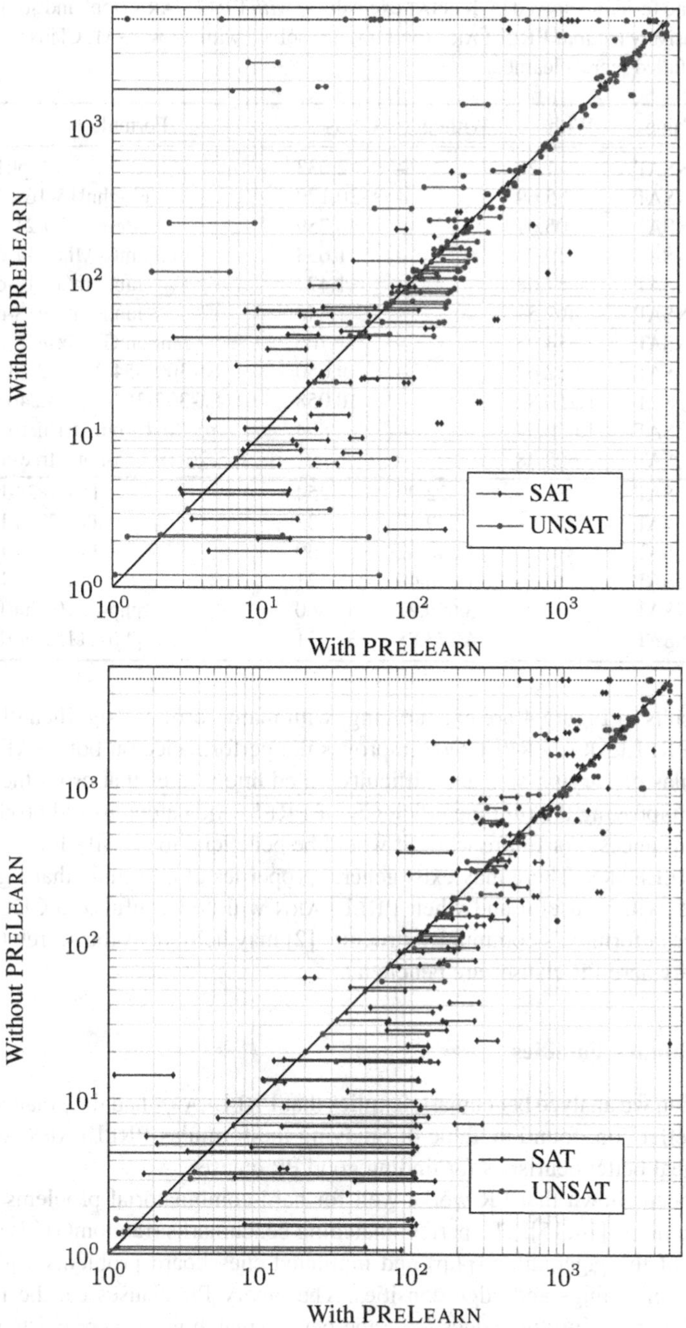

Fig. 4. Execution times w/ and w/o PRELEARN on 0-10k (top) and 10k-50k (bottom) benchmarks. The left-hand point of each segment shows the time for the SAT solver alone; the right-hand point indicates the combined time for preprocessing and solving.

Table 4. Some formulas solved by KISSAT exclusively *with* PRELEARN (top) and some formulas solved exclusively *without* PRELEARN (bottom). (*) solved without KISSAT. Clauses include PR clauses and failed literals learned.

Set	Value	With	Without	Clauses	Formula	Year
0-10k	UNSAT	1.26	–	2,033	ph12*	2013
0-10k	UNSAT	35.69	–	20,179	Pb-chnl15-16_c18*	2019
0-10k	UNSAT	105.01	–	46,759	Pb-chnl20-21_c18	2019
0-10k	UNSAT	59.99	–	1,633	randomG-Mix-n17-d05	2021
0-10k	UNSAT	61.08	–	1,472	randomG-n17-d05	2021
0-10k	UNSAT	407.51	–	1,640	randomG-n18-d05	2021
0-10k	UNSAT	584.95	–	1,706	randomG-Mix-n18-d05	2021
0-10k	SAT	1,082.62	–	9,650	fsf-300-354-2-2-3-2.23.opt	2013
0-10k	SAT	1,250.82	–	10,058	fsf-300-354-2-2-3-2.46.opt	2013
10k-50k	SAT	1,076.34	–	804	sp5-26-19-bin-stri-flat-noid	2021
10k-50k	SAT	608.48	–	901	sp5-26-19-una-nons-tree-noid	2021
10k-50k	SAT	–	22.99	254	Ptn-7824-b13	2016
10k-50k	SAT	–	549.27	133	Ptn-7824-b09	2016
10k-50k	SAT	–	1,246.42	39	Ptn-7824-b02	2016
10k-50k	SAT	–	1,290.49	121	Ptn-7824-b08	2016
10k-50k	UNSAT	–	3,650.21	31,860	rphp4_110_shuffled	2016
10k-50k	UNSAT	–	4,273.88	31,531	rphp4_115_shuffled	2016

The results in Figure 4 are encouraging, with many formulas significantly benefitting from PRELEARN. PRELEARN improves the performance on both SAT and UNSAT formulas of varying size and difficulty. In addition, lines that cross the diagonal imply that improving the runtime efficiency of PRELEARN alone would produce more improved instances. For future work, it would be beneficial to classify formulas before running PRELEARN. There may exist general properties of a formula that signal when PRELEARN will be useful and when PRELEARN will be harmful to a CDCL solver. For instance, a formula's community structure [2] may help focus the search to parts of the formula where PR clauses are beneficial.

7.1 Benchmark Families

In this section we analyze benchmark families that PRELEARN had the greatest positive (negative) effect on, found in Table 4. Studying the formulas PRELEARN works well on may reveal better heuristics for finding good PR clauses.

It has been shown that PR works well for hard combinatorial problems based on perfect matchings [14,15]. The perfect matching benchmarks (randomG) [7] are a generalization of the pigeonhole (php) and mutilated chessboard problems with varying at-most-one encodings and edge densities. The binary PR clauses can be intuitively understood as blocking two edges from the perfect matching if there exists two other edges that match the same nodes. These benchmarks are relatively small but extremely hard for CDCL solvers. Symmetry-breaking with PR clauses greatly reduces the search space and leads KISSAT to a short proof of unsatisfiability. PRELEARN also benefits other hard combinatorial problems that use pseudo-Boolean constraints. The pseudo-Boolean (Pb-chnl) [24] benchmarks are based on at-most-one constraints (using the pairwise encoding) and at-least-one constraints. These formulas have a similar graphical structure to the perfect matching benchmarks. Binary PR clauses block two edges when another set of edges exists that are incident to the same nodes.

For the other two benchmark families that benefited from PRELEARN, the intuition behind PR learning is less clear. The fixed-shape random formulas (fsf) [29] are parameterized non-clausal random formulas built from hyper-clauses. The SAT encoding makes use of the Plaisted-Greenbaum transformation, introducing circuit-like structure to the problem. The superpermutation problem (sp) [22] asks whether a sequence of digits 1–n of length l can contain every permutation of $[1, n]$ as a subsequence, and the optimization variant asks for the smallest such l given n. The sequence of l digits is encoded directly and passed through a multi-layered circuit that checks for the existence of each individual permutation. Digits use the binary (*bin*) or unary (*una*) encoding, are strict *stri* if clauses constrain digit bits to valid encodings and nonstrict *nons* otherwise, and *flat* if the circuit is a large AND or *tree* for prefix recognizing nested circuits. The formulas given ask to find a prefix of a superpermutation for $n = 5$ or length 26 with 19 permutations. The check for 19 permutations was encoded as cardinality constraints in a pseudo-Boolean instance, then converted back to SAT. Each individual permutation is checked by duplicating circuits at each possible starting position of the permutation in l. PR clauses may be pruning certain starting positions for some permutations or affecting the pseudo-Boolean constraints. This cannot be determined without a deeper knowledge of the benchmark generator.

The relativized pigeonhole problem (rphp) [3] involves placing k pigeons in $k - 1$ holes with n nesting places. This problem has polynomial hardness for resolution, unlike the exponential hardness of the classical pigeonhole problem. The symmetry-breaking preprocessor BREAKID [9] generates symmetry-breaking formulas for rphp that are easy for a CDCL solver. PRELEARN can learn many PR clauses but the formula does not become easier. Note PRELEARN can solve the php with $n = 12$ in a second.

One problem is clause and variable permuting (a.k.a. shuffling). The mutilated chessboard problem can still be solved by PRELEARN after permuting variables and clauses. The pigeonhole problem can be solved after permuting clauses but not after permuting variable names. In PRELEARN, PR candidates are sorted by variable name independent of clause ordering, but when the variable names change the order of learned clauses changes. In the mutilated chessboard problem there is local structure, so similar PR clauses are learned under variable renaming. In the pigeonhole problem there is global structure, so a variable renaming can significantly change the binary PR clauses learned and cause earlier saturation with far fewer units.

Another problem is that the addition of PR clauses can change the existing structure of a formula and negatively affect CDCL heuristics. The Pythagorean Triples Problem (Ptn) [19] asks whether monochromatic solutions of the equation $a^2 + b^2 = c^2$ can be avoided. The formulas encode numbers $\{1, \ldots, 7824\}$, for which a valid 2-coloring is possible. In the namings, the *N* in b*N* denotes the number of backbone literals added to the formula. A backbone literal is a literal assigned true in every solution. Adding more than 20 backbone literals makes the problem easy. For each formula KISSAT can find a satisfying assignment, but timeouts with the addition of PR clauses. For one instance, adding only 39 PR clauses will lead to a timeout. In some hard SAT and UNSAT problems solvers require some amount of luck and adding a few clauses or shuffling a formula can cause a CDCL solver's performance to sharply decrease. The Pythagorean Triples problem was originally solved with a local search solver, and local search still performs well after adding PR clauses.

In a straight-forward way, one can avoid the negative effects of adding harmful PR clauses by running two solvers in parallel: one with PRELEARN and one without. This fits with the portfolio approach for solving SAT problems.

8 Conclusion and Future Work

In this paper we presented PRELEARN, a tool built from the SADICAL framework that learns PR clauses in a preprocessing stage. We developed several heuristics for finding PR clauses and multiple configurations for clause learning. In the evaluation we found that PRELEARN improves the performance of the CDCL solver KISSAT on many benchmarks from past SAT competitions.

For future work, quantifying the usefulness of each PR clause in relation to guiding the CDCL solver may lead to better learning heuristics. This is a difficult task that likely requires problem specific information. Separately, failed clause caching can improve performance by remembering and avoiding candidate clauses that fail with unsatisfiable reducts in multiple iterations. This would be most beneficial for problems like the mutilated chessboard that have many conflicting PR clauses. Lastly, incorporating PRELEARN during in-processing may allow for more PR clauses to be learned. This could be implemented with the inner/outer solver framework but would require a significantly narrowed search. CDCL learns many clauses during execution and it would be infeasible to examine binary PR clauses across the entire formula.

Acknowledgements. We thank the community at StarExec for providing computational resources.

References

1. Alekhnovich, M.: Mutilated chessboard problem is exponentially hard for resolution. Theoretical Computer Science **310**(1), 513–525 (2004)
2. Ansótegui, C., Bonet, M.L., Giráldez-Cru, J., Levy, J., Simon, L.: Community structure in industrial SAT instances. Journal of Artificial Intelligence Research (JAR) **66**, 443–472 (2019)
3. Atserias, A., Lauria, M., Nordström, J.: Narrow proofs may be maximally long. ACM Transactions on Computational Logic **17**(3) (2016)
4. Audemard, G., Katsirelos, G., Simon, L.: A restriction of extended resolution for clause learning SAT solvers. In: AAAI Conference on Artificial Intelligence. pp. 15–20. AAAI Press (2010)
5. Biere, A., Fazekas, K., Fleury, M., Heisinger, M.: CaDiCaL, Kissat, Paracooba, Plingeling and Treengeling entering the SAT competition 2020 (2020), unpublished
6. Biere, A., Fröhlich, A.: Evaluating CDCL variable scoring schemes. In: Theory and Applications of Satisfiability Testing (SAT). LNCS, vol. 9340, pp. 405–422 (2015)
7. Codel, C.R., Reeves, J.E., Heule, M.J.H., Bryant, R.E.: Bipartite perfect matching benchmarks. In: Pragmatics of SAT (2021)
8. Cook, S.A.: A short proof of the pigeon hole principle using extended resolution. SIGACT News **8**(4), 28–32 (1976)
9. Devriendt, J., Bogaerts, B., Bruynooghe, M., Denecker, M.: Improved static symmetry breaking for SAT. In: Theory and Applications of Satisfiability Testing (SAT). LNCS, vol. 9710, pp. 104–122. Springer (2016)
10. Eén, N., Biere, A.: Effective preprocessing in SAT through variable and clause elimination. In: Theory and Applications of Satisfiability Testing (SAT). LNCS, vol. 3569, pp. 61–75. Springer (2005)
11. Freeman, J.W.: Improvements to Propositional Satisfiability Search Algorithms. Ph.D. thesis, USA (1995)
12. Haken, A.: The intractability of resolution. Theoretical Computer Science **39**, 297–308 (1985)
13. Heule, M.J.H., Hunt, W.A., Wetzler, N.: Expressing symmetry breaking in DRAT proofs. In: Conference on Automated Deduction (CADE). LNCS, vol. 9195, pp. 591–606. Springer (2015)
14. Heule, M.J.H., Kiesl, B., Biere, A.: Short proofs without new variables. In: Conference on Automated Deduction (CADE). LNCS, vol. 10395, pp. 130–147. Springer (2017)
15. Heule, M.J.H., Kiesl, B., Biere, A.: Clausal proofs of mutilated chessboards. In: NASA Formal Methods. LNCS, vol. 11460, pp. 204–210 (2019)
16. Heule, M.J.H., Kiesl, B., Biere, A.: Encoding redundancy for satisfaction-driven clause learning. In: Tools and Algorithms for the Construction and Analysis of Systems (TACAS). LNCS, vol. 11427, pp. 41–58. Springer (2019)
17. Heule, M.J.H., Kiesl, B., Biere, A.: Strong extension free proof systems. In: Journal of Automated Reasoning. vol. 64, pp. 533–544 (2020)
18. Heule, M.J.H., Kiesl, B., Seidl, M., Biere, A.: PRuning through satisfaction. In: Haifa Verification Conference (HVC). LNCS, vol. 10629, pp. 179–194 (2017)
19. Heule, M.J.H., Kullmann, O., Marek, V.W.: Solving and verifying the boolean pythagorean triples problem via cube-and-conquer. In: Theory and Applications of Satisfiability Testing (SAT). LNCS, vol. 9710, pp. 228–245. Springer (2016)
20. Heule, M.J., Hunt, W.A., Wetzler, N.: Trimming while checking clausal proofs. In: Formal Methods in Computer-Aided Design (FMCAD). pp. 181–188 (2013)
21. Järvisalo, M., Heule, M.J.H., Biere, A.: Inprocessing rules. In: International Joint Conference on Automated Reasoning (IJCAR). LNCS, vol. 7364, pp. 355–370. Springer (2012)

22. Johnston, N.: Non-uniqueness of minimal superpermutations. Discrete Mathematics **313**(14), 1553–1557 (2013)
23. Kiesl, B., Seidl, M., Tompits, H., Biere, A.: Super-blocked clauses. In: International Joint Conference on Automated Reasoning (IJCAR). LNCS, vol. 9706, pp. 45–61 (2016)
24. Lecoutre, C., Roussel, O.: XCSP3 competition 2018 proceedings. pp. 40–41 (2018)
25. Liang, J., Ganesh, V., Poupart, P., Czarnecki, K.: Learning rate based branching heuristic for SAT solvers. In: Theory and Applications of Satisfiability Testing (SAT). LNCS, vol. 9710, pp. 123–140 (2016)
26. Manthey, N., Heule, M.J.H., Biere, A.: Automated reencoding of Boolean formulas. In: Haifa Verification Conference (HVC). LNCS, vol. 7857, pp. 102–117 (2013)
27. Marques-Silva, J., Lynce, I., Malik, S.: Conflict-driven clause learning SAT solvers. In: Handbook of Satisfiability, pp. 131–153. IOS Press (2009)
28. Moskewicz, M.W., Madigan, C.F., Zhao, Y., Zhang, L., Malik, S.: Chaff: Engineering an efficient sat solver. In: Proceedings of the 38th Annual Design Automation Conference. p. 530–535. ACM (2001)
29. Navarro, J.A., Voronkov, A.: Generation of hard non-clausal random satisfiability problems. In: AAAI Conference on Artificial Intelligence. pp. 436–442. The MIT Press (2005)
30. Stump, A., Sutcliffe, G., Tinelli, C.: StarExec: A cross-community infrastructure for logic solving. In: International Joint Conference on Automated Reasoning (IJCAR). LNCS, vol. 8562, pp. 367–373. Springer (2014)
31. Tan, Y.K., Heule, M.J.H., Myreen, M.O.: cake_lpr: Verified propagation redundancy checking in CakeML. In: Tools and Algorithms for the Construction and Analysis of Systems (TACAS), Part II. LNCS, vol. 12652, pp. 223–241 (2021)
32. Tseitin, G.S.: On the Complexity of Derivation in Propositional Calculus, pp. 466–483. Springer (1983)
33. Wetzler, N., Heule, M.J.H., Hunt, W.A.: DRAT-trim: Efficient checking and trimming using expressive clausal proofs. In: Theory and Applications of Satisfiability Testing (SAT). LNCS, vol. 8561, pp. 422–429 (2014)

Reasoning About Vectors Using an SMT Theory of Sequences

Ying Sheng[1(✉)], Andres Nötzli[1], Andrew Reynolds[2], Yoni Zohar[3], David Dill[4], Wolfgang Grieskamp[4], Junkil Park[4], Shaz Qadeer[4], Clark Barrett[1], and Cesare Tinelli[2]

[1] Stanford University, Stanford, USA
ying1123@stanford.edu
[2] The University of Iowa, Iowa City, USA
[3] Bar-Ilan University, Ramat Gan, Israel
[4] Meta Novi, Menlo Park, USA

Abstract. Dynamic arrays, also referred to as vectors, are fundamental data structures used in many programs. Modeling their semantics efficiently is crucial when reasoning about such programs. The theory of arrays is widely supported but is not ideal, because the number of elements is fixed (determined by its index sort) and cannot be adjusted, which is a problem, given that the length of vectors often plays an important role when reasoning about vector programs. In this paper, we propose reasoning about vectors using a theory of sequences. We introduce the theory, propose a basic calculus adapted from one for the theory of strings, and extend it to efficiently handle common vector operations. We prove that our calculus is sound and show how to construct a model when it terminates with a saturated configuration. Finally, we describe an implementation of the calculus in cvc5 and demonstrate its efficacy by evaluating it on verification conditions for smart contracts and benchmarks derived from existing array benchmarks.

1 Introduction

Generic vectors are used in many programming languages. For example, in C++'s standard library, they are provided by `std::vector`. Automated verification of software systems that manipulate vectors requires an efficient and automated way of reasoning about them. Desirable characteristics of any approach for reasoning about vectors include: (i) expressiveness—operations that are commonly performed on vectors should be supported; (ii) generality—vectors are always "vectors of" some type (e.g., vectors of integers), and so it is desirable that vector reasoning be integrated within a more general framework; solvers for satisfiability modulo theories (SMT) provide such a framework and are widely used in verification tools (see [5] for a recent survey); (iii) efficiency—fast and efficient reasoning is essential for usability, especially as verification tools are increasingly used by non-experts and in continuous integration.

This work was funded in part by the Stanford Center for Blockchain Research, NSF-BSF grant numbers 2110397 (NSF) and 2020704 (BSF), and Meta Novi. Part of the work was done when the first author was an intern at Meta Novi.

J. Blanchette et al. (Eds.): IJCAR 2022, LNAI 13385, pp. 125–143, 2022.
https://doi.org/10.1007/978-3-031-10769-6_9

Despite the ubiquity of vectors in software on the one hand and the effectiveness of SMT solvers for software verification on the other hand, there is not currently a clean way to represent vectors using operators from the SMT-LIB standard [3]. While the theory of arrays can be used, it is not a great fit because arrays have a fixed size determined by their index type. Representing a dynamic array thus requires additional modeling work. Moreover, to reach an acceptable level of expressivity, quantifiers are needed, which often makes the reasoning engine less efficient and robust. Indeed, part of the motivation for this work was frustration with array-based modeling in the Move Prover, a verification framework for smart contracts [24] (see Sect. 6 for more information about the Move Prover and its use of vectors). The current paper bridges this gap by studying and implementing a native theory of *sequences* in the SMT framework, which satisfies the desirable properties for vector reasoning listed above.

We present two SMT-based calculi for determining satisfiability in the theory of sequences. Since the decidability of even weaker theories is unknown (see, e.g., [9,15]), we do not aim for a decision procedure. Rather, we prove model and solution soundness (that is, when our procedure terminates, the answer is correct). Our first calculus leverages techniques for the theory of strings. We generalize these techniques, lifting rules specific to string characters to more general rules for arbitrary element types. By itself, this base calculus is already quite effective. However, it misses opportunities to perform high-level vector-based reasoning. For example, both reading from and updating a vector are very common operations in programming, and reasoning efficiently about the corresponding sequence operators is thus crucial. Our second calculus addresses this gap by integrating reasoning methods from array solvers (which handle reads and updates efficiently) into the first procedure. Notice, however, that this integration is not trivial, as it must handle novel combinations of operators (such as the combination of update and read operators with concatenation) as well as out-of-bounds cases that do not occur with ordinary arrays. We have implemented both variants of our calculus in the cvc5 SMT solver [2] and evaluated them on benchmarks originating from the Move prover, as well as benchmarks that were translated from SMT-LIB array benchmarks.

As is typical, both of our calculi are agnostic to the sort of the elements in the sequence. Reasoning about sequences of elements from a particular theory can then be done via theory combination methods such as Nelson-Oppen [18] or polite combination [16,20]. The former can be done for stably infinite theories (and the theory of sequences that we present here is stably infinite), while the latter requires investigating the politeness of the theory, which we expect to do in future work.

The rest of the paper is organized as follows. Section 2 includes basic notions from first-order logic. Section 3 introduces the theory of sequences and shows how it can be used to model vectors. Section 4 presents calculi for this theory and discusses their correctness. Section 5 describes the implementation of these calculi in cvc5. Section 6 presents an evaluation comparing several variations of the sequence solver in cvc5 and Z3. We conclude in Sect. 7 with directions for further research.

Related Work: Our work crucially builds on a proposal by Bjørner et al. [8], but extends it in several key ways. First, their implementation (for a logic they call `QF_BVRE`) restricts the generality of the theory by allowing only bit-vector elements (representing characters) and assuming that sequences are bounded. In contrast, our calculus maintains full generality, allowing unbounded sequences and elements of arbitrary types. Second, while our core calculus focuses only on a subset of the operators in [8], our implementation supports the remaining operators by reducing them to the core operators, and also adds native support for the `update` operator, which is not included in [8].

The base calculus that we present for sequences builds on similar work for the theory of strings [6,17]. We extend our base calculus to support array-like reasoning based on the weak-equivalence approach [10]. Though there exists some prior work on extending the theory of arrays with more operators and reasoning about length [1,12, 14], this work does not include support for most of the of the sequence operators we consider here.

The SMT-solver Z3 [11] also provides a solver for sequences. However, its documentation is limited [7], it does not support `update` directly, and its internal algorithms are not described in the literature. Furthermore, as we show in Sect. 6, the performance of the Z3 implementation is generally inferior to our implementation in cvc5.

2 Preliminaries

We assume the usual notions and terminology of many-sorted first-order logic with equality (see, e.g., [13] for a complete presentation). We consider many-sorted signatures Σ, each containing a set of sort symbols (including a Boolean sort Bool), a family of logical symbols $\approx$ for equality, with sort $\sigma \times \sigma \to \mathsf{Bool}$ for all sorts σ in Σ and interpreted as the identity relation, and a set of interpreted (and sorted) function symbols. We assume the usual definitions of well-sorted terms, literals, and formulas as terms of sort Bool. A literal is *flat* if it has the form $\bot$, $p(x_1, \ldots, x_n)$, $\neg p(x_1, \ldots, x_n)$, $x \approx y$, $\neg x \approx y$, or $x \approx f(x_1, \ldots, x_n)$, where p and f are function symbols and x, y, and $x_1, \ldots, x_n$ are variables. A Σ-interpretation $\mathcal{M}$ is defined as usual, satisfying $\mathcal{M}(\bot) = \text{false}$ and assigns: a set $\mathcal{M}(\sigma)$ to every sort σ of Σ, a function $\mathcal{M}(f) : \mathcal{M}(\sigma_1) \times \ldots \times \mathcal{M}(\sigma_n) \to \mathcal{M}(\sigma)$ to any function symbol f of Σ with arity $\sigma_1 \times \ldots \times \sigma_n \to \sigma$, and an element $\mathcal{M}(x) \in \mathcal{M}(\sigma)$ to any variable x of sort σ. The satisfaction relation between interpretations and formulas is defined as usual and is denoted by $\models$.

A *theory* is a pair $T = (\Sigma, \mathbf{I})$, in which Σ is a signature and $\mathbf{I}$ is a class of Σ-interpretations, closed under variable reassignment. The *models* of T are the interpretations in $\mathbf{I}$ without any variable assignments. A Σ-formula φ is *satisfiable* (resp., *unsatisfiable*) *in* T if it is satisfied by some (resp., no) interpretation in $\mathbf{I}$. Given a (set of) terms S, we write $\mathcal{T}(S)$ to denote the set of all subterms of S. For a theory $T = (\Sigma, \mathbf{I})$, a set S of Σ-formulas and a Σ-formula φ, we write $S \models_T \varphi$ if every interpretation $\mathcal{M} \in \mathbf{I}$ that satisfies S also satisfies φ. By convention and unless otherwise stated, we use letters w, x, y, z to denote variables and s, t, u, v to denote terms.

The theory $T_{\mathsf{LIA}} = (\Sigma_{\mathsf{LIA}}, \mathbf{I}_{T_{\mathsf{LIA}}})$ of *linear integer arithmetic* is based on the signature Σ_{LIA} that includes a single sort Int, all natural numbers as constant symbols, the unary $-$ symbol, the binary $+$ symbol and the binary $\leqslant$ relation. When $k \in \mathbb{N}$, we use the notation $k \cdot x$, inductively defined by $0 \cdot x = 0$ and $(m + 1) \cdot x = x + m \cdot x$. In turn, $\mathbf{I}_{T_{\mathsf{LIA}}}$ consists of all structures $\mathcal{M}$ for Σ_{LIA} in which the domain $\mathcal{M}(\mathsf{Int})$ of Int is the set

Symbol	Arity	SMT-LIB	Description
n	Int	n	All constants n ∈ ℕ
$+$	Int × Int → Int	+	Integer addition
$-$	Int → Int	-	Unary Integer minus
$\leqslant$	Int × Int → Bool	<=	Integer inequality
ϵ	Seq	`seq.empty`	The empty sequence
unit	Elem → Seq	`seq.unit`	Sequence constructor
$\lvert \_ \rvert$	Seq → Int	`seq.len`	Sequence length
nth	Seq × Int → Elem	`seq.nth`	Element access
update	Seq × Int × Elem → Seq	`seq.update`	Element update
extract	Seq × Int × Int → Seq	`seq.extract`	Extraction (subsequence)
$\_ +\!\!+ \cdots +\!\!+ \_$	Seq × ⋯ × Seq → Seq	`seq.concat`	Concatenation

Fig. 1. Signature for the theory of sequences.

of integer numbers, for every constant symbol $n \in \mathbb{N}$, $\mathcal{M}(n) = n$, and $+$, $-$, and $\leqslant$ are interpreted as usual. We use standard notation for integer intervals (e.g., $[a, b]$ for the set of integers i, where $a \leqslant i \leqslant b$ and $[a, b)$ for the set where $a \leqslant i < b$).

3 A Theory of Sequences

We define the theory T_{Seq} of sequences. Its signature Σ_{Seq} is given in Fig. 1. It includes the sorts Seq, Elem, Int, and Bool, intuitively denoting sequences, elements, integers, and Booleans, respectively. The first four lines include symbols of Σ_{LIA}. We write $t_1 \bowtie t_2$, with $\bowtie \in \{>, <, \leqslant\}$, as syntactic sugar for the equivalent literal expressed using $\leqslant$ (and possibly $\neg$). The sequence symbols are given on the remaining lines. Their arities are also given in Fig. 1. Notice that $\_ +\!\!+ \cdots +\!\!+ \_$ is a variadic function symbol.

Interpretations $\mathcal{M}$ of T_{Seq} interpret: Int as the set of integers; Elem as some set; Seq as the set of finite sequences whose elements are from Elem; ϵ as the empty sequence; unit as a function that takes an element from $\mathcal{M}(\mathsf{Elem})$ and returns the sequence that contains only that element; nth as a function that takes an element s from $\mathcal{M}(\mathsf{Seq})$ and an integer i and returns the ith element of s, in case i is non-negative and is smaller than the length of s (we take the first element of a sequence to have index 0). Otherwise, the function has no restrictions; update as a function that takes an element s from $\mathcal{M}(\mathsf{Seq})$, an integer i, and an element a from $\mathcal{M}(\mathsf{Elem})$ and returns the sequence obtained from s by replacing its ith element by a, in case i is non-negative and smaller than the length of s. Otherwise, the returned value is s itself; extract as a function that takes a sequence s and integers i and j, and returns the maximal sub-sequence of s that starts at index i and has length at most j, in case both i and j are non-negative and i is smaller than the length of s. Otherwise, the returned value is the empty sequence;[1] $\lvert \_ \rvert$ as a function that takes a sequence and returns its length; and $\_ +\!\!+ \cdots +\!\!+ \_$ as a function that takes some number of sequences (at least 2) and returns their concatenation.

[1] In [8], the second argument j denotes the end index, while here it denotes the length of the sub-sequence, in order to be consistent with the theory of strings in the SMT-LIB standard.

Notice that the interpretations of Elem and nth are not completely fixed by the theory: Elem can be set arbitrarily, and nth is only defined by the theory for some values of its second argument. For the rest, it can be set arbitrarily.

3.1 Vectors as Sequences

We show the applicability of T_{Seq} by using it for a simple verification task. Consider the C++ function `swap` at the top of Fig. 2. This function swaps two elements in a vector. The comments above the function include a partial specification for it: if both indexes are in-bounds and the indexed elements are equal, then the function should not change the vector (this is expressed by `s_out==s`). We now consider how to encode the verification condition induced by the code and the specification. The function variables a, b, i, and j can be encoded as variables of sort Int with the same names. We include two copies of s: s for its value at the beginning, and s_{out} for its value at the end. But what should be the sorts of s and s_{out}? In Fig. 2 we consider two options: one is based on arrays and the other on sequences.

Example 1 (Arrays). The theory of arrays includes three sorts: index, element (in this case, both are Int), and an array sort Arr, as well as two operators: $x[i]$, interpreted as the ith element of x; and $x[i \leftarrow a]$, interpreted as the array obtained from x by setting the element at index i to a. We declare s and s_{out} as variables of an uninterpreted sort V and declare two functions ℓ and $\mathbf{c}$, which, given v of sort V, return its length (of sort Int) and content (of sort Arr), respectively.[2]

Next, we introduce functions to model vector operations: $\approx_{\mathbf{A}}$ for comparing vectors, $\mathsf{nth}_{\mathbf{A}}$ for reading from them, and $\mathsf{update}_{\mathbf{A}}$ for updating them. These functions need to be axiomatized. We include two axioms (bottom of Fig. 2): Ax_1 states that two vectors are equal iff they have the same length and the same contents. Ax_2 axiomatizes the update operator; the result has the same length, and if the updated index is in bounds, then the corresponding element is updated. These axioms are not meant to be complete, but are rather just strong enough for the example.

The first two lines of the `swap` function are encoded as equalities using $\mathsf{nth}_{\mathbf{A}}$, and the last two lines are combined into one nested constraint that involves $\mathsf{update}_{\mathbf{A}}$. The precondition of the specification is naturally modeled using $\mathsf{nth}_{\mathbf{A}}$, and the post-condition is negated, so that the unsatisfiability of the formula entails the correctness of the function w.r.t. the specification. Indeed, the conjunction of all formulas in this encoding is unsatisfiable in the combined theories of arrays, integers, and uninterpreted functions.

The above encoding has two main shortcomings: It introduces auxiliary symbols, and it uses quantifiers, thus reducing clarity and efficiency. In the next example, we see how using the theory of sequences allows for a much more natural and succinct encoding.

Example 2 (Sequences). In the sequences encoding, s and s_{out} have sort Seq. No auxiliary sorts or functions are needed, as the theory symbols can be used directly. Further,

[2] It is possible to obtain a similar encoding using the theory of datatypes; however, here we use uninterpreted functions which are simpler and better supported by SMT solvers.

```
// @pre: 0 <= i,j < s.size() and s[i] == s[j]
// @post: s out == s
void swap(std::vector<int>& s, int i, int j) {
  int a = s[i];
  int b = s[j];
  s[i] = b;
  s[j] = a;
}
```

	Sequences	Arrays
Problem Variables	a, b, i, j : Int $\quad s, s_{out}$: Seq	a, b, i, j : Int $\quad s, s_{out} : V$
Auxiliary Variables		$\ell : V \to \mathsf{Int} \quad \mathsf{c} : V \to \mathsf{Arr}$ $\approx_{\mathbf{A}}: V \times V \to \mathsf{Bool}$ $\mathsf{nth}_{\mathbf{A}} : V \times \mathsf{Int} \to \mathsf{Int}$ $\mathsf{update}_{\mathbf{A}} : V \times \mathsf{Int} \times \mathsf{Int} \to V$
Axioms		$Ax_1 \wedge Ax_2$
Program	$a \approx \mathsf{nth}(s, i) \wedge b \approx \mathsf{nth}(s, j)$ $s_{out} \approx \mathsf{update}(\mathsf{update}(s, i, b), j, a)$	$a \approx \mathsf{nth}_{\mathbf{A}}(s, i) \wedge b \approx \mathsf{nth}_{\mathbf{A}}(s, j)$ $s_{out} \approx_{\mathbf{A}} \mathsf{update}_{\mathbf{A}}(\mathsf{update}_{\mathbf{A}}(s, i, b), j, a)$
Spec.	$0 \leqslant i, j < \lvert s\rvert \wedge \mathsf{nth}(s, i) \approx \mathsf{nth}(s, j)$ $\neg s_{out} \approx s$	$0 \leqslant i, j < \ell(s) \wedge \mathsf{nth}_{\mathbf{A}}(s, i) \approx \mathsf{nth}_{\mathbf{A}}(s, j)$ $\neg s_{out} \approx_{\mathbf{A}} s$

$$Ax_1 := \forall x, y. x \approx_{\mathbf{A}} y \leftrightarrow (\ell(x) \approx \ell(y) \wedge \forall\, 0 \leqslant i < \ell(x). \mathsf{c}(x)[i] \approx \mathsf{c}(y)[i])$$
$$Ax_2 := \forall x, y, i, a. y \approx_{\mathbf{A}} \mathsf{update}_{\mathbf{A}}(x, i, a) \to (\ell(x) \approx \ell(y) \wedge (0 \leqslant i < \ell(x) \to \mathsf{c}(y) \approx \mathsf{c}(x)[i \leftarrow a]))$$

Fig. 2. An example using T_{Seq}.

these symbols do not need to be axiomatized as their semantics is fixed by the theory. The resulting formula, much shorter than in Exmaple 2 and with no quantifiers, is unsatisfiable in T_{Seq}.

4 Calculi

After introducing some definitions and assumptions, we describe a basic calculus for the theory of sequences, which adapts techniques from previous procedures for the theory of strings. In particular, the basic calculus reduces the operators nth and update by introducing concatenation terms. We then show how to extend the basic calculus by introducing additional rules inspired by solvers for the theory of arrays; the modified calculus can often reason about nth and update terms directly, avoiding the introduction of concatenation terms (which are typically expensive to reason about).

Given a vector of sequence terms $\overline{t} = (t_1, \ldots, t_n)$, we use $\overline{t}$ to denote the term corresponding to the concatenation of $t_1, \ldots, t_n$. If $n = 0$, $\overline{t}$ denotes ϵ, and if $n = 1$, $\overline{t}$ denotes t_1; otherwise (when $n > 1$), $\overline{t}$ denotes a concatenation term having n children. In our calculi, we distinguish between sequence and arithmetic constraints.

Definition 1. *A Σ_{Seq}-formula φ is a sequence constraint if it has the form $s \approx t$ or $s \not\approx t$; it is an arithmetic constraint if it has the form $s \approx t$, $s \geq t$, $s \not\approx t$, or $s < t$ where s, t are terms of sort* Int, *or if it is a disjunction $c_1 \vee c_2$ of two arithmetic constraints.*

Notice that sequence constraints do not have to contain sequence terms (e.g., $x \approx y$ where x, y are Elem-variables). Also, equalities and disequalities between terms of sort Int are both sequence and arithmetic constraints. In this paper we focus on sequence

$$|\epsilon| \to 0 \qquad |\mathsf{unit}(t)| \to 1$$
$$|\mathsf{update}(s,i,t)| \to |s| \qquad |s_1 \mathbin{++} \cdots \mathbin{++} s_n| \to |s_1| + \cdots + |s_n|$$
$$\overline{u} \mathbin{++} \epsilon \mathbin{++} \overline{v} \to \overline{u} \mathbin{++} \overline{v} \qquad \overline{u} \mathbin{++} (s_1 \mathbin{++} \cdots \mathbin{++} s_n) \mathbin{++} \overline{v} \to \overline{u} \mathbin{++} s_1 \mathbin{++} \cdots \mathbin{++} s_n \mathbin{++} \overline{v}$$

Fig. 3. Rewrite rules for the reduced form $t\downarrow$ of a term t, obtained from t by applying these rules to completion.

constraints and arithmetic constraints. This is justified by the following lemma. (Proofs of this lemma and later results can be found in an extended version of this paper [23].)

Lemma 1. *For every quantifier-free Σ_{Seq}-formula φ, there are sets $\mathsf{S}_1, \ldots, \mathsf{S}_n$ of sequence constraints and sets $\mathsf{A}_1, \ldots, \mathsf{A}_n$ of arithmetic constraints such that φ is T_{Seq}-satisfiable iff $\mathsf{S}_i \cup \mathsf{A}_i$ is T_{Seq}-satisfiable for some $i \in [1, n]$.*

Throughout the presentation of the calculi, we will make a few simplifying assumptions.

Assumption 1. *Whenever we refer to a set S of sequence constraints, we assume:*

1. *for every non-variable term $t \in \mathcal{T}(\mathsf{S})$, there exists a variable x such that $x \approx t \in \mathsf{S}$;*
2. *for every Seq-variable x, there exists a variable ℓ_x such that $\ell_x \approx |x| \in \mathsf{S}$;*
3. *all literals in S are flat.*

Whenever we refer to a set of arithmetic constraints, we assume all its literals are flat.

These assumptions are without loss of generality as any set can easily be transformed into an equisatisfiable set satisfying the assumptions by the addition of fresh variables and equalities. Note that some rules below introduce non-flat literals. In such cases, we assume that similar transformations are done immediately after applying the rule to maintain the invariant that all literals in $\mathsf{S} \cup \mathsf{A}$ are flat. Rules may also introduce fresh variables k of sort Seq. We further assume that in such cases, a corresponding constraint $\ell_k \approx |k|$ is added to S with a fresh variable ℓ_k.

Definition 2. *Let C be a set of constraints. We write $\mathsf{C} \models \varphi$ to denote that C entails formula φ in the empty theory, and write $\equiv_{\mathsf{C}}$ to denote the binary relation over $\mathcal{T}(\mathsf{C})$ such that $s \equiv_{\mathsf{C}} t$ iff $\mathsf{C} \models s \approx t$.*

Lemma 2. *For all set S of sequence constraints, $\equiv_{\mathsf{S}}$ is an equivalence relation; furthermore, every equivalence class of $\equiv_{\mathsf{S}}$ contains at least one variable.*

We denote the equivalence class of a term s according to $\equiv_{\mathsf{S}}$ by $[s]_{\equiv_{\mathsf{S}}}$ and drop the $\equiv_{\mathsf{S}}$ subscript when it is clear from the context.

In the presentation of the calculus, it will often be useful to normalize terms to what will be called a *reduced form*.

Definition 3. *Let t be a Σ_{Seq}-term. The* reduced form *of t, denoted by $t\downarrow$, is the term obtained by applying the rewrite rules listed in Fig. 3 to completion.*

Observe that $t\downarrow$ is well defined because the given rewrite rules form a terminating rewrite system. This can be seen by noting that each rule reduces the number of applications of sequence operators in the left-hand side term or keeps that number the same but reduces the size of the term. It is not difficult to show that $\models_{T_{\mathsf{Seq}}} t \approx t\downarrow$.

We now introduce some basic definitions related to concatenation terms.

Definition 4. *A* concatenation term *is a term of the form* $s_1 \mathbin{++} \cdots \mathbin{++} s_n$ *with* $n \geq 2$. *If each* s_i *is a variable, it is a* variable concatenation term. *For a set* S *of sequence constraints, a variable concatenation term* $x_1 \mathbin{++} \cdots \mathbin{++} x_n$ *is* singular *in* S *if* $\mathsf{S} \not\models x_i \approx \epsilon$ *for at most one variable* x_i *with* $i \in [1, n]$. *A sequence variable* x *is* atomic in S *if* $\mathsf{S} \not\models x \approx \epsilon$ *and for all variable concatenation terms* $s \in \mathcal{T}(\mathsf{S})$ *such that* $\mathsf{S} \models x \approx s$, s *is singular in* S.

We lift the concept of atomic variables to atomic representatives of equivalence classes.

Definition 5. *Let* S *be a set of sequence constraints. Assume a choice function* $\alpha : \mathcal{T}(\mathsf{S})/{\equiv_{\mathsf{S}}} \to \mathcal{T}(\mathsf{S})$ *that chooses a variable from each equivalence class of* $\equiv_{\mathsf{S}}$. *A sequence variable* x *is an* atomic representative *in* S *if it is atomic in* S *and* $x = \alpha([x]_{\equiv_{\mathsf{S}}})$.

Finally, we introduce a relation that is the foundation for reasoning about concatenations.

Definition 6. *Let* S *be a set of sequence constraints. We inductively define a relation* $\mathsf{S} \models_{++} x \approx s$, *where* x *is a sequence variable in* S *and* s *is a sequence term whose variables are in* $\mathcal{T}(\mathsf{S})$, *as follows:*

1. $\mathsf{S} \models_{++} x \approx x$ *for all sequence variables* $x \in \mathcal{T}(\mathsf{S})$.
2. $\mathsf{S} \models_{++} x \approx t$ *for all sequence variables* $x \in \mathcal{T}(\mathsf{S})$ *and variable concatenation terms* t, *where* $x \approx t \in \mathsf{S}$.
3. *If* $\mathsf{S} \models_{++} x \approx (\overline{w} \mathbin{++} y \mathbin{++} \overline{z}){\downarrow}$ *and* $\mathsf{S} \models y \approx t$ *and* t *is* ϵ *or a variable concatenation term in* S *that is not singular in* S, *then* $\mathsf{S} \models_{++} x \approx (\overline{w} \mathbin{++} t \mathbin{++} \overline{z}){\downarrow}$.

Let α *be a choice function for* S *as defined in Definition 5. We additionally define the entailment relation* $\mathsf{S} \models^*_{++} x \approx \overline{y}$, *where* $\overline{y}$ *is of length* $n \geq 0$, *to hold if each element of* $\overline{y}$ *is an atomic representative in* S *and there exists* $\overline{z}$ *of length* n *such that* $\mathsf{S} \models_{++} x \approx \overline{z}$ *and* $\mathsf{S} \models y_i \approx z_i$ *for* $i \in [1, n]$.

In other words, $\mathsf{S} \models^*_{++} x \approx t$ holds when t is a concatenation of atomic representatives and is entailed to be equal to x by S. In practice, t is determined by recursively expanding concatenations using equalities in S until a fixpoint is reached.

Example 3. Suppose $\mathsf{S} = \{x \approx y \mathbin{++} z, y \approx w \mathbin{++} u, u \approx v\}$ (we omit the additional constraints required by Assumption 1, part 2 for brevity). It is easy to see that u, v, w, and z are atomic in S, but x and y are not. Furthermore, w and z (and one of u or v) must also be atomic representatives. Clearly, $\mathsf{S} \models_{++} x \approx x$ and $\mathsf{S} \models x \approx y \mathbin{++} z$. Moreover, $y \mathbin{++} z$ is a variable concatenation term that is not singular in S. Hence, we have $\mathsf{S} \models_{++} x \approx (y \mathbin{++} z){\downarrow}$, and so $\mathsf{S} \models_{++} x \approx y \mathbin{++} z$ (by using either Item 2 or Item 3 of Defintion 6, as in fact $x \approx y \mathbin{++} z \in \mathsf{S}$.). Now, since $\mathsf{S} \models_{++} x \approx y \mathbin{++} z$, $\mathsf{S} \models y \approx w \mathbin{++} u$, and $w \mathbin{++} u$ is a variable concatenation term not singular in S, we get that $\mathsf{S} \models_{++} x \approx ((w \mathbin{++} u) \mathbin{++} z){\downarrow}$, and so $\mathsf{S} \models_{++} x \approx w \mathbin{++} u \mathbin{++} z$. Now, assume that $v = \alpha([v]_{\equiv_{\mathsf{S}}}) = \alpha(\{v, u\})$. Then, $\mathsf{S} \models^*_{++} x \approx w \mathbin{++} v \mathbin{++} z$.

Our calculi can be understood as modeling abstractly a cooperation between an *arithmetic subsolver* and a *sequence subsolver*. Many of the derivation rules lift those in the string calculus of Liang et al. [17] to sequences of elements of an arbitrary type. We describe them similarly as rules that modify *configurations*.

Definition 7. *A* configuration *is either the distinguished configuration* unsat *or a pair* (S, A) *of a set* S *of sequence constraints and a set* A *of arithmetic constraints.*

The rules are given in *guarded assignment form*, where the rule premises describe the conditions on the current configuration under which the rule can be applied, and the conclusion is either unsat, or otherwise describes the resulting modifications to the configuration. A rule may have multiple conclusions separated by $\|$. In the rules, some of the premises have the form $\mathsf{S} \models s \approx t$ (see Definition 2). Such entailments can be checked with standard algorithms for congruence closure. Similarly, premises of the form $\mathsf{S} \models_{\mathsf{LIA}} s \approx t$ can be checked by solvers for linear integer arithmetic.

An application of a rule is *redundant* if it has a conclusion where each component in the derived configuration is a subset of the corresponding component in the premise configuration. We assume that for rules that introduce fresh variables, the introduced variables are identical whenever the premises triggering the rule are the same (i.e., we cannot generate an infinite sequence of rule applications by continuously using the same premises to introduce fresh variables).[3] A configuration other than unsat is *saturated* with respect to a set R of derivation rules if every possible application of a rule in R to it is redundant. A *derivation tree* is a tree where each node is a configuration whose children, if any, are obtained by a non-redundant application of a rule of the calculus. A derivation tree is *closed* if all of its leaves are unsat. As we show later, a closed derivation tree with root node (S, A) is a proof that $\mathsf{A} \cup \mathsf{S}$ is unsatisfiable in T_{Seq}. In contrast, a derivation tree with root node (S, A) and a saturated leaf with respect to all the rules of the calculus is a witness that $\mathsf{A} \cup \mathsf{S}$ is satisfiable in T_{Seq}.

4.1 Basic Calculus

Definition 8. *The calculus* BASE *consists of the derivation rules in Figs. 4 and 5.*

Some of the rules are adapted from previous work on string solvers [17,22]. Compared to that work, our presentation of the rules is noticeably simpler, due to our use of the relation $\models^*_{++}$ from Definition 6. In particular, our configurations consist only of pairs of sets of formulas, without any auxiliary data-structures.

Note that judgments of the form $\mathsf{S} \models^*_{++} x \approx t$ are used in premises of the calculus. It is possible to compute whether such a premise holds thanks to the following lemma.

Lemma 3. *Let* S *be a set of sequence constraints and* A *a set of arithmetic constraints. If* (S, A) *is saturated w.r.t.* S-Prop, L-Intro *and* L-Valid, *the problem of determining whether* $\mathsf{S} \models^*_{++} x \approx s$ *for given* x *and* s *is decidable.*

Lemma 3 assumes saturation with respect to certain rules. Accordingly, our proof strategy, described in Sect. 5, will ensure such saturation before attempting to apply rules relying on $\models^*_{++}$. The relation $\models^*_{++}$ induces a normal form for each equivalence class of $\equiv_{\mathsf{S}}$.

[3] In practice, this is implemented by associating each introduced variable with a *witness term* as described in [21].

$$\textsf{A-Conf}\ \frac{\mathsf{A} \models_{\mathsf{LIA}} \bot}{\mathsf{unsat}} \qquad \textsf{A-Prop}\ \frac{\mathsf{A} \models_{\mathsf{LIA}} s \approx t \qquad s, t \in \mathcal{T}(\mathsf{S})}{\mathsf{S} := \mathsf{S}, s \approx t}$$

$$\textsf{S-Conf}\ \frac{\mathsf{S} \models \bot}{\mathsf{unsat}} \qquad \textsf{S-Prop}\ \frac{\mathsf{S} \models s \approx t \qquad s, t \in \mathcal{T}(\mathsf{S}) \qquad s, t \text{ are } \Sigma_{\mathsf{LIA}}\text{-terms}}{\mathsf{A} := \mathsf{A}, s \approx t}$$

$$\textsf{S-A}\ \frac{x, y \in \mathcal{T}(\mathsf{S}) \cap \mathcal{T}(\mathsf{A}) \qquad x, y : \mathsf{Int}}{\mathsf{A} := \mathsf{A}, x \approx y \quad \| \quad \mathsf{A} := \mathsf{A}, x \not\approx y}$$

$$\textsf{L-Intro}\ \frac{s \in \mathcal{T}(\mathsf{S}) \qquad s : \mathsf{Seq}}{\mathsf{S} := \mathsf{S}, |s| \approx (|s|){\downarrow}} \qquad \textsf{L-Valid}\ \frac{x \in \mathcal{T}(\mathsf{S}) \qquad x : \mathsf{Seq}}{\mathsf{S} := \mathsf{S}, x \approx \epsilon \quad \| \quad \mathsf{A} := \mathsf{A}, \ell_x > 0}$$

$$\textsf{U-Eq}\ \frac{\mathsf{S} \models \mathsf{unit}(x) \approx \mathsf{unit}(y)}{\mathsf{S} := \mathsf{S}, x \approx y} \qquad \textsf{C-Eq}\ \frac{\mathsf{S} \models^{*}_{++} x \approx \overline{\boldsymbol{z}} \qquad \mathsf{S} \models^{*}_{++} y \approx \overline{\boldsymbol{z}}}{\mathsf{S} := \mathsf{S}, x \approx y}$$

$$\textsf{C-Split}\ \frac{\mathsf{S} \models^{*}_{++} x \approx (\overline{\boldsymbol{w}} +\!\!+ y +\!\!+ \overline{\boldsymbol{z}}){\downarrow} \qquad \mathsf{S} \models^{*}_{++} x \approx (\overline{\boldsymbol{w}} +\!\!+ y' +\!\!+ \overline{\boldsymbol{z}}'){\downarrow}}{\begin{array}{ll} \mathsf{A} := \mathsf{A}, \ell_y > \ell_{y'} \quad \mathsf{S} := \mathsf{S}, y \approx y' +\!\!+ k & \| \\ \mathsf{A} := \mathsf{A}, \ell_y < \ell_{y'} \quad \mathsf{S} := \mathsf{S}, y' \approx y +\!\!+ k & \| \\ \mathsf{A} := \mathsf{A}, \ell_y \approx \ell_{y'} \quad \mathsf{S} := \mathsf{S}, y \approx y' & \end{array}}$$

$$\textsf{Deq-Ext}\ \frac{x \not\approx y \in \mathsf{S} \qquad x, y : \mathsf{Seq}}{\begin{array}{c} \mathsf{A} := \mathsf{A}, \ell_x \not\approx \ell_y \quad \| \\ \mathsf{A} := \mathsf{A}, \ell_x \approx \ell_y, 0 \leqslant i < \ell_x \quad \mathsf{S} := \mathsf{S}, w_1 \approx \mathsf{nth}(x, i), w_2 \approx \mathsf{nth}(y, i), w_1 \not\approx w_2 \end{array}}$$

Fig. 4. Core derivation rules. The rules use k and i to denote fresh variables of sequence and integer sort, respectively, and w_1 and w_2 for fresh element variables.

Lemma 4. *Let* S *be a set of sequence constraints and* A *a set of arithmetic constraints. Suppose* (S, A) *is saturated w.r.t.* A-Conf, S-Prop, L-Intro, L-Valid, *and* C-Split. *Then, for every equivalence class* e *of* $\equiv_{\mathsf{S}}$ *whose terms are of sort* Seq, *there exists a unique (possibly empty)* $\overline{\boldsymbol{s}}$ *such that whenever* $\mathsf{S} \models^{*}_{++} x \approx \overline{\boldsymbol{s}'}$ *for* $x \in e$, *then* $\overline{\boldsymbol{s}'} = \overline{\boldsymbol{s}}$. *In this case, we call* $\overline{\boldsymbol{s}}$ *the* normal form *of* e *(and of* x*).*

We now turn to the description of the rules in Fig. 4, which form the core of the calculus. For greater clarity, some of the conclusions of the rules include terms before they are flattened. First, either subsolver can report that the current set of constraints is unsatisfiable by using the rules A-Conf or S-Conf. For the former, the entailment $\models_{\mathsf{LIA}}$ (which abbreviates $\models_{\mathcal{T}_{\mathsf{LIA}}}$) can be checked by a standard procedure for linear integer arithmetic, and the latter corresponds to a situation where congruence closure detects a conflict between an equality and a disequality. The rules A-Prop, S-Prop, and S-A correspond to a form of Nelson-Oppen-style theory combination between the two sub-solvers. The first two communicate equalities between the sub-solvers, while the third guesses arrangements for shared variables of sort Int. L-Intro ensures that the length term $|s|$ for each sequence term s is equal to its reduced form $(|s|){\downarrow}$. L-Valid restricts sequence lengths to be non-negative, splitting on whether each sequence is empty or has a length greater than 0. The unit operator is injective, which is captured by U-Eq. C-Eq concludes that two sequence terms are equal if they have the same normal form. If two sequence variables have different normal forms, then C-Split takes the first differing components y and y' from the two normal forms and splits on their length relationship. Note that C-Split is the source for non-termination of the calculus (see, e.g., [17,22]).

$$\text{R-Extract}\ \frac{x \approx \mathsf{extract}(y,i,j) \in \mathsf{S}}{\begin{array}{c}\mathsf{A} := \mathsf{A}, i < 0 \vee i \geqslant \ell_y \vee j \leqslant 0 \quad \mathsf{S} := \mathsf{S}, x \approx \epsilon \quad \| \\ \mathsf{A} := \mathsf{A}, 0 \leqslant i < \ell_y, j > 0, \ell_k \approx i, \ell_x \approx \min(j, \ell_y - i) \\ \mathsf{S} := \mathsf{S}, y \approx k +\!\!+ x +\!\!+ k'\end{array}}$$

$$\text{R-Nth}\ \frac{x \approx \mathsf{nth}(y,i) \in \mathsf{S}}{\begin{array}{c}\mathsf{A} := \mathsf{A}, i < 0 \vee i \geqslant \ell_y \quad \| \\ \mathsf{A} := \mathsf{A}, 0 \leqslant i < \ell_y, \ell_k \approx i \quad \mathsf{S} := \mathsf{S}, y \approx k +\!\!+ \mathsf{unit}(x) +\!\!+ k'\end{array}}$$

$$\text{R-Update}\ \frac{x \approx \mathsf{update}(y,i,z) \in \mathsf{S}}{\begin{array}{c}\mathsf{A} := \mathsf{A}, i < 0 \vee i \geqslant \ell_y \quad \mathsf{S} := \mathsf{S}, x \approx y \quad \| \\ \mathsf{A} := \mathsf{A}, 0 \leqslant i < \ell_y, \ell_k \approx i, \ell_{k'} \approx 1 \\ \mathsf{S} := \mathsf{S}, y \approx k +\!\!+ k' +\!\!+ k'', x \approx k +\!\!+ \mathsf{unit}(z) +\!\!+ k''\end{array}}$$

Fig. 5. Reduction rules for extract, nth, and update. The rules use k, k', and k'' to denote fresh sequence variables. We write $s \approx \min(t, u)$ as an abbreviation for $s \approx t \vee s \approx u, s \leqslant t, s \leqslant u$.

Finally, Deq-Ext handles disequalities between sequences x and y by either asserting that their lengths are different or by choosing an index i at which they differ.

Figure 5 includes a set of reduction rules for handling operators that are not directly handled by the core rules. These reduction rules capture the semantics of these operators by reduction to concatenation. R-Extract splits into two cases: Either the extraction uses an out-of-bounds index or a non-positive length, in which case the result is the empty sequence, or the original sequence can be described as a concatenation that includes the extracted sub-sequence. R-Nth creates an equation between y and a concatenation term with $\mathsf{unit}(x)$ as one of its components, as long as i is not out of bounds. R-Update considers two cases. If i is out of bounds, then the update term is equal to y. Otherwise, y is equal to a concatenation, with the middle component (k') representing the part of y that is updated. In the update term, k' is replaced by $\mathsf{unit}(z)$.

Example 4. Consider a configuration (S, A), where S contains the formulas $x \approx y +\!\!+ z$, $z \approx v +\!\!+ x +\!\!+ w$, and $v \approx \mathsf{unit}(u)$, and A is empty. Hence, $\mathsf{S} \models |x| \approx |y +\!\!+ z|$. By L-Intro, we have $\mathsf{S} \models |y +\!\!+ z| \approx |y| + |z|$. Together with Assumption 1, we have $\mathsf{S} \models \ell_x \approx \ell_y + \ell_z$, and then with S-Prop, we have $\ell_x \approx \ell_y + \ell_z \in \mathsf{A}$. Similarly, we can derive $\ell_z \approx \ell_v + \ell_x + \ell_w, \ell_v \approx 1 \in \mathsf{S}$, and so $(*)\mathsf{A} \models_{\mathsf{LIA}} \ell_z \approx 1 + \ell_y + \ell_z + \ell_w$. Notice that for any variable k of sort Seq, we can apply L-Valid, L-Intro, and S-Prop to add to A either $\ell_k > 0$ or $\ell_k = 0$. Applying this to y, z, w, we have that $\mathsf{A} \models_{\mathsf{LIA}} \bot$ in each branch thanks to $(*)$, and so A-Conf applies and we get unsat.

4.2 Extended Calculus

Definition 9. *The calculus* EXT *is comprised of the derivation rules in Figs. 4 and 6, with the addition of rule R-Extract from Fig. 5.*

Our extended calculus combines array reasoning, based on [10] and expressed by the rules in Fig. 6, with the core rules of Fig. 4 and the R-Extract rule. Unlike in BASE, those rules do not reduce nth and update. Instead, they reason about those operators directly and handle their combination with concatenation. Nth-Concat identifies the ith element of sequence y with the corresponding element selected from its normal form (see Lemma 4). Update-Concat operates similarly, applying update to all the components. Update-Concat-Inv operates similarly on the updated sequence rather than on the original sequence. Nth-Unit captures the semantics of nth when applied to a unit term. Update-Unit is similar and distinguishes an update on an out-of-bounds index (different from 0) from an update within the bound. Nth-Intro is meant to ensure that Nth-Update (explained below) and Nth-Unit (explained above) are applicable whenever an update term exists in the constraints. Nth-Update captures the read-over-write axioms of arrays, adapted to consider their lengths (see, e.g., [10]). It distinguishes three cases: In the first, the update index is out of bounds. In the second, it is not out of bounds, and the corresponding nth term accesses the same index that was updated. In the third case, the index used in the nth term is different from the updated index. Update-Bound considers two cases: either the update changes the sequence, or the sequence remains the same. Finally, Nth-Split introduces a case split on the equality between two sequence variables x and x' whenever they appear as arguments to nth with equivalent second arguments. This is needed to ensure that we detect all cases where the arguments of two nth terms must be equal.

4.3 Correctness

In this section we prove the following theorem:

Theorem 1. *Let $X \in \{\mathsf{BASE}, \mathsf{EXT}\}$ and $(\mathsf{S}_0, \mathsf{A}_0)$ be a configuration, and assume without loss of generality that A_0 contains only arithmetic constraints that are not sequence constraints. Let T be a derivation tree obtained by applying the rules of X with $(\mathsf{S}_0, \mathsf{A}_0)$ as the initial configuration.*

1. *If T is closed, then $\mathsf{S}_0 \cup \mathsf{A}_0$ is T_{Seq}-unsatisfiable.*
2. *If T contains a saturated configuration (S, A) w.r.t. X, then (S, A) is T_{Seq}-satisfiable, and so is $(\mathsf{S}_0, \mathsf{A}_0)$.*

The theorem states that the calculi are correct in the following sense: if a closed derivation tree is obtained for the constraints $\mathsf{S}_0 \cup \mathsf{A}_0$ then those constraints are unsatisfiable in T_{Seq}; if a tree with a saturated leaf is obtained, then they are satisfiable. It is possible, however, that neither kind of tree can be derived by the calculi, making them neither refutation-complete nor terminating. This is not surprising since, as mentioned in the introduction, the decidability of even weaker theories is still unknown.

Proving the first claim in Theorem 1 reduces to a local soundness argument for each of the rules. For the second claim, we sketch below how to construct a satisfying model $\mathcal{M}$ from a saturated configuration for the case of EXT. The case for BASE is similar and simpler.

Model Construction Steps. The full model construction and its correctness are described in a longer version of this paper [23] together with a proof of the theorem above. Here is a summary of the steps needed for the model construction.

1. Sorts: $\mathcal{M}(\mathsf{Elem})$ is interpreted as some arbitrary countably infinite set. $\mathcal{M}(\mathsf{Seq})$ and $\mathcal{M}(\mathsf{Int})$ are then determined by the theory.

$$\text{Nth-Concat}\ \frac{x \approx \mathsf{nth}(y,i) \in \mathsf{S} \qquad \mathsf{S} \models^*_{++} y \approx w_1 ++ \cdots ++ w_n}{\begin{array}{c}\mathsf{A} := \mathsf{A}, i < 0 \vee i \geqslant \ell_y \quad \| \\ \mathsf{A} := \mathsf{A}, 0 \leqslant i < \ell_{w_1} \qquad \mathsf{S} := \mathsf{S}, x \approx \mathsf{nth}(w_1, i) \quad \| \quad \cdots \quad \| \\ \mathsf{A} := \mathsf{A}, \sum_{j=1}^{n-1} \ell_{w_j} \leqslant i < \sum_{j=1}^{n} \ell_{w_j} \qquad \mathsf{S} := \mathsf{S}, x \approx \mathsf{nth}(w_n, i - \sum_{j=1}^{n-1} \ell_{w_j})\end{array}}$$

$$\text{Update-Concat}\ \frac{x \approx \mathsf{update}(y,i,v) \in \mathsf{S} \qquad \mathsf{S} \models^*_{++} y \approx w_1 ++ \cdots ++ w_n}{\begin{array}{c}\mathsf{S} := \mathsf{S}, x \approx z_1 ++ \cdots ++ z_n, \\ z_1 \approx \mathsf{update}(w_1, i, v), \ldots, z_n \approx \mathsf{update}(w_n, i - \sum_{j=1}^{n-1} \ell_{w_j}, v)\end{array}}$$

$$\text{Update-Concat-Inv}\ \frac{x \approx \mathsf{update}(y,i,v) \in \mathsf{S} \qquad \mathsf{S} \models^*_{++} x \approx w_1 ++ \cdots ++ w_n}{\begin{array}{c}\mathsf{S} := \mathsf{S}, y \approx z_1 ++ \cdots ++ z_n, \\ w_1 \approx \mathsf{update}(z_1, i, v), \ldots, w_n \approx \mathsf{update}(z_n, i - \sum_{j=1}^{n-1} \ell_{w_j}, v)\end{array}}$$

$$\text{Nth-Unit}\ \frac{x \approx \mathsf{nth}(y,i) \in \mathsf{S} \qquad \mathsf{S} \models y \approx \mathsf{unit}(u)}{\mathsf{A} := \mathsf{A}, i < 0 \vee i > 0 \quad \| \quad \mathsf{A} := \mathsf{A}, i \approx 0 \qquad \mathsf{S} := \mathsf{S}, x \approx u}$$

$$\text{Update-Unit}\ \frac{x \approx \mathsf{update}(y,i,v) \in \mathsf{S} \qquad \mathsf{S} \models y \approx \mathsf{unit}(u)}{\begin{array}{c}\mathsf{A} := \mathsf{A}, i < 0 \vee i > 0 \qquad \mathsf{S} := \mathsf{S}, x \approx \mathsf{unit}(u) \quad \| \\ \mathsf{A} := \mathsf{A}, i \approx 0 \qquad \mathsf{S} := \mathsf{S}, x \approx \mathsf{unit}(v)\end{array}}$$

$$\text{Nth-Intro}\ \frac{s' \approx \mathsf{update}(s,i,t) \in \mathsf{S}}{\mathsf{S} := \mathsf{S}, e \approx \mathsf{nth}(s,i), e' \approx \mathsf{nth}(s',i)}$$

$$\text{Nth-Update}\ \frac{\mathsf{nth}(x,j) \in \mathcal{T}(\mathsf{S}) \qquad y \approx \mathsf{update}(z,i,v) \in \mathsf{S} \qquad \mathsf{S} \models x \approx y \text{ or } \mathsf{S} \models x \approx z}{\begin{array}{c}\mathsf{A} := \mathsf{A}, j < 0 \vee j \geqslant \ell_x \quad \| \\ \mathsf{A} := \mathsf{A}, i \approx j, 0 \leqslant j < \ell_x \qquad \mathsf{S} := \mathsf{S}, \mathsf{nth}(y,j) \approx v \quad \| \\ \mathsf{A} := \mathsf{A}, i \not\approx j, 0 \leqslant j < \ell_x \qquad \mathsf{S} := \mathsf{S}, \mathsf{nth}(y,j) \approx \mathsf{nth}(z,j)\end{array}}$$

$$\text{Update-Bound}\ \frac{x \approx \mathsf{update}(y,i,v) \in \mathsf{S}}{\mathsf{A} := \mathsf{A}, 0 \leqslant i < \ell_y \qquad \mathsf{S} := \mathsf{S}, \mathsf{nth}(y,i) \not\approx v \quad \| \quad \mathsf{S} := \mathsf{S}, x \approx y}$$

$$\text{Nth-Split}\ \frac{\mathsf{nth}(x,i), \mathsf{nth}(x',i') \in \mathcal{T}(\mathsf{S}) \qquad i \approx i' \in \mathsf{A}}{\mathsf{S} := \mathsf{S}, x \approx x' \quad \| \quad \mathsf{S} := \mathsf{S}, x \not\approx x'}$$

Fig. 6. Extended derivation rules. The rules use $z_1, \ldots, z_n$ to denote fresh sequence variables and e, e' to denote fresh element variables.

2. Σ_{Seq}-symbols: T_{Seq} enforces the interpretation of almost all Σ_{Seq}-symbols, except for nth when the second input is out of bounds. We cover this case below.
3. Integer variables: based on the saturation of A-Conf, we know there is some T_{LIA}-model satisfying A. We set $\mathcal{M}$ to interpret integer variables according to this model.
4. Element variables: these are partitioned into their $\equiv_{\mathsf{S}}$ equivalence classes. Each class is assigned a distinct element from $\mathcal{M}(\mathsf{Elem})$, which is possible since it is infinite.
5. Atomic sequence variables: these are assigned interpretations in several sub-steps:
   (a) length: we first use the assignments to variables ℓ_x to set the length of $\mathcal{M}(x)$, without assigning its actual value.
   (b) unit variables: for variables x with $x \equiv_{\mathsf{S}} \mathsf{unit}(z)$, we set $\mathcal{M}(x)$ to be $[\mathcal{M}(z)]$.

(c) non-unit variables: All other sequence variables are assigned values according to a *weak equivalence graph* we construct in a manner similar to [10]. This construction takes into account constraints that involve update and nth.

6. Non-atomic sequence variables: these are first transformed to their unique normal form (see Lemma 4), consisting of concatenations of atomic variables. Then, the values assigned to these variables are concatenated.
7. nth-terms: for out-of-bounds indices in nth-terms, we rely on $\equiv_S$ to make sure that the assignment is consistent.

We conclude this section with an example of the construction of $\mathcal{M}$.

Example 5. Consider a signature in which Elem is Int, and a saturated configuration $(\mathsf{S}^*, \mathsf{A}^*)$ w.r.t. EXT that includes the following formulas: $y \approx y_1 +\!\!+\, y_2$, $x \approx x_1 +\!\!+\, x_2$, $y_2 \approx x_2$, $y_1 \approx \mathsf{update}(x_1, i, a)$, $|y_1| = |x_1|$, $|y_2| = |x_2|$, $\mathsf{nth}(y, i) \approx a$, $\mathsf{nth}(y_1, i) \approx a$. Following the above construction, a satisfying interpretation $\mathcal{M}$ can be built as follows:

Step 1 Set both $\mathcal{M}(\mathsf{Int})$ and $\mathcal{M}(\mathsf{Elem})$ to be the set of integer numbers. $\mathcal{M}(\mathsf{Seq})$ is fixed by the theory.

Step 3, Step 4 First, find an arithmetic model, $\mathcal{M}(\ell_x) = \mathcal{M}(\ell_y) = 4, \mathcal{M}(\ell_{y_1}) = \mathcal{M}(\ell_{x_1}) = 2, \mathcal{M}(\ell_{y_2}) = \mathcal{M}(\ell_{x_2}) = 2, \mathcal{M}(i) = 0$. Further, set $\mathcal{M}(a) = 0$.

Step 5a Start assigning values to sequences. First, set the lengths of $\mathcal{M}(x)$ and $\mathcal{M}(y)$ to be 4, and the lengths of $\mathcal{M}(x_1), \mathcal{M}(x_2), \mathcal{M}(y_1), \mathcal{M}(y_2)$ to be 2.

Step 5b is skipped as there are no unit terms.

Step 5c Set the 0th element of $\mathcal{M}(y_1)$ to 0 to satisfy $\mathsf{nth}(y_1, i) = a$ (y_1 is atomic, y is not). Assign fresh values to the remaining indices of atomic variables. The result can be, e.g., $\mathcal{M}(y_1) = [0, 2], \mathcal{M}(x_1) = [1, 2], \mathcal{M}(y_2) = \mathcal{M}(x_2) = [3, 4]$.

Step 6 Assign non-atomic sequence variables based on equivalent concatenations: $\mathcal{M}(y) = [0, 2, 3, 4], \mathcal{M}(x) = [1, 2, 3, 4]$.

Step 7 No integer variable in the formula was assigned an out-of-bound value, and so the interpretation of nth on out-of-bounds cases is set arbitrarily.

5 Implementation

We implemented our procedure for sequences as an extension of a previous theory solver for strings [17,22]. This solver is integrated in cvc5, and has been generalized to reason about both strings and sequences. In this section, we describe how the rules of the calculus are implemented and the overall strategy for when they are applied.

Like most SMT solvers, cvc5 is based on the CDCL(T) architecture [19] which combines several subsolvers, each specialized on a specific theory, with a solver for propositional satisfiability (SAT). Following that architecture, cvc5 maintains an evolving set of formulas F. When F starts with quantifier-free formulas over the theory T_{Seq}, the case targeted by this work, the SAT solver searches for a satisfying assignment for F, represented as the set M of literals it satisfies. If none exists, the problem is unsatisfiable at the propositional level and hence T_{Seq}-unsatisfiable. Otherwise, M is partitioned into the arithmetic constraints A and the sequence constraints S and checked for T_{Seq}-satisfiability using the rules of the EXT calculus. Many of those rules, including all those with multiple conclusions, are implemented by adding new formulas to F (following the splitting-on-demand approach [4]). This causes the SAT solver to try to extend its assignment to those formulas, which results in the addition of new literals to M (and thereby also to A and S).

In this setting, the rules of the two calculi are implemented as follows. The effect of rule A-Conf is achieved by invoking cvc5's theory solver for linear integer arithmetic. Rule S-Conf is implemented by the congruence closure submodule of the theory solver for sequences. Rules A-Prop and S-Prop are implemented by the standard mechanism for theory combination. Note that each of these four rules may be applied *eagerly*, that is, before constructing a complete satisfying assignment M for F.

The remaining rules are implemented in the theory solver for sequences. Each time M is checked for satisfiability, cvc5 follows a strategy to determine which rule to apply next. If none of the rules apply and the configuration is different from unsat, then it is saturated, and the solver returns sat. The strategy for EXT prioritizes rules as follows. Only the first applicable rule is applied (and then control goes back to the SAT solver).

1. (Add length constraints) For each sequence term in S, apply L-Intro or L-Valid, if not already done. We apply L-Intro for non-variables, and L-Valid for variables.
2. (Mark congruent terms) For each set of update (resp. nth) terms that are congruent to one another in the current configuration, mark all but one term and ignore the marked terms in the subsequent steps.
3. (Reduce extract) For extract(y, i, j) in S, apply R-Extract if not already done.
4. (Construct normal forms) Apply U-Eq or C-Split. We choose how to apply the latter rule based on constructing normal forms for equivalence classes in a bottom-up fashion, where the equivalence classes of x and y are considered before the equivalence class of $x \mathbin{++} y$. We do this until we find an equivalence class such that S $\models^*_{++} z \approx u_1$ and S $\models^*_{++} z \approx u_2$ for distinct u_1, u_2.
5. (Normal forms) Apply C-Eq if two equivalence classes have the same normal form.
6. (Extensionality) For each disequality in S, apply Deq-Ext, if not already done.
7. (Distribute update and nth) For each term update(x, i, t) (resp. nth(x, j)) such that the normal form of x is a concatenation term, apply Update-Concat and Update-Concat-Inv (resp. Nth-Concat) if not already done. Alternatively, if the normal form of the equivalence class of x is a unit term, apply Update-Unit (resp. Nth-Unit).
8. (Array reasoning on atomic sequences) Apply Nth-Intro and Update-Bound to update terms. For each update term, find the matching nth terms and apply Nth-Update. Apply Nth-Split to pairs of nth terms with equivalent indices.
9. (Theory combination) Apply S-A for all arithmetic terms occurring in both S and A.

Whenever a rule is applied, the strategy will restart from the beginning in the next iteration. The strategy is designed to apply with higher priority steps that are easy to compute and are likely to lead to conflicts. Some steps are ordered based on dependencies from other steps. For instance, Steps 5 and 7 use normal forms, which are computed in Step 4. The strategy for the BASE calculus is the same, except that Steps 7 and 8 are replaced by one that applies R-Update and R-Nth to all update and nth terms in S.

We point out that the C-Split rule may cause non-termination of the proof strategy described above in the presence of *cyclic* sequence constraints, for instance, constraints where sequence variables appear on both sides of an equality. The solver uses methods for detecting some of these cycles, to restrict when C-Split is applied. In particular, when $\mathsf{S} \models^*_{++} x \approx (\overline{u} ++ s ++ \overline{w})\downarrow$, $\mathsf{S} \models^*_{++} x \approx (\overline{u} ++ t ++ \overline{v})\downarrow$, and s occurs in $\overline{v}$, then C-Split is not applied. Instead, other heuristics are used, and in some cases the solver terminates with a response of "unknown" (see e.g., [17] for details). In addition to the version shown here, we also use another variation of the C-Split rule where the normal forms are matched in reverse (starting from the last terms in the concatenations). The implementation also uses fast entailment tests for length inequalities. These tests may allow us to conclude which branch of C-Split, if any, is feasible, without having to branch on cases explicitly.

Although not shown here, the calculus can also accommodate certain *extended* sequence constraints, that is, constraints using a signature with additional functions. For example, our implementation supports sequence containment, replacement, and reverse. It also supports an extended variant of the `update` operator, in which the third argument is a sequence that overrides the sequence being updated starting from the index given in the second argument. Constraints involving these functions are handled by reduction rules, similar to those shown in Fig. 5. The implementation is further optimized by using context-dependent simplifications, which may eagerly infer when certain sequence terms can be simplified to constants based on the current set of assertions [22].

6 Evaluation

We evaluate the performance of our approach, as implemented in cvc5. The evaluation investigates: (i) whether the use of sequences is a viable option for reasoning about vectors in programs, (ii) how our approach compares with other sequence solvers, and (iii) what is the performance impact of our array-style extended rules. As a baseline, we use Version 4.8.14 of the Z3 SMT solver, which supports a theory of sequences without `update`s. For cvc5, we evaluate implementations of both the basic calculus (denoted **cvc5**) and the extended array-based calculus (denoted **cvc5-a**). The benchmarks, solver configurations, and logs from our runs are available for download.[4] We ran all experiments on a cluster equipped with Intel Xeon E5-2620 v4 CPUs. We allocated one physical CPU core and 8 GB of RAM for each solver-benchmark pair and used a time limit of 300 s. We use the following two sets of benchmarks:

Array Benchmarks (ARRAYS). The first set of benchmarks is derived from the `QF_AX` benchmarks in SMT-LIB [3]. To generate these benchmarks, we (i) replace declarations of arrays with declarations of sequences of uninterpreted sorts, (ii) change the sort of index terms to integers, and (iii) replace `store` with `update` and `select` with `nth`. The resulting benchmarks are quantifier-free and do not contain concatenations. Note that the original and the derived benchmarks are not equisatisfiable, because sequences take into account out-of-bounds cases that do not occur in arrays. For the Z3 runs, we add to the benchmarks a definition of `update` in terms of extraction and concatenation.

Smart Contract Verification (DIEM). The second set of benchmarks consists of verification conditions generated by running the Move Prover [24] on smart contracts written for the Diem framework. By default, the encoding does not use the sequence update

[4] http://dx.doi.org/10.5281/zenodo.6146565.

Set		w/ update		
		cvc5	cvc5-a	z3
ARRAYS (551)	Slvd	242	**390**	170
	Time	162	303	4329
DIEM (558)	Slvd	542	**547**	443
	Time	518	440	639

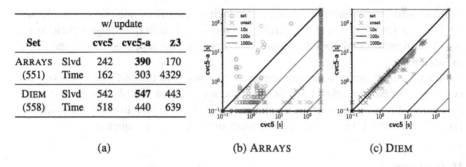

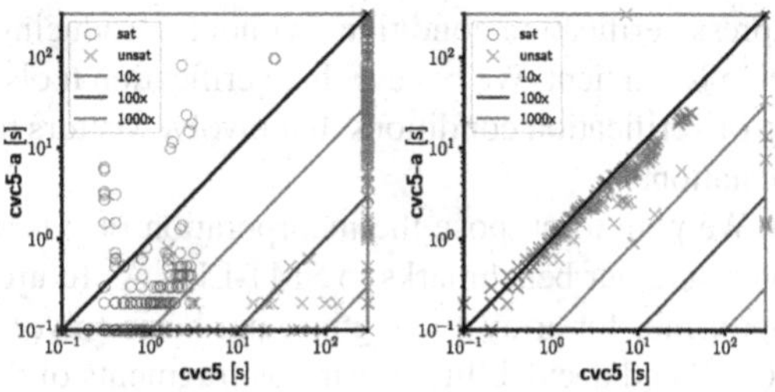

(a) (b) ARRAYS (c) DIEM

Fig. 7. Figure a lists the number of solved benchmarks and total time on commonly solved benchmarks. The scatter plots compare the base solver (**cvc5**) and the extended solver (**cvc5-a**) on ARRAY (Fig. b) and DIEM (Fig. c) benchmarks.

operation, and so Z3 can be used directly. However, we also modified the Move Prover encoding to generate benchmarks that do use the update operator, and ran cvc5 on them. In addition to using the sequence theory, the benchmarks make heavy use of quantifiers and the SMT-LIB theory of datatypes.

Figure 7a summarizes the results in terms of number of solved benchmarks and total time in seconds on commonly solved benchmarks. The configuration that solves the largest number of benchmarks is the implementation of the extended calculus (**cvc5-a**). This approach also successfully solves most of the DIEM benchmarks, which suggests that sequences are a promising option for encoding vectors in programs. The results further show that the sequences solver of cvc5 significantly outperforms Z3 on both the number of solved benchmarks and the solving time on commonly solved benchmarks.

Figures 7b and 7c show scatter plots comparing **cvc5** and **cvc5-a** on the two benchmark sets. We can see a clear trend towards better performance when using the extended solver. In particular, the table shows that in addition to solving the most benchmarks, **cvc5-a** is also fastest on the commonly solved instances from the DIEM benchmark set.

For the ARRAYS set, we can see that some benchmarks are slower with the extended solver. This is also reflected in the table, where **cvc5-a** is slower on the commonly solved instances. This is not too surprising, as the extra machinery of the extended solver can sometimes slow down easy problems. As problems get harder, however, the benefit of the extended solver becomes clear. For example, if we drop Z3 and consider just the commonly solved instances between **cvc5** and **cvc5-a** (of which there are 242), **cvc5-a** is about 2.47× faster (426 vs 1053 s). Of course, further improving the performance of **cvc5-a** is something we plan to explore in future work.

7 Conclusion

We introduced calculi for checking satisfiability in the theory of sequences, which can be used to model the vector data type. We described our implementation in cvc5 and provided an evaluation, showing that the proposed theory is rich enough to naturally express verification conditions without introducing quantifiers, and that our implementation is efficient. We believe that verification tools can benefit by changing their encoding of verification conditions that involve vectors to use the proposed theory and implementation.

We plan to propose the incorporation of this theory in the SMT-LIB standard and contribute our benchmarks to SMT-LIB. As future research, we plan to integrate other approaches for array solving into our basic solver. We also plan to study the politeness [16,20] and decidability of various fragments of the theory of sequences.

References

1. Alberti, F., Ghilardi, S., Pagani, E.: Cardinality constraints for arrays (decidability results and applications). Formal Methods Syst. Des. **51**(3), 545–574 (2017). https://doi.org/10.1007/s10703-017-0279-6
2. Barbosa, H., et al.: cvc5: a versatile and industrial-strength SMT solver. In: Fisman, D., Rosu, G. (eds.) TACAS 2022. LNCS, vol. 13243, pp. 415–442. Springer, Cham (2022). https://doi.org/10.1007/978-3-030-99524-9_24
3. Barrett, C., Fontaine, P., Tinelli, C.: The SMT-LIB Standard: Version 2.6. Technical report, Department of Computer Science, The University of Iowa (2017). www.SMT-LIB.org
4. Barrett, C., Nieuwenhuis, R., Oliveras, A., Tinelli, C.: Splitting on demand in SAT modulo theories. In: Hermann, M., Voronkov, A. (eds.) LPAR 2006. LNCS (LNAI), vol. 4246, pp. 512–526. Springer, Heidelberg (2006). https://doi.org/10.1007/11916277_35
5. Barrett, C., Tinelli, C.: Satisfiability modulo theories. In: Clarke, E., Henzinger, T., Veith, H., Bloem, R. (eds.) Handbook of Model Checking, pp. 305–343. Springer, Cham (2018). https://doi.org/10.1007/978-3-319-10575-8_11
6. Berzish, M., Ganesh, V., Zheng, Y.: Z3str3: a string solver with theory-aware heuristics. In: Stewart, D., Weissenbacher, G. (eds.) 2017 Formal Methods in Computer Aided Design, FMCAD 2017, Vienna, Austria, 2–6 October 2017, pp. 55–59. IEEE (2017)
7. Bjørner, N., de Moura, L., Nachmanson, L., Wintersteiger, C.: Programming Z3 (2018). https://theory.stanford.edu/~nikolaj/programmingz3.html#sec-sequences-and-strings
8. Bjørner, N., Ganesh, V., Michel, R., Veanes, M.: An SMT-LIB format for sequences and regular expressions. SMT **12**, 76–86 (2012)
9. Bjørner, N., Tillmann, N., Voronkov, A.: Path feasibility analysis for string-manipulating programs. In: Kowalewski, S., Philippou, A. (eds.) TACAS 2009. LNCS, vol. 5505, pp. 307–321. Springer, Heidelberg (2009). https://doi.org/10.1007/978-3-642-00768-2_27
10. Christ, J., Hoenicke, J.: Weakly equivalent arrays. In: Lutz, C., Ranise, S. (eds.) FroCoS 2015. LNCS (LNAI), vol. 9322, pp. 119–134. Springer, Cham (2015). https://doi.org/10.1007/978-3-319-24246-0_8
11. de Moura, L., Bjørner, N.: Z3: an efficient SMT solver. In: Ramakrishnan, C.R., Rehof, J. (eds.) TACAS 2008. LNCS, vol. 4963, pp. 337–340. Springer, Heidelberg (2008). https://doi.org/10.1007/978-3-540-78800-3_24
12. Elad, N., Rain, S., Immerman, N., Kovács, L., Sagiv, M.: Summing up smart transitions. In: Silva, A., Leino, K.R.M. (eds.) CAV 2021. LNCS, vol. 12759, pp. 317–340. Springer, Cham (2021). https://doi.org/10.1007/978-3-030-81685-8_15
13. Enderton, H.B.: A Mathematical Introduction to Logic, 2nd edn. Academic Press (2001)
14. Falke, S., Merz, F., Sinz, C.: Extending the theory of arrays: `memset`, `memcpy`, and beyond. In: Cohen, E., Rybalchenko, A. (eds.) VSTTE 2013. LNCS, vol. 8164, pp. 108–128. Springer, Heidelberg (2014). https://doi.org/10.1007/978-3-642-54108-7_6

15. Ganesh, V., Minnes, M., Solar-Lezama, A., Rinard, M.: Word equations with length constraints: what's decidable? In: Biere, A., Nahir, A., Vos, T. (eds.) HVC 2012. LNCS, vol. 7857, pp. 209–226. Springer, Heidelberg (2013). https://doi.org/10.1007/978-3-642-39611-3_21
16. Jovanović, D., Barrett, C.: Polite theories revisited. In: Fermüller, C.G., Voronkov, A. (eds.) LPAR 2010. LNCS, vol. 6397, pp. 402–416. Springer, Heidelberg (2010). https://doi.org/10.1007/978-3-642-16242-8_29
17. Liang, T., Reynolds, A., Tinelli, C., Barrett, C., Deters, M.: A DPLL(T) theory solver for a theory of strings and regular expressions. In: Biere, A., Bloem, R. (eds.) CAV 2014. LNCS, vol. 8559, pp. 646–662. Springer, Cham (2014). https://doi.org/10.1007/978-3-319-08867-9_43
18. Nelson, G., Oppen, D.C.: Simplification by cooperating decision procedures. ACM Trans. Program. Lang. Syst. **1**(2), 245–257 (1979)
19. Nieuwenhuis, R., Oliveras, A., Tinelli, C.: Solving SAT and SAT modulo theories: from an abstract Davis-Putnam-Logemann-Loveland procedure to DPLL(T). J. ACM **53**(6), 937–977 (2006)
20. Ranise, S., Ringeissen, C., Zarba, C.G.: Combining data structures with nonstably infinite theories using many-sorted logic. In: Gramlich, B. (ed.) FroCoS 2005. LNCS (LNAI), vol. 3717, pp. 48–64. Springer, Heidelberg (2005). https://doi.org/10.1007/11559306_3
21. Reynolds, A., Nötzli, A., Barrett, C.W., Tinelli, C.: Reductions for strings and regular expressions revisited. In: 2020 Formal Methods in Computer Aided Design, FMCAD 2020, Haifa, Israel, 21–24 September 2020, pp. 225–235. IEEE (2020)
22. Reynolds, A., Woo, M., Barrett, C., Brumley, D., Liang, T., Tinelli, C.: Scaling up DPLL(T) string solvers using context-dependent simplification. In: Majumdar, R., Kunčak, V. (eds.) CAV 2017. LNCS, vol. 10427, pp. 453–474. Springer, Cham (2017). https://doi.org/10.1007/978-3-319-63390-9_24
23. Sheng, Y.,et al.: Reasoning about vectors using an SMT theory of sequences. CoRR 10.48550/ARXIV.2205.08095 (2022)
24. Zhong, J.E., et al.: The move prover. In: Lahiri, S.K., Wang, C. (eds.) CAV 2020. LNCS, vol. 12224, pp. 137–150. Springer, Cham (2020). https://doi.org/10.1007/978-3-030-53288-8_7

Calculi and Orderings

An Efficient Subsumption Test Pipeline for BS(LRA) Clauses

Martin Bromberger[1], Lorenz Leutgeb[1,2(✉)], and Christoph Weidenbach[1]

[1] Max Planck Institute for Informatics, Saarland Informatics Campus, Saarbrücken, Germany
{mbromber,lorenz,weidenb}@mpi-inf.mpg.de

[2] Graduate School of Computer Science, Saarland Informatics Campus, Saarbrücken, Germany

Abstract. The importance of subsumption testing for redundancy elimination in first-order logic automatic reasoning is well-known. Although the problem is already NP-complete for first-order clauses, the meanwhile developed test pipelines efficiently decide subsumption in almost all practical cases. We consider subsumption between first-oder clauses of the Bernays-Schönfinkel fragment over linear real arithmetic constraints: BS(LRA). The bottleneck in this setup is deciding implication between the LRA constraints of two clauses. Our new *sample point heuristic* preempts expensive implication decisions in about 94% of all cases in benchmarks. Combined with filtering techniques for the first-order BS part of clauses, it results again in an efficient subsumption test pipeline for BS(LRA) clauses.

Keywords: Bernays-Schönfinkel fragment · Linear arithmetic · Redundancy elimination · Subsumption

1 Introduction

The elimination of redundant clauses is crucial for the efficient automatic reasoning in first-order logic. In a resolution [5,50] or superposition setting [4,44], a newly inferred clause might be subsumed by a clause that is already known (*forward subsumption*) or it might subsume a known clause (*backward subsumption*). Although the SCL calculi family [1,11,21] does not require forward subsumption tests, a property also inherent to the propositional CDCL (Conflict Driven Clause Learning) approach [8,34,41,55,63], backward subsumption and hence subsumption remains an important test in order to remove redundant clauses.

In this work we present advances in deciding subsumption for constrained clauses, specifically employing the Bernays-Schönfinkel fragment as foreground logic, and linear real arithmetic as background theory, BS(LRA). BS(LRA) is of particular interest because it can be used to model *supervisors*, i.e., components in technical systems that control system functionality. An example for a supervisor is the electronic control unit of a combustion engine. The logics we use

J. Blanchette et al. (Eds.): IJCAR 2022, LNAI 13385, pp. 147–168, 2022.
https://doi.org/10.1007/978-3-031-10769-6_10

to model supervisors and their properties are called *SupERLogs*—(Sup)ervisor (E)ffective(R)easoning (Log)ics. SupERLogs are instances of function-free first-order logic extended with arithmetic [18], which means BS(LRA) is an example of a SupERLog.

Subsumption is an important redundancy criterion in the context of hierarchic clausal reasoning [6,11,20,35,37]. At the heart of this paper is a new technique to speed up the treatment of linear arithmetic constraints as part of deciding subsumption. For every clause, we store a solution of its associated constraints, which is used to quickly falsify implication decisions, acting as a filter, called the *sample point heuristic.* In our experiments with various benchmarks, the technique is very effective: It successfully preempts expensive implication decisions in about 94% of cases. We elaborate on these findings in Sect. 4.

For example, consider three BS clauses, none of which subsumes another:

$$C_1 := P(a,x) \qquad C_2 := \neg P(y,z) \vee Q(y,z,b) \qquad C_3 := \neg R(b) \vee Q(a,x,b)$$

Let C_4 be the resolvent of C_1 and C_2 upon the atom $P(a,x)$, i.e., $C_4 := Q(a,z,b)$. Now C_4 backward-subsumes C_3 with matcher $\sigma := \{z \mapsto x\}$, i.e. $C_4\sigma \subset C_3$, thus C_3 is redundant and can be eliminated. Now, consider an extension of the above clauses with some simple LRA constraints following the same reasoning:

$$\begin{aligned} C_1' &:= x \geq 1 \parallel P(a,x) \\ C_2' &:= z \geq 0 \parallel \neg P(y,z) \vee Q(y,z,b) \\ C_3' &:= x \geq 0 \parallel \neg R(b) \vee Q(a,x,b) \end{aligned}$$

where $\parallel$ is interpreted as an implication, i.e., clause C_1' stands for $\neg x \geq 1 \vee P(a,x)$ or simply $x < 1 \vee P(a,x)$. The respective resolvent on the constrained clauses is $C_4' := z \geq 0, z \geq 1 \parallel Q(a,z,b)$ or after constraint simplification $C_4' := z \geq 1 \parallel Q(a,z,b)$ because $z \geq 1$ implies $z \geq 0$. For the constrained clauses, C_4' does no longer subsume C_3' with matcher $\sigma := \{z \mapsto x\}$, because $z \geq 0$ does not LRA-imply $z \geq 1$. Now, if we store the sample point $x = 0$ as a solution for the constraint of clause C_3', this sample point already reveals that $z \geq 0$ does not LRA-imply $z \geq 1$. This constitutes the basic idea behind our sample point heuristic. In general, constraints are not just simple bounds as in the above example, and sample points are solutions to the system of linear inequalities of the LRA constraint of a clause.

Please note that our test on LRA constraints is based on LRA theory implication and not on a syntactic notion such as subsumption on the first-order part of the clause. In this sense it is "stronger" than its first-order counterpart. This fact is stressed by the following example, taken from [26, Ex. 2], which shows that first-order implication does not imply subsumption. Let

$$\begin{aligned} C_1 &:= \neg P(x,y) \vee \neg P(y,z) \vee P(x,z) \\ C_2 &:= \neg P(a,b) \vee \neg P(b,c) \vee \neg P(c,d) \vee P(a,d) \end{aligned}$$

Then we have $C_1 \rightarrow C_2$, but again, for all σ we have $C_1\sigma \not\subseteq C_2$: Constructing σ from left to right we obtain $\sigma := \{x \mapsto a, y \mapsto b, z \mapsto c\}$, but $P(a,c) \notin C_2$.

Constructing σ from right to left we obtain $\sigma := \{z \mapsto d, x \mapsto a, y \mapsto c\}$, but $\neg P(a,c) \notin C_2$.

Related Work. Treatment of questions regarding the complexity of deciding subsumption of first-order clauses [27] dates back more than thirty years. Notions of subsumption, varying in generality, are studied in different sub-fields of theorem proving, whereas we restrict our attention to first-order theorem proving. Modern implementations typically decide multiple thousand instances of this problem per second: In [62, Sect. 2], Voronkov states that initial versions of Vampire "seemed to [...] deadlock" without efficient implementations to decide (forward) subsumption.

In order to reduce the number of clauses out of a set of clauses to be considered for pairwise subsumption checking, the best known practice in first-order theorem proving is to use (imperfect) indexing data structures as a means for pre-filtering and research concerning appropriate techniques is plentiful, see [24,25,27–30,33,39,40,43,45–49,52–54,56,59,61] for an evaluation of these techniques. Here we concentrate on the efficiency of a subsumption check between two clauses and therefore do not take indexing techniques into account. Furthermore, the implication test between two linear arithmetic constraints is of a semantic nature and is not related to any syntactic features of the involved constraints and can therefore hardly be filtered by a syntactic indexing approach.

In addition to pre-filtering via indexing, almost all above mentioned implementations of first-order subsumption tests rely on additional filters on the clause level. The idea is to generate an abstraction of clauses together with an ordering relation such that the ordering relation is necessary to hold between two clauses in order for one clause to subsume the other. Furthermore, the abstraction as well as the ordering relation should be efficiently computable. For example, a necessary condition for a first-order clause C_1 to subsume a first-order clause C_2 is $|\operatorname{vars}(C_1)| \geq |\operatorname{vars}(C_2)|$, i.e., the number of different variables in C_1 must be larger or equal than the number of variables in C_2. Further and additional abstractions included by various implementations rely on the size of clauses, number of ground literals, depth of literals and terms, occurring predicate and function symbols. For the BS(LRA) clauses considered here, the structure of the first-order BS part, which consists of predicates and flat terms (variables and constants) only, is not particularly rich.

The exploration of sample points has already been studied in the context of first-order clauses with arithmetic constraints. In [17,36] it was used to improve the performance of iSAT [23] on testing non-linear arithmetic constraints. In general, iSAT tests satisfiability by interval propagation for variables. If intervals get "too small" it typically gives up, however sometimes the explicit generation of a sample point for a small interval can still lead to a certificate for satisfiability. This technique was successfully applied in [17], but was not used for deciding subsumption of constrained clauses.

Motivation. The main motivation for this work is the realization that computing implication decisions required to treat constraints of the background theory presents the bottleneck of an BS(LRA) subsumption check in practice. Inspired by the success of filtering techniques in first-order logic, we devise an exceptionally effective filter for constraints and adopt well-known first-order filters to the BS fragment. Our sample point heuristic for LRA could easily be generalized to other arithmetic theories as well as full first-order logic.

Structure. The paper is structured as follows. After a section defining BS(LRA) and common notions and notation, Sect. 2, we define redundancy notions and our sample point heuristic in Sect. 3. Section 4 justifies the success of the sample point heuristic by numerous experiments in various application domains of BS(LRA). The paper ends with a discussion of the obtained results, Sect. 5. Binaries, utility scripts, benchmarking instances used as input, and the output used for evaluation may be obtained online [13].

2 Preliminaries

We briefly recall the basic logical formalisms and notations we build upon [10]. Our starting point is a standard many-sorted first-order language for BS with *constants* (denoted a, b, c), without non-constant function symbols, with *variables* (denoted w, x, y, z), and *predicates* (denoted P, Q, R) of some fixed *arity*. *Terms* (denoted t, s) are variables or constants. An *atom* (denoted A, B) is an expression $P(t_1, \dots, t_n)$ for a predicate P of arity n. A *positive literal* is an atom A and a *negative literal* is a negated atom $\neg A$. We define $\text{comp}(A) = \neg A$, $\text{comp}(\neg A) = A$, $|A| = A$ and $|\neg A| = A$. Literals are usually denoted L, K, H. Formulas are defined in the usual way using quantifiers $\forall$, $\exists$ and the boolean connectives $\neg$, $\lor$, $\land$, $\rightarrow$, and $\equiv$.

A *clause* (denoted C, D) is a universally closed disjunction of literals $A_1 \lor \dots \lor A_n \lor \neg B_1 \lor \dots \lor \neg B_m$. Clauses are identified with their respective multisets and all standard multiset operations are extended to clauses. For instance, $C \subseteq D$ means that all literals in C also appear in D respecting their number of occurrences. A clause is *Horn* if it contains at most one positive literal, i.e. $n \leqslant 1$, and a *unit clause* if it has exactly one literal, i.e. $n + m = 1$. We write C^+ for the set of positive literals, or *conclusions* of C, i.e. $C^+ := \{A_1, \dots, A_n\}$ and respectively C^- for the set of negative literals, or *premises* of C, i.e. $C^- := \{\neg B_1, \dots, \neg B_m\}$. If Y is a term, formula, or a set thereof, $\text{vars}(Y)$ denotes the set of all variables in Y, and Y is *ground* if $\text{vars}(Y) = \emptyset$.

The *Bernays-Schönfinkel Clause Fragment* (BS) in first-order logic consists of first-order clauses where all involved terms are either variables or constants. The *Horn Bernays-Schönfinkel Clause Fragment* (HBS) consists of all sets of BS Horn clauses.

A *substitution* σ is a function from variables to terms with a finite domain $\text{dom}(\sigma) = \{x \mid x\sigma \neq x\}$ and codomain $\text{codom}(\sigma) = \{x\sigma \mid x \in \text{dom}(\sigma)\}$. We denote substitutions by σ, δ, ρ. The application of substitutions is often written postfix, as in $x\sigma$, and is homomorphically extended to terms, atoms, literals, clauses, and quantifier-free formulas. A substitution σ is *ground* if $\text{codom}(\sigma)$ is ground. Let Y denote some term, literal, clause, or clause set. A substitution σ is a *grounding* for Y if $Y\sigma$ is ground, and $Y\sigma$ is a *ground instance* of Y in this case. We denote by $\text{gnd}(Y)$ the set of all ground instances of Y, and by $\text{gnd}_B(Y)$ the set of all ground instances over a given set of constants B. The *most general unifier* $\text{mgu}(Z_1, Z_2)$ of two terms/atoms/literals Z_1 and Z_2 is defined as usual, and we assume that it does not introduce fresh variables and is idempotent.

We assume a standard many-sorted first-order logic model theory, and write $\mathcal{A} \models \phi$ if an interpretation $\mathcal{A}$ satisfies a first-order formula ϕ. A formula ψ is a logical consequence of ϕ, written $\phi \models \psi$, if $\mathcal{A} \models \psi$ for all $\mathcal{A}$ such that $\mathcal{A} \models \phi$. Sets of clauses are semantically treated as conjunctions of clauses with all variables quantified universally.

2.1 Bernays-Schönfinkel with Linear Real Arithmetic

The extension of BS with linear real arithmetic, BS(LRA), is the basis for the formalisms studied in this paper. We consider a standard *many-sorted* first-order logic with one first-order sort $\mathcal{F}$ and with the sort $\mathcal{R}$ for the real numbers. Given a clause set N, the interpretations $\mathcal{A}$ of our sorts are fixed: $\mathcal{R}^{\mathcal{A}} = \mathbb{R}$ and $\mathcal{F}^{\mathcal{A}} = \mathbb{F}$. This means that $\mathcal{F}^{\mathcal{A}}$ is a Herbrand interpretation, i.e., $\mathbb{F}$ is the set of first-order constants in N, or a single constant out of the signature if no such constant occurs. Note that this is not a deviation from standard semantics in our context as for the arithmetic part the canonical domain is considered and the first-order sort has the finite model property over the occurring constants (note that equality is not part of BS).

Constant symbols, arithmetic function symbols, variables, and predicates are uniquely declared together with their respective sort. The unique sort of a constant symbol, variable, predicate, or term is denoted by the function $\text{sort}(Y)$ and we assume all terms, atoms, and formulas to be well-sorted. We assume *pure* input clause sets, which means the only constants of sort $\mathcal{R}$ are (rational) numbers. This means the only constants that we do allow are rational numbers $c \in \mathbb{Q}$ and the constants defining our finite first-order sort $\mathcal{F}$. Irrational numbers are not allowed by the standard definition of the theory. The current implementation comes with the caveat that only integer constants can be parsed. Satisfiability of pure BS(LRA) clause sets is semi-decidable, e.g., using *hierarchic superposition* [6] or *SCL(T)* [11]. Impure BS(LRA) is no longer compact and satisfiability becomes undecidable, but its restriction to ground clause sets is decidable [22].

All arithmetic predicates and functions are interpreted in the usual way. An interpretation of BS(LRA) coincides with $\mathcal{A}^{\text{LRA}}$ on arithmetic predicates and functions, and freely interprets free predicates. For pure clause sets this is well-defined [6]. Logical satisfaction and entailment is defined as usual, and uses similar notation as for BS.

Example 1. The clause $y < 5 \lor x' \neq x + 1 \lor \neg S_0(x, y) \lor S_1(x', 0)$ is part of a timed automaton with two clocks x and y modeled in BS(LRA). It represents a transition from state S_0 to state S_1 that can be traversed only if clock y is at least 5 and that resets y to 0 and increases x by 1.

Arithmetic terms are constructed from a set $\mathcal{X}$ of *variables*, the set of integer constants $c \in \mathbb{Z}$, and binary function symbols $+$ and $-$ (written infix). Additionally, we allow multiplication $\cdot$ if one of the factors is an integer constant. Multiplication only serves us as syntactic sugar to abbreviate other arithmetic terms, e.g., $x + x + x$ is abbreviated to $3 \cdot x$. Atoms in BS(LRA) are either *first-order atoms* (e.g., $P(13, x)$) or *(linear) arithmetic atoms* (e.g., $x < 42$). Arithmetic atoms are denoted by λ and may use the predicates $\leq, <, \neq, =, >, \geq$, which are written infix and have the expected fixed interpretation. We use $\triangleleft$ as a placeholder for any of these predicates. Predicates used in first-order atoms are called *free*. *First-order literals* and related notation is defined as before. *Arithmetic literals* coincide with arithmetic atoms, since the arithmetic predicates are closed under negation, e.g., $\neg(x \geq 42) \equiv x < 42$.

BS(LRA) clauses are defined as for BS but using BS(LRA) atoms. We often write clauses in the form $\Lambda \parallel C$ where C is a clause solely built of free first-order literals and Λ is a multiset of LRA atoms called the *constraint* of the clause. A clause of the form $\Lambda \parallel C$ is therefore also called a *constrained clause*. The semantics of $\Lambda \parallel C$ is as follows:

$$\Lambda \parallel C \quad \text{iff} \quad \Big(\bigwedge_{\lambda \in \Lambda} \lambda\Big) \rightarrow C \quad \text{iff} \quad \Big(\bigvee_{\lambda \in \Lambda} \neg\lambda\Big) \vee C$$

For example, the clause $x > 1 \vee y \neq 5 \vee \neg Q(x) \vee R(x, y)$ is also written $x \leq 1, y = 5 \| \neg Q(x) \vee R(x, y)$. The negation $\neg(\Lambda \parallel C)$ of a constrained clause $\Lambda \parallel C$ where $C = A_1 \vee \dots \vee A_n \vee \neg B_1 \vee \dots \vee \neg B_m$ is thus equivalent to $(\bigwedge_{\lambda \in \Lambda} \lambda) \wedge \neg A_1 \wedge \dots \wedge \neg A_n \wedge B_1 \wedge \dots \wedge B_m$. Note that since the neutral element of conjunction is $\top$, an empty constraint is thus valid, i.e. equivalent to true.

An *assignment* for a constraint Λ is a substitution (denoted β) that maps all variables in $\text{vars}(\Lambda)$ to real numbers $c \in \mathbb{R}$. An assignment is a *solution* for a constraint Λ if all atoms $\lambda \in (\Lambda\beta)$ evaluate to true. A constraint Λ is *satisfiable* if there exists a solution for Λ. Otherwise it is *unsatisfiable*. Note that assignments can be extended to C by also mapping variables of the first-order sort accordingly.

A clause or clause set is *abstracted* if its first-order literals contain only variables or first-order constants. Every clause C is equivalent to an abstracted clause that is obtained by replacing each non-variable arithmetic term t that occurs in a first-order atom by a fresh variable x while adding an arithmetic atom $x \neq t$ to C. We assume abstracted clauses for theory development, but we prefer non-abstracted clauses in examples for readability, e.g., a unit clause $P(3, 5)$ is considered in the development of the theory as the clause $x = 3, y = 5 \parallel P(x, y)$. In the implementation, we mostly prefer abstracted clauses except that we allow integer constants $c \in \mathbb{Z}$ to appear as arguments of first-order literals. In some cases, this makes it easier to recognize whether two clauses can be matched or not. For instance, we see by syntactic comparison that the two unit clauses $P(3, 5)$ and $P(0, 1)$ have no substitution σ such that $P(3, 5) = P(0, 1)\sigma$. For the abstracted versions on the other hand, $x = 3, y = 5 \parallel P(x,y)$ and $u = 0, v = 1 \parallel P(u,v)$ we can find a matching substitution for the first-order part $\sigma := \{u \mapsto x, v \mapsto y\}$ and would have to check the constraints semantically to exclude the matching.

Hierarchic Resolution. One inference rule, foundational to most algorithms for solving constrained first-order clauses, is *hierarchic resolution* [6]:

$$\frac{\Lambda_1 \parallel L_1 \vee C_1 \quad \Lambda_2 \parallel L_2 \vee C_2 \quad \sigma = \text{mgu}(L_1, \text{comp}(L_2))}{(\Lambda_1, \Lambda_2 \parallel C_1 \vee C_2)\sigma}$$

The conclusion is called *hierarchic resolvent* (of the two clauses in the premise). A *refutation* is the sequence of resolution steps that produces a clause $\Lambda \parallel \bot$ with $\mathcal{A}^{\text{LRA}} \models \Lambda\delta$ for some grounding δ. Hierarchic resolution is sound and refutationally complete for the BS(LRA) clauses considered here, since every set N of BS(LRA) clauses is *sufficiently complete* [6], because all constatnts of the arithemtic sort are numbers. Hence *hierarchic resolution* is sound and refutationally complete for N [6,7]. *Hierarchic unit resolution* is a special case of hierarchic resolution, that only combines two clauses in case one of them is a unit clause. Hierarchic unit resolution is sound and complete for HBS(LRA) [6,7], but not even refutationally complete for BS(LRA).

Most algorithms for Bernays-Schnönfinkel, first-order logic, and beyond utilize resolution. The SCL(T) calculus for HBS(LRA) uses hierarchic resolution in order to learn from the conflicts it encounters during its search. The hierarchic superposition calculus on the other hand derives new clauses via hierarchic resolution based on an ordering. The goal is to either derive the empty clause or a saturation of the clause set, i.e., a state from which no new clauses can be derived. Each of those algorithms must derive new clauses in order to progress, but their subroutines also get progressively slower as more clauses are derived. In order to increase efficiency, it is necessary to eliminate clauses that are obsolete. One measure that determines whether a clause is useful or not is *redundancy*.

Redundancy. In order to define redundancy for constrained clauses, we need an *$\mathcal{H}$-order*, i.e., a well-founded, total, strict ordering $\prec$ on ground literals such that literals in the constraints (in our case arithmetic literals) are always smaller than first-order literals. Such an ordering can be lifted to constrained clauses and sets thereof by its respective multiset extension. Hence, we overload any such order $\prec$ for literals, constrained clauses, and sets of constrained clause if the meaning is clear from the context. We define $\preceq$ as the reflexive closure of $\prec$ and $N^{\preceq \Lambda \parallel C} := \{D \mid D \in N \text{ and } D \preceq \Lambda \parallel C\}$. An instance of an LPO [15] with appropriate precedence can serve as an $\mathcal{H}$-order.

Definition 2 (Clause Redundancy). *A ground clause $\Lambda \parallel C$ is* redundant *with respect to a set N of ground clauses and an $\mathcal{H}$-order $\prec$ if $N^{\preceq \Lambda \parallel C} \models \Lambda \parallel C$. A clause $\Lambda \parallel C$ is* redundant *with respect to a clause set N and an $\mathcal{H}$-order $\prec$ if for all $\Lambda' \parallel C' \in \text{gnd}(\Lambda \parallel C)$ the clause $\Lambda' \parallel C'$ is redundant with respect to* $\text{gnd}(N)$.

If a clause $\Lambda \parallel C$ is redundant with respect to a clause set N, then it can be removed from N without changing its semantics. Determining clause redundancy is an undecidable problem [11,63]. However, there are special cases of redundant clauses that can be easily checked, e.g., tautologies and subsumed clauses. Techniques for tautology deletion and subsumption deletion are the most common elimination techniques in modern first-order provers.

A *tautology* is a clause that evaluates to true independent of the predicate interpretation or assignment. It is therefore redundant with respect to all orders and clause sets; even the empty set.

Corollary 3 (Tautology for Constrained Clauses). *A clause $\Lambda \parallel C$ is a tautology if the existential closure of $\neg(\Lambda \parallel C)$ is unsatisfiable.*

Since $\neg(\Lambda \parallel C)$ is essentially ground (by existential closure and skolemization), it can be solved with an appropriate SMT solver, i.e., an SMT solver that supports unquantified uninterpreted functions coupled with linear real arithmetic. In [2], it is recommended to check only the following conditions for tautology deletion in hierarchic superposition:

Corollary 4 (Tautology Check). *A clause $\Lambda \parallel C$ is a tautology if the existential closure of Λ is unsatisfiable or if C contains two literals L_1 and L_2 with $L_1 = \text{comp}(L_2)$.*

The advantage is that the check on the first-order side of the clause is still purely syntactic and corresponds to the tautology check for pure first-order logic. Nonetheless, there are tautologies that are not captured by Corollary 4, e.g., $x = y \parallel P(x) \vee \neg P(y)$. The SCL(T) calculus on the other hand requires no tautology checks because it never learns tautologies as part of its conflict analysis [1,11,21]. This property is also inherent to the propositional CDCL (Conflict Driven Clause Learning) approach [8,34,41,55,63].

3 Subsumption for Constrained Clauses

A *subsumed* constrained clause is a clause that is redundant with respect to a single clause in our clause set. Formally, subsumption is defined as follows.

Definition 5. (Subsumption for Constrained Clauses [2]). *A constrained clause $\Lambda_1 \parallel C_1$* subsumes *another constrained clause $\Lambda_2 \parallel C_2$ if there exists a substitution σ such that $C_1\sigma \subseteq C_2$, $\text{vars}(\Lambda_1\sigma) \subseteq \text{vars}(\Lambda_2)$, and the universal closure of $\Lambda_2 \rightarrow (\Lambda_1\sigma)$ holds in LRA.*

Eliminating redundant clauses is crucial for the efficient operation of an automatic first-order theorem prover. Although subsumption is considered one of the easier redundancy relationships that we can check in practice, it is still a hard problem in general:

Lemma 6. (Complexity of Subsumption in the BS Fragment). *Deciding subsumption for a pair of BS clauses is NP-complete.*

Proof. Containment in NP follows from the fact that the size of subsumption matchers is limited by the subsumed clause and set inclusion of literals can be decided in polynomial time. For the hardness part, consider the following polynomial-time reduction from 3-SAT. Take a propositional clause set where all clauses have length three. Now introduce a 6-place predicate R and encode each propositional variable P by a first-order variable x_P. Then a propositional clause $L_1 \vee L_2 \vee L_3$ can be encoded by an atom $R(x_{P_1}, p_1, x_{P_2}, p_2, x_{P_3}, p_3)$ where p_i is 0 if L_i is negative and 1 otherwise and P_i is the predicate of L_i. This way the clause set N can be represented by a single BS clause C_N. Now construct a clause D that contains all atoms representing the way a clause of length three can become true by ground atoms over R and constants 0, 1. For example, it contains atoms like $R(0, 0, \ldots)$ and $R(1, 1, \ldots)$ representing that the first literal of a clause is true. Actually, for each such atom $R(0, 0, \ldots)$ the clause D contains $|C_N|$ copies. Finally, C_N subsumes D if and only if N is satisfiable. □

In order to be efficient, modern theorem provers need to decide multiple thousand subsumption checks per second. In the pure first-order case, this is possible because of indexing and filtering techniques that quickly decide most subsumption checks [24,25,27–30,33,39,40,45–49,52–54,56,59,61,62].

For BS(LRA) (and FOL(LRA)), there also exists research on how to perform the subsumption check in general [2,36], but the literature contains no dedicated indexing or filtering techniques for the constraint part of the subsumption check. In this section and as the main contribution of this paper, we present the first such filtering techniques for BS(LRA). But first, we explain how to solve the subsumption check for constrained clauses in general.

First-Order Check. The first step of the subsumption check is exactly the same as in first-order logic without arithmetic. We have to find a substitution σ, also called a *matcher*, such that $C_1\sigma \subseteq C_2$. The only difference is that it is not enough to compute one matcher σ, but we have to compute all matchers for $C_1\sigma \subseteq C_2$ until we find one that satisfies the implication $\Lambda_2 \rightarrow (\Lambda_1\sigma)$. For instance, there are two matchers for the clauses $C_1 := x + y \geq 0 \,\|\, Q(x, y)$ and $C_2 := x < 0, y \geq 0 \,\|\, Q(x, x) \vee Q(y, y)$. The matcher $\{x \mapsto y\}$ satisfies the implication $\Lambda_2 \rightarrow (\Lambda_1\sigma)$ and $\{y \mapsto x\}$ does not. Our own algorithm for finding matchers is in the style of Stillman except that we continue after we find the first matcher [27,58].

Implication Check. The universal closure of the implication $\Lambda_2 \rightarrow (\Lambda_1\sigma)$ can be solved by any SMT solver for the respective theory after we negate it. Note that the resulting formula

$$\exists x_1, \ldots, x_n.\ \Lambda_2 \wedge \neg(\Lambda_1\sigma) \qquad \text{where } \{x_1, \ldots, x_n\} = \text{vars}(\Lambda_2) \tag{1}$$

is already in clause normal form and that the formula can be treated as ground since existential variables can be handled as constants. Intuitively, the universal closure $\Lambda_2 \rightarrow (\Lambda_1\sigma)$ asserts that the set of solutions satisfying Λ_2 is a subset of

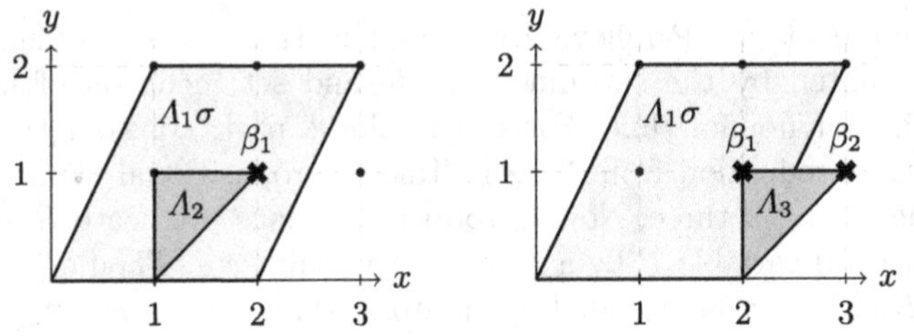

Fig. 1. Solutions of the constraints $\Lambda_1\sigma$, Λ_2, and Λ_3 depicted as polytopes

the set of solutions satisfying $\Lambda_1\sigma$. This means a solution to its negation (1) is a solution for Λ_2, but not for $\Lambda_1\sigma$, thus a counterexample of the subset relation.

Example 7. Let us now look at an example to illustrate the role that formula (1) plays in deciding subsumption. In our example, we have three clauses: $\Lambda_1 \parallel C_1$, $\Lambda_2 \parallel C_2$, and $\Lambda_3 \parallel C_2$, where $C_1 := \neg P(x,y) \vee Q(u,z)$, $C_2 := \neg P(x,y) \vee Q(2,x)$, $\Lambda_1 := y \geq 0\ ,\ y \leq u\ ,\ y \leq x{+}z\ ,\ y \geq x{+}z{-}2{\cdot}u$, $\Lambda_2 := x \geq 1\ ,\ y \leq 1\ ,\ y \geq x{-}1$, and $\Lambda_3 := x \geq 2\ ,\ y \leq 1\ ,\ y \geq x - 2$. Our goal is to test whether $\Lambda_1 \parallel C_1$ subsumes the other two clauses. As our first step, we try to find a substitution σ such that $C_1\sigma \subseteq C_2$. The most general substitution fulfilling this condition is $\sigma := \{z \mapsto x, u \mapsto 2\}$. Next, we check whether $\Lambda_1\sigma$ is implied by Λ_2 and Λ_3. Normally, we would do so by solving the formula (1) with an SMT solver, but to help our intuitive understanding, we instead look at their solution sets depicted in Fig. 1. Note that $\Lambda_1\sigma$ simplifies to $\Lambda_1\sigma := y \geq 0\ ,\ y \leq 2\ ,\ y \leq 2{\cdot}x\ ,\ y \geq 2{\cdot}x{-}4$. Here we see that the solution set for Λ_2 is a subset of $\Lambda_1\sigma$. Hence, Λ_2 implies $\Lambda_1\sigma$, which means that $\Lambda_2 \parallel C_2$ is subsumed by $\Lambda_1 \parallel C_1$. The solution set for Λ_3 is not a subset of $\Lambda_1\sigma$. For instance, the assignment $\beta_2 := \{x \mapsto 3, y \mapsto 1\}$ is a counterexample and therefore a solution to the respective instance of formula (1). Hence, $\Lambda_1 \parallel C_1$ does not subsume $\Lambda_3 \parallel C_2$.

Excess Variables. Note that in general it is not sufficient to find a substitution σ that matches the first-order parts to also match the theory constraints: $C_1\sigma \subseteq C_2$ does not generally imply $\text{vars}(\Lambda_1\sigma) \subseteq \text{vars}(\Lambda_2)$. In particular, if Λ_1 contains variables that do not appear in the first-order part C_1, then these must be projected to Λ_2. We arrive at a variant of (1), that is $\exists x_1,\ldots,x_n \forall y_1,\ldots,y_m.\ \Lambda_2 \wedge \neg(\Lambda_1\sigma)$ where $\{x_1,\ldots,x_n\} = \text{vars}(\Lambda_2)$ and $\{y_1,\ldots,y_m\} = \text{vars}(\Lambda_1) \setminus \text{vars}(C_1)$. Our solution to this problem is to normalize all clauses $\Lambda \parallel C$ by eliminating all *excess variables* $\mathcal{Y} := \text{vars}(\Lambda) \setminus \text{vars}(C)$ such that $\text{vars}(\Lambda) \subseteq \text{vars}(C)$ is guaranteed. For linear real arithmetic this is possible with quantifier elimintation techniques, e.g., Fourier-Motzkin elimination (FME). Although these techniques typically cause the size of Λ to increase exponentially, they often behave well in practice. In fact, we get rid of almost all excess variables in our benchmark examples with simplification techniques based on Gaussian elimination with execution time linear in the number of LRA atoms. Given the precondition $\mathcal{Y} = \emptyset$ achieved by such elimination techniques, we can compute σ as matcher for the first-order parts and then directly use it for testing whether the universal closure of $\Lambda_2 \rightarrow (\Lambda_1\sigma)$ holds. An alternative solution to the issue of excess variables has been proposed: In [2], the substitution σ is decomposed as $\sigma = \delta\tau$, where δ is the first-order matcher and τ is a *theory matcher*, i.e. $\mathrm{dom}(\tau) \subseteq \mathcal{Y}$ and $\mathrm{vars}(\mathrm{codom}(\tau)) \subseteq \mathrm{vars}(\Lambda_2)$. Then, exploiting Farkas' lemma, the computation of τ is reduced to testing the feasibility of a linear program (restricted to matchers that are affine transformations).

The reduction to solving a linear program offers polynomial worst-case complexity but in practice typically behaves worse than solving the variant with quantifier alternations using an SMT solver such as Z3 [36,42].

Filtering First-Order Literals. Even though deciding implication of theory constraints is in practice more expensive than constructing a matcher and deciding inclusion of first-order literals, we still incorporate some lightweight filters for our evaluation. Inspired by Schulz [54] we choose three features, so that every feature f maps clauses to $\mathbb{N}_0$, and $f(C_1) \leqslant f(C_2)$ is necessary for $C_1\sigma \subseteq C_2$.

The features are: $|C^+|$, the number of positive first-order literals in C, $|C^-|$, the number of negative first-order literals in C, and $\lfloor C \rfloor$, the number of occurrences of constants in C.

Sample Point Heuristic. The majority of subsumption tests fail because we cannot find a fitting substitution for their first-order parts. In our experiments, between 66.5% and 99.9% of subsumption tests failed this way. This means our tool only has to check in less than 33.5% of the cases whether one theory constraint implies the other. Despite this, our tool spends more time on implication checks than on the first-order part of the subsumption tests without filtering on the constraint implication tests. The reason is that constraint implication tests are typically much more expensive than the first-order part of a subsumption test. For this reason, we developed the *sample point heuristic* that is much faster to execute than a full constraint implication test, but still filters out the majority of implications that do not hold (in our experiments between 93.8% and 100%).

The idea behind the sample point heuristic is straightforward. We store for each clause $\Lambda \,\|\, C$ a sample solution β for its theory constraint Λ. Before we execute a full constraint implication test, we simply evaluate whether the sample solution β for Λ_2 is also a solution for $\Lambda_1\sigma$. If this is not the case, then β is a solution for (1) and a counterexample for the implication. If β is a solution for $\Lambda_1\sigma$, then the heuristic returns unknown and we have to execute a full constraint implication test, i.e., solve the SMT problem (1).

Often it is possible to get our sample solutions for free. Theorem provers based on hierarchic superposition typically check for every new clause $\Lambda \,\|\, C$ whether Λ is satisfiable in order to eliminate tautologies. This means we can already use this tautology check to compute and store a sample solution for every new clause without extra cost. We only need to pick a solver for the check that returns a solution as a certificate of satisfiability. Although the SCL(T) calculus never learns any tautologies, it is also possible to get a sample solution for free as part of its conflict analysis [11].

Example 8. We revisit Example 7 to illustrate the sample point heuristic. During the tautology check for $\Lambda_2 \,\|\, C_2$ and $\Lambda_3 \,\|\, C_2$, we determined that $\beta_1 := \{x \mapsto 2, y \mapsto 1\}$ is a sample solution for Λ_2 and $\beta_2 := \{x \mapsto 3, y \mapsto 1\}$ a sample solution for Λ_3. Since Λ_2 implies $\Lambda_1\sigma$, all sample solutions for Λ_2 automatically satisfy $\Lambda_1\sigma$. This is the reason why the sample point heuristic never filters out an implication that actually holds, i.e., it returns unknown when we test whether Λ_2 implies $\Lambda_1\sigma$. The assignment β_2 on the other hand does not satisfy $\Lambda_1\sigma$. Hence, the sample point heuristic correctly claims that Λ_3 does not imply $\Lambda_1\sigma$. Note that we could also have chosen β_1 as the sample point for Λ_3. In this case, the sample point heuristic would also return unknown for the implication $\Lambda_3 \rightarrow \Lambda_1\sigma$ although the implication does not hold.

Trivial Cases. Subsumption tests become much easier if the constraint Λ_i of one of the participating clauses is empty. We use two heuristic filters to exploit this fact. We highlight them here because they already exclude some subsumption tests before we reach the sample point heuristic in our implementation.

The *empty conclusion heuristic* exploits that Λ_1 is valid if Λ_1 is empty. In this case, all implications $\Lambda_2 \rightarrow (\Lambda_1\sigma)$ hold because $\Lambda_1\sigma$ evaluates to true under any assignment. So by checking whether $\Lambda_1 = \emptyset$, we can quickly determine whether $\Lambda_2 \rightarrow (\Lambda_1\sigma)$ holds for some pairs of clauses. Note that in contrast to the sample point heuristic, this heuristic is used to find valid implications.

The *empty premise test* exploits that Λ_2 is valid if Λ_2 is empty. In this case, an implication $\Lambda_2 \rightarrow (\Lambda_1\sigma)$ may only hold if $\Lambda_1\sigma$ simplifies to the empty set as well. This is the case because any inequality in the canonical form $\sum_{i=1}^{n} a_i x_i \triangleleft c$ either simplifies to true (because $a_i = 0$ for all $i = 1, \ldots, n$ and $0 \triangleleft c$ holds) and can be removed from $\Lambda_1\sigma$, or the inequality eliminates at least one assignment as a solution for $\Lambda_1\sigma$ [51]. So if $\Lambda_2 = \emptyset$, we check whether $\Lambda_1\sigma$ simplifies to the empty set instead of solving the SMT problem (1).

Pipeline. We call our approach a *pipeline* since it combines multiple procedures, which we call *stages*, that vary in complexity and are independent in principle, for the overall aim of efficiently testing subsumption. Pairs of clauses that "make it through" all stages, are those for which the subsumption relation holds. The pipeline is designed with two goals in mind: (1) To reject as many pairs of clauses as early as possible, and (2) to move stages further towards the end of the pipeline the more expensive they are.

The pipeline consists of six stages, all of which are mentioned above. We divide the pipeline into two phases, the *first-order phase* (FO-phase) consisting of two stages, and the *constraint phase* (C-phase), consisting of four stages. First-order filtering rejects all pairs of clauses for which $f(C_1) > f(C_2)$ holds. Then, matching constructs all matchers σ such that $C_1\sigma \subseteq C_2$. Every matcher is individually tested in the constraint phase. Technically, this means that the input of all following stages is not just a pair of clauses, but a triple of two clauses and a matcher. The constraint phase then proceeds with the empty conclusion heuristic and the empty premise test to accept (resp. reject) all trivial cases of

Algorithm 1: Saturation prover used for evaluation

```
Input  : A set N of clauses.
Output : ⊥ or "unknown".
U := {C ∈ N | |C| = 1}
while U ≠ ∅ do
    M := ∅
    foreach C ∈ U do M := M ∪ resolvents(C, N)
    if ⊥ ∈ M then return ⊥
    reduce M using N          (forward subsumption)
    if M = ∅ then return "unknown"
    reduce N using M          (backward subsumption)
    U := {C ∈ M | |C| = 1}
    N := N ∪ M
end
return "unknown"
```

the constraint implication test. The next stage is the sample point heuristic. If the sample solution β_2 for Λ_2 is no solution for Λ_1 (i.e. $\nvDash \Lambda_1\sigma\beta_2$), then the matcher σ is rejected. Otherwise (i.e. $\vDash \Lambda_1\sigma\beta_2$), the implication test $\Lambda_2 \rightarrow (\Lambda_1\sigma)$ is performed by solving the SMT problem (1) to produce the overall result of the pipeline and finally determine whether subsumption holds.

4 Experimentation

In order to evaluate our new approach on three benchmark instances, derived from BS(LRA) applications, all presented techniques and their combination in form of a pipeline were implemented in the theorem prover SPASS-SPL, a prototype for BS(LRA) reasoning.

Note that SPASS-SPL contains more than one approach for BS(LRA) reasoning, e.g., the Datalog hammer for HBS(LRA) reasoning [10]. These various modes of operation operate independently, and the desired mode is chosen via command-line option. The reasoning approach discussed here is the current default option. On the first-order side, SPASS-SPL consists of a simple saturation prover based on hierarchic unit resolution, see Algorithm 1. It resolves unit clauses with other clauses until either the empty clause is derived or no new clauses can be derived. Note that this procedure is only complete for Horn clauses. For arithmetic reasoning, SPASS-SPL relies on SPASS-SATT, our sound and complete CDCL(LA) solver for quantifier-free linear real and linear mixed/integer arithmetic [12]. SPASS-SATT implements a version of the dual simplex algorithm fine-tuned towards SMT solving [16]. In order to ensure soundness, SPASS-SATT represents all numbers with the help of the *arbitrary-precision arithmetic library* FLINT [31]. This means all calculations, including the implication test and the sample point heuristic, are always exact and thus free of numerical errors. The most relevant part of SPASS-SPL with regards to

Table 1. Overview of how many clause pairs advance in the pipeline (top to bottom)

		lc	bakery, tad	All
	All	1 244 819k	196 437k	1 441 256k
FO	Filtering	61.21%	85.03%	64.45%
	$f(C_1) \leq f(C_2)$	761 905k 61.2061%	167 025k 85.0274%	928 931k 64.4540%
	Matching	0.02%	39.83%	7.18%
	$C_1\sigma \subseteq C_2$	131k 0.0106%	66 531k 33.8694%	66 664k 4.6254%
C	Empty (pre./con.)	44.73%	100.00%	99.89%
	$\not\models \Lambda_1\sigma, \not\models \Lambda_2$	59k 0.0047%	66 531k 33.8694%	66 591k 4.6203%
	Sample point	59.28%	0.12%	0.18%
	$\models \Lambda_1\sigma\beta_2$	35k 0.0028%	82k 0.0416%	117k 0.0081%
	Implication	95.51%	100.00%	98.66%
	Subsumes	33k 0.0027%	82k 0.0416%	115k 0.0080%

Table 2. An overview of the accuracy of non-perfect pipeline stages

Test	Specificity/Sensitivity			Pos./Neg. Predictive Value		
Instances	lc	bakery, tad	All	lc	bakery, tad	All
FO Filtering	0.38797	0.14979	0.35552	0.00013	0.00049	0.00020
FO Matching	0.99996	0.60196	0.92841	0.78456	0.00123	0.00275
Empty Conclusion	0.70973	0.00000	0.00103	0.54474	0.00123	0.00173
Sample Point	0.93864	1.00000	0.99998	0.95510	1.00000	0.98653

this paper is that it performs tautology and subsumption deletion to eliminate redundant clauses. As a preprocessing step, SPASS-SPL eliminates all tautologies from the set of input clauses. Similarly, the function resolvents(C, N) (see Line 4 of Algorithm 1) filters out all newly derived clauses that are tautologies. Note that we also use these tautology checks to eliminate all excess variables and to store sample solutions for all remaining clauses. After each iteration of the algorithm, we also check for subsumed clauses. We first eliminate newly generated clauses by forward subsumption (see Line 6 of Algorithm 1), then use the remaining clauses for backward subsumption (see Line 8 of Algorithm 1).

Benchmarks. Our benchmarking instances come out of three different applications. (1.) A supervisor for an automobile lane change assistant, formulated in the Horn fragment of BS(LRA) [9,10] (five instances, referred to as `lc` in aggregate). (2.) The formalization of reachability for non-deterministic timed automata, formulated in the non-Horn fragment of BS(LRA) [20] (one instance, referred to as `tad`). (3.) Formalizations of variants of mutual exclusion protocols, such as the bakery protocol [38], also formulated in the non-Horn fragment of BS(LRA) [19] (one instance, referred to as `bakery`). The machine used for benchmarking features an Intel Xeon W-1290P CPU (10 cores, 20 threads, up to 5.2 GHz) and 64 GiB DDR4-2933 ECC main memory. Runtime was limited to ten minutes, and memory usage was not limited.

Table 3. Evaluation of the sample point heuristic

Instances		lc	bakery, tad	All
Bottleneck	(C time ÷ FO time)			
	without sample point	127	2757	14867
	with sample point	78	32	89
Avg. pipeline runtime in μs				
	without sample point	0.0315	89.9401	0.5189
	with sample point	0.0311	1.4150	0.2197
Speedup	(C time *with* ÷ *without*)	1.63	137.88	124.16
Benefit-to-cost	(C time *taken* ÷ *saved*)	6.74	181.72	163.72

Evaluation. In Table 1 we give an overview of how many pairs of clauses advance how far in the pipeline (in thousands). Rows with grey background refer to a stage of the pipeline and show which portion of pairs of clauses were kept, relative to the previous stage. Rows with white background refer to (virtual) sets of clauses, their absolute size, and their size relative to the number of attempted tests, as well as the condition(s) established. The three groups of columns refer to groups of benchmark instances. Results vary greatly between lc and the aggregate of bakery and tad. In lc the relative number of subsumed clauses is significantly smaller (0.0027% compared to 0.0416%). FO Matching eliminates a large number of pairs in lc, because the number of predicate symbols, and their arity (lc1, ..., lc4: 36 predicates, arities up to 5; lc5: 53 predicates, arities up to 12) is greater than in bakery (11 predicates, all of arity 2) and tad (4 predicates, all of arity 2).

Binary Classifiers. To evaluate the performance of each stage of the proposed test pipeline, we view each stage individually as a binary classifier on pairs of constrained clauses. The two classes we consider are "subsumes" (positive outcome) and "does not subsume" (negative outcome). Each stage of the pipeline computes a *prediction* on the *actual* result of the overall pipeline. We are thus interested in minimizing two kinds of errors: (1) When one stage of the pipeline predicts that the subsumption test will succeed (the prediciton is positive) but it fails (the actual result is negative), called *false positive* (FP). (2) When one stage of the pipeline predicts that the subsumption test will fail (the prediction is negative) but it succeeds (the actual result is positive), called *false negative* (FN). Dually, a correct prediction is called *true positive* (TP) and *true negative* (TN). For each stage, at least one kind of error is excluded by design: First-order filtering and the sample point heuristic never produce false negatives. The empty conclusion heuristic never produces false positives. The empty premise test is perfect, i.e. it neither produces false positives nor false negatives, with the caveat of not always being applicable. The last stage (implication test) decides the overall result of the pipeline, and thus is also perfect. For evaluation of binary classifiers, we use four different measures (two symmetric pairs):

$$\mathrm{SPC} = \mathrm{TN} \div (\mathrm{TN} + \mathrm{FP}) \qquad \mathrm{PPV} = \mathrm{TP} \div (\mathrm{TP} + \mathrm{FP}) \tag{2}$$

The first pair, *specificity* (SPC) and *positive predictive value*, see (2), is relevant only in presence of false postives (the measures approach 1 as FP approaches 0).

$$\text{SEN} = \text{TP} \div (\text{TP} + \text{FN}) \qquad \text{NPV} = \text{TN} \div (\text{TN} + \text{FN}) \tag{3}$$

The second pair, *sensitivity* (SEN) and *negative predictive value* (NPV), see (3), is relevant only in presence of false negatives (the measures approach 1 as FN approaches 0). Specificity (resp. sensitivity) might be considered the "success rate" in our setup. They answer the question: "Given the *actual* result of the pipeline is 'subsumed' (resp. 'not subsumed'), in how many cases does this stage *predict* correctly?" A specificity (resp. sensitivity) of 0.99 means that the classifier produces a false positive (resp. negative), i.e. a wrong prediction, in one out of one hundred cases. Both measures are independent of the prevalence of particular actual results, i.e. the measures are not biased by instances that feature many (or few) subsumed clauses. On the other hand, positive and negative predictive value are biased by prevalence. They answer the following question: "Given this stage of the pipeline *predicts* 'subsumed' (resp. 'not subsumed'), how likely is it that the *actual* result indeed is 'subsumed' (resp. 'not subsumed')?"

In Table 2 we present for all non-perfect stages of the pipeline specificity (for those that produce false positives) and sensitivity (for those that produce false negatives) as well as the (positive/negative) predictive value. Note that the sample point heuristic has an exceptionally high specificity, still above 93% in the benchmarks where it performed worst. For the benchmarks `bakery` and `tad` it even performs perfectly. Combined, this gives a specificity of above 99.99%. Considering FO Filtering, we expect limited performance, since the structure of terms in BS is flat compared to the rich structure of terms as trees in full first-order logic. This is evidenced by a comparatively low specificity of 35%. However, this classifier is very easy to compute, so pays for itself. FO Matching is a much better classifier, at an aggregate sensitivity of 93%. Even though this classifier is NP-complete, this is not problematic in practice.

Runtime. In Table 3 we focus on the runtime improvement achieved by the sample point heuristic. In the first two lines (Bottleneck), we highlight how much slower testing implication of constraints (the C-phase) is compared to treating the first-order part (the FO-phase). This is equivalent to the time taken for the C-phase per pair of clauses (that reach at least the first C-phase) divided by the time taken for the FO-phase per pair of clauses. We see that without the sample point heuristic, we can expect the constraint implication test to take hundreds to thousands of times longer than the FO-phase. Adding the sample point heuristic decreases this ratio to below one hundred. In the fourth line (avg. pipeline runtime) we do not give a ratio, but the average time it takes to compute the whole pipeline. We achieve millions of subsumption checks per second. In the fifth line (Speedup), we take the time that all C-phases combined take per pair of clauses that reach at least the first C-phase, and take the ratio to the same time without applying the sample point heuristic. In the sixth line (Benefit-to-cost), we consider the time taken to compute the sample point vs. the time it saves. The benefit is about two orders of magnitude greater than the cost.

5 Conclusion

Our next step will be the integration of the subsumption test in the backward subsumption procedure of an SCL based reasoning procedure for BS(LRA) [11] which is currently under development.

There are various ways to improve the sample point heuristic. One improvement would be to store and check multiple sample points per clause. For instance, whenever the sample point heuristic fails and the implication test for $\Lambda_2 \rightarrow (\Lambda_1\sigma)$ also fails, store the solution to (1) as an additional sample point for Λ_2. The new sample point will filter out any future implication tests with $\Lambda_1\sigma$ or similar constraints. However, testing too many sample points might lead to costs outweighing benefits. A potential solution to this problem would be score-based garbage collection, as done in SAT solvers [57]. Another way to store and check multiple sample points per clause is to store a compact description of a set of points that is easy to check against. For instance, we can store the center point and edge length of the largest orthogonal hypercube contained in the solutions of a constraint, which is equivalent to infinitely many sample points. Computing the largest orthogonal hypercube for an LRA constraint is not much harder than finding a sample solution [14]. Checking whether a cube is contained in an LRA constraint works almost the same as evaluating a sample point [14].

Although we developed our sample point technique for the BS(LRA) fragment it is obvious that it will also work for the overall FOL(LRA) clause fragment, because this extension does not affect the LRA constraint part of clauses. From an automated reasoning perspective, satisfiability of the FOL(LRA) and BS(LRA) fragments (clause sets) is undecidable in both cases. Actually, satisfiability of a BS(LRA) clause set is already undecidable if the first-order part is restricted to a single monadic predicate [32]. The first-order part of BS(LRA) is decidable and therefore enables effective guidance for an overall reasoning procedure [11]. Form an application perspective, the BS(LRA) fragment already encompasses a number of used (sub)languages. For example, timed automata [3] and a number of extensions thereof are contained in the BS(LRA) fragment [60].

We also believe that the sample point heuristic will speed up the constraint implication test for FOL(LIA), first-order clauses over linear integer arithmetic, FOL(NRA), i.e., first-order clauses over non-linear real arithmetic, and other combinations of FOL with arithmetic theories. However, the non-linear case will require a more sophisticated setup due to the nature of test points in this case, e.g., a solution may contain root expressions.

Acknowledgments. This work was partly funded by DFG grant 389792660 as part of TRR 248, see https://perspicuous-computing.science. We thank the anonymous reviewers for their thorough reading and detailed constructive comments. Martin Desharnais suggested some textual improvements.

References

1. Alagi, G., Weidenbach, C.: NRCL - a model building approach to the Bernays-Schönfinkel fragment. In: Lutz, C., Ranise, S. (eds.) FroCoS 2015. LNCS (LNAI),

vol. 9322, pp. 69–84. Springer, Cham (2015). https://doi.org/10.1007/978-3-319-24246-0_5
2. Althaus, E., Kruglov, E., Weidenbach, C.: Superposition modulo linear arithmetic SUP(LA). In: Ghilardi, S., Sebastiani, R. (eds.) FroCoS 2009. LNCS (LNAI), vol. 5749, pp. 84–99. Springer, Heidelberg (2009). https://doi.org/10.1007/978-3-642-04222-5_5
3. Alur, R., Dill, D.L.: A theory of timed automata. Theor. Comput. Sci. **126**(2), 183–235 (1994). https://doi.org/10.1016/0304-3975(94)90010-8
4. Bachmair, L., Ganzinger, H.: Rewrite-based equational theorem proving with selection and simplification. J. Log. Comput. **4**(3), 217–247 (1994). https://doi.org/10.1093/logcom/4.3.217
5. Bachmair, L., Ganzinger, H.: Resolution theorem proving. In: Robinson, J.A., Voronkov, A. (eds.) Handbook of Automated Reasoning (in 2 volumes), pp. 19–99. Elsevier and MIT Press, Cambridge (2001). https://doi.org/10.1016/b978-044450813-3/50004-7
6. Bachmair, L., Ganzinger, H., Waldmann, U.: Refutational theorem proving for hierarchic first-order theories. Appl. Algebra Eng. Commun. Comput. **5**, 193–212 (1994). https://doi.org/10.1007/BF01190829
7. Baumgartner, P., Waldmann, U.: Hierarchic superposition revisited. In: Lutz, C., Sattler, U., Tinelli, C., Turhan, A.-Y., Wolter, F. (eds.) Description Logic, Theory Combination, and All That. LNCS, vol. 11560, pp. 15–56. Springer, Cham (2019). https://doi.org/10.1007/978-3-030-22102-7_2
8. Biere, A., Heule, M., van Maaren, H., Walsh, T. (eds.): Handbook of Satisfiability, Frontiers in Artificial Intelligence and Applications, vol. 185. IOS Press, Amsterdam (2009)
9. Bromberger, M., et al.: A sorted datalog hammer for supervisor verification conditions modulo simple linear arithmetic. CoRR abs/2201.09769 (2022). https://arxiv.org/abs/2201.09769
10. Bromberger, M., Dragoste, I., Faqeh, R., Fetzer, C., Krötzsch, M., Weidenbach, C.: A datalog hammer for supervisor verification conditions modulo simple linear arithmetic. In: Konev, B., Reger, G. (eds.) FroCoS 2021. LNCS (LNAI), vol. 12941, pp. 3–24. Springer, Cham (2021). https://doi.org/10.1007/978-3-030-86205-3_1
11. Bromberger, M., Fiori, A., Weidenbach, C.: Deciding the Bernays-Schoenfinkel Fragment over bounded difference constraints by simple clause learning over theories. In: Henglein, F., Shoham, S., Vizel, Y. (eds.) VMCAI 2021. LNCS, vol. 12597, pp. 511–533. Springer, Cham (2021). https://doi.org/10.1007/978-3-030-67067-2_23
12. Bromberger, M., Fleury, M., Schwarz, S., Weidenbach, C.: SPASS-SATT. In: Fontaine, P. (ed.) CADE 2019. LNCS (LNAI), vol. 11716, pp. 111–122. Springer, Cham (2019). https://doi.org/10.1007/978-3-030-29436-6_7
13. Bromberger, M., Leutgeb, L., Weidenbach, C.: An Efficient subsumption test pipeline for BS(LRA) clauses (2022). https://doi.org/10.5281/zenodo.6544456. Supplementary Material
14. Bromberger, M., Weidenbach, C.: Fast cube tests for LIA constraint solving. In: Olivetti, N., Tiwari, A. (eds.) IJCAR 2016. LNCS (LNAI), vol. 9706, pp. 116–132. Springer, Cham (2016). https://doi.org/10.1007/978-3-319-40229-1_9
15. Dershowitz, N.: Orderings for term-rewriting systems. Theor. Comput. Sci. **17**, 279–301 (1982). https://doi.org/10.1016/0304-3975(82)90026-3
16. Dutertre, B., de Moura, L.: A fast linear-arithmetic solver for DPLL(T). In: Ball, T., Jones, R.B. (eds.) CAV 2006. LNCS, vol. 4144, pp. 81–94. Springer, Heidelberg (2006). https://doi.org/10.1007/11817963_11

17. Eggers, A., Kruglov, E., Kupferschmid, S., Scheibler, K., Teige, T., Weidenbach, C.: Superposition Modulo Non-linear Arithmetic. In: Tinelli, C., Sofronie-Stokkermans, V. (eds.) FroCoS 2011. LNCS (LNAI), vol. 6989, pp. 119–134. Springer, Heidelberg (2011). https://doi.org/10.1007/978-3-642-24364-6_9
18. Faqeh, R., Fetzer, C., Hermanns, H., Hoffmann, J., Klauck, M., Köhl, M.A., Steinmetz, M., Weidenbach, C.: towards dynamic dependable systems through evidence-based continuous certification. In: Margaria, T., Steffen, B. (eds.) ISoLA 2020. LNCS, vol. 12477, pp. 416–439. Springer, Cham (2020). https://doi.org/10.1007/978-3-030-61470-6_25
19. Fietzke, A.: Labelled superposition. Ph.D. thesis, Universität des Saarlandes (2014). https://doi.org/10.22028/D291-26569
20. Fietzke, A., Weidenbach, C.: Superposition as a decision procedure for timed automata. Math. Comput. Sci. **6**(4), 409–425 (2012). https://doi.org/10.1007/s11786-012-0134-5
21. Fiori, A., Weidenbach, C.: SCL clause learning from simple models. In: Fontaine, P. (ed.) CADE 2019. LNCS (LNAI), vol. 11716, pp. 233–249. Springer, Cham (2019). https://doi.org/10.1007/978-3-030-29436-6_14
22. Fiori, A., Weidenbach, C.: SCL with theory constraints. CoRR abs/2003.04627 (2020). https://arxiv.org/abs/2003.04627
23. Fränzle, M., Herde, C., Teige, T., Ratschan, S., Schubert, T.: Efficient solving of large non-linear arithmetic constraint systems with complex boolean structure. J. Satisf. Boolean Model. Comput. **1**(3–4), 209–236 (2007). https://doi.org/10.3233/sat190012
24. Ganzinger, H., Nieuwenhuis, R., Nivela, P.: Fast term indexing with coded context trees. J. Autom. Reason. **32**(2), 103–120 (2004). https://doi.org/10.1023/B:JARS.0000029963.64213.ac
25. Gleiss, B., Kovács, L., Rath, J.: Subsumption demodulation in first-order theorem proving. In: Peltier, N., Sofronie-Stokkermans, V. (eds.) IJCAR 2020. LNCS (LNAI), vol. 12166, pp. 297–315. Springer, Cham (2020). https://doi.org/10.1007/978-3-030-51074-9_17
26. Gottlob, G.: Subsumption and implication. Inf. Process. Lett. **24**(2), 109–111 (1987). https://doi.org/10.1016/0020-0190(87)90103-7
27. Gottlob, G., Leitsch, A.: On the efficiency of subsumption algorithms. J. ACM **32**(2), 280–295 (1985). https://doi.org/10.1145/3149.214118
28. Graf, P.: Extended path-indexing. In: Bundy, A. (ed.) CADE 1994. LNCS, vol. 814, pp. 514–528. Springer, Heidelberg (1994). https://doi.org/10.1007/3-540-58156-1_37
29. Graf, P.: Substitution tree indexing. In: Hsiang, J. (ed.) RTA 1995. LNCS, vol. 914, pp. 117–131. Springer, Heidelberg (1995). https://doi.org/10.1007/3-540-59200-8_52
30. Graf, P. (ed.): Term Indexing. LNCS, vol. 1053. Springer, Heidelberg (1995). https://doi.org/10.1007/3-540-61040-5
31. Hart, W.B.: Fast library for number theory: an introduction. In: Fukuda, K., Hoeven, J., Joswig, M., Takayama, N. (eds.) ICMS 2010. LNCS, vol. 6327, pp. 88–91. Springer, Heidelberg (2010). https://doi.org/10.1007/978-3-642-15582-6_18
32. Horbach, M., Voigt, M., Weidenbach, C.: The universal fragment of presburger arithmetic with unary uninterpreted predicates is undecidable. CoRR abs/1703.01212 (2017). http://arxiv.org/abs/1703.01212
33. Purdom, P.W., Brown, C.A.: Fast many-to-one matching algorithms. In: Jouannaud, J.-P. (ed.) RTA 1985. LNCS, vol. 202, pp. 407–416. Springer, Heidelberg (1985). https://doi.org/10.1007/3-540-15976-2_21

34. Bayardo, R.J., Schrag, R.: Using CSP look-back techniques to solve exceptionally hard SAT instances. In: Freuder, E.C. (ed.) CP 1996. LNCS, vol. 1118, pp. 46–60. Springer, Heidelberg (1996). https://doi.org/10.1007/3-540-61551-2_65
35. Korovin, K., Voronkov, A.: Integrating Linear Arithmetic into Superposition Calculus. In: Duparc, J., Henzinger, T.A. (eds.) CSL 2007. LNCS, vol. 4646, pp. 223–237. Springer, Heidelberg (2007). https://doi.org/10.1007/978-3-540-74915-8_19
36. Kruglov, E.: Superposition modulo theory. Ph.D. thesis, Universität des Saarlandes (2013). https://doi.org/10.22028/D291-26547
37. Kruglov, E., Weidenbach, C.: Superposition decides the first-order logic fragment over ground theories. Math. Comput. Sci. **6**(4), 427–456 (2012). https://doi.org/10.1007/s11786-012-0135-4
38. Lamport, L.: A new solution of dijkstra's concurrent programming problem. Commun. ACM **17**(8), 453–455 (1974). https://doi.org/10.1145/361082.361093
39. McCune, W.: Otter 2.0. In: Stickel, M.E. (ed.) CADE 1990. LNCS, vol. 449, pp. 663–664. Springer, Heidelberg (1990). https://doi.org/10.1007/3-540-52885-7_131
40. McCune, W.: Experiments with discrimination-tree indexing and path indexing for term retrieval. J. Autom. Reason. **9**(2), 147–167 (1992). https://doi.org/10.1007/BF00245458
41. Moskewicz, M.W., Madigan, C.F., Zhao, Y., Zhang, L., Malik, S.: Chaff: Engineering an efficient SAT solver. In: Proceedings of the 38th Design Automation Conference, DAC 2001, Las Vegas, NV, USA, 18–22 June 2001, pp. 530–535. ACM (2001). https://doi.org/10.1145/378239.379017
42. de Moura, L., Bjørner, N.: Z3: an efficient SMT solver. In: Ramakrishnan, C.R., Rehof, J. (eds.) TACAS 2008. LNCS, vol. 4963, pp. 337–340. Springer, Heidelberg (2008). https://doi.org/10.1007/978-3-540-78800-3_24
43. Nieuwenhuis, R., Hillenbrand, T., Riazanov, A., Voronkov, A.: On the evaluation of indexing techniques for theorem proving. In: Goré, R., Leitsch, A., Nipkow, T. (eds.) IJCAR 2001. LNCS, vol. 2083, pp. 257–271. Springer, Heidelberg (2001). https://doi.org/10.1007/3-540-45744-5_19
44. Nieuwenhuis, R., Rubio, A.: Paramodulation-based theorem proving. In: Robinson, J.A., Voronkov, A. (eds.) Handbook of Automated Reasoning (in 2 volumes), pp. 371–443. Elsevier and MIT Press, Cambridge (2001). https://doi.org/10.1016/b978-044450813-3/50009-6
45. Ohlbach, H.J.: Abstraction tree indexing for terms. In: 9th European Conference on Artificial Intelligence, ECAI 1990, Stockholm, Sweden, pp. 479–484 (1990)
46. Overbeek, R.A., Lusk, E.L.: Data structures and control architecture for implementation of theorem-proving programs. In: Bibel, W., Kowalski, R. (eds.) CADE 1980. LNCS, vol. 87, pp. 232–249. Springer, Heidelberg (1980). https://doi.org/10.1007/3-540-10009-1_19
47. Ramakrishnan, I.V., Sekar, R.C., Voronkov, A.: Term indexing. In: Robinson, J.A., Voronkov, A. (eds.) Handbook of Automated Reasoning (in 2 volumes), pp. 1853–1964. Elsevier and MIT Press, Cambridge (2001). https://doi.org/10.1016/b978-044450813-3/50028-x
48. Riazanov, A., Voronkov, A.: Partially adaptive code trees. In: Ojeda-Aciego, M., de Guzmán, I.P., Brewka, G., Moniz Pereira, L. (eds.) JELIA 2000. LNCS (LNAI), vol. 1919, pp. 209–223. Springer, Heidelberg (2000). https://doi.org/10.1007/3-540-40006-0_15
49. Riazanov, A., Voronkov, A.: Efficient instance retrieval with standard and relational path indexing. Inf. Comput. **199**(1–2), 228–252 (2005). https://doi.org/10.1016/j.ic.2004.10.012

50. Robinson, J.A.: A machine-oriented logic based on the resolution principle. J. ACM, **12**(1), 23–41 (1965). https://doi.org/10.1145/321250.321253, http://doi.acm.org/10.1145/321250.321253
51. Schrijver, A.: Theory of Linear and Integer Programming. Wiley-Interscience series in discrete mathematics and optimization, Wiley, Hoboken (1999)
52. Schulz, S.: Simple and efficient clause subsumption with feature vector indexing. In: Proceedings of the IJCAR-2004 Workshop on Empirically Successful First-Order Theorem Proving. Elsevier Science (2004)
53. Schulz, S.: Fingerprint Indexing for Paramodulation and Rewriting. In: Gramlich, B., Miller, D., Sattler, U. (eds.) IJCAR 2012. LNCS (LNAI), vol. 7364, pp. 477–483. Springer, Heidelberg (2012). https://doi.org/10.1007/978-3-642-31365-3_37
54. Schulz, S.: Simple and efficient clause subsumption with feature vector indexing. In: Bonacina, M.P., Stickel, M.E. (eds.) Automated Reasoning and Mathematics. LNCS (LNAI), vol. 7788, pp. 45–67. Springer, Heidelberg (2013). https://doi.org/10.1007/978-3-642-36675-8_3
55. Silva, J.P.M., Sakallah, K.A.: GRASP - a new search algorithm for satisfiability. In: Rutenbar, R.A., Otten, R.H.J.M. (eds.) Proceedings of the 1996 IEEE/ACM International Conference on Computer-Aided Design, ICCAD 1996, San Jose, CA, USA, 10–14 November 1996, pp. 220–227. IEEE Computer Society/ACM (1996). https://doi.org/10.1109/ICCAD.1996.569607
56. Socher, R.: A subsumption algorithm based on characteristic matrices. In: Lusk, E., Overbeek, R. (eds.) CADE 1988. LNCS, vol. 310, pp. 573–581. Springer, Heidelberg (1988). https://doi.org/10.1007/BFb0012858
57. Soos, M., Kulkarni, R., Meel, K.S.: CrystalBall: gazing in the black box of SAT solving. In: Janota, M., Lynce, I. (eds.) SAT 2019. LNCS, vol. 11628, pp. 371–387. Springer, Cham (2019). https://doi.org/10.1007/978-3-030-24258-9_26
58. Stillman, R.B.: The concept of weak substitution in theorem-proving. J. ACM **20**(4), 648–667 (1973). https://doi.org/10.1145/321784.321792
59. Tammet, T.: Towards efficient subsumption. In: Kirchner, C., Kirchner, H. (eds.) CADE 1998. LNCS, vol. 1421, pp. 427–441. Springer, Heidelberg (1998). https://doi.org/10.1007/BFb0054276
60. Voigt, M.: Decidable $\exists^*\forall^*$ first-order fragments of linear rational arithmetic with uninterpreted predicates. J. Autom. Reason. **65**(3), 357–423 (2020). https://doi.org/10.1007/s10817-020-09567-8
61. Voronkov, A.: The anatomy of vampire implementing bottom-up procedures with code trees. J. Autom. Reason. **15**(2), 237–265 (1995). https://doi.org/10.1007/BF00881918
62. Voronkov, A.: Algorithms, datastructures, and other issues in efficient automated deduction. In: Goré, R., Leitsch, A., Nipkow, T. (eds.) IJCAR 2001. LNCS, vol. 2083, pp. 13–28. Springer, Heidelberg (2001). https://doi.org/10.1007/3-540-45744-5_3
63. Weidenbach, C.: Automated reasoning building blocks. In: Meyer, R., Platzer, A., Wehrheim, H. (eds.) Correct System Design. LNCS, vol. 9360, pp. 172–188. Springer, Cham (2015). https://doi.org/10.1007/978-3-319-23506-6_12

Ground Joinability and Connectedness in the Superposition Calculus

André Duarte(✉) and Konstantin Korovin(✉)

The University of Manchester, Manchester, UK
{andre.duarte,konstantin.korovin}@manchester.ac.uk

Abstract. Problems in many theories axiomatised by unit equalities (UEQ), such as groups, loops, lattices, and other algebraic structures, are notoriously difficult for automated theorem provers to solve. Consequently, there has been considerable effort over decades in developing techniques to handle these theories, notably in the context of Knuth-Bendix completion and derivatives. The superposition calculus is a generalisation of completion to full first-order logic; however it does not carry over all the refinements that were developed for it, and is therefore not a strict generalisation. This means that (i) as of today, even state of the art provers for first-order logic based on the superposition calculus, while more general, are outperformed in UEQ by provers based on completion, and (ii) the sophisticated techniques developed for completion are not available in any problem which is not in UEQ. In particular, this includes key simplifications such as ground joinability, which have been known for more than 30 years. In fact, all previous completeness proofs for ground joinability rely on proof orderings and proof reductions, which are not easily extensible to general clauses together with redundancy elimination. In this paper we address this limitation and extend superposition with ground joinability, and show that under an adapted notion of redundancy, simplifications based on ground joinability preserve completeness. Another recently explored simplification in completion is connectedness. We extend this notion to "ground connectedness" and show superposition is complete with both connectedness and ground connectedness. We implemented ground joinability and connectedness in a theorem prover, iProver, the former using a novel algorithm which we also present in this paper, and evaluated over the TPTP library with encouraging results.

Keywords: Superposition · Ground joinability · Connectedness · Closure redundancy · First-order theorem proving

1 Introduction

Automated theorem provers based on equational completion [4], such as Waldmeister, MædMax or Twee [13,21,25], routinely outperform superposition-based provers on unit equality problems (UEQ) in competitions such as CASC [22], despite the fact that the superposition calculus was developed as a generalisation

J. Blanchette et al. (Eds.): IJCAR 2022, LNAI 13385, pp. 169–187, 2022.
https://doi.org/10.1007/978-3-031-10769-6_11

of completion to full clausal first-order logic with equality [19]. One of the main ingredients for their good performance is the use of ground joinability criteria for the deletion of redundant equations [1], among other techniques. However, existing proofs of refutational completeness of deduction calculi wrt. these criteria are restricted to unit equalities and rely on proof orderings and proof reductions [1,2,4], which are not easily extensible to general clauses together with redundancy elimination.

Since completion provers perform very poorly (or not at all) on non-UEQ problems (relying at best on incomplete transformations to unit equality [8]), this motivates an attempt to transfer those techniques to the superposition calculus and prove their completeness, so as to combine the generality of the superposition calculus with the powerful simplification rules of completion. To our knowledge, no prover for first-order logic incorporates ground joinability redundancy criteria, except for particular theories such as associativity-commutativity (AC) [20].

For instance, if $f(x, y) \approx f(y, x)$ is an axiom, then the equation $f(x, f(y, z)) \approx f(x, f(z, y))$ is redundant, but this cannot be justified by any simplificaton rule in the superposition calculus. On the other hand, a completion prover which implements ground joinability can easily delete the latter equation wrt. the former. We show that ground joinability can be enabled in the superposition calculus without compromising completeness.

As another example, the simplification rule in completion can use $f(x) \approx s$ (when $f(x) \succ s$) to rewrite $f(a) \approx t$ regardless of how s and t compare, while the corresponding demodulation rule in superposition can only rewrite if $s \prec t$. Our "encompassment demodulation" rule matches the former, while also being complete in the superposition calculus.

In [11] we introduced a novel theoretical framework for proving completeness of the superposition calculus, based on an extension of Bachmair-Ganzinger model construction [5], together with a new notion of redundancy called "closure redundancy". We used it to prove that certain AC joinability criteria, long used in the context of completion [1], could also be incorporated in the superposition calculus for full first-order logic while preserving completeness.

In this paper, we extend this framework to show the completeness of the superposition calculus extended with: (i) a general ground joinability simplification rule, (ii) an improved encompassment demodulation simplification rule, (iii) a connectedness simplification rule extending [3,21], and (iv) a new ground connectedness simplification rule. The proof of completeness that enables these extensions is based on a new encompassment closure ordering. In practice, these extensions help superposition to be competitive with completion in UEQ problems, and improves the performance on non-UEQ problems, which currently do not benefit from these techniques at all.

We also present a novel incremental algorithm to check ground joinability, which is very efficient in practice; this is important since ground joinability can be an expensive criterion to test. Finally, we discuss some of the experimental results we obtained after implementing these techniques in iProver [10,16].

The paper is structured as follows. In Sect. 2 we define some basic notions to be used throughout the paper. In Sect. 3 we define the closure ordering we use to prove redundancies. In Sect. 4 we present redundancy criteria for demodulation, ground joinability, connectedness, and ground connectedness. We prove their completeness in the superposition calculus, and discuss a concrete algorithm for checking ground joinability, and how it may improve on the algorithms used in e.g. Waldmeister [13] or Twee [21]. In Sect. 5 we discuss experimental results.

2 Preliminaries

We consider a signature consisting of a finite set of function symbols and the equality predicate as the only predicate symbol. We fix a countably infinite set of variables. First-order *terms* are defined in the usual manner. Terms without variables are called *ground* terms. A *literal* is an unordered pair of terms with either positive or negative polarity, written $s \approx t$ and $s \not\approx t$ respectively (we write $s \mathrel{\dot{\approx}} t$ to mean either of the former two). A *clause* is a multiset of literals. Collectively terms, literals, and clauses will be called *expressions*.

A *substitution* is a mapping from variables to terms which is the identity for all but finitely many variables. An injective substitution onto variables is called a *renaming*. If e is an expression, we denote application of a substitution σ by $e\sigma$, replacing all variables with their image in σ. Let $\mathrm{GSubs}(e) = \{\sigma \mid e\sigma \text{ is ground}\}$ be the set of *ground substitutions* for e. Overloading this notation for sets we write $\mathrm{GSubs}(E) = \{\sigma \mid \forall e \in E.\ e\sigma \text{ is ground}\}$. Finally, we write e.g. $\mathrm{GSubs}(e_1, e_2)$ instead of $\mathrm{GSubs}(\{e_1, e_2\})$. The identity substitution is denoted by ϵ.

A substitution θ is *more general* than σ if $\theta\rho = \sigma$ for some substitution ρ which is not a renaming. If s and t can be *unified*, that is, if there exists σ such that $s\sigma = t\sigma$, then there also exists the *most general unifier*, written $\mathrm{mgu}(s, t)$. A term s is said to be *more general* than t if there exists a substitution θ that makes $s\theta = t$ but there is no substitution σ such that $t\sigma = s$. Two terms s and t are said to be *equal modulo renaming* if there exist injective θ, σ such that $s\theta = t$ and $t\sigma = s$. The relations "less general than", "equal modulo renaming", and their union are represented respectively by the symbols $\sqsupset$, $\equiv$, and $\sqsupseteq$.

A more refined notion of instance is that of *closure* [6]. Closures are pairs $e \cdot \sigma$ that are said to *represent* the expression $e\sigma$ while retaining information about the original term and its instantiation. Closures where $e\sigma$ is ground are said to be *ground closures*. Let $\mathrm{GClos}(e) = \{e \cdot \sigma \mid e\sigma \text{ is ground}\}$ be the set of ground closures of e. Overloading the notation for sets, if N is a set of clauses then $\mathrm{GClos}(N) = \bigcup_{C \in N} \mathrm{GClos}(C)$.

We write $s[t]$ if t is a *subterm* of s. If also $s \neq t$, then it is a *strict subterm*. We denote these relations by $s \trianglerighteq t$ and $s \triangleright t$ respectively. We write $s[t \mapsto t']$ to denote the term obtained from s by replacing all occurrences of t by t'.

A (strict) partial order is a binary relation which is transitive ($a \succ b \succ c \Rightarrow a \succ c$), irreflexive ($a \not\succ a$), and asymmetric ($a \succ b \Rightarrow b \not\succ a$). A (non-strict) partial preorder (or quasiorder) is any transitive, reflexive relation. A (pre)order is total over X if $\forall x, y \in X.\ x \succeq y \lor y \succeq x$. Whenever a non-strict (pre)order $\succeq$ is given, the induced equivalence relation $\sim$ is $\succeq \cap \preceq$, and the induced strict pre(order) $\succ$ is $\succeq \setminus \sim$. The *transitive closure* of a relation $\succ$, the smallest transitive relation that contains $\succ$, is denoted by $\succ^+$. A *transitive reduction* of a relation $\succ$, the smallest relation whose transitive closure is $\succ$, is denoted by $\succ^-$.

For an ordering $\succ$ over a set X, its *multiset extension* $\succ\!\!\succ$ over multisets of X is given by: $A \succ\!\!\succ B$ iff $A \neq B$ and $\forall x \in B.\ B(x) > A(x)\ \exists y \in A.\ y \succ x \wedge A(y) > B(y)$, where $A(x)$ is the number of occurrences of element x in multiset A (we also use $\succ\!\!\succ\!\!\succ$ for the the multiset extension of $\succ\!\!\succ$). It is well known that the mutltiset extension of a well-founded/total order is also a well-founded/total order, respectively [9]. The *(n-fold) lexicographic extension* of $\succ$ over X is denoted $\succ_{\text{lex}}$ over ordered n-tuples of X, and is given by $\langle x_1, \ldots, x_n\rangle \succ_{\text{lex}} \langle y_1, \ldots, y_n\rangle$ iff $\exists i.\ x_1 = y_1 \wedge \cdots \wedge x_{i-1} = y_{i-1} \wedge x_i \succ y_i$. The lexicographic extension of a well-founded/total order is also a well-founded/total order, respectively.

A binary relation $\to$ over the set of terms is a *rewrite relation* if (i) $l \to r \Rightarrow l\sigma \to r\sigma$ and (ii) $l \to r \Rightarrow s[l] \to s[l \mapsto r]$. The *reflexive-transitive closure* of a relation is the smallest reflexive-transitive relation which contains it. It is denoted by $\xrightarrow{*}$. Two terms are *joinable* ($s \downarrow t$) if $s \xrightarrow{*} u \xleftarrow{*} t$.

If a rewrite relation is also a strict ordering, then it is a *rewrite ordering*. A *reduction ordering* is a rewrite ordering which is well-founded. In this paper we consider reduction orderings which are total on ground terms, such orderings are also *simplification orderings* i.e., satisfy $s \rhd t \Rightarrow s \succ t$.

3 Ordering

In [11] we presented a novel proof of completeness of the superposition calculus based on the notion of closure redundancy, which enables the completeness of stronger redundancy criteria to be shown, including AC normalisation, AC joinability, and encompassment demodulation. In this paper we use a slightly different closure ordering ($\succ_{cc}$), in order to extract better completeness conditions for the redundancy criteria that we present in this paper (the definition of closure redundant clause and closure redundant inference is parametrised by this $\succ_{cc}$).

Let $\succ_t$ be a simplification ordering which is total on ground terms. We extend this first to an ordering on ground term closures, then to an ordering on ground clause closures. Let

$$s \cdot \sigma \succ_{tc'} t \cdot \rho \quad \text{iff} \quad \begin{array}{l} \text{either } s\sigma \succ_t t\rho \\ \text{or else } s\sigma = t\rho \text{ and } s \sqsupset t, \end{array} \tag{1}$$

where $s\sigma$ and $t\rho$ are ground, and let $\succ_{tc}$ be an (arbitrary) total well-founded extension of $\succ_{tc'}$. We extend this to an ordering on clause closures. First let

$$M_{lc}((s \approx t) \cdot \theta) = \{s\theta \cdot \epsilon, t\theta \cdot \epsilon\}, \tag{2}$$

$$M_{lc}((s \not\approx t) \cdot \theta) = \{s\theta \cdot \epsilon, t\theta \cdot \epsilon, s\theta \cdot \epsilon, t\theta \cdot \epsilon\}, \tag{3}$$

and let M_{cc} be defined as follows, depending on whether the clause is unit or non-unit:

$$M_{cc}(\emptyset \cdot \theta) = \emptyset, \tag{4}$$

$$M_{cc}((s \approx t) \cdot \theta) = \{\{s \cdot \theta\}, \{t \cdot \theta\}\}, \tag{5}$$

$$M_{cc}((s \not\approx t) \cdot \theta) = \{\{s \cdot \theta, t \cdot \theta, s\theta \cdot \epsilon, t\theta \cdot \epsilon\}\}, \quad (6)$$
$$M_{cc}((s \dot\approx t \vee \cdots) \cdot \theta) = \{M_{lc}(L \cdot \theta) \mid L \in (s \dot\approx t \vee \cdots)\}, \quad (7)$$

then $\succ_{cc}$ is defined by

$$C \cdot \sigma \succ_{cc} D \cdot \rho \qquad \text{iff} \quad M_{cc}(C \cdot \sigma) \ggg_{tc} M_{cc}(D \cdot \rho). \quad (8)$$

The main purpose of this definition is twofold: (i) that when $s\theta \succ_t t\theta$ and u occurs in a clause D, then $s\theta \lhd u$ or $s \sqsubset s\theta = u$ implies $(s \approx t) \cdot \theta\rho \prec_{cc} D \cdot \rho$, and (ii) that when C is a positive unit clause, D is not, s is the maximal subterm in $C\theta$ and t is the maximal subterm in $D\sigma$, then $s \succeq_t t$ implies $C \cdot \theta \prec_{cc} D \cdot \sigma$. These two properties enable unconditional rewrites via oriented unit equations on positive unit clauses to succeed whenever they would also succeed in unfailing completion [4], and rewrites on negative unit and non-unit clauses to always succeed. This will enable us to prove the correctness of the simplification rules presented in the following section.

4 Redundancies

In this section we present several redundancy criteria for the superposition calculus and prove their completeness. Recall the definitions in [11]: a clause C is redundant in a set S if all its ground closures $C \cdot \theta$ follow from closures in $\mathrm{GClos}(S)$ which are smaller wrt. $\succ_{cc}$; an inference $C_1, \ldots, C_n \vdash D$ is redundant in a set S if, for all $\theta \in \mathrm{GSubs}(C_1, \ldots, C_n, D)$ such that $C_1\theta, \ldots, C_n\theta \vdash D\theta$ is a valid inference, the closure $D \cdot \theta$ follows from closures in $\mathrm{GClos}(S)$ such that each is smaller than some $C_1 \cdot \theta, \ldots, C_n \cdot \theta$. These definitions (in terms of ground closures rather than in terms of ground clauses, as in [19]) arise because they enable us to justify stronger redundancy criteria for application in superposition theorem provers, including the AC criteria developed in [11] and the criteria in this section.

Theorem 1. The superposition calculus [19] is refutationally complete wrt. closure redundancy, that is, if a set of clauses is saturated up to closure redundancy (meaning any inference with non-redundant premises in the set is redundant) and does not contain the empty clause, then it is satisfiable.

Proof. The proof of completeness of the superposition calculus wrt. this closure ordering carries over from [11] with some modifications, which are presented in a full version of this paper [12].

4.1 Encompassment Demodulation

We introduce the following definition, to be re-used throughout the paper.

Definition 1. A rewrite via $l \approx r$ in clause $C[l\theta]$ is *admissible* if one of the following conditions holds: (i) C is not a positive unit, or (let $C = s[l\theta] \approx t$ for some θ) (ii) $l\theta \neq s$, or (iii) $l\theta \sqsupset l$, or (iv) $s \prec_t t$, or (v) $r\theta \prec_t t$.[1]

[1] We note that (iv) is superfluous, but we include it since in practice it is easier to check, as it is local to the clause being rewritten and therefore needs to be checked only once, while (v) needs to be checked with each demodulation attempt.

We then have

$$\text{Encompassment Demodulation}\quad \frac{l \approx r \quad \cancel{C[l\theta]}}{C[l\theta \mapsto r\theta]},\quad \begin{array}{l}\text{where } l\theta \succ_t r\theta \text{, and}\\ \text{rewrite via } l \approx r \text{ in } C \text{ is admissible.}\end{array} \tag{9}$$

In other words, given an equation $l \approx r$, if an instance $l\theta$ is a subterm in C, then the rewrite is admissible (meaning, for example, that an unconditional rewrite is allowed when $l\theta \succ_t r\theta$) if C is not a positive unit, or if $l\theta$ occurs at a strict subterm position, or if $l\theta$ is less general than l, or if $l\theta$ occurs outside a maximal side, or if $r\theta$ is smaller than the other side. This restriction is much weaker than the one given for the usual demodulation rule in superposition [17], and equivalent to the one in equational completion when we restrict ourselves to unit equalities [4].

Example 1. If $f(x) \succ_t s$, we can use $f(x) \approx s$ to rewrite $f(x) \approx t$ when $s \prec_t t$, and $f(a) \approx t$, $f(x) \not\approx t$, or $f(x) \approx t \vee C$ regardless of how s and t compare.

4.2 General Ground Joinability

In [11] we developed redundancy criteria for the theory of AC functions in the superposition calculus. In this section we extend these techniques to develop redundancy criteria for ground joinability in arbitrary equational theories.

Definition 2. Two terms are *strongly joinable* ($s \between t$), in a clause C wrt. a set of equations S, if either $s = t$, or $s \to s[l_1\sigma_1 \mapsto r_1\sigma_1] \xrightarrow{*} t$ via rules $l_i \approx r_i \in S$, where the rewrite via $l_1 \approx r_1$ is admissible in C, or $s \to s[l_1\sigma_1 \mapsto r_1\sigma_1] \downarrow t[l_2\sigma_2 \mapsto r_2\sigma_2] \leftarrow t$ via rules $l_i \approx r_i \in S$, where the rewrites via $l_1 \approx r_1$ and $l_2 \approx r_2$ are admissible in C. To make the ordering explicit, we may write $s \between_{\succ} t$. Two terms are *strongly ground joinable* ($s \Between t$), in a clause C wrt. a set of equations S, if for all $\theta \in \mathrm{GSubs}(s,t)$ we have $s\theta \between t\theta$ in C wrt. S.

We then have:

$$\text{Ground joinability}\quad \frac{\cancel{s \approx t \vee C} \quad S}{},\quad \text{where } s \Between t \text{ in } s \approx t \vee C \text{ wrt. } S, \tag{10a}$$

$$\text{Ground joinability}\quad \frac{\cancel{s \not\approx t \vee C} \quad S}{C},\quad \text{where } s \Between t \text{ in } s \not\approx t \vee C \text{ wrt. } S. \tag{10b}$$

Theorem 2. Ground joinability is a sound and admissible redundancy criterion of the superposition calculus wrt. closure redundancy.

Proof. We will show the positive case first. If $s \Between t$, then for any instance $(s \approx t \vee C) \cdot \theta$ we either have $s\theta = t\theta$, and therefore $\emptyset \models (s \approx t) \cdot \theta$, or we have wlog. $s\theta \succ_t t\theta$, with $s\theta \downarrow t\theta$. Then $s\theta$ and $t\theta$ can be rewritten to the same normal form u by $l_i\sigma_i \to r_i\sigma_i$ where $l_i \approx r_i \in S$. Since $u \prec_t s\theta$ and $u \preceq_t t\theta$, then $(s \approx t \vee C) \cdot \theta$

follows from smaller $(u \approx u \vee C) \cdot \theta^2$ (a tautology, i.e. follows from $\emptyset$) and from the instances of clauses in S used to rewrite $s\theta \rightarrow u \leftarrow t\theta$. It only remains to show that these latter instances are also smaller than $(s \approx t \vee C) \cdot \theta$. Since we have assumed $s\theta \succ_t t\theta$, then at least one rewrite step must be done on $s\theta$. Let $l_1\sigma_1 \rightarrow r_1\sigma_1$ be the instance of the rule used for that step, with $(l_1 \approx r_1) \cdot \sigma_1$ the closure that generates it. By Definition 1 and 2, one of the following holds:

- $C \neq \emptyset$, therefore $(l_1 \approx r_1) \cdot \sigma_1 \prec_{cc} (s \approx t \vee C) \cdot \theta$, or
- $l_1\sigma_1 \lhd s\theta$, therefore $l_1\sigma_1 \prec_t s\theta \Rightarrow l_1 \cdot \sigma_1 \prec_{tc} s \cdot \theta \Rightarrow (l_1 \approx r_1) \cdot \sigma_1 \prec_{cc} (s \approx t) \cdot \theta$, or
- $l_1\sigma_1 = s\theta$ and $s \sqsupset l_1$, therefore $l_1 \cdot \sigma_1 \prec_{tc} s \cdot \theta \Rightarrow (l_1 \approx r_1) \cdot \sigma_1 \prec_{cc} (s \approx t) \cdot \theta$, or
- $l_1\sigma_1 = s\theta$ and $s \equiv l_1$ and $r_1\sigma_1 \prec_t t\theta$, therefore $r_1 \cdot \sigma_1 \prec_{tc} t \cdot \theta \Rightarrow (l_1 \approx r_1) \cdot \sigma_1 \prec_{cc} (s \approx t) \cdot \theta$.

As for the remaining steps, they are done on the smaller side $t\theta$ or on the other side after this first rewrite, which is smaller than $s\theta$. Therefore all subsequent steps done by any $l_j\sigma_j \rightarrow r_j\sigma_j$ will have $r_j \cdot \sigma_j \prec_{tc} l_j \cdot \sigma_j \prec_{tc} s \cdot \theta \Rightarrow (l_j \approx r_j) \cdot \sigma_j \prec_{cc} (s \approx t \vee C) \cdot \theta$. As such, since this holds for all ground closures $(s \approx t \vee C) \cdot \theta$, then $s \approx t \vee C$ is redundant wrt. S.

For the negative case, the proof is similar. We will conclude that $(s \not\approx t \vee C) \cdot \theta$ follows from smaller $(l_i \approx r_i) \cdot \sigma_i \in \mathrm{GClos}(S)$ and smaller $(u \not\approx u \vee C) \cdot \theta$. The latter, of course, follows from smaller $C \cdot \theta$, therefore $s \not\approx t \vee C$ is redundant wrt. $S \cup \{C\}$. □

Example 2. If $S = \{f(x, y) \approx f(y, x)\}$, then $f(x, f(y, z)) \approx f(x, f(z, y))$ is redundant wrt. S. Note that $f(x, y) \approx f(y, x)$ is not orientable by any simplification ordering, therefore this cannot be justified by demodulation alone.

Testing for Ground Joinability. The general criterion presented above begs the question of how to test, in practice, whether $s \Downarrow t$ in a clause $s \approx t \vee C$. Several such algorithms have been proposed [1,18,21]. All of these are based on the observation that if we consider all total preorders $\succeq_v$ on $\mathrm{Vars}(s, t)$ and for all of them show strong joinability with a modified ordering—which we denote $\succ_{t[v]}$—then we have shown strong *ground* joinability in the order $\succ_t$ [18].

Definition 3. A simplification order on terms $\succ_t$ *extended with* a preorder on variables $\succeq_v$, denoted $\succeq_{t[v]}$, is a simplification preorder (i.e. satisfies all the relevant properties in Sect. 2) such that $\succeq_{t[v]} \supseteq \succ_t \cup \succeq_v$.

Example 3. If $x \succ_v y$, then $g(x) \succ_{t[v]} g(y)$, $g(x) \succ_{t[v]} y$, $f(x, y) \succ_{t[v]} f(y, x)$, etc.

The simplest algorithm based on this approach would be to enumerate all possible total preorders $\succeq_v$ over $\mathrm{Vars}(s, t)$, and exhaustively reduce both sides

[2] Wlog. $u\theta = u$, renaming variables in u if necessary.

via equations in S orientable by $\succ_{t[v]}$, checking if the terms can be reduced to the same normal form for all total preorders. This is very inefficient since there are $\mathcal{O}(n!e^n)$ such total preorders [7], where n is the cardinality of Vars(s,t). Another approach is to consider only a smaller number of partial preorders, based on the obvious fact that $s \downarrow_{\succ_{t[v]}} t \Rightarrow \forall \succeq'_v \supseteq \succeq_v .\ s \downarrow_{\succ_{t[v']}} t$, so that joinability under a smaller number of partial preorders can imply joinability under all the total preorders, necessary to prove ground joinability.

However, this poses the question of how to choose which partial preorders to check. Intuitively, for performance, we would like that whenever the two terms are *not* ground joinable, that some total preorder where they are not joinable is found as early as possible, and that whenever the two terms *are* joinable, that all total preorders are covered in as few partial preorders as possible.

Example 4. Let $S = \{f(x, f(y, z)) \approx f(y, f(x, z))\}$. Then $f(x, f(y, f(z, f(w, u))))$ $\approx f(x, f(y, f(w, f(z, u))))$ can be shown to be ground joinable wrt. S by checking just three cases: $\succeq_v \in \{z \succ w\ ,\ z \sim w\ ,\ z \prec w\}$, even though there are 6942 possible preorders.

Waldmeister first tries all partial preorders relating two variables among Vars(s,t), then three, etc. until success, failure (by trying a total order and failing to join) or reaching a predefined limit of attempts [1]. Twee tries an arbitrary total strict order, then tries to weaken it, and repeats until all total preorders are covered [21]. We propose a novel algorithm—incremental ground joinability—whose main improvement is *guiding* the process of picking which preorders to check by finding, during the process of searching for rewrites on subterms of the terms we are attempting to join, minimal extensions of the term order with a variable preorder which allow the rewrite to be done in the $\succ$ direction.

Our algorithm is summarised as follows. We start with an empty queue of variable preorders, V, initially containing only the empty preorder. Then, while V is not empty, we pop a preorder $\succeq_v$ from the queue, and attempt to perform a rewrite via an equation which is newly orientable by some extension $\succeq'_v$ of $\succeq_v$. That is, during the process of finding generalisations of a subterm of s or t among left-hand sides of candidate unoriented unit equations $l \approx r$, when we check that the instance $l\theta \approx r\theta$ used to rewrite is oriented, we try to force this to be true under some minimal extension $\succ_{t[v']}$ of $\succ_{t[v]}$, if possible. If no such rewrite exists, the two terms are not strongly joinable under $\succ_{t[v]}$ or any extension, and so are not strongly ground joinable and we are done. If it exists, we exhaustively rewrite with $\succ_{t[v']}$, and check if we obtain the same normal form. If we do not obtain it yet, we repeat the process of searching rewrites via equations orientable by further extensions of the preorder. But if we do, then we have proven joinability in the extended preorder; now we must add back to the queue a set of preorders O such that all the total preorders which are $\supseteq \succeq_v$ (popped from the queue) but not $\supseteq \succeq'_v$ (minimal extension under which we have proven joinability) are $\supseteq$ of some $\succeq''_v \in O$ (pushed back into the queue to be checked). Obtaining this O is implemented by order_diff$(\succeq_v, \succeq'_v)$, defined below. Whenever there are no more preorders in the queue to check, then we have checked that the terms are strongly joinable under all possible total preorders, and we are done.

Together with this, some book-keeping for keeping track of completeness conditions is necessary. We know that for completeness to be guaranteed, the conditions in Definition 1 must hold. They automatically do if C is not a positive unit or if the rewrite happens on a strict subterm. We also know that after a term has been rewritten at least once, rewrites on that side are always complete (since it was rewritten to a smaller term). Therefore we store in the queue, together with the preorder, a flag in $\mathcal{P}(\{\mathsf{L},\mathsf{R}\})$ indicating on which sides does a top rewrite need to be checked for completeness. Initially the flag is $\{\mathsf{L}\}$ if $s \succ_t t$, $\{\mathsf{R}\}$ if $s \prec_t t$, $\{\mathsf{L},\mathsf{R}\}$ if s and t are incomparable, and $\{\}$ if the clause is not a positive unit. When a rewrite at the top is attempted (say, $l \approx r$ used to rewrite $s = l\theta$ with t being the other side), if the flag for that side is set, then we check if $l\theta \sqsupset l$ or $r\theta \prec t$. If this fails, the rewrite is rejected. Whenever a side is rewritten (at any position), the flag for that side is cleared.

The definition of order_diff is as follows. Let the transitive reduction of $\succeq$ be represented by a set of links of the form $x{\succ}y$ / $x{\sim}y$.

$$\text{order\_diff}(\succeq_1, \succeq_2) = \{\succeq^+ \mid \succeq \in \text{order\_diff}'(\succeq_1, \succeq_2{}^-)\}\,, \tag{11a}$$

$$\text{order\_diff}'(\succeq_1, \succeq_2^-) = \tag{11b}$$

$$\begin{cases} \succeq_2^- = \{x{\succ}y\} \uplus \succeq_2^{-\prime} \Rightarrow \begin{cases} x \succ_1 y \Rightarrow \text{order\_diff}'(\succeq_1, \succeq_2^{-\prime}) \\ x \not\succ_1 y \Rightarrow \begin{array}{l} \{\succeq_1 \cup \{y{\succ}x\},\, \succeq_1 \cup \{x{\sim}y\}\} \\ \cup\, \text{order\_diff}'(\succeq_1 \cup \{x{\succ}y\}, \succeq_2^{-\prime}) \end{array} \end{cases} \\ \succeq_2^- = \{x{\sim}y\} \uplus \succeq_2^{-\prime} \Rightarrow \begin{cases} x \sim_1 y \Rightarrow \text{order\_diff}'(\succeq_1, \succeq_2^{-\prime}) \\ x \not\sim_1 y \Rightarrow \begin{array}{l} \{\succeq_1 \cup \{x{\succ}y\},\, \succeq_1 \cup \{y{\succ}x\}\} \\ \cup\, \text{order\_diff}'(\succeq_1 \cup \{x{\sim}y\}, \succeq_2^{-\prime}) \end{array} \end{cases} \\ \succeq_2^- = \emptyset \qquad\qquad\qquad \Rightarrow \emptyset\,. \end{cases}$$

where $\succeq_1 \subseteq \succeq_2$. In other words, we take a transitive reduction of $\succeq_2$, and for all links ℓ in that reduction which are not part of $\succeq_1$, we return orders $\succeq_1$ augmented with the reverse of ℓ and recurse with $\succeq_1 = \succeq_1 \cup \ell$.

Example 5.

$\succeq_1$	$\succeq_2$	order_diff($\succeq_1, \succeq_2$)
$x \succ y$	$x \succ y \succ z \succ w$	$x \succ y \sim z\,,\, x \succ y \prec z\,,\, x \succ y \succ z \sim w\,,\, x \succ y \succ z \prec w$
$y \prec x \succ z$	$x \succ y \succ z$	$x \succ y \sim z\,,\, x \succ z \succ y$

Theorem 3. For all total $\succeq_v^T \supseteq \succeq_1$, there exists one and only one $\succeq_i \in \{\succeq_2\} \cup$ order_diff($\succeq_1, \succeq_2$) such that $\succeq_v^T \supseteq \succeq_i$. For all $\succeq_v^T \not\supseteq \succeq_1$, there is no $\succeq_i \in \{\succeq_2\} \cup$ order_diff($\succeq_1, \succeq_2$) such that $\succeq_v^T \supseteq \succeq_i$.

Proof. See full version of the paper [12].

An algorithm based on searching for rewrites in minimal extensions of a variable preorder (starting with minimal extensions of the bare term ordering, $\succ_{t[\emptyset]}$), has several advantages. The main benefit of this approach is that, instead of imposing an a priori ordering on variables and then checking joinability under that ordering, we instead build a minimal ordering *while* searching for candidate unit equations to rewrite subterms of s, t. For instance, if two terms are *not* ground joinable, or not even rewritable in any $\succ_{t[v]}$ where it was not rewritable in $\succ_t$, then an approach such as the one used in Avenhaus, Hillenbrand and Löchner [1] cannot detect this until it has extended the preorder arbitrarily to a total ordering, while our incremental algorithm immediately realises this. We should note that empirically this is what happens in most cases: most of the literals we check during a run are *not* ground joinable, so for practical performance it is essential to optimise this case.

Theorem 4. Algorithm 1 returns "Success" only if $s \Downarrow t$ in C wrt. S.[3]

Proof. We will show that Algorithm 1 returns "Success" if and only if $s \downarrow_{\succ_{t[v^T]}} t$ for all total $\succeq_v^T$ over $\mathrm{Vars}(s,t)$, which implies $s \Downarrow_{\succ_t} t$.

When $\langle \succeq_v, s, t, c\rangle$ is popped from V, we exhaustively reduce s, t via equations in S oriented wrt. $\succ_{t[v]}$, obtaining s^r, t^r. If $s^r \sim_{t[v]} t^r$, then $s \downarrow_{\succ_{t[v]}} t$, and so $s \downarrow_{\succ_{t[v^T]}} t$ for all total $\succeq_v^T \supseteq \succeq_v$. If $s^r \not\sim_{t[v]} t^r$, we will attempt to rewrite one of s^r, t^r using *some* extended $\succ_{t[v']}$ where $\succeq_v' \supset \succeq_v$. If this is impossible, then $s \not\downarrow_{\succ_{t[v']}} t$ for any $\succeq_v' \supseteq \succeq_v$, and therefore there exists at least one total $\succeq_v^T$ such that $s \not\downarrow_{\succeq_v^T} t$, and we return "Fail".

If this is possible, then we repeat the process: we exhaustively reduce wrt. $\succ_{t[v']}$, obtaining s', t'. If $s' \not\sim_{t[v']} t'$, then we start again the process from the step where we attempt to rewrite via an extension of $\succeq_v'$: we either find a rewrite with some $\succ_{t[v'']}$ with $\succeq_v'' \supset \succeq_v'$, and exhaustively normalise wrt. $\succ_{t[v'']}$ obtaining s'', t'', etc., or we fail to do so and return "Fail".

If in any such step (after exhaustively normalising wrt. $\succ_{t[v']}$) we find $s' \sim_{t[v']} t'$, then $s \downarrow_{\succ_{t[v']}} t$, and so $s \downarrow_{\succ_{t[v^T]}} t$ for all total $\succeq_v^T \supseteq \succeq_v'$. Now at this point we must add back to the queue a set of preorders $\succeq_{v\,i}''$ such that: for all total $\succeq_v^T \supseteq \succeq_v$, either $\succeq_v^T \supseteq \succeq_v'$ (proven to be $\downarrow$) or $\succeq_v^T \supseteq$ some $\succeq_{v\,i}''$ (added to V to be checked). For efficiency, we would also like for there to be no overlap: no total $\succeq_v^T \supseteq \succeq_v$ is an extension of more than one of $\{\succeq_v', \succeq_{v\,1}'', \ldots\}$.

This is true because of Theorem 3. So we add $\{\langle \succeq_{v\,i}'', s^r, t^r, c^r\rangle \mid \succeq_{v\,i}'' \in \text{order\_diff}(\succeq_v, \succeq_v')\}$ to V, where $c^r = c \setminus (\text{if } s^r \neq s \text{ then } \{\mathsf{L}\} \text{ else } \{\}) \setminus (\text{if } t^r \neq t \text{ then } \{\mathsf{R}\} \text{ else } \{\})$. Note also that $s \downarrow_{\succ_{t[v]}} s^r$ and $t \downarrow_{\succ_{t[v]}} t^r$, therefore also $s \downarrow_{\succ_{t[vi'']}} s^r$ and $t \downarrow_{\succ_{t[vi'']}} t^r$ if $\succeq_{v\,i}'' \supset \succeq_v$.

[3] Note that the other direction may not always hold, there are strongly ground joinable terms which are not detected by this method of analysing all preorders between variables, e.g. $f(x, g(y)) \Downarrow f(g(y), x)$ wrt. $S = \{f(x,y) \approx f(y,x)\}$.

Algorithm 1: Incremental ground joinability test

```
Input: literal s ≈̇ t ∈ C; set of unorientable equations S
Output: whether s ⇓ t in C wrt. S
begin
    c ← ∅ if C is not pos. unit, {L} if s ≻ t, {R} if s ≺ t, {L,R} otherwise
    V ← {⟨∅, s, t, c⟩}
    while V is not empty do
        ⟨⪰_v, s, t, c⟩ ← pop from V
        s, t ← normalise s, t wrt. ≻_t[v], with completeness flag c
        c ← c \ ({L} if s was changed) \ ({R} if t was changed)
        if s ~_t[v] t then
            continue
        else
            s', t', c' ← s, t, c
            while there exists l ≈ r ∈ S that can rewrite s' or t' wrt. some
              ⪰'_v ⊃ ⪰_v, with completeness flag c do
                s', t' ← normalise s', t' wrt. ≻_t[v'], with completeness flag c
                c' ← c' \ ({L} if s' was changed) \ ({R} if t' was changed)
                if s' ~_t[v'] t' then
                    for ⪰''_v in order_diff(⪰_v, ⪰'_v) do push ⟨⪰''_v, s, t, c⟩ to V
                    break
                end
                ⪰_v ← ⪰'_v
            else
                return Fail
            end
        end
    else
        return Success
    end
end
where rewriting u in s, t wrt. ≻ with completeness flag c succeeds if
    (i) u is a strict subterm of s or t,
    (ii) u = s with L ∉ c,
    (iii) u = t with R ∉ c,
    (iv) instance lσ ≈ rσ used to rewrite has l ⊏ u,
    (v) u = s with rσ ≺ t,
    (vi) or u = t with rσ ≺ s.
end
```

During this whole process, any rewrites must pass a completeness test mentioned previously, such that the conditions in the definition of $\Downarrow$ hold. Let s_0, t_0 be the original terms and s, t be the ones being rewritten and c the completeness flag. If the rewrite is at a strict subterm position, it succeeds by Definition 2. If the rewrite is at the top, then we check c. If L is unset ($\mathsf{L} \notin c$), then either $s \succeq s_0 \prec t_0$ or $s \prec s_0$ or the clause is not a positive unit, so we allow a rewrite at the top of s, again by Definition 2. If L is set ($\mathsf{L} \in c$), then an explicit check must be done: we allow a rewrite at the top of s ($= s_0$) iff it is done by $l\sigma \to r\sigma$ with $l\sigma \sqsupset l$ or $r\sigma \prec t_0$. Respectively for R, with the roles of s and t swapped.

In short, we have shown that if $\langle \succeq_v, s', t', c' \rangle$ is popped from V, then V is only ever empty, and so the algorithm only terminates with "Success", if $s' \not\downarrow_{\succ_{t[v^T]}} t'$ for all total $\succeq_v^T \supseteq \succeq_v$. Since V is initialised with $\langle \emptyset, s, t, c \rangle$, then the algorithm only returns "Success" if $s \not\downarrow_{\succ_{t[v^T]}} t$ for all total $\succeq_v^T$. □

Orienting via Extension of Variable Ordering. In order to apply the ground joinability algorithm we need a way to check, for a given $\succ_t$ and $\succeq_v$ and some s, t, whether there exists a $\succeq_v' \supset \succeq_v$ such that $s \succ_{t[v']} t$. Here we show how to do this when $\succ_t$ is a Knuth-Bendix Ordering (KBO) [15].

Recall the definition of KBO. Let $\succ_s$ be a partial order on symbols, w be an $\mathbb{N}$-valued weight function on symbols and variables, with the property that $\exists m\ \forall x \in \mathcal{V}.\ w(x) = m$, $w(c) \geq m$ for all constants c, and there may only exist one unary symbol f with $w(f) = 0$ and in this case $f \succ_s g$ for all other symbols g. For terms, their weight is $w(f(s_1, \ldots)) = w(f) + w(s_1) + \cdots$. Let also $|s|_x$ be the number of occurrences of x in s. Then

$$f(s_1,\ldots) \succ_{\mathrm{KBO}} g(t_1,\ldots) \quad \text{iff} \quad \begin{cases} \text{either } w(f(s_1,\ldots)) > w(g(t_1,\ldots)), \\ \text{or } w(f(s_1,\ldots)) = w(g(t_1,\ldots)) \\ \quad \text{and } f \succ_s g, \\ \text{or } w(f(s_1,\ldots)) = w(g(t_1,\ldots)) \\ \quad \text{and } f = g, \\ \quad \text{and } s_1,\ldots \succ_{\mathrm{KBOlex}} t_1,\ldots; \\ \text{and } \forall x \in \mathcal{V}.\ |f(\ldots)|_x \geq |g(\ldots)|_x. \end{cases} \tag{12a}$$

$$f(s_1,\ldots) \succ_{\mathrm{KBO}} x \quad \text{iff} \quad |f(s_1,\ldots)|_x \geq 1\,. \tag{12b}$$

$$x \succ_{\mathrm{KBO}} y \quad \text{iff} \quad \bot\,. \tag{12c}$$

The conditions on variable occurrences ensure that $s \succ_{\mathrm{KBO}} t \Rightarrow \forall\theta.\ s\theta \succ_{\mathrm{KBO}} t\theta$.

When we extend the order $\succ_{\mathrm{KBO}}$ with a variable preorder $\succeq_v$, the starting point is that $x \succ_v y \Rightarrow x \succ_{\mathrm{KBO}[v]} y$ and $x \sim_v y \Rightarrow x \sim_{\mathrm{KBO}[v]} y$. Then, to ensure that all the properties of a simplification order (included the one mentioned above) hold, we arrive at the following definition (similar to [1]).

$$f(s_1,\ldots) \succ_{\mathrm{KBO}[v]} g(t_1,\ldots) \quad \text{iff} \quad \begin{cases} \text{either } w(f(\ldots)) > w(g(\ldots)), \\ \text{or } w(f(s_1,\ldots)) = w(g(t_1,\ldots)) \\ \quad \text{and } f \succ_s g, \\ \text{or } w(f(s_1,\ldots)) = w(g(t_1,\ldots)) \\ \quad \text{and } f = g, \\ \quad \text{and } s_1,\ldots \succ_{\mathrm{KBO}[v]_{\mathrm{lex}}} t_1,\ldots; \\ \text{and } \forall x \in \mathcal{V}.\ \sum_{y \succeq_v x} |f(\ldots)|_y \\ \quad \geq \sum_{y \succeq_v x} |g(\ldots)|_y. \end{cases} \tag{13a}$$

$$f(s_1,\ldots) \succ_{\mathrm{KBO}[v]} x \quad \text{iff} \quad \exists y \succeq_v x.\ |f(s_1,\ldots)|_y \geq 1\,. \tag{13b}$$

$$x \succ_{\mathrm{KBO}[v]} y \quad \text{iff} \quad x \succ_v y\,. \tag{13c}$$

To check whether there exists a $\succeq'_v \supset \succeq_v$ such that $s \succ_{\mathrm{KBO}[v']} t$, we need to check whether there are some $x \succ y$ or $x = y$ relations that we can add to $\succeq_v$ such that all the conditions above hold (and such that it still remains a valid preorder). Let us denote "there exists a $\succeq'_v \supset \succeq_v$ such that $s \succ_{\mathrm{KBO}[v']} t$" by $s \succ_{\mathrm{KBO}[v,v']} t$. Then the definition is

$$f(s_1,\ldots) \succ_{\mathrm{KBO}[v,v']} g(t_1,\ldots) \quad \text{iff} \quad \begin{cases} \text{either } w(f(\ldots)) > w(g(\ldots)), \\ \text{or } w(f(s_1,\ldots)) = w(g(t_1,\ldots)) \\ \quad \text{and } f \succ_s g, \\ \text{or } w(f(s_1,\ldots)) = w(g(t_1,\ldots)) \\ \quad \text{and } f = g, \\ \quad \text{and } s_1,\ldots \succ_{\mathrm{KBOlex}} t_1,\ldots; \\ \text{and } \exists x_1, y_1, \ldots \\ \quad \succeq'_v = (\succeq_v \cup \{\langle x_1, y_1\rangle, \ldots\})^+ \text{ is a preorder} \\ \quad \text{such that } \forall x \in \mathcal{V}.\ \sum_{y \succeq'_v x} |f(\ldots)|_y \\ \quad\ \geq \sum_{y \succeq'_v x} |g(\ldots)|_y. \end{cases} \tag{14a}$$

$$f(s_1,\ldots) \succ_{\mathrm{KBO}[v,v']} x \quad \text{iff} \quad \begin{cases} \exists y \not\preceq_v x.\ |f(s_1,\ldots)|_y \geq 1, \\ \quad \text{with } \succeq'_v = \succeq_v \cup \{x \succ y\} \\ \quad \text{or } \succeq'_v = \succeq_v \cup \{x = y\}. \end{cases} \tag{14b}$$

$$x \succ_{\mathrm{KBO}[v,v']} y \quad \text{iff} \quad \begin{cases} x \not\preceq_v y \\ \quad \text{with } \succeq'_v = \succeq_v \cup \{x \succ y\}. \end{cases} \tag{14c}$$

This check can be used in Algorithm 1 for finding extensions of variable orderings that orient rewrite rules allowing required normalisations.

4.3 Connectedness

Testing for joinability (i.e. demodulating to $s \approx s$ or $s \not\approx s$) and ground joinability (presented in the previous section) require that each step in proving them is done via an oriented instance of an equation in the set. However, we can weaken this restriction, if we also change the notion of redundancy being used.

As criteria for redundancy of a clause, finding either joinability or ground joinability of a literal in the clause means that the clause can be deleted or the literal removed from the clause (in case of a positive or negative literal, resp.) in any context, that is, we can for example add them to a set of deleted clauses, and for any new clause, if it appears in that set, then immediately remove it since we already saw that it is redundant. The criterion of connectedness [3,21], however, is a criterion for redundancy of *inferences*. This means that a conclusion simplified by this criterion can be deleted (or rather, not added), but in that context only; if it ever comes up again as a conclusion of a different inference, then it is not necessarily also redundant. Connectedness was introduced in the context of equational completion, here we extend it to general clauses and show that it is a redundancy in the superposition calculus.

Definition 4. Terms s and t are *connected* under clauses U and unifier ρ wrt. a set of equations S if there exist terms $v_1, \ldots, v_n$, equations $l_1 \approx r_1, \ldots, l_{n-1} \approx r_{n-1}$, and substitutions $\sigma_1, \ldots, \sigma_{n-1}$ such that:

(i) $v_1 = s$ and $v_n = t$,
(ii) for all $i \in 1, \ldots, n-1$, either $v_{i+1} = v_i[l_i\sigma_i \mapsto r_i\sigma_i]$ or $v_i = v_{i+1}[l_i\sigma_i \mapsto r_i\sigma_i]$, with $l_i \approx r_i \in S$,
(iii) for all $i \in 1, \ldots, n-1$, there exists w in $\bigcup_{C \in U} \bigcup_{p \dot{\approx} q \in C} \{p, q\}$[4] such that for $u_i \in \{l_i, r_i\}$, either (a) $u_i\sigma_i \prec w\rho$, or (b) $u_i\sigma_i = w\rho$ and either $u_i \sqsubset w$ or $w \in C$ such that C is not a positive unit.

Theorem 5. Superposition inferences of the form

$$\frac{\underline{l \approx r} \vee C \quad \underline{s[u] \approx t} \vee D}{(s[u \mapsto r] \approx t \vee C \vee D)\rho}, \quad \begin{array}{l}\text{where } \rho = \text{mgu}(l, u),\\ l\rho \not\preceq r\rho,\ s\rho \not\preceq t\rho,\\ \text{and } u \text{ not a variable,}\end{array} \tag{15}$$

where $s[u \mapsto r]\rho$ and $t\rho$ are connected under $\{l \approx r \vee C,\ s \approx t \vee D\}$ and unifier ρ wrt. some set of clauses S, are redundant inferences wrt. S.

Proof. Let us denote $s' = s[u \mapsto r]$. Let also $U = \{l \approx r \vee C,\ s \approx t \vee D\}$ and $M = \bigcup_{C \in U} \bigcup_{p \dot{\approx} q \in C} \{p, q\}$. We will show that if $s'\rho$ and $t\rho$ are connected under U and ρ, by equations in S, then every instance of that inference obeys the condition for closure redundancy of an inference (see, Sect. 4), wrt. S.

Consider any $(s' \approx t \vee C \vee D)\rho \cdot \theta$ where $\theta \in \text{GSubs}(U\rho)$. Either $s'\rho\theta = t\rho\theta$, and we are done (it follows from $\emptyset$), or $s'\rho\theta \succ t\rho\theta$, or $s'\rho\theta \prec t\rho\theta$.

Consider the case $s'\rho\theta \succ t\rho\theta$. For all $i \in 1, \ldots, n-1$, there exists a $C' \in U$ and a $w \in C'$ such that either (iii.a) $l_i\sigma_i\theta \prec w\rho\theta$, or (iii.b) $l_i\sigma_i\theta = w\rho\theta$ and $l_i \sqsubset v$, or (iii.b) $l_i\sigma_i\theta = w\rho\theta$ and C' is not a positive unit. Likewise for r_i. Therefore, for all $i \in 1, \ldots, n-1$, there exists a $C' \in U$ such that $(l_i \approx r_i) \cdot \sigma_i\theta \prec C' \cdot \rho\theta$. Since $(t \approx t \vee \cdots)\rho \cdot \theta$ is also smaller than $(s' \approx t \vee \cdots)\rho \cdot \theta$ and a tautology, then the instance $(s' \approx t \vee \cdots)\rho \cdot \theta$ of the conclusion follows from closures in $\text{GClos}(S)$ such that each is smaller than one of $(l \approx r \vee C) \cdot \rho\theta$, $(s \approx t \vee D) \cdot \rho\theta$.

In the case that $s'\rho\theta \prec t\rho\theta$, the same idea applies, but now it is $(s' \approx s' \vee \cdots)\rho \cdot \theta$ which is smaller than $(s' \approx t \vee \cdots)\rho \cdot \theta$ and is a tautology.

Therefore, we have shown that for all $\theta \in \text{GSubs}((l \approx r \vee C)\rho,\ (s \approx t \vee D)\rho)$, the instance $(s' \approx t \vee \cdots)\rho \cdot \theta$ of the conclusion follows from closures in $\text{GClos}(S)$ which are all smaller than one of $(l \approx r \vee C) \cdot \rho\theta$, $(s \approx t \vee D) \cdot \rho\theta$. Since any valid superposition inference with ground clauses has to have $l = u$, then any $\theta' \in \text{GSubs}(l \approx r \vee C,\ s \approx t \vee D,\ (s' \approx t \vee C \vee D)\rho)$ such that the inference $(l \approx r \vee C)\theta', (s \approx t \vee D)\theta' \vdash (s' \approx t \vee C \vee D)\rho\theta'$ is valid must have $\theta' = \rho\theta''$, since ρ is the most general unifier. Therefore, we have shown that for all $\theta' \in \text{GSubs}(l \approx r \vee C,\ s \approx t \vee D,\ (s' \approx t \vee C \vee D)\rho)$ for which $(l \approx r \vee C)\theta', (s \approx t \vee D)\theta' \vdash (s' \approx t \vee C \vee D)\rho\theta'$ is a valid superposition inference, the instance $(s' \approx t \vee \cdots)\rho \cdot \theta'$ of the conclusion follows from closures in $\text{GClos}(S)$ which are all smaller than one of $(l \approx r \vee C) \cdot \theta'$, $(s \approx t \vee D) \cdot \theta'$, so the inference is redundant. □

[4] That is, in the set of top-level terms of literals of clauses in U.

Theorem 6. Superposition inferences of the form

$$\frac{\underline{l \approx r} \vee C \quad \underline{s[u] \not\approx t} \vee D}{(s[u \mapsto r] \not\approx t \vee C \vee D)\rho}, \quad \begin{array}{l}\text{where } \rho = \text{mgu}(l, u),\\ l\rho \not\preceq r\rho,\ s\rho \not\preceq t\rho,\\ \text{and } u \text{ not a variable,}\end{array} \tag{16}$$

where $s[u \mapsto r]\rho$ and $t\rho$ are connected under $\{l \approx r \vee C,\ s \not\approx t \vee D\}$ and unifier ρ wrt. some set of clauses S, are redundant inferences wrt. $S \cup \{(C \vee D)\rho\}$.

Proof. Analogously to the previous proof, we find that for all instances of the inference, the closure $(s' \not\approx t \vee \cdots)\rho \cdot \theta$ follows from smaller closure $(t \not\approx t \vee \cdots)\rho \cdot \theta$ or $(s' \not\approx s' \vee \cdots)\rho \cdot \theta$ and closures $(l_i \approx r_i) \cdot \sigma_i \theta$ smaller than $\max\{(l \approx r \vee C) \cdot \theta\ ,$ $(s \not\approx t \vee D) \cdot \theta\ ,\ (s' \not\approx t \vee C \vee D)\rho \cdot \theta\}$. But $(t \not\approx t \vee C \vee D)\rho \cdot \theta$ and $(s' \not\approx s' \vee C \vee D)\rho \cdot \theta$ both follow from smaller $(C \vee D)\rho \cdot \theta$, therefore the inference is redundant wrt. $S \cup \{(C \vee D)\rho\}$. □

4.4 Ground Connectedness

Just as joinability can be generalised to ground joinability, so can connectedness be generalised to ground connectedness. Two terms s, t are *ground connected* under U and ρ wrt. S if, for all $\theta \in \text{GSubs}(s, t)$, $s\theta$ and $t\theta$ are connected under D and ρ wrt. S. Analogously to strong ground joinability, we have that if s and t are connected using $\succ_{t[v]}$ for all total $\succeq_v$ over $\text{Vars}(s, t)$, then s and t are ground connected.

Theorem 7. Superposition inferences of the form

$$\frac{\underline{l \approx r} \vee C \quad \underline{s[u] \approx t} \vee D}{(s[u \mapsto r] \approx t \vee C \vee D)\rho}, \quad \begin{array}{l}\text{where } \rho = \text{mgu}(l, u),\\ l\rho \not\preceq r\rho,\ s\rho \not\preceq t\rho,\\ \text{and } u \text{ not a variable,}\end{array} \tag{17}$$

where $s[u \mapsto r]\rho$ and $t\rho$ are ground connected under $\{l \approx r \vee C,\ s \approx t \vee D\}$ and unifier ρ wrt. some set of clauses S, are redundant inferences wrt. S.

Theorem 8. Superposition inferences of the form

$$\frac{\underline{l \approx r} \vee C \quad \underline{s[u] \not\approx t} \vee D}{(s[u \mapsto r] \not\approx t \vee C \vee D)\rho}, \quad \begin{array}{l}\text{where } \rho = \text{mgu}(l, u),\\ l\rho \not\preceq r\rho,\ s\rho \not\preceq t\rho,\\ \text{and } u \text{ not a variable,}\end{array} \tag{18}$$

where $s[u \mapsto r]\rho$ and $t\rho$ are ground connected under $\{l \approx r \vee C,\ s \not\approx t \vee D\}$ and unifier ρ wrt. some set of clauses S, are redundant inferences wrt. $S \cup \{(C \vee D)\rho\}$.

Proof. The proof of Theorem 7 and 8 is analogous to that of Theorem 5 and 6. The weakening of connectedness to ground connectedness only means that the proof of connectedness (e.g. the v_i, $l_i \approx r_i$, σ_i) may be different for different ground instances. For all the steps in the proof to hold we only need that for all the instances $\theta \in \text{GSubs}(l \approx r \vee C\ ,\ s \dot{\approx} t \vee D\ ,\ (s[u \mapsto r] \dot{\approx} t \vee C \vee D)\rho)$ of the inference, $\theta = \sigma\theta'$ with $\sigma \in \text{GSubs}(s[u \mapsto r]\rho, t\rho)$, which is true. □

Discussion about the strategy for implementation of connectedness and ground connectedness is outside the scope of this paper.

5 Evaluation

We implemented ground joinability in a theorem prover for first-order logic, iProver [10,16].[5] iProver combines superposition, Inst-Gen, and resolution calculi. For superposition, iProver implements a range of simplifications including encompassment demodulation, AC normalisation [10], light normalisation [16], subsumption and subsumption resolution. We run our experiments over FOF problems of the TPTP v7.5 library [23] (17 348 problems) on a cluster of Linux servers with 3 GHz 11 core CPUs, 128 GB memory, with each problem running on a single core with a time limit of 300 s. We used a default strategy (which has not yet been fine-tuned after the introduction of ground joinability), with superposition enabled and the rest of the components disabled. With ground joinability enabled, iProver solved 133 problems more which it did not solve without ground joinability. Note that this excludes the contribution of AC ground joinability or encompassment demodulation [11] (always enabled).

Some of the problems are not interesting for this analysis because ground joinability is not even tried, either because they are solved before superposition saturation begins, or because they are ground. If we exclude these, we are left with 10 005 problems. Ground joinability is successfully used to eliminate clauses in 3057 of them (30.6%, Fig. 1a). This indicates that ground joinability is useful in many classes of problems, including in non-unit problems where it previously had never been used.

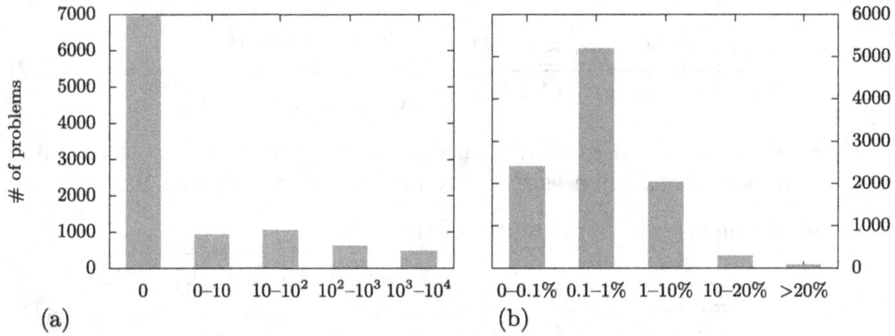

Fig. 1. (a) Clauses simplified by ground joinability. (b) % of runtime spent in gr. joinability

In terms of the performance impact of enabling ground joinability, we measure that among problems whose runtime exceeds 1 s, only in 72 out of 8574 problems does the time spent inside the ground joinability algorithm exceed 20% of runtime, indicating that our incremental algorithm is efficient and suitable for broad application (Fig. 1b).

[5] iProver is available at http://www.cs.man.ac.uk/~korovink/iprover.

TPTP classifies problems by rating in [0,1]. Problems with rating $\geq$0.9 are considered to be very challenging. Problems with rating 1.0 have never been solved by any automated theorem prover. iProver using ground joinability solves 3 previously unsolved rating 1.0 problems, and 7 further problems with rating in [0.9,1.0[(Table 1). We note that some of these latter (e.g. `LAT140-1`, `ROB018-10`, `REL045-1`) were previously only solved by UEQ or SMT provers, but not by any full first-order prover.

Table 1. Hard or unsolved problems in TPTP, solved by iProver with ground joinability.

Name	Rating	Name	Rating
LAT140-1	0.90	ROB018-10	0.95
REL045-1	0.90	LCL477+1	0.97
LCL557+1	0.92	LCL478+1	1.00
LCL563+1	0.92	CSR039+6	1.00
LCL474+1	0.94	CSR040+6	1.00

6 Conclusion and Further Work

In this work we extended the superposition calculus with ground joinability and connectedness, and proved that these rules preserve completness using a modified notion of redundancy, thus bringing for the first time these techniques for use in full first-order logic problems. We have also presented an algorithm for checking ground joinability which attempts to check as few variable preorders as possible.

Preliminary results show three things: (1) ground joinability is applicable in a sizeable number of problems across different domains, including in non-unit problems (where it was never applied before), (2) our proposed algorithm for checking ground joinability is efficient, with over $\frac{3}{4}$ of problems spending less than 1% of runtime there, and (3) application of ground joinability in the superposition calculus of iProver improves overall performance, including discovering solutions to hitherto unsolved problems.

These results are promising, and further optimisations can be done. Immediate next steps include fine-tuning the implementation, namely adjusting the strategies and strategy combinations to make full use of ground joinability and connectedness. iProver uses a sophisticated heuristic system which has not yet been tuned for ground joinability and connectedness [14].

In terms of practical implementation of connectedness and ground connectedness, further research is needed on the interplay between those (criteria for redundancy of inferences) and joinability and ground joinability (criteria for redundancy of clauses).

On the theoretical level, recent work [24] provides a general framework for saturation theorem proving, and we will investigate how techniques developed in this paper can be incorporated into this framework.

References

1. Avenhaus, J., Hillenbrand, T., Löchner, B.: On using ground joinable equations in equational theorem proving. J. Symb. Comput. **36**(1), 217–233 (2003). https://doi.org/10.1016/S0747-7171(03)00024-5
2. Baader, F., Nipkow, T.: Term Rewriting and All That. Cambridge University Press (1998). ISBN 978-0521779203
3. Bachmair, L., Dershowitz, N.: Critical pair criteria for completion. J. Symb. Comput. **6**(1), 1–18 (1988). https://doi.org/10.1016/S0747-7171(88)80018-X
4. Bachmair, L., Dershowitz, N., Plaisted, D.A.: Completion without failure. In: Aït-Kaci, H., Nivat, M. (eds.) Resolution of Equations in Algebraic Structures, vol. II: Rewriting Techniques, pp. 1–30. Academic Press (1989). https://doi.org/10.1016/B978-0-12-046371-8.50007-9
5. Bachmair, L., Ganzinger, H.: Rewrite-based equational theorem proving with selection and simplification. J. Log. Comput. **4**(3), 217–247 (1994). https://doi.org/10.1093/logcom/4.3.217
6. Bachmair, L., Ganzinger, H., Lynch, C.A., Snyder, W.: Basic paramodulation. Inf. Comput. **121**(2), 172–192 (1995). https://doi.org/10.1006/inco.1995.1131. ISSN 0890-5401
7. Barthelemy, J.P.: An asymptotic equivalent for the number of total preorders on a finite set. Discret. Math. **29**(3), 311–313 (1980). https://doi.org/10.1016/0012-365x(80)90159-4
8. Claessen, K., Smallbone, N.: Efficient encodings of first-order horn formulas in equational logic. In: Galmiche, D., Schulz, S., Sebastiani, R. (eds.) IJCAR 2018. LNCS (LNAI), vol. 10900, pp. 388–404. Springer, Cham (2018). https://doi.org/10.1007/978-3-319-94205-6_26
9. Dershowitz, N., Manna, Z.: Proving termination with multiset orderings. Commun. ACM **22**(8), 465–476 (1979). https://doi.org/10.1145/359138.359142
10. Duarte, A., Korovin, K.: Implementing superposition in iProver (system description). In: Peltier, N., Sofronie-Stokkermans, V. (eds.) IJCAR 2020. LNCS (LNAI), vol. 12167, pp. 388–397. Springer, Cham (2020). https://doi.org/10.1007/978-3-030-51054-1_24
11. Duarte, A., Korovin, K.: AC simplifications and closure redundancies in the superposition calculus. In: Das, A., Negri, S. (eds.) TABLEAUX 2021. LNCS (LNAI), vol. 12842, pp. 200–217. Springer, Cham (2021). https://doi.org/10.1007/978-3-030-86059-2_12
12. Duarte, A., Korovin, K.: Ground Joinability and Connectedness in the Superposition Calculus (2022, to appear)
13. Hillenbrand, T., Buch, A., Vogt, R., Löchner, B.: Waldmeister—high-performance equational deduction. J. Autom. Reason. **18**(2), 265–270 (1997). https://doi.org/10.1023/A:1005872405899
14. Holden, E.K., Korovin, K.: Heterogeneous heuristic optimisation and scheduling for first-order theorem proving. In: Kamareddine, F., Sacerdoti Coen, C. (eds.) CICM 2021. LNCS (LNAI), vol. 12833, pp. 107–123. Springer, Cham (2021). https://doi.org/10.1007/978-3-030-81097-9_8
15. Knuth, D.E., Bendix, P.: Simple word problems in universal algebras. In: Leech, J. (ed.) Computational Problems in Abstract Algebra, pp. 263–297. Pergamon (1970). https://doi.org/10.1016/B978-0-08-012975-4.50028-X
16. Korovin, K.: iProver—an instantiation-based theorem prover for first-order logic (system description). In: Armando, A., Baumgartner, P., Dowek, G. (eds.) IJCAR

2008. LNCS (LNAI), vol. 5195, pp. 292–298. Springer, Heidelberg (2008). https://doi.org/10.1007/978-3-540-71070-7_24
17. Kovács, L., Voronkov, A.: First-order theorem proving and VAMPIRE. In: Sharygina, N., Veith, H. (eds.) CAV 2013. LNCS, vol. 8044, pp. 1–35. Springer, Heidelberg (2013). https://doi.org/10.1007/978-3-642-39799-8_1
18. Martin, U., Nipkow, T.: Ordered rewriting and confluence. In: Stickel, M.E. (ed.) CADE 1990. LNCS, vol. 449, pp. 366–380. Springer, Heidelberg (1990). https://doi.org/10.1007/3-540-52885-7_100
19. Nieuwenhuis, R., Rubio, A.: Paramodulation-based theorem proving. In: Robinson, J.A., Voronkov, A. (eds.) Handbook of Automated Reasoning, vol. 2, pp. 371–443. Elsevier and MIT Press (2001). ISBN 0-444-50813-9
20. Schulz, S.: System description: E 1.8. In: McMillan, K., Middeldorp, A., Voronkov, A. (eds.) LPAR 2013. LNCS, vol. 8312, pp. 735–743. Springer, Heidelberg (2013). https://doi.org/10.1007/978-3-642-45221-5_49
21. Smallbone, N.: Twee: an equational theorem prover. In: Platzer, A., Sutcliffe, G. (eds.) CADE 2021. LNCS (LNAI), vol. 12699, pp. 602–613. Springer, Cham (2021). https://doi.org/10.1007/978-3-030-79876-5_35
22. Sutcliffe, G.: The CADE ATP system competition–CASC. AI Mag. **37**(2), 99–101 (2016). https://doi.org/10.1609/aimag.v37i2.2620
23. Sutcliffe, G.: The TPTP problem library and associated infrastructure–from CNF to TH0, TPTP v6.4.0. J. Autom. Reason. **59**(4), 483–502 (2017). https://doi.org/10.1007/s10817-017-9407-7
24. Waldmann, U., Tourret, S., Robillard, S., Blanchette, J.: A comprehensive framework for saturation theorem proving. In: Peltier, N., Sofronie-Stokkermans, V. (eds.) IJCAR 2020. LNCS (LNAI), vol. 12166, pp. 316–334. Springer, Cham (2020). https://doi.org/10.1007/978-3-030-51074-9_18
25. Winkler, S., Moser, G.: MædMax: a maximal ordered completion tool. In: Galmiche, D., Schulz, S., Sebastiani, R. (eds.) IJCAR 2018. LNCS (LNAI), vol. 10900, pp. 472–480. Springer, Cham (2018). https://doi.org/10.1007/978-3-319-94205-6_31

Connection-Minimal Abduction in $\mathcal{EL}$ via Translation to FOL

Fajar Haifani[1,2(✉)], Patrick Koopmann[3(✉)], Sophie Tourret[1,4(✉)], and Christoph Weidenbach[1(✉)]

[1] Max-Planck-Institut für Informatik, Saarland Informatics Campus, Saarbrücken, Germany
{f.haifani,c.weidenbach}@mpi-inf.mpg.de
[2] Graduate School of Computer Science, Saarbrücken, Germany
[3] TU Dresden, Dresden, Germany
patrick.koopmann@tu-dresden.de
[4] Université de Lorraine, CNRS, Inria, LORIA, Nancy, France
sophie.tourret@inria.fr

Abstract. Abduction in description logics finds extensions of a knowledge base to make it entail an observation. As such, it can be used to explain why the observation does not follow, to repair incomplete knowledge bases, and to provide possible explanations for unexpected observations. We consider TBox abduction in the lightweight description logic $\mathcal{EL}$, where the observation is a concept inclusion and the background knowledge is a TBox, i.e., a set of concept inclusions. To avoid useless answers, such problems usually come with further restrictions on the solution space and/or minimality criteria that help sort the chaff from the grain. We argue that existing minimality notions are insufficient, and introduce connection minimality. This criterion follows Occam's razor by rejecting hypotheses that use concept inclusions unrelated to the problem at hand. We show how to compute a special class of connection-minimal hypotheses in a sound and complete way. Our technique is based on a translation to first-order logic, and constructs hypotheses based on prime implicates. We evaluate a prototype implementation of our approach on ontologies from the medical domain.

1 Introduction

Ontologies are used in areas like biomedicine or the semantic web to represent and reason about terminological knowledge. They consist normally of a set of axioms formulated in a description logic (DL), giving definitions of concepts, or stating relations between them. In the lightweight description logic $\mathcal{EL}$ [2], particularly used in the biomedical domain, we find ontologies that contain around a hundred thousand axioms. For instance, SNOMED CT[1] contains over 350,000 axioms, and the Gene Ontology GO[2] defines over 50,000 concepts. A central

[1] https://www.snomed.org/.
[2] http://geneontology.org/.

J. Blanchette et al. (Eds.): IJCAR 2022, LNAI 13385, pp. 188–207, 2022.
https://doi.org/10.1007/978-3-031-10769-6_12

reasoning task for ontologies is to determine whether one concept is subsumed by another, a question that can be answered in polynomial time [1], and rather efficiently in practice using highly optimized description logic reasoners [29]. If the answer to this question is unexpected or hints at an error, a natural interest is in an explanation for that answer—especially if the ontology is complex. But whereas explaining entailments—i.e., explaining why a concept subsumption holds—is well-researched in the DL literature and integrated into standard ontology editors [21,22], the problem of explaining non-entailments has received less attention, and there is no standard tool support. Classical approaches involve counter-examples [5], or *abduction*.

In abduction a non-entailment $\mathcal{T} \not\models \alpha$, for a TBox $\mathcal{T}$ and an observation α, is explained by providing a "missing piece", the *hypothesis*, that, when added to the ontology, would entail α. Thus it provides possible fixes in case the entailment should hold. In the DL context, depending on the shape of the observation, one distinguishes between concept abduction [6], ABox abduction [7–10,12,19,24, 25,30,31], TBox abduction [11,33] or knowledge base abduction [14,26]. We are focusing here on TBox abduction, where the ontology and hypothesis are TBoxes and the observation is a concept inclusion (CI), i.e., a single TBox axiom.

To illustrate this problem, consider the following TBox, about academia,

$$\begin{aligned}\mathcal{T}_{\mathrm{a}} = \{\ & \exists \mathsf{employment.ResearchPosition} \sqcap \exists \mathsf{qualification.Diploma} \sqsubseteq \mathsf{Researcher},\\ & \exists \mathsf{writes.ResearchPaper} \sqsubseteq \mathsf{Researcher},\ \mathsf{Doctor} \sqsubseteq \exists \mathsf{qualification.PhD},\\ & \mathsf{Professor} \equiv \mathsf{Doctor} \sqcap \exists \mathsf{employment.Chair},\\ & \mathsf{FundsProvider} \sqsubseteq \exists \mathsf{writes.GrantApplication}\ \}\end{aligned}$$

that states, in natural language:

- "Being employed in a research position and having a qualifying diploma implies being a researcher."
- "Writing a research paper implies being a researcher."
- "Being a doctor implies holding a PhD qualification."
- "Being a professor is being a doctor employed at a (university) chair."
- "Being a funds provider implies writing grant applications."

The observation $\alpha_{\mathrm{a}} = \mathsf{Professor} \sqsubseteq \mathsf{Researcher}$, "Being a professor implies being a researcher", does not follow from $\mathcal{T}_{\mathrm{a}}$ although it should. We can use TBox abduction to find different ways of recovering this entailment.

Commonly, to avoid trivial answers, the user provides syntactic restrictions on hypotheses, such as a set of abducible axioms to pick from [8,30], a set of abducible predicates [25,26], or patterns on the shape of the solution [11]. But even with those restrictions in place, there may be many possible solutions and, to find the ones with the best explanatory potential, syntactic criteria are usually combined with minimality criteria such as subset minimality, size minimality, or semantic minimality [7]. Even combined, these minimality criteria still retain a major flaw. They allow for explanations that go against the principle of parsimony, also known as Occam's razor, in that they may contain concepts that are completely unrelated to the problem at hands. As an illustration, let us return to our academia example. The TBoxes

$$\mathcal{H}_{a1} = \{\ \mathsf{Chair} \sqsubseteq \mathsf{ResearchPosition},\ \mathsf{PhD} \sqsubseteq \mathsf{Diploma}\} \text{ and}$$
$$\mathcal{H}_{a2} = \{\ \mathsf{Professor} \sqsubseteq \mathsf{FundsProvider},\ \mathsf{GrantApplication} \sqsubseteq \mathsf{ResearchPaper}\}$$

are two hypotheses solving the TBox abduction problem involving $\mathcal{T}_a$ and α_a. Both of them are subset-minimal, have the same size, and are incomparable w.r.t. the entailment relation, so that traditional minimality criteria cannot distinguish them. However, intuitively, the second hypothesis feels more arbitrary than the first. Looking at $\mathcal{H}_{a1}$, Chair and $\mathsf{ResearchPosition}$ occur in $\mathcal{T}_a$ in concept inclusions where the concepts in α_a also occur, and both PhD and $\mathsf{Diploma}$ are similarly related to α_a but via the role $\mathsf{qualification}$. In contrast, $\mathcal{H}_{a2}$ involves the concepts $\mathsf{FundsProvider}$ and $\mathsf{GrantApplication}$ that are not related to α_a in any way in $\mathcal{T}_a$. In fact, any random concept inclusion $A \sqsubseteq \exists \mathsf{writes}.B$ in $\mathcal{T}_a$ would lead to a hypothesis similar to $\mathcal{H}_{a2}$ where A replaces $\mathsf{FundsProvider}$ and B replaces $\mathsf{GrantApplication}$. Such explanations are not parsimonious.

We introduce a new minimality criterion called *connection minimality* that is parsimonious (Sect. 3), defined for the lightweight description logic $\mathcal{EL}$. This criterion characterizes hypotheses for $\mathcal{T}$ and α that connect the left- and right-hand sides of the observation α without introducing spurious connections. To achieve this, every left-hand side of a CI in the hypothesis must follow from the left-hand side of α in $\mathcal{T}$, and, taken together, all the right-hand sides of the CIs in the hypothesis must imply the right-hand side of α in $\mathcal{T}$, as is the case for $\mathcal{H}_{a1}$. To compute connection-minimal hypotheses in practice, we present a technique based on first-order reasoning that proceeds in three steps (Sect. 4). First, we translate the abduction problem into a first-order formula Φ. We then compute the prime implicates of Φ, that is, a set of minimal logical consequences of Φ that subsume all other consequences of Φ. In the final step, we construct, based on those prime implicates, solutions to the original problem. We prove that all hypotheses generated in this way satisfy the connection minimality criterion, and that the method is complete for a relevant subclass of connection-minimal hypotheses. We use the SPASS theorem prover [34] as a restricted SOS-resolution [18,35] engine for the computation of prime implicates in a prototype implementation (Sect. 5), and we present an experimental analysis of its performances on a set of bio-medical ontologies.(Sect. 6). Our results indicate that our method can in many cases be applied in practice to compute connection-minimal hypotheses. A technical report companion of this paper includes all proofs as well as a detailed example of our method as appendices [16].

There are not many techniques that can handle TBox abduction in $\mathcal{EL}$ or more expressive DLs [11,26,33]. In [11], instead of a set of abducibles, a set of *justification patterns* is given, in which the solutions have to fit. An arbitrary oracle function is used to decide whether a solution is admissible or not (which may use abducibles, justification patterns, or something else), and it is shown that deciding the existence of hypotheses is tractable. However, different to our approach, they only consider atomic CIs in hypotheses, while we also allow for hypotheses involving conjunction. The setting from [33] also considers $\mathcal{EL}$, and abduction under various minimality notions such as subset minimality and size minimality. It presents practical algorithms, and an evaluation of an implementation for an always-true informativeness oracle (i.e., limited to subset minimality). Different to our approach, it uses an external DL reasoner to decide entailment relationships. In contrast, we present an approach that directly exploits first-order reasoning, and thus has the potential to be generalisable to more expressive DLs.

While dedicated resolution calculi have been used before to solve abduction in DLs [9,26], to the best of our knowledge, the only work that relies on first-order reasoning for DL abduction is [24]. Similar to our approach, it uses SOS-resolution, but to perform ABox adbuction for the more expressive DL $\mathcal{ALC}$. Apart from the different problem solved, in contrast to [24] we also provide a semantic characterization of the hypotheses generated by our method. We believe this characterization to be a major contribution of our paper. It provides an intuition of what parsimony is for this problem, independently of one's ease with first-order logic calculi, which should facilitate the adoption of this minimality criterion by the DL community. Thanks to this characterization, our technique is calculus agnostic. Any method to compute prime implicates in first-order logic can be a basis for our abduction technique, without additional theoretical work, which is not the case for [24]. Thus, abduction in $\mathcal{EL}$ can benefit from the latest advances in prime implicates generation in first-order logic.

2 Preliminaries

We first recall the descripton logic $\mathcal{EL}$ and its translation to first-order logic [2], as well as TBox abduction in this logic.

Let $\mathsf{N_C}$ and $\mathsf{N_R}$ be pair-wise disjoint, countably infinite sets of unary predicates called *atomic concepts* and of binary predicates called *roles*, respectively. Generally, we use letters A, B, E, F,... for atomic concepts, and r for roles, possibly annotated. Letters C, D, possibly annotated, denote $\mathcal{EL}$ *concepts*, built according to the syntax rule

$$C ::= \top \mid A \mid C \sqcap C \mid \exists r.C\ .$$

We implicitly represent $\mathcal{EL}$ conjunctions as sets, that is, without order, nested conjunctions, and multiple occurrences of a conjunct. We use $\bigsqcap\{C_1, \ldots, C_m\}$ to abbreviate $C_1 \sqcap \ldots \sqcap C_m$, and identify the empty conjunction ($m = 0$) with $\top$. An $\mathcal{EL}$ *TBox* $\mathcal{T}$ is a finite set of *concept inclusions* (CIs) of the form $C \sqsubseteq D$.

$\mathcal{EL}$ is a syntactic variant of a fragment of first-order logic that uses $\mathsf{N_C}$ and $\mathsf{N_R}$ as predicates. Specifically, TBoxes $\mathcal{T}$ and CIs α correspond to closed first-order formulas $\pi(\mathcal{T})$ and $\pi(\alpha)$ resp., while concepts C correspond to open formulas $\pi(C, x)$ with a free variable x. In particular, we have

$$\pi(\top, x) := \mathbf{true}, \qquad \pi(\exists r.C, x) := \exists y.(r(x,y) \wedge \pi(C,y)),$$
$$\pi(A, x) := A(x), \qquad \pi(C \sqsubseteq D) := \forall x.(\pi(C,x) \rightarrow \pi(D,x)),$$
$$\pi(C \sqcap D, x) := \pi(C,x) \wedge \pi(D,x), \qquad \pi(\mathcal{T}) := \bigwedge\{\pi(\alpha) \mid \alpha \in \mathcal{T}\}.$$

As common, we often omit the $\bigwedge$ in conjunctions $\bigwedge \Phi$, that is, we identify sets of formulas with the conjunction over those. The notions of a *term* t; an *atom* $P(\bar{t})$ where $\bar{t}$ is a sequence of terms; a *positive literal* $P(\bar{t})$; a *negative literal* $\neg P(\bar{t})$; and a clause, Horn, definite, positive or negative, are defined as usual for first-order logic, and so are entailment and satisfaction of first-order formulas.

We identify CIs and TBoxes with their translation into first-order logic, and can thus speak of the entailment between formulas, CIs and TBoxes. When $\mathcal{T} \models C \sqsubseteq D$ for some $\mathcal{T}$, we call C a *subsumee* of D and D a *subsumer* of C. We adhere here to the definition of the word "subsume": "to include or contain something else", although the terminology is reversed in first-order logic. We say two TBoxes $\mathcal{T}_1$, $\mathcal{T}_2$ are *equivalent*, denoted $\mathcal{T}_1 \equiv \mathcal{T}_2$ iff $\mathcal{T}_1 \models \mathcal{T}_2$ and $\mathcal{T}_2 \models \mathcal{T}_1$. For example $\{D \sqsubseteq C_1, \ldots, D \sqsubseteq C_n\} \equiv \{D \sqsubseteq C_1 \sqcap \ldots \sqcap C_n\}$. It is well known that, due to the absence of concept negation, every $\mathcal{EL}$ TBox is consistent.

The abduction problem we are concerned with in this paper is the following:

Definition 1. *An $\mathcal{EL}$* TBox abduction problem *(shortened to* abduction problem*) is a tuple $\langle \mathcal{T}, \Sigma, C_1 \sqsubseteq C_2 \rangle$, where $\mathcal{T}$ is a TBox called the* background knowledge*, Σ is a set of atomic concepts called the* abducible signature*, and $C_1 \sqsubseteq C_2$ is a CI called the* observation*, s.t. $\mathcal{T} \not\models C_1 \sqsubseteq C_2$. A solution to this problem is a TBox*

$$\mathcal{H} \subseteq \{A_1 \sqcap \cdots \sqcap A_n \sqsubseteq B_1 \sqcap \cdots \sqcap B_m \mid \{A_1, \ldots, A_n, B_1, \ldots, B_m\} \subseteq \Sigma\}$$

where $m > 0$, $n \geq 0$ and such that $\mathcal{T} \cup \mathcal{H} \models C_1 \sqsubseteq C_2$ and, for all CIs $\alpha \in \mathcal{H}$, $\mathcal{T} \not\models \alpha$. A solution to an abduction problem is called a hypothesis.

For example, $\mathcal{H}_{a1}$ and $\mathcal{H}_{a2}$ are solutions for $\langle \mathcal{T}_a, \Sigma, \alpha_a \rangle$, as long as Σ contains all the atomic concepts that occur in them. Note that in our setting, as in [6, 33], concept inclusions in a hypothesis are *flat*, i.e., they contain no existential role restrictions. While this restricts the solution space for a given problem, it is possible to bypass this limitation in a targeted way, by introducing fresh atomic concepts equivalent to a concept of interest. We exclude the consistency requirement $\mathcal{T} \cup \mathcal{H} \not\models \bot$, that is given in other definitions of DL abduction problem [25], since $\mathcal{EL}$ TBoxes are always consistent. We also allow $m > 1$ instead of the usual $m = 1$. This produces the same hypotheses modulo equivalence.

For simplicity, we assume in the following that the concepts C_1 and C_2 in the abduction problem are atomic. We can always introduce fresh atomic concepts A_1 and A_2 with $A_1 \sqsubseteq C_1$ and $C_2 \sqsubseteq A_2$ to solve the problem for complex concepts.

Common minimality criteria include *subset* minimality, *size* minimality and *semantic* minimality, that respectively favor $\mathcal{H}$ over $\mathcal{H}'$ if: $\mathcal{H} \subsetneq \mathcal{H}'$; the number of atomic concepts in $\mathcal{H}$ is smaller than in $\mathcal{H}'$; and if $\mathcal{H} \models \mathcal{H}'$ but $\mathcal{H}' \not\models \mathcal{H}$.

3 Connection-Minimal Abduction

To address the lack of parsimony of common minimality criteria, illustrated in the academia example, we introduce *connection* minimality, Intuitively, connection minimality only accepts those hypotheses that ensure that every CI in the hypothesis is connected to both C_1 and C_2 in $\mathcal{T}$, as is the case for $\mathcal{H}_{a1}$ in the academia example. The definition of connection minimality is based on the following ideas: 1) Hypotheses for the abduction problem should create a *connection* between C_1 and C_2, which can be seen as a concept D that satisfies $\mathcal{T} \cup \mathcal{H} \models C_1 \sqsubseteq D$, $D \sqsubseteq C_2$. 2) To ensure parsimony, we want this connection to be based on concepts D_1 and D_2 for which we already have $\mathcal{T} \models C_1 \sqsubseteq D_1$, $D_2 \sqsubseteq C_2$. This prevents the introduction of unrelated concepts in the hypothesis. Note however that D_1 and D_2 can be complex, thus the connection from C_1 to D_1 (resp. D_2 to C_2) can be established by arbitrarily long chains of concept inclusions. 3) We additionally want to make sure that the connecting concepts are not more complex than necessary, and that $\mathcal{H}$ only contains CIs that directly connect parts of D_2 to parts of D_1 by closely following their structure.

To address point 1), we simply introduce connecting concepts formally.

Definition 2. *Let C_1 and C_2 be concepts. A concept D* connects *C_1 to C_2 in $\mathcal{T}$ if and only if $\mathcal{T} \models C_1 \sqsubseteq D$ and $\mathcal{T} \models D \sqsubseteq C_2$.*

Note that if $\mathcal{T} \models C_1 \sqsubseteq C_2$ then both C_1 and C_2 are connecting concepts from C_1 to C_2, and if $\mathcal{T} \not\models C_1 \sqsubseteq C_2$, the case of interest, neither of them are.

To address point 2), we must capture *how* a hypothesis creates the connection between the concepts C_1 and C_2. As argued above, this is established via concepts D_1 and D_2 that satisfy $\mathcal{T} \models C_1 \sqsubseteq D_1$, $D_2 \sqsubseteq C_2$. Note that having only two concepts D_1 and D_2 is exactly what makes the approach parsimonious. If there was only one concept, C_1 and C_2 would already be connected, and as soon as there are more than two concepts, hypotheses start becoming more arbitrary: for a very simple example with unrelated concepts, assume given a TBox that entails $\mathsf{Lion} \sqsubseteq \mathsf{Felidae}$, $\mathsf{Mammal} \sqsubseteq \mathsf{Animal}$ and $\mathsf{House} \sqsubseteq \mathsf{Building}$. A possible hypothesis to explain $\mathsf{Lion} \sqsubseteq \mathsf{Animal}$ is $\{\mathsf{Felidae} \sqsubseteq \mathsf{House}, \mathsf{Building} \sqsubseteq \mathsf{Mammal}\}$ but this explanation is more arbitrary than $\{\mathsf{Felidae} \sqsubseteq \mathsf{Mammal}\}$—as is the case when comparing $\mathcal{H}_{a2}$ with $\mathcal{H}_{a1}$ in the academia example—because of the lack of connection of $\mathsf{House} \sqsubseteq \mathsf{Building}$ with both Lion and Animal. Clearly this CI could be replaced by any other CI entailed by $\mathcal{T}$, which is what we want to avoid.

We can represent the structure of D_1 and D_2 in graphs by using *$\mathcal{EL}$ description trees*, originally from Baader et al. [3].

Definition 3. *An $\mathcal{EL}$* description tree *is a finite labeled tree $\mathfrak{T} = (V, E, v_0, l)$ where V is a set of nodes with root $v_0 \in V$, the nodes $v \in V$ are labeled with $l(v) \subseteq \mathsf{N_C}$, and the (directed) edges $vrw \in E$ are such that $v, w \in V$ and are labeled with $r \in \mathsf{N_R}$.*

Given a tree $\mathfrak{T} = (V, E, v_0, l)$ and $v \in V$, we denote by $\mathfrak{T}(v)$ the subtree of $\mathfrak{T}$ that is rooted in v. If $l(v_0) = \{A_1, \ldots, A_k\}$ and $v_1, \ldots, v_n$ are all the children of v_0, we

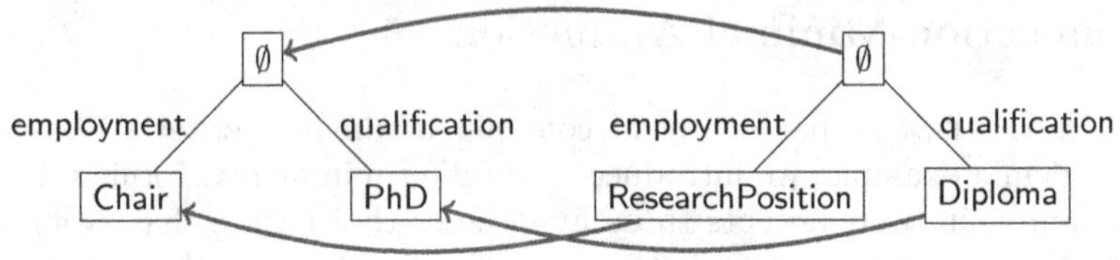

Fig. 1. Description trees of D_1 (left) and D_2 (right).

can define the concept represented by $\mathfrak{T}$ recursively using $C_{\mathfrak{T}} = A_1 \sqcap \ldots \sqcap A_k \sqcap \exists r_1.C_{\mathfrak{T}(v_1)} \sqcap \ldots \sqcap \exists r_l.C_{\mathfrak{T}(v_l)}$ where for $j \in \{1, \ldots, n\}$, $v_0 r_j v_j \in E$. Conversely, we can define $\mathfrak{T}_C$ for a concept $C = A_1 \sqcap \ldots \sqcap A_k \sqcap \exists r_1.C_1 \sqcap \ldots \sqcap \exists r_n.C_n$ inductively based on the pairwise disjoint description trees $\mathfrak{T}_{C_i} = \{V_i, E_i, v_i, l_i\}$, $i \in \{1, \ldots, n\}$. Specifically, $\mathfrak{T}_C = (V_C, E_C, v_C, l_C)$, where

$$V_C = \{v_0\} \cup \bigcup_{i=1}^{n} V_i, \qquad l_C(v) = l_i(v) \text{ for } v \in V_i,$$
$$E_C = \{v_0 r_i v_i \mid 1 \leq i \leq n\} \cup \bigcup_{i=1}^{n} E_i, \qquad l_C(v_0) = \{A_1, \ldots, A_k\}.$$

If $\mathcal{T} = \emptyset$, then subsumption between $\mathcal{EL}$ concepts is characterized by the existence of a homomorphism between the corresponding description trees [3]. We generalise this notion to also take the TBox into account.

Definition 4. *Let $\mathfrak{T}_1 = (V_1, E_1, v_0, l_1)$ and $\mathfrak{T}_2 = (V_2, E_2, w_0, l_2)$ be two description trees and $\mathcal{T}$ a TBox. A mapping $\phi : V_2 \to V_1$ is a* $\mathcal{T}$-homomorphism *from $\mathfrak{T}_2$ to $\mathfrak{T}_1$ if and only if the following conditions are satisfied:*

1. *$\phi(w_0) = v_0$*
2. *$\phi(v) r \phi(w) \in E_1$ for all $vrw \in E_2$*
3. *for every $v \in V_1$ and $w \in V_2$ with $v = \phi(w)$, $\mathcal{T} \models \bigsqcap l_1(v) \sqsubseteq \bigsqcap l_2(w)$*

If only 1 and 2 are satisfied, then ϕ is called a weak *homomorphism.*

$\mathcal{T}$-homomorphisms for a given TBox $\mathcal{T}$ capture subsumption w.r.t. $\mathcal{T}$. If there exists a $\mathcal{T}$-homomorphism ϕ from $\mathfrak{T}_2$ to $\mathfrak{T}_1$, then $\mathcal{T} \models C_{\mathfrak{T}_1} \sqsubseteq C_{\mathfrak{T}_2}$. This can be shown easily by structural induction using the definitions [16]. The weak homomorphism is the structure on which a $\mathcal{T}$-homomorphism can be built by adding some hypothesis $\mathcal{H}$ to $\mathcal{T}$. It is used to reveal missing links between a subsumee D_2 of C_2 and a subsumer D_1 of C_1, that can be added using $\mathcal{H}$.

Example 5. Consider the concepts

$$D_1 = \exists \mathsf{employment}.\mathsf{Chair} \sqcap \exists \mathsf{qualification}.\mathsf{PhD}$$
$$D_2 = \exists \mathsf{employment}.\mathsf{ResearchPosition} \sqcap \exists \mathsf{qualification}.\mathsf{Diploma}$$

from the academia example. Figure 1 illustrates description trees for D_1 (left) and D_2 (right). The curved arrows show a weak homomorphism from $\mathfrak{T}_{D_2}$ to $\mathfrak{T}_{D_1}$ that can be strengthened into a $\mathcal{T}$-homomorphism for some TBox $\mathcal{T}$ that corresponds to the set of CIs in $\mathcal{H}_{a1} \cup \{\top \sqsubseteq \top\}$. The figure can also be used to illustrate what we mean by connection minimality: in order to create a connection between D_1 and D_2, we should *only* add the CIs from $\mathcal{H}_{a1} \cup \{\top \sqsubseteq \top\}$ *unless* they are already entailed by $\mathcal{T}_a$. In practice, this means the weak homomorphism from D_2 to D_1 becomes a $(\mathcal{T}_a \cup \mathcal{H}_{a1})$-homomorphism.

To address point 3), we define a partial order $\preceq_\sqcap$ on concepts, s.t. $C \preceq_\sqcap D$ if we can turn D into C by removing conjuncts in subexpressions, e.g., $\exists r'.B \preceq_\sqcap \exists r.A \sqcap \exists r'.(B \sqcap B')$. Formally, this is achieved by the following definition.

Definition 6. *Let C and D be arbitrary concepts. Then $C \preceq_\sqcap D$ if either:*

- $C = D$,
- $D = D' \sqcap D''$, *and* $C \preceq_\sqcap D'$, *or*
- $C = \exists r.C'$, $D = \exists r.D'$ *and* $C' \preceq_\sqcap D'$.

We can finally capture our ideas on connection minimality formally.

Definition 7 (Connection-Minimal Abduction). *Given an abduction problem $\langle \mathcal{T}, \Sigma, C_1 \sqsubseteq C_2 \rangle$, a hypothesis $\mathcal{H}$ is* connection-minimal *if there exist concepts D_1 and D_2 built over $\Sigma \cup \mathsf{N_R}$ and a mapping ϕ satisfying each of the following conditions:*

1. $\mathcal{T} \models C_1 \sqsubseteq D_1$,
2. D_2 *is a* $\preceq_\sqcap$*-minimal concept s.t.* $\mathcal{T} \models D_2 \sqsubseteq C_2$,
3. ϕ *is a weak homomorphism from the tree* $\mathfrak{T}_{D_2} = (V_2, E_2, w_0, l_2)$ *to the tree* $\mathfrak{T}_{D_1} = (V_1, E_1, v_0, l_1)$, *and*
4. $\mathcal{H} = \{\bigsqcap l_1(\phi(w)) \sqsubseteq \bigsqcap l_2(w) \mid w \in V_2 \wedge \mathcal{T} \not\models \bigsqcap l_1(\phi(w)) \sqsubseteq \bigsqcap l_2(w)\}$.

$\mathcal{H}$ is additionally called packed *if the left-hand sides of the CIs in $\mathcal{H}$ cannot hold more conjuncts than they do, which is formally stated as: for $\mathcal{H}$, there is no $\mathcal{H}'$ defined from the same D_2 and a D_1' and ϕ' s.t. there is a node $w \in V_2$ for which $l_1(\phi(w)) \subsetneq l_1'(\phi'(w))$ and $l_1(\phi(w')) = l_1'(\phi'(w'))$ for $w' \neq w$.*

Straightforward consequences of Definition 7 include that ϕ is a $(\mathcal{T} \cup \mathcal{H})$-homomorphism from $\mathfrak{T}_{D_2}$ to $\mathfrak{T}_{D_1}$ and that D_1 and D_2 are connecting concepts from C_1 to C_2 in $\mathcal{T} \cup \mathcal{H}$ so that $\mathcal{T} \cup \mathcal{H} \models C_1 \sqsubseteq C_2$ as wanted [16]. With the help of Fig. 1 and Example 5, one easily establishes that hypothesis $\mathcal{H}_{a1}$ is connection-minimal—and even packed. Connection-minimality rejects $\mathcal{H}_{a2}$, as a single $\mathcal{T}'$-homomorphism for some $\mathcal{T}'$ between two concepts D_1 and D_2 would be insufficient: we would need two weak homomorphisms, one linking Professor to FundsProvider and another linking $\exists$writes.GrantApplication to $\exists$writes.ResearchPaper.

4 Computing Connection-Minimal Hypotheses Using Prime Implicates

To compute connection-minimal hypotheses in practice, we propose a method based on first-order prime implicates, that can be derived by resolution. We

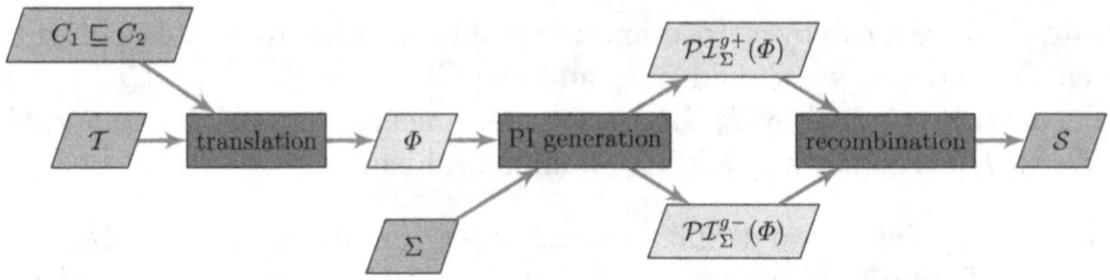

Fig. 2. $\mathcal{EL}$ abduction using prime implicate generation in FOL.

assume the reader is familiar with the basics of first-order resolution, and do not reintroduce notions of clauses, Skolemization and resolution inferences here (for details, see [4]). In our context, every term is built on variables, denoted x, y, a single constant $\mathtt{sk}_0$ and unary Skolem functions usually denoted $\mathtt{sk}$, possibly annotated. Prime implicates are defined as follows.

Definition 8 (Prime Implicate). *Let Φ be a set of clauses. A clause φ is an* implicate *of Φ if $\Phi \models \varphi$. Moreover φ is* prime *if for any other implicate φ' of Φ s.t. $\varphi' \models \varphi$, it also holds that $\varphi \models \varphi'$.*

Let $\Sigma \subseteq \mathsf{N_C}$ be a set of unary predicates. Then $\mathcal{PI}^{g+}_{\Sigma}(\Phi)$ denotes the set of all positive ground prime implicates of Φ that only use predicate symbols from $\Sigma \cup \mathsf{N_R}$, while $\mathcal{PI}^{g-}_{\Sigma}(\Phi)$ denotes the set of all negative ground prime implicates of Φ that only use predicates symbols from $\Sigma \cup \mathsf{N_R}$.

Example 9. Given a set of clauses $\Phi = \{A_1(\mathtt{sk}_0), \neg B_1(\mathtt{sk}_0), \neg A_1(x) \vee r(x, \mathtt{sk}(x)), \neg A_1(x) \vee A_2(\mathtt{sk}(x)), \neg B_2(x) \vee \neg r(x, y) \vee \neg B_3(y) \vee B_1(x)\}$, the ground prime implicates of Φ for $\Sigma = \mathsf{N_C}$ are, on the positive side, $\mathcal{PI}^{g+}_{\Sigma}(\Phi) = \{A_1(\mathtt{sk}_0), A_2(\mathtt{sk}(\mathtt{sk}_0)), r(\mathtt{sk}_0, \mathtt{sk}(\mathtt{sk}_0))\}$ and, on the negative side, $\mathcal{PI}^{g-}_{\Sigma}(\Phi) = \{\neg B_1(\mathtt{sk}_0), \neg B_2(\mathtt{sk}_0) \vee \neg B_3(\mathtt{sk}(\mathtt{sk}_0))\}$. They are implicates because all of them are entailed by Φ. For a ground implicate φ, another ground implicate φ' such that $\varphi' \models \varphi$ and $\varphi \not\models \varphi'$ can only be obtained from φ by dropping literals. Such an operation does not produce another implicate for any of the clauses presented above as belonging to $\mathcal{PI}^{g+}_{\Sigma}(\Phi)$and $\mathcal{PI}^{g-}_{\Sigma}(\Phi)$, thus they really are all prime.

To generate hypotheses, we translate the abduction problem into a set of first-order clauses, from which we can infer prime implicates that we then combine to obtain the result as illustrated in Fig. 2. In more details: We first translate the problem into a set Φ of Horn clauses. Prime implicates can be computed using an off-the-shelf tool [13,28] or, in our case, a slight extension of the resolution-based version of the SPASS theorem prover [34] using the set-of-support strategy and some added features described in Sect. 5. Since Φ is Horn, $\mathcal{PI}^{g+}_{\Sigma}(\Phi)$ contains only unit clauses. A final recombination step looks at the clauses in $\mathcal{PI}^{g-}_{\Sigma}(\Phi)$ one after the other. These correspond to candidates for the connecting concepts D_2 of Definition 7. Recombination attempts to match each literal in one such clause with unit clauses from $\mathcal{PI}^{g+}_{\Sigma}(\Phi)$. If such a match is possible, it produces a suitable D_1 to match D_2, and allows the creation of a solution to the abduction problem. The set $\mathcal{S}$ contains all the hypotheses thus obtained.

In what follows, we present our translation of abduction problems into first-order logic and formalize the construction of hypotheses from the prime implicates of this translation. We then show how to obtain termination for the prime implicate generation process with soundness and completeness guarantees on the solutions computed.

Abduction Method. We assume the $\mathcal{EL}$ TBox in the input is in normal form as defined, e.g., by Baader et al. [2]. Thus every CI is of one of the following forms:

$$A \sqsubseteq B \qquad A_1 \sqcap A_2 \sqsubseteq B \qquad \exists r.A \sqsubseteq B \qquad A \sqsubseteq \exists r.B$$

where A, A_1, A_2, $B \in \mathsf{N_C} \cup \{\top\}$.

The use of normalization is justified by the following lemma.

Lemma 10. *For every $\mathcal{EL}$ TBox $\mathcal{T}$, we can compute in polynomial time an $\mathcal{EL}$ TBox $\mathcal{T}'$ in normal form such that for every other TBox $\mathcal{H}$ and every CI $C \sqsubseteq D$ that use only names occurring in $\mathcal{T}$, we have $\mathcal{T} \cup \mathcal{H} \models C \sqsubseteq D$ iff $\mathcal{T}' \cup \mathcal{H} \models C \sqsubseteq D$.*

After the normalisation, we eliminate occurrences of $\top$, replacing this concept everywhere by the fresh atomic concept $A_\top$. We furthermore add $\exists r.A_\top \sqsubseteq A_\top$ and $B \sqsubseteq A_\top$ in $\mathcal{T}$ for every role r and atomic concept B occurring in $\mathcal{T}$. This simulates the semantics of $\top$ for $A_\top$, namely the implicit property that $C \sqsubseteq \top$ holds for any C no matter what the TBox is. In particular, this ensures that whenever there is a positive prime implicate $B(t)$ or $r(t, t')$, $A_\top(t)$ also becomes a prime implicate. Note that normalisation and $\top$ elimination extend the signature, and thus potentially the solution space of the abduction problem. This is remedied by intersecting the set of abducible predicates Σ with the signature of the original input ontology. We assume that $\mathcal{T}$ is in normal form and without $\top$ in the rest of the paper.

We denote by $\mathcal{T}^-$ the result of renaming all atomic concepts A in $\mathcal{T}$ using fresh *duplicate* symbols A^-. This renaming is done only on concepts but not on roles, and on C_2 but not on C_1 in the observation. This ensures that the literals in a clause of $\mathcal{PI}^{g-}_{\Sigma}(\Phi)$ all relate to the conjuncts of a $\preceq_\sqcap$-minimal subsumee of C_2. Without it, some of these conjuncts would not appear in the negative implicates due to the presence of their positive counterparts as atoms in $\mathcal{PI}^{g+}_{\Sigma}(\Phi)$. The translation of the abduction problem $\langle \mathcal{T}, \Sigma, C_1 \sqsubseteq C_2 \rangle$ is defined as the Skolemization of

$$\pi(\mathcal{T} \uplus \mathcal{T}^-) \wedge \neg\pi(C_1 \sqsubseteq C_2^-)$$

where $\mathbf{sk}_0$ is used as the unique fresh Skolem constant such that the Skolemization of $\neg\pi(C_1 \sqsubseteq C_2^-)$ results in $\{C_1(\mathbf{sk}_0), \neg C_2^-(\mathbf{sk}_0)\}$. This translation is usually denoted Φ and always considered in clausal normal form.

Theorem 11. *Let $\langle \mathcal{T}, \Sigma, C_1 \sqsubseteq C_2 \rangle$ be an abduction problem and Φ be its first-order translation. Then, a TBox $\mathcal{H}'$ is a packed connection-minimal solution to the problem if and only if an equivalent hypothesis $\mathcal{H}$ can be constructed from non-empty sets $\mathcal{A}$ and $\mathcal{B}$ of atoms verifying:*

- $\mathcal{B} = \{B_1(t_1), \ldots, B_m(t_m)\}$ *s.t.* $(\neg B_1^-(t_1) \vee \cdots \vee \neg B_m^-(t_m)) \in \mathcal{PI}_\Sigma^{g-}(\Phi)$,
- *for all* $t \in \{t_1, \ldots, t_m\}$ *there exists an* A *s.t.* $A(t) \in \mathcal{PI}_\Sigma^{g+}(\Phi)$,
- $\mathcal{A} = \{A(t) \in \mathcal{PI}_\Sigma^{g+}(\Phi) \mid t$ *is one of* $t_1, \ldots, t_m\}$, *and*
- $\mathcal{H} = \{C_{\mathcal{A},t} \sqsubseteq C_{\mathcal{B},t} \mid t$ *is one of* $t_1, \ldots, t_m$ *and* $C_{\mathcal{B},t} \not\preceq_\sqcap C_{\mathcal{A},t}\}$, *where* $C_{\mathcal{A},t} = \bigsqcap_{A(t)\in\mathcal{A}} A$ *and* $C_{\mathcal{B},t} = \bigsqcap_{B(t)\in\mathcal{B}} B$.

We call the hypotheses that are constructed as in Theorem 11 *constructible.* This theorem states that every packed connection-minimal hypothesis is equivalent to a constructible hypothesis and vice versa. A constructible hypothesis is built from the concepts in *one* negative prime implicate in $\mathcal{PI}_\Sigma^{g-}(\Phi)$ and *all* matching concepts from prime implicates in $\mathcal{PI}_\Sigma^{g+}(\Phi)$. The matching itself is determined by the Skolem terms that occur in all these clauses. The subterm relation between the terms of the clauses in $\mathcal{PI}_\Sigma^{g+}(\Phi)$ and $\mathcal{PI}_\Sigma^{g-}(\Phi)$ is the same as the ancestor relation in the description trees of subsumers of C_1 and subsumees of C_2 respectively. The terms matching in positive and negative prime implicates allow us to identify where the missing entailments between a subsumer D_1 of C_1 and a subsumee D_2 of C_2 are. These missing entailments become the constructible $\mathcal{H}$. The condition $C_{\mathcal{B},t} \not\preceq_\sqcap C_{\mathcal{A},t}$ is a way to write that $C_{\mathcal{A},t} \sqsubseteq C_{\mathcal{B},t}$ is not a tautology, which can be tested by subset inclusion.

The formal proof of this result is detailed in the technical report [16]. We sketch it briefly here. To start, we link the subsumers of C_1 with $\mathcal{PI}_\Sigma^{g+}(\Phi)$. This is done at the semantics level: We show that all Herbrand models of Φ, i.e., models built on the symbols in Φ, are also models of $\mathcal{PI}_\Sigma^{g+}(\Phi)$, that is itself such a model. Then we show that $C_1(\mathtt{sk}_0)$ as well as the formulas corresponding to the subsumers of C_1 in our translation are satisfied by all Herbrand models. This follows from the fact that Φ is in fact a set of Horn clauses. Next, we show, using a similar technique, how duplicate negative ground implicates, not necessarily prime, relate to subsumees of C_2, with the restriction that there must exist a weak homomorphism from a description tree of a subsumer of C_1 to a description tree of the considered subsumee of C_2. Thus, $\mathcal{H}$ provides the missing CIs that will turn the weak homomorphism into a $(\mathcal{T} \cup \mathcal{H})$-homomorphism. Then, we establish an equivalence between the $\preceq_\sqcap$-minimality of the subsumee of C_2 and the primality of the corresponding negative implicate. Packability is the last aspect we deal with, whose use is purely limited to the reconstruction. It holds because $\mathcal{A}$ contains all $A(t) \in \mathcal{PI}_\Sigma^{g+}(\Phi)$ for all terms t occurring in $\mathcal{B}$.

Example 12. Consider the abduction problem $\langle \mathcal{T}_\mathrm{a}, \Sigma, \alpha_\mathrm{a}\rangle$ where Σ contains all concepts from $\mathcal{T}_\mathrm{a}$. For the translation Φ of this problem, we have

$$\begin{aligned}
\mathcal{PI}_\Sigma^{g+}(\Phi) &= \{\,\mathsf{Professor}(\mathtt{sk}_0), \mathsf{Doctor}(\mathtt{sk}_0), \mathsf{Chair}(\mathtt{sk}_1(\mathtt{sk}_0)), \mathsf{PhD}(\mathtt{sk}_2(\mathtt{sk}_0))\} \\
\mathcal{PI}_\Sigma^{g-}(\Phi) &= \{\,\neg\mathsf{Researcher}^-(\mathtt{sk}_0), \\
&\qquad \neg\mathsf{ResearchPosition}^-(\mathtt{sk}_1(\mathtt{sk}_0)) \vee \neg\mathsf{Diploma}^-(\mathtt{sk}_2(\mathtt{sk}_0))\}
\end{aligned}$$

where $\mathtt{sk}_1$ is the Skolem function introduced for $\mathsf{Professor} \sqsubseteq \exists\mathsf{employment.Chair}$ and $\mathtt{sk}_2$ is introduced for $\mathsf{Doctor} \sqsubseteq \exists\mathsf{qualification.PhD}$. This leads to two constructible solutions: $\{\mathsf{Professor} \sqcap \mathsf{Doctor} \sqsubseteq \mathsf{Researcher}\}$ and $\mathcal{H}_{\mathrm{a}1}$, that are both packed connection-minimal hypotheses if $\Sigma = \mathsf{N_C}$. Another example is presented in full details in the technical report [16].

Termination. If $\mathcal{T}$ contains cycles, there can be infinitely many prime implicates. For example, for $\mathcal{T} = \{C_1 \sqsubseteq A, A \sqsubseteq \exists r.A, \exists r.B \sqsubseteq B, B \sqsubseteq C_2\}$ both the positive and negative ground prime implicates of Φ are unbounded even though the set of constructible hypotheses is finite (as it is for any abduction problem):

$$\mathcal{PI}^{g+}_{\Sigma}(\Phi) = \{C_1(\mathtt{sk_0}), A(\mathtt{sk_0}), A(\mathtt{sk}(\mathtt{sk_0})), A(\mathtt{sk}(\mathtt{sk}(\mathtt{sk_0}))), \ldots\},$$
$$\mathcal{PI}^{g-}_{\Sigma}(\Phi) = \{\neg C_2^-(\mathtt{sk_0}), \neg B^-(\mathtt{sk_0}), \neg B^-(\mathtt{sk}(\mathtt{sk_0})), \ldots\}.$$

To find all constructible hypotheses of an abduction problem, an approach that simply computes all prime implicates of Φ, e.g., using the standard resolution calculus, will never terminate on cyclic problems. However, if we look only for subset-minimal constructible hypotheses, termination can be achieved for cyclic and non-cyclic problems alike, because it is possible to construct all such hypotheses from prime implicates that have a polynomially bounded term depth, as shown below. To obtain this bound, we consider resolution derivations of the ground prime implicates and we show that they can be done under some restrictions that imply this bound.

Before performing resolution, we compute the *presaturation* Φ_p *of the set of clauses* Φ, defined as

$$\Phi_p = \Phi \cup \{\neg A(x) \vee B(x) \mid \Phi \models \neg A(x) \vee B(x)\}$$

where A and B are either both original or both duplicate atomic concepts. The presaturation can be efficiently computed before the translation, using a modern $\mathcal{EL}$ reasoner such as ELK [23], which is highly optimized towards the computation of all entailments of the form $A \sqsubseteq B$. While the presaturation computes nothing a resolution procedure could not derive, it is what allows us to bind the maximal depth of terms in inferences to that in prime implicates. If Φ_p is presaturated, we do not need to perform inferences that produce Skolem terms of a higher nesting depth than what is needed for the prime implicates.

Starting from the presaturated set Φ_p, we can show that all the relevant prime implicates can be computed if we restrict all inferences to those where

R1 at least one premise contains a ground term,
R2 the resolvent contains at most one variable, and
R3 every literal in the resolvent contains Skolem terms of nesting depth at most $n \times m$, where n is the number of atomic concepts in Φ, and m is the number of occurrences of existential role restrictions in $\mathcal{T}$.

The first restriction turns the derivation of $\mathcal{PI}^{g+}_{\Sigma}(\Phi)$ and $\mathcal{PI}^{g-}_{\Sigma}(\Phi)$ into an SOS-resolution derivation [18] with set of support $\{C_1(\mathtt{sk_0}), C_2^-(\mathtt{sk_0})\}$, i.e., the only two clauses with ground terms in Φ. This restriction is a straightforward consequence of our interest in computing only ground implicates, and of the fact that the non-ground clauses in Φ cannot entail the empty clause since every $\mathcal{EL}$ TBox is consistent. The other restrictions are consequences of the following theorems, whose proofs are available in the technical report [16].

Theorem 13. *Given an abduction problem and its translation Φ, every constructible hypothesis can be built from prime implicates that are inferred under restriction 4.*

In fact, for $\mathcal{PI}^{g+}_{\Sigma}(\Phi)$ it is even possible to restrict inferences to generating only ground resolvents, as can be seen in the proof of Theorem 13, that directly looks at the kinds of clauses that are derivable by resolution from Φ.

Theorem 14. *Given an abduction problem and its translation Φ, every subset-minimal constructible hypothesis can be built from prime implicates that have a nesting depth of at most $n \times m$, where n is the number of atomic concepts in Φ, and m is the number of occurrences of existential role restrictions in $\mathcal{T}$.*

The proof of Theorem 14 is based on a structure called a *solution tree*, which resembles a description tree, but with multiple labeling functions. It assigns to each node a Skolem term, a set of atomic concepts called *positive label*, and a single atomic concept called *negative label*. The nodes correspond to matching partners in a constructible hypothesis: The Skolem term is the term on which we match literals. The positive label collects the atomic concepts in the positive prime implicates containing that term. The maximal anti-chains of the tree, i.e., the maximal subsets of nodes s.t. no node is the ancestor of another are such that their negative labels correspond to the literals in a derivable negative implicate. For every solution tree, the Skolem labels and negative labels of the leaves determine a negative prime implicate, and by combining the positive and negative labels of these leaves, we obtain a constructible hypothesis, called the *solution* of the tree. We show that from every solution tree with solution $\mathcal{H}$ we can obtain a solution tree with solution $\mathcal{H}' \subseteq \mathcal{H}$ s.t. on no path, there are two nodes that agree both on the head of their Skolem labeling and on the negative label. Furthermore the number of head functions of Skolem labels is bounded by the total number n of Skolem functions, while the number of distinct negative labels is bounded by the number m of atomic concepts, bounding the depth of the solution tree for $\mathcal{H}'$ at $n \times m$. This justifies the bound in Theorem 14. This bound is rather loose. For the academia example, it is equal to $22 \times 6 = 132$.

5 Implementation

We implemented our method to compute all subset-minimal constructible hypotheses in the tool CAPI.[3] To compute the prime implicates, we used SPASS [34], a first-order theorem prover that includes resolution among other calculi. We implemented everything before and after the prime implicate computation in Java, including the parsing of ontologies, preprocessing (detailed below), clausification of the abduction problems, translation to SPASS input, as well as the parsing and processing of the output of SPASS to build the constructible hypotheses and filter out the non-subset-minimal ones. On the Java side, we used the OWL API for all DL-related functionalities [20], and the $\mathcal{EL}$ reasoner ELK for computing the presaturations [23].

[3] available under https://lat.inf.tu-dresden.de/~koopmann/CAPI.

Preprocessing. Since realistic TBoxes can be too large to be processed by SPASS, we replace the background knowledge in the abduction problem by a subset of axioms relevant to the abduction problem. Specifically, we replace the abduction problem $(\mathcal{T}, \Sigma, C_1 \sqsubseteq C_2)$ by the abduction problem $(\mathcal{M}^{\bot}_{C_1} \cup \mathcal{M}^{\top}_{C_2}, \Sigma, C_1 \sqsubseteq C_2)$, where $\mathcal{M}^{\bot}_{C_1}$ is the $\bot$-*module* of $\mathcal{T}$ for the signature of C_1, and $\mathcal{M}^{\top}_{C_2}$ is the $\top$-*module* of $\mathcal{T}$ for the signature of C_2 [15]. Those notions are explained in the technical report [16]. Their relevant properties are that $\mathcal{M}^{\bot}_{C_1}$ is a subset of $\mathcal{T}$ s.t. $\mathcal{M}^{\bot}_{C_1} \models C_1 \sqsubseteq D$ iff $\mathcal{T} \models C_1 \sqsubseteq D$ for all concepts D, while $\mathcal{M}^{\top}_{C_2}$ is a subset of $\mathcal{T}$ that ensures $\mathcal{M}^{\top}_{C_2} \models D \sqsubseteq C_2$ iff $\mathcal{T} \models D \sqsubseteq C_2$ for all concepts D. It immediately follows that every connection-minimal hypothesis for the original problem $(\mathcal{T}, \Sigma, C_1 \sqsubseteq C_2)$ is also a connection-minimal hypothesis for $(\mathcal{M}^{\bot}_{C_1} \cup \mathcal{M}^{\top}_{C_2}, \Sigma, C_1 \sqsubseteq C_2)$. For the presaturation, we compute with ELK all CIs of the form $A \sqsubseteq B$ s.t. $\mathcal{M}^{\bot}_{C_1} \cup \mathcal{M}^{\top}_{C_2} \models A \sqsubseteq B$.

Prime implicates generation. We rely on a slightly modified version of SPASS v3.9 to compute all ground prime implicates. In particular, we added the possibility to limit the number of variables allowed in the resolvents to enforce **R2**. For each of the restrictions **R1**–**R3** there is a corresponding flag (or set of flags) that is passed to SPASS as an argument.

Recombination. The construction of hypotheses from the prime implicates found in the previous stage starts with a straightforward process of matching negative prime implicates with a set of positive ones based on their Skolem terms. It is followed by subset minimality tests to discard non-subset-minimal hypotheses, since, with the bound we enforce, there is no guarantee that these are valid constructible hypotheses because the negative ground implicates they are built upon may not be prime. If SPASS terminates due to a timeout instead of reaching the bound, then it is possible that some subset-minimal constructible hypotheses are not found, and thus, some non-constructible hypotheses may be kept. Note that these are in any case solutions to the abduction problem.

6 Experiments

There is no benchmark suite dedicated to TBox abduction in $\mathcal{EL}$, so we created our own, using realistic ontologies from the bio-medical domain. For this, we used ontologies from the 2017 snapshot of Bioportal [27]. We restricted each ontology to its $\mathcal{EL}$ fragment by filtering out unsupported axioms, where we replaced domain axioms and n-ary equivalence axioms in the usual way [2]. Note that, even if the ontology contains more expressive axioms, an $\mathcal{EL}$ hypothesis is still useful if found. From the resulting set of TBoxes, we selected those containing at least 1 and at most 50,000 axioms, resulting in a set of 387 $\mathcal{EL}$ TBoxes. Precisely, they contained between 2 and 46,429 axioms, for an average of 3,039 and a median of 569. Towards obtaining realistic benchmarks, we created three different categories of abduction problems for each ontology $\mathcal{T}$, where in each case, we used the signature of the entire ontology for Σ.

- Problems in `ORIGIN` use $\mathcal{T}$ as background knowledge, and as observation a randomly chosen $A \sqsubseteq B$ s.t. A and B are in the signature of $\mathcal{T}$ and $\mathcal{T} \not\models A \sqsubseteq B$. This covers the basic requirements of an abduction problem, but has the disadvantage that A and B can be completely unrelated in $\mathcal{T}$.
- Problems in `JUSTIF` contain as observation a randomly selected CI α s.t., for the original TBox, $\mathcal{T} \models \alpha$ and $\alpha \notin \mathcal{T}$. The background knowledge used is a *justification for* α *in* $\mathcal{T}$ [32], that is, a minimal subset $\mathcal{I} \subseteq \mathcal{T}$ s.t. $\mathcal{I} \not\models \alpha$, from which a randomly selected axiom is removed. The TBox is thus a smaller set of axioms extracted from a real ontology for which we know there is a way of producing the required entailment without adding it explicitly. Justifications were computed using functionalities of the OWL API and ELK.
- Problems in `REPAIR` contain as observation a randomly selected CI α s.t. $\mathcal{T} \models \alpha$, and as background knowledge a *repair for* α *in* $\mathcal{T}$, which is a maximal subset $\mathcal{R} \subseteq \mathcal{T}$ s.t. $\mathcal{R} \not\models \alpha$. Repairs were computed using a justification-based algorithm [32] with justifications computed as for `JUSTIF`. This usually resulted in much larger TBoxes, where more axioms would be needed to establish the entailment.

All experiments were run on Debian Linux (Intel Core i5-4590, 3.30 GHz, 23 GB Java heap size). The code and scripts used in the experiments are available online [17]. The three phases of the method (see Fig. 2) were each assigned a hard time limit of 90 s.

For each ontology, we attempted to create and translate 5 abduction problems of each category. This failed on some ontologies because either there was no corresponding entailment (25/28/25 failures out of the 387 ontologies for `ORIGIN`/`JUSTIF`/`REPAIR`), there was a timeout during the translation (5/5/5 failures for `ORIGIN`/`JUSTIF`/`REPAIR`), or because the computation of justifications caused an exception (-/2/0 failures for `ORIGIN`/`JUSTIF`/`REPAIR`). The final number of abduction problems for each category is in the first column of Table 1.

We then attempted to compute prime implicates for these benchmarks using SPASS. In addition to the hard time limit, we gave a soft time limit of 30 s to SPASS, after which it should stop exploring the search space and return the implicates already found. In Table 1 we show, for each category, the percentage of problems on which SPASS succeeded in computing a non-empty set of clauses (Success) and the percentage of problems on which SPASS terminated within the time limit, where all solutions are computed (Compl.). The high number of CIs in the background knowledge explains most of the cases where SPASS reached the soft time limit. In a lot of these cases, the bound on the term depth goes into the billion, rendering it useless in practice. However, the "Compl." column shows that the bound is reached before the soft time limit in most cases.

The reconstruction never reached the hard time limit. We measured the median, average and maximal number of solutions found ($\#\mathcal{H}$), size of solutions in number of CIs ($|\mathcal{H}|$), size of CIs from solutions in number of atomic concepts ($|\alpha|$), and SPASS runtime (time, in seconds), all reported in Table 1. Except for the simple `JUSTIF` problems, the number of solutions may become very large. At the same time, solutions always contain very few axioms (never

Table 1. Evaluation results.

	#Probl.	Success	Compl.	Median / avg / max #$\mathcal{H}$	$\lvert\mathcal{H}\rvert$	$\lvert\alpha\rvert$	Time (s.)
ORIGIN	1,925	94.7%	61.3%	1/8.51/1850	1/1.00/2	6/7.48/91	0.2/12.4/43.8
JUSTIF	1,803	100.0%	97.2%	1/1.50/5	1/1/1	2/4.21/32	0.2/1.1/34.1
REPAIR	1,805	92.9%	57.0%	43/228.05/6317	1/1.00/2	5/5.09/49	0.6/13.6/59.9

more than 3), though the axioms become large too. We also noticed that highly nested Skolem terms rarely lead to more hypotheses being found: 8/1/15 for ORIGIN/JUSTIF/REPAIR, and the largest nesting depth used was: 3/1/2 for ORIGIN/JUSTIF/REPAIR. This hints at the fact that longer time limits would not have produced more solutions, and motivates future research into redundancy criteria to stop derivations (much) earlier.

7 Conclusion

We have introduced connection-minimal TBox abduction for $\mathcal{EL}$ which finds parsimonious hypotheses, ruling out the ones that entail the observation in an arbitrary fashion. We have established a formal link between the generation of connection-minimal hypotheses in $\mathcal{EL}$ and the generation of prime implicates of a translation Φ of the problem to first-order logic. In addition to obtaining these theoretical results, we developed a prototype for the computation of subset-minimal constructible hypotheses, a subclass of connection-minimal hypotheses that is easy to construct from the prime implicates of Φ. Our prototype uses the SPASS theorem prover as an SOS-resolution engine to generate the needed implicates. We tested this tool on a set of realistic medical ontologies, and the results indicate that the cost of computing connection-minimal hypotheses is high but not prohibitive.

We see several ways to improve our technique. The bound we computed to ensure termination could be advantageously replaced by a redundancy criterion discarding irrelevant implicates long before it is reached, thus greatly speeding computation in SPASS. We believe it should also be possible to further constrain inferences, e.g., to have them produce ground clauses only, or to generate the prime implicates with terms of increasing depth in a controlled incremental way instead of enforcing the soft time limit, but these two ideas remain to be proved feasible. As an alternative to using prime implicates, one may investigate direct method for computing connection-minimal hypotheses in $\mathcal{EL}$.

The theoretical worst-case complexity of connection-minimal abduction is another open question. Our method only gives a very high upper bound: by bounding only the nesting dept of Skolem terms polynomially as we did with Theorem 13, we may still permit clauses with exponentially many literals, and thus double exponentially many clauses in the worst case, which would give us an 2ExpTime upper bound to the problem of computing all subset-minimal constructible hypotheses. Using structure-sharing and guessing, it is likely possible to get a lower bound. We have not looked yet at lower bounds for the complexity either.

While this work focuses on abduction problems where the observation is a CI, we believe that our technique can be generalised to knowledge that also contains ground facts (ABoxes), and to observations that are of the form of conjunctive queries on the ABoxes in such knowledge bases. The motivation for such an extension is to understand why a particular query does not return any results, and to compute a set of TBox axioms that fix this problem. Since our translation already transforms the observation into ground facts, it should be possible to extend it to this setting. We would also like to generalize TBox abduction by finding a reasonable way to allow role restrictions in the hypotheses, and to extend connection-minimality to more expressive DLs such as $\mathcal{ALC}$.

Acknowledgments. This work was supported by the Deutsche Forschungsgemeinschaft (DFG), Grant 389792660 within TRR 248.

References

1. Baader, F., Brandt, S., Lutz, C.: Pushing the $\mathcal{EL}$ envelope. In: Kaelbling, L.P., Saffiotti, A. (eds.) IJCAI-05, Proceedings of the Nineteenth International Joint Conference on Artificial Intelligence, Edinburgh, Scotland, UK, 30 July - 5 August 2005, pp. 364–369. Professional Book Center (2005). http://ijcai.org/Proceedings/05/Papers/0372.pdf
2. Baader, F., Horrocks, I., Lutz, C., Sattler, U.: An Introduction to Description Logic. Cambridge University Press, Cambridge (2017). https://doi.org/10.1017/9781139025355
3. Baader, F., Küsters, R., Molitor, R.: Computing least common subsumers in description logics with existential restrictions. In: Proceedings of IJCAI 1999, pp. 96–103. Morgan Kaufmann (1999)
4. Bachmair, L., Ganzinger, H.: Resolution theorem proving. In: Robinson, J.A., Voronkov, A. (eds.) Handbook of Automated Reasoning (in 2 volumes), pp. 19–99. Elsevier and MIT Press, Cambridge (2001). https://doi.org/10.1016/b978-044450813-3/50004-7
5. Bauer, J., Sattler, U., Parsia, B.: Explaining by example: model exploration for ontology comprehension. In: Grau, B.C., Horrocks, I., Motik, B., Sattler, U. (eds.) Proceedings of the 22nd International Workshop on Description Logics (DL 2009), Oxford, UK, 27–30 July 2009. CEUR Workshop Proceedings, vol. 477. CEUR-WS.org (2009). http://ceur-ws.org/Vol-477/paper_37.pdf
6. Bienvenu, M.: Complexity of abduction in the $\mathcal{EL}$ family of lightweight description logics. In: Proceedings of KR 2008, pp. 220–230. AAAI Press (2008), http://www.aaai.org/Library/KR/2008/kr08-022.php
7. Calvanese, D., Ortiz, M., Simkus, M., Stefanoni, G.: Reasoning about explanations for negative query answers in DL-Lite. J. Artif. Intell. Res. **48**, 635–669 (2013). https://doi.org/10.1613/jair.3870
8. Ceylan, İ.İ., Lukasiewicz, T., Malizia, E., Molinaro, C., Vaicenavicius, A.: Explanations for negative query answers under existential rules. In: Calvanese, D., Erdem, E., Thielscher, M. (eds.) Proceedings of KR 2020, pp. 223–232. AAAI Press (2020). https://doi.org/10.24963/kr.2020/23

9. Del-Pinto, W., Schmidt, R.A.: ABox abduction via forgetting in $\mathcal{ALC}$. In: The Thirty-Third AAAI Conference on Artificial Intelligence, AAAI 2019, pp. 2768–2775. AAAI Press (2019). https://doi.org/10.1609/aaai.v33i01.33012768
10. Du, J., Qi, G., Shen, Y., Pan, J.Z.: Towards practical ABox abduction in large description logic ontologies. Int. J. Semantic Web Inf. Syst. **8**(2), 1–33 (2012). https://doi.org/10.4018/jswis.2012040101
11. Du, J., Wan, H., Ma, H.: Practical TBox abduction based on justification patterns. In: Proceedings of the Thirty-First AAAI Conference on Artificial Intelligence, pp. 1100–1106 (2017). http://aaai.org/ocs/index.php/AAAI/AAAI17/paper/view/14402
12. Du, J., Wang, K., Shen, Y.: A tractable approach to ABox abduction over description logic ontologies. In: Brodley, C.E., Stone, P. (eds.) Proceedings of the Twenty-Eighth AAAI Conference on Artificial Intelligence, pp. 1034–1040. AAAI Press (2014). http://www.aaai.org/ocs/index.php/AAAI/AAAI14/paper/view/8191
13. Echenim, M., Peltier, N., Sellami, Y.: A generic framework for implicate generation modulo theories. In: Galmiche, D., Schulz, S., Sebastiani, R. (eds.) IJCAR 2018. LNCS (LNAI), vol. 10900, pp. 279–294. Springer, Cham (2018). https://doi.org/10.1007/978-3-319-94205-6_19
14. Elsenbroich, C., Kutz, O., Sattler, U.: A case for abductive reasoning over ontologies. In: Proceedings of the OWLED'06 Workshop on OWL: Experiences and Directions (2006). http://ceur-ws.org/Vol-216/submission_25.pdf
15. Grau, B.C., Horrocks, I., Kazakov, Y., Sattler, U.: Modular reuse of ontologies: theory and practice. J. Artif. Intell. Res. **31**, 273–318 (2008). https://doi.org/10.1613/jair.2375
16. Haifani, F., Koopmann, P., Tourret, S., Weidenbach, C.: Connection-minimal abduction in $\mathcal{EL}$ via translation to FOL - technical report (2022). https://doi.org/10.48550/ARXIV.2205.08449, https://arxiv.org/abs/2205.08449
17. Haifani, F., Koopmann, P., Tourret, S., Weidenbach, C.: Experiment data for the paper Connection-minimal Abduction in EL via translation to FOL, May 2022. https://doi.org/10.5281/zenodo.6563656
18. Haifani, F., Tourret, S., Weidenbach, C.: Generalized completeness for SOS resolution and its application to a new notion of relevance. In: Platzer, A., Sutcliffe, G. (eds.) CADE 2021. LNCS (LNAI), vol. 12699, pp. 327–343. Springer, Cham (2021). https://doi.org/10.1007/978-3-030-79876-5_19
19. Halland, K., Britz, K.: ABox abduction in $\mathcal{ALC}$ using a DL tableau. In: 2012 South African Institute of Computer Scientists and Information Technologists Conference, SAICSIT '12, pp. 51–58 (2012). https://doi.org/10.1145/2389836.2389843
20. Horridge, M., Bechhofer, S.: The OWL API: a java API for OWL ontologies. Semant. Web **2**(1), 11–21 (2011). https://doi.org/10.3233/SW-2011-0025
21. Horridge, M., Parsia, B., Sattler, U.: Explanation of OWL entailments in protege 4. In: Bizer, C., Joshi, A. (eds.) Proceedings of the Poster and Demonstration Session at the 7th International Semantic Web Conference (ISWC2008), Karlsruhe, Germany, 28 October 2008. CEUR Workshop Proceedings, vol. 401. CEUR-WS.org (2008). http://ceur-ws.org/Vol-401/iswc2008pd_submission_47.pdf
22. Kazakov, Y., Klinov, P., Stupnikov, A.: Towards reusable explanation services in protege. In: Artale, A., Glimm, B., Kontchakov, R. (eds.) Proceedings of the 30th International Workshop on Description Logics, Montpellier, France, 18–21 July 2017. CEUR Workshop Proceedings, vol. 1879. CEUR-WS.org (2017). http://ceur-ws.org/Vol-1879/paper31.pdf

23. Kazakov, Y., Krötzsch, M., Simancik, F.: The incredible ELK - from polynomial procedures to efficient reasoning with $\mathcal{EL}$ ontologies. J. Autom. Reason. **53**(1), 1–61 (2014). https://doi.org/10.1007/s10817-013-9296-3
24. Klarman, S., Endriss, U., Schlobach, S.: ABox abduction in the description logic $\mathcal{ALC}$. J. Autom. Reason. **46**(1), 43–80 (2011). https://doi.org/10.1007/s10817-010-9168-z
25. Koopmann, P.: Signature-based abduction with fresh individuals and complex concepts for description logics. In: Zhou, Z. (ed.) Proceedings of the Thirtieth International Joint Conference on Artificial Intelligence, IJCAI 2021, Virtual Event/Montreal, Canada, 19–27 August 2021, pp. 1929–1935 (2021). https://doi.org/10.24963/ijcai.2021/266
26. Koopmann, P., Del-Pinto, W., Tourret, S., Schmidt, R.A.: Signature-based abduction for expressive description logics. In: Calvanese, D., Erdem, E., Thielscher, M. (eds.) Proceedings of the 17th International Conference on Principles of Knowledge Representation and Reasoning, KR 2020, pp. 592–602. AAAI Press (2020). https://doi.org/10.24963/kr.2020/59
27. Matentzoglu, N., Parsia, B.: Bioportal snapshot 30.03.2017 (2017). https://doi.org/10.5281/zenodo.439510
28. Nabeshima, H., Iwanuma, K., Inoue, K., Ray, O.: SOLAR: an automated deduction system for consequence finding. AI Commun. **23**(2–3), 183–203 (2010). https://doi.org/10.3233/AIC-2010-0465
29. Parsia, B., Matentzoglu, N., Gonçalves, R.S., Glimm, B., Steigmiller, A.: The owl reasoner evaluation (ORE) 2015 competition report. J. Autom. Reason. **59**(4), 455–482 (2017). https://doi.org/10.1007/s10817-017-9406-8
30. Pukancová, J., Homola, M.: Tableau-based ABox abduction for the $\mathcal{ALCHO}$ description logic. In: Proceedings of the 30th International Workshop on Description Logics (2017). http://ceur-ws.org/Vol-1879/paper11.pdf
31. Pukancová, J., Homola, M.: The AAA Abox abduction solver. KI - Künstliche Intell. **34**(4), 517–522 (2020). https://doi.org/10.1007/s13218-020-00685-4
32. Schlobach, S., Cornet, R.: Non-standard reasoning services for the debugging of description logic terminologies. In: Gottlob, G., Walsh, T. (eds.) Proceedings of the 18th International Joint Conference on Artificial Intelligence (IJCAI 2003), pp. 355–362. Morgan Kaufmann, Acapulco, Mexico (2003). http://ijcai.org/Proceedings/03/Papers/053.pdf
33. Wei-Kleiner, F., Dragisic, Z., Lambrix, P.: Abduction framework for repairing incomplete $\mathcal{EL}$ ontologies: complexity results and algorithms. In: Proceedings of the Twenty-Eighth AAAI Conference on Artificial Intelligence, pp. 1120–1127. AAAI Press (2014). http://www.aaai.org/ocs/index.php/AAAI/AAAI14/paper/view/8239
34. Weidenbach, C., Schmidt, R.A., Hillenbrand, T., Rusev, R., Topic, D.: System description: SPASS version 3.0. In: Pfenning, F. (ed.) CADE 2007. LNCS (LNAI), vol. 4603, pp. 514–520. Springer, Heidelberg (2007). https://doi.org/10.1007/978-3-540-73595-3_38
35. Wos, L., Robinson, G., Carson, D.: Efficiency and completeness of the set of support strategy in theorem proving. J. ACM **12**(4), 536–541 (1965)

Semantic Relevance

Fajar Haifani[1,2] and Christoph Weidenbach[1](✉)

[1] Max Planck Institute for Informatics, Saarland Informatics Campus, Saarbrücken, Germany
{f.haifani,weidenbach}@mpi-inf.mpg.de

[2] Graduate School of Computer Science, Saarbrücken, Germany

Abstract. A clause C is syntactically relevant in some clause set N, if it occurs in every refutation of N. A clause C is syntactically semi-relevant, if it occurs in some refutation of N. While syntactic relevance coincides with satisfiability (if C is syntactically relevant then $N \setminus \{C\}$ is satisfiable), the semantic counterpart for syntactic semi-relevance was not known so far. Using the new notion of a *conflict literal* we show that for independent clause sets N a clause C is syntactically semi-relevant in the clause set N if and only if it adds to the number of conflict literals in N. A clause set is independent, if no clause out of the clause set is the consequence of different clauses from the clause set.

Furthermore, we relate the notion of relevance to that of a minimally unsatisfiable subset (MUS) of some independent clause set N. In propositional logic, a clause C is relevant if it occurs in all MUSes of some clause set N and semi-relevant if it occurs in some MUS. For first-order logic the characterization needs to be refined with respect to ground instances of N and C.

1 Introduction

In our previous work [11], we introduced a notion of syntactic relevance based on refutations while at the same time generalized the completeness result for resolution by the set-of-support strategy (SOS) [28,33] as its test. Our notion of syntactic relevance is useful for explaining why a set of clauses is unsatisfiable. In this paper, we introduce a semantic counterpart of syntactic relevance that sheds further light on the relationship between a clause out of a clause set and the potential refutations of this clause set. In the following Sect. 1.1, we first recall syntactic relevance along with an example and then proceeds to explain it in terms of our new semantic relevance in the later Sect. 1.2.

1.1 Syntactic Relevance

Given an unsatisfiable set of clauses N, $C \in N$ is *syntactically relevant* if it occurs in all refutations, it is *syntactically semi-relevant* if it occurs in some refutation, otherwise it is called *syntactically irrelevant*. The clause-based notion of relevance is useful in relating the contribution of a clause to refutation (goal conjecture).

J. Blanchette et al. (Eds.): IJCAR 2022, LNAI 13385, pp. 208–227, 2022.
https://doi.org/10.1007/978-3-031-10769-6_13

This has in particular been shown in the context of product scenarios built out of construction kits as they are used in the car industry [8,32].

For an illustration of our privous notions and results we now consider the following unsatisfiable first-order clause set N where Fig. 1 presents a refutation of N.

$$\begin{aligned} N = \{&(1)A(f(a)) \lor D(x_3),\\ &(2)\neg D(x_7),\\ &(3)\neg B(c,a) \lor B(b,f(x_6)),\\ &(4)B(x_1,x_2) \lor C(x_1),\\ &(5)\neg C(x_5),\\ &(6)\neg A(x_4) \lor \neg B(b,x_4)\} \end{aligned}$$

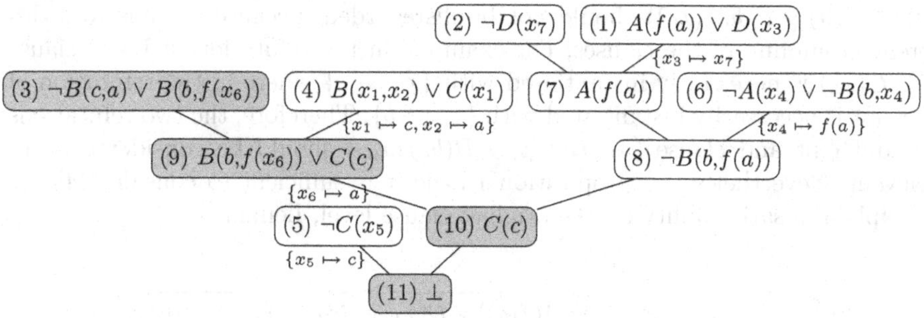

Fig. 1. A refutation of N in tree representation

In essence, inferences in an SOS refutation always involve at least one clause in the SOS and put the resulting clause back in it. So, this refutation is not an SOS refutation from the syntactically semi-relevant clause $(3)\neg B(c,a) \lor B(b,f(x_6))$, because only the shaded part represents an SOS refutation starting with this clause. More specifically, there are two inferences ended in $(8)\neg B(b,f(a))$ which violates the condition for an SOS refutaiton. Nevertheless, it can be transformed into an SOS refutation where the clause $(3)\neg B(c,a) \lor B(b,f(x_6))$ is in the SOS [11], Fig. 2. Please note that $N \setminus \{(3)\neg B(c,a) \lor B(b,f(x_6))\}$ is still unsatisfiable and classical SOS completeness [33] is not sufficient to guarantee the existence of a refutation with SOS $\{(3)\neg B(c,a) \lor B(b,f(x_6))\}$ [11].

In addition, $N \setminus \{(3)\neg B(c,a) \lor B(b,f(x_6))\}$ is also a *minimally unsatisfiable subset* (MUS), where Fig. 3 presents a respective refutation. A MUS is an unsatisfiable clause set such that removing a clause from this set would render it satisfiable. Consequently, a MUS-based defined notion of semi-relevance on the level of the original first-order clauses is not sufficient here. The clause

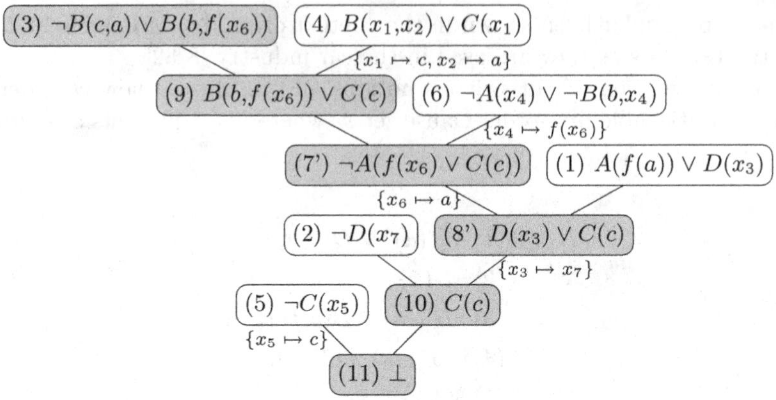

Fig. 2. Semi-relevant clause $(3)\neg B(c,a) \vee B(b, f(x_6))$ in SOS

$(3)\neg B(c,a) \vee B(b, f(x_6))$ should not be disregarded, because it leads to a different grounding of the clauses. For example, in the refutation of Fig. 2 clause $(5)\neg C(x_5)$ is necessarily instantiated with $\{x_5 \mapsto c\}$ where in the refutation of Fig. 3 it is necessarily instantiated with $\{x_5 \mapsto b\}$. Therefore, the two refutations are different and clause $(3)\neg B(c,a) \vee B(b, f(x_6))$ should be considered semi-relevant. Nevertheless, in propositional logic it is sufficient to consider MUSes to explain unsatisfiability on the original clause level, Lemma 18.

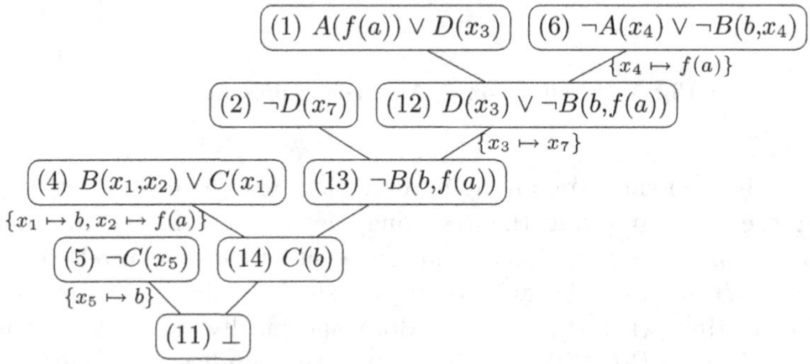

Fig. 3. A refutation of N without $(3)\neg B(c,a) \vee B(b, f(x_6))$

1.2 Semantic Relevance

We now illustrate how our new notion of relevance works on the previous example. First, different from the other works, we propose a way of characterizing semantic relevance by using our novel concept of a *conflict literal*. A ground literal L is a conflict literal in a clause set N if there are some satisfiable sets of instances N_1 and N_2 from N s.t. $N_1 \models L$ and $N_2 \models \text{comp}(L)$. On the one hand, explaining an unsatisfiable clause set as the absence of a model (as it is usually defined) is not that helpful since an absence means there is nothing to discuss in the first place. On the other hand, the contribution of a clause to unsatisfiability of a clause set can only partially be explained using the concept of a MUS which we have discussed before. A conflict literal provides a middle ground to explain the contribution of a clause to unsatisfiability between the absence of a model and MUSes. It also better reflects our intuition that there is a contradiction (in the form of two implied simple facts that cannot be both true at the same time) in an unsatisfiable set of clauses.

From Fig. 1, we can already see that $C(c)$ and its complement $\neg C(c)$ are conflict literals because

$$\begin{aligned} N \setminus \{\neg C(x)\} &\models C(c) \\ \neg C(x) &\models \neg C(c) \end{aligned}$$

Also, in addition to that $\{\neg C(x)\}$ is trivially satisfiable, $N \setminus \{\neg C(x)\}$ is also satisfiable. Based on the refutation in Fig. 3, $\neg C(x)$ is syntactically relevant due to $N \setminus \{(3)\neg B(c,a) \vee B(b, f(x_6))\}$ being a MUS. We will also show that for a ground MUS any ground literal occurring in it is a conflict literal, Lemma 20. For our ongoing example it is still possible to identify the conflict literals by means of ground MUSes by looking into the refutations from Fig. 1 and Fig. 3. This leads to the following conflict literals for N, see Definition 10:

$$\begin{aligned} \text{conflict}(N) = \{&(\neg)A(f(a)), \\ &(\neg)B(b, f(a)), (\neg)B(c,a), \\ &(\neg)C(b), (\neg)C(c)\} \ \cup \\ \{&(\neg)D(t) \mid t \text{ is a ground term}\} \end{aligned}$$

These conflict literals can be identified by pushing the substitutions in the refutations from Fig. 1 and Fig. 3 towards the input clauses. They correspond to two first-order MUSes M_1 and M_2. All the ground literals are conflict literals and all other ground conflict literals can be obtained by grounding the remaining variables.

$$\begin{aligned} M_1 = \{&(5)\neg C(c), (2)\neg D(x_7), \\ &(1)A(f(a)) \vee D(x_3), \\ &(3)\neg B(c,a) \vee B(b, f(a)), \\ &(4)B(c,a) \vee C(c), \\ &(6)\neg A(f((a))) \vee \neg B(b, f(a))\} \\ M_2 = \{&(5)\neg C(b), \\ &(4)B(b, f(a)), (2)\neg D(x_7), \\ &(1)A(f(a)) \vee D(x_3), \\ &(6)\neg A(f(a)) \vee \neg B(b, f(a))\} \end{aligned}$$

One can see that, despite $(3)\neg B(c, a) \vee B(b, f(x_6))$ is outside of the only MUS on the first-order level, an instance of it does occur in some ground MUS, take M_1 and an arbitrary grounding of x_3 and x_7 to the identical term t, and the conflict literal $(\neg)B(c, a)$ depends on clause (3). Nevertheless, determining conflict literals is not so obvious in the general case since we do not necessarily know beforehand which ground terms should substitute the variables in the clauses. Moreover, there can be an infinite number of such ground MUSes of possibly unbounded size.

Based on conflict literals, here we introduce a notion of relevance that is semantic in nature, Definition 16. This will also serve as an alternative characterization to our previous refutation-based syntactic relevance. As redundant clauses, e.g., tautologies, can also be syntactically semi-relevant, we require independent clause sets for the definition of semantic relevance. A clause set is *independent*, if it does not contain clauses with instances implied by satisfiable sets of instances of different clauses out of the set. Given an unsatisfiable independent set of clauses N, a clause C is *relevant* in N if N without C has no conflict literals, it is *semi-relevant* if C is necessary to some conflict literals, and it is *irrelevant* otherwise.

Similar to our previous work, relevant clauses are the obvious ones because removing them would make our set satisfiable. On the other hand, irrelevant clauses can be freely identified once we know the semi-relevant ones. For our running example, in fact $(3)\neg B(c, a) \vee B(b, f(x_6))$ is semi-relevant because it is necessary for the conflict literals $(\neg)C(c)$ and $(\neg)B(c, a)$. More specifically, the set of conflicts for $N \setminus \{\neg B(c, a) \vee B(b, f(x_6))\}$ does not include $(\neg)C(c)$ and $(\neg)B(c, a)$:

$$\begin{aligned}\text{conflict}(N \setminus \{\neg B(c, a) \vee B(b, f(x_6))\}) = {} & \{(\neg)A(f(a)), (\neg)B(b, f(a)), (\neg)C(b)\} \uplus \\ & \{(\neg)D(t) | t \text{ is a ground term}\}\end{aligned}$$

These are conflict literals identifiable from M_2: Assume that the variables x_3 and x_7 in M_2 are both grounded by an identical term t. Take some ground literal, for example, $A(f(a)) \in \text{conflict}(N \setminus \{\neg B(c, a) \vee B(b, f(x_6))\})$, and define

$$\begin{aligned} N_{\emptyset} &= \{C \in M_2 | A(f(a)) \notin C \text{ and } \neg A(f(a)) \notin C\} \\ &= \{(5)\neg C(b), (4)B(b, f(a)), (2)\neg D(t)\} \\ N_{A(f(a))} &= \{C \in M_2 | A(f(a)) \in C\} \\ &= \{(1)A(f(a)) \vee D(t)\} \\ N_{\neg A(f(a))} &= \{C \in M_2 | \neg A(f(a)) \in C\} \\ &= \{(6)\neg A(f(a)) \vee \neg B(b, f(a))\} \end{aligned}$$

$N_{\emptyset} \cup N_A(f(a))$ and $N_{\emptyset} \cup N_{\neg A(f(a))}$ are satisfiable because of the Herbrand model $\{B(b, f(a)), A(f(a))\}$ and $\{B(b, f(a))\}$ respectively. In addition,

$$\begin{aligned} N_{\emptyset} \cup N_A(f(a)) &\models A(f(a)) \\ N_{\emptyset} \cup N_{\neg A(f(a))} &\models \neg A(f(a)) \end{aligned}$$

because $A(f(a))$ can be acquired using resolution between (1) and (2) for $N_\emptyset \cup N_{A(f(a))}$ and $\neg A(f(a))$ can be acquired using resolution between (4) and (6) for $N_\emptyset \cup N_{\neg A(f(a))}$. In a similar manner, we can show that the other ground literals are also conflict literals.

Related Work: Other works which aim to explain unsatisfiability mostly rely on the notion of MUSes, mainly in propositional logic [14–16,21,26]. The complexity of determining whether a clause set is a MUS is D^p-complete for a propositional clause set with at most three literals per clause and at most three occurrences of each propositional variable [25]. In [14], syntactically semi-relevant clauses for propositional logic are called a *plain clause set.* Using the terminology in [16], a clause $C \in N$ is *necessary* if it occurs in all MUSes, it is *potentially necessary* if it occurs in some MUS, otherwise, it is *never necessary.* In addition, a clause is defined to be *usable* if it occurs in some refutation. This is thus similar to our syntactic notion of semi-relevance [11]: Given a clause $C \in N$, C is usable if-and-only-if C is syntactically semi-relevant. It is also argued that a usable clause that is not potentially necessary is semantically superfluous. A different but related notion has also been applied for propositional abduction [7]. The notion of a MUS has also been used for explaining unsatisfiability in first-order logic [20]. There, it has been defined in a more general setting: If a set of clauses N is divided into $N = N' \uplus N''$ with a *non-relaxable* clause set N' and *relaxable* clause set N'' (which must be satisfiable), a MUS is a subset M of N'' s.t. $N' \uplus M$ is unsatisfiable but removing a clause from M would render it satisfiable. There are also some works in satisfiability modulo theory (SMT) [5,6,9,35]. A deletion-based approach well-known in propositional logic has also been used for MUS extraction in SMT [9]. In [5,6], a MUS is extracted by combining an SMT solver with an arbitrary external propositional core extractor. Another approach is to construct some graph representing the subformulas of the problem instance, recursively remove clauses in a depth-first-search manner and additionally use some heuristics to further improve the runtime[35]. For the function-free and equality-free first-order fragment, there is a "decompose-merge" approach to compute all MUSes [19,34]. In description logic, a notion that is related to MUS is called *minimal axiom set* (MinA) usually identified by the problem of axiom pinpointing [1,4,13,30]. Its computation is usually divided into two categories: black-box and white-box. A black-box approach picks some inputs, executes it using some sound and complete reasoner, and then interprets the output [13]. On the other hand, white-box approach takes some reasoner and performs an internal modification for it. In this case, Tableau is mostly used [1,30]. In addition, the concept of a lean kernel has also been used to approximate the union of such MinA's [27]. The way relevance is defined is similar in spirit but usually used for an entailment problem instead of unsatisfiability. The notion of syntactic semi-relevance has also been applied to description logics via a translation scheme to first-order logic [10].

The paper is organized as follows. Section 2 fixes the notations, definitions and existing results in particular from [11]. Section 3 is reserved for our new notion of semantic relevance. Finally, we conclude our work in Sect. 4 with a discussion of our results.

2 Preliminaries

We assume a standard first-order language without equality over a signature $\Sigma = (\Omega, \Pi)$ where Ω is a non-empty set of functions symbols, Π a non-empty set of predicate symbols both coming with their respective fixed arities denoted by the function arity. The set of terms over an infinite set of variables $\mathcal{X}$ is denoted by $T(\Sigma, \mathcal{X})$. Atoms, literals, clauses, and clause sets are defined as usual, e.g., see [24]. We identify a clause with its multiset of literals. Variables in clauses are universally quantified. Then N denotes a clause set; C, D denote clauses; L, K denote literals; A, B denote atoms; P, Q, R, T denote predicates; t, s terms; f, g, h functions; a, b, c, d constants; and x, y, z variables, all possibly indexed. The complement of a literal is denoted by the function comp. Atoms, literals, clauses, and clause sets are *ground* if they do not contain any variable.

An interpretation $\mathcal{I}$ with a nonempty *domain* (or *universe*) $\mathcal{U}$ assigns (i) a total function $f^{\mathcal{I}} : \mathcal{U}^n \mapsto \mathcal{U}$ for each $f \in \Omega$ with $\text{arity}(f) = n$ and (ii) a relation $P \subseteq \mathcal{U}^m$ to every predicate symbol $P^{\mathcal{I}} \in \Pi$ with $\text{arity}(P) = m$. A valuation β is a function $\mathcal{X} \mapsto \mathcal{U}$ where the assignment of some variable x can be modified to $e \in \mathcal{U}$ by $\beta[x \mapsto e]$. It is extended to terms as $\mathcal{I}(\beta) : T(\Sigma, \mathcal{X}) \mapsto \mathcal{U}$. Semantic entailment $\models$ considers variables in clauses to be universally quantified. The extension to atoms, literals, disjunctions, clauses and sets of clauses is as follows: $\mathcal{I}(\beta)(P(t_1, \ldots, t_n)) = 1$ if $(\mathcal{I}(\beta)(t_1), \ldots, \mathcal{I}(\beta)(t_n)) \in P^{\mathcal{I}}$ and 0 otherwise; $\mathcal{I}(\beta)(\neg\phi) = 1 - \mathcal{I}(\beta)(\phi)$; for a disjunction $L_1 \vee \ldots \vee L_k$, $\mathcal{I}(\beta)(L_1 \vee \ldots \vee L_k) = \max(\mathcal{I}(\beta)(L_1), \ldots, \mathcal{I}(\beta)(L_k))$; for a clause C, $\mathcal{I}(\beta)(C) = 1$ if for all valuations $\beta = \{x_1 \mapsto e_1, \ldots, x_n \mapsto e_n\}$ where the x_i are the free variables in C there is a literal $L \in C$ such that $\mathcal{I}(\beta)(L) = 1$; for a set of clauses $N = \{C_1, \ldots, C_k\}$, $\mathcal{I}(\beta)(\{C_1, \ldots, C_k\}) = \min(\mathcal{I}(\beta)(C_1), \ldots, \mathcal{I}(\beta)(C_k))$. A set of clauses N is *satisfiable* if there is an $\mathcal{I}$ of N such that $\mathcal{I}(\beta)(N) = 1$, β arbitrary, (in this case $\mathcal{I}$ is called a *model* of N: $\mathcal{I} \models N$) otherwise N is called *unsatisfiable*.

Substitutions σ, τ are total mappings from variables to terms, where $\text{dom}(\sigma) := \{x \mid x\sigma \neq x\}$ is finite and $\text{codom}(\sigma) := \{t \mid x\sigma = t, x \in \text{dom}(\sigma)\}$. A *renaming* σ is a bijective substitution. The application of substitutions is extended to literals, clauses, and sets/sequences of such objects in the usual way. If $C' = C\sigma$ for some substitution σ, then C' is an *instance* of C. A *unifier* σ for a set of terms $t_1, \ldots, t_k$ satisfies $t_i\sigma = t_j\sigma$ for all $1 \leq i, j \leq k$ and it is called a *most general unifier* if for any unifier σ' of $t_1, \ldots, t_k$ there is a substitution τ s.t. $\sigma' = \sigma\tau$. The function mgu denotes the *most general unifier* of two terms, atoms, literals if it exists. We assume that any mgu of two terms or literals does not introduce any fresh variables and is idempotent.

The resolution calculus consists of two inference rules: Resolution and Factoring [28,29]. The rules operate on a state (N, S) where the initial state for a classical resolution refutation from a clause set N is $(\emptyset, N)$ and for an SOS (Set Of Support) refutation with clause set N and initial SOS clause set S the initial state is (N, S). We describe the rules in the form of abstract rewrite rules operating on states (N, S). As usual we assume for the resolution rule that the involved clauses are variable disjoint. This can always be achieved by applying renamings into fresh variables.

Resolution $(N, S \uplus \{C \vee K\}) \Rightarrow_{\text{RES}} (N, S \cup \{C \vee K, (D \vee C)\sigma\})$
provided $(D \vee L) \in (N \cup S)$ and $\sigma = \text{mgu}(L, \text{comp}(K))$

Factoring $(N, S \uplus \{C \vee L \vee K\}) \Rightarrow_{\text{RES}} (N, S \cup \{C \vee L \vee K\} \cup \{(C \vee L)\sigma\})$
provided $\sigma = \text{mgu}(L, K)$

The clause $(D \vee C)\sigma$ is the result of a *Resolution inference* between its parents and called a *resolvent.* The clause $(C \vee L)\sigma$ is the result of a *Factoring inference* of its parent and called a *factor.* A sequence of rule applications $(N, S) \Rightarrow^*_{RES} (N, S')$ is called a *resolution derivation.* It is called an *SOS resolution derivation* if $N \neq \emptyset$. In case $\bot \in S'$ it is a called a *(SOS) resolution refutation.* If for two clauses C, D there exists a substitution σ such that $C\sigma \subseteq D$, then we say that C *subsumes* D. In this case $C \models D$.

Theorem 1 (Soundness and Refutational Completeness of (SOS) Resolution [11,28,33]**).** *Resolution is sound and refutationally complete [28]. If for some clause set N and initial SOS S, N is satisfiable and $N \cup S$ is unsatisfiable, then there is a (SOS) resolution derivation of $\bot$ from (N, S) [33]. If for some clause set N and clause $C \in N$ there exists a resolution refutation from N using C, then there is an SOS derivation of $\bot$ from $(N \setminus \{C\}, \{C\})$ [11].*

Please note that the recent SOS completeness result of [11] generalizes the classical SOS completeness result by [33].

Theorem 2 (Deductive Completeness of Resolution [17,22]**).** *Given a set of clauses N and a clause D, if $N \models D$, then there is a resolution derivation of some clause C from $(\emptyset, N)$ such that C subsumes D.*

For deductions we require every clause to be used exactly once, so deductions always have a tree form.

Definition 3 (Deduction [11]**).** *A* deduction $\pi_N = [C_1, \dots, C_n]$ *of a clause C_n from some clause set N is a finite sequence of clauses such that for each C_i the following holds:*

1.1 C_i is a renamed, variable-fresh version of a clause in N, or
1.2 there is a clause $C_j \in \pi_N$, $j < i$ s.t. C_i is the result of a Factoring inference from C_j, or
1.3 there are clauses $C_j, C_k \in \pi_N$, $j < k < i$ s.t. C_i is the result of a Resolution inference from C_j and C_k,

and for each $C_i \in \pi_N$, $i < n$:

2.1 there exists exactly one factor C_j of C_i with $j > i$, or
2.2 there exists exactly one C_j and C_k such that C_k is a resolvent of C_i and C_j and $i, j < k$.

We omit the subscript N in π_N if the context is clear.

A deduction π' of some clause $C \in \pi$, where π, π' are deductions from N is a subdeduction of π if $\pi' \subseteq \pi$, where the subset relation is overloaded for sequences. A deduction $\pi_N = [C_1, \ldots, C_{n-1}, \bot]$ is called a *refutation.* While the conditions 3.1.1, 3.1.2, and 3.1.3 are sufficient to represent a resolution derivation, the conditions 3.2.1 and 3.2.2 force deductions to be minimal with respect to C_n.

Note that variable renamings are only applied to clauses from N such that all clauses from N that are introduced in the deduction are variable disjoint. Also recall that our notion of a deduction implies a tree structure. Both assumptions together admit the existence of overall grounding substitutions for a deduction.

Definition 4 (Overall Substitution of a Deduction [11]**).** *Given a deduction π of a clause C_n the* overall substitution *$\tau_{\pi,i}$ of $C_i \in \pi$ is recursively defined by*

1 if C_i is a factor of C_j with $j < i$ and mgu σ, then $\tau_{\pi,i} = \tau_{\pi,j} \circ \sigma$,
2 if C_i is a resolvent of C_j and C_k with $j < k < i$ and mgu σ, then $\tau_{\pi,i} = (\tau_{\pi,j} \circ \tau_{\pi,k}) \circ \sigma$,
3 if C_i is an initial clause, then $\tau_{\pi,i} = \emptyset$,

and the overall substitution of the deduction is $\tau_\pi = \tau_{\pi,n}$. We omit the subscript π if the context is clear.

Overall substitutions are well-defined because clauses introduced from N into the deduction are variable disjoint and each clause is used exactly once in the deduction. A grounding of an overall substitution τ of some deduction π is a substitution $\tau\delta$ such that $\text{codom}(\tau\delta)$ only contains ground terms and $\text{dom}(\delta)$ is exactly the variables from $\text{codom}(\tau)$.

Definition 5 (SOS Deduction [11]**).** *A deduction $\pi_{N \cup S} = [C_1, \ldots, C_n]$ is called an* SOS deduction *with SOS S, if the derivation $(N, S_0) \Rightarrow^*_{RES} (N, S_m)$ is an SOS derivation where $C'_1, \ldots, C'_m$ is the subsequence from $[C_1, \ldots, C_n]$ with input clauses removed, $S_0 = S$, and $S_{i+1} = S_i \cup C'_{i+1}$.*

Oftentimes, it is of particular interest to identify the set of clauses that is minimally unsatisfiable, i.e., removing a clause would make it satisfiable. The earliest mention of such a notion is in [26] where it is introduced via a decision problem. Minimally unsatisfiable sets (MUS) have also gained a lot of attention in practice.

Definition 6 (Minimal Unsatisfiable Subset (MUS) [20]**).** *Given an unsatisfiable set of clauses N, the subset $N' \subseteq N$ is a minimally unsatisfiable subset (MUS) of N if any strict subset of N' is satisfiable.*

In our previous work, we defined a notion of relevance based on how clauses may contribute to unsatisfiability by means of refutations.

Definition 7 (Syntactic Relevance [11]). *Given an unsatisfiable set of clauses N, a clause $C \in N$ is* syntactically relevant *if for all refutations π of N it holds that $C \in \pi$. A clause $C \in N$ is* syntactically semi-relevant *if there exists a refutation π of N in which $C \in \pi$. A clause $C \in N$ is* syntactically irrelevant *if there is no refutation π of N in which $C \in \pi$.*

Syntactic relevance can be identified by using the resolution calculus. A clause $C \in N$ is syntactically semi-relevant if and only if there exists an SOS refutation from SOS $\{C\}$ and $N \setminus \{C\}$.

Theorem 8 (Syntactic Relevance [11]). *Given an unsatisfiable set of clauses N, the clause $C \in N$ is*

1. *syntactically relevant if and only if $N \setminus \{C\}$ is satisfiable,*
2. *syntactically semi-relevant if and only if $(N \setminus \{C\}, \{C\}) \Rightarrow^*_{RES} (N \setminus \{C\}, S \cup \{\bot\})$.*

An open problem from [11] is the question of a semantic counterpart to syntactic semi-relevance. Without any further properties of the clause set N, the notion of semi-relevance can lead to unintuitive results. For example, a tautology could be semi-relevant. Given a refutation showing semi-relevance of some clause C, where, in the refutation, some unary predicate P occurs, the refutation can be immediately extended using the tautology $P(x) \vee \neg P(x)$. We may additionally stumble upon a problem in the case where our set of clauses contains a subsumed clause. For example, if both clauses $Q(a)$ and $Q(x)$ exist in a clause set, they may be both semi-relevant, although from an intuition point of view one may only want to consider $Q(x)$ to be semi-relevant, or even relevant. On the other hand, in some cases, redundant clauses are welcome as semi-relevant clauses.

Example 9 (Redundant Clauses). Given a set of clauses

$$N = \{Q(x), \quad Q(a), \quad \neg Q(a) \vee P(b), \quad \neg P(b), \quad P(x) \vee \neg P(x)\},$$

all clauses are syntactically semi-relevant while $\neg Q(a) \vee P(b)$ and $\neg P(b)$ are syntactically relevant. However, if we disregard the redundant clauses $Q(a)$ and $P(x) \vee \neg P(x)$, then the clause $Q(x)$ becomes a relevant clause. Therefore, for our semantic notion of relevance we will only consider clause sets without clauses implied by other, different clauses from the clause set.

3 Semantic Relevance

Except for the trivially false clause $\bot$, the simplest form of a contradiction is two unit clauses K and L such that K and $\text{comp}(L)$ are unifiable. They will be called *conflict literals*, below. Then the idea for our semantic definition of semi-relevance is to consider clauses that contribute to the number of conflict literals of a clause set. Furthermore, we will show that in any MUS every literal is a conflict literal.

While conflict literals could straightforwardly be defined in propositional logic having the above idea in mind, in first-order logic we have always to relate properties of literals, clauses to their respective ground instances. This is simply due to the fact that unsatisfiability of a first-order clause set is given by unsatisfiability of a finite set of ground instances from this set. Eventually, we will show that for independent clause sets a clause is semi-relevant, if it contributes to the number of conflict literals.

Definition 10 (Conflict Literal). *Given a set of clauses N over some signature Σ, a ground literal L is a* conflict literal *in a clause set N if there are two satisfiable clause sets N_1, N_2 such that*

1. *the clauses in N_1, N_2 are instances of clauses from N and*
2. *$N_1 \models L$ and $N_2 \models \mathrm{comp}(L)$.*

$\mathrm{conflict}(N)$ *denotes the set of conflict literals in N.*

Our notion of a conflict literal generalizes the respective notion in [12] defined for propositional logic.

Example 11 (Conflict Literal). Given an unsatisfiable set of clauses over the signature $\Sigma = (\{a, b, c, d, f\}, \{P\})$:

$$N = \{\neg P(f(a, x)) \vee \neg P(f(c, y)), P(f(x, d)) \vee P(f(y, b))\}$$

Consider the following satisfiable sets of instances from N

$$N_1 = \{\neg P(f(a, d)) \vee \neg P(f(c, y)), P(f(x, d)) \vee P(f(a, b))\}$$
$$N_2 = \{\neg P(f(a, b)) \vee \neg P(f(c, y)), P(f(x, d)) \vee P(f(c, b))\}$$

$P(f(a, b))$ is a conflict literal because $N_1 \models P(f(a, b))$ and $N_2 \models \neg P(f(a, b))$.

We can show that $N_1 \models P(f(a, b))$ because the resolution calculus is sound. Resolving both literals of $\neg P(f(a, d)) \vee \neg P(f(c, y))$ with the first literal of the clause $P(f(x, d)) \vee P(f(a, b))$ results in the clause $P(f(a, b)) \vee P(f(a, b))$ which can be factorized to $P(f(a, b))$. Moreover, N_1 is satisfiable: An interpretation $\mathcal{I}$ with $\mathcal{I}(P(f(a, b))) = 1$ and $\mathcal{I}(P(t)) = 0$ for all terms $t \neq f(a, b)$ satisfies N_1 and $P(f(a, b))$. $N_2 \models \neg P(f(a, b))$ can also be shown in the same manner.

Example 12 (Conflict Literal). Given

$$\begin{aligned} N = \{&\neg R(z), R(c) \vee P(a, y),\\ &Q(a), \neg Q(x) \vee P(x, b),\\ &\neg P(a, b)\} \end{aligned}$$

its conflict literals are

$$\begin{aligned}\text{conflict}(N) = \{&P(a,b), \neg P(a,b),\\ &R(c), \neg R(c),\\ &Q(a), \neg Q(a)\}\end{aligned}$$

In addition to a refutation, the existence of a conflict literal is another way to characterize unsatisfiability of a clause set. Obviously, conflict literals always come in pairs.

Lemma 13 (Minimal Unsatisfiable Ground Clause Sets and Conflict Literals). *If N is a minimally unsatisfiable set of ground clauses (MUS) then any literal occurring in N is a conflict literal.*

Proof Take any ground atom A such that A occurs in N. N can be split into three disjoint clause sets:

$$\begin{aligned}N_\emptyset &= \{C \in N | A \notin C \text{ and } \neg A \notin C\}\\ N_A &= \{C \in N | A \in C\}\\ N_{\neg A} &= \{C \in N | \neg A \in C\}\end{aligned}$$

Since N is minimal, N_A and $N_{\neg A}$ are nonempty, because otherwise A is a pure literal and its corresponding clauses can be removed from N preserving unsatisfiability. Obviously $N_\emptyset \cup N_A$ must be satisfiable, for otherwise the initial choice of N was not minimal. However, $N_\emptyset \cup N'_A$, where N'_A results from all N_A by deleting all A literals from the clauses of N_A, must be unsatisfiable, for otherwise we can construct a satisfying interpretation for N. Thus, every model of $N_\emptyset \cup N_A$ must also be a model of A: $N_\emptyset \cup N_A \models A$. Using the same argument, $N_\emptyset \cup N_{\neg A}$ is satisfiable and $N_\emptyset \cup N_{\neg A} \models \neg A$. Therefore, A is a conflict literal. □

Lemma 14 (Conflict Literals and Unsatisfiability). *Given a set of clauses N, $\text{conflict}(N) \neq \emptyset$ if and only if N is unsatisfiable.*

Proof "⇒" Let $L \in \text{conflict}(N)$. By definition, there are two satisfiable subsets of instances N_1, N_2 from N such that $N_1 \models L$ and $N_2 \models \text{comp}(L)$. Towards contradiction, suppose N is satisfiable. Then, there exists an interpretation $\mathcal{I}$ with $\mathcal{I} \models N$ and therefore it holds that $\mathcal{I} \models N_1$ and $\mathcal{I} \models N_2$. Furthermore, by definition of a conflict literal, $\mathcal{I} \models L$ and $\mathcal{I} \models \text{comp}(L)$, a contradiction.
"⇐" Given an unsatisfiable clause set N, we show that there is a conflict literal in N. Since N is unsatisfiable, by compactness of first-order logic there is a minimal set of ground instances N' from N that is also unsatisfiable. The rest follows from Lemma 13. □

Intuitively, a clause that is implied by other clauses is redundant and can be removed from the set of clauses. However, then applying a calculus generating new clauses, this intuitive notion of redundancy may destroy completeness [2,23]. Still, the detection and elimination of redundant clauses, compatible or incompatible with completeness, is an important concept to the efficiency of automatic reasoning, e.g., in propositional logic [3,18]. It is also apparently important when we try to define a semantic notion of relevance. For example, a syntactically relevant clause would step down to be syntactically semi-relevant if it is duplicated. So, in order to have a semantically robust notion of relevance in first-order logic, we need to use a strong notion of (in)dependency.

Definition 15 (Dependency). *A clause C is* dependent *in N if there exists a satisfiable set of instances N' from $N \setminus \{C\}$ such that $N' \models C\sigma$ for some σ. If C is not dependent in N it is* independent *in N. A clause set N is* independent *if it does not contain any dependent clauses.*

A subsumed clause is obviously a dependent clause. However, there could also be non-subsumed clauses that are dependent. For example, in the set of clauses

$$N = \{P(a, y), P(x, b), \neg P(a, b)\}$$

$P(x, b)$ is dependent because $P(a, b)$ is an instance of $P(x, b)$ and it is entailed by $P(a, y)$. Now, we are ready to define the semantic notion of relevance based on conflict literals and dependency.

In some way, our notion of independence of clause sets is a strong assumption because there might be non-redundant clauses that are considered dependent. While this holds by design in some scenarios (e.g. the mentioned car scenario) in others it is violated by design. In addition, one question that may arise is how to acquire an independent clause set out of a dependent one. For example, in a scenario where some theory is developed out of some independent axioms. Then of course proven lemmas, theorems are dependent with respect to the axioms. In this case one could trace out of the proofs the dependency relations between the intermediate lemmas, theorems and the axioms and this way calculate independent clause sets with respect to some proven conjecture. This would then lead again to independent (sub) clause sets with respect to the proven conjecture where our results are applicable.

Definition 16 (Semantic Relevance). *Given an unsatisfiable set of independent clauses N, a clause $C \in N$ is*

1. relevant, *if* $\text{conflict}(N \setminus \{C\}) = \emptyset$
2. semi-relevant, *if* $\text{conflict}(N \setminus \{C\}) \subsetneq \text{conflict}(N)$
3. irrelevant, *if* $\text{conflict}(N \setminus \{C\}) = \text{conflict}(N)$

Example 17 (Dependent Clauses in Propositional Logic).

$$N = \{P, \neg P, \\ \neg P \vee Q, \neg R \vee P, \\ \neg Q \vee R\}$$

The existence of dependent clauses $\neg P \vee Q$ and $\neg R \vee P$ causes an independent clause $\neg Q \vee R$ to be a semi-relevant clause. However, $\neg Q \vee R$ is not inside the only MUS $\{P, \neg P\}$.

Very often, concepts from propositional logic can be generalized to first-order logic. However, in the context of relevance this is not the case. Our notion of (semi-)relevance can also be characterized by MUSes in propositional logic, but not in first-order logic without considering instances of clauses.

Lemma 18 (Propositional Clause Sets and Relevance). *Given an independent unsatisfiable set of propositional clauses N, the relevant clauses coincide with the intersection of all MUSes and the semi-relevant clauses coincide with the union of all MUSes.*

Proof For the case of relevance: Given $C \in N$, C is relevant if and only if $\text{conflict}(N \setminus \{C\}) = \emptyset$ if and only if $N \setminus \{C\}$ is satisfiable by Lemma 14 if and only if C is contained in all MUSes N' of N.

For the case of semi-relevance: Given $C \in N$, we show C is semi-relevant if and only if C is in some MUS $N' \subseteq N$.

"$\Rightarrow$": Towards contradiction, suppose there is a semi-relevant clause C that is not in any MUS. By definition of semi-relevant clauses, there are satisfiable sets N_1 and N_2 and a propositional variable P such that $N_1 \models P$, $N_2 \models \neg P$ but the MUS M out of $N_1 \cup N_2$ does not contain C. By Theorem 2 there exist deductions π_1 and π_2 of P and $\neg P$ from N_1 and N_2, respectively. Since a deduction is connected, some clauses in M and $(N_1 \cup N_2) \setminus M$ must have some complementary propositional literals Q and $\neg Q$, respectively to be eventually resolved upon in either π_1 or π_2. At least one of these deductions must contain this resolution step between a clause from M and one from $(N_1 \cup N_2) \setminus M$. Now by Lemma 13 the literals Q and $\neg Q$ are conflict literals in M. Thus, there are satisfiable subsets from M which entail Q and $\neg Q$, respectively. Therefore, the clause containing Q or $\neg Q$ in $(N_1 \cup N_2) \setminus M$ is dependent contradicting the assumption that N does not contain dependent clauses.

"$\Leftarrow$": If C is in some MUS $N' \subseteq N$, then, $N' \setminus \{C\}$ is satisfiable. So invoking Lemma 13 any literal $L \in C$ is a conflict literal in N'. In addition, L is not a conflict literal in $N \setminus \{C\}$ for otherwise C is dependent: Suppose L is a conflict literal in $N \setminus \{C\}$ then, by definition, there is satisfiable subset from $N \setminus \{C\}$ which entails L. However, since $L \models C$, it means C is dependent. □

The next example demonstrates that the notion of a MUS cannot be carried over straightforwardly to the level of clauses with variables to characterize semi-relevant clauses in first-order logic.

Example 19 (First-Order Relevant Clauses). Given a set of clauses

$$N = \{P(a, y), \neg P(a, d) \lor Q(b, d), \\ \neg P(x, c), \neg Q(b, d) \lor P(d, c), Q(z, e)\}$$

over $\Sigma = (\{a, b, c, d, e\}, \{P, Q\})$. The conflict literals are

$$\{(\neg)P(a, c), (\neg)Q(b, d), (\neg)P(d, c), (\neg)P(a, d)\}.$$

The clause $P(a, y)$ is relevant. The literals entailed by some satisfiable instances N' from N such that $P(a, y) \notin N'$ are $\{\neg Q(b, d)\} \uplus \{\neg P(t, c), \neg Q(t, e) \mid t \in \{a, b, c, d, e\}\}$ and no two of them are complementary. Thus, $\text{conflict}(N \setminus \{P(a, y)\}) = \emptyset$. The clause $\neg P(a, d) \vee Q(b, d)$ is semi-relevant: $Q(b, d) \notin \text{conflict}(N \setminus \{\neg P(a, d) \vee Q(b, d)\})$. The clause $Q(z, e)$ is irrelevant.

With respect to a MUS, the clause $\neg P(a, d) \vee Q(b, d)$ from Example 19 is irrelevant. The only MUS from N is $\{P(a, y), \neg P(x, c)\}$ with grounding substitution $\{x \mapsto a, y \mapsto c\}$. However, in first-order logic we should not ignore the clauses $\neg P(a, d) \vee Q(b, d)$, $\neg Q(b, d) \vee P(d, c)$, because together with the clauses $P(a, y), \neg P(x, c)$ they result in a different grounding $\{x \mapsto d, y \mapsto d\}$. So, we argue that MUS-based (semi-)relevance on the original clause set is not sufficient to characterize the way clauses are used to derive a contradiction for full first-order logic. However, it does so if ground instances are considered.

Lemma 20 (Relevance and MUSes on First-Order Clauses). *Given an unsatisfiable set of independent first-order clauses N. Then a clause C is relevant in N, if all MUSes of unsatisfiable sets of ground instances from N contain a ground instance of C. The clause C is semi-relevant in N, if there exists a MUS of an unsatisfiable set of ground instances from N that contains a ground instance of C.*

Proof (Relevance) Since all ground instances from N contain a ground instance of C, then, if $N \setminus \{C\}$ contains a ground MUS from N it means that some ground instance of C is entailed by $N \setminus \{C\}$. This violates our assumption that N contains no dependent clauses. Thus, $N \setminus \{C\}$ contains no ground MUSes. This further means that $N \setminus \{C\}$ is satisfiable by the compactness theorem of first-order logic. By Lemma 14 it therefore has no conflict literals and C is relevant. (Semi-Relevance) Take some ground MUS M containing some ground instance C' of C. Due to Lemma 13, any literal $P \in C'$ is a conflict literal in M and consequently also in N. In addition, P is not a conflict literal in $N \setminus \{C\}$ for otherwise C is dependent: Suppose P is a conflict literal in $N \setminus \{C\}$. Then, by definition, there is some satisfiable instances from $N \setminus \{C\}$ which entails P. However, since $P \models C'$, it means C is dependent. In conclusion, $P \in \text{conflict}(N) \setminus \text{conflict}(N \setminus \{C\})$ and thus C is semi-relevant. □

In Example 19, we could identify two ground MUSes:

$$\{P(a, c), \neg P(a, c)\}$$

and

$$\{P(a, d), \neg P(a, d) \vee Q(b, d), \neg P(d, c), \neg Q(b, d) \vee P(d, c)\}$$

Our notion of relevance is thus alternatively explainable using Lemma 20: $P(a, y)$ is relevant because every MUS contains an instance of it ($P(a, c)$ and $P(a, d)$). The clause $\neg P(a, d) \vee Q(b, d)$ is semi-relevant as it is immediately contained in the second MUS. The clause $Q(z, e)$ is irrelevant since no MUS contains any instance of $Q(z, e)$. On the other hand, we may still encounter the case where a dependent clause is actually categorized as syntactically semi-relevant. Therefore, by using the dependency notion while at the same time not restricting a refutation to only use MUS as the input set, we can show that (semi-)relevance actually coincides with the syntactic (semi-)relevance. So, the semi-decidability result also follows.

Theorem 21 (Semantic versus Syntactic Relevance). *Given an independent, unsatisfiable set of clauses N in first-order logic, then (semi)-relevant clauses coincide with syntactically (semi)-relevant clauses.*

Proof We show the following: if N contains no dependent clause, C is (semi-) relevant if and only if C is syntactically (semi-)relevant. The case for relevant clauses is a consequence of Lemma 14. Now, we show it for semi-relevant clauses. "$\Rightarrow$" Let L be a ground literal with $L \in \text{conflict}(N) \setminus \text{conflict}(N \setminus \{C\})$. We can construct a refutation using C. There are two satisfiable subsets of instances N_1, N_2 from N such that $N_1 \models L$ and $N_2 \models \text{comp}(L)$ where $N_1 \cup N_2$ contains at least one instance of C, for otherwise $L \notin \text{conflict}(N) \setminus \text{conflict}(N \setminus \{C\})$. By the deductive completeness, Theorem 2, and the fact that L and $\text{comp}(L)$ are ground literals, there are two variable disjoint deductions π_1 and π_2 of some literals K_1 and K_2 such that $K_1\sigma = L$ and $K_2\sigma = \text{comp}(L)$ for some grounding σ. Obviously, the two variable disjoint deductions can be combined to a refutation $\pi_1.\pi_2.\bot$ containing C. Thus, C is syntactically semi-relevant in N.

"$\Leftarrow$" Given an SOS refutation π using C, i.e., an SOS refutation π from $N \setminus \{C\}$ with SOS $\{C\}$ and overall grounding substitution σ, we show that C is semantically semi-relevant. Let N' be the variable renamed versions of clauses from $N \setminus \{C\}$ used in the refutation and S' be the renamed copies of C used in the refutation. First, we show that $N'\sigma$ is satisfiable. Towards contradiction, suppose $N'\sigma$ is unsatisfiable and let $M\sigma \subseteq N'\sigma$ be its MUS. Since π is connected, some clauses in $M\sigma$ and $S'\sigma \cup (N'\sigma \setminus M\sigma)$ contains literals L and $\text{comp}(L)$ respectively. By Lemma 13, L and $\text{comp}(L)$ are also conflict literals in $M\sigma$. So, by Definition 15, the clause containing $\text{comp}(L)$ in $S'\sigma \cup (N'\sigma \setminus M\sigma)$ is dependent violating our initial assumption.

Now, since $N'\sigma$ is satisfiable, there is a ground MUS from $(N' \cup S')\sigma$ containing some $C'\sigma \in S\sigma$. Due to Lemma 13, any $L \in C'\sigma$ is a conflict literal in N' (and consequently also in N). In addition, L is not a conflict literal in $N \setminus \{C\}$ for otherwise C is dependent: Suppose L is a conflict literal in $N \setminus \{C\}$. Then, by definition, there is some satisfiable instances from $N \setminus \{C\}$ which entails L. However, since $L \models C'\sigma$, it means C is dependent. In conclusion, $L \in \text{conflict}(N) \setminus \text{conflict}(N \setminus \{C\})$ and thus C is semi-relevant. □

When we have a ground MUS, identifification of conflict literals is obvious because all of the literals in it are. However, testing if a literal L is a conflict literal is not trivial, in general. One can try enumerating all MUSes and check if L is contained in some. This definitely works for propositional logic despite being computationally expensive. In first-order logic, this is problematic because there could potentially be an infinite number of MUSes and determining a MUS is not even semi-decidable, in general. The following lemma provides a semi-decidable test using the SOS strategy.

Lemma 22 *Given a ground literal L and an unsatisfiable set of clauses N with no dependent clauses, L is a conflict literal if and only if there is an SOS refutation from $(N, \{L \vee \text{comp}(L)\})$.*

Proof "$\Rightarrow$" By the deductive completeness, Theorem 2, and the fact that L and $\text{comp}(L)$ are ground literals, there are two variable disjoint deductions π_1 and π_2 of some literals K_1 and K_2 such that $K_1\sigma = L$ and $K_2\sigma = \text{comp}(L)$ for some grounding σ. Obviously, the two variable disjoint deductions can be combined to a refutation $\pi_1.\pi_2.\bot$. We can then construct a refutation $\pi_1.\pi_2.(L \vee \neg L).(\text{comp}(L)).\bot$ where K_2 is resolved with $L \vee \text{comp}(L)$ to get $\text{comp}(L)$ which will be resolved with K_1 from π_1 to get $\bot$. By Theorem 7, it means there is an SOS refutation from $(N, \{L \vee \neg L\})$

"$\Leftarrow$" Given an SOS refutation π using $\{L \vee \text{comp}(L)\}$, i.e., an SOS refutation π from $N \setminus \{\{L \vee \text{comp}(L)\}\}$ with SOS $\{\{L \vee \text{comp}(L)\}\}$, Let N' be the variable renamed versions of clauses from N and overall grounding substitution σ. $N'\sigma$ is a MUS for otherwise there is a dependent clause: Suppose $N'\sigma \setminus M$ is an MUS where M is non-empty. Since π is connected, some clause D' in M must be resolved with some $D \in N'\sigma$ upon some literal K. Thus, by Lemma 13, K and $\text{comp}(K)$ are also conflict literals in $N'\sigma \setminus M$. So, by Definition 15, the clause subsuming D' in N is dependent violating our initial assumption. Finally, because L occurs in $N'\sigma$ and $N'\sigma$ is an MUS, by Lemma 13, L is a conflict literal. □

4 Conclusion

The main results of this paper are: (i) a semantic notion of relevance based on the existence of conflict literals, Definition 10, and Definition 16, (ii) its relationship to syntactic relevance, namely, both notions coincide for independent clause sets, Theorem 21, and (iii) the relationship of semantic relevance to minimal unsatisfiable sets, MUSes, both for propositional logic, Lemma 18, and first-order logic, Lemma 20.

The semantic relevance notion sheds some further light on the way clauses may contribute to a refutation beyond what can be offered by the notion of MUSes. While the syntactic notion of semi-relevance also considers redundant clauses such as tautologies to be semi-relevant, the semantic notion rules out redundant clauses. Here, the notions only coincide for independent clause sets. Still, the syntactic notion is "easier" to test and there are applications where clause sets do not contain implied clauses by construction. Hence, the syntactic-relevance coincides with semantic relevance. For example, first-order toolbox formalizations have this property because every tool is formalized by its own distinct predicate. Still a goal, refutation, can be reached by the use of different tools. The classic example is the toolbox for car/truck/tractor building [8,31].

Acknowledgments. This work was partly funded by DFG grant 389792660 as part of TRR 248. We thank Christopher Lynch and David Plaisted for a number of discussions on semantic relevance. We thank the anonymous reviewers for their constructive and detailed comments.

References

1. Baader, F., Peñaloza, R.: Axiom pinpointing in general tableaux. J. Log. Comput. **20**(1), 5–34 (2010)
2. Bachmair, L., Ganzinger, H.: Resolution theorem proving. In: Robinson, A., Voronkov, A. (eds.) Handbook of Automated Reasoning, vol. I, chap. 2, pp. 19–99. Elsevier, Amsterdam (2001)
3. Boufkhad, Y., Roussel, O.: Redundancy in random SAT formulas. In: Kautz, H.A., Porter, B.W. (eds.) Proceedings of the Seventeenth National Conference on Artificial Intelligence and Twelfth Conference on on Innovative Applications of Artificial Intelligence, 30 July - 3 August, 2000, Austin, Texas, USA, pp. 273–278. AAAI Press/The MIT Press (2000)
4. Bourgaux, C., Ozaki, A., Peñaloza, R., Predoiu, L.: Provenance for the description logic ELHr. In: Bessiere, C. (ed.) Proceedings of the Twenty-Ninth International Joint Conference on Artificial Intelligence, IJCAI 2020, pp. 1862–1869. ijcai.org (2020)
5. Cimatti, A., Griggio, A., Sebastiani, R.: A simple and flexible way of computing small unsatisfiable cores in SAT modulo theories. In: Marques-Silva, J., Sakallah, K.A. (eds.) SAT 2007. LNCS, vol. 4501, pp. 334–339. Springer, Heidelberg (2007). https://doi.org/10.1007/978-3-540-72788-0_32
6. Cimatti, A., Griggio, A., Sebastiani, R.: Computing small unsatisfiable cores in satisfiability modulo theories. J. Artif. Intell. Res. **40**, 701–728 (2011)
7. Eiter, T., Gottlob, G.: The complexity of logic-based abduction. J. ACM **42**(1), 3–42 (1995)
8. Fetzer, C., Weidenbach, C., Wischnewski, P.: Compliance, functional safety and fault detection by formal methods. In: Margaria, T., Steffen, B. (eds.) ISoLA 2016. LNCS, vol. 9953, pp. 626–632. Springer, Cham (2016). https://doi.org/10.1007/978-3-319-47169-3_48
9. Guthmann, O., Strichman, O., Trostanetski, A.: Minimal unsatisfiable core extraction for SMT. In: Piskac, R., Talupur, M. (eds.) 2016 Formal Methods in Computer-Aided Design, FMCAD 2016, Mountain View, CA, USA, October 3–6, 2016. pp. 57–64. IEEE (2016)
10. Haifani, F., Koopmann, P., Tourret, S., Weidenbach, C.: On a notion of relevance. In: Borgwardt, S., Meyer, T. (eds.) Proceedings of the 33rd International Workshop on Description Logics (DL 2020) Co-located with the 17th International Conference on Principles of Knowledge Representation and Reasoning (KR 2020), Online Event [Rhodes, Greece], 12th to 14th September 2020. CEUR Workshop Proceedings, vol. 2663. CEUR-WS.org (2020)
11. Haifani, F., Tourret, S., Weidenbach, C.: Generalized completeness for SOS resolution and its application to a new notion of relevance. In: Platzer, A., Sutcliffe, G. (eds.) CADE 2021. LNCS (LNAI), vol. 12699, pp. 327–343. Springer, Cham (2021). https://doi.org/10.1007/978-3-030-79876-5_19
12. Jabbour, S., Ma, Y., Raddaoui, B., Sais, L.: Quantifying conflicts in propositional logic through prime implicates. Int. J. Approx. Reason. **89**, 27–40 (2017)
13. Kalyanpur, A., Parsia, B., Horridge, M., Sirin, E.: Finding all justifications of OWL DL entailments. In: Aberer, K. (ed.) ASWC/ISWC -2007. LNCS, vol. 4825, pp. 267–280. Springer, Heidelberg (2007). https://doi.org/10.1007/978-3-540-76298-0_20
14. Kleine Büning, H., Kullmann, O.: Minimal unsatisfiability and autarkies. In: Biere, A., Heule, M., van Maaren, H., Walsh, T. (eds.) Handbook of Satisfiability, Fron-

tiers in Artificial Intelligence and Applications, vol. 185, pp. 339–401. IOS Press, Amsterdam (2009)
15. Kullmann, O.: Investigations on autark assignments. Discret. Appl. Math. **107**(1–3), 99–137 (2000)
16. Kullmann, O., Lynce, I., Marques-Silva, J.: Categorisation of clauses in conjunctive normal forms: minimally unsatisfiable sub-clause-sets and the lean kernel. In: Biere, A., Gomes, C.P. (eds.) SAT 2006. LNCS, vol. 4121, pp. 22–35. Springer, Heidelberg (2006). https://doi.org/10.1007/11814948_4
17. Lee, C.T.: A Completeness Theorem and a Computer Program for Finding Theorems Derivable from Given Axioms. Ph.D. thesis, University of Berkeley, California, Department of Electrical Engineering (1967)
18. Liberatore, P.: Redundancy in logic I: CNF propositional formulae. Artif. Intell. **163**(2), 203–232 (2005)
19. Liu, S., Luo, J.: FMUS2: an efficient algorithm to compute minimal unsatisfiable subsets. In: Fleuriot, J., Wang, D., Calmet, J. (eds.) AISC 2018. LNCS (LNAI), vol. 11110, pp. 104–118. Springer, Cham (2018). https://doi.org/10.1007/978-3-319-99957-9_7
20. Marques-Silva, J., Mencía, C.: Reasoning about inconsistent formulas. In: Bessiere, C. (ed.) Proceedings of the Twenty-Ninth International Joint Conference on Artificial Intelligence, IJCAI 2020, pp. 4899–4906. ijcai.org (2020)
21. Mencía, C., Kullmann, O., Ignatiev, A., Marques-Silva, J.: On computing the union of MUSes. In: Janota, M., Lynce, I. (eds.) SAT 2019. LNCS, vol. 11628, pp. 211–221. Springer, Cham (2019). https://doi.org/10.1007/978-3-030-24258-9_15
22. Nienhuys-Cheng, S.-H., de Wolf, R.: The equivalence of the subsumption theorem and the refutation-completeness for unconstrained resolution. In: Kanchanasut, K., Lévy, J.-J. (eds.) ACSC 1995. LNCS, vol. 1023, pp. 269–285. Springer, Heidelberg (1995). https://doi.org/10.1007/3-540-60688-2_50
23. Nieuwenhuis, R., Rubio, A.: Paramodulation-based theorem proving. In: Robinson, A., Voronkov, A. (eds.) Handbook of Automated Reasoning, vol. I, chap. 7, pp. 371–443. Elsevier, Amsterdam (2001)
24. Nonnengart, A., Weidenbach, C.: Computing small clause normal forms. In: Robinson, A., Voronkov, A. (eds.) Handbook of Automated Reasoning, vol. 1, chap. 6, pp. 335–367. Elsevier, Amsterdam (2001)
25. Papadimitriou, C.H., Wolfe, D.: The complexity of facets resolved. J. Comput. Syst. Sci. **37**(1), 2–13 (1988)
26. Papadimitriou, C.H., Yannakakis, M.: The complexity of facets (and some facets of complexity). J. Comput. Syst. Sci. **28**(2), 244–259 (1984)
27. Peñaloza, R., Mencía, C., Ignatiev, A., Marques-Silva, J.: Lean kernels in description logics. In: Blomqvist, E., Maynard, D., Gangemi, A., Hoekstra, R., Hitzler, P., Hartig, O. (eds.) ESWC 2017. LNCS, vol. 10249, pp. 518–533. Springer, Cham (2017). https://doi.org/10.1007/978-3-319-58068-5_32
28. Robinson, J.A.: A machine-oriented logic based on the resolution principle. J. ACM **12**(1), 23–41 (1965)
29. Robinson, J.A., Voronkov, A. (eds.): Handbook of Automated Reasoning (in 2 volumes). Elsevier and MIT Press, Cambridge (2001)
30. Schlobach, S., Cornet, R.: Non-standard reasoning services for the debugging of description logic terminologies. In: Gottlob, G., Walsh, T. (eds.) Proceedings of the Eighteenth International Joint Conference on Artificial Intelligence, IJCAI-03, Acapulco, Mexico, 9–15 August 2003, pp. 355–362. Morgan Kaufmann (2003)

31. Sinz, C., Kaiser, A., Küchlin, W.: Formal methods for the validation of automotive product configuration data. Artif. Intell. Eng. Des. Anal. Manuf. **17**(1), 75–97 (2003)
32. Walter, R., Felfernig, A., Küchlin, W.: Constraint-based and SAT-based diagnosis of automotive configuration problems. J. Intell. Inf. Syst. **49**(1), 87–118 (2016). https://doi.org/10.1007/s10844-016-0422-7
33. Wos, L., Robinson, G., Carson, D.: Efficiency and completeness of the set of support strategy in theorem proving. J. ACM **12**(4), 536–541 (1965)
34. Xie, H., Luo, J.: An algorithm to compute minimal unsatisfiable subsets for a decidable fragment of first-order formulas. In: 28th IEEE International Conference on Tools with Artificial Intelligence, ICTAI 2016, San Jose, CA, USA, 6–8 November 2016, pp. 444–451. IEEE Computer Society (2016)
35. Zhang, J., Xu, W., Zhang, J., Shen, S., Pang, Z., Li, T., Xia, J., Li, S.: Finding first-order minimal unsatisfiable cores with a heuristic depth-first-search algorithm. In: Yin, H., Wang, W., Rayward-Smith, V. (eds.) IDEAL 2011. LNCS, vol. 6936, pp. 178–185. Springer, Heidelberg (2011). https://doi.org/10.1007/978-3-642-23878-9_22

SCL(EQ): SCL for First-Order Logic with Equality

Hendrik Leidinger[1,2(✉)] and Christoph Weidenbach[1]

[1] Max-Planck Institute for Informatics, Saarbrücken, Germany
{hleiding,weidenbach}@mpi-inf.mpg.de
[2] Graduate School of Computer Science, Saarbrücken, Germany

Abstract. We propose a new calculus SCL(EQ) for first-order logic with equality that only learns non-redundant clauses. Following the idea of CDCL (Conflict Driven Clause Learning) and SCL (Clause Learning from Simple Models) a ground literal model assumption is used to guide inferences that are then guaranteed to be non-redundant. Redundancy is defined with respect to a dynamically changing ordering derived from the ground literal model assumption. We prove SCL(EQ) sound and complete and provide examples where our calculus improves on superposition.

Keywords: First-order logic with equality · Term rewriting · Model-based reasoning

1 Introduction

There has been extensive research on sound and complete calculi for first-order logic with equality. The current prime calculus is superposition [2], where ordering restrictions guide paramodulation inferences and an abstract redundancy notion enables a number of clause simplification and deletion mechanisms, such as rewriting or subsumption. Still this "syntactic" form of superposition infers many redundant clauses. The completeness proof of superposition provides a "semantic" way of generating only non-redundant clauses, however, the underlying ground model assumption cannot be effectively computed in general [31]. It requires an ordered enumeration of infinitely many ground instances of the given clause set, in general. Our calculus overcomes this issue by providing an effective way of generating ground model assumptions that then guarantee non-redundant inferences on the original clauses with variables.

The underlying ordering is based on the order of ground literals in the model assumption, hence changes during a run of the calculus. It incorporates a standard rewrite ordering. For practical redundancy criteria this means that both rewriting and redundancy notions that are based on literal subset relations are permitted to dynamically simplify or eliminate clauses. Newly generated clauses are non-redundant, so redundancy tests are only needed backwards. Furthermore, the ordering is automatically generated by the structure of the clause set.

J. Blanchette et al. (Eds.): IJCAR 2022, LNAI 13385, pp. 228–247, 2022.
https://doi.org/10.1007/978-3-031-10769-6_14

Instead of a fixed ordering as done in the superposition case, the calculus finds and changes an ordering according to the currently easiest way to make progress, analogous to CDCL (Conflict Driven Clause Learning) [11,21,25,29,34].

Typical for CDCL and SCL (Clause Learning from Simple Models) [1,14,18] approaches to reasoning, the development of a model assumption is done by decisions and propagations. A decision guesses a ground literal to be true whereas a propagation concludes the truth of a ground literal through an otherwise false clause. While propagations in CDCL and propositional logic are restricted to the finite number of propositional variables, in first-order logic there can already be infinite propagation sequences [18]. In order to overcome this issue, model assumptions in SCL(EQ) are at any point in time restricted to a finite number of ground literals, hence to a finite number of ground instances of the clause set at hand. Therefore, without increasing the number of considered ground literals, the calculus either finds a refutation or runs into a *stuck state* where the current model assumption satisfies the finite number of ground instances. In this case one can check whether the model assumption can be generalized to a model assumption of the overall clause set or the information of the stuck state can be used to appropriately increase the number of considered ground literals and continue search for a refutation. SCL(EQ) does not require exhaustive propagation, in general, it just forbids the decision of the complement of a literal that could otherwise be propagated.

For an example of SCL(EQ) inferring clauses, consider the three first-order clauses

$$C_1 := h(x) \approx g(x) \lor c \approx d \qquad C_2 := f(x) \approx g(x) \lor a \approx b$$
$$C_3 := f(x) \not\approx h(x) \lor f(x) \not\approx g(x)$$

with a Knuth-Bendix Ordering (KBO), unique weight 1, and precedence $d \prec c \prec b \prec a \prec g \prec h \prec f$. A Superposition Left [2] inference between C_2 and C_3 results in

$$C_4' := h(x) \not\approx g(x) \lor f(x) \not\approx g(x) \lor a \approx b.$$

For SCL(EQ) we start by building a partial model assumption, called a *trail*, with two decisions

$$\Gamma := [h(a) \approx g(a)^{1:(h(x) \approx g(x) \lor h(x) \not\approx g(x)) \cdot \sigma}, f(a) \approx g(a)^{2:(f(x) \approx g(x) \lor f(x) \not\approx g(x)) \cdot \sigma}]$$

where $\sigma := \{x \mapsto a\}$. Decisions and propagations are always ground instances of literals from the first-order clauses, and are annotated with a level and a justification clause, in case of a decision a tautology. Now with respect to Γ clause C_3 is false with grounding σ, and rule Conflict is applicable; see Sect. 3.1 for details on the inference rules. In general, clauses and justifications are considered variable disjoint, but for simplicity of the presentation of this example, we repeat variable names here as long as the same ground substitution is shared. The maximal literal in $C_3\sigma$ is $(f(x) \not\approx h(x))\sigma$ and a rewrite refutation using the ground equations from the trail results in the justification clause

$$(g(x) \not\approx g(x) \lor f(x) \not\approx g(x) \lor f(x) \not\approx g(x) \lor h(x) \not\approx g(x)) \cdot \sigma$$

where for the refutation justification clauses and all otherwise inferred clauses we use the grounding σ for guidance, but operate on the clauses with variables. The respective ground clause is smaller than $(f(x) \not\approx h(x))\sigma$, false with respect to Γ and becomes our new conflict clause by an application of our inference rule Explore-Refutation. It is simplified by our inference rules Equality-Resolution and Factorize, resulting in the finally learned clause

$$C_4 := h(x) \not\approx g(x) \vee f(x) \not\approx g(x)$$

which is then used to apply rule Backtrack to the trail. Observe that C_4 is strictly stronger than C_4' the clause inferred by superposition and that C_4 cannot be inferred by superposition. Thus SCL(EQ) can infer stronger clauses than superposition for this example.

Related Work: SCL(EQ) is based on ideas of SCL [1,14,18] but for the first time includes a native treatment of first-order equality reasoning. Similar to [14] propagations need not to be exhaustively applied, the trail is built out of decisions and propagations of ground literals annotated by first-order clauses, SCL(EQ) only learns non-redundant clauses, but for the first time conflicts resulting out of a decision have to be considered, due to the nature of the equality relation.

There have been suggested several approaches to lift the idea of an inference guiding model assumption from propositional to full first-order logic [6,12,13,18]. They do not provide a native treatment of equality, e.g., via paramodulation or rewriting.

Baumgartner et al. describe multiple calculi that handle equality by using unit superposition style inference rules and are based on either hyper tableaux [5] or DPLL [15,16]. Hyper tableaux fix a major problem of the well-known free variable tableaux, namely the fact that free variables within the tableau are rigid, i.e., substitutions have to be applied to all occurrences of a free variable within the entire tableau. Hyper tableaux with equality [7] in turn integrates unit superposition style inference rules into the hyper tableau calculus.

Another approach that is related to ours is the model evolution calculus with equality ($\mathcal{ME}_\mathcal{E}$) by Baumgartner et al. [8,9] which lifts the DPLL calculus to first-order logic with equality. Similar to our approach, $\mathcal{ME}_\mathcal{E}$ creates a candidate model until a clause instance contradicts this model or all instances are satisfied by the model. The candidate model results from a so-called context, which consists of a finite set of non-ground rewrite literals. Roughly speaking, a context literal specifies the truth value of all its ground instances unless a more specific literal specifies the complement. Initially the model satisfies the identity relation over the set of all ground terms. Literals within a context may be universal or parametric, where universal literals guarantee all its ground instances to be true. If a clause contradicts the current model, it is repaired by a non-deterministic split which adds a parametric literal to the current model. If the added literal does not share any variables in the contradictory clause it is added as a universal literal.

Another approach by Baumgartner and Waldmann [10] combined the superposition calculus with the Model Evolution calculus with equality. In this calculus the atoms of the clauses are labeled as "split atoms" or "superposition atoms". The superposition part of the calculus then generates a model for the superposition atoms while the model evolution part generates a model for the split atoms. Conversely, this means that if all atoms are labeled as "split atom", the calculus behaves similar to the model evolution calculus. If all atoms are labeled as "superposition atom", it behaves like the superposition calculus.

Both the hyper tableaux calculus with equality and the model evolution calculus with equality allow only unit superposition applications, while SCL(EQ) inferences are guided paramodulation inferences on clauses of arbitrary length. The model evolution calculus with equality was revised and implemented in 2011 [8] and compares its performance with that of hyper tableaux. Model evolution performed significantly better, with more problems solved in all relevant TPTP [30] categories, than the implementation of the hyper tableaux calculus.

Plaisted et al. [27] present the Ordered Semantic Hyper-Linking (OSHL) calculus. OSHL is an instantiation based approach that repeatedly chooses ground instances of a non-ground input clause set such that the current model does not satisfy the current ground clause set. A further step repairs the current model such that it satisfies the ground clause set again. The algorithm terminates if the set of ground clauses contains the empty clause. OSHL supports rewriting and narrowing, but only with unit clauses. In order to handle non-unit clauses it makes use of other mechanisms such as Brand's Transformation [3].

Inst-Gen [22] is an instantiation based calculus, that creates ground instances of the input first-order formulas which are forwarded to a SAT solver. If a ground instance is unsatisfiable, then the first-order set is as well. If not then the calculus creates more instances. The Inst-Gen-EQ calculus [23] creates instances by extracting instantiations of unit superposition refutations of selected literals of the first-order clause set. The ground abstraction is then extended by the extracted clauses and an SMT solver then checks the satisfiability of the resulting set of equational and non-equational ground literals.

In favor of examples and explanations we omit all proofs. They are available in an extended version published as a research report [24]. The rest of the paper is organized as follows. Section 2 provides basic formalisms underlying SCL(EQ). The rules of the calculus are presented in Sect. 3. Soundness and completeness results are provided in Sect. 4. We end with a discussion of obtained results and future work, Sect. 5. The main contribution of this paper is the SCL(EQ) calculus that only learns non-redundant clauses, permits subset based redundancy elimination and rewriting, and its soundness and completeness.

2 Preliminaries

We assume a standard first-order language with equality and signature $\Sigma = (\Omega, \emptyset)$ where the only predicate symbol is equality $\approx$. N denotes a set of clauses, C, D denote clauses, L, K, H denote equational literals, A, B denote equational atoms, t, s terms from $T(\Omega, \mathcal{X})$ for an infinite set of variables $\mathcal{X}$, f, g, h function symbols from Ω, a, b, c constants from Ω and x, y, z variables from $\mathcal{X}$. The function *comp* denotes the complement of a literal. We write $s \not\approx t$ as a shortcut for

$\neg(s \approx t)$. The literal $s \# t$ may denote both $s \approx t$ and $s \not\approx t$. The semantics of first-order logic and semantic entailment $\models$ is defined as usual.

By σ, τ, δ we denote substitutions, which are total mappings from variables to terms. Let σ be a substitution, then its finite domain is defined as $dom(\sigma) := \{x \mid x\sigma \neq x\}$ and its codomain is defined as $codom(\sigma) = \{t \mid x\sigma = t, x \in dom(\sigma)\}$. We extend their application to literals, clauses and sets of such objects in the usual way. A term, literal, clause or sets of these objects is ground if it does not contain any variable. A substitution σ is *ground* if $codom(\sigma)$ is ground. A substitution σ is *grounding* for a term t, literal L, clause C if $t\sigma$, $L\sigma$, $C\sigma$ is ground, respectively. By $C \cdot \sigma$, $L \cdot \sigma$ we denote a closure consisting of a clause C, literal L and a grounding substitution σ, respectively. The function *gnd* computes the set of all ground instances of a literal, clause, or clause set. The function mgu denotes the most general unifier of terms, atoms, literals, respectively. We assume that mgus do not introduce fresh variables and that they are idempotent.

The set of positions $\text{pos}(L)$ of a literal (term $\text{pos}(t)$) is inductively defined as usual. The notion $L|_p$ denotes the subterm of a literal L ($t|_p$ for term t) at position $p \in \text{pos}(L)$ ($p \in \text{pos}(t)$). The replacement of a subterm of a literal L (term t) at position $p \in \text{pos}(L)$ ($p \in \text{pos}(t)$) by a term s is denoted by $L[s]_p$ ($t[s]_p$). For example, the term $f(a, g(x))$ has the positions $\{\epsilon, 1, 2, 21\}$, $f(a, g(x))|_{21} = x$ and $f(a, g(x))[b]_2$ denotes the term $f(a, b)$.

Let R be a set of rewrite rules $l \to r$, called a *term rewrite system* (TRS). The rewrite relation $\to_R \subseteq T(\Omega, \mathcal{X}) \times T(\Omega, \mathcal{X})$ is defined as usual by $s \to_R t$ if there exists $(l \to r) \in R$, $p \in \text{pos}(s)$, and a matcher σ, such that $s|_p = l\sigma$ and $t = s[r\sigma]_p$. We write $s = t{\downarrow_R}$ if s is the normal form of t in the rewrite relation $\to_R$. We write $s \# t = (s' \# t'){\downarrow_R}$ if s is the normal form of s' and t is the normal form of t'. A rewrite relation is terminating if there is no infinite descending chain $t_0 \to t_1 \to ...$ and confluent if $t \mathrel{{}^*{\leftarrow}} s \to^* t'$ implies $t \leftrightarrow^* t'$. A rewrite relation is convergent if it is terminating and confluent. A rewrite order is a irreflexive and transitive rewrite relation. A TRS R is terminating, confluent, convergent, if the rewrite relation $\to_R$ is terminating, confluent, convergent, respectively. A term t is called irreducible by a TRS R if no rule from R rewrites t. Otherwise it is called reducible. A literal, clause is irreducible if all of its terms are irreducible, and reducible otherwise. A substitution σ is called irreducible if any $t \in codom(\sigma)$ is irreducible, and reducible otherwise.

Let $\prec_T$ denote a well-founded rewrite ordering on terms which is total on ground terms and for all ground terms t there exist only finitely many ground terms $s \prec_T t$. We call $\prec_T$ a *desired* term ordering. We extend $\prec_T$ to equations by assigning the multiset $\{s, t\}$ to positive equations $s \approx t$ and $\{s, s, t, t\}$ to inequations $s \not\approx t$. Furthermore, we identify $\prec_T$ with its multiset extension comparing multisets of literals. For a (multi)set of terms $\{t_1, \ldots, t_n\}$ and a term t, we define $\{t_1, \ldots, t_n\} \prec_T t$ if $\{t_1, \ldots, t_n\} \prec_T \{t\}$. For a (multi)set of Literals $\{L_1, \ldots, L_n\}$ and a term t, we define $\{L_1, \ldots, L_n\} \prec_T t$ if $\{L_1, \ldots, L_n\} \prec_T \{\{t\}\}$. Given a ground term β then $\text{gnd}_{\prec_T \beta}$ computes the set of all ground instances of a literal, clause, or clause set where the groundings are smaller than β according to the ordering $\prec_T$. Given a set (sequence) of ground literals Γ let conv(Γ) be a convergent rewrite system out of the positive equations in Γ using $\prec_T$.

Let $\prec$ be a well-founded, total, strict ordering on ground literals, which is lifted to clauses and clause sets by its respective multiset extension. We overload $\prec$ for literals, clauses, clause sets if the meaning is clear from the context. The ordering is lifted to the non-ground case via instantiation: we define $C \prec D$ if for all grounding substitutions σ it holds $C\sigma \prec D\sigma$. Then we define $\preceq$ as the reflexive closure of $\prec$ and $N^{\preceq C} := \{D \mid D \in N \text{ and } D \preceq C\}$ and use the standard superposition style notion of redundancy [2].

Definition 1 (Clause Redundancy). *A ground clause C is* redundant *with respect to a set N of ground clauses and an ordering $\prec$ if $N^{\preceq C} \models C$. A clause C is* redundant *with respect to a clause set N and an ordering $\prec$ if for all $C' \in gnd(C)$, C' is redundant with respect to $gnd(N)$.*

3 The SCL(EQ) Calculus

We start the introduction of the calculus by defining the ingredients of an SCL(EQ) state.

Definition 2 (Trail). *A* trail *$\Gamma := [L_1^{i_1:C_1\cdot\sigma_1}, ..., L_n^{i_n:C_n\cdot\sigma_n}]$ is a consistent sequence of ground equations and inequations where L_j is annotated by a level i_j with $i_{j-1} \leq i_j$, and a closure $C_j\cdot\sigma_j$. We omit the annotations if they are not needed in a certain context. A ground literal L is true in Γ if $\Gamma \models L$. A ground literal L is false in Γ if $\Gamma \models comp(L)$. A ground literal L is undefined in Γ if $\Gamma \not\models L$ and $\Gamma \not\models comp(L)$. Otherwise it is defined. For each literal L_j in Γ it holds that L_j is undefined in $[L_1, ..., L_{j-1}]$ and irreducible by* conv$(\{L_1, ..., L_{j-1}\})$.

The above definition of truth and undefinedness is extended to clauses in the obvious way. The notions of true, false, undefined can be parameterized by a ground term β by saying that L is β-undefined in a trail Γ if $\beta \prec_T L$ or L is undefined. The notions of a β-true, β-false term are restrictions of the above notions to literals smaller β, respectively. All SCL(EQ) reasoning is layered with respect to a ground term β.

Definition 3. *Let Γ be a trail and L a ground literal such that L is defined in Γ. By* core$(\Gamma; L)$ *we denote a minimal subsequence $\Gamma' \subseteq \Gamma$ such that L is defined in Γ'. By* cores$(\Gamma; L)$ *we denote the set of all cores.*

Note that *core*$(\Gamma; L)$ is not necessarily unique. There can be multiple cores for a given trail Γ and ground literal L.

Definition 4 (Trail Ordering). *Let $\Gamma := [L_1, ..., L_n]$ be a trail. The (partial) trail ordering $\prec_\Gamma$ is the sequence ordering given by Γ, i.e., $L_i \prec_\Gamma L_j$ if $i < j$ for all $1 \leq i, j \leq n$.*

Definition 5 (Defining Core and Defining Literal). *For a trail Γ and a sequence of literals $\Delta \subseteq \Gamma$ we write $max_{\prec_\Gamma}(\Delta)$ for the largest literal in Δ according to the trail ordering $\prec_\Gamma$. Let Γ be a trail and L a ground literal such that L is defined in Γ. Let $\Delta \in cores(\Gamma; L)$ be a sequence of literals where $max_{\prec_\Gamma}(\Delta) \preceq_\Gamma max_{\prec_\Gamma}(\Lambda)$ for all $\Lambda \in cores(\Gamma; L)$, then* $\max_\Gamma(L) := \max_{\prec_\Gamma}(\Delta)$ *is called the* defining literal *and Δ is called a* defining core *for L in Γ. If $cores(\Gamma; L)$ contains only the empty core, then L has no* defining literal *and no* defining core.

Note that there can be multiple defining cores but only one defining literal for any defined literal L. For example, consider a trail $\Gamma := [f(a) \approx f(b)^{1:C_1 \cdot \sigma_1}, a \approx b^{2:C_2 \cdot \sigma_2}, b \approx c^{3:C_3 \cdot \sigma_3}]$ with an ordering $\prec_T$ that orders the terms of the equations from left to right, and a literal $g(f(a)) \approx g(f(c))$. Then the defining cores are $\Delta_1 := [a \approx b, b \approx c]$ and $\Delta_2 := [f(a) \approx f(b), b \approx c]$. The defining literal, however, is in both cases $b \approx c$. Defined literals that have no defining core and therefore no defining literal are literals that are trivially false or true. Consider, for example, $g(f(a)) \approx g(f(a))$. This literal is trivially true in Γ. Thus an empty subset of Γ is sufficient to show that $g(f(a)) \approx g(f(a))$ is defined in Γ.

Definition 6 (Literal Level). *Let Γ be a trail. A ground literal $L \in \Gamma$ is of* level *i if L is annotated with i in Γ. A defined ground literal $L \notin \Gamma$ is of level i if the defining literal of L is of level i. If L has no defining literal, then L is of level 0. A ground clause D is of level i if i is the maximum level of a literal in D.*

The restriction to minimal subsequences for the defining literal and definition of a level eventually guarantee that learned clauses are smaller in the trail ordering. This enables completeness in combination with learning non-redundant clauses as shown later.

Lemma 7. *Let Γ_1 be a trail and K a defined literal that is of level i in Γ_1. Then K is of level i in a trail $\Gamma := \Gamma_1, \Gamma_2$.*

Definition 8. *Let Γ be a trail and $L \in \Gamma$ a literal. L is called a* decision literal *if $\Gamma = \Gamma_0, K^{i:C \cdot \tau}, L^{i+1:C' \cdot \tau'}, \Gamma_1$. Otherwise L is called a* propagated literal.

In our above example $g(f(a)) \approx g(f(c))$ is of level 3 since the defining literal $b \approx c$ is annotated with 3. $a \not\approx b$ on the other hand is of level 2.

We define a well-founded total strict ordering which is induced by the trail and with which non-redundancy is proven in Sect. 4. Unlike SCL [14,18] we use this ordering for the inference rules as well. In previous SCL calculi, conflict resolution automatically chooses the greatest literal and resolves with this literal. In SCL(EQ) this is generalized. Coming back to our running example above, suppose we have a conflict clause $f(b) \not\approx f(c) \vee b \not\approx c$. The defining literal for both inequations is $b \approx c$. So we could do paramodulation inferences with both literals. The following ordering makes this non-deterministic choice deterministic.

Definition 9 (Trail Induced Ordering). *Let $\Gamma := [L_1^{i_1:C_1\cdot\sigma_1}, ..., L_n^{i_n:C_n\cdot\sigma_n}]$ be a trail, β a ground term such that $\{L_1, ..., L_n\} \prec_T \beta$ and $M_{i,j}$ all β-defined ground literals not contained in $\Gamma \cup comp(\Gamma)$: for a defining literal $max_\Gamma(M_{i,j}) = L_i$ and for two literals $M_{i,j}$, $M_{i,k}$ we have $j < k$ if $M_{i,j} \prec_T M_{i,k}$. The trail induces a total well-founded strict order $\prec_{\Gamma^*}$ on β-defined ground literals $M_{k,l}, M_{m,n}, L_i, L_j$ of level greater than zero, where*

1. $M_{i,j} \prec_{\Gamma^*} M_{k,l}$ *if* $i < k$ *or* $(i = k$ *and* $j < l)$
2. $L_i \prec_{\Gamma^*} L_j$ *if* $L_i \prec_\Gamma L_j$
3. $comp(L_i) \prec_{\Gamma^*} L_j$ *if* $L_i \prec_\Gamma L_j$
4. $L_i \prec_{\Gamma^*} comp(L_j)$ *if* $L_i \prec_\Gamma L_j$ *or* $i = j$
5. $comp(L_i) \prec_{\Gamma^*} comp(L_j)$ *if* $L_i \prec_\Gamma L_j$
6. $L_i \prec_{\Gamma^*} M_{k,l}$, $\mathrm{comp}(L_i) \prec_{\Gamma^*} M_{k,l}$ *if* $i \leq k$
7. $M_{k,l} \prec_{\Gamma^*} L_i$, $M_{k,l} \prec_{\Gamma^*} \mathrm{comp}(L_i)$ *if* $k < i$

and for all β-defined literals L of level zero:

8. $\prec_{\Gamma^*} := \prec_T$
9. $L \prec_{\Gamma^*} K$ *if K is of level greater than zero and K is β-defined*

and can eventually be extended to β-undefined ground literals K, H by

10. $K \prec_{\Gamma^*} H$ *if* $K \prec_T H$
11. $L \prec_{\Gamma^*} H$ *if L is β-defined*

The literal ordering $\prec_{\Gamma^*}$ is extended to ground clauses by multiset extension and identified with $\prec_{\Gamma^*}$ as well.

Lemma 10 (Properties of $\prec_{\Gamma^*}$).

1. *$\prec_{\Gamma^*}$ is well-defined.*
2. *$\prec_{\Gamma^*}$ is a total strict order, i.e. $\prec_{\Gamma^*}$ is irreflexive, transitive and total.*
3. *$\prec_{\Gamma^*}$ is a well-founded ordering.*

Example 11. Assume a trail $\Gamma := [a \approx b^{1:C_0\cdot\sigma_0}, c \approx d^{1:C_1\cdot\sigma_1}, f(a') \not\approx f(b')^{1:C_2\cdot\sigma_2}]$, select KBO as the term ordering $\prec_T$ where all symbols have weight one and $a \prec a' \prec b \prec b' \prec c \prec d \prec f$ and a ground term $\beta := f(f(a))$. According to the trail induced ordering we have that $a \approx b \prec_{\Gamma^*} c \approx d \prec_{\Gamma^*} f(a') \not\approx f(b')$ by 9.2. Furthermore we have that

$$a \approx b \prec_{\Gamma^*} a \not\approx b \prec_{\Gamma^*} c \approx d \prec_{\Gamma^*} c \not\approx d \prec_{\Gamma^*} f(a') \not\approx f(b') \prec_{\Gamma^*} f(a') \approx f(b')$$

by 9.3 and 9.4. Now for any literal L that is *β-defined* in Γ and the defining literal is $a \approx b$ it holds that $a \not\approx b \prec_{\Gamma^*} L \prec_{\Gamma^*} c \approx d$ by 9.6 and 9.7. This holds analogously for all literals that are *β-defined* in Γ and the defining literal is $c \approx d$ or $f(a') \not\approx f(b')$. Thus we get:

$$\begin{gathered} L_1 \prec_{\Gamma^*} ... \prec_{\Gamma^*} a \approx b \prec_{\Gamma^*} a \not\approx b \prec_{\Gamma^*} f(a) \approx f(b) \prec_{\Gamma^*} f(a) \not\approx f(b) \prec_{\Gamma^*} \\ c \approx d \prec_{\Gamma^*} c \not\approx d \prec_{\Gamma^*} f(c) \approx f(d) \prec_{\Gamma^*} f(c) \not\approx f(d) \prec_{\Gamma^*} \\ f(a') \not\approx f(b') \prec_{\Gamma^*} f(a') \approx f(b') \prec_{\Gamma^*} a' \approx b' \prec_{\Gamma^*} a' \not\approx b' \prec_{\Gamma^*} K_1 \prec_{\Gamma^*} \dots \end{gathered}$$

where K_i are the *β-undefined* literals and L_j are the trivially defined literals.

Definition 12 (Rewrite Step). *A* rewrite step *is a five-tuple* $(s\#t\cdot\sigma, s\#t \vee C\cdot\sigma, R, S, p)$ *and inductively defined as follows. The tuple* $(s\#t\cdot\sigma, s\#t \vee C\cdot\sigma, \epsilon, \epsilon, \epsilon)$ *is a rewrite step. Given rewrite steps* R, S *and a position* p *then* $(s\#t\cdot\sigma, s\#t \vee C\cdot\sigma, R, S, p)$ *is a* rewrite step. *The literal* $s\#t$ *is called the* rewrite literal. *In case* R, S *are not* ϵ*, the rewrite literal of* R *is an equation.*

Rewriting is one of the core features of our calculus. The following definition describes a rewrite inference between two clauses. Note that unlike the superposition calculus we allow rewriting below variable level.

Definition 13 (Rewrite Inference). *Let* $I_1 := (l_1 \approx r_1\cdot\sigma_1, l_1 \approx r_1 \vee C_1\cdot\sigma_1, R_1, L_1, p_1)$ *and* $I_2 := (l_2\#r_2\cdot\sigma_2, l_2\#r_2 \vee C_2\cdot\sigma_2, R_2, L_2, p_2)$ *be two variable disjoint rewrite steps where* $r_1\sigma_1 \prec_T l_1\sigma_1$*,* $(l_2\#r_2)\sigma_2|_p = l_1\sigma_1$ *for some position* p*. We distinguish two cases:*

1. *if* $p \in \text{pos}(l_2\#r_2)$ *and* $\mu := \text{mgu}((l_2\#r_2)|_p, l_1)$ *then* $(((l_2\#r_2)[r_1]_p)\mu\cdot\sigma_1\sigma_2, ((l_2\#r_2)[r_1]_p)\mu \vee C_1\mu \vee C_2\mu\cdot\sigma_1\sigma_2, I_1, I_2, p)$ *is the result of a rewrite inference.*
2. *if* $p \notin \text{pos}(l_2\#r_2)$ *then let* $(l_2\#r_2)\delta$ *be the most general instance of* $l_2\#r_2$ *such that* $p \in \text{pos}((l_2\#r_2)\delta)$*,* δ *introduces only fresh variables and* $(l_2\#r_2)\delta\sigma_2\rho = (l_2\#r_2)\sigma_2$ *for some minimal* ρ*. Let* $\mu := \text{mgu}((l_2\#r_2)\delta|_p, l_1)$*. Then* $((l_2\#r_2)\delta[r_1]_p\mu\cdot\sigma_1\sigma_2\rho, (l_2\#r_2)\delta[r_1]_p\mu \vee C_1\mu \vee C_2\delta\mu\cdot\sigma_1\sigma_2\rho, I_1, I_2, p)$ *is the result of a rewrite inference.*

Lemma 14. *Let* $I_1 := (l_1 \approx r_1\cdot\sigma_1, l_1 \approx r_1 \vee C_1\cdot\sigma_1, R_1, L_1, p_1)$ *and* $I_2 := (l_2\#r_2\cdot\sigma_2, l_2\#r_2 \vee C_2\cdot\sigma_2, R_2, L_2, p_2)$ *be two variable disjoint rewrite steps where* $r_1\sigma_1 \prec_T l_1\sigma_1$*,* $(l_2\#r_2)\sigma_2|_p = l_1\sigma_1$ *for some position* p*. Let* $I_3 := (l_3\#r_3\cdot\sigma_3, l_3\#r_3 \vee C_3\cdot\sigma_3, I_1, I_2, p)$ *be the result of a rewrite inference. Then:*

1. $C_3\sigma_3 = (C_1 \vee C_2)\sigma_1\sigma_2$ *and* $l_3\#r_3\sigma_3 = (l_2\#r_2)\sigma_2[r_1\sigma_1]_p$*.*
2. $(l_3\#r_3)\sigma_3 \prec_T (l_2\#r_2)\sigma_2$
3. *If* $N \models (l_1 \approx r_1 \vee C_1) \wedge (l_2\#r_2 \vee C_2)$ *for some set of clauses* N*, then* $N \models l_3\#r_3 \vee C_3$

Now that we have defined rewrite inferences we can use them to define a *reduction chain application* and a *refutation*, which are sequences of rewrite steps. Intuitively speaking, a *reduction chain application* reduces a literal in a clause with literals in $conv(\Gamma)$ until it is irreducible. A *refutation* for a literal L that is *β-false* in Γ for a given β, is a sequence of rewrite steps with literals in Γ, L such that $\bot$ is inferred. Refutations for the literals of the conflict clause will be examined during conflict resolution by the rule Explore-Refutation.

Definition 15 (Reduction Chain). *Let* Γ *be a trail. A* reduction chain $\mathcal{P}$ *from* Γ *is a sequence of rewrite steps* $[I_1, ..., I_m]$ *such that for each* $I_i = (s_i\#t_i\cdot\sigma_i, s_i\#t_i \vee C_i\cdot\sigma_i, I_j, I_k, p_i)$ *either*

1. $s_i\#t_i^{n_i:s_i\#t_i\vee C_i\cdot\sigma}$ *is contained in* Γ *and* $I_j = I_k = p_i = \epsilon$ *or*
2. I_i *is the result of a rewriting inference from rewrite steps* I_j, I_k *out of* $[I_1, ..., I_m]$ *where* $j, k < i$*.*

Let $(l \# r)\delta^{o:l \# r \vee C \cdot \delta}$ *be an annotated ground literal. A* reduction chain application *from* Γ *to* $l \# r$ *is a reduction chain* $[I_1, ..., I_m]$ *from* $\Gamma, (l \# r)\delta^{o:l \# r \vee C \cdot \delta}$ *such that* $l\delta\downarrow_{\text{conv}(\Gamma)} = s_m\sigma_m$ *and* $r\delta\downarrow_{\text{conv}(\Gamma)} = t_m\sigma_m$*. We assume reduction chain applications to be minimal, i.e., if any rewrite step is removed from the sequence it is no longer a reduction chain application.*

Definition 16 (Refutation). *Let* Γ *be a trail and* $(l \# r)\delta^{o:l \# r \vee C \cdot \delta}$ *an annotated ground literal that is* β*-false in* Γ *for a given* β*. A* refutation $\mathcal{P}$ *from* Γ *and* $l \# r$ *is a reduction chain* $[I_1, ..., I_m]$ *from* $\Gamma, (l \# r)\delta^{o:l \# r \vee C \cdot \delta}$ *such that* $(s_m \# t_m)\sigma_m = s \not\approx s$ *for some* s*. We assume refutations to be minimal, i.e., if any rewrite step* I_k*,* $k < m$ *is removed from the refutation, it is no longer a refutation.*

3.1 The SCL(EQ) Inference Rules

We can now define the rules of our calculus based on the previous definitions. A *state* is a six-tuple $(\Gamma; N; U; \beta; k; D)$ similar to the SCL calculus, where Γ a sequence of annotated ground literals, N and U the sets of initial and learned clauses, β is a ground term such that for all $L \in \Gamma$ it holds $L \prec_T \beta$, k is the decision level, and D a status that is $\top$, $\bot$ or a closure $C \cdot \sigma$. Before we propagate or decide any literal, we make sure that it is irreducible in the current trail. Together with the design of $\prec_{\Gamma^*}$ this eventually enables rewriting as a simplification rule.

Propagate

$(\Gamma; N; U; \beta; k; \top) \Rightarrow_{\text{SCL(EQ)}} \quad (\Gamma, s_m \# t_m \sigma_m^{k:(s_m \# t_m \vee C_m)\cdot\sigma_m}; N; U; \beta; k; \top)$
provided there is a $C \in (N \cup U)$, σ grounding for C, $C = C_0 \vee C_1 \vee L$, $\Gamma \models \neg C_0\sigma$, $C_1\sigma = L\sigma \vee ... \vee L\sigma$, $C_1 = L_1 \vee ... \vee L_n$, $\mu = mgu(L_1, ..., L_n, L)$ $L\sigma$ is *β-undefined* in Γ, $(C_0 \vee L)\mu\sigma \prec_T \beta$, σ is irreducible by *conv*(Γ), $[I_1, \ldots, I_m]$ is a reduction chain application from Γ to $L\sigma^{k:(L \vee C_0)\mu\cdot\sigma}$ where $I_m = (s_m \# t_m \cdot \sigma_m, s_m \# t_m \vee C_m \cdot \sigma_m, I_j, I_k, p_m)$.

Note that the definition of Propagate also includes the case where $L\sigma$ is irreducible by Γ. In this case $L = s_m \# t_m$ and $m = 1$. The rule Decide below, is similar to Propagate, except for the subclause C_0 which must be *β-undefined* or *β-true* in Γ, i.e., Propagate cannot be applied and the decision literal is annotated by a tautology.

Decide

$(\Gamma; N; U; \beta; k; \top) \Rightarrow_{\text{SCL(EQ)}} \quad (\Gamma, s_m \# t_m \sigma_m^{k+1:(s_m \# t_m \vee comp(s_m \# t_m))\cdot\sigma_m}; N; U; \beta; k+1; \top)$
provided there is a $C \in (N \cup U)$, σ grounding for C, $C = C_0 \vee L$, $C_0\sigma$ is *β-undefined* or *β-true* in Γ, $L\sigma$ is *β-undefined* in Γ, $(C_0 \vee L)\sigma \prec_T \beta$, σ is irreducible by *conv*(Γ), $[I_1, \ldots, I_m]$ is a reduction chain application from Γ to $L\sigma^{k+1:L \vee C_0 \cdot \sigma}$ where $I_m = (s_m \# t_m \cdot \sigma_m, s_m \# t_m \vee C_m \cdot \sigma_m, I_j, I_k, p_m)$.

Conflict
$(\Gamma; N; U; \beta; k; \top) \Rightarrow_{\mathrm{SCL(EQ)}} \quad (\Gamma; N; U; \beta; k; D)$
provided there is a $D' \in (N \cup U)$, σ grounding for D', $D'\sigma$ is β-*false* in Γ, σ is irreducible by *conv*(Γ), $D = \bot$ if $D'\sigma$ is of level 0 and $D = D' \cdot \sigma$ otherwise.

For the non-equational case, when a conflict clause is found by an SCL calculus [14,18], the complements of its first-order ground literals are contained in the trail. For equational literals this is not the case, in general. The proof showing D to be β-*false* with respect to Γ is a rewrite proof with respect to conv(Γ). This proof needs to be analyzed to eventually perform paramodulation steps on D or to replace D by a $\prec_{\Gamma^*}$ smaller β-*false* clause showing up in the proof.

Skip
$(\Gamma, K^{l:C\cdot\tau}, L^{k:C'\cdot\tau'}; N; U; \beta; k; D\cdot\sigma) \Rightarrow_{\mathrm{SCL(EQ)}} \quad (\Gamma, K^{l:C\cdot\tau}; N; U; \beta; l; D\cdot\sigma)$ if $D\sigma$ is β-*false* in $\Gamma, K^{l:C\cdot\tau}$.

The Explore-Refutation rule is the FOL with Equality counterpart to the resolve rule in CDCL or SCL. While in CDCL or SCL complementary literals of the conflict clause are present on the trail and can directly be used for resolution steps, this needs a generalization for FOL with Equality. Here, in general, we need to look at (rewriting) refutations of the conflict clause and pick an appropriate clause from the refutation as the next conflict clause.

Explore-Refutation
$(\Gamma, L; N; U; \beta; k; (D \vee s \# t)\cdot\sigma)) \Rightarrow_{\mathrm{SCL(EQ)}} \quad (\Gamma, L; N; U; \beta; k; (s_j \# t_j \vee C_j)\cdot\sigma_j)$
if $(s \# t)\sigma$ is strictly $\prec_{\Gamma^*}$ maximal in $(D \vee s \# t)\sigma$, L is the defining literal of $(s \# t)\sigma$, $[I_1, ..., I_m]$ is a refutation from Γ and $(s \# t)\sigma$, $I_j = (s_j \# t_j \cdot \sigma_j, (s_j \# t_j \vee C_j)\cdot\sigma_j, I_l, I_k, p_j)$, $1 \le j \le m$, $(s_j \# t_j \vee C_j)\sigma_j \prec_{\Gamma^*} (D \vee s \# t)\sigma$, $(s_j \# t_j \vee C_j)\sigma_j$ is β-*false* in Γ.

Factorize
$(\Gamma; N; U; \beta; k; (D \vee L \vee L')\cdot\sigma) \Rightarrow_{\mathrm{SCL(EQ)}} \quad (\Gamma; N; U; \beta; k; (D \vee L)\mu\cdot\sigma)$
provided $L\sigma = L'\sigma$, and $\mu = \mathrm{mgu}(L, L')$.

Equality-Resolution
$(\Gamma; N; U; \beta; k; (D \vee s \not\approx s')\cdot\sigma) \Rightarrow_{\mathrm{SCL(EQ)}} \quad (\Gamma; N; U; \beta; k; D\mu\cdot\sigma)$
provided $s\sigma = s'\sigma$, $\mu = mgu(s, s')$.

Backtrack
$(\Gamma, K, \Gamma'; N; U; \beta; k; (D \vee L)\cdot\sigma) \Rightarrow_{\mathrm{SCL(EQ)}} \quad (\Gamma; N; U \cup \{D \vee L\}; \beta; j - i; \top)$
provided $D\sigma$ is of level i' where $i' < k$, K is of level j and Γ, K the minimal trail subsequence such that there is a grounding substitution τ with $(D \vee L)\tau$ β-false in Γ, K but not in Γ; $i = 1$ if K is a decision literal and $i = 0$ otherwise.

Grow
$(\Gamma; N; U; \beta; k; \top) \Rightarrow_{\mathrm{SCL(EQ)}} \quad (\epsilon; N; U; \beta'; 0; \top)$
provided $\beta \prec_T \beta'$.

In addition to soundness and completeness of the SCL(EQ) rules their tractability in practice is an important property for a successful implementation. In particular, finding propagating literals or detecting a false clause under some grounding. It turns out that these operations are NP-complete, similar to first-order subsumption which has been shown to be tractable in practice.

Lemma 17. *Assume that all ground terms t with $t \prec_T \beta$ for any β are polynomial in the size of β. Then testing Propagate (Conflict) is NP-Complete, i.e., the problem of checking for a given clause C whether there exists a grounding substitution σ such that $C\sigma$ propagates (is false) is NP-Complete.*

Example 18 (SCL(EQ) vs. Superposition: Saturation). Consider the following clauses:

$$N := \{C_1 := c \approx d \vee D, C_2 := a \approx b \vee c \not\approx d, C_3 := f(a) \not\approx f(b) \vee g(c) \not\approx g(d)\}$$

where again we assume a KBO with all symbols having weight one, precedence $d \prec c \prec b \prec a \prec g \prec f$ and $\beta := f(f(g(a)))$. Suppose that we first decide $c \approx d$ and then propagate $a \approx b$: $\Gamma = [c \approx d^{1:c\approx d \vee c \not\approx d}, a \approx b^{1:C_2}]$. Now we have a conflict with C_3. Explore-Refutation applied to the conflict clause C_3 results in a paramodulation inference between C_3 and C_2. Another application of Equality-Resolution gives us the new conflict clause $C_4 := c \not\approx d \vee g(c) \not\approx g(d)$. Now we can Skip the last literal on the trail, which gives us $\Gamma = [c \approx d^{1:c\approx d \vee c \not\approx d}]$. Another application of the Explore-Refutation rule to C_4 using the decision justification clause followed by Equality-Resolution and Factorize gives us $C_5 := c \not\approx d$. Thus with SCL(EQ) the following clauses remain:

$$\begin{array}{cc} C_1' = D & C_5 = c \not\approx d \\ \multicolumn{2}{c}{C_3 = f(a) \not\approx f(b) \vee g(c) \not\approx g(d)} \end{array}$$

where we derived C_1' out of C_1 by subsumption resolution [33] using C_5. Actually, subsumption resolution is compatible with the general redundancy notion of SCL(EQ), see Lemma 25. Now we consider the same example with superposition and the very same ordering (N_i is the clause set of the previous step and N_0 the initial clause set N).

$$\begin{aligned} N_0 &\Rightarrow_{Sup(C_2,C_3)} N_1 \cup \{C_4 := c \not\approx d \vee g(c) \not\approx g(d)\} \\ &\Rightarrow_{Sup(C_1,C_4)} N_2 \cup \{C_5 := c \not\approx d \vee D\} \Rightarrow_{Sup(C_1,C_5)} N_3 \cup \{C_6 := D\} \end{aligned}$$

Thus superposition ends up with the following clauses:

$$\begin{array}{ll} C_2 = a \approx b \vee c \not\approx d & C_3 = f(a) \not\approx f(b) \vee g(c) \not\approx g(d) \\ C_4 = c \not\approx d \vee g(c) \not\approx g(d) & C_6 = D \end{array}$$

The superposition calculus generates more and larger clauses.

Example 19 (SCL(EQ) vs. Superposition: Refutation). Suppose the following set of clauses: $N := \{C_1 := f(x) \not\approx a \vee f(x) \approx b, C_2 := f(f(y)) \approx y, C_3 := a \not\approx b\}$ where again we assume a KBO with all symbols having weight one, precedence

$b \prec a \prec f$ and $\beta := f(f(f(a)))$. A long refutation by the superposition calculus results in the following (N_i is the clause set of the previous step and N_0 the initial clause set N):

$$\begin{aligned} N_0 &\Rightarrow_{Sup(C_1,C_2)} N_1 \cup \{C_4 := y \not\approx a \vee f(f(y)) \approx b\} \\ &\Rightarrow_{Sup(C_1,C_4)} N_2 \cup \{C_5 := a \not\approx b \vee f(f(y)) \approx b \vee y \not\approx a\} \\ &\Rightarrow_{Sup(C_2,C_5)} N_3 \cup \{C_6 := a \not\approx b \vee b \approx y \vee y \not\approx a\} \\ &\Rightarrow_{Sup(C_2,C_4)} N_4 \cup \{C_7 := y \approx b \vee y \not\approx a\} \\ &\Rightarrow_{EqRes(C_7)} N_5 \cup \{C_8 := a \approx b\} \Rightarrow_{Sup(C_3,C_8)} N_6 \cup \{\bot\} \end{aligned}$$

The shortest refutation by the superposition calculus is as follows:

$$\begin{aligned} N_0 &\Rightarrow_{Sup(C_1,C_2)} N_1 \cup \{C_4 := y \not\approx a \vee f(f(y)) \approx b\} \\ &\Rightarrow_{Sup(C_2,C_4)} N_2 \cup \{C_5 := y \approx b \vee y \not\approx a\} \\ &\Rightarrow_{EqRes(C_5)} N_3 \cup \{C_6 := a \approx b\} \Rightarrow_{Sup(C_3,C_6)} N_4 \cup \{\bot\} \end{aligned}$$

In SCL(EQ) on the other hand we would always first propagate $a \not\approx b$, $f(f(a)) \approx a$ and $f(f(b)) \approx b$. As soon as $a \not\approx b$ and $f(f(a)) \approx a$ are propagated we have a conflict with $C_1\{x \rightarrow f(a)\}$. So suppose in the worst case we propagate:

$$\Gamma := [a \not\approx b^{0:a \not\approx b}, f(f(b)) \approx b^{0:(f(f(y)) \approx y)\{y \rightarrow b\}}, f(f(a)) \approx a^{0:(f(f(y)) \approx y)\{y \rightarrow a\}}]$$

Now we have a conflict with $C_1\{x \rightarrow f(a)\}$. Since there is no decision literal on the trail, *Conflict* rule immediately returns $\bot$ and we are done.

4 Soundness and Completeness

In this section we show soundness and refutational completeness of SCL(EQ) under the assumption of a regular run. We provide the definition of a regular run and show that for a regular run all learned clauses are non-redundant according to our trail induced ordering. We start with the definition of a sound state.

Definition 20. *A state $(\Gamma; N; U; \beta; k; D)$ is sound if the following conditions hold:*

1. *Γ is a consistent sequence of annotated literals,*
2. *for each decomposition $\Gamma = \Gamma_1, L\sigma^{i:(C \vee L)\cdot\sigma}, \Gamma_2$ where $L\sigma$ is a propagated literal, we have that $C\sigma$ is β-false in Γ_1, $L\sigma$ is β-undefined in Γ_1 and irreducible by $conv(\Gamma_1)$, $N \cup U \models (C \vee L)$ and $(C \vee L)\sigma \prec_T \beta$,*
3. *for each decomposition $\Gamma = \Gamma_1, L\sigma^{i:(L \vee comp(L))\cdot\sigma}, \Gamma_2$ where $L\sigma$ is a decision literal, we have that $L\sigma$ is β-undefined in Γ_1 and irreducible by $conv(\Gamma_1)$, $N \cup U \models (L \vee comp(L))$ and $(L \vee comp(L))\sigma \prec_T \beta$,*
4. *$N \models U$,*
5. *if $D = C \cdot \sigma$, then $C\sigma$ is β-false in Γ, $N \cup U \models C$,*

Lemma 21. *The initial state $(\epsilon; N; \emptyset; \beta; 0; \top)$ is sound.*

Definition 22. *A run is a sequence of applications of SCL(EQ) rules starting from the initial state.*

Theorem 23. *Assume a state $(\Gamma; N; U; \beta; k; D)$ resulting from a run. Then $(\Gamma; N; U; \beta; k; D)$ is sound.*

Next, we give the definition of a regular run. Intuitively speaking, in a regular run we are always allowed to do decisions except if

1. a literal can be propagated before the first decision and
2. the negation of a literal can be propagated.

To ensure non-redundant learning we enforce at least one application of Skip during conflict resolution except for the special case of a conflict after a decision.

Definition 24 (Regular Run). *A run is called* regular *if*

1. *the rules Conflict and Factorize have precedence over all other rules,*
2. *If $k = 0$ in a state $(\Gamma; N; U; \beta; k; D)$, then Propagate has precedence over Decide,*
3. *If an annotated literal $L^{k:C\cdot\sigma}$ could be added by an application of Propagate on Γ in a state $(\Gamma; N; U; \beta; k; D)$ and $C \in N \cup U$, then the annotated literal $comp(L)^{k+1:C'\cdot\sigma'}$ is not added by Decide on Γ,*
4. *during conflict resolution Skip is applied at least once, except if Conflict is applied immediately after an application of Decide.*
5. *if Conflict is applied immediately after an application of Decide, then Backtrack is only applied in a state $(\Gamma, L'; N; U; \beta; k; D\cdot\sigma)$ if $L\sigma = comp(L')$ for some $L \in D$.*

Now we show that any learned clause in a regular run is non-redundant according to our trail induced ordering.

Lemma 25 (Non-Redundant Clause Learning). *Let N be a clause set. The clauses learned during a regular run in SCL(EQ) are not redundant with respect to $\prec_{\Gamma^*}$ and $N \cup U$. For the trail only non-redundant clauses need to be considered.*

The proof of Lemma 25 is based on the fact that conflict resolution eventually produces a clause smaller then the original conflict clause with respect to $\prec_{\Gamma^*}$. All simplifications, e.g., contextual rewriting, as defined in [2,20,33,35–37], are therefore compatible with Lemma 25 and may be applied to the newly learned clause as long as they respect the induced trail ordering. In detail, let Γ be the trail before the application of rule Backtrack. The newly learned clause can be simplified according to the induced trail ordering $\prec_{\Gamma^*}$ as long as the simplified clause is smaller with respect to $\prec_{\Gamma^*}$.

Another important consequence of Lemma 25 is that newly learned clauses need not to be considered for redundancy. Furthermore, the SCL(EQ) calculus always terminates, Lemma 33, because there only finitely many non-redundant clauses with respect to a fixed β.

For dynamic redundancy, we have to consider the fact that the induced trail ordering changes. At this level, only redundancy criteria and simplifications that are compatible with *all* induced trail orderings may be applied. Due to the construction of the induced trail ordering, it is compatible with $\prec_T$ for unit clauses.

Lemma 26 (Unit Rewriting). *Assume a state* $(\Gamma; N; U; \beta; k; D)$ *resulting from a regular run where the current level* $k > 0$ *and a unit clause* $l \approx r \in N$. *Now assume a clause* $C \vee L[l']_p \in N$ *such that* $l' = l\mu$ *for some matcher* μ. *Now assume some arbitrary grounding substitutions* σ' *for* $C \vee L[l']_p$, σ *for* $l \approx r$ *such that* $l\sigma = l'\sigma'$ *and* $r\sigma \prec_T l\sigma$. *Then* $(C \vee L[r\mu\sigma\sigma']_p)\sigma' \prec_{\Gamma^*} (C \vee L[l']_p)\sigma'$.

In addition, any notion that is based on a literal subset relationship is also compatible with ordering changes. The standard example is subsumption.

Lemma 27. *Let* C, D *be two clauses. If there exists a substitution* σ *such that* $C\sigma \subset D$, *then* D *is redundant with respect to* C *and any* $\prec_{\Gamma^*}$.

The notion of redundancy, Definition 1, only supports a strict subset relation for Lemma 27, similar to the superposition calculus. However, the newly generated clauses of SCL(EQ) are the result of paramodulation inferences [28]. In a recent contribution to dynamic, abstract redundancy [32] it is shown that also the non-strict subset relation in Lemma 27, i.e., $C\sigma \subseteq D$, preserves completeness.

If all stuck states, see below Definition 28, with respect to a fixed β are visited before increasing β then this provides a simple dynamic fairness strategy.

When unit reduction or any other form of supported rewriting is applied to clauses smaller than the current β, it can be applied independently from the current trail. If, however, unit reduction is applied to clauses larger than the current β then the calculus must do a restart to its initial state, in particular the trail must be emptied, as for otherwise rewriting may result generating a conflict that did not exist with respect to the current trail before the rewriting. This is analogous to a restart in CDCL once a propositional unit clause is derived and used for simplification. More formally, we add the following new Restart rule to the calculus to reset the trail to its initial state after a unit reduction.

Restart
$(\Gamma; N; U; \beta; k; \top) \Rightarrow_{\text{SCL(EQ)}} (\epsilon; N; U; \beta; 0; \top)$

Next we show refutation completeness of SCL(EQ). To achieve this we first give a definition of a stuck state. Then we show that stuck states only occur if all ground literals $L \prec_T \beta$ are *β-defined* in Γ and not during conflict resolution. Finally we show that conflict resolution will always result in an application of Backtrack. This allows us to show termination (without application of Grow) and refutational completeness.

Definition 28 (Stuck State). *A state* $(\Gamma; N; U; \beta; k; D)$ *is called* stuck *if* $D \neq \bot$ *and none of the rules of the calculus, except for Grow, is applicable.*

Lemma 29 (Form of Stuck States). *If a regular run (without rule Grow) ends in a stuck state* $(\Gamma; N; U; \beta; k; D)$, *then* $D = \top$ *and all ground literals* $L\sigma \prec_T \beta$, *where* $L \vee C \in (N \cup U)$ *are β-defined in* Γ.

Lemma 30. *Suppose a sound state $(\Gamma; N; U; \beta; k; D)$ resulting from a regular run where $D \notin \{\top, \bot\}$. If Backtrack is not applicable then any set of applications of Explore-Refutation, Skip, Factorize, Equality-Resolution will finally result in a sound state $(\Gamma'; N; U; \beta; k; D')$, where $D' \prec_{\Gamma^*} D$. Then Backtrack will be finally applicable.*

Corollary 31 (Satisfiable Clause Sets). *Let N be a satisfiable clause set. Then any regular run without rule Grow will end in a stuck state, for any β.*

Thus a stuck state can be seen as an indication for a satisfiable clause set. Of course, it remains to be investigated whether the clause set is actually satisfiable. Superposition is one of the strongest approaches to detect satisfiability and constitutes a decision procedure for many decidable first-order fragments [4,19]. Now given a stuck state and some specific ordering such as KBO, LPO, or some polynomial ordering [17], it is decidable whether the ordering can be instantiated from a stuck state such that Γ coincides with the superposition model operator on the ground terms smaller than β. In this case it can be effectively checked whether the clauses derived so far are actually saturated by the superposition calculus with respect to this specific ordering. In this sense, SCL(EQ) has the same power to decide satisfiability of first-order clause sets than superposition.

Definition 32. *A regular run terminates in a state $(\Gamma; N; U; \beta; k; D)$ if $D = \top$ and no rule is applicable, or $D = \bot$.*

Lemma 33. *Let N be a set of clauses and β be a ground term. Then any regular run that never uses Grow terminates.*

Lemma 34. *If a regular run reaches the state $(\Gamma; N; U; \beta; k; \bot)$ then N is unsatisfiable.*

Theorem 35 (Refutational Completeness). *Let N be an unsatisfiable clause set, and $\prec_T$ a desired term ordering. For any ground term β where $gnd_{\prec_T \beta}(N)$ is unsatisfiable, any regular SCL(EQ) run without rule Grow will terminate by deriving $\bot$.*

5 Discussion

We presented SCL(EQ), a new sound and complete calculus for reasoning in first-order logic with equality. We will now discuss some of its aspects and present ideas for future work beyond the scope of this paper.

The trail induced ordering, Definition 9, is the result of letting the calculus follow the logical structure of the clause set on the literal level and at the same time supporting rewriting at the term level. It can already be seen by examples on ground clauses over (in)equations over constants that this combination requires a layered approach as suggested by Definition 9, see [24].

In case the calculus runs into a stuck state, i.e., the current trail is a model for the set of considered ground instances, then the trail information can be effectively used for a guided continuation. For example, in order to use the trail to certify a model, the trail literals can be used to guide the design of a lifted ordering for the clauses with variables such that propagated trail literals are maximal in respective clauses. Then it could be checked by superposition, if the current clause is saturated by such an ordering. If this is not the case, then there must be a superposition inference larger than the current β, thus giving a hint on how to extend β. Another possibility is to try to extend the finite set of ground terms considered in a stuck state to the infinite set of all ground terms by building extended equivalence classes following patterns that ensure decidability of clause testing, similar to the ideas in [14]. If this fails, then again this information can be used to find an appropriate extension term β for rule Grow.

In contrast to superposition, SCL(EQ) does also inferences below variable level. Inferences in SCL(EQ) are guided by a false clause with respect to a partial model assumption represented by the trail. Due to this guidance and the different style of reasoning this does not result in an explosion in the number of possibly inferred clauses but also rather in the derivation of more general clauses, see [24].

Currently, the reasoning with solely positive equations is done on and with respect to the trail. It is well-known that also inferences from this type of reasoning can be used to speed up the overall reasoning process. The SCL(EQ) calculus already provides all information for such a type of reasoning, because it computes the justification clauses for trail reasoning via rewriting inferences. By an assessment of the quality of these clauses, e.g., their reduction potential with respect to trail literals, they could also be added, independently from resolving a conflict.

The trail reasoning is currently defined with respect to rewriting. It could also be performed by congruence closure [26].

Towards an implementation, the aspect of how to find interesting ground decision or propagation literals for the trail can be treated similar to CDCL [11, 21,25,29]. A simple heuristic may be used from the start, like counting the number of instance relationships of some ground literal with respect to the clause set, but later on a bonus system can focus the search towards the structure of the clause sets. Ground literals involved in a conflict or the process of learning a new clause get a bonus or preference. The regular strategy requires the propagation of all ground unit clauses smaller than β. For an implementation a propagation of the (explicit and implicit) unit clauses with variables to the trail will be a better choice. This complicates the implementation of refutation proofs and rewriting (congruence closure), but because every reasoning is layered by a ground term β this can still be efficiently done.

Acknowledgments. This work was partly funded by DFG grant 389792660 as part of TRR 248, see https://perspicuous-computing.science. We thank the anonymous reviewers and Martin Desharnais for their thorough reading, detailed comments, and corrections.

References

1. Alagi, G., Weidenbach, C.: NRCL - a model building approach to the Bernays-Schönfinkel fragment. In: Lutz, C., Ranise, S. (eds.) FroCoS 2015. LNCS (LNAI), vol. 9322, pp. 69–84. Springer, Cham (2015). https://doi.org/10.1007/978-3-319-24246-0_5
2. Bachmair, L., Ganzinger, H.: Rewrite-based equational theorem proving with selection and simplification. J. Log. Comput. **4**(3), 217–247 (1994)
3. Bachmair, L., Ganzinger, H., Voronkov, A.: Elimination of equality via transformation with ordering constraints. In: Kirchner, C., Kirchner, H. (eds.) CADE 1998. LNCS, vol. 1421, pp. 175–190. Springer, Heidelberg (1998). https://doi.org/10.1007/BFb0054259
4. Bachmair, L., Ganzinger, H., Waldmann, U.: Superposition with simplification as a decision procedure for the monadic class with equality. In: Gottlob, G., Leitsch, A., Mundici, D. (eds.) KGC 1993. LNCS, vol. 713, pp. 83–96. Springer, Heidelberg (1993). https://doi.org/10.1007/BFb0022557
5. Baumgartner, P.: Hyper tableau — the next generation. In: de Swart, H. (ed.) TABLEAUX 1998. LNCS (LNAI), vol. 1397, pp. 60–76. Springer, Heidelberg (1998). https://doi.org/10.1007/3-540-69778-0_14
6. Baumgartner, P., Fuchs, A., Tinelli, C.: Lemma learning in the model evolution calculus. In: Hermann, M., Voronkov, A. (eds.) LPAR 2006. LNCS (LNAI), vol. 4246, pp. 572–586. Springer, Heidelberg (2006). https://doi.org/10.1007/11916277_39
7. Baumgartner, P., Furbach, U., Pelzer, B.: Hyper tableaux with equality. In: Pfenning, F. (ed.) CADE 2007. LNCS (LNAI), vol. 4603, pp. 492–507. Springer, Heidelberg (2007). https://doi.org/10.1007/978-3-540-73595-3_36
8. Baumgartner, P., Pelzer, B., Tinelli, C.: Model evolution with equality-revised and implemented. J. Symb. Comput. **47**(9), 1011–1045 (2012)
9. Baumgartner, P., Tinelli, C.: The model evolution calculus with equality. In: Nieuwenhuis, R. (ed.) CADE 2005. LNCS (LNAI), vol. 3632, pp. 392–408. Springer, Heidelberg (2005). https://doi.org/10.1007/11532231_29
10. Baumgartner, P., Waldmann, U.: Superposition and model evolution combined. In: Schmidt, R.A. (ed.) CADE 2009. LNCS (LNAI), vol. 5663, pp. 17–34. Springer, Heidelberg (2009). https://doi.org/10.1007/978-3-642-02959-2_2
11. Biere, A., Heule, M., van Maaren, H., Walsh, T. (eds.): Handbook of Satisfiability, Frontiers in Artificial Intelligence and Applications, vol. 185. IOS Press, Amsterdam (2009)
12. Bonacina, M.P., Furbach, U., Sofronie-Stokkermans, V.: On First-Order Model-Based Reasoning. In: Martí-Oliet, N., Ölveczky, P.C., Talcott, C. (eds.) Logic, Rewriting, and Concurrency. LNCS, vol. 9200, pp. 181–204. Springer, Cham (2015). https://doi.org/10.1007/978-3-319-23165-5_8
13. Bonacina, M.P., Plaisted, D.A.: SGGS theorem proving: an exposition. In: Schulz, S., Moura, L.D., Konev, B. (eds.) PAAR-2014. 4th Workshop on Practical Aspects of Automated Reasoning. EPiC Series in Computing, vol. 31, pp. 25–38. EasyChair (2015)
14. Bromberger, M., Fiori, A., Weidenbach, C.: Deciding the Bernays-Schoenfinkel fragment over bounded difference constraints by simple clause learning over theories. In: Henglein, F., Shoham, S., Vizel, Y. (eds.) VMCAI 2021. LNCS, vol. 12597, pp. 511–533. Springer, Cham (2021). https://doi.org/10.1007/978-3-030-67067-2_23

15. Davis, M., Logemann, G., Loveland, D.: A machine program for theorem-proving. Commun. ACM **5**(7), 394–397 (1962)
16. Davis, M., Putnam, H.: A computing procedure for quantification theory. J. ACM (JACM) **7**(3), 201–215 (1960)
17. Dershowitz, N., Plaisted, D.A.: Rewriting. In: Robinson, A., Voronkov, A. (eds.) Handbook of Automated Reasoning, vol. I, chap. 9, pp. 535–610. Elsevier (2001)
18. Fiori, A., Weidenbach, C.: SCL clause learning from simple models. In: Fontaine, P. (ed.) CADE 2019. LNCS (LNAI), vol. 11716, pp. 233–249. Springer, Cham (2019). https://doi.org/10.1007/978-3-030-29436-6_14
19. Ganzinger, H., de Nivelle, H.: A superposition decision procedure for the guarded fragment with equality. In: LICS, pp. 295–304 (1999)
20. Gleiss, B., Kovács, L., Rath, J.: Subsumption demodulation in first-order theorem proving. In: Peltier, N., Sofronie-Stokkermans, V. (eds.) IJCAR 2020. LNCS (LNAI), vol. 12166, pp. 297–315. Springer, Cham (2020). https://doi.org/10.1007/978-3-030-51074-9_17
21. Bayardo, R.J., Schrag, R.: Using CSP look-back techniques to solve exceptionally hard SAT instances. In: Freuder, E.C. (ed.) CP 1996. LNCS, vol. 1118, pp. 46–60. Springer, Heidelberg (1996). https://doi.org/10.1007/3-540-61551-2_65
22. Korovin, K.: Inst-Gen – a modular approach to instantiation-based automated reasoning. In: Voronkov, A., Weidenbach, C. (eds.) Programming Logics. LNCS, vol. 7797, pp. 239–270. Springer, Heidelberg (2013). https://doi.org/10.1007/978-3-642-37651-1_10
23. Korovin, K., Sticksel, C.: iProver-Eq: an instantiation-based theorem prover with equality. In: Giesl, J., Hähnle, R. (eds.) IJCAR 2010. LNCS (LNAI), vol. 6173, pp. 196–202. Springer, Heidelberg (2010). https://doi.org/10.1007/978-3-642-14203-1_17
24. Leidinger, H., Weidenbach, C.: SCL(EQ): SCL for first-order logic with equality (2022). arXiv: 2205.08297
25. Moskewicz, M.W., Madigan, C.F., Zhao, Y., Zhang, L., Malik, S.: Chaff: engineering an efficient SAT solver. In: Proceedings of the Design Automation Conference, pp. 530–535. ACM (2001)
26. Nelson, G., Oppen, D.C.: Fast decision procedures based on congruence closure. J. ACM **27**(2), 356–364 (1980)
27. Plaisted, D.A., Zhu, Y.: Ordered semantic hyper-linking. J. Autom. Reason. **25**(3), 167–217 (2000)
28. Robinson, G., Wos, L.: Paramodulation and theorem-proving in first-order theories with equality. In: Meltzer, B., Michie, D. (eds.) Machine Intelligence 4, pp. 135–150 (1969)
29. Silva, J.P.M., Sakallah, K.A.: GRASP - a new search algorithm for satisfiability. In: International Conference on Computer Aided Design, ICCAD, pp. 220–227. IEEE Computer Society Press (1996)
30. Sutcliffe, G.: The TPTP problem library and associated infrastructure - from CNF to th0, TPTP v6.4.0. J. Autom. Reasoning **59**(4), 483–502 (2017)
31. Teucke, A.: An approximation and refinement approach to first-order automated reasoning. Doctoral thesis, Saarland University (2018)
32. Waldmann, U., Tourret, S., Robillard, S., Blanchette, J.: A comprehensive framework for saturation theorem proving. In: Peltier, N., Sofronie-Stokkermans, V. (eds.) IJCAR 2020. LNCS (LNAI), vol. 12166, pp. 316–334. Springer, Cham (2020). https://doi.org/10.1007/978-3-030-51074-9_18

33. Weidenbach, C.: Combining superposition, sorts and splitting. In: Robinson, A., Voronkov, A. (eds.) Handbook of Automated Reasoning, vol. 2, chap. 27, pp. 1965–2012. Elsevier (2001)
34. Weidenbach, C.: Automated reasoning building blocks. In: Meyer, R., Platzer, A., Wehrheim, H. (eds.) Correct System Design. LNCS, vol. 9360, pp. 172–188. Springer, Cham (2015). https://doi.org/10.1007/978-3-319-23506-6_12
35. Weidenbach, C., Wischnewski, P.: Contextual rewriting in SPASS. In: PAAR/ESHOL. CEUR Workshop Proceedings, vol. 373, pp. 115–124. Australien, Sydney (2008)
36. Weidenbach, C., Wischnewski, P.: Subterm contextual rewriting. AI Commun. **23**(2–3), 97–109 (2010)
37. Wischnewski, P.: Effcient Reasoning Procedures for Complex First-Order Theories. Ph.D. thesis, Saarland University, November 2012

Term Orderings for Non-reachability of (Conditional) Rewriting

Akihisa Yamada(✉)

National Institute of Advanced Industrial Science and Technology, Tokyo, Japan
akihisa.yamada@aist.go.jp

Abstract. We propose generalizations of reduction pairs, well-established techniques for proving termination of term rewriting, in order to prove unsatisfiability of reachability (infeasibility) in plain and conditional term rewriting. We adapt the weighted path order, a merger of the Knuth–Bendix order and the lexicographic path order, into the proposed framework. The proposed approach is implemented in the termination prover NaTT, and the strength of our approach is demonstrated through examples and experiments.

1 Introduction

In the research area of term rewriting, among the most well-studied topics are termination, confluence, and reachability analyses.

In termination analysis, a crucial task used to be to design *reduction orders*, well-founded orderings over terms that are closed under contexts and substitutions. Well-known examples of such orderings include the *Knuth–Bendix ordering* [14], *polynomial interpretations* [18], *multiset/lexicographic path ordering* [4,13], and *matrix interpretations* [5]. The *dependency pair framework* generalized reduction orders into *reduction pairs* [2,9,12], and there are a number of implementations that automatically find reduction pairs, e.g., AProVE [7], $\mathsf{T_TT_2}$ [16], MU-TERM [11], NaTT [35], competing in the International Termination Competition [8].

Traditional reachability analysis (cf. [6]) has been concerned with the possibility of rewriting a given source term s to a target t, where variables in the terms are treated as constants. There is an increasing need for solving a more general question: is it possible to instantiate variables so that the instance of s rewrites to the instance of t? Let us illustrate the problem with an elementary example.

Example 1. Consider the following TRS encoding addition of natural numbers:

$$\mathcal{R}_{\mathsf{add}} := \{\ \mathsf{add}(0, y) \to y,\ \ \mathsf{add}(\mathsf{s}(x), y) \to \mathsf{s}(\mathsf{add}(x, y))\ \}$$

The reachability constraint $\mathsf{add}(\mathsf{s}(x), y) \twoheadrightarrow y$ represents the possibility of rewriting from $\mathsf{add}(\mathsf{s}(x), y)$ to y, where variables x and y can be arbitrary terms.

J. Blanchette et al. (Eds.): IJCAR 2022, LNAI 13385, pp. 248–267, 2022.
https://doi.org/10.1007/978-3-031-10769-6_15

This (un)satisfiability problem of reachability, also called (in)feasibility, plays important roles in termination [24] and confluence analyses of (conditional) rewriting [21]. A tool competition dedicated for this problem has been founded as the infeasibility (INF) category in the International Confluence Competition (CoCo) since 2019 [25].

In this paper, we propose a new method for proving unsatisfiability of reachability, using the term ordering techniques developed for termination analysis. Specifically, in Sect. 3, we first generalize reduction pairs to *rewrite pairs*, and show that they can be used for proving unsatisfiability of reachability. We further generalize the notion to *co-rewrite pairs*, yielding a sound and complete method. The power of the proposed method is demonstrated by importing (relaxed) *semantic* term orderings from termination analysis.

In order to import also *syntactic* term orderings, in Sect. 4 we identify a condition when the *weighted path order (WPO)* [36] forms a rewrite pair. Since KBO and LPO are instances of WPO, we see that these orderings can also be used in our method. In Sect. 5 we also present how to derive co-rewrite pairs from WPO.

In Sect. 6, we adapt the approach into conditional rewriting. Section 7 reports on the implementation and experiments conducted on examples in the paper and the benchmark set of CoCo 2021.

Related Work Our rewrite pairs are essentially Aoto's *discrimination pairs* [1] which are closed under substitutions. On way of disproving confluence, Aoto introduced discrimination pairs and used them in proving non-joinability. The *joinability* of terms s and t is expressed as $\exists u.\ s \to_{\mathcal{R}}^* u \leftarrow_{\mathcal{R}}^* t$, while the current paper is concerned with $\exists\theta.\ s\theta \to_{\mathcal{R}}^* t\theta$. As substitutions are not considered, discrimination pairs do not need closure under substitutions, and Aoto's insights are mainly for dealing with the reverse rewriting $\leftarrow_{\mathcal{R}}^*$.

Lucas and Gutiérrez [19] proposed reducing infeasibility to the model finding of first-order logic. Our formulations especially in Sect. 6 are similar to theirs. A crucial difference is that, while they encode the closure properties and order properties into logical formulas and delegate these tasks to the background theory solvers, we ensure these properties by means of reduction pairs, for which well-established techniques exist in the literature.

Sternagel and Yamada [30] proposed a framework for analyzing reachability by combining basic logical manipulations, and Gutiérrez and Lucas [10] proposed another framework, similar to the dependency pair framework. The present work focuses on atomic analysis techniques, and is orthogonal to these efforts of combining techniques.

2 Preliminaries

We assume familiarity with term rewriting, cf. [3] or [32]. For a binary relation denoted by a symbol like $\sqsupset$, we denote its dual relation by $\sqsubset$ and the negated relation by $\not\sqsupset$. Relation composition is denoted by $\circ$.

Throughout the paper we fix a set $\mathcal{V}$ of *variable symbols*. A *signature* is a set $\mathcal{F}$ of function symbols, where each $f \in \mathcal{F}$ is associated with its *arity*, the number of arguments. The set of *terms* built from $\mathcal{F}$ and $\mathcal{V}$ is denoted by $\mathcal{T}(\mathcal{F}, \mathcal{V})$, where a term is either in $\mathcal{V}$ or of form $f(s_1, \ldots, s_n)$ where $f \in \mathcal{F}$ is n-ary and $s_1, \ldots, s_n \in \mathcal{T}(\mathcal{F}, \mathcal{V})$. Given a term $s \in \mathcal{T}(\mathcal{F}, \mathcal{V})$ and a *substitution* $\theta : \mathcal{V} \to \mathcal{T}(\mathcal{F}, \mathcal{V})$, $s\theta$ denotes the term obtained from s by replacing every variable x by $\theta(x)$. A *context* is a term $C \in \mathcal{T}(\mathcal{F}, \mathcal{V} \cup \{\square\})$ where a special variable $\square$ occurs exactly once. Given $s \in \mathcal{T}(\mathcal{F}, \mathcal{V})$, we denote by $C[s]$ the term obtained by replacing $\square$ by s in C.

A relation $\sqsupset$ over terms is *closed under substitutions (resp. contexts)* iff $s \sqsupset t$ implies $s\theta \sqsupset t\theta$ for any substitution θ (resp. $C[s] \sqsupset C[t]$ for any context C). Relations over terms that are closed under contexts and substitutions are called *rewrite relations*. Rewrite relations which are also preorders are called *rewrite preorders*, and those which are strict orders are *rewrite orders*. Well-founded rewrite orders are called *reduction orders*.

A *term rewrite system (TRS)* $\mathcal{R}$ is a (usually finite) relation over terms, where each $\langle l, r\rangle \in \mathcal{R}$ is called a *rewrite rule* and written $l \to r$. We do not require the usual assumption that $l \notin \mathcal{V}$ and variables occurring in r must occur in l. The *rewrite step* $\to_{\mathcal{R}}$ induced by TRS $\mathcal{R}$ is the least rewrite relation containing $\mathcal{R}$. Its reflexive transitive closure is denoted by $\to_{\mathcal{R}}^*$, which is the least rewrite preorder containing $\mathcal{R}$.

A *reachability atom* is a pair of terms s and t, written $s \twoheadrightarrow t$. We say that $s \twoheadrightarrow t$ is *$\mathcal{R}$-satisfiable* iff $s\theta \to_{\mathcal{R}}^* t\theta$ for some θ, and *$\mathcal{R}$-unsatisfiable* otherwise.

3 Term Orderings for Non-reachability

Reduction pairs constitute the core ingredient in proving termination with dependency pairs. Just as rewrite orders generalize reduction orders, we first introduce the notion of "rewrite pairs" by removing the well-foundedness assumption of reduction pairs.

Definition 1 (rewrite pair). *We call a pair $\langle \sqsupseteq, \sqsupset\rangle$ of relations an* order pair *if $\sqsupseteq$ is a preorder, $\sqsupset$ is irreflexive, $\sqsupset \subseteq \sqsupseteq$, and $\sqsupseteq \circ \sqsupset \circ \sqsupseteq \subseteq \sqsupset$. A* rewrite pair *is an order pair $\langle \sqsupseteq, \sqsupset\rangle$ over terms such that both $\sqsupseteq$ and $\sqsupset$ are closed under substitutions and $\sqsupseteq$ is closed under contexts. It is called a* reduction pair *if moreover $\sqsupset$ is well-founded.*

Standard definitions of reduction pairs put less order-like assumptions than the above definition, but the above (more natural) assumptions do not lose the generality of previous definitions [34]. Due to these assumptions, our rewrite pair satisfies the assumption of discrimination pairs [1].

The following statement is our first observation: a rewrite pair can prove non-reachability.

Theorem 1. *If $\langle \sqsupseteq, \sqsupset\rangle$ is a rewrite pair, $\mathcal{R} \subseteq \sqsupseteq$ and $s \sqsubset t$, then $s \twoheadrightarrow t$ is $\mathcal{R}$-unsatisfiable.*

A similar observation has been made [20, Theorem 11], where well-foundedness is assumed instead of irreflexivity. Note that irreflexivity is essential: if $s \sqsubset s$ for some s, then we have $s \sqsubset s$ but $s \twoheadrightarrow s$ is $\mathcal{R}$-satisfiable.

The proof of Theorem 1 will be postponed until more general Theorem 2 will be obtained. Instead, we start with utilizing Theorem 1 by generalizing a classical way of defining reduction pairs: the semantic approach [23].

Definition 2 (model). *An $\mathcal{F}$-algebra $\mathcal{A} = \langle A, [\cdot]\rangle$ specifies a set A called the* carrier *and an* interpretation $[f] : A^n \to A$ *to each n-ary $f \in \mathcal{F}$. The evaluation of a term s under assignment $\alpha : \mathcal{V} \to A$ is defined as usual and denoted by $[s]\alpha$.*

A related/preordered $\mathcal{F}$-algebra $\langle \mathcal{A}, \sqsupset\rangle = \langle A, [\cdot], \sqsupset\rangle$ *consists of an $\mathcal{F}$-algebra and a relation/preorder $\sqsupset$ on A. Given $\alpha : \mathcal{V} \to A$, we write $[s \sqsupset t]\alpha$ to mean $[s]\alpha \sqsupset [t]\alpha$. We write $\mathcal{A} \models s \sqsupset t$ if $[s \sqsupset t]\alpha$ holds for every $\alpha : \mathcal{V} \to A$. We say $\langle \mathcal{A}, \sqsupset\rangle$ is a* (relational) model *of a TRS $\mathcal{R}$ if $\mathcal{A} \models l \sqsupset r$ for every $l \to r \in \mathcal{R}$. We say $\langle \mathcal{A}, \sqsupset\rangle$ is* monotone *if $a_i \sqsupset a_i'$ implies $[f](a_1, \dots, a_i, \dots, a_n) \sqsupset [f](a_1, \dots, a_i', \dots, a_n)$ for arbitrary $a_1, \dots, a_n, a_i' \in A$ and n-ary $f \in \mathcal{F}$.*

The notion of relational models is due to van Oostrom [28]. In this paper, we simply call them models. Models in terms of equational theory are models $\langle \mathcal{A}, =\rangle$ in the above definition, where monotonicity is inherent. *Quasi-models* of Zantema [37] are preordered (or partially ordered) monotone models. Theorem 1 can be reformulated in the semantic manner as follows:

Corollary 1. *If $\langle \geq, >\rangle$ is an order pair, $\langle \mathcal{A}, \geq\rangle$ is a monotone model of $\mathcal{R}$, and $\mathcal{A} \models s < t$, then $s \twoheadrightarrow t$ is $\mathcal{R}$-unsatisfiable.*

Note that Corollary 1 does not demand well-foundedness on $>$. In particular, one can employ models over negative numbers (or equivalently, positive numbers with the order pair $\langle \leq, <\rangle$).

Example 2. Consider again the TRS $\mathcal{R}_{\mathsf{add}}$ of Example 1. The monotone ordered $\mathcal{F}$-algebra $\langle \mathbb{Z}_{\leq 0}, [\cdot], \geq\rangle$ defined by

$$[\mathsf{add}](x, y) = x + y \qquad [\mathsf{s}](x) = x - 1 \qquad [\mathsf{0}] = 0$$

is a model of $\mathcal{R}_{\mathsf{add}}$: Whenever $x, y \in \mathbb{Z}_{\leq 0}$, we have

$$[\mathsf{add}]([\mathsf{0}], y) = y \qquad [\mathsf{add}]([\mathsf{s}](x), y) = x + y - 1 = [\mathsf{s}]([\mathsf{add}](x, y))$$

Now we can conclude that the reachability constraint $\mathsf{add}(\mathsf{s}(x), y) \twoheadrightarrow y$ is $\mathcal{R}_{\mathsf{add}}$-unsatisfiable by $\langle \mathbb{Z}_{\leq 0}, [\cdot]\rangle \models \mathsf{add}(\mathsf{s}(x), y) < y$: Whenever $x, y \in \mathbb{Z}_{\leq 0}$, we have

$$[\mathsf{add}]([\mathsf{s}](x), y) = x + y - 1 < y$$

Observe that in Theorem 1, $\sqsupset$ occurs only in the dual form $\sqsubset$. Hence we now directly analyze the condition which $\sqsupseteq$ and $\sqsubset$ should satisfy to prove non-reachability, and this gives a sound and complete method.

Definition 3 (co-rewrite pair). *We call a pair $\langle \sqsupseteq, \sqsubset \rangle$ of relations over terms a* co-rewrite pair, *if $\sqsupseteq$ is a rewrite preorder, $\sqsubset$ is closed under substitutions, and $\sqsupseteq \cap \sqsubset = \emptyset$.*

Theorem 2. *$s \twoheadrightarrow t$ is $\mathcal{R}$-unsatisfiable if and only if there exists a co-rewrite pair $\langle \sqsupseteq, \sqsubset \rangle$ such that $\mathcal{R} \subseteq \sqsupseteq$ and $s \sqsubset t$.*

Proof. For the "if" direction, suppose on the contrary that $s\theta \rightarrow^*_{\mathcal{R}} t\theta$ for some θ. Since $\sqsupseteq$ is a rewrite preorder containing $\mathcal{R}$ and $\rightarrow^*_{\mathcal{R}}$ is the least of such, we must have $s\theta \sqsupseteq t\theta$. On the other hand, since $s \sqsubset t$ and $\sqsubset$ is closed under substitutions, we have $s\theta \sqsubset t\theta$. This is not possible since $\sqsupseteq \cup \sqsubset = \emptyset$.

For the "only if" direction, take $\rightarrow^*_{\mathcal{R}}$ as $\sqsupseteq$ and define $\sqsubset$ by $s \sqsubset t$ iff $s \twoheadrightarrow t$ is $\mathcal{R}$-unsatisfiable. Then clearly $\sqsubset$ is closed under substitutions, $\rightarrow^*_{\mathcal{R}} \cap \sqsubset = \emptyset$, and $\mathcal{R} \subseteq \rightarrow^*_{\mathcal{R}}$. □

Theorem 2 can be more concisely reformulated in the model-oriented manner, as the greatest choice of $\sqsubset$ can be specified: $s \sqsubset t$ iff $\mathcal{A} \models s \not\gtrsim t$.

Corollary 2. *$s \twoheadrightarrow t$ is $\mathcal{R}$-unsatisfiable if and only if there exists a monotone preordered model $\langle \mathcal{A}, \gtrsim \rangle$ of $\mathcal{R}$ such that $\mathcal{A} \models s \not\gtrsim t$.*

Corollary 2 is useful when models over non-totally ordered carriers are considered. There are important methods (for termination) that crucially rely on such carriers: the *matrix interpretations* [5], or more generally the *tuple interpretations* [15,34].

Example 3. Consider the following TRS, where the first rule is from [5]:

$$\mathcal{R}_{mat} = \{\ \mathsf{f}(\mathsf{f}(x)) \rightarrow \mathsf{f}(\mathsf{g}(\mathsf{f}(x))),\ \mathsf{f}(x) \rightarrow x\ \}$$

The preordered $\{\mathsf{f}, \mathsf{g}\}$-algebra $\langle \mathbb{N}^2, [\cdot], \geq \rangle$ defined by

$$[\mathsf{f}]\begin{pmatrix} x \\ y \end{pmatrix} = \begin{pmatrix} x + y + 1 \\ y + 1 \end{pmatrix} \qquad [\mathsf{g}]\begin{pmatrix} x \\ y \end{pmatrix} = \begin{pmatrix} x + 1 \\ 0 \end{pmatrix}$$

is a model of $\mathcal{R}_{mat}$, where $\geq$ is extended pointwise over $\mathbb{N}^2$. Indeed, the first rule is oriented as the following calculation demonstrates:

$$[\mathsf{f}]\left([\mathsf{f}]\begin{pmatrix} x \\ y \end{pmatrix}\right) = \begin{pmatrix} x + 2y + 3 \\ y + 2 \end{pmatrix} \begin{matrix} \geq \\ \geq \end{matrix} \begin{pmatrix} x + y + 3 \\ 1 \end{pmatrix} = [\mathsf{f}]\left([\mathsf{g}]\left([\mathsf{f}]\begin{pmatrix} x \\ y \end{pmatrix}\right)\right)$$

and the second rule can be easily checked. Now we prove that $x \twoheadrightarrow \mathsf{g}(x)$ is $\mathcal{R}_{mat}$-unsatisfiable by Corollary 2. Indeed, $\langle \mathbb{N}^2, [\cdot] \rangle \models x \not\geq \mathsf{g}(x)$ is shown as follows:

$$\begin{pmatrix} x \\ y \end{pmatrix} \begin{matrix} \not\geq \\ \geq \end{matrix} \begin{pmatrix} x + 1 \\ 0 \end{pmatrix} = [\mathsf{g}]\begin{pmatrix} x \\ y \end{pmatrix}$$

for any $x, y \in \mathbb{N}$. Note also that Theorem 1 is not applicable, since $\langle \mathbb{N}^2, [\cdot] \rangle \not\models x < \mathsf{g}(x)$ due to the second coordinate.

We conclude the section by proving Theorem 1 via Theorem 2.

Proof (of Theorem 1). We show that $\langle \sqsupseteq, \sqsubset \rangle$ form a co-rewrite pair when $\langle \sqsupseteq, \sqsupset \rangle$ is a rewrite pair. It suffices to show that $\sqsupseteq \cap \sqsubset = \emptyset$. To this end, suppose on the contrary that $s \sqsupseteq t \sqsupset s$. By compatibility, we have $s \sqsupset s$, which contradicts the irreflexivity of $\sqsupset$. □

4 Weighted Path Order for Non-reachability

The previous section was concerned with the semantic approach towards obtaining (co-)rewrite pairs. In this section we focus on the syntactic approach. We choose the weighted path order (WPO), which subsumes both the lexicographic path order (LPO) and the Knuth-Bendix order (KBO), so the result of this section applies to these more well-known methods. The *multiset path order* [4] can also be subsumed [29], but we omit this extension to keep the presentation simple. WPO is induced by three ingredients: an $\mathcal{F}$-algebra; a *precedence* ordering over function symbols; and a *(partial) status*, which controls the recursive behavior of the ordering.

Definition 4 (partial status). *A* partial status π *specifies for each n-ary $f \in \mathcal{F}$ a list $\pi(f) \in \{1, \dots, n\}^*$, also seen as a set, of its argument positions. We say π is* total *if $1, \dots, n \in \pi(f)$ whenever f is n-ary. When $\pi(f) = [i_1, \dots, i_m]$, we denote $[s_{i_1}, \dots, s_{i_m}]$ by $\pi_f(s_1, \dots, s_n)$.*

For instance, the *empty* status $\pi(f) = []$ allows WPO to subsume weakly monotone interpretations [36, Section 4.1]. We allow positions to be duplicated, following [33].

Definition 5 (WPO [36]). *Let π be a partial status, $\mathcal{A}$ an $\mathcal{F}$-algebra, and $\langle \geq, > \rangle$ and $\langle \succsim, \succ \rangle$ be pairs of relations on $\mathcal{A}$ and $\mathcal{F}$, respectively. The* weighted path order $\mathsf{WPO}(\pi, \mathcal{A}, \geq, >, \succsim, \succ)$, *or* $\mathsf{WPO}(\mathcal{A})$ *or even* WPO *for short, is the pair $\langle \sqsupseteq_{\mathsf{WPO}}, \sqsupset_{\mathsf{WPO}} \rangle$ of relations over terms defined as follows: $s \sqsupset_{\mathsf{WPO}} t$ iff*

1. $\mathcal{A} \models s > t$ *or*
2. $\mathcal{A} \models s \geq t$ *and*
   (a) $s = f(s_1, \dots, s_n)$, $s_i \sqsupseteq_{\mathsf{WPO}} t$ *for some* $i \in \pi(f)$;
   (b) $s = f(s_1, \dots, s_n)$, $t = g(t_1, \dots, t_m)$, $s \sqsupset_{\mathsf{WPO}} t_j$ *for every* $j \in \pi(g)$ *and*
      i. $f \succ g$, *or*
      ii. $f \succsim g$ *and* $\pi_f(s_1, \dots, s_n) \sqsupset^{\mathsf{lex}}_{\mathsf{WPO}} \pi_g(t_1, \dots, t_m)$.

The relation $\sqsupseteq_{\mathsf{WPO}}$ is defined similarly, but with $\sqsupseteq^{\mathsf{lex}}_{\mathsf{WPO}}$ instead of $\sqsupset^{\mathsf{lex}}_{\mathsf{WPO}}$ in (2b-ii) and the following subcase is added in case 2:

   (c) $s = t \in \mathcal{V}$.

Here $\langle \sqsupseteq^{\mathsf{lex}}_P, \sqsupset^{\mathsf{lex}}_P \rangle$ denotes the lexicographic extension *of a pair $P = \langle \sqsupseteq_P, \sqsupset_P \rangle$ of relations, defined by: $[s_1, \dots, s_n] \mathrel{{}_{(\sqsupseteq)}\!\sqsupset^{\mathsf{lex}}_P} [t_1, \dots, t_m]$ iff*

- $m = 0$ *and* $n \mathrel{{}_{(}\geq_{)}} 0$*, or*
- $m, n > 0$ *and* $s_1 > t_1$ *or both* $s_1 \sqsupseteq_P t_1$ *and* $[s_2, \ldots, s_n] \mathrel{{}_{(}\sqsupseteq_{)}{}_P^{\mathsf{lex}}} [t_2, \ldots, t_m]$.

LPO is WPO induced by a total status π and a trivial $\mathcal{F}$-algebra as $\mathcal{A}$, and is written LPO. Allowing partial statuses corresponds to applying *argument filters* [2,17] (except for collapsing ones). KBO is a special case of WPO where π is total and $\mathcal{A}$ is induced by an admissible weight function.

For termination analysis, a precondition for WPO to be a reduction pair is crucial. In this work, we only need it to be a rewrite pair; that is, well-foundedness is not necessary. Thus, for instance, it is possible to have $x \sqsupset_{\mathsf{WPO}} f(x)$ by $[f](x) = x - 1$. This explains why $s \in \mathcal{V}$ is permitted in case 1, which might look useless to those who are already familiar with termination analysis.

We formulate the main claim of this section as follows.

Definition 6 (π-simplicity). *We say a related $\mathcal{F}$-algebra $\langle A, [\cdot], \geq\rangle$ is π-*simple[1] *for a partial status π iff $[f](a_1, \ldots, a_n) \geq a_i$ for arbitrary n-ary $f \in \mathcal{F}$, $a_1, \ldots, a_n \in A$, and $i \in \pi(f)$.*

Proposition 1. *If $\langle \geq, > \rangle$ and $\langle \succsim, \succ \rangle$ are order pairs on $\mathcal{A}$ and $\mathcal{F}$, and $\langle \mathcal{A}, \geq \rangle$ is monotone and π-simple, then $\langle \sqsupseteq_{\mathsf{WPO}}, \sqsupset_{\mathsf{WPO}} \rangle$ is a rewrite pair.*

Under these conditions, it is known that $\sqsupseteq_{\mathsf{WPO}}$ is closed under contexts and $\sqsupset_{\mathsf{WPO}}$ is compatible with $\sqsupseteq_{\mathsf{WPO}}$ [36, Lemmas 7, 10, 13]. Later in this section we prove other properties necessary for Proposition 1, for which the claims in [36] must be generalized for the purpose of this paper.

The benefit of having syntax-aware methods can be easily observed by recalling why we have them in termination analysis.

Example 4 ([13]). Consider the TRS $\mathcal{R}_{\mathsf{A}}$ consisting of the following rules:

$$\mathsf{A}(\mathsf{0}, y) \to \mathsf{s}(y) \quad \mathsf{A}(\mathsf{s}(x), \mathsf{0}) \to \mathsf{A}(x, \mathsf{s}(\mathsf{0})) \quad \mathsf{A}(\mathsf{s}(x), \mathsf{s}(y)) \to \mathsf{A}(x, \mathsf{A}(\mathsf{s}(x), y))$$

and suppose that a monotone $\{\mathsf{A}, \mathsf{s}, \mathsf{0}\}$-algebra $\langle \mathbb{N}, [\cdot], \geq \rangle$ is a model of $\mathcal{R}_{\mathsf{A}}$. Then, denoting the Ackermann function by A, we have

$$[\mathsf{A}]([\mathsf{s}]^m(\mathsf{0}), [\mathsf{s}]^n(\mathsf{0})) \geq [\mathsf{s}]^{A(m,n)}(\mathsf{0}) \tag{1}$$

Now consider proving the obvious fact that $x \twoheadrightarrow \mathsf{s}(x)$ is $\mathcal{R}_{\mathsf{A}}$-unsatisfiable. This requires $\langle \mathbb{N}, [\cdot] \rangle \models x < \mathsf{s}(x)$, and then $[\mathsf{s}]^n(0) \geq n$ by an inductive argument. This is not possible if $[\mathsf{A}]$ is primitive recursive (e.g., a polynomial), since (1) with $[\mathsf{s}]^{A(m,n)}(0) \geq A(m, n)$ contradicts the well-known fact that the Ackermann function has no primitive-recursive bound.

On the other hand, LPO with $\mathsf{A} \succ \mathsf{s}$ satisfies $\mathcal{R}_{\mathsf{A}} \subseteq \sqsupset_{\mathsf{LPO}}$ ($\subseteq \sqsupseteq_{\mathsf{LPO}}$) and $x \sqsubset_{\mathsf{LPO}} \mathsf{s}(x)$. Thus Theorem 1 with $\langle \sqsupseteq, \sqsupset \rangle = \langle \sqsupseteq_{\mathsf{LPO}}, \sqsupset_{\mathsf{LPO}} \rangle$ proves that $x \twoheadrightarrow \mathsf{s}(x)$ is $\mathcal{R}_{\mathsf{A}}$-unsatisfiable, thanks to Proposition 1 and Theorem 1.

[1] Such a property would be called *inflationary* in the mathematics literature. In the term rewriting, the word *simple* has been used (see, e.g., [32]) in accordance with *simplification orders*.

Example 5. Consider the TRS consisting of the following rules:

$$\mathcal{R}_{kbo} := \{\ \mathsf{f}(\mathsf{g}(x)) \to \mathsf{g}(\mathsf{f}(\mathsf{f}(x))),\ \mathsf{g}(x) \to x\ \}$$

WPO (or KBO) induced by $\mathcal{A} = \langle \mathbb{N}, [\cdot] \rangle$ and precedence $\langle \succsim, \succ \rangle$ such that

$$[\mathsf{f}](x) = x \qquad [\mathsf{g}](x) = x + 1 \qquad \mathsf{f} \succ \mathsf{g}$$

satisfies $\mathcal{R}_{kbo} \subseteq \sqsupseteq_{\mathsf{WPO}}$. Thus, for instance $\mathsf{g}(x) \twoheadrightarrow \mathsf{g}(\mathsf{f}(x))$ is $\mathcal{R}_{kbo}$-unsatisfiable by Theorem 1. On the other hand, let $\langle A, [\cdot], \geq \rangle$ with $A \subseteq \mathbb{Z}$ be a model of $\mathcal{R}_{kbo}$. Using the idea of [38, Proposition 11], one can show $[\mathsf{f}](x) \leq x$. Hence, Corollary 2 with models over a subset of integers cannot handle the problem. LPO orients the first rule from right to left and hence cannot handle the problem either.

The power of WPO can also be easily verified, by considering

$$\mathcal{R}_{wpo} := \mathcal{R}_{kbo} \cup \{\ \mathsf{f}(\mathsf{h}(x)) \to \mathsf{h}(\mathsf{h}(\mathsf{f}(x))),\ \mathsf{f}(x) \to x\ \}$$

By extending the above WPO with $[\mathsf{h}](x) = x$ and $\mathsf{f} \succ \mathsf{h}$, which does not fall into the class of KBO anymore,[2] we can prove, e.g., that $\mathsf{f}(x) \twoheadrightarrow \mathsf{f}(\mathsf{h}(x))$ is $\mathcal{R}$-unsatisfiable. None of the above mentioned methods can handle this problem.

The rest of this section is dedicated for proving Proposition 1. Similar results are present in [36], but they make implicit assumptions such as that $\geq$ and $\succsim$ are preorders. In this paper we need more essential assumptions as we will consider non-transitive relations in the next section.

First we reprove the reflexivity of $\sqsupseteq_{\mathsf{WPO}}$. The proof also serves as a basis for the more complicated irreflexivity proof.

Lemma 1. *If both $\geq$ and $\succsim$ are reflexive and $\langle \mathcal{A}, \geq \rangle$ is π-simple, then*

1. *$i \in \pi(f)$ implies $f(s_1, \ldots, s_n) \sqsupset_{\mathsf{WPO}} s_i$, and*
2. *$s \sqsupseteq_{\mathsf{WPO}} s$, i.e., $\sqsupseteq_{\mathsf{WPO}}$ is reflexive.*

Proof. As $s \sqsupseteq_{\mathsf{WPO}} s$ is trivial when $s \in \mathcal{V}$, we assume $s = f(s_1, \ldots, s_n)$ and prove the two claims by induction on the structure of s. For the first claim, by π-simplicity, for any α we have $[s]\alpha = [f]([s_1]\alpha, \ldots, [s_n]\alpha) \geq [s_i]\alpha$, and hence $\mathcal{A} \models s \geq s_i$. By the second claim of induction hypothesis we have $s_i \sqsupseteq_{\mathsf{WPO}} s_i$, and thus $s \sqsupset_{\mathsf{WPO}} s_i$ follows by (2a) of Definition 5. Next we show $s \sqsupseteq_{\mathsf{WPO}} s$ holds by (2b-ii). Indeed, $\mathcal{A} \models s \geq s$ follows from the reflexivity of $\geq$; $s \sqsupset_{\mathsf{WPO}} s_i$ for every $i \in \pi(f)$ as shown above; $f \succsim f$ as $\succsim$ is reflexive; and finally, $\pi_f(s_1, \ldots, s_n) \sqsupseteq^{\mathsf{lex}}_{\mathsf{WPO}} \pi_f(s_1, \ldots, s_n)$ is due to induction hypothesis and the fact that lexicographic extension preserves reflexivity. □

Using reflexivity, we can show that both $\sqsupseteq_{\mathsf{WPO}}$ and $\sqsupset_{\mathsf{WPO}}$ are closed under substitutions. This result will be reused in Sect. 5, where it will be essential that neither $\geq$ nor $>$ need be transitive.

---

[2] When $[\mathsf{h}]$ is the identity. KBO requires $\mathsf{h} \succsim f$ for any f.

Lemma 2. *If both $\geq$ and $\gtrsim$ are reflexive and $\langle \mathcal{A}, \geq \rangle$ is π-simple, then both $\sqsupseteq_{\mathsf{WPO}}$ and $\sqsupset_{\mathsf{WPO}}$ are closed under substitutions.*

Proof. We prove by induction on s and t that $s \sqsupseteq_{\mathsf{WPO}} t$ implies $s\theta \sqsupseteq_{\mathsf{WPO}} t\theta$ and that $s \sqsupset_{\mathsf{WPO}} t$ implies $s\theta \sqsupset_{\mathsf{WPO}} t\theta$. We prove the first claim by case analysis on how $s \sqsupseteq_{\mathsf{WPO}} t$ is derived. The other claim is analogous, without case (2c) below.

1. $\mathcal{A} \models s > t$: Then we have $\mathcal{A} \models s\theta > t\theta$ and thus $s\theta \sqsupseteq_{\mathsf{WPO}} t\theta$ by case 1.
2. $\mathcal{A} \models s \geq t$: Then we have $\mathcal{A} \models s\theta \geq t\theta$. There are the following subcases.
   (a) $s = f(s_1, \dots, s_n)$ and $s_i \sqsupseteq_{\mathsf{WPO}} t$ for some $i \in \pi(f)$: In this case, we know $s_i\theta \sqsupseteq_{\mathsf{WPO}} t\theta$ by induction hypothesis on s. Thus (2a) concludes $s\theta \sqsupseteq_{\mathsf{WPO}} t\theta$.
   (b) $s = f(s_1, \dots, s_n)$, $t = g(t_1, \dots, t_m)$, and $s \sqsupset_{\mathsf{WPO}} t_j$ for every $j \in \pi(g)$: By induction hypothesis on t, we have $s\theta \sqsupset_{\mathsf{WPO}} t_j\theta$. So the precondition of (2b) for $s\theta \sqsupseteq_{\mathsf{WPO}} t\theta$ is satisfied. There are the following subcases:
      i. $f \succ g$: Then (2b-i) concludes.
      ii. $f \gtrsim g$ and $\pi_f(s_1, \dots, s_n) \sqsupseteq^{\mathsf{lex}}_{\mathsf{WPO}} \pi_g(t_1, \dots, t_m)$: Then by induction hypothesis we have $\pi_f(s_1\theta, \dots, s_n\theta) \sqsupseteq^{\mathsf{lex}}_{\mathsf{WPO}} \pi_g(t_1\theta, \dots, t_m\theta)$, and thus (2b-ii) concludes.
   (c) $s = t \in \mathcal{V}$: Then we have $s\theta \sqsupseteq_{\mathsf{WPO}} t\theta$ by Lemma 1. □

Irreflexivity of $\sqsupset_{\mathsf{WPO}}$ is less obvious to have. In fact, [36] uses well-foundedness to claim it. Here we identify more essential conditions.

Lemma 3. *If $\langle \geq, > \rangle$ is an order pair on $\mathcal{A}$, and $\succ$ is irreflexive on $\mathcal{F}$, and $\langle \mathcal{A}, \geq \rangle$ is π-simple, then $\sqsupset_{\mathsf{WPO}}$ is irreflexive.*

Proof. We show $s \not\sqsupset_{\mathsf{WPO}} s$ for every s by induction on the structure of s. This is clear if $s \in \mathcal{V}$, so consider $s = f(s_1, \dots, s_n)$. Since $>$ is irreflexive, we have $\mathcal{A} \not\models s > s$, and thus $s \sqsupset_{\mathsf{WPO}} s$ cannot be due to case 1 of Definition 5. As $\succ$ is irreflexive on $\mathcal{F}$, $f \not\succ f$ and thus (2b-i) is not possible, either. Thanks to induction hypothesis and the fact that lexicographic extension preserves irreflexivity, we have $\pi_f(s_1, \dots, s_n) \not\sqsupset^{\mathsf{lex}}_{\mathsf{WPO}} \pi_f(s_1, \dots, s_n)$, and thus (2b-ii) is not possible either.

The remaining (2a) is more involving. To show $s_i \not\sqsupseteq_{\mathsf{WPO}} f(s_1, \dots, s_n)$ for any $i \in \pi(f)$, we prove the following more general claim: $s' \lhd^+_\pi s$ implies $s' \not\sqsupseteq_{\mathsf{WPO}} s$, where $\lhd_\pi$ denotes the least relation such that $s_i \lhd_\pi f(s_1, \dots, s_n)$ if $i \in \pi(f)$. This claim is proved by induction on s'. Due to the simplicity assumption, we have $\mathcal{A} \models s \geq s'$ for every $s' \lhd_\pi s$, and this generalizes for every $s' \lhd^+_\pi s$ by easy induction and the transitivity of $\geq$. Thus we cannot have $\mathcal{A} \models s' > s$, since $\mathcal{A} \models s \geq s' > s$ contradicts the assumption that $\langle \geq, > \rangle$ is an order pair. This tells us that $s' \sqsupseteq_{\mathsf{WPO}} s$ cannot be due to case 1. Case (2a) is not applicable thanks to (inner) induction hypothesis on s'. Case (2b) is not possible either, since $s' \not\sqsupset_{\mathsf{WPO}} s'$ thanks to (outer) induction hypothesis on s. This concludes $s' \not\sqsupseteq_{\mathsf{WPO}} s$ for any $s' \lhd^+_\pi s$, and in particular $s_i \not\sqsupseteq_{\mathsf{WPO}} s$ for any $i \in \pi(f)$, refuting the last possibility for $s \sqsupset_{\mathsf{WPO}} s$ to hold. □

5 Co-WPO

The preceding section demonstrated how to use WPO as a rewrite pair in Theorem 1. In this section we show how to use WPO in combination with Theorem 2, that is, when $\sqsupseteq = \sqsupseteq_{\mathsf{WPO}}$, what $\sqsubset$ should be. We show that $\sqsubset_{\overline{\mathsf{WPO}}}$, where $\overline{\mathsf{WPO}} := \mathsf{WPO}(\pi, \mathcal{A}, \not<, \not\leq, \not\prec, \not\preceq)$, serves the purpose.

Proposition 2. *If $\langle \geq, > \rangle$ and $\langle \succeq, \succ \rangle$ are order pairs on $\mathcal{A}$ and $\mathcal{F}$, $\langle \mathcal{A}, \geq \rangle$ is π-simple and monotone, then $\langle \sqsupseteq_{\mathsf{WPO}}, \sqsubset_{\overline{\mathsf{WPO}}} \rangle$ is a co-rewrite pair.*

When $\langle \mathcal{A}, \geq \rangle$ is not total, Example 3 also demonstrates that using Proposition 2 with Theorem 2 is more powerful than using Proposition 1 in combination with Theorem 1, by taking $\pi(f) = []$ for every f. At the time of writing, however, it is unclear to the author if the difference still exists when $\langle \mathcal{A}, \geq \rangle$ is totally ordered but $\langle \mathcal{F}, \succeq \rangle$ is not. Nevertheless we will clearly see the merit of Proposition 2 under the setting of conditional rewriting in the next section.

The remainder of this section proves Proposition 2. Unfortunately, $\overline{\mathsf{WPO}}$ does not satisfy many important properties of WPO, mostly due to the fact that $\langle \not<, \not\leq \rangle$ is not even an order pair. Nevertheless, Lemma 2 is applicable to $\sqsupset_{\overline{\mathsf{WPO}}}$ and gives the following fact:

Lemma 4. *If $\langle \geq, > \rangle$ is an order pair on $\mathcal{A}$, $\langle \mathcal{A}, \geq \rangle$ is π-simple, and $\succ$ is irreflexive, then $\sqsupset_{\overline{\mathsf{WPO}}}$ is closed under substitutions.*

Proof. We apply Lemma 2 to $\overline{\mathsf{WPO}}$. To this end, we need to prove the following:

- $\langle \mathcal{A}, \not< \rangle$ is π-simple: Suppose on the contrary one had $[f](a_1, \dots, a_n) < a_i$ with $i \in \pi(f)$. Due to the simplicity assumption, we have $[f](a_1, \dots, a_n) \geq a_i$. By compatibility we must have $a_i < a_i$, contradicting irreflexivity.
- $\not<$ and $\not\prec$ are reflexive: This follows from the irreflexivity of $<$ and $\prec$. □

The remaining task is to show that $\sqsupseteq_{\mathsf{WPO}} \cap \sqsubset_{\overline{\mathsf{WPO}}} = \emptyset$. Due to the mutual inductive definition of WPO, we need to simultaneously prove the property for the other combination: $\sqsupseteq_{\overline{\mathsf{WPO}}} \cap \sqsubset_{\mathsf{WPO}} = \emptyset$.

Definition 7. *We say that two pairs $P = \langle \sqsupseteq_P, \sqsupset_P \rangle$ and $Q = \langle \sqsupseteq_Q, \sqsupset_Q \rangle$ of relations are* co-compatible *iff $\sqsupseteq_P \cap \sqsubset_Q = \sqsupset_P \cap \sqsubseteq_Q = \emptyset$.*

The next claim is a justification for the word "compatible" in Definition 7. Here the compatibility assumption of order pairs is crucial.

Proposition 3. *An order pair $\langle \sqsupseteq, \sqsupset \rangle$ is co-compatible with itself.*

Proof. Suppose on the contrary that $a \sqsupseteq b$ and $b \sqsupset a$. Then we have $a \sqsupset a$ by compatibility, contradicting the irreflexivity of $\sqsupset$. □

Lemma 5. *If $P = \langle \sqsupseteq_P, \sqsupset_P \rangle$ and $Q = \langle \sqsupseteq_Q, \sqsupset_Q \rangle$ are co-compatible pairs of relations, then $\langle \sqsupseteq_P^{\mathsf{lex}}, \sqsupset_P^{\mathsf{lex}} \rangle$ and $\langle \sqsupseteq_Q^{\mathsf{lex}}, \sqsupset_Q^{\mathsf{lex}} \rangle$ are co-compatible.*

Proof. Let us assume that both

$$[s_1, \ldots, s_n] \sqsupseteq_P^{\mathsf{lex}} [t_1, \ldots, t_m] \tag{2}$$

$$[s_1, \ldots, s_n] \sqsubset_Q^{\mathsf{lex}} [t_1, \ldots, t_m] \tag{3}$$

hold and derive a contradiction. The other part $\sqsupset_P^{\mathsf{lex}} \cap \sqsubseteq_Q^{\mathsf{lex}}$ is analogous. We proceed by induction on the length of $[s_1, \ldots, s_n]$. If $n = 0$, then (2) demands $m = 0$ but (3) demands $m > 0$. Hence we have $n > 0$, and then (3) demands $m > 0$. If $s_1 \sqsupset_P t_1$ then by assumption we have $s_1 \not\sqsubseteq_Q t_1$ but (3) demands $s_1 \sqsubseteq_Q t_1$ (or $s_1 \sqsubset_Q t_1$). Hence (2) is due to $s_1 \sqsupseteq_P t_1$ and $[s_2, \ldots, s_n] \sqsupseteq_P^{\mathsf{lex}} [t_2, \ldots, t_m]$. By assumption we have $s_1 \not\sqsubset_Q t_1$, so (3) is due to $s_1 \sqsubseteq_Q t_1$ and $[s_2, \ldots, s_n] \sqsubset_Q^{\mathsf{lex}} [t_2, \ldots, t_m]$. We derive a contradiction by induction hypothesis. □

We arrive at the main lemma for $\overline{\mathsf{WPO}}$.

Lemma 6. *If $\langle \geq, > \rangle$ and $\langle \succsim, \succ \rangle$ are order pairs on $\mathcal{A}$ and $\mathcal{F}$, and $\langle \mathcal{A}, \geq \rangle$ is π-simple, then* WPO *and* $\overline{\mathsf{WPO}}$ *are co-compatible.*

Proof. We show that neither $s \sqsupseteq_{\mathsf{WPO}} t \wedge s \sqsubset_{\overline{\mathsf{WPO}}} t$ nor $s \sqsupset_{\mathsf{WPO}} t \wedge s \sqsubseteq_{\overline{\mathsf{WPO}}} t$ hold for any s and t, by induction on the structure of s and then t. Let us assume $s \sqsupseteq_{\mathsf{WPO}} t$ and prove $s \not\sqsubset_{\overline{\mathsf{WPO}}} t$. The other claim is analogous. We proceed by case analysis on the derivation of $s \sqsupseteq_{\mathsf{WPO}} t$.

1. $\mathcal{A} \models s > t$: Then $s \sqsubset_{\overline{\mathsf{WPO}}} t$ cannot hold as it demands $\mathcal{A} \models s \not> t$ (or $s \not\geq t$).
2. $\mathcal{A} \models s \geq t$: Then $\mathcal{A} \models s \not\geq t$ cannot happen and thus $s \sqsubset_{\overline{\mathsf{WPO}}} t$ must be due to case 2 of Definition 5. There are the following subcases for $s \sqsupseteq_{\mathsf{WPO}} t$:
   (a) $s = f(s_1, \ldots, s_n)$, $s_i \sqsupseteq_{\mathsf{WPO}} t$ for some $i \in \pi(f)$: By induction hypothesis on s, we have $s_i \not\sqsubset_{\overline{\mathsf{WPO}}} t$, and thus $s \sqsubset_{\overline{\mathsf{WPO}}} t$ can only be due to (2a). So $t = g(t_1, \ldots, t_m)$ and $s \sqsubseteq_{\overline{\mathsf{WPO}}} t_j$ for some $j \in \pi(g)$. Then $s \not\sqsupset_{\mathsf{WPO}} t_j$ by induction hypothesis on t. On the contrary we must have $s \sqsupset_{\mathsf{WPO}} t_j$: By Lemma 1–1. we have $s \sqsupset_{\mathsf{WPO}} s_i \sqsupseteq_{\mathsf{WPO}} t \sqsupset_{\mathsf{WPO}} t_j$ and hence $s \sqsupset_{\mathsf{WPO}} t_j$ as $\langle \sqsupseteq_{\mathsf{WPO}}, \sqsupset_{\mathsf{WPO}} \rangle$ is an order pair.
   (b) $s = f(s_1, \ldots, s_n)$, $t = g(t_1, \ldots, t_m)$, and $s \sqsupset_{\mathsf{WPO}} t_j$ for every $j \in \pi(g)$: By induction hypothesis on t, we have $s \not\sqsubseteq_{\overline{\mathsf{WPO}}} t_j$ for any $j \in \pi(g)$. Thus $s \sqsubset_{\overline{\mathsf{WPO}}} t$ must be due to (2b). We proceed by further considering the following two possibilities.
      i. $f \succ g$: As neither $f \not\succ g$ nor $f \not\succsim g$ hold, $s \sqsubseteq_{\overline{\mathsf{WPO}}} t$ is not possible.
      ii. $f \succsim g$ and $\pi_f(s_1, \ldots, s_n) \sqsupseteq_{\mathsf{WPO}}^{\mathsf{lex}} \pi_g(t_1, \ldots, s_m)$: As $f \not\succsim g$ does not hold, (2b-i) is not applicable to have $s \sqsubset_{\overline{\mathsf{WPO}}} t$. By Lemma 5 and induction hypothesis, we have $\pi_f(s_1, \ldots, s_n) \not\sqsubset_{\overline{\mathsf{WPO}}}^{\mathsf{lex}} \pi_g(t_1, \ldots, t_m)$ and thus (2b-ii) is also not applicable, either.
   (c) $s = t \in \mathcal{V}$: Then clearly $s \sqsubset_{\overline{\mathsf{WPO}}} t$ cannot hold. □

6 Conditional Rewriting

Conditional term rewriting (cf. [27]) is an extension of term rewriting so that rewrite rules can be guarded by conditions. We are interested in the "oriented" variants, as they naturally correspond to functional programming concepts such as `where` clauses of Haskell or `when` clauses of OCaml.

A *conditional rewrite rule* $l \to r \Leftarrow \phi$ consists of terms l and r, and a list ϕ of pairs of terms. We may omit "$\Leftarrow []$" and write $s_1 \twoheadrightarrow t_1, \ldots, s_n \twoheadrightarrow t_n$ for $[\langle s_1, t_1\rangle, \ldots, \langle s_n, t_n\rangle]$. A *conditional TRS (CTRS)* $\mathcal{R}$ is a set of conditional rewrite rules. A CTRS $\mathcal{R}$ yields the rewrite preorder $\to_{\mathcal{R}}^*$ by the following derivation rules [22]:

$$\frac{}{s \to_{\mathcal{R}}^* s}\ \textsc{Refl} \qquad \frac{s \to_{\mathcal{R}} t \quad t \to_{\mathcal{R}}^* u}{s \to_{\mathcal{R}}^* u}\ \textsc{Trans}$$

$$\frac{s_i \to_{\mathcal{R}} s_i'}{f(s_1, \ldots, s_i, \ldots, s_n) \to_{\mathcal{R}} f(s_1, \ldots, s_i', \ldots, s_n)}\ \textsc{Mono}$$

$$\frac{s_1\theta \to_{\mathcal{R}}^* t_1\theta \quad \cdots \quad s_n\theta \to_{\mathcal{R}}^* t_n\theta}{l\theta \to_{\mathcal{R}} r\theta}\ \textsc{Rule} \quad \text{if } (l \to r \Leftarrow s_1 \twoheadrightarrow t_1, \ldots, s_n \twoheadrightarrow t_n) \in \mathcal{R}$$

To approximate reachability with respect to CTRSs by means of (co-)rewrite pairs, one needs to be careful when dealing with conditions.

Example 6. Consider the following CTRS:

$$\mathcal{R}_{\mathsf{fg}} := \{\ \mathsf{f}(x) \to x,\ \ \mathsf{g}(x) \to y \Leftarrow \mathsf{f}(x) \twoheadrightarrow y\ \}$$

and a reachability atom $\mathsf{g}(x) \twoheadrightarrow \mathsf{f}(x)$. One might expect that a rewrite preorder $\sqsupseteq$ such that

$$\mathsf{f}(x) \sqsupseteq x \qquad\qquad \mathsf{g}(x) \sqsupseteq y \ \text{ if }\ \mathsf{f}(x) \sqsupseteq y$$

can over-approximate $\to_{\mathcal{R}_{\mathsf{fg}}}^*$, but this is unfortunately false. For instance, any LPO satisfies the above constraints: $\mathsf{f}(x) \sqsupseteq_{\mathsf{LPO}} x$ as LPO is a simplification order, and the second constraints also vacuously holds as the condition $\mathsf{f}(x) \sqsupseteq_{\mathsf{LPO}} y$ is false. However, it is unsound to conclude that $\mathsf{g}(x) \twoheadrightarrow \mathsf{f}(x)$ is $\mathcal{R}_{\mathsf{fg}}$-unsatisfiable even if $\mathsf{g}(x) \sqsubset_{\mathsf{LPO}} \mathsf{f}(x)$: by setting $\mathsf{f} \succ \mathsf{g}$ one can have $\mathsf{g}(x) \sqsubset_{\mathsf{LPO}} \mathsf{f}(x)$ and $\mathsf{g}(x) \sqsubset_{\overline{\mathsf{LPO}}} \mathsf{f}(x)$, but $\mathsf{g}(x) \to_{\mathcal{R}_{\mathsf{fg}}} \mathsf{f}(x)$.

A solution is to use co-rewrite pairs already for dealing with conditions.

Proposition 4. *If $\langle \sqsupseteq, \sqsubset \rangle$ is a co-rewrite pair, $(l \to r \Leftarrow \phi) \in \mathcal{R}$ implies $l \sqsupseteq r$ or $u \sqsubset v$ for some $u \twoheadrightarrow v \in \phi$, and $s \sqsubset t$, then $s \twoheadrightarrow t$ is $\mathcal{R}$-unsatisfiable.*

Proof. We show that $s \to_{\mathcal{R}}^* t$ implies $s \sqsupseteq t$. This is sufficient, since, then $s\theta \to_{\mathcal{R}}^* t\theta$ implies $s\theta \sqsupseteq t\theta$, while $s \sqsubset t$ demands $s\theta \sqsubset t\theta$, which is not possible since $\sqsupseteq \cap \sqsubset = \emptyset$. The claim is proved by induction on the derivation of $s \to_{\mathcal{R}}^* t$.

– Refl: Since $\sqsupseteq$ is reflexive, we have $s \sqsupseteq s$.

- TRANS: We have $s \to_{\mathcal{R}} t$ and $t \to^*_{\mathcal{R}} u$ as premises, and $s \sqsupseteq t$ and $t \sqsupseteq u$ by induction hypothesis. Since $\sqsupseteq$ is transitive we conclude $s \sqsupseteq u$.
- MONO: We have $s_i \to_{\mathcal{R}} s'_i$ as a premise and $s_i \sqsupseteq s'_i$ by induction hypothesis. Since $\sqsupseteq$ is closed under contexts, we get $f(s_1, \dots, s_i, \dots, s_n) \sqsupseteq f(s_1, \dots, s'_i, \dots, s_n)$.
- RULE: We have $(l \to r \Leftarrow s_1 \twoheadrightarrow t_1, \dots, s_n \twoheadrightarrow t_n) \in \mathcal{R}$, and for every $i \in \{1, \dots, n\}$ have $s_i\theta \to^*_{\mathcal{R}} t_i\theta$ as a premise and $s_i\theta \sqsupseteq t_i\theta$ by induction hypothesis. Since $\sqsupseteq \cap \sqsubset = \emptyset$, we get $s_i\theta \not\sqsubset t_i\theta$. Since $\sqsubset$ is closed under substitutions, we conclude $s_i \not\sqsubset t_i$ for every $i \in \{1, \dots, n\}$. By assumption, this entails $l \sqsupseteq r$, and since $\sqsupseteq$ is closed under substitution, we conclude $l\theta \sqsupseteq r\theta$. □

Example 7. Consider the following singleton CTRS:

$$\mathcal{R}_{\mathsf{ab}} := \{\ \mathsf{a} \to \mathsf{b} \Leftarrow \mathsf{b} \twoheadrightarrow \mathsf{a}\ \}$$

Proposition 4 combined with LPO or WPO induced by a partial precedence such that $\mathsf{a} \not\gtrsim \mathsf{b}$ and $\mathsf{b} \not\gtrsim \mathsf{a}$ proves that $\mathsf{a} \twoheadrightarrow \mathsf{b}$ is $\mathcal{R}_{\mathsf{ab}}$-unsatisfiable: Clearly $\mathsf{b} \not\sqsubseteq_{\mathsf{LPO}} \mathsf{a}$ and $\mathsf{a} \not\sqsubseteq_{\mathsf{LPO}} \mathsf{b}$ by case (2b-i) of Definition 5. On the other hand, Proposition 4 with the term ordering induced by a totally ordered algebra $\langle \mathcal{A}, \geq \rangle$ cannot solve the problem, since $\mathcal{A} \models \mathsf{a} \not\geq \mathsf{b}$ implies $\mathcal{A} \models \mathsf{b} \geq \mathsf{a}$ by totality, which then demands $\mathcal{A} \models \mathsf{a} \geq \mathsf{b}$ to satisfy the assumption of Proposition 4. For the same reason, WPO induced by a totally ordered algebra and a total precedence cannot handle the problem either.

Note that the condition of the rule in $\mathcal{R}_{\mathsf{ab}}$ is unsatisfiable, and this is one of the two cases where Proposition 4 is effective. The other case is when a condition can be ignored. Proposition 4 is incomplete when conditions are essential, as in Example 6. For dealing with essential conditional rules, the variable binding in a rule should be taken into account. At this point, a model-oriented formulation (*a la* [19]) seems more suitable.

Definition 8 (model of CTRS). *We extend the notation $[s \sqsupset t]\alpha$ of Definition 2 to $[\phi]\alpha$ for an arbitrary Boolean formula ϕ with the single binary predicate $\sqsupset$ in the obvious manner. We say $\mathcal{A} = \langle A, [\cdot] \rangle$* validates *$\phi$, written $\mathcal{A} \models \phi$, iff $[\phi]\alpha$ for every $\alpha : \mathcal{V} \to A$. We say a related $\mathcal{F}$-algebra $\langle \mathcal{A}, \sqsupset \rangle$ is a* model *of a CTRS $\mathcal{R}$ iff*[3] *$\mathcal{A} \models l \sqsupset r \vee s_1 \not\sqsupset t_1 \vee \dots \vee s_n \not\sqsupset t_n$ for every $(l \to r \Leftarrow s_1 \twoheadrightarrow t_1, \dots, s_n \twoheadrightarrow t_n) \in \mathcal{R}$.*

Besides minor simplifications (e.g., we do not need two predicates as we are only concerned with reachability in many steps in this paper), the major difference with [19] is that here we do not encode the monotonicity or order axioms into logical formulas (using $\overline{\mathcal{R}}$ of [19]). Instead, we impose these properties as meta-level assumptions over models.

Theorem 3. *For a CTRS $\mathcal{R}$, $s \twoheadrightarrow t$ is $\mathcal{R}$-unsatisfiable if and only if there exists a monotone preordered model $\langle \mathcal{A}, \geq \rangle$ of $\mathcal{R}$ such that $\mathcal{A} \models s \not\geq t$.*

---

[3] Here the formula $s \not\sqsupset t$ is a shorthand for $\neg\ s \sqsupset t$.

Proof. We start with the "if" direction. Let $\langle \mathcal{A}, \geq \rangle$ be a monotone preordered model of $\mathcal{R}$. As in Proposition 4, it suffices to show that $s \to_{\mathcal{R}}^* t$ implies $\mathcal{A} \models s \geq t$. The claim is proved by induction on the derivation of $s \to_{\mathcal{R}}^* t$.

- Refl: Since $\geq$ is reflexive, we have $\mathcal{A} \models s \geq s$.
- Trans: We have $s \to_{\mathcal{R}} t$ and $t \to_{\mathcal{R}}^* u$ as premises, and $\mathcal{A} \models s \geq t$ and $\mathcal{A} \models t \geq u$ by induction hypothesis. Since $\geq$ is transitive we conclude $\mathcal{A} \models s \geq u$.
- Mono: We have $s_i \to_{\mathcal{R}} s_i'$ as a premise and $\mathcal{A} \models s_i \geq s_i'$ by induction hypothesis. Since $\langle \mathcal{A}, \geq \rangle$ is monotone, we get $\mathcal{A} \models f(s_1, \dots, s_i, \dots, s_n) \geq f(s_1, \dots, s_i', \dots, s_n)$.
- Rule: We have $(l \to r \Leftarrow s_1 \twoheadrightarrow t_1, \dots, s_n \twoheadrightarrow t_n) \in \mathcal{R}$, and for every $i \in \{1, \dots, n\}$ have $s_i\theta \to_{\mathcal{R}}^* t_i\theta$ as a premise and $\mathcal{A} \models s_i\theta \geq t_i\theta$ by induction hypothesis. Since $\langle \mathcal{A}, \geq \rangle$ is a model of $\mathcal{R}$, and by the fact that validity is closed under substitutions, we have $\mathcal{A} \models l\theta \geq r\theta \vee s_1\theta \not\geq t_1\theta \vee \dots \vee s_n\theta \not\geq t_n\theta$. Together with the induction hypotheses we conclude $\mathcal{A} \models l\theta \geq r\theta$.

Next consider the "only if" direction. We show that $\langle \mathcal{T}(\mathcal{F}, \mathcal{V}), \to_{\mathcal{R}}^* \rangle$ is a model of $\mathcal{R}$, that is, for every $(l \to r \Leftarrow s_1 \twoheadrightarrow t_1, \dots, s_n \twoheadrightarrow t_n) \in \mathcal{R}$, we show $\mathcal{T}(\mathcal{F}, \mathcal{V}) \models l \to_{\mathcal{R}}^* r \vee s_1 \not\to_{\mathcal{R}}^* t_1 \vee \dots \vee s_n \not\to_{\mathcal{R}}^* t_n$. This means $l\theta \to_{\mathcal{R}}^* r\theta$ for every $\theta : \mathcal{V} \to \mathcal{T}(\mathcal{F}, \mathcal{V})$ such that $s_1\theta \to_{\mathcal{R}}^* t_1\theta, \dots, s_n\theta \to_{\mathcal{R}}^* t_n\theta$, which is immediate by Rule. The fact that $\to_{\mathcal{R}}^*$ is a preorder and closed under contexts is also immediate. Finally, $s \twoheadrightarrow t$ being $\mathcal{R}$-unsatisfiable means that $s\theta \not\to_{\mathcal{R}}^* t\theta$ for any $\theta : \mathcal{V} \to \mathcal{T}(\mathcal{F}, \mathcal{V})$, that is, $\mathcal{T}(\mathcal{F}, \mathcal{V}) \models s \not\to_{\mathcal{R}}^* t$. □

Putting implementation issues aside, it is trivial to use semantic (termination) methods in Theorem 3.

Example 8. Consider again the CTRS $\mathcal{R}_{\mathsf{fg}}$ of Example 6. The monotone ordered $\{\mathsf{f}, \mathsf{g}\}$-algebra $\langle \mathbb{N}, [\cdot], \geq \rangle$ defined by

$$[\mathsf{f}](x) = x \qquad\qquad [\mathsf{g}](x) = x + 1$$

is a model of $\mathcal{R}_{\mathsf{fg}}$, since for arbitrary $x, y \in \mathbb{N}$, we have

$$[\mathsf{f}](x) \geq x \qquad\qquad [\mathsf{g}](x) = x + 1 \geq y \vee [\mathsf{f}](x) = x \not\geq y$$

Then, with Theorem 3 we can show that $\mathsf{f}(x) \twoheadrightarrow \mathsf{g}(x)$ is $\mathcal{R}_{\mathsf{fg}}$-unsatisfiable, as $[\mathsf{f}](x) = x \not\geq x + 1 = [\mathsf{g}](x)$ for every $x \in \mathbb{N}$.

To use $\mathsf{WPO}(\mathcal{A})$ in combination with Theorem 3, we need to validate formulas with predicate $\sqsupseteq_{\mathsf{WPO}(\mathcal{A})}$ in the term algebra $\mathcal{T}(\mathcal{F}, \mathcal{V})$. We encode these formulas into formulas with predicates $\geq$ and $>$, which are then interpreted in $\mathcal{A}$.

Definition 9 (formal WPO). *Let $\langle \geq, > \rangle$ and $\langle \succsim, \succ \rangle$ be pairs of relations over some set and over $\mathcal{F}$, respectively, and let π be a partial status. We define $\mathtt{wpo}(\pi, \geq, >, \succsim, \succ)$ or $\mathtt{wpo}$ for short to be the pair $\langle \sqsupseteq_{\mathtt{wpo}}, \sqsupset_{\mathtt{wpo}} \rangle$, where for terms $s, t \in \mathcal{T}(\mathcal{F}, \mathcal{V})$, $s \sqsupseteq_{\mathtt{wpo}} t$ and $s \sqsupset_{\mathtt{wpo}} t$ are Boolean formulas defined as follows:*

$$s \sqsupset_{\mathtt{wpo}} t \;:=\; s > t \vee (s \geq t \wedge \phi)$$

where ϕ *is* FALSE *if* $s \in \mathcal{V}$ *and is* $\bigvee_{i\in\pi(f)} s_i \sqsupseteq_{\mathtt{wpo}} t \vee \psi$ *if* $s = f(s_1,\dots,s_n)$, *and* ψ *is* FALSE *if* $t \in \mathcal{V}$ *and is*

$$\bigwedge_{j\in\pi(g)} s \sqsupset_{\mathtt{wpo}} t_j \wedge (f \succ g \vee (f \succsim g \wedge \pi_f(s_1,\dots,s_n) \sqsupset^{\mathsf{lex}}_{\mathtt{wpo}} \pi_g(t_1,\dots,t_m)))$$

if $t = g(t_1,\dots,t_m)$. *Formula* $s \sqsupseteq_{\mathtt{wpo}} t$ *is defined analogously, except that* ϕ *is* TRUE *if* $s = t \in \mathcal{V}$, *and* $\sqsupset^{\mathsf{lex}}_{\mathtt{wpo}}$ *in formula* ψ *is replaced by* $\sqsupseteq^{\mathsf{lex}}_{\mathtt{wpo}}$.

We omit an easy proof that verifies that wpo encodes WPO:

Lemma 7. $s \mathrel{_{(}\sqsupseteq_{)}}_{\mathsf{WPO}(\mathcal{A})} t$ *iff* $\mathcal{A} \models s \mathrel{_{(}\sqsupseteq_{)}}_{\mathtt{wpo}} t$.

Note carefully that $s \not\sqsupseteq_{\mathsf{WPO}(\mathcal{A})} t$ is $\mathcal{A} \not\models s \sqsupseteq_{\mathtt{wpo}} t$ but not $\mathcal{A} \models s \not\sqsupseteq_{\mathtt{wpo}} t$. Hence we ensure $s \not\sqsupseteq_{\mathsf{WPO}(\mathcal{A})} t$ by $\mathcal{A} \models s \sqsubset_{\overline{\mathtt{wpo}}} t$, where $\overline{\mathtt{wpo}}$ denotes $\mathtt{wpo}(\pi, \not<, \not\leq, \not\succ, \not\succsim)$.

Theorem 4. *If* $\mathcal{R}$ *is a CTRS,* $\langle \geq, > \rangle$ *and* $\langle \succsim, \succ \rangle$ *are order pairs on* $\mathcal{A}$ *and* $\mathcal{F}$, $\langle \mathcal{A}, \geq \rangle$ *is* π*-simple and monotone,* $\mathcal{A} \models l \sqsupseteq_{\mathtt{wpo}} r \vee u_1 \sqsubset_{\overline{\mathtt{wpo}}} v_1 \vee \dots \vee u_n \sqsubset_{\overline{\mathtt{wpo}}} v_n$ *for every* $(l \to r \Leftarrow u_1 \twoheadrightarrow v_1, \dots, u_n \twoheadrightarrow v_n) \in \mathcal{R}$, *and* $\mathcal{A} \models s \sqsubset_{\overline{\mathtt{wpo}}} t$, *then* $s \twoheadrightarrow t$ *is* $\mathcal{R}$*-unsatisfiable.*

Proof. We apply Theorem 3. To this end, we first show that $\langle \mathcal{T}(\mathcal{F},\mathcal{V}), \sqsupseteq_{\mathsf{WPO}(\mathcal{A})} \rangle$ is a monotone preordered model of $\mathcal{R}$. Monotonicity and preorderedness are due to Proposition 1. For being a model, let $(l \to r \Leftarrow u_1 \twoheadrightarrow v_1, \dots, u_n \twoheadrightarrow v_n) \in \mathcal{R}$. Due to assumption and Lemma 7, we have $l \sqsupseteq_{\mathsf{WPO}(\mathcal{A})} r \vee u_1 \sqsubset_{\overline{\mathsf{WPO}}(\mathcal{A})} v_1 \vee \dots \vee u_n \sqsubset_{\overline{\mathsf{WPO}}(\mathcal{A})} v_n$. Due to Lemmas 2 and 4, we get $l\theta \sqsupseteq_{\mathsf{WPO}(\mathcal{A})} r\theta \vee u_1\theta \sqsubset_{\overline{\mathsf{WPO}}(\mathcal{A})} v_1\theta \vee \dots \vee u_n\theta \sqsubset_{\overline{\mathsf{WPO}}(\mathcal{A})} v_n\theta$ for every $\theta : \mathcal{V} \to \mathcal{T}(\mathcal{F},\mathcal{V})$. With Proposition 2 we conclude $\mathcal{T}(\mathcal{F},\mathcal{V}) \models l \sqsupseteq_{\mathsf{WPO}(\mathcal{A})} r \vee u_1 \not\sqsupseteq_{\mathsf{WPO}(\mathcal{A})} v_1 \vee \dots \vee u_n \not\sqsupseteq_{\mathsf{WPO}(\mathcal{A})} v_n$. Finally, we need $\mathcal{T}(\mathcal{F},\mathcal{V}) \models s \not\sqsupseteq_{\mathsf{WPO}(\mathcal{A})} t$, i.e., $s\theta \not\sqsupseteq_{\mathsf{WPO}(\mathcal{A})} t\theta$ for any $\theta : \mathcal{V} \to \mathcal{T}(\mathcal{F},\mathcal{V})$. As we assume $s \sqsubset_{\overline{\mathsf{WPO}}(\mathcal{A})} t$, by Lemma 4 we have $s\theta \sqsubset_{\overline{\mathsf{WPO}}(\mathcal{A})} t\theta$. By Proposition 2 we conclude $s\theta \not\sqsupseteq_{\mathsf{WPO}(\mathcal{A})} t\theta$. □

7 Experiments

The proposed methods are implemented in the termination prover NaTT [35], available at https://www.trs.cm.is.nagoya-u.ac.jp/NaTT/.

Internally, NaTT reduces the problem of finding an algebra $\mathcal{A}$ that make $\langle \mathcal{A}, \geq \rangle$ a model of a TRS $\mathcal{R}$ (or $\sqsupseteq_{\mathsf{WPO}(\mathcal{A})} \subseteq \mathcal{R}$) into a satisfiability modulo theory (SMT) problem, which is then solved by the backend SMT solver z3 [26]. The implementation of Theorem 1 and Corollary 1 is a trivial adaptation from the termination methods. Cororllary 2 is also trivial for totally ordered carriers, since $\mathcal{A} \models s \not\geq t$ is equivalent to $\mathcal{A} \models s < t$. Matrix/tuple interpretations are also easy, since $\mathcal{A} \models (a_1,\dots,a_n) \not\geq (b_1,\dots,b_n)$ is equivalent to $\mathcal{A} \models a_1 < b_1 \vee \dots \vee a_n < b_n$. Theorem 2 with $\overline{\mathsf{WPO}}$ is obtained by parametrizing WPO.

Table 1. Experimental results.

	TRS						CTRS		
Method	$\mathcal{R}_{\mathsf{add}}$	$\mathcal{R}_{mat}$	$\mathcal{R}_{\mathsf{A}}$	$\mathcal{R}_{kbo}$	$\mathcal{R}_{wpo}$	COPS(15)	$\mathcal{R}_{\mathsf{ab}}$	$\mathcal{R}_{\mathsf{fg}}$	COPS(126)
$\mathcal{S}um$						6		✓	15
$\mathcal{S}um^+$						6		✓	24
$\mathcal{S}um^-$	✓					5			28
$\mathcal{M}at$		✓				6	✓	✓	25 (TO:88)
LPO			✓			5	✓		19
WPO($\mathcal{S}um$)				✓	✓	6	✓	✓	15
WPO($\mathcal{S}um^+$)				✓	✓	6	✓	✓	25 (TO:29)
WPO($\mathcal{S}um^-$)				✓	✓	6	✓		15
infChecker	✓		✓			13	✓	✓	51+42 (TO:25)
CO3		✓				5	✓		20
NaTT 2.1						3			19
NaTT 2.2	✓	✓	✓	✓	✓	6 (TO:4)	✓	✓	31 (TO:79)

Theorem 3 needs some tricks. In the unconditional case, finding a desired algebra $\mathcal{A}$ can be encoded into SMT over quantifier-free linear arithmetic for a large class of $\mathcal{A}$ [36]. For the conditional case, we need to find ($\exists$) parameters that validates ($\forall$) a disjunctive clause. Farkas' lemma would reduce such a problem into quantifier-free SMT, but then the resulting problem is nonlinear. Experimentally, we observe that our backend z3 performs better on quantified linear arithmetic than quantifier-free nonlinear arithmetic, and hence we choose to leave the $\forall$ quantifiers.

We conducted experiments using the examples presented in the paper and the examples in the INF category of the standard benchmark set COPS. The execution environment is StarExec [31] with the same settings as CoCo 2019.

Many COPS examples contain conjunctive reachability constraints of form $s_1 \twoheadrightarrow t_1 \wedge \cdots \wedge s_n \twoheadrightarrow t_n$. In this experiment we naively collapsed such a constraint into $\mathsf{tp}(s_1, \ldots, s_n) \twoheadrightarrow \mathsf{tp}(t_1, \ldots, t_n)$ by introducing a fresh function symbol tp. Two benchmarks exceed the scope of oriented CTRSs, on which NaTT immediately gives up.

As co-rewrite pairs we tested algebras $\mathcal{S}um$, $\mathcal{S}um^+$, $\mathcal{S}um^-$, $\mathcal{M}at$, LPO, and WPO. The basic algebra $\mathcal{S}um = \langle \mathbb{Z}, [\cdot] \rangle$ is given by $[f](x_1, \ldots, x_n) = c_0 + \sum_{i=1}^{n} c_i \cdot x_i$, where $c_0 \in \mathbb{Z}, c_1, \ldots, c_n \in \{0, 1\}$. Similarly $\mathcal{S}um^+$ and $\mathcal{S}um^-$ are defined, where the ranges of c_0, which also determine the carrier, are $\mathbb{N}$ and $\mathbb{Z}_{\leq 0}$, respectively. The algebra $\mathcal{M}at$ represents the 2D matrix interpretations.

Table 1 presents the results. For TRSs, we can observe that our proposed methods advance the state of the art, in the sense that they prove new examples that no tool previously participated in CoCo could handle. As there are only 15 TRS examples in the INF category of COPS 2021, we could not derive interesting observations there. Taking CTRS examples into account, we see $\mathcal{S}um$ is not as good as $\mathcal{S}um^+$ or $\mathcal{S}um^-$, while the carrier is bigger ($\mathbb{Z}$ versus $\mathbb{N}$ or $\mathbb{Z}_{\leq 0}$). This phenomenon is explained as follows: For the latter two one knows variables are bounded by 0 (from below or above), and hence one can have $\mathcal{S}um^+ \models x \geq \mathsf{a}$ or $\mathcal{S}um^- \models \mathsf{a} \geq x$ by $[\mathsf{a}] = 0$. Neither is possible when the carrier is unbounded. This observation also suggests another choice of carriers that are bounded from below and above, which is left for future work.

From the figures in CTRS examples, $\mathcal{S}um^-$ performs the best among our methods. However, $\mathcal{M}at$ and $\mathsf{WPO}(\mathcal{S}um^+)$ solve more examples if TRS examples are counted. It does not seem appropriate yet to judge practical significance from these experiments.

Finally, we implemented as the default strategy of NaTT 2.2 the sequential application of $\mathcal{S}um^-$, LPO, $\mathsf{WPO}(\mathcal{S}um^+)$, and $\mathcal{M}at$ after the test NaTT already have implemented. There improvement over previous NaTT 2.1 should be clear, although the number of timeouts (indicated by "TO:") is significant.

8 Conclusion

We proposed generalizations of termination techniques that can prove unsatisfiability of reachability, both for term rewriting and for conditional term rewriting. We implemented the approach in the termination prover NaTT, and experimentally evaluated the significance of the proposed approach.

The implementation focused on evaluating the proposed methods separately. The only implemented way of combining their power is a naive one: apply the tests one by one while they fail. For future work, it will be interesting to incorporate the proposed method into the existing frameworks [10,30].

Acknowledgments. The author would like to thank Kiraku Shintani for the technical help with the COPS database system. I would also like to thank Nao Hirokawa, Salvador Lucas, Naoki Nishida, and Sarah Winkler for discussions, and the anonymous reviewers for their detailed comments that improved the presentation of the paper.

References

1. Aoto, T.: Disproving confluence of term rewriting systems by interpretation and ordering. In: Fontaine, P., Ringeissen, C., Schmidt, R.A. (eds.) FroCoS 2013. LNCS (LNAI), vol. 8152, pp. 311–326. Springer, Heidelberg (2013). https://doi.org/10.1007/978-3-642-40885-4_22
2. Arts, T., Giesl, J.: Termination of term rewriting using dependency pairs. Theor. Compt. Sci. **236**(1–2), 133–178 (2000). https://doi.org/10.1016/S0304-3975(99)00207-8
3. Baader, F., Nipkow, T.: Term Rewriting and All That. Cambridge University Press, Cambridge (1998)
4. Dershowitz, N.: Orderings for term-rewriting systems. Theor. Compt. Sci. **17**(3), 279–301 (1982). https://doi.org/10.1016/0304-3975(82)90026-3
5. Endrullis, J., Waldmann, J., Zantema, H.: Matrix interpretations for proving termination of term rewriting. J. Autom. Reason. **40**(2–3), 195–220 (2008). https://doi.org/10.1007/s10817-007-9087-9

6. Feuillade, G., Genet, T., Tong, V.V.T.: Reachability analysis over term rewriting systems. J. Autom. Reasoning **33**, 341–383 (2004). https://doi.org/10.1007/s10817-004-6246-0
7. Giesl, J., et al.: Proving termination of programs automatically with AProVE. In: Demri, S., Kapur, D., Weidenbach, C. (eds.) IJCAR 2014. LNCS (LNAI), vol. 8562, pp. 184–191. Springer, Cham (2014). https://doi.org/10.1007/978-3-319-08587-6_13
8. Giesl, J., Rubio, A., Sternagel, C., Waldmann, J., Yamada, A.: The termination and complexity competition. In: Beyer, D., Huisman, M., Kordon, F., Steffen, B. (eds.) TACAS 2019. LNCS, vol. 11429, pp. 156–166. Springer, Cham (2019). https://doi.org/10.1007/978-3-030-17502-3_10
9. Giesl, J., Thiemann, R., Schneider-Kamp, P., Falke, S.: Mechanizing and improving dependency pairs. J. Autom. Reason. **37**(3), 155–203 (2006). https://doi.org/10.1007/s10817-006-9057-7
10. Gutiérrez, R., Lucas, S.: Automatically proving and disproving feasibility conditions. In: Peltier, N., Sofronie-Stokkermans, V. (eds.) IJCAR 2020. LNCS (LNAI), vol. 12167, pp. 416–435. Springer, Cham (2020). https://doi.org/10.1007/978-3-030-51054-1_27
11. Gutiérrez, R., Lucas, S.: MU-TERM: verify termination properties automatically (system description). In: Peltier, N., Sofronie-Stokkermans, V. (eds.) IJCAR 2020. LNCS (LNAI), vol. 12167, pp. 436–447. Springer, Cham (2020). https://doi.org/10.1007/978-3-030-51054-1_28
12. Hirokawa, N., Middeldorp, A.: Automating the dependency pair method. Inf. Comput. **199**(1–2), 172–199 (2005). https://doi.org/10.1016/j.ic.2004.10.004
13. Kamin, S., Lévy, J.J.: Two generalizations of the recursive path ordering (1980). unpublished note
14. Knuth, D.E., Bendix, P.: Simple word problems in universal algebras. In: Computational Problems in Abstract Algebra, pp. 263–297. Pergamon Press, New York (1970). https://doi.org/10.1016/B978-0-08-012975-4.50028-X
15. Kop, C., Vale, D.: Tuple interpretations for higher-order complexity. In: Kobayashi, N. (ed.) FSCD 2021. LIPIcs, vol. 195, pp. 31:1–31:22. Schloss Dagstuhl - Leibniz-Zentrum für Informatik (2021). https://doi.org/10.4230/LIPIcs.FSCD.2021.31
16. Korp, M., Sternagel, C., Zankl, H., Middeldorp, A.: Tyrolean termination tool 2. In: Treinen, R. (ed.) RTA 2009. LNCS, vol. 5595, pp. 295–304. Springer, Heidelberg (2009). https://doi.org/10.1007/978-3-642-02348-4_21
17. Kusakari, K., Nakamura, M., Toyama, Y.: Argument filtering transformation. In: Nadathur, G. (ed.) PPDP 1999. LNCS, vol. 1702, pp. 47–61. Springer, Heidelberg (1999). https://doi.org/10.1007/10704567_3
18. Lankford, D.: Canonical algebraic simplification in computational logic. Technical report ATP-25, University of Texas (1975)
19. Lucas, S., Gutiérrez, R.: Use of logical models for proving infeasibility in term rewriting. Inf. Process. Lett. **136**, 90–95 (2018). https://doi.org/10.1016/j.ipl.2018.04.002
20. Lucas, S., Meseguer, J.: 2D dependency pairs for proving operational termination of CTRSs. In: Escobar, S. (ed.) WRLA 2014. LNCS, vol. 8663, pp. 195–212. Springer, Cham (2014). https://doi.org/10.1007/978-3-319-12904-4_11
21. Lucas, S., Meseguer, J.: Dependency pairs for proving termination properties of conditional term rewriting systems. J. Log. Algebraic Methods Program. **86**(1), 236–268 (2017). https://doi.org/10.1016/j.jlamp.2016.03.003

22. Lucas, S., Meseguer, J., Gutiérrez, R.: The 2D dependency pair framework for conditional rewrite systems. part I: definition and basic processors. J. Comput. Syst. Sci. **96**, 74–106 (2018). https://doi.org/10.1016/j.jcss.2018.04.002
23. Manna, Z., Ness, S.: Termination of Markov algorithms (1969). unpublished manuscript
24. Middeldorp, A.: Approximating dependency graphs using tree automata techniques. In: Goré, R., Leitsch, A., Nipkow, T. (eds.) IJCAR 2001. LNCS, vol. 2083, pp. 593–610. Springer, Heidelberg (2001). https://doi.org/10.1007/3-540-45744-5_49
25. Middeldorp, A., Nagele, J., Shintani, K.: CoCo 2019: report on the eighth confluence competition. Int. J. Softw. Tools Technol. Transfer **23**(6), 905–916 (2021). https://doi.org/10.1007/s10009-021-00620-4
26. de Moura, L., Bjørner, N.: Z3: an efficient SMT solver. In: Ramakrishnan, C.R., Rehof, J. (eds.) TACAS 2008. LNCS, vol. 4963, pp. 337–340. Springer, Heidelberg (2008). https://doi.org/10.1007/978-3-540-78800-3_24
27. Ohlebusch, E.: Advanced Topics in Term Rewriting. Springer, New York (2002). https://doi.org/10.1007/978-1-4757-3661-8
28. Oostrom, V.: Sub-Birkhoff. In: Kameyama, Y., Stuckey, P.J. (eds.) FLOPS 2004. LNCS, vol. 2998, pp. 180–195. Springer, Heidelberg (2004). https://doi.org/10.1007/978-3-540-24754-8_14
29. Sternagel, C., Thiemann, R., Yamada, A.: A formalization of weighted path orders and recursive path orders. Arch. Formal Proofs (2021). https://isa-afp.org/entries/Weighted_Path_Order.html, Formal proof development
30. Sternagel, C., Yamada, A.: Reachability analysis for termination and confluence of rewriting. In: Vojnar, T., Zhang, L. (eds.) TACAS 2019. LNCS, vol. 11427, pp. 262–278. Springer, Cham (2019). https://doi.org/10.1007/978-3-030-17462-0_15
31. Stump, A., Sutcliffe, G., Tinelli, C.: StarExec: a cross-community infrastructure for logic solving. In: Demri, S., Kapur, D., Weidenbach, C. (eds.) IJCAR 2014. LNCS (LNAI), vol. 8562, pp. 367–373. Springer, Cham (2014). https://doi.org/10.1007/978-3-319-08587-6_28
32. TeReSe: Term Rewriting Systems, Cambridge Tracts in Theoretical Computer Science, vol. 55. Cambridge University Press, Cambridge (2003)
33. Thiemann, R., Schöpf, J., Sternagel, C., Yamada, A.: Certifying the weighted path order (invited talk). In: Ariola, Z.M. (ed.) FSCD 2020. LIPIcs, vol. 167, pp. 4:1–4:20. Schloss Dagstuhl-Leibniz-Zentrum für Informatik, Dagstuhl, Germany (2020). https://doi.org/10.4230/LIPIcs.FSCD.2020.4
34. Yamada, A.: Multi-dimensional interpretations for termination of term rewriting. In: Platzer, A., Sutcliffe, G. (eds.) CADE 2021. LNCS (LNAI), vol. 12699, pp. 273–290. Springer, Cham (2021). https://doi.org/10.1007/978-3-030-79876-5_16
35. Yamada, A., Kusakari, K., Sakabe, T.: Nagoya termination tool. In: Dowek, G. (ed.) RTA 2014. LNCS, vol. 8560, pp. 466–475. Springer, Cham (2014). https://doi.org/10.1007/978-3-319-08918-8_32
36. Yamada, A., Kusakari, K., Sakabe, T.: A unified ordering for termination proving. Sci. Comput. Program. **111**, 110–134 (2015). https://doi.org/10.1016/j.scico.2014.07.009
37. Zantema, H.: Termination of term rewriting by semantic labelling. Fundam. Informaticae **24**(1/2), 89–105 (1995). https://doi.org/10.3233/FI-1995-24124
38. Zantema, H.: The termination hierarchy for term rewriting. Appl. Algebr. Eng. Comm. Compt. **12**(1/2), 3–19 (2001). https://doi.org/10.1007/s002000100061

Knowledge Representation and Justification

Evonne: Interactive Proof Visualization for Description Logics (System Description)

Christian Alrabbaa[1(✉)], Franz Baader[1(✉)], Stefan Borgwardt[1(✉)], Raimund Dachselt[2(✉)], Patrick Koopmann[1(✉)], and Julián Méndez[2(✉)]

[1] Institute of Theoretical Computer Science, TU Dresden, Dresden, Germany
{christian.alrabbaa,franz.baader,stefan.borgwardt, patrick.koopmann}@tu-dresden.de

[2] Interactive Media Lab Dresden, TU Dresden, Dresden, Germany
{raimund.dachselt,julian.mendez2}@tu-dresden.de

Abstract. Explanations for description logic (DL) entailments provide important support for the maintenance of large ontologies. The "justifications" usually employed for this purpose in ontology editors pinpoint the parts of the ontology responsible for a given entailment. Proofs for entailments make the intermediate reasoning steps explicit, and thus explain how a consequence can actually be derived. We present an interactive system for exploring description logic proofs, called Evonne, which visualizes proofs of consequences for ontologies written in expressive DLs. We describe the methods used for computing those proofs, together with a feature called *signature-based proof condensation*. Moreover, we evaluate the quality of generated proofs using real ontologies.

1 Introduction

Proofs generated by Automated Reasoning (AR) systems are sometimes presented to humans in textual form to convince them of the correctness of a theorem [9,11], but more often employed as certificates that can automatically be checked [20]. In contrast to the AR setting, where very long proofs may be needed to derive a deep mathematical theorem from very few axioms, DL-based ontologies are often very large, but proofs of a single consequence are usually of a more manageable size. For this reason, the standard method of explanation in description logic [8] has long been to compute so-called *justifications*, which point out a minimal set of source statements responsible for an entailment of interest. For example, the ontology editor Protégé[1] supports the computation of justifications since 2008 [12], which is very useful when working with large DL ontologies. Nevertheless, it is often not obvious why a given consequence actually follows from such a justification [13]. Recently, this explanation capability has been extended towards showing full *proofs* with intermediate reasoning steps, but this is restricted to ontologies written in the lightweight DLs supported by the Elk reasoner [15,16], and the graphical presentation of proofs is very basic.

[1] https://protege.stanford.edu/.

J. Blanchette et al. (Eds.): IJCAR 2022, LNAI 13385, pp. 271–280, 2022.
https://doi.org/10.1007/978-3-031-10769-6_16

In this paper, we present EVONNE as an interactive system, for exploring DL proofs for description logic entailments, using the methods for computing small proofs presented in [3,5]. Initial prototypes of EVONNE were presented in [6,10], but since then, many improvements were implemented. While EVONNE does more than just visualizing proofs, this paper focuses on the proof component of EVONNE: specifically, we give a brief overview of the interface for exploring proofs, describe the proof generation methods implemented in the back-end, and present an experimental evaluation of these proofs generation methods in terms of proof size and run time. The improved back-end uses Java libraries that extract proofs using various methods, such as from the ELK calculus, or *forgetting-based proofs* [3] using the forgetting tools LETHE [17] and FAME [21] in a black-box fashion. The new front-end is visually more appealing than the prototypes presented in [6,10], and allows to inspect and explore proofs using various interaction techniques, such as zooming and panning, collapsing and expanding, text manipulation, and compactness adjustments. Additional features include the minimization of the generated proofs according to various measures and the possibility to select a *known signature* that is used to automatically hide parts of the proofs that are assumed to be obvious for users with certain previous knowledge. Our evaluation shows that proof sizes can be significantly reduced in this way, making the proofs more user-friendly. EVONNE can be tried and downloaded at https://imld.de/evonne. The version of EVONNE described here, as well as the data and scripts used in our experiments, can be found at [2].

2 Preliminaries

We recall some relevant notions for DLs; for a detailed introduction, see [8]. DLs are decidable fragments of first-order logic (FOL) with a special, variable-free syntax, and that use only unary and binary predicates, called *concept names* and *role names*, respectively. These can be used to build complex *concepts*, which correspond to first-order formulas with one free variable, and *axioms* corresponding to first-order sentences. Which kinds of concepts and axioms can be built depends on the expressivity of the used DL. Here we mainly consider the light-weight DL $\mathcal{ELH}$ and the more expressive $\mathcal{ALCH}$. We have the usual notion of FOL *entailment* $\mathcal{O} \models \alpha$ of an axiom α from a finite set of axioms $\mathcal{O}$, called an ontology. of special interest are entailments of *atomic CIs* (concept inclusions) of the form $A \sqsubseteq B$, where A and B are concept names. Following [3], we define *proofs* of $\mathcal{O} \models \alpha$ as finite, acyclic, directed hypergraphs, where vertices v are labeled with axioms $\ell(v)$ and hyperedges are of the form (S, d), with S a set of vertices and d a vertex such that $\{\ell(v) \mid v \in S\} \models \ell(d)$; the leaves of a proof must be labeled by elements of $\mathcal{O}$ and the root by α. In this paper, all proofs are *trees*, i.e. no vertex can appear in the first component of multiple hyperedges (see Fig. 1).

3 The Graphical User Interface

The user interface of EVONNE is implemented as a web application. To support users in understanding large proofs, they are offered various layout options and

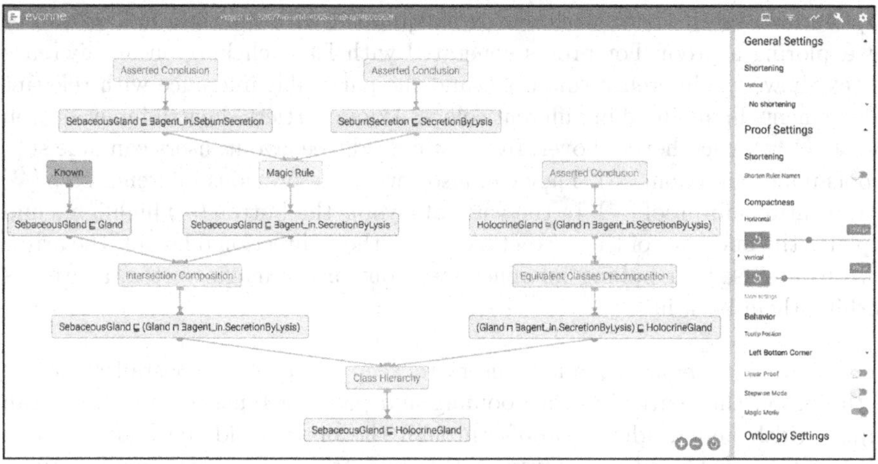

Fig. 1. Overview of EVONNE - a condensed proof in the bidirectional layout

interaction components. The proof visualization is linked to a second view showing the context of the proof in a relevant subset of the ontology. In this ontology view, interactions between axioms are visualized, so that users can understand the context of axioms occurring in the proof. The user can also examine possible ways to eliminate unwanted entailments in the ontology view. The focus of this system description, however, is on the proof component: we describe how the proofs are generated and how users can interact with the proof visualization. For details on the ontology view, we refer the reader to the workshop paper [6], where we also describe how EVONNE supports ontology repair.

Initialization. After starting EVONNE for the first time, users create a new project, for which they specify an ontology file. They can then select an entailed atomic CI to be explained. The user can choose between different proof methods, and optionally select a signature of *known terms* (cf. Sect. 4), which can be generated using the term selection tool Protégé-TS [14].

Layout. Proofs are shown as graphs with two kinds of vertices: colored vertices for axioms, gray ones for inference steps. By default, proofs are shown using a *tree layout.* To take advantage of the width of the display when dealing with long axioms, it is possible to show proofs in a *vertical layout*, placing axioms linearly below each other, with inferences represented through edges on the side (without the inference vertices). It is possible to automatically re-order vertices to minimize the distance between conclusion and premises in each step. The third layout option is the *bidirectional layout* (see Fig. 1), a tree layout where, initially, the entire proof is collapsed into a *magic vertex* that links the conclusion directly to its justification, and from which individual inference steps can be pulled out and pushed back from both directions.

Exploration. In all views, each vertex is equipped with multiple functionalities for exploring a proof. For proofs generated with ELK, clicking on an inference vertex shows the inference rule used, and the particular inference with relevant sub-elements highlighted in different colors. Axiom vertices show different button (⬆, ⬇, ◀, ◉) when hovered over. In the standard tree layout, users can hide sub-proofs under an axiom (⬇). They can also reveal the previous inference step (◀) or the entire-sub-proof (⬆). In the vertical layout, the button (◉) highlights and explains the inference of the current axiom. In the bidirectional layout, the arrow buttons are used for pulling inference steps out of the magic vertex, as well as pushing them back in.

Presentation. A *minimap* allows users to keep track of the overall structure of the proof, thus enriching the zooming and panning functionality. Users can adjust width and height of proofs through the options side-bar. Long axiom labels can be *shortened* in two ways: either by setting a fixed size to all vertices, or by abbreviating names based on capital letters. Afterwards, it is possible to restore the original labels individually.

4 Proof Generation

To obtain the proofs that are shown to the user, we implemented different proof generation techniques, some of which were initially described in [3]. For $\mathcal{ELH}$ ontologies, proofs can be generated natively by the DL reasoner ELK [16]. These proofs use rules from the calculus described in [16]. We apply the Dijkstra-like algorithm introduced in [4,5] to compute a *minimized proof* from the ELK output. This minimization can be done w.r.t. different measures, such as the size, depth, or weighted sum (where each axiom is weighted by its size), as long as they are *monotone* and *recursive* [5]. For ontologies outside of the $\mathcal{ELH}$ fragment, we use the forgetting-based approach originally described in [3], for which we now implemented two alternative algorithms for computing more compact proofs (Sect. 4.1). Finally, independently of the proof generation method, one can specify a signature of known terms. This signature contains terminology that the user is familiar with, so that entailments using only those terms do not need to be explained. The condensation of proofs w.r.t. signatures is described in Sect. 4.2.

4.1 Forgetting-Based Proofs

In a forgetting-based proof, proof steps represent inferences on concept or role names using a *forgetting* operation. Given an ontology $\mathcal{O}$ and a predicate name x, the result $\mathcal{O}^{-x}$ of forgetting x in $\mathcal{O}$ does not contain any occurrences of x, while still capturing all entailments of $\mathcal{O}$ that do not use x [18]. In a forgetting-based proof, an inference takes as premises a set $\mathcal{P}$ of axioms and has as conclusion some axiom $\alpha \in \mathcal{P}^{-x}$ (where a particular forgetting operation is used to compute $\mathcal{P}^{-x}$). Intuitively, α is obtained from $\mathcal{P}$ by performing inferences on x. To compute a forgetting-based proof, we have to forget the names occuring in the ontology one after the other, until only the names occurring in the statement to be proved are left. For the forgetting operation, the user can select between two implementations: LETHE [17] (using the method supporting $\mathcal{ALCH}$) and FAME [21] (using the method supporting $\mathcal{ALCOI}$). Since the space of possible inference steps is exponentially large, it is not feasible to minimize proofs after their computation, as we do for $\mathcal{EL}$ entailments, which is why we rely on heuristics and search algorithms to generate small proofs. Specifically, we implemented three methods for computing forgetting-based proofs: `HEUR` tries to find proofs fast, `SYMB` tries to minimize the number of predicates forgotten in a proof, with the aim of obtaining proofs of small depth, and `SIZE` tries to optimize the size of the proof. The heuristic method `HEUR` is described in [3], and its implementation has not been changed since then. The search methods `SYMB` and `SIZE` are new (details can be found in the extended version [1]).

4.2 Signature-Based Proof Condensation

When inspecting a proof over a real-world ontology, different parts of the proof will be more or less familiar to the user, depending on their knowledge about the involved concepts or their experience with similar inference steps in the past. For CIs between concepts for which a user has application knowledge, they may not need to see a proof, and consequently, sub-proofs for such axioms can be automatically hidden. We assume that the user's knowledge is given in the form of a *known signature* Σ and that axioms that contain only symbols from Σ do not need to be explained. The effect can be seen in Fig. 1 through the "known"-inference on the left, where Σ contains SebaceousGland and Gland. The known signature is taken into consideration when minimizing the proofs, so that proofs are selected for which more of the known information can be used if convenient. This can be easily integrated into the Dijsktra approach described in [3], by initially assigning to each axiom covered by Σ a proof with a single vertex.

5 Evaluation

For EVONNE to be usable in practice, it is vital that proofs are computed efficiently and that they are not too large. An experimental evaluation of minimized proofs for $\mathcal{EL}$ and forgetting-based proofs obtained with FAME and LETHE is provided in [3]. We here present an evaluation of additional aspects: 1) a comparison of the three methods for computing forgetting-based proofs, and 2) an evaluation on the impact of signature-based proof condensation. All experiments were performed on Debian Linux (Intel Core i5-4590, 3.30 GHz, 23 GB Java heap size).

5.1 Minimal Forgetting-Based Proofs

To evaluate forgetting-based proofs, we extracted $\mathcal{ALCH}$ "proof tasks" from the ontologies in the 2017 snapshot of BioPortal [19]. We restricted all ontologies

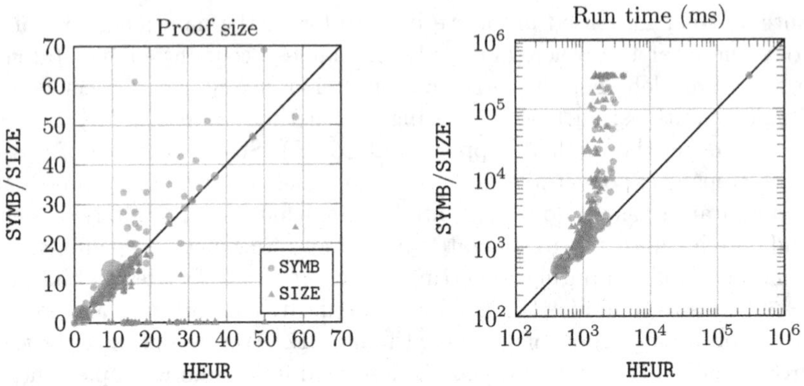

Fig. 2. Run times and proof sizes for different forgetting-based proof methods. Marker size indicates how often each pattern occurred in the BioPortal snapshot. Instances that timed out were assigned size 0.

to $\mathcal{ALCH}$ and collected all entailed atomic CIs α, for each of which we computed the union $\mathcal{U}$ of all their justifications. We identified pairs $(\alpha, \mathcal{U})$ that were isomorphic modulo renaming of predicates, and kept only those patterns $(\alpha, \mathcal{U})$ that contained at least one axiom not expressible in $\mathcal{ELH}$. This was successful in 373 of the ontologies[2] and resulted in 138 distinct *justification patterns* $(\alpha, \mathcal{U})$, representing 327 different entailments in the BioPortal snapshot. We then computed forgetting-based proofs for $\mathcal{U} \models \alpha$ with our three methods using LETHE, with a 5-minute timeout. This was successful for 325/327 entailments for the heuristic method (`HEUR`), 317 for the symbol-minimizing method (`SYMB`), and 279 for the size-minimizing method (`SIZE`). In Fig. 2 we compare the resulting *proof sizes* (left) and the *run times* (right), using `HEUR` as baseline (x-axis). `HEUR` is indeed faster in most cases, but `SIZE` reduces proof size by 5% on average compared to `HEUR`, which is not the case for `SYMB`. Regarding *proof depth* (not shown in the figure), `SYMB` did not outperform `HEUR` on average, while `SIZE` surprisingly yielded an average reduction of 4% compared to `HEUR`. Despite this good performance of `SIZE` for proof size and depth, for entailments that depend on many or complex axioms, computation times for both `SYMB` and `SIZE` become unacceptable, while proof generation with `HEUR` mostly stays in the area of seconds.

5.2 Signature-Based Proof Condensation

To evaluate how much hiding proof steps in a known signature decreases proof size in practice, we ran experiments on the large medical ontology SNOMED CT (International Edition, July 2020) that is mostly formulated in $\mathcal{ELH}$.[3] As signatures we used SNOMED CT *Reference Sets*,[4] which are restricted vocabularies

[2] The other ontologies could not be processed in this way within the memory limit.
[3] https://www.snomed.org/.
[4] https://confluence.ihtsdotools.org/display/DOCRFSPG/2.3.+Reference+Set.

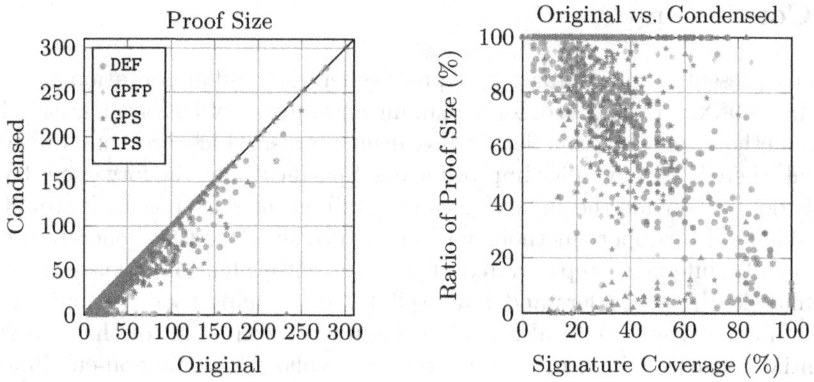

Fig. 3. Size of original and condensed proofs (left). Ratio of proof size depending on the signature coverage (right).

for specific use cases. We extracted justifications similarly to the previous experiment, but did not rename predicates and considered only proof tasks that use at least 5 symbols from the signature, since otherwise no improvement can be expected by using the signatures. For each signature, we randomly selected 500 out of 6.689.452 *proof tasks* (if at least 500 existed). This left the 4 reference sets *General Practitioner/Family Practitioner* (GPFP), *Global Patient Set* (GPS), *International Patient Summary* (IPS), and the one included in the SNOMED CT distribution (DEF). For each of the resulting 2.000 proof tasks, we used ELK [16] and our proof minimization approach to obtain (a) a proof of minimal size and (b) a proof of minimal size after hiding the selected signature. The distribution of proof sizes can be seen in Fig. 3. In 770/2.000 cases, a smaller proof was generated when using the signature. In 91 of these cases, the size was even be reduced to 1, i.e. the target axiom used only the given signature and therefore nothing else needed to be shown. In the other 679 cases with reduced size, the average *ratio* of reduced size to original size was 0.68–0.93 (depending on the signature). One can see that this ratio is correlated with the *signature coverage* of the original proof (i.e. the ratio of signature symbols to total symbols in the proof), with a weak or strong correlation depending on the signature (r between -0.26 and -0.74). However, a substantial number of proofs with relatively high signature coverage could still not be reduced in size at all (see the top right of the right diagram). In summary, we can see that signature-based condensation can be useful, but this depends on the proof task and the signature. We also conducted experiments on the Galen ontology,[5] with comparable results (see the extended version of this paper [1]).

[5] https://bioportal.bioontology.org/ontologies/GALEN.

6 Conclusion

We have presented and compared the proof generation and presentation methods used in Evonne, a visual tool for explaining entailments of DL ontologies. While these methods produce smaller or less deep proofs, which are thus easier to present, there is still room for improvements. Specifically, as the forgetting-based proofs do not provide the same degree of detail as the Elk proofs, it would be desirable to also support methods for more expressive DLs that generate proofs with smaller inference steps. Moreover, our current evaluation focuses on proof size and depth—to understand how well Evonne helps users to understand DL entailments, we would also need a qualitative evaluation of the tool with potential end-users. We are also working on explanations for non-entailments using countermodels [7] and a plugin for the ontology editor Protégé that is compatible with the PULi library and Proof Explanation plugin presented in [15], which will support all proof generation methods discussed here and more.[6]

Acknowledgements. This work was supported by the German Research Foundation (DFG) in Germany's Excellence Strategy: EXC-2068, 390729961 - Cluster of Excellence "Physics of Life" and EXC 2050/1, 390696704 - Cluster of Excellence "Centre for Tactile Internet" (CeTI) of TU Dresden, by DFG grant 389792660 as part of TRR 248 - CPEC, by the AI competence center ScaDS.AI Dresden/Leipzig, and the DFG Research Training Group QuantLA, GRK 1763.

References

1. Alrabbaa, C., Baader, F., Borgwardt, S., Dachselt, R., Koopmann, P., Méndez, J.: Evonne: interactive proof visualization for description logics (system description) - extended version (2022). https://doi.org/10.48550/ARXIV.2205.09583
2. Alrabbaa, C., Baader, F., Borgwardt, S., Dachselt, R., Koopmann, P., Méndez, J.: Evonne: interactive proof visualization for description logics (system description) - IJCAR22 - resources, May 2022. https://doi.org/10.5281/zenodo.6560603
3. Alrabbaa, C., Baader, F., Borgwardt, S., Koopmann, P., Kovtunova, A.: Finding small proofs for description logic entailments: theory and practice. In: Albert, E., Kovács, L. (eds.) Proceedings of the 23rd International Conference on Logic for Programming, Artificial Intelligence and Reasoning (LPAR 2020). EPiC Series in Computing, vol. 73, pp. 32–67. EasyChair (2020). https://doi.org/10.29007/nhpp
4. Alrabbaa, C., Baader, F., Borgwardt, S., Koopmann, P., Kovtunova, A.: On the complexity of finding good proofs for description logic entailments. In: Borgwardt, S., Meyer, T. (eds.) Proceedings of the 33rd International Workshop on Description Logics (DL 2020). CEUR Workshop Proceedings, vol. 2663. CEUR-WS.org (2020). http://ceur-ws.org/Vol-2663/paper-1.pdf
5. Alrabbaa, C., Baader, F., Borgwardt, S., Koopmann, P., Kovtunova, A.: Finding good proofs for description logic entailments using recursive quality measures. In: Platzer, A., Sutcliffe, G. (eds.) CADE 2021. LNCS (LNAI), vol. 12699, pp. 291–308. Springer, Cham (2021). https://doi.org/10.1007/978-3-030-79876-5_17

[6] https://github.com/de-tu-dresden-inf-lat/evee.

6. Alrabbaa, C., Baader, F., Dachselt, R., Flemisch, T., Koopmann, P.: Visualising proofs and the modular structure of ontologies to support ontology repair. In: Borgwardt, S., Meyer, T. (eds.) Proceedings of the 33rd International Workshop on Description Logics (DL 2020). CEUR Workshop Proceedings, vol. 2663. CEUR-WS.org (2020). http://ceur-ws.org/Vol-2663/paper-2.pdf
7. Alrabbaa, C., Hieke, W., Turhan, A.: Counter model transformation for explaining non-subsumption in $\mathcal{EL}$. In: Beierle, C., Ragni, M., Stolzenburg, F., Thimm, M. (eds.) Proceedings of the 7th Workshop on Formal and Cognitive Reasoning. CEUR Workshop Proceedings, vol. 2961, pp. 9–22. CEUR-WS.org (2021). http://ceur-ws.org/Vol-2961/paper_2.pdf
8. Baader, F., Horrocks, I., Lutz, C., Sattler, U.: An Introduction to Description Logic. Cambridge University Press, Cambridge (2017). https://doi.org/10.1017/9781139025355
9. Fiedler, A.: Natural language proof explanation. In: Hutter, D., Stephan, W. (eds.) Mechanizing Mathematical Reasoning. LNCS (LNAI), vol. 2605, pp. 342–363. Springer, Heidelberg (2005). https://doi.org/10.1007/978-3-540-32254-2_20
10. Flemisch, T., Langner, R., Alrabbaa, C., Dachselt, R.: Towards designing a tool for understanding proofs in ontologies through combined node-link diagrams. In: Ivanova, V., Lambrix, P., Pesquita, C., Wiens, V. (eds.) Proceedings of the Fifth International Workshop on Visualization and Interaction for Ontologies and Linked Data (VOILA 2020). CEUR Workshop Proceedings, vol. 2778, pp. 28–40. CEUR-WS.org (2020). http://ceur-ws.org/Vol-2778/paper3.pdf
11. Horacek, H.: Presenting proofs in a human-oriented way. In: CADE 1999. LNCS (LNAI), vol. 1632, pp. 142–156. Springer, Heidelberg (1999). https://doi.org/10.1007/3-540-48660-7_10
12. Horridge, M., Parsia, B., Sattler, U.: Explanation of OWL entailments in Protege 4. In: Bizer, C., Joshi, A. (eds.) Proceedings of the Poster and Demonstration Session at the 7th International Semantic Web Conference (ISWC 2008). CEUR Workshop Proceedings, vol. 401. CEUR-WS.org (2008). http://ceur-ws.org/Vol-401/iswc2008pd_submission_47.pdf
13. Horridge, M., Parsia, B., Sattler, U.: Justification oriented proofs in OWL. In: Patel-Schneider, P.F., Pan, Y., Hitzler, P., Mika, P., Zhang, L., Pan, J.Z., Horrocks, I., Glimm, B. (eds.) ISWC 2010. LNCS, vol. 6496, pp. 354–369. Springer, Heidelberg (2010). https://doi.org/10.1007/978-3-642-17746-0_23
14. Hyland, I., Schmidt, R.A.: Protégé-TS: An OWL ontology term selection tool. In: Borgwardt, S., Meyer, T. (eds.) Proceedings of the 33rd International Workshop on Description Logics (DL 2020). CEUR Workshop Proceedings, vol. 2663. CEUR-WS.org (2020). http://ceur-ws.org/Vol-2663/paper-12.pdf
15. Kazakov, Y., Klinov, P., Stupnikov, A.: Towards reusable explanation services in protege. In: Artale, A., Glimm, B., Kontchakov, R. (eds.) Proceedings of the 30th International Workshop on Description Logics (DL 2017). CEUR Workshop Proceedings, vol. 1879. CEUR-WS.org (2017). http://ceur-ws.org/Vol-1879/paper31.pdf
16. Kazakov, Y., Krötzsch, M., Simancik, F.: The incredible ELK - from polynomial procedures to efficient reasoning with $\mathcal{EL}$ ontologies. J. Autom. Reason. **53**(1), 1–61 (2014). https://doi.org/10.1007/s10817-013-9296-3
17. Koopmann, P.: LETHE: forgetting and uniform interpolation for expressive description logics. Künstliche Intell. **34**(3), 381–387 (2020). https://doi.org/10.1007/s13218-020-00655-w

18. Koopmann, P., Schmidt, R.A.: Forgetting concept and role symbols in $\mathcal{ALCH}$-ontologies. In: McMillan, K., Middeldorp, A., Voronkov, A. (eds.) LPAR 2013. LNCS, vol. 8312, pp. 552–567. Springer, Heidelberg (2013). https://doi.org/10.1007/978-3-642-45221-5_37
19. Matentzoglu, N., Parsia, B.: Bioportal snapshot 30.03.2017, March 2017. https://doi.org/10.5281/zenodo.439510
20. Reger, G., Suda, M.: Checkable proofs for first-order theorem proving. In: Reger, G., Traytel, D. (eds.) 1st International Workshop on Automated Reasoning: Challenges, Applications, Directions, Exemplary Achievements (ARCADE 2017). EPiC Series in Computing, vol. 51, pp. 55–63. EasyChair (2017). https://doi.org/10.29007/s6d1
21. Zhao, Y., Schmidt, R.A.: FAME: an automated tool for semantic forgetting in expressive description logics. In: Galmiche, D., Schulz, S., Sebastiani, R. (eds.) IJCAR 2018. LNCS (LNAI), vol. 10900, pp. 19–27. Springer, Cham (2018). https://doi.org/10.1007/978-3-319-94205-6_2

Actions over Core-Closed Knowledge Bases

Claudia Cauli[1,2(✉)], Magdalena Ortiz[3], and Nir Piterman[1]

[1] University of Gothenburg, Gothenburg, Sweden
claudiacauli@gmail.com
[2] Amazon Web Services, Seattle, USA
[3] TU Wien, Vienna, Austria

Abstract. We present new results on the application of semantic- and knowledge-based reasoning techniques to the analysis of cloud deployments. In particular, to the security of *Infrastructure as Code* configuration files, encoded as description logic knowledge bases. We introduce an action language to model *mutating actions*; that is, actions that change the structural configuration of a given deployment by adding, modifying, or deleting resources. We mainly focus on two problems: the problem of determining whether the execution of an action, no matter the parameters passed to it, will not cause the violation of some security requirement (*static verification*), and the problem of finding sequences of actions that would lead the deployment to a state where (un)desirable properties are (not) satisfied (*plan existence* and *plan synthesis*). For all these problems, we provide definitions, complexity results, and decision procedures.

1 Introduction

The use of automated reasoning techniques to analyze the properties of cloud infrastructure is gaining increasing attention [4–7,18]. Despite that, more effort needs to be put into the modeling and verification of generic security requirements over cloud infrastructure pre-deployment. The availability of formal techniques, providing strong security guarantees, would assist complex system-level analyses such as threat modeling and data flow, which now require considerable time, manual intervention, and expert domain knowledge.

We continue our research on the application of semantic-based and knowledge-based reasoning techniques to cloud deployment *Infrastructure as Code* configuration files. In [14], we reported on our experience using expressive description logics to model and reason about Amazon Web Services' proprietary Infrastructure as Code framework (AWS CloudFormation). We used the rich constructs of these logics to encode domain knowledge, simulate closed-world reasoning, and express mitigations and exposures to security threats. Due to the high complexity of basic tasks [3,26], we found reasoning in such a framework to be not efficient at cloud scale. In [15], we introduced *core-closed knowledge*

C. Cauli—This work was done prior to joining Amazon.

J. Blanchette et al. (Eds.): IJCAR 2022, LNAI 13385, pp. 281–299, 2022.
https://doi.org/10.1007/978-3-031-10769-6_17

bases—a lightweight description logic combining closed- and open-world reasoning that is tailored to model cloud infrastructure and efficiently query its security properties. Core-closed knowledge bases enable partially-closed predicates whose interpretation is closed over a *core* part of the knowledge base but open elsewhere. To encode potential exposure to security threats, we studied the query satisfiability problem and (together with the usual query entailment problem) applied it to a new class of conjunctive queries that we called MUST/MAY queries. We were able to answer such queries over core-closed knowledge bases in LOGSPACE in data complexity and NP in combined complexity, improving the required NEXPTIME complexity for satisfiability over $\mathcal{ALCOIQ}$ (used in [14]).

Here, we enhance the quality of the analyses done over pre-deployment artifacts, giving users and practitioners additional precise insights on the impact of potential changes, fixes, and general improvements to their cloud projects. We enrich core-closed knowledge bases with the notion of *core-completeness*, which is needed to ensure that updates are consistent. We define the syntax and semantics of an action language that is expressive enough to encode *mutating* API calls, i.e., operations that change a cloud deployment configuration by creating, modifying, or deleting existing resources. As part of our effort to improve the quality of automated analysis, we also provide relevant reasoning tools to identify and predict the consequences of these changes. To this end, we consider procedures that determine whether the execution of a mutating action always preserves given properties (*static verification*); determine whether there exists a sequence of operations that would lead a deployment to a configuration meeting certain requirements (*plan existence*); and find such sequences of operations (*plan synthesis*).

The paper is organized as follows. In Sect. 2, we provide background on core-closed knowledge bases, conjunctive queries, and MUST/MAY queries. In Sect. 3, we motivate and introduce the notion of *core-completeness*. In Sect. 4, we define the action language. In Sect. 5, we describe the static verification problem and characterize its complexity. In Sect. 6, we address the planning problem and concentrate on the synthesis of minimal plans satisfying a given requirement expressed using MUST/MAY queries. We discuss related works in Sect. 7 and conclude in Sect. 8. Results and proofs that are omitted in this paper are found in the full version [16].

2 Background

Description logics (DLs) are a family of logics for encoding knowledge in terms of concepts, roles, and individuals; analogous to first-order logic unary predicates, binary predicates, and constants, respectively. Standard DL knowledge bases (KBs) have a set of axioms, called *TBox*, and a set of assertions, called *ABox*. The TBox contains axioms that relate to concepts and roles. The ABox contains assertions that relate individuals to concepts and pairs of individuals to roles. KBs are usually interpreted under the open-world assumption, meaning that the asserted facts are not assumed to be complete.

Core-Closed Knowledge Bases. In [15], we introduced core-closed knowledge bases (ccKBs) as a suitable description logic formalism to encode cloud deployments. The main characteristic of ccKBs is to allow for a combination of open- and closed-world reasoning that ensures tractability. A DL-Lite$^{\mathcal{F}}$ ccKB is the tuple $\mathcal{K} = \langle \mathcal{T}, \mathcal{A}, \mathcal{S}, \mathcal{M} \rangle$ built from the standard knowledge base $\langle \mathcal{T}, \mathcal{A} \rangle$ and the *core* system $\langle \mathcal{S}, \mathcal{M} \rangle$. The former encodes incomplete terminological and assertional knowledge. The latter is, in turn, composed of two parts: $\mathcal{S}$ (also called the *SBox*), containing axioms that encode the core structural specifications, and $\mathcal{M}$ (also called the *MBox*), containing positive concept and role assertions that encode the core configuration. Syntactically, $\mathcal{M}$ is similar to an ABox but, semantically, is assumed to be complete with respect to the specifications in $\mathcal{S}$.

The ccKB $\mathcal{K}$ is defined over the alphabets $\mathbf{C}$ (of concepts), $\mathbf{R}$ (of roles), and $\mathbf{I}$ (of individuals), all partitioned into an open subset and a partially-closed subset. That is, the set of concepts is partitioned into the open concepts $\mathbf{C}^{\mathcal{K}}$ and the closed (specification) concepts $\mathbf{C}^{\mathcal{S}}$; the set of roles is partitioned into open roles $\mathbf{R}^{\mathcal{K}}$ and closed (specification) roles $\mathbf{R}^{\mathcal{S}}$; and the set of individuals is partitioned into open individuals $\mathbf{I}^{\mathcal{K}}$ and closed (model) individuals $\mathbf{I}^{\mathcal{M}}$. We call $\mathbf{C}^{\mathcal{S}}$ and $\mathbf{R}^{\mathcal{S}}$ core-closed predicates, or partially-closed predicates, as their extension is closed over the core domain $\mathbf{I}^{\mathcal{M}}$ and open otherwise. In contrast, we call $\mathbf{C}^{\mathcal{K}}$ and $\mathbf{R}^{\mathcal{K}}$ open predicates. The syntax of concept and role expressions in DL-Lite$^{\mathcal{F}}$ [2,8] is as follows:

$$\mathsf{B} ::= \bot \mid \mathsf{A} \mid \exists \mathsf{p}$$

where A denotes a concept name and p is either a role name r or its inverse r^-. The syntax of axioms provides for the three following axioms:

$$\mathsf{B}^1 \sqsubseteq \mathsf{B}^2, \qquad \mathsf{B}^1 \sqsubseteq \neg \mathsf{B}^2, \qquad (\mathsf{funct}\ \mathsf{p}),$$

respectively called: *positive inclusion* axioms, *negative inclusion* axioms, and *functionality* axioms. These axioms are contained in the sets $\mathcal{S}$ and $\mathcal{T}$. To precisely denote the subsets of $\mathcal{S}$ and $\mathcal{T}$ having only axioms of a given type we use the notation $PI_{\mathcal{X}}$, $NI_{\mathcal{X}}$, and $F_{\mathcal{X}}$, for $\mathcal{X} \in \{\mathcal{S}, \mathcal{T}\}$, which respectively contain only positive inclusion axioms, negative inclusion axioms, and functionality axioms. From now on, we denote symbols from the alphabet $\mathbf{X}^{\mathcal{X}}$ with the subscript $\mathcal{X}$, and symbols from the generic alphabet $\mathbf{X}$ with no subscript. In core-closed knowledge bases, axioms and assertions fall into the scope of a different set depending on the predicates and individuals that they refer to, according to the set definitions below.

$$\begin{aligned}
\mathcal{M} &\subseteq \{\mathsf{A}_{\mathcal{S}}(a_{\mathcal{M}}),\ \mathsf{R}_{\mathcal{S}}(a_{\mathcal{M}}, a),\ \mathsf{R}_{\mathcal{S}}(a, a_{\mathcal{M}})\} \\
\mathcal{A} &\subseteq \{\mathsf{A}_{\mathcal{K}}(a_{\mathcal{K}}),\ \mathsf{R}_{\mathcal{K}}(a_{\mathcal{K}}, b_{\mathcal{K}}),\ \mathsf{A}_{\mathcal{S}}(a_{\mathcal{K}}),\ \mathsf{R}_{\mathcal{S}}(a_{\mathcal{K}}, b_{\mathcal{K}})\} \\
\mathcal{S} &\subseteq \{\mathsf{B}^1_{\mathcal{S}} \sqsubseteq \mathsf{B}^2_{\mathcal{S}},\ \mathsf{B}^1_{\mathcal{S}} \sqsubseteq \neg \mathsf{B}^2_{\mathcal{S}},\ \mathsf{Func}(\mathsf{P}_{\mathcal{S}})\} \\
\mathcal{T} &\subseteq \{\mathsf{B}^1 \sqsubseteq \mathsf{B}^2_{\mathcal{K}},\ \mathsf{B}^1 \sqsubseteq \neg \mathsf{B}^2_{\mathcal{K}},\ \mathsf{Func}(\mathsf{P}_{\mathcal{K}})\}
\end{aligned}$$

In the above definition of the set $\mathcal{M}$, role assertions link at least one individual from the core domain $\mathbf{I}^{\mathcal{M}}$ (denoted as $a_{\mathcal{M}}$) to one individual from the general set

$\mathbf{I}$ (denoted as a). Node a could either be an individual from the open partition $\mathbf{I}^{\mathcal{K}}$ or the closed partition $\mathbf{I}^{\mathcal{M}}$. When a is an element from the set $\mathbf{I}^{\mathcal{K}}$, we refer to it as a "boundary node", as it sits at the boundary between the core and the open parts of the knowledge base. As mentioned earlier, $\mathcal{M}$-assertions are assumed to be complete and consistent with respect to the terminological knowledge given in $\mathcal{S}$; whereas the usual open-world assumption is made for $\mathcal{A}$-assertions. The semantics of a DL-Lite$^{\mathcal{F}}$ core-closed KB is given in terms of interpretations $\mathcal{I}$, consisting of a non-empty domain $\Delta^{\mathcal{I}}$ and an interpretation function $\cdot^{\mathcal{I}}$. The latter assigns to each concept A a subset $\mathsf{A}^{\mathcal{I}}$ of $\Delta^{\mathcal{I}}$, to each role r a subset $\mathsf{r}^{\mathcal{I}}$ of $\Delta^{\mathcal{I}} \times \Delta^{\mathcal{I}}$, and to each individual a a node $a^{\mathcal{I}}$ in $\Delta^{\mathcal{I}}$, and it is extended to concept expressions in the usual way. An interpretation $\mathcal{I}$ is a model of an inclusion axiom $\mathsf{B}_1 \sqsubseteq \mathsf{B}_2$ if $\mathsf{B}_1^{\mathcal{I}} \subseteq \mathsf{B}_2^{\mathcal{I}}$. An interpretation $\mathcal{I}$ is a model of a membership assertion $\mathsf{A}(a)$, (resp. $\mathsf{r}(a, b)$) if $a^{\mathcal{I}} \in \mathsf{A}^{\mathcal{I}}$ (resp. $(a^{\mathcal{I}}, b^{\mathcal{I}}) \in \mathsf{r}^{\mathcal{I}}$). We say that $\mathcal{I}$ models $\mathcal{T}$, $\mathcal{S}$, and $\mathcal{A}$ if it models all axioms or assertions contained therein. We say that $\mathcal{I}$ models $\mathcal{M}$, denoted $\mathcal{I} \models^{\mathsf{CWA}} \mathcal{M}$, when it models an $\mathcal{M}$-assertion f *if and only if* $f \in \mathcal{M}$. Finally, $\mathcal{I}$ models $\mathcal{K}$ if it models $\mathcal{T}$, $\mathcal{S}$, $\mathcal{A}$, and $\mathcal{M}$. When $\mathcal{K}$ has at least one model, we say that $\mathcal{K}$ is satisfiable.

In the remainder of this paper, we will sometimes refer to the *lts* interpretation of $\mathcal{M}$. The *lts* interpretation of $\mathcal{M}$, denoted $lts(\mathcal{M})$, is the interpretation $(\Delta^{lts(\mathcal{M})}, \cdot^{lts(\mathcal{M})})$ defined only over concept and role names from the set $\mathbf{C}^{\mathcal{S}}$ and $\mathbf{R}^{\mathcal{S}}$, respectively, and over individual names from $\mathbf{I}^{\mathcal{K}}$ that appear in the scope of $\mathcal{M}$-assertions. The interpretation $lts(\mathcal{M})$ is the *unique* model of $\mathcal{M}$ such that $lts(\mathcal{M}) \models^{\mathsf{CWA}} \mathcal{M}$.

In the application presented in [14], description logic KBs are used to encode machine-readable deployment files containing multiple resource declarations. Every resource declaration has an underlying tree structure, whose leaves can potentially link to the roots of other resource declarations. Let $\mathbf{I}^r \subseteq \mathbf{I}^{\mathcal{M}}$ be the set of all resource nodes, we encode their resource declarations in $\mathcal{M}$, and formalize the resulting forest structure by partitioning $\mathcal{M}$ into multiple subsets $\{\mathcal{M}_i\}_{i \in \mathbf{I}^r}$, each representing a tree of assertions rooted at a resource node i (we generally refer to constants in $\mathcal{M}$ as nodes). For the purpose of this work, we will refer to core-closed knowledge bases where $\mathcal{M}$ is partitioned as described; that is, ccKBs such that $\mathcal{K} = \langle \mathcal{T}, \mathcal{A}, \mathcal{S}, \{\mathcal{M}_i\}_{i \in \mathbf{I}^r} \rangle$.

Conjunctive Queries. A *conjunctive query* (CQ) is an existentially-quantified formula $q[\vec{x}]$ of the form $\exists \vec{y}.conj(\vec{x}, \vec{y})$, where *conj* is a conjunction of positive atoms and potentially inequalities. A *union of conjunctive queries* (UCQ) is a disjunction of CQs. The variables in $\vec{x}$ are called *answer variables*, those in $\vec{y}$ are the existentially-quantified *query variables.* A tuple $\vec{c}$ of constants appearing in the knowledge base $\mathcal{K}$ is an answer to q if for all interpretations $\mathcal{I}$ model of $\mathcal{K}$ we have $\mathcal{I} \models q[\vec{c}]$. We call these tuples the *certain answers* of q over $\mathcal{K}$, denoted $ans(\mathcal{K}, q)$, and the problem of testing whether a tuple is a certain answer *query entailment.* A tuple $\vec{c}$ of constants appearing in $\mathcal{K}$ satisfies q if there exists an interpretation $\mathcal{I}$ model of $\mathcal{K}$ such that $\mathcal{I} \models q[\vec{c}]$. We call these tuples the *sat answers* of q over $\mathcal{K}$, denoted $sat{-}ans(\mathcal{K}, q)$, and the problem of testing whether a given tuple is a sat answer *query satisfiability.*

MUST/MAY *Queries.* A MUST/MAY query ψ [15] is a Boolean combination of nested UCQs in the scope of a MUST or a MAY operator as follows:

$$\psi ::= \neg\psi \mid \psi_1 \wedge \psi_2 \mid \psi_1 \vee \psi_2 \mid \text{MUST } \varphi \mid \text{MAY } \varphi_{\not\approx}$$

where φ and $\varphi_{\not\approx}$ are unions of conjunctive queries potentially containing inequalities. The reasoning needed for answering the nested queries can be decoupled from the reasoning needed to answer the higher-level formula: nested queries MUST φ are reduced to conjunctive query entailment, and nested queries MAY $\varphi_{\not\approx}$ are reduced to conjunctive query satisfiability. We denote by $\mathsf{ANS}(\psi, \mathcal{K})$ the answers of a MUST/MAY query ψ over the core-closed knowledge base $\mathcal{K}$.

3 Core-Complete Knowledge Bases

The algorithm Consistent presented in [15] computes satisfiability of DL-Lite$^{\mathcal{F}}$ core-closed knowledge bases relying on the assumption that $\mathcal{M}$ is complete and consistent with respect to $\mathcal{S}$. Such an assumption effectively means that the information contained in $\mathcal{M}$ is *explicitly* present and *cannot be completed by inference.* The algorithm relies on the existence of a theoretical object, the canonical interpretation, in which missing assertions can always be introduced when they are logically implied by the positive inclusion axioms. As a matter of fact, positive inclusion axioms are not even included in the inconsistency formula built for the satisfiability check, as it is proven that the canonical interpretation always satisfies them ([15], Lemma 3). When the assumption that $\mathcal{M}$ is consistent with respect to $\mathcal{S}$ is dropped, the algorithm Consistent becomes insufficient to check satisfiability. We illustrate this with an example.

Example 1 (Required Configuration). Let us consider the axioms constraining the AWS resource type S3::Bucket. In particular, the $\mathcal{S}$-axiom S3::Bucket $\sqsubseteq$ $\exists$loggingConfiguration prescribing that all buckets must have a *required* logging configuration. For a set $\mathcal{M} = \{\text{S3::Bucket}(b)\}$, according to the partially-closed semantics of core-closed knowledge bases, the absence of an assertion loggingConfiguration(b, x), for some x, is interpreted as the assertion being false in $\mathcal{M}$, which is therefore not consistent with respect to $\mathcal{S}$. However, the algorithm Consistent will check the *lts* interpretation of $\mathcal{M}$ for an empty formula (as there are no negative inclusion or functionality axioms) and return *true.*

In essence, the algorithm Consistent does not compute the full satisfiability of the whole core-closed knowledge base, but only of its open part. Satisfiability of $\mathcal{M}$ with respect to the positive inclusion axioms in $\mathcal{S}$ needs to be checked separately. We introduce a new notion to denote when a set $\mathcal{M}$ is complete with respect to $\mathcal{S}$ that is distinct from the notion of consistency. Let $\mathcal{K} = \langle \mathcal{T}, \mathcal{A}, \mathcal{S}, \mathcal{M} \rangle$ be a DL-Lite$^{\mathcal{F}}$ core-closed knowledge base; we say that $\mathcal{K}$ is *core-complete* when $\mathcal{M}$ models *all* positive inclusion axioms in $\mathcal{S}$ under a closed-world assumption; we say that $\mathcal{K}$ is *open-consistent* when $\mathcal{M}$ and $\mathcal{A}$ model all negative inclusion and functionality axioms in $\mathcal{K}$'s negative inclusion closure. Finally, we say that $\mathcal{K}$ is *fully satisfiable* when is both *core-complete* and *open-consistent.*

Lemma 1. *In order to check* full satisfiability *of a DL-Lite$^{\mathcal{F}}$ core-closed KB, one simply needs to check if $\mathcal{K}$ is* core-complete *(that is, if $\mathcal{M}$ models all* positive axioms *in $\mathcal{S}$ under a closed-world assumption) and if $\mathcal{K}$ is* open-consistent *(that is, to run the algorithm* Consistent*).*

Proof. Dropping the assumption that $\mathcal{M}$ is consistent w.r.t. $\mathcal{S}$ causes Lemma 3 from [15] to fail. In particular, the canonical interpretation of $\mathcal{K}$, $can(\mathcal{K})$, would still be a model of $PI_{\mathcal{T}}$, $\mathcal{A}$, and $\mathcal{M}$, but may *not* be a model of $PI_{\mathcal{S}}$. This is due to the construction of the canonical model that is based on the notion of applicable axioms. In rules **c5-c8** of [15] Definition 1, axioms in $PI_{\mathcal{S}}$ are defined as applicable to assertions involving open nodes $a_{\mathcal{K}}$ but *not* to model nodes $a_{\mathcal{M}}$ in $\mathbf{I}^{\mathcal{M}}$. As a result, if the implications of such axioms on model nodes are not included in $\mathcal{M}$ itself, then they will not be included in $can(\mathcal{K})$ either, and $can(\mathcal{K})$ will not be a model of $PI_{\mathcal{S}}$. On the other hand, one can easily verify that Lemmas 1,2,4,5,6,7 and Corollary 1 would still hold as they do not rely on the assumption. However, since it is not guaranteed anymore that $\mathcal{M}$ satisfies all positive inclusion axioms from $\mathcal{S}$, the *if* direction of [15] Theorem 1 does not hold anymore: there can be an unsatisfiable ccKB $\mathcal{K}$ such that $db(\mathcal{A}) \cup lts(\mathcal{M}) \models cln(\mathcal{T} \cup \mathcal{S}), \mathcal{A}, \mathcal{M}$. For instance, the knowledge base from Example 1. We also note that the negative inclusion and functionality axioms from $\mathcal{S}$ will be checked anyway by the consistency formula, both on $db(\mathcal{A})$ and on $lts(\mathcal{M})$.

Lemma 2. *Checking whether a DL-Lite$^{\mathcal{F}}$ core-closed knowledge base is* core-complete *can be done in polynomial time in $\mathcal{M}$. As a consequence, checking full satisfiability is also done in polynomial time in $\mathcal{M}$.*

Proof. One can write an algorithm that checks *core-completeness* by searching for the existence of a positive inclusion axiom $\mathsf{B}^1_{\mathcal{S}} \sqsubseteq \mathsf{B}^2_{\mathcal{S}} \in PI_{\mathcal{S}}$ such that $\mathcal{M} \models \mathsf{B}^1_{\mathcal{S}}(a_{\mathcal{M}})$ and $\mathcal{M} \not\models \mathsf{B}^2_{\mathcal{S}}(a_{\mathcal{M}})$, where the relation $\models$ is defined over DL-Lite$^{\mathcal{F}}$ concept expressions as follows:

$$
\begin{aligned}
\mathcal{M} &\models \bot(a_{\mathcal{M}}) &&\leftrightarrow\quad \mathit{false}\\
\mathcal{M} &\models \mathsf{A}_{\mathcal{S}}(a_{\mathcal{M}}) &&\leftrightarrow\quad \mathsf{A}_{\mathcal{S}}(a_{\mathcal{M}}) \in \mathcal{M}\\
\mathcal{M} &\models \exists \mathsf{r}_{\mathcal{S}}(a_{\mathcal{M}}) &&\leftrightarrow\quad \exists b.\ \mathsf{r}_{\mathcal{S}}(a_{\mathcal{M}}, b) \in \mathcal{M}\\
\mathcal{M} &\models \exists \mathsf{r}^-_{\mathcal{S}}(a_{\mathcal{M}}) &&\leftrightarrow\quad \exists b.\ \mathsf{r}_{\mathcal{S}}(b, a_{\mathcal{M}}) \in \mathcal{M}.
\end{aligned}
$$

The knowledge base is *core-complete* if such a node cannot be found.

4 Actions

We now introduce a formal language to encode mutating actions. Let us remind ourselves that, in our application of interest, the execution of a mutating action modifies the configuration of a deployment by either adding new resource instances, deleting existing ones, or modifying their settings. Here, we introduce a framework for DL-Lite$^{\mathcal{F}}$ core-closed knowledge base updates, triggered by the execution of an action that enables all the above mentioned effects. The only component of the core-closed knowledge base that is modified by the action execution is $\mathcal{M}$; while $\mathcal{T}$, $\mathcal{S}$, and $\mathcal{A}$ remain unchanged. As a consequence of updating $\mathcal{M}$, actions can introduce new individuals and delete old ones, thus updating the set $\mathbf{I}^{\mathcal{M}}$ as well. Note that this may force changes outside $\mathbf{I}^{\mathcal{M}}$ due to the axioms in $\mathcal{T}$ and $\mathcal{S}$. The effects of applying an action over $\mathcal{M}$ depend on a set of input parameters that will be instantiated at execution time, resulting in different assertions being added or removed from $\mathcal{M}$. As a consequence of assertions being added, fresh individuals might be introduced in the active domain of $\mathcal{M}$, including both model nodes from $\mathbf{I}^{\mathcal{M}}$ and boundary nodes from $\mathbf{I}^{B}$. Differently, as a consequence of assertions being removed, individuals might be removed from the active domain of $\mathcal{M}$, including model nodes from $\mathbf{I}^{\mathcal{M}}$ but *not* including boundary nodes from $\mathbf{I}^{B}$. In fact, boundary nodes are owned by the open portion of the knowledge base and are known to exist regardless of them being used in $\mathcal{M}$. We invite the reader to review the set definitions for $\mathcal{A}$- and $\mathcal{M}$-assertions (Sect. 2) to note that it is indeed possible for a generic boundary individual a involved in an $\mathcal{M}$-assertion to also be involved in an $\mathcal{A}$-assertion.

4.1 Syntax

An action is defined by a signature and a body. The signature consists of an action name and a list of formal parameters, which will be replaced with actual parameters at execution time. The body, or action effect, can include conditional statements and concatenation of atomic operations over $\mathcal{M}$-assertions. For example, let α be the action $act(\vec{x}) = \gamma$; that is, the action denoted by signature $act(\vec{x})$ and body γ, with signature name act, signature parameters $\vec{x}$, and body effect γ. Since it contains unbound parameters, or free variables, action α is ungrounded and needs to be instantiated with actual values in order to be executed over a set $\mathcal{M}$. In the following, we assume the existence of a set Var, of variable names, and consider a generic input parameters substitution $\vec{\theta} : \mathsf{Var} \to \mathbf{I}$, which replaces each variable name by an individual node. For simplicity, we will denote an ungrounded action by its effect γ, and a grounded action by the composition of its effect with an input parameter substitution $\gamma\vec{\theta}$. Action effects can either be *complex* or *basic*. The syntax of complex action effects γ and basic effects β is constrained by the following grammar.

$$\gamma ::= \epsilon \mid \beta \cdot \gamma \mid [\varphi \rightsquigarrow \beta] \cdot \gamma$$
$$\beta ::= \oplus_x S \mid \ominus_x S \mid \odot_{x_{new}} S \mid \ominus_x$$

The complex action effects γ include: the empty effect (ϵ), the execution of a basic effect followed by a complex one ($\beta \cdot \gamma$), and the conditional execution of a basic effect upon evaluation of a formula φ over the set $\mathcal{M}$ ($[\varphi \rightsquigarrow \beta] \cdot \gamma$). The basic action effects β include: the addition of a set S of $\mathcal{M}$-assertions to the subset $\mathcal{M}_x$ ($\oplus_x S$), the removal of a set S of $\mathcal{M}$-assertions from the subset $\mathcal{M}_x$ ($\ominus_x S$), the addition of a fresh subset $\mathcal{M}_{x_{new}}$ containing all the $\mathcal{M}$-assertions in the set S ($\odot_{x_{new}} S$), and the removal of an existing $\mathcal{M}_x$ subset in its entirety ($\ominus_x$). The set S, the formula φ, and the operators $\oplus/\ominus$ might contain *free variables*. These variables are of two types: *(1)* variables that are replaced by the grounding of the action input parameters, and *(2)* variables that are the answer variables of the formula φ and appear in the nested effect β.

Example 2. The following is the definition of the action createBucket from the API reference of the AWS resource type S3::Bucket. The input parameters are two: the new bucket name "*name*" and the canned access control list "*acl*" (one of *Private*, *PublicRead*, *PublicReadWrite*, *AuthenticatedRead*, etc.). The effect of the action is to add a fresh subset $\mathcal{M}_x$ for the newly introduced individual x containing the two assertions $\mathsf{S3::Bucket}(x)$ and $\mathsf{accessControl}(x, y)$.

$$\mathsf{createBucket}(x : name, y : acl) = \odot_x\{\mathsf{S3::Bucket}(x), \mathsf{accessControl}(x, y)\} \cdot \epsilon$$

The action needs to be instantiated by a specific parameter assignment, for example the substitution $\theta = [\ x \leftarrow DataBucket,\ \ y \leftarrow Private\ ]$, which binds the variable x to the node $DataBucket$ and the variable y to the node $Private$, both taken from a pool of inactive nodes in $\mathbf{I}$.

Action Query φ. The syntax introduced in the previous paragraph allows for complex actions that conditionally execute a basic effect β depending on the evaluation of a formula φ over $\mathcal{M}$. This is done via the construct $[\,\varphi \rightsquigarrow \beta\,] \cdot \gamma$. The formula φ might have a set $\vec{y}$ of answer variables that appear free in its body and are then bound to concrete tuples of nodes during evaluation. The answer tuples are in turn used to instantiate the free variables in the nested effect β. We call φ the *action query* since we use it to select all the nodes that will be involved in the action effect. According to the grammar below, φ is a boolean combination of $\mathcal{M}$-assertions potentially containing free variables.

$$\varphi ::= \mathsf{A}_{\mathcal{S}}(t) \ \mid\ \mathsf{R}_{\mathcal{S}}(t_1, t_2) \mid \varphi_1 \wedge \varphi_2 \ \mid\ \varphi_2 \vee \varphi_2 \ \mid\ \neg\varphi$$

In particular, $\mathsf{A}_{\mathcal{S}}$ is a symbol from the set $\mathbf{C}^{\mathcal{S}}$ of partially-closed concepts; $\mathsf{R}_{\mathcal{S}}$ is a symbol from the set $\mathbf{R}^{\mathcal{S}}$ of partially-closed roles; and t, t_1, t_2 are either individual or variable names from the set $\mathbf{I} \uplus \mathsf{Var}$, chosen in such a way that the resulting assertion is an $\mathcal{M}$-assertion. Since the formula φ can only refer to $\mathcal{M}$-assertions, which are interpreted under a closed semantics, its evaluation requires looking at the content of the set $\mathcal{M}$. A formula φ with no free variables is a boolean formula and evaluates to either true or false. A formula φ with answer variables $\vec{y}$ and arity $ar(\varphi)$ evaluates to all the tuples $\vec{t}$, of size equal the arity of φ, that make the formula true in $\mathcal{M}$. The free variables of φ can only appear in the action β such that $\varphi \rightsquigarrow \beta$. We denote by $\mathsf{ANS}(\varphi, \mathcal{M})$ the set of answers to the action query φ over $\mathcal{M}$. It is easy to see that the maximum number of tuples that could be returned by the evaluation (that is, the size of the set $\mathsf{ANS}(\varphi, \mathcal{M})$) is bounded by $|\mathbf{I}^{\mathcal{M}} \uplus \mathbf{I}^{B}|^{ar(\varphi)}$, in turn bounded by $(\,2|\mathcal{M}|\,)^{2|\varphi|}$.

Example 3. The following example shows the encoding of the S3 API operation called deleteBucketEncryption, which requires as unique input parameter the name of the bucket whose encryption configuration is to be deleted. Since a bucket can have multiple encryption configuration rules (each prescribing different encryption keys and algorithms to be used) we use an action query φ to select *all* the nodes that match the assertions structure to be removed.

$$\varphi[y,k,z](x) = \mathsf{S3::Bucket}(x) \wedge \mathsf{encrRule}(x,y) \wedge \mathsf{SSEKey}(y,k) \wedge \mathsf{SSEAlgo}(y,z)$$

The query φ is instantiated by the specific bucket instance (which will replace the variable x) and returns all the triples (y,k,z) of encryption rule, key, and algorithm, respectively, which identify the assertions corresponding to the different encryption configurations that the bucket has. The answer variables are then used in the action effect to instantiate the assertions to remove from $\mathcal{M}_x$:

$$\begin{aligned}&\mathsf{deleteBucketEncryption}(x:\mathit{name})\\&= [\varphi[y,k,z](x) \rightsquigarrow \ominus_x\{\mathsf{encrRule}(x,y), \mathsf{SSEKey}(y,k), \mathsf{SSEAlgo}(y,z)\}] \cdot \epsilon\end{aligned}$$

4.2 Semantics

So far, we have described the syntax of our action language and provided two examples that showcase the encoding of real-world API calls. Now, we define the semantics of action effects with respect to the changes that they induce over a knowledge base. Let us recall that given a substitution $\vec{\theta}$ for the input parameters of an action γ, we denote by $\gamma\vec{\theta}$ the grounded action where all the input variables are replaced according to what prescribed by $\vec{\theta}$. Let us also recall that the effects of an action apply only to assertions in $\mathcal{M}$ and individuals from $\mathbf{I}^{\mathcal{M}}$, and cannot affect nodes and assertions from the open portion of the knowledge base.

The execution of a grounded action $\gamma\vec{\theta}$ over a DL-Lite$^{\mathcal{F}}$ core-closed knowledge base $\mathcal{K} = (\mathcal{T}, \mathcal{A}, \mathcal{S}, \mathcal{M})$, defined over the set $\mathbf{I}^{\mathcal{M}}$ of partially-closed individuals, generates a new knowledge base $\mathcal{K}^{\gamma\vec{\theta}} = (\mathcal{T}, \mathcal{A}, \mathcal{S}, \mathcal{M}^{\gamma\vec{\theta}})$, defined over an updated set of partially-closed individuals $\mathbf{I}^{\mathcal{M}^{\gamma\vec{\theta}}}$. Let S be a set of $\mathcal{M}$-assertions, γ a complex action, $\vec{\theta}$ an input parameter substitution, and $\vec{\rho}$ a generic substitution that potentially replaces all free variables in the action γ. Let $\vec{\rho_1}$ and $\vec{\rho_2}$ be two substitutions with signature $\mathsf{Var} \rightarrow \mathbf{I}$ such that $dom(\vec{\rho_1}) \cap dom(\vec{\rho_2}) = \emptyset$; we denote their composition by $\vec{\rho_1}\vec{\rho_2}$ and define it as the new substitution such that $\vec{\rho_1}\vec{\rho_2}(x) = a$ if $\vec{\rho_1}(x) = a \vee \vec{\rho_2}(x) = a$, and $\vec{\rho_1}\vec{\rho_2}(x) = \perp$ if $\vec{\rho_1}(x) = \perp \wedge \vec{\rho_2}(x) = \perp$. We formalize the application of the grounded action $\gamma\vec{\theta}$ as the transformation $T_{\gamma\vec{\theta}}$ that maps the pair $\langle \mathcal{M}, \mathbf{I}^{\mathcal{M}} \rangle$ into the new pair $\left\langle \mathcal{M}', \mathbf{I}^{\mathcal{M}'} \right\rangle$. We sometimes use the notation $T_{\gamma\vec{\theta}}(\mathcal{M})$ or $T_{\gamma\vec{\theta}}(\mathbf{I}^{\mathcal{M}})$ to refer to the updated MBox or to the updated set of model nodes, respectively. The rules for applying the transformation depend on the structure of the action γ and are reported in Fig. 1. The transformation starts with an initial generic substitution $\vec{\rho} = \vec{\theta}$. As the transformation progresses, the generic substitution $\vec{\rho}$ can be updated only as a result of the evaluation of an action query φ over $\mathcal{M}$. Precisely, all the tuples $\vec{t_1}, ..., \vec{t_n}$ making φ true in $\mathcal{M}$ will be considered and composed with the current substitution $\vec{\rho}$ generating n fresh substitutions $\rho\vec{t_1}, ..., \rho\vec{t_n}$ which are used in the subsequent application of the nested effect β. Since the core $\mathcal{M}$ of the knowledge base $\mathcal{K}$ changes at every action execution, its domain of model nodes $\mathbf{I}^{\mathcal{M}}$ changes as well. The execution of an action $\gamma\vec{\theta}$ over the knowledge base $\mathcal{K} = (\mathcal{T}, \mathcal{A}, \mathcal{S}, \mathcal{M})$ with set of model nodes $\mathbf{I}^{\mathcal{M}}$ could generate a new $\mathcal{K}^{\gamma\vec{\theta}} = (\mathcal{T}, \mathcal{A}, \mathcal{S}, \mathcal{M}^{\gamma\vec{\theta}})$ with a new set of model nodes $\mathbf{I}^{\mathcal{M}'}$ that is not *core-complete* or not *open-consistent* (see Sect. 3 for the corresponding definitions). We illustrate two examples next.

$$
\begin{aligned}
T_{\epsilon\vec{\rho}}(\mathcal{M},\mathbf{I}^{\mathcal{M}}) &= (\mathcal{M},\mathbf{I}^{\mathcal{M}})\\
T_{\beta\cdot\gamma\vec{\rho}}(\mathcal{M},\mathbf{I}^{\mathcal{M}}) &= T_{\gamma\vec{\rho}}(\ T_{\beta\vec{\rho}}(\mathcal{M},\mathbf{I}^{\mathcal{M}})\ )\\
T_{[\varphi\rightsquigarrow\beta]\cdot\gamma\vec{\rho}}(\mathcal{M},\mathbf{I}^{\mathcal{M}}) &= \begin{cases} T_{\gamma\vec{\rho}}(T_{\beta\vec{\rho}}(\mathcal{M},\mathbf{I}^{\mathcal{M}})\ ) & \text{if } \mathsf{ANS}(\varphi,\mathcal{M}) = tt\\ T_{\gamma\vec{\rho}}(\mathcal{M},\mathbf{I}^{\mathcal{M}}) & \text{if } \mathsf{ANS}(\varphi,\mathcal{M}) = \emptyset \text{ or } f\!f\\ T_{\gamma\vec{\rho}}(T_{\beta\vec{\rho}\vec{t}_1\cdot\ldots\cdot\beta\vec{\rho}\vec{t}_n}(\mathcal{M},\mathbf{I}^{\mathcal{M}})) & \text{if } \mathsf{ANS}(\varphi,\mathcal{M}) = \{\vec{t}_1,..,\vec{t}_n\}\end{cases}\\
T_{\oplus_x S\vec{\rho}}(\mathcal{M},\mathbf{I}^{\mathcal{M}}) &= (\ \{\mathcal{M}_i\}_{i\neq\vec{\rho}(x)} \cup \{\mathcal{M}_{\vec{\rho}(x)} \cup S_{\vec{\rho}}\}\ ,\ \mathbf{I}^{\mathcal{M}} \cup ind(S_{\vec{\rho}})\ )\\
T_{\ominus_x S\vec{\rho}}(\mathcal{M},\mathbf{I}^{\mathcal{M}}) &= (\ \{\mathcal{M}_i\}_{i\neq\vec{\rho}(x)} \cup \{\mathcal{M}_{\vec{\rho}(x)} \setminus S_{\vec{\rho}}\}\ ,\ \mathbf{I}^{\mathcal{M}} \setminus ind(S_{\vec{\rho}})\ )\\
T_{\odot_x S\vec{\rho}}(\mathcal{M},\mathbf{I}^{\mathcal{M}}) &= (\ \mathcal{M} \cup \{\mathcal{M}_{\vec{\rho}(x)} = S_{\vec{\rho}}\}\ ,\ \mathbf{I}^{\mathcal{M}} \cup ind(S_{\vec{\rho}})\ )\\
T_{\ominus_x\vec{\rho}}(\mathcal{M},\mathbf{I}^{\mathcal{M}}) &= (\ \mathcal{M} \setminus \mathcal{M}_{\vec{\rho}(x)}\ ,\ \mathbf{I}^{\mathcal{M}} \setminus ind(\mathcal{M}_{\vec{\rho}(x)})\ )
\end{aligned}
$$

Fig. 1. Semantic of the action language defined over the MBox $\mathcal{M}$ and set $\mathbf{I}^{\mathcal{M}}$.

Example 4 (Violation of core-completeness). Consider the case where the general specifications of the system require all objects of type bucket to have a logging configuration, and an action that removes the logging configuration from a bucket. Consider the core-closed knowledge base $\mathcal{K}$ where $\mathcal{S} = \{\mathsf{S3::Bucket} \sqsubseteq \exists\mathsf{loggingConfiguration}\}$ and $\mathcal{M} = \{\mathsf{S3::Bucket}(b), \mathsf{loggingConfiguration}(b, c)\}$ (consistent wrt $\mathcal{S}$) and the action γ defined as

$$
\begin{aligned}
&\mathsf{deleteLoggingConfiguration}(x : name)\\
&\quad = [(\varphi[y](x) = \mathsf{S3::Bucket}(x) \wedge \mathsf{loggingConfiguration}(x, y))\\
&\qquad \rightsquigarrow \ominus_x\{\mathsf{loggingConfiguration}(x, y)\}] \cdot \epsilon
\end{aligned}
$$

For the input parameter substitution $\vec{\theta} = [x \leftarrow b]$, it is easy to see that the transformation $T_{\gamma\vec{\theta}}$ applied to $\mathcal{M}$ results in the update $\mathcal{M}^{\gamma\vec{\theta}} = \{\mathsf{S3::Bucket}(b)\}$, which is *not* core-complete.

Example 5 (Violation of open-consistency). Consider the case where an action application indirectly affects boundary nodes and their properties, leading to inconsistencies in the open portion of the knowledge base. For example, when the knowledge base prescribes that buckets used to store logs cannot be public; however, a change in the configuration of a bucket instance causes a second bucket (initially known to be public) to also become a log store. In particular, this happens when the knowledge base $\mathcal{K}$ contains the $\mathcal{T}$-axiom $\exists\mathsf{loggingDestination}^{-} \sqsubseteq \neg\mathsf{PublicBucket}$ and the $\mathcal{A}$-assertion $\mathsf{PublicBucket}(b)$, and we apply an action that introduces a new bucket storing its logs to b, defined as follows:

$$\begin{aligned}&\mathsf{createBucketWithLogging}(x : \mathit{name}, y : \mathit{log})\\&\quad = \odot_x\{\mathsf{S3{::}Bucket}(x), \mathsf{loggingDestination}(x, y)\}\end{aligned}$$

For the input parameter substitution $\vec{\theta} = [x \leftarrow \mathit{newBucket}, y \leftarrow b]$, the result of applying the transformation $T_{\gamma\vec{\theta}}$ is the set $\mathcal{M} = \{\mathsf{S3{::}Bucket}(\mathit{newBucket}), \mathsf{loggingDestination}(\mathit{newBucket}, b)\}$ which, combined with the pre-existing and unchanged sets $\mathcal{T}$ and $\mathcal{A}$, causes the updated $\mathcal{K}^{\gamma\vec{\theta}}$ to be *not* open-consistent.

From a practical point of view, the examples highlight the need to re-evaluate core-completeness and open-consistency of a core-closed knowledge base after each action execution. Detecting a violation to core-completeness signals that we have modeled an action that is inconsistent with respect to the systems specifications, which most likely means that the action is missing something and needs to be revised. Detecting a violation to open-consistency signals that our action, even when consistent with respect to the specifications, introduces a change that conflicts with other assumptions that we made about the system, and generally indicates that we should either revise the assumptions or forbid the application of the action. Both cases are important to consider in the development life cycle of the core-closed KB and the action definitions.

5 Static Verification

In this section, we investigate the problem of computing whether the execution of an action, no matter the specific instantiation, always preserves given properties of core-closed knowledge bases. We focus on properties expressed as MUST/MAY queries and define the static verification problem as follows.

Definition 1 (Static Verification). *Let $\mathcal{K}$ be a DL-Lite$^{\mathcal{F}}$ core-closed knowledge base, q be a* MUST/MAY *query, and γ be an action with free variables from the language presented above. Let $\vec{\theta}$ be an assignment for the input variables of γ that transforms γ into the* grounded *action $\gamma\vec{\theta}$. Let $\mathcal{K}^{\gamma\vec{\theta}}$ be the DL-Lite$^{\mathcal{F}}$ core-closed knowledge base resulting from the application of the grounded action $\gamma\vec{\theta}$ onto $\mathcal{K}$. We say that the action γ "*preserves q over $\mathcal{K}$*" iff for every grounded instance $\gamma\vec{\theta}$ we have that $\mathsf{ANS}(q, \mathcal{K}) = \mathsf{ANS}(q, \mathcal{K}^{\gamma\vec{\theta}})$. The static verification problem is that of determining whether an action γ is q-preserving over $\mathcal{K}$.*

An action γ is *not* q-preserving over $\mathcal{K}$ iff there exists a grounding $\vec{\theta}$ for the input variables of γ such that $\mathsf{ANS}(q, \mathcal{K}) \neq \mathsf{ANS}(q, \mathcal{K}^{\gamma\vec{\theta}})$; that is, fixed the grounding $\vec{\theta}$ there exists a tuple $\vec{t}$ for q's answer variables such that $\vec{t} \in \mathsf{ANS}(q, \mathcal{K}) \setminus \mathsf{ANS}(q, \mathcal{K}^{\gamma\vec{\theta}})$ or $\vec{t} \in \mathsf{ANS}(q, \mathcal{K}^{\gamma\vec{\theta}}) \setminus \mathsf{ANS}(q, \mathcal{K})$.

Theorem 1 (Complexity of the Static Verification Problem). *The static verification problem, i.e. deciding whether an action γ is q-preserving over $\mathcal{K}$, can be decided in* PTIME *in data complexity and* EXPTIME *in the arities of γ and q.*

Proof. The proof relies on the fact that one could: enumerate all possible assignments $\vec{\theta}$; compute the updated knowledge bases $\mathcal{K}^{\gamma\vec{\theta}}$; check whether these are fully satisfiable; enumerate all tuples $\vec{t}$ for the query q; and, finally, check whether there exists at least one such tuple that satisfies q over $\mathcal{K}$ but not $\mathcal{K}^{\gamma\vec{\theta}}$ or vice versa. The number of assignments $\vec{\theta}$ is bounded by $\left(|\mathbf{I}^{\mathcal{M}} \uplus \mathbf{I}^{\mathcal{K}}| + ar(\gamma)\right)^{ar(\gamma)}$ as it is sufficient to replace each variable appearing in the action γ either by a known object from $\mathbf{I}^{\mathcal{M}} \uplus \mathbf{I}^{\mathcal{K}}$ or by a fresh one. The computation of the updated $\mathcal{K}^{\gamma\vec{\theta}}$ is done in polynomial time in $\mathcal{M}$ (and is exponential in the size of the action γ) as it may require the evaluation of an internal action query φ and the consecutive re-application of the transformation for a number of tuples that is bounded by a polynomial over the size of $\mathcal{M}$. As explained in Sect. 3, checking full satisfiability of the resulting core-closed knowledge base is also polynomial in $\mathcal{M}$. The number of tuples $\vec{t}$ is bounded by $\left(|\mathbf{I}^{\mathcal{M}} \uplus \mathbf{I}^{\mathcal{K}}| + ar(\gamma)\right)^{ar(q)}$ as it is enough to consider all those tuples involving known objects plus the fresh individuals introduced by the assignment $\vec{\theta}$. Checking whether a tuple $\vec{t}$ satisfies the query q over a core-closed knowledge base is decided in LogSpace in the size of $\mathcal{M}$ [15] which is, thus, also polynomial in $\mathcal{M}$.

6 Planning

As discussed throughout the paper, the execution of a mutating action modifies the configuration of a deployment and potentially changes its posture with respect to a given set of requirements. In the previous two sections, we introduced a language to encode mutating actions and we investigated the problem of checking whether the application of an action preserves the properties of a core-closed knowledge base. In this section, we investigate the plan existence and synthesis problems; that is, the problem of deciding whether there exists a sequence of grounded actions that leads the knowledge base to a state where a certain requirement is met, and the problem of finding a set of such plans, respectively. We start by defining a notion of transition system that is generated by applying actions to a core-closed knowledge base and then use this notion to focus on the mentioned planning problems. As in classical planning, the plan existence problem for plans computed over unbounded domains is undecidable [17,19]. The undecidability proof is done via reduction from the Word problem. The problem of deciding whether a deterministic Turing machine M accepts a word $w \in \{0,1\}^*$ is reduced to the plan existence problem. Since undecidability holds even for basic action effects, we can show undecidability over an unbounded domain by using the same encoding of [1].

Transition Systems. In the style of the work done in [10,21], the combination of a DL-Lite$^{\mathcal{F}}$ core-closed knowledge base and a set of actions can be viewed as the transition system it generates. Intuitively, the states of the transition system correspond to MBoxes and the transitions between states are labeled by grounded actions. A DL-Lite$^{\mathcal{F}}$ core-closed knowledge base $\mathcal{K} = (\mathcal{T}, \mathcal{A}, \mathcal{S}, \mathcal{M}_0)$, defined over the possibly infinite set of individuals $\mathbf{I}$ (and model nodes $\mathbf{I}_0^{\mathcal{M}} \subseteq \mathbf{I}$)

and the set Act of ungrounded actions, generates the transition system (TS) $\Upsilon_{\mathcal{K}} = (\mathbf{I}, \mathcal{T}, \mathcal{A}, \mathcal{S}, \Sigma, \mathcal{M}_0, \rightarrow)$ where Σ is a set of *fully satisfiable* (i.e., *core-complete* and *open-consistent*) MBoxes; $\mathcal{M}_0$ is the initial MBox; and $\rightarrow \subseteq \Sigma \times L_{\mathsf{Act}} \times \Sigma$ is a labeled transition relation with L_{Act} the set of all possible *grounded actions*. The sets Σ and $\rightarrow$ are defined by mutual induction as the smallest sets such that: if $\mathcal{M}_i \in \Sigma$ then for every grounded action $\gamma\vec{\theta} \in L_{\mathsf{Act}}$ such that the fresh MBox $\mathcal{M}_{i+1}$ resulting from the transformation $T_{\gamma\vec{\theta}}$ is core-complete and open-consistent, we have that $\mathcal{M}_{i+1} \in \Sigma$ and $(\mathcal{M}_i, \gamma\vec{\theta}, \mathcal{M}_{i+1}) \in \rightarrow$.

Since we assume that actions have input parameters that are replaced during execution by values from $\mathbf{I}$, which contains both known objects from $\mathbf{I}^{\mathcal{M}} \uplus \mathbf{I}^{\mathcal{K}}$ and possibly infinitely many fresh objects, the generated transition system $\Upsilon_{\mathcal{K}}$ is generally infinite. To keep the planning problem decidable, we concentrate on a known finite subset $\mathcal{D} \subset \mathbf{I}$ containing all the fresh nodes and value assignments to action variables that are of interest for our application. In the remainder of this paper, we discuss the plan existence and synthesis problem for finite transition systems $\Upsilon_{\mathcal{K}} = (\mathcal{D}, \mathcal{T}, \mathcal{A}, \mathcal{S}, \Sigma, \mathcal{M}_0, \rightarrow)$, whose states in Σ have a domain that is also bounded by $\mathcal{D}$.

The Plan Existence Problem. A plan is a sequence of grounded actions whose execution leads to a state satisfying a given property. Let $\mathcal{K} = (\mathcal{T}, \mathcal{A}, \mathcal{S}, \mathcal{M}_0)$ be a DL-Lite$^{\mathcal{F}}$ core-closed knowledge base; Act be a set of ungrounded actions; and let $\Upsilon_{\mathcal{K}} = (\mathcal{D}, \mathcal{T}, \mathcal{A}, \mathcal{S}, \Sigma, \mathcal{M}_0, \rightarrow)$ be its generated finite TS. Let π be a finite sequence $\gamma_1\vec{\theta}_1 \cdots \gamma_n\vec{\theta}_n$ of grounded actions taken from the set L_{Act}. We call the sequence π *consistent* iff there exists a run $\rho = \mathcal{M}_0 \xrightarrow{\gamma_1\vec{\theta}_1} \mathcal{M}_1 \xrightarrow{\gamma_2\vec{\theta}_2} \cdots \xrightarrow{\gamma_n\vec{\theta}_n} \mathcal{M}_n$ in $\Upsilon_{\mathcal{K}}$. Let q be a MUST/MAY query mentioning objects from $adom(\mathcal{K})$ and $\vec{t}$ a tuple from the set $adom(\mathcal{K})^{ar(q)}$. A consistent sequence π of grounded actions is a *plan* from $\mathcal{K}$ to $(\vec{t}, q)$ iff $\vec{t} \in \mathsf{ANS}(q, \mathcal{K}_n = (\mathcal{T}, \mathcal{A}, \mathcal{S}, \mathcal{M}_n))$ with $\mathcal{M}_n$ the final state of the run induced by π.

Definition 2 (Plan Existence). *Given a DL-Lite$^{\mathcal{F}}$ core-closed knowledge base $\mathcal{K}$, a tuple $\vec{t}$, and a* MUST/MAY *query q, the* plan existence *problem is that of deciding whether there exists a plan from $\mathcal{K}$ to $(\vec{t}, q)$.*

Example 6. Let us consider the transition system $\Upsilon_{\mathcal{K}}$ generated by the core-closed knowledge base $\mathcal{K} = (\mathcal{T}, \mathcal{A}, \mathcal{S}, \mathcal{M}_0)$ having the set of partially-closed assertions $\mathcal{M}_0$ defined as

$$\{\mathsf{S3::Bucket}(b),\ \mathsf{KMS::Key}(k),\ \mathsf{bucketEncryptionRule}(b, r),\ \mathsf{bucketKey}(r, k),\ \mathsf{bucketKeyEnabled}(r, true),\ \mathsf{enableKeyRotation}(k, false)\}$$

and the set of action labels Act containing the actions deleteBucket, createBucket, deleteKey, createKey, enableKeyRotation, putBucketEncryption, and deleteBucketEncryption. Let us assume that we are interested in verifying the existence of a sequence of grounded actions that when applied onto the knowledge base would configure the bucket node b to be encrypted with a rotating key. Formally, this is equivalent to checking the existence of a consistent plan π that when executed on the transition system $\Upsilon_{\mathcal{K}}$ leads to a state $\mathcal{M}_n$ such that the tuple $\vec{t} = b$ is in the set $\mathsf{ANS}(q, \mathcal{K}_n = (\mathcal{T}, \mathcal{A}, \mathcal{S}, \mathcal{M}_n))$ for q the query

$$q[x] = \mathsf{S3::Bucket}(x) \wedge \textsc{Must}\ (\exists y, z.\ \mathsf{bucketSSEncryption}(x, y) \wedge \mathsf{bucketKey}(y, z) \wedge \mathsf{enableKeyRotation}(z, \mathit{true}))$$

It is easy to see that the following three sequences of grounded actions are valid plans from $\mathcal{K}$ to (b, q):

$$\pi_1 = \mathsf{enableKeyRotation}(k)$$
$$\pi_2 = \mathsf{createKey}(k_1) \cdot \mathsf{enableKeyRotation}(k_1) \cdot \mathsf{putBucketEncryption}(b, k_1)$$
$$\pi_3 = \mathsf{deleteBucketEncryption}(b, k) \cdot \mathsf{createKey}(k_1) \cdot \mathsf{enableKeyRotation}(k_1) \cdot \mathsf{putBucketEncryption}(b, k_1)$$

If, for example, a bucket was only allowed to have one encryption (by means of a functional axiom in $\mathcal{S}$), then π_2 would not be a valid plan, as it would generate an inconsistent run leading to a state $\mathcal{M}_i$ that is not open-consistent w.r.t. $\mathcal{S}$.

Lemma 3. *The plan existence problem for a finite transition system $\Upsilon_{\mathcal{K}}$ generated by a DL-Lite$^{\mathcal{F}}$ core-closed knowledge base $\mathcal{K}$ and a set of actions* Act, *over a finite domain of objects $\mathcal{D}$, reduces to graph reachability over a graph whose number of states is at most exponential in the size of $\mathcal{D}$.*

The Plan Synthesis Problem. We now focus on the problem of finding plans that satisfy a given condition. As discussed in the previous paragraph, we are mostly driven by query answering; in particular, by conditions corresponding to a tuple (of objects from our starting deployment configuration) satisfying a given requirement expressed as a Must/May query. Clearly, this problem is meaningful in our application of interest because it corresponds to finding a set of potential sequences of changes that would allow one to reach a configuration satisfying (resp., not satisfying) one, or more, security mitigations (resp., vulnerabilities). We concentrate on DL-Lite$^{\mathcal{F}}$ core-closed knowledge bases and their generated finite transition systems, where potential fresh objects are drawn from a fixed set $\mathcal{D}$. We are interested in sequences of grounded actions that are minimal and ignore sequences that extend these. We sometimes call such minimal sequences *simple plans.* A plan π from an initial core-closed knowledge base $\mathcal{K}$ to a goal condition b is minimal (or simple) *iff* there does not exist a plan π' (from the same initial $\mathcal{K}$ to the same goal condition b) s.t. $\pi = \pi' \cdot \sigma$, for σ a non-empty suffix of grounded actions.

In Algorithm 1, we present a depth-first search algorithm that, starting from $\mathcal{K}$, searches for all simple plans that achieve a given target query membership condition. The transition system $\Upsilon_{\mathcal{K}}$ is computed, and stored, on the fly in the Successors sub-procedure and the graph is explored in a depth-first search traversal fashion.

Algorithm 1: $\mathsf{FindPlans}(\mathcal{K}, \mathcal{D}, \mathsf{Act}, \langle \vec{t}, q\rangle)$

Inputs : A ccKB $\mathcal{K} = (\mathcal{T}, \mathcal{A}, \mathcal{S}, \mathcal{M}_0)$, a domain $\mathcal{D}$, a set of actions Act and a pair $\langle \vec{t}, q\rangle$ of an answer tuple and a Must/May query

Output: A possibly empty set Π of consistent simple plans

1 **def** `FindPlans` *(* $\mathcal{K}, \mathcal{D}, \mathsf{Act}, \langle \vec{t}, q\rangle$ *):*
2   $\Pi := \emptyset$;
3   $S := \bot$;
4   $\mathsf{AllPlanSearch}(\mathcal{M}_0, \epsilon, \emptyset, \mathcal{K}, \mathcal{D}, \mathsf{Act}, \langle \vec{t}, q\rangle)$;
5   **return** Π;
6 **def** `AllPlanSearch` *(* $\mathcal{M}, \pi, V, \mathcal{K}, \mathcal{D}, \mathsf{Act}, \langle \vec{t}, q\rangle$ *):*
7   **if** $\mathcal{M} \in V$ **then**
8   **return**;
9   **if** $\vec{t} \in \mathsf{ANS}(q, \langle \mathcal{T}, \mathcal{A}, \mathcal{S}, \mathcal{M}\rangle)$ **then**
10     $\Pi := \Pi \cup \{\pi\}$;
11   **return**;
12   $Q := \emptyset$;
13   **foreach** $\left\langle \gamma\vec{\theta}, \mathcal{M}' \right\rangle \in \mathsf{Successors}(\mathcal{M}, \mathsf{Act}, \mathcal{D})$ **do**
14     $Q.push(\left\langle \gamma\vec{\theta}, \mathcal{M}' \right\rangle)$;
15   $V := V \cup \{\mathcal{M}\}$;
16   **while** $Q \neq \emptyset$ **do**
17     $\left\langle \gamma\vec{\theta}, \mathcal{M}' \right\rangle = Q.pop()$;
18     $\mathsf{AllPlanSearch}(\mathcal{M}', \pi \cdot \gamma\vec{\theta}, V, \mathcal{K}, \mathcal{D}, \mathsf{Act}, \langle \vec{t}, q\rangle)$;
19   $V := V \setminus \{\mathcal{M}\}$;
20   **return**;
21 **def** `Successors` *(* $\mathcal{M}, \mathsf{Act}, \mathcal{D}$ *):*
22   **if** $S[\mathcal{M}]$ *is defined* **then**
23   **return** $S[\mathcal{M}]$;
24   $N := \emptyset$;
25   **foreach** $\gamma \in \mathsf{Act}$, $\vec{\theta} \in \mathcal{D}^{ar(\gamma)}$ **do**
26     $\mathcal{M}' := T_{\gamma\vec{\theta}}(\mathcal{M})$;
27   **if** $\mathcal{M}'$ *is fully satisfiable* **then**
28       $N := N \cup \{\left\langle \gamma\vec{\theta}, \mathcal{M}' \right\rangle\}$
29   $S[\mathcal{M}] := N$;
30   **return** N;

We note that the condition $\vec{t} \in \mathsf{ANS}(q, \langle \mathcal{T}, \mathcal{A}, \mathcal{S}, \mathcal{M}\rangle)$ (line 9) could be replaced by any other query satisfiability condition and that one could easily rewrite the algorithm to be parameterized by a more general boolean goal. For example, the condition that a given tuple $\vec{t}$ is *not* an answer to a query q over the analyzed state, with the query q representing an undesired configuration, or a boolean formula over multiple query membership assertions. We also note that Algorithm 1 could be simplified to return only one simple plan, if a plan exists, or NULL, if a plan does not exist, thus solving the so-called *plan generation problem*. We refer the reader to the full version of this paper [16] containing the plan generation algorithm (full version, Appendix A.1) and the proofs of Theorem 2 and 3 below (full version, Appendices A.2 and A3, respectively).

Theorem 2 (Minimal Plan Synthesis Correctness). *Let $\mathcal{K}$ be a DL-Lite$^{\mathcal{F}}$ core-closed knowledge base, $\mathcal{D}$ be a fixed finite domain,* Act *be a set of ungrounded action labels, and $\langle \vec{t}, q \rangle$ be a goal. Then a plan π is returned by the algorithm* FindPlans($\mathcal{K}, \mathcal{D}$, Act, $\langle \vec{t}, q \rangle$) *if and only if π is a minimal plan from $\mathcal{K}$ to $\langle \vec{t}, q \rangle$.*

Theorem 3 (Minimal Plan Synthesis Complexity). *The* FindPlans *algorithm runs in polynomial time in the size of $\mathcal{M}$ and exponential time in size of $\mathcal{D}$.*

7 Related Work

The syntax of the action language that we presented in this paper is similar to that of [1,12,13]. Differently from their work, we disallow complex action effects to be nested inside conditional statements, and we define basic action effects that consist purely in the addition and deletion of concept and role $\mathcal{M}$-assertions. Thus, our actions are much less general than those used in their framework. The semantics of their action language is defined in terms of changes applied to instances, and the action effects are captured and encoded through a variant of $\mathcal{ALCHOIQ}$ called $\mathcal{ALCHOIQ}_{br}$. In our work, instead, the execution of an action updates a portion of the core-closed knowledge base $\mathcal{K}$—the core $\mathcal{M}$, which is interpreted under a close-world assumption and can be seen as a partial assignment for the interpretations that are models of $\mathcal{K}$. Since we directly manipulate $\mathcal{M}$, the semantics of our actions is more similar to that of [21] and, in general, to ABox updates [22,23]. Like the frameworks introduced in [9–11,20], our actions are parameterized and when combined with a core-closed knowledge base generate a transition system. In [11], the authors focus on a variant of *Knowledge and Action Bases* [21] called *Explicit-Input KABs* (eKABs); in particular, on finite and on state-bounded eKABs, for which planning existence is decidable. Our generated transition systems are an adaptation of the work done in *Description Logic based Dynamic Systems*, *KABs*, and *eKABs* to our setting of core-closed knowledge bases. In [24], the authors address decidability of the plan existence problem for logics that are subset of $\mathcal{ALCOI}$. Their action language is similar to the one presented in this paper; including pre-conditions, in the form of a set of ABox assertions, post-conditions, in the form of basic addition or removal of assertions, concatenation, and input parameters. In [11], the plan synthesis problem is discussed also for lightweight description logics. Relying on the FOL-reducibility of DL-Lite$^{\mathcal{A}}$, it is shown that plan synthesis over DL-Lite$^{\mathcal{A}}$ can be compiled into an ADL planning problem [25]. This does not seem possible in our case, as not all necessary tests over core-closed knowledge bases are known to be FOL-reducible. In [10] and [9], the authors concentrate on verifying and synthesizing temporal properties expressed in a variant of μ-calculus over description logic based dynamic systems, both problems are relevant in our application scenario and we will consider them in future works.

8 Conclusion

We focused on the problem of analyzing cloud infrastructure encoded as description logic knowledge bases combining complete and incomplete information. From a practical standpoint, we concentrated on formalizing and foreseeing the impact of potential changes pre-deployment. We introduced an action language to encode mutating actions, whose semantics is given in terms of changes induced to the complete portion of the knowledge base. We defined the static verification problem as the problem of deciding whether the execution of an action, no matter the specific parameters passed, always preserves a set of properties of the knowledge base. We characterized the complexity of the problem and provided procedural steps to solve it. We then focused on three formulations of the classical AI planning problem: namely, plan existence, generation, and synthesis. In our setting, the planning problem is formulated with respect to the transition system arising from the combination of a core-closed knowledge base and a set of actions; goals are given in terms of one, or more, MUST/MAY conjunctive query membership assertion; and plans of interest are simple sequences of parameterized actions.

Acknowledgments. This work is supported by the ERC Consolidator grant D-SynMA (No. 772459).

References

1. Ahmetaj, S., Calvanese, D., Ortiz, M., Simkus, M.: Managing change in graph-structured data using description logics. ACM Trans. Comput. Log. **18**(4), 27:1–27:35 (2017)
2. Artale, A., Calvanese, D., Kontchakov, R., Zakharyaschev, M.: The DL-lite family and relations. J. Artif. Intell. Res. **36**, 1–69 (2009)
3. Baader, F., Horrocks, I., Lutz, C., Sattler, U.: An Introduction to Description Logic. Cambridge University Press, Cambridge (2017)
4. Backes, J., et al.: Reachability analysis for AWS-based networks. In: Dillig, I., Tasiran, S. (eds.) CAV 2019. LNCS, vol. 11562, pp. 231–241. Springer, Cham (2019). https://doi.org/10.1007/978-3-030-25543-5_14
5. Backes, J., et al.: Stratified abstraction of access control policies. In: Lahiri, S.K., Wang, C. (eds.) CAV 2020. LNCS, vol. 12224, pp. 165–176. Springer, Cham (2020). https://doi.org/10.1007/978-3-030-53288-8_9

6. Backes, J., et al.: Semantic-based automated reasoning for AWS access policies using SMT. In: Bjørner, N., Gurfinkel, A. (eds.) 2018 Formal Methods in Computer Aided Design, FMCAD 2018, Austin, TX, USA, 30 October–2 November 2018, pp. 1–9. IEEE (2018). https://doi.org/10.23919/FMCAD.2018.8602994
7. Bouchet, M., et al.: Block public access: trust safety verification of access control policies. In: Devanbu, P., Cohen, M.B., Zimmermann, T. (eds.) ESEC/FSE 2020: 28th ACM Joint European Software Engineering Conference and Symposium on the Foundations of Software Engineering, Virtual Event, USA, 8–13 November 2020, pp. 281–291. ACM (2020). https://doi.org/10.1145/3368089.3409728
8. Calvanese, D., Giacomo, G.D., Lembo, D., Lenzerini, M., Rosati, R.: EQL-lite: effective first-order query processing in description logics. In: Veloso, M.M. (ed.) Proceedings of the 20th International Joint Conference on Artificial Intelligence, IJCAI 2007, Hyderabad, India, 6–12 January 2007, pp. 274–279 (2007). http://ijcai.org/Proceedings/07/Papers/042.pdf
9. Calvanese, D., De Giacomo, G., Montali, M., Patrizi, F.: Verification and synthesis in description logic based dynamic systems. In: Faber, W., Lembo, D. (eds.) RR 2013. LNCS, vol. 7994, pp. 50–64. Springer, Heidelberg (2013). https://doi.org/10.1007/978-3-642-39666-3_5
10. Calvanese, D., Montali, M., Patrizi, F., Giacomo, G.D.: Description logic based dynamic systems: modeling, verification, and synthesis. In: Yang, Q., Wooldridge, M.J. (eds.) Proceedings of the Twenty-Fourth International Joint Conference on Artificial Intelligence, IJCAI 2015, Buenos Aires, Argentina, 25–31 July 2015, pp. 4247–4253. AAAI Press (2015). http://ijcai.org/Abstract/15/604
11. Calvanese, D., Montali, M., Patrizi, F., Stawowy, M.: Plan synthesis for knowledge and action bases. In: Kambhampati, S. (ed.) Proceedings of the Twenty-Fifth International Joint Conference on Artificial Intelligence, IJCAI 2016, New York, NY, USA, 9–15 July 2016, pp. 1022–1029. IJCAI/AAAI Press (2016). http://www.ijcai.org/Abstract/16/149
12. Calvanese, D., Ortiz, M., Simkus, M.: Evolving graph databases under description logic constraints. In: Eiter, T., Glimm, B., Kazakov, Y., Krötzsch, M. (eds.) Informal Proceedings of the 26th International Workshop on Description Logics, Ulm, Germany, 23–26 July 2013. CEUR Workshop Proceedings, vol. 1014, pp. 120–131. CEUR-WS.org (2013). http://ceur-ws.org/Vol-1014/paper_82.pdf
13. Calvanese, D., Ortiz, M., Simkus, M.: Verification of evolving graph-structured data under expressive path constraints. In: Martens, W., Zeume, T. (eds.) 19th International Conference on Database Theory, ICDT 2016, Bordeaux, France, 15–18 March 2016. LIPIcs, vol. 48, pp. 15:1–15:19. Schloss Dagstuhl - Leibniz-Zentrum für Informatik (2016). https://doi.org/10.4230/LIPIcs.ICDT.2016.15
14. Cauli, C., Li, M., Piterman, N., Tkachuk, O.: Pre-deployment security assessment for cloud services through semantic reasoning. In: Silva, A., Leino, K.R.M. (eds.) CAV 2021. LNCS, vol. 12759, pp. 767–780. Springer, Cham (2021). https://doi.org/10.1007/978-3-030-81685-8_36
15. Cauli, C., Ortiz, M., Piterman, N.: Closed- and open-world reasoning in dl-lite for cloud infrastructure security. In: Proceedings of the 18th International Conference on Principles of Knowledge Representation and Reasoning, KR 2021, Hanoi, Vietnam (2021)
16. Cauli, C., Ortiz, M., Piterman, N.: Actions over core-closed knowledge bases (2022). https://doi.org/10.48550/ARXIV.2202.12592. https://arxiv.org/abs/2202.12592
17. Chapman, D.: Planning for conjunctive goals. Artif. Intell. **32**(3), 333–377 (1987). https://doi.org/10.1016/0004-3702(87)90092-0

18. Cook, B.: Formal reasoning about the security of amazon web services. In: Chockler, H., Weissenbacher, G. (eds.) CAV 2018. LNCS, vol. 10981, pp. 38–47. Springer, Cham (2018). https://doi.org/10.1007/978-3-319-96145-3_3
19. Erol, K., Nau, D.S., Subrahmanian, V.S.: Complexity, decidability and undecidability results for domain-independent planning. Artif. Intell. **76**(1–2), 75–88 (1995). https://doi.org/10.1016/0004-3702(94)00080-K
20. Giacomo, G.D., Masellis, R.D., Rosati, R.: Verification of conjunctive artifact-centric services. Int. J. Cooperative Inf. Syst. **21**(2), 111–140 (2012). https://doi.org/10.1142/S0218843012500025
21. Hariri, B.B., Calvanese, D., Montali, M., Giacomo, G.D., Masellis, R.D., Felli, P.: Description logic knowledge and action bases. J. Artif. Intell. Res. **46**, 651–686 (2013)
22. Kharlamov, E., Zheleznyakov, D., Calvanese, D.: Capturing model-based ontology evolution at the instance level: the case of dl-lite. J. Comput. Syst. Sci. **79**(6), 835–872 (2013). https://doi.org/10.1016/j.jcss.2013.01.006
23. Liu, H., Lutz, C., Milicic, M., Wolter, F.: Foundations of instance level updates in expressive description logics. Artif. Intell. **175**(18), 2170–2197 (2011). https://doi.org/10.1016/j.artint.2011.08.003
24. Milicic, M.: Planning in action formalisms based on DLS: first results. In: Calvanese, D., et al. (eds.) Proceedings of the 2007 International Workshop on Description Logics (DL2007), Brixen-Bressanone, near Bozen-Bolzano, Italy, 8–10 June 2007. CEUR Workshop Proceedings, vol. 250. CEUR-WS.org (2007). http://ceur-ws.org/Vol-250/paper_59.pdf
25. Pednault, E.P.D.: ADL and the state-transition model of action. J. Logic Comput. **4**(5), 467–512 (1994). https://doi.org/10.1093/logcom/4.5.467
26. Tobies, S.: A NExpTime-complete description logic strictly contained in C^2. In: Flum, J., Rodriguez-Artalejo, M. (eds.) CSL 1999. LNCS, vol. 1683, pp. 292–306. Springer, Heidelberg (1999). https://doi.org/10.1007/3-540-48168-0_21

GK: Implementing Full First Order Default Logic for Commonsense Reasoning (System Description)

Tanel Tammet[1](✉), Dirk Draheim[2], and Priit Järv[1]

[1] Applied Artificial Intelligence Group, Tallinn University of Technology, Tallinn, Estonia
{tanel.tammet,priit.jarv1}@taltech.ee

[2] Information Systems Group, Tallinn University of Technology, Tallinn, Estonia
dirk.draheim@taltech.ee

Abstract. Our goal is to develop a logic-based component for hybrid – machine learning plus logic – commonsense question answering systems. The paper presents an implementation GK of default logic for handling rules with exceptions in unrestricted first order knowledge bases. GK is built on top of our existing automated reasoning system with confidence calculation capabilities. To overcome the problem of undecidability of checking potential exceptions, GK performs delayed recursive checks with diminishing time limits. These are combined with the taxonomy-based priorities for defaults and numerical confidences.

1 Introduction

The problem of handling uncertainty is one of the critical issues when considering the use of logic for automating commonsense reasoning. Most of the facts and rules people use in their daily lives are uncertain. There are many types of uncertainty, like fuzziness (is a person somewhat tall or very tall), confidence (how certain does some fact seem) and exceptions (birds can typically fly, but penguins, ostriches etc., can not). Some of these uncertainties, like fuzziness and confidence, can be represented numerically, while others, like rules with exceptions, are discrete. In [18] we present the design and implementation of the CONFER framework for extending existing automated reasoning systems with confidence calculation capabilities. In the current paper we present the implementation called GK for default logic [13], built by further extending the CONFER implementation. Importantly, we design a novel practical framework for implementing default logic for the full, undecidable first order logic on the basis of a conventional resolution prover.

1.1 Default Logic

Default logic was introduced in 1980 by R. Reiter [13] to model one aspect of common-sense reasoning: rules with exceptions. It has remained one of the most

J. Blanchette et al. (Eds.): IJCAR 2022, LNAI 13385, pp. 300–309, 2022.
https://doi.org/10.1007/978-3-031-10769-6_18

well-known logic-based mechanisms devoted to this goal, with the *circumscription* by J. McCarthy and the *autoepistemic logic* being the early alternatives. Several similar systems have been proposed later, like defeasible logic [11].

Default logic [13] extends classical logic with default rules of the form

$$\frac{\alpha(x) : \beta_1(x), ...\beta_n(x)}{\gamma(x)}$$

where a *precondition* α(x), *justifications* $\beta_1(x), ...\beta_n(x)$ and a *consequent* $\gamma(x)$ are first order predicate calculus formulas whose free variables are among $x = x_1, ..., x_m$. For every tuple of individuals $t = t_1, ..., t_n$, if the precondition $\alpha(t)$ is derivable and none of the *negated* justifications $\neg\beta(t)$ are derivable from a given knowledge base KB, then the consequent $\gamma(t)$ can be derived from KB. Differently from classical and most other logics, default logic is *non-monotonic*: adding new assumptions can make some previously derivable formulas non-derivable.

As investigated in [7], the interpretation of quantifiers in default rules can lead to several versions of default logic. We follow the original interpretation of Reiter in [13] which requires the use of Skolemization in a specific manner over default rules. For example, a default rule: $\exists x P(x) \vdash \exists x P(x)$ should be interpreted as : $P(c) \vdash P(c)$, where c is a Skolem constant.

Consider a typical example for default logic: birds can normally fly, but penguins cannot fly. The classical logic part

$$penguin(p) \,\&\, bird(b) \,\&\, \forall x.penguin(x) \Rightarrow bird(x) \,\&\, \forall x.penguin(x) \Rightarrow \neg fly(x).$$

is extended with the default rule $bird(x) : fly(x) \vdash fly(x)$. From here we can derive that an arbitrary bird b can fly, but a penguin p cannot. The default rule cannot be applied to p, since a contradiction is derivable from $fly(p)$. This argument cannot be easily modelled using numerical confidences: the probability of an arbitrary living bird being able to fly is relatively high, while the penguins form a specific subset of birds, for which this probability is zero.

Another well-known example – Nixon's triangle – introduces the problem of multiple extensions and *sceptical* vs *credulous* entailment: the classical facts $republican(nixon) \,\&\, quaker(nixon)$ extended with two mutually excluding default rules $republican(x) : \neg pacifist(x) \vdash \neg pacifist(x)$ and $quaker(x) : pacifist(x) \vdash pacifist(x)$. The credulous entailment allows giving different priorities to the default rules and accepts different sets (*extensions*) of consequences, if there is a way to assign priorities so that all the consequences in an extension can be derived. The sceptical entailment requires that a consequence is present in all extensions. GK follows the latter interpretation, but allows explicit priorities to be assigned to the default rules.

The concept of *priorities* for default rules has been well investigated, with several mechanisms proposed. G. Brewka argues in [4] that "for realistic applications involving default reasoning it is necessary to reason about the priorities of defaults" and introduces an ordering of defaults based on specificity: default rules for a more specific class of objects should take priority over rules for more general classes. For example, since birds (who typically do fly) are physical objects and physical objects typically do not fly, we have contradictory default rules describing the flying capability of arbitrary birds. Since birds are a subset of physical objects, the flying rule of birds should have a higher priority than the non-flying rule of physical objects.

1.2 Undecidability, Grounding and Implementations

Perhaps the most significant problem standing in the way of automating default logic is undecidability of the applicability of rules. Indeed, in order to apply a default rule, we must prove that the justifications do not lead to a contradiction with the rest of the knowledge base KB. For full first order logic this is undecidable. Hence, the standard approach for handling default logic has been creating a large ground instance KB_g of the KB, and then performing decidable propositional reasoning on the KB_g.

Almost all the existing implementations of default logic like DeReS [5], DLV2 [1] or CLINGO [8], with the noteworthy exception of s(CASP) [2], follow the same principle. More generally, the field of *Answer Set Programming* (ASP), see [10], is devoted to this approach. As an exception, the s(CASP) system [2] solves queries without the grounding step and is thus better suited for large domains. It is noteworthy that the s(CASP) system has been used in [9] for automating common sense reasoning for autonomous driving with the help of default rules. However, s(CASP) is a logic programming system, not a universal automated reasoner. For example, when we add a rule `bird(father(X)) :- bird(X)` to the formulation of the above birds example in s(CASP), the search does not terminate, apparently due to the infinitely growing nesting of terms.

While ASP systems are very well suited for specific kinds of problems over a small finite domain, grounding becomes infeasible for large first order knowledge bases (*KB* in the following), in particular when the domain is infinite and nested terms can be derived from the KB. The approach described in this paper accepts the lack of logical omniscience and performs delayed recursive checking of exceptions with diminishing time limits directly on non-grounded clauses, combined with the taxonomy-based priorities for defaults and numerical confidences.

2 Algorithms

Our approach of implementing default rules in GK for first order logic is to delay justification checking until a first-order proof is found and then perform recursively deepening checks with diminishing time limits. Thus, our system first produces a potentially large number of different candidate proofs and then enters a recursive checking phase. The idea of delaying justification checking is already present in the original paper of R. Reiter [13], where he uses linear resolution and delayed checks as the main machinery of his proofs. The results produced by GK thus depend on the time limits and are not stable. Showing specific fixpoint properties of the algorithm is not in the scope of our paper.

A practical question for implementation is the actual representation of default rules and making the rules fit the first-order proof search machinery. To this end we introduce *blocker atoms* which are similar to the justification indexes of Reiter.

In the following we will assume that the underlying first order reasoner uses the resolution method, see [3] for details. The rest of the paper assumes familiarity with the basic concepts, terminology and algorithms of the resolution method.

2.1 Background: Queries and Answers

We assume our system is presented with a question in one of two forms: *(1)* Is the statement Q true? *(2)* Find values V for existentially bound variables in Q so that Q is true. For simplicity's sake we will assume that the statement Q is in the prefix form, i.e., no quantifiers occur in the scope of other logical connectives.

In the second case, it could be that several different value vectors can be assigned to the variables, essentially giving different answers. We also note that an answer could be a disjunction, giving possible options instead of a single definite answer.

A widely used machinery in resolution-based theorem provers for extracting values of existentially bound variables in Q is to use a special *answer predicate*, converting a question statement Q to a formula $\exists X(Q(X)\&\neg answer(X))$ for a tuple of existentially quantified variables X in Q [6]. Whenever a clause is derived which consists of only answer predicates, it is treated as a contradiction (essentially, answer) and the arguments of the answer predicate are returned as the values looked for. A common convention is to call such clauses *answer clauses.* We will require that the proof search does not stop whenever an answer clause is found, but will continue to look for new answer clauses until a predetermined time limit is reached. See [16] for a framework of extracting multiple answers.

We also assume that queries take a general form $(KB\&A) \Rightarrow Q$ where KB is a commonsense knowledge base, A is an optional set of precondition statements for this particular question and Q is a question statement. The whole general query form is negated and converted to clauses, i.e., disjunctions of literals (positive or negative atoms). We will call the clauses stemming from the question statement *question clauses.*

2.2 Blocker Atoms and Justification Checking

Without loss of generality we assume that the precondition and consequent formulas α and γ in default rules are clauses and justifications $\beta_1, ..., \beta_n$ are literals, i.e. positive or negative atoms: $\alpha : \beta_1, ...\beta_n \vdash \gamma$. Complex formulas can be encoded with a new predicate over the free variables of the formula and an equivalence of the new atom with the formula. Recall that Reiter assumes that the default rules are Skolemized.

We encode a default rule as a clause by concatenating into one clause the precondition and consequent clauses $\alpha(x)$ and $\gamma(x)$ and blocker atoms $block(\neg\beta_1)$,

..., $block(\neg\beta_n)$ where each justification β_i is either a positive or a negative atom. The negation $\neg$ is used since we prefer to speak about *blockers* and not *justificatons*. For example, the "birds can fly" default rule is represented as a clause

$$\neg\texttt{bird(X)} \vee \texttt{fly(X)} \vee \texttt{block(0,neg(fly(X)))}$$

where X is a variable and neg(fly(X)) encodes the negated justification. The first argument of the blocker (0 above) encodes priority information covered in the next section.

A proof of a question clause is a clause containing only answer atoms and blocker atoms. In the justification checking phase the system attempts to prove each decoded second blocker argument $\neg\beta_i$ in turn: the proof is considered invalid if some of $\neg\beta_i$ can be proved and this checking-proof itself is valid. If we pose a question fly(X) $\Rightarrow$ answer(X) to the system to be proved (see the earlier example), we get two different answers: answer(p) $\vee$ block(neg(fly(p)) and answer(b) $\vee$ block(neg(fly(b)). Checking the first of these means trying to prove $\neg$fly(p) which succeeds, hence the first answer is invalid. Checking the second answer we try to prove $\neg$fly(b) which fails, hence the answer is valid.

Notice that the contents $\neg\beta_i$ of blockers, just like answer clauses, have a role of collecting substitutions during the proof search: this enables us to disregard the order in which the clauses are used, i.e. both top-down, bottom-up and mixed proof search strategies can be used.

Importantly, blockers are used during the subsumption checks similarly to ordinary literals. A clause C_1 with fewer or more general literals than C_2 is hence always preferred to C_2, given that (a) the literals of C_1 subsume C_2, disregarding the priority arguments of blockers, and (b) the priority arguments of corresponding blocker literals in C_1 are equal or stronger than these of C_2. When combined with the uncertainty and inconsistency handling mechanisms of CONFER, the subsumption restrictions of the latter also apply. There are also other differences to ordinary literals. First, we prohibit the application of equality (demodulation or paramodulation) to the contents of blocker atoms during proof search. Second, we discard clauses containing mutually contradictory blockers (assuming the decoding of the second argument) like we would discard ordinary tautologies.

2.3 Priorities, Recursion and Infinite Branches

Default rule priorities are critical for the practical encoding of commonsense knowledge. The usage of priorities in proof search is simple: when checking a blocker with a given priority, it is not allowed to use default rules with a lower priority. We encode priority information as a first argument of the blocker literal, offering several ways to determine priority: either as an integer, a taxonomy class number, a string in a taxonomy or a combination of these with an integer.

For automatically using specificity we employ taxonomy classes: a class has a higher prirority than those above it on the taxonomy branch. We have built a topologically sorted acyclic graph of English words using the WordNet taxonomy along with an efficient algorithm for quick priority checks during proof search. Taxonomy classes are indicated with a special term like `$(61598)`. Alternatively one can use an actual English word like `$("bird")` which is automatically recognized to be more specific than, say, `$("object")`. To enable more fine-grained priorities, an integer can be added to the term like `$("bird",2)` generating a lexicographic order.

The recursive check for the non-provability of blockers can go arbitrarily deep, except for the time limits. Our algorithm allocates N seconds for the whole proof search and spends half of N for looking for different proofs and answers for the query, with the other half split evenly for each answer. Again, the time allocated for checking an answer is split evenly between the blockers in the answer. Each such time snippet is again split between a search for the proof of the blocker, and if found, for recursively checking the validity of this proof. Once the allocated time is below a given threshold (currently one millisecond) the proof is assumed to be not found.

Answers given by the system depend on the amount of time given, the search strategy chosen etc. For example, consider the Nixon triangle presented earlier, with two contradictory default rules. In case the priorities of these rules are equal and we allow defaults with the same priority to be used for checking an answer containing a blocker, the recursive check terminates only because of a time limit, which is unpredictable. Hence, we may sometimes get one answer and sometimes another. In order to increase both stability and efficiency, GK checks the blockers in the search nodes above, and terminates with failure in cases nonterminating loops are detected. Therefore GK always gives a sceptical result to the Nixon triangle: neither $\mathit{pacifist}(\mathit{nixon})$ nor $\neg \mathit{pacifist}(\mathit{nixon})$ is proven.

3 Confidences and Inconsistencies

GK integrates the exception-handling algorithms described in the previous chapter with the algorithms designed for handling inconsistent KB-s and numeric confidences assigned to clauses, previously presented as a CONFER framework in [18]. The framework is built on the resolution method. It calculates the estimates for the confidences of derived clauses, using both (a) the decreasing confidence of a conjunction of clauses as performed by the resolution and paramodulation rule, and (b) the increasing confidence of a disjunction of clauses for cumulating evidence. CONFER handles inconsistent KB-s by requiring the proofs of answers to contain the clauses stemming from the question posed. It performs searches both for the question and its negation and returns the resulting confidence calculated as a difference of the confidences found by these two searches.

The integrated algorithm is more complex than the one we previously described. Whenever the algorithms of the previous chapter speak about "proving", the system actually performs two independent searches – one for the positive and one for the negated goal – with the confidences calculated for both of these. A blocker is considered to be proved in case the resulting confidence is over a pre-determined configurable threshold, by default 0.5. Blocker proofs must also contain the clause built from the blocker. Thus, the whole search tree for a query consists of two types of interleaved layers: positive/negative confidence searches and blocker checking searches, the latter type potentially making the tree arbitrarily deep up to the minimal time limit threshold.

4 Implementation and Experiments

The described algorithms are implemented by the first author as a software system GK available at https://logictools.org/gk/. GK is written in C on top of our implementation of the CONFER framework [18] which is built on top of a high-performance resolution prover GKC [17] (see https://github.com/tammet/gkc) for conventional first order logic. Thus GK inherits most of the capabilities and algorithms of GKC.

A tutorial and a set of default logic example problems along with proofs from GK are also available at http://logictools.org/gk. GK is able to quickly solve nontrivial problems built by extending classic default logic examples. It is also able to solve classification problems combining exception and cumulative evidence and problems with dynamic situations using fluents, including planning problems. We have built a very large integrated knowledge base from the Quasimodo [14] and ConceptNet [15] knowledge bases, converting these to default logic plus confidences. GK is able to solve simple problems using this large knowledge base along with the Wordnet taxonomy for specificity: see the referenced web page for examples.

The following small example illustrates the fundamental difference of GK from the existing ASP systems for default logic. The standard penguins and birds example presented above in the ASP syntax is

```
bird(b1).
penguin(p1).
bird(X) :- penguin(X).
flies(X) :- bird(X), not -flies(X).
-flies(X) :- penguin(X).
```

Both GK and the ASP systems clingo 5.4.0, dlv 2.1.1 and s(CASP) 0.21.10.09 give an expected answer to the queries `flies(b1)` and `flies(p1)`. However, when we add the rules

```
bird(father(X)) :- bird(X).
penguin(father(X)) :- penguin(X).
```

none of these ASP systems terminate for these queries, while GK does solve the queries as expected. Notably, as pointed out by the author of s(CASP), this system does terminate for the reformulation of the same problem with the two replacement rules

```
flies(X) :- bird(X), not abs(X).
abs(X) :- penguin(X).
```

while clingo and dlv do not terminate. When we instead add the facts and rules

```
father(b1,b2).
father(p1,p2).
...
father(bN-1,bN).
father(pN-1,pN).

ancestor(X,Y):- father(X,Y).
ancestor(X,Y) :- ancestor(X,Z), ancestor(Z,Y).
```

for a large N, s(CASP) does not terminate and clingo and dlv become slow for `flies(b1)`: ca 8 s for $N = 500$ and ca 1 min for $N = 1000$ on a laptop with a 10-th generation i7 processor. GK solves the same question with $N = 1000$ under half a second and with $N = 100000$ under three seconds: the latter problem size is clearly out of scope of the capabilities of existing ASP systems.

We have previously shown that the confidence handling mechanisms in CONFER may slow down proof search for certain types of problems, but do not have a strong negative effect on very large commonsense CYC [12] problems in the TPTP problem collection. Differently from CONFER, the algorithms for default logic described above do not substantially modify the resolution method implementation of pure first order logic search, thus the performance of these parts of GK are mostly the same as of GKC. The ability to give a correct answer to a query during a given time limit depends on the performance of these components, and not on the overall recursively branching algorithm.

5 Summary and Future Work

We have presented algorithms and an implementation of an automated reasoning system for default logic on the basis of unrestricted first order logic and a resolution method. While there are several systems able to solve default logic or similar nonmonotonic logic problems, these are built on the basis of answer set programming and are normally based on grounding. We are not aware of other full first order logic reasoning systems for default logic, and neither of systems integrating confidences and inconsistency-handling with rules with exceptions.

Future work is planned on three directions: adding features to the solver, proving several useful properties of the algorithms and incorporating the solver into a commonsense reasoning system able to handle nontrivial tasks posed in natural language. The work on incorporating similarity-based reasoning into GK and building a suitable semantic parser for natural language is currently ongoing. We are particularly interested in exploring practical ways to integrate GK with the machine learning techniques for natural language.

References

1. Alviano, M., et al.: The ASP system DLV2. In: Balduccini, M., Janhunen, T. (eds.) LPNMR 2017. LNCS (LNAI), vol. 10377, pp. 215–221. Springer, Cham (2017). https://doi.org/10.1007/978-3-319-61660-5_19

2. Arias, J., Carro, M., Salazar, E., Marple, K., Gupta, G.: Constraint answer set programming without grounding. Theor. Pract. Logic Program. **18**(3–4), 337–354 (2018)
3. Bachmair, L., Ganzinger, H.: Resolution theorem proving. In: Robinson, A., Voronkov, A. (eds.) Handbook of Automated Reasoning, vol. I, ch. 2, pp. 19–99. Elsevier, Amsterdam (2001)
4. Brewka, G.: Adding priorities and specificity to default logic. In: MacNish, C., Pearce, D., Pereira, L.M. (eds.) JELIA 1994. LNCS, vol. 838, pp. 247–260. Springer, Heidelberg (1994). https://doi.org/10.1007/BFb0021977
5. Cholewinski, P., Marek, V.W., Truszczynski, M.: Default reasoning system deres. KR **96**, 518–528 (1996)
6. Green, C.: Theorem proving as a basis for question-answering systems. Mach. Intell. **4**, 183–205 (1969)
7. Kaminski, M.: A comparative study of open default theories. Artif. Intell. **77**(2), 285–319 (1995)
8. Kaminski, R., Schaub, T., Wanko, P.: A tutorial on hybrid answer set solving with *clingo*. In: Ianni, G. (ed.) Reasoning Web 2017. LNCS, vol. 10370, pp. 167–203. Springer, Cham (2017). https://doi.org/10.1007/978-3-319-61033-7_6
9. Kothawade, S., Khandelwal, V., Basu, K., Wang, H., Gupta, G.: Auto-discern: Autonomous driving using common sense reasoning (2021). arXiv preprint. arXiv:2110.13606
10. Lifschitz, V.: Answer Set Programming. Springer, Berlin (2019). https://doi.org/10.1007/978-3-030-24658-7
11. Nute, D.: Defeasible Logic, vol. 3. Oxford University Press, Oxford (1994)
12. Ramachandran, D., Reagan, P., Goolsbey, K.: First-orderized researchcyc: expressivity and efficiency in a common-sense ontology. In: AAAI Workshop on Contexts and Ontologies: Theory, Practice and Applications, pp. 33–40 (2005)
13. Reiter, R.: A logic for default reasoning. Artif. Intell. **13**(1–2), 81–132 (1980)
14. Romero, J., Razniewski, S., Pal, K., Pan, J.Z., Sakhadeo, A., Weikum, G.: Commonsense properties from query logs and question answering forums. In: Zhu, W. (eds.) Proceedings of the 28th ACM International Conference on Information and Knowledge Management, CIKM'19, pp. 1411–1420. ACM (2019)
15. Speer, R., Chin, J., Havasi, C.: ConceptNet 5.5: An open multilingual graph of general knowledge. In: Singh, S.P., Markovitch, S. (eds.) Proceedings of the 31st AAAI Conference on Artificial Intelligence, pp. 4444–4451. AAAI (2017)
16. Sutcliffe, G., Yerikalapudi, A., Trac, S.: Multiple answer extraction for question answering with automated theorem proving systems. In: Lane, H.C., Guesgen, H.W. (eds.) Proceedings of the 22nd International Florida Artificial Intelligence Research Society Conference, FLAIRS'22. AAAI (2009)
17. Tammet, T.: GKC: a reasoning system for large knowledge bases. In: Fontaine, P. (ed.) CADE 2019. LNCS (LNAI), vol. 11716, pp. 538–549. Springer, Cham (2019). https://doi.org/10.1007/978-3-030-29436-6_32
18. Tammet, T., Draheim, D., Järv, P.: Confidences for Commonsense Reasoning. In: Platzer, A., Sutcliffe, G. (eds.) CADE 2021. LNCS (LNAI), vol. 12699, pp. 507–524. Springer, Cham (2021). https://doi.org/10.1007/978-3-030-79876-5_29

Hypergraph-Based Inference Rules for Computing $\mathcal{EL}^+$-Ontology Justifications

Hui Yang([✉]), Yue Ma, and Nicole Bidoit

LISN, CNRS, Université Paris-Saclay, Gif-sur-Yvette, France
{yang,ma,nicole.bidoit}@lisn.fr

Abstract. To give concise explanations for a conclusion obtained by reasoning over ontologies, *justifications* have been proposed as minimal subsets of an ontology that entail the given conclusion. Even though computing one justification can be done in polynomial time for tractable Description Logics such as $\mathcal{EL}^+$, computing all justifications is complicated and often challenging for real-world ontologies. In this paper, based on a graph representation of $\mathcal{EL}^+$-ontologies, we propose a new set of *inference rules* (called H-rules) and take advantage of them for providing a new method of computing all justifications for a given conclusion. The advantage of our setting is that most of the time, it reduces the number of *inferences* (generated by H-rules) required to derive a given conclusion. This accelerates the enumeration of justifications relying on these inferences. We validate our approach by running real-world ontology experiments. Our graph-based approach outperforms PULi [14], the state-of-the-art algorithm, in most of cases.

1 Introduction

Ontologies provide structured representations of domain knowledge that are suitable for AI reasoning. They are used in various domains, including medicine, biology, and finance. In the domain of ontologies, one of the interesting topics is to provide explanations of reasoning conclusions. To this end, *justifications* have been proposed to offer users a brief explanation for a given conclusion. Computing justifications has been widely explored for different tasks, for instance for debugging ontologies [1,9,11] and computing ontology modules [6]. Extracting just one justification can be easy for tractable ontologies, such as $\mathcal{EL}^+$ [17]. For instance, we can find one justification by deleting unnecessary axioms one by one. However, there may exist more than one justification for a given conclusion. Computing all such justifications is computationally complex and reveals itself to be a challenging problem [18].

There are mainly two different approaches [17] to compute all justifications for a given conclusion, the *black-box* approach and the *glass-box* approach. The *black-box* approach [11] relies only on a reasoner and, as such, can be

This work is funded by the BPI-France (PSPC AIDA: 2019-PSPC-09).

J. Blanchette et al. (Eds.): IJCAR 2022, LNAI 13385, pp. 310–328, 2022.
https://doi.org/10.1007/978-3-031-10769-6_19

used for ontologies in any existing Description Logics. For example, a simple (naive) *black-box* approach would check all the subsets of the ontology using an existing reasoner and then filter the subset-minimal ones (i.e., justifications). Many advanced and optimized black-box algorithms have been proposed since 2007 [10]. Meanwhile, the glass-box approaches have achieved better performances over certain specific ontology languages (such as $\mathcal{EL}^+$-ontology) by going deep into the reasoning process. Among them, the class of SAT-based methods [1–3,14,16] performs the best. The main idea developed by SAT-based methods is to trace, in a first step, a *complete set of inferences* (*complete set* for short) that contribute to the derivation of a given conclusion, and then, in a second step, to use SAT-tools or resolution to extract all justifications from these inferences. A detailed example is provided in Sect. 4.1.

In the real world, ontologies are always huge. For instance, the SnomedCT ontology contains more than 300,000 axioms. Thus, the traced *complete set* can be large, which could make it challenging to extract the justifications over them. Several techniques could be applied to reduce the size of the traced *complete set*, like the *locality-based modules* [8] and the *goal-directed tracing algorithm* [12]. One of their shared ideas is to identify, for a given conclusion, a particular part of the ontology relevant for the extraction of justifications. For example, the state-of-the-art algorithm, PULi [14], uses a *goal-directed tracing algorithm*. However, even for PULi, a simple ontology $\mathcal{O} = \{A_i \sqsubseteq A_{i+1} \mid 1 \leq i \leq n-1\}$ with the conclusion $A_0 \sqsubseteq A_n$ leads to a *complete set* containing $n-1$ inferences. This set can not be reduced further even with the previously mentioned optimizations. From this observation, we decided to explore a new SAT-based glass-box method to handle such situations better.

Now, let us look carefully at the ontology $\mathcal{O}$ above, and let us regard each A_i as a graph node N_{A_i}. Then we are able to construct, for $\mathcal{O}$, a directed graph whose edges are of the form $N_{A_i} \rightarrow N_{A_{i+1}}$. It turns out that all the justifications for the conclusion $A_0 \sqsubseteq A_n$ are extracted from all the paths from N_{A_0} to N_{A_n}, and here we have only one such path. We can easily extend this idea on $\mathcal{EL}^+$-ontology because most of the $\mathcal{EL}^+$-axioms can be interpreted as direct edges except one case (i.e., $A \equiv B_1 \sqcap \cdots \sqcap B_n$), for which we need a hyperedge (for more details see Definition 3). However, for more expressive ontologies, this translation becomes more complicated. For example, it is hard to map $\mathcal{ALC}$-axioms to edges as those axioms may contain negation or disjunction of concepts.

This example inspired us to explore a hypergraph representation of the ontology and reformulate inferences and justifications. Roughly, our inferences are built from elementary paths of the hypergraph and lead to particular paths called H-paths. Then, computing all the justifications for a given conclusion is made using such H-paths. For the previous ontology $\mathcal{O}$ and the conclusion $A_0 \sqsubseteq A_n$, our *complete set* is reduced to only two inferences (no matter the value of n) corresponding to the unique path from N_{A_0} to N_{A_n}. The source of improvement provided by our method is twofold. On the one hand, it comes from the fact that elementary paths are pre-computed while extracting the inferences and that existing algorithms like depth-first search can efficiently compute such paths. On the other hand, yet as a consequence, decreasing the size of the *complete sets* of inferences leads to smaller inputs for the SAT-based algorithm extracting justifications from the *complete set* (recall here that our method is a SAT-based glass-box method).

The paper is organized as follows. Section 2 introduces preliminary definitions and notions. In Sect. 3, we associate a hypergraph representation to $\mathcal{EL}^+$-ontology and introduce a new set of inference rules, called H-rules, that generate our inferences. In Sect. 4, we develop the algorithm `minH`, which compute justifications based on our inferences. Section 5 shows experimental results and Sect. 6 summarizes our work.

2 Preliminaries

2.1 $\mathcal{EL}^+$-Ontology

Given sets of atomic concepts $\mathsf{N}_C = \{A, B, \cdots\}$ and atomic roles $\mathsf{N}_R = \{r, s, t, \cdots\}$, the set of $\mathcal{EL}^+$concepts C and axioms α are built by the following grammar rules:

$$C ::= \top \mid A \mid C \sqcap C \mid \exists r.C, \quad a ::= C \sqsubseteq C \mid C \equiv C \mid r \sqsubseteq s \mid r_1 \circ \cdots \circ r_n \sqsubseteq s.$$

A $\mathcal{EL}^+$-ontology $\mathcal{O}$ is a finite set of $\mathcal{EL}^+$-axioms. An **interpretation** $\mathcal{I} = (\Delta^\mathcal{I}, \cdot^\mathcal{I})$ of $\mathcal{O}$ consists of a non-empty set $\Delta^\mathcal{I}$ and a mapping from atomic concepts $A \in \mathsf{N}_C$ to a subset $A^\mathcal{I} \subseteq \Delta^\mathcal{I}$ and from roles $r \in \mathsf{N}_R$ to a subset $r^\mathcal{I} \subseteq \Delta^\mathcal{I} \times \Delta^\mathcal{I}$. For a concept C built from the grammar rules, we define $C^\mathcal{I}$ inductively by: $(\top)^\mathcal{I} = \Delta^\mathcal{I}, (C \sqcap D)^\mathcal{I} = C^\mathcal{I} \cap D^\mathcal{I}$, $(\exists r.C)^\mathcal{I} = \{a \in \Delta^\mathcal{I} \mid \exists b, (a,b) \in r^\mathcal{I}, b \in C^\mathcal{I}\}$, $(r \circ s)^\mathcal{I} = \{(a,b) \in \Delta^\mathcal{I} \times \Delta^\mathcal{I} \mid \exists c, (a,c) \in r^\mathcal{I}, (c,b) \in s^\mathcal{I}\}$. An interpretation is a **model** of $\mathcal{O}$ if it is compatible with all axioms in $\mathcal{O}$, i.e., for all $C \sqsubseteq D, C \equiv D, r \sqsubseteq s, r_1 \circ \cdots \circ r_n \sqsubseteq s \in \mathcal{O}$, we have $C^\mathcal{I} \subseteq D^\mathcal{I}, C^\mathcal{I} = D^\mathcal{I}, r^\mathcal{I} \subseteq s^\mathcal{I}, (r_1 \circ \cdots \circ r_n)^\mathcal{I} \subseteq s^\mathcal{I}$, respectively. We say $\mathcal{O} \models a$ where α is an axiom iff each model of $\mathcal{O}$ is compatible with α. A concept A is **subsumed** by B w.r.t. $\mathcal{O}$ if $\mathcal{O} \models A \sqsubseteq B$.

Next, we use $A, B, \cdots, G$ (possibly with subscripts) to denote atomic concepts and we use X, Y, Z (possibly with subscripts) to denote atomic concepts $A, \cdots, G$, or complex concepts $\exists r.A$, $\cdots$, $\exists r.G$.

We assume that ontologies are normalized. A $\mathcal{EL}^+$-ontology $\mathcal{O}$ is **normalized** if all its axioms are of the form $A \equiv B_1 \sqcap \cdots \sqcap B_m, A \sqsubseteq B_1 \sqcap \cdots \sqcap B_m, A \equiv \exists r.B, A \sqsubseteq \exists r.B, r \sqsubseteq s$, or $r \circ s \sqsubseteq t$, where $A, B, B_i \in \mathsf{N}_C$, and $r, s, t \in \mathsf{N}_R$. Every $\mathcal{EL}^+$-ontology can be normalised in polynomial time by introducing new atomic concepts and atomic roles.

Example 1. *The following set of axioms is a $\mathcal{EL}^+$-ontology:*
$\mathcal{O} = \{\ a_1{:}A \sqsubseteq D,\ a_2{:}D \sqsubseteq \exists r.E,\ a_3{:}E \sqsubseteq F, a_4{:}B \equiv \exists t.F,\ a_5{:}r \sqsubseteq t,\ a_6{:}G \equiv C \sqcap B\ , a_7{:}C \sqsubseteq A\}$.

It is clear that $\mathcal{O} \models A \sqsubseteq \exists r.E$ as for all models $\mathcal{I}$, we have $A^\mathcal{I} \subseteq D^\mathcal{I}$ by the axiom a_1 and $D^\mathcal{I} \subseteq (\exists r.E)^\mathcal{I}$ by a_2.

Table 1. Inference rules over $\mathcal{EL}^+$-ontology.

$$\mathcal{R}_1: \frac{A \sqsubseteq A_1, \cdots, A \sqsubseteq A_n, \quad A_1 \sqcap A_2 \sqcap \cdots \sqcap A_n \sqsubseteq B}{A \sqsubseteq B}$$

$$\mathcal{R}_2: \frac{A \sqsubseteq A_1, \quad A_1 \sqsubseteq \exists r.B}{A \sqsubseteq \exists r.B} \qquad \mathcal{R}_3: \frac{A \sqsubseteq \exists r.B_1, \quad B_1 \sqsubseteq B_2, \quad \exists r.B_2 \sqsubseteq B}{A \sqsubseteq B}$$

$$\mathcal{R}_4: \frac{A_0 \sqsubseteq \exists r_1.A_1, \quad \cdots, A_{n-1} \sqsubseteq \exists r_n.A_n, \quad r_1 \circ \cdots \circ r_n \sqsubseteq r}{A_0 \sqsubseteq \exists r.A_n}$$

2.2 Inference, Support and Justification

Given a $\mathcal{EL}^+$-ontology $\mathcal{O}$, a major reasoning task over $\mathcal{O}$ is *classification*, which aims at finding all subsumptions $\mathcal{O} \models A \sqsubseteq B$ for atomic concepts A, B occurring in $\mathcal{O}$. Generally, it can be solved by applying *inferences* recursively over $\mathcal{O}$ [5].

An **inference** ρ is a pair $\langle \rho_{pre}, \rho_{con} \rangle$ whose *premise* set ρ_{pre} consists of $\mathcal{EL}^+$-axioms and *conclusion* ρ_{con} is a single $\mathcal{EL}^+$-axiom. As usual, a sequence of inferences $\rho^1, \cdots, \rho^n$ is a **derivation** of an axiom α from $\mathcal{O}$ if $\rho^n_{con} = \alpha$ and for any $\beta \in \rho^i_{pre}$, $1 \leq i \leq n$, we have $\beta \in \mathcal{O}$ or $\beta = \rho^j_{con}$ for some $j < i$.

As usual, **inference rules** are used to generate inferences. For instance, Table 1 [1,5] shows a set of inference rules for $\mathcal{EL}^+$-ontologies. Next, we use $\mathcal{O} \vdash A \sqsubseteq B$ to denote that $A \sqsubseteq B$ is derivable from $\mathcal{O}$ using inferences generated by the rules in Table 1. The set of inference rules in Table 1 is *sound* and *complete* for classification [5], i.e., $\mathcal{O} \models A \sqsubseteq B$ iff $\mathcal{O} \vdash A \sqsubseteq B$ for any $A, B \in \mathsf{N}_C$.

A **support** of $A \sqsubseteq B$ over $\mathcal{O}$ is a sub-ontology $\mathcal{O}' \subseteq \mathcal{O}$ such that $\mathcal{O}' \models A \sqsubseteq B$. The **justifications** for $A \sqsubseteq B$ are subset-minimal supports of $A \sqsubseteq B$. We denote the collection of all justifications for $A \sqsubseteq B$ w.r.t. $\mathcal{O}$ by $J_{\mathcal{O}}(A \sqsubseteq B)$.

We say S is a **complete set** (of inferences) for $A \sqsubseteq B$ if for any justifications $\mathcal{O}'$ of $A \sqsubseteq B$, we can derive $A \sqsubseteq B$ from $\mathcal{O}'$ using only the inferences in S.

Example 2 (Example 1 cont'd). *Before applying inference rules, axioms in $\mathcal{O}$ are preprocessed in order to be compatible with Table 1. For example, a_4 is replaced by $B \sqsubseteq \exists t.F$ and $\exists t.F \sqsubseteq B$. Then, according to the inference rules of Table 1, we may produce the following inferences: $\rho = \langle \{A \sqsubseteq D, D \sqsubseteq \exists r.E\}, A \sqsubseteq \exists r.E \rangle$, $\rho' = \langle \{A \sqsubseteq \exists r.E, r \sqsubseteq t\}, A \sqsubseteq \exists t.E \rangle$ and $\rho'' = \langle \{A \sqsubseteq \exists t.E, E \sqsubseteq F, \exists t.F \sqsubseteq B\}, A \sqsubseteq B \rangle$ generated by rule $\mathcal{R}_2$, $\mathcal{R}_4$ and $\mathcal{R}_3$ respectively. Then $\mathcal{O} \vdash A \sqsubseteq B$ since $A \sqsubseteq B$ is derivable from $\mathcal{O}$ by the sequence ρ, ρ', ρ''.*

Notice that $\mathcal{O}' = \{a_1, a_2, a_3, a_4, a_5\}$ is a support for $A \sqsubseteq B$, and thus, any superset $\mathcal{O}''$ of $\mathcal{O}'$ is a support of $A \sqsubseteq B$. $\mathcal{O}'$ is also one of the justifications for $A \sqsubseteq B$ as for any $\mathcal{O}''' \subset \mathcal{O}'$, we have $\mathcal{O}''' \not\models A \sqsubseteq B$. Moreover, here the three inferences ρ, ρ', ρ'' provide a complete set for $A \sqsubseteq B$.

3 Hypergraph-Based Inference Rules

3.1 H-Inferences

In general, a (directed) hypergraph $\mathcal{G} = (\mathcal{V}, \mathcal{E})$ is defined by a set of nodes $\mathcal{V}$ and a set of hyperedges $\mathcal{E}$ [4,7]. A hyperedge is of the form $e = (S_1, S_2), S_1, S_2 \subseteq \mathcal{V}$. In this paper, a hypergraph is associated to an ontology as follows:

Definition 3. *For a given $\mathcal{EL}^+$-ontology $\mathcal{O}$, the associated hypergraph is $\mathcal{G}_\mathcal{O} = (\mathcal{V}_\mathcal{O}, \mathcal{E}_\mathcal{O})$ where (i) the set of nodes $\mathcal{V}_\mathcal{O} = \{N_A, N_r, N_{\exists r.A} \mid A \in \mathsf{N_C}, r \in \mathsf{N_R}\}$ and (ii) the set of edges $\mathcal{E}_\mathcal{O}$ is defined by $f(\mathcal{O})$ where f is the multi-valued mapping shown in Fig. 1. Given a hyperedge e of $\mathcal{E}_\mathcal{O}$, the inverse image of e, $f^{-1}(e)$, is defined in the obvious manner. For a set E of hyperedges, $f^{-1}(E) = \cup_{e \in E} f^{-1}(e)$.*

	α	$f(\alpha)$
1.	$A \sqsubseteq B_1 \sqcap \cdots B_n$	$(\{N_A\}, \{N_{B_i}\}), 1 \leq i \leq n$
2.	$A \equiv B_1 \sqcap \cdots B_m$	$(\{N_A\}, \{N_{B_i}\}), 1 \leq i \leq m$ $(\{N_{B_1}, \cdots, N_{B_m}\}, \{N_A\})$
3.	$A \sqsubseteq \exists r.B$	$(\{N_A\}, \{N_{\exists r.B}\})$
4.	$A \equiv \exists r.B$	$(\{N_{\exists r.B}\}, \{N_A\})$ and $(\{N_A\}, \{N_{\exists r.B}\})$
5.	$r \sqsubseteq s$	$(\{N_r\}, \{N_s\})$ and $(\{N_{\exists r.A}\}, \{N_{\exists s.A}\})$ for all A
6.	$r \circ s \sqsubseteq t$	$(\{N_r, N_s\}, \{N_s, N_t\})$

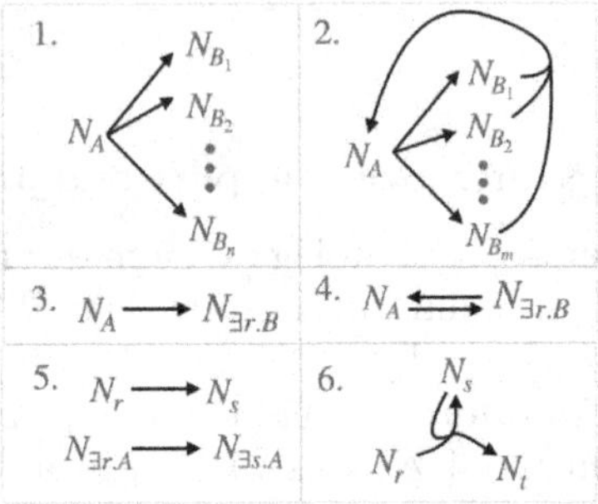

Fig. 1. Definition of f (left) and graphical illustrations of $f(\alpha)$ (right)

Notice that, the hyperedges associated with $A \equiv B_1 \sqcap \cdots \sqcap B_m$ are (i) the hyperedge $(\{N_{B_1}, \cdots, N_{B_m}\}, \{N_A\})$ and (2) of course, the edges corresponding to $A \sqsubseteq B_1 \sqcap \cdots \sqcap B_m$.

Example 4 (Example 1 cont'd). *The hypergraph $\mathcal{G}_\mathcal{O}$ for $\mathcal{O}$ is shown in Fig. 2, where $e_0 = (\{N_C\}, \{N_A\})$, $e_1 = (\{N_A\}, \{N_D\})$, $e_2 = (\{N_D\}, \{N_{\exists r.E}\})$, etc. Also, $f^{-1}(e_0) = C \sqsubseteq A$, $f^{-1}(e_1) = A \sqsubseteq D$, and $f^{-1}(e_2) = D \sqsubseteq \exists r.E$, etc.*

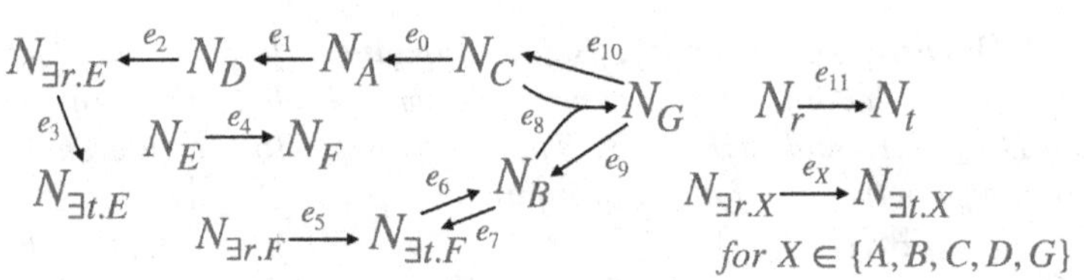

Fig. 2. The hypergraph associated with the ontology $\mathcal{O}$.

As for graphs, a path (next called **regular path**) from nodes N_1 to N_2 in a hypergraph is a sequence of edges:

$$e_0 = (S_1^0, S_2^0), e_1 = (S_1^1, S_2^1), \cdots, e_n = (S_1^n, S_2^n) \tag{1}$$

where $N_1 \in S_1^0, N_2 \in S_2^n$ and $S_2^{i-1} = S_1^i, 1 \leq i \leq n$. Next, the **existence** of a regular path from N_X to N_Y in a hypergraph $\mathcal{G}_\mathcal{O}$ is denoted $N_X \rightsquigarrow N_Y$. Now, we introduce hypergraph-based inferences which are based on the existence of regular paths as follows:

Table 2. H-rules over $\mathcal{G}_\mathcal{O} = (\mathcal{V}_\mathcal{O}, \mathcal{E}_\mathcal{O})$.

$$\mathcal{H}_0 : \frac{N_X \rightsquigarrow N_Y}{N_X \overset{h}{\rightsquigarrow} N_Y} \qquad \mathcal{H}_2 : \frac{N_X \overset{h}{\rightsquigarrow} N_{\exists r.B_1},\ N_{B_1} \overset{h}{\rightsquigarrow} N_{B_2},\ N_{\exists r.B_2} \rightsquigarrow N_Y}{N_X \overset{h}{\rightsquigarrow} N_Y}$$

$$\mathcal{H}_1 : \frac{N_X \overset{h}{\rightsquigarrow} N_{B_1}, \cdots, N_X \overset{h}{\rightsquigarrow} N_{B_m},\ N_A \rightsquigarrow N_Y,\ e}{N_X \overset{h}{\rightsquigarrow} N_Y} : e{=}(\{N_{B_1}, \cdots, N_{B_m}\}, \{N_A\}){\in}\mathcal{E}_\mathcal{O}$$

$$\mathcal{H}_3 : \frac{N_X \overset{h}{\rightsquigarrow} N_{\exists r.A_1},\ N_{A_1} \overset{h}{\rightsquigarrow} N_{\exists s.A_2},\ N_{\exists t.A_2} \rightsquigarrow N_Y,\ e}{N_X \overset{h}{\rightsquigarrow} N_Y} : e{=}(\{N_r, N_s\}, \{N_s, N_t\}){\in}\mathcal{E}_\mathcal{O}$$

Definition 5. *Given a hypergraph $\mathcal{G}_\mathcal{O}$, Table 2 gives a set of inference rules called **H-rules**. Inferences based on H-rules are called **H-inferences**. Next, we denote by $\mathcal{O} \vdash_h N_X \overset{h}{\rightsquigarrow} N_Y$ (or simply $N_X \overset{h}{\rightsquigarrow} N_Y$) the fact that $N_X \overset{h}{\rightsquigarrow} N_Y$ can be derived from $\mathcal{G}_\mathcal{O}$ using the H-inferences.*

Example 6 (Example 4 cont'd). *As shown in Fig. 2, we have $N_A \rightsquigarrow N_{\exists r.E}$, $N_E \rightsquigarrow N_F$, $N_{\exists r.F} \rightsquigarrow N_B$ from the existence of regular paths. Then we can derive $N_A \overset{h}{\rightsquigarrow} N_B$ from $\mathcal{G}_\mathcal{O}$ by the H-rules $\mathcal{H}_0$, $\mathcal{H}_0$ and $\mathcal{H}_2$ which generate the H-inferences ρ^1, ρ^2, ρ^3, where $\rho^1 = \langle\{N_A \rightsquigarrow N_{\exists r.E}\}, N_A \overset{h}{\rightsquigarrow} N_{\exists r.E}\rangle$, $\rho^2 = \langle\{N_E \rightsquigarrow N_F\}, N_E \overset{h}{\rightsquigarrow} N_F\rangle$ and $\rho^3 = \langle\{N_A \overset{h}{\rightsquigarrow} N_{\exists r.E}, N_E \overset{h}{\rightsquigarrow} N_F, N_{\exists r.F} \rightsquigarrow N_B\}, N_A \overset{h}{\rightsquigarrow} N_B\rangle$, respectively.*

Note that the first rule $\mathcal{H}_0$, the initialization rule, makes regular paths the elementary components of H-rules. Moreover, Proposition 7 formally states that, in our H-inference system, we do not need to add the transitive inference rule:

$$\frac{N_X \overset{h}{\rightsquigarrow} N_Z, N_Z \overset{h}{\rightsquigarrow} N_Y}{N_X \overset{h}{\rightsquigarrow} N_Y}.$$

Proposition 7. *If $\mathcal{O} \vdash_h N_X \overset{h}{\rightsquigarrow} N_Z$ and $\mathcal{O} \vdash_h N_Z \overset{h}{\rightsquigarrow} N_Y$ then $\mathcal{O} \vdash_h N_X \overset{h}{\rightsquigarrow} N_Y$.*

3.2 Completeness and Soundness of H-Inferences

The following result is the main result of this section. It states the equivalence of $N_X \overset{h}{\rightsquigarrow} N_Y$ derivation (by Table 2) and ontology entailment for $X \sqsubseteq Y$, and thus states that our H-rules are sound and complete for $\mathcal{EL}^+$-ontology.

Theorem 8. *If $\mathcal{O}$ is an $\mathcal{EL}^+$-ontology, then $\mathcal{O} \models X \sqsubseteq Y$ iff $\mathcal{O} \vdash_h N_X \overset{h}{\rightsquigarrow} N_Y$, where X, Y are concepts of either form A or $\exists r.B$.*

Proof. "$\Leftarrow$" is obvious by induction over Table 2 and the fact that $N_X \rightsquigarrow N_Y$ implies $\mathcal{O} \models X \sqsubseteq Y$, so we only need to prove the direction "$\Rightarrow$".

Assume that $\mathcal{O} \models X \sqsubseteq Y$. We consider two cases:

Case 1. We assume $\mathcal{O} \vdash X \sqsubseteq Y$[1]. Let $d(X,Y)$ be the length of one shortest derivation of $X \sqsubseteq Y$ from $\mathcal{O}$ using Table 1. We prove "⇒" by induction on $d(X,Y)$.

- **Assume** $d(X,Y) = 0$**.** In this case $\mathcal{O}$ must contain axioms of the form $X \equiv Y \sqcap \cdots$ or $X \sqsubseteq Y \sqcap \cdots$. Clearly we have $N_X \rightsquigarrow N_Y$ thus $\mathcal{O} \vdash_h N_X \overset{h}{\rightsquigarrow} N_Y$.
- **Assuming "⇒" holds when** $d(X,Y) < k$**, let us prove "⇒" holds for** $d(X,Y) = k$**.** Suppose ρ^{last} is the last inference in one shortest derivation of $X \sqsubseteq Y$ using Table 1. Two cases arise:
  1. Assume ρ^{last} is generated by $\mathcal{R}_1(n > 1), \mathcal{R}_3$ or $\mathcal{R}_4(n = 2)$. For example, assume $\rho^{last} = \langle\{X \sqsubseteq \exists r.B_1, B_1 \sqsubseteq B_2, \exists r.B_2 \sqsubseteq Y\}, X \sqsubseteq Y\rangle$ comes from $\mathcal{R}_3$. We have $d(X, \exists r.B_1), d(B_1, B_2), d(\exists r.B_2, Y) < k$ because their corresponding subsumptions can be derived without ρ^{last}. By the assumption $\mathcal{O} \vdash_h N_X \overset{h}{\rightsquigarrow} N_{\exists r.B_1}, N_{B_1} \overset{h}{\rightsquigarrow} N_{B_2}, N_{\exists r.B_2} \overset{h}{\rightsquigarrow} N_Y$. Then we have $\mathcal{O} \vdash_h N_X \overset{h}{\rightsquigarrow} N_{\exists r.B_2}$ by first deriving $N_X \overset{h}{\rightsquigarrow} N_{\exists r.B_1}, N_{B_1} \overset{h}{\rightsquigarrow} N_{B_2}$, and then applying H-inference:

     $$\rho^{new} = \langle\{N_X \overset{h}{\rightsquigarrow} N_{\exists r.B_1}, N_{B_1} \overset{h}{\rightsquigarrow} N_{B_2}, N_{\exists r.B_2} \rightsquigarrow N_{\exists r.B_2}\}, N_X \overset{h}{\rightsquigarrow} N_{\exists r.B_2}\rangle.$$

     Then $\mathcal{O} \vdash_h N_X \overset{h}{\rightsquigarrow} N_Y$ by Proposition 7 since $\mathcal{O} \vdash_h N_X \overset{h}{\rightsquigarrow} N_{\exists r.B_2}, N_{\exists r.B_2} \overset{h}{\rightsquigarrow} N_B$. The argument also holds for $\mathcal{R}_1(n > 1)$(or $\mathcal{R}_4(n = 2)$) by applying $\mathcal{H}_1$ (or $\mathcal{H}_3$) instead of $\mathcal{H}_2$.
  2. Assume ρ^{last} is generated by $\mathcal{R}_1(n = 1), \mathcal{R}_2$ or $\mathcal{R}_4(n = 1)$. Then, in each case, we have ρ^{last} has the form $\langle\{X \sqsubseteq Z, Z \sqsubseteq Y\}, X \sqsubseteq Y\rangle$. As in case 1, we have $d(X, Z), d(Z, Y) < k$. By the assumption, $\mathcal{O} \vdash_h N_X \overset{h}{\rightsquigarrow} N_Z, N_Z \overset{h}{\rightsquigarrow} N_Y$, then $\mathcal{O} \vdash_h N_X \overset{h}{\rightsquigarrow} N_Y$ by Proposition 7.

Case 2. If $\mathcal{O} \vdash X \sqsubseteq Y$ does not hold, then X or Y is not atomic. In this case, we introduce new axioms $A \equiv X$, $B \equiv Y$ with new atomic concepts A, B and denote the extended ontology by $\mathcal{O}'$. Clearly, $\mathcal{O}' \models A \sqsubseteq B$ and thus $\mathcal{O}' \vdash A \sqsubseteq B$ since Table 1 is sound and complete. Therefore, we have $\mathcal{O}' \vdash_h N_A \overset{h}{\rightsquigarrow} N_B$ by the same arguments as above. Now, notice that $\mathcal{G}_{\mathcal{O}'}$ is obtained from $\mathcal{G}_{\mathcal{O}}$ by adding 4 edges: $(\{N_A\}, \{N_X\}), (\{N_X\}, \{N_A\}), (\{N_B\}, \{N_Y\})$ and $(\{N_Y\}, \{N_B\})$, thus we have $\mathcal{O}' \vdash_h N_A \overset{h}{\rightsquigarrow} N_B$ iff $\mathcal{O} \vdash_h N_X \overset{h}{\rightsquigarrow} N_Y$.

3.3 Extracting Justifications from $\mathcal{G}_{\mathcal{O}}$

Now, we formally define H-paths as a hypergraph representation of classical derivations based on H-rules. The reader should pay attention to the fact that H-paths are not classical hyperpaths [7]. Next, for the sake of homogeneity, we consider a regular path from N_X to N_Y as the set of its edges and denote it as $P_{X,Y}$.

[1] The reader should recall that the equivalence ($\mathcal{O} \models X \sqsubseteq Y$ iff $\mathcal{O} \vdash X \sqsubseteq Y$) only holds when X and Y are atomic concepts wrt. the inference system presented in Table 1.

Definition 9 (H-paths). *In the hypergraph $\mathcal{G}_\mathcal{O}$, an H-path $H_{X,Y}$ from N_X to N_Y is a set of edges recursively generated by the following composition rules:*

0. *A regular path $P_{X,Y}$ is an H-path from N_X to N_Y;*
1. *If $e = (\{N_{B_1}, \cdots, N_{B_m}\}, \{N_A\}) \in \mathcal{V}_\mathcal{O}$, if H_{X,B_i} are H-paths for $i = 1..m$, and if $P_{A,Y}$ is a regular path, then $H_{X,B_1} \cup \cdots \cup H_{X,B_m} \cup P_{A,Y} \cup \{e\}$ is an H-path from N_X to N_Y;*
2. *If $H_{X,\exists r.B_1}, H_{B_1,B_2}$ are H-paths and $P_{\exists r.B_2,Y}$ is a regular path, then $H_{X,\exists r.B_1} \cup H_{B_1,B_2} \cup P_{\exists r.B_2,Y}$ is an H-path from N_X to N_Y;*
3. *If $e = (\{N_r, N_s\}, \{N_s, N_t\}) \in \mathcal{V}_\mathcal{O}$, if $H_{X,\exists r.A_1}, H_{A_1,\exists s.A_2}$ are H-paths and if $P_{\exists t.A_2,B}$ is a regular path, then $H_{X,\exists r.A_1} \cup H_{A_1,\exists s.A_2} \cup P_{\exists t.A_2,B} \cup \{e\}$ is an H-path from N_X to N_Y.*

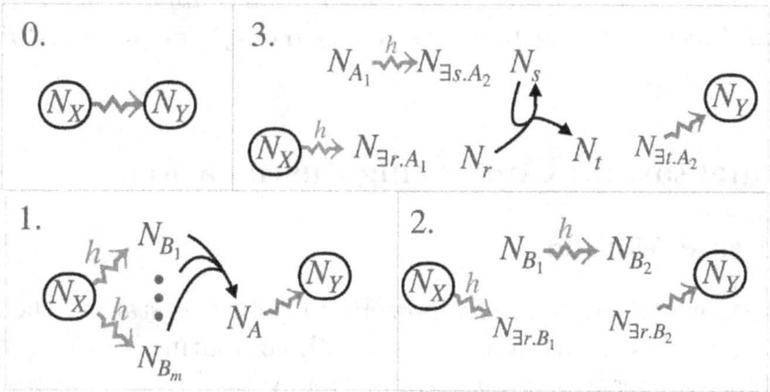

Fig. 3. Structure of H-paths from N_X to N_Y

Figure 3 gives an illustration of H-paths: the blue arrows $\rightsquigarrow$ correspond to regular paths, and the red ones $\overset{h}{\rightsquigarrow}$ to H-paths. It is straightforward to compare composition rules building H-paths with H-rules building derivations in Table 2. One may also consider H-paths as deviation-trees with leaves corresponding to the edges in $\mathcal{G}_\mathcal{O}$. However, our approach provides a more direct characterization of justifications as shown in Theorem 10.

We say that an H-path $H_{X,Y}$ is **minimal** if there is no H-path $H'_{X,Y}$ such that $H'_{X,Y} \subset H_{X,Y}$.

Now, we are ready to explain how H-paths and justifications are related. We can compute justifications from minimal H-paths as stated below:

Theorem 10. *Given X, Y of either form A or $\exists r.B$. Let*

$$\mathcal{S} = \{f^{-1}(H_{X,Y}) \mid H_{X,Y} \text{ is a minimal H-path from } N_X \text{ to } N_Y\}.$$

Then $\mathcal{J}_\mathcal{O}(X \sqsubseteq Y) = \{s \in \mathcal{S} \mid s' \not\subset s, \forall s' \in \mathcal{S}\}$. That is, all justifications for $X \sqsubseteq Y$ are the minimal subsets in $\mathcal{S}$.

Proof. For any justification $\mathcal{O}'$ of $X \sqsubseteq Y$, there exists a minimal H-path $H_{X,Y}$ such that $\mathcal{O}' = f^{-1}(H_{X,Y})$. The reason is that, since $\mathcal{O}' \models X \sqsubseteq Y$, there exists an H-path $H_{X,Y}$ from N_X to N_Y on $\mathcal{G}_{\mathcal{O}'}$ by Theorem 8. Without loss of generality, we can assume $H_{X,Y}$ is minimal on $\mathcal{G}_{\mathcal{O}'}$, then it is also minimal on $\mathcal{G}_{\mathcal{O}}$ since $\mathcal{G}_{\mathcal{O}'}$ is a sub-graph of $\mathcal{G}_{\mathcal{O}}$. We have $\mathcal{O}' = f^{-1}(H_{X,Y})$ because otherwise there exists $\mathcal{O}'' \subsetneq \mathcal{O}'$ such that $\mathcal{O}'' = f^{-1}(H_{X,Y})$, and thus $\mathcal{O}'' \models X \sqsubseteq Y$ by Theorem 8 again. Therefore, $\mathcal{O}'$ is not a justification. Contradiction.

Now, we know $\mathcal{S}$ contains all justifications for $X \sqsubseteq Y$. Moreover, $f^{-1}(H_{X,Y}) \models X \sqsubseteq Y$ for any H-path $H_{X,Y}$. Therefore, we have $\mathcal{J}_{\mathcal{O}}(X \sqsubseteq Y) = \{s \in \mathcal{S} \mid s' \not\subset s, \forall s' \in \mathcal{S}\}$ by the definition of justifications.

Example 11. (Example 4 cont'd). *The regular paths from N_A to $N_{\exists r.E}$ and from N_E to N_F produce two H-paths $H_{A,\exists rE} = \{e_1, e_2, e_3\}$ and $H_{E,F} = \{e_4\}$. Then, applying the third composition rule with $H_{A,\exists rE}, H_{E,F}$ and $P_{\exists r.F,B} = \{e_6\}$, we get $H_{A,B} = \{e_1, e_2, e_3, e_4, e_6\}$, which is the unique H-path from N_A to N_B. Thus, by Theorem 10, we have $\{\alpha_1, \alpha_2, \alpha_3, \alpha_4, \alpha_5\}$, the only justification for $A \sqsubseteq B$.*

4 Implementation: Computing Justifications

4.1 SAT-Based Method

In this section, we describe briefly how PULi [14], the state-of-the-art *glass-box* algorithm, proceeds. Given an ontology $\mathcal{O}$, computing $J_{\mathcal{O}}(X \sqsubseteq Y)$ is done through 2 steps: (1) tracing *a complete set* for $X \sqsubseteq Y$, (2) using resolution to extract the justifications from *the complete set*. The following example illustrates both steps:

Example 12 (Example 1 cont'd). *Let us compute $J_{\mathcal{O}}(G \sqsubseteq D)$ using PULi's method.*

1. *Using the goal-directed tracing algorithm in [12], the first step produces a complete set of inferences*[2] *$\{\rho_1, \rho_2\}$ for $G \sqsubseteq D$, where $\rho_1 = \langle\{G \sqsubseteq C, C \sqsubseteq A\}, G \sqsubseteq A\rangle, \rho_2 = \langle\{G \sqsubseteq A, A \sqsubseteq D\}, G \sqsubseteq D\rangle$.*
2. *This step is again composed of two parts:*
   - *(a) The first part proceeds to the translation of the inferences into clauses. Let us denote $\overline{p}_1{:}G \sqsubseteq C$, $\overline{p}_2{:}C \sqsubseteq A$, $\overline{p}_3{:}A \sqsubseteq D$, $p_4{:}G \sqsubseteq A$, $p_5{:}G \sqsubseteq D$. Here the literals $\overline{p}_1, \overline{p}_2, \overline{p}_3$ (with a bar) are called* **answer literals** *as they correspond to the axioms a_6, a_7, a_1 in $\mathcal{O}$. Thus, we obtain $\mathcal{C} = \{\neg\overline{p}_1 \vee \neg\overline{p}_2 \vee p_4, \neg p_4 \vee \neg\overline{p}_3 \vee p_5\}$ by rewriting the inferences ρ_1, ρ_2 as clauses.*
   - *(b) Secondly, a new clause $\neg p_5$ is added to $\mathcal{C}$, where p_5 corresponds to the conclusion $G \sqsubseteq D$, and resolution is applied over $\mathcal{C}$. The set of all justifications $J_{\mathcal{O}}(G \sqsubseteq D)$ is obtained by considering (i) the clauses formed of*

[2] For the sake of simplicity, we use the inference rules in Table 1 although PULi uses a slightly different set of inference rules [13].

Algorithm 1: minH

input : $X \sqsubseteq Y$
output: J: $\mathcal{J}_{\mathcal{O}}(X \sqsubseteq Y)$.
1 J $\leftarrow \emptyset$;
2 $\mathcal{U} \leftarrow$ `CompleteH(`$N_X \overset{h}{\rightsquigarrow} N_Y$`)`;
3 min_hpaths $\leftarrow$ `resolution(clauses(`$\mathcal{U}$`))`;
4 **for** h $\in$ min_hpaths **do**
5 　**if** f^{-1}(h') $\not\subset f^{-1}$(h) *for any* h' $\in$ min_hpaths **then**
6 　　J.*add*(f^{-1}(h))
7 　**end**
8 **end**

answer literals only and (ii) among them keeping the minimal ones[3]. *In this example, after the resolution phase, the only clause that consists of merely answer literals is* $\neg\overline{p}_1 \vee \neg\overline{p}_2 \vee \neg\overline{p}_3$. *Thus, the set of all justifications is* $J_{\mathcal{O}}(G \sqsubseteq D) = \{\{a_1, a_6, a_7\}\}$.

Our method for computing justifications follows the same steps as PULi although here the major difference is that the first step computes a complete set of H-inferences instead of a complete set of inferences wrt. Table 1.

4.2 Computing Justification by Minimal H-Paths

In this section, given an ontology $\mathcal{O}$ and its associated hypergraph $\mathcal{G}_{\mathcal{O}}$, we present `minH` (Algorithm 1) that computes all justifications for $X_0 \sqsubseteq Y_0$ using the minimal H-paths from N_{X_0} to N_{Y_0} over $\mathcal{G}_{\mathcal{O}}$. The algorithm `minH` proceeds in two steps described below.

Step 1. First, at Line 2, `minH` computes a *complete set* of inferences $\mathcal{U}$ for $N_{X_0} \overset{h}{\rightsquigarrow} N_{Y_0}$ using `CompleteH` (See Algorithm 2). Here, $\mathcal{U}$ is complete in the sense that for any H-path $H_{X,Y}$, we can derive $N_X \overset{h}{\rightsquigarrow} N_Y$ using inferences in $\mathcal{U}$ from the edge set $H_{X,Y}$. `CompleteH` computes $\mathcal{U}$ as follows:

- **Line 3–12 of Algorithm 2:** The recursive application of `trace_one_turn` (See Algorithm 3) outputs the set of all H-inferences whose conclusion is the given input $N_{X_1} \overset{h}{\rightsquigarrow} N_{Y_1}$;
- **Line 13–17 of Algorithm 2:** Let `path` be the depth-first search algorithm that computes all regular paths from N_X to N_Y in $\mathcal{G}_{\mathcal{O}}$ with input (N_X, N_Y). Intuitively, the purpose is to shift inferences from regular paths to edges.

Step 2. Then Algorithm `minH` computes all justifications for $X_0 \sqsubseteq Y_0$ as follows:

[3] Here a clause c is smaller than c_1 if all the literals of c are in c_1.

Algorithm 2: CompleteH

input : $N_X \overset{h}{\rightsquigarrow} N_Y$
output: $\mathcal{U}$: a complete set of inferences for $N_X \overset{h}{\rightsquigarrow} N_Y$.

1 $\mathcal{U}$, history, Q $\leftarrow \emptyset$; // Q is a queue
2 Q.*add*($N_X \overset{h}{\rightsquigarrow} N_Y$);
3 **while** Q $\neq \emptyset$ **do**
4     $N_{X_1} \overset{h}{\rightsquigarrow} N_{Y_1} \leftarrow$ Q.*takeNext*();
5   history.*add*($N_{X_1} \overset{h}{\rightsquigarrow} N_{Y_1}$);
6     $\mathcal{U} \leftarrow \mathcal{U} \bigcup$ `trace_one_turn`($N_{X_1} \overset{h}{\rightsquigarrow} N_{Y_1}$);
7   **for** $N_{X_2} \overset{h}{\rightsquigarrow} N_{Y_2}$ *appearing in* `trace_one_turn`($N_{X_1} \overset{h}{\rightsquigarrow} N_{Y_1}$) **do**
8   **if** $N_{X_2} \overset{h}{\rightsquigarrow} N_{Y_2} \notin$ history *and* $N_{X_2} \overset{h}{\rightsquigarrow} N_{Y_2} \notin$ Q **then**
9   Q.add($N_{X_2} \overset{h}{\rightsquigarrow} N_{Y_2}$)
10   **end**
11   **end**
12 **end**
13 **for** $N_{X_2} \rightsquigarrow N_{Y_2}$ *appearing in* $\mathcal{U}$ **do**
14   **for** $p=\{e_1, e_2, \cdots, e_n\} \in$ `path`(N_{X_2}, N_{Y_2}) **do**
15         $\mathcal{U}$.*add*($\langle\{e_1, e_2, \cdots, e_n\}, N_{X_2} \rightsquigarrow N_{Y_2}\rangle$);
16   **end**
17 **end**

- **Line 3 of Algorithm 1:** It computes all minimal H-paths from N_{X_0} to N_{Y_0} using resolution, which is developed by PULi[4], over the clauses generated from $\mathcal{U}$ as illustrated in Sect. 4.1. Here, a literal p is associated with each edge e, each $N_X \overset{h}{\rightsquigarrow} N_Y$, and each $N_X \rightsquigarrow N_Y$ in $\mathcal{U}$. The answer literals are those associated with edges.
- **Line 4–8 of Algorithm 1:** It computes justifications by mapping back all the minimal H-paths and select the subset-minimal sets as stated in Theorem 10.

Example 13 (Example 4 cont'd). *Assume $X_0 = G$ and $Y_0 = D$ are the input of* **minH**. *Then at line 2 of* **minH**, *we have $\mathcal{U} = \{\rho^1, \rho^2\}$, where $\rho^1 = \langle\{N_G \rightsquigarrow N_D\}, N_G \overset{h}{\rightsquigarrow} N_D\rangle$ is H-inference obtained by* **CompleteH** *(line 3–12) and $\rho^2 = \langle\{e_0, e_1, e_8\}, N_G \rightsquigarrow N_D\rangle$ is produced from regular paths obtained by* **CompleteH** *(line 13–17). Let us denote $\overline{p}_0{:}e_0$, $\overline{p}_1{:}e_1$, $\overline{p}_2{:}e_8$ as answer literals and $p_3{:}N_G \rightsquigarrow N_D$, $p_4{:}N_G \overset{h}{\rightsquigarrow} N_D$. Then* **clauses**$(\mathcal{U}) = \{\neg p_3 \vee p_4,\ \neg\overline{p}_0 \vee \neg\overline{p}_1 \vee \neg\overline{p}_2 \vee p_3\}$.

By resolution over **clauses**$(\mathcal{U})$, *we obtain* **min_hpaths** $=\{\{e_0, e_1, e_8\}\}$ *at line 3 of* **minH**. *Then the output of* **minH** *is $J=\{\{a_1, a_6, a_7\}\}$, which is the set of all justifications for $G \sqsubseteq D$.*

[4] Available at https://github.com/liveontologies/pinpointing-experiments.

Algorithm 3: `trace_one_turn`

input : $N_X \overset{h}{\rightsquigarrow} N_Y$

output: the set result of all H-inferences whose conclusion is $N_X \overset{h}{\rightsquigarrow} N_Y$.

1 result $\leftarrow \emptyset$;
2 $\mathcal{P}_1(X,Y) \leftarrow \{(\{N_{B_1},\cdots,N_{B_m}\},\{N_A\}) \in \mathcal{E}^{\mathcal{O}} \mid \mathcal{O} \models X \sqsubseteq A \sqsubseteq Y\}$;
3 **for** $(\{N_{B_1},\cdots,N_{B_m}\},\{N_A\}) \in \mathcal{P}_1(X,Y)$ **do**
4   **if** **path**$(N_A, N_Y) \neq \emptyset$ *or* $Y{=}A$ **then**
5   result.*add*$(\langle\{N_X \overset{h}{\rightsquigarrow} N_{B_1},\cdots,N_X \overset{h}{\rightsquigarrow} N_{B_m}, N_A \rightsquigarrow N_Y\}, N_X \overset{h}{\rightsquigarrow} N_Y\rangle)$;
6   **end**
7 **end**
8 $\mathcal{P}_2(X,Y) \leftarrow \{(r,B_1,B_2) \mid \mathcal{O} \models X \sqsubseteq \exists r.B_1,\ B_1 \sqsubseteq B_2, \exists r.B_2 \sqsubseteq Y\}$;
9 **for** $(r,B_1,B_2) \in \mathcal{P}_2(X,Y)$ **do**
10   **if** **path**$(N_{\exists r.B_2}, N_Y) \neq \emptyset$ *or* $Y{=}\exists r.B_2$ **then**
11   result.*add*$(\langle\{N_X \overset{h}{\rightsquigarrow} N_{\exists r.B_1}, N_{B_1} \overset{h}{\rightsquigarrow} N_{B_2}, N_{\exists r.B_2} \rightsquigarrow N_Y\}, N_X \overset{h}{\rightsquigarrow} N_Y\rangle)$;
12   **end**
13 **end**
14 $\mathcal{P}_3(X,Y) \leftarrow \{(r,s,t,A_1,A_2) \mid r{\circ}s \sqsubseteq t \in \mathcal{O},\ \mathcal{O} \models X \sqsubseteq \exists r.A_1, A_1 \sqsubseteq \exists s.A_2, \exists t.A_2 \sqsubseteq Y\}$;
    for $(r,s,t,A_1,A_2) \in \mathcal{P}_3(X,Y)$ **do**
15   **if** **path**$(N_{\exists t.A_2}, N_Y) \neq \emptyset$ *or* $Y{=}\exists t.A_2$ **then**
16   result.*add*$(\langle\{N_X \overset{h}{\rightsquigarrow} N_{\exists r.A_1}, N_{A_1} \overset{h}{\rightsquigarrow} N_{\exists s.A_2}, N_{\exists t.A_2} \rightsquigarrow N_Y, (\{N_r, N_t\}, \{N_s, N_t\})\},$
          $\{N_s, N_t\})\}, N_X \overset{h}{\rightsquigarrow} N_Y\rangle)$;
17   **end**
18 **end**

4.3 Optimization

Below we present two optimizations that have been implemented in order to accelerate the computation of all justifications.

1. In Algorithm 3, for the H-inference added at Line 5, we require that there exists at least one regular path from N_A to N_Y that does not contain an edge $e_i = (\{N_A\},\{N_{B_i}\})$ for some $1 \leq i \leq m$. Otherwise, as shown in Fig. 4, H-paths corresponding to this H-inference are not minimal, as they all contain one H-path from N_X to N_Y of the form $H_{X,B_i} \cup (P_{A,Y} - \{e_i\})$. In the same spirit, we require that the H-path from N_X to N_{B_i} does not pass by N_A.
2. If we have an H-path $H_{A,B} = H_{A,\exists r.B_1} \cup H_{B_1,B_2} \cup P_{\exists r.B_2,B}$ where

$$H_{A,\exists r.B_1} = H_{A,\exists r.C} \cup H_{C,B_1}. \tag{2}$$

then $H_{C,B_2} = H_{C,B_1} \cup H_{B_1,B_2}$ is also an H-path and $H_{A,B} = H_{A,\exists r.C} \cup H_{C,B_2} \cup P_{\exists r.B_2,B}$. The two different ways to decompose $H_{A,B}$ above are already considered in Line 8 when executing Algorithm 3 with the input $N_A \overset{h}{\rightsquigarrow} N_B$. It means that the decomposition (2) is redundant. We can avoid such redundancy by requiring $\exists r.B_2 \neq Y$ at Line 11.

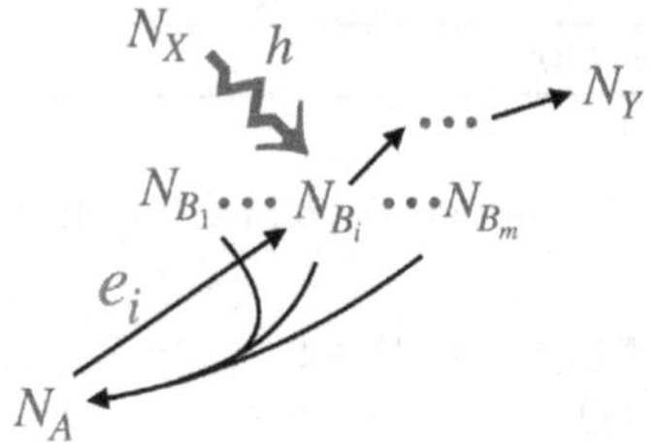

Fig. 4. Illustration of Optimization 1

5 Experiments

To evaluate and validate our approach, we compare `minH`[5] with PULi [14], the state-of-the-art algorithm for computing justifications at this moment. Both methods compute all justifications based on resolution but with different inference rules generated in different ways. PULi uses a complete set (next denoted by *elk*) generated by the ELK reasoner [13], which uses inference rules slightly different from those in Table 1. Our method uses the complete set $\mathcal{U}$ generated by Step 1 of `minH`, described in Sect. 4.2. To analyze the performance of our setting, we make the following two measures: (1) we compare the size of *elk* with that of $\mathcal{U}$, (2) we compare the time cost of PULi with that of `minH`. All the experiments were conducted on a machine with an INTEL Xeon 2.6 GHz and 128 GiB of RAM.

The experiments were processed with four different ontologies[6]: go-plus, galen7, SnomedCT (version Jan. 2015 and Jan. 2021). All the non-$\mathcal{EL}^+$ axioms are deleted. Here, go-plus, galen7 are the same ontologies used in [14]. We denote the four ontologies above by go-plus, galen7, snt2015 and snt2021. The number of axioms, concepts, relations, and queries for each ontology are shown in Table 3.

Next a **query** refers to a *direct subsumption*[7] $A \sqsubseteq B$. In our experiments, for the four ontologies, the set of all justifications $J_{\mathcal{O}}(A \sqsubseteq B)$ is computed for each query $A \sqsubseteq B$. A query $A \sqsubseteq B$ is called **trivial** iff all minimal H-paths from N_A to N_B are regular paths, otherwise, the query is **non-trivial**.

Comparing Complete Sets: $\mathcal{U}$ vs. *elk*. We summarize our results in Table 4 and Fig. 5. Table 4 shows that on all four ontologies, $\mathcal{U}$ is much smaller than *elk* on average. Especially on galen7, the difference between *elk* and $\mathcal{U}$ is even up to 50 times. The gap is even more significant for the median value since a large part of the queries is trivial. However, the gap is much smaller for the maximal number. On snt2021, the largest $\mathcal{U}$ in size is three times larger than that of *elk*.

[5] A prototype is available at https://gitlab.lisn.upsaclay.fr/yang/minH.

[6] Available at https://osf.io/9sj8n/, https://www.snomed.org/.

[7] i.e., $\mathcal{O} \models A \sqsubseteq B$ and there is no other atomic concept A' such that $\mathcal{O} \models A \sqsubseteq A', A' \sqsubseteq B$. Direct subsumptions can be computed by a reasoner supporting ontology classification.

Table 3. Summary of sizes of the input ontologies.

	go-plus	galen7	snt2015	snt2021
#axioms	105557	44475	311466	362638
#concepts	57173	28482	311480	361226
#roles	157	964	58	132
#queries	90443	91332	461854	566797

Table 4. Summary of size of *elk*, $\mathcal{U}$.

		go-plus	galen7	snt2015	snt2021
elk	average	166.9	3602.0	114.7	67.3
	median	43.0	3648.0	10.0	31.0
	max	7919.0	81501.0	2357	2226
$\mathcal{U}$	average	34.2	74.6	29.4	19.4
	median	4.0	5.0	1.0	3.0
	max	7772	24103	2002	6452
#non-trivial query		50272	62470	195082	304321

In Fig. 5, for a given query, if the complete set *elk* contains fewer inference rules than $\mathcal{U}$, the corresponding blue point is below the red line. The percentage of such cases are: 0.34% for go-plus, 0.066% for galen7, 0.79% for snt2015, and 1.01% for snt2021. It means that for most of the queries, the corresponding $\mathcal{U}$ is smaller than *elk*.

As shown in Table 4 and in Fig. 5, sometimes `minH` generates bigger complete set $\mathcal{U}$ than PULi. It may happen when, for example, there might be exponentially many different regular paths occurring in the computation process of `minH`. Therefore, `minH` could produce a huge complete set. Also, $\mathcal{U}$ can be bigger than *elk* when all the regular paths involved are simple. For example, if all regular paths contain only one edge, then the complete set $\mathcal{U}$ includes many clauses of the form $\neg p_e \vee p_{N_A \rightsquigarrow N_B}$, which happens because H-rules use regular paths. Indeed, the clause $\neg p_e \vee p_{N_A \rightsquigarrow N_B}$ is redundant since we can omit this clause by replacing $p_{N_A \rightsquigarrow N_B}$ by p_e. For *elk*, this does not happen.

Comparing Time Cost: `minH` vs. PULi. In the following, we only compare the time cost on non-trivial queries. For trivial queries, all H-path are regular paths. Thus all the justifications have already been enumerated by `path` in `minH`. It is also easy to compute all the justifications for trivial queries for PULi.

We set a limit of 60 s for each query. The timed-out queries contribute of 60 s to the total time cost. To compare `minH` with PULi, we test all three different strategies, *threshold, top down, bottom up* of the resolution algorithm proposed in [14]. We summarize in Table 5 the total time cost (top) and the timed-out queries (bottom). Figure 6 gives the comparisons over queries that are successful for both `minH` and PULi.

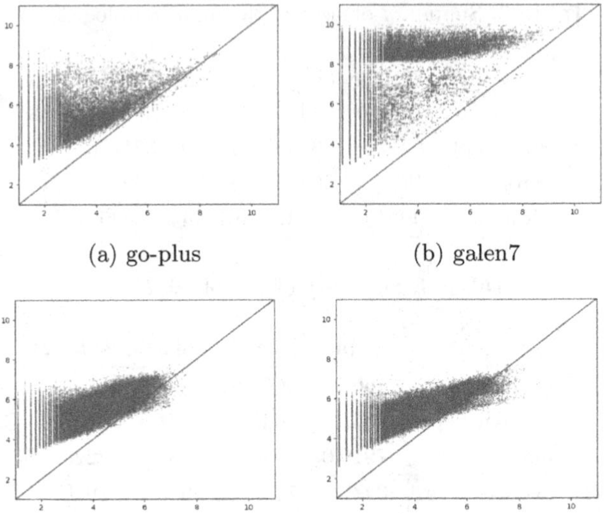

(a) go-plus (b) galen7

(c) snt2015 (d) snt2021

Fig. 5. Each blue point has coordinate $(\log(\#|\mathcal{U}|), \log(\#|elk|))$, where $\mathcal{U}$, *elk* are generated from a non-trivial query, the red line is $x = y$. (Color figure online)

As shown in Table 5, when using the threshold strategy, `minH` is more time consuming in total (+5%) on snt2021, and `minH` has more timed-out queries than PULi on snt2015 and snt2021. This is in part due to the fact that $\mathcal{U}$ is larger than *elk* for relatively many queries on snt2015 and snt2021 as shown in Fig. 5. For the remaining 11 cases, `minH` performs better than PULi in terms of total time cost and the number of timed-out queries. Especially on galen7, the gap between the two methods is even up to ten times for the total time cost. We can see from Table 5 that the threshold strategy performs the best for PULi on all four ontologies. This strategy is also the best strategy for `minH` except for galen7, for which the *bottom up* strategy is the best with `minH`.

For each strategy detailed in Fig. 6, the black curve (the ordered time costs of `minH` on successful queries) is always below the red curve (the ordered time costs of PULi on successful queries) for all the ontologies. This suggests that `minH` spends less time over successful queries. Also, most of the green points are below the red lines, which suggests that `minH` performs better than PULi most of the time for a given query. In some cases, we can see that PULi is more efficient than `minH`. One of the reasons might be as follows. Note that when computing justifications by resolution, we have to compare two different clauses and delete the redundant one (i.e., the non-minimal one). When regular paths are big, `minH` might be time consuming because of these comparisons.

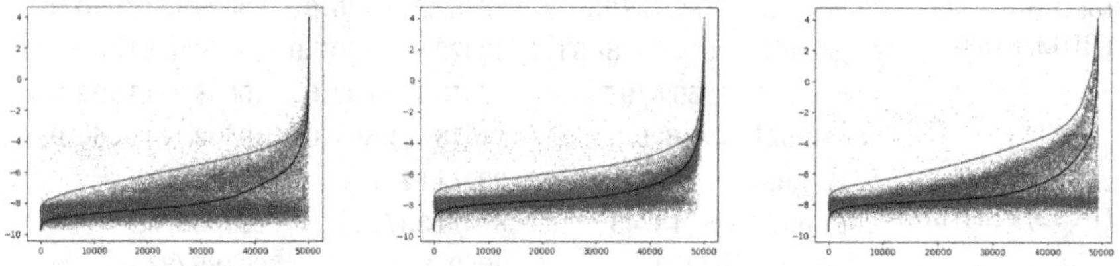

(a) go-plus

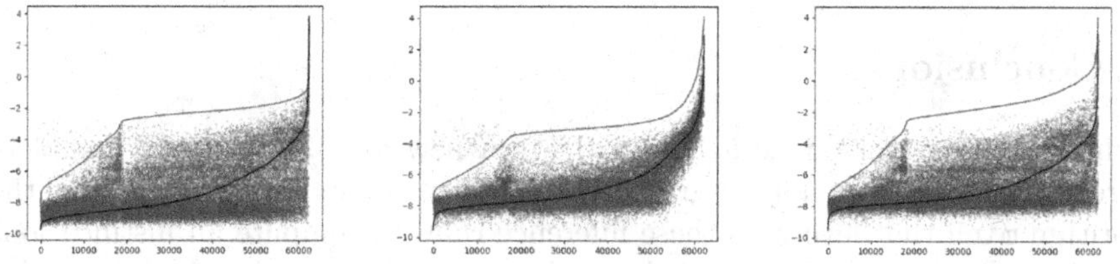

(b) galen7

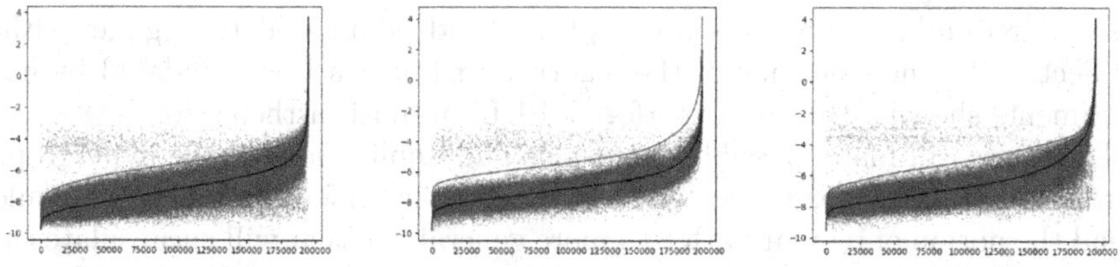

(c) snt2015

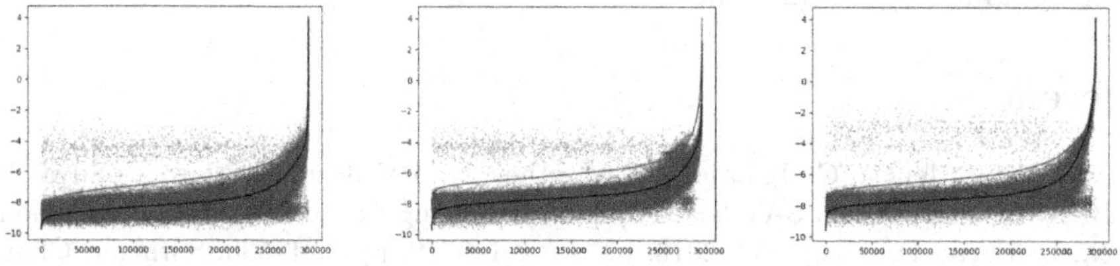

(d) snt2021

Fig. 6. For each line, the left, middle and right charts correspond to *threshold, top down, bottom up* strategies respectively. The y-axis is the log value of time(s). The red (resp. black) curve presents the ascending ordered (log value of) time cost of PULi (resp. `minH`). For a green point (x, y), e^y is the time cost of `minH` for the query corresponding to the red line point (x, y'). (Color figure online)

Table 5. Total time cost and number of timed-out queries.

		threshold	*top down*	*bottom up*
total times(s) (PULi/minH)	go-plus	8482.7/**7350.3**	16352.3/**8935.6**	73629.1/**17950.9**
	galen7	10796.2/**3681.4**	43372.9/**10607.9**	36300.9/**3156.3**
	snt2015	1956.8/**973.5**	13650.7/**1107.6**	15058.3/**11392.2**
	snt2021	**2116.1**/2222.6	11573.9/**2361.6**	19402.1/**17154.9**
timed-out queries (PULi/minH/both)	go-plus	116/**103** /93	202/**117**/114	935/**223**/223
	galen7	48/**43**/43	370/**123**/120	228/**38**/38
	snt2015	**0**/3/0	49/**3**/3	96/**88**/83
	snt2021	**2**/8/1	39/**9**/9	144/**133**/128

6 Conclusion

In this paper, we introduce and investigate a new set of sound and complete inference rules based on a hypergraph representation of ontologies. We design the algorithm `minH` that leverages these inference rules to compute all justifications for a given conclusion. The key of the performance of our method is that regular paths are used as elementary components of H-paths and this leads to reducing the size of complete sets because (1) rules are more compact than standard ones, (2) redundant inferences are captured and eliminated by regular paths (see Sect. 4.3). The efficiency of the algorithm `minH` has been validated by our experiments showing that it outperforms PULi in most of the cases.

There are still many possible extensions and applications of the hypergraph approach. For instance, to get even more compact inference rules, we could extend the notion of regular path to a more general one that will encapsulate the inference rule $\mathcal{H}_2$ in the same way as regular paths are encapsulated in H-rules. Moreover, we will try to apply our approach for other tasks like classification and to compute logical differences [15].

References

1. Arif, M.F., Mencía, C., Ignatiev, A., Manthey, N., Peñaloza, R., Marques-Silva, J.: BEACON: an efficient SAT-based tool for debugging $\mathcal{EL}^+$ ontologies. In: Creignou, N., Le Berre, D. (eds.) SAT 2016. LNCS, vol. 9710, pp. 521–530. Springer, Cham (2016). https://doi.org/10.1007/978-3-319-40970-2_32
2. Arif, M.F., Mencía, C., Marques-Silva, J.: Efficient axiom pinpointing with EL2MCS. In: Hölldobler, S., Krötzsch, M., Peñaloza, R., Rudolph, S. (eds.) KI 2015. LNCS (LNAI), vol. 9324, pp. 225–233. Springer, Cham (2015). https://doi.org/10.1007/978-3-319-24489-1_17
3. Arif, M.F., Mencía, C., Marques-Silva, J.: Efficient MUS enumeration of horn formulae with applications to axiom pinpointing. In: Heule, M., Weaver, S. (eds.) SAT 2015. LNCS, vol. 9340, pp. 324–342. Springer, Cham (2015). https://doi.org/10.1007/978-3-319-24318-4_24
4. Ausiello, G., Laura, L.: Directed hypergraphs: introduction and fundamental algorithms-a survey. Theor. Comput. Sci. **658**, 293–306 (2017)

5. Baader, F., Brandt, S., Lutz, C.: Pushing the EL envelope. In: IJCAI, vol. 5, pp. 364–369 (2005)
6. Chen, J., Ludwig, M., Ma, Y., Walther, D.: Zooming in on ontologies: minimal modules and best excerpts. In: d'Amato, C., et al. (eds.) ISWC 2017. LNCS, vol. 10587, pp. 173–189. Springer, Cham (2017). https://doi.org/10.1007/978-3-319-68288-4_11
7. Gallo, G., Longo, G., Pallottino, S., Nguyen, S.: Directed hypergraphs and applications. Discret. Appl. Math. **42**(2–3), 177–201 (1993)
8. Grau, B.C., Horrocks, I., Kazakov, Y., Sattler, U.: Modular reuse of ontologies: theory and practice. J. Artif. Intell. Res. **31**, 273–318 (2008)
9. Ignatiev, A., Marques-Silva, J., Mencía, C., Peñaloza, R.: Debugging EL+ ontologies through horn MUS enumeration. In: Artale, A., Glimm, B., Kontchakov, R. (eds.) Proceedings of the 30th International Workshop on Description Logics, Montpellier, France, 18–21 July 2017. CEUR Workshop Proceedings, vol. 1879. CEUR-WS.org (2017). http://ceur-ws.org/Vol-1879/paper54.pdf
10. Kalyanpur, A., Parsia, B., Horridge, M., Sirin, E.: Finding all justifications of OWL DL entailments. In: Aberer, K., et al. (eds.) ASWC/ISWC -2007. LNCS, vol. 4825, pp. 267–280. Springer, Heidelberg (2007). https://doi.org/10.1007/978-3-540-76298-0_20
11. Kalyanpur, A., Parsia, B., Sirin, E., Hendler, J.: Debugging unsatisfiable classes in OWL ontologies. J. Web Semant. **3**(4), 268–293 (2005)
12. Kazakov, Y., Klinov, P.: Goal-directed tracing of inferences in EL ontologies. In: Mika, P., et al. (eds.) ISWC 2014. LNCS, vol. 8797, pp. 196–211. Springer, Cham (2014). https://doi.org/10.1007/978-3-319-11915-1_13
13. Kazakov, Y., Krötzsch, M., Simancik, F.: ELK reasoner: architecture and evaluation. In: ORE (2012)
14. Kazakov, Y., Skočovský, P.: Enumerating justifications using resolution. In: Galmiche, D., Schulz, S., Sebastiani, R. (eds.) IJCAR 2018. LNCS (LNAI), vol. 10900, pp. 609–626. Springer, Cham (2018). https://doi.org/10.1007/978-3-319-94205-6_40
15. Ludwig, M., Walther, D.: The logical difference for $\mathcal{ELH}^r$-terminologies using hypergraphs. In: Schaub, T., Friedrich, G., O'Sullivan, B. (eds.) 21st European Conference on Artificial Intelligence, ECAI 2014, Prague, Czech Republic, 18–22 August 2014 - Including Prestigious Applications of Intelligent Systems (PAIS 2014). Frontiers in Artificial Intelligence and Applications, vol. 263, pp. 555–560. IOS Press (2014). https://doi.org/10.3233/978-1-61499-419-0-555
16. Manthey, N., Peñaloza, R., Rudolph, S.: Efficient axiom pinpointing in EL using sat technology. In: Description Logics (2016)
17. Peñaloza, R.: Axiom pinpointing. In: Cota, G., Daquino, M., Pozzato, G.L. (eds.) Applications and Practices in Ontology Design, Extraction, and Reasoning, Studies on the Semantic Web, vol. 49, pp. 162–177. IOS Press (2020). https://doi.org/10.3233/SSW200042
18. Penaloza, R., Sertkaya, B.: Understanding the complexity of axiom pinpointing in lightweight description logics. Artif. Intell. **250**, 80–104 (2017)

Choices, Invariance, Substitutions, and Formalizations

Sequent Calculi for Choice Logics

Michael Bernreiter[1(✉)], Anela Lolic[1], Jan Maly[2], and Stefan Woltran[1]

[1] Institute of Logic and Computation, TU Wien, Vienna, Austria
{mbernrei,alolic,woltran}@dbai.tuwien.ac.at
[2] Institute for Logic, Language and Computation, University of Amsterdam, Amsterdam, The Netherlands
j.f.maly@uva.nl

Abstract. Choice logics constitute a family of propositional logics and are used for the representation of preferences, with especially *qualitative choice logic* (QCL) being an established formalism with numerous applications in artificial intelligence. While computational properties and applications of choice logics have been studied in the literature, only few results are known about the proof-theoretic aspects of their use. We propose a sound and complete sequent calculus for preferred model entailment in QCL, where a formula F is entailed by a QCL-theory T if F is true in all preferred models of T. The calculus is based on labeled sequent and refutation calculi, and can be easily adapted for different purposes. For instance, using the calculus as a cornerstone, calculi for other choice logics such as *conjunctive choice logic* (CCL) can be obtained in a straightforward way.

1 Introduction

Choice logics are propositional logics for the representation of alternative options for problem solutions [4]. These logics add new connectives to classical propositional logic that allow for the formalization of ranked options. A prominent example is *qualitative choice logic* (QCL for short) [7], which adds the connective *ordered disjunction* $\vec{\times}$ to classical propositional logic. Intuitively, $A \vec{\times} B$ means that if possible A, but if A is not possible than at least B. The semantics of a choice logic induce a preference ordering over the models of a formula.

As choice logics are well suited for preference handling, they have a multitude of applications in AI such as logic programming [8], alert correlation [3], or database querying [13]. But while computational properties and applications of choice logics have been studied in the literature, only few results are known about the proof-theoretic aspects of their use. In particular, there is no proof system capable of deriving valid sentences containing choice operators. In this paper we propose a sound and complete calculus for preferred model entailment in QCL that can easily be generalized to other choice logics.

Entailment in choice logics is non-monotonic: conclusions that have been drawn might not be derivable in light of new information. It is therefore not surprising that choice logics are related to other non-monotonic formalisms. For

J. Blanchette et al. (Eds.): IJCAR 2022, LNAI 13385, pp. 331–349, 2022.
https://doi.org/10.1007/978-3-031-10769-6_20

instance, it is known [7] that QCL can capture propositional circumscription and that, if additional symbols in the language are admitted, circumscription can be used to generate models corresponding to the inclusion-preferred QCL models up to the additional atoms. We do not intend to use this translation of our choice logic formulas (or sequents) in order to employ an existing calculus for circumscription, for instance [5].

Instead, we define calculi in sequent format directly for choice logics, which are different from existing non-monotonic logics in the way non-monotonicity is introduced. Specifically, the non-standard part of our logics is a new logical connective which is fully embedded in the logical language. For this reason, calculi for choice logics also differ from most other calculi for non-monotonic logics: our calculi do not use non-standard inference rules as in default logic, modal operators expressing consistency or belief as in autoepistemic logic, or predicates whose extensions are minimized as in circumscription. However, one method that can also be applied to choice logics is the use of a refutation calculus (also known as rejection or antisequent calculus) axiomatising invalid formulas, i.e., non-theorems. Refutation calculi for non-monotonic logics were used in [5]. Specifically, by combining a refutation calculus with an appropriate sequent calculus, elegant proof systems for the central non-monotonic formalisms of default logic [16], autoepistemic logic [15], and circumscription [14] were obtained. However, to apply this idea to choice logics, we have to take another facet of their semantics into account.

With choice logics, we are working in a setting similar to many-valued logics. Interpretations ascribe a natural number called satisfaction degree to choice logic formulas. Preferred models of a formula are then those models with the least degree. There are several kinds of sequent calculus systems for many-valued logics, where the representation as a hypersequent calculus [1,10] plays a prominent role. However, there are crucial differences between choice logics and many-valued logics in the usual sense. Firstly, choice logic interpretations are classical, i.e., they set propositional variables to either true or false. Secondly, non-classical satisfaction degrees only arise when choice connectives, e.g. ordered disjunction in QCL, occur in a formula. Thirdly, when applying a choice connective $\circ$ to two formulas A and B, the degree of $A \circ B$ does not only depend on the degrees of A and B, but also on the maximum degrees that A and B can possibly assume. Therefore, techniques used in proof systems for conventional many-valued logics can not be applied directly to choice logics.

In [11] a sequent calculus based system for reasoning with contrary-to-duty obligations was introduced, where a non-classical connective was defined to capture the notion of reparational obligation, which is in force only when a violation of a norm occurs. This is related to the ordered disjunction in QCL, however, based on the intended use in [11] the system was defined only for the occurrence of the new connective on the right side of the sequent sign. We aim for a proof system for reasoning with choice logic operators, and to deduce formulas from choice logic formulas. Thus, we need a calculus with left and right inference rules.

To obtain such a calculus we combine the idea of a refutation calculus with methods developed for multi-valued logics in a novel way. First, we develop a (monotonic) sequent calculus for reasoning about satisfaction degrees using a labeled calculus, a method developed for (finite) many-valued logics [2,9,12]. Secondly, we define a labeled refutation calculus for reasoning about invalidity in terms of satisfaction degrees. Finally, we join both calculi to obtain a sequent calculus for the non-monotonic entailment of QCL. To this end, we introduce a new, non-monotonic inference rule that has sequents of the two labeled calculi as premises and formalizes degree minimization.

The rest of this paper is organized as follows. In the next section we present the basic notions of choice logics and introduce the most prominent choice logics QCL and CCL (*conjunctive choice logic*). In Sect. 3 we develop a labeled sequent calculus for propositional logic extended by the QCL connective $\vec{\times}$. This calculus is shown to be sound and complete and already can be used to derive interesting sentences containing choice operators. In Sect. 4 we extend the previously defined sequent calculus with an appropriate refutation calculus and non-monotonic reasoning, to capture entailment in QCL. The developed methodology for QCL can be extended to other choice logics as well. In particular we show in Sect. 5 how the calculi can be adapted for CCL.

2 Choice Logics

First, we formally define the notion of choice logics in accordance with the choice logic framework of [4] before giving concrete examples in the form of QCL and CCL. Finally, we define preferred model entailment.

Definition 1. *Let $\mathcal{U}$ denote the alphabet of propositional variables. The set of choice connectives $\mathcal{C}_\mathcal{L}$ of a choice logic $\mathcal{L}$ is a finite set of symbols such that $\mathcal{C}_\mathcal{L} \cap \{\neg, \wedge, \vee\} = \emptyset$. The set $\mathcal{F}_\mathcal{L}$ of formulas of $\mathcal{L}$ is defined inductively as follows: (i) $a \in \mathcal{F}_\mathcal{L}$ for all $a \in \mathcal{U}$; (ii) if $F \in \mathcal{F}_\mathcal{L}$, then $(\neg F) \in \mathcal{F}_\mathcal{L}$; (iii) if $F, G \in \mathcal{F}_\mathcal{L}$, then $(F \circ G) \in \mathcal{F}_\mathcal{L}$ for $\circ \in (\{\wedge, \vee\} \cup \mathcal{C}_\mathcal{L})$.*

For example, $\mathcal{C}_{\text{QCL}} = \{\vec{\times}\}$ and $((a \vec{\times} c) \wedge (b \vec{\times} c)) \in \mathcal{F}_{\text{QCL}}$. Formulas that do not contain a choice connective are referred to as classical formulas.

The semantics of a choice logic is given by two functions, satisfaction degree and optionality. The satisfaction degree of a formula given an interpretation is either a natural number or ∞. The lower this degree, the more preferable the interpretation. The optionality of a formula describes the maximum finite satisfaction degree that this formula can be ascribed, and is used to penalize non-satisfaction.

Definition 2. *The optionality of a choice connective $\circ \in \mathcal{C}_\mathcal{L}$ in a choice logic $\mathcal{L}$ is given by a function $opt^\circ_\mathcal{L} \colon \mathbb{N}^2 \to \mathbb{N}$ such that $opt^\circ_\mathcal{L}(k, \ell) \leq (k+1) \cdot (\ell+1)$ for all $k, \ell \in \mathbb{N}$. The optionality of an $\mathcal{L}$-formula is given via $opt_\mathcal{L} \colon \mathcal{F}_\mathcal{L} \to \mathbb{N}$ with (i) $opt_\mathcal{L}(a) = 1$ for every $a \in \mathcal{U}$; (ii) $opt_\mathcal{L}(\neg F) = 1$; (iii) $opt_\mathcal{L}(F \wedge G) = opt_\mathcal{L}(F \vee G) = max(opt_\mathcal{L}(F), opt_\mathcal{L}(G))$; (iv) $opt_\mathcal{L}(F \circ G) = opt^\circ_\mathcal{L}(opt_\mathcal{L}(F), opt_\mathcal{L}(G))$ for every choice connective $\circ \in \mathcal{C}_\mathcal{L}$.*

The optionality of a classical formula is always 1. Note that, for any choice connective $\circ$, the optionality of $F \circ G$ is bounded such that $opt_{\mathcal{L}}(F \circ G) \leq (opt_{\mathcal{L}}(F) + 1) \cdot (opt_{\mathcal{L}}(G) + 1)$. In the following, we write $\overline{\mathbb{N}}$ for $(\mathbb{N} \cup \{\infty\})$.

Definition 3. *The satisfaction degree of a choice connective $\circ \in \mathcal{C}_{\mathcal{L}}$ in a choice logic $\mathcal{L}$ is given by a function $deg^{\circ}_{\mathcal{L}} \colon \mathbb{N}^2 \times \overline{\mathbb{N}}^2 \to \overline{\mathbb{N}}$ such that $deg^{\circ}_{\mathcal{L}}(k, \ell, m, n) \leq opt^{\circ}_{\mathcal{L}}(k, \ell)$ or $deg^{\circ}_{\mathcal{L}}(k, \ell, m, n) = \infty$ for all $k, \ell \in \mathbb{N}$ and all $m, n \in \overline{\mathbb{N}}$. The satisfaction degree of an $\mathcal{L}$-formula under an interpretation $\mathcal{I} \subseteq \mathcal{U}$ is given via the function $deg_{\mathcal{L}} \colon 2^{\mathcal{U}} \times \mathcal{F}_{\mathcal{L}} \to \overline{\mathbb{N}}$ with*

1. *$deg_{\mathcal{L}}(\mathcal{I}, a) = 1$ if $a \in \mathcal{I}$, $deg_{\mathcal{L}}(\mathcal{I}, a) = \infty$ otherwise for every $a \in \mathcal{U}$;*
2. *$deg_{\mathcal{L}}(\mathcal{I}, \neg F) = 1$ if $deg_{\mathcal{L}}(\mathcal{I}, F) = \infty$, $deg_{\mathcal{L}}(\mathcal{I}, \neg F) = \infty$ otherwise;*
3. *$deg_{\mathcal{L}}(\mathcal{I}, F \wedge G) = max(deg_{\mathcal{L}}(\mathcal{I}, F), deg_{\mathcal{L}}(\mathcal{I}, G))$;*
4. *$deg_{\mathcal{L}}(\mathcal{I}, F \vee G) = min(deg_{\mathcal{L}}(\mathcal{I}, F), deg_{\mathcal{L}}(\mathcal{I}, G))$;*
5. *$deg_{\mathcal{L}}(\mathcal{I}, F \circ G) = deg^{\circ}_{\mathcal{L}}(opt_{\mathcal{L}}(F), opt_{\mathcal{L}}(G), deg_{\mathcal{L}}(\mathcal{I}, F), deg_{\mathcal{L}}(\mathcal{I}, G))$, $\circ \in \mathcal{C}_{\mathcal{L}}$.*

We also write $\mathcal{I} \models^{\mathcal{L}}_{m} F$ for $deg_{\mathcal{L}}(\mathcal{I}, F) = m$. If $m < \infty$, we say that $\mathcal{I}$ satisfies F (to a finite degree), and if $m = \infty$, then $\mathcal{I}$ does not satisfy F. If F is a classical formula, then $\mathcal{I} \models^{\mathcal{L}}_{1} F \iff \mathcal{I} \models F$ and $\mathcal{I} \models^{\mathcal{L}}_{\infty} F \iff \mathcal{I} \not\models F$. The symbols $\top$ and $\bot$ are shorthand for the formulas $(a \vee \neg a)$ and $(a \wedge \neg a)$, where a can be any variable. We have $opt_{\mathcal{L}}(\top) = opt_{\mathcal{L}}(\bot) = 1$, $deg_{\mathcal{L}}(\mathcal{I}, \top) = 1$ and $deg_{\mathcal{L}}(\mathcal{I}, \bot) = \infty$ for any interpretation $\mathcal{I}$ in every choice logic.

Models and preferred models of formulas are defined in the following way:

Definition 4. *Let $\mathcal{L}$ be a choice logic, $\mathcal{I}$ an interpretation, and F an $\mathcal{L}$-formula. $\mathcal{I}$ is a model of F, written as $\mathcal{I} \in Mod_{\mathcal{L}}(F)$, if $deg_{\mathcal{L}}(\mathcal{I}, F) < \infty$. $\mathcal{I}$ is a preferred model of F, written as $\mathcal{I} \in Prf_{\mathcal{L}}(F)$, if $\mathcal{I} \in Mod_{\mathcal{L}}(F)$ and $deg_{\mathcal{L}}(\mathcal{I}, F) \leq deg_{\mathcal{L}}(\mathcal{J}, F)$ for all other interpretations $\mathcal{J}$.*

Moreover, we define the notion of classical counterparts for choice connectives.

Definition 5. *Let $\mathcal{L}$ be a choice logic. The classical counterpart of a choice connective $\circ \in \mathcal{C}_{\mathcal{L}}$ is the classical binary connective $\circledast$ such that, for all atoms a and b, $deg_{\mathcal{L}}(\mathcal{I}, a \circ b) < \infty \iff \mathcal{I} \models a \circledast b$. The classical counterpart of an $\mathcal{L}$-formula F is denoted as $cp(F)$ and is obtained by replacing all occurrences of choice connectives in F by their classical counterparts.*

A natural property of known choice logics is that choice connectives can be replaced by their classical counterpart without affecting satisfiability, meaning that $deg_{\mathcal{L}}(\mathcal{I}, F) < \infty \iff \mathcal{I} \models cp(F)$ holds for all $\mathcal{L}$-formulas F.

So far we introduced choice logics in a quite abstract way. We now introduce two particular instantiations, namely QCL, the first and most prominent choice logic in the literature, and CCL, which introduces a connective $\vec{\odot}$ called ordered conjunction in place of QCL's ordered disjunction.

Definition 6. QCL *is the choice logic such that* $\mathcal{C}_{\text{QCL}} = \{\vec{\times}\}$, *and, if* $k = opt_{\text{QCL}}(F)$, $\ell = opt_{\text{QCL}}(G)$, $m = deg_{\text{QCL}}(\mathcal{I}, F)$, *and* $n = deg_{\text{QCL}}(\mathcal{I}, G)$, *then*

$$opt_{\text{QCL}}(F\vec{\times}G) = opt^{\vec{\times}}_{\text{QCL}}(k,\ell) = k+\ell, \text{ and}$$

$$deg_{\text{QCL}}(\mathcal{I}, F\vec{\times}G) = deg^{\vec{\times}}_{\text{QCL}}(k,\ell,m,n) = \begin{cases} m & \text{if } m < \infty; \\ n+k & \text{if } m = \infty, n < \infty; \\ \infty & \text{otherwise.} \end{cases}$$

In the above definition, we can see how optionality is used to penalize non-satisfaction: given a QCL-formula $F\vec{\times}G$ and an interpretation $\mathcal{I}$, if $\mathcal{I}$ satisfies F (to some finite degree), then $deg_{\text{QCL}}(\mathcal{I}, F\vec{\times}G) = deg_{\text{QCL}}(\mathcal{I}, F)$; if $\mathcal{I}$ does not satisfy F, then $deg_{\text{QCL}}(\mathcal{I}, F\vec{\times}G) = opt_{\text{QCL}}(F) + deg_{\text{QCL}}(\mathcal{I}, G)$. Since $deg_{\text{QCL}}(\mathcal{I}, F) \leq opt_{\text{QCL}}(F)$, interpretations that satisfy F result in a lower degree, i.e., are more preferable, compared to interpretations that do not satisfy F. Let us take a look at a concrete example:

Example 1. Consider the QCL-formula $F = (a\vec{\times}c) \wedge (b\vec{\times}c)$. Note that the classical counterpart of $\vec{\times}$ is $\vee$, i.e., $cp(F) = (a \vee c) \wedge (b \vee c)$. Thus, $\{c\}, \{a,b\}, \{a,c\}, \{b,c\}, \{a,b,c\} \in Mod_{\text{QCL}}(F)$. Of these models, $\{a,b\}$ and $\{a,b,c\}$ satisfy F to a degree of 1 while $\{c\}$, $\{a,c\}$, and $\{b,c\}$ satisfy F to a degree of 2. Therefore, $\{a,b\}, \{a,b,c\} \in Prf_{\text{QCL}}(F)$.

Next, we define CCL. Note that we follow the revised definition of CCL [4], which differs from the initial specification[1]. Intuitively, given a CCL-formula $F\vec{\odot}G$ it is best to satisfy both F and G, but also acceptable to satisfy only F.

Definition 7. CCL *is the choice logic such that* $\mathcal{C}_{\text{CCL}} = \{\vec{\odot}\}$, *and, if* $k = opt_{\text{CCL}}(F)$, $\ell = opt_{\text{CCL}}(G)$, $m = deg_{\text{CCL}}(\mathcal{I}, F)$, *and* $n = deg_{\text{CCL}}(\mathcal{I}, G)$, *then*

$$opt_{\text{CCL}}(F\vec{\odot}G) = k+\ell, \text{ and}$$

$$deg_{\text{CCL}}(\mathcal{I}, F\vec{\odot}G) = \begin{cases} n & \text{if } m = 1, n < \infty; \\ m+\ell & \text{if } m < \infty \text{ and } (m > 1 \text{ or } n = \infty); \\ \infty & \text{otherwise.} \end{cases}$$

Example 2. Consider the CCL-formula $G = (a\vec{\odot}c) \wedge (b\vec{\odot}c)$. Note that the classical counterpart of $\vec{\odot}$ is the first projection, i.e., $cp(G) = a \wedge b$. Thus, $\{a,b\}$, $\{a,b,c\} \in Mod_{\text{CCL}}(G)$. Of these models, $\{a,b,c\}$ satisfies G to a degree of 1 while $\{a,b\}$ satisfies G to a degree of 2. Therefore, $\{a,b,c\} \in Prf_{\text{CCL}}(G)$.

If $\mathcal{L}$ is a choice logic, then a set of $\mathcal{L}$-formulas is called an $\mathcal{L}$-theory. An $\mathcal{L}$-theory T entails a classical formula F, written as $T \mathrel{|\!\sim} F$, if F is true in all preferred models of T. However, we first need to define what the preferred models of a choice logic theory are. There are several approaches for this. In the original QCL paper [7], a lexicographic and an inclusion-based approach were introduced.

[1] It seems that, under the initial definition of CCL, $a\vec{\odot}b$ is always ascribed a degree of 1 or ∞, i.e., non-classical degrees can not be obtained (cf. Definition 8 in [6]).

Definition 8. *Let $\mathcal{L}$ be a choice logic, $\mathcal{I}$ an interpretation, and T an $\mathcal{L}$-theory. $\mathcal{I} \in Mod_{\mathcal{L}}(T)$ if $deg_{\mathcal{L}}(\mathcal{I}, F) < \infty$ for all $F \in T$. $\mathcal{I}^k_{\mathcal{L}}(T)$ denotes the set of formulas in T satisfied to a degree of k by $\mathcal{I}$, i.e., $\mathcal{I}^k_{\mathcal{L}}(T) = \{F \in T \mid deg_{\mathcal{L}}(\mathcal{I}, F) = k\}$.*

- *$\mathcal{I}$ is a lexicographically preferred model of T, written as $\mathcal{I} \in Prf^{lex}_{\mathcal{L}}(T)$, if $\mathcal{I} \in Mod_{\mathcal{L}}(T)$ and if there is no $\mathcal{J} \in Mod_{\mathcal{L}}(T)$ such that, for some $k \in \mathbb{N}$ and all $l < k$, $|\mathcal{I}^k_{\mathcal{L}}(T)| < |\mathcal{J}^k_{\mathcal{L}}(T)|$ and $|\mathcal{I}^l_{\mathcal{L}}(T)| = |\mathcal{J}^l_{\mathcal{L}}(T)|$ holds.*
- *$\mathcal{I}$ is an inclusion-based preferred model of T, written as $\mathcal{I} \in Prf^{inc}_{\mathcal{L}}(T)$, if $\mathcal{I} \in Mod_{\mathcal{L}}(T)$ and if there is no $\mathcal{J} \in Mod_{\mathcal{L}}(T)$ such that, for some $k \in \mathbb{N}$ and all $l < k$, $\mathcal{I}^k_{\mathcal{L}}(T) \subset \mathcal{J}^k_{\mathcal{L}}(T)$ and $\mathcal{I}^l_{\mathcal{L}}(T) = \mathcal{J}^l_{\mathcal{L}}(T)$ holds.*

In our calculus for preferred model entailment we focus on the lexicographic approach, but it will become clear how it can be adapted to other preferred model semantics (see Sect. 4). We now formally define preferred model entailment:

Definition 9. *Let $\mathcal{L}$ be a choice logic, T an $\mathcal{L}$-theory, S a classical theory, and $\sigma \in \{lex, inc\}$. $T \mathrel{|\!\!\sim}^{\sigma}_{\mathcal{L}} S$ if for all $\mathcal{I} \in Prf^{\sigma}_{\mathcal{L}}(T)$ there is $F \in S$ such that $\mathcal{I} \models F$.*

Example 3. Consider the QCL-theory $T = \{\neg(a \wedge b), a \vec{\times} c, b \vec{\times} c\}$. Then $\{c\}, \{a, c\}, \{b, c\} \in Mod_{\mathrm{QCL}}(T)$. Note that, because of $\neg(a \wedge b)$, a model of T can not satisfy both $a \vec{\times} c$ and $b \vec{\times} c$ to a degree of 1. Specifically,

$$\{a, c\}^1_{\mathrm{QCL}}(T) = \{\neg(a \wedge b), a \vec{\times} c\} \text{ and } \{a, c\}^2_{\mathrm{QCL}}(T) = \{b \vec{\times} c\},$$
$$\{b, c\}^1_{\mathrm{QCL}}(T) = \{\neg(a \wedge b), b \vec{\times} c\} \text{ and } \{b, c\}^2_{\mathrm{QCL}}(T) = \{a \vec{\times} c\},$$
$$\{c\}^1_{\mathrm{QCL}}(T) = \{\neg(a \wedge b)\} \text{ and } \{c\}^2_{\mathrm{QCL}}(T) = \{a \vec{\times} c, b \vec{\times} c\}.$$

Thus, $\{a, c\}, \{b, c\} \in Prf^{lex}_{\mathrm{QCL}}(T)$ but $\{c\} \notin Prf^{lex}_{\mathrm{QCL}}(T)$. It can be concluded that $T \mathrel{|\!\!\sim}^{lex}_{\mathrm{QCL}} c \wedge (a \vee b)$. However, $T \not\mathrel{|\!\!\sim}^{lex}_{\mathrm{QCL}} a$ and $T \not\mathrel{|\!\!\sim}^{lex}_{\mathrm{QCL}} b$.

It is easy to see that preferred model entailment is non-monotonic. For example, $\{a \vec{\times} b\} \mathrel{|\!\!\sim}^{lex}_{\mathrm{QCL}} a$ but $\{a \vec{\times} b, \neg a\} \not\mathrel{|\!\!\sim}^{lex}_{\mathrm{QCL}} a$.

3 The Sequent Calculus L[QCL]

As a first step towards a calculus for preferred model entailment, we propose a labeled calculus [2,12] for reasoning about the satisfaction degrees of QCL formulas in sequent format and prove its soundness and completeness. One advantage of the sequent calculus format is having symmetrical left and right rules for all connectives, in particular for the choice connectives. This is in contrast to the representation of ordered disjunction in the calculus for deontic logic [11], in which only right-hand side rules are considered.

As the calculus will be concerned with satisfaction degrees rather than preferred models, we need to define entailment in terms of satisfaction degrees. To this end, the formulas occurring in the sequents of our calculus are labeled with natural numbers, i.e., they are of the form $(A)_k$, where A is a choice logic formula and $k \in \mathbb{N}$. $(A)_k$ is satisfied by those interpretations that satisfy A to a degree of

k. Instead of labeling formulas with degree ∞ we use the negated formula, i.e., instead of $(A)_\infty$ we use $(\neg A)_1$. We observe that $(A)_k$ for $opt_{\mathcal{L}}(A) > k$ can never have a model. We will deal with such formulas by replacing them with $(\bot)_1$. For classical formulas, we may write A for $(A)_1$.

Definition 10. *Let $(A_1)_{k_1}, \ldots, (A_m)_{k_m}$ and $(B_1)_{l_1}, \ldots, (B_n)_{l_n}$ be labeled* QCL-*formulas. $(A_1)_{k_1}, \ldots, (A_m)_{k_m} \vdash (B_1)_{l_1}, \ldots, (B_n)_{l_n}$ is a labeled* QCL-*sequent. $\Gamma \vdash \Delta$ is valid iff every interpretation that satisfies all labeled formulas in Γ to the degree specified by the label also satisfies at least one labeled formula in Δ to the degree specified by the label.*

Note that entailment in terms of satisfaction degrees, as defined above, is montonic. Frequently we will write $(A)_{<k}$ as shorthand for $(A)_1, \ldots, (A)_{k-1}$ and $(A)_{>k}$ for $(A)_{k+1}, \ldots, (A)_{opt_{\mathrm{QCL}}(A)}, (\neg A)_1$. Moreover, $\langle \Gamma, (A)_i \vdash \Delta \rangle_{i<k}$ denotes the sequence of sequents

$$\Gamma, (A)_1 \vdash \Delta \ldots \Gamma, (A)_{k-1} \vdash \Delta.$$

Analogously, $\langle \Gamma, (A)_i \vdash \Delta \rangle_{i>k}$ stands for the sequence of sequents $\Gamma, (A)_{k+1} \vdash \Delta \ldots \Gamma, (A)_{opt_{\mathrm{QCL}}(A)} \vdash \Delta \quad \Gamma, (\neg A)_1 \vdash \Delta$.

We define the sequent calculus **L**[QCL] over labeled sequents below. In addition to introducing inference rules for $\vec{\times}$ we have to modify the inference rules for conjunction and disjunction of propositional **LK**. The idea behind the $\lor$-left rule is that a model M of $(A)_k$ is only a model of $(A \lor B)_k$ if there is no $l < k$ s.t. M is a model of $(B)_l$. Therefore, every model of $(A \lor B)_k$ is a model of Δ iff

- every model of $(A)_k$ is a model of Δ or of some $(B)_l$ with $l < k$,
- every model of $(B)_k$ is a model of Δ or of some $(A)_l$ with $l < k$.

Essentially the same idea works for $\land$-left but with $l > k$. For the $\lor$-right rule, in order for every model of Γ to be a model of $(A \lor B)_k$, every model of Γ must either be a model of $(A)_k$ or of $(B)_k$ and no model of Γ can be a model of $(A)_l$ for $l < k$, i.e., $\Gamma, (A)_l \vdash \bot$. Similarly for $\land$-right.

Definition 11 (L[QCL]). *The axioms of* **L**[QCL] *are of the form $(p)_1 \vdash (p)_1$ for propositional variables p. The inference rules are given below. For the structural and logical rules, whenever a labeled formula $(F)_k$ appears in the conclusion of an inference rule it holds that $k \leq opt_{\mathcal{L}}(F)$.*

The structural rules are:

$$\frac{\Gamma \vdash \Delta}{\Gamma, (A)_k \vdash \Delta}\,wl \qquad \frac{\Gamma \vdash \Delta}{\Gamma \vdash (A)_k, \Delta}\,wr \qquad \frac{\Gamma, (A)_k, (A)_k \vdash \Delta}{\Gamma, (A)_k \vdash \Delta}\,cl \qquad \frac{\Gamma \vdash (A)_k, (A)_k, \Delta}{\Gamma \vdash (A)_k, \Delta}\,cr$$

The logical rules are:

$$\frac{\Gamma \vdash (cp(A))_1, \Delta}{\Gamma, (\neg A)_1 \vdash \Delta}\,\neg l \qquad \frac{\Gamma, (cp(A))_1 \vdash \Delta}{\Gamma \vdash (\neg A)_1, \Delta}\,\neg r$$

$$\frac{\Gamma, (A)_k \vdash (B)_{<k}, \Delta \qquad \Gamma, (B)_k \vdash (A)_{<k}, \Delta}{\Gamma, (A \lor B)_k \vdash \Delta}\,\lor l \qquad \frac{\langle \Gamma, (A)_i \vdash \Delta \rangle_{i<k} \quad \langle \Gamma, (B)_i \vdash \Delta \rangle_{i<k} \quad \Gamma \vdash (A)_k, (B)_k, \Delta}{\Gamma \vdash (A \lor B)_k, \Delta}\,\lor r$$

$$\frac{\Gamma,(A)_k \vdash (B)_{>k},\Delta \qquad \Gamma,(B)_k \vdash (A)_{>k},\Delta}{\Gamma,(A\wedge B)_k \vdash \Delta}\wedge l \qquad \frac{\langle \Gamma,(A)_i \vdash \Delta\rangle_{i>k} \quad \langle \Gamma,(B)_i \vdash \Delta\rangle_{i>k} \quad \Gamma \vdash (A)_k,(B)_k,\Delta}{\Gamma \vdash (A\wedge B)_k,\Delta}\wedge r$$

The rules for ordered disjunction, with $k \leq opt_{\mathcal{L}}(A)$ and $l \leq opt_{\mathcal{L}}(B)$, are:

$$\frac{\Gamma,(A)_k \vdash \Delta}{\Gamma,(A\vec{\times}B)_k \vdash \Delta}\vec{\times}l_1 \qquad \frac{\Gamma,(B)_l,(\neg A)_1 \vdash \Delta}{\Gamma,(A\vec{\times}B)_{opt_{\mathrm{QCL}}(A)+l} \vdash \Delta}\vec{\times}l_2$$

$$\frac{\Gamma \vdash (A)_k,\Delta}{\Gamma \vdash (A\vec{\times}B)_k,\Delta}\vec{\times}r_1 \qquad \frac{\Gamma \vdash (\neg A)_1,\Delta \qquad \Gamma \vdash (B)_l,\Delta}{\Gamma \vdash (A\vec{\times}B)_{opt_{\mathrm{QCL}}(A)+l},\Delta}\vec{\times}r_2$$

The degree overflow rules[2]*, with $k \in \mathbb{N}$, are:*

$$\frac{\Gamma,\bot \vdash \Delta}{\Gamma,(A)_{opt_{\mathrm{QCL}}(A)+k} \vdash \Delta}dol \qquad \frac{\Gamma \vdash \Delta}{\Gamma \vdash (A)_{opt_{\mathrm{QCL}}(A)+k},\Delta}dor$$

Observe that the modified $\wedge$ and $\vee$ inference rules correspond to the $\wedge$ and $\vee$ inference rules of propositional **LK** in case we are dealing only with classical formulas. Our $\wedge$-left rule splits the proof-tree unnecessarily for classical theories, and the $\wedge$-right rule adds an unnecessary third condition $\Gamma \vdash A, B, \Delta$. These additional conditions are necessary when dealing with non-classical formulas.

The intuition behind the degree overflow rules is that we sometimes need to fix invalid sequences, i.e., sequences in which a formula F is assigned a label k with $opt_{\mathrm{QCL}}(F) < k < \infty$.

Example 4. The following is an **L**[QCL]-proof of a valid sequent.[3]

$$\frac{\dfrac{\dfrac{\vdots}{b\vee c,\neg a,b \vdash a\wedge b,a\wedge c,b}}{\dfrac{b\vee c,(a\vec{\times}b)_2 \vdash a\wedge b,a\wedge c,b}{(a\vec{\times}b)_2 \vdash \neg(b\vec{\times}c),a\wedge b,a\wedge c,b}\neg r}\vec{\times}l_2 \qquad \dfrac{\dfrac{\dfrac{\vdots}{a\vee b,\neg b,c \vdash a\wedge b,a\wedge c,b}}{a\vee b,(b\vec{\times}c)_2 \vdash a\wedge b,a\wedge c,b}\vec{\times}l_2}{(b\vec{\times}c)_2 \vdash \neg(a\vec{\times}b),a\wedge b,a\wedge c,b}\neg r}{\dfrac{((a\vec{\times}b)\wedge(b\vec{\times}c))_2 \vdash a\wedge b,a\wedge c,b}{\neg(a\wedge b),((a\vec{\times}b)\wedge(b\vec{\times}c))_2 \vdash a\wedge c,b}\neg l}\wedge l$$

Example 5. The following proof shows how the $\wedge r$-rule can introduce more than three premises. Note that we make use of the *dol*-rule in the leftmost branch.

$$\frac{\dfrac{\dfrac{\vdots}{a,b,\bot \vdash}}{a,b,(a)_2 \vdash}dol \qquad \dfrac{\vdots}{a,b,\neg a \vdash} \qquad \dfrac{\dfrac{\vdots}{a,b,c,\neg b \vdash}}{a,b,(b\vec{\times}c)_2 \vdash}\vec{\times}l_2 \qquad \dfrac{\dfrac{\vdots}{a,b \vdash b\vee c}}{a,b,\neg(b\vec{\times}c) \vdash}\neg l \qquad \dfrac{\dfrac{\vdots}{a,b \vdash a,b}}{a,b \vdash a,(b\vec{\times}c)_1}\vec{\times}r_1}{a,b \vdash (a\wedge(b\vec{\times}c))_1}\wedge r$$

We now show soundness and completeness of **L**[QCL].

---

[2] *dol*/*dor* stands for degree overflow left/right.

[3] Note that, once we reach sequents containing only classical formulas, we do not continue the proof. However, it can be verified that the classical sequents on the left and right branch are provable in this case. Moreover, given a formula $(A)_1$ with a label of 1, the label is often omitted for readability.

Proposition 1. **L**[QCL] *is sound.*

Proof. We show for all rules that they are sound.

- For (ax) and the structural rules this is clearly the case.
- $(\neg r)$ and $(\neg l)$: follows from the fact that $deg_{\mathrm{QCL}}(\mathcal{I}, F) < \infty \iff \mathcal{I} \models cp(F)$ for all QCL-formulas F.
- $(\vee l)$: Assume that the conclusion of the rule is not valid, i.e., there is a model M of Γ and $(A \vee B)_k$ that is not a model of Δ. Then, M satisfies either A or B to degree k and neither to a degree smaller than k. Assume M satisfies A to a degree of k, the other case is symmetric. Then M is a model of Γ and $(A)_k$ but, by assumption, neither of Δ nor of $(B)_j$ for any $j < k$. Hence at least one of the premises is not valid. Analogously for $(\wedge l)$.
- $(\vee r)$: Assume there is a model M of Γ that is not a model of Δ or of $(A \vee B)_k$. There are two possible cases why M is not a model of $(A \vee B)_k$: (1) M satisfies neither A nor B to degree k. But then the premise $\Gamma \vdash (A)_k, (B)_k, \Delta$ is not valid as M is also not a model of Δ by assumption. (2) M satisfies either A or B to a degree smaller than k. Assume that M satisfies A to degree $j < k$ (the other case is symmetric). Then the premise $\Gamma, (A)_j \vdash \Delta$ is not valid. Indeed, M is a model of Γ and $(A)_j$ but not of Δ. Analogously for $(\wedge r)$.
- $(\vec{\times} l_1)$ and $(\vec{\times} r_1)$: follows from the fact that $(A)_k$ has the same models as $(A \vec{\times} B)_k$ for $k \leq opt_{\mathcal{L}}(A)$.
- $(\vec{\times} l_2)$: Assume the conclusion of the rule is not valid and let M be the model witnessing this. Then M is a model of $(A \vec{\times} B)_{opt_{QCL}(A)+l}$. By definition, M satisfies B to degree l and is not a model of A. However, then it is also a model of Γ, $(B)_l$ and $(\neg A)_1$, which means that the premise is not valid.
- $(\vec{\times} r_2)$. Assume that both premises are valid, i.e., every model of Γ is either a model of Δ or of $(\neg A)_1$ and $(B)_l$ with $l \leq opt_{\mathcal{L}}(B)$. Now, by definition, any model that is not a model of A (and hence a model of $(\neg A)_1$) and of $(B)_l$ satisfies $A \vec{\times} B$ to degree $opt_{QCL}(A) + l$. Therefore, every model of Γ is either a model of Δ or of $(A \vec{\times} B)_{opt_{QCL}(A)+l}$, which means that the conclusion of the rule is valid.
- (dol): $\Gamma, \bot$ has no models, i.e., the premise $\Gamma, \bot \vdash \Delta$ is valid. Crucially, the sequent $\Gamma, (A)_{opt_{\mathrm{QCL}}(A)+k}$ has no models as well since A cannot be satisfied to a degree m with $opt_{\mathcal{L}}(A) < m < \infty$. (dor) is clearly sound. □

Proposition 2. **L**[QCL] *is complete.*

Proof. We show this by induction over the (aggregated) formula complexity of the non-classical formulas.

- For the base case, we observed that if all formulas are classical and labeled with 1, then all our rules reduce to the classical sequent calculus, which is known to be complete. Moreover, we observe that $(A)_1$ is equivalent to A. Hence, we can turn labeled atoms into classical atoms.
- Assume that a sequent of the form $\Gamma, (A)_{opt_{\mathrm{QCL}}(A)+k} \vdash \Delta$ with $k \in \mathbb{N}$ is valid. Since $\Gamma, \bot$ has no models, $\Gamma, \bot \vdash \Delta$ is valid and, by the induction hypothesis, provable. Thus, $\Gamma, (A)_{opt_{\mathrm{QCL}}(A)+k} \vdash \Delta$ is provable using the (dol) rule.

- Assume that a sequent $\Gamma \vdash (A)_{opt_{\text{QCL}}(A)+k}, \Delta$ is valid. $(A)_{opt_{\text{QCL}}(A)+k}$ can not be satisfied, i.e., $\Gamma \vdash \Delta$ is valid and, by the induction hypothesis, provable. Therefore, $\Gamma \vdash (A)_{opt_{\text{QCL}}(A)+k}, \Delta$ is provable using the (dor) rule.
- Assume that a sequent of the form $\Gamma \vdash (\neg A)_1, \Delta$ is valid. Then every model of Γ is either a model of $(\neg A)_1$ or of Δ. In other words, every model of Γ that is not a model of $(\neg A)_1$ (i.e., is model of $cp(A)$) is a model of Δ. Therefore, every interpretation that is a model of both Γ and $cp(A)$ must be a model of Δ. It follows that $\Gamma, cp(A) \vdash \Delta$ is valid and, by the induction hypothesis, provable. Thus, $\Gamma \vdash (\neg A)_1, \Delta$ is provable using the $(\neg r)$ rule. Similarly for $\Gamma, (\neg A)_1 \vdash \Delta$.
- Assume that a sequent of the form $\Gamma, (A \vee B)_k \vdash \Delta$ is valid, with $k \leq opt_{\mathcal{L}}(A \vee B)$. We claim that then both $\Gamma, (A)_k \vdash (B)_{<k}, \Delta$ and $\Gamma, (B)_k \vdash (A)_{<k}, \Delta$ are valid. Assume to the contrary that $\Gamma, (A)_k \vdash (B)_{<k}, \Delta$ is not valid (the other case is symmetric). Then, there is a model M of Γ and $(A)_k$ that is neither a model of $(B)_{<k}$ nor of Δ. But then M is also a model of Γ and $(A \vee B)_k$, but not of Δ, which contradicts the assumption that $\Gamma, (A \vee B)_k \vdash \Delta$ is valid. Therefore, both $\Gamma, (A)_k \vdash (B)_{<k}, \Delta$ and $\Gamma, (B)_k \vdash (A)_{<k}, \Delta$ are valid and, by the induction hypothesis, provable. This means that $\Gamma, (A \vee B)_k \vdash \Delta$ is provable by $(\vee l)$. Similarly for a sequent of the form $\Gamma, (A \wedge B)_k \vdash \Delta$.
- Assume that a sequent of the form $\Gamma \vdash (A \vee B)_k, \Delta$ is valid, with $k \leq opt_{\mathcal{L}}(A \vee B)$. We claim that then for all $i < k$ the sequents $\Gamma, (A)_i \vdash \Delta$ and $\Gamma, (B)_i \vdash \Delta$ and $\Gamma \vdash (A)_k, (B)_k, \Delta$ are valid. Assume by contradiction that there is an $i < k$ s.t. $\Gamma, (A)_i \vdash \Delta$ is not valid. Then, there is a model M of Γ and $(A)_i$ that is not a model of Δ. However, then M is a model of Γ but neither of Δ nor of $(A \vee B)_k$ (as M satisfies $A \vee B$ to degree $i \neq k$), which contradicts our assumption that $\Gamma \vdash (A \vee B)_k, \Delta$ is valid. The case that there is an $i < k$ s.t. $\Gamma, (B)_i \vdash \Delta$ is not valid is symmetric. Finally, we assume that $\Gamma \vdash (A)_k, (B)_k, \Delta$ is not valid. Then, there is a model M of Γ that is not a model of $(A)_k$, $(B)_k$ or Δ. Then, M is model of Γ but neither of Δ nor of $(A \vee B)_k$, contradicting our assumption. Therefore, all sequents listed above must be valid, and, by the induction hypothesis, $\Gamma \vdash (A \vee B)_k, \Delta$ is provable. Similarly for a sequent of the form $\Gamma \vdash (A \wedge B)_k, \Delta$.
- Assume that a sequent of the form $\Gamma, (A \vec{\times} B)_k \vdash \Delta$ with $k \leq opt_{\text{QCL}}(A)$ is valid. Then $\Gamma, (A)_k \vdash \Delta$ is also valid since $(A \vec{\times} B)_k$ and $(A)_k$ have the same models if $k \leq opt_{\text{QCL}}(A)$. By the induction hypothesis $\Gamma, (A \vec{\times} B)_k \vdash \Delta$ is provable. Analogously for sequents of the form $\Gamma \vdash (A \vec{\times} B)_k, \Delta$.
- Assume that a sequent of the form $\Gamma, (A \vec{\times} B)_{opt_{\text{QCL}}(A)+l} \vdash \Delta$ is valid, with $l \leq opt_{\mathcal{L}}(B)$. We claim that the sequent $\Gamma, (B)_l, \neg A \vdash \Delta$ is then also valid. Indeed, if M is a model of Γ, $(B)_l$ and $\neg A$, then it is also a model of Γ and $(A \vec{\times} B)_{opt_{\text{QCL}}(A)+l}$. Hence, by assumption, M must be a model of Δ. From this, we can conclude as before that $\Gamma, (A \vec{\times} B)_{opt_{\text{QCL}}(A)+l} \vdash \Delta$ is provable.
- Assume that a sequent of the form $\Gamma \vdash (A \vec{\times} B)_{opt_{QCL}(A)+l}, \Delta$ is valid, with $l \leq opt_{\mathcal{L}}(B)$. We claim that then also the sequents $\Gamma \vdash \neg A, \Delta$ and $\Gamma \vdash (B)_l, \Delta$ are valid. Assume by contradiction that the first sequent is not valid. This means that there is a model M of Γ that is not a model of either $\neg A$

nor of Δ. However, then M is a model of A and therefore satisfies $A \vec{\times} B$ to a degree smaller than $opt_{\text{QCL}}(A)$. This contradicts our assumption that $\Gamma \vdash (A \vec{\times} B)_{opt_{\text{QCL}}(A)+l}, \Delta$ is valid. Assume now that the second sequent is not valid, i.e., that there is a model M of Γ that is neither a model of $(B)_l$ nor of Δ. Then, M cannot be a model of $(A \vec{\times} B)_{opt_{\text{QCL}}(A)+l}$ and we again have a contradiction to our assumption. As before, it follows by the induction hypothesis that $\Gamma \vdash (A \vec{\times} B)_{opt_{\text{QCL}}(A)+l}, \Delta$ is provable. □

So far we have not introduced a cut rule, and as we have shown our calculus is complete without such a rule. However, it is easy to see that we have cut-admissibility, i.e., **L**[QCL] can be extended by:

$$\frac{\Gamma \vdash (A)_k, \Delta \qquad \Gamma', (A)_k \vdash \Delta'}{\Gamma, \Gamma' \vdash \Delta, \Delta'} \; cut$$

Another aspect of our calculus that should be mentioned is that, although **L**[QCL] is cut-free, we do not have the subformula property. This is especially obvious when looking at the rules for negation, where we use the classical counterpart $cp(A)$ of QCL-formulas. For example, $\neg(a \vec{\times} b)$ in the conclusion of the $\neg$-left rule becomes $cp(a \vec{\times} b) = a \vee b$ in the premise.

While we believe that **L**[QCL] is interesting in its own right, the question of how we can use it to obtain a calculus for preferred model entailment arises. Essentially, we have to add a rule that allows us to go from standard to preferred model inferences. As a first approach we consider theories $\Gamma \cup \{A\}$ with Γ consisting only of classical formulas and A being a QCL-formula. In this simple case, preferred models of $\Gamma \cup \{A\}$ are those models of $\Gamma \cup \{A\}$ that satisfy A to the smallest possible degree. One might add the following rule to **L**[QCL]:

$$\frac{\langle \Gamma, (A)_i \vdash \bot \rangle_{i<k} \qquad \Gamma, (A)_k \vdash \Delta}{\Gamma, A \mid\!\sim^{lex}_{\text{QCL}} \Delta} \; \mid\!\sim_{naive}$$

Intuitively, the above rule states that, if there are no interpretations that satisfy Γ while also satisfying A to a degree lower than k, and if Δ follows from all models of $\Gamma, (A)_k$, then Δ is entailed by the preferred models of $\Gamma \cup \{A\}$. However, the obtained calculus **L**[QCL] + $\mid\!\sim_{naive}$ derives invalid sequents.

Example 6. The invalid entailment $\neg a, a \vec{\times} b \mid\!\sim^{lex}_{\text{QCL}} a$ can be derived via $\mid\!\sim_{naive}$.

$$\cfrac{\cfrac{\cfrac{a \vdash a}{\neg a, a \vdash a}\; wl}{\neg a, (a \vec{\times} b)_1 \vdash a}\; \vec{\times} l_1}{\neg a, a \vec{\times} b \mid\!\sim^{lex}_{\text{QCL}} a}\; \mid\!\sim_{naive}$$

What is missing is an assertion that $\Gamma, (A)_k$ is satisfiable. Unfortunately, this can not be formulated in **L**[QCL]. A way of addressing this problem is to define a refutation calculus, as has been done for other non-monotonic logics [5].

4 Calculus for Preferred Model Entailment

We now introduce a calculus for preferred model entailment. However, as argued above, we first need to introduce the refutation calculus $\mathbf{L}[\mathrm{QCL}]^-$. In the literature, a rejection method for first-order logic with equality was first introduced in [17] and proved complete w.r.t. finite model theory. Our refutation calculus is based on a simpler rejection method for propositional logic defined in [5]. Using the refutation calculus, we prove that $(A)_k$ is satisfiable by deriving the antisequent $(A)_k \nvdash \bot$.

Definition 12. *A labeled* QCL*-antisequent is denoted by* $\Gamma \nvdash \Delta$ *and it is valid if and only if the corresponding labeled* QCL*-sequent* $\Gamma \vdash \Delta$ *is not valid, i.e., if at least one model that satisfies all formulas in* Γ *to the degree specified by the label satisfies no formula in* Δ *to the degree specified by the label.*

Below we give a definition of the refutation calculus $\mathbf{L}[\mathrm{QCL}]^-$. Note that most rules coincide with their counterparts in $\mathbf{L}[\mathrm{QCL}]$. Binary rules are translated into two rules; one inference rule per premise. $(\vee r)$ and $(\wedge l)$ in $\mathbf{L}[\mathrm{QCL}]$ have an unbounded number of premises, but due to their structure they can be translated into three inference rules. For $(\wedge r)$ we need to introduce two extra rules for the case that either A or B is not satisfied.

Definition 13 ($\mathbf{L}[\mathrm{QCL}]^-$). *The axioms of* $\mathbf{L}[\mathrm{QCL}]^-$ *are of the form* $\Gamma \nvdash \Delta$*, where* Γ *and* Δ *are disjoint sets of atoms and* $\bot \notin \Gamma$*. The inference rules of* $\mathbf{L}[\mathrm{QCL}]^-$ *are given below. Whenever a labeled formula* $(F)_k$ *appears in the conclusion of an inference rule it holds that* $k \leq opt_{\mathcal{L}}(F)$*.*

The logical rules are:

$$\frac{\Gamma, (cp(A))_1 \nvdash \Delta}{\Gamma \nvdash (\neg A)_1, \Delta}\nvdash \neg r \qquad \frac{\Gamma \nvdash (cp(A))_1, \Delta}{\Gamma, (\neg A)_1 \nvdash \Delta}\nvdash \neg l$$

$$\frac{\Gamma, (A)_k \nvdash (B)_{<k}, \Delta}{\Gamma, (A \vee B)_k \nvdash \Delta}\nvdash \vee l_1 \qquad \frac{\Gamma, (B)_k \nvdash (A)_{<k}, \Delta}{\Gamma, (A \vee B)_k \nvdash \Delta}\nvdash \vee l_2$$

$$\frac{\Gamma, (A)_i \nvdash \Delta}{\Gamma \nvdash (A \vee B)_k, \Delta}\nvdash \vee r_1 \qquad \frac{\Gamma, (B)_i \nvdash \Delta}{\Gamma \nvdash (A \vee B)_k, \Delta}\nvdash \vee r_2 \qquad \frac{\Gamma \nvdash (A)_k, (B)_k, \Delta}{\Gamma \nvdash (A \vee B)_k, \Delta}\nvdash \vee r_3$$

where $i < k$*.*

$$\frac{\Gamma, (A)_k \nvdash (B)_{>k}, \Delta}{\Gamma, (A \wedge B)_k \nvdash \Delta}\nvdash \wedge l_1 \qquad \frac{\Gamma, (B)_k \nvdash (A)_{>k}, \Delta}{\Gamma, (A \wedge B)_k \nvdash \Delta}\nvdash \wedge l_2$$

$$\frac{\Gamma, (A)_i \nvdash \Delta}{\Gamma \nvdash (A \wedge B)_k, \Delta}\nvdash \wedge r_1 \qquad \frac{\Gamma, (\neg A)_1 \nvdash \Delta}{\Gamma \nvdash (A \wedge B)_k, \Delta}\nvdash \wedge r_2 \qquad \frac{\Gamma, (B)_i \nvdash \Delta}{\Gamma \nvdash (A \wedge B)_k, \Delta}\nvdash \wedge r_3$$

$$\frac{\Gamma, (\neg B)_1 \nvdash \Delta}{\Gamma \nvdash (A \wedge B)_k, \Delta}\nvdash \wedge r_4 \qquad \frac{\Gamma \nvdash (A)_k, (B)_k, \Delta}{\Gamma \nvdash (A \wedge B)_k, \Delta}\nvdash \wedge r_5$$

where $i > k$*.*

The rules for ordered disjunction, with $k \leq opt_{\mathcal{L}}(A)$ *and* $l \leq opt_{\mathcal{L}}(B)$*, are:*

$$\frac{\Gamma, (A)_k \nvdash \Delta}{\Gamma, (A \vec{\times} B)_k \nvdash \Delta}\nvdash \vec{\times} l_1 \qquad \frac{\Gamma, (B)_l, (\neg A)_1 \nvdash \Delta}{\Gamma, (A \vec{\times} B)_{opt_{\mathrm{QCL}}(A)+l} \nvdash \Delta}\nvdash \vec{\times} l_2$$

$$\frac{\Gamma \nvdash (A)_k, \Delta}{\Gamma \nvdash (A \vec{\times} B)_k, \Delta}\nvdash \vec{\times} r_1 \qquad \frac{\Gamma \nvdash (\neg A)_1, \Delta}{\Gamma \nvdash (A \vec{\times} B)_{opt_{\mathrm{QCL}}(A)+l}, \Delta}\nvdash \vec{\times} r_2 \qquad \frac{\Gamma \nvdash (B)_l, \Delta}{\Gamma \nvdash (A \vec{\times} B)_{opt_{\mathrm{QCL}}(A)+l}, \Delta}\nvdash \vec{\times} r_3$$

The degree overflow rules, with $k \in \mathbb{N}$, are:

$$\frac{\Gamma, \bot \nvdash \Delta}{\Gamma, (A)_{opt_{\mathrm{QCL}}(A)+k} \nvdash \Delta} \nvdash dol \qquad \frac{\Gamma \nvdash \Delta}{\Gamma \nvdash (A)_{opt_{\mathrm{QCL}}(A)+k}, \Delta} \nvdash dor$$

Example 7. The following is related to Example 4 and shows that the sequent $\neg(a \wedge b), ((a \vec{\times} b) \wedge (b \vec{\times} c))_2$ is satisfiable.

$$\dfrac{\dfrac{\dfrac{\dfrac{\vdots \atop (a \vee b), c, \neg b \nvdash a \wedge b, \bot}{(a \vee b), (b \vec{\times} c)_2 \nvdash a \wedge b, \bot} \nvdash \vec{\times} l_2}{(b \vec{\times} c)_2 \nvdash \neg(a \vec{\times} b), a \wedge b, \bot} \nvdash \neg r}{((a \vec{\times} b) \wedge (b \vec{\times} c))_2 \nvdash a \wedge b, \bot} \nvdash \wedge l_2}{\neg(a \wedge b), ((a \vec{\times} b) \wedge (b \vec{\times} c))_2 \nvdash \bot} \nvdash \neg l$$

Note that the interpretation $\{a, c\}$ witnesses $(a \vee b), c, \neg b \nvdash a \wedge b, \bot$.

Proposition 3. $\mathbf{L}[\mathrm{QCL}]^-$ *is sound.*

Proof. The soundness of the negation rules is straightforward. The soundness of the rules $(\vec{\times} l_1)$, $(\vec{\times} l_2)$ and $(\vec{\times} r_1)$ follows by the same argument as for $\mathbf{L}[\mathrm{QCL}]$. For the remaining rules, it is easy to check that the same model witnessing the validity of the premise also witnesses the validity of the conclusion. □

Proposition 4. $\mathbf{L}[\mathrm{QCL}]^-$ *is complete.*

Proof. We show completeness by an induction over the (aggregated) formula complexity. Assume $\Gamma \nvdash \Delta$ is valid, i.e. $\Gamma \vdash \Delta$ is not valid. Now, there must be a rule in $\mathbf{L}[\mathrm{QCL}]$ for which $\Gamma \vdash \Delta$ is the conclusion. By the soundness of $\mathbf{L}[\mathrm{QCL}]$, this implies that at least one of the premises $\Gamma^* \vdash \Delta^*$ is not valid. However, then $\Gamma^* \nvdash \Delta^*$ is valid and, by induction, also provable. Now, by the construction of $\mathbf{L}[\mathrm{QCL}]^-$, there is a rule that allows us to derive $\Gamma \nvdash \Delta$ from $\Gamma^* \nvdash \Delta^*$. □

So far no cut-rule has been introduced for $\mathbf{L}[\mathrm{QCL}]^-$, and indeed, a counterpart of the cut rule would not be sound. One possibility is to introduce a contrapositive of cut as described by Bonatti and Olivetti [5]. Again, it is easy to see that this rule is admissible in our calculus:

$$\frac{\Gamma \nvdash \Delta \qquad \Gamma, (A)_k \vdash \Delta}{\Gamma \nvdash (A)_k, \Delta} \, cut2$$

We are now ready to combine $\mathbf{L}[\mathrm{QCL}]$ and $\mathbf{L}[\mathrm{QCL}]^-$ by defining an inference rule that allows us to go from labeled sequents to non-monotonic inferences. Again, we first consider the case where Γ is classical and A is a choice logic formula. The preferred model inference rule is:

$$\frac{\langle \Gamma, (A)_i \vdash \bot \rangle_{i<k} \qquad \Gamma, (A)_k \nvdash \bot \qquad \Gamma, (A)_k \vdash \Delta}{\Gamma, A \mid\!\sim^{lex}_{\mathrm{QCL}} \Delta} \mid\!\sim_{simple}$$

Intuitively, the premises $\langle \Gamma, (A)_i \vdash \bot \rangle_{i<k}$ along with $\Gamma, (A)_k \nvdash \bot$ ensure that models satisfying A to a degree of k are preferred, while the premise $\Gamma, (A)_k \vdash \Delta$ ensures that Δ is entailed by those preferred models.

Example 8. The valid entailment $\neg(a \wedge b), (a \vec{\times} b) \wedge (b \vec{\times} c) \mid\!\sim^{lex}_{\text{QCL}} a \wedge c, b$ is provable by choosing $k = 2$:

$$\dfrac{\overset{(\varphi_1)}{\Gamma, ((a \vec{\times} b) \wedge (b \vec{\times} c))_1 \vdash \bot} \quad \overset{(\varphi_2)}{\Gamma, ((a \vec{\times} b) \wedge (b \vec{\times} c))_2 \nvdash \bot} \quad \overset{(\varphi_3)}{\Gamma, ((a \vec{\times} b) \wedge (b \vec{\times} c))_2 \vdash \Delta}}{\Gamma, (a \vec{\times} b) \wedge (b \vec{\times} c) \mid\!\sim^{lex}_{\text{QCL}} \Delta} \mid\!\sim_{simple}$$

with $\Gamma = \neg(a \wedge b)$ and $\Delta = a \wedge c, b$. φ_3 is the **L**[QCL]-proof from Example 4 and φ_2 is the **L**[QCL]$^-$-proof from Example 7. φ_1 is not shown explicitly, but it can be verified that the corresponding sequent is provable.

We extend $\mid\!\sim_{simple}$ to the more general case, where more than one non-classical formula may be present, to obtain a calculus for preferred model entailment. An additional rule $\mid\!\sim_{unsat}$ is needed in case a theory is classically unsatisfiable.

Definition 14 (L[QCL]$^{lex}_{\mid\sim}$). *Let $\leq_l$ be the order on vectors in $\mathbb{N}^k$ defined by*

- *$\boldsymbol{v} <_l \boldsymbol{w}$ if there is some $n \in \mathbb{N}$ such that $\boldsymbol{v}$ has more entries of value n and for all $1 \leq m < n$ both vectors have the same number of entries of value m.*
- *$\boldsymbol{v} =_l \boldsymbol{w}$ if, for all $n \in \mathbb{N}$, $\boldsymbol{v}$ and $\boldsymbol{w}$ have the same number of entries of value n.*

L[QCL]$^{lex}_{\mid\sim}$ *consists of the axioms and rules of* **L**[QCL] *and* **L**[QCL]$^-$ *plus the following rules, where $\boldsymbol{v}, \boldsymbol{w} \in \mathbb{N}^k$, Γ consists of only classical formulas, and every A_i with $1 \leq i \leq k$ is a* QCL*-formula:*

$$\dfrac{\langle \Gamma, (A_1)_{w_1}, \ldots, (A_k)_{w_k} \vdash \bot \rangle_{\boldsymbol{w} < \boldsymbol{v}} \quad \Gamma, (A_1)_{v_1}, \ldots, (A_k)_{v_k} \nvdash \bot \quad \langle \Gamma, (A_1)_{w_1}, \ldots, (A_k)_{w_k} \vdash \Delta \rangle_{\boldsymbol{w} = \boldsymbol{v}}}{\Gamma, A_1, \ldots, A_k \mid\!\sim^{lex}_{\text{QCL}} \Delta} \mid\!\sim_{lex}$$

$$\dfrac{\Gamma, cp(A_1), \ldots, cp(A_k) \vdash \bot}{\Gamma, A_1, \ldots, A_k \mid\!\sim^{lex}_{\text{QCL}} \Delta} \mid\!\sim_{unsat}$$

We first provide a small example and then show soundness and completeness.

Example 9. Consider the valid entailment $\neg(a \wedge b), (a \vec{\times} b), (b \vec{\times} c) \mid\!\sim^{lex}_{\text{QCL}} a \wedge c, b$ similar to Example 8, but with the information that we require $(a \vec{\times} b)$ and $(b \vec{\times} c)$ encoded as separate formulas. It is not possible to satisfy all formulas on the left to a degree of 1. Rather, it is optimal to either satisfy $(\neg(a \wedge b))_1, (a \vec{\times} b)_1, (b \vec{\times} c)_2$ or, alternatively, $(\neg(a \wedge b))_1, (a \vec{\times} b)_2, (b \vec{\times} c)_1$. We choose $\boldsymbol{v} = (1, 1, 2)$, with $\boldsymbol{w} = (1, 1, 1)$ being the only vector $\boldsymbol{w}$ s.t. $\boldsymbol{w} < \boldsymbol{v}$. Thus, we get

$$\dfrac{\overset{\vdots}{\Gamma, (a \vec{\times} b)_1, (b \vec{\times} c)_1 \vdash \bot} \quad \overset{\vdots}{\Gamma, (a \vec{\times} b)_1, (b \vec{\times} c)_2 \nvdash \bot} \quad \overset{\vdots}{\Gamma, (a \vec{\times} b)_1, (b \vec{\times} c)_2 \vdash \Delta} \quad \overset{\vdots}{\Gamma, (a \vec{\times} b)_2, (b \vec{\times} c)_1 \vdash \Delta}}{\Gamma, (a \vec{\times} b), (b \vec{\times} c) \mid\!\sim^{lex}_{\text{QCL}} \Delta} \mid\!\sim_{lex}$$

with $\Gamma = \neg(a \wedge b)$ and $\Delta = a \wedge c, b$. It can be verified that indeed all branches are provable, but we do not show this explicitly here.

Proposition 5. $\mathbf{L}[\mathrm{QCL}]^{lex}_{\mid\sim}$ *is sound.*

Proof. Consider the first the $\mid\sim_{lex}$-rule and assume that all premises are derivable. By the soundness of $\mathbf{L}[\mathrm{QCL}]$ and $\mathbf{L}[\mathrm{QCL}]^-$ they are also valid. From the first set of premises $\langle \Gamma, (A_1)_{w_1}, \ldots, (A_k)_{w_k} \vdash \bot \rangle_{\boldsymbol{w} < \boldsymbol{v}}$ we can conclude that if there is some model M of Γ that satisfies A_i to a degree of v_i for all $1 \leq i \leq k$, then $M \in Prf^{lex}_{\mathrm{QCL}}(\Gamma \cup \{A_1, \ldots, A_k\})$. The premise $\Gamma, (A_1)_{v_1}, \ldots, (A_k)_{v_k} \nvdash \bot$ ensures that there is such a model M. By the last set of premises $\langle \Gamma, (A_1)_{w_1}, \ldots, (A_k)_{w_k} \vdash \Delta \rangle_{\boldsymbol{w} = \boldsymbol{v}}$, we can conclude that all models of $\Gamma \cup \{A_1, \ldots, A_k\}$ that are equally as preferred as M, i.e., all $M' \in Prf^{lex}_{\mathrm{QCL}}(\Gamma \cup \{A_1, \ldots, A_k\})$, satisfy at least one formula in Δ. Therefore, $\Gamma, A_1, \ldots, A_k \mid\sim^{lex}_{\mathrm{QCL}} \Delta$ is valid.

Now consider the $\mid\sim_{unsat}$-rule and assume that $\Gamma, cp(A_1), \ldots, cp(A_k) \vdash \bot$ is derivable and therefore valid. Thus, $\Gamma \cup \{A_1, \ldots, A_k\}$ has no models and therefore also no preferred models. Then $\Gamma, A_1, \ldots, A_k \mid\sim^{lex}_{\mathrm{QCL}} \Delta$ is valid. □

Proposition 6. $\mathbf{L}[\mathrm{QCL}]^{lex}_{\mid\sim}$ *is complete.*

Proof. Assume that $\Gamma, A_1, \ldots, A_k \mid\sim^{lex}_{\mathrm{QCL}} \Delta$ is valid. If $\Gamma \cup \{A_1, \ldots, A_k\}$ is unsatisfiable then $\Gamma, cp(A_1), \ldots, cp(A_k) \vdash \bot$ is valid, i.e., we can apply the $\mid\sim_{unsat}$-rule. Now consider the case that $\Gamma \cup \{A_1, \ldots, A_k\}$ is satisfiable and assume that some preferred model M of $\Gamma \cup \{A_1, \ldots, A_k\}$ satisfies A_i to a degree of v_i for all $1 \leq i \leq k$. Then, we claim that all premises of the rule are valid and, by the completeness of $\mathbf{L}[\mathrm{QCL}]$ and $\mathbf{L}[\mathrm{QCL}]^-$, also derivable.

Assume by contradiction that one of the premises is not valid. First, consider the case that $\Gamma, (A_1)_{w_1}, \ldots, (A_k)_{w_k} \vdash \bot$ is not valid for some $\boldsymbol{w} < \boldsymbol{w}$. Then there is a model M' of Γ that satisfies A_i to a degree of w_i for all $1 \leq i \leq k$. However, this contradicts the assumption that M is a preferred model of $\Gamma \cup \{A_1, \ldots, A_k\}$.

Next, assume that $\Gamma, (A_1)_{v_1}, \ldots, (A_k)_{v_k} \nvdash \bot$ is not valid. However, M satisfies $\Gamma, (A_1)_{v_1}, \ldots, (A_k)_{v_k}$ and does not satisfy $\bot$. Contradiction.

Finally, we assume that $\Gamma, (A_1)_{w_1}, \ldots, (A_k)_{w_k} \vdash \Delta$ is not valid for some $\boldsymbol{w} = \boldsymbol{v}$. Then, there is a model M' of Γ that satisfies A_i to a degree of w_i for all $1 \leq i \leq k$ but does not satisfy any formula in Δ. But M' is a preferred model of $\Gamma \cup \{A_1, \ldots, A_k\}$, which contradicts $\Gamma, A_1, \ldots, A_k \mid\sim^{lex}_{\mathrm{QCL}} \Delta$ being valid. □

In this paper, we focused on the lexicographic semantics for preferred models of choice logic theories. However, rules for other semantics, e.g. a rule $\mid\sim_{inc}$ for the inclusion based approach (cf. Definition 8), can be obtained by simply adapting the way in which vectors over $\mathbb{N}^k$ are compared (cf. Definition 14).

5 Beyond QCL

QCL was the first choice logic to be described [7], and applications concerned with QCL and ordered disjunction have been discussed in the literature [3,8,13]. For this reason, the main focus in this paper lies with QCL. However, as we have seen in Sect. 2, CCL and its ordered conjunction show that interesting logics similar to QCL exist. We will now demonstrate that $\mathbf{L}[\mathrm{QCL}]$ can easily be adapted for other choice logics. In particular, we introduce **L**[CCL] in which the rules of **L**[QCL] for the classical connectives can be retained. All that is needed is to replace the $\vec{\times}$-rules by appropriate rules for the choice connective $\vec{\odot}$ of CCL.

Definition 15 (L[CCL]). **L**[CCL] *is* **L**[QCL], *except that the* $\vec{\times}$*-rules are replaced by the following* $\vec{\odot}$*-rules:*

$$\frac{\Gamma, (A)_1, (B)_k \vdash \Delta}{\Gamma, (A\vec{\odot}B)_k \vdash \Delta}\ \vec{\odot}l_1 \qquad \frac{\Gamma, (A)_l, (\neg B)_1 \vdash \Delta}{\Gamma, (A\vec{\odot}B)_{opt_{\mathrm{CCL}}(B)+l} \vdash \Delta}\ \vec{\odot}l_2 \qquad \frac{\Gamma, (A)_m \vdash \Delta}{\Gamma, (A\vec{\odot}B)_{opt_{\mathrm{CCL}}(B)+m} \vdash \Delta}\ \vec{\odot}l_3$$

$$\frac{\Gamma \vdash (A)_1, \Delta \qquad \Gamma \vdash (B)_k, \Delta}{\Gamma \vdash (A\vec{\odot}B)_k, \Delta}\ \vec{\odot}r_1 \quad \frac{\Gamma \vdash (A)_l, \Delta \qquad \Gamma \vdash (\neg B)_1, \Delta}{\Gamma \vdash (A\vec{\odot}B)_{opt_{\mathrm{CCL}}(B)+l}, \Delta}\ \vec{\odot}r_2 \quad \frac{\Gamma \vdash (A)_m, \Delta}{\Gamma \vdash (A\vec{\odot}B)_{opt_{\mathrm{CCL}}(B)+m}, \Delta}\ \vec{\odot}r_3$$

where $k \leq opt_{\mathrm{CCL}}(B)$, $l \leq opt_{\mathrm{CCL}}(A)$, *and* $1 < m \leq opt_{\mathrm{CCL}}(A)$.

Note that, given $\Gamma, (A\vec{\odot}B)_{opt_{\mathrm{CCL}}(B)+m} \vdash \Delta$ with $1 < m \leq opt_{\mathrm{CCL}}(A)$, we need to guess whether $\vec{\odot}l_2$ or $\vec{\odot}l_3$ has to be applied. We do not define **L**[CCL]$^-$ here, but the necessary rules for $\vec{\odot}$ can be inferred from the $\vec{\odot}$-rules of **L**[CCL] in a similar way to how **L**[QCL]$^-$ was derived from **L**[QCL].

Proposition 7. **L**[CCL] *is sound.*

Proof. We consider the newly introduced rules.

- For $\vec{\odot}l_1$, $\vec{\odot}l_2$, and $\vec{\odot}l_3$ this follows directly from the definition of CCL.
- ($\vec{\odot}r_1$). Assume both premises are valid, i.e., every model of Γ is a model of Δ or of $(A)_1$ and $(B)_k$ with $k \leq opt_{\mathcal{L}}(B)$. By definition, any model that satisfies $(A)_1$ and $(B)_k$ satisfies $A\vec{\odot}B$ to degree k. Thus, every model of Γ is a model of Δ or of $(A\vec{\odot}B)_k$, which means the conclusion of the rule is valid.
- ($\vec{\odot}r_2$). Assume both premises are valid, i.e., every model of Γ is either a model of Δ or of $(A)_l$ and $(\neg B)_1$ with $l \leq opt_{\mathrm{CCL}}(A)$. By definition, any model that satisfies $(A)_l$ and does not satisfy B (and hence satisfies $(\neg B)_1$) satisfies $A\vec{\odot}B$ to degree $opt_{\mathrm{CCL}}(B) + l$.
- ($\vec{\odot}r_3$). Assume that the premise is valid, i.e., every model of Γ is either a model of Δ or of $(A)_m$ with $1 < m \leq opt_{\mathrm{CCL}}(A)$. By definition, any model that satisfies $(A)_m$, regardless of what degree this model ascribes to B, satisfies $A\vec{\odot}B$ to degree $opt_{\mathrm{CCL}}(B) + m$. □

Proposition 8. **L**[CCL] *is complete.*

Proof. We adapt the induction of the proof of Proposition 2:

- Assume that a sequent of the form $\Gamma, (A\vec{\odot}B)_k \vdash \Delta$ is valid, with $k \leq opt_{\mathcal{L}}(B)$. All models that satisfy $(A\vec{\odot}B)_k$ must satisfy A to a degree of 1 and B to a degree of k. Thus, $\Gamma, (A)_1, (B)_k \vdash \Delta$ is valid, and, by the induction hypothesis, $\Gamma, (A\vec{\odot}B)_k \vdash \Delta$ is provable. Similarly for the cases $\Gamma, (A\vec{\odot}B)_{opt_{\mathrm{CCL}}(B)+l} \vdash \Delta$ with $l \leq opt_{\mathrm{CCL}}(A)$, and $\Gamma, (A\vec{\odot}B)_{opt_{\mathrm{CCL}}(B)+m} \vdash \Delta$ with $1 < m \leq opt_{\mathrm{CCL}}(A)$.

- Assume that a sequent of the form $\Gamma \vdash (A\vec{\odot}B)_k, \Delta$ is valid, with $k \leq opt_{\mathcal{L}}(B)$. We claim that then $\Gamma \vdash (A)_1, \Delta$ and $\Gamma \vdash (B)_k, \Delta$ are valid. Assume, for the sake of a contradiction, that the first sequent is not valid. This means that there is a model M of Γ that is neither a model of $(A)_1$ nor of Δ. However, then M satisfies $A\vec{\odot}B$ to a degree higher than $opt_{\mathrm{CCL}}(B)$. This contradicts the assumption that $\Gamma \vdash (A\vec{\odot}B)_k, \Delta$ is valid. Assume now that the second sequent is not valid, i.e., that there is a model M of Γ that is neither a model of $(B)_k$ nor of Δ. Then M cannot be a model of $(A\vec{\odot}B)_k$, contradicting the assumption. As before, it follows by the induction hypothesis that $\Gamma \vdash (A\vec{\odot}B)_k, \Delta$ is provable. Similarly for the cases $\Gamma \vdash (A\vec{\odot}B)_{opt_{\mathrm{CCL}}(B)+l}, \Delta$ with $l \leq opt_{\mathrm{CCL}}(A)$, and $\Gamma \vdash (A\vec{\odot}B)_{opt_{\mathrm{CCL}}(B)+m}, \Delta$ with $1 < m \leq opt_{\mathrm{CCL}}(A)$. □

We are confident that our methods can be adapted not only for QCL and CCL, but for numerous other instantiations of the choice logic framework defined in Sect. 2. We mention here *lexicographic choice logic* (LCL) [4], in which $A\vec{\diamond}B$ expresses that it is best to satisfy A and B, second best to satisfy only A, third best to satisfy only B, and unacceptable to satisfy neither.

Moreover, note that the inference rules $\mid\!\sim_{lex}$ and $\mid\!\sim_{unsat}$ (cf. Definition 14) do not depend on any specific choice logic. Thus, once labeled calculi are developed for a choice logic, a calculus for preferred model entailment follows immediately.

6 Conclusion

In this paper we introduce a sound and complete sequent calculus for preferred model entailment in QCL. This non-monotonic calculus is built on two calculi: a monotonic labeled sequent calculus and a corresponding refutation calculus.

Our systems are modular and can easily be adapted: on the one hand, calculi for choice logics other than QCL can be obtained by introducing suitable rules for the choice connectives of the new logic, as exemplified with our calculus for CCL; on the other hand, a non-monotonic calculus for preferred model semantics other than the lexicographic semantics can be obtained by adapting the inference rule $\mid\!\sim_{lex}$ which transitions from preferred model entailment to the labeled calculi.

Our work contributes to the line of research on non-monotonic sequent calculi that make use of refutation systems [5]. Our system is the first proof calculus for choice logics, which have been studied mainly from the viewpoint of their computational properties [4] and their potential applications [3,8,13] so far.

Regarding future work, we aim to investigate the proof complexity of our calculi, and how this complexity might depend on which choice logic or preferred model semantics is considered. Also, calculi for other choice logics such as LCL could be explicitly defined, as was done with CCL in Sect. 5.

Acknowledgments. We thank the anonymous reviewers for their valuable feedback. This work was funded by the Austrian Science Fund (FWF) under the grants Y698 and J4581.

References

1. Avron, A.: The method of hypersequents in the proof theory of propositional non-classical logics. In: Hodges, W., Hyland, M., Steinhorn, C., Truss, J. (eds.) Logic: From Foundations to Applications, pp. 1–32. Oxford Science Publications, Oxford (1996)
2. Baaz, M., Lahav, O., Zamansky, A.: Finite-valued semantics for canonical labelled calculi. J. Autom. Reason. **51**(4), 401–430 (2013)
3. Benferhat, S., Sedki, K.: Alert correlation based on a logical handling of administrator preferences and knowledge. In: SECRYPT, pp. 50–56. INSTICC Press (2008)
4. Bernreiter, M., Maly, J., Woltran, S.: Choice logics and their computational properties. In: IJCAI, pp. 1794–1800. ijcai.org (2021)
5. Bonatti, P.A., Olivetti, N.: Sequent calculi for propositional nonmonotonic logics. ACM Trans. Comput. Log. **3**(2), 226–278 (2002)
6. Boudjelida, A., Benferhat, S.: Conjunctive choice logic. In: ISAIM (2016)
7. Brewka, G., Benferhat, S., Berre, D.L.: Qualitative choice logic. Artif. Intell. **157**(1–2), 203–237 (2004)
8. Brewka, G., Niemelä, I., Syrjänen, T.: Logic programs with ordered disjunction. Comput. Intell. **20**(2), 335–357 (2004)
9. Carnielli, W.A.: Systematization of finite many-valued logics through the method of tableaux. J. Symb. Log. **52**(2), 473–493 (1987)
10. Geibinger, T., Tompits, H.: Sequent-type calculi for systems of nonmonotonic paraconsistent logics. In: ICLP Technical Communications. EPTCS, vol. 325, pp. 178–191 (2020)
11. Governatori, G., Rotolo, A.: Logic of violations: a Gentzen system for reasoning with contrary-to-duty obligations. Australas. J. Logic **4**, 193–215 (2006)
12. Kaminski, M., Francez, N.: Calculi for many-valued logics. Log. Univers. **15**(2), 193–226 (2021)
13. Liétard, L., Hadjali, A., Rocacher, D.: Towards a gradual QCL model for database querying. In: Laurent, A., Strauss, O., Bouchon-Meunier, B., Yager, R.R. (eds.) IPMU 2014. CCIS, vol. 444, pp. 130–139. Springer, Cham (2014). https://doi.org/10.1007/978-3-319-08852-5_14
14. McCarthy, J.: Circumscription - a form of non-monotonic reasoning. Artif. Intell. **13**(1–2), 27–39 (1980)
15. Moore, R.C.: Semantical considerations on nonmonotonic logic. Artif. Intell. **25**(1), 75–94 (1985)
16. Reiter, R.: A logic for default reasoning. Artif. Intell. **13**(1–2), 81–132 (1980)
17. Tiomkin, M.L.: Proving unprovability. In: LICS, pp. 22–26. IEEE Computer Society (1988)

Lash 1.0 (System Description)

Chad E. Brown[1] and Cezary Kaliszyk[2](✉)

[1] Czech Technical University in Prague, Prague, Czech Republic
[2] University of Innsbruck, Innsbruck, Austria
cezary.kaliszyk@uibk.ac.at

Abstract. Lash is a higher-order automated theorem prover created as a fork of the theorem prover Satallax. The basic underlying calculus of Satallax is a ground tableau calculus whose rules only use shallow information about the terms and formulas taking part in the rule. Lash uses new, efficient C representations of vital structures and operations. Most importantly, Lash uses a C representation of (normal) terms with perfect sharing along with a C implementation of normalizing substitutions. We describe the ways in which Lash differs from Satallax and the performance improvement of Lash over Satallax when used with analogous flag settings. With a 10 s timeout Lash outperforms Satallax on a collection TH0 problems from the TPTP. We conclude with ideas for continuing the development of Lash.

Keywords: Higher-order logic · Automated reasoning · TPTP

1 Introduction

Satallax [4,7] is an automated theorem prover for higher-order logic that was a top competitor in the THF division of CASC [10] for most of the 2010s. The basic calculus of Satallax is a complete ground tableau calculus [2,5,6]. In recent years the top systems of the THF division of CASC are primarily based on resolution and superposition [3,8,11]. At the moment it is an open question whether there is a research and development path via which a tableau based prover could again become competitive. As a first step towards answering this question we have created a fork of Satallax, called Lash, focused on giving efficient C implementations of data structures and operations needed for search in the basic calculus.

Satallax was partly competitive due to (optional) additions that went beyond the basic calculus. Three of the most successful additions were the use of higher-order pattern clauses during search, the use of higher-order unification as a heuristic to suggest instantiations at function types and the use of the first-order theorem prover E as a backend to try to prove the first-order part of the current state is already unsatisfiable. Satallax includes flags that can be used to activate or deactivate such additions so that search only uses the basic calculus. They are deactivated by default. Satallax has three representations of terms in Ocaml. The basic calculus rules use the primary representation. Higher-order

J. Blanchette et al. (Eds.): IJCAR 2022, LNAI 13385, pp. 350–358, 2022.
https://doi.org/10.1007/978-3-031-10769-6_21

unification and pattern clauses make use of a representation that includes a case for metavariables to be instantiated. Communication with E uses a third representation restricted to first-order terms and formulas. When only the basic calculus is used, only the primary representation is needed.

Assuming only the basic calculus is used only limited information about (normal) terms is needed during the search. Typically we only need to know the outer structure of the principal formulas of each rule, and so the full term does not need to be traversed. In some cases Satallax either implicitly or explicitly traverses the term. The implicit cases are when a rule needs to know if two terms are equal. In Satallax, Ocaml's equality is used to test for equality of terms, implicitly relying on a recursion over the term. The explicit cases are quantifier rules that instantiate with either a term or a fresh constant. In the former case we may also need to normalize the result after instantiating with a term.

In order to give an optimized implementation of the basic calculus we have created a new theorem prover, Lash[1], by forking a recent version of Satallax (Satallax 3.4), the last version that won the THF division of CASC (in 2019). Generally speaking, we have removed all the additional code that goes beyond the basic calculus. In particular we do not need terms with metavariables since we support neither pattern clauses nor higher-order unification in Lash. Likewise we do not need a special representation for first-order terms and formulas since Lash does not communicate with E. We have added efficient C implementations of (normal) terms with perfect sharing. Additionally we have added new efficient C implementations of priority queues and the association of formulas with integers (to communicate with MiniSat). To measure the speedup given by the new parts of the implementation we have run Satallax 3.4 using flag settings that only use the basic calculus and Lash 1.0 using the same flag settings. We have also compared Lash to Satallax 3.4 using Satallax's default strategy with a timeout of 10 s, and have found that Lash 1.0 outperforms Satallax with this short timeout even when Satallax is using the optional additions (including calling E). We describe the changes and present a number of examples for which the changes lead to a significant speedup.

2 Preliminaries

We will presume a familiarity with simple type theory and only give a quick description to make our use of notation clear, largely following [6]. We assume a set of base types, one of which is the type o of propositions (also called booleans), and the rest we refer to as sorts. We use α, β to range over sorts and σ, τ to range over types. The only types other than base types are function types $\sigma\tau$, which can be thought of as the type of functions from σ to τ.

All terms have a unique type and are inductively defined as (typed) variables, (typed) constants, well-typed applications $(t\ s)$ and λ-abstractions $(\lambda x.t)$. We

[1] Lash 1.0 along with accompanying material is available at http://grid01.ciirc.cvut.cz/~chad/ijcar2022lash/.

also include the logical constant $\bot$ as a term of type o, terms (of type o) of the form $(s \Rightarrow t)$ (implications) and $(\forall x.t)$ (universal quantifiers) where s, t have type o and terms (of type o) of the form $(s =_\sigma t)$ where s, t have a common type σ. We also include choice constants ε_σ of $(\sigma o)\sigma$ at each type σ. We write $\neg t$ for $t \Rightarrow \bot$ and $(s \neq_\sigma t)$ for $(s =_\sigma t \Rightarrow \bot)$. We omit type parentheses and type annotations except where they are needed for clarity. Terms of type o are also called propositions. We also use $\top$, $\vee$, $\wedge$, $\exists$ with the understanding that these are notations for equivalent propositions in the set of terms above.

We assume terms are equal if they are the same up to α-conversion of bound variables (using de Bruijn indices in the implementation). We write $[s]$ for the $\beta\eta$-normal form of s.

The tableau calculi of [6] (without choice) and [2] (with choice) define when a branch is refutable. A branch is a finite set of normal propositions. We let A range over branches and write A, s for the branch $A \cup \{s\}$. We will not give a full calculus, but will instead discuss a few of the rules with surprising properties. Before doing so we emphasize rules that are *not* in the calculus. There is no cut rule stating that if A, s and $A, \neg s$ are refutable, then A is refutable. (During search such a rule would require synthesizing the cut formula s.) There is also no rule stating that if the branch $A, (s = t), [ps], [pt]$ is refutable, then $A, (s = t), [ps]$ is refutable (where s, t have type σ and p is a term of type σo). That is, there is no rule for rewriting into arbitrarily deep positions using equations.

All the tableau rules only need to examine the outer structure to test if they apply (when searching backwards for a refutation). When applying the rule, new formulas are constructed and added to the branch (or potentially multiple branches, each a subgoal to be refuted). An example is the confrontation rule, the only rule involving positive equations. The confrontation rule states that if $s =_\alpha t$ and $u \neq_\alpha v$ are on a branch A (where α is a sort), then we can refute A by refuting $A, s \neq u, t \neq u$ and $A, s \neq v, t \neq v$. A similar rule is the mating rule, which states that if $ps_1 \ldots s_n$ and $\neg pt_1 \ldots t_n$ are on a branch A (where p is a constant of type $\sigma_1 \cdots \sigma_n o$), then we can refute A by refuting each of the branches $A, s_i \neq t_i$ for each $i \in \{1, \ldots, n\}$. The mating rule demonstrates how disequations can appear on a branch even if the original branch to refute contained no reference to equality at all. One way a branch can be closed is if $s \neq s$ is on the branch. In an implementation, this means an equality check is done for s and t whenever a disequation $s \neq t$ is added to the branch. In Satallax this requires Ocaml to traverse the terms. In Lash this only requires comparing the unique integer ids the implementation assigns to the terms.

The disequations generated on a branch play an important role. Terms (of sort α) occuring on one side of a disequation on a branch are called *discriminating terms*. The rule for instantiating a quantified formula $\forall x.t$ (where x has sort α) is restricted to instantiating with discriminating terms (or a default term if no terms of sort α are discriminating). During search in Satallax this means there is a finite set of permitted instantiations (at sort α) and this set grows as disequations are produced. Note that, unlike most automated theorem provers, the instantiations do not arise from unification. In Satallax (and Lash) when

$\forall x.t$ is being processed it is instantiated with all previously processed instantiations. When a new instantiation is produced, previously processed universally quantified propositions are instantiated with it. When $\forall x.t$ is instantiated with s, then $[(\lambda x.t)s]$ is added to the branch. Such an instantiation is the important case where the new formula involves term traversals: both for substitution and normalization. In Satallax the substitution and normalization require multiple term traversals. In Lash we have used normalizing substitutions and memorized previous computations, minimizing the number of term traversals. The need to instantiate arises when processing either a universally quantified proposition (giving a new quantifier to instantiate) or a disequation at a sort (giving new discriminating terms).

We discuss a small example both Satallax and Lash can easily prove. We briefly describe what both do in order to give the flavor of the procedure and (hopefully) prevent readers from assuming the provers behave too similarly from readers based on other calculi (e.g., resolution).

Example `SEV241^5` from TPTP v7.5.0 [9] (`X5201A` from TPS [1]) contains a minor amount of features going beyond first-order logic. The statement to prove is

$$\forall x.U\ x \wedge W\ x \Rightarrow \forall S.(S = U \vee S = W) \Rightarrow Sx.$$

Here U and W are constants of type αo, x is a variable of type α and S is a variable of type αo. The higher-order aspects of this problem are the quantifier for S (though this could be circumvented by making S a constant like U and W) and the equations between predicates (though these could be circumvented by replacing $S = U$ by $\forall y.Sy \Leftrightarrow Uy$ and replacing $S = W$ similarly). The tableau rules effectively do both during search.

Satallax never clausifies. The formula above is negated and assumed. We will informally describe tableau rules as splitting the problem into subgoals, though this is technically mediated through MiniSat (where the set of MiniSat clauses is unsatisfiable when all branches are closed). Tableau rules are applied until the problem involves a constant c (for x), a constant S' for S and assumptions $U\ c$, $W\ c$, $S' = U \vee S' = W$ and $\neg S'c$ on the branch. The disjunction is internally $S' \neq U \Rightarrow S' = W$ and the implication rule splits the problem into two branches, one with $S' = U$ and one with $S' = W$. Both branches are solved in analogous ways and we only describe the $S' = U$ branch. Since $S' = U$ is an equation at function type, the relevant rule adds $\forall y.S'y = Uy$ to the branch. Since there are no disequations on the branch, there is no instantiation available for $\forall y.S'y = Uy$. In such a case, a default instantiation is created and used. That is, a default constant d (of sort α) is generated and we instantiate with this d, giving $S'd =_o Ud$. The rule for equations at type o splits into two subgoals: one branch with $S'd$ and Ud and another with $\neg S'd$ and $\neg Ud$. On the first branch we mate $S'd$ with $\neg S'c$ adding the disequation $d \neq c$ to the branch. This makes c available as an instantiation for $\forall y.S'y = Uy$. After instantiating with c the rest of the subcase is straightforward. In the other subgoal we mate $U\ c$ with $\neg Ud$ giving the disequation $c \neq d$. Again, c becomes available as an instantiation and the rest of the subcase is straightforward.

3 Terms with Perfect Sharing

Lash represents normal terms as C structures, with a unique integer id assigned to each term. The structure contains a tag indicating which kind of term is represented, a number that is used to either indicate the de Bruijn index (for a variable), the name (for a constant), or the type (for a λ-abstraction, a universal quantifier, a choice operator, or an equation). Two pointers (optionally) point to relevant subterms in each case. In addition the structure maintains the information of which de Bruijn indices are free in the term (with de Bruijn indices limited to a maximum of 255). Knowing the free de Bruijn indices of terms makes recognizing potential η-redexes possible without traversing the λ-abstraction. Likewise it is possible to determine when shifting and substitution of de Bruijn indices would not affect a term, avoiding the need to traverse the term.

In Ocaml only the unique integer id is directly revealed and this is sufficient to test for equality of terms. Hash tables are used to uniquely assign types to integers and strings (for names) to integers and these integers are used to interface with the C code. Various functions are used in the Ocaml-C interface to request the construction of (normal) terms. For example, given the two Ocaml integer ids i and j corresponding to terms s and t, the function `mk_norm_ap` given i and j will return an integer k corresponding to the normal term $[s\ t]$. The C implementation recognizes if s is a λ-abstraction and performs all $\beta\eta$-reductions to obtain a normal term. Additionally, the C implementation treats terms as graphs with perfect sharing, and additionally caches previous operations (including substitutions and de Bruijn shifting) to prevent recomputation.

In addition to the low-level C term reimplementation, we have also provided a number of other low-level functionalities replacing the slower parts of the Ocaml code. This includes low-level priority queues, as well as C code used to associate the integers representing normal propositions with integers that are used to communicate with MiniSat. The MiniSat integers are nonzero and satisfy the property that minus on integers corresponds to negation of propositions.

4 Results and Examples

The first mode in the default schedule for Satallax 3.1 is MODE213. This mode activates one feature that goes beyond the basic calculus: pattern clauses. Additionally the mode sets a flag that tries to split the initial goal into several independent subgoals before beginning the search proper. Through experimentation we have found that setting a flag (common to both Satallax and Lash) to essentially prevent MiniSat from searching (i.e., only using MiniSat to recognize contradictions that are evident without search) often improves the performance. We have created a modified mode MODE213D that deactivates these additions (and delays the use of MiniSat) so that Satallax and Lash will have a similar (and often the same) search space. (Sometimes the search spaces differ due to differences in the way Satallax and Lash enumerate instantiations for function types, an issue we

Table 1. Lash vs. Satallax on 2053 TH0 Problems.

Prover	Problems Solved
Lash	1501 (73%)
Satallax (with E)	1487 (72%)
Satallax (without E)	1445 (70%)
Satallax (Lash Schedule)	1412 (69%)

will not focus on here.) We have also run Lash with many variants of Satallax modes with similar modifications. From such test runs we have created a 10 s schedule consisting of 5 modes.

To give a general comparison of Satallax and Lash we have run both on 2053 TH0 problems from a recent release of the TPTP [9] (7.5.0). We initially selected all problems with TPTP status of Theorem or Unsatisfiable (so they should be provable in principle) without polymorphism (or similar extensions of TH0). We additionally removed a few problems that could not be parsed by Satallax 3.4 and removed a few hundred problems big enough to activate SINE in Satallax 3.4.

We ran Lash for 10 s with its default schedule over this problem set. For comparison, we have run Satallax 3.4 for 10 s in three different ways: using the Lash schedule (since the flag settings make sense for both systems) and using Satallax 3.4's default schedule both with and without access to E [12]. The results are reported in Table 1. It is already promising that Lash has the ability to slightly outperform Satallax even when Satallax is allowed to call E.

To get a clearer view of the improvement we discuss a few specific examples.

TPTP problem `NUM638^1` (part of Theorem 3 from the AUTOMATH formalization of Landau's book) is about the natural numbers (starting from 1). The problem assumes a successor function s is injective and that every number other than 1 has a predecessor. An abstract notion of existence is used by having a constant some of type $(\iota o)o$ about which no extra assumptions are made, so the assumption is formally $\forall x.x \neq 1 \Rightarrow \mathsf{some}(\lambda u.x = su)$. For a fixed n, $n \neq 1$ is assumed and the conjecture to prove is the negation of the implication $(\forall xy.n = sx \Rightarrow n = sy \Rightarrow x = y) \Rightarrow \neg(\mathsf{some}(\lambda u.n = su))$. The implication is assumed and the search must rule out the negation of the antecedent (i.e., that n has two predecessors) and the succedent (that n has no predecessor). Satallax and Lash both take 3911 steps to prove this example. With MODE213D, Lash completes the search in 0.4 s while Satallax requires almost 29 s.

TPTP problem `SEV108^5` (`SIX_THEOREM` from TPS [1]) corresponds to proving the Ramsey number R(3,3) is at most 6. The problem assumes there is a symmetric binary relation R (the edge relation of a graph with the sort as vertices) and there are (at least) 6 distinct elements. The conclusion is that there are either 3 distinct elements all of which are R-related or 3 distinct elements none of which are R-related. Satallax and Lash can solve the problem in 14129 steps with mode MODE213D. Satallax proves the theorem in 0.153 s while Lash proves the theorem in the same number of steps but in 0.046 s.

The difference is more impressive if we consider the modified problem of proving R(3, 4) is at most 9. That is, we assume there are (at least) 9 distinct elements and modify the second disjunct of the conclusion to be that there are 4 distinct elements none of which are R-related. Satallax and Lash both use 186127 steps to find the proof. For Satallax this takes 44 s while for Lash this takes 5.5 s.

The TPTP problem `SYO506^1` is about an if-then-else operator. The problem has a constant c of type $o\iota\iota\iota$. Instead of giving axioms indicating c behaves as an if-then-else operator, the conjecture is given as a disjunction:

$$(\forall xy.c\ (x = y)\ x\ y\ =\ y) \vee \neg(\forall xy.c\ \top\ x\ y\ =\ x) \vee \neg(\forall xy.c\ \bot\ x\ y\ =\ y).$$

After negating the conjecture and applying the first few tableau rules the branch will contain the propositions $\forall xy.c\ \top\ x\ y\ =\ x$, $\forall xy.c\ \bot\ x\ y\ =\ y$ and the disequation $c\ (d = e)\ d\ e \neq e$ for fresh d and e of type ι. In principle the rules for if-then-else given in [2] could be used to solve the problem without using the universally quantified formulas (other than to justify that c *is* an if-then-else operator). However, these are not implemented in Satallax or Lash. Instead search proceeds as usual via the basic underlying procedure. Both Satallax and Lash can prove the example using modes MODE0C1 in 32704 steps. Satallax performs the search in 9.8 s while Lash completes the search in 0.2 s.

In addition to the examples considered above, we have constructed a family of examples intended to demonstrate the power of the shared term representation and caching of operations. Let cons have type $\iota\iota\iota$ and nil have type ι. For each natural number n, consider the proposition C^n given by

$$\overline{n}\ (\lambda x.\mathsf{cons}\ x\ x)\ (\mathsf{cons}\ \mathsf{nil}\ \mathsf{nil}) = \mathsf{cons}\ (\overline{n}\ (\lambda x.\mathsf{cons}\ x\ x)\ \mathsf{nil})\ (\overline{n}\ (\lambda x.\mathsf{cons}\ x\ x)\ \mathsf{nil})$$

where $\overline{n}$ is the appropriately typed Church numeral. Proving the proposition C^n does not require any search and merely requires the prover to normalize the conjecture and note the two sides have the same normal form. However, this normal form on both sides will be a complete binary tree of depth $n+1$. We have run Lash and Satallax on C^n with $n \in \{20, 21, 22, 23, 24\}$ using mode MODE213D. Lash solves all five problems in the same amount of time, less than 0.02 s for each. Satallax takes 4 s, 8 s, 16 s, 32 s and 64 s. As expected, since Satallax is not using a shared representation, the computation time exponentially increases with respect to n.

5 Conclusion and Future Work

We have used Lash as a vehicle to demonstrate that giving a more efficient implementation of the underlying tableau calculus of Satallax can lead to significant performance improvements. An obvious possible extension of Lash would be to implement pattern clauses, higher-order unification and the ability to call E.

While we may do this, our current plans are to focus on directions that further diverge from the development path followed by Satallax.

Interesting theoretical work would be to modify the underlying calculus (while maintaining completeness). For example the rules of the calculus might be able to be further restricted based on orderings of ground terms. On the other hand, new rules might be added to support a variety of constants with special properties. This was already done for constants that satisfy axioms indicating the constant is a choice, description or if-then-else operator [2]. Suppose a constant r of type $\iota\iota o$ is known to be reflexive due to a formula $\forall x.r\ x\ x$ being on the branch. One could avoid ever instantiating this universally quantified formula by simply including a tableau rule that extends a branch with $s \neq t$ whenever $\neg r\ s\ t$ is on the branch. Similar rules could operationalize other special cases of universally quantified formulas, e.g., formulas giving symmetry or transitivity of a relation. A modification of the usual completeness proof would be required to prove completeness of the calculus with these additional rules (and with the restriction disallowing instantiating the corresponding universally quantified formulas).

Finally the C representation of terms could be extended to include precomputed special features. Just as the current implementation knows which de Bruijns are free in the term (without traversing the term), a future implementation could know other features of the term without requiring traversal. Such features could be used to guide the search.

Acknowledgements. The results were supported by the Ministry of Education, Youth and Sports within the dedicated program ERC CZ under the project POSTMAN no. LL1902 and the ERC starting grant no. 714034 SMART.

References

1. Andrews, P.B., Bishop, M., Issar, S., Nesmith, D., Pfenning, F., Xi, H.: TPS: a theorem-proving system for classical type theory. J. Autom. Reason. **16**(3), 321–353 (1996). https://doi.org/10.1007/BF00252180
2. Backes, J., Brown, C.E.: Analytic tableaux for higher-order logic with choice. J. Autom. Reason. **47**(4), 451–479 (2011). https://doi.org/10.1007/s10817-011-9233-2
3. Bhayat, A., Reger, G.: A combinator-based superposition calculus for higher-order logic. In: Peltier, N., Sofronie-Stokkermans, V. (eds.) IJCAR 2020. LNCS (LNAI), vol. 12166, pp. 278–296. Springer, Cham (2020). https://doi.org/10.1007/978-3-030-51074-9_16
4. Brown, C.E.: Satallax: an automatic higher-order prover. In: Gramlich, B., Miller, D., Sattler, U. (eds.) IJCAR 2012. LNCS (LNAI), vol. 7364, pp. 111–117. Springer, Heidelberg (2012). https://doi.org/10.1007/978-3-642-31365-3_11
5. Brown, C.E., Smolka, G.: Extended first-order logic. In: Berghofer, S., Nipkow, T., Urban, C., Wenzel, M. (eds.) TPHOLs 2009. LNCS, vol. 5674, pp. 164–179. Springer, Heidelberg (2009). https://doi.org/10.1007/978-3-642-03359-9_13
6. Brown, C.E., Smolka, G.: Analytic tableaux for simple type theory and its first-order fragment. Logical Methods Comput. Sci. **6**(2) (2010). https://doi.org/10.2168/LMCS-6(2:3)2010

7. Färber, M., Brown, C.: Internal guidance for Satallax. In: Olivetti, N., Tiwari, A. (eds.) IJCAR 2016. LNCS (LNAI), vol. 9706, pp. 349–361. Springer, Cham (2016). https://doi.org/10.1007/978-3-319-40229-1_24
8. Steen, A., Benzmüller, C.: Extensional higher-order paramodulation in Leo-III. J. Autom. Reason. **65**(6), 775–807 (2021). https://doi.org/10.1007/s10817-021-09588-x
9. Sutcliffe, G.: The TPTP problem library and associated infrastructure. From CNF to TH0, TPTP v6.4.0. J. Autom. Reason. **59**(4), 483–502 (2017)
10. Sutcliffe, G.: The 10th IJCAR automated theorem proving system competition - CASC-J10. AI Commun. **34**(2), 163–177 (2021). https://doi.org/10.3233/AIC-201566
11. Vukmirović, P., Bentkamp, A., Blanchette, J., Cruanes, S., Nummelin, V., Tourret, S.: Making higher-order superposition work. In: Platzer, A., Sutcliffe, G. (eds.) CADE 2021. LNCS (LNAI), vol. 12699, pp. 415–432. Springer, Cham (2021). https://doi.org/10.1007/978-3-030-79876-5_24
12. Vukmirović, P., Blanchette, J.C., Cruanes, S., Schulz, S.: Extending a Brainiac prover to lambda-free higher-order logic. In: Vojnar, T., Zhang, L. (eds.) TACAS 2019. LNCS, vol. 11427, pp. 192–210. Springer, Cham (2019). https://doi.org/10.1007/978-3-030-17462-0_11

Goéland: A Concurrent Tableau-Based Theorem Prover (System Description)

Julie Cailler, Johann Rosain, David Delahaye, Simon Robillard(✉), and Hinde Lilia Bouziane

LIRMM, Univ Montpellier, CNRS, Montpellier, France
{julie.cailler,johann.rosain,david.delahaye,simon.robillard,
hinde.bouziane}@lirmm.fr

Abstract. We describe Goéland, an automated theorem prover for first-order logic that relies on a concurrent search procedure to find tableau proofs, with concurrent processes corresponding to individual branches of the tableau. Since branch closure may require instantiating free variables shared across branches, processes communicate via channels to exchange information about substitutions used for closure. We present the proof search procedure and its implementation, as well as experimental results obtained on problems from the TPTP library.

Keywords: Automated Theorem Proving · Tableaux · Concurrency

1 Introduction

Although clausal proof techniques have enjoyed success in automated theorem proving, some applications benefit from reasoning on unaltered formulas (rather than Skolemized clauses), while others require the production of proofs in a sequent calculus. These roles are fulfilled by provers based on the tableau method [17], as initially designed by Beth and Hintikka [2,13]. For first-order logic, efficient handling of universal formulas is typically achieved with free variables that are instantiated only when needed to close a branch. This step is said to be *destructive* because it may affect open branches sharing variables. This causes fairness (and consequently, completeness) issues, as illustrated in Fig. 1. In this example, exploring the left branch produces a substitution that prevents direct closure of the right branch. Reintroducing the original quantified formula with a different free variable is not sufficient to close the right branch, because an applicable δ-rule creates a new Skolem symbol that will result in a different but equally problematic substitution every time a left branch is explored. Thus, systematically exploring the left branch before the right leads to non-termination of the search. Conversely, exploring the right branch first produces a substitution (which instantiates the free variable X with a rather than b) that closes both branches.

Concurrent computing offers a way to implement a proof search procedure that explores branches simultaneously. Such a procedure can compare closing

J. Blanchette et al. (Eds.): IJCAR 2022, LNAI 13385, pp. 359–368, 2022.
https://doi.org/10.1007/978-3-031-10769-6_22

$$\frac{\dfrac{P(a) \wedge \neg P(b) \wedge \forall x. \ (P(x) \Leftrightarrow \forall y. \ P(y))}{P(a), \neg P(b), \forall x. \ (P(x) \Leftrightarrow \forall y. \ P(y))} \ \alpha_\wedge}{P(X/b) \Leftrightarrow \forall y. \ P(y)} \ \gamma_\forall$$

$$\frac{P(X/b), \forall y. \ P(y)}{\sigma = \{X \mapsto b\}} \ \odot_\sigma \qquad \frac{\dfrac{\neg P(X/b), \neg \forall y. \ P(y)}{\neg P(sk_1)} \ \delta_{\neg\forall}}{P(X'/b) \Leftrightarrow \forall y. \ P(y)} \ \gamma_\forall \ \beta_\Leftrightarrow$$

$$\frac{p(X'/b), \forall y. \ P(y)}{\begin{array}{c} \sigma = \{X' \mapsto b\} \\ \sigma' = \{X' \mapsto sk_1\} \end{array}} \ \odot_\sigma \qquad \frac{\neg P(X'/b), \neg \forall y. \ P(y)}{\dfrac{\neg P(sk_2)}{\cdots} \ \gamma_\forall} \ \delta_{\neg\forall} \ \beta_\Leftrightarrow$$

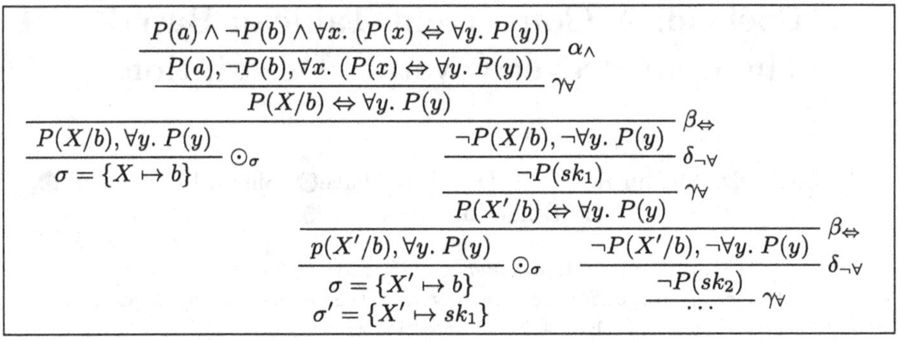

Fig. 1. Incompleteness caused by unfair selection of branches

substitutions to detect (dis)agreements between branches, and consequently either close branches early, or restart proof attempts with limited backtracking. The simultaneous exploration of branches is handled by the concurrency system, either by interleaving computations through scheduling, or by executing tasks in parallel if the hardware resources allow it. A concurrent procedure naturally lends itself to parallel execution, allowing us to take advantage of multi-core architectures for efficient first-order theorem proving. Thus, concurrency provides an elegant and efficient solution to proof search with free variable tableaux.

In this paper, we describe a concurrent destructive proof search procedure for first-order analytic tableaux (Sect. 2) and its implementation in a tool called Goéland, as well as its evaluation on problems from the TPTP library [19] and comparison to other state-of-the-art provers (Sect. 3).

Related Work. A lot of research has been carried out on the parallelization of proof search procedures [4], often focusing primarily on parallel execution and performance. In contrast, we use concurrency not only as a way to take advantage of multi-core architectures, but also as an algorithmic device that is useful even for sequential execution (with interleaved threads). Some concurrent and parallel approaches focus more distinctly on the exploration of the search space, either by dividing the search space between processes (*distributed search*) or by using processes with different search plans on the same space (*multi search*) [3]. These approaches can be performed either by *heterogeneous systems* that rely on cooperation between systems with different inference systems [1,8,12], or *homogeneous systems* where all deductive processes use the same inference system. According to this classification, the technique presented here is a homogeneous system that performs a distributed search. Concurrent tableaux provers include the model-elimination provers CPTheo [12] and Partheo [18], and the higher-order prover Hot [15], which notably uses concurrency to deal with fairness issues arising from the non-terminating nature of higher-order unification. Lastly, concurrency has been used as the basis of a generic framework to present various proof strategies [10] or allow distributed calculations over a network [21].

2 Concurrent Proof Search

Free Variable Tableaux. Goéland attempts to build a refutation proof for a first-order formula, i.e., a closed tableau for its negation, using a standard free-variable tableau calculus [11]. The calculus is composed of α-, γ- and δ-rules that extend a branch with one formula, β-rules that divide a branch by extending it with two formulas, and a $\odot$-rule that closes a branch. γ-rules deal with universally-quantified formulas by introducing a formula with a free variable. A free variable is not universally quantified, but is instead a placeholder for some term instantiation, typically determined upon branch closure. δ-rules deal with existentially-quantified formulas by introducing a formula with a Skolem function symbol that takes as arguments the free variables in the branch. This ensures freshness of the Skolem symbol independently of variable instantiation.

The branch closure rule applies to a branch carrying atomic formulas P and Q such that, for some substitution σ, $\sigma(P) = \sigma(\neg Q)$. In that case, σ is applied to all branches. That rule is consequently *destructive*: applying a substitution to close one branch may modify another, removing the possibility to close it immediately. A tableau is closed when all its branches are closed. Closing a tableau can thus be seen as providing a global unifier that closes all branches.

Semantics for Concurrency. Goéland relies on a concurrent search procedure. In order to present this procedure, we use a simple WHILE language augmented with instructions for concurrency, in the style of CSP [14]. Each process has its own variable store, as well as a collection of process identifiers used for communication: π_{parent} denotes the identifier of a process's parent, while Π_{children} denotes the collection of identifiers of active children of that process. Given a process identifier π and an expression e, the command $\pi\,!\,e$ is used to send an asynchronous message with the value e to the process identified by π. Conversely, the command $\pi\,?\,x$ blocks the execution until the process identified by π sends a message, which is stored in the variable x. Lastly, the instruction **start** creates a new process that executes a function with some given arguments, while the instruction **kill** interrupts the execution of a process according to its identifier.

Proof Search Procedure. The proof search is carried out concurrently by processes corresponding to branches of the tableau. Processes are started upon application of a β-rule, one for each new branch. Communications between processes take two forms: a process may send a set of closing substitutions for its branch to its parent, or a parent may send a substitution (that closes one of its children's branch) to the other children. The proof search is performed by the $proofSearch$, $waitForParent$, and $waitForChildren$ procedures (described in Procedures 1, 2, and 3, respectively).

The $proofSearch$ procedure initiates the proof search for a branch. It first attempts to apply the closure rule. A closing substitution is called *local* to a process if its domain includes only free variables introduced by this process or one of its descendants (i.e., if the variables do not occur higher in the proof tree). If one of the closing substitutions is local to the process, it is reported and the

---

Procedure 1: *proofSearch*

Data: a tableau T
1 **begin**
2   **var** $\Theta \leftarrow applyClosingRule(T)$;
3   **for** $\theta \in \Theta$ **do**
4   **if** $isLocal(\theta)$ **then**
5   π_{parent} ! $\{\theta\}$
6   **return**
7   **if** $\Theta \neq \emptyset$ **then**
8   π_{parent} ! Θ
9          $waitForParent(T, \Theta)$
10   **else if** $applicableAlphaRule(T)$ **then**
11          $proofSearch(applyAlphaRule(T))$
12   **else if** $applicableDeltaRule(T)$ **then**
13          $proofSearch(applyDeltaRule(T))$
14   **else if** $applicableBetaRule(T)$ **then**
15   **for** $T' \in applyBetaRule(T)$ **do**
16   **start** $proofSearch(T')$
17          $waitForChildren(T, \emptyset, \emptyset)$
18   **else if** $applicableGammaRule(T)$ **then**
19          $proofSearch(applyGammaRule(T))$
20   **else**
21   π_{parent} ! $\emptyset$

---

process terminates. If only non-local closing substitutions are found, they are reported and the process executes *waitForParent*. Otherwise, the procedure applies tableau expansion rules according to the priority: $\alpha \prec \delta \prec \beta \prec \gamma$. If a β-rule is applied, new processes are started, and each of them executes *proofSearch* on the newly created branch, while the current process executes *waitForChildren*.

The *waitForParent* procedure is executed by a process after it has found closing non-local substitutions. Such substitutions may prevent closure in other branches. In these cases, the parent will eventually send another candidate substitution. *waitForParent* waits until such a substitution is received, and triggers a new step of proof search. The process may also be terminated by its parent (via the **kill** instruction) during the execution of this procedure, if one of the substitutions previously sent by the process leads to closing the parent's branch.

The *waitForChildren* procedure is executed by a process after the application of a β-rule and the creation of child processes. The set of substitutions sent by each child is stored in a map *subst* (Line 2), initially undefined everywhere $(f_\perp)$. This procedure closes the branch (Line 13) if there exists a substitution θ that agrees with one closing substitution of each child process, i.e., for each child process, the process has reported a substitution σ such that $\sigma(X) = \theta(X)$ for any variable X in the domain of σ. If no such substitution can be found

Procedure 2: *waitForParent*

Data: a tableau T, a set Θ_{sent} of substitutions sent by this process to its parent

begin
- $\pi_{\text{parent}}\ ?\ \sigma$
- **if** $\sigma \in \Theta_{sent}$ **then**
  - $\pi_{\text{parent}}\ !\ \sigma$
  - $waitForParent(T, \Theta_{\text{sent}})$
- **else**
  - $proofSearch(\sigma(T))$

after all the children have closed their branches, then one closing substitution $\sigma \in subst$ is picked arbitrarily (Line 18) and sent to all the children (which are at that point executing *waitForParent*) to restart their proof attempts. With the additional constraint of the substitution σ, the new proof attempts may fail, hence the necessity for backtracking among candidate substitutions $\Theta_{\text{backtrack}}$ (Line 5 and 6). At the end, if all the substitutions were tried and failed, the process sends a failure message (symbolized by $\emptyset$) to its parent.

Thus, concurrency and backtracking are used to prevent incompleteness resulting from unfair instantiation of free variables. Another potential source of unfairness is the γ-rule, when applied more than once to a universal formula (reintroduction). This may be needed to find a refutation, but unbounded reintroductions would lead to unfairness. Iterative deepening [16] is used to guard against this: a bound limits the number of reintroductions on any single branch, and if no proof is found, the bound is increased and the proof search restarted.

Figure 2 illustrates the interactions between processes for the problem in Fig. 1, and shows how concurrency helps ensure fairness. It describes the parent process, in the top box, and below, the two children processes created upon application of the β-rule. Dotted lines separate successive states of a process (i.e., Procedures 1, 2 and 3 seen above), while arrows and boxes represent substitution exchanges. The number above each arrow indicates the chronology of the interactions. After both children have returned a substitution (1), the parent arbitrarily chooses one of them, starting with $X \mapsto b$, and sends it to the children (2). Since this substitution prevents closure in the right branch (3), the parent later backtracks and sends the other substitution $X \mapsto a$ (4), allowing both children (5) and then the parent to close successfully.

3 Implementation and Experimental Results

Implementation. The procedures presented in Sect. 2 are implemented in the Goéland prover[1] using the Go language. Go supports concurrency and parallelism, based on lightweight execution threads called *goroutines* [20]. Goroutines

[1] Available at: https://github.com/GoelandProver/Goeland/releases/tag/v1.0.0-beta.

Procedure 3: *waitForChildren*

```
Data: a tableau T, a set Θ_sent of substitutions sent by this process to its
      parent, a set Θ_backtrack of substitutions used for backtracking
begin
    var subst ← f_⊥
    while ∃π ∈ Π_children. subst[π] = ⊥ do
        π ? subst[π]
        if subst[π] = ∅ then
            if ∃θ ∈ Θ_backtrack then
                for π ∈ Π_children do π ! θ;
                waitForChildren(T, Θ_sent, Θ_backtrack \ {θ})
            else
                for π ∈ Π_children do kill π;
                π_parent ! ∅
                return
    if ∃θ, agreement(θ, subst) then
        π_parent ! {θ}
        for π ∈ Π_children do kill π;
        waitForParent(T, Θ_sent ∪ {θ})
    else
        σ ← choice(subst)
        for π ∈ Π_children do π ! σ;
        waitForChildren(T, Θ_sent, Θ_backtrack ∪ ⋃_π subst[π] \ {σ})
```

are executed according to a so-called *hybrid threading* (or $M : N$) model: M goroutines are executed over N effective threads and scheduling is managed by both the Go runtime and the operating system. This threading model allows the execution of a large number of goroutines with a reasonable consumption of system resources. Goroutines use channels to exchange messages, so that the implementation is close to the presentation of Sect. 2.

Goéland has, for the time being, no dedicated mechanism for equality reasoning. However, we have implemented an extension that implements deduction modulo theory [9], i.e., transforms axioms into rewrite rules over propositions and terms. Deduction modulo theory has proved very useful to improve proof search when integrated into usual automated proof techniques [5], and also produces excellent results with manually-defined rewrite rules [6, 7]. In Goéland, deduction modulo theory selects some axioms on the basis of a simple syntactic criterion and replaces them by rewrite rules.

Experimental Results. We evaluated Goéland on two problems categories with FOF theorems in the TPTP library (v7.4.0): syntactic problems without equality (SYN) and problems of set theory (SET). The former was chosen for its elementary nature, whereas the latter was picked primarily to evaluate the performance of the deduction modulo theory, as the axioms of set theory are good

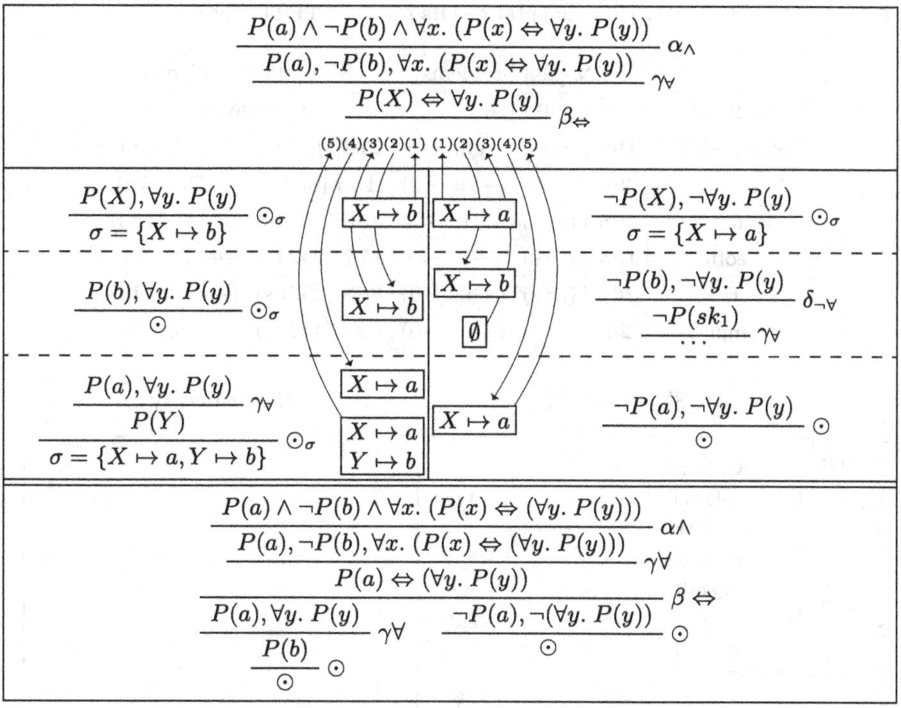

Fig. 2. Proof search and resulting proof for $P(a) \wedge \neg P(b) \wedge \forall x.(P(x) \Leftrightarrow \forall y.P(y))$

targets for rewriting. We compared the results with those of five other provers: tableau-based provers Zenon (v0.8.5), Princess (v2021-05-10) and LeoIII (v1.6), as well as saturation-based provers E (v2.6) and Vampire (v4.6.1). Experiments were executed on a computer equipped with an Intel Xeon E5-2680 v4 2.4GHz 2×14-core processor and 128 GB of memory. Each proof attempt was limited to 300 s. Table 1 and Fig. 3 report the results. Table 1 shows the number of problems solved by each prover, the cumulative time, and the number of problems solved by a given prover but not by Goéland (+) and conversely (−). Figure 3 presents the cumulative time required to solve the number of problems.

As can be observed, the results of Goéland are comparable to, or slightly better than those of other tableau-based provers on problems from SYN, while saturation theorem provers achieve the best results. On this category, the axioms do not trigger deduction modulo theory rewriting rules, hence the similar results of Goéland and Goéland+DMT. On SET, Goéland+DMT obtains significantly better results than other tableau-based provers. This confirms the previous results on the performance of deduction modulo theory for set theory [6,7].

Table 1. Experimental results over the TPTP library

	SYN (263 problems)	SET (464 problems)
Goéland	**199** (190 s)	**150** (4659 s)
Goéland+DMT	**199** (196 s) (+0, −0)	**278** (1292 s) (+142, −14)
Zenon	**256** (67 s) (+60, −3)	**150** (562 s) (+75, −75)
Princess	**195** (189 s) (+1, −5)	**258** (1168 s) (+141, −33)
LeoIII	**195** (268 s) (+1, −5)	**177** (2925 s) (+77, −50)
E	**261** (168 s) (+62, −0)	**363** (2377 s) (+223, −10)
Vampire	**262** (13 s) (+63, −0)	**321** (4122 s) (+188, −17)

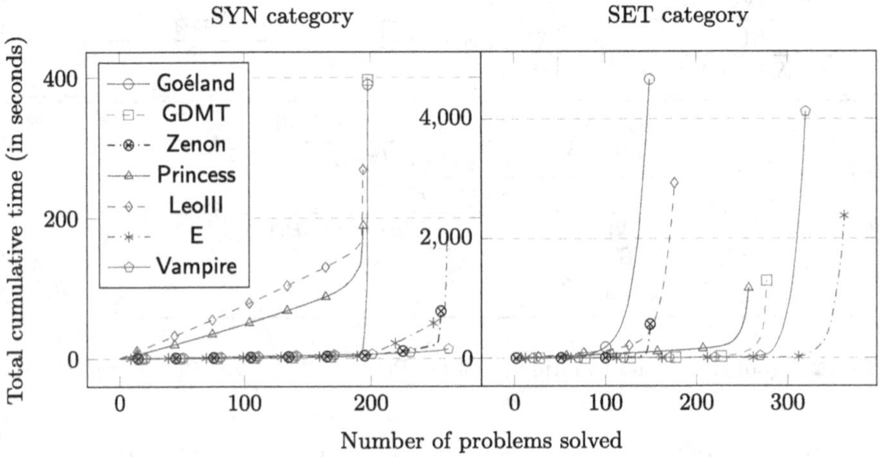

Fig. 3. Cumulative time per problem solved between Goéland, Goéland+DMT(GDMT), Zenon, Princess, LeoIII, E, and Vampire

4 Conclusion

We have presented a concurrent proof search procedure for tableaux in first-order logic with the aim of ensuring a fair exploration of the search space. This procedure has been implemented in the prover Goéland. This tool is still in an early stage, and (with the exception of deduction modulo theory) implements only the most basic functionalities, yet empirical results are encouraging. We plan on adding functionalities such as equality reasoning, arithmetic reasoning, and support for polymorphism to Goéland, which should increase its usability and performance. The integration of these functionalities in the context of a concurrent prover seems to be a promising line of research. Further investigation is also needed to prove the fairness, and therefore completeness, of our procedure.

References

1. Benzmüller, C., Kerber, M., Jamnik, M., Sorge, V.: Experiments with an agent-oriented reasoning system. In: Baader, F., Brewka, G., Eiter, T. (eds.) KI 2001. LNCS (LNAI), vol. 2174, pp. 409–424. Springer, Heidelberg (2001). https://doi.org/10.1007/3-540-45422-5_29
2. Beth, E.W.: Formal Methods: An Introduction to Symbolic Logic and to the Study of Effective Operations in Arithmetic and Logic, Synthese Library, vol. 4. D. Reidel Pub. Co. (1962)
3. Bonacina, M.P.: A taxonomy of parallel strategies for deduction. Ann. Math. Artif. Intell. **29**(1), 223–257 (2000)
4. Bonacina, M.P.: Parallel theorem proving. In: Hamadi, Y., Sais, L. (eds.) Handbook of Parallel Constraint Reasoning, pp. 179–235. Springer, Cham (2018). https://doi.org/10.1007/978-3-319-63516-3_6
5. Burel, G., Bury, G., Cauderlier, R., Delahaye, D., Halmagrand, P., Hermant, O.: First-order automated reasoning with theories: when deduction modulo theory meets practice. J. Autom. Reason. **64**(6), 1001–1050 (2019). https://doi.org/10.1007/s10817-019-09533-z
6. Bury, G., Cruanes, S., Delahaye, D., Euvrard, P.-L.: An automation-friendly set theory for the B method. In: Butler, M., Raschke, A., Hoang, T.S., Reichl, K. (eds.) ABZ 2018. LNCS, vol. 10817, pp. 409–414. Springer, Cham (2018). https://doi.org/10.1007/978-3-319-91271-4_32
7. Bury, G., Delahaye, D., Doligez, D., Halmagrand, P., Hermant, O.: Automated deduction in the B set theory using typed proof search and deduction modulo. In: Fehnker, A., McIver, A., Sutcliffe, G., Voronkov, A. (eds.) Logic for Programming, Artificial Intelligence and Reasoning (LPAR). EPiC Series in Computing, vol. 35, pp. 42–58. EasyChair (2015)
8. Denzinger, J., Kronenburg, M., Schulz, S.: DISCOUNT - a distributed and learning equational prover. J. Autom. Reason. **18**(2), 189–198 (1997)
9. Dowek, G., Hardin, T., Kirchner, C.: Theorem proving modulo. J. Autom. Reason. (JAR) **31**(1), 33–72 (2003)
10. Fisher, M.: An open approach to concurrent theorem proving. Parallel Process. Artif. Intell. **3**, 80011 (1997)
11. Fitting, M.: First-Order Logic and Automated Theorem Proving. Springer, Heidelberg (1990)
12. Fuchs, M., Wolf, A.: System description: cooperation in model elimination: CPTHEO. In: Kirchner, C., Kirchner, H. (eds.) CADE 1998. LNCS, vol. 1421, pp. 42–46. Springer, Heidelberg (1998). https://doi.org/10.1007/BFb0054245
13. Hintikka, J.: Two papers on symbolic logic: form and content in quantification theory and reductions in the theory of types. Societas Philosophica, Acta philosophica Fennica **8**, 7–55 (1955)
14. Hoare, C.A.R.: Communicating sequential processes. Commun. ACM **21**(8), 666–677 (1978)
15. Konrad, K.: Hot: a concurrent automated theorem prover based on higher-order tableaux. In: Grundy, J., Newey, M. (eds.) TPHOLs 1998. LNCS, vol. 1479, pp. 245–261. Springer, Heidelberg (1998). https://doi.org/10.1007/BFb0055140
16. Korf, R.E.: Depth-first iterative-deepening: an optimal admissible tree search. Artif. Intell. **27**(1), 97–109 (1985)

17. Letz, R.: First-order tableau methods. In: D'Agostino, M., Gabbay, D.M., Hähnle, R., Posegga, J. (eds.) Handbook of Tableau Methods, pp. 125–196. Springer, Heidelberg (1999). https://doi.org/10.1007/978-94-017-1754-0_3. ISBN 978-94-017-1754-0
18. Schumann, J., Letz, R.: Partheo: a high-performance parallel theorem prover. In: Stickel, M.E. (ed.) CADE 1990. LNCS, vol. 449, pp. 40–56. Springer, Heidelberg (1990). https://doi.org/10.1007/3-540-52885-7_78
19. Sutcliffe, G.: The TPTP problem library and associated infrastructure. From CNF to TH0, TPTP v6.4.0. J. Autom. Reason. (JAR) **59**(4), 483–502 (2017)
20. Tsoukalos, M.: Mastering Go: Create Golang Production Applications Using Network Libraries, Concurrency, Machine Learning, and Advanced Data Structures, pp. 439–463. Packt Publishing Ltd. (2019)
21. Wu, C.H.: A multi-agent framework for distributed theorem proving. Expert Syst. Appl. **29**(3), 554–565 (2005)

Binary Codes that Do Not Preserve Primitivity

Štěpán Holub[1(✉)], Martin Raška[1], and Štěpán Starosta[2(✉)]

[1] Faculty of Mathematics and Physics, Charles University, Prague, Czech Republic
holub@karlin.mff.cuni.cz

[2] Faculty of Information Technology, Czech Technical University in Prague, Prague, Czech Republic
stepan.starosta@fit.cvut.cz

Abstract. A code X is not primitivity preserving if there is a primitive list $\mathbf{w} \in \texttt{lists}\, X$ whose concatenation is imprimitive. We formalize a full characterization of such codes in the binary case in the proof assistant Isabelle/HOL. Part of the formalization, interesting on its own, is a description of $\{x, y\}$-interpretations of the square xx if $|y| \leq |x|$. We also provide a formalized parametric solution of the related equation $x^j y^k = z^\ell$.

1 Introduction

Consider two words abba and b. It is possible to concatenate (several copies of) them as b·abba·b, and obtain a power of a third word, namely a square bab·bab of bab. In this paper, we completely describe all ways how this can happen for two words, and formalize it in Isabelle/HOL.

The corresponding theory has a long history. The question can be formulated as solving equations in three variables of the special form $W(x, y) = z^\ell$ where the left hand side is a sequence of x's and y's, and $\ell \geq 2$. The seminal result in this direction is the paper by R. C. Lyndon and M.-P. Schützenberger [10] from 1962, which solves in a more general setting of free groups the equation $x^j y^k = z^\ell$ with $2 \leq j, k, \ell$. It was followed, in 1967, by a partial answer to our question by A. Lentin and M.-P. Schützenberger [9]. A complete characterization of monoids generated by three words was provided by L. G. Budkina and Al. A. Markov in 1973 [4]. The characterization was later, in 1976, reproved in a different way by Lentin's student J.-P. Spehner in his Ph.D. thesis [14], which even explicitly mentions the answer to the present question. See also a comparison of the two classifications by T. Harju and D. Nowotka [7]. In 1985, the result was again reproved by E. Barbin-Le Rest and M. Le Rest [1], this time specifically focusing on our question. Their paper contains a characterization of binary interpretations of a square as a crucial tool. The latter combinatorial result is interesting on its own, but is very little known. In addition to the fact that, as far as we know, the proof is not available in English, it has to be reconstructed from Théorème 2.1 and Lemme 3.1 in [1], it is long, technical and little structured, with many

J. Blanchette et al. (Eds.): IJCAR 2022, LNAI 13385, pp. 369–387, 2022.
https://doi.org/10.1007/978-3-031-10769-6_23

intuitive steps that have to be clarified. It is symptomatic, for example, that Maňuch [11] cites the claim as essentially equivalent to his desired result but nevertheless provides a different, shorter but similarly technical proof.

The fact that several authors opted to provide their own proof of the already known result, and that even a weaker result was republished as new shows that the existing proof was not considered sufficiently convincing and approachable. This makes the topic a perfect candidate for formalization. The proof we present here naturally contains some ideas of the proof from [1] but is significantly different. Our main objective was to follow the basic methodological requirement of a good formalization, namely to identify claims that are needed in the proof and formulate them as separate lemmas and as generally as possible so that they can be reused not only in the proof but also later. Moreover, the formalization naturally forced us to consider carefully the overall strategy of the proof (which is rather lost behind technical details of published works on this topic). Under Isabelle's pressure we eventually arrived at a hopefully clear proof structure which includes a simple, but probably innovative use of the idea of "gluing" words. The analysis of the proof is therefore another, and we believe the most important contribution of our formalization, in addition to the mere certainty that there are no gaps in the proof.

In addition, we provide a complete parametric solution of the equation $x^k y^j = z^\ell$ for arbitrary j, k and ℓ, a classification which is not very difficult, but maybe too complicated to be useful in a mere unverified paper form.

The formalization presented here is an organic part of a larger project of formalization of combinatorics of words (see an introductory description in [8]). We are not aware of a similar formalization project in any proof assistant. The existence of the underlying library, which in turn extends the theories of "List" and "HOL-Library.Sublist" from the standard Isabelle distribution, critically contributes to a smooth formalization which is getting fairly close to the way a human paper proof would look like, outsourcing technicalities to the (reusable) background. We accompany claims in this text with names of their formalized counterparts.

2 Basic Facts and Notation

Let Σ be an arbitrary set. Lists (i.e. finite sequences) $[x_1, x_2, \ldots, x_n]$ of elements $x_i \in \Sigma$ are called *words* over Σ. The set of all words over Σ is usually denoted as Σ^*, using the Kleene star. A notorious ambivalence of this notation is related to the situation when we consider a set of words $X \subset \Sigma^*$, and are interested in lists over X. They should be denoted as elements of X^*. However, X^* usually means something else (in the theory of rational languages), namely the set of all words in Σ^* generated by the set X. To avoid the confusion, we will therefore follow the notation used in the formalization in Isabelle, and write `lists` X instead, to make clear that the entries of an element of `lists` X are themselves words. In order to further help to distinguish words over the basic alphabet from lists over a set of words, we shall use boldface variables for the latter.

In particular, it is important to keep in mind the difference between a letter a and the word $[a]$ of length one, the distinction which is usually glossed over lightly in the literature on combinatorics on words. The set of words over Σ generated by X is then denoted as $\langle X \rangle$. The (associative) binary operation of concatenation of two words u and v is denoted by $u \cdot v$. We prefer this algebraic notation to the Isabelle's original `@`. Moreover, we shall often omit the dot as usual. If $\mathbf{u} = [x_1, x_2, \ldots, x_n] \in \texttt{lists}\, X$ is a list of words, then we write $\texttt{concat}\, \mathbf{u}$ for $x_1 \cdot x_2 \cdots x_n$. We write ε for the empty list, and u^k for the concatenation of k copies of u (we use $u^{@}k$ in the formalization). We write $u \leq_p v$, $u <_p v$, $u \leq_s v$, $u <_s v$, and $u \leq_f v$ to denote that u is a *prefix*, a *strict prefix*, *suffix*, *strict suffix* and *factor* (that is, a contiguous sublist) respectively. A word is *primitive* if it is nonempty and not a power of a shorter word. Otherwise, we call it *imprimitive*. Each nonempty word w is a power of a unique primitive word $\rho\, w$, its *primitive root*. A nonempty word r is a *periodic root* of a word w if $w \leq_p r \cdot w$. This is equivalent to w being a prefix of the right infinite power of r, denoted r^ω. Note that we deal with finite words only, and we use the notation r^ω only as a convenient shortcut for "a sufficiently long power of r". Two words u and v are *conjugate*, we write $u \sim v$, if $u = rq$ and $v = qr$ for some words r and q. Note that conjugation is an equivalence whose classes are also called *cyclic words*. A word u is a *cyclic factor* of w if it is a factor of some conjugate of w. A set of words X is a *code* if its elements do not satisfy any nontrivial relation, that is, they are a basis of a free semigroup. For a two-element set $\{x, y\}$, this is equivalent to x and y being non-commuting, i.e., $xy \neq yx$, and/or to $\rho\, x \neq \rho\, y$. An important characterization of a semigroup S of words to be free is the *stability condition* which is the implication $u, v, uz, zv \in S \implies z \in S$. The longest common prefix of u and v is denoted by $u \wedge_p v$. If $\{x, y\}$ is a (binary) code, then $(x \cdot w) \wedge_p (y \cdot w') = xy \wedge_p yx$ for any $w, w' \in \langle\{x, y\}\rangle$ sufficiently long. We explain some elementary facts from combinatorics on words used in this article in more detail in Sect. 8.

3 Main Theorem

Let us introduce the central definition of the paper.

Definition 1. *We say that a set X of words is* primitivity preserving *if there is no word* $\mathbf{w} \in \texttt{lists}\, X$ *such that*

- $|\mathbf{w}| \geq 2$*;*
- $\mathbf{w}$ *is primitive; and*
- $\texttt{concat}\, \mathbf{w}$ *is imprimitive.*

Note that our definition does not take into account singletons $\mathbf{w} = [x]$. In particular, X can be primitivity preserving even if some $x \in X$ is imprimitive. Nevertheless, in the binary case, we will also provide some information about the cases when one or both elements of the code have to be primitive.

In [12], V. Mitrana formulates the primitivity of a set in terms of morphisms, and shows that X is primitivity preserving if and only if it is the minimal set of generators of a "pure monoid", cf. [3, p. 276]. This brings about a wider concept of morphisms preserving a given property, most classically square-freeness, see for example a characterization of square-free morphisms over three letters by M. Crochemore [5].

The target claim of our formalization is the following characterization of words witnessing that a binary code is not primitivity preserving:

Theorem 1 (`bin_imprim_code`). *Let* $B = \{x, y\}$ *be a code that is not primitivity preserving. Then there are integers* $j \geq 1$ *and* $k \geq 1$*, with* $k = 1$ *or* $j = 1$*, such that the following conditions are equivalent for any* $\mathbf{w} \in \texttt{lists}\, B$ *with* $|\mathbf{w}| \geq 2$*:*

- $\mathbf{w}$ *is primitive, and* $\texttt{concat}\,\mathbf{w}$ *is imprimitive*
- $\mathbf{w}$ *is conjugate with* $[x]^j[y]^k$.

Moreover, assuming $|y| \leq |x|$,

- *if* $j \geq 2$*, then* $j = 2$ *and* $k = 1$*, and both* x *and* y *are primitive;*
- *if* $k \geq 2$*, then* $j = 1$ *and* x *is primitive.*

Proof. Let $\mathbf{w}$ be a word witnessing that B is not primitivity preserving. That is, $|\mathbf{w}| \geq 2$, $\mathbf{w}$ is primitive, and $\texttt{concat}\,\mathbf{w}$ is imprimitive. Since $[x]^j[y]^k$ and $[y]^k[x]^j$ are conjugate, we can suppose, without loss of generality, that $|y| \leq |x|$.

First, we want to show that $\mathbf{w}$ is conjugate with $[x]^j[y]^k$ for some $j, k \geq 1$ such that $k = 1$ or $j = 1$. Since $\mathbf{w}$ is primitive and of length at least two, it contains both x and y. If it contains one of these letters exactly once, then $\mathbf{w}$ is clearly conjugate with $[x]^j[y]^k$ for $j = 1$ or $k = 1$. Therefore, the difficult part is to show that no primitive $\mathbf{w}$ with $\texttt{concat}\,\mathbf{w}$ imprimitive can contain both letters at least twice. This is the main task of the rest of the paper, which is finally accomplished by Theorem 4 claiming that words that contain at least two occurrences of x are conjugate with $[x, x, y]$. To complete the proof of the first part of the theorem, it remains to show that j and k do not depend on $\mathbf{w}$. This follows from Lemma 1.

Note that the imprimitivity of $\texttt{concat}\,\mathbf{w}$ induces the equality $x^j y^k = z^\ell$ for some z and $\ell \geq 2$. The already mentioned seminal result of Lyndon and Schützenberger shows that j and k cannot be simultaneously at least two, since otherwise x and y commute. For the same reason, considering its primitive root, the word y is primitive if $j \geq 2$. Similarly, x is primitive if $k \geq 2$. The primitivity of x when $j = 2$ is a part of Theorem 4. □

We start by giving a complete parametric solution of the equation $x^j y^k = z^\ell$ in the following theorem. This will eventually yield, after the proof of Theorem 1 is completed, a full description of not primitivity preserving binary codes. Since the equation is mirror symmetric, we omit symmetric cases by assuming $|y| \leq |x|$.

Theorem 2 (`LS_parametric_solution`). *Let* $\ell \geq 2$*,* $j, k \geq 1$ *and* $|y| \leq |x|$.

The equality $x^j y^k = z^\ell$ *holds if and only if one of the following cases takes place:*

A. There exists a word r, *and integers* $m, n, t \geq 0$ *such that*

$$mj + nk = t\ell, \quad \textit{and}$$
$$x = r^m, \quad y = r^n, \quad z = r^t;$$

B. $j = k = 1$ *and there exist non-commuting words* r *and* q, *and integers* $m, n \geq 0$ *such that*

$$m + n + 1 = \ell, \quad \textit{and}$$
$$x = (rq)^m r, \quad y = q(rq)^n, \quad z = rq;$$

C. $j = \ell = 2$, $k = 1$ *and there exist non-commuting words* r *and* q *and an integer* $m \geq 2$ *such that*

$$x = (rq)^m r, \quad y = qrrq, \quad z = (rq)^m rrq;$$

D. $j = 1$ *and* $k \geq 2$ *and there exist non-commuting words* r *and* q *such that*

$$x = (qr^k)^{\ell-1} q, \quad y = r, \quad z = qr^k;$$

E. $j = 1$ *and* $k \geq 2$ *and there are non-commuting words* r *and* q, *an integer* $m \geq 1$ *such that*

$$x = (qr(r(qr)^m)^{k-1})^{\ell-2} qr(r(qr)^m)^{k-2} rq, \quad y = r(qr)^m, \quad z = qr(r(qr)^m)^{k-1}.$$

Proof. If x and y commute, then all three words commute, hence they are a power of a common word. A length argument yields the solution A.

Assume now that $\{x, y\}$ is a code. Then no pair of words x, y and z commutes. We have shown in the overview of the proof of Theorem 1 that $j = 1$ or $k = 1$ by the Lyndon-Schützenberger theorem. The solution is then split into several cases.

Case 1: $j = k = 1$.
Let m and r be such that $z^m r = x$ with r a strict prefix of z. By setting $z = rq$, we obtain the solution B with $n = \ell - m - 1$.

Case 2: $j \geq 2, k = 1$.
Since $|y| \leq |x|$ and $\ell \geq 2$, we have

$$2|z| \leq |z^\ell| = |x^j| + |y| < 2|x^j|,$$

so z is a strict prefix of x^j.

As x^j has periodic roots both z and x, and z does not commute with x, the Periodicity lemma implies $|x^j| < |z| + |x|$. That is, $z = x^{j-1}u$, $x^j = zv$ and $x = uv$ for some nonempty words u and v. As v is a prefix of z, it is also a prefix of x. Therefore, we have

$$x = uv = vu'$$

for some word u'. This is a well known conjugation equality which implies $u = rq$, $u' = qr$ and $v = (rq)^n r$ for some words r, q and an integer $n \geq 0$.

We have
$$j|x| + |y| = |x^j y| = |z^\ell| = \ell(j-1)|x| + \ell|u|,$$
and thus $|y| = (\ell j - \ell - j)|x| + \ell|u|$. Since $|y| \leq |x|$, $|u| > 0$, $j \geq 2$, and $\ell \geq 2$, it follows that $\ell j - \ell - j = 0$, which implies $j = l = 2$. We therefore have $x^2 y = z^2$ and $x^2 = zv$, hence $vy = z$.

Combining $u = rq$, $u' = qr$, and $v = (rq)^n r$ with $x = vu'$, $z = x^{j-1}u = xu = vu'u$, and $vy = z$, we obtain the solution C with $m = n + 1$. The assumption $|y| \leq |x|$ implies $m \geq 2$.

Case 3: $j = 1, k \geq 2, y^k \leq_s z$.
We have $z = qy^k$ for some word q. Noticing that $x = z^{\ell-1}q$ yields the solution D.

Case 4: $j = 1, k \geq 2, z <_s y^k$.
This case is analogous to the second part of Case 2. Using the Periodicity lemma, we obtain $uy^{k-1} = z$, $y^k = vz$, and $y = vu$ with nonempty u and v. As v is a suffix of z, it is also a suffix of y, and we have $y = vu = u'v$ for some u'. Plugging the solution of the last conjugation equality, namely $u' = rq$, $u = qr$, $v = (rq)^n r$, into $y = u'v$, $z = uy^{k-1}$ and $z^{\ell-1} = xv$ gives the solution E with $m = n + 1$.

Finally, the words r and q do not commute since x and y, which are generated by r and q, do not commute.

The proof is completed by a direct verification of the converse. □

We now show that, for a given not primitivity preserving binary code, there is a unique pair of exponents (j, k) such that $x^j y^k$ is imprimitive.

Lemma 1 (`LS_unique`). *Let $B = \{x, y\}$ be a code. Assume $j, k, j', k' \geq 1$. If both $x^j y^k$ and $x^{j'} y^{k'}$ are imprimitive, then $j = j'$ and $k = k'$.*

Proof. Let z_1, z_2 be primitive words and $\ell, \ell' \geq 2$ be such that
$$x^j y^k = z_1^\ell \quad \text{and} \quad x^{j'} y^{k'} = z_2^{\ell'}. \tag{1}$$

Since B is a code, the words x and y do not commute. We proceed by contradiction.

Case 1: First, assume that $j = j'$ and $k \neq k'$.
Let, without loss of generality, $k < k'$. From (1) we obtain $z_1^\ell y^{k'-k} = z_2^{\ell'}$. The case $k' - k \geq 2$ is impossible due to the Lyndon-Schützenberger theorem. Hence $k' - k = 1$. This is another place where the formalization triggered a simple and nice general lemma (easily provable by the Periodicity lemma) which will turn out to be useful also in the proof of Theorem 4. Namely, the lemma `imprim_ext_suf_comm` claims that if both uv, and uvv are imprimitive, then u and v commute. We apply this lemma to $u = x^j y^{k-1}$ and $v = y$, obtaining a contradiction with the assumption that x and y do not commute.

Case 2. The case $k = k'$ and $j \neq j'$ is symmetric to Case 1.

Case 3. Let finally $j \neq j'$ and $k \neq k'$. The Lyndon-Schützenberger theorem implies that either j or k is one, and similarly either j' or k' is one. We can therefore assume that $k = j' = 1$ and $k', j \geq 2$. Moreover, we can assume that $|y| \leq |x|$. Indeed, in the opposite case, we can consider the words $y^k x^j$ and $y^{k'} x^{j'}$ instead, which are also both imprimitive.

Theorem 2 now allows only the case C for the equality $x^j y = z_1^\ell$. We therefore have $j = \ell = 2$ and $x = (rq)^m r$, $y = qrrq$ for an integer $m \geq 2$ and some non-commuting words r and q. Since $y = qrrq$ is a suffix of z_2^ℓ, this implies that z_2 and rq do not commute. Consider the word $x \cdot qr = (rq)^m rqr$, which is a prefix of xy, and therefore also of z_2^ℓ. This means that $x \cdot qr$ has two periodic roots, namely rq and z_2, and the Periodicity lemma implies that $|x \cdot qr| < |rq| + |z_2|$. Hence x is shorter than z_2. The equality $xy^{k'} = z_2^{\ell'}$, with $\ell' \geq 2$, now implies on one hand that $rqrq$ is a prefix of z_2, and on the other hand that z_2 is a suffix of $y^{k'}$. It follows that $rqrq$ is a factor of $(qrrq)^k$. Hence $rqrq$ and $qrrq$ are conjugate, thus they both have a period of length $|rq|$, which implies $qr = rq$. This is a contradiction. □

The rest of the paper, and therefore also of the proof of Theorem 1, is organized as follows. In Sect. 4, we introduce a general theory of interpretations, which is behind the main idea of the proof, and apply it to the (relatively simple) case of a binary code with words of the same length. In Sect. 5 we characterize the unique disjoint extendable $\{x, y\}$-interpretation of the square of the longer word x. This is a result of independent interest, and also the cornerstone of the proof of Theorem 1 which is completed in Sect. 6 by showing that a word containing at least two x's witnessing that $\{x, y\}$ is not primitivity preserving is conjugate with $[x, x, y]$.

4 Interpretations and the Main Idea

Let X be a code, let u be a factor of $\texttt{concat}\, \mathbf{w}$ for some $\mathbf{w} \in \texttt{lists}\, X$. The natural question is to decide how u can be produced as a factor of words from X, or, in other words, how it can be interpreted in terms of X. This motivates the following definition.

Definition 2. *Let X be a set of words over Σ. We say that the triple $(p, s, \mathbf{w}) \in \Sigma^* \times \Sigma^* \times \texttt{lists}\, X$ is an* X-interpretation *of a word $u \in \Sigma^*$ if*

- $\mathbf{w}$ *is nonempty;*
- $p \cdot u \cdot s = \texttt{concat}\, \mathbf{w}$*;*
- $p <_p \texttt{hd}\, \mathbf{w}$ *and*
- $s <_s \texttt{last}\, \mathbf{w}$.

The definition is illustrated by the following figure, where $\mathbf{w} = [w_1, w_2, w_3, w_4]$:

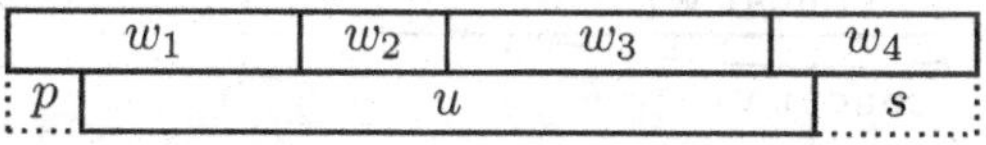

The second condition of the definition motivates the notation $p\, u\, s \sim_{\mathcal{I}} \mathbf{w}$ for the situation when $(p, s, \mathbf{w})$ is an X-interpretation of u.

Remark 1. For sake of historical reference, we remark that our definition of X-interpretation differs from the one used in [1]. Their formulation of the situation depicted by the above figure would be that u is interpreted by the triple $(s', w_2 \cdot w_3, p')$ where $p \cdot s' = w_1$ and $p' \cdot s = w_4$. This is less convenient for two reasons. First, the decomposition of $w_2 \cdot w_3$ into $[w_2, w_3]$ is only implicit here (and even ambiguous if X is not a code). Second, while it is required that the the words p' and s' are a prefix and a suffix, respectively, of an element from X, the identity of that element is left open, and has to be specified separately.

If u is a nonempty element of $\langle X \rangle$ and $u = \texttt{concat}\,\mathbf{u}$ for $\mathbf{u} \in \texttt{lists}\,X$, then the X-interpretation $\varepsilon\, u\, \varepsilon \sim_{\mathcal{I}} \mathbf{u}$ is called *trivial.* Note that the trivial X-interpretation is unique if X is a code.

As nontrivial X-interpretations of elements from $\langle X \rangle$ are of particular interest, the following two concepts are useful.

Definition 3. *An X-interpretation $p\, u\, s \sim_{\mathcal{I}} \mathbf{w}$ of $u = \texttt{concat}\,\mathbf{u}$ is called*

- disjoint *if* $\texttt{concat}\,\mathbf{w}' \neq p \cdot \texttt{concat}\,\mathbf{u}'$ *whenever* $\mathbf{w}' \leq_p \mathbf{w}$ *and* $\mathbf{u}' \leq_p \mathbf{u}$.
- extendable *if* $p \leq_s w_p$ *and* $s \leq_p w_s$ *for some elements* $w_p, w_s \in \langle X \rangle$.

Note that a disjoint X-interpretation is not trivial, and that being disjoint is relative to a chosen factorization $\mathbf{u}$ of u (which is nevertheless unique if X is a code).

The definitions above are naturally motivated by **the main idea** of the characterization of sets X that do not preserve primitivity, which dates back to Lentin and Schützenberger [9]. If $\mathbf{w}$ is primitive, while $\texttt{concat}\,\mathbf{w}$ is imprimitive, say $\texttt{concat}\,\mathbf{w} = z^k$, $k \geq 2$, then the shift by z provides a nontrivial and extendable X-interpretation of $\texttt{concat}\,\mathbf{w}$. (In fact, $k-1$ such nontrivial interpretations). Moreover, the following lemma, formulated in a more general setting of two words $\mathbf{w}_1$ and $\mathbf{w}_2$, implies that the X-interpretation is disjoint if X is a code.

Lemma 2 (shift_interpret, shift_disjoint). *Let X be a code. Let $\mathbf{w}_1, \mathbf{w}_2 \in \texttt{lists}\,X$ be such that $z \cdot \texttt{concat}\,\mathbf{w}_1 = \texttt{concat}\,\mathbf{w}_2 \cdot z$ where $z \notin \langle X \rangle$. Then $z \cdot \texttt{concat}\,\mathbf{v}_1 \neq \texttt{concat}\,\mathbf{v}_2$, whenever $\mathbf{v}_1 \leq_p \mathbf{w}_1^n$ and $\mathbf{v}_2 \leq_p \mathbf{w}_2^n$, $n \in \mathbb{N}$.*

In particular, $\texttt{concat}\,\mathbf{u}$ has a disjoint extendable X-interpretation for any prefix $\mathbf{u}$ of $\mathbf{w}_1$.

The excluded possibility is illustrated by the following figure.

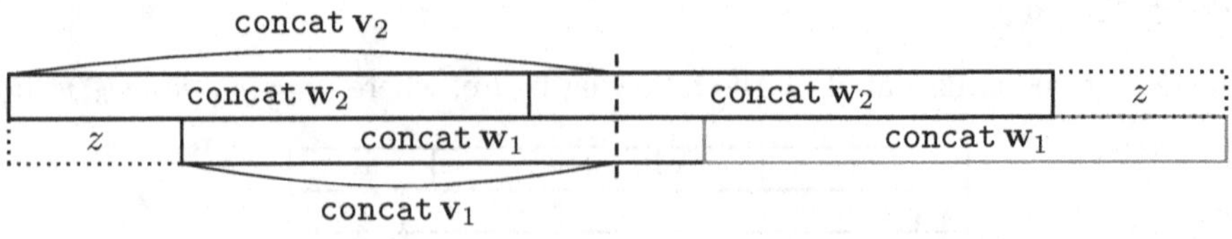

Proof. First, note that $z \cdot \texttt{concat}\, \mathbf{w}_1^n = \texttt{concat}\, \mathbf{w}_2^n \cdot z$ for any n. Let $\mathbf{w}_1^n = \mathbf{v}_1 \cdot \mathbf{v}_1'$ and $\mathbf{w}_2^n = \mathbf{v}_2 \cdot \mathbf{v}_2'$. If $z \cdot \texttt{concat}\, \mathbf{v}_1 = \texttt{concat}\, \mathbf{v}_2$, then also $\texttt{concat}\, \mathbf{v}_2' \cdot z = \texttt{concat}\, \mathbf{v}_1'$. This contradicts $z \notin \langle X \rangle$ by the stability condition.

An extendable X-interpretation of $\mathbf{u}$ is induced by the fact that $\texttt{concat}\, \mathbf{u}$ is covered by $\texttt{concat}(\mathbf{w}_2 \cdot \mathbf{w}_2)$. The interpretation is disjoint by the first part of the proof. □

In order to apply the above lemma to the imprimitive $\texttt{concat}\, \mathbf{w} = z^k$ of a primitive $\mathbf{w}$, set $\mathbf{w}_1 = \mathbf{w}_2 = \mathbf{w}$. The assumption $z \notin \langle X \rangle$ follows from the primitivity of $\mathbf{w}$: indeed, if $z = \texttt{concat}\, \mathbf{z}$, with $\mathbf{z} \in \texttt{lists}\, X$, then $\mathbf{w} = \mathbf{z}^k$ since B is a code.

We first apply the main idea to a relatively simple case of nontrivial $\{x, y\}$-interpretation of the word $x \cdot y$ where x and y are of the same length.

Lemma 3 (uniform_square_interp). *Let $B = \{x, y\}$ be a code with $|x| = |y|$. Let $p\ (x \cdot y)\ s \sim_{\mathcal{I}} \mathbf{v}$ be a nontrivial B-interpretation. Then $\mathbf{v} = [x, y, x]$ or $\mathbf{v} = [y, x, y]$ and $x \cdot y$ is imprimitive.*

Proof. From $p \cdot x \cdot y \cdot s = \texttt{concat}\, \mathbf{v}$, it follows, by a length argument, that $|\mathbf{v}|$ is three. A straightforward way to prove the claim is to consider all eight possible candidates. In each case, it is then a routine few line proof that shows that $x = y$, unless $\mathbf{v} = [x, y, x]$ or $\mathbf{v} = [y, x, y]$, which we omit. In the latter cases, $x \cdot y$ is a nontrivial factor of its square $(x \cdot y) \cdot (x \cdot y)$, which yields the imprimitivity of $x \cdot y$. □

The previous (sketch of the) proof nicely illustrates on a small scale the advantages of formalization. It is not necessary to choose between a tedious elementary proof for sake of completeness on one hand, and the suspicion that something was missed on the other hand (leaving aside that the same suspicion typically remains even after the tedious proof). A bit ironically, the most difficult part of the formalization is to show that $\mathbf{v}$ is indeed of length three, which needs no further justification in a human proof.

We have the following corollary which is a variant of Theorem 4, and also illustrates the main idea of its proof.

Lemma 4 (bin_imprim_not_conjug). *Let $B = \{x, y\}$ be a binary code with $|x| = |y|$. If $\mathbf{w} \in \texttt{lists}\, B$ is such that $|\mathbf{w}| \geq 2$, $\mathbf{w}$ is primitive, and $\texttt{concat}\, \mathbf{w}$ is imprimitive, then x and y are not conjugate.*

Proof. Since $\mathbf{w}$ is primitive and of length at least two, it contains both letters x and y. Therefore, it has either $[x, y]$ or $[y, x]$ as a factor. The imprimitivity of $\texttt{concat}\, \mathbf{w}$ yields a nontrivial B-interpretation of $x \cdot y$, which implies that $x \cdot y$ is not primitive by Lemma 3.

Let x and y be conjugate, and let $x = r \cdot q$ and $y = q \cdot r$. Since $x \cdot y = r \cdot q \cdot q \cdot r$ is imprimitive, also $r \cdot r \cdot q \cdot q$ is imprimitive. Then r and q commute by the theorem of Lyndon and Schützenberger, a contradiction with $x \neq y$. □

5 Binary Interpretation of a Square

Let $B = \{x, y\}$ be a code such that $|y| \leq |x|$. In accordance with the main idea, the core technical component of the proof is the description of the disjoint extendable B-interpretations of the square x^2. This is a very nice result which is relatively simple to state but difficult to prove, and which is valuable on its own. As we mentioned already, it can be obtained from Théorème 2.1 and Lemme 3.1 in [1].

Theorem 3 (square_interp_ext.sq_ext_interp). *Let $B = \{x, y\}$ be a code such that $|y| \leq |x|$, both x and y are primitive, and x and y are not conjugate. Let $p(x \cdot x)s \sim_\mathcal{I} \mathbf{w}$ be a disjoint extendable B-interpretation. Then*

$$\mathbf{w} = [x, y, x], \qquad s \cdot p = y, \qquad p \cdot x = x \cdot s.$$

In order to appreciate the theorem, note that the definition of interpretation implies

$$p \cdot x \cdot x \cdot s = x \cdot y \cdot x,$$

hence $x \cdot y \cdot x = (p \cdot x)^2$. This will turn out to be the only way how primitivity may not be preserved if x occurs at least twice in $\mathbf{w}$. Here is an example with $x = 01010$ and $y = 1001$:

0	1	0	1	0	1	0	0	1	0	1	0	1	0
0	1	0	1	0	1	0	0	1	0	1	0	1	0

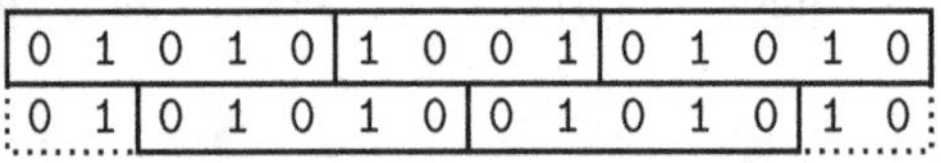

Proof. By the definition of a disjoint interpretation, we have $p \cdot x \cdot x \cdot s = \texttt{concat}\, \mathbf{w}$, where $p \neq \varepsilon$ and $s \neq \varepsilon$. A length argument implies that $\mathbf{w}$ has length at least three. Since a primitive word is not a nontrivial factor of its square, we have $\mathbf{w} = [\texttt{hd}\, \mathbf{w}] \cdot [y]^k \cdot [\texttt{last}\, \mathbf{w}]$, with $k \geq 1$. Since the interpretation is disjoint, we can split the equality into $p \cdot x = \texttt{hd}\, \mathbf{w} \cdot y^m \cdot u$ and $x \cdot s = v \cdot y^\ell \cdot \texttt{last}\, \mathbf{w}$, where $y = u \cdot v$, both u and v are nonempty, and $k = \ell + m + 1$. We want to show $\texttt{hd}\, \mathbf{w} = \texttt{last}\, \mathbf{w} = x$ and $m = \ell = 0$. The situation is mirror symmetric so we can solve cases two at a time.

If $\texttt{hd}\, \mathbf{w} = \texttt{last}\, \mathbf{w} = y$, then powers of x and y share a factor of length at least $|x| + |y|$. Since they are primitive, this implies that they are conjugate, a contradiction. The same argument applies when $\ell \geq 1$ and $\texttt{hd}\, \mathbf{w} = y$ (if $m \geq 1$ and $\texttt{last}\, \mathbf{w} = y$ respectively). Therefore, in order to prove $\texttt{hd}\, \mathbf{w} = \texttt{last}\, \mathbf{w} = x$, it remains to exclude the case $\texttt{hd}\, \mathbf{w} = y$, $\ell = 0$ and $\texttt{last}\, \mathbf{w} = x$ ($\texttt{last}\, \mathbf{w} = y$, $m = 0$ and $\texttt{hd}\, \mathbf{w} = x$ respectively). This is covered by one of the technical lemmas that we single out:

Lemma 5 (pref_suf_pers_short). *Let $x \leq_p v \cdot x$, $x \leq_s p \cdot u \cdot v \cdot u$ and $|x| > |v \cdot u|$ with $p \in \langle \{u, v\} \rangle$. Then $u \cdot v = v \cdot u$.*

This lemma indeed excludes the case we wanted to exclude, since the conclusion implies that y is not primitive. We skip the proof of the lemma here and make instead an informal comment. Note that v is a period root of x. In other words, x is a factor of v^ω. Therefore, with the stronger assumption that $v \cdot u \cdot v$ is a factor of x, the conclusion follows easily by the familiar principle that v being a factor of v^ω "synchronizes" primitive roots of v. Lemma 5 then exemplifies one of the virtues of formalization, which makes it easy to generalize auxiliary lemmas, often just by following the most natural proof and checking its minimal necessary assumptions.

Now we have $\texttt{hd}\,\mathbf{w} = \texttt{last}\,\mathbf{w} = x$, hence $p \cdot x = x \cdot y^m \cdot u$ and $x \cdot s = v \cdot y^\ell \cdot x$. The natural way to describe this scenario is to observe that x has both the (prefix) period root $v \cdot y^\ell$, and the suffix period root $y^m \cdot u$. Using again Lemma 5, we exclude situations when $\ell = 0$ and $m \geq 1$ ($m = 0$ and $\ell \geq 1$ resp.). It therefore remains to deal with the case when both m and ℓ are positive. We divide this into four lemmas according to the size of the overlap the prefix $v \cdot y^\ell$ and the suffix $y^m \cdot u$ have in x. More exactly, the cases are:

- $\left|v \cdot y^\ell\right| + |y^m \cdot u| \leq |x|$
- $|x| < \left|v \cdot y^\ell\right| + |y^m \cdot u| \leq |x| + |u|$
- $|x| + |u| < \left|v \cdot y^\ell\right| + |y^m \cdot u| < |x| + |u \cdot v|$
- $|x| + |u \cdot v| \leq \left|v \cdot y^\ell\right| + |y^m \cdot u|$

and they are solved by an auxiliary lemma each. The first three cases yield that u and v commute, the first one being a straightforward application of the Periodicity lemma. The last one is also straightforward application of the "synchronization" idea. It implies that $x \cdot x$ is a factor of y^ω, a contradiction with the assumption that x and y are primitive and not conjugate. Consequently, the technical, tedious part of the whole proof is concentrated in lemmas dealing with the second, and the third case (see lemmas `short_overlap` and `medium_overlap` in the theory `Binary_Square_Interpretation.thy`). The corresponding proofs are further analyzed and decomposed into more elementary claims in the formalization, where further details can be found.

This completes the proof of $\mathbf{w} = [x, y, x]$. A byproduct of the proof is the description of words x, y, p and s. Namely, there are non-commuting words r and t, and integers m, k and ℓ such that

$$x = (rt)^{m+1} \cdot r, \qquad y = (tr)^{k+1} \cdot (rt)^{\ell+1}, \qquad p = (rt)^{k+1}, \qquad s = (tr)^{\ell+1}.$$

The second claim of the present theorem, that is, $y = s \cdot p$ is then equivalent to $k = \ell$, and it is an easy consequence of the assumption that the interpretation is extendable. □

6 The Witness with Two x's

In this section, we characterize words witnessing that $\{x, y\}$ is not primitivity preserving and containing at least two x's.

Theorem 4 (bin_imprim_longer_twice). *Let $B = \{x, y\}$ be a code such that $|y| \leq |x|$. Let* $\mathbf{w} \in$ lists $\{x, y\}$ *be a primitive word which contains x at least twice such that* concat $\mathbf{w}$ *is imprimitive.*

Then $\mathbf{w} \sim [x, x, y]$ and both x and y are primitive.

We divide the proof in three steps.

The Core Case. We first prove the claim with two additional assumptions which will be subsequently removed. Namely, the following lemma shows how the knowledge about the B-interpretation of $x \cdot x$ from the previous section is used. The additional assumptions are displayed as items.

Lemma 6 (bin_imprim_primitive). *Let $B = \{x, y\}$ be a code with $|y| \leq |x|$ where*

- *both x and y are primitive,*

and let $\mathbf{w} \in$ lists B *be primitive such that* concat $\mathbf{w}$ *is imprimitive, and*

- *$[x, x]$ is a cyclic factor of* $\mathbf{w}$.

Then $\mathbf{w} \sim [x, x, y]$.

Proof. Choosing a suitable conjugate of $\mathbf{w}$, we can suppose, without loss of generality, that $[x, x]$ is a prefix of $\mathbf{w}$. Now, we want to show $\mathbf{w} = [x, x, y]$. Proceed by contradiction and assume $\mathbf{w} \neq [x, x, y]$. Since $\mathbf{w}$ is primitive, this implies $\mathbf{w} \cdot [x, x, y] \neq [x, x, y] \cdot \mathbf{w}$.

By Lemma 4, we know that x and y are not conjugate. Let concat $\mathbf{w} = z^k$, $2 \leq k$ and z primitive. Lemma 2 yields a disjoint extendable B-interpretation of $(\text{concat}\ \mathbf{w})^2$. In particular, the induced disjoint extendable B-interpretation of the prefix $x \cdot x$ is of the form $p\,(x \cdot x)\,s \sim_{\mathcal{I}} [x, y, x]$ by Theorem 3:

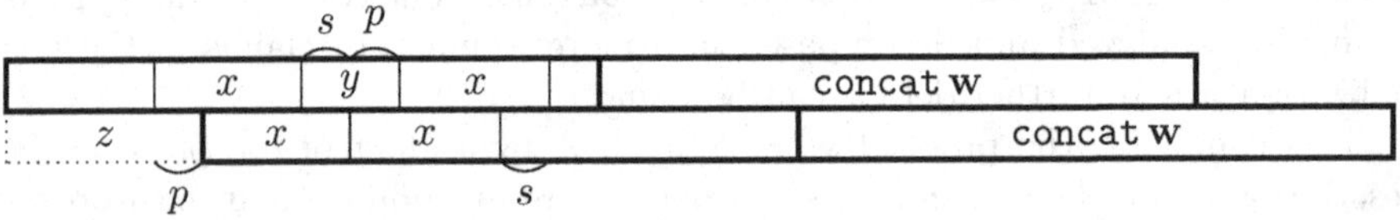

Let $\mathbf{p}$ be the prefix of $\mathbf{w}$ such that concat $\mathbf{p} \cdot p = z$. Then

$$\text{concat}(\mathbf{p} \cdot [x, y]) = z \cdot (x \cdot p), \quad \text{concat}\,[x, x, y] = (x \cdot p)^2, \quad \text{concat}\,\mathbf{w} = z^k,$$

and we want to show $z = xp$, which will imply $\text{concat}([x, x, y] \cdot \mathbf{w}) = \text{concat}(\mathbf{w} \cdot [x, x, y])$, hence $\mathbf{w} = [x, x, y]$ since $\{x, y\}$ is a code, and both $\mathbf{w}$ and $[x, x, y]$ are primitive.

Again, proceed by contradiction, and assume $z \neq xp$. Then, since both z and $x \cdot p$ are primitive, they do not commute. We now have two binary codes, namely $\{\mathbf{w}, [x, x, y]\}$ and $\{z, xp\}$. The following two equalities, (2) and (3) exploit the fundamental property of longest common prefixes of elements of binary codes mentioned in Sect. 2. In particular, we need the following lemma:

Lemma 7 (`bin_code_lcp_concat`). *Let* $X = \{u_0, u_1\}$ *be a binary code, and let* $\mathbf{z}_0, \mathbf{z}_1 \in$ lists X *be such that* concat $\mathbf{z}_0$ *and* concat $\mathbf{z}_1$ *are not prefix-comparable. Then*

$$(\texttt{concat}\,\mathbf{z}_0) \wedge_p (\texttt{concat}\,\mathbf{z}_1) = \texttt{concat}(\mathbf{z}_0 \wedge_p \mathbf{z}_1) \cdot (u_0 \wedge u_1).$$

See Sect. 8 for more comments on this property. Denote $\alpha_{z,xp} = z \cdot xp \wedge_p xp \cdot z$. Then also $\alpha_{z,xp} = z^k \cdot (xp)^2 \wedge_p (xp)^2 \cdot z^k$. Similarly, let $\alpha_{x,y} = x \cdot y \wedge_p y \cdot x$. Then Lemma 7 yields

$$\begin{aligned}\alpha_{z,xp} &= \texttt{concat}(\mathbf{w} \cdot [x,x,y]) \wedge_p \texttt{concat}([x,x,y] \cdot \mathbf{w}) \\ &= \texttt{concat}(\mathbf{w} \cdot [x,x,y] \wedge_p [x,x,y] \cdot \mathbf{w}) \cdot \alpha_{x,y}\end{aligned} \tag{2}$$

and also

$$\begin{aligned}z \cdot \alpha_{z,xp} =& \texttt{concat}(\mathbf{w} \cdot \mathbf{p} \cdot [x,y]) \wedge_p \texttt{concat}(\mathbf{p} \cdot [x,y] \cdot \mathbf{w}) \\ =& \texttt{concat}(\mathbf{w} \cdot \mathbf{p} \cdot [x,y] \wedge_p \mathbf{p} \cdot [x,y] \cdot \mathbf{w}) \cdot \alpha_{x,y}.\end{aligned} \tag{3}$$

Denote

$$\mathbf{v}_1 = \mathbf{w} \cdot [x,x,y] \wedge_p [x,x,y] \cdot \mathbf{w}, \qquad \mathbf{v}_2 = \mathbf{w} \cdot \mathbf{p} \cdot [x,y] \wedge_p \mathbf{p} \cdot [x,y] \cdot \mathbf{w}.$$

From (2) and (3) we now have $z \cdot \texttt{concat}\,\mathbf{v}_1 = \texttt{concat}\,\mathbf{v}_2$. Since $\mathbf{v}_1$ and $\mathbf{v}_2$ are prefixes of some $\mathbf{w}^n$, we have a contradiction with Lemma 2. □

Dropping the Primitivity Assumption. We first deal with the situation when x and y are not primitive. A natural idea is to consider the primitive roots of x and y instead of x and y. This means that we replace the word $\mathbf{w}$ with $\mathcal{R}\mathbf{w}$, where $\mathcal{R}$ is the morphism mapping $[x]$ to $[\rho\, x]^{e_x}$ and $[y]$ to $[\rho\, y]^{e_y}$ where $x = (\rho\, x)^{e_x}$ and $y = (\rho\, y)^{e_y}$. For example, if $x = abab$ and $y = aa$, and $\mathbf{w} = [x, y, x] = [abab, aa, abab]$, then $\mathcal{R}\mathbf{w} = [ab, ab, a, a, ab, ab]$.

Let us check which hypotheses of Lemma 6 are satisfied in the new setting, that is, for the code $\{\rho\, x, \rho\, y\}$ and the word $\mathcal{R}\mathbf{w}$. The following facts are not difficult to see.

- $\texttt{concat}\,\mathbf{w} = \texttt{concat}(\mathcal{R}\mathbf{w})$;
- if $[c, c]$, $c \in \{x, y\}$, is a cyclic factor $\mathbf{w}$, then $[\rho\, c, \rho\, c]$ is a cyclic factor of $\mathcal{R}\mathbf{w}$.

The next required property:

- if $\mathbf{w}$ is primitive, then $\mathcal{R}\mathbf{w}$ is primitive;

deserves more attention. It triggered another little theory of our formalization which can be found in locale `sings_code`. Note that it fits well into our context, since the claim is that $\mathcal{R}$ is a primitivity preserving morphism, which implies that its image on the singletons $[x]$ and $[y]$ forms a primitivity preserving set of words, see theorem `code.roots_prim_morph`.

Consequently, the only missing hypothesis preventing the use of Lemma 6 is $|y| \le |x|$ since it may happen that $|\rho\, x| < |\rho\, y|$. In order to solve this difficulty, we shall ignore for a while the length difference between x and y, and obtain the following intermediate lemma.

Lemma 8 (bin_imprim_both_squares, bin_imprim_both_squares_prim). *Let $B = \{x, y\}$ be a code, and let* $\mathbf{w} \in$ lists B *be a primitive word such that* concat $\mathbf{w}$ *is imprimitive. Then* $\mathbf{w}$ *cannot contain both* $[x, x]$ *and* $[y, y]$ *as cyclic factors.*

Proof. Assume that $\mathbf{w}$ contains both $[x, x]$ and $[y, y]$ as cyclic factors.

Consider the word $\mathcal{R}\mathbf{w}$ and the code $\{\rho\, x, \rho\, y\}$. Since $\mathcal{R}\mathbf{w}$ contains both $[\rho\, x, \rho\, x]$ and $[\rho\, y, \rho\, y]$, Lemma 6 implies that $\mathcal{R}\mathbf{w}$ is conjugate either with the word $[\rho\, x, \rho\, x, \rho\, y]$ or with $[\rho\, y, \rho\, y, \rho\, x]$, which is a contradiction with the assumed presence of both squares. □

Concluding the Proof by Gluing. It remains to deal with the existence of squares. We use an idea that is our main innovation with respect to the proof from [1], and contributes significantly to the reduction of length of the proof, and hopefully also to its increased clarity. Let $\mathbf{w}$ be a list over a set of words X. The idea is to choose one of the words, say $u \in X$, and to concatenate (or "glue") blocks of u's to words following them. For example, if $\mathbf{w} = [u, v, u, u, z, u, z]$, then the resulting list is $[uv, uuz, uz]$. This procedure is in the general case well defined on lists whose last "letter" is not the chosen one and it leads to a new alphabet $\{u^i \cdot v \mid v \neq u\}$ which is a code if and only if X is. This idea is used in an elegant proof of the Graph lemma (see [8] and [2]). In the binary case, which is of interest here, if $\mathbf{w}$ in addition does not contain a square of a letter, say $[x, x]$, then the new code $\{x \cdot y, y\}$ is again binary. Moreover, the resulting glued list $\mathbf{w}'$ has the same concatenation, and it is primitive if (and only if) $\mathbf{w}$ is. Note that gluing is in this case closely related to the Nielsen transformation $y \mapsto x^{-1}y$ known from the theory of automorphisms of free groups.

Induction on $|\mathbf{w}|$ now easily leads to the proof of Theorem 4.

Proof (of Theorem 4). If $\mathbf{w}$ contains y at most once, then we are left with the equation $x^j \cdot y = z^\ell$, $\ell \geq 2$. The equality $j = 2$ follows from the Periodicity lemma, see Case 2 in the proof of Theorem 2.

Assume for contradiction that y occurs at least twice in $\mathbf{w}$. Lemma 8 implies that at least one square, $[x, x]$ or $[y, y]$ is missing as a cyclic factor. Let $\{x', y'\} = \{x, y\}$ be such that $[x', x']$ is not a cyclic factor of $\mathbf{w}$. We can therefore perform the gluing operation, and obtain a new, strictly shorter word $\mathbf{w}' \in$ lists $\{x' \cdot y', y'\}$. The longer element $x' \cdot y'$ occurs at least twice in $\mathbf{w}'$, since the number of its occurrences in $\mathbf{w}'$ is the same as the number of occurrences of x' in $\mathbf{w}$, the latter word containing both letters at least twice by assumption. Moreover, $\mathbf{w}'$ is primitive, and concat $\mathbf{w}'$ = concat $\mathbf{w}$ is imprimitive. Therefore, by induction on $|\mathbf{w}|$, we have $\mathbf{w}' \sim [x' \cdot y', x' \cdot y', y']$. In order to show that this is not possible we can successfully reuse the lemma imprim_ext_suf_comm mentioned in the proof of Lemma 1, this time for $u = x'y'x'$ and $v = y'$. The words u and v do not commute because x' and y' do not commute. Since uv is imprimitive, the word $uvv \sim$ concat $\mathbf{w}'$ is primitive. □

This also completes the proof of our main target, Theorem 1.

7 Additional Notes on the Formalization

The formalization is a part of an evolving combinatorics on words formalization project. It relies on its backbone session, called CoW, a version of which is also available in the Archive of Formal Proofs [15]. This session covers basics concepts in combinatorics on words including the Periodicity lemma. An overview is available in [8].

The evolution of the parent session CoW continued along with the presented results and its latest stable version is available at our repository [16]. The main results are part of another Isabelle session CoW_Equations, which, as the name suggests, aims at dealing with word equations. We have greatly expanded its elementary theory `Equations_Basic.thy` which provides auxiliary lemmas and definitions related to word equations. Noticeably, it contains the definition `factor_interpretation` (Definition 2) and related facts.

Two dedicated theories were created: `Binary_Square_Interpretation.thy` and `Binary_Code_Imprimitive.thy`. The first contains lemmas and locales dealing with $\{x, y\}$-interpretation of the square xx (for $|y| \leq |x|$), culminating in Theorem 3. The latter contains Theorems 1 and 4.

Another outcome was an expansion of formalized results related to the Lyndon-Schützenberger theorem. This result, along with many useful corollaries, was already part of the backbone session CoW, and it was newly supplemented with the parametric solution of the equation $x^j y^k = z^\ell$, specifically Theorem 2 and Lemma 1. This formalization is now part of CoW_Equations in the theory `Lyndon_Schutzenberger.thy`.

Similarly, the formalization of the main results triggered a substantial expansion of existing support for the idea of gluing as mentioned in Sect. 6. Its reworked version is now in a separate theory called `Glued_Codes.thy` (which is part of the session CoW_Graph_Lemma).

Let us give a few concrete highlights of the formalization. A very useful tool, which is part of the CoW session, is the `reversed` attribute. The attribute produces a symmetrical fact where the symmetry is induced by the mapping **rev**, i.e., the mapping which reverses the order of elements in a list. For instance, the fact stating that if p is a prefix of v, then p a prefix of $v \cdot w$, is transformed by the reversed attribute into the fact saying that if s is suffix of v, then s is a suffix of $w \cdot v$. The attribute relies on ad hoc defined rules which induce the symmetry. In the example, the main reversal rule is

$$(\text{rev u} \leq \text{p rev v}) = \text{u} \leq \text{s v}.$$

The attribute is used frequently in the present formalization. For instance, Fig. 1 shows the formalization of the proof of Cases 1 and 2 of Theorem 1. Namely, the proof of Case 2 is smoothly deduced from the lemma that deals with Case 1, avoiding writing down the same proof again up to symmetry. See [13] for more details on the symmetry and the attribute `reversed`.

To be able to use this attribute fully in the formalization of main results, it needed to be extended to be able to deal with elements of type `'a list list`, as the constant `factor_interpretation` is of the function type over this exact

```
proof(cases)
  case 1
  then show ?thesis
    using LS-unique-same
      assms(1, 4–8) by blast
next
  case 2
  then show ?thesis
    using LS-unique-same[reversed]
      assms(1, 3, 5–8) by blast
```

(a) Using the `reversed` attribute to solve symmetric cases.

```
have primitive [x,x,y]
  using ‹x ≠ y›
  by primitivity-inspection

from ‹|ws| = 3› ‹ws ∈ lists {x,y}›
‹x ≠ y› ‹[x, x] ≤f ws · ws›
‹[y, y] ≤f ws · ws›
show False
  by list-inspection simp-all

from ‹p · t · s = t · t · p›
have p · t = t · p
  by mismatch
```

(b) Methods `primitivity_inspection`, `list_inspection` and `mismatch`.

Fig. 1. Highlights from the formalization in Isabelle/HOL.

type. The new theories of the session CoW_Equations contain almost 50 uses of this attribute.

The second highlight of the formalization is the use of simple but useful proof methods. The first method, called `primitivity_inspection`, is able to show primitivity or imprimitivity of a given word.

Another method named `list_inspection` is used to deal with claims that consist of straightforward verification of some property for a set of words given by their length and alphabet. For instance, this method painlessly concludes the proof of lemma `bin_imprim_both_squares_prim`. The method divides the goal into eight easy subgoals corresponding to eight possible words. All goals are then discharged by `simp_all`.

The last method we want to mention is `mismatch`. It is designed to prove that two words commute using the property of a binary code mentioned in Sect. 2 and explained in Sect. 8. Namely, if a product of words from $\{x, y\}$ starting with x shares a prefix of length at least $|xy|$ with another product of words from $\{x, y\}$, this time starting with y, then x and y commute. Examples of usage of the attribute `reversed` and all three methods are given in Fig. 1.

8 Appendix: Background Results in Combinatorics on Words

A periodic root r of w need not be primitive, but it is always possible to consider the corresponding primitive root $\rho\, r$, which is also a periodic root of w. Note that any word has infinitely many periodic roots since we allow r to be longer than w. Nevertheless, a word can have more than one period even if we consider only periods shorter than $|w|$. Such a possibility is controlled by the Periodicity lemma, often called the Theorem of Fine and Wilf (see [6]):

Lemma 9 (per_lemma_comm). *If w has a period u and v, i.e., $w \leq_p uw$ and $w \leq_p vw$, with $|u| + |v| - \gcd(|u|, |v|) \leq |w|$, then $uv = vu$.*

Usually, the weaker test $|u| + |v| \leq |w|$ is sufficient to indicate that u and v commute.

Conjugation $u \sim v$ is characterized as follows:

Lemma 10 (conjugation). *If $uz = zv$ for nonempty u, then there exists words r and q and an integer k such that $u = rq$, $v = qr$ and $z = (rq)^k r$.*

We have said that w has a periodic root r if it is a prefix of r^ω. If w is a factor, not necessarily a prefix, of r^ω, then it has a periodic root which is a conjugate of r. In particular, if $|u| = |v|$, then $u \sim v$ is equivalent to u and v being mutually factors of a power of the other word.

Commutation of two words is characterized as follows:

Lemma 11 (comm). *$xy = yx$ if and only if $x = t^k$ and $y = t^m$ for some word t and some integers $k, m \geq 0$.*

Since every nonempty word has a (unique) primitive root, the word t can be chosen primitive (k or m can be chosen 0 if x or y is empty).

We often use the following theorem, called "the theorem of Lyndon and Schützenberger":

Theorem 5 (Lyndon_Schutzenberger). *If $x^j y^k = z^\ell$ with $j \geq 2$, $k \geq 2$ and $\ell \geq 2$, then the words x, y and z commute.*

A crucial property of a primitive word t is that it cannot be a nontrivial factor of its own square. For a general word u, the equality $u \cdot u = p \cdot u \cdot s$ with nonempty p and s implies that all three words p, s, u commute, that is, have a common primitive root t. This can be seen by writing $u = t^k$, and noticing that the presence of a nontrivial factor u inside uu can be obtained exclusively by a shift by several t's. This idea is often described as "synchronization".

Let x and y be two words that do not commute. The longest common prefix of xy and yx is denoted α. Let c_x and c_y be the letter following α in xy and yx respectively. A crucial property of α is that it is a prefix of any sufficiently long word in $\langle\{x, y\}\rangle$. Moreover, if $\mathbf{w} = [u_1, u_2, \ldots, u_n] \in \texttt{lists}\,\{x, y\}$ is such that concat $\mathbf{w}$ is longer than α, then $\alpha \cdot [c_x]$ is a prefix of concat $\mathbf{w}$ if $u_1 = x$ and $\alpha \cdot [c_y]$ is a prefix of concat $\mathbf{w}$ if $u_1 = y$. That is why the length of α is sometimes called "the decoding delay" of the binary code $\{x, y\}$. Note that the property indeed in particular implies that $\{x, y\}$ is a code, that is, it does not satisfy any nontrivial relation. It is also behind our method mismatch. Finally, using this property, the proof of Lemma 7 is straightforward.

Acknowledgments. The authors acknowledge support by the Czech Science Foundation grant GAČR 20-20621S.

References

1. Barbin-Le Rest, E., Le Rest, M.: Sur la combinatoire des codes à deux mots. Theor. Comput. Sci. **41**, 61–80 (1985)
2. Berstel, J., Perrin, D., Perrot, J.F., Restivo, A.: Sur le théorème du défaut. J. Algebra **60**(1), 169–180 (1979). http://www.sciencedirect.com/science/article/pii/0021869379901133. https://doi.org/10.1016/0021-8693(79)90113-3
3. Berstel, J., Perrin, D., Reutenauer, C.: Codes and Automata. Cambridge (2010). https://www.ebook.de/de/product/8629820/jean_berstel_dominique_perrin_christophe_reutenauer_codes_and_automata.html
4. Budkina, L.G., Markov, Al.A.: F-semigroups with three generators. Mat. Zametki **14**, 267–277 (1973). Translated from Mat. Zametki **14**(2), 267–277 (1973)
5. Crochemore, M.: Sharp characterizations of squarefree morphisms. Theor. Comput. Sci. **18**(2), 221–226 (1982). https://doi.org/10.1016/0304-3975(82)90023-8
6. Fine, N.J., Wilf, H.S.: Uniqueness theorems for periodic functions. Proc. Am. Math. Soc. **16**(1), 109–109 (1965). https://doi.org/10.1090/S0002-9939-1965-0174934-9
7. Harju, T., Nowotka, D.: On the independence of equations in three variables. Theoret. Comput. Sci. **307**(1), 139–172 (2003). https://doi-org.ezproxy.is.cuni.cz/10.1016/S0304-3975(03)00098-7. https://doi.org/10.1016/S0304-3975(03)00098-7
8. Holub, Š., Starosta, Š.: Formalization of basic combinatorics on words. In: Cohen, L., Kaliszyk, C. (eds.) 12th International Conference on Interactive Theorem Proving (ITP 2021). Leibniz International Proceedings in Informatics (LIPIcs), vol. 193, pp. 22:1–22:17, Dagstuhl, Germany. Schloss Dagstuhl - Leibniz-Zentrum für Informatik (2021). https://drops.dagstuhl.de/opus/volltexte/2021/13917. https://doi.org/10.4230/LIPIcs.ITP.2021.22
9. Lentin, A., Schützenberger, M.-P.: A combinatorial problem in the theory of free monoids. In: Combinatorial Mathematics and its Applications (Proc. Conf., Univ. North Carolina, Chapel Hill, N.C., 1967), pp. 128–144. University North Carolina Press, Chapel Hill (1969)
10. Lyndon, R.C., Schützenberger, M.-P.: The equation $a^m = b^n c^p$ in a free group. Michigan Math. J. **9**(4), 289–298 (1962). https://doi.org/10.1307/mmj/1028998766
11. Manuch, J.: Defect effect of bi-infinite words in the two-element case. Discret. Math. Theor. Comput. Sci. **4**(2), 273–290 (2001). http://dmtcs.episciences.org/279
12. Mitrana, V.: Primitive morphisms. Inform. Process. Lett. **64**(6), 277–281 (1997). https://doi.org/10.1016/s0020-0190(97)00178-6
13. Raška, M., Starosta, Š.: Producing symmetrical facts for lists induced by the list reversal mapping in Isabelle/HOL (2021). https://arxiv.org/abs/2104.11622
14. Spehner, J.-P.: Quelques problèmes d'extension, de conjugaison et de presentation des sous-monoïdes d'un monoïde libre. Ph.D. thesis, Université Paris VII, Paris (1976)
15. Holub, Š., Raška, M., Starosta, Š.: Combinatorics on words basics. Archive of Formal Proofs, May 2021. https://isa-afp.org/entries/Combinatorics_Words.html. Formal proof development
16. Holub, Š., Raška, M., Starosta, Š.: Combinatorics on words formalized (release v1.6) (2022). https://gitlab.com/formalcow/combinatorics-on-words-formalized

Formula Simplification via Invariance Detection by Algebraically Indexed Types

Takuya Matsuzaki(✉) and Tomohiro Fujita

Tokyo University of Science, 1-3 Kagurazaka, Shinjuku-ku, Tokyo 162-8601, Japan
matuzaki@rs.tus.ac.jp, 1418097@ed.tus.ac.jp

Abstract. We describe a system that detects an invariance in a logical formula expressing a math problem and simplifies it by eliminating variables utilizing the invariance. Pre-defined function and predicate symbols in the problem representation language are associated with algebraically indexed types, which signify their invariance property. A Hindley-Milner style type reconstruction algorithm is derived for detecting the invariance of a problem. In the experiment, the invariance-based formula simplification significantly enhanced the performance of a problem solver based on quantifier-elimination for real-closed fields, especially on the problems taken from the International Mathematical Olympiads.

1 Introduction

It is very common to find an argument marked by the phrase "without loss of generality" (w.l.o.g.) in a mathematical proof by human. An argument of this kind is most often based on the symmetry or the invariance in the problem [9].

Suppose that we are going to prove, by an algebraic method, that the three median lines of a triangle meet at a point (Fig. 1). Six real variables are needed to represent three points on a plane. Since the concepts of 'median lines' and 'meeting at a point' are translation-invariant, we may fix one of the corners at the origin. Furthermore, because these concepts are also invariant under any invertible linear map, we may fix the other two points to, e.g., $(1, 0)$ and $(0, 1)$. Thus, all six variables were eliminated and the task of proof became much easier.

W.l.o.g. arguments may thus have strong impact on the efficiency of inference. It has drawn attention in several research areas including the relative strength of proof systems (e.g., [2,3,12,20]), propositional SAT (e.g., [1,6,8,17,19]), proof assistants [9], and algebraic methods for geometry problem solving [7,10].

Among others, Iwane and Anai [10] share exactly the same objective with us; both aim at solving geometry problems stated in natural language, using an algebraic method as the backend. Logical formulas resulted from mechanical translation of problem text tend to be huge and very redundant, while the computational cost of algebraic methods is generally quite sensitive to the size of the input measured by, e.g., the number of variables. Simplification of the input formula is hence a mandatory part of such a problem-solving system.

J. Blanchette et al. (Eds.): IJCAR 2022, LNAI 13385, pp. 388–406, 2022.
https://doi.org/10.1007/978-3-031-10769-6_24

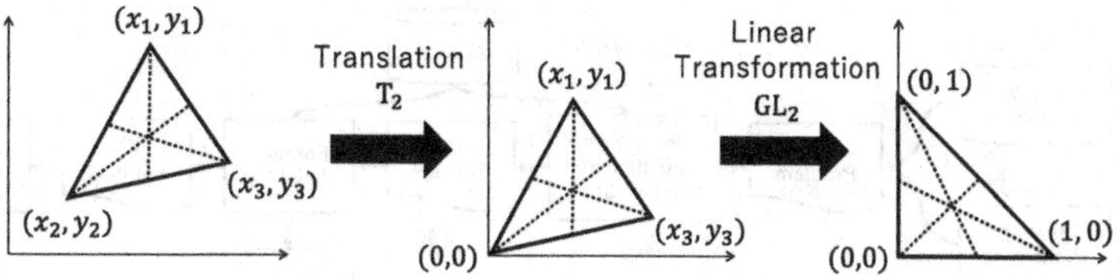

Fig. 1. Variable Elimination w.l.o.g. by Invariance

Iwane and Anai's method operates on the first-order formula of real-closed fields (RCFs), i.e., a quantified boolean combination of equalities and inequalities between polynomials. They proposed to detect the invariance of a problem by testing the invariance of the polynomials under translation, scaling, and rotation. While being conceptually simple, it amounts to discover the geometric property of the problem solely by its algebraic representation. The detection of rotational invariance is especially problematic because, to test that on a system of polynomials, one needs to identify all the pairs (or triples) of variables that originate from the x and y (and z) coordinates of the same points. Thus their algorithm for 2D rotational invariance already incurs a search among a large number of possibilities and they left the detection of 3D rotational invariance untouched. Davenport [7] also suggests essentially the same method.

In this paper, we propose to detect the invariance in a more high-level language than that of RCF. We use algebraically indexed types (AITs) proposed by Atkey et al. [4] as the representation language. In AIT, each symbol in a formula has a type with indices. An indexed-type of a function indicates that its output undergoes the same or a related transformation as the input. The invariances of the functions are combined via type reconstruction and an invariance in a problem is detected.

The contribution of the current paper is summarized as follows:

1. A type reconstruction algorithm for AIT is derived. Atkey et al. [4] laid out the formalism of AIT but did not provide a type inference/reconstruction algorithm. We devised, for a version of AIT, a type reconstruction algorithm that is based on semantic unification in the theory of transformation groups.
2. A set of variable elimination rules are worked out. Type reconstruction in AIT discerns a more fine-grained notion of invariance than previous approaches. We derived a set of elimination rules that covers all cases.
3. The practicality of the proposed method is verified; it significantly enhanced the performance of a problem solver based on quantifier elimination for RCF, especially on the problems from International Mathematical Olympiads.

In the rest of the paper, we first introduce a math problem solver, on which the proposed method was implemented, and summarize the formalism of AIT. We then detail the type reconstruction procedure and the variable elimination rules. We finally present the experimental results and conclude the paper.

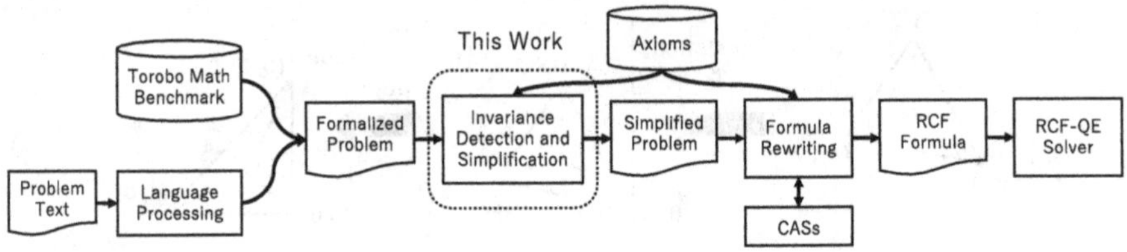

Fig. 2. Overview of Todai Robot Math Problem Solver

```
(Show (forall (A B C D X K L M )
        (-> (&& (is-triangle A B C)
                (= (rad-of-angle (angle B C A)) (* 90 (Degree)))
                (= D (foot-of-perp-line-from-to C (line A B)))
                (on X (seg C D)) (! (= X C)) (! (= X D))
                (on K (seg A X))
                (= (length-of (seg B K)) (length-of (seg B C)))
                (on L (seg B X))
                (= (length-of (seg A L)) (length-of (seg A C)))
                (intersect (seg A L) (seg B K) M))
            (= (length-of (seg M L)) (length-of (seg M K))))))
```

Fig. 3. Example of Manually Formalized Problem (IMO 2012, Problem 5)

2 Todai Robot Math Solver and Problem Library

This work is a part of the development of the Todai Robot Math Problem Solver (henceforth ToroboMath) [13–16]. Figure 2 presents an overview of the system. ToroboMath is targeted at solving pre-university math problems. Our long-term goal is to develop a system that solves problems stated in natural language.

The natural language processing (NLP) module of the system accepts a problem text and derives its logical representation through syntactic analysis. Currently, it produces a correct logical form for around 50% of sentences [13], which is not high enough to cover a wide variety of problems. Although the motivation behind the current work is to cope with the huge formulas produced by the NLP module, we instead used a library of *manually* formalized problems for the evaluation of the formula simplification procedure.

The problem library has been developed along with the ToroboMath system. It contains approximately one thousand math problems collected from several sources including the International Mathematical Olympiads (IMOs). Figure 3 presents a problem that was taken from IMO 2012.

The problems in the library are manually encoded in a polymorphic higher-order language, which is the same language as the output of the NLP module. Table 1 lists some of its primitive types. The language includes a large set of predicate and function symbols that are tailored for formalizing pre-university math problems. Currently, 1387 symbols are defined using 2808 axioms. Figure 4 provides an example of the axioms that defines the predicate `maximum`.

Table 1. Example of Primitive Types

truth values	`Bool`
numbers	`Z` (integers), `Q` (rationals), `R` (reals), `C` (complex)
vectors	`2d.Vec`, `3d.Vec`
geometric objects	`2d.Shape`, `3d.Shape`
angles	`2d.Angle`, `3d.Angle`
sets and lists	`SetOf`(α), `ListOf`(α)

```
(axiom def_maximum
  (forall (set max)
    (<-> (maximum set max)
      (& (elem max set)
        (forall (v)
          (-> (elem v set)
            (<= v max)))))))
```

Fig. 4. Example of Axiom

The problem solving module of the ToroboMath accepts a formalized problem and iteratively rewrites it using: (1) basic transformations such as $\forall x.(x = \alpha \rightarrow \phi(x)) \Leftrightarrow \phi(\alpha)$ and beta-reduction, (2) simplification of expressions such as polynomial division and integration by computer algebra systems (CASs), and (3) the axioms that define the predicate and function symbols.

Once the rewritten formula is in the language of real-closed fields (RCFs) or Peano arithmetic, it is handed to a solver for the theory. For RCF formulas, we use an implementation of the quantifier-elimination (QE) procedure for RCF based on cylindrical algebraic decomposition. Finally, we solve the resulting quantifier-free formula with CASs and obtain the answer. The time complexity of RCF-QE is quite high; it is doubly exponential in the number of variables [5]. Hence, the simplification of the formula *before* RCF-QE is a crucial step.

3 Algebraically Indexed Types

This section summarizes the framework of AIT. We refrain from presenting it in full generality and describe its application to geometry ([4, §2]) with the restriction we made on it in incorporating it into the type system of ToroboMath.

In AIT, some of the primitive types have associated *indices*. An index represents a transformation on the object of that type. For instance, in $\texttt{Vec}\langle B, t\rangle$, the index B stands for an invertible linear transformation and t stands for a translation. The index variables bound by universal quantifiers signify that a function of that type is invariant under any transformations indicated by the indices, e.g.,

$$\texttt{midpoint} : \forall B{:}\mathsf{GL}_2.\forall t{:}\mathsf{T}_2.\ \texttt{Vec}\langle B, t\rangle \rightarrow \texttt{Vec}\langle B, t\rangle \rightarrow \texttt{Vec}\langle B, t\rangle.$$

The type of `midpoint` certifies that, when two points P and Q undergo an arbitrary affine transformation, the midpoint of P and Q moves accordingly.

3.1 Sort and Index Expression

The *sort* of an index signifies the kind of transformations represented by the index. We assume the set Sort of index sorts includes GL_k ($k = 1, 2, 3$) (general linear transformations), O_k ($k = 2, 3$) (orthogonal transformations), and T_k ($k = 2, 3$) (translations). In the type of `midpoint`, B is of sort GL_2 and t is of sort T_2.

An *index expression* is composed of index variables and index operators. In the current paper, we use the following operators: $\langle +, -, 0 \rangle$ are addition, negation, and unit of T_k $(k = 2, 3)$; $\langle \cdot\,, {}^{-1}, 1 \rangle$ are multiplication, inverse, and unit of GL_k and O_k; det is the determinant; $|\cdot|$ is the absolute value. An *index context* Δ is a list of index variables paired with their sorts: $\Delta = i_1{:}S_1, i_2{:}S_2, \ldots, i_n{:}S_n$. The well-sortedness of an index expression e of sort S, written $\Delta \vdash e : S$, is defined analogously to the well-typedness in simple type theory.

3.2 Type, Term, and Typing Judgement

The set of primitive types, $\textsc{PrimType} = \{\texttt{Bool}, \texttt{R}, \texttt{2d.Vec}, \texttt{3d.Vec}, \texttt{2d.Shape}, \ldots\}$, is the same as that in the language of ToroboMath. A function tyArity: $\textsc{PrimType} \to \textsc{Sort}^*$ specifies the number and sorts of indices appropriate for the primitive types: e.g., $\text{tyArity}(\texttt{2d.Vec}) = (\mathsf{GL}_2, \mathsf{T}_2)$.

A judgement $\Delta \vdash A$ type means that type A is well-formed and well-indexed with respect to an index context Δ. Here are the derivation rules:

$$\frac{\texttt{X} \in \textsc{PrimType} \quad \text{tyArity}(\texttt{X}) = (S_1, \ldots, S_m) \quad \{\Delta \vdash e_j : S_j\}_{1 \le j \le m}}{\Delta \vdash \texttt{X}\langle e_1, \ldots, e_m \rangle \ \mathsf{type}} \ \textsc{TyPrim}$$

$$\frac{\Delta \vdash A \ \mathsf{type} \quad \Delta \vdash B \ \mathsf{type}}{\Delta \vdash A \to B \ \mathsf{type}} \ \textsc{TyArr} \qquad \frac{\Delta, i{:}S \vdash A \ \mathsf{type}}{\Delta \vdash \forall i{:}S.A \ \mathsf{type}} \ \textsc{TyForall}$$

While Atkey et al.'s system is formulated in the style of System F, we allow the quantifiers only at the outermost (prenex) position. The restriction permits an efficient type reconstruction algorithm analogous to Hindley-Milner's, while being expressive enough to capture the invariance of the pre-defined functions in ToroboMath and the invariance in the majority of math problems.

The well-typedness of a term M, written $\Delta; \Gamma \vdash M : A$, is judged with respect to an index context Δ and a typing context $\Gamma = x_1 : A_1, \ldots, x_n : A_n$. A typing context is a list of variables with their types. A special context Γ_{ops} consists of the pre-defined symbols and their types, e.g., $+ : \forall s{:}\mathsf{GL}_1.\, \texttt{R}\langle s \rangle \to \texttt{R}\langle s \rangle \to \texttt{R}\langle s \rangle \in \Gamma_{\text{ops}}$. We assume Γ_{ops} is always available in the typing derivation and suppress it in a judgement. The typing rules are analogous to those for lambda calculus with rank-1 polymorphism except for TyEq:

$$\frac{x : A \in \Gamma}{\Delta; \Gamma \vdash x : A} \ \textsc{Var} \qquad \frac{\Delta; \Gamma \vdash M : \forall i{:}S.A \quad \Delta \vdash e{:}S}{\Delta; \Gamma \vdash M : A\{i \mapsto e\}} \ \textsc{UnivInst} \qquad \frac{\Delta; \Gamma, x : A \vdash M : B}{\Delta; \Gamma \vdash \lambda x.M : A \to B} \ \textsc{Abs}$$

$$\frac{\Delta; \Gamma \vdash M : A \to B \quad \Delta; \Gamma \vdash N : A}{\Delta; \Gamma \vdash MN : B} \ \textsc{App} \qquad \frac{\Delta; \Gamma \vdash M : A \quad \Delta \vdash A \equiv B}{\Delta; \Gamma \vdash M : B} \ \textsc{TyEq}$$

In the Abs and App rules, the meta-variables A and B only designate a type without quantifiers. In the UnivInst rule, $A\{i \mapsto e\}$ is the result of substituting e for i in A. The 'polymorphism' of the types with quantifiers hence takes place only when a pre-defined symbol (e.g., `midpoint`) enters a derivation via the Var rule and then the bound index variable is instantiated via the UnivInst rule.

The type equivalence judgement $\Delta \vdash A \equiv B$ in the TyEQ rule equates two types involving *semantically* equivalent index expressions; thus, e.g., $s{:}\mathsf{GL}_1 \vdash \mathtt{R}\langle s \cdot s^{-1}\rangle \equiv \mathtt{R}\langle 1\rangle$ and $O{:}\mathsf{O}_2 \vdash \mathtt{R}\langle|\det O|\rangle \equiv \mathtt{R}\langle 1\rangle$.

3.3 Index Erasure Semantics and Transformational Interpretation

The abstraction theorem for AIT [4] enables us to know the invariance of a term by its type. The theorem relates two kinds of interpretations of types and terms: index erasure semantics and relational interpretations. We will restate the theorem with what we here call *transformational* interpretations (t-interpretations hereafter), instead of the relational interpretations. It suffices for the purpose of justifying our algorithm and makes it easier to grasp the idea of the theorem.

The index-erasure semantics of a primitive type $\mathtt{X}\langle e_1, \dots, e_n\rangle$ is determined only by $\mathtt{X}$. We thus write $\lfloor \mathtt{X}\langle e_1, \dots, e_n\rangle \rfloor = \lfloor \mathtt{X} \rfloor$. The interpretation $\lfloor \mathtt{X} \rfloor$ is the set of mathematical objects intended for the type: e.g., $\lfloor \mathtt{2d.Vec}\langle B, t\rangle \rfloor = \lfloor \mathtt{2d.Vec} \rfloor = \mathbb{R}^2$ and $\lfloor \mathtt{R}\langle s\rangle \rfloor = \lfloor \mathtt{R} \rfloor = \mathbb{R}$. The index-erasure semantics of a non-primitive type is determined by the type structure: $\lfloor A \to B \rfloor = \lfloor A \rfloor \to \lfloor B \rfloor$ and $\lfloor \forall i{:}S.\, T \rfloor = \lfloor T \rfloor$.

The index-erasure semantics of a typing context $\Gamma = x_1{:}\mathtt{T}_1, \dots, x_n{:}\mathtt{T}_n$ is the direct product of the domains of the variables: $\lfloor \Gamma \rfloor = \lfloor \mathtt{T}_1 \rfloor \times \cdots \times \lfloor \mathtt{T}_n \rfloor$. The erasure semantics of a term $\Delta; \Gamma \vdash M : A$ is a function of the values assigned to its free variables: $\lfloor M \rfloor : \lfloor \Gamma \rfloor \to \lfloor A \rfloor$ and defined as usual (see, e.g., [18,21]).

The t-interpretation of a type $\mathtt{T}$, denoted by $[\![\mathtt{T}]\!]$, is a function from the assignments to the index variables to a transformation on $\lfloor \mathtt{T} \rfloor$. To be precise, we first define the semantics of index context $\Delta = i_1{:}S_1, \dots, i_n{:}S_n$ as the direct product of the interpretation of the sorts: $[\![\Delta]\!] = [\![S_1]\!] \times \cdots \times [\![S_n]\!]$, where $[\![S_1]\!], \dots, [\![S_n]\!]$ are the intended sets of transformations: e.g., $[\![\mathsf{GL}_2]\!] = \mathrm{GL}_2$ and $[\![\mathsf{T}_2]\!] = \mathrm{T}_2$. The interpretation of an index expression e of sort S is a function $[\![e]\!] : [\![\Delta]\!] \to [\![S]\!]$ that is determined by the structure of the expression; for $\rho \in [\![\Delta]\!]$,

$$[\![\mathtt{f}(e_1, \dots, e_n)]\!](\rho) = [\![\mathtt{f}]\!]([\![e_1]\!](\rho), \dots, [\![e_n]\!](\rho)), \quad [\![i_k]\!](\rho) = \rho(i_k),$$

where, in the last equation, we regard $\rho \in [\![\Delta]\!]$ as a function from index variables to their values. The index operations det and $|\cdot|$ are interpreted as intended.

The t-interpretation of a primitive type $\mathtt{X}\langle e_1, \dots, e_n\rangle$ is then determined by $\mathtt{X}$ and the structures of the index expressions $e_1, \dots, e_n$. The t-interpretation of $\mathtt{Vec}$ and $\mathtt{Shape}$ is the affine transformation of vectors and geometric objects parametrized by $\rho \in [\![\Delta]\!]$; for index expressions $\beta{:}\mathsf{GL}_2$ and $\tau{:}\mathsf{T}_2$,

$$[\![\mathtt{Vec}\langle \beta, \tau\rangle]\!](\rho) : \mathbb{R}^2 \ni x \mapsto M_{[\![\beta]\!](\rho)} x + v_{[\![\tau]\!](\rho)} \in \mathbb{R}^2$$
$$[\![\mathtt{Shape}\langle \beta, \tau\rangle]\!](\rho) : \mathcal{P}(\mathbb{R}^2) \ni S \mapsto \{M_{[\![\beta]\!](\rho)} x + v_{[\![\tau]\!](\rho)} \mid x \in S\} \in \mathcal{P}(\mathbb{R}^2),$$

where $M_{[\![\beta]\!](\rho)}$ and $v_{[\![\tau]\!](\rho)}$ are the representation matrix and vector of $[\![\beta]\!](\rho)$ and $[\![t]\!](\rho)$, and $\mathcal{P}(\mathbb{R}^2)$ denotes the power set of $\mathbb{R}^2$. Similarly, for the real numbers,

$$[\![\mathtt{R}\langle \sigma\rangle]\!](\rho) : \mathbb{R} \ni x \mapsto [\![\sigma]\!](\rho) x \in \mathbb{R}.$$

That is, $[\![\mathsf{R}\langle\sigma\rangle]\!](\rho)$ is a change of scale with the scaling factor determined by the expression $\sigma{:}\mathsf{GL}_1$ and the assignment ρ. For a primitive type $\mathtt{X}$ with no indices, its t-interpretation is the identity map on $\lfloor \mathtt{X} \rfloor$: i.e., $[\![\mathtt{X}]\!](\rho) = \mathrm{id}_{\lfloor X \rfloor}$.

The t-interpretation of a function type $A \to B$ is a higher-order function that maps a (mathematical) function $f : \lfloor A \rfloor \to \lfloor B \rfloor$ to another function on the same domain and codomain such that: $[\![A \to B]\!](\rho)(f) = [\![B]\!](\rho) \circ f \circ ([\![A]\!](\rho))^{-1}$. It is easy to check that this interpretation is compatible with currying. Equivalently, we may say that if $g = [\![A \to B]\!](\rho)(f)$, then f and g are in the commutative relation $g \circ [\![A]\!](\rho) = [\![B]\!](\rho) \circ f$. The typing derivation in AIT is a way to 'pull out' the effect of transformation $[\![A]\!](\rho)$ on a free variable deep inside a term by combining such commutative relations.

The t-interpretation of a fully-quantified type is the identity map on its erasure semantics: $[\![\forall i_1{:}S_1. \ldots \forall i_n{:}S_n.\ T]\!] = \mathrm{id}_{\lfloor T \rfloor}$. We don't define that of partially-quantified types because we don't need it to state the abstraction theorem.

3.4 Abstraction Theorem

The abstraction theorem for AIT enables us to detect the invariance of (the erasure-semantics of) a term under a certain set of transformations on its free variables. We first define the t-interpretation of the typing context $\Gamma = x_1 : T_1, \ldots, x_n : T_n$ as a simultaneous transformation of $\eta = (v_1, \ldots, v_n) \in \lfloor \Gamma \rfloor$:

$$[\![\Gamma]\!](\rho) : \lfloor \Gamma \rfloor \ni \eta \mapsto [\![\Gamma]\!](\rho) \circ \eta = ([\![T_1]\!](\rho) \circ v_1, \ldots, [\![T_n]\!](\rho) \circ v_n) \in \lfloor \Gamma \rfloor .$$

We now present a version of the abstraction theorem, restricted to the case of a term of quantifier-free type and restated with the t-interpretation:

Theorem 1 *(Abstraction [4], restated using transformational interpretation). If A is a quantifier-free type and $\Delta; \Gamma \vdash M : A$, then for all $\rho \in [\![\Delta]\!]$ and all $\eta \in \lfloor \Gamma \rfloor$, we have $[\![A]\!](\rho) \circ \lfloor M \rfloor(\eta) = \lfloor M \rfloor([\![\Gamma]\!](\rho) \circ \eta)$.*

Here we provide two easy corollaries of the theorem. The first one is utilized to eliminate variables from a formula while preserving the equivalence.

Corollary 1. *If $\Delta; x_1 : \mathtt{T}_1, \ldots, x_n : \mathtt{T}_n \vdash \phi(x_1, \ldots, x_n) : \mathtt{Bool}$, then for all $\rho \in [\![\Delta]\!]$, we have $\phi(x_1, \ldots, x_n) \Leftrightarrow \phi([\![\mathtt{T}_1]\!](\rho) \circ x_1, \ldots, [\![\mathtt{T}_n]\!](\rho) \circ x_n)$.*

This is by the abstraction theorem and the fact $[\![\mathtt{Bool}]\!](\rho) = \mathrm{id}_{\lfloor \mathtt{Bool} \rfloor}$ for any ρ. It indicates that, without loss of generality, we may 'fix' some of the variables to, e.g., zeros by appropriately choosing ρ.

The second corollary is for providing more intuition about the theorem.

Corollary 2. *If $\epsilon; \epsilon \vdash \lambda x_1. \ \ldots .\lambda x_n.\ f(x_1, \ldots, x_n) : \forall \Delta.\ \mathtt{T}_1 \to \cdots \to \mathtt{T}_n \to \mathtt{T}_0$ then, for all $\rho \in [\![\Delta]\!]$ and all $v_i \in \lfloor \mathtt{T}_i \rfloor$ $(i = 1, \ldots, n)$,*

$$[\![\mathtt{T}_0]\!](\rho) \circ \lfloor f \rfloor(v_1, \ldots, v_n) = \lfloor f \rfloor([\![\mathtt{T}_1]\!](\rho) \circ v_1, \ldots, [\![\mathtt{T}_n]\!](\rho) \circ v_n).$$

In the statement, $\forall \Delta$ signifies the universal quantification over all index variables in Δ. By this corollary, for instance, we can tell from the type of `midpoint` that, for all $x_1, x_2 \in \mathbb{R}^2$ and for all $g \in \mathrm{GL}_2$ and $t \in \mathrm{T}_2$,

$$\lfloor \mathtt{midpoint} \rfloor (M_g x_1 + v_t, M_g x_2 + v_t) = M_g \lfloor \mathtt{midpoint} \rfloor (x_1, x_2) + v_t.$$

3.5 Restriction on the Index Expressions of Sort $\mathsf{GL}_k/\mathsf{O}_k$ ($k \geq 2$)

We found that the type reconstruction in AIT is far more straightforward when we assume an index expression of sort GL_k or O_k ($k \geq 2$) includes at most one index variable of sort GL_k or O_k that is not in the determinant operator. Assuming this, any expression e of sort GL_k or O_k can be written in the form of

$$e = \prod_{i \in I} s_i^{w_i} \cdot \prod_{i \in I} |s_i|^{x_i} \cdot \prod_{j \in J} \det(B_j)^{y_j} \cdot \prod_{j \in J} |\det(B_j)|^{z_j} \cdot B_0^{\delta},$$

where $\{s_i\}_{i \in I}$ are of sort GL_1, $\{B_0\} \cup \{B_j\}_{j \in J}$ are of sort GL_k or O_k, $w_i, x_i, y_j, z_j \in \mathbb{Z}$, and $\delta \in \{0, 1\}$. We henceforth say an expression e in the above form satisfies *the head variable property* and call B_0 the *head variable* of e.

Empirically, this restriction is not too restrictive; as far as we are aware of, the invariance of all the pre-defined functions and predicates in TOROBOMATH is expressible with an indexed-type satisfying this.

4 Invariance Detection Through Type Reconstruction

We need type reconstruction in AIT for two purposes: to infer the invariance of the pre-defined symbols in TOROBOMATH and to infer the invariance in a math problem. To this end, we only have to derive the judgement $\Delta; \Gamma \vdash \phi : \texttt{Bool}$ where ϕ is either a defining axiom of a symbol or a formula of a problem. For a pre-defined symbol s, by a judgement $\Delta; s : T, \cdots \vdash \phi : \texttt{Bool}$, we know s is of type T and it has the invariance signified by T. For a problem ϕ, by the judgement $\Delta; x_1 : T_1, \ldots, x_n : T_n \vdash \phi : \texttt{Bool}$, we know the invariance of ϕ under the transformation on the free variables $x_1, \ldots, x_n$ according to $[\![T_1]\!], \ldots, [\![T_n]\!]$.

Since all types are in prenex form, we can find the typing derivation by a procedure analogous to the Hindley-Milner (H-M) algorithm. It consists of two steps: deriving equations among index expressions, and solving them. The procedure for solving the equations in $\mathsf{T}_2/\mathsf{T}_3$ is essentially the same as in the type inference for Kennedy's unit-of-measure types [11], which is a precursor of AIT. Further development is required to solve the equations in $\mathsf{GL}_2/\mathsf{GL}_3$, even under the restriction on the form of index expressions mentioned in Sect. 3.5, due to the existence of the index operations $|\cdot|$ and det.

4.1 Equation Derivation

We first assign a type variable α_i for each subterm t_i in ϕ. Then, for a subterm t_i in the form $t_j t_k$ (i.e., application of t_j to t_k), we have the equation $\alpha_j = \alpha_k \to \alpha_i$. The case for a subterm t_i in the form of $\lambda x.t_j$ is also analogous to H-M and we omit it here. For a leaf term (i.e., a variable) t_i, if it is one of the pre-defined symbols and $t_i : \forall i_1{:}S_1. \ldots \forall i_n{:}S_n.T \in \Gamma_{\text{ops}}$, we set $\alpha_i = T\{i_1 \mapsto \beta_1, \ldots, i_n \mapsto \beta_n\}$, where $\{i_1 \mapsto \beta_1, \ldots, i_n \mapsto \beta_n\}$ stands for the substitution of fresh variables $\beta_1, \ldots, \beta_n$ for $i_1, \ldots, i_n$. By solving the equations for the type and index variables $\{\alpha_i\}$ and $\{\beta_j\}$, we reconstruct the most general indexed-types of all the subterms.

For example, consider the following axiom defining perpendicular:

$$\forall v_1.\forall v_2.(\texttt{perpendicular}(v_1, v_2) \longleftrightarrow \texttt{inner-prod}(v_1, v_2) = 0),$$

and suppose that inner-prod is in Γ_{ops}. We are going to reconstruct the type of perpendicular. The type of inner-prod is

$$\texttt{inner-prod} : \forall s_1, s_2{:}\mathsf{GL}_1.\ \forall O{:}\mathsf{O}_2.\ \texttt{Vec}\langle s_1 O, 0\rangle \rightarrow \texttt{Vec}\langle s_2 O, 0\rangle \rightarrow \texttt{R}\langle s_1 \cdot s_2\rangle$$

and it is instantiated as $\texttt{inner-prod} : \texttt{Vec}\langle s_1 O, 0\rangle \rightarrow \texttt{Vec}\langle s_2 O, 0\rangle \rightarrow \texttt{R}\langle s_1 \cdot s_2\rangle$ where s_1, s_2, and O are fresh variables. Since the type of perpendicular in the non-AIT version of our language is $\texttt{Vec} \rightarrow \texttt{Vec} \rightarrow \texttt{Bool}$, we set fresh variables to all indices in the primitive types and have:

$$\texttt{perpendicular} : \texttt{Vec}\langle \beta_1, \tau_1\rangle \rightarrow \texttt{Vec}\langle \beta_2, \tau_2\rangle \rightarrow \texttt{Bool}.$$

Since perpendicular is applied to v_1 and v_2, the types of v_1 and v_2 are equated to $\texttt{Vec}\langle \beta_1, \tau_1\rangle$ and $\texttt{Vec}\langle \beta_2, \tau_2\rangle$. Additionally, since inner-prod is also applied to v_1 and v_2, we have the following equations:

$$\texttt{Vec}\langle s_1 O, 0\rangle = \texttt{Vec}\langle \beta_1, \tau_1\rangle, \quad \texttt{Vec}\langle s_2 O, 0\rangle = \texttt{Vec}\langle \beta_2, \tau_2\rangle \tag{4.1}$$

If we have an equation between the same primitive type, by unifying both sides of the equation, in turn we have one or more equations between index expressions, i.e., if we have $\texttt{X}\langle e_1, \ldots, e_m\rangle = \texttt{X}\langle e'_1, \ldots, e'_m\rangle$, then we have: $e_1 = e'_1, \ldots, e_m = e'_m$. For Eq. (4.1), we hence have $s_1 O = \beta_1, s_2 O = \beta_2, 0 = \tau_1$, and $0 = \tau_2$. Thus, by recursively unifying all the equated types, we are left with a system of equations between index expressions.

4.2 Equation Solving

To solve the derived equations between index expressions, we need to depart from the analogy with the H-M algorithm. Namely, instead of applying syntactic unification, we need semantic unification, i.e., we solve the equations as simultaneous equations in the transformation groups.

We first order the equations with respect to the sort of the equated expressions. We then process them in the order $\mathsf{T}_2/\mathsf{T}_3 \rightarrow \mathsf{GL}_2/\mathsf{GL}_3 \rightarrow \mathsf{GL}_1$ as follows.[1]

First, since equations of sort $\mathsf{T}_2/\mathsf{T}_3$ are always in the form of $\sum_i a_i t_i = 0$ ($a_i \in \mathbb{Z}$), where $\{t_i\}$ are variables of sort T_k ($k \in \{2, 3\}$), we can solve the equations as is the case with a linear homogeneous system. Although the solution may involve rational coefficients as in $t_i = \sum_j \frac{n_{ij}}{m_{ij}} t_j$ ($n_{ij}, m_{ij} \in \mathbb{Z}$), we can clear the denominators by introducing new variables t'_j such that $t_j = \mathrm{lcm}\{m_{ij}\}_i \cdot t'_j$.

Next, by the head variable property, equations of sort $\mathsf{GL}_2/\mathsf{GL}_3$ (henceforth $\mathsf{GL}_{\geq 2}$) are always in the form of $\sigma_1 B_1 = \sigma_2 B_2$, where σ_1 and σ_2 are index

[1] In this subsection, $\mathsf{GL}_2, \mathsf{GL}_3, \mathsf{O}_2$, and O_3 are collectively denoted as $\mathsf{GL}_2/\mathsf{GL}_3$ or $\mathsf{GL}_{\geq 2}$.

expressions of sort GL_1, and B_1 and B_2 are the head variables of sort $\mathsf{GL}_{\geq 2}$. We decompose these equations according to Table 2, which summarizes the following argument: Let E denote the identity transformation. Since $\sigma_1 B_1 = \sigma_2 B_2 \iff \sigma_1^{-1}\sigma_2 E = B_1 B_2^{-1}$, there must be some $s \in \mathsf{GL}_1$ such that $B_1 B_2^{-1} = sE$ and $\sigma_1^{-1}\sigma_2 = s$. Furthermore, by the superset-subset relation between the sorts of B_1 and B_2, e.g., $\mathsf{O}_2 \subset \mathsf{GL}_2$ for $B_1 : \mathsf{O}_2$ and $B_2 : \mathsf{GL}_2$, we can express one of the broader sort with the other as a parameter.

The algorithm for $\mathsf{GL}_{\geq 2}$ equations works as follows. First, we initialize the set of solution with the empty substitution: $S \leftarrow \{\}$. For each $\mathsf{GL}_{\geq 2}$ equation $\sigma_1 B_1 = \sigma_2 B_2$, we look up Table 2 and find the $\mathsf{GL}_{\geq 2}$ solution $B_i \mapsto sB_j$ and one or more new GL_1 equations. We populate the current set of GL_1 equations with the new ones, and apply the solution $B_i \mapsto sB_j$ to all the remaining GL_1 and $\mathsf{GL}_{\geq 2}$ equations. We also compose the GL_2 solution $B_i \mapsto sB_j$ with the current solution set: $S \leftarrow S \circ \{B_i \mapsto sB_j\}$.

By processing all $\mathsf{GL}_{\geq 2}$ equations as above, we are left with a partial solution S and a system of GL_1 equations, each of which is in the following form:

$$\prod_{i \in I} s_i^{w_i} \cdot \prod_{i \in I} |s_i|^{x_i} \cdot \prod_{j \in J} \det(B_j)^{y_j} \cdot \prod_{j \in J} |\det(B_j)|^{z_j} = 1 \quad (w_i, x_i, y_j, z_j \in \mathbb{Z}),$$

where we assume about I and J that $\{s_i\}_{i \in I}$ are all the GL_1 variables, $\{B_j\}_{j \in J}$ are all the remaining $\mathsf{GL}_{\geq 2}$ variables, and $I \cap J = \emptyset$. Letting $u_i = s_i \cdot |s_i|^{-1}$, $v_i = |s_i|$, $u_j = \det(B_j) \cdot |\det(B_j)|^{-1}$, and $v_i = |\det(B_j)|$, we have $s_i = u_i v_i$ and $\det(B_j) = u_j v_j$ for all $i \in I$ and $j \in J$. By using them, we have

$$\prod_i u_i^{w_i} \cdot \prod_i v_i^{w_i + x_i} \cdot \prod_j u_j^{y_j} \cdot \prod_j v_j^{y_j + z_j} = 1.$$

Since $u_i, u_j \in \{+1, -1\}$ and $v_i, v_j > 0$ for all i and j, we know the above equation is equivalent to the following two equations:

$$\prod_i u_i^{w_i} \cdot \prod_j u_j^{y_j} = 1, \quad \prod_i v_i^{w_i + x_i} \cdot \prod_j v_j^{y_j + z_j} = 1.$$

We thus have two systems of equations, one in $\{+1, -1\}$ and the other in $\mathbb{R}_{>0}$. Now we temporarily rewrite the solution with u_i and v_i: $S \leftarrow S \circ \{s_i \mapsto u_i v_i\}_{i \in I}$.

First consider the system in $\mathbb{R}_{>0}$. As long as there remains an equation involving a variable v_i, which originates from a GL_1 variable, we solve it for v_i and compose the solution $v_i \mapsto \prod_{i' \neq i} v_{i'}^{p_{i'}} \cdot \prod_j v_j^{q_j}$ with S while applying it to the remaining equations. The denominators of fractional exponents (i.e., $p_{i'}, q_j \in \mathbb{Q} \backslash \mathbb{Z}$) can be cleared similarly to the case of T_k equations. If all the equations in $\mathbb{R}_{>0}$ are solved this way, then S is the most general solution. Otherwise, there remain one or more equations of the form $\prod_{j \in J'} |\det B_j|^{d_j} = 1$ for some $J' \subset J$ and $\{d_j\}_{j \in J'}$. This is the only case where we may miss some invariance of a formula; in general, we cannot express the most general solution to this equation only with the index variables of sort GL_k and O_k. We make a compromise here and are satisfied with a less general solution $S \circ \{B_j \mapsto E\}_{j \in J'}$. Fortunately, this

Table 2. Decomposition of $\mathsf{GL}_2/\mathsf{GL}_3$ equation $\sigma_i B_i = \sigma_j B_j$ (s: a fresh variable)

Combination of head variables	Solution in GL_k	Equations in GL_1
$B_i = B_j$	none	$\sigma_i = \sigma_j$
$B_i : \mathsf{O}_k \ \wedge \ B_j = E$	$B_i \mapsto sE$	$s\sigma_i = \sigma_j, \ \lvert s \rvert = 1$
$B_i : \mathsf{GL}_k \ \wedge \ B_j = E$	$B_i \mapsto sE$	$s\sigma_i = \sigma_j$
$B_i : \mathsf{O}_k \ \wedge \ B_j : \mathsf{O}_k$	$B_i \mapsto sB_j$	$s\sigma_i = \sigma_j, \ \lvert s \rvert = 1$
$B_i : \mathsf{GL}_k \ \wedge \ B_j : \mathsf{O}_k$	$B_i \mapsto sB_j$	$s\sigma_i = \sigma_j$
$B_i : \mathsf{GL}_k \ \wedge \ B_j : \mathsf{GL}_k$	$B_i \mapsto sB_j$	$s\sigma_i = \sigma_j$

does not frequently happen in practice. We made this compromise only on three out of 533 problems used in the experiment. We expect that having more sorts, e.g., $\mathrm{SL}_k^{\pm} = \{M \in \mathrm{GL}_k \mid |\det M| = 1\}$, in the language of index expressions might be of help here, but leave it as a future work.

The system in $\{+1, -1\}$ is processed analogously to that in $\mathbb{R}_{>0}$. Finally, by restoring $\{u_i, v_i\}_{i \in I}$ and $\{u_j, v_j\}_{j \in J}$ in the solution S to their original forms, e.g., $u_i \mapsto s_i \cdot |s_i|^{-1}$, we have a solution to the initial set of equations in terms of the variables of sort GL_k and O_k.

4.3 Type Reconstruction for Pre-defined Symbols with Axioms

We incrementally determined the indexed-types of the pre-defined symbols according to the hierarchy of their definitions. We first constructed a directed acyclic graph wherein the nodes are the pre-defined symbols and the edges represent the dependency between their definitions. We manually assigned an indexed-type to the symbols without defining axioms (e.g., $+ : \mathtt{R} \to \mathtt{R} \to \mathtt{R}$) and initialized Γ_{ops} with them. We then reconstructed the indexed-types of other symbols in a topological order of the graph. After the reconstruction of the type of each symbol, we added the symbol with its inferred type to Γ_{ops}.

For some of the symbols, type reconstruction does not go as well as we hope. For example, the following axiom defines the symbol `midpoint`:

$$\forall p_1, p_2.(\mathtt{midpoint}(p_1, p_2) = \frac{1}{2} \cdot (p_1 + p_2)).$$

At the beginning of the type reconstruction of `midpoint`, the types of the symbols in the axiom are instantiated as follows:

$$\begin{aligned} \mathtt{midpoint} &: \mathtt{Vec}\langle \beta_1, \tau_1 \rangle \to \mathtt{Vec}\langle \beta_2, \tau_2 \rangle \to \mathtt{Vec}\langle \beta_3, \tau_3 \rangle \\ \cdot &: \mathtt{R}\langle s_1 \rangle \to \mathtt{Vec}\langle B_1, 0 \rangle \to \mathtt{Vec}\langle s_1 B_1, 0 \rangle \\ + &: \mathtt{Vec}\langle B_2, t_1 \rangle \to \mathtt{Vec}\langle B_2, t_2 \rangle \to \mathtt{Vec}\langle B_2, t_1 + t_2 \rangle. \end{aligned}$$

The derived equations between the index expressions are as follows:

$$\{B_2 = \beta_1, B_2 = \beta_2, B_1 = B_2, \beta_3 = s_1 B_1, s_1 = 1, t_1 = \tau_1, t_2 = \tau_2, 0 = t_1 + t_2, \tau_3 = 0\}.$$

By solving these equations, we obtain the indexed-type of `midpoint` as follows:

$$\texttt{midpoint} : \forall B_1{:}\mathsf{GL}_2.\ \forall t_1{:}\mathrm{T}_2.\ \texttt{Vec}\langle B_1, t_1\rangle \to \texttt{Vec}\langle B_1, -t_1\rangle \to \texttt{Vec}\langle B_1, 0\rangle.$$

This type indicates that the midpoint of any two points P and Q remains the same when we move P and Q respectively to $P + t_1$ and $Q - t_1$ for any $t_1 \in \mathbb{R}^2$. While it is *not wrong*, the following type is more useful for our purpose:

$$\texttt{midpoint} : \forall B{:}\mathsf{GL}_2.\ \forall t{:}\mathrm{T}_2.\ \texttt{Vec}\langle B, t\rangle \to \texttt{Vec}\langle B, t\rangle \to \texttt{Vec}\langle B, t\rangle. \qquad (1)$$

To such symbols, we manually assigned a more appropriate type.[2]

In the current system, 945 symbols have a type that includes indices. We manually assigned the types to 255 symbols that have no defining axioms. For 203 symbols we manually overwrote the inferred type as in the case of `midpoint`. The types of the remaining 487 symbols were derived through the type reconstruction.

5 Variable Elimination Based on Invariance

In this section, we first provide an example of the variable elimination procedure based on invariance. We then describe the top-level algorithm of the variable elimination, which takes a formula as input and eliminates some of the quantified variables in it by utilizing the invariance indicated by an index variable. We finally list the elimination rule for each sort of index variable.

5.1 Example of Variable Elimination Based on Invariance

Let us consider again the proof of the existence of the centroid of a triangle. For triangle ABC, the configuration of the midpoints P, Q, R of the three sides and the centroid G is described by the following formula:

$$\psi(A, B, C, P, Q, R, G) := \begin{pmatrix} P = \texttt{midpoint}(B, C) \land \texttt{on}(G, \texttt{segment}(A, P)) \land \\ Q = \texttt{midpoint}(C, A) \land \texttt{on}(G, \texttt{segment}(B, Q)) \land \\ R = \texttt{midpoint}(A, B) \land \texttt{on}(G, \texttt{segment}(C, R)) \end{pmatrix}$$

where $\texttt{on}(X, Y)$ stands for the inclusion of point X in a geometric object Y, and $\texttt{segment}(X, Y)$ stands for the line segment between points X and Y. Let ϕ denote the existence of the centroid (and the three midpoints):

$$\phi(A, B, C) := \exists G.\ \exists P.\ \exists Q.\ \exists R.\ \psi(A, B, C, P, Q, R, G).$$

Our goal is to prove $\forall A.\ \forall B.\ \forall C.\ \phi(A, B, C)$.

[2] The awkwardness of the type inferred for `midpoint` is a price for the efficiency of type reconstruction; it is due to the fact that we ignore the linear space structure of T_2 (and also, we do not posit $\mathrm{T}_1 (\simeq \mathbb{R})$ as the second index of type `R`). Otherwise, the type reconstruction comes closer to a search for an invariance on the algebraic representation of the problems and the defining axioms. Hence $1/2 * (t + t) = t$ is not deduced for $t : \mathsf{T}_2$, which is necessary to infer the type in Eq. (1).

The functions `midpoint`, `on`, and `segment` are invariant under translations and general linear transformations. The reconstruction algorithm hence derives

$$\beta : \mathsf{GL}_2, \tau : \mathsf{T}_2 \;;\; A : \mathtt{Vec}\langle\beta,\tau\rangle, B : \mathtt{Vec}\langle\beta,\tau\rangle, C : \mathtt{Vec}\langle\beta,\tau\rangle \vdash \phi(A,B,C) : \mathtt{Bool}.$$

By the abstraction theorem, this judgement implies the invariance of the proposition $\phi(A,B,C)$ under arbitrary affine transformations:

$$\forall g \in \mathrm{GL}_2.\ \forall t \in \mathrm{T}_2.\ \forall A,B,C.\ \phi(A,B,C) \Leftrightarrow \phi(t \circ g \circ A, t \circ g \circ B, t \circ g \circ C).$$

First, by considering the case of g being identity, we have

$$\forall t \in \mathrm{T}_2.\ \forall A,B,C.\ \phi(A,B,C) \Leftrightarrow \phi(t \circ A, t \circ B, t \circ C). \tag{2}$$

By using this, we are going to verify $\forall B,C.\ \phi(\mathbf{0},B,C) \Leftrightarrow \forall A,B,C.\ \phi(A,B,C)$, by which we know that we only have to prove $\forall B,C.\ \phi(\mathbf{0},B,C)$.

Suppose that $\forall B,C.\ \phi(\mathbf{0},B,C)$ holds. Since T_2 acts transitively on $\mathbb{R}^2$, for any $A \in \mathbb{R}^2$, there exists $t \in \mathrm{T}_2$ such that $t \circ \mathbf{0} = A$. Furthermore, for any $B,C \in \mathbb{R}^2$, by instantiating $\forall B,C.\ \phi(\mathbf{0},B,C)$ with $B \mapsto t^{-1}\circ B$ and $C \mapsto t^{-1}\circ C$, we have $\phi(\mathbf{0}, t^{-1}\circ B, t^{-1}\circ C)$. By Eq. (2), we obtain $\phi(t\circ\mathbf{0}, t\circ t^{-1}\circ B, t\circ t^{-1}\circ C)$, which is equivalent to $\phi(A,B,C)$. Since A,B,C were arbitrary, we proved

$$\forall B,C.\ \phi(\mathbf{0},B,C) \Rightarrow \forall A,B,C.\ \phi(A,B,C).$$

The converse is trivial. We thus proved $\forall B,C.\ \phi(\mathbf{0},B,C) \Leftrightarrow \forall A,B,C.\ \phi(A,B,C)$.

The simplified formula, $\forall B,C.\ \phi(\mathbf{0},B,C)$, is still invariant under the simultaneous action of GL_2 on B and C. Hence, by applying the type reconstruction again, we have $\beta : \mathsf{GL}_2 \;;\; B : \mathtt{Vec}\langle\beta,0\rangle, C : \mathtt{Vec}\langle\beta,0\rangle \vdash \phi(\mathbf{0},B,C) : \mathtt{Bool}$. It implies the following invariance: $\forall g \in \mathrm{GL}_2.\ \forall B,C.\ \phi(\mathbf{0},B,C) \Leftrightarrow \phi(\mathbf{0}, g\circ B, g\circ C)$.

We now utilize it to eliminate the remaining variables B and C. Although it is tempting to 'fix' B and C respectively at, e.g., $\mathbf{e}_1 := (1,0)$ and $\mathbf{e}_2 := (0,1)$, it incurs some *loss* of generality. For instance, when B is at the origin, there is no way to move B to $\mathbf{e}_1$ by any $g \in \mathrm{GL}_2$. We consider four cases:

1. B and C are linearly independent,
2. $B \neq \mathbf{0}$, and B and C are linearly dependent,
3. $C \neq \mathbf{0}$, and B and C are linearly dependent, and
4. B and C are both at the origin.

For each of these cases, we can find a suitable transformation in GL_2 as follows:

1. There exists $g_1 \in \mathrm{GL}_2$ s.t. $g_1 \circ B = \mathbf{e}_1$ and $g_1 \circ C = \mathbf{e}_2$,
2. There exist $g_2 \in \mathrm{GL}_2$ and $r \in \mathbb{R}$ s.t. $g_2 \circ B = \mathbf{e}_1$ and $g_2 \circ C = r\mathbf{e}_1$,
3. There exist $g_3 \in \mathrm{GL}_2$ and $r' \in \mathbb{R}$ s.t. $g_3 \circ C = \mathbf{e}_1$ and $g_3 \circ B = r'\mathbf{e}_1$, and
4. We only have to know whether or not $\phi(\mathbf{0},\mathbf{0},\mathbf{0})$ holds.

By a similar argument to the one for the translation-invariance, we have

$$\forall B,C.\ \phi(\mathbf{0},B,C) \Leftrightarrow \phi(\mathbf{0},\mathbf{e}_1,\mathbf{e}_2) \wedge \forall r.\ \phi(\mathbf{0},\mathbf{e}_1,r\mathbf{e}_1) \wedge \forall r'.\ \phi(\mathbf{0},r'\mathbf{e}_1,\mathbf{e}_1) \wedge \phi(\mathbf{0},\mathbf{0},\mathbf{0}).$$

Thus, we eliminated all four coordinate values (i.e., x and y coordinates for B and C) in the first and the last case and three of them in the other two cases.

5.2 Variable Elimination Algorithm

The variable elimination algorithm works as follows. We traverse the formula of a problem in a top-down order and, for each subformula in the form of

$$Qx_1.Qx_2.\cdots Qx_n.\ \phi(x_1, x_2, \ldots, x_n, \mathbf{y}) \quad (Q \in \{\forall, \exists\})$$

where $\mathbf{y} = y_1, \ldots, y_m$ are the free variables, we apply the type reconstruction procedure to $\phi(x_1, x_2, \ldots, x_n, \mathbf{y})$ and derive a judgement $\Delta; \Gamma, x_1{:}\mathtt{T}_1, \ldots, x_n{:}\mathtt{T}_n \vdash \phi(x_1, \ldots, x_n, \mathbf{y}) : \mathtt{Bool}$. We then choose an index variable i that appears at least once in $\mathtt{T}_1, \ldots, \mathtt{T}_n$ but in none of the types of $\mathbf{y}$. It means the transformation signified by i acts on some of $\{x_1, \ldots, x_n\}$ but on none of $\mathbf{y}$. We select from $\{x_1, \ldots, x_n\}$ one or more variables whose types include i and are of the form $\mathtt{R}\langle\sigma\rangle$ or $\mathtt{Vec}\langle\beta, \tau\rangle$. Suppose that we select $x_1, \ldots, x_l$. Then we know the judgement $\Delta; \Gamma, x_1{:}\mathtt{T}_1, \ldots, x_l{:}\mathtt{T}_l \vdash Qx_{l+1}.\cdots Qx_n.\ \phi(x_1, \ldots, x_n, \mathbf{y}) : \mathtt{Bool}$ also holds. We then eliminate (or add restriction on) the bound variables $x_1, \ldots, x_l$ by one of the lemmas in Sect. 5.3 according to the sort of i. After the elimination, the procedure is recursively applied to the resulting formula and its subformulas.

5.3 Variable Elimination Rules

We now present how to eliminate variables based on a judgement of the form

$$\Delta; \Gamma,\ x_1 : \mathtt{T}_1, \ldots, x_n : \mathtt{T}_n \vdash \psi(x_1, \ldots, x_n, \mathbf{y}) : \mathtt{Bool}$$

where $\mathtt{T}_1, \ldots, \mathtt{T}_n$ include no other variables than i; $\Gamma = y_1{:}\mathtt{U}_1, \ldots, y_m{:}\mathtt{U}_m$ is a typing context for $\mathbf{y} = y_1, \ldots, y_m$; and $\mathtt{U}_1 \ldots, \mathtt{U}_m$ do not include i. Note that we can obtain a judgement of this form by the procedure in Sect. 5.2 and by substituting the unity of appropriate sorts for all index variables other than i in $\mathtt{T}_1, \ldots, \mathtt{T}_n$.

We provide the variable elimination rules as lemmas, one for each sort of i. They state the rules for variables bound by $\forall$. The rules for $\exists$ are analogous. In stating the lemma, we suppress Δ and Γ in the judgement and $\mathbf{y}$ in ψ for brevity but we still assume the above-mentioned condition hold.

Some complication arises due to the fact that if $k \neq l$, then $\mathtt{T}_k$ and $\mathtt{T}_l$ may be indexed with *different* expressions of i. We thus need to consider potentially different transformations $[\![\mathtt{T}_1]\!](i), \ldots, [\![\mathtt{T}_n]\!](i)$ applied simultaneously on $x_1, \ldots, x_n$. Please refer to supplementary material on the first author's web page for a general argument behind the rules and the proofs of the lemmas (https://researchmap.jp/mtzk/?lang=en).

T_k: The following lemma states that, as we saw in Sect. 5.1, we have only to consider the truth of a formula $\psi(x)$ at $x = \mathbf{0}$ if $\psi(x)$ is translation-invariant.

Lemma 1. *If* $x : \mathtt{Vec}\langle 1, \tau(t)\rangle \vdash \psi(x) : \mathtt{Bool}$ *holds for* $t : \mathsf{T}_k$ $(t \in \{2, 3\})$, *then* $\forall x.\ \psi(x) \Leftrightarrow \psi(\mathbf{0})$.

O_2: The following lemma means that we may assume x is on the x-axis if $\psi(x)$ is invariant under rotation and reflection.

Lemma 2. *If* $x : \mathtt{Vec}\langle\beta(O), 0\rangle \vdash \psi(x) : \mathtt{Bool}$ *holds for* $O : \mathsf{O}_2$, *then* $\forall x.\ \psi(x) \Leftrightarrow \forall r.\ \psi(r\mathbf{e}_1)$.

O_3: A judgement in the following form implies different kinds of invariance according to β_1 and β_2:

$$x_1 : \mathtt{Vec}\langle\beta_1(O), 0\rangle, x_2 : \mathtt{Vec}\langle\beta_2(O), 0\rangle \vdash \psi(x_1, x_2) : \mathtt{Bool}. \tag{3}$$

In any case, we may assume x_1 is on the x-axis and x_2 is on the xy-plane for proving $\forall x_1, x_2.\ \psi(x_1, x_2)$, as stated in the following lemma.

Lemma 3. *If judgement (3) holds for* $O : \mathsf{O}_3$, *then*

$$\forall x_1.\ \forall x_2.\ \psi(x_1, x_2) \Leftrightarrow \forall p, q, r \in \mathbb{R}.\ \psi(p\mathbf{e}_1, q\mathbf{e}_1 + r\mathbf{e}_2).$$

GL_1: For $s : \mathsf{GL}_1$, a judgement $x : \mathtt{R}\langle\sigma(s)\rangle \vdash \psi(x) : \mathtt{Bool}$ implies, either

- $\psi(x)$ is invariant under change of sign, i.e., $\psi(x) \Leftrightarrow \psi(-x)$,
- $\psi(x)$ is invariant under positive scaling, i.e., $\psi(x) \Leftrightarrow \psi(fx)$ for all $f > 0$, or
- $\psi(x)$ is invariant under arbitrary scaling, i.e., $\psi(x) \Leftrightarrow \psi(fx)$ for all $f \neq 0$.

The form of σ determines the type of invariance. The following lemma summarizes how we can eliminate or restrict a variable for these cases.

Lemma 4. *Let* $\sigma(s) = s^e \cdot |s|^f \quad (e \neq 0$ *or* $f \neq 0)$ *and suppose a judgement* $x : \mathtt{R}\langle\sigma(s)\rangle \vdash \psi(s) : \mathtt{Bool}$ *holds for* $s : \mathsf{GL}_1$. *We have three cases:*

1. *if* $e + f = 0$, *then* $\forall x.\ \psi(x) \Leftrightarrow \forall x \geq 0.\ \psi(x)$, *otherwise,*
2. *if* e *is an even number, then* $\forall x.\ \psi(x) \Leftrightarrow \psi(1) \wedge \psi(0) \wedge \psi(-1)$, *and*
3. *if* e *is an odd number, then* $\forall x.\ \psi(x) \Leftrightarrow \psi(1) \wedge \psi(0)$.

GL_2 For $B : \mathsf{GL}_2$, a judgement in the following form implies different kinds of invariance of $\psi(x_1, x_2)$ depending on the form of β_1 and β_2:

$$x_1 : \mathtt{Vec}\langle\beta_1(B), 0\rangle, x_2 : \mathtt{Vec}\langle\beta_2(B), 0\rangle \vdash \psi(x_1, x_2). \tag{4}$$

The following lemma summarizes how we eliminate the variables in each case.

Lemma 5. *Let* $\beta_j(B) = \det(B)^{e_j} \cdot |\det(B)|^{f_j} \cdot B$ *and* $g_j = e_j + f_j$ $(j \in \{1, 2\})$. *If judgement (4) holds, then, letting* $\psi_0 := \psi(\mathbf{0}, \mathbf{0}) \wedge \forall r.\ \psi(r\mathbf{e}_1, \mathbf{e}_1) \wedge \forall r.\ \psi(\mathbf{e}_1, r\mathbf{e}_1)$ *and* $\Psi := \forall x_1.\ \forall x_2.\ \psi(x_1, x_2)$, *the following equivalences hold:*

1. *If* $g_1 + g_2 + 1 = 0$ *and*
   - *if* $e_1 + e_2$ *is an even number, then* $\Psi \Leftrightarrow \psi_0 \wedge \psi(\mathbf{e}_1, \mathbf{e}_2)$
   - *if* $e_1 + e_2$ *is an odd number, then* $\Psi \Leftrightarrow \psi_0 \wedge \psi(\mathbf{e}_1, \mathbf{e}_2) \wedge \psi(\mathbf{e}_1, -\mathbf{e}_2)$
2. *If* $g_1 + g_2 + 1 \neq 0$, *then* $\Psi \Leftrightarrow \psi_0 \wedge \forall r.\ \psi(r\mathbf{e}_1, \mathbf{e}_2)$.

A similar lemma holds for the invariances indicated by an index variable of sort GL_3. We refrain from presenting it for space reasons.

Table 3. Results on All RCF Problems in TOROBOMATH Benchmark

Division/#Prblms		ALGIDX		BASELINE	
		Solved	Time	Solved	Time
IMO	116	28%	51.7s	16%	19.7s
Univ	243	69%	22.1s	62%	26.8s
Chart	174	68%	9.7s	62%	12.0s
All	533	60%	20.5s	52%	20.6s

Table 4. Results on RCF Problems with Invariance Detected and Variable Eliminated

Division/#Prblms		ALGIDX		BASELINE		Speed
		Solved	Time	Solved	Time	up
IMO	77	19%	91.3s	1%	3.6s	23%
Univ	49	57%	31.0s	33%	62.7s	495%
Chart	77	49%	14.3s	36%	26.0s	529%
All	203	40%	34.3s	22%	38.5s	505%

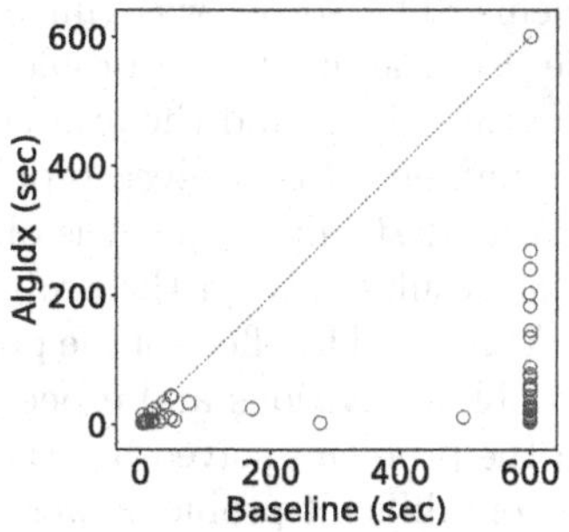

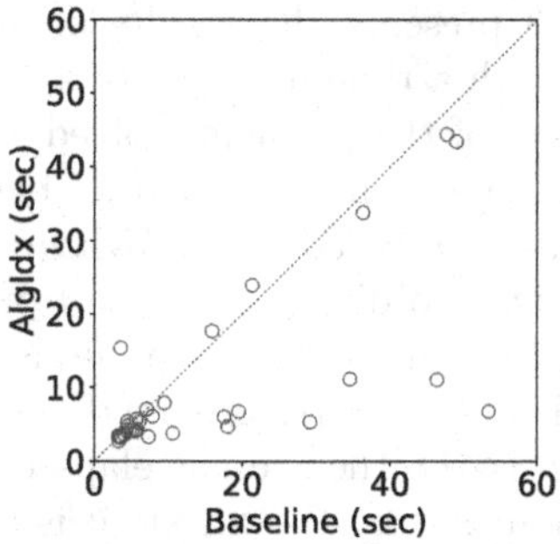

Fig. 5. Comparison of Elapsed Time with and without the Invariance Detection based on AITs (Left: All Problems; Right: Problems Solved within 60 s)

6 Experiment

We evaluated the effectiveness of the proposed method on the pre-university math problems in the TOROBOMATH benchmark. We used a subset of the problems that can be naturally expressible (by human) in the language of RCF. Most of them are either in geometry or algebra. Note that the formalization was done in the language introduced in Sect. 2 but not directly in the language of RCF. The problems are divided according to the source of the problems; **IMO** problems were taken from past International Mathematical Olympiads, **Univ** problems were from entrance exams of Japanese universities, and **Chart** problems were from a popular math practice book series. Please refer to another paper [16] on the TOROBOMATH benchmark for the details of the problems.

The type reconstruction and formula simplification procedures presented in Sect. 4 and Sect. 5 were implemented as a pre-processor of the formalized problems. The time spent for the preprocessing was almost negligible (0.76 s per problem on average) compared to that for solving the problems.

We compared the TOROBOMATH system with and without the pre-processor (respectively called ALGIDX and BASELINE below). The BASELINE system *is equipped with* Iwane and Anai's invariance detection and simplification algorithm [10] that operates on the language of RCF while ALGIDX *is not with it.* Thus, our evaluation shall reveal the advantage of detecting and exploiting the invariance of the problem expressed in a language that directly encodes its geometric meaning.

Table 5. Percentage of Problems from which one or more Variables are Eliminated by the Rule for each Sort

GL_1	T_2	O_2	GL_2	T_3	O_3	GL_3	any
22.3	27.4	26.1	1.7	6.6	7.5	0.0	38.1

Table 6. Most Frequent Invariance Types Detected and Eliminated

Invariance	(%)	Invariance	(%)
$\mathsf{GL}_1, \mathsf{T}_2, \mathsf{O}_2$	17.1	GL_1	2.4
$\mathsf{T}_2, \mathsf{O}_2$	8.1	$\mathsf{GL}_1, \mathsf{T}_3, \mathsf{O}_3$	2.1
$\mathsf{T}_3, \mathsf{O}_3$	4.5	$\mathsf{T}_2, \mathsf{GL}_2$	1.1

Table 3 presents the results on all problems. The solver was run on each problem with a time limit of 600 s. The table lists the number of problems, the percentages of the problems solved within the time limit, and the average wall-clock time spent on the solved problems. The number of the solved problems is significantly increased in the **IMO** division. A modest improvement is observed in the other two divisions. Table 4 presents the results only on the problems in which at least one variable was eliminated by AlgIdx. The effect of the proposed method is quite clearly observed across all problem divisions and especially on **IMO**. On **IMO**, the average elapsed time on the problems solved by AlgIdx is longer than that by Baseline; it is because more difficult problems were solved by AlgIdx within the time limit. In fact, the average speed-up by AlgIdx (last column in Table 4) is around 500% on **Univ** and **Chart**; i.e., on the problems solved by both, AlgIdx output the answer five times faster than Baseline.

A curious fact is that both AlgIdx and Baseline tended to need more time to solve the problems on which an invariance was detected and eliminated by AlgIdx (i.e., Time in Table 4) than the average over all solved problems (Time in Table 3). It suggests that a problem having an invariance, or equivalently a symmetry, is harder for automatic solvers than those without it.

Figure 5 shows a comparison of the elapsed time for each problem. Each point represents a problem, and the x and y coordinates respectively indicate the elapsed time to solve (or to timeout) by Baseline and AlgIdx. We can see many problems that were not solved by Baseline within 600 s were solved within 300 s by AlgIdx. The speed-up is also observed on easier problems (those solved in 60 s) as can be seen in the right panel of Fig. 5.

Table 5 lists the fraction of problems on which one or more variables are eliminated based on the invariance indicated by an index variable of each sort. Table 6 provides the distribution of the combination of the sorts of invariances detected and eliminated by AlgIdx.

7 Conclusion

A method for automating w.l.o.g. arguments on geometry problems has been presented. It detects an invariance in a problem through type reconstruction in AIT and simplifies the problem utilizing the invariance. It was especially effective on harder problems including past IMO problems. Our future work includes the exploration for a more elaborate language of the index expressions that captures various kind of invariance while keeping the type inference amenable.

References

1. Aloul, F.A., Sakallah, K.A., Markov, I.L.: Efficient symmetry breaking for boolean satisfiability. In: Proceedings of the 18th International Joint Conference on Artificial Intelligence, IJCAI 2003, pp. 271–276 (2003)
2. Arai, N.H.: Tractability of cut-free Gentzen type propositional calculus with permutation inference. Theoret. Comput. Sci. **170**(1), 129–144 (1996)
3. Arai, N.H., Urquhart, A.: Local symmetries in propositional logic. In: Dyckhoff, R. (ed.) TABLEAUX 2000. LNCS (LNAI), vol. 1847, pp. 40–51. Springer, Heidelberg (2000). https://doi.org/10.1007/10722086_3
4. Atkey, R., Johann, P., Kennedy, A.: Abstraction and invariance for algebraically indexed types. In: Proceedings of the 40th Annual ACM SIGPLAN-SIGACT Symposium on Principles of Programming Languages, POPL 2013, pp. 87–100 (2013)
5. Brown, C.W., Davenport, J.H.: The complexity of quantifier elimination and cylindrical algebraic decomposition. In: Proceedings of the 2007 International Symposium on Symbolic and Algebraic Computation, ISSAC 2007, pp. 54–60 (2007)
6. Crawford, J.M., Ginsberg, M.L., Luks, E.M., Roy, A.: Symmetry-breaking predicates for search problems. In: Proceedings of the Fifth International Conference on Principles of Knowledge Representation and Reasoning, KR 1996, pp. 148–159 (1996)
7. Davenport, J.H.: What does "without loss of generality" mean, and how do we detect it. Math. Comput. Sci. **11**(3), 297–303 (2017)
8. Devriendt, J., Bogaerts, B., Bruynooghe, M., Denecker, M.: Improved static symmetry breaking for SAT. In: Creignou, N., Le Berre, D. (eds.) SAT 2016. LNCS, vol. 9710, pp. 104–122. Springer, Cham (2016). https://doi.org/10.1007/978-3-319-40970-2_8
9. Harrison, J.: Without loss of generality. In: Berghofer, S., Nipkow, T., Urban, C., Wenzel, M. (eds.) TPHOLs 2009. LNCS, vol. 5674, pp. 43–59. Springer, Heidelberg (2009). https://doi.org/10.1007/978-3-642-03359-9_3
10. Iwane, H., Anai, H.: Formula simplification for real quantifier elimination using geometric invariance. In: Proceedings of the 2017 ACM on International Symposium on Symbolic and Algebraic Computation, ISSAC 2017, pp. 213–220 (2017)
11. Kennedy, A.: Types for units-of-measure: theory and practice. In: Horváth, Z., Plasmeijer, R., Zsók, V. (eds.) CEFP 2009. LNCS, vol. 6299, pp. 268–305. Springer, Heidelberg (2010). https://doi.org/10.1007/978-3-642-17685-2_8
12. Krishnamurthy, B.: Short proofs for tricky formulas. Acta Inform. **22**(3), 253–275 (1985)
13. Matsuzaki, T., Ito, T., Iwane, H., Anai, H., Arai, N.H.: Semantic parsing of pre-university math problems. In: Proceedings of the 55th Annual Meeting of the Association for Computational Linguistics (Volume 1: Long Papers), ACL 2017, pp. 2131–2141 (2017)
14. Matsuzaki, T., Iwane, H., Anai, H., Arai, N.H.: The most uncreative examinee: a first step toward wide coverage natural language math problem solving. In: Proceedings of the Twenty-Eighth AAAI Conference on Artificial Intelligence, AAAI 2014, pp. 1098–1104 (2014)

15. Matsuzaki, T., et al.: Can an A.I. win a medal in the mathematical olympiad? - Benchmarking mechanized mathematics on pre-university problems. AI Communications **31**(3), 251–266 (2018)
16. Matsuzaki, T., et al.: Race against the teens – benchmarking mechanized math on pre-university problems. In: Olivetti, N., Tiwari, A. (eds.) IJCAR 2016. LNCS (LNAI), vol. 9706, pp. 213–227. Springer, Cham (2016). https://doi.org/10.1007/978-3-319-40229-1_15
17. Metin, H., Baarir, S., Colange, M., Kordon, F.: CDCLSym: introducing effective symmetry breaking in SAT solving. In: Beyer, D., Huisman, M. (eds.) TACAS 2018. LNCS, vol. 10805, pp. 99–114. Springer, Cham (2018). https://doi.org/10.1007/978-3-319-89960-2_6
18. Reynolds, J.C.: Types, abstraction and parametric polymorphism. In: Information Processing 83, Proceedings of the IFIP 9th World Computer Congress, pp. 513–523 (1983)
19. Sabharwal, A.: SymChaff: exploiting symmetry in a structure-aware satisfiability solver. Constraints **14**(4), 478–505 (2009)
20. Szeider, S.: Homomorphisms of conjunctive normal forms. Discrete Appl. Math. **130**(2), 351–365 (2003)
21. Wadler, P.: Theorems for free! In: Proceedings of the Fourth International Conference on Functional Programming Languages and Computer Architecture, FPCA 1989, pp. 347–359 (1989)

Synthetic Tableaux: Minimal Tableau Search Heuristics

Michał Sochański(✉), Dorota Leszczyńska-Jasion(✉), Szymon Chlebowski, Agata Tomczyk, and Marcin Jukiewicz

Adam Mickiewicz University, ul. Wieniawskiego 1, 61-712 Poznań, Poland
{Michal.Sochanski,Dorota.Leszczynska,Szymon.Chlebowski, Agata.Tomczyk,Marcin.Jukiewicz}@amu.edu.pl

Abstract. We discuss the results of our work on heuristics for generating minimal synthetic tableaux. We present this proof method for classical propositional logic and its implementation in Haskell. Based on mathematical insights and exploratory data analysis we define heuristics that allows building a tableau of optimal or nearly optimal size. The proposed heuristics has been first tested on a data set with over 200,000 short formulas (length 12), then on 900 formulas of length 23. We describe the results of data analysis and examine some tendencies. We also confront our approach with the pigeonhole principle.

Keywords: Synthetic tableau · Minimal tableau · Data analysis · Proof-search heuristics · Haskell · Pigeonhole principle

1 Introduction

The method of *synthetic tableaux* (ST, for short) is a proof method based entirely on direct reasoning but yet designed in a tableau format. The basic idea is that all the laws of logic, and only laws of logic, can be derived directly by cases from parts of some partition of the whole logical space. Hence an ST-proof of a formula typically starts with a division between 'p-cases' and '$\neg p$-cases' and continues with further divisions, if necessary. Further process of derivation consists in applying the so-called *synthesizing* rules that build complex formulas from their parts—subformulas and/or their negations. For example, if p holds, then every implication with p in the succedent holds, '$q \rightarrow p$' in particular; then also '$p \rightarrow (q \rightarrow p)$' holds by the same argument. If $\neg p$ is the case, then every implication with p in the antecedent holds, thus '$p \rightarrow (q \rightarrow p)$' is settled. This kind of reasoning *proves* that '$p \rightarrow (q \rightarrow p)$' holds in every possible case (unless we reject *tertium non datur* in the partition of the logical space). There are no indirect assumptions, no *reductio ad absurdum*, no assumptions that need to be discharged. The ST method needs no labels, no derivation of a normal form (clausal form) is required.

This work was supported financially by National Science Centre, Poland, grant no 2017/26/E/HS1/00127.

J. Blanchette et al. (Eds.): IJCAR 2022, LNAI 13385, pp. 407–425, 2022.
https://doi.org/10.1007/978-3-031-10769-6_25

In the case of Classical Propositional Logic (CPL, for short) the method may be viewed as a formalization of the truth-tables method. The assumption that p amounts to considering all Boolean valuations that make p true; considering $\neg p$ exhausts the logical space. The number of cases to be considered corresponds to the number of branches of an ST, and it clearly depends on the number of distinct propositional variables in a formula, thus the upper bound for complexity of an ST-search is the complexity of the truth-tables method. In the worst case this is exponential with respect to the number of variables, but for some classes of formulas truth-tables behave better than standard analytic tableaux (see [4–7] for this diagnosis). However, the method of ST can perform better than truth-tables, as shown by the example of '$p \to (q \to p)$', where we do not need to partition the space of valuations against the $q/\neg q$ cases.[1] The question, obviously, is *how much* better? The considerations presented in this paper aim at developing a *quasi*-experimental framework for answering it.

The ST method was introduced in [19], then extended to some non-classical logics in [20,22]. An adjustment to the first-order level was presented in [14]. There were also interesting applications of the method in the domain of abduction: [12,13]. On the propositional level, the ST method is both a proof- and model-checking method, which means that one can examine satisfiability of a formula A (equivalently, validity of $\neg A$) and its falsifiability (equivalently, inconsistency of $\neg A$) at the same time. Normally, one needs to derive a clausal form of both A and $\neg A$ to check the two dual semantic cases (satisfiability and validity) with one of the quick methods, while the ST-system is designed to examine both of them. Wisely used, this property can contribute to limiting the increase in complexity in verification of semantic properties.

For the purpose of optimization of the ST method we created a heuristics that leads to construction of a variable ordering—a task similar to the one performed in research on Ordered Binary Decision Diagrams (OBDDs), and, generally, in Boolean satisfiability problem (SAT) [8,15]. In Sect. 3 we sketch a comparison of STs to OBDDs. Let us stress at this point, however, that the aim of our analysis remains proof-theoretical—the ST method is a 'full-blooded' proof method working on formulas of arbitrary representation. It was already adjusted to first-order and to some non-classical logics, and has a large scope of applications beyond satisfiability checking of clausal forms.

The optimization methods that we present are based on exploratory data analysis performed on millions of tableaux. Some aspects of the analysis are also discussed in the paper. The data are available on https://ddsuam.wordpress.com/software-and-data/.

Here is a plan of what follows. The next section introduces the ST method, Sect. 3 compares STs with analytic tableaux and with BDDs, and Sect. 4 presents the implementation in Haskell. In Sect. 5 we introduce the mathematical concepts

---

[1] On a side note, it is easy to show that the ST system is polynomially equivalent to system **KE** introduced in [4], as both systems contain cut. What is more, there is a strict analogy between the ST method and the inverse method (see [4,16]). The relation between ST and **KI** was examined by us in detail in Sect. 2 of [14].

needed to analyse heuristics of small tableaux generation. In Sect. 6 we describe the analysed data, and in Sect. 7—the obtained results. Section 8 confronts our approach with the pigeonhole principle, and Sect. 9 indicates plans for further research.

2 The Method of Synthetic Tableaux

Language. Let $\mathcal{L}_{\mathsf{CPL}}$ stand for the language of CPL with negation, $\neg$, and implication, $\rightarrow$. $\mathsf{Var} = \{p, q, r, \ldots, p_i, \ldots\}$ is the set of propositional variables and 'Form' stands for the set of all formulas of the language, where the notion of formula is understood in a standard way. $A, B, C \ldots$ will be used for formulas of $\mathcal{L}_{\mathsf{CPL}}$. Propositional variables and their negations are called *literals. Length* of a formula A is understood as the number of occurrences of characters in A, parentheses excluded.

Let $A \in \mathsf{Form}$. We define the notion of a *component* of A as follows. (i) A is a component of A. (ii) If A is of the form '$\neg\neg B$', then B is a component of A. (iii) If A is of the form '$B \rightarrow C$', then '$\neg B$' and C are components of A. (iv) If A is of the form '$\neg(B \rightarrow C)$', then B and '$\neg C$' are components of A. (v) If C is a component of B and B is a component of A, then C is a component of A. (vi) Nothing else is a component of A. By '$\mathsf{Comp}(A)$' we mean the set of all components of A. For example, $\mathsf{Comp}(\, p \rightarrow (q \rightarrow p)\,) = \{p \rightarrow (q \rightarrow p), \neg p, q \rightarrow p, \neg q, p\}$. As we can see, *component of a formula* is not the same as *subformula of a formula*; $\neg q$ is not a subformula of the law of antecedent, q is, but it is not its component. Components refer to *uniform notation* as defined by Smullyan (see [18]) which is very convenient to use with a larger alphabet. Let us also observe that the association of $\mathsf{Comp}(A)$ with a Hintikka set is quite natural, although $\mathsf{Comp}(A)$ need not be consistent. In the sequel we shall also use '$\mathsf{Comp}^{\pm}(A)$' as a short for '$\mathsf{Comp}(A) \cup \mathsf{Comp}(\neg A)$'.

Rules. The system of ST consists of the set of rules (see Table 1) and the notion of proof (see Definition 2). The rules can be applied in the construction of an ST for a formula A on the proviso that (a) the premises already occur on a given branch, (b) the conclusion (conclusions, in the case of (*cut*)) of a particular application of the rule belongs (both belong) to $\mathsf{Comp}^{\pm}(A)$. The only branching rule, called (*cut*) by analogy to its famous sequent-calculus formulation, is at the same time the only rule that needs no premises, hence every ST starts with an application of this rule. If its application creates branches with p_i and $\neg p_i$, then we say that the rule was *applied with respect to* p_i.

One of the nice properties of this method is that it is easy to keep every branch *consistent*: it is sufficient to restrict the applications of (*cut*), so that on every branch (*cut*) is applied with respect to a given variable p_i at most once. This warrants that $p_i, \neg p_i$ never occur together on the same branch.

The notion of a proof is formalized by that of a tree. If $\mathcal{T}$ is a labelled tree, then by $X_{\mathcal{T}}$ we mean the set of its nodes, and by $r_{\mathcal{T}}$ we mean its root. Moreover, $\eta_{\mathcal{T}}$ is used for a function assigning labels to the nodes in $X_{\mathcal{T}}$.

Table 1. Rules of the ST system for $\mathcal{L}_{\mathsf{CPL}}$

$(\mathbf{r}^1_{\rightarrow})$	$(\mathbf{r}^2_{\rightarrow})$	$(\mathbf{r}^3_{\rightarrow})$	$(\mathbf{r}_{\neg})$	(cut)
$\dfrac{\neg A}{A \rightarrow B}$	$\dfrac{B}{A \rightarrow B}$	$\dfrac{A \quad \neg B}{\neg(A \rightarrow B)}$	$\dfrac{A}{\neg\neg A}$	$p_i \quad \mid \quad \neg p_i$

Definition 1 (synthetic tableau). *A synthetic tableau for a formula A is a finite labelled tree $\mathcal{T}$ generated by the above rules, such that* $\eta_{\mathcal{T}} : X \backslash \{r_{\mathcal{T}}\} \longrightarrow \mathsf{Comp}^{\pm}(A)$ *and each leaf is labelled with A or with ¬A.*

$\mathcal{T}$ is called consistent *if the applications of (cut) are subject to the restriction defined above: there are no two applications of (cut) on the same branch with respect to the same variable.*

$\mathcal{T}$ is called regular *provided that literals are introduced in the same order on each branch, otherwise $\mathcal{T}$ is called* irregular.

Finally, $\mathcal{T}$ is called canonical, *if, first, it is consistent and regular, and second, it starts with an introduction of all possible literals by (cut) and only after that the other rules are applied on the created branches.*

In the above definition we have used the notion of *literals introduced in the same order on each branch.* It seems sufficiently intuitive at the moment, so we postpone the clarification of this notion until the end of this section.

Definition 2 (proof in ST system). *A synthetic tableau $\mathcal{T}$ for a formula A is a* proof of *A* in the ST system *iff each leaf of $\mathcal{T}$ is labelled with A.*

Theorem 1. (soundness and completeness, see [21]**).** *A formula A is valid in* CPL *iff A has a proof in the ST-system.*

Example 1. Below we present two different STs for one formula: $B = p \rightarrow (q \rightarrow p)$. Each of them is consistent and regular. Also, each of them is a proof of the formula in the ST system.

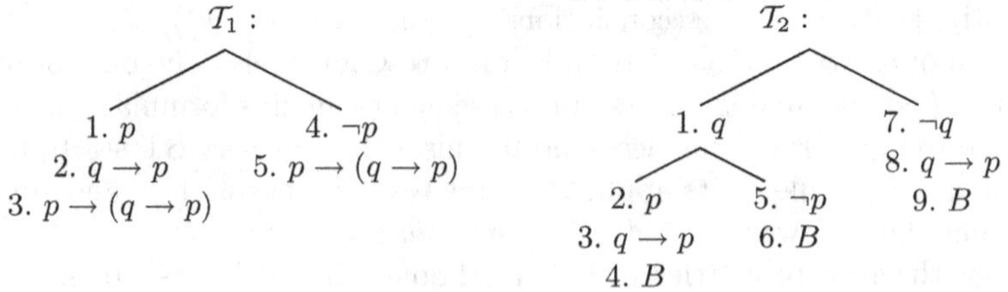

In $\mathcal{T}_1$: 2 comes from 1 by $\mathbf{r}^2_{\rightarrow}$, similarly 3 comes from 2 by $\mathbf{r}^2_{\rightarrow}$. 5 comes from 4 by $\mathbf{r}^1_{\rightarrow}$. In $\mathcal{T}_2$: nothing can be derived from 1, hence the application of (*cut*) wrt p is the only possible move. The numbering of the nodes is not part of the ST.

There are at least two important size measures used with respect to trees: the number of nodes and the number of branches. As witnessed by our data, there is a very high overall correlation between the two measures, we have thus used only one of them—the number of branches—in further analysis. Among various STs for the same formula there can be those of smaller, and those of bigger size. An ST of a minimal size is called *optimal*. In the above example, $\mathcal{T}_1$ is an optimal ST for B. Let us also observe that there can be many STs for a formula of the same size, in particular, there can be many optimal STs.

Example 2. Two possible canonical synthetic tableaux for $B = p \to (q \to p)$. Each of them is regular, consistent, but clearly not optimal (*cf.* $\mathcal{T}_1$).

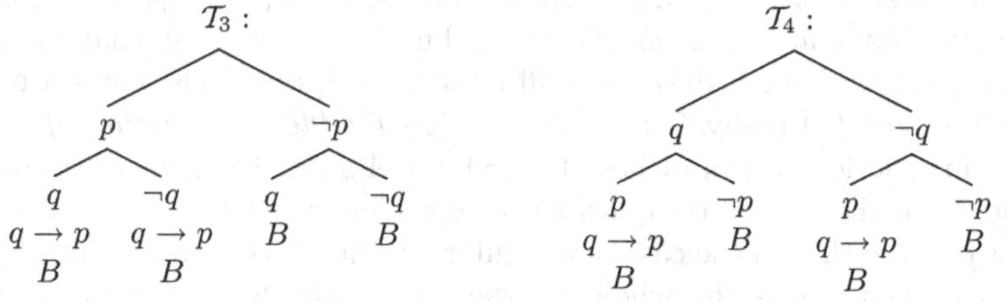

In the case of formulas with at most two distinct variables regularity is a trivial property. Here comes an example with three variables.

Example 3. $\mathcal{T}_5$ is an irregular ST for formula $C = (p \to \neg q) \to \neg(r \to p)$, i.e. variables are introduced in various orders on different branches. $\mathcal{T}_6$ is an example of an inconsistent ST for C, i.e. there are two applications of (*cut*) on one branch with respect to p, which results in a branch carrying both p and $\neg p$ (the blue one). The whole right subtree of $\mathcal{T}_5$, starting with $\neg p$, is repeated twice in $\mathcal{T}_6$, where it is symbolized with letter $\mathcal{T}^*$. Let us observe that $\neg\neg(r \to p)$ is a component of $\neg C$ due to clause (iv) defining the concept of component.

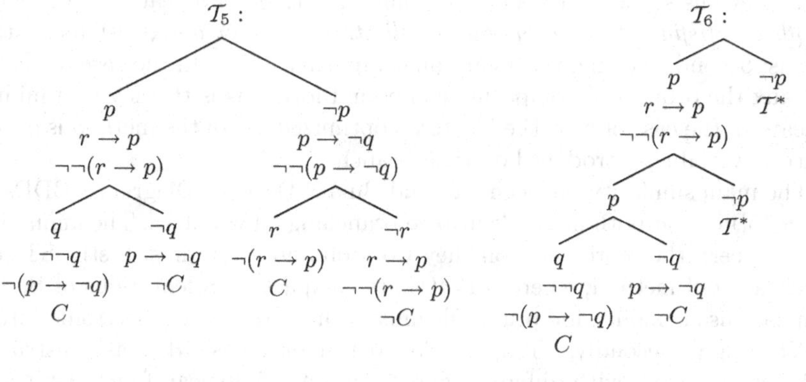

On the level of CPL we can use only consistent STs while still having a complete calculus (for details see [19,21]). An analogue of closing a branch of an analytic tableau for formula A is, in the case of an ST, ending a branch with

A synthesized. And the fact that an ST for A has a consistent branch ending with $\neg A$ witnesses satisfiability of $\neg A$. The situation concerning consistency of branches is slightly different, however, in the formalization of first-order logic presented in [14], as a restriction of the calculus to consistent STs produces an incomplete formalization.

Finally, let us introduce some auxiliary terminology to be used in the sequel. Suppose $\mathcal{T}$ is an ST for a formula A and $\mathcal{B}$ is a branch of $\mathcal{T}$. Literals occur on $\mathcal{B}$ in an order set by the applications of (cut), suppose that it is $\langle \pm p_1, \ldots, \pm p_n \rangle$, where '$\pm$' is a negation sign or no sign. In this situation we call sequence $o = \langle p_1, \ldots, p_n \rangle$ the *order on* $\mathcal{B}$. It can happen that o contains *all* variables that occur in A, or that some of them are missing. Suppose that $q_1, \ldots, q_m$ are all of (and only) the distinct variables occurring in A. Each permutation of $q_1, \ldots, q_m$ will be called *an instruction for a branch of an ST for A*. Further, we will say that the order o on $\mathcal{B}$ *complies with an instruction* I iff either $o = I$, or o constitutes a proper initial segment of I. Finally, $\mathcal{I}$ is an *instruction for the construction of* $\mathcal{T}$, if $\mathcal{I}$ is a set of instructions for branches of an ST for A such that for each branch of $\mathcal{T}$, the order on the branch complies with some element of $\mathcal{I}$.

Let us observe that in the case of a regular ST the set containing one instruction for a branch makes the whole instruction for the ST, as the instruction describes all the branches. Let us turn to examples. $\mathcal{T}_5$ from Example 3 has four branches with the following orders (from the left): $\langle p, q \rangle$, $\langle p, q \rangle$, $\langle p, r \rangle$, $\langle p, r \rangle$. On the other hand, there are six permutations of p, q, r, and hence six possible instructions for branches of an arbitrary ST for the discussed formula. Order $\langle p, q \rangle$ complies with instruction $\langle p, q, r \rangle$, and order $\langle p, r \rangle$ complies with instruction $\langle p, r, q \rangle$. The set $\{\langle p, q, r \rangle, \langle p, r, q \rangle\}$ is an instruction for the construction of an ST for C, more specifically, it is an instruction for the construction of $\mathcal{T}_5$.

3 ST, Analytic Tableaux, BDDs, and SAT Solvers

The analogy between STs and analytic tableaux sketched in the last paragraph of the previous section breaks in two points. First, let us repeat: the ST method is *both a satisfiability checker and a validity checker at once*, just like a truth table is. Second, the analogy breaks on complexity issues. In the case of analytic tableaux the order of decomposing compound formulas is the key to a minimal tableau. In the case of STs, the key to an optimized use of the method is a clever choice of variables introduced on each branch.

The main similarity between STs and Binary Decision Diagrams (BDDs, see e.g. [8,15]) is that both methods involve branching on variables. The main differences concern the representation they work on and their aims: firstly, STs constitute a proof method, whereas BDDs are compact representations of Boolean formulas, used mainly for practical aims such as design of electronic circuits (VLSI design); secondly, ST applies to logical formulas, whereas construction of BDDs may start with different representations of Boolean functions, usually circuits or Boolean formulas.

The structure of the constructed tree is also slightly different in the two approaches: in BDDs the inner nodes correspond to variables with outgoing edges labelled with 1 or 0; in STs, on the other hand, inner nodes are labelled with literals or more complex formulas. The terminal nodes of a BDD (also called sinks, labelled with 1 or 0) indicate the value of a Boolean function calculated for the arguments introduced along the path from the root, whereas the leaves of an ST carry a synthesized formula (the initial one or its negation). In addition to that, the methods differ in terms of the construction process: in case of BDDs, tree structures are first generated and then reduced to a more compact form using the elimination and merging rules; the STs, in turn, are built 'already reduced'. However, the interpretation of the outcome of both constructions is analogous. Firstly, for a formula A with n distinct variables $p_1, \ldots, p_n$ and the associated Boolean function $f_A = f_A(x_1, \ldots, x_n)$, the following fact holds: If a branch of an ST containing literals from a set L ends with A or $\neg A$ synthesized (which means that assuming that the literals from L are true is sufficient to calculate the value of A), then the two mentioned reduction rules can be used in a BDD for f_A, so that the route that contains the variables occurring in L followed by edges labelled according to the signs in L can be directed to a terminal node (sink). For example, if A can be synthesized on a branch with literals $\neg p_1$, p_2 and $\neg p_3$, then $f_A(0, 1, 0, x_4, \ldots, x_n) = 1$ for all values of the variables $y \in \{x_4, \ldots, x_n\}$ and so the route in the associated BDD containing the variables x_1, x_2 and x_3 followed by the edges labelled with 0, 1 and 0, respectively, leads directly to the sink labelled with 1.

However, possibility of applying the reduction procedures for a BDD does not always correspond to the possibility of reducing an ST. For example, the reduced BDD for formula $p \vee (q \wedge \neg q)$ consists of the single node labelled with p with two edges directed straight to the sinks 1 and 0; on the other hand, construction of an ST for the formula requires introducing q following the literal $\neg p$. This observation suggests that ST, in general, have greater size than the reduced BDDs.

Strong similarity of the two methods is also illustrated by the fact that they both allow the construction of a disjunctive normal form (DNF) of the logical or Boolean formula to which they were applied. In the case of ST, DNF is the disjunction of conjunctions of literals that appear on branches finished with the formula synthesized. The smaller the ST, the smaller the DNF. Things are analogous with BDDs.

Due to complexity issues, research on BDDs centers on ordered binary decision diagrams (OBDDs), in which different variables appear in the same order on all paths from the root. A number of heuristics have been proposed in order to construct a variable ordering that will lead to the smallest OBDDs, using characteristics of the different types of representation of Boolean function (for example, for circuits, topological characteristics have been used for that purpose). OBDDs are clearly analogous to regular STs, the construction of which also requires finding a good variable ordering, leading to a smaller ST. We suppose that our methodology can also be used to find orderings for OBDDs by expressing Boolean functions as logical formulas. It is not clear to us whether the OBDDs methodology can be used in our framework.

Let us move on to other comparisons, this time with a lesser degree of detail. It is very instructive to compare the ST method to SAT-solvers, as their effectiveness is undeniably impressive nowadays[2]. The ST method does not aim at challenging this effectiveness. Let us explain, however, in what aspect the ST method can still be viewed as a computationally attractive alternative to a SAT solver. The latter produces an answer to question about satisfiability, sometimes producing also examples of satisfying valuations and/or counting the satisfying valuations. In order to obtain an answer to another question—that about validity—one needs to ask about satisfiability of the initial problem negated. As we stressed above, the ST method answers the two questions at once, providing at the same time a description of classes of valuations satisfying and not satisfying the initial formula. Hence one ST is worth two SAT-checks together with a rough model counting.

Another interesting point concerns clausal forms. The method of ST does not require derivation of clausal form, but the applications of the rules of the system, defined via α-, β-notation, reflects the breaking of a formula into its components, and thus, in a way, leads to a definition of a normal form (a DNF, as we mentioned above). But this is not to say that an ST needs a full conversion to DNF. In this respect the ST method is rather similar to non-clausal theorem provers (*e.g.* non-clausal resolution, see [9,17]).

Let us finish this section with a summary of the ST method. Formally, it is a proof method with many applications beyond the realm of CPL. In the area of CPL, semantically speaking, it is both satisfiability and validity checker, displaying semantic properties of a formula like a truth table does, but amenable to work more efficiently (in terms of the number of branches) than the latter method. The key to this efficiency is in the order of variables introduced in an ST. In what follows we present a method of construction of such variable orders and examine our approach in an experimental setting.

4 Implementation

The main functionality of the implementation described in this section is a construction of an ST for a formula according to an instruction provided by the user. If required, it can also produce *all possible instructions* for a given formula and build all STs according to them. In our research we have mainly used the second possibility.

The implemented algorithm generates non-canonical, possibly irregular STs. Let us start with some basics. There are three main datatypes employed. Standard, recursively defined formula type, `For`, used to represent propositional formulas; Monad `Maybe Formula`, `MF`, consisting of `Just Formula` and `Nothing`—used to express the fact that the synthesis of a given formula on a given branch was successful (`Just`) or not (`Nothing`). To represent an ST we use type of trees imported from `Data.Tree`. Thus every ST can be represented as `Tree [MF]`

[2] See [23, p. 2021]: *contemporary SAT solvers can often handle practical instances with millions of variables and constraints.*

`[Tree [MF]]`, that is a tree labelled by lists of `MF`. We employed such a general structure having in mind possible extensions to non-classical logics (for CPL a binary tree is sufficient). The algorithm generating all possible ST for a given formula consists of the following steps:

1. We start by performing a few operations on the goal-formula A:
   (a) a list of all components of A and all components of $\neg A$, and a separate list of the variables occurring in A (`atoms A`) is generated;
   (b) the first list is sorted in such a way that all components of a given formula in that list precede it (`sort A`).
2. After this initial step, based on the list `atoms A`, all possible instructions for the construction of an ST for A are generated (`allRules (atoms A)`).
3. For each instruction from `allRules (atoms A)` we build an ST using the following strategy, called 'compulsory':
   (a) after each introduction of a literal (by (*cut*)) we try to synthesize (by the other rules) as many formulas from `sort A` as possible;
   (b) if no synthesizing rule is applicable we look into the instruction to introduce an appropriate literal and we go back to (a). Let us note that $\mathcal{T}_1$, $\mathcal{T}_2$, $\mathcal{T}_5$ are constructed according to this strategy.
4. Lastly, we generate a `CSV` file containing some basic information about each generated tree: *int.al.* the number of nodes and whether the tree is a proof.

Please observe that the length of a single branch is linear in the size of a formula; this follows from the fact that `sort A` contains only the components of A. On the other hand, an 'outburst' of computational complexity enters on the level of the number of STs. In general, if k is the number of distinct variables in a formula A, then for $k = 3$ there are 12 different canonical STs, for $k = 4$ and $k = 5$ this number is, respectively, 576 and 1,688,800. In the case of $k = 6$ the number of canonical STs *per* formula exceeds 10^{12} and this approach is no longer feasible[3].

The Haskell implementation together with necessary documentation is available on https://ddsuam.wordpress.com/software-and-data/.

5 dp-Measure and the Rest of Our Toolbox

As we have already observed, in order to construct an optimal ST for a given formula one needs to make a clever choice of the literals to start with. The following function was defined to facilitate the smart choices. It assigns a rational value from the interval $\langle 0; 1\rangle$ to each occurrence of a literal in a syntactic tree for

[3] It can be shown (*e.g.* by mathematical induction) that for formulas with k different variables, the total number of canonical STs is given by the following explicit formula:

$$\prod_{i=1}^{k} (k - i + 1)^{2^{i-1}}.$$

formula A (in fact, it assigns the values to all elements of $\mathsf{Comp}(A)$). Intuitively, the value reflects the *derivative power* of the literal in synthesizing A.

The first case of the equation in Definition 3 is to make the function full (=total) on $\mathsf{Form} \times \mathsf{Form}$, it also corresponds with the intended meaning of the defined measure: if $B \notin \mathsf{Comp}(A)$, then B is of no use in deriving A. The second case expresses the starting point: to calculate the values of $dp(A, B)$ for atomic B, one needs to assign $1 = dp(A, A)$; then the value is propagated down along the branches of a formula's syntactic tree. Dividing the value a by 2 in the fourth line reflects the fact that both components of an α-formula are needed to synthesize the formula. In order to use the measure, we need to calculate it for both A and $\neg A$; this follows from the fact that we do not know whether A or $\neg A$ will be synthesized on a given branch.

Definition 3. $dp : \mathsf{Form} \times \mathsf{Form} \longrightarrow \langle 0; 1\rangle$

$$dp(\, A, B\,) = \begin{cases} 0 & \textit{if } B \notin \mathsf{Comp}(A), \\ 1 & \textit{if } B = A, \\ a & \textit{if } dp(\, A, \neg\neg B\,) = a, \\ \frac{a}{2} & \textit{if } B \in \{C, \neg D\} \textit{ and } dp(\, A, \neg(C \to D)\,) = a, \\ a & \textit{if } B \in \{\neg C, D\} \textit{ and } dp(\, A, C \to D\,) = a. \end{cases}$$

Example 4. A visualization of calculating dp for formulas B, C from Examples 2, 3 and for $D = (p \to \neg p) \to p$.

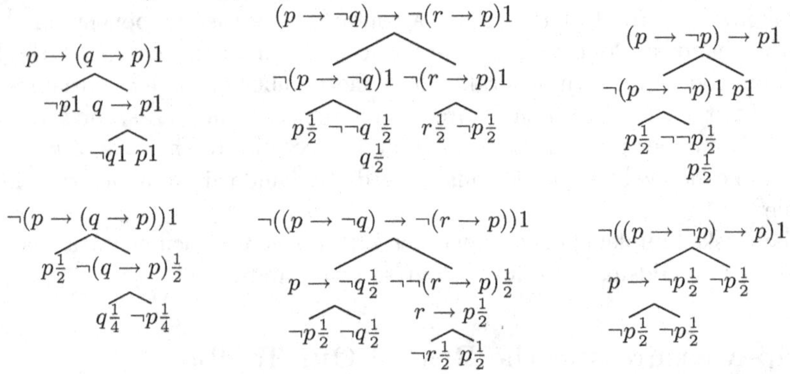

As one can see from Example 4, the effect of applying the dp measure to a formula and its negation is a number of values that need to be aggregated in order to obtain a clear instruction for an ST construction. However, some conclusions can be drawn already from the above example. It seems clear that the value $dp(\, p \to (q \to p), p\,) = 1$ corresponds to the fact that p is sufficient to synthesize the whole formula (as witnessed by $\mathcal{T}_1$, see Example 1). So is the case with $\neg p$. On the other hand, even if $\neg q$ is sufficient to synthesize the formula, q is not (see $\mathcal{T}_2$, Example 1), hence the choice between p and q is plain. But it seems to be the only obvious choice at the moment. In the case of the second formula, every literal gets the same value: 0.5. What is more, in the case of longer formulas a situation depicted by the rightmost syntactic trees is very likely to happen:

we obtain $dp(D, p) = 0.5$ *twice* (since dp works on *occurrences* of literals), and $dp(\neg D, \neg p) = 0.5$ *three times.*

In the aggregation of the dp-values we use the parametrised Hamacher s-norm, defined for $a, b \in \langle 0; 1 \rangle$ as follows:

$$a\, \mathbf{s}_\lambda\, b = \frac{a + b - ab - (1 - \lambda)ab}{1 - (1 - \lambda)ab}$$

for which we have taken $\lambda = 0.1$, as the value turned out to give the best results. Hamacher s-norm can be seen as a fuzzy alternative; it is commutative and associative, hence it is straightforward to extend its application to an arbitrary finite number of arguments. For $a = b = c = 0.5$ we obtain:

$$a\, \mathbf{s}_\lambda\, b \approx 0.677, \quad \text{and } (a\, \mathbf{s}_\lambda\, b)\, \mathbf{s}_\lambda\, c \approx 0.768$$

The value of this norm is calculated for a formula A and a literal l by taking the dp-values $dp(A, l)$ for each occurrence l in the syntactic tree of A. This value will be denoted as '$h(A, l)$'; in case there is only one value $dp(A, l)$, we take $h(A, l) = dp(A, l)$. Hence, referring to the above Example 4, we have *e.g.* $h(B, p) = 1$, $h(\neg B, \neg p) = 0.25$, $h(\neg D, \neg p) \approx 0.768$.

Finally, function H is defined for *variables*, not their occurrences, in formula A as follows:

$$H(A, p_i) = \frac{\max(h(A, p_i), h(\neg A, p_i)) + \max(h(A, \neg p_i), h(\neg A, \neg p_i))}{2}$$

The important property of this apparatus is that for $a, b < 1$ we have $a\, \mathbf{s}_{0.1}\, b > \max\{a, b\}$, and thus $h(A, l)$ and $H(A, p_i)$ are sensitive to the number of aggregated elements. Another desirable feature of the introduced functions is that $h(A, p_i) = 1$ indicates that one can synthesize A on a branch starting with p_i without further applications of (*cut*); furthermore, $H(A, p_i) = 1$ indicates that both p_i and $\neg p_i$ have this property.

Let us stress that the values of dp, h and H are very easy to calculate. Given a formula A, we need to assign a dp-value to each of its components, and the number of components is linear in the length of A. On the other hand, the information gained by these calculations is sometimes not sufficient. The assignment $dp(A, p_i) = 2^{-m}$ says only that A can be built from p_i and m other components of A, but it gives us no clue as to which components are needed. In Example 4, H works perfectly, as we have $H(B, p) = 1$ and $H(B, q) = 0.625$, hence H indicates the following instruction of construction of an ST: $\{\langle p, q \rangle\}$. Unfortunately, in the case of formula C we have $H(C, p) = H(C, q) = H(C, r) = 0.5$, hence a more sophisticated solution is needed.

6 Data

At the very beginning of the process of data generation we faced the following general problem: how to make any *conclusive* inferences about an infinite population (all Form) on the basis of finite data? Considering the methodological problems connected with applying classical statistical inference methods in this context, we limited our analysis to descriptive statistics, exploratory analysis and testing. To make this as informative as possible, we took a 'big data' approach: for every formula we generated all possible STs, differing in the order of applications of (*cut*) on particular branches. In addition to that, where it was feasible, we generated all possible formulas falling under some syntactical specifications. The approach is aimed at testing different optimisation methods as well as exploring data in search for patterns and new hypotheses. The knowledge gained in this way is further used on samples of longer formulas to examine tendencies.

From now on we use l for the length of a formula, k for the number of distinct variables occurring in a formula, and n for the number of all occurrences of variables (leaves, if we think of formulas as trees). On the first stage we examined a dataset containing all possible STs for formulas with $l = 12$ and $k \leqslant 4$. There are over 33 million of different STs already for these modest values; for larger k the data to analyse was simply too big. We generated 242,265 formulas, from which we have later removed those with $k \leqslant 2$ and/or $k = n$, as the results for them where not interesting. In the case of further datasets we also generated all possible STs, but the formulas were longer and they were randomly generated[4]. And so we considered (i) 400 formulas with $l = 23, k = 3$, (ii) 400 formulas with $l = 23, k = 4$, (iii) 100 formulas with $l = 23, k = 5$. In all cases $9 \leqslant n \leqslant 12$; this value is to be combined with the occurrences of negations in a formula—the smaller n, the more occurrences of negation.

Having all possible STs for a formula generated, we could simply check what is the optimal ST' size for this formula. The idea was to look for possible relations between, on the one hand, instructions producing the small STs, and, on the other hand, properties of formulas that are easy to calculate, like *dp* or numbers of occurrences of variables. The first dataset included only relatively small formulas; however, with all possible formulas of a given type available, it was possible *e.g.* to track various types of 'unusual' behaviour of formulas and all possible problematic issues regarding the optimisation methods, which could remain unnoticed if only random samples of formulas were generated. In case of randomly generated formulas the 'special' or 'difficult' types of formulas may not be tracked (as the probability of drawing them may be small), but instead we have an idea of an 'average' formula, or average behaviour of the optimisation methods. By generating all the STs, in turn, we gained access to full information not only about the regular but also irregular STs, which is the basis for indicating the set of optimal STs and the evaluation of the optimisation methods.

7 Data Analysis and a Discussion of Results

In this section we present some results of analyses performed on our data. The main purpose of the analyses is to test the effectiveness of the function H in terms

[4] The algorithm of generating random formulas is described in [11]. The author prepared also the Haskell implementation of the algorithm. See https://github.com/kiryk/random-for.

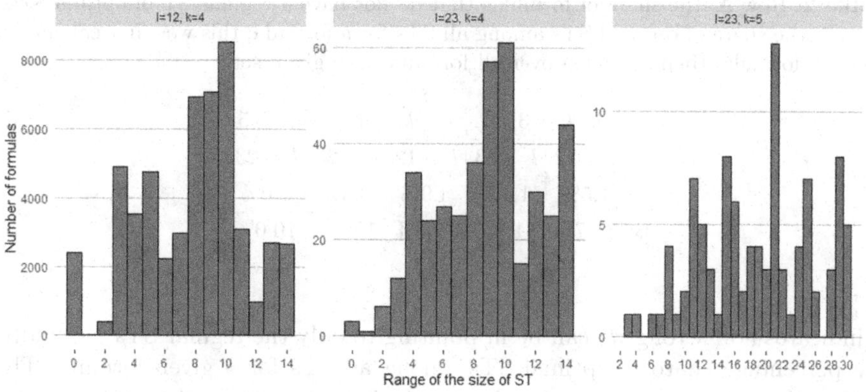

Fig. 1. Distribution of the difference between the size of a maximal and that of a minimal ST for formulas with $k = 4, 5$.

of indicating a small ST. Moreover, we performed different types of exploratory analysis on the data, aiming at understanding the variation of size among all STs for different formulas, and how it relates to the effectiveness of H.

Most results will be presented for the five combinations of the values of l and k in our data, that is, $l = 12, k \in \{3, 4\}$ and $l = 23, k \in \{3, 4, 5\}$; however, some results will be presented with the values of $k = 3$ and $k = 4$ grouped together (where the difference between them is insignificant) and the charts are presented only for $k \geqslant 4$.

We will examine the variation of size among STs using a range statistic: by *range of the size of ST for a formula A* (ST range, for short) we mean the difference between an ST of maximal and minimal size; this value indicates the possible room for optimization. The maximal-size ST is bounded by the size of a canonical ST for a given formula; its size depends only on k. For $k = 4$ a canonical ST has 16 branches, for $k = 5$ it is 32 branches.

The histograms on Fig. 1 present the distributions of ST range for formulas with $k = 4$ and $k = 5$. The rightmost bar in the histogram for $l = 23, k = 5$ says that for 5 (among 100) formulas there are STs with only two branches, where the maximal STs for these formulas have 32 branches. We can also read from the histograms that for formulas with $k = 4$ the ST range of some formulas is equal to 0 (7.9% of formulas with $l = 12$ and 3.5% with $l = 23$), which means that all STs have the same size. We have decided to exclude these formulas from the results of tests of efficiency of H, as the formulas leave no room for optimization. However, as can be seen on the histogram, there were no formulas of this kind among those with $k = 5$. This indicates that with the increase of k the internal differentiation of the set of STs for a formula increases as well, leading to a smaller share of formulas with small ST range.

Two more measures relating to the distribution of the size of ST may be of interest. Firstly, the share of formulas for which no regular ST is of optimal size—

Table 2. Row A: the share of formulas that do not have a regular ST of optimal size. Row B: the share of optimal STs among all STs for a formula; this was first calculated for each formula, then averaged over all formulas in a given set.

	$k = 3$		$k = 4$		$k = 5$
	$l = 12$	$l = 23$	$l = 12$	$l = 23$	$l = 23$
A	1.5%	1.1%	4.9%	3.3%	8.0%
B	31.7%	31.8%	17.3%	17.4%	10.0%

it indicates how wrong we can be in pointing to only the regular STs. Secondly, the percentage share of optimal STs among all STs for a given formula. The latter gives an idea what is the chance of picking an optimal ST at random. Table 2 presents both values for formulas depending on k and l (let us recall that formulas with ST range equal to 0 are excluded from the analysis). In both cases we can see clearly a tendency with growing k. As was to be expected, the table shows that the average share of optimal STs depends on the value of k rather than the size of the formula. This is understandable—as the number of branches depends on k only, the length of a formula translates to the length of branches, and the latter is linear in the former. In a way, this explains why the results are almost identical when the size of STs is calculated in terms of nodes rather than branches (as we mentioned above, the overall correlation between the two measures makes the choice between them irrelevant).

We can categorise the output of the function H into three main classes. In the first case, the values assigned to variables by H strictly order the variables, which results in one specific instruction of construction of a regular ST. The general score of such unique indications was very high: 70.9% for formulas with $l = 12$, 92.0% for $l = 23, k = 3, 4$, and 72.0% for $k = 5$. The second possibility is when H assigns the same value to each variable; in this case we gain no information at all (let us recall that we have excluded the only cases that could justify such assignments, that is, the formulas for which each ST is of the same size). The share of such formulas in our datasets was small: 0.6% for $l = 12$, 0.1% for $l = 23, k = 3, 4$ and 0% for $k = 5$, suggesting that it tends to fall with k rising. The third possibility is that the ordering is not strict, yet some information is gained. In this case for some, but not all, variables the value of H is the same.

The methodology used to asses effectiveness of H is quite simple. We assume that every indication must be a single regular instruction, hence we use additional criteria in case of formulas of the second and third kind described, in order to obtain a strict ordering. If H outputs the same value for some variables, we first order the variables by the number of occurrences in the formula; if the ordering is still not strict, we give priority to variables for which the sum of depths for all occurrences of literals in the syntactic tree is smaller; finally, where the above criteria do not provide a strict ordering, the order is chosen at random.

We used three evaluating functions to asses the quality of indications. Each function takes as arguments a formula and the ST for this formula indicated by

Table 3. The third column gives the number of formulas satisfying the characteristic presented in the first and the second column. The further three columns display values averaged on the sets. F_1 indicates how often we indicate an optimal ST. F_2 reports the mistake of our indication calculated as the difference of sizes between the indicated ST and an optimal one. Finally, POT indicates proximity to an optimal ST in a standardized way.

k	l	no of formulas	F_1	F_2	POT
3	12	113,190	0.935	0.089	0.974
	23	400	0.923	0.104	0.966
4	12	53,130	0.859	0.286	0.966
	23	400	0.836	0.297	0.960
5	23	100	0.75	0.52	0.971

our heuristics. The first function (F_1 in Table 3) outputs 1 if the indicated ST is of optimal size, 0 otherwise. The second function (F_2 in Table 3) outputs the difference between the size of the indicated ST and the optimal size. The third function is called *proximity to optimal tableau*, POT_A in symbols:

$$\mathsf{POT}_A(\mathcal{T}) = 1 - \frac{|\mathcal{T}| - min_A}{max_A - min_A}$$

where $\mathcal{T}$ is the ST for formula A indicated by H, $|\mathcal{T}|$ is the size of $\mathcal{T}$, max_A is the size of an ST for A of maximal size, and min_A is the size of an optimal ST for A. Later on we skip the relativization to A. Let us observe that the value $\frac{|\mathcal{T}|-min}{max-min}$ represents a mistake in indication relative to the ST range of a formula, and in this sense POT_A can be considered as a standardized measure of the quality of indication. Finally, values of each of the three evaluating functions were calculated for sets of formulas, by taking average values over all formulas in the set.

The results of the three functions presented in Table 3 show that optimal STs are indicated less often for formulas with greater k; however, the POT values seem to remain stable across all data, indicating that, on average, proximity of the indicated ST to the optimal ones does not depend on k or l.

Further analysis showed that the factor that most influenced the efficiency of our methodology was whether there is at least one value 1 among the *dp*-values of literals for a formula A. We shall write 'Max(dp) $= 1$' if this is the case, and 'Max(dp) < 1' otherwise (we skip the relativisation to A for simplicity). For formulas with Max(dp) $= 1$, results of the evaluating functions were much better; for example, the value of the POT function for formulas with $l = 12$ was 0.979 if Max(dp) $= 1$, and 0.814 for those with Max(dp) < 1; in case of formulas with $l = 23, k = 3, 4$ those values were 0.968 and 0.869, respectively, and for formulas with $l = 23, k = 5$ it was 0.974 and 0.901, respectively. This shows that our methodology works significantly worse if Max(dp) < 1; on the other hand, if Max(dp) $= 1$, the *dp* measure works very well. It should also be pointed

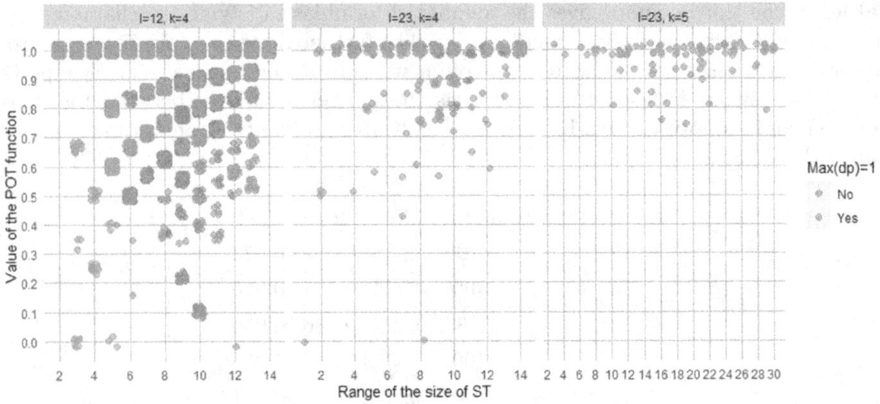

Fig. 2. Distribution of the difference between indicated and optimal ST in relation to ST-range. Every point corresponds to a formula, the points are slightly jittered in order to improve readibility. Each chart corresponds to different data, formulas $k = 3$ are excluded; additionally the colour indicates whether Max(dp) $= 1$ for a formula.

out that the difference between the POT values for both groups is smaller for formulas with greater l and k. Figure 2 presents a scatter plot that gives an idea of the whole distribution of the values of the POT function in relation to the ST range. Each formula on the plot is represented by a point, the colours additionally indicating whether Max(dp) < 1. The chart suggests, similarly as Table 3, that the method works well as the values of l and k rise for formulas, indicating STs that are on average equally close to the optimal ones.

One can point at two possible explanations of the fact that our methodology works worse for formulas with Max(dp) < 1. Firstly, if *e.g.*, $dp(A, p) = 2^{-m}$, we only obtain the information that, except for p, m more occurrences of components of A are required in order to synthesize the whole formula. Secondly, the function H neglects the complex dependencies between the various aggregated occurrences of a given variable, taking into account only the number of occurrences of literals in an aggregated group. However, considering very low computational complexity of the method based on the dp values and the function H, the outlined framework seems to provide good heuristics for indicating small STs. Methods that would reflect more aspects of the complex structure of logical formulas would likely require much more computational resources.

On a final note, we would like to add that exploration of the data allowed us to study properties of formulas that went beyond the scope of the optimisation of ST. The data was used in a similar way as in so called Experimental Mathematics, where numerous instances are analysed and visualized in order to *e.g.* gain insight, search for new patterns and relationships, test conjectures and introduce new concepts (see *e.g.* [1]).

Table 4. The pigeonhole principle

PHP_m			the size of ST		
m	k	l	indicated by H	minimal	canonical ST
1	2	7	3	3	4
2	6	34	15	11	64
3	12	90	99	43	4096
4	20	184	783	189	2^{20}

8 The Pigeonhole Principle

At the end we consider the propositional version of the principle introduced by Cook and Reckhow in [3, p. 43]. In the field of proof complexity the principle was used to prove that resolution is *intractable*, that is, any resolution proof of the propositional pigeonhole principle must be of exponential size (wrt the size of the formula). This has been proved by Haken in [10], see also[2].

Here is PHP_m in the propositional version:

$$\bigwedge_{0\leqslant i\leqslant m}\ \bigvee_{0\leqslant j<m} p_{i,j} \rightarrow \bigvee_{0\leqslant i<n\leqslant m}\ \bigvee_{0\leqslant j<m} (p_{i,j} \wedge p_{n,j})$$

where $\bigwedge$ and $\bigvee$ stand for generalized conjunction, disjunction (respectively) with the range indicated beneath.

The pigeonhole principle is constructed in a perfect symmetry of the roles played by the consecutive variables. Each variable has the same number of occurrences in the formula, and each of them gets the same value under H, they also have occurrences at the same depth of a syntactic tree. All this means that in our account we can only suggest a random, regular ST. However, it is worth noticing that, first, H behaves consistently with the structure of the formula, and second, the result is still attractive. In Table 4 the fourth column presents the size of the ST indicated by our heuristics, that is, in fact, generated by random ordering of variables. It is to be contrasted with the number 2^k in the last column describing the size of a canonical ST for the formula, which is at the same time the number of rows in a truth table for the formula. The minimal STs for the formulas were found with pen and paper and they are irregular.

9 Summary and Further Work

We presented a proof method of Synthetic Tableaux for CPL and explained how the efficiency of tableau construction depends on the choices of variables to apply (*cut*) to. We defined possible algorithms to choose the variables and experimentally tested their efficiency.

Our plan for the next research is well defined and it is to implement heuristics amenable to produce instructions for irregular STs. We have an algorithm, yet untested.

As far as proof-theoretical aims are concerned, the next task is to extend and adjust the framework to the first-order level based on the already described ST system for first-order logic [14]. We also wish to examine the efficiency of our indications on propositional non-classical logics for which the ST method exists (see [20,22]). In the area of data analysis another possible step would be to perform more complex statistical analysis using e.g. machine learning methods.

References

1. Borwein, J., Bailey, D.: Mathematics by Experiment: Plausible Reasoning in the 21st Century. A K Peters, Ltd., Natick (2004)
2. Buss, S.R.: Polynomial size proofs of the propositional pigeonhole principle. J. Symb. Log. **52**(4), 916–927 (1987)
3. Cook, S.A., Reckhow, R.A.: The relative efficiency of propositional proof systems. J. Symb. Log. **44**(1), 36–50 (1979)
4. D'Agostino, M.: Investigations into the complexity of some propositional calculi. Technical Monograph. Oxford University Computing Laboratory, Programming Research Group, November 1990
5. D'Agostino, M.: Are tableaux an improvement on truth-tables? Cut-free proofs and bivalence. J. Log. Lang. Comput. **1**, 235–252 (1992)
6. D'Agostino, M.: Tableau methods for classical propositional logic. In: D'Agostino, M., Gabbay, D.M., Hähnle, R., Posegga, J. (eds.) Handbook of Tableau Methods, pp. 45–123. Kluwer Academic Publishers (1999)
7. D'Agostino, M., Mondadori, M.: The taming of the cut. Classical refutations with analytic cut. J. Log. Comput. **4**(3), 285–319 (1994)
8. Ebendt, R., Fey, G., Drechsler, R.: Advanced BDD Opimization. Springer, Heidelberg (2005). https://doi.org/10.1007/b107399
9. Fitting, M.: First-Order Logic and Automated Theorem Proving, 2nd edn. Springer, New York (1996). https://doi.org/10.1007/978-1-4612-2360-3
10. Haken, A.: The intractability of resolution. Theoret. Comput. Sci. **39**, 297–308 (1985)
11. Kiryk, A.: A modified Korsh algorithm for random trees with various arity (manuscript) (2022)
12. Komosinski, M., Kups, A., Leszczyńska-Jasion, D., Urbański, M.: Identifying efficient abductive hypotheses using multi-criteria dominance relation. ACM Trans. Comput. Log. **15**(4), 1–20 (2014)
13. Komosinski, M., Kups, A., Urbański, M.: Multi-criteria evaluation of abductive hypotheses: towards efficient optimization in proof theory. In: Proceedings of the 18th International Conference on Soft Computing, Brno, Czech Republic, pp. 320–325 (2012)
14. Leszczyńska-Jasion, D., Chlebowski, S.: Synthetic tableaux with unrestricted cut for first-order theories. Axioms **8**(4), 133 (2019)
15. Meinel, C., Theobald, T.: Algorithms and Data Structures in VLSI Design. OBDD - Foundations and Applications, Springer, Heidelberg (1998). https://doi.org/10.1007/978-3-642-58940-9
16. Mondadori, M.: Efficient inverse tableaux. J. IGPL **3**(6), 939–953 (1995)
17. Murray, N.V.: Completely non-clausal theorem proving. Artif. Intell. **18**, 67–85 (1982)

18. Smullyan, R.M.: First-Order Logic. Springer, Berlin, Heidelberg, New York (1968). https://doi.org/10.1007/978-3-642-86718-7
19. Urbański, M.: Remarks on synthetic tableaux for classical propositional calculus. Bull. Sect. Log. **30**(4), 194–204 (2001)
20. Urbański, M.: Synthetic tableaux for Łukasiewicz' calculus Ł3. Logique Anal. (N.S.) **177–178**, 155–173 (2002)
21. Urbański, M.: Tabele syntetyczne a logika pytań (Synthetic Tableaux and the Logic of Questions). Wydawnictwo UMCS, Lublin (2002)
22. Urbański, M.: How to synthesize a paraconsistent negation. The case of CLuN. Logique Anal. **185–188**, 319–333 (2004)
23. Vizel, Y., Weissenbacher, G., Malik, S.: Boolean satisfiability solvers and their applications in model checking. Proc. IEEE **103**, 2021–2035 (2015)

Modal Logics

Paraconsistent Gödel Modal Logic

Marta Bílková[1], Sabine Frittella[2], and Daniil Kozhemiachenko[2](✉)

[1] The Czech Academy of Sciences, Institute of Computer Science, Prague, Czech Republic
bilkova@cs.cas.cz

[2] INSA Centre Val de Loire, Univ. Orléans, LIFO EA 4022, Bourges, France
{sabine.frittella,daniil.kozhemiachenko}@insa-cvl.fr

Abstract. We introduce a paraconsistent modal logic $\mathbf{K}\mathsf{G}^2$, based on Gödel logic with coimplication (bi-Gödel logic) expanded with a De Morgan negation $\neg$. We use the logic to formalise reasoning with graded, incomplete and inconsistent information. Semantics of $\mathbf{K}\mathsf{G}^2$ is two-dimensional: we interpret $\mathbf{K}\mathsf{G}^2$ on crisp frames with two valuations v_1 and v_2, connected via $\neg$, that assign to each formula two values from the real-valued interval $[0, 1]$. The first (resp., second) valuation encodes the positive (resp., negative) information the state gives to a statement. We obtain that $\mathbf{K}\mathsf{G}^2$ is strictly more expressive than the classical modal logic $\mathbf{K}$ by proving that finitely branching frames are definable and by establishing a faithful embedding of $\mathbf{K}$ into $\mathbf{K}\mathsf{G}^2$. We also construct a constraint tableau calculus for $\mathbf{K}\mathsf{G}^2$ over finitely branching frames, establish its decidability and provide a complexity evaluation.

Keywords: Constraint tableaux · Gödel logic · Two-dimensional logics · Modal logics

1 Introduction

People believe in many things. Sometimes, they even have contradictory beliefs. Sometimes, they believe in one statement more than in the other. However, if a person has contradictory beliefs, they are not bound to believe in anything. Likewise, believing in ϕ *strictly more than* in χ makes one believe in ϕ *completely*. These properties of beliefs are natural, and yet hardly expressible in the classical modal logic. In this paper, we present a two-dimensional modal logic based on Gödel logic that can formalise beliefs taking these traits into account.

Two-Dimensional Treatment of Uncertainty. Belnap-Dunn four-valued logic (BD, or First Degree Entailment—FDE) [4,16,34] can be used to formalise

The research of Marta Bílková was supported by the grant 22-01137S of the Czech Science Foundation. The research of Sabine Frittella and Daniil Kozhemiachenko was funded by the grant ANR JCJC 2019, project PRELAP (ANR-19-CE48-0006). This research is part of the MOSAIC project financed by the European Union's Marie Skłodowska-Curie grant No. 101007627.

J. Blanchette et al. (Eds.): IJCAR 2022, LNAI 13385, pp. 429–448, 2022.
https://doi.org/10.1007/978-3-031-10769-6_26

reasoning with both incomplete and inconsistent information. In BD, formulas are evaluated on the De Morgan algebra **4** (Fig. 1, left) where the four values $\{t, f, b, n\}$ encode the information available about the formula: true, false, both true and false, neither true nor false. b and n thus represent inconsistent and incomplete information, respectively. It is important to note that the values represent the available information about the statement, not its intrinsic truth or falsity. Furthermore, this approach essentially treats *evidence for* a statement (its positive support) as being independent of *evidence against* it (negative support) which allows to differentiate between 'absence of evidence' and the 'evidence of absence'. The BD negation $\neg$ then swaps positive and negative supports.

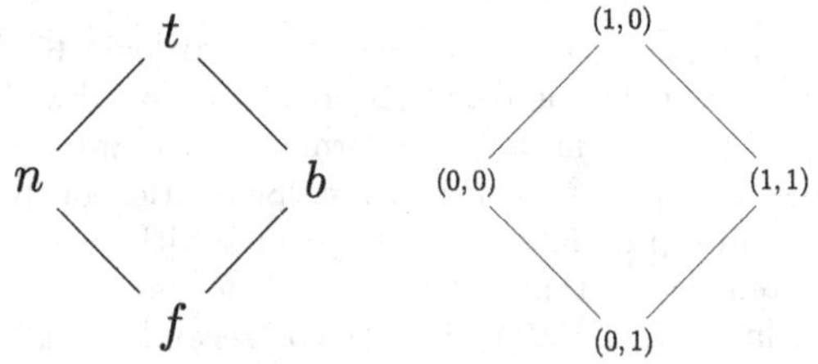

Fig. 1. **4** (left) and its continuous extension $[0,1]^{\bowtie}$ (right). $(x, y) \leq_{[0,1]^{\bowtie}} (x', y')$ iff $x \leq x'$ and $y \geq y'$.

The information regarding a statement, however, might itself be not crisp—after all, our sources are not always completely reliable. Thus, to capture the uncertainty, we extend **4** to the lattice $[0,1]^{\bowtie}$ (Fig. 1, right). $[0,1]^{\bowtie}$ is a twist product (cf, [37] for definitions) of $[0,1]$ with itself: the order on the second coordinate is reversed w.r.t. the order on the first coordinate. This captures the intuition behind the usual 'truth' (upwards) order: an agent is more certain in χ than in ϕ when the evidence for χ is stronger than the evidence for ϕ while the evidence against χ is weaker than the evidence against ϕ.

Note that $[0,1]^{\bowtie}$ is a bilattice whose left-to-right order can be interpreted as the information order. This links the logics we consider to bilattice logics applied to reasoning in AI in [19] and then studied further in [24,35].

Comparing Beliefs. Uncertainty is manifested not only in the non-crisp character of the information. An agent might often lack the capacity to establish the concrete numerical value that represents their certainty in a given statement. Indeed, 'I am 43% certain that the wallet is Paula's' does not sound natural. On the other hand, it is reasonable to assume that the agents' beliefs can be compared in most contexts: neither 'I am more confident that the wallet is Paula's than that the wallet is Quentin's', nor 'Alice is more certain than Britney that Claire loves pistachio ice cream' require us to give a concrete numerical representation to the (un)certainty.

These considerations lead us to choosing the two-dimensional relative of the Gödel logic dubbed G^2 as the propositional fragment of our logic. G^2 was introduced in [5] and is, in fact, an extension of Moisil's logic[1] from [31] with the prelinearity axiom $(p \to q) \vee (q \to p)$. As in the original Gödel logic G, the validity of a formula in G^2 depends not on the values of its constituent variables but on the relative order between them. In this sense, G is a logic of comparative truth. Thus, as we treat positive and negative supports of a given statement independently, G^2 is a logic of comparative truth and falsity. Note that while the values of two statements may not be comparable (say, p is evaluated as $(0.5, 0.3)$ and q as $(0, 0)$), the coordinates of the values always are. We will see in Sect. 2, how we can formalise statements comparing agents' beliefs.

The sources available to the agents as well as the references between these sources can be represented as states in a Kripke model and its accessibility relation, respectively. It is important to mention that we account for the possibility that a source can give us contradictory information regarding some statement. Still, we want our reasoning with such information to be non-trivial. This is reflected by the fact that $(p \wedge \neg p) \to q$ is not valid in G^2. Thus, the logic (treated as a set of valid formulas) lacks the explosion principle. In this sense, we call G^2 and its modal expansions 'paraconsistent'. This links our approach to other paraconsistent fuzzy logics such as the ones discussed in [17].

To reason with the information provided by the sources, we introduce two interdefinable modalities—$\Box$ and $\Diamond$—interpreted as infima and suprema w.r.t. the upwards order on $[0,1]^{\bowtie}$. We mostly assume (unless stated otherwise) that accessibility relations in models are crisp. Intuitively, it means that the sources are either accessible or not (and, likewise, either refer to the other ones, or not).

Broader Context. This paper is a part of the project introduced in [6] and carried on in [5] aiming to develop a modular logical framework for reasoning based on uncertain, incomplete and inconsistent information. We model agents who build their epistemic attitudes (like beliefs) based on information aggregated from multiple sources. $\Box$ and $\Diamond$ can be then viewed as two simple aggregation strategies: a pessimistic one (the infimum of positive support and the supremum of the negative support), and an optimistic one (the dual strategy), respectively. They can be defined via one another using $\neg$ in the expected manner: $\Box\phi$ stands for $\neg\Diamond\neg\phi$ and $\Diamond\phi$ for $\neg\Box\neg\phi$. In this paper, in contrast to [15] and [6], we do allow for modalities to nest.

The other part of our motivation comes from the work on modal Gödel logic ($\mathfrak{GK}$—in the notation of [36]) equipped with relational semantics [12,13, 36]. There, the authors develop proof and model theory of modal expansions of G interpreted over frames with both crisp and fuzzy accessibility relations. In particular, it was shown that the $\Box$-fragment[2] of $\mathfrak{GK}$ lacks the finite model property (FMP) w.r.t. fuzzy frames while the $\Diamond$-fragment has FMP[3] only w.r.t. fuzzy (but not crisp) frames. Furthermore, both $\Box$ and $\Diamond$ fragments of $\mathfrak{GK}$ are PSPACE-complete [28,29].

[1] This logic was introduced several times: by Wansing [38] as $\mathsf{I_4C_4}$ and then by Leitgeb [27] as HYPE. Cf. [33] for a recent and more detailed discussion.

[2] Note that $\Box$ and $\Diamond$ are not interdefinable in $\mathfrak{GK}$—cf. [36, Lemma 6.1] for details.

[3] There is, however, a semantics in [11] w.r.t. which bi-modal $\mathfrak{GK}$ has FMP.

Description Gödel logics, a notational version of modal logics, have found their use the field of knowledge representation [8–10], in particular, in the representation of vague or uncertain data which is not possible in the classical ontologies. In this respect, our paper provides a further extension of representable data types as we model not only vague reasoning but also non-trivial reasoning with inconsistent information.

In the present paper, we are expanding the language with the Gödel coimplication $\prec$ to allow for the formalisation of statements expressing that an agent is *strictly more confident* in one statement than in another one (cf. Sect. 2 for the details). Furthermore, the presence of $\neg$ will allow us to simplify the frame definability. Still, we will show that our logic is a conservative extension of $\mathfrak{G}\mathfrak{K}^c$—the modal Gödel logic of crisp frames from [36] in the language with both $\Box$ and $\Diamond$.

Logics. We are discussing many logics obtained from the propositional Gödel logic G. Our main interest is in the logic we denote $\mathbf{K}\mathsf{G}^2$. It can be produced from G in several ways: (1) adding De Morgan negation $\neg$ to obtain G^2 (in which case $\phi \prec \phi'$ can be defined as $\neg(\neg\phi' \rightarrow \neg\phi)$) and then further expanding the language with $\Box$ or $\Diamond$; (2) adding $\prec$ or Δ (Baaz' delta) to G, then both $\Box$ and $\Diamond$ thus acquiring $\mathbf{K}\mathsf{biG}$[4] (modal bi-Gödel logic) which is further enriched with $\neg$. These and other relations are given on Fig. 2.

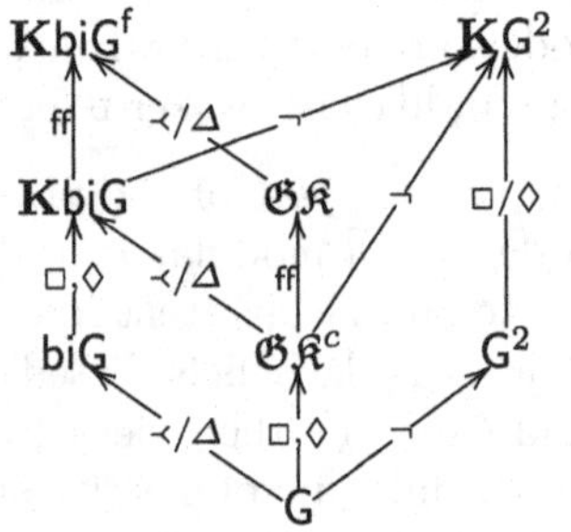

Fig. 2. Logics in the article. ff stands for 'permitting fuzzy frames'. Subscripts on arrows denote language expansions. / stands for 'or' and comma for 'and'.

Plan of the Paper. The remainder of the paper is structured as follows. In Sect. 2, we define bi-Gödel algebras and use them to present $\mathbf{K}\mathsf{biG}$ (on both fuzzy and crisp frames) and then $\mathbf{K}\mathsf{G}^2$ (on crisp frames), show how to formalise statements where beliefs of agents are compared, and prove some semantical properties. In Sect. 3, we show that $\Diamond$ fragment of $\mathbf{K}\mathsf{biG}^{\mathsf{f}}$ ($\mathbf{K}\mathsf{biG}$ on fuzzy frames) lacks finite model property. We then present a finitely branching fragment of $\mathbf{K}\mathsf{G}^2$ ($\mathbf{K}\mathsf{G}^2_{\mathsf{fb}}$) and argue for its use in representation of agents' beliefs. In Sect. 4, we design a constraint tableaux calculus for $\mathbf{K}\mathsf{G}^2_{\mathsf{fb}}$ which we use to obtain the complexity results. Finally, in Sect. 5 we discuss further lines of research.

[4] To the best of our knowledge, the only work on bi-Gödel (symmetric Gödel) modal logic is [20]. There, the authors propose an expansion of biG with $\Box$ and $\Diamond$ equipped with proof-theoretic interpretation and provide its algebraic semantics.

2 Language and Semantics

In this section, we present semantics for **K**biG (modal bi-Gödel logic) over both fuzzy and crisp frames and the one for $\mathbf{KG}^2$ over crisp frames. Let Var be a countable set of propositional variables. The language $\mathsf{bi}\mathcal{L}^{\neg}_{\Box,\Diamond}$ is defined via the following grammar.

$$\phi := p \in \mathsf{Var} \mid \neg\phi \mid (\phi \wedge \phi) \mid (\phi \vee \phi) \mid (\phi \rightarrow \phi) \mid (\phi \prec \phi) \mid \Box\phi \mid \Diamond\phi$$

Two constants, $\mathbf{0}$ and $\mathbf{1}$, can be introduced in the traditional fashion: $\mathbf{0} := p \prec p$, $\mathbf{1} := p \rightarrow p$. Likewise, the Gödel negation can be also defined as expected: $\sim\phi := \phi \rightarrow \mathbf{0}$. The $\neg$-less fragment of $\mathsf{bi}\mathcal{L}^{\neg}_{\Box,\Diamond}$ is denoted with $\mathsf{bi}\mathcal{L}_{\Box,\Diamond}$.

To facilitate the presentation, we introduce bi-Gödel algebras.

Definition 1. *The bi-Gödel algebra* $[0,1]_{\mathsf{G}} = ([0,1], 0, 1, \wedge_{\mathsf{G}}, \vee_{\mathsf{G}}, \rightarrow_{\mathsf{G}}, \prec_{\mathsf{G}})$ *is defined as follows: for all* $a, b \in [0,1]$, *the standard operations are given by* $a \wedge_{\mathsf{G}} b := \min(a,b)$, $a \vee_{\mathsf{G}} b := \max(a,b)$,

$$a \rightarrow_G b = \begin{cases} 1, & \text{if } a \leq b \\ b & \text{else,} \end{cases} \qquad b \prec_G a = \begin{cases} 0, & \text{if } b \leq a \\ b & \text{else.} \end{cases}$$

Definition 2.

- *A* fuzzy frame *is a tuple* $\mathfrak{F} = \langle W, R\rangle$ *with* $W \neq \varnothing$ *and* $R : W \times W \rightarrow [0,1]$.
- *A* crisp frame *is a tuple* $\mathfrak{F} = \langle W, R\rangle$ *with* $W \neq \varnothing$ *and* $R \subseteq W \times W$.

Definition 3 (KbiG models). *A* **K**biG model *is a tuple* $\mathfrak{M} = \langle W, R, v\rangle$ *with* $\langle W, R\rangle$ *being a (crisp or fuzzy) frame, and* $v : \mathsf{Var} \times W \rightarrow [0,1]$. v *(a valuation) is extended on complex* $\mathsf{bi}\mathcal{L}_{\Box,\Diamond}$ *formulas as follows:*

$$v(\phi \circ \phi', w) = v(\phi, w) \circ_{\mathsf{G}} v(\phi', w). \qquad (\circ \in \{\wedge, \vee, \rightarrow, \prec\})$$

The interpretation of modal formulas on fuzzy *frames is as follows:*

$$v(\Box\phi, w) = \inf_{w' \in W} \{wRw' \rightarrow_{\mathsf{G}} v(\phi, w')\}, \quad v(\Diamond\phi, w) = \sup_{w' \in W} \{wRw' \wedge_{\mathsf{G}} v(\phi, w')\}.$$

On crisp *frames, the interpretation is simpler (here,* $\inf(\varnothing) = 1$ *and* $\sup(\varnothing) = 0$*):*

$$v(\Box\phi, w) = \inf\{v(\phi, w') : wRw'\}, \qquad v(\Diamond\phi, w) = \sup\{v(\phi, w') : wRw'\}.$$

We say that $\phi \in \mathsf{bi}\mathcal{L}_{\Box,\Diamond}$ *is* **K**biG valid *on frame* $\mathfrak{F}$ *(denote,* $\mathfrak{F} \models_{\mathbf{K}\mathsf{biG}} \phi$*) iff for any* $w \in \mathfrak{F}$, *it holds that* $v(\phi, w) = 1$ *for any model* $\mathfrak{M}$ *on* $\mathfrak{F}$.

Note that the definitions of validity in $\mathfrak{GK}^c$ and $\mathfrak{GK}$ coincide with those in **K**biG and $\mathbf{K}\mathsf{biG}^{\mathsf{f}}$ if we consider the $\prec$-free fragment of $\mathsf{bi}\mathcal{L}_{\Box,\Diamond}$.

As we have already mentioned, on *crisp* frames, the accessibility relation can be understood as availability of (trusted or reliable) sources. In *fuzzy* frames, it can be thought of as the degree of trust one has in a source. Then, $\Diamond\phi$ represents the search for evidence from trusted sources that supports ϕ: $v(\Diamond\phi, t) > 0$ iff there is t' s.t. $tRt' > 0$ and $v(\phi, t') > 0$, i.e., there must be a source t' to which t has positive degree of trust and that has at least some certainty in ϕ. On the other hand, if no source is trusted by t (i.e., $tRu = 0$ for all u), then $v(\Diamond\phi, t) = 0$. Likewise, $\Box\chi$ can be construed as the search of evidence against χ given by trusted sources: $v(\Box\chi, t) < 1$ iff there is a source t' that gives to χ less certainty than t gives trust to t'. In other words, if t trusts no sources, or if all sources have at least as high confidence in χ as t has in them, then t fails to find a trustworthy enough counterexample.

Definition 4 ($\mathbf{KG}^2$ models). *A* $\mathbf{KG}^2$ model *is a tuple* $\mathfrak{M} = \langle W, R, v_1, v_2\rangle$ *with* $\langle W, R\rangle$ *being a* crisp *frame, and* $v_1, v_2 : \mathsf{Var} \times W \to [0, 1]$. *The valuations which we interpret as support of truth and support of falsity, respectively, are extended on complex formulas as expected.*

$$
\begin{array}{rlrl}
v_1(\neg\phi, w) &= v_2(\phi, w) & v_2(\neg\phi, w) &= v_1(\phi, w)\\
v_1(\phi \wedge \phi', w) &= v_1(\phi, w) \wedge_{\mathsf{G}} v_1(\phi', w) & v_2(\phi \wedge \phi', w) &= v_2(\phi, w) \vee_{\mathsf{G}} v_2(\phi', w)\\
v_1(\phi \vee \phi', w) &= v_1(\phi, w) \vee_{\mathsf{G}} v_1(\phi', w) & v_2(\phi \vee \phi', w) &= v_2(\phi, w) \wedge_{\mathsf{G}} v_2(\phi', w)\\
v_1(\phi \to \phi', w) &= v_1(\phi, w) \to_{\mathsf{G}} v_1(\phi', w) & v_2(\phi \to \phi', w) &= v_2(\phi', w) \prec_{\mathsf{G}} v_2(\phi, w)\\
v_1(\phi \prec \phi', w) &= v_1(\phi, w) \prec_{\mathsf{G}} v_1(\phi', w) & v_2(\phi \prec \phi', w) &= v_2(\phi', w) \to_{\mathsf{G}} v_2(\phi, w)\\
v_1(\Box\phi, w) &= \inf\{v_1(\phi, w') : wRw'\} & v_2(\Box\phi, w) &= \sup\{v_2(\phi, w') : wRw'\}\\
v_1(\Diamond\phi, w) &= \sup\{v_1(\phi, w') : wRw'\} & v_2(\Diamond\phi, w) &= \inf\{v_2(\phi, w') : wRw'\}
\end{array}
$$

We say that $\phi \in \mathsf{bi}\mathcal{L}^{\neg}_{\Box,\Diamond}$ *is* $\mathbf{KG}^2$ valid on frame $\mathfrak{F}$ *(*$\mathfrak{F} \models_{\mathbf{KG}^2} \phi$*) iff for any* $w \in \mathfrak{F}$, *it holds that* $v_1(\phi, w) = 1$ *and* $v_2(\phi, w) = 0$ *for any model* $\mathfrak{M}$ *on* $\mathfrak{F}$.

Convention 1. *In what follows, we will denote a pair of valuations* $\langle v_1, v_2\rangle$ *just with* v *if there is no risk of confusion. Furthermore, for each frame* $\mathfrak{F}$ *and each* $w \in \mathfrak{F}$, *we denote*

$$
\begin{array}{ll}
R(w) = \{w' : wRw' = 1\}, & \text{(for fuzzy frames)}\\
R(w) = \{w' : wRw'\}. & \text{(for crisp frames)}
\end{array}
$$

Convention 2. *We will further denote with* $\mathbf{K}\mathsf{biG}$ *the set of all formulas* $\mathbf{K}\mathsf{biG}$*-valid on all* crisp *frames;* $\mathbf{K}\mathsf{biG}^{\mathsf{f}}$ *the set of all formulas* $\mathbf{K}\mathsf{biG}$*-valid on all* fuzzy *frames; and* $\mathbf{KG}^2$*—the set of all formulas* $\mathbf{KG}^2$ *valid on all* crisp *frames.*

Before proceeding to establish some semantical properties, let us make two remarks. First, neither $\Box$ nor $\Diamond$ are trivialised by contradictions: in contrast to $\mathbf{K}$, $\Box(p \wedge \neg p) \to \Box q$ is not $\mathbf{KG}^2$ valid, and neither is $\Diamond(p \wedge \neg p) \to \Diamond q$. Intuitively, this means that one can have contradictory but non-trivial beliefs. Second, we can formalise statements of comparative belief such as the ones we have already given before:

wallet: *I am more confident that the wallet is Paula's than that the wallet is Quentin's.*
ice cream: *Alice is more certain than Britney that Claire loves pistachio ice cream.*

For this, consider the following defined operators.

$$\Delta\tau := {\sim}(\mathbf{1} \prec \tau) \tag{1}$$

$$\Delta^{\neg}\phi := {\sim}(\mathbf{1} \prec \phi) \wedge \neg{\sim}{\sim}(\mathbf{1} \prec \phi) \tag{2}$$

It is clear that for any $\tau \in \mathsf{bi}\mathcal{L}_{\Box,\Diamond}$ and $\phi \in \mathsf{bi}\mathcal{L}^{\neg}_{\Box,\Diamond}$ interpreted on $\mathbf{K}\mathsf{biG}$ and $\mathbf{KG}^2$ models, respectively, it holds that

$$v(\Delta\tau, w) = \begin{cases} 1 & \text{if } v(\tau, w) = 1 \\ 0 & \text{otherwise,} \end{cases} \quad v(\Delta^{\neg}\phi, w) = \begin{cases} (1,0) & \text{if } v(\phi, w) = (1,0) \\ (0,1) & \text{otherwise.} \end{cases} \tag{3}$$

Now we can define formulas that express order relations between values of two formulas both for $\mathbf{K}\mathsf{biG}$ and $\mathbf{KG}^2$.

For $\mathbf{K}\mathsf{biG}$ they look as follows:

$$v(\tau, w) \leq v(\tau', w) \text{ iff } v(\Delta(\tau \to \tau'), w) = 1,$$
$$v(\tau, w) > v(\tau', w) \text{ iff } v\left({\sim}\Delta(\tau' \to \tau), w\right) = 1.$$

In $\mathbf{KG}^2$, the orders are defined in a more complicated way:

$$v(\phi, w) \leq v(\phi', w) \text{ iff } v(\Delta^{\neg}(\phi \to \phi'), w) = (1,0),$$
$$v(\phi, w) > v(\phi', w) \text{ iff } v(\Delta^{\neg}(\phi' \to \phi) \wedge {\sim}\Delta^{\neg}(\phi \to \phi'), w) = (1,0).$$

Observe, first, that both in $\mathbf{K}\mathsf{biG}$ and $\mathbf{KG}^2$ the relation 'the value of τ (ϕ) is less or equal to the value of τ' (ϕ')' is defined as '$\tau \to \tau'$ ($\phi \to \phi'$) has the designated value'. In $\mathbf{K}\mathsf{biG}$, the strict order is just a negation of the non-strict order since all values are comparable. On the other hand, in contrast to $\mathbf{K}\mathsf{biG}$, the strict order in $\mathbf{KG}^2$ is not a simple negation of the non-strict order since $\mathbf{KG}^2$ is essentially two-dimensional. We provide further details in Remark 2.

Finally, we can formalise wallet as follows. We interpret 'I am confident' as $\Box$ and substitute 'the wallet is Paula's' with p, and 'the wallet is Quentin's' with q. Now, we just use the definition of $>$ in $\mathsf{bi}\mathcal{L}^{\neg}_{\Box,\Diamond}$ to get

$$\Delta^{\neg}(\Box p \to \Box q) \wedge {\sim}\Delta^{\neg}(\Box q \to \Box p). \tag{4}$$

For $\mathsf{ice\ cream}$, we need two different modalities: $\Box_a$ and $\Box_b$ for Alice and Brittney, respectively. Replacing 'Alice loves pistachio ice cream' with p, we get

$$\Delta^{\neg}(\Box_a p \to \Box_b p) \wedge {\sim}\Delta^{\neg}(\Box_b p \to \Box_a p). \tag{5}$$

Remark 1. Δ is called Baaz' delta (cf., e.g. [3] for more details). Intuitively, $\Delta\tau$ can be interpreted as 'τ has the designated value' and acts much like a necessity modality: if τ is $\mathbf{K}\mathsf{biG}$ valid, then so is $\Delta\tau$; moreover, $\Delta(p \to q) \to (\Delta p \to \Delta q)$ is valid. Furthermore, Δ and $\prec$ can be defined via one another in $\mathbf{K}\mathsf{biG}$, thus the addition of Δ to G makes it more expressive and allows to define both strict and non-strict orders.

Remark 2. Recall that we mentioned in Sect. 1 that an agent should usually be able to compare their beliefs in different statements: this is reflected by the fact that $\Delta(p \to q) \vee \Delta(q \to p)$ is **K**biG valid. It can be counter-intuitive if the contents of beliefs have nothing in common, however.

This drawback is avoided if we treat support of truth and support of falsity independently. Here is where a difference between **K**biG and $\mathbf{KG}^2$ lies. In $\mathbf{KG}^2$, we can *only compare the values of formulas coordinate-wise*, whence $\Delta^{\neg}(p \to q) \vee \Delta^{\neg}(q \to p)$ is not $\mathbf{KG}^2$ valid. E.g., if we set $v(p, w) = (0.7, 0.6)$ and $v(q, w) = (0.4, 0.2)$, $v(p, w)$ and $v(q, w)$ will not be comparable w.r.t. the truth (upward) order on $[0, 1]^{\bowtie}$.

We end this section with establishing some useful semantical properties.

Proposition 1. *$\mathfrak{F} \models_{\mathbf{KG}^2} \phi$ iff for any model $\mathfrak{M}$ on $\mathfrak{F}$ and any $w \in \mathfrak{F}$, $v_1(\phi, w) = 1$.*

Proof. The 'if' direction is evident from the definition of validity. We show the 'only if' part. It suffices to show that the following statement holds for any ϕ and $w \in \mathfrak{F}$:

for any $v(p, w) = (x, y)$, let $v^*(p, w) = (1-y, 1-x)$. Then $v(\phi, w) = (x, y)$ iff $v^*(\phi, w) = (1-y, 1-x)$.

We proceed by induction on ϕ. The proof of propositional cases is identical to the one in [5, Proposition 5]. We consider only the case of $\phi = \Box\psi$ since $\Box$ and $\Diamond$ are interdefinable.

Let $v(\Box\psi, w) = (x, y)$. Then $\inf\{v_1(\psi, w') : wRw'\} = x$, and $\sup\{v_2(\psi, w') : wRw'\} = y$. Now, we apply the induction hypothesis to ψ, and thus if $v(\psi, s) = (x', y')$, then $v^*(\psi, s) = (1-y', 1-x')$ for any $s \in R(w)$. But then $\inf\{v_1^*(\psi, w') : wRw'\} = 1 - y$, and $\sup\{v_2^*(\psi, w') : wRw'\} = 1 - x$ as required.

Now, assume that $v_1(\phi, w) = 1$ for any v_1 and w. We can show that $v_2(\phi, w) = 0$ for any w and v_2. Assume for contradiction that $v_2(\phi, w) = y > 0$ but $v_1(\phi, w) = 1$. Then, $v^*(\phi) = (1-y, 1-1) = (1-y, 0)$. But since $y > 0$, $v^*(\phi) \neq (1, 0)$.

Proposition 2.

1. *Let ϕ be a formula over $\{\mathbf{0}, \wedge, \vee, \to, \Box, \Diamond\}$. Then, $\mathfrak{F} \models_{\mathfrak{G}\mathfrak{K}} \phi$ iff $\mathfrak{F} \models_{\mathbf{K}\mathsf{biG}^{\mathsf{f}}} \phi$ and $\mathfrak{F} \models_{\mathfrak{G}\mathfrak{K}^c} \phi$ iff $\mathfrak{F} \models_{\mathbf{K}\mathsf{biG}} \phi$, for any $\mathfrak{F}$.*
2. *Let $\phi \in \mathsf{bi}\mathcal{L}_{\Box,\Diamond}$. Then, $\mathfrak{F} \models_{\mathbf{K}\mathsf{biG}} \phi$ iff $\mathfrak{F} \models_{\mathbf{KG}^2} \phi$, for any crisp $\mathfrak{F}$.*

Proof. 1. follows directly from the semantic conditions of Definition 3. We consider 2. The 'only if' direction is straightforward since the semantic conditions of v_1 in $\mathbf{KG}^2$ models and v in **K**biG models coincide. The 'if' direction follows from Proposition 1: if ϕ is valid on $\mathfrak{F}$, then $v(\phi, w) = 1$ for any $w \in \mathfrak{F}$ and any v on $\mathfrak{F}$. But then, $v_1(\phi, w) = 1$ for any $w \in \mathfrak{F}$. Hence, $\mathfrak{F} \models_{\mathbf{KG}^2} \phi$.

3 Model-Theoretic Properties of $\mathbf{KG}^2$

In the previous section, we have seen how the addition of $\prec$ allowed us to formalise statements considering comparison of beliefs. Here, we will show that both $\Box$ and $\Diamond$ fragments of **K**biG, and hence $\mathbf{KG}^2$, are strictly more expressive than the classical modal logic **K**, i.e. that they can define all classically definable classes of crisp frames as well as some undefinable ones.

Definition 5 (Frame definability). *Let Σ be a set of formulas. Σ* defines *a class of frames $\mathbb{K}$ in a logic* **L** *iff it holds that $\mathfrak{F} \in \mathbb{K}$ iff $\mathfrak{F} \models_{\mathbf{L}} \Sigma$.*

The next statement follows from Proposition 2 since **K** can be faithfully embedded in $\mathfrak{GK}^c$ by substituting each variable p with $\sim\sim p$ (cf. [28,29] for details).

Theorem 1. *Let $\mathbb{K}$ be a class of frames definable in* **K**. *Then, $\mathbb{K}$ is definable in* $\mathbf{K}\mathsf{biG}$ *and* $\mathbf{K}\mathsf{G}^2$.

Theorem 2. *1. Let $\mathfrak{F}$ be* crisp. *Then $\mathfrak{F}$ is finitely branching (i.e., $R(w)$ is finite for every $w \in \mathfrak{F}$) iff $\mathfrak{F} \models_{\mathbf{K}\mathsf{biG}} \mathbf{1} \prec \lozenge((p \prec q) \wedge q)$.*
2. Let $\mathfrak{F}$ be fuzzy. *Then $\mathfrak{F}$ is finitely branching and* $\sup\{wRw' : wRw' < 1\} < 1$ *for all $w \in \mathfrak{F}$ iff $\mathfrak{F} \models_{\mathbf{K}\mathsf{biG}} \mathbf{1} \prec \lozenge((p \prec q) \wedge q)$.*

Proof. We show the case of fuzzy frames since the crisp ones can be tackled in the same manner. Assume that $\mathfrak{F}$ is finitely branching and that $\sup\{wRw' : wRw' < 1\} < 1$ for all $w \in \mathfrak{F}$. It suffices to show that $v(\lozenge((p \prec q) \wedge q), w) < 1$ for all $w \in \mathfrak{F}$. First of all, observe that there is no $w' \in \mathfrak{F}$ s.t. $v((p \prec q) \wedge q, w') = 1$. It is clear that $\sup\limits_{wRw'<1} \{v((p \prec q) \wedge q, w') \wedge_{\mathsf{G}} wRw'\} < 1$ and that

$$\sup\{v((p \prec q) \wedge q, w') : wRw' = 1\} = \max\{v((p \prec q) \wedge q, w') : wRw' = 1\} < 1$$

since $R(w)$ is finite. But then $v(\lozenge((p \prec q) \wedge q), w) < 1$ as required.

For the converse, either (1) $R(w)$ is infinite for some w, or (2) $\sup\{wRw' : wRw' < 1\} = 1$ for some w. For (1), set $v(p, w') = 1$ for every $w' \in R(w)$. Now let $W' \subseteq R(w)$ and $W' = \{w_i : i \in \{1, 2, \ldots\}\}$. We set $v(q, w_i) = \frac{i}{i+1}$. It is easy to see that $\sup\{v(q, w_i) : w_i \in W'\} = 1$ and that $v((p \prec q) \wedge q, w_i) = v(q, w_i)$. Therefore, $v(\mathbf{1} \prec \lozenge((p \prec q) \wedge q), w) = 0$.

For (2), we let $v(p, w') = 1$ and further, $v(q, w') = wRw'$ for all $w' \in \mathfrak{F}$. Now since $\sup\{wRw' : wRw' < 1\} = 1$ and $v(((p \prec q) \wedge q), w') = v(q, w')$ for all $w' \in \mathfrak{F}$, it follows that $v(\lozenge((p \prec q) \wedge q), w) = 1$, whence $v(\mathbf{1} \prec \lozenge((p \prec q) \wedge q), w) = 0$.

Remark 3. The obvious corollary of Theorem 2 is the lack of FMP for the $\lozenge$-fragment of $\mathbf{K}\mathsf{biG}^{\mathsf{f}}$[5] since $\lozenge((p \prec q) \wedge q)$ in never true in a finite model. This differentiates $\mathbf{K}\mathsf{biG}^{\mathsf{f}}$ from $\mathfrak{GK}$ since the $\lozenge$-fragment of $\mathfrak{GK}$ *has* FMP [12, Theorem 7.1]. Moreover, one can define finitely branching frames in $\Box$ fragments of $\mathfrak{GK}$ and $\mathfrak{GK}^c$. Indeed, $\sim\sim\Box(p \vee \sim p)$ serves as such definition.

Corollary 1. $\mathbf{K}\mathsf{G}^2$ *and both $\Box$ and $\lozenge$ fragments of* $\mathbf{K}\mathsf{biG}$ *are strictly more expressive than* **K**.

Proof. From Theorems 1 and 2 since **K** is complete both w.r.t. all frames and all finitely branching frames. The result for $\mathbf{K}\mathsf{G}^2$ follows since it is conservative over $\mathbf{K}\mathsf{biG}$ (Proposition 2).

[5] Bi-modal $\mathbf{K}\mathsf{biG}^{\mathsf{f}}$ lacks have FMP since it is a conservative extension of $\mathfrak{GK}$.

These results show us that addition of $\prec$ greatly enhances the expressive power of our logic. Here it is instructive to remind ourselves that classical epistemic logics are usually complete w.r.t. finitely branching frames (cf. [18] for details). It is reasonable since for practical reasoning, agents cannot consider infinitely many alternatives. In our case, however, if we wish to use $\mathbf{K}\mathsf{biG}$ and $\mathbf{K}\mathsf{G}^2$ for knowledge representation, we need to *impose* finite branching explicitly.

Furthermore, allowing for infinitely branching frames in $\mathbf{K}\mathsf{biG}$ or $\mathbf{K}\mathsf{G}^2$ leads to counter-intuitive consequences. In particular, it is possible that $v(\Box\phi, w) = (0, 1)$ even though there are no $w', w'' \in R(w)$ s.t. $v_1(\phi, w') = 0$ or $v_2(\phi, w'') = 1$. In other words, there is no source that decisively falsifies ϕ, furthermore, all sources have some evidence *for* ϕ, and yet we somehow believe that ϕ is completely false and untrue. Dually, it is possible that $v(\Diamond\phi, w) = (1, 0)$ although there are no $w', w'' \in R(w)$ s.t. $v_1(\phi, w') = 1$ or $v_2(\phi, w'') = 0$. Even though $\Diamond$ is an 'optimistic' aggregation, it should not ignore the fact that *all* sources have some evidence *against* ϕ but *none* supports it completely.

Of course, this situation is impossible if we consider only finitely branching frames for infima and suprema will become minima and maxima. There, all values of modal formulas will be *witnessed* by some accessible states in the following sense. For $\heartsuit \in \{\Box, \Diamond\}$, $i \in \{1, 2\}$, if $v_i(\heartsuit\phi, w) = x$, then there is $w' \in R(w)$ s.t. $v_i(\phi, w') = x$. Intuitively speaking, finitely branching frames represent the situation when our degree of certainty in some statement is based uniquely on the data given by the sources.

Convention 3. *We will further use* $\mathbf{K}\mathsf{biG}_{\mathsf{fb}}$ *and* $\mathbf{K}\mathsf{G}^2_{\mathsf{fb}}$ *to denote the sets of all* $\mathsf{bi}\mathcal{L}_{\Box,\Diamond}$ *and* $\mathsf{bi}\mathcal{L}^{\neg}_{\Box,\Diamond}$ *formulas valid on finitely branching crisp frames.*

Observe, moreover, that $\Box$ and $\Diamond$ are still undefinable via one another in $\mathsf{bi}\mathcal{L}_{\Box,\Diamond}$. The proof is the same as that of [36, Lemma 6.1].

Proposition 3. $\Box$ *and* $\Diamond$ *are not interdefinable in* $\mathbf{K}\mathsf{biG}_{\mathsf{fb}}$.

Corollary 2.

1. $\Box$ *and* $\Diamond$ *are not interdefinable in* $\mathbf{K}\mathsf{biG}$, $\mathbf{K}\mathsf{biG}^{\mathsf{f}}_{\mathsf{fb}}$, *and* $\mathbf{K}\mathsf{biG}^{\mathsf{f}}$.
2. *Both* $\Box$ *and* $\Diamond$ *fragments of* $\mathbf{K}\mathsf{biG}$ *are more expressive than* $\mathbf{K}$.

In the remainder of the paper, we are going to provide a complete proof system for $\mathbf{K}\mathsf{G}^2_{\mathsf{fb}}$ (and hence, $\mathbf{K}\mathsf{biG}_{\mathsf{fb}}$), and establish its decidability and complexity as well as finite model property. Note, however, that the latter is not entirely for granted. In fact, several expected ways of defining filtration (cf. [7,14] for more details thereon) fail.

Let $\Sigma \subseteq \mathsf{bi}\mathcal{L}_{\Box,\Diamond}$ be closed under subformulas. If we want to have filtration for $\mathbf{K}\mathsf{biG}_{\mathsf{fb}}$, there are three intuitive ways to define $\sim_\Sigma$ on the carrier of a model that is supposed to relate states satisfying the same formulas.

1. $w \sim^1_\Sigma w'$ iff $v(\phi, w) = v(\phi, w')$ for all $\phi \in \Sigma$.
2. $w \sim^2_\Sigma w'$ iff $v(\phi, w) = 1 \Leftrightarrow v(\phi, w') = 1$ for all $\phi \in \Sigma$.
3. $w \sim^3_\Sigma w'$ iff $v(\phi, w) \leq v(\phi', w) \Leftrightarrow v(\phi, w') \leq v(\phi', w')$ for all $\phi, \phi' \in \Sigma \cup \{\mathbf{0}, \mathbf{1}\}$.

Consider the model on Fig. 3 and two formulas:

$$\phi^{\leq} := {\sim\sim}(p \to \Diamond p) \qquad\qquad \phi^{>} := {\sim\sim}(p \prec \Diamond p)$$

Now let Σ to be the set of all subformulas of $\phi^{\leq} \wedge \phi^{>}$.

First of all, it is clear that $v(\phi^{\leq} \wedge \phi^{>}, w) = 1$ for any $w \in \mathfrak{M}$. Observe now that all states in $\mathfrak{M}$ are *distinct* w.r.t. $\sim^1_\Sigma$. Thus, the first way of constructing the carrier of the new model does not give the FMP.

$$\mathfrak{M} : w_1 \longrightarrow w_2 \longrightarrow \ldots \longrightarrow w_n \longrightarrow \ldots$$

Fig. 3. $v(p, w_n) = \frac{1}{n+1}$

As regards to $\sim^2_\Sigma$ and $\sim^3_\Sigma$, one can check that for any $w, w' \in \mathfrak{M}$, it holds that $w \sim^2_\Sigma w'$ and $w \sim^3_\Sigma w'$. So, if we construct a filtration of $\mathfrak{M}$ using equivalence classes of either of these two relations, the carrier of the resulting model is going to be finite. Even more so, it is going to be a singleton.

However, we can show that there is *no finite model* $\mathfrak{N} = \langle U, S, e\rangle$ s.t.

$$\forall s \in \mathfrak{N} : v(\phi^{\leq} \wedge \phi^{>}, s) = 1.$$

Indeed, $e(\phi^{\leq}, t) = 1$ iff $e(p, t') > 0$ for some $t' \in S(t)$, while $e(\phi^{>}, t) = 1$ iff $v(p, t) > v(p, t')$ for any $t' \in S(t)$. Now, if U is finite, we have two options: either (1) there is $u \in U$ s.t. $R(u) = \varnothing$, or (2) U contains a finite S-cycle.

For (1), note that $v(\Diamond p, u) = 0$, and we have two options: if $e(p, u) = 0$, then $e(\phi^{>}, u) = 0$; if, on the other hand, $e(p, u) > 0$, then $e(\phi^{\leq}, u) = 0$. For (2), assume w.l.o.g. that the S-cycle looks as follows: $u_0 S u_1 S u_2 \ldots S u_n S u_0$.

If $e(p, u_0) = 0$, $e(\phi^{>}, u_0) = 0$, so $e(p, u_0) > 0$. Furthermore, $e(p, u_i) > e(p, u_{i+1})$. Otherwise, again, $e(\phi^{>}, u_i) = 0$. But then we have $e(\phi^{>}, u_i) = 0$.

But this means that $\sim^2_\Sigma$ and $\sim^3_\Sigma$ do not preserve truth of formulas from w to $[w]_\Sigma$, i.e., neither of these two relations can be used to define filtration. Thus, in order to explicitly prove the finite model property and establish complexity evaluations for $\mathbf{K}\mathsf{biG}_{\mathsf{fb}}$ and $\mathbf{K}\mathsf{G}^2_{\mathsf{fb}}$, we will provide a tableaux calculus. It will also serve as a decision procedure for satisfiability and validity of formulas.

4 Tableaux for $\mathbf{K}\mathsf{G}^2_{\mathsf{fb}}$

Usually, proof theory for modal and many-valued logics is presented in one of the following several forms. The first one is a Hilbert-style axiomatisation as given in e.g. [23] for the propositional Gödel logic and in [12,13,36] for its modal expansions. Hilbert calculi are useful for establishing frame correspondence results as well as for showing that one logic extends another one in the same language. On the other hand, their completeness proofs might be quite complicated, and the proof-search not at all straightforward. Second, there are non-labelled sequent and hyper-sequent calculi (cf. [30] for the propositional proof systems and [28,29] for the modal hypersequent calculi). With regards to modal logics, completeness proofs of (hyper)sequent calculi often provide the answer for the decidability problem. Furthermore, the proof search can be quite straightforwardly automatised provided that the calculus is *cut-free.*

Finally, there are proof systems that directly incorporate semantics: in particular, tableaux (e.g., the ones for Gödel logics [2] and tableaux for Łukasiewicz description logic [25]) and labelled sequent calculi (cf., e.g. [32] for labelled sequent calculi for classical modal logics). Because of the calculi's nature, their completeness proofs are usually simple. Besides, the calculi serve as a decision procedure that either establishes that the given formula is valid or provides an explicit countermodel.

Our tableaux system $\mathcal{T}(\mathbf{KG}^2_{\mathsf{fb}})$ is a straightforward modal expansion of constraint tableaux for G^2 presented in [5]. It is inspired by constraint tableaux for Łukasiewicz logics from [21,22] (but cf. [26] for an approach similar to ours) which we modify with two-sorted labels corresponding to the support of truth and support of falsity in the model. This idea comes from tableaux for the Belnap—Dunn logic by D'Agostino [1]. Moreover, since $\mathbf{KG}^2_{\mathsf{fb}}$ is a conservative extension of $\mathbf{K}\mathsf{biG}_{\mathsf{fb}}$, our calculus can be used for that logic as well if we apply only the rules that govern the support of truth of $\mathsf{bi}\mathcal{L}_{\Box,\Diamond}$ formulas.

Definition 6 $(\mathcal{T}(\mathbf{KG}^2_{\mathsf{fb}}))$**.** *We fix a set of state-labels* W *and let* $\lesssim\in\{<,\leqslant\}$ *and* $\gtrsim\in\{>,\geqslant\}$. *Let further* $w\in\mathsf{W}$, $\mathbf{x}\in\{1,2\}$, $\phi\in\mathsf{bi}\mathcal{L}^{\neg}_{\Box,\Diamond}$, *and* $c\in\{0,1\}$. *A structure is either* $w{:}\mathbf{x}{:}\phi$ *or* c. *We denote the set of structures with* Str.

We define a constraint tableau *as a downward branching tree whose branches are sets containing the following types of entries:*

- relational constraints *of the form* $w\mathsf{R}w'$ *with* $w,w'\in\mathsf{W}$*;*
- structural constraints *of the form* $\mathfrak{X}\lesssim\mathfrak{X}'$ *with* $\mathfrak{X},\mathfrak{X}'\in\mathsf{Str}$.

Each branch can be extended by an application of a rule[6] *from Fig. 4 or Fig. 5.*

A tableau's branch $\mathcal{B}$ *is* closed *iff one of the following conditions applies:*

- *the transitive closure of* $\mathcal{B}$ *under* $\lesssim$ *contains* $\mathfrak{X}<\mathfrak{X}$*;*
- $0\geqslant1\in\mathcal{B}$*, or* $\mathfrak{X}>1\in\mathcal{B}$*, or* $\mathfrak{X}<0\in\mathcal{B}$.

A tableau is closed *iff all its branches are closed. We say that there is a* tableau proof *of* ϕ *iff there is a closed tableau starting from the constraint* $w{:}1{:}\phi<1$.

An open branch $\mathcal{B}$ *is* complete *iff the following condition is met.*

* If all premises of a rule occur on $\mathcal{B}$, then its one conclusion[7] occurs on $\mathcal{B}$.

Remark 4. Note that due to Proposition 1, we need to check only one valuation of ϕ to verify its validity.

Convention 4 (Interpretation of constraints). *The following table gives the interpretations of structural constraints on the example of* $\leqslant$.

[6] If $\mathfrak{X}<1$ and $\mathfrak{X}<\mathfrak{X}'$ (or $0<\mathfrak{X}'$ and $\mathfrak{X}<\mathfrak{X}'$) occur on $\mathcal{B}$, then the rules are applied only to $\mathfrak{X}<\mathfrak{X}'$.

[7] Note that branching rules have *two* conclusions.

$$\neg_1\lesssim\frac{w:1:\neg\phi\lesssim\mathfrak{X}}{w:2:\phi\lesssim\mathfrak{X}}\quad\neg_2\lesssim\frac{w:2:\neg\phi\lesssim\mathfrak{X}}{w:1:\phi\lesssim\mathfrak{X}}\quad\neg_1\gtrsim\frac{w:1:\neg\phi\gtrsim\mathfrak{X}}{w:2:\phi\gtrsim\mathfrak{X}}\quad\neg_2\gtrsim\frac{w:2:\neg\phi\gtrsim\mathfrak{X}}{w:1:\phi\gtrsim\mathfrak{X}}$$

$$\wedge_1\gtrsim\frac{w:1:\phi\wedge\phi'\gtrsim\mathfrak{X}}{\begin{array}{c}w:1:\phi\gtrsim\mathfrak{X}\\ w:1:\phi'\gtrsim\mathfrak{X}\end{array}}\quad\wedge_2\lesssim\frac{w:2:\phi\wedge\phi'\lesssim\mathfrak{X}}{\begin{array}{c}w:2:\phi\lesssim\mathfrak{X}\\ w:2:\phi'\lesssim\mathfrak{X}\end{array}}\quad\vee_1\lesssim\frac{w:1:\phi\vee\phi'\lesssim\mathfrak{X}}{\begin{array}{c}w:1:\phi\lesssim\mathfrak{X}\\ w:1:\phi'\lesssim\mathfrak{X}\end{array}}\quad\vee_2\gtrsim\frac{w:2:\phi\vee\phi'\gtrsim\mathfrak{X}}{\begin{array}{c}w:2:\phi\gtrsim\mathfrak{X}\\ w:2:\phi'\gtrsim\mathfrak{X}\end{array}}$$

$$\wedge_1\lesssim\frac{w:1:\phi\wedge\phi'\lesssim\mathfrak{X}}{w:1:\phi\lesssim\mathfrak{X}\mid w:1:\phi'\lesssim\mathfrak{X}}\qquad\wedge_2\gtrsim\frac{w:2:\phi\wedge\phi'\gtrsim\mathfrak{X}}{w:2:\phi\gtrsim\mathfrak{X}\mid w:2:\phi'\gtrsim\mathfrak{X}}$$

$$\vee_1\gtrsim\frac{w:1:\phi\vee\phi'\gtrsim\mathfrak{X}}{w:1:\phi\gtrsim\mathfrak{X}\mid w:1:\phi'\gtrsim\mathfrak{X}}\qquad\vee_2\lesssim\frac{w:2:\phi\vee\phi'\lesssim\mathfrak{X}}{w:2:\phi\lesssim\mathfrak{X}\mid w:2:\phi'\lesssim\mathfrak{X}}$$

$$\to_1\leqslant\frac{w:1:\phi\to\phi'\leqslant\mathfrak{X}}{\begin{array}{c|c} & \mathfrak{X}<1\\ \mathfrak{X}\geqslant 1 & w:1:\phi'\leqslant\mathfrak{X}\\ & w:1:\phi>w:1:\phi'\end{array}}\qquad\to_1\gtrsim\frac{w:1:\phi\to\phi'\gtrsim\mathfrak{X}}{w:1:\phi\leqslant w:1:\phi'\mid w:1:\phi'\gtrsim\mathfrak{X}}$$

$$\to_2\lesssim\frac{w:2:\phi\to\phi'\lesssim\mathfrak{X}}{w:2:\phi'\leqslant w:2:\phi\mid w:2:\phi'\lesssim\mathfrak{X}}\qquad\to_2\geqslant\frac{w:2:\phi\to\phi'\geqslant\mathfrak{X}}{\begin{array}{c|c} & \mathfrak{X}>0\\ \mathfrak{X}\leqslant 0 & w:2:\phi'\geqslant\mathfrak{X}\\ & w:2:\phi'>w:2:\phi\end{array}}$$

$$\prec_1\lesssim\frac{w:1:\phi\prec\phi'\lesssim\mathfrak{X}}{w:1:\phi\leqslant w:1:\phi'\mid w:1:\phi\lesssim\mathfrak{X}}\qquad\prec_1\geqslant\frac{w:1:\phi\prec\phi'\geqslant\mathfrak{X}}{\begin{array}{c|c} & \mathfrak{X}>0\\ \mathfrak{X}\leqslant 0 & w:1:\phi\geqslant\mathfrak{X}\\ & w:1:\phi>w:1:\phi'\end{array}}$$

$$\prec_2\gtrsim\frac{w:2:\phi\prec\phi'\gtrsim\mathfrak{X}}{w:2:\phi\gtrsim\mathfrak{X}\mid w:2:\phi'\leqslant w:2:\phi}\qquad\prec_2\leqslant\frac{w:2:\phi\prec\phi'\leqslant\mathfrak{X}}{\begin{array}{c|c} & \mathfrak{X}<1\\ \mathfrak{X}\geqslant 1 & w:2:\phi\leqslant\mathfrak{X}\\ & w:2:\phi'>w:2:\phi\end{array}}$$

$$\to_1<\frac{w:1:\phi\to\phi'<\mathfrak{X}}{\begin{array}{c}w:1:\phi'<\mathfrak{X}\\ w:1:\phi>w:1:\phi'\end{array}}\quad\to_2>\frac{w:2:\phi\to\phi'>\mathfrak{X}}{\begin{array}{c}w:2:\phi'>\mathfrak{X}\\ w:2:\phi'>w:2:\phi\end{array}}$$

$$\prec_1>\frac{w:1:\phi\prec\phi'>\mathfrak{X}}{\begin{array}{c}w:1:\phi>\mathfrak{X}\\ w:1:\phi>w:1:\phi'\end{array}}\quad\prec_2<\frac{w:2:\phi\prec\phi'<\mathfrak{X}}{\begin{array}{c}w:2:\phi<\mathfrak{X}\\ w:2:\phi<w:2:\phi'\end{array}}$$

Fig. 4. Propositional rules of $\mathcal{T}\left(\mathbf{KG}^2_{\mathsf{fb}}\right)$. Bars denote branching.

entry	interpretation
$w:1:\phi \leqslant w':2:\phi'$	$v_1(\phi, w) \leq v_2(\phi', w')$
$w:2:\phi \leqslant c$	$v_2(\phi, w) \leq c$ with $c \in \{0,1\}$

As one can see from Fig. 4 and Fig. 5, the rules follow the semantical conditions from Definition 4. Let us discuss $\rightarrow_1\!\leqslant$ and $\Box_1\!\lesssim$ in more details.

The premise of $\rightarrow_1\!\leqslant$ is interpreted as $v_1(\phi \rightarrow \phi', w) \leqslant x$. To decompose the implication, we check two options: either $x = 1$ (then, the value of $\phi \rightarrow \phi'$ is arbitrary) or $x < 1$. In the second case, we use the semantics to obtain that $v_1(\phi', w) \leqslant x$ and $v_1(\phi, w) > v_1(\phi', w)$.

$$\Box_1\!\gtrsim\frac{\begin{array}{c}w:1:\Box\phi \gtrsim \mathfrak{X}\\ wRw'\end{array}}{w':1:\phi \gtrsim \mathfrak{X}} \quad \Box_1\!\lesssim\frac{w:1:\Box\phi \lesssim \mathfrak{X}}{\begin{array}{c}wRw''\\ w'':1:\phi \lesssim \mathfrak{X}\end{array}} \quad \Box_2\!\gtrsim\frac{w:2:\Box\phi \gtrsim \mathfrak{X}}{\begin{array}{c}wRw''\\ w'':2:\phi \gtrsim \mathfrak{X}\end{array}} \quad \Box_2\!\lesssim\frac{\begin{array}{c}w:2:\Box\phi \lesssim \mathfrak{X}\\ wRw'\end{array}}{w':2:\phi \lesssim \mathfrak{X}}$$

$$\Diamond_1\!\lesssim\frac{\begin{array}{c}w:1:\Diamond\phi \lesssim \mathfrak{X}\\ wRw'\end{array}}{w':1:\phi \lesssim \mathfrak{X}} \quad \Diamond_1\!\gtrsim\frac{w:1:\Diamond\phi \gtrsim \mathfrak{X}}{\begin{array}{c}wRw''\\ w'':1:\phi \gtrsim \mathfrak{X}\end{array}} \quad \Diamond_2\!\lesssim\frac{w:2:\Diamond\phi \lesssim \mathfrak{X}}{\begin{array}{c}wRw''\\ w'':2:\phi \lesssim \mathfrak{X}\end{array}} \quad \Diamond_2\!\gtrsim\frac{\begin{array}{c}w:2:\Diamond\phi \gtrsim \mathfrak{X}\\ wRw'\end{array}}{w':2:\phi \gtrsim \mathfrak{X}}$$

Fig. 5. Modal rules of $\mathcal{T}\left(\mathbf{KG}^2_{\mathsf{fb}}\right)$. w'' is fresh on the branch.

In order to apply $\Box_1\!\lesssim$ to $w:1:\Box\phi \lesssim \mathfrak{X}$, we introduce a new state w'' that is seen by w. Since we work in a finite branching model, w'' can witness the value of $\Box\phi$. Thus, we add $w'':1:\phi \lesssim \mathfrak{X}$.

We also provide an example of how our tableaux work. On Fig. 6, one can see a successful proof on the left and a failed proof on the right.

$w_0:1:\mathbf{1}\prec\Diamond((p\prec q)\wedge q)<1$

Left branch:

$w_0:1:\mathbf{1} \leqslant w_0:1:\Diamond((p\prec q)\wedge q)$
$w_0:1:\Diamond((p\prec q)\wedge q)\geqslant 1$
$w_0 R w_1$
$w_1:1:(p\prec q)\wedge q\geqslant 1$
$w_1:1:p\prec q \geqslant 1$

- $1 \leqslant 0$ ×
- $w_1:1:q \geqslant 1$, $w_1:1:p \geqslant 1$, $w_1:1:q < 1$ ×

Right branch:

$w_0:1:\mathbf{1} < 1$ ×

$w_0:1:\Box p \rightarrow \Box\Box p < 1$
$w_0:1:\Box\Box p<1$
$w_0:1:\Box p > w_0:1:\Box\Box p$
$w_0 R w_1$
$w_0:1:\Box p > w_1:1:\Box p$
$w_1:1:p > w_1:1:\Box p$
$w_1 R w_2$
$w_1:1:p > w_2:1:p$
☹

Fig. 6. × indicates closed branches; ☹ indicates complete open branches.

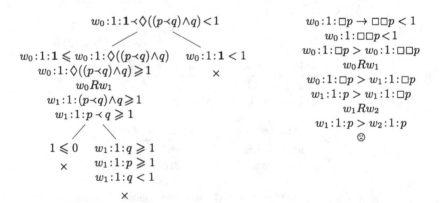

Definition 7 (Branch realisation). *We say that a model* $\mathfrak{M} = \langle W, R, v_1, v_2 \rangle$ *with* $W = \{w : w \text{ occurs on } \mathcal{B}\}$ *and* $R = \{\langle w, w' \rangle : wRw' \in \mathcal{B}\}$ realises a branch $\mathcal{B}$ *of a tree iff the following conditions are met.*

- $v_{\mathbf{x}}(\phi, w) \leq v_{\mathbf{x}'}(\phi', w')$ *for any* $w : \mathbf{x} : \phi \leqslant w' : \mathbf{x}' : \phi' \in \mathcal{B}$ *with* $\mathbf{x}, \mathbf{x}' \in \{1, 2\}$.
- $v_{\mathbf{x}}(\phi, w) \leq c$ *for any* $w : \mathbf{x} : \phi \leqslant c \in \mathcal{B}$ *with* $c \in \{0, 1\}$.

Theorem 3 (Completeness). ϕ *is* $\mathbf{KG}^2_{\mathsf{fb}}$ *valid iff it has a* $\mathcal{T}(\mathbf{KG}^2_{\mathsf{fb}})$ *proof.*

Proof. We consider only the $\mathbf{KG}^2_{\mathsf{fb}}$ case since $\mathbf{KbiG}_{\mathsf{fb}}$ can be handled the same way. For soundness, we check that if the premise of the rule is realised, then so is at least one of its conclusions. We consider the cases of $\rightarrow_1\leqslant$ and $\Box_1\lesssim$. Assume that $w : 1 : \phi \rightarrow \phi' \leqslant \mathfrak{X}$ is realised and assume w.l.o.g. that $\mathfrak{X} = u : 2 : \psi$. It is clear that either $v_2(\psi, u) = 1$ or $v_2(\psi, u) < 1$. In the first case, $\mathfrak{X} \geqslant 1$ is realised. In the second case, we have that $v_1(\phi, w) > v_1(\phi', w)$ and $v_1(\phi', w) \leqslant v_2(\psi, u)$. Thus, $\mathfrak{X} < 1$, $w : 1 : \phi > w : 1 : \phi'$, and $w : 1 : \phi' \leqslant u : 1 : \psi$ are realised as well, as required.

For $\Box_1\lesssim$, assume that $w : 1 : \Box\phi \leqslant \mathfrak{X}$ is realised and assume w.l.o.g. that $\mathfrak{X} = u : 2 : \psi$. Thus, $v_1(\Box\phi, w) \leqslant v_2(\psi, u)$ Then, since the model is finitely branching, there is an accessible state w'' s.t. $v_1(\phi, w) \leqslant v_2(\psi, u)$. Thus, $w'' : 1 : \phi \leqslant \mathfrak{X}$ is realised too.

As no closed branch is realisable, the result follows.

For completeness, we show that every complete open branch $\mathcal{B}$ is realisable. We construct the model as follows. We let $W = \{w : w \text{ occurs in } \mathcal{B}\}$, and set $R = \{\langle w, w' \rangle : wRw' \in \mathcal{B}\}$. Now, it remains to construct the suitable valuations.

For $i \in \{1, 2\}$, if $w : i : p \geqslant 1 \in \mathcal{B}$, we set $v_i(p, w) = 1$. If $w : i : p \leqslant 0 \in \mathcal{B}$, we set $v_i(p, w) = 0$. To set the values of the remaining variables $q_1, \ldots, q_n$, we proceed as follows. Denote $\mathcal{B}^+$ the transitive closure of $\mathcal{B}$ under $\lesssim$ and let

$$[w : \mathbf{x} : q_i] = \left\{ w' : \mathbf{x}' : q_j \;\middle|\; \begin{array}{c} w : \mathbf{x} : q_i \leqslant w' : \mathbf{x}' : q_j \in \mathcal{B}^+ \text{ and } w : \mathbf{x} : q_i < w' : \mathbf{x}' : q_j \notin \mathcal{B}^+ \\ \text{or} \\ w : \mathbf{x} : q_i \geqslant w' : \mathbf{x}' : q_j \in \mathcal{B}^+ \text{ and } w : \mathbf{x} : q_i > w' : \mathbf{x}' : q_j \notin \mathcal{B}^+ \end{array} \right\}$$

It is clear that there are at most $2 \cdot n \cdot |W|$ $[w : \mathbf{x} : q_i]$'s since the only possible loop in $\mathcal{B}^+$ is $w_{i_1} : \mathbf{x} : r \leqslant \ldots \leqslant w_{i_1} : \mathbf{x} : r$, but in such a loop all elements belong to $[w_{i_1} : \mathbf{x} : r]$. We put $[w : \mathbf{x} : q_i] \prec [w' : \mathbf{x}' : q_j]$ iff there are $w_k : \mathbf{x} : r \in [w : \mathbf{x} : q_i]$ and $w'_k : \mathbf{x}' : r' \in [w' : \mathbf{x}' : q_j]$ s.t. $w_k : \mathbf{x} : r < w'_k : \mathbf{x}' : r' \in \mathcal{B}^+$.

We now set the valuation of these variables as follows

$$v_{\mathbf{x}}(q_i, w) = \frac{|\{[w' : \mathbf{x}' : q'] \mid [w' : \mathbf{x}' : q'] \prec [w : \mathbf{x} : q_i]\}|}{2 \cdot n \cdot |W|}$$

Note that if some ϕ contains s but $\mathcal{B}^+$ contains no inequality with it, the above definition ensures that s is going to be evaluated at 0. Thus, all constraints containing only variables are satisfied.

It remains to show that all other constraints are satisfied. For that, we prove that if at least one conclusion of the rule is satisfied, then so is the premise. The propositional cases are straightforward and can be tackled in the same manner as in [5, Theorem 2]. We consider only the case of $\Diamond_2 \gtrsim$. Assume w.l.o.g. that $\gtrsim = \geqslant$ and $\mathfrak{X} = u:1:\psi$. Since $\mathcal{B}$ is complete, if $w: 2:\Diamond\phi \geqslant u:1:\psi \in \mathcal{B}$, then for any w' s.t. $w\mathsf{R}w' \in \mathcal{B}$, we have $w': 2:\phi \geqslant u:1:\psi \in \mathcal{B}$, and all of them are realised by $\mathfrak{M}$. But then $w: 2:\Diamond\phi \geqslant u:1:\psi$ is realised too, as required.

Theorem 4.

1. *Let* $\phi \in \mathsf{bi}\mathcal{L}^{\neg}_{\Box,\Diamond}$ *be* not $\mathbf{KG}^2_{\mathsf{fb}}$ valid, *and let* $|\phi|$ *denote the number of symbols in it. Then there is a model* $\mathfrak{M}$ *of the size* $O(|\phi|^{|\phi|})$ *and depth* $O(|\phi|)$ *and* $w \in \mathfrak{M}$ *s.t.* $v_1(\phi, w) \neq 1$.
2. $\mathbf{KG}^2_{\mathsf{fb}}$ *validity and satisfiability*[8] *are* PSPACE*-complete.*

Proof. We begin with 1. By Theorem 3, if ϕ is *not* $\mathbf{KG}^2_{\mathsf{fb}}$ *valid*, we can build a falsifying model using tableaux. It is also clear from the rules on Fig. 5 that the depth of the constructed model is bounded from above by the maximal number of nested modalities in ϕ. The width of the model is bounded by the maximal number of modalities on the same level of nesting. The sharpness of the bound is obtained using the embedding of $\mathbf{K}$ into $\mathbf{KG}^2_{\mathsf{fb}}$ since $\mathbf{K}$ is complete w.r.t. finitely branching models and it is possible to force shallow trees of exponential size in $\mathbf{K}$ (cf., e.g. [7, §6.7]). The embedding also entails PSPACE-hardness. It remains to tackle membership.

First, observe from the proof of Theorem 3 that $\phi(p_1, \ldots, p_n)$ is satisfiable (falsifiable) on $\mathfrak{M} = \langle W, R, v_1, v_2\rangle$ iff there are v_1 and v_2 that give variables values from $\mathsf{V} = \left\{0, \frac{1}{2\cdot n\cdot|W|}, \ldots, \frac{2\cdot n\cdot|W|-1}{2\cdot n\cdot|W|}, 1\right\}$ under which ϕ is satisfied (falsified).

As we mentioned, $|W|$ is bounded from above by k^{k+1} with k being the number of modalities in ϕ. Therefore, we replace structural constraints with labelled formulas of the form $w:i:\phi = \mathsf{v}$ ($\mathsf{v} \in \mathsf{V}$) avoiding comparisons of values of formulas in different states. As expected, we close the branch if it contains $w:i:\psi = \mathsf{v}$ and $w:i:\psi = \mathsf{v}'$ for $\mathsf{v} \neq \mathsf{v}'$.

Now we replace the rules with the new ones that work with labelled formulas instead of structural constraints. Below, we give as an example new rules for $\rightarrow$ and $\Diamond$[9] (with $|\mathsf{V}| = m + 1$):

$$\frac{w:1:\phi \rightarrow \phi' = 1}{w:1:\phi = 0 \left| \begin{array}{l} w:1:\phi = \frac{1}{m+1} \\ w:1:\phi' = \frac{1}{m+1} \end{array} \right| \begin{array}{l} w:1:\phi = \frac{1}{m+1} \\ w:1:\phi' = \frac{2}{m+1} \end{array} \left| \ldots \right| \begin{array}{l} w:1:\phi = \frac{m-1}{m+1} \\ w:1:\phi' = \frac{m}{m+1} \end{array} \left| w:1:\phi' = 1 \right.}$$

$$\frac{w:1:\Diamond\phi = \frac{r}{m+1}}{w\mathsf{R}w''; w'':1:\phi = \frac{r}{m+1}} \qquad \frac{w:1:\Diamond\phi = \frac{r}{m+1}; w\mathsf{R}w'}{w':1:\phi = 0 \mid \ldots \mid w':1:\phi = \frac{r-1}{m+1}}$$

[8] Satisfiability and falsifiability (non-validity) are reducible to each other using $\prec$: ϕ is satisfiable iff $\sim\!\sim(\phi \prec \mathbf{0})$ is falsifiable; ϕ is falsifiable iff $\sim\!\sim(\mathbf{1} \prec \phi)$ is satisfiable.

[9] Intuitively, for a value $1 > \mathsf{v} > 0$ of $\Diamond\phi$ at w, we add a new state that witnesses v, and for a state on the branch, we guess a value smaller than v. Other modal rules can be rewritten similarly.

We now show how to build a satisfying model for ϕ using polynomial space. We begin with $w_0 : 1 : \phi = 1$ and start applying propositional rules (first, those that do not require branching). If we implement a branching rule, we pick one branch and work only with it: either until the branch is closed, in which case we pick another one; until no more rules are applicable (then, the model is constructed); or until we need to apply a modal rule to proceed. At this stage, we need to store only the subformulas of ϕ with labels denoting their value at w_0.

Now we guess a modal formula (say, $w_0 : 2 : \Box\chi = \frac{1}{m+1}$) whose decomposition requires an introduction of a new state (w_1) and apply this rule. Then we apply all modal rules that use $w_0 \mathsf{R} w_1$ as a premise (again, if those require branching, we guess only one branch) and start from the beginning with the propositional rules. If we reach a contradiction, the branch is closed. Again, the only new entries to store are subformulas of ϕ (now, with fewer modalities), their values at w_1, and a relational term $w_0 \mathsf{R} w_1$. Since the depth of the model is $O(|\phi|)$ and since we work with modal formulas one by one, we need to store subformulas of ϕ with their values $O(|\phi|)$ times, so, we need only $O(|\phi|^2)$ space.

Finally, if no rule is applicable and there is no contradiction, we mark $w_0 : 2 : \Box\chi = \frac{1}{m+1}$ as 'safe'. Now we *delete all entries of the tableau below it* and pick another unmarked modal formula that requires an introduction of a new state. Dealing with these one by one allows us to construct the model branch by branch. But since the length of each branch of the model is bounded by $O(|\phi|)$ and since we delete *branches of the model* once they are shown to contain no contradictions, we need only polynomial space.

We end the section with two simple observations. First, Theorems 3 and 4 are applicable both to $\mathbf{K}\mathsf{biG}_{\mathsf{fb}}$ and $\mathbf{K}\mathsf{G}^2_{\mathsf{fb}}$ because the latter is conservative over the former. Secondly, since $\mathbf{K}\mathsf{G}^2$ and $\mathbf{K}\mathsf{biG}$ are conservative over $\mathfrak{GK}^c$ and since $\mathbf{K}$ can be embedded in $\mathfrak{GK}^c$, the lower bounds on complexity of a classical modal logic of some class of frames $\mathbb{K}$ and G^2 modal logic of $\mathbb{K}$ will coincide.

5 Concluding Remarks

In this paper, we developed a crisp modal expansion of the two-dimensional Gödel logic G^2 as well as an expansion of bi-Gödel logic with $\Box$ and $\Diamond$ both for crisp and fuzzy frames. We also established their connections with modal Gödel logics, and gave a complexity analysis of their finitely branching fragments.

The following steps are: to study the proof theory of $\mathbf{K}\mathsf{G}^2$ and $\mathbf{K}\mathsf{G}^2_{\mathsf{fb}}$: both in the form of Hilbert-style and sequent calculi; establish the decidability (or lack thereof) for the case of $\mathbf{K}\mathsf{G}^2$. Moreover, two-dimensional treatment of information invites for different modalities, e.g. those formalising aggregation strategies given in [6]—in particular, the cautious one (where the agent takes minima/infima of *both* positive and negative supports of a given statement) and the confident one (whereby the maxima/suprema are taken). Last but not least, while in this paper we assumed that our *access* to sources is crisp, one can argue that the *degree of our bias* towards the given source can be formalised via *fuzzy* frames. Thus, it would be instructive to construct a fuzzy version of $\mathbf{K}\mathsf{G}^2$.

In a broader perspective, we plan to provide a general treatment of two-dimensional modal logics of uncertainty. Indeed, within our project [5,6], we are formalising reasoning with heterogeneous and possibly incomplete and inconsistent information (such as crisp or fuzzy data, personal beliefs, etc.) in a modular fashion. This modularity is required because different contexts should be treated with different logics—indeed, not only the information itself can be of various nature but the reasoning strategies of different agents even applied to the same data are not necessarily the same either. Thus, since we wish to account for this diversity, we should be able to combine different logics in our approach.

References

1. Agostino, M.D.: Investigations into the complexity of some propositional calculi. Oxford University Computing Laboratory, Oxford (1990)
2. Avron, A., Konikowska, B.: Decomposition proof systems for Gödel-Dummett logics. Stud. Logica **69**(2), 197–219 (2001)
3. Baaz, M.: Infinite-valued Gödel logics with 0-1-projections and relativizations. In: Gödel 1996: Logical Foundations of Mathematics, Computer Science and Physics–Kurt Gödel's legacy, Brno, Czech Republic, August 1996, Proceedings, pp. 23–33. Association for Symbolic Logic (1996)
4. Belnap, Nuel D..: How a computer should think. In: Omori, H., Wansing, H. (eds.) New Essays on Belnap-Dunn Logic. SL, vol. 418, pp. 35–53. Springer, Cham (2019). https://doi.org/10.1007/978-3-030-31136-0_4
5. Bílková, M., Frittella, S., Kozhemiachenko, D.: Constraint tableaux for two-dimensional fuzzy logics. In: Das, A., Negri, S. (eds.) TABLEAUX 2021. LNCS (LNAI), vol. 12842, pp. 20–37. Springer, Cham (2021). https://doi.org/10.1007/978-3-030-86059-2_2
6. Bílková, M., Frittella, S., Majer, O., Nazari, S.: Belief based on inconsistent information. In: Martins, M.A., Sedlár, I. (eds.) Dynamic Logic. New Trends and Applications, pp. 68–86. Springer, Cham (2020). https://doi.org/10.1007/978-3-030-65840-3_5
7. Blackburn, P., Rijke, M.D., Venema, Y.: Modal logic. Cambridge Tracts in Theoretical Computer Science, vol. 53. Cambridge University Press, 4. print. with corr. edn. (2010)
8. Bobillo, F., Delgado, M., Gómez-Romero, J., Straccia, U.: Fuzzy description logics under Gödel semantics. Int. J. Approx. Reason. **50**(3), 494–514 (2009)
9. Bobillo, F., Delgado, M., Gómez-Romero, J., Straccia, U.: Joining Gödel and Zadeh fuzzy logics in fuzzy description logics. Int. J. Uncertain. Fuzziness Knowledge-Based Syst. **20**(04), 475–508 (2012)
10. Borgwardt, S., Distel, F., Peñaloza, R.: Decidable Gödel description logics without the finitely-valued model property. In: Fourteenth International Conference on the Principles of Knowledge Representation and Reasoning (2014)
11. Caicedo, X., Metcalfe, G., Rodríguez, R., Rogger, J.: A finite model property for Gödel modal logics. In: Libkin, L., Kohlenbach, U., de Queiroz, R. (eds.) WoLLIC 2013. LNCS, vol. 8071, pp. 226–237. Springer, Heidelberg (2013). https://doi.org/10.1007/978-3-642-39992-3_20
12. Caicedo, X., Rodriguez, R.: Standard Gödel modal logics. Stud. Logica **94**(2), 189–214 (2010). https://doi.org/10.1007/s11225-010-9230-1

13. Caicedo, X., Rodríguez, R.: Bi-modal Gödel logic over [0,1]-valued Kripke frames. J. Log. Comput. **25**(1), 37–55 (2015)
14. Chagrov, A., Zakharyaschev, M.: Modal Logic. Clarendon Press, Oxford (1997)
15. Cintula, P., Noguera, C.: Modal logics of uncertainty with two-layer syntax: a general completeness theorem. In: Kohlenbach, U., Barceló, P., de Queiroz, R. (eds.) WoLLIC 2014. LNCS, vol. 8652, pp. 124–136. Springer, Heidelberg (2014). https://doi.org/10.1007/978-3-662-44145-9_9
16. Dunn, J.M.: Intuitive semantics for first-degree entailments and 'coupled trees'. Philos. Stud. **29**(3), 149–168 (1976)
17. Ertola, R., Esteva, F., Flaminio, T., Godo, L., Noguera, C.: Paraconsistency properties in degree-preserving fuzzy logics. Soft Comput. **19**, 1–16 (2014). https://doi.org/10.1007/s00500-014-1489-0
18. Fagin, R., Halpern, J.Y., Moses, Y., Vardi, M.Y.: Reasoning About Knowledge. MIT Press, Cambridge (2003)
19. Ginsberg, M.: Multivalued logics: a uniform approach to reasoning in AI. Comput. Intell **4**, 256–316 (1988)
20. Grigolia, R., Kiseliova, T., Odisharia, V.: Free and projective bimodal symmetric gödel algebras. Stud. Logica **104**(1), 115–143 (2016)
21. Hähnle, R.: A new translation from deduction into integer programming. In: Calmet, J., Campbell, J.A. (eds.) AISMC 1992. LNCS, vol. 737, pp. 262–275. Springer, Heidelberg (1993). https://doi.org/10.1007/3-540-57322-4_18
22. Hähnle, R.: Many-valued logic and mixed integer programming. Ann. Math. Artif. Intell. **12**(3–4), 231–263 (1994)
23. Hájek, P.: Metamathematics of Fuzzy Logic. Trends in Logic, 4th edn. Springer, Dordrecht (1998). https://doi.org/10.1007/978-94-011-5300-3
24. Jansana, R., Rivieccio, U.: Residuated bilattices. Soft. Comput. **16**(3), 493–504 (2012)
25. Kułacka, A., Pattinson, D., Schröder, L.: Syntactic labelled tableaux for Łukasiewicz fuzzy ALC. In: Twenty-Third International Joint Conference on Artificial Intelligence. AAAI Press (2013)
26. Lascio, L.D., Gisolfi, A.: Graded tableaux for rational Pavelka logic. Int. J. Intell. Syst. **20**(12), 1273–1285 (2005)
27. Leitgeb, H.: Hype: a system of hyperintensional logic (with an application to semantic paradoxes). J. Philos. Log. **48**(2), 305–405 (2019)
28. Metcalfe, G., Olivetti, N.: Proof systems for a Gödel modal logic. In: Giese, M., Waaler, A. (eds.) TABLEAUX 2009. LNCS (LNAI), vol. 5607, pp. 265–279. Springer, Heidelberg (2009). https://doi.org/10.1007/978-3-642-02716-1_20
29. Metcalfe, G., Olivetti, N.: Towards a proof theory of Gödel modal logics. Log. Methods Comput. Sci. **7** (2011)
30. Metcalfe, G., Olivetti, N., Gabbay, D.: Proof Theory for Fuzzy Logics. Applied Logic Series, vol. 36. Springer, Dordrecht (2008). https://doi.org/10.1007/978-1-4020-9409-5
31. Moisil, G.: Logique modale. Disquisitiones mathematicae et physicae **2**, 3–98 (1942)
32. Negri, S.: Proof analysis in modal logic. J. Philos. Log. **34**(5–6), 507–544 (2005)
33. Odintsov, S., Wansing, H.: Routley star and hyperintensionality. J. Philos. Log. **50**, 33–56 (2021)
34. Omori, H., Wansing, H.: 40 years of FDE: an introductory overview. Stud. Logica. **105**(6), 1021–1049 (2017). https://doi.org/10.1007/s11225-017-9748-6
35. Rivieccio, U.: An algebraic study of bilattice-based logics. Ph.D. thesis, University of Barcelona – University of Genoa (2010)

36. Rodriguez, R.O., Vidal, A.: Axiomatization of Crisp Gödel Modal Logic. Stud. Logica **109**(2), 367–395 (2020). https://doi.org/10.1007/s11225-020-09910-5
37. Vakarelov, D.: Notes on N-lattices and constructive logic with strong negation. Stud. Logica **36**(1–2), 109–125 (1977)
38. Wansing, H.: Constructive negation, implication, and co-implication. J. Appl. Non-Classical Logics **18**(2–3), 341–364 (2008). https://doi.org/10.3166/jancl.18.341-364

Non-associative, Non-commutative Multi-modal Linear Logic

Eben Blaisdell[1], Max Kanovich[2], Stepan L. Kuznetsov[3,4], Elaine Pimentel[2(✉)], and Andre Scedrov[1]

[1] Department of Mathematics, University of Pennsylvania, Philadelphia, USA
[2] Department of Computer Science, University College London, London, UK
elaine.pimentel@gmail.com
[3] Steklov Mathematical Institute of RAS, Moscow, Russia
[4] Faculty of Computer Science, HSE University, Moscow, Russia

Abstract. Adding multi-modalities (called *subexponentials*) to linear logic enhances its power as a logical framework, which has been extensively used in the specification of *e.g.* proof systems, programming languages and bigraphs. Initially, subexponentials allowed for classical, linear, affine or relevant behaviors. Recently, this framework was enhanced so to allow for commutativity as well. In this work, we close the cycle by considering associativity. We show that the resulting system (acLL_Σ) admits the (multi)cut rule, and we prove two undecidability results for fragments/variations of acLL_Σ.

1 Introduction

Resource aware logics have been object of passionate study for quite some time now. The motivations for this passion vary: resource consciousness are adequate for modeling steps of computation; logics have interesting algebraic semantics; calculi have nice proof theoretic properties; multi-modalities allow for the specification of several behaviors; there are many interesting applications in linguistics, etc.

With this variety of subjects, applications and views, it is not surprising that different groups developed different systems based on different principles. For example, the Lambek calculus (L) [29] was introduced for mathematical modeling of natural language syntax, and it extends a basic categorial grammar [3,4] by a concatenation operator. Linear logic (LL) [16], originally discovered by Girard from a semantical analysis of the models of polymorphic λ-calculus, turned out to be a refinement of classical and intuitionistic logic, having the dualities of the former and constructive properties of the

The work of Max Kanovich was partially supported by EPSRC Programme Grant EP/R006865/1: "Interface Reasoning for Interacting Systems (IRIS)."
The work of Stepan L. Kuznetsov was supported by the Theoretical Physics and Mathematics Advancement Foundation "BASIS" and partially performed within the framework of HSE University Basic Research Program and within the project MK-1184.2021.1.1 "Algorithmic and Semantic Questions of Modal Extensions of Linear Logic" funded by the Ministry of Science and Higher Education of Russia.
Elaine Pimentel acknowledges partial support by the MOSAIC project (EU H2020-MSCA-RISE-2020 Project 101007627).

J. Blanchette et al. (Eds.): IJCAR 2022, LNAI 13385, pp. 449–467, 2022.
https://doi.org/10.1007/978-3-031-10769-6_27

latter. The key point is the presence of the *modalities* !, ?, called *exponentials* in LL. In the intuitionistic version of LL, denoted by ILL, only the ! exponential is present.

L and LL were compared in [2], when Abrusci showed that Lambek calculus coincides with a variant of the non-commutative, multiplicative version of ILL [41]. This correspondence can be lifted for considering also the additive connectives: Full (multiplicative-additive) Lambek calculus FL relates to non-commutative multiplicative-additive version of ILL, here denoted by cLL.

In this paper we propose the sequent based system acLL_Σ, a conservative extension of cLL, where associativity is allowed only for formulas marked with a special kind of modality, determined by a *subexponential signature* Σ. The notation adopted is modular, uniform and scalable, in the sense that many well known systems will appear as fragments or special cases of acLL_Σ, by only modifying the signature Σ. The core fragment of acLL_Σ (*i.e.*, without the subexponentials) corresponds to the non-associative version of full Lambek calculus, FNL [8].[1]

The language of acLL_Σ consists of a denumerable infinite set of propositional variables $\{p, q, r, \ldots\}$, the unities $\{1, \top\}$, the binary connectives for additive conjunction and disjunction $\{\&, \oplus\}$, the non-commutative multiplicative conjunction $\otimes$, the non-commutative linear implications $\{\rightarrow, \leftarrow\}$, and the unary subexponentials $!^i$, with i belonging to a pre-ordered set of labels $(I, \preceq)$.

Roughly speaking, subexponentials [13] are substructural multi-modalities. In LL, $!A$ indicates that the linear formula A behaves *classically*, that is, it can be contracted *and* weakened. Labeling ! with indices allows moving one step further: The set I can be partitioned so that, in $!^i A$, A can be contracted *and/or* weakened. This allows for two other types of behavior (other than classical or linear): affine (only weakening) or relevant (only contraction). Pre-ordering the labels (together with an upward closeness requirement) guarantees cut-elimination [42]. But then, why consider only weakening and contraction? Why not also take into account other structural properties, like commutativity or associativity? In [20,21] commutativity was added to the picture, so that in $!^i A$, A can be contracted, weakened, classical or linear, but it may also commute with the neighbor formula. In this work we consider the last missing part: Associativity.

Smoothly extending cLL to allow consideration of the non-associative case is non trivial. This requires a structural recasting/reframing of sequents: we pass from sets/multisets to lists in the non-commutative case, onto trees in the case of non-associativity [28]. As a consequence, the inference rules should act deeply over formulas in tree-structured sequents, which can be tricky in the presence of modalities [17].

On the other side, the multi-modal Lambek calculus introduced in [35,45] and extended/compiled/implemented in [18,36–38][2] use different *families of connectives and contexts*, distinguished by means of indices, or *modes*. Contexts are indexed binary trees, with formulas built from the indexed adjoint connectives $\{\rightarrow_i, \leftarrow_i\}$ and $\otimes_i$ (*e.g.*

[1] The multiplicative fragment of acLL_Σ is the non-associative version of Lambek's calculus, NL, introduced by Lambek himself in [30]. Both the associative calculus L and the non-associative calculus NL have their advantages and disadvantages for the analysis of natural language syntax, as we discuss in more detail in Sect. 2.2.

[2] The Grail family of theorem provers [37] works with a variety of modern type-logical frameworks, including multimodal type-logical grammars.

$(A \rightarrow_i B, (C \otimes_j D, H)^k)^i$). Each mode has its own set of logical rules (following the same rule scheme), and different structural features can be combined via the mode information on the formulas. This gives to the resulting system a multi-modal flavor, but it also results in a language of *binary connectives*, determined by the modes. This forces an unfortunate second level synchronization between implications and tensor, and modalities act over whole *sequents*, not on single *formulas*.

In order to attribute particular resource management properties to individual resources, in [27,33] explicit (classical) multi-modalities $\Diamond_i, \Box_i$ were proposed. While such unary modalities were inspired in LL exponentials, the resemblance stops there. First of all, the logical connectives come together with structural constructors for contexts, which turns $\Diamond_i, \Box_i$ into truncated forms of product and implication.

Second, $\Diamond_i, \Box_i$ have a *temporal behavior*, in the sense that $\Diamond\Box F \Rightarrow F$ and $F \Rightarrow \Box\Diamond F$, which are not provable in LL using the "natural interpretation" $\Diamond = ?, \Box = !$.

In this paper, multi-modality is *totally local*, given by the subexponentials. The signature Σ contains the pre-ordered set of labels, together with a function stating which axioms, among weakening, contraction, exchange and associativity, are assumed for each label. Sequents will have a *nested structure*, corresponding to trees of formulas. And rules will be applied deeply in such structures. This not only gives the LL based system a more modern presentation (based on nested systems, like *e.g.* in [10,15]), but it also brings the notation closer to the one adopted by the Lambek community, like in [25]. Finally, it also uniformly extends several LL based systems present in the literature, as Example 8 in the next section shows.

Designing a good system serves more than simple pure proof-theoretic interests: Well behaved, neat proof systems can be used in order to approach several important problems, such as interpolation, complexity and decidability. And decidability of extensions/variants/fragments of L and LL is a fascinating subject of study, since the presence or absence of substructural properties/connectives may completely change the outcome. Indeed, it is well known that LL is undecidable [32], but adding weakening (affine LL) turns the system decidable [24], while removing the additives (MELL – multiplicative, exponential LL) reaches the border of knowledge: It is a long standing open problem [50]. Non-associativity also alters decidability and complexity: L is NP-complete [47], while NL is decidable in polynomial time [1,6]. Finally, the number of subexponentials also plays a role in decision problems: MELL with two subexponentials is undecidable [9].

In this work, we will present two undecidability results, all orbiting (but not encompassing) MELL/FNL. First, we show that acLL_Σ containing the multiplicatives $\otimes, \rightarrow$, the additive $\oplus$ and one classical subexponential (allowing contraction and weakening) is undecidable. This is a refinement of the unpublished result by Tanaka [51], which states that FNL plus one fully-powered subexponential is undecidable.

In the second undecidability result, we keep two subexponentials, but with a minimalist configuration: the implicational fragment of the logic plus two subexponentials: the "main" one allowing for contraction, exchange, and associativity (weakening is optional), and an "auxiliary" one allowing only associativity. This is a variation of Chaudhuri's result (in the non-associative, non-commutative case), making use of fewer connectives (tensor is not needed) and less powerful subexponentials.

Table 1. Acronyms/decidability of systems mentioned in the paper.

Acronym	System	Decidable?
L	Lambek calculus	✓
LL	(propositional) linear logic	✗
ILL	intuitionistic LL	✗
MALL	multiplicative-additive LL	✓
iMALL	intuitionistic MALL	✓
FL	full (multiplicative-additive) L	✓
cLL	non-commutative iMALL	✓
acLL_Σ	non-commutative, non-associative ILL with subexponentials	–
NL	non-associative L	✓
FNL	full (multiplicative-additive) NL	✓
MELL	multiplicative-exponential LL	unknown
SDML	simply dependent multimodal linear logics	–
SMALC_Σ	FL with subexponentials	–

The rest of the paper is organized as follows: Sect. 2 presents the system acLL_Σ, showing that it has the cut-elimination property and presenting an example in linguistics; Sect. 3 shows the undecidability results; and Sect. 4 concludes the paper.

We have placed, in Table 1, the acronyms for and decidability of all considered systems. Decidability for the cases marked with "–" depends on the signature Σ.

2 A Nested System for Non-associativity

Similar to modal connectives, the exponential ! in ILL is not *canonical* [13], in the sense that if $i \neq j$ then $!^i F \not\equiv !^j F$. Intuitively, this means that we can mark the exponential with *labels* taken from a set I organized in a pre-order $\preceq$ (*i.e.*, reflexive and transitive), obtaining (possibly infinitely-many) exponentials ($!^i$ for $i \in I$). Also as in multi-modal systems, the pre-order determines the provability relation: for a general formula F, $!^b F$ *implies* $!^a F$ iff $a \preceq b$.

The algebraic structure of subexponentials, combined with their intrinsic structural property allow for the proposal of rich linear logic based frameworks. This opened a venue for proposing different multi-modal substructural logical systems, that encountered a number of different applications. Originally [42], subexponentials could assume only weakening and contraction axioms:

$$\mathsf{C}: \quad !^i F \rightarrow !^i F \otimes !^i F \qquad \mathsf{W}: \quad !^i F \rightarrow 1$$

This allows the specification of systems with multiple contexts, which may be represented by sets or multisets of formulas [44], as well as the specification and verification of concurrent systems [43], and biological systems [46]. In [20,21], non-commutative systems allowing commutative subexponentials were presented:

$$\mathsf{E}: \quad (!^i F) \otimes G \equiv G \otimes (!^i F)$$

and this has many applications, *e.g.*, in linguistics [21].

In this work, we will present a non-commutative, non-associative linear logic based system, and add the possibility of assuming associativity[3]

$$\mathsf{A1}: \ !^i F \otimes (G \otimes H) \equiv (!^i F \otimes G) \otimes H \qquad \mathsf{A2}: \ (G \otimes H) \otimes !^i F \equiv G \otimes (H \otimes !^i F)$$

as well as commutativity and other structural properties.

We start by presenting an adaption of simply dependent multimodal linear logics (SDML) appearing in [31] to the non-associative/commutative case.

The language of non-commutative SDML is that of (propositional intuitionistic) linear logic with subexponentials [21] supplied with the *left residual*; or similarly, that of FL with subexponentials. Non-associative contexts will be organized via binary trees, here called *structures*.

Definition 1 (Structured sequents). Structures *are formulas or pairs containing structures:*

$$\Gamma, \Delta := F \mid (\Gamma, \Gamma)$$

where the constructors may be empty but never a singleton.

An n-ary context $\Gamma\left\{^1\right\}\ldots\left\{^n\right\}$ *is a context that contains n pairwise distinct numbered* holes $\{\,\}$ *wherever a formula may otherwise occur. Given n contexts $\Gamma_1, \ldots, \Gamma_n$, we write $\Gamma\{\Gamma_1\}\cdots\{\Gamma_n\}$ for the context where the k-th hole in $\Gamma\left\{^1\right\}\ldots\left\{^n\right\}$ has been replaced by Γ_k (for $1 \leq k \leq n$). If $\Gamma_k = \varnothing$ the hole is removed.*

A structured sequent *(or simply* sequent*) has the form $\Gamma \Rightarrow F$ where Γ is a structure and F is a formula.*

Example 2. Structures are binary trees, with formulas as leaves and commas as nodes. The structure $!^i A, (B, C)$ represents the tree below left, while $(!^i A, B), C$ represents the tree below right

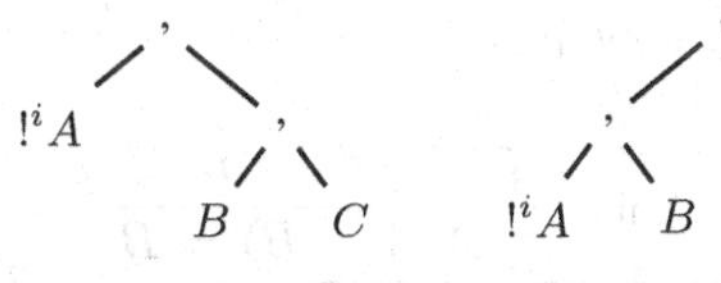

Definition 3 (SDML). *Let $\mathcal{A}$ be a set of axioms. A (non-associative/commutative)* simply dependent multimodal logical system *(*SDML*) is given by a triple $\Sigma = (I, \preccurlyeq, f)$, where I is a set of indices, $(I, \preccurlyeq)$ is a pre-order, and f is a mapping from I to $2^{\mathcal{A}}$.*

If Σ is a SDML, *then the* logic described by Σ *has the modality $!^i$ for every $i \in I$, with the rules of* FNL *depicted in Fig. 1, together with rules for the axioms $f(i)$ and the interaction axioms $!^j A \rightarrow !^i A$ for every $i, j \in I$ with $i \preccurlyeq j$. Finally, every* SDML *is assumed to be upwardly closed w.r.t. $\preceq$, that is, if $i \preceq j$ then $f(i) \subseteq f(j)$ for all $i, j \in I$.*

[3] Note that the implemented rules in Fig. 2 reflect the left to right direction of such axioms only.

Figure 2 presents the structured system acLL_Σ, for the logic described by the SDML determined by Σ, with $\mathcal{A} = \{\mathsf{C}, \mathsf{W}, \mathsf{A1}, \mathsf{A2}, \mathsf{E}\}$ where, in the subexponential rule for $\mathsf{S} \in \mathcal{A}$, the respective $s \in I$ is such that $\mathsf{S} \in f(s)$ (*e.g.* the subexponential symbol e indicates that $\mathsf{E} \in f(e)$). We will denote by $!^{\mathsf{Ax}}\Delta$ the fact that the structure Δ contains only banged formulas as leaves, each of them assuming the axiom Ax.

As an economic notation, we will write $\uparrow i$ for the *upset* of the index i, *i.e.*, the set $\{j \in I : i \preccurlyeq j\}$. We extend this notation to structures in the following way. Let Γ be a structure containing only banged formulas as leaves. If such formulas admit the multiset partition

$$\{!^j F \in \Gamma : i \preccurlyeq j\} \cup \{!^k F \in \Gamma : i \not\preccurlyeq k \text{ and } \mathsf{W} \in f(k)\}$$

then $\Gamma^{\uparrow i}$ is the structure obtained from Γ by easing the formulas in the second component of the partition (equivalently, the substructure of Γ formed with all and only formulas of the first component of the partition). Otherwise, $\Gamma^{\uparrow i}$ is undefined.

Example 4. Let $\Gamma = (!^i A, (!^j B, !^k C))$ be represented below left, $i \preceq j$ but $i \not\preceq k$, and $\mathsf{W} \in f(k)$. Then $\Gamma^{\uparrow i} = (!^i A, !^j B)$ is depicted below right

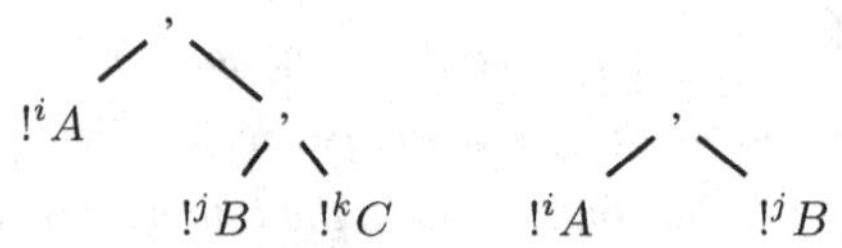

Observe that, if $\mathsf{W} \notin f(k)$, then $\Gamma^{\uparrow i}$ cannot be built. In this case, any derivation of $\Gamma \Rightarrow !^i(A \otimes B)$ cannot start with an application of the promotion rule $!^i R$ (similarly to how promotion in ILL cannot be applied in the presence of non-classical contexts). In this case, if A, B are atomic, this sequent would not be provable.

Example 5. The use of subexponentials to deal with associativity can be illustrated by the prefixing sequent $A \to B \Rightarrow (C \to A) \to (C \to B)$: It is not provable for an arbitrary formula C, but if $C = !^a C'$, then

$$\dfrac{\dfrac{\dfrac{\dfrac{\dfrac{}{!^aC' \Rightarrow !^aC'}\,\mathsf{init} \quad \dfrac{\dfrac{}{A \Rightarrow A}\,\mathsf{init} \quad \dfrac{}{B \Rightarrow B}\,\mathsf{init}}{(A, A \to B) \Rightarrow B}\,{\to}L}{((!^aC', (!^aC' \to A)), (A \to B)) \Rightarrow B}\,{\to}L}{(!^aC', ((!^aC' \to A), (A \to B))) \Rightarrow B}\,\mathsf{A1}}{((!^aC' \to A), (A \to B)) \Rightarrow !^aC' \to B}\,{\to}R}{A \to B \Rightarrow (!^aC' \to A) \to (!^aC' \to B)}\,{\to}R$$

2.1 Cut-Elimination

When it comes to the proof of cut-elimination for acLL_Σ, the cut reductions for the propositional connectives follow the standard steps for similar systems such as, *e.g.*, Moot and Retoré's system $\mathsf{NL}\Diamond$ in [38, Chapter 5.2.2]. The case of structural rules, on the other hand, should be treated with care.

Propositional rules

$$\frac{\Gamma\{(F,G)\} \Rightarrow H}{\Gamma\{F \otimes G\} \Rightarrow H}\ \otimes L \qquad \frac{\Gamma_1 \Rightarrow F \quad \Gamma_2 \Rightarrow G}{(\Gamma_1,\Gamma_2) \Rightarrow F \otimes G}\ \otimes R \qquad \frac{\Gamma\{F\} \Rightarrow H \quad \Gamma\{G\} \Rightarrow H}{\Gamma\{F \oplus G\} \Rightarrow H}\ \oplus L$$

$$\frac{\Gamma \Rightarrow F_i}{\Gamma \Rightarrow F_1 \oplus F_2}\ \oplus R_i \qquad \frac{\Gamma\{F_i\} \Rightarrow G}{\Gamma\{F_1 \,\&\, F_2\} \Rightarrow G}\ \& L_i \qquad \frac{\Gamma \Rightarrow F \quad \Gamma \Rightarrow G}{\Gamma \Rightarrow F \,\&\, G}\ \& R$$

$$\frac{\Delta \Rightarrow F \quad \Gamma\{G\} \Rightarrow H}{\Gamma\{(\Delta, F \to G)\} \Rightarrow H}\ \to L \qquad \frac{(F,\Gamma) \Rightarrow G}{\Gamma \Rightarrow F \to G}\ \to R \qquad \frac{\Delta \Rightarrow F \quad \Gamma\{G\} \Rightarrow H}{\Gamma\{(G \leftarrow F, \Delta)\} \Rightarrow H}\ \leftarrow L$$

$$\frac{(\Gamma,F) \Rightarrow G}{\Gamma \Rightarrow G \leftarrow F}\ \leftarrow R \qquad \frac{\Gamma\{\} \Rightarrow F}{\Gamma\{1\} \Rightarrow F}\ 1L \qquad \frac{}{\Rightarrow 1}\ 1R \qquad \frac{}{\Gamma \Rightarrow \top}\ \top R$$

Initial and cut rules

$$\frac{}{F \Rightarrow F}\ \mathsf{init} \qquad \frac{\Delta \Rightarrow F \quad \Gamma\{^1F\}\ldots\{^nF\} \Rightarrow G}{\Gamma\{^1\Delta\}\ldots\{^n\Delta\} \Rightarrow G}\ \mathsf{mcut}$$

Fig. 1. Structured system FNL for non-associative, full Lambek calculus.

Subexponential rules

$$\frac{\Gamma^{\uparrow i} \Rightarrow F}{\Gamma \Rightarrow\, !^i F}\ !^i R \qquad \frac{\Gamma\{F\} \Rightarrow G}{\Gamma\{!^i F\} \Rightarrow G}\ \mathsf{der}$$

Structural rules

$$\frac{\Gamma\{((!^a\Delta_1,\Delta_2),\Delta_3)\} \Rightarrow G}{\Gamma\{(!^a\Delta_1,(\Delta_2,\Delta_3))\} \Rightarrow G}\ \mathsf{A1} \qquad \frac{\Gamma\{(\Delta_1,(\Delta_2,!^a\Delta_3))\} \Rightarrow G}{\Gamma\{((\Delta_1,\Delta_2),!^a\Delta_3)\} \Rightarrow G}\ \mathsf{A2} \qquad \frac{\Gamma\{(\Delta_2,!^e\Delta_1)\} \Rightarrow G}{\Gamma\{(!^e\Delta_1,\Delta_2)\} \Rightarrow G}\ \mathsf{E1}$$

$$\frac{\Gamma\{(!^e\Delta_2,\Delta_1)\} \Rightarrow G}{\Gamma\{(\Delta_1,!^e\Delta_2)\} \Rightarrow G}\ \mathsf{E2} \qquad \frac{\Gamma\{\} \Rightarrow G}{\Gamma\{!^w\Delta\} \Rightarrow G}\ \mathsf{W} \qquad \frac{\Gamma\{^1\,!^c\Delta\}\ldots\{^n\,!^c\Delta\} \Rightarrow G}{\Gamma\{^1\}\ldots\{^k\,!^c\Delta\}\ldots\{^n\} \Rightarrow G}\ \mathsf{C}$$

Fig. 2. Structured system acLL_Σ for the logic described by Σ.

Theorem 6. *If the sequent $\Gamma \Rightarrow F$ is provable in acLL_Σ, then it has a proof with no instances of the rule* mcut.

Proof. The most representative cases of cut reductions involving subexponentials are detailed next. In order to simplify the notation, when possible, the mcut rule is presented in its simple form, with an 1-ary context.

Case $!^a$: Suppose that

$$\dfrac{\dfrac{\overset{\pi_1}{\Delta_1^{\uparrow a} \Rightarrow F}}{\Delta_1 \Rightarrow\, !^a F}\ !^a R \qquad \dfrac{\overset{\pi_2}{\Gamma\{((!^a F,\Delta_2),\Delta_3)\} \Rightarrow G}}{\Gamma\{(!^a F,(\Delta_2,\Delta_3))\} \Rightarrow G}\ \mathsf{A1}}{\Gamma\{(\Delta_1,(\Delta_2,\Delta_3))\} \Rightarrow G}\ \mathsf{mcut}$$

Since axioms are upwardly closed w.r.t. $\preceq$, it must be the case that $\Delta_1^{\uparrow a}$ contains only formulas marked with subexponentials allowing associativity. All the other formulas in Δ_1 can be weakened; this is guaranteed by the application of the rule $!^a R$ in π_1. Hence the derivation above reduces to

$$\dfrac{\dfrac{\dfrac{\dfrac{\overset{\pi_1}{\Delta_1^{\uparrow a} \Rightarrow F}}{\Delta_1^{\uparrow a} \Rightarrow !^a F}\ !^a R \qquad \overset{\pi_2}{\Gamma\{(!^a F, \Delta_2), \Delta_3)\} \Rightarrow G}}{\Gamma\left\{((\Delta_1^{\uparrow a}, \Delta_2), \Delta_3)\right\} \Rightarrow G}\ \mathsf{mcut}}{\Gamma\left\{(\Delta_1^{\uparrow a}, (\Delta_2, \Delta_3))\right\} \Rightarrow G}\ \mathsf{A1}}{\Gamma\{(\Delta_1, (\Delta_2, \Delta_3))\} \Rightarrow G}\ \mathsf{W}$$

Case $!^c$: Suppose that

$$\dfrac{\dfrac{\overset{\pi_1}{\Delta^{\uparrow c} \Rightarrow F}}{\Delta \Rightarrow !^c F}\ !^c R \qquad \dfrac{\overset{\pi_2}{\Gamma\{!^c F\}\dots\{!^c F\}\dots\{!^c F\} \Rightarrow G}}{\Gamma\{\}\dots\{!^c F\}\dots\{\} \Rightarrow G}\ \mathsf{C}}{\Gamma\{\}\dots\{\Delta\}\dots\{\} \Rightarrow G}\ \mathsf{mcut}$$

Since $\Delta^{\uparrow c}$ contains only formulas marked with subexponentials allowing contraction, the derivation above reduces to

$$\dfrac{\dfrac{\dfrac{\dfrac{\overset{\pi_1}{\Delta^{\uparrow c} \Rightarrow F}}{\Delta^{\uparrow c} \Rightarrow !^c F}\ !^c R \qquad \overset{\pi_2}{\Gamma\{!^c F\}\dots\{!^c F\}\dots\{!^c F\} \Rightarrow G}}{\Gamma\{\Delta^{\uparrow c}\}\dots\{\Delta^{\uparrow c}\}\dots\{\Delta^{\uparrow c}\} \Rightarrow G}\ \mathsf{mcut}}{\Gamma\{\}\dots\{\Delta^{\uparrow c}\}\dots\{\} \Rightarrow G}\ \mathsf{C}}{\Gamma\{\}\dots\{\Delta\}\dots\{\} \Rightarrow G}\ \mathsf{W}$$

Observe that here, as usual, the multicut rule is needed in order to reduce the cut complexity.

Case $!^i R$: Suppose that

$$\dfrac{\dfrac{\overset{\pi_1}{\Delta^{\uparrow i} \Rightarrow F}}{\Delta \Rightarrow !^i F}\ !^i R \qquad \dfrac{\overset{\pi_2}{(\Gamma\{!^i F\})^{\uparrow j} \Rightarrow G}}{\Gamma\{!^i F\} \Rightarrow !^j G}\ !^j R}{\Gamma\{\Delta\} \Rightarrow !^j G}\ \mathsf{mcut}$$

If $j \not\preceq i$, then it should be the case that $\mathsf{W} \in f(i)$ and $\left(\Gamma\{!^i F\}\right)^{\uparrow j} = \Gamma\{\}^{\uparrow j}$, since $!^i F$ will be weakened in the application of rule $!^j R$. Hence, all formulas in Δ can be weakened as well and the reduction is

$$\dfrac{\dfrac{\overset{\pi_2}{\Gamma\{\}^{\uparrow j} \Rightarrow G}}{\Gamma\{\} \Rightarrow !^j G}\ !^j R}{\Gamma\{\Delta\} \Rightarrow !^j G}\ \mathsf{W}$$

On the other hand, if $j \preceq i$, by transitivity all the formulas in $\Delta^{\uparrow i}$ also have this property (implying that $\Delta^{\uparrow i}$ is a substructure of $\Delta^{\uparrow j}$), and the rest of formulas of Δ can be weakened. Hence the derivation above reduces to

$$\dfrac{\dfrac{\dfrac{\dfrac{\pi_1}{\Delta^{\uparrow i} \Rightarrow F}}{\Delta^{\uparrow j} \Rightarrow !^i F}\ !^i R \qquad \dfrac{\pi_2}{(\Gamma\{!^i F\})^{\uparrow j} \Rightarrow G}}{(\Gamma\{\Delta\})^{\uparrow j} \Rightarrow G}}{\Gamma\{\Delta\} \Rightarrow !^j G}\ !^j R$$

The other cases for subexponentials are similar or simpler. □

The next examples illustrate what we mean by acLL_Σ being a "conservative extension" of subsystems and variants. Indeed, although we remove structural properties of the core LL, subexponentials allow them to be added back, either locally or globally.

Example 7 (Structural variants of iMALL*).* Adding combinations of contraction C and / or weakening W for *arbitrary formulas* to additive-multiplicative intuitionistic linear logic (iMALL) yields, respectively, propositional intuitionistic logic ILP = iMALL + {C, W}, and the intuitionistic versions of affine linear logic aLL = iMALL + W and relevant logic R = iMALL + C. For the sake of presentation we overload the notation and use the connectives of linear logic also for these logics. In order to embed the logics above into acLL_Σ, let $\alpha \in \{\mathsf{ILP}, \mathsf{aLL}, \mathsf{R}\}$ and consider modalities $!^\alpha$ with $f(\alpha) = \{\mathsf{E}, \mathsf{A1}, \mathsf{A2}\} \cup \mathcal{A}$ where $\mathcal{A} \subseteq \{\mathsf{C}, \mathsf{W}\}$ is the set of axioms whose corresponding rules are in α. The translation τ_α prefixes *every subformula* with the modality $!^\alpha$. For $\mathcal{L} \in \{\mathsf{ILP}, \mathsf{aLL}, \mathsf{R}\}$ it is then straightforward to show that a structured sequent S is cut-free derivable in $\mathcal{L}$ iff its translation $\tau_\alpha(S)$ is cut-free derivable in the logic described by $(\{\alpha\}, \preccurlyeq, f)$ with $\preccurlyeq$ the obvious relation, and f as given above.

Example 8 (Structural variants of FNL*).* Following the same script as above and starting from FNL:

- considering $f(\alpha) = \mathcal{A} \subseteq \{\mathsf{E}, \mathsf{A1}, \mathsf{A2}\}$;
  - If $\mathcal{A} = \{\mathsf{A1}, \mathsf{A2}\}$, then we obtain the system FL;
  - If $\mathcal{A} = \{\mathsf{E}, \mathsf{A1}, \mathsf{A2}\}$ then the resulting system corresponds to iMALL.
  - Adding C, W as options to $\mathcal{A}$ will result the affine/relevant versions of the systems above.
- in a pre-order $(I, \preceq)$, if $f(i) = \{\mathsf{A1}, \mathsf{A2}\} \cup \mathcal{A}_i$ where $\mathcal{A}_i \subseteq \{\mathsf{E}, \mathsf{C}, \mathsf{W}\}$ for each $i \in I$, then the resulting system corresponds to SMALC_Σ in [21] (that is, the extension of FL with subexponentials).

2.2 An Example in Linguistics

Since its inception, Lambek calculus [29] has been applied to the modeling of natural language syntax by means of categorial grammars. In a categorial grammar, each word is assigned one or several Lambek formulas, which serve as syntactic categories. For a simple example, *John* and *Mary* are assigned np ("noun phrase") and *loves* gets

$(np \to s) \leftarrow np$. Here s stands for "sentence", and *loves* is a transitive verb, which lacks noun phrases on both sides to become a sentence. Grammatical validity of *"John loves Mary"* is supported by derivability of the sequent $np, (np \to s) \leftarrow np, np \Rightarrow s$. Notice that this derivability keeps valid also in the non-associative setting, if the correct nested structure is provided: $(np, ((np \to s) \leftarrow np, np)) \Rightarrow s$.

The original Lambek calculus L is associative. In some cases, however, associativity leads to over-generation, *i.e.*, validation of grammatically incorrect sentences. Lambek himself realized this and proposed the non-associative calculus NL in [30]. We will illustrate this issue with the example given in [38, Sect. 4.2.2]. The syntactic category assignment is as follows (where n stands for "noun"):

Words	Types
the	$np \leftarrow n$
Hulk	n
is	$(np \to s) \leftarrow (n \leftarrow n)$
green, incredible	$n \leftarrow n$

With this assignment, sentences *"The Hulk is green"* and *"The Hulk is incredible"* are correctly marked as valid, by deriving the sequent

$$(np \leftarrow n, n), ((np \to s) \leftarrow (n \leftarrow n), n \leftarrow n) \Rightarrow s$$

However, in the associative setting the sequent for the phrase *"The Hulk is green incredible,"* which is grammatically incorrect, also becomes derivable:

$$np \leftarrow n, n, (np \to s) \leftarrow (n \leftarrow n), n \leftarrow n, n \leftarrow n \Rightarrow s,$$

essentially due to derivability of $n \leftarrow n, n \leftarrow n \Rightarrow n \leftarrow n$.

In other situations, however, associativity is useful. Standard examples include handling of dependent clauses, *e.g.*, *"the girl whom John loves,"* which is validated as a noun phrase by the following derivable sequent:

$$np \leftarrow n, n, (n \to n) \leftarrow (s \leftarrow np), np, (np \to s) \leftarrow np \Rightarrow np$$

Here $(n \to n) \leftarrow (s \leftarrow np)$ is the syntactic category for *who.*

Our subexponential extension of NL, however, handles this case using local associativity instead of the global one. Namely, the category for *whom* now becomes $(n \to n) \leftarrow (s \leftarrow !^a np)$, where $!^a$ is a subexponential which allows the A2 rule, and the following sequent is happily derivable:

$$np \leftarrow n, (n, ((n \to n) \leftarrow (s \leftarrow !^a np), (np, (np \to s) \leftarrow np))) \Rightarrow np$$

The necessity of this more fine-grained control of associativity, instead of a global associativity rule, is seen via a combination of these examples. Namely, we talk about sentences like *"The superhero whom Hawkeye killed was incredible"* and *"... was green"*. With $!^a$, each of them is handled in the same way as the previous examples:

$$(np \leftarrow n, (n, ((n \to n) \leftarrow (s \leftarrow !^a np), (np, (np \to s) \leftarrow np)))), \\ ((np \to s) \leftarrow (n \leftarrow n), n \leftarrow n) \Rightarrow s.$$

On one hand, without $!^a$ this sequent cannot be derived in the non-associative system. On the other hand, if we make the system globally associative, it would validate incorrect sentences like *"The superhero whom Hawkeye killed was green incredible."*

3 Some Undecidability Results

Non-associativity makes a significant difference in decidability and complexity matters. For example, while L is NP-complete [47], NL is decidable in polynomial time [1,14].

For our system acLL_Σ, its decidability or undecidability depends on its signature Σ. In fact, we have a family of different systems acLL_Σ, with Σ as a parameter. Recall that the subexponential signature Σ controls not just the number of subexponentials and the preorder among them. More importantly, it dictates, for each subexponential, which structural rules this subexponential licenses. If for every $i \in I$ we have $\mathsf{C} \notin f(s)$, that is, no subexponential allows contraction, then acLL_Σ is clearly decidable, since the cut-free proof search space is finite. Therefore, for undecidability it is necessary to have at least one subexponential which allows contraction.

For a non-associative system with only one fully-powered exponential modality s (that is, $f(s) = \{\mathsf{E}, \mathsf{C}, \mathsf{W}, \mathsf{A1}, \mathsf{A2}\}$), undecidability was proven in a preprint by Tanaka [51], based on Chvalovský's [11] result on undecidability of the finitary consequence relation in FNL.

In this section, we prove two undecidability results. The first one is a refinement of Tanaka's result: We establish undecidability with at least one subexponential which allows contraction and weakening (commutativity/associativity are optional), in a subsystem containing only the additive connective $\oplus$ and the multiplicatives $\otimes$ and $\multimap$.

The second undecidability result is for the minimalistic, purely multiplicative fragment, which includes only $\multimap$ (not even $\otimes$). As a trade-off, however, it requires two subexponentials: the "main" one, which allows contraction, exchange, and associativity (weakening is optional), and an "auxiliary" one, which allows only associativity.

It should be noted that this undecidability result is orthogonal to Tanaka's [51], and the proof technique is essentially different. Indeed, Chvalovský's undecidability theorem does not hold for the non-associative Lambek calculus without additives, where the consequence relation is decidable [7].

Finally, we observe that *if* the intersection of these systems is decidable (which is still an open question), then our two undecidability results are *incomparable:* we have two undecidable fragments of acLL_Σ, but their common part, which includes only divisions and one exponential, would be decidable.

3.1 Undecidability with Additives and One Subexponential

We are going to derive the next theorem from undecidability of the finitary consequence relation in FNL [11]. Recall that FNL is, in fact, the fragment of acLL_Σ without subexponentials (that is, with an empty I).

Theorem 9. *If there exists such $s \in I$ that $f(s) \supseteq \{\mathsf{C}, \mathsf{W}\}$, then the derivability problem in acLL_Σ is undecidable. Moreover, this holds for the fragment with only $\otimes$, $\multimap$, $\oplus$, $!^s$.*

In fact, using C and W, one can also derive A1, A2, E1, and E2. Therefore, if $f(s) \supseteq \{\mathsf{C}, \mathsf{W}\}$, then $!^s$ is actually a full-power exponential modality. (In the proof of Theorem 9 below, we use only W and C rules, in order to avoid confusion.) However, Theorem 9 does not directly follow from undecidability of propositional linear logic [32], because here the basic system is non-associative and non-commutative, while linear logic is both associative and commutative. Thus, we need a different encoding for undecidability.

Let $\mathbf{\Phi}$ be a finite set of FNL sequents. By $\mathsf{FNL}(\mathbf{\Phi})$ let us denote FNL extended by adding sequents from $\mathbf{\Phi}$ as additional (non-logical) axioms. In general, $\mathsf{FNL}(\mathbf{\Phi})$ does not enjoy cut-elimination, so mcut is kept as a rule of inference in $\mathsf{FNL}(\mathbf{\Phi})$. A sequent $\Gamma \Rightarrow F$ is called *a consequence of* $\mathbf{\Phi}$ if this sequent is derivable in $\mathsf{FNL}(\mathbf{\Phi})$.

Theorem 10 (Chvalovský [11]**).** *The consequence relation in* FNL *is undecidable, that is, there exists no algorithm which, given* $\mathbf{\Phi}$ *and* $\Gamma \Rightarrow F$*, determines whether* $\Gamma \Rightarrow F$ *is a consequence of* $\mathbf{\Phi}$*. Moreover, undecidability keeps valid when* $\mathbf{\Phi}$ *and* $\Gamma \Rightarrow F$ *are built from variables using only* $\otimes$ *and* $\oplus$*.*

Now, in order to prove Theorem 9, we internalize $\mathbf{\Phi}$ into the sequent using $!^s$, assuming $f(s) \supseteq \{\mathsf{C}, \mathsf{W}\}$.

First we notice that we may suppose, without loss of generality, that all sequents in $\mathbf{\Phi}$ are of the form $\Rightarrow A$, that is, have empty antecedents. Namely, each sequent of the form $\Pi \Rightarrow B$ can be replaced by $\Rightarrow (\bigotimes \Pi) \to B$, where $\bigotimes \Pi$ is obtained from Π by replacing each comma with $\otimes$. Indeed, these sequents are derivable from one another: from $\Pi \Rightarrow B$ to $\Rightarrow (\bigotimes \Pi) \to B$ we apply a sequence of $\otimes L$ followed by $\to R$, and for the other direction we apply a series of cuts, first with $(\bigotimes \Pi, (\bigotimes \Pi) \to B) \Rightarrow B$, and then with $(F, G) \Rightarrow F \otimes G$ several times, for the corresponding subformulas of $\bigotimes \Pi$. The following embedding lemma ("modalized deduction theorem") holds.

Lemma 11. *The sequent* $\Gamma \Rightarrow F$ *is a consequence of* $\mathbf{\Phi} = \{\Rightarrow A_1, \ldots, \Rightarrow A_n\}$ *if and only if the sequent* $\big((\ldots((!^s A_1, !^s A_2), !^s A_3), \ldots, !^s A_n), \Gamma\big) \Rightarrow F$ *is derivable in* acLL_Σ.

Proof. Let us denote $(\ldots((!^s A_1, !^s A_2), !^s A_3), \ldots, !^s A_n)$ by $!\Phi$. Notice that C and W can be applied to $!\Phi$ as a whole; this is easily proven by induction on n.

For the "only if" direction let us take the derivation of $\Gamma \Rightarrow F$ in $\mathsf{FNL}(\mathbf{\Phi})$ (with cuts) and replace each sequent of the form $\Delta \Rightarrow G$ in it with $(!\Phi, \Delta) \Rightarrow G$, and each sequent of the form $\Rightarrow G$ with $!\Phi \Rightarrow G$. The translations of non-logical axioms from $\mathbf{\Phi}$ are derived as follows:

$$\dfrac{\dfrac{\dfrac{}{A_i \Rightarrow A_i}\ \mathsf{init}}{!^s A_i \Rightarrow A_i}\ \mathsf{der}}{!\Phi \Rightarrow A_i}\ \mathsf{W},\ n-1 \text{ times}$$

Translations of axioms init and $1R$ are derived from the corresponding original axioms by W, n times; $\top R$ remains valid.

Rules $\otimes L$, $\oplus L$, $\oplus R_i$, $\& L_i$, $\& R$, and $1L$ remain valid. For $\to L$, $\leftarrow L$, and mcut we contract $!\Phi$ as a whole:

$$\dfrac{\dfrac{(!\Phi, \Delta) \Rightarrow F \quad (!\Phi, \Gamma\{G\}) \Rightarrow H}{(!\Phi, \Gamma\{((!\Phi, \Delta), F \to G)\}) \Rightarrow H} \to L}{(!\Phi, \Gamma\{(\Delta, F \to G)\}) \Rightarrow H} \mathsf{C} \qquad \dfrac{\dfrac{(!\Phi, \Delta) \Rightarrow F \quad (!\Phi, \Gamma\{F\}\ldots\{F\}) \Rightarrow C}{(!\Phi, \Gamma\{(!\Phi, \Delta)\}\ldots\{(!\Phi, \Delta)\}) \Rightarrow C} \mathsf{mcut}}{(!\Phi, \Gamma\{\Delta\}\ldots\{\Delta\}) \Rightarrow C} \mathsf{C}$$

For $\otimes R$, $\to R$, and $\leftarrow R$, we combine contraction and weakening:

$$\dfrac{\dfrac{\dfrac{(!\Phi, \Gamma_1) \Rightarrow F \quad (!\Phi, \Gamma_2) \Rightarrow G}{((!\Phi, \Gamma_1), (!\Phi, \Gamma_2)) \Rightarrow F \otimes G} \otimes R}{(!\Phi, ((!\Phi, \Gamma_1), (!\Phi, \Gamma_2))) \Rightarrow F \otimes G} \mathsf{W}}{(!\Phi, (\Gamma_1, \Gamma_2)) \Rightarrow F \otimes G} \mathsf{C} \qquad \dfrac{\dfrac{\dfrac{(!\Phi, (F, \Gamma)) \Rightarrow G}{(!\Phi, (F, (!\Phi, \Gamma)) \Rightarrow G} \mathsf{W}}{(F, (!\Phi, \Gamma)) \Rightarrow G} \mathsf{C}}{(!\Phi, \Gamma) \Rightarrow F \to G} \to R$$

Notice that our original derivation was in $\mathsf{FNL}(\mathbf{\Phi})$, so it does not include rules operating subexponentials.

For the "if" direction we take a cut-free proof of $(!\Phi, \Gamma) \Rightarrow F$ in acLL_Σ and erase all formulas which include the subexponential. In the resulting derivation tree all rules and axioms, except those which operate $!^s$, remain valid. Structural rules for $!^s$ trivialize (since the !-formula was erased). The $!^s R$ rule could not have been used, since we do not have positive occurrences of $!^s F$, and our proof is cut-free.

Finally, der translates into

$$\dfrac{\Gamma\{A_i\} \Rightarrow G}{\Gamma\{\} \Rightarrow G}$$

This is modeled by cut with one of the sequents from $\mathbf{\Phi}$:

$$\dfrac{\Rightarrow A_i \quad \Gamma\{A_i\} \Rightarrow G}{\Gamma\{\} \Rightarrow G} \mathsf{mcut}$$

Thus, we get a correct derivation in $\mathsf{FNL}(\mathbf{\Phi})$. □

Theorem 10 and Lemma 11 immediately yield Theorem 9.

3.2 Undecidability Without Additives and with Two Subexponentials

Theorem 12. *If there are $a, c \in I$ such that $f(a) = \{\mathsf{A1}, \mathsf{A2}\}$ and $f(c) \supseteq \{\mathsf{C}, \mathsf{E}, \mathsf{A1}, \mathsf{A2}\}$, then the derivability problem in acLL_Σ is undecidable. Moreover, this holds for the fragment with only $\to$, $!^a$, and $!^c$.*

Remember from Example 8 that SMALC_Σ [21] denotes the extension of FL with subexponentials. The undecidability theorem above is proved by encoding the one-division fragment of SMALC_Σ containing one exponential c such that $f(c) \supseteq \{\mathsf{C}, \mathsf{E}\}$. It turns out that that such a system is undecidable.

Theorem 13 (Kanovich et al. [22,23]**).** *If there exists such $c \in I$ that $f(c) \supseteq \{\mathsf{C}, \mathsf{E}\}$, then the derivability problem in SMALC_Σ is undecidable. Moreover, this holds for the fragment with only $\to$ and $!^c$.*

Observe that SMALC_Σ can be obtained from acLL_Σ by adding "global" associativity rules:

$$\frac{\Gamma\{((\Delta_1, \Delta_2), \Delta_3)\} \Rightarrow G}{\Gamma\{(\Delta_1, (\Delta_2, \Delta_3))\} \Rightarrow G} \qquad \frac{\Gamma\{(\Delta_1, (\Delta_2, \Delta_3))\} \Rightarrow G}{\Gamma\{((\Delta_1, \Delta_2), \Delta_3)\} \Rightarrow G}$$

The usual formulation of SMALC_Σ, of course, uses sequences of formulas instead of nested structures as antecedents. The alternative formulation, however, would be more convenient for us now. It will be also convenient for us to regard all subexponentials in SMALC_Σ to be associative, that is, $f(s) \supseteq \{\mathsf{A1}, \mathsf{A2}\}$ for each $s \in I$.

In order to embed SMALC_Σ into acLL_Σ, we define two translations, $A^{!-}$ and $A^{!+}$, by mutual recursion:

$$\begin{array}{lll}
z^{!-} = {!^a} z & z^{!+} = z & \text{where } z \text{ is a variable,1, or } \top \\
(A \to B)^{!-} = {!^a}(A^{!+} \to B^{!-}) & (A \to B)^{!+} = A^{!-} \to B^{!+} & \\
(B \leftarrow A)^{!-} = {!^a}(B^{!-} \leftarrow A^{!+}) & (B \leftarrow A)^{!+} = B^{!+} \leftarrow A^{!-} & \\
(A \circledast B)^{!-} = {!^a}(A^{!-} \circledast B^{!-}) & (A \circledast B)^{!+} = A^{!+} \circledast B^{!+} & \text{where } \circledast \in \{\otimes, \oplus, \&\} \\
({!^s} A)^{!-} = {!^s}(A^{!-}) & ({!^s} A)^{!+} = {!^s}(A^{!+}) &
\end{array}$$

Informally, our translation adds a $!^a$ over any formula (not only over atoms) of negative polarity, unless this formula was already marked with a $!^s$. Thus, all formulae in antecedents would begin with either the new subexponential $!^a$ or one of the old subexponentials $!^s$, and all these subexponentials allow associativity rules $\mathsf{A1}$ and $\mathsf{A2}$.

Lemma 14. *A sequent $A_1, \ldots, A_n \Rightarrow B$ is derivable in SMALC_Σ if and only if its translation $(\ldots(A_1^{!-}, A_2^{!-}), \ldots, A_n^{!-}) \Rightarrow B^{!+}$ is derivable in acLL_Σ.*

Proof. For the "only if" part, let us first note that each formula $A_i^{!-}$ is of the form $!^s F$ and $\mathsf{A1}, \mathsf{A2} \in f(s)$. Indeed, either s is an "old" subexponential label (for which we added $\mathsf{A1}, \mathsf{A2}$) or $s = a$. Thus brackets can be freely rearranged in the antecedent.

Now we take a cut-free proof of $A_1, \ldots, A_n \Rightarrow B$ in SMALC_Σ and replace each sequent in it with its translation. Right rules for connectives other than subexponentials, i.e., $\otimes R$, $\oplus R_i$, $\& R$, $\to R$, and $\leftarrow R$, remain valid as they are, up to rearranging brackets in antecedents. For $!^i R$, we notice that the translation of a formula of the form $!^j F$, where $j \preceq i$, is also a formula of the form $!^j F'$. Thus, this rule also remains valid. The same holds for the dereliction rule der, because $(!^i F)^{!-}$ is exactly $!^i(F^{!-})$. Finally, the "old" structural rules (exchange, contraction, weakening) also remain valid (up to rearranging of brackets), since $!^i F$ gets translated into $!^i(F^{!-})$, which enjoys the same structural rules.

For the other left rules, we need to derelict $!^a$ first, and then perform the corresponding rule application. Rearrangement of brackets, if needed, is performed below dereliction or above the application of the rule in question.

The "if" part is easier. Given a derivation of $(\ldots(A_1^{!-}, A_2^{!-}), \ldots, A_n^{!-}) \Rightarrow B^{!+}$ in acLL_Σ, we erase $!^a$ everywhere, and consider it as a derivation in SMALC_Σ. Associativity rules for the erased $!^a$ (which are the only structural rules for this subexponential) keep valid, because now associativity is global. Dereliction and right introduction for $!^a$ trivialize. All other rules, which do not operate $!^a$, remain as they are. Thus, we get a derivation of $A_1, \ldots, A_n \Rightarrow B$ in SMALC_Σ, since erasing $!^a$ makes our translations just identical. □

4 Related Work and Conclusion

In this paper, we have presented acLL_Σ, a sequent-based system for non-associative, non-commutative linear logic with subexponentials. Starting form FNL, we modularly and uniformly added rules for exchange, associativity, weakening and contraction, which can be applied with the subexponentials having with the respective features. This allows for the application of structural rules locally, and it conservatively extends well known systems in the literature, continuing the path of controlling structural properties started by Girard himself [16].

Another approach to combining associative and non-associative behavior in Lambek-style grammars is the framework of *the Lambek calculus with brackets* by Morrill [39,40] and Moortgat [34]. The bracket approach is dual to ours: there the base system is associative, and brackets, which are controlled by bracket modalities, introduce local non-associativity. Both the associative Lambek calculus and the non-associative Lambek calculus can be embedded into the Lambek calculus with brackets: the former is just by design of the system and the latter was shown by Kurtonina [26] by constructing a translation.

From the point of view of generative power, however, the (associative) Lambek calculus with brackets is weaker than the non-associative system with subexponentials, which is presented in this paper. Namely, as shown by Kanazawa [19], grammars based on the Lambek calculus with brackets can generate only context-free languages. In contrast, grammars based on our system with subexponentials go beyond context-free languages, even when no subexponential allows contraction (subexponentials allowing contraction may lead to undecidability, as shown in the last section).

As a quick example, let us consider a subexponential $!^{ae}$ which allows both associativity (A1 and A2) and exchange (E). If we put this subexponential over any (sub)formula, the system becomes associative and commutative. Using this system, one can describe the non context-free language MIX_3, which contains all non-empty words over $\{a, b, c\}$, in which the numbers of a, b, and c are equal. Indeed, MIX_3 is the permutation closure of the language $\{(abc)^n \mid n \geq 1\}$. The latter is regular, therefore context-free, and therefore definable by a Lambek grammar. The ability of our system to go beyond context-free languages is important from the point of view of applications, since there are known linguistic phenomena which are essentially non-context-free [49].

Regarding decidability, let us compare our results with the more well-known associative non-commutative and associative commutative cases.

In the associative and commutative case the situation is as follows. In the presence of additives, the system is known to be undecidable with one exponential modality [32]. Without additives, we get MELL, the (un)decidability of which is a well-known open problem [50]. However, with two subexponentials MELL again becomes undecidable [9]. Thus, we have the same trade-off as in our non-associative non-commutative case: for undecidability one needs either additives, or two subexponentials.

Our results help to shed some light in the (un)decidability problem for the spectrum of logical systems surrounding MELL/FNL, allowing for a fine-grained analysis of the problem, specially the trade-offs on connectives and subexponentials for guaranteeing (un)decidability.

There is a lot to be done from now on. First of all, we would like to analyze better the minimalist fragment of acLL_Σ containing only implication and one fully-powered subexponential, as it seems to be crucial for understanding the lower bound of undecidability (or the upper bound of decidability). Second, one should definitely explore more the use of acLL_Σ in modeling natural language syntax. The examples in Sect. 2.2 show how to locally combine sentences with different grammatical characteristics, and the MIX_3 example above illustrates how that can be of importance. That is, it would be interesting to have a formal study about acLL_Σ and categorial grammars. Third, we plan to investigate the connections between our work and Adjoint logic [48] as well as with Display calculus [5, 12]. Finally, we intend to study proof-theoretic properties of acLL_Σ, such as normalization of proofs (*e.g.* via focusing) and interpolation.

Acknowledgements. We are grateful for the useful suggestions from the anonymous referees. We would like to thank L. Beklemishev, M. Moortgat, and C. Retoré for their inspiring and helpful comments regarding an approach based on non-associativity.

References

1. Aarts, E., Trautwein, K.: Non-associative Lambek categorial grammar in polynomial time. Math. Log. Q. **41**(4), 476–484 (1995)
2. Abrusci, V.M.: A comparison between Lambek syntactic calculus and intuitionistic linear logic. Zeitschr. Math. Logik Grundl. Math. (Math. Logic Q.) **36**, 11–15 (1990)
3. Ajdukiewicz, K.: Die syntaktische Konnexität. Studia Philosophica **1**, 1–27 (1935)
4. Bar-Hillel, Y.: A quasi-arithmetical notation for syntactic description. Language **29**, 47–58 (1953)
5. Belnap, N.: Display logic. J. Philos. Log. **11**(4), 375–417 (1982). https://doi.org/10.1007/BF00284976
6. Bulinska, M.: On the complexity of nonassociative Lambek calculus with unit. Stud. Logica **93**(1), 1–14 (2009). https://doi.org/10.1007/s11225-009-9205-2
7. Buszkowski, W.: Lambek calculus with nonlogical axioms. In: Language and Grammar, Studies in Mathematical Linguistics and Natural Language, pp. 77–93. CSLI Publications (2005)
8. Buszkowski, W., Farulewski, M.: Nonassociative Lambek calculus with additives and context-free languages. In: Grumberg, O., Kaminski, M., Katz, S., Wintner, S. (eds.) Languages: From Formal to Natural. LNCS, vol. 5533, pp. 45–58. Springer, Heidelberg (2009). https://doi.org/10.1007/978-3-642-01748-3_4
9. Chaudhuri, K.: Undecidability of multiplicative subexponential logic. In: Alves, S., Cervesato, I. (eds.) Proceedings Third International Workshop on Linearity, LINEARITY 2014, Vienna, Austria, 13th July 2014. EPTCS, vol. 176, pp. 1–8 (2014). https://doi.org/10.4204/EPTCS.176.1
10. Chaudhuri, K., Marin, S., Straßburger, L.: Modular focused proof systems for intuitionistic modal logics. In: FSCD, pp. 16:1–16:18 (2016)
11. Chvalovský, K.: Undecidability of consequence relation in full non-associative Lambek calculus. J. Symb. Logic **80**(2), 567–586 (2015)
12. Clouston, R., Dawson, J., Goré, R., Tiu, A.: Annotation-free sequent calculi for full intuitionistic linear logic. In: Rocca, S.R.D. (ed.) Computer Science Logic 2013 (CSL 2013), CSL 2013, Torino, Italy, 2–5 September 2013. LIPIcs, vol. 23, pp. 197–214. Schloss Dagstuhl - Leibniz-Zentrum für Informatik (2013). https://doi.org/10.4230/LIPIcs.CSL.2013.197

13. Danos, V., Joinet, J.-B., Schellinx, H.: The structure of exponentials: uncovering the dynamics of linear logic proofs. In: Gottlob, G., Leitsch, A., Mundici, D. (eds.) KGC 1993. LNCS, vol. 713, pp. 159–171. Springer, Heidelberg (1993). https://doi.org/10.1007/BFb0022564
14. de Groote, P., Lamarche, F.: Classical non-associative Lambek calculus. Stud. Logica **71**(3), 355–388 (2002). https://doi.org/10.1023/A:1020520915016
15. Gheorghiu, A., Marin, S.: Focused proof-search in the logic of bunched implications. In: FOSSACS 2021. LNCS, vol. 12650, pp. 247–267. Springer, Cham (2021). https://doi.org/10.1007/978-3-030-71995-1_13
16. Girard, J.-Y.: Linear logic. Theor. Comput. Sci. **50**, 1–102 (1987). https://doi.org/10.1016/0304-3975(87)90045-4
17. Guglielmi, A., Straßburger, L.: Non-commutativity and MELL in the calculus of structures. In: Fribourg, L. (ed.) CSL 2001. LNCS, vol. 2142, pp. 54–68. Springer, Heidelberg (2001). https://doi.org/10.1007/3-540-44802-0_5
18. Hepple, M.: A general framework for hybrid substructural categorial logics. Technical report 94–14, IRCS (1994)
19. Kanazawa, M.: On the recognizing power of the Lambek calculus with brackets. J. Logic Lang. Inform. **27**(4), 295–312 (2018)
20. Kanovich, M., Kuznetsov, S., Nigam, V., Scedrov, A.: A logical framework with commutative and non-commutative subexponentials. In: Galmiche, D., Schulz, S., Sebastiani, R. (eds.) IJCAR 2018. LNCS (LNAI), vol. 10900, pp. 228–245. Springer, Cham (2018). https://doi.org/10.1007/978-3-319-94205-6_16
21. Kanovich, M., Kuznetsov, S., Nigam, V., Scedrov, A.: Subexponentials in non-commutative linear logic. Math. Struct. Comput. Sci. **29**(8), 1217–1249 (2019). https://doi.org/10.1017/S0960129518000117
22. Kanovich, M., Kuznetsov, S., Scedrov, A.: Undecidability of the Lambek calculus with a relevant modality. In: Foret, A., Morrill, G., Muskens, R., Osswald, R., Pogodalla, S. (eds.) FG 2015-2016. LNCS, vol. 9804, pp. 240–256. Springer, Heidelberg (2016). https://doi.org/10.1007/978-3-662-53042-9_14
23. Kanovich, M., Kuznetsov, S., Scedrov, A.: The multiplicative-additive Lambek calculus with subexponential and bracket modalities. J. Log. Lang. Inform. **30**, 31–88 (2021)
24. Kopylov, A.: Decidability of linear affine logic. Inf. Comput. **164**(1), 173–198 (2001). https://doi.org/10.1006/inco.1999.2834
25. Kozak, M.: Cyclic involutive distributive full Lambek calculus is decidable. J. Log. Comput. **21**(2), 231–252 (2011). https://doi.org/10.1093/logcom/exq021
26. Kurtonina, N.: Frames and labels. A modal analysis of categorial inference. Ph.D. thesis, Universiteit Utrecht, ILLC (1995)
27. Kurtonina, N., Moortgat, M.: Structural control. In: Blackburn, P., de Rijke, M. (eds.) Specifying Syntactic Structures, CSLI, Stanford, pp. 75–113 (1997)
28. Lamarche, F.: On the Algebra of Structural Contexts. Mathematical Structures in Computer Science, p. 51 (2003). Article dans revue scientifique avec comité de lecture. https://hal.inria.fr/inria-00099461
29. Lambek, J.: The mathematics of sentence structure. Am. Math. Monthly **65**(3), 154–170 (1958)
30. Lambek, J.: On the calculus of syntactic types. In: Jakobson, R. (ed.) Structure of Language and Its Mathematical Aspects, pp. 166–178. American Mathematical Society (1961)
31. Lellmann, B., Olarte, C., Pimentel, E.: A uniform framework for substructural logics with modalities. In: LPAR-21, pp. 435–455 (2017)
32. Lincoln, P., Mitchell, J., Scedrov, A., Shankar, N.: Decision problems for propositional linear logic. Ann. Pure Appl. Logic **56**(1–3), 239–311 (1992)
33. Moortgat, M.: Multimodal linguistic inference. Log. J. IGPL **3**(2–3), 371–401 (1995). https://doi.org/10.1093/jigpal/3.2-3.371

34. Moortgat, M.: Multimodal linguistic inference. J. Logic Lang. Inform. **5**(3–4), 349–385 (1996)
35. Moortgat, M., Morrill, G.: Heads and phrases: type calculus for dependency and constituent structure. Technical report (1991)
36. Moortgat, M., Oehrle, R.: Logical parameters and linguistic variation. In: Fifth European Summer School in Logic, Language and Information. Lecture Notes on Categorial Grammar (1993)
37. Moot, R.: The grail theorem prover: type theory for syntax and semantics. CoRR, abs/1602.00812 (2016). arXiv:1602.00812
38. Moot, R., Retoré, C.: The Logic of Categorial Grammars. LNCS, vol. 6850. Springer, Heidelberg (2012). https://doi.org/10.1007/978-3-642-31555-8
39. Morrill, G.: Categorial formalisation of relativisation: Pied piping, islands, and extraction sites. Technical report LSI-92-23-R, Universitat Politècnica de Catalunya (1992)
40. Morrill, G.: Parsing/theorem-proving for logical grammar CatLog3. J. Log. Lang. Inf. **28**(2), 183–216 (2019). https://doi.org/10.1007/s10849-018-09277-w
41. Morrill, G., Leslie, N., Hepple, M., Barry, G.: Categorial deductions and structural operations. In: Studies in Categorial Grammar, Edinburgh Working Paper in Cognitive Science, vol. 5, pp. 1–21 (1990)
42. Nigam, V., Miller, D.: A framework for proof systems. J. Autom. Reason. **45**(2), 157–188 (2010). https://doi.org/10.1007/s10817-010-9182-1
43. Nigam, V., Olarte, C., Pimentel, E.: On subexponentials, focusing and modalities in concurrent systems. Theor. Comput. Sci. **693**, 35–58 (2017). https://doi.org/10.1016/j.tcs.2017.06.009
44. Nigam, V., Pimentel, E., Reis, G.: An extended framework for specifying and reasoning about proof systems. J. Log. Comput. **26**(2), 539–576 (2016). https://doi.org/10.1093/logcom/exu029
45. Oehrle, R., Zhang, S.: Lambek calculus and preposing of embedded subjects. Coyote Papers (1989). http://hdl.handle.net/10150/226572
46. Olarte, C., Chiarugi, D., Falaschi, M., Hermith, D.: A proof theoretic view of spatial and temporal dependencies in biochemical systems. Theor. Comput. Sci. **641**, 25–42 (2016). https://doi.org/10.1016/j.tcs.2016.03.029
47. Pentus, M.: Lambek calculus is NP-complete. Theor. Comput. Sci. **357**, 186–201 (2006)
48. Pruiksma, K., Chargin, W., Pfenning, F., Reed, J.: Adjoint logic (2018, Unpublished manuscript)
49. Shieber, S.: Evidence against the context-freeness of natural languages. Linguist. Philos. **8**, 333–343 (1985)
50. Straßburger, L.: On the decision problem for MELL. Theor. Comput. Sci. **768**, 91–98 (2019). https://doi.org/10.1016/j.tcs.2019.02.022
51. Tanaka, H.: A note on undecidability of propositional non-associative linear logics (2019). arXiv preprint arXiv:1909.13444

Effective Semantics for the Modal Logics K and KT via Non-deterministic Matrices

Ori Lahav[1] and Yoni Zohar[2](✉)

[1] Tel Aviv University, Tel Aviv-Yafo, Israel
[2] Bar Ilan University, Ramat Gan, Israel
yoni.zohar@biu.ac.il

Abstract. A four-valued semantics for the modal logic K is introduced. Possible worlds are replaced by a hierarchy of four-valued valuations, where the valuations of the first level correspond to valuations that are legal w.r.t. a basic non-deterministic matrix, and each level further restricts its set of valuations. The semantics is proven to be effective, and to precisely capture derivations in a sequent calculus for K of a certain form. Similar results are then obtained for the modal logic KT, by simply deleting one of the truth values.

1 Introduction

Propositional modal logics extend classical logic with *modalities*, intuitively interpreted as necessity, knowledge, or temporal operators. Such extensions have several applications in computer science and artificial intelligence (see, e.g., [7,9,13]).

The most common and successful semantic framework for modal logics is the so called *possible worlds semantics*, in which each world is equipped with a two-valued valuation, and the semantic constraints regarding the modal operators consider the valuations in *accessible* worlds. While this has been the gold standard for modal logic semantics for many years, alternative semantic frameworks have been proposed. One of these approaches, initiated by Kearns [10], is based on an infinite sequence of sets of valuations in a non-deterministic many-valued semantics. Since then, several non-deterministic many-valued semantics, without possible worlds, were developed for modal logics (see, e.g., [4,8,12,14]). The current paper is a part of that body of work. Having an alternative semantic framework for modal logics, different than the common possible worlds semantics, has the potential of exposing new intuitions and understandings of modal logics, and also to form the basis to new decision procedures.

Our main contribution is a four-valued semantics for the modal logic K. The key characteristic of the semantics that we present is *effectiveness*: when checking

We thank the anonymous reviewers for their useful feedback. This research was supported by NSF-BSF (grant number 2020704), ISF (grant numbers 619/21 and 1566/18), and the Alon Young Faculty Fellowship.

J. Blanchette et al. (Eds.): IJCAR 2022, LNAI 13385, pp. 468–485, 2022.
https://doi.org/10.1007/978-3-031-10769-6_28

for the entailment of a formula φ from a set Γ of formulas in K, it suffices to only consider *partial* models, defined over the subformulas of Γ and φ. To the best of our knowledge, this is the first effective Nmatrices-based semantics for K. Such a semantics has the potential of being subject to reductions to classical satisfiability [3], as it is based on finite-valued truth tables, and thus improving the performance of solvers for modal logic by utilizing off-the-shelf SAT solvers. Another advantage of this semantics is that it precisely captures derivations in a sequent calculus for K that admit a certain property. Following Kearns, models of this semantics are based on the concept of *levels*—valuations of level 0 are the ordinary valuations of Nmatrices, while each level $m > 0$ introduces more constraints. We show that valuations of level m correspond to derivations in the calculus whose largest number of applications of the rule that correspond to the axiom (K) in any branch of the derivation is at most m. Our restrictions between the levels are more complex than the original restrictions in Kearns' work, in order to obtain effectiveness. Another precise correspondence between the semantics and the proof system that we prove, is between the domains of valuations and the formulas allowed to be used in derivations.

Finally, we observe that by deleting one of the truth values, a three-valued semantics for the modal logic KT is obtained, which is similar to the one presented in [8]. Like the case of K, the resulting semantics is effective, and tightly correspond to derivations in a sequent calculus for KT.

Outline. The paper is organized as follows: Sect. 2 reviews standard notions in non-deterministic matrices. In Sect. 3, we present our semantics for the modal logic K, as well as the sequent calculus our investigation will be based on, which is coupled with the notion of (K)-depth of derivations. In Sect. 4, we prove soundness and completeness theorems between the sequent calculus and the semantics. In Sect. 5, we prove that the semantics that we provide is effective, not only for deciding entailment, but also for producing countermodels when an entailment does not hold. In Sect. 6 we establish similar results for the modal logic KT. We conclude with §7, where directions for future research are outlined.

Related Work. In [10], Kearns initiated the study of modal semantics without possible worlds. This work was recently revisited by Skurt and Omori [14], who generalized Kearns' work and reframed his framework within the framework of logical Non-deterministic matrices. As indicated in [14], it was not clear how to make this semantics effective, as it requires checking truth values of infinitely many formulas when considering the validity of a given formula (see, e.g., Remark 42 of [14]). In [4], Coniglio et al. develop a similar framework for modal logics, and some bound over the formulas that need to be considered was achieved. However, in [5], the authors clarified that it is unclear how to effectively use the resulting semantics. A semantics based on Nmatrices for the modal logics KT and S4 was presented in [8] by Grätz, that includes a method to extend a partial model in that semantics into a total one, which results in an effective semantics. We chose here to focus on K, which is a weaker logic, forming a common basis to all other normal modal logics. By deleting one out of four truth values, we obtain corresponding results for KT as well. The semantics that we present here is similar in nature to the one presented in [8], however: (*i*) the truth tables are different, as we intentionally enforced the many-valued tables of the classical connectives to be obtained by a straightforward duplication of truth values from the original two-valued truth tables; and (*ii*) the semantic condition for levels of valuations that we define here is inductive, where each level relies on lower levels (thus refraining from a definition of a more cyclic nature as the one in [8], that is better understood operationally). A variant of the semantics from [14] was also introduced and studied in [12], but without considering the ability to perform effective automated reasoning but instead focusing on infinite valuations rather than on partial ones. A complete proof theoretic characterization in terms of sequent calculi to the various levels of valuations was not given in any of the above works. Also, an effective semantics for K, which is the most basic modal logic, was not given in any of the above works.

Non-deterministic matrices were introduced in [2], and have since became a useful tool for investigating non-classical logics and proof systems (see [1] for a survey). They generalize (deterministic) matrices [15] by allowing a non-deterministic choice of truth values in the truth tables. Like matrices, Nmatrices enjoy the semantic *analyticity* property, which allows one to extend a partial valuation into a full one. Our semantic framework can be viewed as a further refinement of non-deterministic matrices, namely *restricted* non-deterministic matrices, introduced in [6].

2 Preliminaries

In this section we provide the necessary definitions about Nmatrices following [1]. We assume a propositional language $\mathcal{L}$ with countably infinitely many atomic variables $p_1, p_2, \ldots$. When there is no room for confusion, we identify $\mathcal{L}$ with its set of well-formed formulas (e.g., when writing $\varphi \in \mathcal{L}$). We write $sub(\varphi)$ for the set of subformulas of a formula φ. This notation is extended to sets of formulas in the natural way.

Valuations. In the context of a set $\mathcal{V}$ of "truth values", a *valuation* is a function v from some domain $\mathsf{Dom}(v) \subseteq \mathcal{L}$ to $\mathcal{V}$. For a set $\mathcal{F} \subseteq \mathcal{L}$, an *$\mathcal{F}$-valuation* is a valuation with domain $\mathcal{F}$. (In particular, an $\mathcal{L}$-valuation is defined on all formulas.) For $X \subseteq \mathcal{V}$, we write $v^{-1}[X]$ for the set $\{\varphi \mid v(\varphi) \in X\}$. For $x \in \mathcal{V}$, we also write $v^{-1}[x]$ for the set $\{\varphi \mid v(\varphi) = x\}$.

Definition 1. Let $\mathcal{D} \subseteq \mathcal{V}$ be a set of "designated truth values". A valuation v *$\mathcal{D}$-satisfies* a formula φ, denoted by $v \models_{\mathcal{D}} \varphi$, if $v(\varphi) \in \mathcal{D}$. For a set Σ of formulas, we write $v \models_{\mathcal{D}} \Sigma$ if $v \models_{\mathcal{D}} \varphi$ for every $\varphi \in \Sigma$.

Notation 2. Let $\mathcal{D} \subseteq \mathcal{V}$ be a set of designated truth values and $\mathbb{V}$ be a set of valuations. For sets L, R of formulas, we write $L \vdash^{\mathbb{V}}_{\mathcal{D}} R$ if for every $v \in \mathbb{V}$, $v \models_{\mathcal{D}} L$ implies that $v \models_{\mathcal{D}} \varphi$ for some $\varphi \in R$. We omit L or R in this notation when they are empty (e.g., when writing $\vdash^{\mathbb{V}}_{\mathcal{D}} R$), and set parentheses for singletons (e.g., when writing $L \vdash^{\mathbb{V}}_{\mathcal{D}} \varphi$).

Nmatrices. An Nmatrix M for $\mathcal{L}$ is a triple of the form $\langle \mathcal{V}, \mathcal{D}, \mathcal{O} \rangle$, where $\mathcal{V}$ is a set of *truth values*, $\mathcal{D} \subseteq \mathcal{V}$ is a set of *designated truth values*, and $\mathcal{O}$ is a function assigning a *truth table* $\mathcal{V}^n \to P(\mathcal{V}) \setminus \{\emptyset\}$ to every n-ary connective $\diamond$ of $\mathcal{L}$ (which assigns a set of possible values to each tuple of values). In the context of an Nmatrix $M = \langle \mathcal{V}, \mathcal{D}, \mathcal{O} \rangle$, we often denote $\mathcal{O}(\diamond)$ by $\tilde{\diamond}$.

An $\mathcal{F}$-valuation v is *M-legal* if $v(\varphi) \in \textsf{pos-val}(\varphi, M, v)$ for every formula $\varphi \in \mathcal{F}$ whose immediate subformulas are contained in $\mathcal{F}$, where $\textsf{pos-val}(\varphi, M, v)$ is defined by:

1. $\textsf{pos-val}(p, M, v) = \mathcal{V}$ for every atomic formula p.
2. $\textsf{pos-val}(\diamond(\psi_1, \ldots, \psi_n), M, v) = \tilde{\diamond}(v(\psi_1), \ldots, v(\psi_n))$ for every non-atomic formula $\diamond(\psi_1, \ldots, \psi_n)$.

In other words, there is no restriction regarding the values assigned to atomic formulas, whereas the values of compound formulas should respect the truth tables.

Lemma 1 ([1]). *Let $\mathcal{F} \subseteq \mathcal{L}$ be a set closed under subformulas and M an Nmatrix for $\mathcal{L}$. Then every M-legal $\mathcal{F}$-valuation v can be extended to an M-legal $\mathcal{L}$-valuation.*

3 The Modal Logic K

In this section we introduce a novel effective semantics for the model logic K. We first present a known proof system for this logic (Sect. 3.1), and then our semantics (Sect. 3.2). From here on, we assume that the language $\mathcal{L}$ consists of the connectives $\supset$, $\wedge$, $\vee$, $\neg$ and $\Box$ with their usual arities. The standard $\Diamond$ operator can be defined as a macro $\Diamond\varphi \stackrel{\text{def}}{=} \neg\Box\neg\varphi$. Obviously, using De-Morgan rules, fewer connectives can be used. However, we chose this set of connectives in order to have a primitive language rich enough for the examples that we include along the paper.

3.1 Proof System

Figure 1 presents a Gentzen-style calculus, denoted by $\mathsf{G_K}$, for the modal logic K that was proven to be equivalent to the original formulation of the logic as a Hilbert system (see, e.g., [16]). We take *sequents* to be pairs $\langle \Gamma, \Delta \rangle$ of finite *sets* of formulas. For readability, we write $\Gamma \Rightarrow \Delta$ instead of $\langle \Gamma, \Delta \rangle$ and use standard notations such as $\Gamma, \varphi \Rightarrow \psi$ instead of $(\Gamma \cup \{\varphi\}) \Rightarrow \{\psi\}$.

The (CUT) rule is included in $\mathsf{G_K}$ for convenience, but applications of (CUT) can be eliminated from derivations (see, e.g., [11]). Since the focus of this paper is semantics rather than cut-elimination, we allow ourselves to use cut freely and do not distinguish derivations that use it from derivations that do not. We write $\vdash_{\mathsf{G_K}} \Gamma \Rightarrow \Delta$ if there is a derivation of a sequent $\Gamma \Rightarrow \Delta$ in the calculus $\mathsf{G_K}$.

In the sequel, we provide a semantic characterization of $\vdash_{\mathsf{G_K}}$. It is based on a more refined notion of derivability that takes into account: (*i*) the set $\mathcal{F}$ of formulas used in the derivation; and (*ii*) the (K)-*depth* of the derivation, as defined next.

$$(\text{WEAK})\ \frac{\Gamma \Rightarrow \Delta}{\Gamma, \Gamma' \Rightarrow \Delta, \Delta'} \qquad (\text{ID})\ \frac{}{\Gamma, \varphi \Rightarrow \varphi, \Delta} \qquad (\text{CUT})\ \frac{\Gamma, \varphi \Rightarrow \Delta \quad \Gamma \Rightarrow \varphi, \Delta}{\Gamma \Rightarrow \Delta} \qquad (K)\ \frac{\Gamma \Rightarrow \varphi}{\Box\Gamma \Rightarrow \Box\varphi}$$

$$(\neg\Rightarrow)\ \frac{\Gamma \Rightarrow \varphi, \Delta}{\Gamma, \neg\varphi \Rightarrow \Delta} \qquad (\Rightarrow\neg)\ \frac{\Gamma, \varphi \Rightarrow \Delta}{\Gamma \Rightarrow \neg\varphi, \Delta} \qquad (\supset\Rightarrow)\ \frac{\Gamma \Rightarrow \varphi, \Delta \quad \Gamma, \psi \Rightarrow \Delta}{\Gamma, \varphi \supset \psi, \Delta} \qquad (\Rightarrow\supset)\ \frac{\Gamma, \varphi \Rightarrow \psi, \Delta}{\Gamma \Rightarrow \varphi \supset \psi, \Delta}$$

$$(\wedge\Rightarrow)\ \frac{\Gamma, \varphi, \psi \Rightarrow \Delta}{\Gamma, \varphi \wedge \psi \Rightarrow \Delta} \qquad (\Rightarrow\wedge)\ \frac{\Gamma \Rightarrow \varphi, \Delta \quad \Gamma \Rightarrow \psi, \Delta}{\Gamma \Rightarrow \varphi \wedge \psi, \Delta} \qquad (\vee\Rightarrow)\ \frac{\Gamma, \varphi \Rightarrow \Delta \quad \Gamma, \psi \Rightarrow \Delta}{\Gamma, \varphi \vee \psi \Rightarrow \Delta} \qquad (\Rightarrow\vee)\ \frac{\Gamma \Rightarrow \varphi, \psi, \Delta}{\Gamma \Rightarrow \varphi \vee \psi, \Delta}$$

Fig. 1. The sequent calculus $\mathsf{G_K}$

Definition 3. A *derivation* of a sequent $\Gamma \Rightarrow \Delta$ in $\mathsf{G_K}$ is a tree in which the nodes are labeled with sequents, the root is labeled with $\Gamma \Rightarrow \Delta$, and every node is the result of an application of some rule of $\mathsf{G_K}$ where the premises are the labels of its children in the tree. A derivation is called an $\mathcal{F}$-*derivation* if it employs only sequents composed of formulas from $\mathcal{F}$. The (K)-*depth* of a derivation is the maximal number of applications of rule (K) in any of the branches of the derivation.

Notation 4. We write $\vdash^{\mathcal{F},m}_{\mathsf{G_K}} \Gamma \Rightarrow \Delta$ if there is a derivation of $\Gamma \Rightarrow \Delta$ in $\mathsf{G_K}$ in which only $\mathcal{F}$-sequents occur and that has (K)-depth at most m. We drop $\mathcal{F}$ from this notation when $\mathcal{F} = \mathcal{L}$; and drop m to dismiss the restriction regarding the (K)-depth.

Example 1. Let $\varphi \stackrel{\text{def}}{=} \Box(p_1 \wedge p_2) \supset (\Box p_1 \wedge \Box p_2)$ and $\mathcal{F} = sub(\varphi)$. The following is a derivation of $\Rightarrow \varphi$ in $\mathsf{G_K}$ that only uses $\mathcal{F}$-formulas and has (K)-depth of 1 (though the number of applications of (K) in the derivation is 2):

$$\dfrac{\dfrac{\dfrac{\dfrac{\dfrac{}{p_1, p_2 \Rightarrow p_1}(\text{ID})}{p_1 \wedge p_2 \Rightarrow p_1}(\wedge\Rightarrow)}{\Box(p_1 \wedge p_2) \Rightarrow \Box p_1}(\text{K}) \quad \dfrac{\dfrac{\dfrac{}{p_1, p_2 \Rightarrow p_2}(\text{ID})}{p_1 \wedge p_2 \Rightarrow p_2}(\wedge\Rightarrow)}{\Box(p_1 \wedge p_2) \Rightarrow \Box p_2}(\text{K})}{\Box(p_1 \wedge p_2) \Rightarrow \Box p_1 \wedge \Box p_2}(\Rightarrow\wedge)}{\Rightarrow \Box(p_1 \wedge p_2) \supset \Box p_1 \wedge \Box p_2}(\Rightarrow\supset)$$

3.2 Semantics

The semantics is based on a four-valued Nmatrix stratified with "levels", where for every m, legal valuations of level $m+1$ are a subset of legal valuations of level m. The underlying Nmatrix, denoted by $\mathsf{M_K}$, is obtained by duplicating the classical truth values. Thus, the sets of truth values and of designated truth values are given by:

$$\mathcal{V}_4 \stackrel{\text{def}}{=} \{\mathsf{T}, \mathsf{t}, \mathsf{f}, \mathsf{F}\} \qquad \mathcal{D} \stackrel{\text{def}}{=} \{\mathsf{T}, \mathsf{t}\}$$

The truth tables are as follows (we have $\overline{\mathcal{D}} = \{\mathsf{f}, \mathsf{F}\}$):

$x \tilde{\supset} y$	T	t	F	f
T	$\mathcal{D}$	$\mathcal{D}$	$\overline{\mathcal{D}}$	$\overline{\mathcal{D}}$
t	$\mathcal{D}$	$\mathcal{D}$	$\overline{\mathcal{D}}$	$\overline{\mathcal{D}}$
F	$\mathcal{D}$	$\mathcal{D}$	$\mathcal{D}$	$\mathcal{D}$
f	$\mathcal{D}$	$\mathcal{D}$	$\mathcal{D}$	$\mathcal{D}$

$x \tilde{\wedge} y$	T	t	F	f
T	$\mathcal{D}$	$\mathcal{D}$	$\overline{\mathcal{D}}$	$\overline{\mathcal{D}}$
t	$\mathcal{D}$	$\mathcal{D}$	$\overline{\mathcal{D}}$	$\overline{\mathcal{D}}$
F	$\overline{\mathcal{D}}$	$\overline{\mathcal{D}}$	$\overline{\mathcal{D}}$	$\overline{\mathcal{D}}$
f	$\overline{\mathcal{D}}$	$\overline{\mathcal{D}}$	$\overline{\mathcal{D}}$	$\overline{\mathcal{D}}$

$x \tilde{\vee} y$	T	t	F	f
T	$\mathcal{D}$	$\mathcal{D}$	$\mathcal{D}$	$\mathcal{D}$
t	$\mathcal{D}$	$\mathcal{D}$	$\mathcal{D}$	$\mathcal{D}$
F	$\mathcal{D}$	$\mathcal{D}$	$\overline{\mathcal{D}}$	$\overline{\mathcal{D}}$
f	$\mathcal{D}$	$\mathcal{D}$	$\overline{\mathcal{D}}$	$\overline{\mathcal{D}}$

x	$\tilde{\neg} x$
T	$\overline{\mathcal{D}}$
t	$\overline{\mathcal{D}}$
F	$\mathcal{D}$
f	$\mathcal{D}$

x	$\tilde{\Box} x$
T	$\mathcal{D}$
t	$\overline{\mathcal{D}}$
F	$\mathcal{D}$
f	$\overline{\mathcal{D}}$

We employ the following notations for subsets of truth values:

$$\mathsf{TF} \stackrel{\text{def}}{=} \{\mathsf{T}, \mathsf{F}\} \qquad \mathsf{tf} \stackrel{\text{def}}{=} \{\mathsf{t}, \mathsf{f}\}$$

For the classical connectives, the truth tables of $\mathsf{M_K}$ treat t just like T, and f just like F, and are essentially two-valued—the result is either $\mathcal{D}$ or $\overline{\mathcal{D}}$, and it depends solely on whether the inputs are elements of $\mathcal{D}$ or $\overline{\mathcal{D}}$. Thus, for the language without $\Box$, this Nmatrix provides a (non-economic) four-valued semantics for classical logic.

While the output for $\Box$ is also always $\mathcal{D}$ or $\overline{\mathcal{D}}$, it differentiates between T (that results in $\mathcal{D}$) and t (that results in $\overline{\mathcal{D}}$), and similarly between F and f. In fact, this table is captured by the condition: $\tilde{\Box}(x) \in \mathcal{D}$ iff $x \in \mathsf{TF}$.

Example 2. Let $\mathcal{F} = sub(\varphi)$ where φ is the formula from Example 1. The following valuation v is an $\mathcal{F}$-valuation that is $\mathsf{M_K}$-legal:

$$v(p_1) = v(p_2) = \mathsf{f} \quad v(p_1 \wedge p_2) = \mathsf{F} \quad v(\Box p_1) = v(\Box p_2) = v(\Box p_1 \wedge \Box p_2) = \mathsf{F}$$

$$v(\Box(p_1 \wedge p_2)) = \mathsf{T} \quad v(\Box(p_1 \wedge p_2) \supset (\Box p_1 \wedge \Box p_2)) = \mathsf{F}$$

To show that it is $\mathsf{M_K}$-legal, one needs to verify that $v(\psi) \in \textsf{pos-val}(\psi, \mathsf{M_K}, v)$ for each $\psi \in \mathcal{F}$. For example, $v(p_1) = \mathsf{f} \in \mathcal{V}_4 = \textsf{pos-val}(p_1, \mathsf{M_K}, v)$. As another example, since $v(p_1) = \mathsf{f}$, we have that $\textsf{pos-val}(\Box p_1, \mathsf{M_K}, v) = \tilde{\Box}(\mathsf{f}) = \{\mathsf{F}, \mathsf{f}\}$, and hence $v(\Box p_1) = \mathsf{F} \in \textsf{pos-val}(\Box p_1, \mathsf{M_K}, v)$. Notice that v does not satisfy φ.

The truth table for $\Box$ can be understood via "possible worlds" intuition. Our four truth values are intuitively captured as follows, assuming a given formula ψ and a world w:

- T: ψ holds in w and in every world accessible from w;
- t: ψ holds in w but it does not hold in some world accessible from w;
- F: ψ does not hold in w but does hold in every world accessible from w; and
- f: ψ does not hold in w and it does not hold in some world accessible from w.

In the possible worlds semantics, $\Box\psi$ holds in some world w iff ψ holds in every world that is accessible from w, which intuitively explains the table for $\Box$. Note that non-determinism is inherent here. For example, if ψ holds in w and in every world accessible from w (i.e., ψ has value T), we know that $\Box\psi$ holds in w, but we do not know whether $\Box\psi$ holds in every world accessible from w (thus $\Box\psi$ has value T or t).

Now, the Nmatrix M_{K} by itself is not adequate for the modal logic K (as Examples 1 and 2 demonstrate). What is missing is the relation between the choices we make to resolve non-determinism for different formulas. Continuing with the possible worlds intuition, we observe that if a formula φ follows from a set of formulas Σ that hold in all accessible worlds (i.e., φ follows from formulas whose truth value is T or F), then φ itself should hold in all accessible worlds (i.e., φ's truth value should be T or F). Directly encoding this condition requires us to consider a set $\mathbb{V}$ of M_{K}-legal $\mathcal{F}$-valuations for which the following holds (recall Notation 2 from Sect. 2):

$$\forall v \in \mathbb{V}. \forall \varphi \in \mathcal{F}. (v^{-1}[\mathsf{TF}] \vdash^{\mathbb{V}}_{\mathcal{D}} \varphi \implies v(\varphi) \in \mathsf{TF}) \qquad (\textit{necessitation})$$

In turn, to obtain completeness we take a maximal set $\mathbb{V}$ that satisfies the *necessitation* condition. While it is possible to define this set of valuations as the greatest fixpoint of *necessitation*, following previous work, we find it convenient to reach this set using "levels":

Definition 5. The set $\mathbb{V}^{\mathcal{F},m}_{\mathsf{K}}$ is inductively defined as follows:

- $\mathbb{V}^{\mathcal{F},0}_{\mathsf{K}}$ is the set of M_{K}-legal $\mathcal{F}$-valuations.
- $\mathbb{V}^{\mathcal{F},m+1}_{\mathsf{K}} \stackrel{\text{def}}{=} \left\{ v \in \mathbb{V}^{\mathcal{F},m}_{\mathsf{K}} \mid \forall \varphi \in \mathcal{F}. v^{-1}[\mathsf{TF}] \vdash^{\mathbb{V}^{\mathcal{F},m}_{\mathsf{K}}}_{\mathcal{D}} \varphi \implies v(\varphi) \in \mathsf{TF} \right\}$

We also define:

$$\mathbb{V}^{\mathcal{F}}_{\mathsf{K}} \stackrel{\text{def}}{=} \bigcap_{m \geq 0} \mathbb{V}^{\mathcal{F},m}_{\mathsf{K}} \qquad \mathbb{V}^{m}_{\mathsf{K}} \stackrel{\text{def}}{=} \mathbb{V}^{\mathcal{L},m}_{\mathsf{K}} \qquad \mathbb{V}_{\mathsf{K}} \stackrel{\text{def}}{=} \bigcap_{m \geq 0} \mathbb{V}^{\mathcal{L},m}_{\mathsf{K}}$$

Similarly to the idea originated by Kearns in [10], valuations are partitioned into *levels*, which are inductively defined. The first level, $\mathbb{V}^{\mathcal{F},0}_{\mathsf{K}}$, consists solely of the M_{K}-legal valuations with domain $\mathcal{F}$. For each $m > 0$, the m'th level is defined as a subset of the $(m-1)$'th level, with an additional constraint: a valuation v from level $m-1$ remains in level m, only if every formula $\varphi \in \mathcal{F}$ entailed (at the $m-1$ level) from the set of formulas that were assigned a value from TF by v, is itself assigned a value from TF by v. As we show below, in the "end" of this process, by taking $\bigcap_{m\geq 0} \mathbb{V}^{\mathcal{F},m}_{\mathsf{K}}$, one obtains the greatest set $\mathbb{V}$ satisfying the *necessitation* condition

Remark 1. The *necessitation* condition is similar to the one provided in [8] to the modal logics KT and $\mathsf{S4}$. In contrast, the condition from [4,10,14] is simpler and does not involve $v^{-1}[\mathsf{TF}]$ at all, but also does not give rise to decision procedures.

Example 3. Following Example 2, while the formula φ is not satisfied by all valuations in $\mathbb{V}^{\mathcal{F},0}_{\mathsf{K}}$, it is satisfied by all valuations in $\mathbb{V}^{\mathcal{F},m}_{\mathsf{K}}$ for every $m > 0$. In particular, the valuation v from Example 2 is not in $\mathbb{V}^{\mathcal{F},1}_{\mathsf{K}}$: we have $p_1 \wedge p_2 \vdash^{\mathbb{V}^{\mathcal{F},0}_{\mathsf{K}}}_{\mathcal{D}} p_1$ and $v(p_1 \wedge p_2) = \mathsf{F}$ (so $p_1 \wedge p_2 \in v^{-1}[\mathsf{TF}]$), but $v(p_1) = \mathsf{f} \notin \mathsf{TF}$.

For each set $\mathcal{F} \subseteq \mathcal{L}$ and $m \geq 0$, we obtain a consequence relation $\vdash_{\mathcal{D}}^{\mathbb{V}_{\mathsf{K}}^{\mathcal{F},m}}$ between sets of $\mathcal{F}$-formulas. Disregarding m, we also obtain the relation $\vdash_{\mathcal{D}}^{\mathbb{V}_{\mathsf{K}}^{\mathcal{F}}}$ (for every $\mathcal{F}$), which we will show to be sound and complete for K. We note that all these relations are compact. The proof of the following theorem relies on the completeness theorems that we prove in Sect. 4.

Theorem 1 (Compactness).

1. *For every $m \geq 0$, if $L \vdash_{\mathcal{D}}^{\mathbb{V}_{\mathsf{K}}^{\mathcal{F},m}} R$, then $\Gamma \vdash_{\mathcal{D}}^{\mathbb{V}_{\mathsf{K}}^{\mathcal{F},m}} \Delta$ for some finite $\Gamma \subseteq L$ and $\Delta \subseteq R$.*
2. *If $L \vdash_{\mathcal{D}}^{\mathbb{V}_{\mathsf{K}}^{\mathcal{F}}} R$, then $\Gamma \vdash_{\mathcal{D}}^{\mathbb{V}_{\mathsf{K}}^{\mathcal{F}}} \Delta$ for some finite $\Gamma \subseteq L$ and $\Delta \subseteq R$.*

Now, to show that $\mathbb{V}_{\mathsf{K}}^{\mathcal{F}}$ is indeed the largest set $\mathbb{V}$ of M_{K}-legal $\mathcal{F}$-valuations that satisfies *necessitation*, we use the following two lemmas. The first is a general construction that relies only on the use of *finite-valued* valuation functions.

Lemma 2. *Let $v_0, v_1, v_2, \ldots$ be an infinite sequence of valuations over a common domain $\mathcal{F}$. Then, there exists some v such that for every finite set $\mathcal{F}' \subseteq \mathcal{F}$ of formulas and $m \geq 0$, we have $v|_{\mathcal{F}'} = v_k|_{\mathcal{F}'}$ for some $k \geq m$.*

Proof (Outline). First, if $\mathcal{F}$ is finite, then there is only a finite number of $\mathcal{F}$-valuations, and there must exists some $\mathcal{F}$-valuation v_m that occurs infinitely often in the sequence $v_0, v_1, \ldots$. We take $v = v_m$, and the required property trivially holds. Now, assume that $\mathcal{F}$ is infinite, and let $\varphi_0, \varphi_1, \ldots$ be an enumeration of the formulas in $\mathcal{F}$. For every $i \geq 0$, let $\mathcal{F}_i = \{\varphi_0, \ldots, \varphi_i\}$. We construct a sequence of infinite sets $A_0, A_1, \ldots \subseteq \mathbb{N}$ such that:

- For every $i \geq 0$, $A_{i+1} \subseteq A_i$.
- For every $0 \leq j \leq i$, $a \in A_j$, and $b \in A_i$, $v_a(\varphi_j) = v_b(\varphi_j)$.

To do so, take some infinite set $A_0 \subseteq \mathbb{N}$ such that $v_a(\varphi_0) = v_b(\varphi_0)$ for every $a, b \in A_0$ (such set must exist since we have a finite number of truth values). Then, given A_i, we let A_{i+1} be some infinite subset of A_i such that $v_a(\varphi_{i+1}) = v_b(\varphi_{i+1})$ for every $a, b \in A_{i+1}$. The valuation v is defined by $v(\varphi_i) = v_a(\varphi_i)$ for some $a \in A_i$. The properties of the A_i's ensure that v is well defined, and it can be shown that it also satisfies the required property. □

Using Lemma 2 and the compactness property, we can show the following:

Lemma 3. *Let $v_0, v_1, \ldots$ be a sequence of valuations over a common domain $\mathcal{F}$ such that $v_m \in \mathbb{V}_{\mathsf{K}}^{\mathcal{F},m}$ for every $m \geq 0$. Then, there exists some $v \in \mathbb{V}_{\mathsf{K}}^{\mathcal{F}}$ such that for every $\varphi \in \mathcal{F}$, $v(\varphi) = v_m(\varphi)$ for some $m \geq 0$.*

Proof (Outline). By Lemma 2, there exists some v such that for every finite set $\mathcal{F}'$ of formulas, $v|_{\mathcal{F}'} = v_m|_{\mathcal{F}'}$ for some $m \geq 0$. It is easy to verify that v satisfies the required properties. In particular, one shows that $v \in \mathbb{V}_{\mathsf{K}}^{\mathcal{F},m}$ for every $m \geq 0$ by induction on m. In that proof we use Theorem 1 to obtain a finite $\Gamma \subseteq v^{-1}[\mathsf{TF}]$ such that $\Gamma \vdash_{\mathcal{D}}^{\mathbb{V}_{\mathsf{K}}^{\mathcal{F},m-1}} \varphi$ from the assumption that $v^{-1}[\mathsf{TF}] \vdash_{\mathcal{D}}^{\mathbb{V}_{\mathsf{K}}^{\mathcal{F},m-1}} \varphi$. Then, the above property of v is applied with $\mathcal{F}' = \Gamma \cup \{\varphi\}$. □

Now, our characterization theorem easily follows:

Theorem 2. *The set $\mathbb{V}^{\mathcal{F}}_{\mathsf{K}}$ is the largest set $\mathbb{V}$ of M_{K}-legal $\mathcal{F}$-valuations that satisfies necessitation.*

Proof (Outline). To prove that $\mathbb{V}^{\mathcal{F}}_{\mathsf{K}}$ satisfies *necessitation*, one needs to prove that if $v^{-1}[\mathsf{TF}] \vdash_{\mathcal{D}}^{\mathbb{V}^{\mathcal{F}}_{\mathsf{K}}} \varphi$, then also $v^{-1}[\mathsf{TF}] \vdash_{\mathcal{D}}^{\mathbb{V}^{\mathcal{F},m}_{\mathsf{K}}} \varphi$ for some $m \geq 0$. This is done using Lemma 3. For maximality, given a set $\mathbb{V}$, we assume by contradiction that there is some m such that $\mathbb{V} \not\subseteq \mathbb{V}^{\mathcal{F},m}_{\mathsf{K}}$, take a minimal such m, and show that it cannot be 0. Then, from $\mathbb{V} \subseteq \mathbb{V}^{\mathcal{F}}_{\mathsf{K}} m-1$, it follows that actually $\mathbb{V} \subseteq \mathbb{V}^{\mathcal{F},m}_{\mathsf{K}}$, and thus we obtain a contradiction. □

Finite Domain. By definition we have $\mathbb{V}^{\mathcal{F},0}_{\mathsf{K}} \supseteq \mathbb{V}^{\mathcal{F},1}_{\mathsf{K}} \supseteq \mathbb{V}^{\mathcal{F},2}_{\mathsf{K}} \supseteq \ldots$ (and so, $\vdash_{\mathcal{D}}^{\mathbb{V}^{\mathcal{F},0}_{\mathsf{K}}} \subseteq \vdash_{\mathcal{D}}^{\mathbb{V}^{\mathcal{F},1}_{\mathsf{K}}} \subseteq \vdash_{\mathcal{D}}^{\mathbb{V}^{\mathcal{F},2}_{\mathsf{K}}} \subseteq \ldots$). Next, we show that when $\mathcal{F}$ is finite, then this sequence must converge.

Lemma 4. *Suppose that $\mathbb{V}^{\mathcal{F},m}_{\mathsf{K}} = \mathbb{V}^{\mathcal{F},m+1}_{\mathsf{K}}$ for some $m \geq 0$. Then, $\mathbb{V}^{\mathcal{F}}_{\mathsf{K}} = \mathbb{V}^{\mathcal{F},m}_{\mathsf{K}}$.*

Lemma 5. *For a* finite *set $\mathcal{F}$ of formulas, $\mathbb{V}^{\mathcal{F}}_{\mathsf{K}} = \mathbb{V}^{\mathcal{F},4^{|\mathcal{F}|}}_{\mathsf{K}}$.*

Proof. The left-to-right inclusion follows from our definitions. For the right-to-left inclusion, note that by Lemma 4, $\mathbb{V}^{\mathcal{F},m}_{\mathsf{K}} = \mathbb{V}^{\mathcal{F},m+1}_{\mathsf{K}}$ implies that $\mathbb{V}^{\mathcal{F},m}_{\mathsf{K}} = \mathbb{V}^{\mathcal{F},k}_{\mathsf{K}}$ for every $k \geq m$. Thus, it suffices to show that $\mathbb{V}^{\mathcal{F},m}_{\mathsf{K}} = \mathbb{V}^{\mathcal{F},m+1}_{\mathsf{K}}$ for some $0 \leq m \leq 4^{|\mathcal{F}|}+1$. Indeed, otherwise we have $\mathbb{V}^{\mathcal{F},0}_{\mathsf{K}} \supset \mathbb{V}^{\mathcal{F},1}_{\mathsf{K}} \supset \mathbb{V}^{\mathcal{F},2}_{\mathsf{K}} \supset \ldots \supset \mathbb{V}^{\mathcal{F},4^{|\mathcal{F}|}+1}_{\mathsf{K}}$, but this is impossible since there are only $4^{|\mathcal{F}|}$ functions from $\mathcal{F}$ to $\mathcal{V}_4$. □

Optimized Tables. Starting from level 1, the condition on valuations allows us to refine the truth tables of M_{K}, and reduce the search space for countermodels. For instance, since $\psi \vdash_{\mathcal{D}}^{\mathbb{V}^{\mathcal{F},0}_{\mathsf{K}}} \varphi \supset \psi$ (for every $\mathcal{F}$ with $\{\psi, \varphi, \varphi \supset \psi\} \subseteq \mathcal{F}$), at level 1 we have that if $\psi \in v^{-1}[\mathsf{TF}]$, then $v(\varphi \supset \psi) \in \mathsf{TF}$. This allows us to remove t and f from the first and third columns (when $y \in \mathsf{TF}$) in the table presenting $\tilde{\supset}$. The following entailments (at level 0), all with a single occurrence of some connective, lead to similar refinements, resulting in the optimized tables below for $\supset$, $\wedge$ and $\vee$:

$$\varphi, \varphi \supset \psi \vdash_{\mathcal{D}}^{\mathbb{V}^{\mathcal{F},0}_{\mathsf{K}}} \psi \qquad \varphi, \psi \vdash_{\mathcal{D}}^{\mathbb{V}^{\mathcal{F},0}_{\mathsf{K}}} \varphi \wedge \psi \qquad \varphi \wedge \psi \vdash_{\mathcal{D}}^{\mathbb{V}^{\mathcal{F},0}_{\mathsf{K}}} \varphi \qquad \varphi \wedge \psi \vdash_{\mathcal{D}}^{\mathbb{V}^{\mathcal{F},0}_{\mathsf{K}}} \psi$$

$$\varphi \vdash_{\mathcal{D}}^{\mathbb{V}^{\mathcal{F},0}_{\mathsf{K}}} \varphi \vee \psi \qquad \psi \vdash_{\mathcal{D}}^{\mathbb{V}^{\mathcal{F},0}_{\mathsf{K}}} \varphi \vee \psi$$

$x \tilde{\supset} y$	T	t	F	f
T	{T}	{t}	{F}	{f}
t	{T}	$\mathcal{D}$	{F}	$\overline{\mathcal{D}}$
F	{T}	{t}	{T}	{t}
f	{T}	$\mathcal{D}$	{T}	$\mathcal{D}$

$x \tilde{\wedge} y$	T	t	F	f
T	{T}	{t}	{F}	{f}
t	{t}	{t}	{f}	{f}
F	{F}	{f}	{F}	{f}
f	{f}	{f}	{f}	{f}

$x \tilde{\vee} y$	T	t	F	f
T	{T}	{T}	{T}	{T}
t	{T}	$\mathcal{D}$	{T}	$\mathcal{D}$
F	{T}	{T}	{F}	{F}
f	{T}	$\mathcal{D}$	{F}	$\overline{\mathcal{D}}$

We note that level 1 valuations are not fully captured by these tables. For example, they must assign T to every formula of the form $\varphi \supset \varphi$, while the table above allows also t when $v(\varphi) \in \mathsf{tf}$. A decision procedure for K can benefit from relying on these optimized tables instead of the original ones, starting from level 1.

4 Soundness and Completeness

In this section we establish the soundness and completeness of the proposed semantics. For that matter, we first extend the notion of satisfaction to sequents:

Definition 6. An $\mathcal{F}$-valuation v *$\mathcal{D}$-satisfies* an $\mathcal{F}$-sequent $\Gamma \Rightarrow \Delta$, denoted by $v \models_{\mathcal{D}} \Gamma \Rightarrow \Delta$, if $v \not\models_{\mathcal{D}} \varphi$ for some $\varphi \in \Gamma$ or $v \models_{\mathcal{D}} \varphi$ for some $\varphi \in \Delta$.

To prove soundness, we first note that except for (K), the soundness of each derivation rule easily follows from the Nmatrix semantics:

Lemma 6 (Local Soundness). *Consider an application of a rule of G_K other than* (K) *deriving a sequent $\Gamma \Rightarrow \Delta$ from sequents $\Gamma_1 \Rightarrow \Delta_1, \ldots, \Gamma_n \Rightarrow \Delta_n$, such that $\Gamma \cup \Gamma_1 \cup \ldots \cup \Gamma_n \cup \Delta \cup \Delta_1 \cup \ldots \cup \Delta_n \subseteq \mathcal{F}$. Let $v \in \mathbb{V}_K^{\mathcal{F},m}$ for some $m \geq 0$. If $v \models_{\mathcal{D}} \Gamma_i \Rightarrow \Delta_i$ for every $1 \leq i \leq n$, then $v \models_{\mathcal{D}} \Gamma \Rightarrow \Delta$.*

For (K), we make use of the level requirement, and prove the following lemma.

Lemma 7 (Soundness of (K)). *Suppose that $\Gamma \cup \Box\Gamma \cup \{\varphi, \Box\varphi\} \subseteq \mathcal{F}$, and $\Gamma \vdash_{\mathcal{D}}^{\mathbb{V}_K^{\mathcal{F},m-1}} \varphi$. Then, $\Box\Gamma \vdash_{\mathcal{D}}^{\mathbb{V}_K^{\mathcal{F},m}} \Box\varphi$.*

Proof. Let $v \in \mathbb{V}_{\mathsf{K}}^{\mathcal{F},m}$ such that $v \models_{\mathcal{D}} \Box\Gamma$. We prove that $v \models_{\mathcal{D}} \Box\varphi$. By the truth table of $\Box$, we have that $v(\psi) \in \mathbb{TF}$ for every $\psi \in \Gamma$, and we need to show that $v(\varphi) \in \mathbb{TF}$. Since $v(\psi) \in \mathbb{TF}$ for every $\psi \in \Gamma$, we have $\Gamma \subseteq v^{-1}[\mathbb{TF}]$. Since $\Gamma \vdash_{\mathcal{D}}^{\mathbb{V}_{\mathsf{K}}^{\mathcal{F},m-1}} \varphi$, we have $v^{-1}[\mathbb{TF}] \vdash_{\mathcal{D}}^{\mathbb{V}_{\mathsf{K}}^{\mathcal{F},m-1}} \varphi$. Since $v \in \mathbb{V}_{\mathsf{K}}^{\mathcal{F},m}$, it follows that $v(\varphi) \in \mathbb{TF}$. □

The above two lemmas together establish soundness, and from soundness for each level, we easily derive soundness for arbitrary (K)-depth.

Theorem 3 (Soundness for m). *If $\vdash_{\mathsf{G}_K}^{\mathcal{F},m} \Gamma \Rightarrow \Delta$, then $\Gamma \vdash_{\mathcal{D}}^{\mathbb{V}_K^{\mathcal{F},m}} \Delta$.*

Theorem 4 (Soundness without m). *If $\vdash_{\mathsf{G}_K}^{\mathcal{F}} \Gamma \Rightarrow \Delta$, then $\Gamma \vdash_{\mathcal{D}}^{\mathbb{V}_K^{\mathcal{F}}} \Delta$.*

By taking $\mathcal{F} = \mathcal{L}$ in Theorem 4 we get that if $\vdash_{\mathsf{G}_{\mathsf{K}}} \Gamma \Rightarrow \Delta$, then $\Gamma \vdash_{\mathcal{D}}^{\mathbb{V}_{\mathsf{K}}} \Delta$.

Next, we prove the following two completeness theorems:

Theorem 5 (Completeness for m). *Let $\mathcal{F} \subseteq \mathcal{L}$ closed under subformulas and $\Gamma \Rightarrow \Delta$ an $\mathcal{F}$-sequent. If $\Gamma \vdash_{\mathcal{D}}^{\mathbb{V}_K^{\mathcal{F},m}} \Delta$, then $\vdash_{\mathsf{G}_K}^{\mathcal{F},m} \Gamma \Rightarrow \Delta$.*

Theorem 6 (Completeness without m**).** *Let* $\mathcal{F} \subseteq \mathcal{L}$ *closed under subformulas and* $\Gamma \Rightarrow \Delta$ *an* $\mathcal{F}$*-sequent. If* $\Gamma \vdash_{\mathcal{D}}^{\mathbb{V}_{K}^{\mathcal{F}}} \Delta$*, then* $\vdash_{\mathsf{G}_K}^{\mathcal{F}} \Gamma \Rightarrow \Delta$.

In fact, since $\mathcal{F}$ may be infinite, we need to prove stronger theorems than Theorems 5 and 6, that incorporate infinite sequents.

Definition 7. An ω-sequent is a pair $\langle L, R\rangle$, denoted by $L \Rightarrow R$, such that L and R are (possibly infinite) sets of formulas. We write $\Vdash_{\mathsf{G}_K}^{\mathcal{F},m} L \Rightarrow R$ if $\vdash_{\mathsf{G}_K}^{\mathcal{F},m} \Gamma \Rightarrow \Delta$ for some finite $\Gamma \subseteq L$ and $\Delta \subseteq R$.

Other notions for sequents (e.g., being an $\mathcal{F}$-sequent) are extended to ω-sequents in the obvious way. In particular, $v \models_{\mathcal{D}} L \Rightarrow R$ if $v(\psi) \notin \mathcal{D}$ for some $\psi \in L$ or $v(\psi) \in \mathcal{D}$ for some $\psi \in R$.

Theorem 7 (ω**-Completeness for** m**).** *Let* $\mathcal{F} \subseteq \mathcal{L}$ *closed under subformulas and* $L \Rightarrow R$ *an* ω*-*$\mathcal{F}$*-sequent. If* $L \vdash_{\mathcal{D}}^{\mathbb{V}_{K}^{\mathcal{F},m}} R$*, then* $\Vdash_{\mathsf{G}_K}^{\mathcal{F},m} L \Rightarrow R$.

Theorem 8 (ω**-Completeness without** m**).** *Let* $\mathcal{F} \subseteq \mathcal{L}$ *closed under subformulas and* $L \Rightarrow R$ *an* ω*-*$\mathcal{F}$*-sequent. If* $L \vdash_{\mathcal{D}}^{\mathbb{V}_{K}^{\mathcal{F}}} R$*, then* $\Vdash_{\mathsf{G}_K}^{\mathcal{F}} L \Rightarrow R$.

Theorem 5 is a consequence of Theorem 7. Indeed, by Theorem 7, $\Gamma \vdash_{\mathcal{D}}^{\mathbb{V}_{K}^{\mathcal{F},m}} \Delta$ implies that $\vdash_{\mathsf{G}_K}^{\mathcal{F},m} \Gamma' \Rightarrow \Delta'$ for some (finite) $\Gamma' \subseteq \Gamma$ and $\Delta' \subseteq \Delta$. Using (WEAK), we obtain that $\vdash_{\mathsf{G}_K}^{\mathcal{F},m} \Gamma \Rightarrow \Delta$. Similarly, Theorem 6 is a consequence of Theorem 8. Also, using Lemma 3, we obtain Theorem 8 from Theorem 7. Hence in the remainder of this section we focus on the proof of Theorem 7.

Proof of Theorem 7. We start by defining maximal and consistent ω-sequents, and proving their existence.

Definition 8 (Maximal and consistent ω**-sequent).** Let $\mathcal{F} \subseteq \mathcal{L}$ and $m \geq 0$. An $\mathcal{F}$-ω-sequent $L \Rightarrow R$ is called:

1. $\mathcal{F}$*-maximal* if $\mathcal{F} \subseteq L \cup R$.
2. $\langle \mathsf{G}_K, \mathcal{F}, m\rangle$*-consistent* if $\nVdash_{\mathsf{G}_K}^{\mathcal{F},m} L \Rightarrow R$.
3. $\langle \mathsf{G}_K, \mathcal{F}, m\rangle$*-maximal-consistent* (in short, $\langle \mathsf{G}_K, \mathcal{F}, m\rangle$*-max-con*) if it is $\mathcal{F}$-maximal and $\langle \mathsf{G}_K, \mathcal{F}, m\rangle$-consistent.

Lemma 8. *Let* $\mathcal{F} \subseteq \mathcal{L}$ *and* $L \Rightarrow R$ *an* $\mathcal{F}$*-*ω*-sequent. Suppose that* $\nVdash_{\mathsf{G}_K}^{\mathcal{F},m} L \Rightarrow R$. *Then, there exist sets* $L_{MC(\mathsf{G}_K,\mathcal{F},m,L\Rightarrow R)}$ *and* $R_{MC(\mathsf{G}_K,\mathcal{F},m,L\Rightarrow R)}$ *such that the following hold:*

- $L \subseteq L_{MC(\mathsf{G}_K,\mathcal{F},m,L\Rightarrow R)}$ *and* $R \subseteq R_{MC(\mathsf{G}_K,\mathcal{F},m,L\Rightarrow R)}$.
- $L_{MC(\mathsf{G}_K,\mathcal{F},m,L\Rightarrow R)} \cup R_{MC(\mathsf{G}_K,\mathcal{F},m,L\Rightarrow R)} \subseteq \mathcal{F}$.
- $L_{MC(\mathsf{G}_K,\mathcal{F},m,L\Rightarrow R)} \Rightarrow R_{MC(\mathsf{G}_K,\mathcal{F},m,L\Rightarrow R)}$ *is* $\langle \mathsf{G}_K, \mathcal{F}, m\rangle$*-max-con.*

Thus, given an underivable ω-sequent, we can extend it to a $\langle \mathsf{G}_K, \mathcal{F}, m\rangle$-max-con ω-sequent. This ω-sequent induces the canonical countermodel, as defined next.

Algorithm 1. Deciding $\Gamma \vdash_{\mathcal{D}}^{\mathbb{V}_{\mathsf{K}}} \varphi$.

$\mathcal{F} \leftarrow sub(\Gamma \cup \{\varphi\})$
$m \leftarrow 4^{|\mathcal{F}|}$
for $v \in \mathbb{V}_{\mathsf{K}}^{\mathcal{F},m}$ **do**
  if $v \models_{\mathcal{D}} \Gamma$ and $v \not\models_{\mathcal{D}} \varphi$ **then**
    return ("NO", v)
return "YES"

Notation 9. We denote the set $\{\psi \in \mathcal{F} \mid \Box\psi \in X\}$ by $\mathbb{B}_{\mathcal{F}}^{X}$.

Definition 10. Suppose that $L \uplus R = \mathcal{F}$. The *canonical model w.r.t.* $L \Rightarrow R$, $\mathcal{F}$, *and* m, denoted by $v(\mathcal{F}, L \Rightarrow R, m)$, is the $\mathcal{F}$-valuation defined as follows (in λ notation):

For $m = 0$:

$$\lambda\varphi \in \mathcal{F}. \begin{cases} \mathsf{T} & \varphi \in L \text{ and } \Box\varphi \in L \\ \mathsf{t} & \varphi \in L \text{ and } \Box\varphi \notin L \\ \mathsf{F} & \varphi \in R \text{ and } \Box\varphi \in L \\ \mathsf{f} & \varphi \in R \text{ and } \Box\varphi \notin L \end{cases}$$

For $m > 0$:

$$\lambda\varphi \in \mathcal{F}. \begin{cases} \mathsf{T} & \varphi \in L \text{ and } \Vdash_{\mathsf{G_K}}^{\mathcal{F},m-1} \mathbb{B}_{\mathcal{F}}^{L} \Rightarrow \varphi \\ \mathsf{t} & \varphi \in L \text{ and } \nVdash_{\mathsf{G_K}}^{\mathcal{F},m-1} \mathbb{B}_{\mathcal{F}}^{L} \Rightarrow \varphi \\ \mathsf{F} & \varphi \in R \text{ and } \Vdash_{\mathsf{G_K}}^{\mathcal{F},m-1} \mathbb{B}_{\mathcal{F}}^{L} \Rightarrow \varphi \\ \mathsf{f} & \varphi \in R \text{ and } \nVdash_{\mathsf{G_K}}^{\mathcal{F},m-1} \mathbb{B}_{\mathcal{F}}^{L} \Rightarrow \varphi \end{cases}$$

Clearly, $v(\mathcal{F}, L \Rightarrow R, m) \not\models_{\mathcal{D}} L \Rightarrow R$. The proof of Theorem 7 is done by induction on m, and then carries on by showing that if $L \Rightarrow R$ is $\langle \mathsf{G_K}, \mathcal{F}, m\rangle$-max-con, then $v(\mathcal{F}, L \Rightarrow R, m)$ belongs to $\mathbb{V}_{\mathsf{K}}^{\mathcal{F},m}$ for every m.

Concretely, let $v \overset{\text{def}}{=} v(\mathcal{F}, L \Rightarrow R, m)$. We show that $v \in \mathbb{V}_{\mathsf{K}}^{\mathcal{F},k}$ for every $k \leq m$ by induction on k. The base case $k = 0$ is straightforward. For $k > 0$, we have $v \in \mathbb{V}_{\mathsf{K}}^{\mathcal{F},k-1}$ by the induction hypothesis. Let $\varphi \in \mathcal{F}$, and suppose that $v^{-1}[\mathbb{TF}] \vdash_{\mathcal{D}}^{\mathbb{V}_{\mathsf{K}}^{\mathcal{F},k-1}} \varphi$. To show that $v(\varphi) \in \mathbb{TF}$, we prove that $\Vdash_{\mathsf{G_K}}^{\mathcal{F},m-1} \mathbb{B}_{\mathcal{F}}^{L} \Rightarrow \varphi$. By the outer induction hypothesis (regarding the completeness theorem itself), $v^{-1}[\mathbb{TF}] \vdash_{\mathcal{D}}^{\mathbb{V}_{\mathsf{K}}^{\mathcal{F},k-1}} \varphi$ implies that $\Vdash_{\mathsf{G_K}}^{\mathcal{F},k-1} v^{-1}[\mathbb{TF}] \Rightarrow \varphi$, which implies that $\Vdash_{\mathsf{G_K}}^{\mathcal{F},m-1} v^{-1}[\mathbb{TF}] \Rightarrow \varphi$. Hence, there is a finite set $\{\varphi_1, \ldots, \varphi_n\} \subseteq v^{-1}[\mathbb{TF}]$ such that $\vdash_{\mathsf{G_K}}^{\mathcal{F},m-1} \{\varphi_1, \ldots, \varphi_n\} \Rightarrow \varphi$. For every $1 \leq i \leq n$, since $\varphi_i \in v^{-1}[\mathbb{TF}]$, we have that $\Vdash_{\mathsf{G_K}}^{\mathcal{F},m-1} \mathbb{B}_{\mathcal{F}}^{L} \Rightarrow \varphi_i$ and hence $\vdash_{\mathsf{G_K}}^{\mathcal{F},m-1} \Gamma_i \Rightarrow \varphi_i$ for some $\Gamma_i \subseteq \mathbb{B}_{\mathcal{F}}^{L}$. Using n applications of (CUT) on these sequents and $\vdash_{\mathsf{G_K}}^{\mathcal{F},m-1} \{\varphi_1, \ldots, \varphi_n\} \Rightarrow \varphi$, we obtain that $\vdash_{\mathsf{G_K}}^{\mathcal{F},m-1} \Gamma_1, \ldots, \Gamma_n \Rightarrow \varphi$, and so $\Vdash_{\mathsf{G_K}}^{\mathcal{F},m-1} \mathbb{B}_{\mathcal{F}}^{L} \Rightarrow \varphi$.

5 Effectiveness of the Semantics

In this section we study the effectiveness of the semantics introduced in Definition 5 for deciding $\vdash_{\mathsf{M_K}}$. Roughly speaking, a semantic framework is said to be *effective* if it induces a decision procedure that decides its underlying logic.

Consider Algorithm 1. Given a finite set Γ of formulas and a formula φ, it checks whether any valuations in $\mathbb{V}_{\mathsf{K}}^{\mathcal{F},m}$ is a countermodel. The correctness of this algorithm relies on the analyticity of $\mathsf{G_K}$, namely:

Lemma 9 ([11]). *If* $\vdash_{\mathsf{G}_{\mathcal{K}}} \Gamma \Rightarrow \Delta$, *then* $\vdash_{\mathsf{G}_{\mathcal{K}}}^{sub(\Gamma \cup \{\varphi\})} \Gamma \Rightarrow \Delta$.

Using Lemma 9, we show that the algorithm is correct.

Lemma 10. *Algorithm 1 always terminates, and returns "YES" iff* $\Gamma \vdash_{\mathcal{D}}^{\mathbb{V}_{\mathsf{K}}} \varphi$.

Proof. Termination follows from the fact that $\mathbb{V}_{\mathsf{K}}^{\mathcal{F},m}$ is finite. Suppose that the result is "YES" and assume for contradiction that $\Gamma \not\vdash_{\mathcal{D}}^{\mathbb{V}_{\mathsf{K}}} \varphi$. Hence, there exists some $u \in \mathbb{V}_{\mathsf{K}}$ such that $u \models_{\mathcal{D}} \Gamma$ and $u \not\models_{\mathcal{D}} \varphi$. Consider $v \stackrel{\mathsf{def}}{=} u|_{\mathcal{F}}$. Then, $v \in \mathbb{V}_{\mathsf{K}}^{\mathcal{F}} \subseteq \mathbb{V}_{\mathsf{K}}^{\mathcal{F},m}$, which contradicts the fact that the algorithm returns "YES". Now, suppose that the result is "NO". Then, there exists some $v \in \mathbb{V}_{\mathsf{K}}^{\mathcal{F},m}$ such that $v \models_{\mathcal{D}} \Gamma$ and $v \not\models_{\mathcal{D}} \varphi$. By Lemma 5, $v \in \mathbb{V}_{\mathsf{K}}^{\mathcal{F}}$. Hence, $\Gamma \not\vdash_{\mathcal{D}}^{\mathbb{V}_{\mathsf{K}}^{\mathcal{F}}} \varphi$. By Theorem 3, we have $\not\vdash_{\mathsf{G}_{\mathsf{K}}}^{\mathcal{F}} \Gamma \Rightarrow \varphi$. By Lemma 9, we have $\not\vdash_{\mathsf{G}_{\mathsf{K}}} \Gamma \Rightarrow \varphi$. By Theorem 6, we have $\Gamma \not\vdash_{\mathcal{D}}^{\mathbb{V}_{\mathsf{K}}} \varphi$. □

Lemma 10 shows that Algorithm 1 is a decision procedure for $\vdash_{\mathsf{M}_{\mathsf{K}}}$, when ignoring the additional output provided in Line 5. However, it is typical in applications that a "YES" or "NO" answer is not enough, and often it is expected that a "NO" result is accompanied with a countermodel. Algorithm 1 returns a valuation v in case the answer is "NO", but Lemma 10 does not ensure that v is indeed a countermodel for $\Gamma \vdash_{\mathcal{D}}^{\mathbb{V}_{\mathsf{K}}} \varphi$. The issue is that the valuation v from the proof of Lemma 10 witnesses the fact that $\not\vdash_{\mathcal{D}}^{\mathbb{V}_{\mathsf{K}}}$ only in a non-constructive way. Indeed, using the soundness and completeness theorems, we are able to deduce that $v' \models_{\mathcal{D}} \Gamma$ and $v' \not\models_{\mathcal{D}} \varphi$ for some $v' \in \mathbb{V}_{\mathsf{K}}$, but the relation between v and v' is unclear. Most importantly, it is not clear whether v and v' agree on $\mathcal{F}$-formulas. In the remainder of this section we prove that v' extends v, and so the returned countermodel of Line 5 can be trusted.

We say that a valuation v' *extends* a valuation v if $\mathsf{Dom}(v) \subseteq \mathsf{Dom}(v')$ and $v'(\varphi) = v(\varphi)$ for every $\varphi \in \mathsf{Dom}(v)$ (identifying functions with sets of pairs, this means $v \subseteq v'$). Clearly, for a $\mathsf{Dom}(v)$-formula ψ we have that $v' \models_{\mathcal{D}} \psi$ iff $v \models_{\mathcal{D}} \psi$. We first show how to extend a given valuation $v \in \mathbb{V}_{\mathsf{K}}^{\mathcal{F},m}$ by a single formula ψ such that $sub(\psi) \setminus \{\psi\} \subseteq \mathcal{F}$, obtaining a valuation $v' \in \mathbb{V}_{\mathsf{K}}^{\mathcal{F} \cup \{\psi\},m}$ that agrees with v on all formulas in $\mathcal{F}$.

Lemma 11. *Let* $m \geq 0$, $\mathcal{F} \subseteq \mathcal{L}$, *and* $v \in \mathbb{V}_{K}^{\mathcal{F},m}$. *Let* $\psi \in \mathcal{L} \setminus \mathcal{F}$ *such that* $sub(\psi) \setminus \{\psi\} \subseteq \mathcal{F}$. *Then,* v *can be extended to some* $v' \in \mathbb{V}_{K}^{\mathcal{F} \cup \{\psi\},m}$.

We sketch the proof of Lemma 11.

When $m = 0$, v' exists from Lemma 1. For $m > 0$, we define v' as follows:[1]

$$v' \stackrel{\mathsf{def}}{=} \lambda \varphi \in \mathcal{F} \cup \{\psi\} . \begin{cases} v(\varphi) & \varphi \in \mathcal{F} \\ \min(\mathsf{pos\text{-}val}(\psi, \mathsf{M}_{\mathsf{K}}, v) \cap \mathsf{TF}) & \varphi = \psi \wedge v^{-1}[\mathsf{TF}] \vdash_{\mathcal{D}}^{\mathbb{V}_{\mathsf{K}}^{\mathcal{F} \cup \{\psi\}, m-1}} \psi \\ \min(\mathsf{pos\text{-}val}(\psi, \mathsf{M}_{\mathsf{K}}, v) \cap \mathsf{tf}) & \text{otherwise} \end{cases}$$

[1] The use of min here assumes an arbitrary order on truth values. It is used here only to choose *some* element from a non-empty set of truth values.

The proof of Lemma 11 then carries on by showing that $v' \in \mathbb{V}_{\mathsf{K}}^{\mathcal{F}\cup\{\psi\},m}$.
Next, Lemma 11 is used in order to extend partial valuations into total ones.

Lemma 12. *Let $v \in \mathbb{V}_K^{\mathcal{F},m}$ for some $\mathcal{F}$ closed under subformulas. Then, v can be extended to some $v' \in \mathbb{V}_K^m$.*

Finally, Lemmas 3 and 12 can be used in order to extend any partial valuation in $\mathbb{V}_{\mathsf{K}}^{\mathcal{F}}$ into a total one.

Lemma 13. *Let $v \in \mathbb{V}_K^{\mathcal{F}}$ for some set $\mathcal{F}$ closed under subformulas. Then, v can be extended to some $v' \in \mathbb{V}_K$.*

We conclude by showing that when Algorithm 1 returns ("NO", v), then v is a finite representation of a true countermodel for $\Gamma \vdash_{\mathsf{M}_{\mathsf{K}}} \varphi$.

Corollary 1. *If $\Gamma \nvdash_{\mathcal{D}}^{\mathbb{V}_K} \varphi$. Then Algorithm 1 returns ("NO", v) for some v for which there exists $v' \in \mathbb{V}_K$ such that $v = v'|_{sub(\Gamma\cup\{\varphi\})}$, $v' \models_{\mathcal{D}} \Gamma$, and $v' \not\models_{\mathcal{D}} \varphi$.*

Proof. Suppose that $\Gamma \nvdash_{\mathcal{D}}^{\mathbb{V}_{\mathsf{K}}} \varphi$. Then by Lemma 10, Algorithm 1 does not return "YES". Therefore, it returns ("NO", v) for some $v \in \mathbb{V}_{\mathsf{K}}^{\mathcal{F},m}$ such that $v \models_{\mathcal{D}} \Gamma$ and $v \not\models_{\mathcal{D}} \varphi$, where $\mathcal{F} = sub(\Gamma \cup \{\varphi\})$ and $m = 4^{|\mathcal{F}|}$. By Lemma 5, $v \in \mathbb{V}_{\mathsf{K}}\mathcal{F}$. By Lemma 13, v can be extended to some $v' \in \mathbb{V}_{\mathsf{K}}$. Therefore, $v = v'|_{sub(\Gamma\cup\{\varphi\})}$, $v' \models_{\mathcal{D}} \Gamma$, and $v' \not\models_{\mathcal{D}} \varphi$. □

Remark 2. Notice that in scenarios where model generation is not important, m can be set to a much smaller number in Line 2 of Algorithm 1, namely, the "modal depth" of the input.[2] The reason for that is that for such m, it can be shown that $\vdash_{\mathsf{G_K}}^{\mathcal{F},m} \Gamma \Rightarrow \varphi$ iff $\vdash_{\mathsf{G_K}}^{\mathcal{F}} \Gamma \Rightarrow \varphi$, by reasoning about the applications of rule (K). Using the soundness and completeness theorems, we can get $\Gamma \vdash_{\mathcal{D}}^{\mathbb{V}_{\mathsf{K}}^{\mathcal{F},m}} \varphi$ iff $\Gamma \vdash_{\mathcal{D}}^{\mathbb{V}_{\mathsf{K}}^{\mathcal{F}}} \varphi$, and so limiting to such m is enough. Notice however, that we do not necessarily get $\mathbb{V}_{\mathsf{K}}^{\mathcal{F},m} = \mathbb{V}_{\mathsf{K}}^{\mathcal{F}}$ for such m, and so the valuation returned in Line 5 might not be an element of $\mathbb{V}_{\mathsf{K}}^{\mathcal{F}}$.

6 The Modal Logic KT

In this section we obtain similar results for the modal logic KT. First, the calculus $\mathsf{G_{KT}}$ is obtained from $\mathsf{G_K}$ by adding the following rule (see, e.g., [16]):

$$(T)\ \frac{\Gamma, \varphi \Rightarrow \Delta}{\Gamma, \Box\varphi \Rightarrow \Delta}$$

Derivations are defined as before. (In particular, the (K)-depth of a derivation still depends on applications of rule (K), not of rule (T).) We write $\vdash_{\mathsf{G_{KT}}}^{\mathcal{F},m} \Gamma \Rightarrow \Delta$

[2] The *modal depth* of an atomic formula p is 0. The modal depth of $\Box\varphi$ is the modal depth of φ plus 1. The modal depth of $\diamond(\varphi_1, \ldots, \varphi_n)$ for $\diamond \neq \Box$ is the maximum among the modal depths of $\varphi_1, \ldots, \varphi_n$.

if there is a derivation of $\Gamma \Rightarrow \Delta$ in G_{KT} in which only $\mathcal{F}$-sequents occur and that has (K)-depth at most m.

Next, we consider the semantics. For a valuation $v \in \mathbb{V}_{\mathsf{K}}$ to respect rule (T), we must have that if $v \models_{\mathcal{D}} \Gamma, \varphi \Rightarrow \Delta$, then $v \models_{\mathcal{D}} \Gamma, \Box\varphi \Rightarrow \Delta$. In particular, when $v \not\models_{\mathcal{D}} \Gamma \Rightarrow \Delta$, we get that if $v(\varphi) \notin \mathcal{D}$, then $v(\Box\varphi) \notin \mathcal{D}$. Now, if $v(\varphi) = \mathsf{F}$, then $v(\Box\varphi) \in \mathcal{D}$ according to the truth table of $\Box$ in M_{K}. But, we must have $v(\Box\varphi) \notin \mathcal{D}$. This leads us to remove F from M_{K}.

We thus obtain the following Nmatrix M_{KT}: The sets of truth values and of designated truth values are given by[3]

$$\mathcal{V}_3 \stackrel{\text{def}}{=} \{\mathsf{T}, \mathsf{t}, \mathsf{f}\} \qquad \mathcal{D} \stackrel{\text{def}}{=} \{\mathsf{T}, \mathsf{t}\}$$

and the truth tables are as follows:

$x \tilde{\supset} y$	T	t	f
T	$\mathcal{D}$	$\mathcal{D}$	{f}
t	$\mathcal{D}$	$\mathcal{D}$	{f}
f	$\mathcal{D}$	$\mathcal{D}$	$\mathcal{D}$

$x \tilde{\wedge} y$	T	t	f
T	$\mathcal{D}$	$\mathcal{D}$	{f}
t	$\mathcal{D}$	$\mathcal{D}$	{f}
f	{f}	{f}	{f}

$x \tilde{\vee} y$	T	t	f
T	$\mathcal{D}$	$\mathcal{D}$	$\mathcal{D}$
t	$\mathcal{D}$	$\mathcal{D}$	$\mathcal{D}$
f	$\mathcal{D}$	$\mathcal{D}$	{f}

x	$\tilde{\neg} x$
T	{f}
t	{f}
f	$\mathcal{D}$

x	$\tilde{\Box} x$
T	$\mathcal{D}$
t	{f}
f	{f}

Again, one may gain intuition from the possible worlds semantics. There, the logic KT is characterized by frames with *reflexive* accessibility relation. Thus, for instance, if ψ holds in w but not in some world accessible from w (i.e., ψ has value t), we know that $\Box\psi$ does not hold in w, and the reflexivity of the accessibility relation implies that $\Box\psi$ does not hold in some world accessible from w (thus $\Box\psi$ has value f).

Example 4. Let $\varphi \stackrel{\text{def}}{=} \Box\Box(p_1 \wedge p_2) \supset \Box p_1$ and $\mathcal{F} \stackrel{\text{def}}{=} sub(\varphi)$. The sequent $\Rightarrow \varphi$ has a derivation in G_{KT} using only $\mathcal{F}$ formulas of (K)-depth of 1. However, it is not satisfied by all M_{KT}-legal $\mathcal{F}$-valuations. For example, the following valuation is an M_{KT}-legal valuation that does not satisfy φ:

$$v(p_1) = v(p_2) = \mathsf{t} \quad v(\Box p_1) = \mathsf{f}$$

$$v(p_1 \wedge p_2) = v(\Box(p_1 \wedge p_2)) = v(\Box\Box(p_1 \wedge p_2)) = \mathsf{T} \quad v(\varphi) = \mathsf{f}$$

Next, we define the levels of valuations for M_{KT}. These are obtained from Definition 5 by removing the value F:

Definition 11. The set $\mathbb{V}_{\mathsf{KT}}^{\mathcal{F},m}$ is recursively defined as follows:

- $\mathbb{V}_{\mathsf{KT}}^{\mathcal{F},0}$ is the set of M_{KT}-legal $\mathcal{F}$-valuations.
- $\mathbb{V}_{\mathsf{KT}}^{\mathcal{F},m+1} \stackrel{\text{def}}{=} \left\{ v \in \mathbb{V}_{\mathsf{KT}}^{\mathcal{F},m} \mid \forall \varphi \in \mathcal{F}.\, v^{-1}[\mathsf{T}] \vdash_{\mathcal{D}}^{\mathbb{V}_{\mathsf{KT}}^{\mathcal{F},m}} \varphi \implies v(\varphi) = \mathsf{T} \right\}$

We also define:

$$\mathbb{V}_{\mathsf{KT}}^{\mathcal{F}} \stackrel{\text{def}}{=} \bigcap_{m \geq 0} \mathbb{V}_{\mathsf{KT}}^{\mathcal{F},m} \qquad \mathbb{V}_{\mathsf{KT}}^{m} \stackrel{\text{def}}{=} \mathbb{V}_{\mathsf{KT}}^{\mathcal{L},m} \qquad \mathbb{V}_{\mathsf{KT}} \stackrel{\text{def}}{=} \bigcap_{m \geq 0} \mathbb{V}_{\mathsf{KT}}^{\mathcal{L},m}$$

[3] In this section we denote the set $\{\mathsf{T}\}$ by $\mathbb{T}\mathsf{F}$.

Example 5. Following Example 4, we note that for every $v \in \mathbb{V}_{\mathsf{KT}}^{\mathcal{F},m}$ with $m > 0$, we have $v \models_{\mathcal{D}} \varphi$. In particular, the valuation v from Example 4 does not belong to $\mathbb{V}_{\mathsf{KT}}^{\mathcal{F},m}$: $\Box(p_1 \wedge p_2) \in v^{-1}[\mathsf{T}]$, $\Box(p_1 \wedge p_2) \vdash_{\mathcal{D}}^{\mathbb{V}_{\mathsf{KT}}^{\mathcal{F},0}} p_1$, but $v(p_1) = \mathsf{t}$.

Similarly to Theorem 2, the levels of valuations converge to a maximal set that satisfies the following condition:

$$\forall v \in \mathbb{V}. \forall \varphi \in \mathcal{F}. v^{-1}[\mathsf{T}] \vdash_{\mathcal{D}}^{\mathbb{V}} \varphi \implies v(\varphi) = \mathsf{T} \qquad (\textit{necessitation}_{\mathsf{KT}})$$

Theorem 9. *The set* $\mathbb{V}_{KT}^{\mathcal{F}}$ *is the largest set* $\mathbb{V}$ *of* M_{KT}*-legal* $\mathcal{F}$*-valuations that satisfies necessitation*$_{KT}$.

The proof of Theorem 9 is analogous to that of Theorem 2.

Remark 3. The *necessitation*$_{\mathsf{KT}}$ condition is equivalent to the one given in [8], except that the underlying truth table is different. Theorem 9 proves that our gradual way of defining $\mathbb{V}_{\mathsf{KT}}^{\mathcal{F}}$ via levels coincides with the semantic condition from [8].

As we demonstrated for K, starting from level 1, the condition on valuations allows us to refine the truth tables of M_{KT}, and reduce the search space. Simple entailments (at level 0) lead to the optimized tables below for $\supset$, $\wedge$ and $\vee$:

$x \tilde{\supset} y$	T	t	f
T	{T}	{t}	{f}
t	{T}	$\mathcal{D}$	{f}
f	{T}	$\mathcal{D}$	$\mathcal{D}$

$x \tilde{\wedge} y$	T	t	f
T	{T}	{t}	{f}
t	{t}	{t}	{f}
f	{f}	{f}	{f}

$x \tilde{\vee} y$	T	t	f
T	{T}	{T}	{T}
t	{T}	$\mathcal{D}$	$\mathcal{D}$
f	{T}	$\mathcal{D}$	{f}

Soundness and completeness for G_{KT} are obtained analogously to G_{K}, keeping in mind that M_{KT} is obtained from M_{K} by deleting the value F. For soundness, this is captured by the rule (T). For completeness, the same construction of a countermodel is performed , while rule (T) ensures that it is three-valued.

Theorem 10 (Soundness and Completeness). *Let* $\mathcal{F} \subseteq \mathcal{L}$ *closed under subformulas and* $\Gamma \Rightarrow \Delta$ *an* $\mathcal{F}$*-sequent.*

1. *For every* $m \geq 0$, $\Gamma \vdash_{\mathcal{D}}^{\mathbb{V}_{KT}^{\mathcal{F},m}} \Delta$ *iff* $\vdash_{\mathsf{G}_{KT}}^{\mathcal{F},m} \Gamma \Rightarrow \Delta$.
2. $\Gamma \vdash_{\mathcal{D}}^{\mathbb{V}_{KT}^{\mathcal{F}}} \Delta$ *iff* $\vdash_{\mathsf{G}_{KT}}^{\mathcal{F}} \Gamma \Rightarrow \Delta$.

Effectiveness is also shown similarly to K. For that matter, we use the following main lemma, whose proof is similar to Lemma 13. The only component that is added to that proof is making sure that the constructed model is three-valued.

Lemma 14. *Let* $v \in \mathbb{V}_{KT}^{\mathcal{F}}$ *for some set* $\mathcal{F}$ *closed under subformulas. Then,* v *can be extended to some* $v' \in \mathbb{V}_{KT}$.

Let Algorithm 2 be obtained from Algorithm 1 by setting m to $3^{|\mathcal{F}|}$ in Line 2, and taking $v \in \mathbb{V}_{\mathsf{KT}}^{\mathcal{F},m}$ in Line 3. Similarly to Lemma 10 and Corollary 1, we get that Algorithm 2 is a model-producing decision procedure for $\vdash_{\mathsf{M}_{\mathsf{KT}}}$.

Lemma 15. *Algorithm 2 always terminates, and returns "YES" iff* $\Gamma \vdash_{\mathcal{D}}^{\mathbb{V}_{KT}} \varphi$. *Further, if* $\Gamma \nvdash_{\mathcal{D}}^{\mathbb{V}_{KT}} \varphi$, *then it returns ("NO", v) for some v for which there exists* $v' \in \mathbb{V}_{KT}$ *such that* $v = v'|_{sub(\Gamma\cup\{\varphi\})}$, $v' \models_{\mathcal{D}} \Gamma$, *and* $v' \not\models_{\mathcal{D}} \varphi$.

7 Future Work

We have introduced a new semantics for the modal logic K, based on levels of valuations in many-valued non-deterministic matrices. Our semantics is effective, and was shown to tightly correspond to derivations in a sequent calculus for K. We also adapted these results for the modal logic KT.

There are two main directions for future work. The first is to establish similar semantics for other normal modal logics, such as KD, K4, S4 and S5, and to investigate ◊ as an independent modality. The second is to analyze the complexity, implement and experiment with decision procedures for K and KT based on the proposed semantics. In particular, we plan to consider SAT-based decision procedures that would encode this semantics in SAT, directly or iteratively.

References

1. Avron, A., Zamansky, A.: Non-deterministic semantics for logical systems - a survey. In: Gabbay, D., Guenther, F. (eds.) Handbook of Philosophical Logic, vol. 16, pp. 227–304. Springer, Dordrecht (2011). https://doi.org/10.1007/978-94-007-0479-4_4
2. Avron, A., Lev, I.: Non-deterministic multi-valued structures. J. Log. Comput. **15**, 241–261 (2005). Conference version: Avron, A., Lev, I.: Canonical propositional Gentzen-type systems. In: International Joint Conference on Automated Reasoning, IJCAR 2001. Proceedings, LNAI, vol. 2083, pp. 529–544. Springer (2001)
3. Biere, A., Heule, M., van Maaren, H., Walsh, T. (eds.): Handbook of Satisfiability. Frontiers in Artificial Intelligence and Applications, 2nd edn., vol. 336. IOS Press (2021)
4. Coniglio, M.E., del Cerro, L.F., Peron, N.M.: Finite non-deterministic semantics for some modal systems. J. Appl. Non Class. Log. **25**(1), 20–45 (2015)
5. Coniglio, M.E., del Cerro, L.F., Peron, N.M.: Errata and addenda to 'finite non-deterministic semantics for some modal systems'. J. Appl. Non Class. Log. **26**(4), 336–345 (2016)
6. Coniglio, M.E., Toledo, G.V.: Two decision procedures for da costa's Cn logics based on restricted Nmatrix semantics. Stud. Log. **110**, 1–42 (2021)
7. Fagin, R., Halpern, J.Y., Moses, Y., Vardi, M.: Reasoning About Knowledge. MIT Press, Cambridge (2004)
8. Grätz, L.: Truth tables for modal logics T and S4, by using three-valued non-deterministic level semantics. J. Log. Comput. **32**(1), 129–157 (2022)
9. Halpern, J., Manna, Z., Moszkowski, B.: A hardware semantics based on temporal intervals. In: Diaz, J. (ed.) ICALP 1983. LNCS, vol. 154, pp. 278–291. Springer, Heidelberg (1983). https://doi.org/10.1007/BFb0036915
10. Kearns, J.T.: Modal semantics without possible worlds. J. Symb. Log. **46**(1), 77–86 (1981)

11. Lahav, O., Avron, A.: A unified semantic framework for fully structural propositional sequent systems. ACM Trans. Comput. Log. **14**(4), 271–273 (2013)
12. Pawlowski, P., La Rosa, E.: Modular non-deterministic semantics for T, TB, S4, S5 and more. J. Log. Comput. **32**(1), 158–171 (2022)
13. Pratt, V.R.: Application of modal logic to programming. Stud. Log.: Int. J. Symb. Log. **39**(2/3), 257–274 (1980)
14. Skurt, D., Omori, H.: More modal semantics without possible worlds. FLAP **3**(5), 815–846 (2016)
15. Urquhart, A.: Many-valued logic. In: Gabbay, D., Guenthner, F. (eds.) Handbook of Philosophical Logic, vol. II, 2nd edn., pp. 249–295. Kluwer (2001)
16. Wansing, H.: Sequent systems for modal logics. In: Gabbay, D.M., Guenthner, F. (eds.) Handbook of Philosophical Logic, vol. 8, 2nd edn., pp. 61–145. Springer, Dordrecht (2002). https://doi.org/10.1007/978-94-010-0387-2_2

Local Reductions for the Modal Cube

Cláudia Nalon[1], Ullrich Hustadt[2(✉)], Fabio Papacchini[3], and Clare Dixon[4]

[1] Department of Computer Science, University of Brasília, Brasília, Brazil
nalon@unb.br

[2] Department of Computer Science, University of Liverpool, Liverpool, UK
U.Hustadt@liverpool.ac.uk

[3] School of Computing and Communications, Lancaster University in Leipzig, Leipzig, Germany
f.papacchini@lancaster.ac.uk

[4] Department of Computer Science, University of Manchester, Manchester, UK
clare.dixon@manchester.ac.uk

Abstract. The modal logic K is commonly used to represent and reason about necessity and possibility and its extensions with combinations of additional axioms are used to represent knowledge, belief, desires and intentions. Here we present local reductions of all propositional modal logics in the so-called modal cube, that is, extensions of K with arbitrary combinations of the axioms B, D, T, 4 and 5 to a normal form comprising a formula and the set of modal levels it occurs at. Using these reductions we can carry out reasoning for all these logics with the theorem prover K_SP. We define benchmarks for these logics and experiment with the reduction approach as compared to an existing resolution calculus with specialised inference rules for the various logics.

1 Introduction

Modal logics have been used to represent and reason about mental attitudes such as knowledge, belief, desire and intention, see for example [17,20,31]. These can be represented using extensions of the basic modal logic K with one or more of the axioms B (symmetry), D (seriality), T (reflexivity), 4 (transitivity) and 5 (Euclideaness). The logic K and these extensions form the so-called *modal cube*, see Fig. 1. In the diagram, a line from a logic L_1 to a logic L_2 to its right and/or above means that all theorems of L_1 are also theorems of L_2, but not vice versa. As indicated in Fig. 1, some of the logics have the same theorems, e.g., KB5 and KB4. Also, all logics not explicitly listed have the same theorems as KT5 aka S5. In total there are 15 distinct logics.

While these modal logics are well-studied and a multitude of calculi and translations to other logics exist, see, e.g., [1,3–6,9,13,14,16,18,22,41], fully

C. Dixon was partially supported by the EPSRC funded RAI Hubs FAIR-SPACE (EP/R026092/1) and RAIN (EP/R026084/1), and the EPSRC funded programme Grant S4 (EP/N007565/1).

J. Blanchette et al. (Eds.): IJCAR 2022, LNAI 13385, pp. 486–505, 2022.
https://doi.org/10.1007/978-3-031-10769-6_29

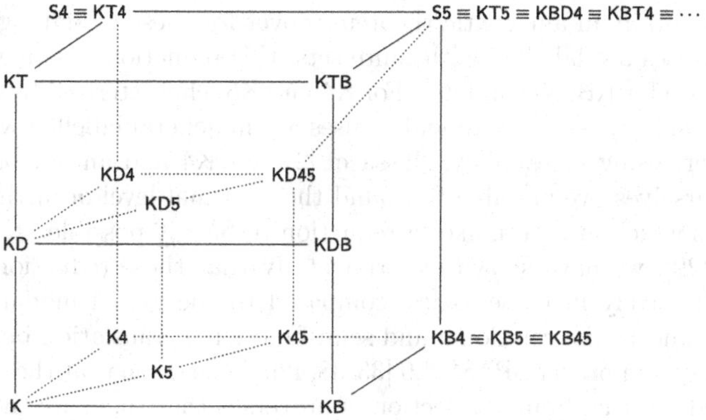

Fig. 1. Modal Cube: Relationships between modal logics

automatic support by provers is still lacking. Early implementations covering the full modal cube, such as Catach's TABLEAUX system [7], are no longer available. LoTREC 2.0 [10] supports a wide range of logics but is not intended as an automatic theorem prover. MOIN [11] supports all the logics but the focus is on producing human-readable proofs and countermodels for small formulae. Other provers that go beyond just K, like MleanCoP [28] and CEGARBox [15] only support a small subset of the 15 logics. There are also a range of translations from modal logics to first-order and higher-order logics [13,18,19,27,33]. Regarding implementations of those, SPASS [33,43] is limited to a subset of the 15 logics, while LEO-III [13,36] supports all the logics in the modal cube, but can only solve very few of the available benchmark formulae.

K$_S$P [23] is a modal logic theorem prover that implements both the modal-layered resolution (MLR) calculus [25] for the modal logic K and the global resolution (GMR) calculus [24] for all the 15 logics considered here. It also supports several refinements of resolution and a range of simplification rules. In this paper, we give reductions of all logics of the modal cube into a normal form for the basic modal logic K. We then compare the performance of the combination of these reductions with the modal-layered resolution calculus to that of the global resolution calculus on a new benchmark collection for the modal cube.

In [29] we have presented new reductions[1] of the propositional modal logics KB, KD, KT, K4, and K5 to Separated Normal Form with Sets of Modal Levels SNF_{sml}. SNF_{sml} is a generalisation of the Separated Normal Form with Modal Level, SNF_{ml}. In the latter, labelled modal clauses are used where a natural number label refers to a particular level within a tree Kripke structure at which a modal clause holds. In the former, a finite or infinite set of natural numbers labels each modal clause with the intended meaning that such a modal clause is true at every level of a tree Kripke structure contained in that set. As our prover K$_S$P and the modal-layered resolution calculus it implements currently only support sets of modal clauses in SNF_{ml}, we then use a further reduction from SNF_{sml}

[1] A *reduction* here is a satisfiability preserving mapping between logics.

to SNF_{ml} to obtain an automatic theorem prover for these modal logics. Where all modal clauses are labelled with finite sets, this reduction is straightforward. This is the case for KB, KD and KT. For K4 and K5, characterised by the axioms $\Box\varphi \rightarrow \Box\Box\varphi$ and $\Diamond\varphi \rightarrow \Box\Diamond\varphi$, modal clauses are in general labelled with infinite sets. However, using a result by Massacci [21] for K4 and an analogous result for K5 by ourselves, we are able to bound the maximal level occurring in those labelling sets which in turn makes a reduction to SNF_{ml} possible.

Also in [29], we have shown experimentally that these reductions allow us to reason effectively in these logics, compared to the global modal resolution calculus [24] and to the relational and semi-functional translation built into the first-order theorem prover SPASS 3.9 [33,38,42]. The reason that the comparison only included a rather limited selection of provers is that these are the only ones with built-in support for all six logics our reductions covered.

Unfortunately, we cannot simply combine our reductions for single axioms to obtain satisfiability preserving reductions for their combinations. There are two main reasons for this. First, our calculus does not use an explicit representation of the accessibility relationship within a Kripke structure, which would make it possible to reflect modal axioms via corresponding properties of that accessibility relationship. Instead, we add labelled modal clauses based on instances of the modal axioms for $\Box$-formulae occurring in the modal formula we want to check for satisfiability. However, if we deal with multiple modal axioms, then these axioms might interact making it necessary to add instances that are not necessary for each individual axiom. For instance, consider, the converse of axiom B, $\Diamond\Box\varphi \rightarrow \varphi$, and axiom 4, $\Box\varphi \rightarrow \Box\Box\varphi$. Together they imply $\Diamond\Box\varphi \rightarrow \Box\varphi$. Instances of this derived axiom are necessary for completeness of a reduction from KB4 to K, but are unsound for KB and K4 separately.

Second, our reductions attempt to keep the labelling sets minimal in size in order to decrease the number of inferences that can be performed. Again, taking axioms B and 4 as examples, in KB, a $\Box$-formula $\Box\psi$ true at level ml in a tree-like Kripke structure M forces ψ to be true at level $ml - 1$, while in K4, $\Box\psi$ true at level ml in M forces ψ to be true all levels ml' with $ml' > ml$. This is reflected in the labelling sets we use for these two logics. However, for KB4, $\Box\psi$ true at level ml forces ψ to be true at every level in a tree-like Kripke structure M (unless M consists only of a single world).

Since we intend to maintain these two properties of our reductions, we have to consider each modal logic individually. As we will see, for some logics a reduction can be obtained as the union of the existing reductions while for others we need a logic-specific reduction to accommodate the interaction of axioms.

The structure of the paper is as follows. In Sect. 2 we recall common concepts of propositional modal logic and the definition of our normal form SNF_{ml}. Section 3 introduces our reduction for extensions of the basic modal logic K with combinations of the axioms B, D, T, 4, and 5. Section 4 presents a transformation from SNF_{sml} to SNF_{ml} which allows us to use the modal resolution prover KSP to reason in all the modal logics. In Sect. 5 we compare the performance of a combination of our reductions and the modal-layered resolution calculus implemented in the prover KSP with resolution calculi specifically designed for the logics under consideration as well as the prover LEO-III.

2 Preliminaries

The language of modal logic is an extension of the language of propositional logic with a unary modal operator $\Box$ and its dual $\Diamond$. More precisely, given a denumerable set of *propositional symbols*, $P = \{p, p_0, q, q_0, t, t_0, \ldots\}$ as well as propositional *constants* **true** and **false**, *modal formulae* are inductively defined as follows: constants and propositional symbols are modal formulae. If φ and ψ are modal formulae, then so are $\neg\varphi$, $(\varphi \wedge \psi)$, $(\varphi \vee \psi)$, $(\varphi \rightarrow \psi)$, $\Box\varphi$, and $\Diamond\varphi$. We also assume that $\wedge$ and $\vee$ are associative and commutative operators and consider, e.g., $(p \vee (q \vee r))$ and $(r \vee (q \vee p))$ to be identical formulae. We often omit parentheses if this does not cause confusion. By $\mathsf{var}(\varphi)$ we denote the set of all propositional symbols occurring in φ. This function straightforwardly extends to finite sets of modal formulae. A *modal axiom (schema)* is a modal formula ψ representing the set of all instances of ψ.

A *literal* is either a propositional symbol or its negation; the set of literals is denoted by L_P. By $\neg l$ we denote the *complement* of the literal $l \in L_P$, that is, if l is the propositional symbol p then $\neg l$ denotes $\neg p$, and if l is the literal $\neg p$ then $\neg l$ denotes p. By $|l|$ for $l \in L_P$ we denote p if $l = p$ or $l = \neg p$. A *modal literal* is either $\Box l$ or $\Diamond l$, where $l \in L_P$.

A *(normal) modal logic* is a set of modal formulae which includes all propositional tautologies, the axiom schema $\Box(\varphi \rightarrow \psi) \rightarrow (\Box\varphi \rightarrow \Box\psi)$, called the *axiom* K, it is closed under modus ponens (if $\vdash \varphi$ and $\vdash \varphi \rightarrow \psi$ then $\vdash \psi$) and the rule of necessitation (if $\vdash \varphi$ then $\vdash \Box\varphi$).

K is the weakest modal logic, that is, the logic given by the smallest set of modal formulae constituting a normal modal logic. By $\mathsf{K}\Sigma$ we denote an *extension* of K by a set Σ of axioms.

The standard semantics of modal logics is the *Kripke semantics* or *possible world semantics*. A *Kripke frame* F is an ordered pair $\langle W, R\rangle$ where W is a non-empty set of *worlds* and R is a binary (accessibility) relation over W. A *Kripke structure* M over P is an ordered pair $\langle F, V\rangle$ where F is a Kripke frame and the *valuation* V is a function mapping each propositional symbol in P to a subset $V(p)$ of W. A *rooted Kripke structure* is an ordered pair $\langle M, w_0\rangle$ with $w_0 \in W$. To simplify notation, in the following we write $\langle W, R, V\rangle$ and $\langle W, R, V, w_0\rangle$ instead of $\langle\langle W, R\rangle, V\rangle$ and $\langle\langle\langle W, R\rangle, V\rangle, w_0\rangle$, respectively.

Satisfaction (or truth) of a formula at a world w of a Kripke structure $M = \langle W, R, V\rangle$ is inductively defined by:

$$
\begin{array}{ll}
\langle M, w\rangle \models \mathbf{true}; & \langle M, w\rangle \not\models \mathbf{false}; \\
\langle M, w\rangle \models p & \text{iff } w \in V(p), \text{ where } p \in P; \\
\langle M, w\rangle \models \neg\varphi & \text{iff } \langle M, w\rangle \not\models \varphi; \\
\langle M, w\rangle \models (\varphi \wedge \psi) & \text{iff } \langle M, w\rangle \models \varphi \text{ and } \langle M, w\rangle \models \psi; \\
\langle M, w\rangle \models (\varphi \vee \psi) & \text{iff } \langle M, w\rangle \models \varphi \text{ or } \langle M, w\rangle \models \psi; \\
\langle M, w\rangle \models (\varphi \rightarrow \psi) & \text{iff } \langle M, w\rangle \models \neg\varphi \text{ or } \langle M, w\rangle \models \psi; \\
\langle M, w\rangle \models \Box\varphi & \text{iff for every } v, w\,R\,v \text{ implies } \langle M, v\rangle \models \varphi; \\
\langle M, w\rangle \models \Diamond\varphi & \text{iff there is } v, w\,R\,v \text{ and } \langle M, v\rangle \models \varphi.
\end{array}
$$

Table 1. Modal axioms and relational frame properties

Name	Axiom	Frame Property	
D	$\Box\varphi \to \Diamond\varphi$	Serial	$\forall v \exists w. v\, R\, w$
T	$\Box\varphi \to \varphi$	Reflexive	$\forall w. w\, R\, w$
B	$\varphi \to \Box\Diamond\varphi$	Symmetric	$\forall vw. v\, R\, w \to w\, R\, v$
4	$\Box\varphi \to \Box\Box\varphi$	Transitive	$\forall uvw.(u\, R\, v \land v\, R\, w) \to u\, R\, w$
5	$\Diamond\varphi \to \Box\Diamond\varphi$	Euclidean	$\forall uvw.(u\, R\, v \land u\, R\, w) \to v\, R\, w$

Table 2. Rewriting Rules for Simplification

$\varphi \land \varphi \Rightarrow \varphi$	$\varphi \land \neg\varphi \Rightarrow \mathbf{false}$	$\Box\mathbf{true} \Rightarrow \mathbf{true}$	$\neg\mathbf{true} \Rightarrow \mathbf{false}$	$\neg\neg\varphi \Rightarrow \varphi$
$\varphi \lor \varphi \Rightarrow \varphi$	$\varphi \lor \neg\varphi \Rightarrow \mathbf{true}$	$\Diamond\mathbf{false} \Rightarrow \mathbf{false}$	$\neg\mathbf{false} \Rightarrow \mathbf{true}$	
$\varphi \land \mathbf{true} \Rightarrow \varphi$	$\varphi \land \mathbf{false} \Rightarrow \mathbf{false}$	$\varphi \lor \mathbf{false} \Rightarrow \varphi$	$\varphi \lor \mathbf{true} \Rightarrow \mathbf{true}$	

If $\langle M, w\rangle \models \varphi$ holds then M is a *model* of φ, φ is *true at w in M* and M *satisfies* φ. A modal formula φ is *satisfiable* iff there exists a Kripke structure M and a world w in M such that $\langle M, w\rangle \models \varphi$.

We are interested in extensions of K with the modal axioms shown in Table 1 and their combinations. Each of these axioms defines a class of Kripke frames where the accessibility relation R satisfies the first-order property stated in the table. Combinations of axioms then define a class of Kripke frames where the accessibility relation satisfies the combination of their corresponding properties.

Given a normal modal logic L with corresponding class of frames $\mathfrak{F}$, we say a modal formula φ is *L-satisfiable* iff there exists a frame $F \in \mathfrak{F}$, a valuation V and a world $w \in F$ such that $\langle F, V, w\rangle \models \varphi$. It is *$L$-valid* or *valid in L* iff for every frame $F \in \mathfrak{F}$, every valuation V and every world $w \in F$, $\langle F, V, w\rangle \models \varphi$. A normal modal logic L_2 is *an extension* of a normal modal logic L_1 iff all L_1-valid formulae are also L_2-valid.

A rooted Kripke structure $M = \langle W, R, V, w_0\rangle$ is a *rooted tree Kripke structure* iff R is a tree, that is, a directed acyclic connected graph where each node has at most one predecessor, with *root* w_0. It is a *rooted tree Kripke model* of a modal formula φ iff $\langle W, R, V, w_0\rangle \models \varphi$. In a rooted tree Kripke structure with root w_0 for every world $w_k \in W$ there is exactly one path connecting w_0 and w_k, the length of that path is the *modal level of w_k (in M), denoted by* $\mathsf{ml}_M(w_k)$.

It is well-known [17] that a modal formula φ is K-satisfiable iff there is a finite rooted tree Kripke structure $M = \langle F, V, w_0\rangle$ such that $\langle M, w_0\rangle \models \varphi$.

For the reductions presented in the next section we assume that any modal formula φ has been simplified by exhaustively applying the rewrite rules in Table 2, and it is in Negation Normal Form (NNF). That is, a formula where only propositional symbols are allowed in the scope of negations. We say that such a formula is in *simplified NNF*.

The reductions produce formulae in a clausal normal form, called *Separated Normal Form with Sets of Modal Levels* SNF_{sml}, introduced in [29]. The language of SNF_{sml} extends that of the basic modal logic K with sets of modal levels as labels. Clauses in SNF_{sml} have one of the following forms:

$S : \bigvee_{i=1}^{n} l_i$	$S : l' \rightarrow \Box l$	$S : l' \rightarrow \Diamond l$
(literal clause)	(positive modal clause)	(negative modal clause)

where $S \subseteq \mathbb{N}$ and l, l', l_i are propositional literals with $1 \leq i \leq n$, $n \in \mathbb{N}$. We write $\star : \varphi$ instead of $\mathbb{N} : \varphi$ and such clauses are called *global clauses*. Positive and negative modal clauses are together known as *modal clauses*.

Given a rooted tree Kripke structure M and a set S of natural numbers, by $M[S]$ we denote the set of worlds that are at a modal level in S, that is, $M[S] = \{w \in W \mid \mathsf{ml}_M(w) \in S\}$. Then

$$M \models S : \varphi \text{ iff } \langle M, w\rangle \models \varphi \text{ for every world } w \in M[S].$$

The motivation for using a set S to label clauses is that in our reductions the formula φ may hold at several levels, possibly an infinite number of levels. It therefore makes sense to label such formulae not with just a single level, but a set of levels. The Separated Normal Form with Modal Level, SNF_{ml}, can be seen as the special case of SNF_{sml} where all labelling sets are singletons.

Note that if $S = \emptyset$, then $M \models S : \varphi$ trivially holds. Also, a Kripke structure M can satisfy S : **false** if there is no world w with $\mathsf{ml}_M(w) \in S$. On the other hand, S : **false** with $0 \in S$ is unsatisfiable as a rooted tree Kripke structure always has a world with modal level 0.

If $M \models S : \varphi$, then we say that $S : \varphi$ *holds in* M or *is true in* M. For a set Φ of labelled formulae, $M \models \Phi$ iff $M \models S : \varphi$ for every $S : \varphi$ in Φ, and we say Φ is K-*satisfiable*.

We introduce some notation that will be used in the following. Let $S^+ = \{l+1 \in \mathbb{N} \mid l \in S\}$, $S^- = \{l-1 \in \mathbb{N} \mid l \in S\}$, and $S^{\geq} = \{n \mid n \geq \min(S)\}$, where $\min(S)$ is the least element in S. Note that the restriction of the elements being in $\mathbb{N}$ implies that S^- cannot contain negative numbers.

3 Extensions of K

In this section we define reductions from all the logics in the modal cube to SNF_{sml}. We assume that the set P of propositional symbols is partitioned into two infinite sets Q and T such that Q contains the propositional symbols of the modal formula φ under consideration, and T surrogate symbols t_ψ for every subformula ψ of φ and supplementary propositional symbols. In particular, for every modal formula ψ we have $\mathsf{var}(\psi) \subset Q$ and there exists a propositional symbol $t_\psi \in T$ uniquely associated with ψ. These surrogate symbols serve the same purpose as Tseitin variables [40] and Skolem predicates [30,39] in the transformation of propositional and first-order formulae, respectively, to clausal form via structural transformation.

It turns out that given a reduction $\rho_{\mathsf{K}\Sigma}$ for $\mathsf{K}\Sigma$ with $\{\mathsf{D}, \mathsf{T}\} \cap \Sigma = \emptyset$, there is a uniform and straightforward way we can obtain a reduction for $\mathsf{KD}\Sigma$ and $\mathsf{KT}\Sigma$ from $\rho_{\mathsf{K}\Sigma}$. Also, the valid formulae of $\mathsf{KDT}\Sigma$ are the same as those of

Table 3. Categorisation of modal logics in the modal cube

'Base logics'	K	KB	K4	K5	KB4	K45
Extensions with D	KD	KDB	KD4	KD5		KD45
Extensions with T	KT	KTB	KT4	KT5		

$\mathsf{KT}\Sigma$, so we do not need to consider the case of adding both axioms to $\mathsf{K}\Sigma$. Similarly, the logics KT45, KDB4, KTB4 and KT5 all have the same set of valid formulae. Therefore, as shown in Table 3, we can divide the 15 modal logics into three categories: Six 'base logics', five modal logics obtained by extending a 'base logic' with D, and a further four modal logics obtained by extending a 'base logic' with T. For four of the six 'base logics' (namely, K, KB, K4, and K5) we have already devised reductions in [29], so only two (i.e., KB4 and K45) remain.

Given a modal formula φ in simplified NNF and $L = \mathsf{K}\Sigma$ with $\Sigma \subseteq \{\mathsf{B}, \mathsf{D}, \mathsf{T}, 4, 5\}$, we can obtain a set Φ_L of clauses in SNF_{sml} such that φ is L-satisfiable iff Φ_L is K-satisfiable with $\Phi_L = \rho_L^{sml}(\varphi) = \{\{0\} : t_\varphi\} \cup \rho_L(\{0\} : t_\varphi \rightarrow \varphi)$, where ρ_L is defined as follows:

$$\begin{aligned}
\rho_L(S : t \rightarrow \mathbf{true}) &= \emptyset \\
\rho_L(S : t \rightarrow \mathbf{false}) &= \{S : \neg t\} \\
\rho_L(S : t \rightarrow (\psi_1 \wedge \psi_2)) &= \{S : \neg t \vee \eta(\psi_1), S : \neg t \vee \eta(\psi_2)\} \cup \delta_L(S, \psi_1) \cup \delta_L(S, \psi_2) \\
\rho_L(S : t \rightarrow \psi) &= \{S : \neg t \vee \psi\} \\
&\quad \text{if } \psi \text{ is a disjunction of literals} \\
\rho_L(S : t \rightarrow (\psi_1 \vee \psi_2)) &= \{S : \neg t \vee \eta(\psi_1) \vee \eta(\psi_2)\} \cup \delta_L(S, \psi_1) \cup \delta_L(S, \psi_2) \\
&\quad \text{if } \psi_1 \vee \psi_2 \text{ is not a disjunction of literals} \\
\rho_L(S : t \rightarrow \Diamond\psi) &= \{S : t \rightarrow \Diamond\eta(\psi)\} \cup \delta_L(S^+, \psi) \\
\rho_L(S : t \rightarrow \Box\psi) &= P_L(S : t \rightarrow \Box\psi) \cup \Delta_L(S : t \rightarrow \Box\psi)
\end{aligned}$$

η and δ_L are defined as follows:

$$\eta(\psi) = \begin{cases} \psi, & \text{if } \psi \text{ is a literal} \\ t_\psi, & \text{otherwise} \end{cases} \qquad \delta_L(S, \psi) = \begin{cases} \emptyset, & \text{if } \psi \text{ is a literal} \\ \rho_L(S : t_\psi \rightarrow \psi), & \text{otherwise} \end{cases}$$

and functions P_L and Δ_L, are defined as shown in Table 4.

We can see in Table 4 that the reduction for KB4 has an additional SNF_{sml} clause $\star : t_{\Box\psi} \vee t_{\Box\neg t_{\Box\psi}}$ that occurs neither in the reduction for KB nor in that for K4. It can be seen as an encoding of the derived axiom $\Diamond\Box\psi \rightarrow \Box\psi$ that follows from the contrapositive $\Diamond\Box\psi \rightarrow \psi$ of B and 4 $\Box\psi' \rightarrow \Box\Box\psi'$.

For K45 we see that all the SNF_{sml} clauses in the reduction for K5 carry over. These clauses are already sufficient to ensure that, semantically, if $t_{\Box\psi}$ is true at any world at a level other than 0, then $t_{\Box\psi}$ is true at every world. Consequently, to accommodate axiom 4, it suffices to add the SNF_{sml} clause $\{0\} : t_{\Box\psi} \rightarrow \Box t_{\Box\psi}$ to ensure that this also holds for the root world at level 0.

L	$P_L(S : t_{\Box\psi} \rightarrow \Box\psi)$	$\Delta_L(S : t_{\Box\psi} \rightarrow \Box\psi)$
K	$S : t_{\Box\psi} \rightarrow \Box\eta(\psi)$	$\delta_L(S^+, \psi)$
KB	$S : t_{\Box\psi} \rightarrow \Box\eta(\psi)$, $S^- : \eta(\psi) \vee t_{\Box\neg t_{\Box\psi}}$, $S^- : t_{\Box\neg t_{\Box\psi}} \rightarrow \Box\neg t_{\Box\psi}$	$\delta_L(S^- \cup S^+, \psi)$
K4	$S^{\geq} : t_{\Box\psi} \rightarrow \Box\eta(\psi)$, $S^{\geq} : t_{\Box\psi} \rightarrow \Box t_{\Box\psi}$	$\delta_L((S^+)^{\geq}, \psi)$
K5	$\star : t_{\Box\psi} \rightarrow \Box\eta(\psi)$, $\star : \neg t_{\Diamond t_{\Box\psi}} \vee t_{\Box\psi}$, $\star : t_{\Diamond t_{\Box\psi}} \rightarrow \Diamond t_{\Box\psi}$, $\star : \neg t_{\Diamond t_{\Box\psi}} \rightarrow \Box\neg t_{\Box\psi}$, $\star : t_{\Diamond t_{\Box\psi}} \rightarrow \Box t_{\Diamond t_{\Box\psi}}$	$\delta_L(\star, \psi)$
KB4	$\star : t_{\Box\psi} \rightarrow \Box\eta(\psi)$, $\star : \eta(\psi) \vee t_{\Box\neg t_{\Box\psi}}$, $\star : t_{\Box\psi} \vee t_{\Box\neg t_{\Box\psi}}$, $\star : t_{\Box\neg t_{\Box\psi}} \rightarrow \Box\neg t_{\Box\psi}$, $\star : t_{\Box\psi} \rightarrow \Box t_{\Box\psi}$	$\delta_L(\star, \psi)$
K45	$\star : t_{\Box\psi} \rightarrow \Box\eta(\psi)$, $\{0\} : t_{\Box\psi} \rightarrow \Box t_{\Box\psi}$ iff $0 \in S$, $\star : \neg t_{\Diamond t_{\Box\psi}} \vee t_{\Box\psi}$, $\star : t_{\Diamond t_{\Box\psi}} \rightarrow \Diamond t_{\Box\psi}$, $\star : \neg t_{\Diamond t_{\Box\psi}} \rightarrow \Box\neg t_{\Box\psi}$, $\star : t_{\Diamond t_{\Box\psi}} \rightarrow \Box t_{\Diamond t_{\Box\psi}}$	$\delta_L(\star, \psi)$
KDΣ	$\{lb^P_{K\Sigma}(S) : t_{\Box\psi} \rightarrow \Diamond\eta(\psi)\} \cup P_{K\Sigma}(S : t_{\Box\psi} \rightarrow \Box\psi)$	$\delta_L(lb^\delta_{K\Sigma}(S), \psi)$
KTΣ	$\{lb^P_{K\Sigma}(S) : \neg t_{\Box\psi} \vee \eta(\psi)\} \cup P_{K\Sigma}(S : t_{\Box\psi} \rightarrow \Box\psi)$	$\delta_L(lb^\delta_{K\Sigma}(S) \cup S, \psi)$

where $lb^P_{K\Sigma}$ and $lb^\delta_{K\Sigma}$ are defined as follows

Table 4. Reduction of $\Box$-formulae, $\Sigma \subseteq \{\mathsf{B}, 4, 5\}$.

L	K	KB	K4	K5	KB4	K45
$lb^P_L(S)$	S	S	$S^{\geq}$	$\star$	$\star$	$\star$
$lb^\delta_L(S)$	S^+	$S^- \cup S^+$	$(S^+)^{\geq}$	$\star$	$\star$	$\star$

For reductions of KDΣ and KTΣ we have favoured the reuse of reductions for KΣ, KD and KT over optimisation for specific logics. For example, take KBD. Given that in a symmetric model, every world w except the root world w_0 has an R-successor, the axiom D only 'enforces' that w_0 also has an R-successor. So, instead of adding a clause $S : t_{\Box\psi} \rightarrow \Diamond\psi$ for every clause $S : t_{\Box\psi} \rightarrow \Box\eta(\psi)$ we could just add $\{0\} : t_{\Box\psi} \rightarrow \Diamond\psi$ iff $0 \in S$. Similarly, in KT5, because of 5, for all worlds w except w_0 we already have $w\,R\,w$. So, we could again $\{0\} : \neg t_{\Box\psi} \vee \eta(\psi)$ for every clause $S : t_{\Box\psi} \rightarrow \Box\eta(\psi)$ iff $0 \in S$.

For the KB4-unsatisfiable formula $\psi_1 = (\neg p \wedge \Diamond\Diamond\Box p)$, if we were to independently apply the reductions for KB and K4, that is, we compute $\{\{0\} : t_{\psi_1}\} \cup \rho_{\mathsf{KB}}(\{0\} : t_{\psi_1} \rightarrow \psi_1) \cup \rho_{\mathsf{K4}}(\{0\} : t_{\psi_1} \rightarrow \psi_1)$, then the result is the following set of clauses Φ_1:

(1) $\{0\} : t_{\psi_1}$
(2) $\{0\} : \neg t_{\psi_1} \vee \neg p$
(3) $\{0\} : \neg t_{\psi_1} \vee t_{\Diamond\Diamond\Box p}$
(4) $\{0\} : t_{\Diamond\Diamond\Box p} \rightarrow \Diamond t_{\Diamond\Box p}$
(5) $\{1\} : t_{\Diamond\Box p} \rightarrow \Diamond t_{\Box p}$
(6) $\{2\}^{\geq} : t_{\Box p} \rightarrow \Box p$
(7) $\{2\}^{\geq} : t_{\Box p} \rightarrow \Box t_{\Box p}$
(8) $\{1\} : p \vee t_{\Box\neg t_{\Box p}}$
(9) $\{1\} : t_{\Box\neg t_{\Box p}} \rightarrow \Box\neg t_{\Box p}$

Clauses (1) to (5) stem from the transformation of ψ_1 to SNF_{sml} for K, Clauses (6) and (7) stem from the reduction for 4 and Clauses (8) and (9) stem from the reduction for B. This set of SNF_{sml} clauses is K-satisfiable. The clauses imply $\{1\} : p$, but neither $\{1\} : \Box p$ nor $\{0\} : p$ which we need to obtain a contradiction. Part of the reason is that we would need to apply the reduction for 4 and B recursively to newly introduced surrogates for $\Box$-formulae which in turn leads to the introduction of further surrogates and problems with the termination of the reduction.

In contrast, the clause set Φ_2 obtained by our reduction for KB4 is:

(10) $\{0\} : t_{\psi_1}$
(11) $\{0\} : \neg t_{\psi_1} \vee \neg p$
(12) $\{0\} : \neg t_{\psi_1} \vee t_{\Diamond\Diamond\Box p}$
(13) $\{0\} : t_{\Diamond\Diamond\Box p} \to \Diamond t_{\Diamond\Box p}$
(14) $\{1\} : t_{\Diamond\Box p} \to \Diamond t_{\Box p}$
(15) $\star : t_{\Box p} \to \Box p$
(16) $\star : t_{\Box p} \to \Box t_{\Box p}$
(17) $\star : p \vee t_{\Box\neg t_{\Box p}}$
(18) $\star : t_{\Box\neg t_{\Box p}} \to \Box\neg t_{\Box p}$
(19) $\star : t_{\Box p} \vee t_{\Box\neg t_{\Box p}}$
(20) $\star : t_{\Box\neg t_{\Box p}} \to \Box t_{\Box\neg t_{\Box p}}$

Note Clauses (19) and (20) in Φ_2 for which there are no corresponding clauses in Φ_1. Also, the set of labels of Clauses (15) to (18) are strict supersets of those of the corresponding Clauses (6) to (9). Φ_2 implies both $\{1\} : \Box p$ and $\{0\} : p$. The latter, together with Clauses (10) and (11), means Φ_2 is K-unsatisfiable.

Theorem 1. *Let φ be a modal formula in simplified NNF, $\Sigma \subseteq \{\mathsf{B}, \mathsf{D}, \mathsf{T}, \mathsf{4}, \mathsf{5}\}$, and $\Phi_{\mathsf{K}\Sigma} = \rho^{sml}_{\mathsf{K}\Sigma}(\varphi)$. Then φ is $\mathsf{K}\Sigma$-satisfiable iff $\Phi_{\mathsf{K}\Sigma}$ is K-satisfiable.*

Proof (Sketch). For $|\Sigma| \leq 1$ this follows from Theorem 5 in [29].

For K45, KB4, $\mathsf{KD}\Sigma'$, and $\mathsf{KT}\Sigma'$ with $\Sigma' \subseteq \{\mathsf{B}, \mathsf{4}, \mathsf{5}\}$ we proceed in analogy to the proofs of Theorems 3 and 4 in [29]. Let L be one of these logics.

To show that if φ is L-satisfiable then Φ_L is K-satisfiable, we show that given a rooted L-model M of φ a small variation of the unravelling of M is a rooted tree K-model $\vec{M}_L$ of Φ_L. The main step is to define the valuation of the additional propositional symbols t_ψ so that we can prove that all clauses in Φ_L hold in $\vec{M}_L$. To show that if Φ_L is K-satisfiable then φ is L-satisfiable, we take a rooted tree K-model $M = \langle W, R, V, w_0\rangle$ of Φ_L and construct a Kripke structure $M_L = \langle W, R^L, V, w_0\rangle$. The relation R^L is the closure of R under the relational properties associated with the axioms of L. The proof that M_L is a model of φ relies on the fact that the clauses in Φ_L ensure that for subformulae $\Box\psi$ of φ, ψ will be true at all worlds reachable via R^L from a world where $\Box\psi$ is true. □

4 From SNF_{sml} to SNF_{ml}

As K$_S$P does not support SNF_{sml}, in our evaluation of the effectiveness of the reductions defined in Sect. 3, we have used a transformation from SNF_{sml} to SNF_{ml}. An alternative approach would be to reflect the use of SNF_{sml} in the calculus and re-implement the prover. Whilst we believe that redesigning the calculus presents few problems, re-implementing K$_S$P needs more thought in particular how to represent infinite sets. The route we adopt here allows us to experiment with the approach in general without having to change the prover. For extensions of K with one or more of the axioms B, D, T such a transformation

Table 5. Bounds on the length of prefixes in SST tableaux

Logic L	Bound db^{φ}_{L}
K,KD,KT, KB,KDB,KTB	$1 + d^{\varphi}_{m}$
K4,S4	$2 + d^{\varphi}_{\Diamond} + n^{\varphi}_{\Diamond} \times n^{\varphi}_{\Box}$
KD4	$2 + d^{\varphi}_{\Diamond} + (\max(1, n^{\varphi}_{\Diamond}) \times n^{\varphi}_{\Box})$
KB4,KTB4, K5,S5,K45	$2 + d^{\varphi}_{\Diamond} + n^{\varphi}_{\Diamond}$
KD5	$2 + d^{\varphi}_{\Diamond} + \max(1, n^{\varphi}_{\Diamond})$

is straightforward as the sets of modal levels occurring in the normal form of modal formulae are all finite. Thus, instead of a single SNF_{sml} clause $S : \neg t_{\psi} \vee \eta_f(\psi)$ we can use the finite set of SNF_{ml} clauses $\{ml : \neg t_{\psi} \vee \eta_f(\psi) \mid ml \in S\}$.

For extensions of K with at least one of the axioms 4 and 5, potentially together with other axioms, the sets of modal levels labelling clauses are in general infinite. For each logic L it is, however, possible to define a computable function that maps the modal formula φ under consideration onto a bound db^{φ}_{L} such that, restricting the modal levels in the normal form of φ by db^{φ}_{L}, preserves satisfiability equivalence.

To establish the bound and prove satisfiability equivalence, we need to introduce the basic notions of Single Step Tableaux (SST) calculi for a modal logic L [14,21], which uses sequences of natural numbers to prefix modal formulae in a tableau. The SST calculus consists of a set of rules, with the (π) rule being the only rule increasing prefixes' lengths (i.e., $\sigma : \Diamond\varphi / \sigma.n : \varphi$ with $\sigma.n$ new on the branch). For a logic L, an *L-tableau* $\mathcal{T}$ in the SST calculus for a modal formula φ is a (binary) tree where the root of $\mathcal{T}$. is labelled with $1 : \varphi$, and every other node is labelled with a prefixed formula $\sigma : \psi$ obtained by application of a rule of the calculus. A *branch* $\mathcal{B}$ is a path from the root to a leaf. A *branch* $\mathcal{B}$ is *closed* if it contains either **false** or a propositional contradiction at the same prefix. A *tableau* "$\mathcal{T}$ is *closed* if all its branches are closed. A prefixed formula $\sigma : \psi$ is *reduced for rule* (r) *in* $\mathcal{B}$ if the branch $\mathcal{B}$ already contains the conclusion of such rule application. By a *systematic tableau construction* we mean an application of the procedure in [14, p. 374] adapted to SST rules.

For each logic L, we establish its bound by considering an L-SST calculus, where a modal level in an SNF_{sml} clause corresponds to the length of a prefix in an SST tableau. The bound then either follows from an already known bound on the length of prefixes in an SST tableau preserving correctness of the SST calculus, or we establish such a bound ourselves. To prove satisfiability equivalence, we show that, for a closed SST tableau with such a bound on the length of prefixes in place, we can construct a resolution refutation of a set of SNF_{sml} or SNF_{ml} clauses with a corresponding bound on modal levels in those clauses.

For a modal formula φ in simplified NNF let d^{φ}_{m} be the modal depth of φ, $d^{\varphi}_{\Diamond}$ be the maximal nesting of $\Diamond$-operators not under the scope of any $\Box$ operators in φ, $n^{\varphi}_{\Box}$ be the number of $\Box$-subformulae in φ, and $n^{\varphi}_{\Diamond}$ be the number of

$\Diamond$-subformulae below $\Box$-operators in φ. Our results for the bounds on the length of prefixes in SST tableaux can then be summarised by the following theorem.

Theorem 2. *Let $L = \mathsf{K}\Sigma$ with $\Sigma \subseteq \{\mathsf{B}, \mathsf{D}, \mathsf{T}, \mathsf{4}, \mathsf{5}\}$. A systematic tableau construction of an L-tableau for a modal formula φ in simplified NNF under the following Constraints (TC1) and (TC2)*

(TC1) a rule (r) of the SST calculus is only applicable to a prefixed formula $\sigma : \psi$ in a branch $\mathcal{B}$ if the formula is not already reduced for (r) in $\mathcal{B}$;
(TC2) rule (π) of the SST calculus is only applicable to prefixed formulae $\sigma : \Diamond\psi$ with $|\sigma| < db^{\varphi}_{L}$ for db^{φ}_{L} as defined in Table 5

terminates in one of following states:

(1) all branches of the constructed tableau are closed and φ is L-unsatisfiable or
(2) at least one branch $\mathcal{B}$ is not closed, no rule is still applicable to a labelled formula in $\mathcal{B}$, and φ is L-satisfiable.

The proof is analogous to Massacci's [21, Section B.2]. Note that for logics $\mathsf{KD4}$ and $\mathsf{KD5}$, we use $\max(1, n^{\varphi}_{\Diamond})$ in the calculation of the bound. That is, if $n^{\varphi}_{\Diamond} \geq 1$ then $\max(1, n^{\varphi}_{\Diamond}) = n^{\varphi}_{\Diamond}$ and the bound is the same as for $\mathsf{K4}$ and $\mathsf{K5}$. Otherwise $\max(1, n^{\varphi}_{\Diamond}) = 1$, that is, the bound is the same as for a formula with a single $\Diamond$-subformula below $\Box$-operators in φ.

For K, KD, KT, KB and KDB these bounds were already stated in [21, Tables III and IV]. The bound for KTB follows straightforwardly from that for KB and KDB. For $\mathsf{KD4}$, Massacci [21, Tables III and IV] states the bound to be the same as for $\mathsf{K4}$. However, this is not correct for the case that the formula φ contains no $\Diamond$-formulae, where its bound would simply be 2, independent of φ. For example, the formula $\Box\Box\Box\mathbf{false}$ which is $\mathsf{KD4}$-unsatisfiable, does not have a closed $\mathsf{KD4}$-tableau with this bound. For the other logics the bounds are new. As argued in [21], the bounds allow tableau decision procedures for extensions of K with axioms $\mathsf{4}$ and $\mathsf{5}$ that do not require a loop check and are therefore of wider interest.

Note that in $\mathsf{KT4}$, $\Box\Box\psi$ and $\Box\psi$ are equivalent and so are $\Box(\psi\wedge\Box\vartheta)$ and $\Box(\psi\wedge \vartheta)$. So, it makes sense to further simplify $\mathsf{KT4}$ formulae using such equivalences before computing the normal form and the bound with the benefit that it may not only reduce the bound but also the size of the normal form. Similar equivalences that can be used to reduce the number of modal operators in a formula also exist for other logics, see, e.g., [8, Chapter 4].

To establish a relationship between closed tableaux and resolution refutations of a set of SNF_{ml} clauses, we formally define the modal layered resolution calculus. Table 6 shows the inference rules of the calculus restricted to labels occurring in our normal form. For GEN1 and GEN3, if the modal clauses in the premises occur at the modal level ml, then the literal clause in the premises occurs at modal level $ml + 1$.

Let Φ be a set of SNF_{ml} clauses. A *(resolution) derivation from* Φ is a sequence of sets $\Phi_0, \Phi_1, \ldots$ where $\Phi_0 = \Phi$ and, for each $i > 0$, $\Phi_{i+1} = \Phi_i \cup \{D\}$, where $D \notin \Phi_i$ is the resolvent obtained from Φ_i by an application of one of the inference rules to premises in Φ_i. A *(resolution) refutation of* Φ is a derivation $\Phi_0, \ldots, \Phi_k$, $k \in \mathbb{N}$, where $0 : \mathbf{false} \in \Phi_k$.

To map a set of SNF_{sml} clauses to a set of SNF_{ml} clauses, using a bound $n \in \mathbb{N}$ on the modal levels, we define a function db_n on clauses and sets of clauses in SNF_{sml} as follows:

$$\mathrm{db}_n(S : \varphi) = \{ml : \varphi \mid ml \in S \text{ and } ml \leq n\}$$
$$\mathrm{db}_n(\Phi) = \bigcup_{S:\varphi\in\Phi} \mathrm{db}_n(S : \varphi)$$

Note that prefixes in SST-tableaux have a minimal length of 1 while the minimal modal level in SNF_{ml} clauses is 0. So, a prefix of length n in a prefixed formula corresponds to a modal level $n-1$ in an SNF_{ml} clause.

The proof of the following theorem then takes advantage of the fact that we have surrogates and associated clauses for each subformula of φ and proceeds by induction over applications of rule (π).

Theorem 3. *Let* $L = \mathsf{K}\Sigma$ *with* $\Sigma \subseteq \{\mathsf{B}, \mathsf{D}, \mathsf{T}, \mathsf{4}, \mathsf{5}\}$, φ *be a* $\mathsf{K}\Sigma$*-unsatisfiable formula in simplified NNF,* db^{φ}_L *be as defined in Table 5, and* $\Phi_L = \rho^{ml}_L(\varphi) = \mathrm{db}_{db^{\varphi}_L-1}(\rho^{sml}_L(\varphi))$. *Then there is a resolution refutation of* Φ_L.

Regarding the size of the encoding, we note that, ignoring the labelling sets, the reduction ρ^{sml}_L into SNF_{sml} is linear with respect to the size of the original formula. The size including the labelling sets would depend on the exact representation of those sets, in particular, of infinite sets. As those are not arbitrary, there is still an overall polynomial bound on the size of the sets of SNF_{sml} clauses produced by ρ^{sml}_L. When transforming clauses from SNF_{sml} into SNF_{ml}, we may need to add every clause to all levels within the bounds provided by Theorem 3. The parameters for calculating those bounds, d^{φ}_m, $d^{\varphi}_{\Diamond}$, $n^{\varphi}_{\Diamond}$, and $n^{\varphi}_{\Box}$, are all themselves linearly bound by the size of the formula. Thus, in the worst case, which is $\mathsf{S4}$, the size of the clause set produced by ρ^{ml}_L is bounded by a polynomial of degree 3 with respect to the size of the original formula.

It is worth pointing out that both the reduction ρ^{sml}_L of a modal formula to SNF_{sml} and the reduction ρ^{ml}_L to SNF_{ml} are also reversible, that is, we can reconstruct the original formula from the SNF_{sml} and from the SNF_{ml} clause set obtained by ρ^{sml}_L or ρ^{ml}_L, respectively. This reconstruction can also be performed in polynomial time. Thus the reduction itself does not affect the complexity of the satisfiability problem. For instance, the satisfiability problem for $\mathsf{S5}$ is NP-complete and so is the satisfiability problem of the subclass $\mathbb{C}_{\mathsf{S5}}$ of SNF_{ml} clause sets that can be obtained as the result of an application of ρ^{ml}_{S5} to a modal formula. However, a generic decision procedure for K will not be a complexity-optimal decision procedure for $\mathbb{C}_{\mathsf{S5}}$.

Table 6. Inference rules of the MLR calculus

$$\mathsf{LRES}: \frac{\begin{array}{l} ml : D \vee l \\ ml : D' \vee \neg l \end{array}}{ml : D \vee D'} \qquad \mathsf{MRES}: \frac{\begin{array}{l} ml : l_1 \rightarrow \Box l \\ ml : l_2 \rightarrow \Diamond \neg l \end{array}}{ml : \neg l_1 \vee \neg l_2} \qquad \mathsf{GEN2}: \frac{\begin{array}{l} ml : l'_1 \rightarrow \Box l_1 \\ ml : l'_2 \rightarrow \Box \neg l_1 \\ ml : l'_3 \rightarrow \Diamond l_2 \end{array}}{ml : \neg l'_1 \vee \neg l'_2 \vee \neg l'_3}$$

$$\mathsf{GEN1}: \frac{\begin{array}{l} ml : l'_1 \rightarrow \Box \neg l_1 \\ \vdots \\ ml : l'_m \rightarrow \Box \neg l_m \\ ml : l' \rightarrow \Diamond \neg l \\ ml+1 : l_1 \vee \ldots \vee l_m \vee l \end{array}}{ml : \neg l'_1 \vee \ldots \vee \neg l'_m \vee \neg l'} \qquad \mathsf{GEN3}: \frac{\begin{array}{l} ml : l'_1 \rightarrow \Box \neg l_1 \\ \vdots \\ ml : l'_m \rightarrow \Box \neg l_m \\ ml : l' \rightarrow \Diamond l \\ ml+1 : l_1 \vee \ldots \vee l_m \end{array}}{ml : \neg l'_1 \vee \ldots \vee \neg l'_m \vee \neg l'}$$

5 Evaluation

An empirical evaluation of the practical usefulness of the reductions we presented in Sects. 3 and 4 faces the challenge that there is no substantive collection of benchmark formulae for the 15 logics of the modal cube except for basic modal logic. Catach [7] evaluates his prover on 31 modal formulae with a maximal length of 22 and maximal modal depth of 4. They are not sufficiently challenging. The QMLTP Problem Library for First-Order Modal Logics [32] focuses on quantified formulae and contains only a few formulae taken from the research literature that are purely propositional and were not written for the basic modal logic K. The Logics Workbench (LWB) benchmark collection [2] contains formulae for K, KT and S4 but not for any of the other logics we consider. For each of these three logics, the collection consists of 18 parameterised classes with 21 formulae each, plus scripts with which further formulae could be generated if needed. All formulae in 9 classes are satisfiable and all formulae in the other 9 classes are unsatisfiable in the respective logic.

In [29] we have used the 18 classes of the LWB benchmark collection for K to evaluate our approach for the six logics consisting of K and its extensions with a single axiom. One drawback of using these 18 classes for other modal logics is that formulae that are K-satisfiable are not necessarily KΣ-satisfiable for non-empty sets Σ of additional axioms. For example, for K5, only 60 out of 180 K-satisfiable formulae were K5-satisfiable. Another drawback is that while K-unsatisfiable formulae are also KΣ-unsatisfiable, a resolution refutation would not necessarily involve any of the additional clauses introduced by our reduction for KΣ. It may be that the additional clauses allow us to find a shorter refutation, but it may just be a case of finding the same refutation in a larger search space. It is also worth recalling that simplification alone is sufficient to determine that all formulae in the class `k_lin_p` are K-unsatisfiable while pure literal elimination can be used to reduce all formulae in `k_grz_p` to the same simple formula [26].

Table 7. Logic-specific modification of unsatisfiable benchmark formulae

Logic L	ψ_l^p	Logic L	ψ_l^p
K	**false**	KD4	$(\Box q_p \wedge \Diamond\Diamond\Box\Diamond\neg q_p)$
KB	$(\neg q_p \wedge \Diamond\Box q_p)$	K5	$(\Diamond\neg q_p \wedge \Diamond\Box q_p)$
KDB	$(\neg q_p \wedge \Diamond\Box((\Box\neg q'_p \wedge \Box q'_p) \vee q_p))$	KD5	$((\Box\neg q_p \wedge \Box q_p) \vee (\Diamond\Box q'_p \wedge \Diamond\neg q'_p)$
KTB	$(\neg q_p \wedge \Diamond\Box((\neg q'_p \wedge \Box q'_p) \vee q_p))$	K45	$(\Box q_p \wedge \Diamond\Box q'_p \wedge \Diamond\Diamond(\neg q_p \vee \neg q'_p))$
KD	$(\Box\neg q_p \wedge \Box q_p)$	KD45	$((\Box\neg q'_p \wedge \Box q'_p) \wedge$ $(\Box q_p \wedge \Diamond\Box q'_p \wedge \Diamond\Diamond(\neg q_p \vee \neg q'_p))$
KT	$(\neg q_p \wedge \Box q_p)$		
K4	$(\Box q_p \wedge \Diamond\Diamond\neg q_p)$	S4	$(\neg q'_p \wedge \Box(\neg q'_p \vee \Box q_p) \wedge \Diamond\Diamond\neg q_p)$
K4B	$(\neg q_p \wedge \Diamond\Diamond\Box q_p)$	S5	$((\neg q_p \wedge \Box q_p) \vee (\neg q'_p \wedge \Diamond\Diamond\Diamond\Box q'_p)$

Thus, some of the classes evaluate the preprocessing capabilities of a prover but not the actual calculus and its implementation.

We therefore propose a different approach here. The principles underlying our approach are that (i) there should be the same number of formulae for each logic though not necessarily the same formulae across all logics; (ii) there should be an equal number of satisfiable and unsatisfiable formulae for each logic; (iii) a formula that is L-unsatisfiable should only be L'-unsatisfiable for every extension L' of L; (iv) a formula that is L'-satisfiable should be L-satisfiable for every extension L' of L; (v) the formulae should belong to parameterised classes of formulae of increasing difficulty. Note that Principles (iii) and (iv) are intentionally not symmetric. For L-unsatisfiable formulae it should be necessary for a prover to use the rules or clauses specific to L instead of being able to find a refutation without those. For L-satisfiable formulae we want to maximise the search space for a model.

For unsatisfiable formulae, we take the five LWB classes `k_branch_p`, `k_path_p`, `k_ph_p`, `k_poly_p`, `k_t4p_p` and for each logic L in the modal cube transform each formula in a class so that is L-unsatisfiable, but L'-satisfiable for any logic L' that is not an extension of L. The transformation proceeds by first converting a formula φ to simplified NNF. Then for each propositional literal l it replaces all its occurrences by $(l \vee \psi_L^p)$ where $|l| = p$ and ψ_L^p is a modal formula uniquely associated with p and L, resulting in a formula φ'. Finally, for logics KD4 and KDB we need to add a disjunct $(\Box q \wedge \Box\neg q)$ to φ', while for logics S4 and KTB we need to add a disjunct $(q \wedge \Box\neg q)$, where q is a propositional symbol not occurring in φ'. These disjuncts are unsatisfiable in the respective logics but satisfiable in logics where D, or T, do not hold. Table 7 shows the formulae ψ_L^p that we use in our evaluation. In the table, q_p and q'_p are propositional variables uniquely associated with p that do not occur in φ. The overall effect of this transformation is that the resulting classes of formulae satisfy Principles (iii) and (v).

For satisfiable formulae, we use the five classes `k_poly_n`, `s4_md_n`, `s4_ph_n`, `s4_path_n`, `s4_s5_n` without modification. Although the first of these classes was designed to be K-satisfiable and the other four to be S4-satisfiable, the formulae in those classes are satisfiable in all the logics we consider. `s4_ipc_n` also consists

Table 8. Benchmarking results

Logic	Status	Total	GMR (cneg)	GMR (cord)	GMR (cplain)	R+MLR (cneg)	R+MLR (cord)	R+MLR (cplain)	LEO-III+E
K	S	100	84	85	77	**100**	**100**	**100**	0
KD	S	100	84	85	77	96	**100**	93	0
KT	S	100	70	**81**	50	66	68	61	0
KB	S	100	58	58	29	51	**64**	51	0
K4	S	100	83	**85**	77	56	57	50	0
K5	S	100	**67**	60	45	36	37	26	0
KDB	S	100	63	70	40	56	**73**	55	0
KTB	S	100	58	**59**	38	52	57	31	0
KD4	S	100	83	**85**	77	52	53	46	0
KD5	S	100	**73**	70	61	46	47	38	0
K45	S	100	45	**53**	34	36	37	25	0
K4B	S	100	18	19	11	23	**38**	15	0
KD45	S	100	**67**	66	56	46	47	38	0
S4	S	100	66	**76**	48	45	44	33	0
S5	S	100	32	28	32	32	**35**	24	0
All	S	1500	951	**980**	752	793	857	686	0
K	U	100	74	76	71	**79**	78	77	21
KD	U	100	74	**76**	71	73	75	62	13
KT	U	100	74	**77**	70	71	74	67	30
KB	U	100	71	**78**	68	71	52	55	10
K4	U	100	55	52	**57**	41	29	35	4
K5	U	100	74	46	**75**	50	30	48	8
KDB	U	100	73	**77**	71	73	52	56	8
KTB	U	100	72	**77**	69	67	50	53	9
KD4	U	100	**70**	59	67	40	32	39	1
KD5	U	100	75	46	**77**	51	40	46	3
K45	U	100	**51**	37	49	16	12	8	3
K4B	U	100	47	52	46	**53**	30	49	5
KD45	U	100	**64**	43	55	33	22	28	1
S4	U	100	47	**68**	66	45	21	23	4
S5	U	100	47	51	**52**	36	13	29	2
All	U	1500	**968**	915	964	799	610	675	122

only of S5-satisfiable formulae but these appear to be insufficiently challenging and have not been included in our benchmark set. All other classes of the LWB benchmark classes for K and S4 are satisfiable in some of the logics, but not in all. The five classes satisfy Principles (iv) and (v). The benchmark collection consisting of all ten classes together then also satisfies Principles (i) and (ii).

Another challenge for an empirical evaluation is the lack of available fully automatic theorem provers for all 15 logics that we have already discussed in Sect. 1. This leaves us with just three different approaches we can compare (i) the higher-order logic prover LEO-III [12,37], with **E** 2.6 as external reasoner, *LEO-III+E for short*, that supports a wide range of logics via semantic embedding into higher-order logic (ii) the combination of our reductions with the modal-layered resolution (MLR) calculus for SNF_{ml} clauses [25], *R+MLR calculus for short*, implemented in the modal theorem prover K$_\mathrm{S}$P (iii) the global modal resolution (GMR) calculus, implemented in K$_\mathrm{S}$P, which has resolution rules for all 15 logics [24]. For R+MLR and GMR calculi, resolution inferences between literal clauses can either be unrestricted (`cplain` option), restricted by negative resolution (`cneg` option), or restricted by an ordering (`cord` option). It is worth pointing out that negative and ordered resolution require slightly different transformations to the normal form that introduce additional clauses (`snf+` and `snf++` options, respectively). Also, the ordering cannot be arbitrary [25]. For the experiments, we have used the following options: (i) input processing: prenexing, together with simplification and pure literal elimination (`bnfsimp`, `prenex`, `early_ple`); (ii) preprocessing of clauses: renaming reuses symbols (`limited_reuse_renaming`), forward and backward subsumption (`fsub`, `bsub`) are enabled; the usable is populated with clauses whose maximal literal is positive (`populate_usable, max_lit_positive`); pure literal elimination is set for GMR (`ple`) and modal level ple is set for MLR (`mlple`); (iii) processing: inference rules not required for completeness are also used (`unit, lhs_unit,mres`), the options for preprocessing of clauses are kept and clause selection takes the shortest clause by level (`shortest`).

For LEO-III we provide the prover with a modal formula in the syntax it expects plus a logic specification that tells the prover in which modal logic the formula is meant to be solved, for example, `$modal_system_S4`. LEO-III can collaborate with external reasoners during proof search and we have used **E** 2.6 [34,35] as external reasoner and restricted LEO-III to one instance of **E** running in parallel. LEO-III is implemented in Java and we have set the maximum heap size to 1 GB and the thread stack size to 64 MB for the JVM.

Table 8 shows our benchmarking results. The first three columns of the table show the logic in which we determine the satisfiability status of each formula, the satisfiability status of the formulae, and their number. The next six columns then show how many of those formulae were solved by K$_\mathrm{S}$P with a particular calculus and refinement. The last column shows the result for LEO-III. The highest number or numbers are highlighted in bold. A time limit of 100 CPU seconds was set for each formula. Benchmarking was performed on a PC with an AMD Ryzen 5 5600X CPU @ 4.60 GHz max and 64 GB main memory using Fedora release 34 as operating system.

While the R+MLR calculus is competitive with GMR on extensions of K with axioms D, T and, possibly, B, the GMR calculus has better performance on extensions with axioms 4 and 5.

On satisfiable formulae, where for all logics we use exactly the same formulae and both resolution calculi have to saturate the set of clauses up to redundancy, the number of formulae solved is directly linked to the number of inferences necessary to do so. The fact that we reduce SNF_{sml} clauses to SNF_{ml} clauses via the introduction of multiple copies of the same clausal formulae with different labels clearly leads to a corresponding multiplication of the inferences that need to be performed. LEO-III+E does not solve any of the satisfiable formulae. This can be seen as an illustration of how important the use of additional techniques is that can turn resolution into a decision procedure on embeddings of modal logics into first-order logic [18,33].

On unsatisfiable formulae, where we use different formulae for each logic, the number of formulae solved is linked to the number of inferences it takes to find a refutation. For instance, on K it takes the GMR calculus on average 6.2 times the number of inferences to find a refutation than the R+MLR calculus. However, for all other logics the opposite is true. On the remaining 14 logics, the R+MLR calculus on average requires 6.5 times the number of inferences to find a refutation than the GMR calculus. Given that the R+MLR calculus currently uses a reduction from a modal logic to SNF_{sml} followed by a transformation from SNF_{sml} to SNF_{ml}, it is difficult to discern which of the two is the major problem. It is clear that multiple copies of the same clausal formulae are also detrimental to proof search. LEO-III+E does reasonably well on unsatisfiable formulae and the results clearly show the impact that additional axioms have on its performance. It performs best for KT and K but for logics involving axioms 4 and 5 very few formulae can be solved. The external prover **E** finds the proof for 121 out of the 122 modal formulae LEO-III+E can solve.

6 Conclusions

We have presented novel reductions of extensions of the modal logic K with arbitrary combinations of the axioms B, D, T, 4, 5 to clausal normal forms SNF_{sml} and SNF_{ml} for K. The implementation of those reductions combined with K_SP [26], allows us to reason in all 15 logics of the modal cube in a fully automatic way. Such support was so far extremely limited.

The transformation of sets of SNF_{sml} to sets of SNF_{ml} relies on new results that show that non-clausal closed tableaux in the Single Step Tableaux calculus [14,21] can be simulated by refutations in the modal-layered resolution (MLR) calculus for SNF_{ml} clauses [25].

We have also developed a new collection of benchmark formulae that covers all 15 logics of the modal cube. The collection consists of classes of parameterised and therefore scalable formulae. It contains an equal number of satisfiable and unsatisfiable formulae for each logic and the satisfiability status of each formula is known in advance. So far extensive collections of benchmark formulae were only available for K with smaller collections available for KT and S4. A key feature of the approach is that it uses the systematic modification of K-unsatisfiable formulae to obtain unsatisfiable formulae in other logics. Thus, we could obtain a more extensive collection by applying this approach to further collections of benchmark formulae for K.

The evaluation we presented shows that on most of the 15 modal logics the combination of our reduction to SNF_{ml} with the MLR calculus does not perform as well as the global modal resolution (GMR) calculus, also implemented in K$_\mathsf{S}$P. This contrasts with the evaluation in [29], where we only considered six logics and used a different collection of benchmarks. We believe that the new benchmark collection more clearly indicates weaknesses in the current approach, in particular, the reduction from SNF_{sml} to SNF_{ml}. It is possible that the implementation of a calculus that operates directly on sets of SNF_{sml} clauses would perform considerably better as it avoids the repetition of clauses with different labels. However, it does so by using potentially infinite sets of labels which makes an implementation challenging. We intend to explore this possibility in future work.

References

1. Areces, C., de Rijke, M., de Nivelle, H.: Resolution in modal, description and hybrid logic. J. Log. Comput. **11**(5), 717–736 (2001)
2. Balsiger, P., Heuerding, A., Schwendimann, S.: A benchmark method for the propositional modal logics K, KT, S4. J. Autom. Reasoning **24**(3), 297–317 (2000). https://doi.org/10.1023/A:1006249507577
3. Basin, D., Matthews, S., Vigano, L.: Labelled propositional modal logics: theory and practice. J. Log. Comput. **7**(6), 685–717 (1997)
4. Blackburn, P., van Benthem, J., Wolter, F. (eds.): Handbook of Modal Logic. Elsevier (2006)
5. Blackburn, P., de Rijke, M., Venema, Y.: Modal Logic. Cambridge Tracts in Theoretical Computer Science. Cambridge University Press (2002)
6. Brünnler, K.: Deep sequent systems for modal logic. Arch. Math. Log. **48**(6), 551–577 (2009). https://doi.org/10.1007/s00153-009-0137-3
7. Catach, L.: TABLEAUX: a general theorem prover for modal logics. J. Autom. Reason. **7**(4), 489–510 (1991). https://doi.org/10.1007/BF01880326
8. Chellas, B.F.: Modal Logic: An Introduction. Cambridge University Press, Cambridge (1980)
9. Fitting, M.: Prefixed tableaus and nested sequents. Ann. Pure Appl. Log. **163**(3), 291–313 (2012)
10. Gasquet, O., Herzig, A., Longin, D., Sahade, M.: LoTREC: logical tableaux research engineering companion. In: Beckert, B. (ed.) TABLEAUX 2005. LNCS (LNAI), vol. 3702, pp. 318–322. Springer, Heidelberg (2005). https://doi.org/10.1007/11554554_25
11. Girlando, M., Straßburger, L.: MOIN: a nested sequent theorem prover for intuitionistic modal logics (system description). In: Peltier, N., Sofronie-Stokkermans, V. (eds.) IJCAR 2020. LNCS (LNAI), vol. 12167, pp. 398–407. Springer, Cham (2020). https://doi.org/10.1007/978-3-030-51054-1_25
12. Gleißner, T., Steen, A.: LEO-III (2022). https://github.com/leoprover/Leo-III
13. Gleißner, T., Steen, A., Benzmüller, C.: Theorem provers for every normal modal logic. In: Eiter, T., Sands, D. (eds.) LPAR 2017. EPiC Series in Computing, vol. 46, pp. 14–30. EasyChair (2017). https://doi.org/10.29007/jsb9
14. Goré, R.: Tableau methods for modal and temporal logics. In: D'Agostino, M., Gabbay, D., Hähnle, R., Posegga, J. (eds.) Handbook of Tableau Methods, pp.

297–396. Springer, Heidelberg (1999). https://doi.org/10.1007/978-94-017-1754-0_6
15. Goré, R., Kikkert, C.: CEGAR-Tableaux: improved modal satisfiability via modal clause-learning and SAT. In: Das, A., Negri, S. (eds.) TABLEAUX 2021. LNCS (LNAI), vol. 12842, pp. 74–91. Springer, Cham (2021). https://doi.org/10.1007/978-3-030-86059-2_5
16. Governatori, G.: Labelled modal tableaux. In: Areces, C., Goldblatt, R. (eds.) AiML 2008, pp. 87–110. College Publications (2008)
17. Halpern, J.Y., Moses, Y.: A guide to completeness and complexity for modal logics of knowledge and belief. Artif. Intell. **54**(3), 319–379 (1992)
18. Horrocks, I., Hustadt, U., Sattler, U., Schmidt, R.A.: Computational modal logic. In: Blackburn, P., van Benthem, J., Wolter, F. (eds.) Handbook of Modal Logic, chap. 4, pp. 181–245. Elsevier (2006)
19. Hustadt, U., de Nivelle, H., Schmidt, R.A.: Resolution-based methods for modal logics. Log. J. IGPL **8**(3), 265–292 (2000)
20. van Linder, B., van der Hoek, W., Meyer, J.J.C.: Formalising abilities and opportunities of agents. Fundamenta Informaticae **34**(1–2), 53–101 (1998)
21. Massacci, F.: Single step tableaux for modal logics. J. Autom. Reason. **24**, 319–364 (2000). https://doi.org/10.1023/A:1006155811656
22. Mayer, M.C.: Herbrand style proof procedures for modal logics. J. Appl. Non Class. Logics **3**(2), 205–223 (1993)
23. Nalon, C.: K$_S$P (2022). https://www.nalon.org/#software
24. Nalon, C., Dixon, C.: Clausal resolution for normal modal logics. J. Algorithms **62**, 117–134 (2007)
25. Nalon, C., Dixon, C., Hustadt, U.: Modal resolution: proofs, layers, and refinements. ACM Trans. Comput. Log. **20**(4), 23:1–23:38 (2019)
26. Nalon, C., Hustadt, U., Dixon, C.: K$_S$P: architecture, refinements, strategies and experiments. J. Autom. Reason. **64**(3), 461–484 (2020). https://doi.org/10.1007/s10817-018-09503-x
27. Ohlbach, H.J., Nonnengart, A., de Rijke, M., Gabbay, D.M.: Encoding two-valued nonclassical logics in classical logic. In: Robinson, J.A., Voronkov, A. (eds.) Handbook of Automated Reasoning, chap. 21, pp. 1403–1485. Elsevier (2001)
28. Otten, J.: MleanCoP: a connection prover for first-order modal logic. In: Demri, S., Kapur, D., Weidenbach, C. (eds.) IJCAR 2014. LNCS (LNAI), vol. 8562, pp. 269–276. Springer, Cham (2014). https://doi.org/10.1007/978-3-319-08587-6_20
29. Papacchini, F., Nalon, C., Hustadt, U., Dixon, C.: Efficient local reductions to basic modal logic. In: Platzer, A., Sutcliffe, G. (eds.) CADE 2021. LNCS (LNAI), vol. 12699, pp. 76–92. Springer, Cham (2021). https://doi.org/10.1007/978-3-030-79876-5_5
30. Plaisted, D.A., Greenbaum, S.: A structure-preserving clause form translation. J. Symb. Comput. **2**(3), 293–304 (1986)
31. Rao, A.S., Georgeff, M.P.: Modeling rational agents within a BDI-architecture. In: KR 1991, pp. 473–484. Morgan Kaufmann (1991)
32. Raths, T., Otten, J.: The QMLTP problem library for first-order modal logics. In: Gramlich, B., Miller, D., Sattler, U. (eds.) IJCAR 2012. LNCS (LNAI), vol. 7364, pp. 454–461. Springer, Heidelberg (2012). https://doi.org/10.1007/978-3-642-31365-3_35
33. Schmidt, R.A., Hustadt, U.: First-order resolution methods for modal logics. In: Voronkov, A., Weidenbach, C. (eds.) Programming Logics. LNCS, vol. 7797, pp. 345–391. Springer, Heidelberg (2013). https://doi.org/10.1007/978-3-642-37651-1_15

34. Schulz, S.: E 2.6 (2022). http://wwwlehre.dhbw-stuttgart.de/~sschulz/E/Download.html
35. Schulz, S., Cruanes, S., Vukmirović, P.: Faster, higher, stronger: E 2.3. In: Fontaine, P. (ed.) CADE 2019. LNCS (LNAI), vol. 11716, pp. 495–507. Springer, Cham (2019). https://doi.org/10.1007/978-3-030-29436-6_29
36. Steen, A., Benzmüller, C.: The higher-order prover Leo-III. In: Giacomo, G.D., et al. (eds.) ECAI 2020. Frontiers in Artificial Intelligence and Applications, vol. 325, pp. 2937–2938. IOS Press (2020). https://doi.org/10.3233/FAIA200462
37. Steen, A., Benzmüller, C.: Extensional higher-order paramodulation in Leo-III. J. Autom. Reason. **65**(6), 775–807 (2021). https://doi.org/10.1007/s10817-021-09588-x
38. The SPASS Team: SPASS 3.9 (2016). http://www.spass-prover.org/
39. Boy de la Tour, T.: An optimality result for clause form translation. J. Symb. Comput. **14**(4), 283–301 (1992)
40. Tseitin, G.S.: On the complexity of derivation in propositional calculus. In: Siekmann, J.H., Wrightson, G. (eds.) Automation of Reasoning: Classical Papers on Computational Logic 1967–1970, vol. 2, pp. 466–483. Springer, Heidelberg (1983). https://doi.org/10.1007/978-3-642-81955-1_28. Original paper (in Russian) appeared in 1968
41. Wansing, H.: Sequent calculi for normal modal proposisional logics. J. Log. Comput. **4**(2), 125–142 (1994)
42. Weidenbach, C.: Combining superposition, sorts and splitting. In: Robinson, J.A., Voronkov, A. (eds.) Handbook of Automated Reasoning, pp. 1965–2013. Elsevier and MIT Press (2001)
43. Weidenbach, C., Dimova, D., Fietzke, A., Kumar, R., Suda, M., Wischnewski, P.: SPASS version 3.5. In: Schmidt, R.A. (ed.) CADE 2009. LNCS (LNAI), vol. 5663, pp. 140–145. Springer, Heidelberg (2009). https://doi.org/10.1007/978-3-642-02959-2_10

Proof Systems and Proof Search

Cyclic Proofs, Hypersequents, and Transitive Closure Logic

Anupam Das(✉) and Marianna Girlando

University of Birmingham, Birmingham, UK
{a.das,m.girlando}@bham.ac.uk

Abstract. We propose a cut-free cyclic system for Transitive Closure Logic (TCL) based on a form of *hypersequents*, suitable for automated reasoning via proof search. We show that previously proposed sequent systems are cut-free incomplete for basic validities from Kleene Algebra (KA) and Propositional Dynamic Logic (PDL), over standard translations. On the other hand, our system faithfully simulates known cyclic systems for KA and PDL, thereby inheriting their completeness results. A peculiarity of our system is its richer correctness criterion, exhibiting 'alternating traces' and necessitating a more intricate soundness argument than for traditional cyclic proofs.

Keywords: Cyclic proofs · Transitive Closure Logic · Hypersequents · Propositional Dynamic Logic

1 Introduction

Transitive Closure Logic (TCL) is the extension of first-order logic by an operator computing the transitive closure of definable binary relations. It has been studied by numerous authors, e.g. [15–17], and in particular has been proposed as a foundation for the mechanisation and automation of mathematics [1].

Recently, Cohen and Rowe have proposed *non-wellfounded* and *cyclic* systems for TCL [9,11]. These systems differ from usual ones by allowing proofs to be infinite (finitely branching) trees, rather than finite ones, under some appropriate global correctness condition (the 'progressing criterion'). One particular feature of the cyclic approach to proof theory is the facilitation of automation, since complexity of inductive invariants is effectively traded off for a richer proof structure. In fact this trade off has recently been made formal, cf. [3,12], and has led to successful applications to automated reasoning, e.g. [6,7,24,26,27].

In this work we investigate the capacity of cyclic systems to automate reasoning in TCL. Our starting point is the demonstration of a key shortfall of Cohen and Rowe's system: its cut-free fragment, here called TC_G, is unable to cyclically prove even standard theorems of relational algebra, e.g. $(a \cup b)^* = a^*(ba^*)^*$ and

This work was supported by a UKRI Future Leaders Fellowship, 'Structure vs Invariants in Proofs', project reference MR/S035540/1.

J. Blanchette et al. (Eds.): IJCAR 2022, LNAI 13385, pp. 509–528, 2022.
https://doi.org/10.1007/978-3-031-10769-6_30

$(aa \cup aba)^+ \leq a^+((ba^+)^+ \cup a))$ (Theorem 12). An immediate consequence of this is that cyclic proofs of TC_G do not enjoy cut-admissibility (Corollary 13). On the other hand, these (in)equations are theorems of Kleene Algebra (KA) [18,19], a decidable theory which admits automation-via-proof-search thanks to the recent cyclic system of Das and Pous [14].

What is more, TCL is well-known to interpret Propositional Dynamic Logic (PDL), a modal logic whose modalities are just terms of KA, by a natural extension of the 'standard translation' from (multi)modal logic to first-order logic (see, e.g., [4,5]). Incompleteness of cyclic-TC_G for PDL over this translation is inherited from its incompleteness for KA. This is in stark contrast to the situation for modal logics without fixed points: the standard translation from K (and, indeed, all logics in the 'modal cube') to first-order logic actually *lifts* to cut-free proofs for a wide range of modal logic systems, cf. [21,22].

A closer inspection of the systems for KA and PDL reveals the stumbling block to any simulation: these systems implicitly conduct a form of 'deep inference', by essentially reasoning underneath $\exists$ and $\wedge$. Inspired by this observation, we propose a form of *hypersequents* for predicate logic, with extra structure admitting the deep reasoning required. We present the cut-free system HTC and a novel notion of cyclic proof for these hypersequents. In particular, the incorporation of some deep inference at the level of the rules necessitates an 'alternating' trace condition corresponding to *alternation* in automata theory.

Our first main result is the Soundness Theorem (Theorem 23): non-wellfounded proofs of HTC are sound for *standard semantics*. The proof is rather more involved than usual soundness arguments in cyclic proof theory, due to the richer structure of hypersequents and the corresponding progress criterion. Our second main result is the Simulation Theorem (Theorem 28): HTC is complete for PDL over the standard translation, by simulating a cut-free cyclic system for the latter. This result can be seen as a formal interpretation of cyclic modal proof theory within cyclic predicate proof theory, in the spirit of [21,22].

To simplify the exposition, we shall mostly focus on equality-free TCL and 'identity-free' PDL in this paper, though all our results hold also for the 'reflexive' extensions of both logics. We discuss these extensions in Sect. 7, and present further insights and conclusions in Sect. 8. Full proofs and further examples not included here (due to space constraints) can be found in [13].

2 Preliminaries

We shall work with a fixed first-order vocabulary consisting of a countable set Pr of unary *predicate* symbols, written p, q, etc., and of a countable set Rel of binary *relation* symbols, written a, b, etc. We shall generally reserve the word 'predicate' for unary and 'relation' for binary. We could include further relational symbols too, of higher arity, but choose not to in order to calibrate the semantics of both our modal and predicate settings.

We build formulas from this language differently in the modal and predicate settings, but all our formulas may be formally evaluated within *structures*:

Definition 1 (Structures). *A* structure $\mathcal{M}$ *consists of a set* D*, called the* domain *of* $\mathcal{M}$*, which we sometimes denote by* $|\mathcal{M}|$*; a subset* $p^{\mathcal{M}} \subseteq D$ *for each* $p \in \mathsf{Pr}$*; and a subset* $a^{\mathcal{M}} \subseteq D \times D$ *for each* $a \in \mathsf{Rel}$.

2.1 Transitive Closure Logic

In addition to the language introduced at the beginning of this section, in the predicate setting we further make use of a countable set of *function* symbols, written f^i, g^j, etc. where the superscripts $i, j \in \mathbb{N}$ indicate the *arity* of the function symbol and may be omitted when it is not ambiguous. Nullary function symbols (aka *constant* symbols), are written c, d etc. We shall also make use of *variables*, written x, y, etc., typically bound by quantifiers. *Terms*, written s, t, etc., are generated as usual from variables and function symbols by function application. A term is *closed* if it has no variables.

We consider the usual syntax for first-order logic formulas over our language, with an additional operator for transitive closure (and its dual). Formally, TCL formulas, written A, B, etc., are generated as follows:

$$\begin{aligned} A, B ::= {} & p(t) \mid \bar{p}(t) \mid a(s,t) \mid \bar{a}(s,t) \mid (A \wedge B) \mid (A \vee B) \mid \forall x A \mid \exists x A \mid \\ & TC(\lambda x, y.A)(s,t) \mid \overline{TC}(\lambda x, y.A)(s,t) \end{aligned}$$

When variables x, y are clear from context, we may write $TC(A(x,y))(s,t)$ or $TC(A)(s,t)$ instead of $TC(\lambda x, y.A)(s,t)$, as an abuse of notation, and similarly for $\overline{TC}$. We may write $A[t/x]$ for the formula obtained from A by replacing every free occurrence of the variable x by the term t. We have included both TC and $\overline{TC}$ as primitive operators, so that we can reduce negation to atomic formulas, shown below. This will eventually allow a one-sided formulation of proofs.

Definition 2 (Duality). *For a formula* A *we define its* complement*,* $\bar{A}$*, by:*

$$\begin{array}{llllll} \overline{p(t)} := \bar{p}(t) & \overline{\bar{p}(t)} := p(t) & & & & \\ \overline{a(s,t)} := \bar{a}(s,t) & \overline{\forall x A} := \exists x \bar{A} & \overline{A \wedge B} := \bar{A} \vee \bar{B} & \overline{TC(A)(s,t)} := \overline{TC}(\bar{A})(s,t) \\ \overline{\bar{a}(s,t)} := a(s,t) & \overline{\exists x A} := \forall x \bar{A} & \overline{A \vee B} := \bar{A} \wedge \bar{B} & \overline{\overline{TC}(A)(s,t)} := TC(\bar{A})(s,t) \end{array}$$

We shall employ standard logical abbreviations, e.g. $A \supset B$ for $\bar{A} \vee B$.

We may evaluate formulas with respect to a structure, but we need additional data for interpreting function symbols:

Definition 3 (Interpreting function symbols). *Let* $\mathcal{M}$ *be a structure with domain* D*. An* interpretation *is a map* ρ *that assigns to each function symbol* f^n *a function* $D^n \to D$*. We may extend any interpretation* ρ *to an action on (closed) terms by setting recursively* $\rho(f(t_1, \dots, t_n)) := \rho(f)(\rho(t_1), \dots, \rho(t_n))$.

We only consider *standard* semantics in this work: TC (and $\overline{TC}$) is always interpreted as the *real* transitive closure (and its dual) in a structure, rather than being axiomatised by some induction (and coinduction) principle.

Definition 4 (Semantics). *Given a structure $\mathcal{M}$ with domain D and an interpretation ρ, the judgement $\mathcal{M}, \rho \models A$ is defined as usual for first-order logic with the following additional clauses for TC and $\overline{TC}$:*[1]

- *$\mathcal{M}, \rho \models TC(A(x,y))(s,t)$ if there are $v_0, \dots, v_{n+1} \in D$ with $\rho(s) = v_0$, $\rho(t) = v_{n+1}$, such that for every $i \leq n$ we have $\mathcal{M}, \rho \models A(v_i, v_{i+1})$.*
- *$\mathcal{M}, \rho \models \overline{TC}(A(x,y))(s,t)$ if for all $v_0, \dots, v_{n+1} \in D$ with $\rho(s) = v_0$ and $\rho(t) = v_{n+1}$, there is some $i \leq n$ such that $\mathcal{M}, \rho \models A(v_i, v_{i+1})$.*

If $\mathcal{M}, \rho \models A$ for all $\mathcal{M}$ and ρ, we simply write $\models A$.

Remark 5 (TC and $\overline{TC}$ as least and greatest fixed points). As expected, we have $\mathcal{M}, \rho \not\models TC(A)(s,t)$ just if $\mathcal{M}, \rho \models \overline{TC}(\bar{A})(s,t)$, and so the two operators are semantically dual. Thus, TC and $\overline{TC}$ duly correspond to least and greatest fixed points, respectively, satisfying in any model:

$$TC(A)(s,t) \iff A(s,t) \vee \exists x(A(s,x) \wedge TC(A)(x,t)) \tag{1}$$
$$\overline{TC}(A)(s,t) \iff A(s,t) \wedge \forall x(A(s,x) \vee \overline{TC}(A)(x,t)) \tag{2}$$

Let us point out that our $\overline{TC}$ operator is not the same as Cohen and Rowe's transitive 'co-closure' operator TC^{op} in [10], but rather the De Morgan dual of TC. In the presence of negation, TC and $\overline{TC}$ are indeed interdefinable, cf. Definition 2.

2.2 Cohen-Rowe Cyclic System for TCL

Cohen and Rowe proposed in [9,11] a non-wellfounded system for TCL that extends a usual sequent calculus $\mathsf{LK}_=$ for first-order logic with equality and substitution by rules for TC inspired by its characterisation as a least fixed point, cf. (1).[2] Note that the presence of the substitution rule is critical for the notion of 'regularity' in predicate cyclic proof theory. The resulting notions of non-wellfounded and cyclic proofs are formulated similarly to those for first-order logic with (ordinary) inductive definitions [8]:

Definition 6 (Sequent system). TC_G *is the extension of* $\mathsf{LK}_=$ *by the rules:*

$$TC_0 \frac{\Gamma, A(s,t)}{\Gamma, TC(A)(s,t)} \qquad TC_1 \frac{\Gamma, A(s,r) \quad \Gamma, TC(A)(r,t)}{\Gamma, TC(A)(s,t)}$$
$$\overline{TC} \frac{\Gamma, A(s,t) \quad \Gamma, A(s,c), \overline{TC}(A)(c,t)}{\Gamma, \overline{TC}(A)(s,t)} \; c \text{ fresh} \tag{3}$$

TC_G-preproofs *are possibly infinite trees of sequents generated by the rules of* TC_G*. A preproof is* regular *if it has only finitely many distinct sub-preproofs.*

[1] Note that we are including 'parameters from the model' in formulas here. Formally, this means each $v \in D$ is construed as a constant symbol for which $\rho(v) = v$.

[2] Cohen and Rowe's system is originally called RTC_G, rather using a 'reflexive' version RTC of the TC operator. However this (and its rules) can be encoded (and simulated) by defining $RTC(\lambda x, y.A)(s,t) := TC(\lambda x, y(x = y \vee A))(s,t)$.

The notion of 'correct' non-wellfounded proof is obtained by a standard *progressing criterion* in cyclic proof theory. We shall not go into details here, being beyond the scope of this work, but refer the reader to those original works (as well as [13] for our current variant). Let us write $\vdash_{cyc}$ for their notion of cyclic provability using the above rules, cf. [9,11]. A standard infinite descent countermodel argument yields:

Proposition 7 (Soundness, [9,11]**).** *If* $\mathsf{TC}_G \vdash_{cyc} A$ *then* $\models A$.

In fact, this result is subsumed by our main soundness result for HTC (Theorem 23) and its simulation of TC_G (Theorem 19). In the presence of cut, a form of converse of Proposition 7 holds: cyclic TC_G proofs are 'Henkin complete', i.e. complete for all models of a particular axiomatisation of TCL based on (co)induction principles for TC (and $\overline{TC}$) [9,11]. However, the counterexample we present in the next section implies that cut is not eliminable (Corollary 13).

3 Interlude: Motivation from PDL and Kleene Algebra

Given the TCL sequent system proposed by Cohen and Rowe, why do we propose a hypersequential system? Our main argument is that proof search in TC_G is rather weak, to the extent that cut-free cyclic proofs are unable to simulate a basic (cut-free) system for modal logic PDL (regardless of proof search strategy). At least one motivation here is to 'lift' the *standard translation* from cut-free cyclic proofs for PDL to cut-free cyclic proofs in an adequate system for TCL.

3.1 Identity-Free PDL

Identity-free propositional dynamic logic (PDL$^+$) is a version of the modal logic PDL without tests or identity, thereby admitting an 'equality-free' standard translation into predicate logic. Formally, PDL$^+$ *formulas*, written A, B, etc., and *programs*, written α, β, etc., are generated by the following grammars:

$$A, B ::= p \mid \overline{p} \mid (A \wedge B) \mid (A \vee B) \mid [\alpha]A \mid \langle\alpha\rangle A$$
$$\alpha, \beta ::= a \mid (\alpha;\beta) \mid (\alpha \cup \beta) \mid \alpha^+$$

We sometimes simply write $\alpha\beta$ instead of $\alpha;\beta$, and $(\alpha)A$ for a formula that is either $\langle\alpha\rangle A$ or $[\alpha]A$.

Definition 8 (Duality). *For a formula A we define its* complement, $\bar{A}$, *by:*

$$\bar{\overline{p}} := p \qquad \begin{array}{rl} \overline{A \wedge B} & := \bar{A} \vee \bar{B} \\ \overline{A \vee B} & := \bar{A} \wedge \bar{B} \end{array} \qquad \begin{array}{rl} \overline{[\alpha]A} & := \langle\alpha\rangle\bar{A} \\ \overline{\langle\alpha\rangle A} & := [\alpha]\bar{A} \end{array}$$

We *evaluate* PDL$^+$ formulas using the traditional relational semantics of modal logic, by associating each program with a binary relation in a structure. Again, we only consider standard semantics, in the sense that the $+$ operator is interpreted as the real transitive closure within a structure.

Definition 9 (Semantics). *For structures $\mathcal{M}$ with domain D, elements $v \in D$, programs α and formulas A, we define $\alpha^{\mathcal{M}} \subseteq D \times D$ and the judgement $\mathcal{M}, v \models A$ as follows:*

- *($a^{\mathcal{M}}$ is already given in the specification of $\mathcal{M}$, cf. Definition 1).*
- *$(\alpha;\beta)^{\mathcal{M}} := \{(u,v) :$ there is $w \in D$ s.t. $(u,w) \in \alpha^{\mathcal{M}}$ and $(w,v) \in \beta^{\mathcal{M}}\}$.*
- *$(\alpha \cup \beta)^{\mathcal{M}} := \{(u,v) : (u,v) \in \alpha^{\mathcal{M}}$ or $(u,v) \in \beta^{\mathcal{M}}\}$.*
- *$(\alpha^+)^{\mathcal{M}} := \{(u,v)$: there are $w_0, \dots, w_{n+1} \in D$ s.t. $u = w_0, v = w_{n+1}$ and, for every $i \leq n, (w_i, w_{i+1}) \in \alpha^{\mathcal{M}}\}$.*
- *$\mathcal{M}, v \models p$ if $v \in p^{\mathcal{M}}$.*
- *$\mathcal{M}, v \models \bar{p}$ if $v \notin p^{\mathcal{M}}$.*
- *$\mathcal{M}, v \models A \wedge B$ if $\mathcal{M}, v \models A$ and $\mathcal{M}, v \models B$.*
- *$\mathcal{M}, v \models A \vee B$ if $\mathcal{M}, v \models A$ or $\mathcal{M}, v \models B$.*
- *$\mathcal{M}, v \models [\alpha]A$ if $\forall (v,w) \in \alpha^{\mathcal{M}}$ we have $\mathcal{M}, w \models A$.*
- *$\mathcal{M}, v \models \langle\alpha\rangle A$ if $\exists (v,w) \in \alpha^{\mathcal{M}}$ with $\mathcal{M}, w \models A$.*

If $\mathcal{M}, v \models A$ for all $\mathcal{M}$ and $v \in |\mathcal{M}|$, then we write $\models A$.

Note that we are overloading the satisfaction symbol $\models$ here, for both PDL$^+$ and TCL. This should never cause confusion, in particular since the two notions of satisfaction are 'compatible' as we shall now see.

3.2 The Standard Translation

The so-called 'standard translation' of modal logic into predicate logic is induced by reading the semantics of modal logic as first-order formulas. We now give a natural extension of this that interprets PDL$^+$ into TCL. At the logical level our translation coincides with the usual one for basic modal logic; our translation of programs, as expected, requires the TC operator to interpret the $+$ of PDL$^+$.

Definition 10. *For* PDL$^+$ *formulas A and programs α, we define the* standard translations *$\mathsf{ST}(A)(x)$ and $\mathsf{ST}(\alpha)(x,y)$ as* TCL*-formulas with free variables x and x, y, resp., inductively as follows:*

$$\begin{aligned}
\mathsf{ST}(p)(x) &:= p(x) & \mathsf{ST}(a)(x,y) &:= a(x,y)\\
\mathsf{ST}(\bar{p})(x) &:= \bar{p}(x) & \mathsf{ST}(\alpha \cup \beta)(x,y) &:= \mathsf{ST}(\alpha)(x,y) \vee \mathsf{ST}(\beta)(x,y)\\
\mathsf{ST}(A \vee B)(x) &:= \mathsf{ST}(A)(x) \vee \mathsf{ST}(B)(x) & \mathsf{ST}(\alpha;\beta)(x,y) &:= \exists z(\mathsf{ST}(\alpha)(x,z) \wedge \mathsf{ST}(\beta)(z,y))\\
\mathsf{ST}(A \wedge B)(x) &:= \mathsf{ST}(A)(x) \wedge \mathsf{ST}(B)(x) & \mathsf{ST}(\alpha^+)(x,y) &:= TC(\mathsf{ST}(\alpha))(x,y)\\
\mathsf{ST}(\langle\alpha\rangle A)(x) &:= \exists y(\mathsf{ST}(\alpha)(x,y) \wedge \mathsf{ST}(A)(y))\\
\mathsf{ST}([\alpha]A)(x) &:= \forall y(\overline{\mathsf{ST}(\alpha)(x,y)} \vee \mathsf{ST}(A)(y))
\end{aligned}$$

where $TC(\mathsf{ST}(\alpha))$ is shorthand for $TC(\lambda x, y.\mathsf{ST}(\alpha)(x,y))$.

It is routine to show that $\overline{\mathsf{ST}(A)(x)} = \mathsf{ST}(\bar{A})(x)$, by structural induction on A, justifying our overloading of the notation $\bar{A}$, in both TCL and PDL$^+$. Yet another advantage of using the same underlying language for both the modal and predicate settings is that we can state the following (expected) result without the need for encodings, following by a routine structural induction (see, e.g., [5]):

Theorem 11. *For* PDL$^+$ *formulas A, we have $\mathcal{M}, v \models A$ iff $\mathcal{M} \models \mathsf{ST}(A)(v)$.*

3.3 Cohen-Rowe System is not Complete for PDL$^+$

PDL$^+$ admits a standard cut-free cyclic proof system LPD^+ (see Sect. 6.1) which is both sound and complete (cf. Theorem 30). However, a shortfall of TC_G is that it is unable to cut-free simulate LPD^+. In fact, we can say something stronger:

Theorem 12 (Incompleteness). *There exist a* PDL$^+$ *formula A such that $\models A$ but $\mathsf{TC}_G \not\vdash_{cyc} \mathsf{ST}(A)(x)$ (in the absence of cut).*

This means not only that TC_G is unable to locally cut-free simulate the rules of LPD^+, but also that there are some validities for which there are no cut-free cyclic proofs at all in TC_G. One example of such a formula is:

$$\langle (aa \cup aba)^+ \rangle p \supset \langle a^+((ba^+)^+ \cup a) \rangle p \tag{4}$$

A detailed proof of this is found in [13], but let us briefly discuss it here. First, the formula above is not artificial: it is derived from the well-known PDL validity $\langle (a \cup b)^* \rangle p \supset \langle a^*(ba^*)^* \rangle p$ by identity-elimination. This in turn is essentially a theorem of relational algebra, namely $(a \cup b)^* \leq a^*(ba^*)^*$, which is often used to eliminate $\cup$ in (sums of) regular expressions. The same equation was (one of those) used by Das and Pous in [14] to show that the sequent system LKA for Kleene Algebra is cut-free cyclic incomplete.

The argument that $\mathsf{TC}_G \not\vdash_{cyc} \mathsf{ST}(4)(x)$ is much more involved than the one from [14], due to the fact we are working in predicate logic, but the underlying basic idea is similar. At a very high level, the RHS of (4) (viewed as a relational inequality) is translated to an existential formula $\exists z(\mathsf{ST}(a^+)(x,z) \wedge \mathsf{ST}((ba^+)^+ \cup a)(z,y)$ that, along some branch (namely the one that always chooses aa when decomposing the LHS of (4)) can never be instantiated while remaining valid. This branch witnesses the non-regularity of any proof. However $\mathsf{ST}(4)(x)$ is cyclically provable in TC_G with cut, so an immediate consequence of Theorem 12 is:

Corollary 13. *The class of cyclic proofs of TC_G does not enjoy cut-admissibility.*

4 Hypersequent Calculus for TCL

Let us take a moment to examine why any 'local' simulation of LPD^+ by TC_G fails, in order to motivate the main system that we shall present. The program rules, in particular the $\langle\,\rangle$-rules, require a form of *deep inference* to be correctly simulated, over the standard translation. For instance, let us consider the action of the standard translation on two rules we shall see later in LPD^+ (cf. Sect. 6.1):

$$\langle\cup\rangle_0 \frac{\Gamma, \langle a_0 \rangle p}{\Gamma, \langle a_0 \cup a_1 \rangle p} \quad \rightsquigarrow \quad \frac{\mathsf{ST}(\Gamma)(c), \exists x(a_0(c,x) \wedge p(x))}{\mathsf{ST}(\Gamma)(c), \exists x((a_0(c,x) \vee a_1(c,x)) \wedge p(x))}$$

$$\langle;\rangle \frac{\Gamma, \langle a \rangle \langle b \rangle p}{\Gamma, \langle a; b \rangle p} \quad \rightsquigarrow \quad \frac{\mathsf{ST}(\Gamma)(c), \exists y(a(c,y) \wedge \exists x(b(y,x) \wedge p(x)))}{\mathsf{ST}(\Gamma)(c), \exists x(\exists y(a(c,y) \wedge b(y,x)) \wedge p(x))}$$

$$\mathsf{init}\,\frac{}{\{\}^{\varnothing}} \qquad \mathsf{wk}\,\frac{\mathbf{S}}{\mathbf{S},\mathbf{S}'} \qquad \sigma\,\frac{\mathbf{S}}{\sigma(\mathbf{S})} \qquad \cup\,\frac{\mathbf{S},\{\Gamma\}^{\mathbf{x}} \quad \mathbf{S},\{\Delta\}^{\mathbf{y}}}{\mathbf{S},\{\Gamma,\Delta\}^{\mathbf{x},\mathbf{y}}}\ \begin{matrix}\mathsf{fv}(\Delta)\cap\mathbf{x}=\varnothing\\ \mathsf{fv}(\Gamma)\cap\mathbf{y}=\varnothing\end{matrix}$$

$$\mathsf{id}\,\frac{\mathbf{S},\{\Gamma\}^{\mathbf{x}}}{\mathbf{S},\{\Gamma,A\}^{\mathbf{x}},\{\bar{A}\}^{\varnothing}}\ A\text{ closed} \qquad \wedge\,\frac{\mathbf{S},\{\Gamma,A,B\}^{\mathbf{x}}}{\mathbf{S},\{\Gamma,A\wedge B\}^{\mathbf{x}}} \qquad \vee\,\frac{\mathbf{S},\{\Gamma,A_i\}^{\mathbf{x}}}{\mathbf{S},\{\Gamma,A_0\vee A_1\}^{\mathbf{x}}}\ i\in\{0,1\}$$

$$\mathsf{inst}\,\frac{\mathbf{S},\{\Gamma(t)\}^{\mathbf{x}}}{\mathbf{S},\{\Gamma(y)\}^{\mathbf{x},y}} \qquad \exists\,\frac{\mathbf{S},\{\Gamma,A(y)\}^{\mathbf{x},y}}{\mathbf{S},\{\Gamma,\exists x(A(x))\}^{\mathbf{x}}}\ y\text{ fresh} \qquad \forall\,\frac{\mathbf{S},\{\Gamma,A(f(\mathbf{x}))\}^{\mathbf{x}}}{\mathbf{S},\{\Gamma,\forall x(A(x))\}^{\mathbf{x}}}\ f\text{ fresh}$$

$$TC\,\frac{\mathbf{S},\{\Gamma,A(s,t)\}^{\mathbf{x}},\{\Gamma,A(s,z),TC(A)(z,t)\}^{\mathbf{x},z}}{\mathbf{S},\{\Gamma,TC(A)(s,t)\}^{\mathbf{x}}}\ z\text{ fresh}$$

$$\overline{TC}\,\frac{\mathbf{S},\{\Gamma,A(s,t),A(s,f(\mathbf{x}))\}^{\mathbf{x}},\{\Gamma,A(s,t),\overline{TC}(A)(f(\mathbf{x}),t)\}^{\mathbf{x}}}{\mathbf{S},\{\Gamma,\overline{TC}(A)(s,t)\}^{\mathbf{x}}}\ f\text{ fresh}$$

Fig. 1. Hypersequent calculus HTC. σ is a 'substitution' map from constants to terms and a renaming of other function symbols and variables.

The first case above suggests that any system to which the standard translation lifts must be able to reason *underneath* $\exists$ *and* $\wedge$, so that the inference indicated in blue is 'accessible' to the prover. The second case above suggests that the existential-conjunctive meta-structure necessitated by the first case should admit basic equivalences, in particular certain *prenexing*. This section is devoted to the incorporation of these ideas (and necessities) into a bona fide proof system.

4.1 A System for Predicate Logic via Annotated Hypersequents

An *annotated cedent*, or simply *cedent*, written S, S' etc., is an expression $\{\Gamma\}^{\mathbf{x}}$, where Γ is a set of formulas and the *annotation* $\mathbf{x}$ is a set of variables. We sometimes construe annotations as lists rather than sets when it is convenient, e.g. when taking them as inputs to a function.

Each cedent may be intuitively read as a TCL formula, under the following interpretation: $fm(\{\Gamma\}^{x_1,\dots,x_n}) := \exists x_1 \dots \exists x_n \bigwedge \Gamma$. When $\mathbf{x} = \varnothing$ then there are no existential quantifiers above, and when $\Gamma = \varnothing$ we simply identify $\bigwedge \Gamma$ with $\top$. We also sometimes write simply A for the annotated cedent $\{A\}^{\varnothing}$.

A *hypersequent*, written $\mathbf{S}, \mathbf{S}'$ etc., is a set of annotated cedents. Each hypersequent may be intuitively read as the disjunction of its cedents. Namely we set: $fm(\{\Gamma_1\}^{\mathbf{x}_1}, \dots, \{\Gamma_n\}^{\mathbf{x}_n}) := fm(\{\Gamma_1\}^{\mathbf{x}_1}) \vee \dots \vee fm(\{\Gamma_n\}^{\mathbf{x}_n})$.

Definition 14 (System). *The rules of* HTC *are given in Fig. 1. A* HTC *pre-proof is a (possibly infinite) derivation tree generated by the rules of* HTC. *A preproof is* regular *if it has only finitely many distinct subproofs.*

Our hypersequential system is somewhat more refined than usual sequent systems for predicate logic. E.g., the usual $\exists$ rule is decomposed into $\exists$ and inst, whereas the usual $\wedge$ rule is decomposed into $\wedge$ and $\cup$. The rules for TC and $\overline{TC}$ are induced directly from their characterisations as fixed points in (1).

Note that the rules $\overline{TC}$ and $\forall$ introduce, bottom-up, the fresh function symbol f, which plays the role of the *Herbrand function* of the corresponding $\forall$ quantifier: just as $\forall \mathbf{x} \exists x A(x)$ is equisatisfiable with $\forall \mathbf{x} A(f(\mathbf{x}))$, when f is fresh, by Skolemisation, by duality $\exists \mathbf{x} \forall x A(x)$ is equivalid with $\exists \mathbf{x} A(f(\mathbf{x}))$, when f is fresh, by Herbrandisation. The usual $\forall$ rule of the sequent calculus corresponds to the case when $\mathbf{x} = \varnothing$.

4.2 Non-wellfounded Hypersequent Proofs

Our notion of ancestry, as compared to traditional sequent systems, must account for the richer structure of hypersequents:

Definition 15 (Ancestry). *Fix an inference step* r*, as typeset in Fig. 1. A formula C in the premiss is an* immediate ancestor *of a formula C' in the conclusion if they have the same colour; if $C, C' \in \Gamma$ then we further require $C = C'$, and if C, C' occur in $\mathbf{S}$ then $C = C'$ occur in the same cedent. A cedent S in the premiss is an immediate ancestor of a cedent S' in the conclusion if some formula in S is an* immediate ancestor *of some formula in S'.*

Immediate ancestry on both formulas and cedents is a binary relation, inducing a directed graph whose paths form the basis of our correctness condition:

Definition 16 ((Hyper)traces). *A* hypertrace *is a maximal path in the graph of immediate ancestry on cedents. A* trace *is a maximal path in the graph of immediate ancestry on formulas.*

Definition 17 (Progress and proofs). *Fix a preproof $\mathcal{D}$. A (infinite) trace $(F_i)_{i\in\omega}$ is* progressing *if there is k such that, for all $i > k$, F_i has the form $\overline{TC}(A)(s_i, t_i)$ and is infinitely often principal.*[3] *A (infinite) hypertrace $\mathcal{H}$ is* progressing *if every infinite trace within it is progressing. A (infinite) branch is progressing if it has a progressing hypertrace. $\mathcal{D}$ is a* proof *if every infinite branch is progressing. If, furthermore, $\mathcal{D}$ is regular, we call it a* cyclic proof.

We write HTC $\vdash_{nwf}$ $\mathbf{S}$ *(or* HTC $\vdash_{cyc}$ $\mathbf{S}$*) if there is a proof (or cyclic proof, respectively) of* HTC *of the hypersequent* $\mathbf{S}$.

In usual cyclic systems, checking that a regular preproof is progressing is decidable by straightforward reduction to the universality of nondeterministic ω-automata, with runs 'guessing' a progressing trace along an infinite branch. Our notion of progress exhibits an extra quantifier alternation: we must *guess* an infinite hypertrace in which *every* trace is progressing. Nonetheless, by appealing to determinisation or alternation, we can still decide our progressing condition:

Proposition 18. *Checking whether a* HTC *preproof is a proof is decidable by reduction to universality of ω-regular languages.*

[3] In fact, by a simple well-foundedness argument, it is equivalent to say that $(F_i)_{i<\omega}$ is progressing if it is infinitely often principal for a $\overline{TC}$-formula.

As we mentioned earlier, cyclic proofs of HTC indeed are at least as expressive as those of Cohen and Rowe's system by a routine local simulation of rules:

Theorem 19 (Simulating Cohen-Rowe). *If* $\mathsf{TC}_G \vdash_{cyc} A$ *then* $\mathsf{HTC} \vdash_{cyc} A$.

4.3 Some Examples

Example 20 (Fixed point identity). The sequent $\{\overline{TC}(a)(c,d)\}^{\varnothing}, \{TC(\bar{a})(c,d)\}^{\varnothing}$ is finitely derivable using rule id on $TC(a)(c,d)$ and the init rule. However we can also cyclically reduce it to a simpler instance of id. Due to the granularity of the inference rules of HTC, we actually have some liberty in how we implement such a derivation. E.g., the HTC-proof below applies TC rules below $\overline{TC}$ ones, and delays branching until the 'end' of proof search, which is impossible in TC_G. The only infinite branch, looping on $\bullet$, is progressing by the blue hypertrace.

$$
\mathit{TC}\,\dfrac{\overline{\mathit{TC}}\,\dfrac{\mathsf{inst}\,\dfrac{2\cup\,\dfrac{\mathsf{id}\,\dfrac{\mathsf{init}\,\dfrac{}{\{\}^{\varnothing}}}{\{a(c,d)\}^{\varnothing},\{\bar{a}(c,d)\}^{\varnothing}} \qquad \cup\,\dfrac{\mathsf{id}\,\dfrac{\mathsf{init}\,\dfrac{}{\{\}^{\varnothing}}}{\{a(c,e)\}^{\varnothing},\{\bar{a}(c,e)\}^{\varnothing}} \qquad \mathit{TC}\,\dfrac{\vdots}{\{\overline{TC}(a)(e,d)\}^{\varnothing},\{TC(\bar{a})(e,d)\}^{\varnothing}}\,\bullet}{\{a(c,e)\}^{\varnothing},\{\overline{TC}(a)(e,d)\}^{\varnothing},\{\bar{a}(c,e),TC(\bar{a})(e,d)\}^{\varnothing}}}{\{a(c,d),a(c,e)\}^{\varnothing},\{a(c,d),\overline{TC}(a)(e,d)\}^{\varnothing},\{\bar{a}(c,d)\}^{\varnothing},\{\bar{a}(c,e),TC(\bar{a})(e,d)\}^{\varnothing}}}{\{a(c,d),a(c,e)\}^{\varnothing},\{a(c,d),\overline{TC}(a)(e,d)\}^{\varnothing},\{\bar{a}(c,d)\}^{\varnothing},\{\bar{a}(c,x),TC(\bar{a})(x,d)\}^{x}}}{\{\overline{TC}(a)(c,d)\}^{\varnothing},\{\bar{a}(c,d)\}^{\varnothing},\{\bar{a}(c,x),TC(\bar{a})(x,d)\}^{x}}}{\{\overline{TC}(a)(c,d)\}^{\varnothing},\{TC(\bar{a})(c,d)\}^{\varnothing}}\,\bullet
$$

This is an example of the more general 'rule permutations' available in HTC, hinting at a more flexible proof theory (we discuss this further in Sect. 8).

Example 21 (Transitivity). TC can be proved transitive by way of a cyclic proof in TC_G of the sequent $\overline{TC}(a)(c,d), \overline{TC}(a)(d,e), TC(\bar{a})(c,e)$. As in the previous example we may mimic that proof line by line, but we give a slightly different one that cannot directly be interpreted as a TC_G proof:

$$
\mathit{TC}\,\dfrac{\overline{\mathit{TC}}\,\dfrac{2\cup\,\dfrac{\mathsf{inst}\,\dfrac{\cup\,\dfrac{\mathsf{id}\,\dfrac{\mathsf{init}\,\dfrac{}{\{\}^{\varnothing}}}{a(c,d),\bar{a}(c,d)} \qquad \dfrac{\text{Ex. 20}}{TC(\bar{a})(d,e),\overline{TC}(a)(d,e)}}{a(c,d),\overline{TC}(a)(d,e),\bar{a}(c,d),TC(\bar{a})(d,e)}}{a(c,d),\overline{TC}(a)(d,e),\{\bar{a}(c,x),TC(\bar{a})(x,e)\}^{x}} \qquad \mathsf{inst}\,\dfrac{\cup\,\dfrac{\mathsf{id}\,\dfrac{\mathsf{init}\,\dfrac{}{\{\}^{\varnothing}}}{a(c,c'),\bar{a}(c,c')} \qquad \mathit{TC}\,\dfrac{\vdots}{\overline{TC}(a)(c',d),\overline{TC}(a)(d,e),TC(\bar{a})(c',e)}\,\circ}{a(c,c'),\overline{TC}(a)(c',d),\overline{TC}(a)(d,e),\bar{a}(c,e),\bar{a}(c,c'),TC(\bar{a})(c',e)}}{a(c,c'),\overline{TC}(a)(c',d),\overline{TC}(a)(d,e),\bar{a}(c,e),\{\bar{a}(c,x),TC(\bar{a})(x,e)\}^{x}}}{a(c,d),a(c,c'),a(c,d),\overline{TC}(a)(c',d),\overline{TC}(a)(d,e),\bar{a}(c,e),\{\bar{a}(c,x),TC(\bar{a})(x,e)\}^{x}}}{\overline{TC}(a)(c,d),\overline{TC}(a)(d,e),\bar{a}(c,e),\{\bar{a}(c,x),TC(\bar{a})(x,e)\}^{x}}}{\overline{TC}(a)(c,d),\overline{TC}(a)(d,e),TC(\bar{a})(c,e)}\,\circ
$$

The only infinite branch (except for that from Example 20), looping on $\circ$, is progressing by the red hypertrace.

Finally, it is pertinent to revisit the 'counterexample' (4) that witnessed incompleteness of TC_G for PDL$^+$. The following result is, in fact, already implied by our later completeness result, Theorem 28, but we shall present it nonetheless:

Proposition 22. $\mathsf{HTC} \vdash_{cyc} \mathsf{ST}((aa \cup aba)^+)(c,d) \supset \mathsf{ST}(a^+((ba^+)^+ \cup a))(c,d)$.

Proof. We give the required cyclic proof in Fig. 2, using the abbreviations: $\alpha(c,d) = \mathsf{ST}(aa \cup aba)(c,d)$ and $\beta(c,d) = \mathsf{ST}((ba^+)^+ \cup a)(c,d)$. The only infinite branch (looping on $\bullet$) has progressing hypertrace is marked in blue. Hypersequents $\mathbf{R} = \{\overline{\alpha}(c,d)\}^{\varnothing}, \{\overline{\alpha}(c,d), \overline{TC}(\overline{\alpha})(e,d)\}^{\varnothing}, \{TC(a)(c,y), \beta(y,d)\}^{y}$ and $\mathbf{R}' = \{\overline{\alpha}(c,d)\}^{\varnothing}, \{\overline{\alpha}(c,d)\}^{\varnothing}, \{TC(a)(c,y), \beta(y,d)\}^{y}$ have finitary proofs, while $\mathbf{P} = \{\overline{aba}(c,e)\}^{\varnothing}, \{\overline{TC}(\overline{\alpha})(e,d)\}^{\varnothing}, \{TC(a)(c,y), \beta(y,d)\}^{y}$ has a cyclic proof.

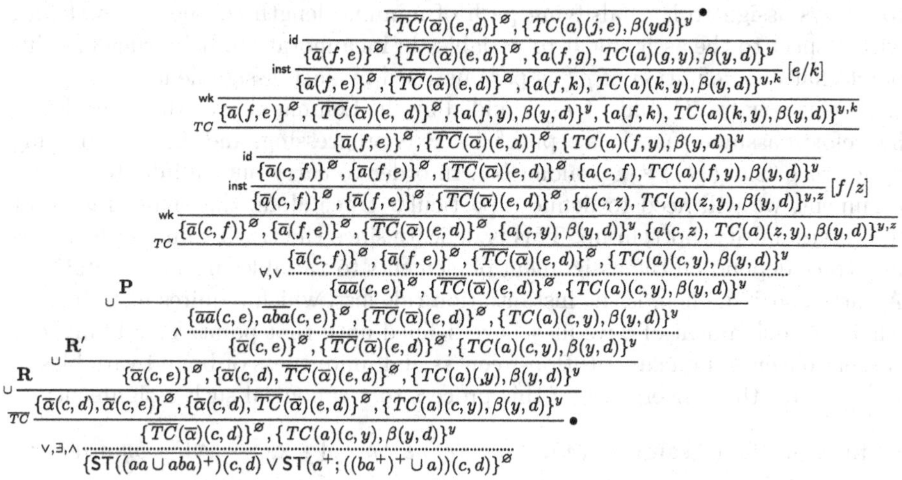

Fig. 2. Cyclic proof for sequent not cyclically provable by TC_G.

5 Soundness of HTC

This section is devoted to the proof of the first of our main results:

Theorem 23 (Soundness). *If* $\mathsf{HTC} \vdash_{nwf} \mathbf{S}$ *then* $\models \mathbf{S}$.

The argument is quite technical due to the alternating nature of our progress condition. In particular the treatment of traces within hypertraces requires a more fine grained argument than usual, bespoke to our hypersequential structure.

Throughout this section, we shall fix a HTC preproof $\mathcal{D}$ of a hypersequent $\mathbf{S}$. For practical reasons we shall assume that $\mathcal{D}$ is substitution-free (at the cost of regularity) and that each quantifier in $\mathbf{S}$ binds a distinct variable.[4] We further assume some structure $\mathcal{M}^{\times}$ and an interpretation ρ_0 such that $\rho_0 \not\models \mathbf{S}$ (within $\mathcal{M}^{\times}$). Since each rule is locally sound, by contraposition we can continually choose 'false premisses' to construct an infinite 'false branch':

Lemma 24 (Countermodel branch). *There is a branch* $\mathcal{B}^{\times} = (\mathbf{S}_i)_{i<\omega}$ *of* $\mathcal{D}$ *and an interpretation* $\rho^{\times}$ *such that, with respect to* $\mathcal{M}^{\times}$*:*

[4] Note that this convention means we can simply take $y = x$ in the $\exists$ rule in Fig. 1.

1. $\rho^\times \not\models \mathbf{S}_i$, *for all* $i < \omega$;
2. *Suppose that* $\mathbf{S}_i$ *concludes a* $\overline{TC}$ *step, as typeset in Fig. 1, and* $\rho^\times \models TC(\bar{A})(s,t)\,[\mathbf{d}/\mathbf{x}]$. *If* n *is minimal such that* $\rho^\times \models \bar{A}(d_i, d_{i+1})$ *for all* $i \le n$, $\rho^\times(s) = d_0$ *and* $\rho^\times(t) = d_n$, *and* $n > 1$, *then* $\rho^\times(f)(\mathbf{d}) = d_1$[5] *so that* $\rho_{i+1} \models \bar{A}(s, f(\mathbf{x}))[\mathbf{d}/\mathbf{x}]$ *and* $\rho^\times \models TC(\bar{A})(f(\mathbf{x}), t)[\mathbf{d}/\mathbf{x}]$.

Unpacking this a little, our interpretation $\rho^\times$ is actually defined as the limit of a chain of 'partial' interpretations $(\rho_i)_{i<\omega}$, with each $\rho_i \not\models \mathbf{S}_i$ (within $\mathcal{M}^\times$). Note in particular that, by 2, whenever some $\overline{TC}$-formula is principal, we choose ρ_{i+1} to always assign to it a falsifying path of minimal length (if one exists at all), with respect to the assignment to variables in its annotation. It is crucial at this point that our definition of $\rho^\times$ is parametrised by such assignments.

Let us now fix $\mathcal{B}^\times$ and $\rho^\times$ as provided by the Lemma above. Moreover, let us henceforth assume that $\mathcal{D}$ is a proof, i.e. it is progressing, and fix a progressing hypertrace $\mathcal{H} = (\{\Gamma_i\}^{\mathbf{x}_i})_{i<\omega}$ along $\mathcal{B}^\times$. In order to carry out an infinite descent argument, we will need to define a particular trace along this hypertrace that 'preserves' falsity, bottom-up. This is delicate since the truth values of formulas in a trace depend on the assignment of elements to variables in the annotations. A particular issue here is the instantiation rule inst, which requires us to 'revise' whatever assignment of y we may have defined until that point. Thankfully, our earlier convention on substitution-freeness and uniqueness of bound variables in $\mathcal{D}$ facilitates the convergence of this process to a canonical such assignment:

Definition 25 (Assignment). *We define* $\delta_\mathcal{H} : \bigcup_{i<\omega} \mathbf{x}_i \to |\mathcal{M}^\times|$ *by* $\delta_\mathcal{H}(x) := \rho(t)$ *if* x *is instantiated by* t *in* $\mathcal{H}$; *otherwise* $\delta_\mathcal{H}(x)$ *is some arbitrary* $d \in |\mathcal{M}^\times|$.

Note that $\delta_\mathcal{H}$ is indeed well-defined, thanks to the convention that each quantifier in $\mathbf{S}$ binds a distinct variable. In particular we have that each variable x is instantiated at most once along a hypertrace. Henceforth we shall simply write $\rho, \delta_\mathcal{H} \models A(\mathbf{x})$ instead of $\rho \models A(\delta_\mathcal{H}(\mathbf{x}))$. Working with such an assignment ensures that false formulas along $\mathcal{H}$ always have a false immediate ancestor:

Lemma 26 (Falsity through $\mathcal{H}$). *If* $\rho^\times, \delta_\mathcal{H} \not\models F$ *for some* $F \in \Gamma_i$, *then* F *has an immediate ancestor* $F' \in \Gamma_{i+1}$ *with* $\rho^\times, \delta_\mathcal{H} \not\models F'$.

In particular, regarding the inst rule of Fig. 1, note that if $F \in \Gamma(y)$ then we can choose $F' = F[t/y]$ which, by definition of $\delta_\mathcal{H}$, has the same truth value. By repeatedly applying this Lemma we obtain:

Proposition 27 (False trace). *There exists an infinite trace* $\tau^\times = (F_i)_{i<\omega}$ *through* $\mathcal{H}$ *such that, for all* i, *it holds that* $\mathcal{M}^\times, \rho^\times, \delta_\mathcal{H} \not\models F_i$.

We are now ready to prove our main soundness result.

Proof (of Theorem 23, sketch). Fix the infinite trace $\tau^\times = (F_i)_{i<\omega}$ through $\mathcal{H}$ obtained by Proposition 27. Since $\tau^\times$ is infinite, by definition of HTC proofs, it

[5] To be clear, we here choose an arbitrary such minimal '$\bar{A}$-path'.

needs to be progressing, i.e., it is infinitely often $\overline{TC}$-principal and there is some $k \in \mathbb{N}$ s.t. for $i > k$ we have that $F_i = \overline{TC}(A)(s_i, t_i)$ for some terms s_i, t_i.

To each F_i, for $i > k$, we associate the natural number n_i measuring the '$\bar{A}$-distance between s_i and t_i'. Formally, $n_i \in \mathbb{N}$ is least such that there are $d_0, \dots, d_{n_i} \in |\mathcal{M}^\times|$ with $\rho^\times(s) = d_0, \rho^\times(t) = d_{n_i}$ and, for all $i < n_i$, $\rho^\times, \delta_{\mathcal{H}} \models \bar{A}(d_i, d_{i+1})$. Our aim is to show that $(n_i)_{i>k}$ has no minimal element, contradicting wellfoundness of $\mathbb{N}$. For this, we establish the following two local properties:

$$\text{id}\ \frac{}{p, \bar{p}} \qquad \text{wk}\ \frac{\Gamma}{\Gamma, A} \qquad \mathsf{k}_a\ \frac{\Gamma, A}{\langle a\rangle\Gamma, [a]A} \qquad \wedge\ \frac{\Gamma, A \quad \Gamma, B}{\Gamma, A \wedge B} \qquad \vee_0\ \frac{\Gamma, A_0}{\Gamma, A_0 \vee A_1} \qquad \vee_1\ \frac{\Gamma, A_1}{\Gamma, A_0 \vee A_1}$$

$$\langle;\rangle\ \frac{\Gamma, \langle\alpha\rangle\langle\beta\rangle A}{\Gamma, \langle\alpha;\beta\rangle A} \qquad \langle\cup\rangle_0\ \frac{\Gamma, \langle\alpha_0\rangle A}{\Gamma, \langle\alpha_0 \cup \alpha_1\rangle A} \qquad \langle\cup\rangle_1\ \frac{\Gamma, \langle\alpha_1\rangle A}{\Gamma, \langle\alpha_0 \cup \alpha_1\rangle A} \qquad [\cup]\ \frac{\Gamma, [\alpha]A \quad \Gamma, [\beta]A}{\Gamma, [\alpha \cup \beta]A}$$

$$[;]\ \frac{\Gamma, [\alpha][\beta]A}{\Gamma, [\alpha;\beta]A} \qquad \langle+\rangle_0\ \frac{\Gamma, \langle\alpha\rangle A}{\Gamma, \langle\alpha^+\rangle A} \qquad \langle+\rangle_1\ \frac{\Gamma, \langle\alpha\rangle\langle\alpha^+\rangle A}{\Gamma, \langle\alpha^+\rangle A} \qquad [+]\ \frac{\Gamma, [\alpha]A \quad \Gamma, [\alpha][\alpha^+]A}{\Gamma, [\alpha^+]A}$$

Fig. 3. Rules of LPD^+.

1. $(n_i)_{i>k}$ is *monotone decreasing*, i.e., for all $i > k$, we have $n_{i+1} \leq n_i$;
2. Whenever F_i is principal, we have $n_{i+1} < n_i$.

So $(n_i)_{i>k}$ is monotone decreasing, by 1, but cannot converge, by 2 and the definition of progressing trace. Thus $(n_i)_{k<i}$ has no minimal element, yielding the required contradiction.

6 HTC is Complete for PDL$^+$, Over Standard Translation

In this section we give our next main result:

Theorem 28 (Completeness for PDL$^+$**).** *For a* PDL$^+$ *formula A, if $\models A$ then* $\mathsf{HTC} \vdash_{cyc} \mathsf{ST}(A)(c)$.

The proof is by a direct simulation of a cut-free cyclic system for PDL$^+$ that is complete. We shall briefly sketch this system below.

6.1 Circular System for PDL$^+$

The system LPD^+, given in Fig. 3, is the natural extension of the usual sequent calculus for basic multimodal logic K by rules for programs. In Fig. 3, $\langle a\rangle\Gamma$ is shorthand for $\{\langle a\rangle B : B \in \Gamma\}$. (Regular) preproofs for this system are defined just like for HTC or TC_G. The notion of 'immediate ancestor' is induced by the indicated colouring: a formula C in a premiss is an immediate ancestor of a formula C' in the conclusion if they have the same colour; if $C, C' \in \Gamma$ then we furthermore require $C = C'$.

Definition 29 (Non-wellfounded proofs). *Fix a preproof $\mathcal{D}$ of a sequent Γ. A* thread *is a maximal path in its graph of immediate ancestry. We say a thread is* progressing *if it has a smallest infinitely often principal formula of the form $[\alpha^+]A$. $\mathcal{D}$ is a* proof *if every infinite branch has a progressing thread. If $\mathcal{D}$ is regular, we call it a* cyclic proof *and we may write $\mathsf{LPD}^+ \vdash_{cyc} \Gamma$.*

Soundness of cyclic-LPD^+ is established by a standard infinite descent argument, but is also implied by the soundness of cyclic-HTC (Theorem 23) and the simulation we are about to give (Theorem 28), though this is somewhat overkill. Completeness may be established by the game theoretic approach of Niwinskí and Walukiewicz [23], as done by Lange [20] for PDL (with identity), or by purely proof theoretic techniques of Studer [25]. Either way, both results follow from a standard embedding of PDL$^+$ into the μ-calculus and its known completeness results [23,25], by way of a standard 'proof reflection' argument: μ-calculus proofs of the embedding are 'just' step-wise embeddings of LPD^+ proofs:

Theorem 30 (Soundness and completeness, [20]). *Let A be a PDL$^+$ formula. $\models A$ iff $\mathsf{LPD}^+ \vdash_{cyc} A$.*

6.2 A 'Local' Simulation of LPD^+ by HTC

In this subsection we show that LPD^+-preproofs can be stepwise transformed into HTC-proofs, with respect to the standard translation. In order to produce this local simulation, we need a more refined version of the standard translation that incorporates the structural elements of hypersequents.

Fix a PDL$^+$ formula $A = [\alpha_1] \dots [\alpha_n]\langle\beta_1\rangle \dots \langle\beta_m\rangle B$, for $n, m \geq 0$. The *hypersequent translation* of A, written $\mathsf{HT}(A)(c)$, is defined as:

$$\{\overline{\mathsf{ST}(\alpha_1)(c, d_1)}\}^{\varnothing}, \{\overline{\mathsf{ST}(\alpha_2)(d_1, d_2)}\}^{\varnothing}, \dots, \{\overline{\mathsf{ST}(\alpha_n)(d_{n-1}, d_n)}\}^{\varnothing},$$
$$\{\mathsf{ST}(\beta_1)(d_n, y_1), \mathsf{ST}(\beta_2)(y_2, y_3), \dots, \mathsf{ST}(\beta_m)(y_{m-1}, y_m), \mathsf{ST}(B)(y_m)\}^{y_1, \dots, y_m}$$

For $\Gamma = A_1, \dots, A_k$, we write $\mathsf{HT}(\Gamma)(c) := \mathsf{HT}(A_1)(c), \dots, \mathsf{HT}(A_k)(c)$.

Definition 31 (HT-translation). *Let $\mathcal{D}$ be a PDL$^+$ preproof. We shall define a HTC preproof $\mathsf{HT}(\mathcal{D})(c)$ of the hypersequent $\mathsf{HT}(A)(c)$ by a local translation of inference steps. We give only a few of the important cases here, but a full definition can be found in [13].*

– *A step* $\mathsf{k}_a \dfrac{B_1, \dots, B_k, A}{\langle a\rangle B_1, \dots, \langle a\rangle B_k, [a]A}$ *is translated to:*

$$
\begin{array}{rc}
 & \mathsf{HT}(B_1)(c), \dots, \mathsf{HT}(B_k)(c), \mathsf{HT}(A)(c) \\ \hline
{\scriptstyle [d/c]} & \mathsf{HT}(B_1)(d), \dots, \mathsf{HT}(B_k)(d), \mathsf{HT}(A)(d) \\ \hline\hline
{\scriptstyle \vee,\forall} & \{\mathsf{CT}(B_1)(d)\}^{\mathbf{x}_{B_1}}, \dots, \{\mathsf{CT}(B_k)(d)\}^{\mathbf{x}_{B_k}}, \mathsf{HT}(A)(d) \\ \hline
{\scriptstyle \mathsf{wk}} & \{\mathsf{CT}(B_1)(d)\}^{\mathbf{x}_{B_1}}, \dots, \{\mathsf{CT}(B_k)(d)\}^{\mathbf{x}_{B_k}}, \{\overline{\mathsf{ST}(a)(c, d)}\}^{\varnothing}, \mathsf{HT}(A)(d) \\ \hline\hline
{\scriptstyle \cup} & \{\mathsf{CT}(B_1)(d)\}^{\mathbf{x}_{B_1}}, \dots, \{\mathsf{ST}(a)(c, d), \mathsf{CT}(B_k)(d)\}^{\mathbf{x}_{B_k}}, \{\overline{\mathsf{ST}(a)(c, d)}\}^{\varnothing}, \mathsf{HT}(A)(d) \\ \hline\hline
{\scriptstyle \mathsf{inst}} & \{\mathsf{ST}(a)(c, y), \mathsf{CT}(B_1)(y)\}^{\mathbf{x}_{B_1}, y}, \dots, \{\mathsf{ST}(a)(c, y), \mathsf{CT}(B_k)(y)\}^{\mathbf{x}_{B_k}, y}, \{\overline{\mathsf{ST}(a)(c, d)}\}^{\varnothing}, \mathsf{HT}(A)(d) \\
{\scriptstyle =} & \cdots\cdots\cdots\cdots\cdots\cdots\cdots\cdots\cdots\cdots\cdots\cdots\cdots\cdots\cdots\cdots \\
 & \mathsf{HT}(\langle a\rangle B_1)(c), \dots, \mathsf{HT}(\langle a\rangle B_k)(c), \mathsf{HT}([a]A)(c)
\end{array}
$$

where (omitted) left-premisses of $\cup$ *steps are simply proved by* wk, id, init. *In this and the following cases, we use the notation* $\mathsf{CT}(A)(c)$ *and* $\mathbf{x}_A$ *for the appropriate sets of formulas and variables forced by the definition of* HT *(again, see [13] for further details).*

– *A* $\langle \cup \rangle_i$ *step (for* $i = 0, 1$*), as typeset in Fig. 3, is translated to:*

$$
\begin{array}{rc}
 & \mathsf{HT}(\Gamma)(c), \mathsf{HT}(\langle \alpha_i \rangle A)(c) \\ \hline
= & \mathsf{HT}(\Gamma)(c), \{\mathsf{ST}(\alpha_i)(c, y), \mathsf{CT}(A)(y)\}^{\mathbf{x}_B, y} \\ \hline
\vee & \mathsf{HT}(\Gamma)(c), \{\mathsf{ST}(\alpha_0)(c, y) \vee \mathsf{ST}(\alpha_1)(c, y), \mathsf{CT}(A)(y)\}^{\mathbf{x}_A, y} \\ \hline
= & \mathsf{HT}(\Gamma)(c), \mathsf{HT}(\langle \alpha_0 \cup \alpha_1 \rangle A)(c)
\end{array}
$$

– *A* $\langle ; \rangle$ *step, as typeset in Fig. 3, is translated to:*

$$
\begin{array}{rc}
 & \mathsf{HT}(\Gamma)(c), \mathsf{HT}(\langle \alpha \rangle \langle \beta \rangle A)(c) \\ \hline
= & \mathsf{HT}(\Gamma)(c), \{\mathsf{ST}(\alpha)(c, z), \mathsf{ST}(\alpha)(z, y), \mathsf{CT}(A)(y)\}^{\mathbf{x}_A, y, z} \\ \hline
\wedge & \mathsf{HT}(\Gamma)(c), \{\mathsf{ST}(\alpha)(c, z) \wedge \mathsf{ST}(\alpha)(z, y), \mathsf{CT}(A)(y)\}^{\mathbf{x}_A, y, z} \\ \hline
\exists & \mathsf{HT}(\Gamma)(c), \{\exists z(\mathsf{ST}(\alpha)(c, z) \wedge \mathsf{ST}(\alpha)(z, y)), \mathsf{CT}(A)(y)\}^{\mathbf{x}_A, y} \\ \hline
= & \mathsf{HT}(\Gamma)(c), \mathsf{HT}(\langle \alpha ; \beta \rangle A)(c)
\end{array}
$$

– *A* $[+]$ *step, as typeset in Fig. 3, is translated to:*

$$
\begin{array}{rcc}
 & & \mathsf{HT}(\Gamma)(c), \mathsf{HT}([\alpha][\alpha^+]A)(c) \\ \hline
= & \mathcal{E}' & \mathsf{HT}(\Gamma)(c), \{\overline{\mathsf{ST}(\alpha)(c, f)}\}^{\varnothing}, \{\overline{TC}(\overline{\mathsf{ST}(\alpha)})(f, d)\}^{\varnothing}, \mathsf{HT}(A)(d) \\ \hline
\cup & \mathcal{E} & \mathsf{HT}(\Gamma)(c), \{\overline{\mathsf{ST}(\alpha)(c, f)}\}^{\varnothing}, \{\overline{\mathsf{ST}(\alpha)(c, d)}, \overline{TC}(\overline{\mathsf{ST}(\alpha)})(f, d)\}^{\varnothing}, \mathsf{HT}(A)(d) \\ \hline
\cup & & \mathsf{HT}(\Gamma)(c), \{\overline{\mathsf{ST}(\alpha)(c, d)}, \overline{\mathsf{ST}(\alpha)(c, f)}\}^{\varnothing}, \{\overline{\mathsf{ST}(\alpha)(c, d)}, \overline{TC}(\overline{\mathsf{ST}(\alpha)})(f, d)\}^{\varnothing}, \mathsf{HT}(A)(d) \\ \hline
\overline{TC} & & \mathsf{HT}(\Gamma)(c), \{\overline{TC}(\overline{\mathsf{ST}(\alpha)})(c, d)\}^{\varnothing}, \mathsf{HT}(A)(d) \\ \hline
= & & \mathsf{HT}(\Gamma)(c), \mathsf{HT}([\alpha^+]A)(c)
\end{array}
$$

where $\mathcal{E}$ *and* $\mathcal{E}'$ *derive* $\mathsf{HT}(\Gamma)(c)$ *and* $\mathsf{HT}([\alpha]A)(c)$*, resp., using* wk*-steps.*

Note that, formally speaking, the well-definedness of $\mathsf{HT}(\mathcal{D})(c)$ in the definition above is guaranteed by coinduction: each rule of $\mathcal{D}$ is translated into a (nonempty) derivation.

Remark 32 (Deeper inference). Observe that HTC can also simulate 'deeper' program rules than are available in LPD^+. E.g. a rule $\dfrac{\Gamma, \langle \alpha \rangle \langle \beta_i \rangle A}{\Gamma, \langle \alpha \rangle \langle \beta_0 \cup \beta_1 \rangle A}$ may be simulated too (similarly for $[\,]$). E.g. $\langle a^+ \rangle \langle b \rangle p \supset \langle a^+ \rangle \langle b \cup c \rangle p$ admits a *finite* proof in HTC (under ST), rather than a necessarily infinite (but cyclic) one in LPD^+.

6.3 Justifying Regularity and Progress

Proposition 33. *If* $\mathcal{D}$ *is regular, then so is* $\mathsf{HT}(\mathcal{D})(c)$.

Proof. Notice that each rule in $\mathcal{D}$ is translated to a finite derivation in $\mathsf{HT}(\mathcal{D})(c)$. Thus, if $\mathcal{D}$ has only finitely many distinct subproofs, then also $\mathsf{HT}(\mathcal{D})(c)$ has only finitely many distinct subproofs.

Proposition 34. *If $\mathcal{D}$ is progressing, then so is* $\mathsf{HT}(\mathcal{D})(c)$.

Proof (sketch). We need to show that every infinite branch of $\mathsf{HT}(\mathcal{D})(c)$ has a progressing hypertrace. Since the HT translation is defined stepwise on the individual steps of $\mathcal{D}$, we can associate to each infinite branch $\mathcal{B}$ of $\mathsf{HT}(\mathcal{D})(c)$ a unique infinite branch $\mathcal{B}'$ of $\mathcal{D}$. Since $\mathcal{D}$ is progressing, let $\tau = (F_i)_{i<\omega}$ be a progressing thread along $\mathcal{B}'$. By inspecting the rules of LPD^+ (and by definition of progressing thread), for some $k \in \mathbb{N}$, each F_i for $i > k$ has the form: $[\alpha_{i,1}] \cdots [\alpha_{i,n_i}][\alpha^+]A$, for some $n_i \geq 0$. So, for $i > k$, $\mathsf{HT}(F_i)(d_i)$ has the form:

$$\{\overline{\mathsf{ST}(\alpha_{i,1})(c, d_{i,1})}\}^{\varnothing}, \ldots, \{\overline{\mathsf{ST}(\alpha_{i,n_i})(d_{i,n_i-1}, d_{i,n_i})}\}^{\varnothing}, \{\overline{TC}(\overline{\mathsf{ST}(\alpha)})(d_{i,n_i}, d_i)\}^{\varnothing}, \mathsf{HT}(A)(d_i)$$

By inspection of the HT-translation (Definition 31) whenever F_{i+1} is an immediate ancestor of F_i in $\mathcal{B}'$, there is a path from the cedent $\{\overline{TC}(\overline{\mathsf{ST}(\alpha)})(d_{i+1,n_{i+1}}, d_{i+1})\}^{\varnothing}$ to the cedent $\{\overline{TC}(\overline{\mathsf{ST}(\alpha)})(d_{i,n_i}, d_i)\}^{\varnothing}$ in the graph of immediate ancestry along $\mathcal{B}$. Thus, since $\tau = (F_i)_{i<\omega}$ is a trace along $\mathcal{B}'$, we have a (infinite) hypertrace of the form $\mathcal{H}_\tau := (\{\Delta_i, \overline{TC}(\overline{\mathsf{ST}(\alpha)})(d_{i,n_i}, d_i)\}^{\varnothing})_{i>k'}$ along $\mathcal{B}$. By construction $\Delta_i = \varnothing$ for infinitely many $i > k'$, and so $\mathcal{H}_\tau$ has just one infinite trace. Moreover, by inspection of the $[+]$ step in Definition 31, this trace progresses in $\mathcal{B}$ every time τ does in $\mathcal{B}'$, and so progresses infinitely often. Thus, $\mathcal{H}$ is a progressing hypertrace. Since the choice of the branch $\mathcal{B}$ of $\mathcal{D}$ was arbitrary, we are done.

6.4 Putting it all Together

We can now finally conclude our main simulation theorem:

Proof (of Theorem 28, sketch). Let A be a PDL$^+$ formula s.t. $\models A$. By the completeness result for LPD^+, Theorem 30, we have that $\mathsf{LPD}^+ \vdash_{cyc} A$, say by a cyclic proof $\mathcal{D}$. From here we construct the HTC preproof $\mathsf{HT}(\mathcal{D})(c)$ which, by Propositions 33 and 34, is in fact a cyclic proof of $\mathsf{HT}(A)(c)$. Finally, we apply some basic $\vee, \wedge, \exists, \forall$ steps to obtain a cyclic HTC proof of $\mathsf{ST}(A)(c)$.

7 Extension by Equality and Simulating Full PDL

We now briefly explain how our main results are extended to the 'reflexive' version of TCL. The language of $\mathsf{HTC}_=$ allows further atomic formulas of the form $s = t$ and $s \neq t$. The calculus $\mathsf{HTC}_=$ extends HTC by the rules:

$$= \frac{\mathbf{S}, \{\Gamma\}^{\mathsf{x}}}{\mathbf{S}, \{t = t, \Gamma\}^{\mathsf{x}}} \qquad \neq \frac{\mathbf{S}, \{\Gamma(s), \Delta(s)\}^{\mathsf{x}}}{\mathbf{S}, \{\Gamma(s), s \neq t\}^{\mathsf{x}}, \{\Delta(t)\}^{\mathsf{x}}}$$

The notion of immediate ancestry is colour-coded as in Definition 15, and the resulting notions of (pre)proof, (hyper)trace and progress are as in Definition 17. The simulation of Cohen and Rowe's system TC_G extends to their reflexive system, RTC_G, by defining their operator $RTC(\lambda x, y.A)(s,t) := TC(\lambda x, y.(x = y \vee A))(s,t)$. Note that, while it is semantically correct to set $RTC(A)(s,t)$ to be $s = t \vee TC(A)(s,t)$, this encoding does not lift to the Cohen-Rowe rules for RTC. Understanding that structures interpret $=$ as true equality, a modular adaptation of the soundness argument for HTC, cf. Sect. 5, yields:

Theorem 35 (Soundness of $\mathsf{HTC}_=$). *If* $\mathsf{HTC}_= \vdash_{nwf} \mathbf{S}$ *then* $\models \mathbf{S}$.

Turning to the modal setting, PDL may be defined as the extension of PDL$^+$ by including a program $A?$ for each formula A. Semantically, we have $(A?)^{\mathcal{M}} = \{(v, v) : \mathcal{M}, v \models A\}$. From here we may define $\varepsilon := \top?$ and $\alpha^* := (\varepsilon \cup \alpha)^+$; again, while it is semantically correct to set $\alpha^* = \varepsilon \cup \alpha^+$, this encoding does not lift to the standard sequent rules for $*$. The system LPD is obtained from LPD^+ by including the rules:

$$\langle ? \rangle \frac{\Gamma, A \quad \Gamma, B}{\Gamma, \langle A? \rangle B} \qquad [?] \frac{\Gamma, \bar{A}, B}{\Gamma, [A?]B}$$

Again, the notion of immediate ancestry is colour-coded as for LPD^+; the resulting notions of (pre)proof, thread and progress are as in Definition 29. Just like for LPD^+, a standard encoding of LPD into the μ-calculus yields its soundness and completeness, thanks to known sequent systems for the latter, cf. [23,25], but has also been established independently [20]. Again, a modular adaptation of the simulation of LPD^+ by HTC, cf. Sect. 6, yields:

Theorem 36 (Completeness for PDL). *Let A be a PDL formula. If* $\models A$ *then* $\mathsf{HTC}_= \vdash_{cyc} \mathsf{ST}(A)(c)$.

8 Conclusions

In this work we proposed a novel cyclic system HTC for Transitive Closure Logic (TCL) based on a form of hypersequents. We showed a soundness theorem for standard semantics, requiring an argument bespoke to our hypersequents. Our system is cut-free, rendering it suitable for automated reasoning via proof search. We showcased its expressivity by demonstrating completeness for PDL, over the standard translation. In particular, we demonstrated formally that such expressivity is not available in the previously proposed system TC_G of Cohen and Rowe (Theorem 12). Our system HTC locally simulates TC_G too (Theorem 19).

As far as we know, HTC is the first cyclic system employing a form of *deep inference* resembling *alternation* in automata theory, e.g. wrt. proof checking, cf. Proposition 18. It would be interesting to investigate the structural proof theory that emerges from our notion of hypersequent. As hinted at in Examples 20 and 21, our hypersequential system exhibits more liberal rule permutations than usual sequents, so we expect their *focussing* and *cut-elimination* behaviours to similarly be richer, cf. [21,22]. Note however that such investigations are rather pertinent for pure predicate logic (without TC): focussing and cut-elimination arguments do not typically preserve regularity of non-wellfounded proofs, cf. [2].

Finally, our work bridges the cyclic proof theories of (identity-free) PDL and (reflexive) TCL. With increasing interest in both modal and predicate cyclic proof theory, it would be interesting to further develop such correspondences.

Acknowledgements. The authors would like to thank Sonia Marin, Jan Rooduijn and Reuben Rowe for helpful discussions on matters surrounding this work.

References

1. Avron, A.: Transitive closure and the mechanization of mathematics. In: Kamareddine, F.D. (eds) Thirty Five Years of Automating Mathematics. Applied Logic Series, vol. 28, pp. 149–171. Springer, Dordrecht (2003). https://doi.org/10.1007/978-94-017-0253-9_7
2. Baelde, D., Doumane, A., Saurin, A.: Infinitary proof theory: the multiplicative additive case. In: Talbot, J., Regnier, L. (eds.) 25th EACSL Annual Conference on Computer Science Logic, CSL 2016, 29 August–1 September 2016, Marseille, France. LIPIcs, vol. 62, pp. 42:1–42:17. Schloss Dagstuhl - Leibniz-Zentrum für Informatik (2016). https://doi.org/10.4230/LIPIcs.CSL.2016.42
3. Berardi, S., Tatsuta, M.: Classical system of martin-lof's inductive definitions is not equivalent to cyclic proofs. CoRR abs/1712.09603 (2017). http://arxiv.org/abs/1712.09603
4. Blackburn, P., van Benthem, J.: Modal logic: a semantic perspective. In: Blackburn, P., van Benthem, J.F.A.K., Wolter, F. (eds.) Handbook of Modal Logic, Studies in Logic and Practical Reasoning, vol. 3, pp. 1–84. North-Holland (2007). https://doi.org/10.1016/s1570-2464(07)80004-8
5. Blackburn, P., De Rijke, M., Venema, Y.: Modal Logic, vol. 53. Cambridge University Press (2002)
6. Brotherston, J., Distefano, D., Petersen, R.L.: Automated cyclic entailment proofs in separation logic. In: Bjørner, N., Sofronie-Stokkermans, V. (eds.) CADE 2011. LNCS (LNAI), vol. 6803, pp. 131–146. Springer, Heidelberg (2011). https://doi.org/10.1007/978-3-642-22438-6_12
7. Brotherston, J., Gorogiannis, N., Petersen, R.L.: A generic cyclic theorem prover. In: Jhala, R., Igarashi, A. (eds.) APLAS 2012. LNCS, vol. 7705, pp. 350–367. Springer, Heidelberg (2012). https://doi.org/10.1007/978-3-642-35182-2_25
8. Brotherston, J., Simpson, A.: Sequent calculi for induction and infinite descent. J. Log. Comput. **21**(6), 1177–1216 (2011)
9. Cohen, L., Rowe, R.N.S.: Uniform inductive reasoning in transitive closure logic via infinite descent. In: Ghica, D.R., Jung, A. (eds.) 27th EACSL Annual Conference on Computer Science Logic, CSL 2018, 4–7 September 2018, Birmingham, UK. LIPIcs, vol. 119, pp. 17:1–17:16. Schloss Dagstuhl - Leibniz-Zentrum für Informatik (2018). https://doi.org/10.4230/LIPIcs.CSL.2018.17
10. Cohen, L., Rowe, R.N.S.: Integrating induction and coinduction via closure operators and proof cycles. In: Peltier, N., Sofronie-Stokkermans, V. (eds.) IJCAR 2020. LNCS (LNAI), vol. 12166, pp. 375–394. Springer, Cham (2020). https://doi.org/10.1007/978-3-030-51074-9_21
11. Cohen, L., Rowe, R.N.: Non-well-founded proof theory of transitive closure logic. ACM Trans. Comput. Log. **21**(4), 1–31 (2020)
12. Das, A.: On the logical complexity of cyclic arithmetic. Log. Methods Comput. Sci. **16**(1) (2020). https://doi.org/10.23638/LMCS-16(1:1)2020

13. Das, A., Girlando, M.: Cyclic proofs, hypersequents, and transitive closure logic (2022). https://doi.org/10.48550/ARXIV.2205.08616
14. Das, A., Pous, D.: A cut-free cyclic proof system for Kleene algebra. In: Schmidt, R.A., Nalon, C. (eds.) TABLEAUX 2017. LNCS (LNAI), vol. 10501, pp. 261–277. Springer, Cham (2017). https://doi.org/10.1007/978-3-319-66902-1_16
15. Grädel, E.: On transitive closure logic. In: Börger, E., Jäger, G., Kleine Büning, H., Richter, M.M. (eds.) CSL 1991. LNCS, vol. 626, pp. 149–163. Springer, Heidelberg (1992). https://doi.org/10.1007/BFb0023764
16. Gurevich, Y.: Logic and the Challenge of Computer Science, pp. 1–57. Computer Science Press (1988). https://www.microsoft.com/en-us/research/publication/logic-challenge-computer-science/
17. Immerman, N.: Languages that capture complexity classes. SIAM J. Comput. **16**(4), 760–778 (1987). https://doi.org/10.1137/0216051
18. Kozen, D.: A completeness theorem for Kleene algebras and the algebra of regular events. In: Proceedings of the Sixth Annual Symposium on Logic in Computer Science (LICS 1991), Amsterdam, The Netherlands, 15–18 July 1991, pp. 214–225. IEEE Computer Society (1991). https://doi.org/10.1109/LICS.1991.151646
19. Krob, D.: Complete systems of b-rational identities. Theor. Comput. Sci. **89**(2), 207–343 (1991). https://doi.org/10.1016/0304-3975(91)90395-I
20. Lange, M.: Games for modal and temporal logics. Ph.D. thesis (2003)
21. Marin, S., Miller, D., Volpe, M.: A focused framework for emulating modal proof systems. In: Beklemishev, L.D., Demri, S., Maté, A. (eds.) Advances in Modal Logic 11, Proceedings of the 11th Conference on "Advances in Modal Logic," held in Budapest, Hungary, 30 August–2 September 2016, pp. 469–488. College Publications (2016). http://www.aiml.net/volumes/volume11/Marin-Miller-Volpe.pdf
22. Miller, D., Volpe, M.: Focused labeled proof systems for modal logic. In: Davis, M., Fehnker, A., McIver, A., Voronkov, A. (eds.) LPAR 2015. LNCS, vol. 9450, pp. 266–280. Springer, Heidelberg (2015). https://doi.org/10.1007/978-3-662-48899-7_19
23. Niwiński, D., Walukiewicz, I.: Games for the mu-calculus. Theor. Comput. Sci. **163**(1), 99–116 (1996). https://doi.org/10.1016/0304-3975(95)00136-0
24. Rowe, R.N.S., Brotherston, J.: Automatic cyclic termination proofs for recursive procedures in separation logic. In: Bertot, Y., Vafeiadis, V. (eds.) Proceedings of the 6th ACM SIGPLAN Conference on Certified Programs and Proofs, CPP 2017, Paris, France, 16–17 January 2017, pp. 53–65. ACM (2017). https://doi.org/10.1145/3018610.3018623
25. Studer, T.: On the proof theory of the modal mu-calculus. Stud. Logica. **89**(3), 343–363 (2008)
26. Tellez, G., Brotherston, J.: Automatically verifying temporal properties of pointer programs with cyclic proof. In: de Moura, L. (ed.) CADE 2017. LNCS (LNAI), vol. 10395, pp. 491–508. Springer, Cham (2017). https://doi.org/10.1007/978-3-319-63046-5_30
27. Tellez, G., Brotherston, J.: Automatically verifying temporal properties of pointer programs with cyclic proof. J. Autom. Reason. **64**(3), 555–578 (2020). https://doi.org/10.1007/s10817-019-09532-0

Equational Unification and Matching, and Symbolic Reachability Analysis in Maude 3.2 (System Description)

Francisco Durán[1], Steven Eker[2], Santiago Escobar[3](✉), Narciso Martí-Oliet[4], José Meseguer[5], Rubén Rubio[4], and Carolyn Talcott[2]

[1] Universidad de Málaga, Málaga, Spain
duran@lcc.uma.es
[2] SRI International, Menlo Park, CA, USA
eker@csl.sri.com, clt@cs.stanford.edu
[3] VRAIN, Universitat Politècnica de València, Valencia, Spain
sescobar@upv.es
[4] Universidad Complutense de Madrid, Madrid, Spain
{narciso,rubenrub}@ucm.es
[5] University of Illinois at Urbana-Champaign, Urbana, IL, USA
meseguer@illinois.edu

Abstract. Equational unification and matching are fundamental mechanisms in many automated deduction applications. Supporting them efficiently for as wide as possible a class of equational theories, and in a typed manner supporting type hierarchies, benefits many applications; but this is both challenging and nontrivial. We present Maude 3.2's efficient support of these features as well as of symbolic reachability analysis of infinite-state concurrent systems based on them.

1 Introduction

Unification is a key mechanism in resolution [41] and paramodulation-based [36] theorem proving. Since Plotkin's work [40] on *equational unification*, i.e.,

Durán was supported by the grant UMA18-FEDERJA-180 funded by J. Andalucía/FEDER and the grant PGC2018-094905-B-I00 funded by MCIN/AEI/10.13039/501100011033 and ERDF A way of making Europe. Escobar was supported by the EC H2020-EU grant 952215, by the grant RTI2018-094403-B-C32 funded by MCIN/AEI/10.13039/501100011033 and ERDF A way of making Europe, by the grant PROMETEO/2019/098 funded by Generalitat Valenciana, and by the grant PCI2020-120708-2 funded by MICIN/AEI/10.13039/501100011033 and by the European Union NextGenerationEU/PRTR. Martí-Oliet and Rubio were supported by the grant PID2019-108528RB-C22 funded by MCIN/AEI/10.13039/501100011033 and ERDF A way of making Europe. Talcott was partially supported by the U. S. Office of Naval Research under award numbers N00014-15-1-2202 and N00014-20-1-2644, and NRL grant N0017317-1-G002.

J. Blanchette et al. (Eds.): IJCAR 2022, LNAI 13385, pp. 529–540, 2022.
https://doi.org/10.1007/978-3-031-10769-6_31

E-unification modulo an equational theory E, it is widely used for increased effectiveness. Since Walther's work [47] it has been well understood that *typed* E-unification, exploiting types and subtype hierarchies, can drastically reduce a prover's search space. Many other automated deduction applications use typed E-unification as a key mechanism, including, inter alia: (i) constraint logic programming, e.g., [12,23]; (ii) narrowing-based infinite-state reachability analysis and model checking, e.g., [6,35]; (iii) cryptographic protocol analysis modulo algebraic properties, e.g., [8,19,28]; (iv) partial evaluation, e.g., [4,5]; and (v) SMT solving, e.g., [32,48]. The special case of typed E-*matching* is also a key component in all the above areas as well as in: (vi) E-generalization (also called anti-unification), e.g., [1,2]; and (vii) E-homeomorphic embedding, e.g., [3].

Maximizing the scope and effectiveness of typed E-unification and E-matching means efficiently supporting as wide a class of theories E as possible. Such efficiency crucially depends on both efficient algorithms (and their combinations) and —since the number of E-unifiers may be large— on computing complete *minimal* sets of solutions to reduce the search space. The recent Maude 3.2 release[1] provides this kind of efficient support for typed E-unification and E-matching in three, increasingly more general classes of theories E:

1. Typed B-unification and B-matching, where B is any combination of associativity (A) and/or commutativity (C) and/or unit element (U) axioms.
2. Typed $E \cup B$-unification and matching in the *user-definable* infinite class of theories $E \cup B$ with B as in (1), and $E \cup B$ having the *finite variant property* (FVP) [13,21].
3. Typed $E \cup B$-unification for the infinite class of *user-definable* theories $E \cup B$ with B as in (1), and E confluent, terminating, and coherent modulo B.

For classes (1) and (2) the set of B- (resp. $E \cup B$-) unifiers is always *complete, minimal and finite*, except for the *AwoC* case when B contains an A but not C axiom for some binary symbol f.[2] The typing is order-sorted [22,29] and thus contains many-sorted and unsorted B- (resp. $E \cup B$-) unification as special cases. For class (3), Maude enumerates a possibly infinite complete set of $E \cup B$-unifiers, with the same AwoC exception on B. We discuss new features for classes (1)–(2), and a new narrowing modulo $E \cup B$-based *symbolic reachability analysis* feature for infinite-state systems specified in Maude as rewrite theories $(\Sigma, E \cup B, R)$ with equations $E \cup B$ in class (2) and concurrent transition rules R. In Sect. 5 we discuss various applications that can benefit from these new features.

In comparison with previous Maude tool papers reporting on new features —the last one was [16]— the new features reported here include: (i) computing minimal complete sets of most general B- (resp. $E \cup B$-) unifiers for classes (1) and (2) except for the AwoC case; (ii) a new $E \cup B$-matching algorithm for class (2); and (iii) a new symbolic reachability analysis for concurrent systems

[1] Publicly available at http://maude.cs.illinois.edu.

[2] In the AwoC case, Maude's algorithms are optimized to favor many commonly occurring cases where typed A-unification is finitary, and provides a finite set of solutions and an incompleteness warning outside such cases (see [18]).

based on narrowing with transition rules modulo equations $E \cup B$ in class (2) enjoying powerful state-space reduction capabilities based on the minimality and completeness feature (i) and on "folding" less general symbolic states into more general ones through subsumption. Section 3.1 shows the importance of the new $E \cup B$-matching algorithm for efficient computation of minimal $E \cup B$-unifiers.

Notation, Strict-B-Coherence, and FVP. For notation involving either term positions, $p \in pos(t)$, $t|_p$, $t[t']_p$, or substitutions, $t\theta$, $\theta\mu$, see [14]. Equations $(u = v) \in E$ oriented as rules $(u \to v) \in \vec{E}$ are *strictly coherent* modulo axioms B iff $(t =_B t' \wedge t \to_{\vec{E},B} w) \Rightarrow \exists w'(t \to_{\vec{E},B} w' \wedge w =_B w')$, where $t \to_{\vec{E},B} w$ iff $\exists (u \to v) \in \vec{E}$, $\exists \theta$, $\exists p \in pos(t)(u\theta =_B t|_p \wedge w = t[v\theta]_p)$. For $(\Sigma, E \cup B)$ an equational theory with $\vec{E}$ confluent, terminating and strictly coherent modulo B, (1) an $\vec{E}, B$-*t-variant* is a pair (v, θ) s.t. $v = (t\theta)!_{\vec{E},B} \wedge \theta = \theta!_{\vec{E},B}$, where $u!_{\vec{E},B}$ (resp. $\theta!_{\vec{E},B}$) denotes the $\vec{E}, B$-normal form of u, resp. θ; (2) for $\vec{E}, B$-t-variants (v, θ), (u, μ), the *more general relation* $(v, \theta) \sqsupseteq_B (u, \mu)$ holds iff $\exists \gamma (u =_B v\gamma \wedge \theta\gamma =_B \mu)$; (3) $(\Sigma, E \cup B)$ is *FVP* [13,21] iff any Σ-term t has a *finite* set of most general $\vec{E}, B$-t-variants. Footnote 5 explains how FVP can be checked.

2 Complete and Minimal Order-Sorted B-Unifiers

Throughout the paper we use the following equational theory $E \cup B$ of the Booleans as a running example (with self-explanatory, user-definable syntax[3]):

```
fmod BOOL-FVP is protecting TRUTH-VALUE .
   op _and_ : Bool Bool -> Bool [assoc comm] .
   op _xor_ : Bool Bool -> Bool [assoc comm] .
   op not_ : Bool -> Bool .
   op _or_ : Bool Bool -> Bool .
   op _<=>_ : Bool Bool -> Bool .
   vars X Y Z W : Bool .

   eq X and true = X [variant] .
   eq X and false = false [variant] .
   eq X and X = X  [variant] .
   eq X and X and Y = X and Y [variant] .    *** AC extension
   eq X xor false = X [variant] .
   eq X xor X = false [variant] .
   eq X xor X xor Y = Y [variant] .          *** AC extension
   eq not X = X xor true [variant] .
   eq X or Y = (X and Y) xor X xor Y [variant] .
   eq X <=> Y = true xor X xor Y [variant] .
endfm
```

[3] This module imports Maude's `TRUTH-VALUE` module and the command "`set include BOOL off .`" must be typed before the module to avoid default importation of `BOOL`.

The axioms B are the associativity-commutativity (AC) axioms for `xor` and `and` (specified with the `assoc comm` attributes). The equations E are terminating and confluent modulo B [42]. To achieve strict *B-coherence* [30], the needed AC-extensions [39] are added —for example, the AC-extension of `X xor X = false` is `X xor X xor Y = Y`. The equations E for `xor` and `and` define the theory of *Boolean rings, except for the missing*[4] *distributivity equation* `X and (Y xor Z) = (X and Y) xor (X and Z)`. The remaining equations in E define `or`, `not` and `<=>` as definitional extensions. The `variant` attribute declares that the equation will be used for folding variant narrowing [21]. The theory is FVP,[5] in class (2). In this section we will consider B-unification (for $B = AC$) using this example. $E \cup B$-unification for the same example will be discussed in Sect. 3.

For B any combination of associativity and/or commutativity and/or identity axioms, Maude's `unify` command computes a complete finite set of most general B-unifiers, except for the AwoC case. The new `irredundant unify` command always returns[6] a *finite, complete and minimal* set of B-unifiers, except for the AwoC case. The output of `unify` for the equation below can be found in [10, §13].

```
Maude> irredundant unify X and not Y and not Z =? W and Y and not X .
Decision time: 0ms cpu (0ms real)

Unifier 1                               Unifier 2
X --> #1:Bool and #2:Bool               X --> #2:Bool
Z --> #1:Bool and #2:Bool               Z --> #1:Bool
Y --> #1:Bool                           Y --> #2:Bool
W --> #2:Bool and not #1:Bool           W --> not #1:Bool
```

3 $E \cup B$-Unification and Matching for FVP Theories

It is a general result from [21] that if $E \cup B$ is FVP and B-unification is finitary, then $E \cup B$-unification is *finitary* and a complete finite set of $E \cup B$-unifiers can be computed by *folding variant narrowing* [21]. Furthermore, assuming that $T_{\Sigma/E,s}$ is non-empty for each sort s, a finitary $E \cup B$-unification algorithm automatically provides a decision procedure for *satisfiability* of any *positive* (the $\wedge, \vee$-fragment) quantifier-free formula φ in the initial algebra $T_{\Sigma/E}$, since φ can be put in DNF, and a conjunction of equalities Γ is satisfiable in $T_{\Sigma/E}$ iff Γ is $E \cup B$-unifiable.

Since for our running example `BOOL-FVP` the equations $E \cup B$ are FVP and B-unification (in this case $B = AC$) is finitary, all this has useful consequences for

[4] By missing distributivity, this theory is *weaker* than the theory of Boolean rings. Nevertheless, its *initial algebra* $T_{\Sigma/E \cup B}$ is exactly the Booleans on {`true`,`false`} with the standard truth tables for all connectives. Thus, all equations provable in Boolean algebra hold in $T_{\Sigma/E \cup B}$, including the missing distributivity equation.

[5] This can be easily checked in Maude by checking the finiteness of the variants for each $f(X)$, resp. $f(X, Y)$, for each unary, resp. binary, symbol f in `BOOL-FVP` using the `get variants` command; see [9] for a theoretical justification of this check.

[6] Fresh variables follow the form `#1:Bool`.

BOOL-FVP. Indeed, $T_{\Sigma/E \cup B}$ is exactly the Booleans[7] on {true,false} with the well-known truth tables for and, xor, not, or and <=>. This means that $E \cup B$-unification provides a Boolean *satisfiability decision procedure* for a Boolean expression u on such symbols, namely, u is Boolean satisfiable iff the equation $u = \texttt{true}$ is $E \cup B$-unifiable. Furthermore, a ground assignment ρ to the variables of u is a satisfying assignment for u iff there exists an $E \cup B$-unifier α of $u = \texttt{true}$ and a ground substitution δ such that $\rho = \alpha\delta$. For the same reasons, u is a Boolean *tautology* iff the equation $u = \texttt{false}$ has no $E \cup B$-unifiers.

A complete, finite set of $E \cup B$-unifiers can be computed with Maude's variant unify command whenever $E \cup B$ is FVP, except for the AwoC case. Instead, the new[8] filtered variant unify command computes a *finite, complete and minimal* set of $E \cup B$-unifiers, which can be considerably smaller than that computed by variant unify. For our BOOL-FVP example, filtered variant unify gives us a Boolean satisfiability decision procedure plus a symbolic specification of satisfying assignments. Such a procedure is not practical: it cannot compete with standard SAT-solvers; but that was never our purpose: our purpose here is to illustrate with simple examples how $E \cup B$-unification works for the *infinite* class of *user-definable* FVP theories $E \cup B$, of which BOOL-FVP is just a simple example; dozens of other examples can be found in [32].

The difference between the variant unify and the new filtered variant unify command is illustrated with the following example; its unfiltered output can be found in [10, §14]. Note that the single $E \cup B$-unifier gives us a compact symbolic description of this Boolean expression's satisfying assignments.

```
Maude> filtered variant unify (X or Y) <=> Z =? true .
rewrites: 3224 in 12765ms cpu (14776ms real) (252 rewrites/second)

Unifier 1
X --> #1:Bool xor #2:Bool
Y --> #1:Bool
Z --> #2:Bool xor (#1:Bool and (#1:Bool xor #2:Bool))

No more unifiers.
Advisory: Filtering was complete.
```

The computation of a minimal set of $E \cup B$-unifiers relies on filtering by $E \cup B$-matching between two $E \cup B$-unifiers, as explained in the following section.

3.1 FVP $E \cup B$-Matching and Minimality of $E \cup B$-Unifiers

By definition, a term u $E \cup B$-*matches* another term v iff there is a substitution γ such that $u =_{E \cup B} v\gamma$. Besides the existing match command modulo axioms

[7] Each connective's truth table can be checked with Maude's reduce command. Actually, need only check and and xor (other connectives are definitional extensions).

[8] In Maude, different command names are used to emphasize different algorithms. The word 'filtered' is used instead of 'irredundant' because irredundancy is not guaranteed in the AwoC case.

B, Maude's new `variant match` command computes a complete, minimal set of $E \cup B$-*matching substitutions* for any FVP theory $E \cup B$ in class (2), except for the AwoC case. Such an algorithm could always be derived from an $E \cup B$-unification algorithm by replacing u by $\overline{u}$, where all variables in u are replaced by fresh constants in $\overline{u}$, and computing the $E \cup B$-unifiers of $\overline{u} = v$. But a more efficient special-purpose algorithm has been designed and implemented for this purpose. $E \cup B$-matching algorithms are automatically provided by Maude for any *user-definable* theory in class (2) with the `variant match` command.

```
Maude> variant match in BOOL-FVP : Z and W <=? X .
rewrites: 12 in 21ms cpu (27ms real) (545 rewrites/second)

Matcher 1            Matcher 2            Matcher 3
Z --> true           Z --> X              Z --> X
W --> X              W --> true           W --> X
```

This is a good moment to ask and answer a relevant question: Why is computing a complete *minimal* set of $E \cup B$-unifiers for a unification problem Γ, where $E \cup B$ is an FVP theory in class (2) except for the AwoC case, *nontrivial*? We first need to explain how minimality is achieved. Suppose that α and β are two $E \cup B$-unifiers of a system of equations Γ with, say, typed variables $x_1, \ldots, x_n$. We then say that α *is more general than* β modulo $E \cup B$, denoted $\alpha \sqsupseteq_{E \cup B} \beta$, iff there is a substitution γ such that for each x_i, $1 \leq i \leq n$, $\gamma(\alpha(x_i)) =_{E \cup B} \beta(x_i)$. But this exactly means that the vector $[\beta(x_1), \ldots, \beta(x_n)]$ $E \cup B$-matches the vector $[\alpha(x_1), \ldots, \alpha(x_n)]$ with $E \cup B$-matching substitution γ. A complete set of $E \cup B$-unifiers of Γ is by definition *minimal* iff for any two different unifiers α and β in it we have $\alpha \not\sqsupseteq_{E \cup B} \beta$ *and* $\beta \not\sqsupseteq_{E \cup B} \alpha$, i.e., the two associated $E \cup B$-matching problems fail.

What is *nontrivial* is computing a minimal complete set of $E \cup B$-unifiers *efficiently*. One could do so inefficiently by simulating $E \cup B$-matching with $E \cup B$-unification, and more efficiently by using an $E \cup B$-matching algorithm. Maude achieves still greater efficiency by directly computing the $\alpha \sqsupseteq_{E \cup B} \beta$ relation. The key difference between the `variant unify` command and the new **`filtered variant unify`** command is that the second computes a $E \cup B$-minimal set of $E \cup B$-unifiers of Γ using the $\alpha \sqsupseteq_{E \cup B} \beta$ relation, whereas the first only computes a set of B-minimal $E \cup B$-unifiers of Γ using the cheaper $\alpha \sqsupseteq_B \beta$ relation. There are three ideas we use to make it fast in practice: (i) variant matching is faster than variant unification because one side is variable-free; (ii) enumerating the variant matchers between two variant unifiers is far more expensive than checking existence of a matcher; and (iii) variant unifiers are discarded on-the-fly avoiding further narrowing steps and computation.

4 Narrowing-Based Symbolic Reachability Analysis

In Maude, concurrent systems are specified in so-called *system modules* as *rewrite theories* of the form: $\mathcal{R} = (\Sigma, G, R)$, where G is an equational theory either of the form B in class (1), or $E \cup B$ in classes (2) or (3), and R are the *system transition rules*, specified as rewrite rules. When the theory $\mathcal{R}$ is *topmost*, meaning that the rules R rewrite the entire state, narrowing with rules R modulo the equations G is a *complete* symbolic reachability analysis method for *infinite-state systems* [35]. That is, given a term u with variables $\overrightarrow{x}$, representing a typically infinite set of initial states, and another term v with variables $\overrightarrow{y}$, representing a possibly infinite set of target states, narrowing can answer the question: *can an instance of u reach an instance of v?* That is, does the formula $\exists \overrightarrow{x}, \overrightarrow{y} \;\; u \rightarrow^* v$ hold in $\mathcal{R}$? Note that, if the *complement* of a system invariant I can be symbolically described as the set of ground instances of terms in a set $\{v_1, \ldots, v_n\}$ of pattern terms, then narrowing provides a semi-decision procedure for verifying whether the system specified by $\mathcal{R}$ fails to satisfy I starting from an initial set of states specified by u. Namely, I holds iff no instance of any v_i can be reached from some instance of u.

Assuming G is in class (1) or (2), Maude's `vu-narrow` command implements narrowing with R modulo G by performing G-unification at each narrowing step. However, the number of symbolic states that need to be explored can be *infinite*. This means that if no solution exists for the narrowing search, Maude will search forever, so that only *depth-bounded searches* will terminate. The great advantage of the new `{fold} vu-narrow {filter,delay}` command is that it performs a powerful *symbolic state space reduction* by: (i) removing a newly explored symbolic state v' if it $E \cup B$-matches a previously explored state v and replacing transition with target v' by transitions with target v; and (ii) using minimal sets of $E \cup B$-unifiers for each narrowing step and for checking common instances between a newly explored state and the target term (ensured by words `filter` and `delay`). This can make the entire search space finite and allow full verification of invariants for some infinite-state systems. Consider the following Maude specification of Lamport's bakery protocol.

```
mod BAKERY is
  sorts Nat LNat Nat? State WProcs Procs .
  subsorts Nat LNat < Nat? .  subsort WProcs < Procs .
  op 0 : -> Nat .
  op s : Nat -> Nat .
  op [_] : Nat -> LNat .            *** number-locking operator
  op < wait,_> : Nat -> WProcs .
  op < crit,_> : Nat -> Procs .
  op mt : -> WProcs .               *** empty multiset
  op __ : Procs Procs -> Procs [assoc comm id: mt] .    *** union
  op __ : WProcs WProcs -> WProcs [assoc comm id: mt] . *** union
  op _|_|_ : Nat Nat? Procs -> State .
  vars n m i j k : Nat . var x? : Nat? . var PS : Procs . var WPS : WProcs .

  rl [new]:  m | n | PS => s(m) | n | < wait,m > PS [narrowing] .
  rl [enter]: m | n | < wait,n > PS => m | [n] | < crit,n > PS [narrowing] .
  rl [leave]: m | [n] | < crit,n > PS => m | s(n) | PS [narrowing] .
endm
```

The states of BAKERY have the form "m | x? | PS" with m the ticket-dispensing counter, x? the (possibly locked) counter to access the critical section, and PS a multiset of processes either waiting or in the critical section. BAKERY is infinite-state: [new] creates new processes, and the counters can grow unboundedly. When a waiting process enters the critical section with [enter], the second counter n is locked as [n]; and it is unlocked and incremented when it leaves it with [leave]. The key invariant is *mutual exclusion*. Note that the term "i | x? | < crit, j > < crit, k > PS" describes all states in the *complement* of mutual exclusion states. Without the fold option, narrowing does not terminate, but with the following command we can verify that BAKERY satisfies mutual exclusion, not just for the initial state "0 | 0 | mt", but for the much more general infinite set of initial states with waiting processes only "m | n | WPS".

```
Maude> {fold} vu-narrow {filter,delay}
        m | n | WPS =>* i | x? | < crit, j > < crit, k > PS .

No solution.
rewrites: 4 in 1ms cpu (1ms real) (2677 rewrites/second)
```

The new vu-narrow {filter,delay} command can achieve dramatic state space reductions over the previous vu-narrow command by filtering $E \cup B$-unifiers. This is illustrated by a simple cryptographic protocol example in [10, §15] exploiting the unitary nature of unification in the exclusive-or theory [24].

5 Applications and Conclusion

Maude can be used as a meta-tool to develop new formal tools because: (i) its underlying equational and rewriting logics are logical —and reflective meta-logical— frameworks [7,27,46]; (ii) Maude's efficient support of logical reflection through its META-LEVEL module; (iii) Maude's rewriting, search, model checking, and strategy language features [11,15]; and (iv) Maude's symbolic reasoning features [15,33], the latest reported here. We refer to [11,15,31,33] for references on various Maude-based tools. Many of them can benefit from these new features.

By way of example we mention some areas ready to reap such benefits: (1) *Formal Analysis of Cryptographic Protocols.* The new features can yield substantial improvements to tools such as Maude-NPA [19], Tamarin [28] and AKISS [8]. (2) *Model Checking of Infinite-State Systems.* The narrowing-based LTL symbolic model checker reported in [6,20], and the addition of new symbolic capabilities to Real-Time Maude [37,38] can both benefit from the new features. (3) *SMT Solving.* In Sect. 3 we noted that FVP $E \cup B$-unification makes satisfiability of positive QF formulas in $T_{\Sigma/E \cup B}$ decidable. Under mild conditions, this has been extended in [32,44] to a procedure for satisfiability in $T_{\Sigma/E \cup B}$ of all QF formulas which will also benefit from the new features. (4) *Theorem Proving.* The new Maude Inductive Theorem Prover under construction [34], as well as

Maude's Invariant Analyzer [43] and Reachability Logic Theorem Prover [45] all use equational unification and narrowing modulo equations; so all will benefit from the new features. (5) *Theory Transformations* based on equational unification, e.g., partial evaluation [4], ground confluence methods [17] or program termination methods [25,26] could likewise become more efficient.

In conclusion, we have presented and illustrated with examples new equational unification and matching, and symbolic reachability analysis features in Maude 3.2. Thanks to the above-mentioned properties (i)–(iv) of Maude as a meta-tool, we hope that this work will encourage other researchers to use Maude and its symbolic features to develop new tools in many different logics.

References

1. Aït-Kaci, H., Sasaki, Y.: An axiomatic approach to feature term generalization. In: De Raedt, L., Flach, P. (eds.) ECML 2001. LNCS (LNAI), vol. 2167, pp. 1–12. Springer, Heidelberg (2003). https://doi.org/10.1007/3-540-44795-4_1
2. Alpuente, M., Ballis, D., Cuenca-Ortega, A., Escobar, S., Meseguer, J.: ACUOS^2: a high-performance system for modular ACU generalization with subtyping and inheritance. In: Calimeri, F., Leone, N., Manna, M. (eds.) JELIA 2019. LNCS (LNAI), vol. 11468, pp. 171–181. Springer, Cham (2019). https://doi.org/10.1007/978-3-030-19570-0_11
3. Alpuente, M., Cuenca-Ortega, A., Escobar, S., Meseguer, J.: Order-sorted homeomorphic embedding modulo combinations of associativity and/or commutativity axioms. Fundam. Inform. **177**(3–4), 297–329 (2020)
4. Alpuente, M., Cuenca-Ortega, A., Escobar, S., Meseguer, J.: A partial evaluation framework for order-sorted equational programs modulo axioms. J. Log. Algebraic Methods Program. **110**, 100501 (2020)
5. Alpuente, M., Falaschi, M., Vidal, G.: Partial evaluation of functional logic programs. ACM Trans. Program. Lang. Syst. **20**(4), 768–844 (1998)
6. Bae, K., Escobar, S., Meseguer, J.: Abstract logical model checking of infinite-state systems using narrowing. In: RTA 2013. LIPIcs, vol. 21, pp. 81–96. Schloss Dagstuhl - Leibniz-Zentrum fuer Informatik (2013)
7. Basin, D., Clavel, M., Meseguer, J.: Rewriting logic as a metalogical framework. ACM Trans. Comput. Log. **5**, 528–576 (2004)
8. Chadha, R., Cheval, V., Ciobâcă, Ş, Kremer, S.: Automated verification of equivalence properties of cryptographic protocols. ACM Trans. Comput. Log. **17**(4), 23:1–23:32 (2016)
9. Cholewa, A., Meseguer, J., Escobar, S.: Variants of variants and the finite variant property. Technical report, CS Dept. University of Illinois at Urbana-Champaign, February 2014. http://hdl.handle.net/2142/47117
10. Clavel, M., et al.: Maude manual (version 3.2.1). SRI International, February 2022. http://maude.cs.illinois.edu
11. Clavel, M., et al.: All About Maude, A High-Performance Logical Framework. Lecture Notes in Computer Science, vol. 4350. Springer, Heidelberg (2007). https://doi.org/10.1007/978-3-540-71999-1
12. Colmerauer, A.: An introduction to Prolog III. Commun. ACM **33**(7), 69–90 (1990)
13. Comon-Lundh, H., Delaune, S.: The finite variant property: how to get rid of some algebraic properties. In: Giesl, J. (ed.) RTA 2005. LNCS, vol. 3467, pp. 294–307. Springer, Heidelberg (2005). https://doi.org/10.1007/978-3-540-32033-3_22

14. Dershowitz, N., Jouannaud, J.-P.: Rewrite systems. In: van Leeuwen, J. (ed.) Handbook of Theoretical Computer Science, Volume B: Formal Models and Semantics, pp. 243–320. North-Holland (1990)
15. Durán, F., et al.: Programming and symbolic computation in Maude. J. Log. Algebraic Methods Program. **110**, 100497 (2020)
16. Durán, F., Eker, S., Escobar, S., Martí-Oliet, N., Meseguer, J., Talcott, C.: Associative unification and symbolic reasoning modulo associativity in Maude. In: Rusu, V. (ed.) WRLA 2018. LNCS, vol. 11152, pp. 98–114. Springer, Cham (2018). https://doi.org/10.1007/978-3-319-99840-4_6
17. Durán, F., Meseguer, J., Rocha, C.: Ground confluence of order-sorted conditional specifications modulo axioms. J. Log. Algebraic Methods Program. **111**, 100513 (2020)
18. Eker, S.: Associative unification in Maude. J. Log. Algebraic Methods Program. **126**, 100747 (2022)
19. Escobar, S., Meadows, C., Meseguer, J.: Maude-NPA: cryptographic protocol analysis modulo equational properties. In: Aldini, A., Barthe, G., Gorrieri, R. (eds.) FOSAD 2007-2009. LNCS, vol. 5705, pp. 1–50. Springer, Heidelberg (2009). https://doi.org/10.1007/978-3-642-03829-7_1
20. Escobar, S., Meseguer, J.: Symbolic model checking of infinite-state systems using narrowing. In: Baader, F. (ed.) RTA 2007. LNCS, vol. 4533, pp. 153–168. Springer, Heidelberg (2007). https://doi.org/10.1007/978-3-540-73449-9_13
21. Escobar, S., Sasse, R., Meseguer, J.: Folding variant narrowing and optimal variant termination. J. Algebraic Log. Program. **81**, 898–928 (2012)
22. Goguen, J., Meseguer, J.: Order-sorted algebra I: equational deduction for multiple inheritance, overloading, exceptions and partial operations. Theoret. Comput. Sci. **105**, 217–273 (1992)
23. Jaffar, J., Maher, M.J.: Constraint logic programming: a survey. J. Log. Program. **19**(20), 503–581 (1994)
24. Kapur, D., Narendran, P.: Matching, unification and complexity. SIGSAM Bull. **21**(4), 6–9 (1987)
25. Lucas, S., Meseguer, J., Gutiérrez, R.: The 2D dependency pair framework for conditional rewrite systems. Part I: definition and basic processors. J. Comput. Syst. Sci. **96**, 74–106 (2018)
26. Lucas, S., Meseguer, J., Gutiérrez, R.: The 2D dependency pair framework for conditional rewrite systems - Part II: advanced processors and implementation techniques. J. Autom. Reason. **64**(8), 1611–1662 (2020)
27. Martí-Oliet, N., Meseguer, J.: Rewriting logic as a logical and semantic framework. In: Gabbay, D., Guenthner, F. (eds.) Handbook of Philosophical Logic, 2nd. Edition, pages 1–87. Kluwer Academic Publishers (2002). First published as SRI Technical report SRI-CSL-93-05, August 1993
28. Meier, S., Schmidt, B., Cremers, C., Basin, D.: The TAMARIN prover for the symbolic analysis of security protocols. In: Sharygina, N., Veith, H. (eds.) CAV 2013. LNCS, vol. 8044, pp. 696–701. Springer, Heidelberg (2013). https://doi.org/10.1007/978-3-642-39799-8_48
29. Meseguer, J.: Membership algebra as a logical framework for equational specification. In: Presicce, F.P. (ed.) WADT 1997. LNCS, vol. 1376, pp. 18–61. Springer, Heidelberg (1998). https://doi.org/10.1007/3-540-64299-4_26
30. Meseguer, J.: Strict coherence of conditional rewriting modulo axioms. Theor. Comput. Sci. **672**, 1–35 (2017)

31. Meseguer, J.: Symbolic reasoning methods in rewriting logic and Maude. In: Moss, L.S., de Queiroz, R., Martinez, M. (eds.) WoLLIC 2018. LNCS, vol. 10944, pp. 25–60. Springer, Heidelberg (2018). https://doi.org/10.1007/978-3-662-57669-4_2
32. Meseguer, J.: Variant-based satisfiability in initial algebras. Sci. Comput. Program. **154**, 3–41 (2018)
33. Meseguer, J.: Symbolic computation in Maude: some tapas. In: LOPSTR 2020. LNCS, vol. 12561, pp. 3–36. Springer, Cham (2021). https://doi.org/10.1007/978-3-030-68446-4_1
34. Meseguer, J., Skeirik, S.: Inductive reasoning with equality predicates, contextual rewriting and variant-based simplification. In: Escobar, S., Martí-Oliet, N. (eds.) WRLA 2020. LNCS, vol. 12328, pp. 114–135. Springer, Cham (2020). https://doi.org/10.1007/978-3-030-63595-4_7
35. Meseguer, J., Thati, P.: Symbolic reachability analysis using narrowing and its application to verification of cryptographic protocols. High.-Order Symb. Comput. **20**(1–2), 123–160 (2007)
36. Nieuwenhuis, R., Rubio, A.: Paramodulation-based theorem proving. In: Robinson, J.A., Voronkov, A. (eds.) Handbook of Automated Reasoning (in 2 volumes), pp. 371–443. Elsevier and MIT Press (2001)
37. Ölveczky, P.C.: Real-time Maude and its applications. In: Escobar, S. (ed.) WRLA 2014. LNCS, vol. 8663, pp. 42–79. Springer, Cham (2014). https://doi.org/10.1007/978-3-319-12904-4_3
38. Ölveczky, P.C., Meseguer, J.: Semantics and pragmatics of real-time Maude. High.-Order Symb. Comput. **20**(1–2), 161–196 (2007)
39. Peterson, G.E., Stickel, M.E.: Complete sets of reductions for some equational theories. J. Assoc. Comput. Mach. **28**(2), 233–264 (1981)
40. Plotkin, G.: Building-in equational theories. In: Meltzer, B., Michie, D. (eds.) 1971 Proceedings of the Seventh Annual Machine Intelligence Workshop on Machine Intelligence 7, Edinburgh, pp. 73–90. Edinburgh University Press (1972)
41. Robinson, J.A.: A machine-oriented logic based on the resolution principle. J. Assoc. Comput. Mach. **12**(1), 23–41 (1965)
42. Rocha, C., Meseguer, J.: Five isomorphic Boolean theories and four equational decision procedures. Technical report UIUCDCS-R-2007-2818, CS Department, University of Illinois at Urbana-Champaign, February 2007. http://hdl.handle.net/2142/11295
43. Rocha, C., Meseguer, J.: Proving safety properties of rewrite theories. In: Corradini, A., Klin, B., Cîrstea, C. (eds.) CALCO 2011. LNCS, vol. 6859, pp. 314–328. Springer, Heidelberg (2011). https://doi.org/10.1007/978-3-642-22944-2_22
44. Skeirik, S., Meseguer, J.: Metalevel algorithms for variant satisfiability. J. Log. Algebr. Meth. Program. **96**, 81–110 (2018)
45. Skeirik, S., Stefanescu, A., Meseguer, J.: A constructor-based reachability logic for rewrite theories. Fundam. Inform. **173**(4), 315–382 (2020)
46. Stehr, M.-O., Meseguer, J.: Pure type systems in rewriting logic: specifying typed higher-order languages in a first-order logical framework. In: Owe, O., Krogdahl, S., Lyche, T. (eds.) From Object-Orientation to Formal Methods. LNCS, vol. 2635, pp. 334–375. Springer, Heidelberg (2004). https://doi.org/10.1007/978-3-540-39993-3_16
47. Walther, C.: A mechanical solution of Schubert's steamroller by many-sorted resolution. Artif. Intell. **26**(2), 217–224 (1985)
48. Zheng, Y., et al.: Z3str2: an efficient solver for strings, regular expressions, and length constraints. Formal Methods Syst. Design **50**(2–3), 249–288 (2017)

Leśniewski's Ontology – Proof-Theoretic Characterization

Andrzej Indrzejczak(✉)

Department of Logic, University of Lodz, Łódź, Poland
andrzej.indrzejczak@filhist.uni.lodz.pl

Abstract. The ontology of Leśniewski is commonly regarded as the most comprehensive calculus of names and the theoretical basis of mereology. However, ontology was not examined by means of proof-theoretic methods so far. In the paper we provide a characterization of elementary ontology as a sequent calculus satisfying desiderata usually formulated for rules in well-behaved systems in modern structural proof theory. In particular, the cut elimination theorem is proved and the version of subformula property holds for the cut-free version.

Keywords: Leśniewski · Ontology · Calculus of Names · Sequent Calculus · Cut Elimination

1 Introduction

The ontology of Leśniewski is a kind of calculus of names proposed as a formalization of logic alternative to Fregean paradigm. Basically, it is a theory of the binary predicate ε understood as the formalization of the Greek 'esti'. Informally a formula $a\varepsilon b$ is to be read as "(the) a is (a/the) b", so in order to be true a must be an individual name whereas b can be individual or general name. In the original formulation Leśniewski's ontology is the middle part of the hierarchical structure involving also the protothetics and mereology (see the presentation in Urbaniak [20]). Protothetics, a very general form of propositional logic, is the basis of the overall construction. Its generality follows from the fact that, in addition to sentence variables, arbitrary sentence-functors (connectives) are allowed as variables, and quantifiers binding all these kinds of variables are involved. Similarly in Leśniewski's ontology, we have a quantification over name variables but also over arbitrary name-functors creating complex names. In consequence we obtain very expressive logic which is then extended to mereology. The latter, which is the most well-known ingredient of Leśniewski's construction, is a theory of parthood relation, which provides an alternative formalization of the theory of classes and foundations of mathematics.

Despite of the dependence of Leśniewski's ontology on his protothetics, we can examine this theory, in particular its part called elementary ontology, in isolation, as a kind of first-order theory of ε based on classical first-order logic (FOL). Elementary ontology, in this sense, was investigated, among others, by

J. Blanchette et al. (Eds.): IJCAR 2022, LNAI 13385, pp. 541–558, 2022.
https://doi.org/10.1007/978-3-031-10769-6_32

Słupecki [17] and Iwanuś [7], and we follow this line here. The expressive power of such an approach is strongly reduced, in particular, quantifiers apply only to name variables. One should note however that, despite of the appearances, it is not just another elementary theory in the standard sense, since the range of variables is not limited to individual names but admits general and even empty names. Thus, name variables may represent not only 'Napoleon Bonaparte' but also 'an emperor' and 'Pegasus'. This leads to several problems concerning the interpretation of quantifiers in ontology, encountered in the semantical treatment (see e.g. Küng and Canty [8] or Rickey [16]). However, for us the problems of proper interpretation are not important here, since we develop purely syntactical formulation, which is shown to be equivalent to Leśniewski's axiomatic formulation.

Taking into account the importance and originality of Leśniewski's ontology it is interesting, if not surprising, that so far no proof-theoretic study was offered, in particular, in terms of sequent calculus (SC). In fact, a form of natural deduction proof system was applied by many authors following the original way of presenting proofs by Leśniewski (see, e.g. his [9–11]). However this can hardly be treated as a proof-theoretic study of Leśniewski's ontology but only as a convenient way of simplifying presentation of axiomatic proofs. Ishimoto and Kobayashi [6] introduced also a tableau system for part of (quantifier-free) ontology – we will say more about this system later.

In this paper we present a sequent calculus for elementary ontology and focus on its most important properties. More specifically, in Sect. 2 we briefly characterise elementary ontology which will be the object of our study. In Sect. 3 we present an adequate sequent calculus for the basic part of elementary ontology and prove that it is equivalent with the axiomatic formulation. Then we prove the cut elimination theorem for this calculus in Sect. 4. In the next section we focus on the problem of extensionality and discuss some alternative formulations of ontology and some of its parts, as well as the intuitionistic version of it. Section 6 shows how the basic system can be extended with rules for new predicate constants which preserve cut elimination. The problem of extension with rules for term constants is discussed briefly in Sect. 7. A summary of obtained results and open problems closes the paper.

2 Elementary Ontology

Roughly, in this article, by Leśniewski's elementary ontology we mean standard FOL (in some chosen adequate formalization) with Leśniewski's axiom LA added. For more detailed general presentation of Leśniewski's systems one may consult Urbaniak [20] and for a detailed study of Leśniewski's ontology see Iwanuś [7] or Słupecki [17]. In the next section we will select a particular sequent system as representing FOL and investigate several ways of possible representation of LA in this framework.

We will consider two languages for ontology. In both we assume a denumerable set of name variables. Following the well-known Gentzen's custom we apply a graphical distinction between the bound variables, which will be denoted by $x, y, z, ...$ (possibly with subscripts), and the free variables usually called parameters, which will be denoted by $a, b, c, ...$ These are the only terms we admit, and both kinds will be called simply name variables. The basic language L_o consists of the following vocabulary:

- connectives: $\neg, \wedge, \vee, \rightarrow$;
- first-order quantifiers: $\forall, \exists$;
- predicate: ε.

As we can see, in addition to the standard logical vocabulary of FOL, the only specific constant is a binary predicate ε with the formation rule: $t \varepsilon t'$ is an atomic formula, for any terms t, t'. In what follows we will use a convention: instead of $t \varepsilon t'$ we will write tt'. The complexity of formulae of L_o is defined as the number of occurrences of logical constants, i.e. connectives and quantifiers. Hence the complexity of atomic formulae is 0.

The language L_p, considered in Sect. 6, adds to this vocabulary a number of unary and binary predicates: $D, V, S, G, U, =, \equiv, \approx, \bar{\varepsilon}, \subset, \not\subset, A, E, I, O$.

In L_o and L_p we have name variables, which range over all names (individual, general and empty), as the only terms. However Leśniewski considered also complex terms built with the help of specific term-forming functors. We will discuss briefly such extensions in the setting of sequent calculus in Sect. 7 and notice important problems they generate for decent proof-theoretic treatment.

The only specific axiom of elementary ontology is Leśniewski's axiom LA:

$$\forall xy(xy \leftrightarrow \exists z(zx) \wedge \forall z(zx \rightarrow zy) \wedge \forall zv(zx \wedge vx \rightarrow zv))$$

$\mathrm{LA}^{\rightarrow}$, $\mathrm{LA}^{\leftarrow}$ will be used to refer to the respective implications forming LA, with dropped outer universal quantifier. Note that:

Lemma 1. *The following formulae are equivalent to LA:*

1. $\forall xy(xy \leftrightarrow \exists z(zx \wedge zy) \wedge \forall zv(zx \wedge vx \rightarrow zv))$
2. $\forall xy(xy \leftrightarrow \exists z(zx \wedge zy \wedge \forall v(vx \rightarrow vz)))$
3. $\forall xy(xy \leftrightarrow \exists z(\forall v(vx \leftrightarrow vz) \wedge zy))$

We start with the system in the language L_o, i.e. with ε (conventionally omitted) as the only specific predicate constant added to the standard language of FOL.

3 Sequent Calculus

Elementary ontology will be formalised as a sequent calculus with sequents $\Gamma \Rightarrow \Delta$ which are ordered pairs of finite multisets of formulae called the antecedent and the succedent, respectively. We will use the calculus G (after Gentzen) which is essentially the calculus G1 of Troelstra and Schwichtenberg [19]. All necessary

$$(AX)\ \varphi \Rightarrow \varphi \qquad (Cut)\ \frac{\Gamma \Rightarrow \Delta, \varphi \quad \varphi, \Pi \Rightarrow \Sigma}{\Gamma, \Pi \Rightarrow \Delta, \Sigma}$$

$$(W\Rightarrow)\ \frac{\Gamma \Rightarrow \Delta}{\varphi, \Gamma \Rightarrow \Delta} \qquad (\Rightarrow W)\ \frac{\Gamma \Rightarrow \Delta}{\Gamma \Rightarrow \Delta, \varphi}$$

$$(C\Rightarrow)\ \frac{\varphi, \varphi, \Gamma \Rightarrow \Delta}{\varphi, \Gamma \Rightarrow \Delta} \qquad (\Rightarrow C)\ \frac{\Gamma \Rightarrow \Delta, \varphi, \varphi}{\Gamma \Rightarrow \Delta, \varphi}$$

$$(\neg\Rightarrow)\ \frac{\Gamma \Rightarrow \Delta, \varphi}{\neg\varphi, \Gamma \Rightarrow \Delta} \qquad (\Rightarrow\neg)\ \frac{\varphi, \Gamma \Rightarrow \Delta}{\Gamma \Rightarrow \Delta, \neg\varphi}$$

$$(\wedge\Rightarrow)\ \frac{\varphi, \psi, \Gamma \Rightarrow \Delta}{\varphi \wedge \psi, \Gamma \Rightarrow \Delta} \qquad (\Rightarrow\wedge)\ \frac{\Gamma \Rightarrow \Delta, \varphi \quad \Gamma \Rightarrow \Delta, \psi}{\Gamma \Rightarrow \Delta, \varphi \wedge \psi}$$

$$(\vee\Rightarrow)\ \frac{\varphi, \Gamma \Rightarrow \Delta \quad \psi, \Gamma \Rightarrow \Delta}{\varphi \vee \psi, \Gamma \Rightarrow \Delta} \qquad (\Rightarrow\vee)\ \frac{\Gamma \Rightarrow \Delta, \varphi, \psi}{\Gamma \Rightarrow \Delta, \varphi \vee \psi}$$

$$(\rightarrow\Rightarrow)\ \frac{\Gamma \Rightarrow \Delta, \varphi \quad \psi, \Gamma \Rightarrow \Delta}{\varphi \rightarrow \psi, \Gamma \Rightarrow \Delta} \qquad (\Rightarrow\rightarrow)\ \frac{\varphi, \Gamma \Rightarrow \Delta, \psi}{\Gamma \Rightarrow \Delta, \varphi \rightarrow \psi}$$

$$(\leftrightarrow\Rightarrow)\ \frac{\Gamma \Rightarrow \Delta, \varphi, \psi \quad \varphi, \psi, \Gamma \Rightarrow \Delta}{\varphi \leftrightarrow \psi, \Gamma \Rightarrow \Delta} \qquad (\forall\Rightarrow)\ \frac{\varphi[x/b], \Gamma \Rightarrow \Delta}{\forall x\varphi, \Gamma \Rightarrow \Delta}$$

$$(\Rightarrow\leftrightarrow)\ \frac{\varphi, \Gamma \Rightarrow \Delta, \psi \quad \psi, \Gamma \Rightarrow \Delta, \varphi}{\Gamma \Rightarrow \Delta, \varphi \leftrightarrow \psi} \qquad (\Rightarrow\forall)\ \frac{\Gamma \Rightarrow \Delta, \varphi[x/a]}{\Gamma \Rightarrow \Delta, \forall x\varphi}$$

$$(\exists\Rightarrow)\ \frac{\varphi[x/a], \Gamma \Rightarrow \Delta}{\exists x\varphi, \Gamma \Rightarrow \Delta} \qquad (\Rightarrow\exists)\ \frac{\Gamma \Rightarrow \Delta, \varphi[x/b]}{\Gamma \Rightarrow \Delta, \exists x\varphi}$$

where a is a fresh parameter (eigenvariable), not present in Γ, Δ and φ, whereas b is an arbitrary parameter.

Fig. 1. Calculus G

structural rules, including cut, weakening and contraction are primitive. The calculus G consists of the rules from Fig. 1:

Let us recall that formulae displayed in the schemata are active, whereas the remaining ones are parametric, or form a context. In particular, all active formulae in the premisses are called side formulae, and the one in the conclusion is the principal formula of the respective rule application. Proofs are defined in a standard way as finite trees with nodes labelled by sequents. The height of a proof $\mathcal{D}$ of $\Gamma \Rightarrow \Delta$ is defined as the number of nodes of the longest branch in $\mathcal{D}$. $\vdash_k \Gamma \Rightarrow \Delta$ means that $\Gamma \Rightarrow \Delta$ has a proof of the height at most k.

G provides an adequate formalization of the classical pure FOL (i.e. with no terms other than variables). However, we should remember that here terms in quantifier rules are restricted to variables ranging over arbitrary names (including empty and general). This means, in particular, that quantifiers do not have an existential import, like in standard FOL.

Let us call G+LA an extension of G with LA as an additional axiomatic sequent. The following hold:

Lemma 2. *The following sequents are provable in G+LA:*
$ab \Rightarrow \exists x(xa)$
$ab \Rightarrow \forall x(xa \rightarrow xb)$
$ab \Rightarrow \forall xy(xa \wedge ya \rightarrow xy)$
$\exists x(xa), \forall x(xa \rightarrow xb), \forall xy(xa \wedge ya \rightarrow xy) \Rightarrow ab$

The proof is obvious. In fact, these sequents together allow us to derive LA so we could use them alternatively in a characterization of elementary ontology on the basis of G.

G+LA is certainly an adequate formalization of elementary ontology in the sense of Słupecki and Iwanuś. However, from the standpoint of proof theoretic analysis it is not an interesting form of sequent calculus and it will be used only for showing the adequacy of our main system called GO.

To obtain the basic GO we add the following four rules to G:

$$(R)\ \frac{aa, \Gamma \Rightarrow \Delta}{ab, \Gamma \Rightarrow \Delta} \qquad (T)\ \frac{ac, \Gamma \Rightarrow \Delta}{ab, bc, \Gamma \Rightarrow \Delta} \qquad (S)\ \frac{ba, \Gamma \Rightarrow \Delta}{ab, bb, \Gamma \Rightarrow \Delta}$$

$$(E)\ \frac{da, \Gamma \Rightarrow \Delta, dc \quad dc, \Gamma \Rightarrow \Delta, da \quad ab, \Gamma \Rightarrow \Delta}{cb, \Gamma \Rightarrow \Delta}$$

where d in (E) is a new parameter (eigenvariable), and a, b, c are arbitrary.

The names of rules come from reflexivity, transitivity, symmetry and extensionality. In case of (R) and (S) it is a kind of prefixed reflexivity and symmetry $(ab \rightarrow aa, bb \rightarrow (ab \rightarrow ba))$. Why (E) comes from extensionality will be explained later.

We can show that GO is an adequate characterization of elementary ontology.

Theorem 1. *If G+LA* $\vdash \Gamma \Rightarrow \Delta$, *then GO* $\vdash \Gamma \Rightarrow \Delta$.

Proof. It is sufficient to prove that the axiomatic sequent LA is provable in GO.

$$\dfrac{\dfrac{\dfrac{aa \Rightarrow aa}{ab \Rightarrow aa}\,(R)}{ab \Rightarrow \exists x(xa)}\,(\Rightarrow \exists) \qquad \dfrac{\dfrac{\dfrac{cb \Rightarrow cb}{ca, ab \Rightarrow cb}\,(T)}{ab \Rightarrow ca \rightarrow cb}\,(\Rightarrow\rightarrow)}{ab \Rightarrow \forall x(xa \rightarrow xb)}\,(\Rightarrow \forall)}{ab \Rightarrow \exists x(xa) \wedge \forall x(xa \rightarrow xb)}\,(\Rightarrow \wedge)$$

$(\Rightarrow \wedge)$ with:

$$\dfrac{\dfrac{\dfrac{\dfrac{\dfrac{\dfrac{cd \Rightarrow cd}{ca, ad \Rightarrow cd}\,(T)}{ca, da, aa \Rightarrow cd}\,(S)}{ca, da, ab \Rightarrow cd}\,(R)}{ab, ca \wedge da \Rightarrow cd}\,(\wedge \Rightarrow)}{ab \Rightarrow ca \wedge da \rightarrow cd}\,(\Rightarrow\rightarrow)}{ab \Rightarrow \forall xy(xa \wedge ya \rightarrow xy)}\,(\Rightarrow \forall)$$

yields LA$^{\rightarrow}$ after ($\Rightarrow\rightarrow$). A proof of the converse is more complicated (for readability and space-saving we ommited all applications of weakening rules necessary for the application of two- and three-premiss rules; this convention will be applied hereafter with no comments):

$$\dfrac{ca \Rightarrow ca \qquad \dfrac{\dfrac{\dfrac{\dfrac{da \Rightarrow da \qquad ca \Rightarrow ca}{da, ca \Rightarrow da \wedge ca}\,(\Rightarrow\wedge) \qquad dc \Rightarrow dc}{da, ca, da \wedge ca \rightarrow dc \Rightarrow dc}\,(\rightarrow\Rightarrow)}{da, ca, \forall xy(xa \wedge ya \rightarrow xy) \Rightarrow dc}\,(\forall\Rightarrow) \qquad \dfrac{da \Rightarrow da}{dc, ca \Rightarrow da}\,(T) \qquad ab \Rightarrow ab}{cb, ca, \forall xy(xa \wedge ya \rightarrow xy) \Rightarrow ab}\,(E)}{\dfrac{\dfrac{ca, ca \rightarrow cb, \forall xy(xa \wedge ya \rightarrow xy) \Rightarrow ab}{ca, \forall x(xa \rightarrow xb), \forall xy(xa \wedge ya \rightarrow xy) \Rightarrow ab}\,(\forall\Rightarrow)}{\exists x(xa), \forall x(xa \rightarrow xb), \forall xy(xa \wedge ya \rightarrow xy) \Rightarrow ab}\,(\exists\Rightarrow)}\,(\rightarrow\Rightarrow)$$

It is routine to prove LA. □

Note that to prove LA$^{\rightarrow}$ the rules $(R), (T), (S)$ were sufficient, whereas in order to derive the converse, (E) alone is not sufficient - we need (T) again.

Theorem 2. *If GO* $\vdash \Gamma \Rightarrow \Delta$*, then G+LA* $\vdash \Gamma \Rightarrow \Delta$.

Proof. It is sufficient to prove that the four rules of GO are derivable in G+LA.
For (T):

$$\dfrac{\dfrac{bc \Rightarrow \forall x(xb \rightarrow xc) \qquad \dfrac{\dfrac{ab \Rightarrow ab \qquad ac \Rightarrow ac}{ab \rightarrow ac, ab \Rightarrow ac}\,(\rightarrow\Rightarrow)}{\forall x(xb \rightarrow xc), ab \Rightarrow ac}\,(\forall\Rightarrow)}{ab, bc \Rightarrow ac}\,(Cut) \qquad ac, \Gamma \Rightarrow \Delta}{ab, bc, \Gamma \Rightarrow \Delta}\,(Cut)$$

where the leftmost leaf is provable in G+LA (Lemma 2).

For (S):

$$\dfrac{\dfrac{bb \Rightarrow \forall xy(xb \wedge yb \rightarrow xy) \qquad \dfrac{\dfrac{\dfrac{bb \Rightarrow bb \qquad ab \Rightarrow ab}{bb, ab \Rightarrow bb \wedge ab}\,(\Rightarrow\wedge) \qquad ba \Rightarrow ba}{bb \wedge ab \rightarrow ba, bb, ab \Rightarrow ba}\,(\rightarrow\Rightarrow)}{\forall xy(xb \wedge yb \rightarrow xy), bb, ab \Rightarrow ba}\,(\forall\Rightarrow)}{bb, bb, ab \Rightarrow ba}\,(Cut)}{bb, ab \Rightarrow ba}\,(C\Rightarrow)$$

where the leftmost leaf is provable in G+LA (Lemma 2). By cut with the premiss of (S) we obtain its conclusion.
For (R):

$$\dfrac{\dfrac{ab \Rightarrow \forall xy(xa \wedge ya \rightarrow xy) \qquad \dfrac{ab \Rightarrow \exists x(xa) \qquad S}{\forall xy(xa \wedge ya \rightarrow xy), \forall x(xa \rightarrow xa), ab \Rightarrow aa}\,(Cut)}{\forall x(xa \rightarrow xa), ab, ab \Rightarrow aa}\,(Cut)}{\forall x(xa \rightarrow xa), ab \Rightarrow aa}\,(C\Rightarrow)$$

where $S := \exists x(xa), \forall xy(xa \wedge ya \rightarrow xy), \forall x(xa \rightarrow xa) \Rightarrow aa$ and all leaves are provable in G+LA (Lemma 2); in particular S is the fourth sequent with b replaced with a. By cut with $\Rightarrow \forall x(xa \rightarrow xa)$ and the premiss of (R) we obtain its conclusion.

Since $(R), (T), (S)$ are all derivable in G+LA we use them in the proof of the derivability of (E) to simplify matters. Note first the following three proofs with weakenings omitted:

$$\dfrac{\dfrac{\dfrac{\dfrac{cc \Rightarrow cc}{cb \Rightarrow cc}\,(R) \qquad ca \Rightarrow ca}{ca \leftrightarrow cc, cb \Rightarrow ca}\,(\leftrightarrow\Rightarrow)}{ca \leftrightarrow cc, cb \Rightarrow \exists x(xa)}\,(\Rightarrow \exists)}{\forall x(xa \leftrightarrow xc), cb \Rightarrow \exists x(xa)}\,(\forall \Rightarrow)$$

$$\dfrac{\dfrac{\dfrac{\dfrac{da \Rightarrow da \qquad \dfrac{db \Rightarrow db}{dc, cb \Rightarrow db}\,(T)}{da \leftrightarrow dc, cb, da \Rightarrow db}\,(\leftrightarrow\Rightarrow)}{\forall x(xa \leftrightarrow xc), cb, da \Rightarrow db}\,(\forall \Rightarrow)}{\forall x(xa \leftrightarrow xc), cb \Rightarrow da \rightarrow db}\,(\Rightarrow\rightarrow)}{\forall x(xa \leftrightarrow xc), cb \Rightarrow \forall x(xa \rightarrow xb)}\,(\Rightarrow \forall)$$

and

$$\dfrac{\dfrac{\dfrac{\dfrac{\dfrac{\dfrac{da \Rightarrow da \qquad \dfrac{\dfrac{ea \Rightarrow ea \qquad \dfrac{\dfrac{\dfrac{de \Rightarrow de}{ce, dc \Rightarrow de}\,(T)}{ec, dc, cc \Rightarrow de}\,(S)}{ec, dc, cb \Rightarrow de}\,(R \Rightarrow)}{dc, ea \leftrightarrow ec, cb, ea \Rightarrow de}\,(\leftrightarrow\Rightarrow)}{dc, \forall x(xa \leftrightarrow xc), cb, ea \Rightarrow de}\,(\forall \Rightarrow)}{da \leftrightarrow dc, \forall x(xa \leftrightarrow xc), cb, da, ea \Rightarrow de}\,(\leftrightarrow\Rightarrow)}{\forall x(xa \leftrightarrow xc), \forall x(xa \leftrightarrow xc), cb, da, ea \Rightarrow de}\,(\forall \Rightarrow)}{\forall x(xa \leftrightarrow xc), cb, da, ea \Rightarrow de}\,(C \Rightarrow)}{\forall x(xa \leftrightarrow xc), cb, da \wedge ea \Rightarrow de}\,(\wedge \Rightarrow)}{\forall x(xa \leftrightarrow xc), cb \Rightarrow da \wedge ea \rightarrow de}\,(\Rightarrow\rightarrow)}{\forall x(xa \leftrightarrow xc), cb \Rightarrow \forall xy(xa \wedge ya \rightarrow xy)}\,(\Rightarrow \forall)$$

By three cuts with $\exists x(xa), \forall x(xa \rightarrow xb), \forall xy(xa \wedge ya \rightarrow xy) \Rightarrow ab$ and contractions we obtain a proof of $S := \forall x(xa \leftrightarrow xc), cb \Rightarrow ab$. Then we finish in the following way:

$$\dfrac{\dfrac{\dfrac{\dfrac{da, \Gamma \Rightarrow \Delta, dc \qquad dc, \Gamma \Rightarrow \Delta, da}{\Gamma \Rightarrow \Delta, da \leftrightarrow dc}\,(\Rightarrow\leftrightarrow)}{\Gamma \Rightarrow \Delta, \forall x(xa \leftrightarrow xc)}\,(\Rightarrow \forall) \qquad S}{cb, \Gamma \Rightarrow \Delta, ab}\,(Cut) \qquad ab, \Gamma \Rightarrow \Delta}{cb, \Gamma \Rightarrow \Delta}\,(Cut)$$

Note that to prove derivability of (E) we need in fact the whole LA. We elaborate on the strength of this rule in Sect. 5. □

4 Cut Elimination

The possibility of representing LA by means of these four rules makes GO a calculus with desirable proof-theoretic properties. First of all note that for G the cut elimination theorem holds. Since the only primitive rules for ε are all one-sided, in the sense that principal formulae occur in the antecedents only, we can easily extend this result to GO. We follow the general strategy of cut elimination proofs applied originally for hypersequent calculi in Metcalfe, Olivetti and Gabbay [13] but which works well also in the context of standard sequent calculi (see Indrzejczak [5]). Such a proof has a particularly simple structure and allows us to avoid many complexities inherent in other methods of proving cut elimination. In particular, we avoid well known problems with contraction, since two auxiliary lemmata deal with this problem in advance. Note first that for GO the following result holds:

Lemma 3 (Substitution). *If* $\vdash_k \Gamma \Rightarrow \Delta$, *then* $\vdash_k \Gamma[a/b] \Rightarrow \Delta[a/b]$.

Proof. By induction on the height of a proof. Note that (E) may require similar relettering like $(\exists \Rightarrow)$ and $(\Rightarrow \forall)$. Note that the proof provides the height-preserving admissibility of substitution. □

Let us assume that all proofs are regular in the sense that every parameter a which is fresh by side condition on the respective rule must be fresh in the entire proof, not only on the branch where the application of this rule takes place. There is no loss of generality since every proof may be systematically transformed into a regular one by the substitution lemma. The following notions are crucial for the proof:

1. The cut-degree is the complexity of cut-formula φ, i.e. the number of connectives and quantifiers occurring in φ; it is denoted as $d\varphi$.
2. The proof-degree ($d\mathcal{D}$) is the maximal cut-degree in $\mathcal{D}$.

Remember that the complexity of atomic formulae, and consequently of cut- and proof-degree in case of atomic cuts, is 0. The proof of the cut elimination theorem is based on two lemmata which successively make a reduction: first on the height of the right, and then on the height of the left premiss of cut. φ^k, Γ^k denote $k > 0$ occurrences of φ, Γ, respectively.

Lemma 4 (Right reduction). *Let* $\mathcal{D}_1 \vdash \Gamma \Rightarrow \Delta, \varphi$ *and* $\mathcal{D}_2 \vdash \varphi^k, \Pi \Rightarrow \Sigma$ *with* $d\mathcal{D}_1, d\mathcal{D}_2 < d\varphi$, *and* φ *principal in* $\Gamma \Rightarrow \Delta, \varphi$, *then we can construct a proof* $\mathcal{D}$ *such that* $\mathcal{D} \vdash \Gamma^k, \Pi \Rightarrow \Delta^k, \Sigma$ *and* $d\mathcal{D} < d\varphi$.

Proof. By induction on the height of $\mathcal{D}_2$. The basis is trivial, since $\Gamma \Rightarrow \Delta, \varphi$ is identical with $\Gamma^k, \Pi \Rightarrow \Delta^k, \Sigma$. The induction step requires examination of all cases of possible derivations of $\varphi^k, \Pi \Rightarrow \Sigma$, and the role of the cut-formula in the transition. In cases where all occurrences of φ are parametric we simply apply the induction hypotheses to the premisses of $\varphi^k, \Pi \Rightarrow \Sigma$ and then apply the respective rule – it is essentially due to the context independence of almost all rules and the regularity of proofs, which together prevent violation of side conditions on eigenvariables. If one of the occurrences of φ in the premiss(es) is a side formula of the last rule we must additionally apply weakening to restore the missing formula before the application of the relevant rule.

In cases where one occurrence of φ in $\varphi^k, \Pi \Rightarrow \Sigma$ is principal we make use of the fact that φ in the left premiss is also principal; for the cases of contraction and weakening it is trivial. Note that due to condition that φ is principal in the left premiss it must be compound, since all rules introducing atomic formulae as principal are working only in the antecedents. Hence all cases where one occurrence of atomic φ in the right premiss would be introduced by means of $(R), (S), (T), (E)$ are not considered in the proof of this lemma. The only exceptions are axiomatic sequents $\Gamma \Rightarrow \Delta, \varphi$ with principal atomic φ, but they do not make any harm. □

Lemma 5 (Left reduction). *Let* $\mathcal{D}_1 \vdash \Gamma \Rightarrow \Delta, \varphi^k$ *and* $\mathcal{D}_2 \vdash \varphi, \Pi \Rightarrow \Sigma$ *with* $d\mathcal{D}_1, d\mathcal{D}_2 < d\varphi$, *then we can construct a proof* $\mathcal{D}$ *such that* $\mathcal{D} \vdash \Gamma, \Pi^k \Rightarrow \Delta, \Sigma^k$ *and* $d\mathcal{D} < d\varphi$.

Proof. By induction on the height of $\mathcal{D}_1$ but with some important differences. First note that we do not require φ to be principal in $\varphi, \Pi \Rightarrow \Sigma$ so it includes the case with φ atomic. In all these cases we just apply the induction hypothesis. This guarantees that even if an atomic cut formula was introduced in the right premiss by one of the rules $(R), (S), (T), (E)$ the reduction of the height is done only on the left premiss, and we always obtain the expected result. Now, in cases where one occurrence of φ in $\Gamma \Rightarrow \Delta, \varphi^k$ is principal we first apply the induction hypothesis to eliminate all other $k-1$ occurrences of φ in premisses and then we apply the respective rule. Since the only new occurrence of φ is principal we can make use of the right reduction lemma again and obtain the result, possibly after some applications of structural rules. □

Now we are ready to prove the cut elimination theorem:

Theorem 3. *Every proof in GO can be transformed into cut-free proof.*

Proof. By double induction: primary on $d\mathcal{D}$ and subsidiary on the number of maximal cuts (in the basis and in the inductive step of the primary induction). We always take the topmost maximal cut and apply Lemma 5 to it. By successive repetition of this procedure we diminish either the degree of a proof or the number of cuts in it until we obtain a cut-free proof. □

As a consequence of the cut elimination theorem for GO we obtain:

Corollary 1. *If* $\vdash \Gamma \Rightarrow \Delta$*, then it is provable in a proof which is closed under subformulae of* $\Gamma \cup \Delta$ *and atomic formulae.*

So cut-free GO satisfies the form of the subformula property which holds for several elementary theories as formalised by Negri and von Plato [14].

5 Modifications

Construction of rules which are deductively equivalent to axioms may be to some extent automatised (see e.g. Negri and von Plato [14], Braüner [1], or Marin, Miller, Pimentel and Volpe [12]). Still, even the choice of the version of (equivalent) axiom which will be used for transformation, may have an impact on the quality of obtained rules. Moreover, very often some additional tuning is necessary to obtain rules, which are well-behaved from the proof-theoretic point of view. In this section we will focus briefly on this problem and sketch some alternatives.

In our adequacy proofs we referred to the original formulation of LA, since rules $(R), (T), (S)$ correspond directly in a modular way to three conjuncts of $LA^{\rightarrow}$. Our rule (E) however, is modelled not on $LA^{\leftarrow}$ but rather on the suitable implication of variant 3 of LA from Lemma 1. As a first approximation we can obtain the rule:

$$\frac{\Gamma \Rightarrow \Delta, \exists z(\forall v(va \leftrightarrow vz) \wedge zb)}{\Gamma \Rightarrow \Delta, ab}$$

which after further decomposition and quantifier elimination yields:

$$\frac{da, \Gamma \Rightarrow \Delta, dc \quad dc, \Gamma \Rightarrow \Delta, da \quad \Gamma \Rightarrow \Delta, cb}{\Gamma \Rightarrow \Delta, ab}$$

(where d is a new parameter) which is very similar to (E) but with some active atoms in the succedents. This is troublesome for proving cut elimination if ab is a cut formula and a principal formula of $(R), (S)$ or (T) in the right premiss of cut. Fortunately, (E) is interderivable with this rule (it follows from the rule generation theorem in Indrzejczak [5]) and has the principal formula in the antecedent.

It is clear that if we focus on other variants then we can obtain different rules by their decomposition. In effect note that instead of (E) we may equivalently use the following rules based directly on LA, or on variants 2 and 1 respectively:

$$(E_{LA}) \ \frac{da, \Gamma \Rightarrow \Delta, db \quad da, ea, \Gamma \Rightarrow \Delta, de \quad ab, \Gamma \Rightarrow \Delta}{ca, \Gamma \Rightarrow \Delta}$$

$$(E_2) \ \frac{da, \Gamma \Rightarrow \Delta, dc \quad da, \Gamma \Rightarrow \Delta, cd \quad ab, \Gamma \Rightarrow \Delta}{ca, cb, \Gamma \Rightarrow \Delta}$$

$$(E_1) \ \frac{da, ea, \Gamma \Rightarrow \Delta, de \quad ab, \Gamma \Rightarrow \Delta}{ca, cb, \Gamma, \Rightarrow \Delta}$$

where d, e are new parameters (eigenvariables).

Note, that each of these rules, used instead of (E), yields a variant of GO for which we can also prove cut elimination. However, as we will show by the end of this section, (E) seems to be optimal. Perhaps, the last one is the most economical in the sense of branching factor. However, since its left premiss directly corresponds to the condition $\forall xy(xa \wedge ya \rightarrow xy)$ it introduces two different new parameters to premisses which makes it more troublesome in some respects. In fact, if we want to reduce the branching factor it is possible to replace all these rules by the following variants:

$$(E') \quad \frac{da, \Gamma \Rightarrow \Delta, dc \quad dc, \Gamma \Rightarrow \Delta, da}{cb, \Gamma \Rightarrow \Delta, ab}$$

$$(E'_{LA}) \quad \frac{da, \Gamma \Rightarrow \Delta, db \quad da, ea, \Gamma \Rightarrow \Delta, de}{ca, \Gamma \Rightarrow \Delta, ab}$$

$$(E'_2) \quad \frac{da, \Gamma \Rightarrow \Delta, dc \quad da, \Gamma \Rightarrow \Delta, cd}{ca, cb, \Gamma \Rightarrow \Delta, ab}$$

$$(E'_1) \quad \frac{da, ea, \Gamma \Rightarrow \Delta, de}{ca, cb, \Gamma \Rightarrow \Delta, ab}$$

with the same proviso on eigenvariables d, e. Their interderivability with the rules stated first is easily obtained by means of the rule generation theorem too. These rules seem to be more convenient for proof search. However, for these primed rules cut elimination cannot be proved in the constructive way, for the reasons mentioned above, and it is an open problem if cut-free systems with these rules as primitive are complete.

We finish this section with stating the last reason for choosing (E). Let us explain why (E), the most complicated specific rule of GO, was claimed to be connected with extensionality. Consider the following two principles:

WE $\forall x(xa \leftrightarrow xb) \rightarrow \forall x(ax \leftrightarrow bx)$
$WExt$ $\forall x(xa \leftrightarrow xb) \rightarrow \forall x(\varphi(x, a) \leftrightarrow \varphi(x, b))$

where $\varphi(x, a)$ denotes arbitrary formula with at least one occurrence of x (not bound by any quantifier within φ) and a.

Lemma 6. *WE is equivalent to $WExt$.*

Proof. That WE follows from $WExt$ is obvious since the former is a specific instance of the latter. The other direction is by induction on the complexity of φ. In the basis there are just two cases: $\varphi(x, a)$ is either xa or ax; the former is trivial and the latter is just WE. The induction step goes like an ordinary proof of the extensionality principle in FOL. □

Lemma 7. *In G (E) is equivalent to (WE).*

Proof. Note first that in G the following sequents are provable:

- $\forall x(ax \leftrightarrow cx), cb \Rightarrow ab$
- $\forall x(xa \leftrightarrow xc), da \Rightarrow dc$
- $\forall x(xa \leftrightarrow xc), dc \Rightarrow da$

we will use them in the proofs to follow.

For derivability of (E):

$$\dfrac{\dfrac{\dfrac{da, \Gamma \Rightarrow \Delta, dc \quad dc, \Gamma \Rightarrow \Delta, da}{\Gamma \Rightarrow \Delta, da \leftrightarrow dc}(\Rightarrow\leftrightarrow)}{\Gamma \Rightarrow \Delta, \forall x(xa \leftrightarrow xc)}(\Rightarrow \forall) \quad \mathcal{D}}{\Gamma \Rightarrow \Delta, \forall x(ax \leftrightarrow cx)}(Cut) \quad \forall x(ax \leftrightarrow cx), cb \Rightarrow ab}{cb, \Gamma \Rightarrow \Delta, ab}(Cut)$$

where $\mathcal{D}$ is a proof of $\forall x(xa \leftrightarrow xc) \Rightarrow \forall x(ax \leftrightarrow cx)$ from WE and the rightmost sequent is provable. The endsequent by cut with $ab, \Gamma \Rightarrow \Delta$ yields the conclusion of (E).

Provability of WE in G with (E):

$$(E)\ \dfrac{\forall x(xa \leftrightarrow xc), da \Rightarrow dc \quad \forall x(xa \leftrightarrow xc), dc \Rightarrow da \quad ab \Rightarrow ab}{\forall x(xa \leftrightarrow xc), cb \Rightarrow ab}$$

In the same way we prove $\forall x(xa \leftrightarrow xc), ab \Rightarrow cb$ which by $(\Rightarrow\leftrightarrow)$, $(\Rightarrow \forall)$ and $(\Rightarrow\rightarrow)$ yields WE.

□

This shows that we can obtain the axiomatization of elementary ontology by means of $\text{LA}^{\rightarrow}$ and WE (or $WExt$). Also instead of $\text{LA}^{\rightarrow}$ we can use three axioms corresponding to our three rules $(R), (S), (T)$. Note that if we get rid of (E) (or WE) we obtain a weaker version of ontology investigated by Takano [18]. If we get rid of quantifier rules we obtain a quantifier-free version of this system investigated by Ishimoto and Kobayashi [6].

On the basis of the specific features of sequent calculus we can obtain here for free also the intuitionistic version of ontology. As is well known it is sufficient to restrict the rules of G to sequents having at most one formula in the succedent (which requires small modifications like replacement of $(\leftrightarrow\Rightarrow)$ and $(\Rightarrow \vee)$ with two variants having always one side formula in the succedent) to obtain the version adequate for the intuitionistic FOL. Since all specific rules for ε can be restricted in a similar way, we can obtain the calculus GIO for the intuitionistic version of elementary ontology. One can easily check that all proofs showing the adequacy of GO and the cut elimination theorem are either intuitionistically correct or can be easily changed into such proofs. The latter remark concerns these proofs in which the classical version of $(\leftrightarrow\Rightarrow)$ required the introduction of the second side formula into succedent by $(\Rightarrow W)$; the intuitionistic two versions of $(\leftrightarrow\Rightarrow)$ do not require this step.

6 Extensions

Leśniewski and his followers were often working on ontology enriched with definitions of special predicates and name-creating functors. In this section we focus on a number of unary and binary predicates which are popular ontological constants. Instead of adding these definitions to GO we will introduce predicates by means of sequent rules satisfying conditions formulated for well-behaved SC rules. Let us call L_p the language of L_o enriched with all these predicates and GOP, the calculus with the additional rules for predicates. The definitions of the most important unary predicates are:

$$Da := \exists x(xa) \quad Va := \neg\exists x(xa)$$
$$Sa := \exists x(ax) \quad Ga := \exists xy(xa \wedge ya \wedge \neg xy)$$

D, V, S, G are unary predicates informing that a is denoting, empty (or void), singular or general. D and S are Leśniewski's *ex* and *ob* respectively. He preferred also to apply $sol(a)$ which we symbolize with U (for unique):

$$Ua := \forall xy(xa \wedge ya \rightarrow xy) \text{ [or simply } \neg Ga]$$

The additional rules for these predicates are of the form:

$$(D\Rightarrow)\ \frac{ba, \Gamma \Rightarrow \Delta}{Da, \Gamma \Rightarrow \Delta} \quad (\Rightarrow D)\ \frac{\Gamma \Rightarrow \Delta, ca}{\Gamma \Rightarrow \Delta, Da} \quad (S\Rightarrow)\ \frac{ab, \Gamma \Rightarrow \Delta}{Sa, \Gamma \Rightarrow \Delta}$$
$$(\Rightarrow S)\ \frac{\Gamma \Rightarrow \Delta, ac}{\Gamma \Rightarrow \Delta, Sa} \quad (V\Rightarrow)\ \frac{\Gamma \Rightarrow \Delta, ca}{Va, \Gamma \Rightarrow \Delta} \quad (\Rightarrow V)\ \frac{ba, \Gamma \Rightarrow \Delta}{\Gamma \Rightarrow \Delta, Va}$$

where b is new and c arbitrary in all schemata.

$$(G\Rightarrow)\ \frac{ba, ca, \Gamma \Rightarrow \Delta, bc}{Ga, \Gamma \Rightarrow \Delta} \quad (\Rightarrow G)\ \frac{\Gamma \Rightarrow \Delta, da \quad \Gamma \Rightarrow \Delta, ea \quad de, \Gamma \Rightarrow \Delta}{\Gamma \Rightarrow \Delta, Ga}$$
$$(\Rightarrow U)\ \frac{ba, ca, \Gamma \Rightarrow \Delta, bc}{\Gamma \Rightarrow \Delta, Ua} \quad (U\Rightarrow)\ \frac{\Gamma \Rightarrow \Delta, da \quad \Gamma \Rightarrow \Delta, ea \quad de, \Gamma \Rightarrow \Delta}{Ua, \Gamma \Rightarrow \Delta}$$

where b, c are new, and d, e are arbitrary parameters.

The binary predicates of identity, (weak and strong) coextensiveness, nonbeing b, subsumption and antysubsumption are defined in the following way:

$$a = b := ab \wedge ba \qquad a\bar{\varepsilon}b := aa \wedge \neg ab$$
$$a \equiv b := \forall x(xa \leftrightarrow xb) \quad a \subset b := \forall x(xa \rightarrow xb)$$
$$a \approx b := a \equiv b \wedge Da \qquad a \not\subset b := \forall x(xa \rightarrow \neg xb)$$

Finally note that Aristotelian categorical sentences can be also defined in Leśniewski's ontology:

$$aAb := a \subset b \wedge Da \quad aEb := a \not\subset b \wedge Da$$
$$aIb := \exists x(xa \wedge xb) \quad aOb := \exists x(xa \wedge \neg xb)$$

The rules for binary predicates:

$$(=\Rightarrow)\ \frac{ab, ba, \Gamma \Rightarrow \Delta}{a = b, \Gamma \Rightarrow \Delta} \qquad (\Rightarrow=)\ \frac{\Gamma \Rightarrow \Delta, ab \quad \Gamma \Rightarrow \Delta, ba}{\Gamma \Rightarrow \Delta, a = b}$$
$$(\equiv\Rightarrow)\ \frac{\Gamma \Rightarrow \Delta, ca, cb \quad ca, cb, \Gamma \Rightarrow \Delta}{a \equiv b, \Gamma \Rightarrow \Delta} \quad (\Rightarrow\equiv)\ \frac{da, \Gamma \Rightarrow \Delta, db \quad db, \Gamma \Rightarrow \Delta, da}{\Gamma \Rightarrow \Delta, a \equiv b}$$
$$(\approx\Rightarrow)\ \frac{da, \Gamma \Rightarrow \Delta, ca, cb \quad ca, cb, da, \Gamma \Rightarrow \Delta}{a \approx b, \Gamma \Rightarrow \Delta}$$

$$(\Rightarrow\approx)\ \frac{da, \Gamma \Rightarrow \Delta, db \quad db, \Gamma \Rightarrow \Delta, da \quad \Gamma \Rightarrow \Delta, ca}{\Gamma \Rightarrow \Delta, a \approx b}$$

$$(\bar{\varepsilon}\Rightarrow)\ \frac{aa, \Gamma \Rightarrow \Delta, ab}{a\bar{\varepsilon}b, \Gamma \Rightarrow \Delta} \qquad (\Rightarrow\bar{\varepsilon})\ \frac{\Gamma \Rightarrow \Delta, aa \quad ab, \Gamma \Rightarrow \Delta}{\Gamma \Rightarrow \Delta, a\bar{\varepsilon}b}$$

$$(\subset\Rightarrow)\ \frac{\Gamma \Rightarrow \Delta, ca \quad cb, \Gamma \Rightarrow \Delta}{a \subset b, \Gamma \Rightarrow \Delta} \qquad (\Rightarrow\subset)\ \frac{da, \Gamma \Rightarrow \Delta, db}{\Gamma \Rightarrow \Delta, a \subset b}$$

$$(\not\subset\Rightarrow)\ \frac{\Gamma \Rightarrow \Delta, ca \quad \Gamma \Rightarrow \Delta, cb}{a \not\subset b, \Gamma \Rightarrow \Delta} \qquad (\Rightarrow\not\subset)\ \frac{da, db, \Gamma \Rightarrow \Delta}{\Gamma \Rightarrow \Delta, a \not\subset b}$$

$$(A\Rightarrow)\ \frac{da, \Gamma \Rightarrow \Delta, ca \quad cb, da, \Gamma \Rightarrow \Delta}{aAb, \Gamma \Rightarrow \Delta} \qquad (\Rightarrow A)\ \frac{da, \Gamma \Rightarrow \Delta, db \quad \Gamma \Rightarrow \Delta, ca}{\Gamma \Rightarrow \Delta, aAb}$$

$$(E\Rightarrow)\ \frac{da, \Gamma \Rightarrow \Delta, ca \quad da, \Gamma \Rightarrow \Delta, cb}{aEb, \Gamma \Rightarrow \Delta} \qquad (\Rightarrow E)\ \frac{da, db, \Gamma \Rightarrow \Delta \quad \Gamma \Rightarrow \Delta, ca}{\Gamma \Rightarrow \Delta, aEb}$$

$$(I\Rightarrow)\ \frac{da, db, \Gamma \Rightarrow \Delta}{aIb, \Gamma \Rightarrow \Delta} \qquad (\Rightarrow I)\ \frac{\Gamma \Rightarrow \Delta, ca \quad \Gamma \Rightarrow \Delta, cb}{\Gamma \Rightarrow \Delta, aIb}$$

$$(O\Rightarrow)\ \frac{da, \Gamma \Rightarrow \Delta, db}{aOb, \Gamma \Rightarrow \Delta} \qquad (\Rightarrow O)\ \frac{\Gamma \Rightarrow \Delta, ca \quad cb, \Gamma \Rightarrow \Delta}{\Gamma \Rightarrow \Delta, aOb}$$

where d is new and c arbitrary (but c can be identical to d in rules for $\approx, A, E$).

Proofs of interderivability with equivalences corresponding to suitable definitions are trivial in most cases. We provide only one for the sake of illustration. The hardest case is $\approx$.

$$(\Rightarrow\forall)\ \dfrac{(\Rightarrow\leftrightarrow)\ \dfrac{(\approx\Rightarrow)\ \dfrac{da, ca \Rightarrow ca, cb \quad da, ca, cb \Rightarrow cb}{a \approx b, ca \Rightarrow cb} \qquad \dfrac{da, ca \Rightarrow ca, cb \quad da, ca, cb \Rightarrow ca}{a \approx b, cb \Rightarrow ca}}{a \approx b \Rightarrow ca \leftrightarrow cb}}{a \approx b \Rightarrow \forall x(xa \leftrightarrow xb)}$$

and

$$(\approx\Rightarrow)\ \dfrac{(\Rightarrow\exists)\ \dfrac{ca \Rightarrow ca, aa, ab}{ca \Rightarrow \exists x(xa), aa, ab} \qquad (\Rightarrow\forall)\ \dfrac{ca, aa, ab \Rightarrow ca}{ca, aa, ab \Rightarrow \exists x(xa)}}{a \approx b \Rightarrow \exists x(xa)}$$

by $(\Rightarrow\wedge)$ yield one part. For the second:

$$(\wedge\Rightarrow)\ \dfrac{(\exists\Rightarrow)\ \dfrac{(\Rightarrow\approx)\ \dfrac{\forall x(xa \leftrightarrow xb), da \Rightarrow db \quad \forall x(xa \leftrightarrow xb), db \Rightarrow da \quad ca \Rightarrow ca}{\forall x(xa \leftrightarrow xb), ca \Rightarrow a \approx b}}{\forall x(xa \leftrightarrow xb), \exists x(xa) \Rightarrow a \approx b}}{\forall x(xa \leftrightarrow xb) \wedge \exists x(xa) \Rightarrow a \approx b}$$

where the left and the middle premiss are obviously provable by means of $(\forall\Rightarrow)$, $(\leftrightarrow\Rightarrow)$. We omit proofs of the derivability of both rules in GO enriched with the axiom $\Rightarrow \forall x(xa \leftrightarrow xb) \wedge \exists x(xa) \leftrightarrow a \approx b$.

We treat all these predicates as new constants hence their complexity is fixed as 1, in contrast to atomic formulae, which are of complexity 0. Of course we can consider ontology with an arbitrary selection of these predicates according to the needs. Accordingly we can enrich GO also with arbitrary selection of suitable rules for predicates. All the results holding for GOP are correct for any subsystem. Let us list some important features of these rules and enriched GO:

1. All rules for predicates are explicit, separate and symmetric, which are usual requirements for well-behaved rules in sequent calculi (see e.g. [5]). In this respect they are similar to the rules for logical constants and differ from specific rules for ε which are one-sided (in the sense of having principal formulae always in the antecedent).
2. All these new rules satisfy the subformula property in the sense that side formulae are only atomic.
3. The substitution lemma holds for GO with any combination of the above rules.
4. All rules are pairwise reductive, modulo substitution of terms,

We do not prove the substitution lemma, since the proof is standard, but we comment on the last point, since cut elimination holds due to 3 and 4. The notion of reductivity for sequent rules was introduced by Ciabattoni [2] and it may be roughly defined as follows: A pair of introduction rules $(\Rightarrow \star)$, $(\star \Rightarrow)$ for a constant $\star$ is reductive if an application of cut on cut formulae introduced by these rules may be replaced by the series of cuts made on less complex formulae, in particular on their subformulae. Basically it enables the reduction of cut-degree in the proof of cut elimination. Again we illustrate the point with respect to the most complicated case. Let us consider the application of cut with the cut formula $a \approx b$, then the left premiss of this cut was obtained by:

$$(\Rightarrow\approx)\ \frac{ca, \Gamma \Rightarrow \Delta, cb \qquad cb, \Gamma \Rightarrow \Delta, ca \qquad \Gamma \Rightarrow \Delta, da}{\Gamma \Rightarrow \Delta, a \approx b}$$

where c is new and d is arbitrary. And the right premiss was obtained by:

$$(\approx\Rightarrow)\ \frac{ea, \Pi \Rightarrow \Sigma, fa, fb \qquad ea, fa, fb, \Pi \Rightarrow \Sigma}{a \approx b, \Pi \Rightarrow \Sigma}$$

where e is new and f is arbitrary.

By the substitution lemma on the premisses of $(\Rightarrow\approx), (\approx\Rightarrow)$ we obtain:

1. $fa, \Gamma \Rightarrow \Delta, fb$
2. $fb, \Gamma \Rightarrow \Delta, fa$
3. $da, \Pi \Rightarrow \Sigma, fa, fb$
4. $da, fa, fb, \Pi \Rightarrow \Sigma$

and we can derive:

$$\dfrac{(Cut)\ \dfrac{(C)\ \dfrac{(Cut)\ \dfrac{(Cut)\ \dfrac{\Gamma \Rightarrow \Delta, da \qquad da, \Pi \Rightarrow \Sigma, fa, fb}{\Gamma, \Pi \Rightarrow \Delta, \Sigma, fa, fb} \qquad fb, \Gamma \Rightarrow \Delta, fa}{\Gamma, \Gamma, \Pi \Rightarrow \Delta, \Delta, fa, fa}}{\Gamma, \Pi \Rightarrow \Delta, \Sigma, fa} \qquad \mathcal{D}}{(C)\ \dfrac{\Gamma, \Gamma, \Pi, \Pi \Rightarrow \Delta, \Delta, \Sigma, \Sigma}{\Gamma, \Pi \Rightarrow \Delta, \Sigma}}$$

where $\mathcal{D}$ is a similar proof of $fa, \Gamma, \Pi \Rightarrow \Delta, \Sigma$ from $\Gamma \Rightarrow \Delta, da$, 4 and 1 by cuts and contractions. All cuts are of lower degree than the original cut. It is routine exercise to check that all rules for predicates are reductive and this is sufficient for proving Lemma 4 and 5 for GOP. As a consequence we obtain:

Theorem 4. *Every proof in GOP can be transformed into cut-free proof.*

Since the rules are modular this holds for every subsystem based on a selection of the above rules.

7 Conclusion

Both the basic system GO and its extension GOP are cut-free and satisfy a form of the subformula property. It shows that Leśniewski's ontology admits standard proof-theoretical study and allows us to obtain reasonable results. In particular, we can prove for GO the interpolation theorem using the Maehara strategy (see e.g. [19]) and this implies for GO other expected results like e.g. Beth's definability theorem. Space restrictions forbid to present it here. On the other hand, we restricted our study to the system with simple names only, whereas fuller study should cover also complex names built with the help of several name-forming functors. The typical ones are the counterparts of the well-known class operations definable in Leśniewski's ontology in the following way:

$$a\bar{b} := aa \wedge \neg ab \quad a(b \cap c) := ab \wedge ac \quad a(b \cup c) := ab \vee ac$$

It is not a problem to provide suitable rules corresponding to these definitions:

$$(- \Rightarrow)\ \frac{aa, \Gamma \Rightarrow \Delta, ab}{a\bar{b}, \Gamma \Rightarrow \Delta} \qquad (\Rightarrow -)\ \frac{ab, \Gamma \Rightarrow \Delta \quad \Gamma \Rightarrow \Delta, aa}{\Gamma \Rightarrow \Delta, a\bar{b}}$$

$$(\cap \Rightarrow)\ \frac{ab, ac, \Gamma \Rightarrow \Delta}{a(b \cap c), \Gamma \Rightarrow \Delta} \qquad (\Rightarrow \cap)\ \frac{\Gamma \Rightarrow \Delta, ab \quad \Gamma \Rightarrow \Delta, ac}{\Gamma \Rightarrow \Delta, a(b \cap c)}$$

$$(\cup \Rightarrow)\ \frac{ab, \Gamma \Rightarrow \Delta \quad ac, \Gamma \Rightarrow \Delta}{a(b \cup c), \Gamma \Rightarrow \Delta} \qquad (\Rightarrow \cup)\ \frac{\Gamma \Rightarrow \Delta, ab, ac}{\Gamma \Rightarrow \Delta, a(b \cup c)}$$

Although their structure is similar to the rules provided for predicates in the last section, their addition raises important problems. One is of a more general nature and well-known: definitions of term-forming operations in ontology are creative. Although it was intended in the original architecture of Leśniewski's systems, in the modern approach this is not welcome. Iwanuś [7] has shown that the problem can be overcome by enriching elementary ontology with two axioms corresponding to special versions of the comprehension axiom but this opens a problem of derivability of these axioms in GO enriched with special rules.

There is also a specific problem with cut elimination for GO with added complex terms and suitable rules. Even if they are reductive (and the rules stated above are reductive, as a reader can check), we run into a problem with quantifier rules. If unrestricted instantiation of terms is admitted in $(\Rightarrow \exists)$, $(\forall \Rightarrow)$ the subformula property is lost. One can find some solutions for this problem, for example by using two separated measures of complexity for formula-makers and term-makers (see e.g. [3]), or by restricting in some way the instantiation of terms in respective quantifier rules (see e.g. [4]). The examination of these possibilities is left for further study.

The last open problem deserving careful study is the possibility of application for automated proof-search and obtaining semi-decision procedures (or decision procedures for quantifier-free subsystems) on the basis of the provided sequent calculus. In particular, due to modularity of provided rules, one could obtain in this way decision procedures for several quantifier-free subsystems investigated by Pietruszczak [15], or by Ishimoto and Kobayashi [6].

Acknowledgements. The research is supported by the National Science Centre, Poland (grant number: DEC-2017/25/B/HS1/01268). I am also greatly indebted to Nils Kürbis for his valuable comments.

References

1. Braüner, T.: Hybrid Logic and its Proof-Theory. Springer, Cham (2011). https://doi.org/10.1007/978-94-007-0002-4
2. Ciabattoni, A.: Automated generation of analytic calculi for logics with linearity. In: Marcinkowski, J., Tarlecki, A. (eds.) CSL 2004. LNCS, vol. 3210, pp. 503–517. Springer, Heidelberg (2004). https://doi.org/10.1007/978-3-540-30124-0_38
3. Indrzejczak, A.: Fregean description theory in proof-theoretic setting. Logic Log. Philos. **28**(1), 137–155 (2019)
4. Indrzejczak, A.: Free logics are cut-free. Stud. Log. **109**(4), 859–886 (2021)
5. Indrzejczak, A.: Sequents and Trees. An Introduction to the Theory and Applications of Propositional Sequent Calculi. Birkhäuser (2021)
6. Ishimoto, A., Kobayashi, M.: A propositional fragment of Leśniewski's ontology and its formulation by the tableau method. Stud. Log. **41**(2–3), 181–196 (1982)
7. Iwanuś, B.: On Leśniewski's elementary ontology. Stud. Log. **31**(1), 73–119 (1973)
8. Küng, G., Canty, J.T.: Substitutional quantification and Leśniewskian quantifiers. Theoria **36**, 165–182 (1970)
9. Leśniewski, S.: Über Funktionen, deren Felder Gruppen mit Rücksicht auf diese Funktionen sind. Fundam. Math. **13**, 319–332 (1929). [English translation. In: Leśniewski [11]]
10. Leśniewski, S.: Über Funktionen, deren Felder Abelsche Gruppen in Bezug auf diese Funktionen sind. Fundam. Math. **14**, 242–251 (1929). [English translation. In: Leśniewski [11]]
11. Leśniewski, S.: Collected Works, vol. II. Surma, S., Srzednicki, J., Barnett, D.I. Kluwer/PWN (1992)
12. Marin, S., Miller, D., Pimentel, E., Volpe, M.: From axioms to synthetic inference rules via focusing. Ann. Pure Appl. Log. **173**(5), 103091 (2022)
13. Metcalfe, G., Olivetti, N., Gabbay, D.: Proof Theory for Fuzzy Logics. Springer, Heidelberg (2008). https://doi.org/10.1007/978-1-4020-9409-5
14. Negri, S., von Plato, J.: Structural Proof Theory. Cambridge University Press, Cambridge (2001)
15. Pietruszczak, A.: Quantifier-free Calculus of Names. Systems and Metatheory, Toruń (1991). [in Polish]
16. Rickey, F.: Interpretations of Leśniewski's ontology. Dialectica **39**(3), 181–192 (1985)

17. Słupecki, J.: S. Leśniewski's calculus of names. Stud. Log. **3**(1), 7–72 (1955)
18. Takano, M.: A semantical investigation into Leśniewski's axiom and his ontology. Stud. Log. **44**(1), 71–78 (1985)
19. Troelstra, A.S., Schwichtenberg, H.: Basic Proof Theory. Oxford University Press, Oxford (1996)
20. Urbaniak, R.: Leśniewski's Systems of Logic and Foundations of Mathematics. Springer, Cham (2014). https://doi.org/10.1007/978-3-319-00482-2

Bayesian Ranking for Strategy Scheduling in Automated Theorem Provers

Chaitanya Mangla(✉), Sean B. Holden, and Lawrence C. Paulson

Computer Laboratory, University of Cambridge, Cambridge, England
{cm772,sbh11,lp15}@cl.cam.ac.uk

Abstract. A *strategy schedule* allocates time to proof strategies that are used in sequence in a theorem prover. We employ Bayesian statistics to propose alternative sequences for the strategy schedule in each proof attempt. Tested on the TPTP problem library, our method yields a time saving of more than 50%. By extending this method to optimize the fixed time allocations to each strategy, we obtain a notable increase in the number of theorems proved.

Keywords: Bayesian machine learning · Strategy scheduling · Automated theorem proving

1 Introduction

Theorem provers have wide-ranging applications, including formal verification of large mathematical proofs [9] and reasoning in knowledge-bases [37]. Thus, improvements in provers that lead to more successful proofs, and savings in the time taken to discover proofs, are desirable.

Automated theorem provers generate proofs by utilizing inference procedures in combination with heuristic search. A specific configuration of a prover, which may be specialized for a certain class of problems, is termed a *strategy*. Provers such as E [27] can select from a portfolio of strategies to solve the goal theorem. Furthermore, certain provers hedge their allocated proof time across a number of proof strategies by use of a *strategy schedule*, which specifies a time allocation for each strategy and the sequence in which they are used until one proves the goal theorem. This method was pioneered in the Gandalf prover [33].

Prediction of the effectiveness of a strategy prior to a proof attempt is usually intractable or undecidable [12]. A practical implementation must infer such a prediction by tractable approximations. Therefore, machine learning methods for strategy invention, selection and scheduling are actively researched. Machine learning methods for strategy selection conditioned on the proof goal have shown promising results [3]. Good results have also been reported for strategy synthesis using machine learning [1]. Work on machine learning for algorithm portfolios—which allocate resources to multiple solvers simultaneously—is also relevant to strategy scheduling because of its similar goals. For this purpose, Silverthorn and Miikkulainen propose latent class models [31] .

J. Blanchette et al. (Eds.): IJCAR 2022, LNAI 13385, pp. 559–577, 2022.
https://doi.org/10.1007/978-3-031-10769-6_33

In this work, we present a method for generating strategy schedules using Bayesian learning with two primary goals: to reduce proving time or to prove more theorems. We have evaluated this method for both purposes using iLeanCoP, an intuitionistic first-order logic prover with a compact implementation and good performance [18]. Intuitionistic logic is a non-standard form of first-order logic, of which relatively little is known with regard to automation. It is of interest in theoretical computer science and philosophy of mathematics [7]. Among intuitionistic provers, iLeanCoP is seen as impressive and is able to prove a sufficient number of theorems in our benchmarks for significance testing. Its core is implemented in around thirty lines of Prolog; such simplicity adds clarity to interpretations of our results. Our method was benchmarked on the Thousands of Problems for Theorem Provers (TPTP) problem library [32], in which we are able to save more than 50% on proof time when aiming for the former goal. Towards the latter goal, we are able to prove notably more theorems.

Our two primary, complementary, contributions presented here are: first, a Bayesian machine learning model for strategy scheduling; and second, engineered features for use in that model. The text below is organized as follows. In Sect. 2, we introduce preliminary material used subsequently to construct a machine learning model for strategy scheduling, described in Sects. 3–7. The data used to train and evaluate this model are described in Sect. 8, followed by experiments, results and conclusions in Sects. 9–12.

2 Distribution of Permutations

We model a strategy schedule using a vector of strategies, and thus all schedules are *permutations* of the same.

Definition 1 (Permutation). *Let $M \in \mathbb{N}$. A permutation $\boldsymbol{\pi} \in \mathbb{N}^M$ is a vector of indices, with $\pi_i \in \{1, \ldots, M\}$ and $\forall i \neq j : \pi_i \neq \pi_j$, representing a reordering of the components of an M-dimensional vector $\boldsymbol{s}$ to $[s_{\pi_1}, s_{\pi_2}, \ldots, s_{\pi_M}]^\mathsf{T}$.*

In this text, vector-valued variables, such as $\boldsymbol{\pi}$ above, are in boldface, which must change when they are indexed, like π_1 for example. For probabilistic modelling of schedules represented using permutations, we use the Plakett-Luce model [14,21] to define a parametric probability distribution over permutations.

Definition 2 (Plakett-Luce distribution). *The Plakett-Luce distribution* $\mathrm{Perm}(\boldsymbol{\lambda})$ *with parameter $\boldsymbol{\lambda} \in \mathbb{R}^M_{>0}$, has support over permutations of indices $\{1, \ldots, M\}$. For permutation $\boldsymbol{\Pi}$ distributed as* $\mathrm{Perm}(\boldsymbol{\lambda})$,

$$\Pr(\boldsymbol{\Pi} = \boldsymbol{\pi}; \boldsymbol{\lambda}) = \prod_{j=1}^{M} \frac{\lambda_{\pi_j}}{\sum_{u=j}^{M} \lambda_{\pi_u}}.$$

In latter sections, we use the parameter $\boldsymbol{\lambda}$ to assign an abstract 'score' to strategies when modelling distributions over schedules. This score is particularly useful due to the following theorem.

Theorem 1. *Let* $\boldsymbol{\pi}^*$ *be a mode of the distribution* $\mathrm{Perm}(\boldsymbol{\lambda})$, *that is*

$$\boldsymbol{\pi}^* = \operatorname*{argmax}_{\boldsymbol{\pi}} \Pr(\boldsymbol{\pi}; \boldsymbol{\lambda}).$$

Then, $\lambda_{\pi_1^*} \geqslant \lambda_{\pi_2^*} \geqslant \lambda_{\pi_3^*} \geqslant \ldots \geqslant \lambda_{\pi_M^*}$.

Thus, assuming $\boldsymbol{\lambda}$ is a vector of the score of each strategy, the highest probability permutation indexes the strategies in decreasing order of scores. Conversely, the highest probability permutation can be obtained efficiently by sorting the indices of $\boldsymbol{\lambda}$ with respect to their corresponding values in decreasing order. Cao et al. [4] have presented a proof of Theorem 1, and Cheng et al. [5] have discussed some further interesting details.

Example 1. Let $\boldsymbol{\lambda} = [1, 9]^\intercal$, $\boldsymbol{\pi}^{(1)} = [1, 2]^\intercal$ and $\boldsymbol{\pi}^{(2)} = [2, 1]^\intercal$. Then,

$$\Pr(\boldsymbol{\Pi} = \boldsymbol{\pi}^{(1)}; \boldsymbol{\lambda}) = \frac{\lambda_{\pi_1^{(1)}}}{\lambda_{\pi_1^{(1)}} + \lambda_{\pi_2^{(1)}}} \cdot \frac{\lambda_{\pi_2^{(1)}}}{\lambda_{\pi_2^{(1)}}} = \frac{1}{1+9} \cdot \frac{9}{9} = \frac{1}{10}.$$

Similarly, $\Pr(\boldsymbol{\Pi} = \boldsymbol{\pi}^{(2)}; \boldsymbol{\lambda}) = 9/10$. □

Theorem 2. $\mathrm{Perm}(c\boldsymbol{\lambda}) = \mathrm{Perm}(\boldsymbol{\lambda})$, *for any scalar constant* $c > 0$.

In other words, the Plakett-Luce distribution is invariant to the scale of the parameter vector.

Lemma 1. $\mathrm{Perm}(\exp(\boldsymbol{\lambda} + c)) = \mathrm{Perm}(\exp(\boldsymbol{\lambda}))$, *for any scalar constant* $c \in \mathbb{R}$.

Lemma 1 follows from Theorem 2, and shows the same distribution is translation invariant if the parameter is exponentiated. Cao et al. [4] give proofs of both.

3 A Maximum Likelihood Model

We model a strategy schedule as a ranking of known strategies, where each strategy is constructed by a parameter setting and time allocation. A ranking therein is a permutation of strategies, with each strategy retaining its time allocation irrespective of the ordering. We construct, in this section, a model for inference of such permutations that is linear in the parameters.

Suppose we have a repository of N theorems which we test against each of our M known strategies to build a data-set $\mathcal{D} = \{(\boldsymbol{\pi}^{(i)}, \boldsymbol{x}^{(i)})\}_{i=1}^{N}$, where $\boldsymbol{\pi}^{(i)}$ is a desirable ordering of strategies for theorem i and $\boldsymbol{x}^{(i)}$ is a feature vector representation of the theorem. In Sect. 9, we detail how we instantiated $\mathcal{D}$ for our experiments, which serves as an example for any other implementation. We assume that $\boldsymbol{\pi}^{(i)}$ has Plakett-Luce distribution conditioned on $\boldsymbol{x}^{(i)}$ such that

$$\Pr(\boldsymbol{\pi}; \boldsymbol{x}, \boldsymbol{\omega}) = \mathrm{Perm}(\boldsymbol{\Lambda}(\boldsymbol{x}, \boldsymbol{\omega})), \quad (1)$$

where $\boldsymbol{\omega}$ is a parameter the model must learn and $\boldsymbol{\Lambda}(\cdot)$ is a vector-valued function of range $\mathbb{R}^M_{>0}$. We use the notation $\Lambda(\cdot)_i$ to index into the value of $\boldsymbol{\Lambda}(\cdot)$. We represent our prover strategies with feature vectors $\{\boldsymbol{d}^{(j)}\}_{j=1}^{M}$. To calculate the score of strategy j using $\Lambda(\cdot)_j$, we specify

$$\Lambda(\boldsymbol{x}^{(i)}, \boldsymbol{\omega})_j = \exp\left(\boldsymbol{\phi}(\boldsymbol{x}^{(i)}, \boldsymbol{d}^{(j)})^{\mathsf{T}} \boldsymbol{\omega}\right) \tag{2}$$

to ensure that the scores are positive valued, where $\boldsymbol{\phi}$ is a suitable basis expansion function. Assuming the data is i.i.d, the likelihood of the parameter vector is given by

$$\mathcal{L}(\boldsymbol{\omega}) = p(\mathcal{D}; \boldsymbol{\omega}) = \prod_{i=1}^{N} \Pr(\boldsymbol{\pi}^{(i)}; \boldsymbol{\Lambda}(\boldsymbol{x}^{(i)}, \boldsymbol{\omega})). \tag{3}$$

An $\hat{\boldsymbol{\omega}}$ that maximizes this likelihood can then be used to forecast the distribution over permutations for a new theorem $\boldsymbol{x}^*$ by evaluating $\mathrm{Perm}(\boldsymbol{\Lambda}(\boldsymbol{x}^*, \hat{\boldsymbol{\omega}}))$ for all permutations. This would incur factorial complexity; however, we are often only interested in the most likely permutation, which can be retrieved in polynomial time. Specifically for strategy scheduling the permutation with the highest predicted probability should reflect the orderings in the data. For this purpose, we use Theorem 1 to find the highest probability permutation $\boldsymbol{\pi}^*$ by sorting the values of $\{\boldsymbol{\Lambda}(\boldsymbol{x}^*, \hat{\boldsymbol{\omega}})_j\}_{j=1}^{M}$ in descending order.

Remark 1. A method named ListNet designed to rank documents for search queries using the Plakett-Luce distribution is evaluated by Cao et al. [4]. Their evaluation uses a linear basis expansion. We can derive a similar construction in our model by setting

$$\boldsymbol{\phi}(\boldsymbol{x}^{(i)}, \boldsymbol{d}^{(j)}) = [\boldsymbol{x}^{(i)\mathsf{T}}, \boldsymbol{d}^{(j)\mathsf{T}}]^{\mathsf{T}}. \tag{4}$$

Remark 2. The likelihood in Equation (3) can be maximized by minimizing the negative log likelihood $\ell(\boldsymbol{\omega}) = -\log \mathcal{L}(\boldsymbol{\omega})$, which (as shown by Schäfer and Hüllermeier [26]) is convex and therefore can be minimized using gradient-based methods. The minima may, however, be unidentifiable due to translation invariance, as demonstrated by Lemma 1. This problem is eliminated in our Bayesian model by the use of a Gaussian prior, as explained in Sect. 4.

Example 2. Let there be $N = 2$ theorems and $M = 2$ strategies. Let the theorems and strategies be characterized by univariate values such that $x^{(1)} = 1$, $x^{(2)} = 2$, $d^{(1)} = 1$ and $d^{(2)} = 2$.

Suppose strategy $d^{(1)}$ is ideal for theorem $x^{(1)}$ and strategy $d^{(2)}$ for $x^{(2)}$, as shown on the right, where a + indicates the preferred strategy.

	$d^{(1)}$	$d^{(2)}$
$x^{(1)}$	+	−
$x^{(2)}$	−	+

This is evidently an example of a parity problem [34], and hence cannot be modelled by a simple linear expansion using the basis function mentioned in Remark 1. A solution in this instance is to use

$$\phi(x^{(i)}, d^{(j)}) = x^{(i)} \cdot d^{(j)}.$$

The parameter ω is then one-dimensional, and the required training data takes the form $\mathcal{D} = \{([1,2]^{\mathsf{T}}, 1), ([2,1]^{\mathsf{T}}, 2)\}$. We find that $\mathcal{L}(w)$ is convex, with maxima at $\hat{\omega} = 0.42$ as shown in Fig. 1.

□

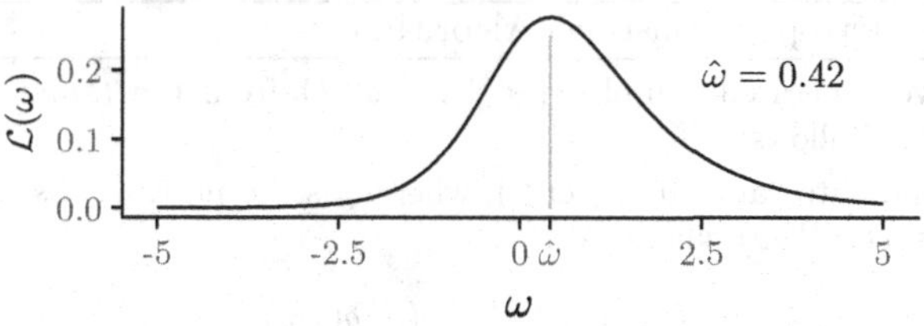

Fig. 1. The likelihood function in Example 2.

4 Bayesian Inference

We place a Gaussian prior distribution on the parameter $\boldsymbol{\omega}$ of the model described in Sect. 3. This has two advantages: first, the posterior mode is identifiable, as noted by Johnson et al. [11] and demonstrated in Example 3 on page 7; second, the parameter is regularized. With this prior specified as the normal distribution

$$\boldsymbol{\omega} \sim \mathcal{N}(\boldsymbol{m}_0, \boldsymbol{S}_0), \tag{5}$$

and assuming $\boldsymbol{\pi}$ is independent of $\mathcal{D}$ given $(\boldsymbol{x}, \boldsymbol{\omega})$, the posterior predictive distribution is

$$p(\boldsymbol{\pi}|\boldsymbol{x}^*, \mathcal{D}) = \int p(\boldsymbol{\pi}|\boldsymbol{x}^*, \boldsymbol{\omega})p(\boldsymbol{\omega}|\mathcal{D})\mathrm{d}\boldsymbol{\omega},$$

which may be approximated by sampling from the posterior,

$$\boldsymbol{\omega}^s \sim p(\boldsymbol{\omega}|\mathcal{D}), \tag{6}$$

to obtain

$$p(\boldsymbol{\pi}|\boldsymbol{x}^*, \mathcal{D}) \approx \frac{1}{S}\sum_{s=1}^{S} p(\boldsymbol{\pi}|\boldsymbol{x}^*, \boldsymbol{\omega}^s). \tag{7}$$

Given a new theorem $\boldsymbol{x}^*$, to find the permutation of strategies with the highest probability of success, using the approximation above would require its evaluation for every permutation of $\boldsymbol{\pi}$. This process incurs factorial complexity. We instead make a Bayes point approximation [16] using the mean values of the samples such that,

$$\begin{aligned} p(\boldsymbol{\pi}|\boldsymbol{x}^*, \mathcal{D}) &\approx p(\boldsymbol{\pi}|\boldsymbol{x}^*, \langle\boldsymbol{\omega}^s\rangle) && \text{using Eq. (7)} \\ &= \Pr(\boldsymbol{\pi}|\boldsymbol{\Lambda}(\boldsymbol{x}^*, \langle\boldsymbol{\omega}^s\rangle)) && \text{using Eq. (1)}, \end{aligned}$$

where $\langle\cdot\rangle$ denotes mean value. The mean of the Plakett-Luce parameter for Bayesian inference has been used in prior work [8] to obtain good results. Furthermore, using that, the highest probability permutation can be obtained by using Theorem 1, thereby incurring only the cost of sorting the items. This saving is substantial when generating a strategy schedule, because it saves on prediction time, which is important for the following reason.

Algorithm 1. Metropolis-Hastings Algorithm

Suppose we have generated samples $\{\boldsymbol{\omega}^{(1)}, \ldots, \boldsymbol{\omega}^{(i)}\}$ from the *target distribution* p. Generate $\boldsymbol{\omega}^{(i+1)}$ as follows.

1: Generate candidate value $\dot{\boldsymbol{\omega}} \sim q(\boldsymbol{\omega}^{(i)})$, where q is the *proposal distribution.*
2: Evaluate $r \equiv r(\boldsymbol{\omega}^{(i)}, \dot{\boldsymbol{\omega}})$ where

$$r(x, y) = \min\left\{\frac{p(y)}{p(x)}\frac{q(x|y)}{q(y|x)}, 1\right\}.$$

3: Set

$$\boldsymbol{\omega}^{(i+1)} = \begin{cases} \dot{\boldsymbol{\omega}} & \text{with probability } r \\ \boldsymbol{\omega}^{(i)} & \text{with probability } 1-r. \end{cases}$$

Remark 3. While benchmarking and in typical use, a prover is allocated a fixed amount of time for a proof attempt, and any time taken to predict a strategy schedule must be accounted for within this allocation. Time taken for this prediction is time taken away from the prover itself which could have been invested in the proof search. Therefore, it is essential to minimize schedule prediction time. It is particularly wise to favour a saving in prediction time at the cost of model optimization and training time.

Remark 4. In our implementation we set $\boldsymbol{m}_0 = 0$. This has the effect of prioritizing smaller weights $\boldsymbol{\omega}$ in the posterior. Furthermore, we set $\boldsymbol{S}_0 = \eta\boldsymbol{I}, \eta \in \mathbb{R}$, where $\boldsymbol{I}$ is the identity matrix. Consequently, the hyperparameter η controls the strength of the prior, since the entropy of the Gaussian prior scales linearly by $\log|\boldsymbol{S}_0|$.

Remark 5. A specialization of the Plakett-Luce distribution using the Thurstonian interpretation admits a Gamma distribution conjugate prior [8]. That, however, is unavailable to our model when parametrized as shown in Eq. (1).

5 Sampling

We use the Markov chain Monte Carlo (MCMC) Metropolis-Hastings algorithm [38] to generate samples from the posterior distribution. In MCMC sampling, one constructs a Markov chain whose stationary distribution matches the target distribution p. For the Metropolis-Hastings algorithm, stated in Algorithm 1, this chain is constructed using a proposal distribution $y|x \sim q$, where q is set to a distribution that can be conveniently sampled from.

Note that while calculating r in Algorithm 1, the normalization constant of the target density p cancels out. This is to our advantage; to generate samples $\boldsymbol{\omega}^s$ from the posterior, which is, by Eq. (3) and Eq. (5),

$$\begin{aligned} p(\boldsymbol{\omega}|\mathcal{D}) &\propto p(\mathcal{D}|\boldsymbol{\omega})p(\boldsymbol{\omega}) \\ &= \mathcal{L}(\boldsymbol{\omega})\mathcal{N}(\boldsymbol{m}_0, \boldsymbol{S}_0), \end{aligned} \tag{8}$$

the posterior only needs to be computed in this unnormalized form.

In this work, we choose a random walk proposal of the form

$$q(\boldsymbol{\omega}'|\boldsymbol{\omega}) = \mathcal{N}(\boldsymbol{\omega}'|\boldsymbol{\omega}, \Sigma_q), \tag{9}$$

and tune Σ_q for efficient sampling simulation. We start the simulation at a local mode $\hat{\boldsymbol{\omega}}$, and set $\mathcal{N}(\hat{\boldsymbol{\omega}}, \Sigma_q)$ to approximate the local curvature of the posterior at that point using methods by Rossi [25]. Specifically, our procedure for computing Σ_q is as follows.

1. First, writing the posterior from Eq. (8) as

$$p(\boldsymbol{\omega}|\mathcal{D}) = \frac{1}{Z}e^{-E(\boldsymbol{\omega})},$$

   where Z is the normalization constant, we have

$$E(\boldsymbol{\omega}) = -\log \mathcal{L}(\boldsymbol{\omega}) - \log \mathcal{N}(\boldsymbol{m}_0, \boldsymbol{S}_0). \tag{10}$$

   We find a local mode $\hat{\boldsymbol{\omega}}$ by optimizing $E(\boldsymbol{\omega})$ using a gradient-based method.
2. Then, using a Laplace approximation [2], we approximate the posterior in the locality of this mode to

$$\mathcal{N}(\hat{\boldsymbol{\omega}}, H^{-1}), \text{ where } H = \nabla\nabla E(\boldsymbol{\omega})|_{\hat{\boldsymbol{\omega}}}$$

   is the Hessian matrix of $E(\boldsymbol{\omega})$ evaluated at that local mode.
3. Finally, we set

$$\Sigma_q = s^2 H^{-1}$$

   in Eq. (9), where s is used to tune all the length scales. We set this value to $s^2 = 2.38$ based on the results by Roberts and Rosenthal [24].

Remark 6. When calculating r in Algorithm 1 during sampling, to evaluate the unnormalized posterior at any point $\boldsymbol{\omega}^s$ we compute it from Equation (10) as $\exp(-E(\boldsymbol{\omega}^s))$—it is therefore the only form in which the posterior needs to be coded in the implementation.

Example 3 (Gaussian Prior). To demonstrate the effect of using a Gaussian prior, we build upon Example 2, with the data taking the form

$$\mathcal{D} = \{([1,2]^\intercal, 1), ([2,1]^\intercal, 2)\}.$$

We perform basis expansion as explained in Sect. 6 with prior parameter $\eta = 1.0$, kernel $\sigma = 0.1$ and $\varsigma = 2$ centres. Thus, the model parameter is

$$\boldsymbol{\omega} = [\omega_1, \omega_2]^\intercal, \quad \boldsymbol{\omega} \in \mathbb{R}^2.$$

The unnormalized negative log posterior $E(\omega_1, \omega_2)$, as defined in Eq. (10), is shown in Fig. 2b; and the negative log likelihood $\ell(\omega_1, \omega_2) = -\log \mathcal{L}(\omega_1, \omega_2)$ as mentioned in Remark 2, is shown in Fig. 2a. Note the contrast in the shape of the two surfaces. The minimum is along the top-right portion in Fig. 2a, which is flat and leads to an unidentifiable point estimate, whereas in Fig. 2b, the minimum is in a narrow region near the centre. The Gaussian prior, in informal terms, has lifted the surface up, with an effect that increases in proportion to the distance from the origin.

□

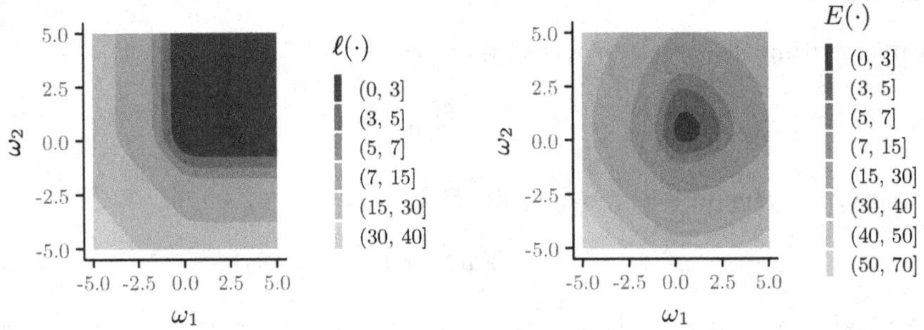

(a) Likelihood function $\ell(\omega_1, \omega_2)$ (b) Posterior function $E(\omega_1, \omega_2)$

Fig. 2. Comparison of the shape of the likelihood and the posterior functions.

6 Basis Expansion

Example 2 shows how the linear expansion in Remark 1 is ineffective even in very simple problem instances. The maximum likelihood bilinear model presented by Schäfer and Hüllermeier [26] is related to our model defined in Sect. 2 with the basis performing the Kronecker (tensor) product $\phi(x, d) = x \otimes d$. Their results show such an expansion produces a competitive model, but falls behind in comparison to their non-linear model.

To model non-linear interactions between theorems and strategies, we use a *Gaussian kernel* for the basis expansion.

Definition 3 (Gaussian Kernel). *A Gaussian kernel* κ *is defined by*

$$\kappa(\boldsymbol{y}, \boldsymbol{z}) = \exp\left(-\frac{\|\boldsymbol{y} - \boldsymbol{z}\|^2}{2\sigma^2}\right), \quad \textit{for } \sigma > 0.$$

The Gaussian kernel $\kappa(\boldsymbol{y}, \boldsymbol{z})$ effectively represents the inner product of $\boldsymbol{y}$ and $\boldsymbol{z}$ in a Hilbert space whose bandwidth is controlled by σ. Smaller values of σ correspond to a higher bandwidth, more flexible, inner product space. Larger values of σ will reduce the kernel to a constant function, as detailed in [30]. For our ranking model, we must tune σ to balance between over-fitting and under-performance.

We use the Gaussian kernel for basis expansion by setting

$$\phi(\boldsymbol{x}, \boldsymbol{d}) = \left[\kappa([\boldsymbol{x}^\intercal, \boldsymbol{d}^\intercal]^\intercal, \boldsymbol{c}^{(1)}), \ldots, \kappa([\boldsymbol{x}^\intercal, \boldsymbol{d}^\intercal]^\intercal, \boldsymbol{c}^{(C)})\right]^\intercal,$$

where $\{\boldsymbol{c}^{(i)}\}_{i=1}^{C}$ is a collection of *centres*. By choosing centres to be themselves composed of theorems $\boldsymbol{x}^{(\cdot)}$ and strategies $\boldsymbol{d}^{(\cdot)}$, such that $\boldsymbol{c}^{(\cdot)} = [\boldsymbol{x}^{(\cdot)\intercal}, \boldsymbol{d}^{(\cdot)\intercal}]^\intercal$, the basis expansion above represents each data item with a non-linear inner product against other known items.

To find the relevant subset of $\mathcal{D}$ from which centres should be formed, we follow the method described in the steps below.

1. Initially, we set the collection of centres to every possible centre. That is, for N theorems and M strategies, we produce a centre for every combination of the two, thereby producing $C = N \cdot M$ centres.
2. Next, we use ϕ to expand every centre to produce the $C \times C$ matrix $\boldsymbol{\Gamma}$ such that
$$\boldsymbol{\Gamma}_{i,j} = \phi(\boldsymbol{c}^{(i)})_j = \kappa(\boldsymbol{c}^{(i)}, \boldsymbol{c}^{(j)}).$$
3. Then, we generate a vector $\boldsymbol{\gamma}$ such that γ_i represents a *score* for centre $\boldsymbol{c}^{(i)}$. Since each centre is a combination of a theorem and a strategy, we set the score to signify how well the strategy performs for that theorem, as detailed in Remark 7 below.
4. Finally, we use Automatic Relevance Determination (ARD) [17] with $\boldsymbol{\Gamma}$ as input and $\boldsymbol{\gamma}$ as the response variable. The result is a weight assignment to each centre to signify its relevance. The highest absolute-weighted ς centres are chosen, where ς is a parameter which decides the total number of centres.

This method is inspired by the procedure used in Relevance Vector Machines [35] for a similar purpose.

Remark 7 (score). For a strategy that succeeds in proving a theorem, the score for the pair is the fraction of the time allocation left unconsumed by the prover. For an unsuccessful strategy-theorem combination, we set the score to a value close to zero.

Remark 8 (ς). The parameter ς is another tunable parameter which, in similar fashion to the parameter σ earlier in this section, controls the model complexity introduced by the basis expansion. Both variables must be tuned together.

7 Model Selection and Time Allocations

From Remark 8, ς and σ are hyperparameters that control the complexity introduced into our model through the Gaussian basis expansion; and Remark 4 introduces η, the hyperparameter that controls the strength of the prior. The final model is selected by tuning them. Tuning must aim to avoid overfitting to the training data; and to maximize, during testing, either the savings in proof-search time or the number of theorems proved. However, we do not have a closed-form expression relating these parameters to this aim, thus any combination of the parameters can be judged only by testing them.

In this work we have used *Bayesian optimization* [29] to optimize these hyperparameters. Bayesian optimization is a black-box parameter optimization method that attempts to search for a global optimum within the scope of a set resource budget. It models the optimization target as a user-specified *objective function*, which maps from the parameter space to a loss metric. This model of the objective function is constructed using *Gaussian Process* (GP) regression [22], using data generated by repeatedly testing the objective function.

Our specified objective function maps from the hyperparameters $(\varsigma, \sigma, \eta)$ to a loss metric ξ. We use cross-validation within the training data while calculating ξ to penalize hyperparameters that over-fit. Hyperparameters are tuned at training time only, after which they are fixed for subsequent testing. The final test set is never used for any hyperparameter optimization.

In the method presented thus far we are only permuting strategies with fixed time allocations to build a sequence for a strategy schedule. In this setting, the number of theorems proved cannot change, but the time taken to prove theorems can be reduced. Therefore, with this aim, a useful metric for ξ is the total time taken by the theorem prover to prove the theorems in the cross-validation test set.

However, we can take further advantage of the hyperparameter tuning phase to additionally tune the times allocated to each strategy, by treating these times as hyperparameters. Therefore, for each strategy $\boldsymbol{d}^{(i)}$ we create a hyperparameter $\nu^{(i)} \in (0, 1)$ which sets the proportion of the proof time allocated to that strategy. We can then optimize our model to maximize the number of theorems proved; a count of the remaining theorems is then a viable metric for ξ. Note that once the $\nu^{(\cdot)}$ are set, time allocation for $\boldsymbol{d}^{(i)}$ is fixed to $\nu^{(i)}$, irrespective of its order in the strategy schedule.

Remark 9. Our results include two types of experiment:

- one where the time allocations for each strategy are set to the defaults shipped with our reference theorem prover, and so we optimize for saving proof time; and
- another wherein we allocate time to each strategy during the hyperparameter tuning phase, and so we optimize for proving the maximum number of theorems.

8 Training Data and Feature Extraction

Our chosen theorem prover, iLeanCoP, is shipped with a fixed strategy schedule consisting of 5 strategies. It splits the allocated proof time across the first four strategies by 2%, 60%, 20% and 10%. However, only the first strategy is complete and therefore usually expected to take up its entire time allocation. The remaining strategies are incomplete, and may exit early on failure. Therefore, the fifth and final strategy, which we refer to as the fallback strategy, is allocated all the remaining time.

Emulating iLeanCop. We have constructed a dataset by attempting to prove every theorem in our problem library using each of these strategies individually. With this information, the result of any proof attempt can be calculated by emulating the behaviour of iLeanCoP. This is how we evaluate the predicted schedules—we emulate a proof attempt by iLeanCoP using that schedule for each theorem in the test set. For a faithful emulation of the fallback strategy, it is always attempted last, and therefore any new schedule is only a permutation of the first four strategies. Our experiments allocate a time of 600 s per theorem. The dataset is built to ensure that, within this proof time, any such strategy permutation can be emulated. We kept a timeout of 1200 s per strategy per theorem when building the dataset, which is more than sufficient for current experiments and gives us headroom for future experiments with longer proof times.

Strategy Features. Each strategy in iLeanCoP consists of a time allocation and parameter settings; the parameters are described by Otten [19]. We use a one-hot encoding feature representation for strategies based on the parameter setting as shown in Table 1. Another feature noting the completeness of each strategy is also shown. Another feature (not shown in the table) contains the time allocated to each strategy. Note the fallback strategy is used in prover emulation but not in the schedule prediction.

Table 1. Features of the four main strategies.

Strategy	Parameter					Completeness
	`def`	`scut`	`cut`	`comp(7)`	`conj`	
`def,scut,cut,comp(7)`	1	1	1	1	0	1
`def,scut,cut`	1	1	1	0	0	0
`conj,scut,cut`	0	1	1	0	1	0
`def,conj,cut`	1	0	1	0	1	0

Theorem Features. The TPTP problem library contains a large, comprehensive collection of theorems and is designed for testing automated theorem provers. The problems are taken from a range of domains such as Logic Calculi, Algebra, Software Verification, Biology and Philosophy, and presented in multiple logical forms. For iLeanCoP, we select the subset in first-order form, denoted there as `FOF`. In version 7.1.0, there are 8157 such problems covering 43 domains. Each problem consists of a set of formulae and a goal theorem. The problems are of varying sizes. For example, the problem named `HWV134+1` from the Hardware Verification domain contains 128975 formulae, whilst `SET703+4` from the Set Theory domain contains only 12.

We have constructed a dataset containing features extracted from the first-order logic problems in TPTP (see Appendix A). Here, we describe how those features were developed.

In deployment, a prover using our method to generate strategy schedules would have to extract features from the goal theorem at the beginning of a proof attempt. To minimize the computational overhead of feature extraction, in keeping with our goal noted in Remark 3, we use features that can be collected when the theorem is parsed by the prover. The collection of features developed in this work is based on the authors' prior experience, and later we will briefly examine the quality of each feature to discard the uninformative ones. We extract the following features, which are all considered candidates for the subsequent feature selection process.

Symbol Counts: A count of the logical connectives and quantifiers. We extract one feature per symbol by tracking lexical symbols encountered while parsing.
Quantifier Rank: The maximum depth of nesting of quantifiers.
Quantifier Count: A count of the number of quantifiers.
Mean and Maximum Function Arity: Obtained by keeping track of functions during parsing.
Number of Functions: A count of the number of functions.
Quantifier Alternations: A count of the number of times the quantifiers flip between the existential and universal. When calculated by examining only the sequence of lexical symbols, the count may be inaccurate. An accurate count is obtained by tracking negations during parsing while collecting quantifiers. We extract both as candidates.

Feature Selection and Pre-processing. We examine the degree of association between the individual theorem features described above and the speed with which the strategies solve each theorem; for this we use the Maximal Information Coefficient (MIC) measure [23]. For every theorem we calculate the score, as defined in Remark 7, averaged over all strategies. This score is paired with each feature to calculate its MIC. Most lexical symbols achieve an MIC close to zero. We selected the features with relatively high MIC for the presented work, and these are shown in Fig. 3.

The two features based on quantifier alternations are clearly correlated, but both meet the above criterion for selection. Correlations can also be expected between the other features. Furthermore, our features range over different scales. For example, the maximal function arity in TPTP averages 2, whereas the number of predicate symbols averages 2097. It is desirable to remove these correlations to alleviate any burden on the subsequent modelling phase, and to standardize the features to zero mean and unit variance to create a feature space with similar length-scales in all dimensions. The former is achieved by *decorrelation*, the latter by *standardization*, and both together by a *sphering transformation*.

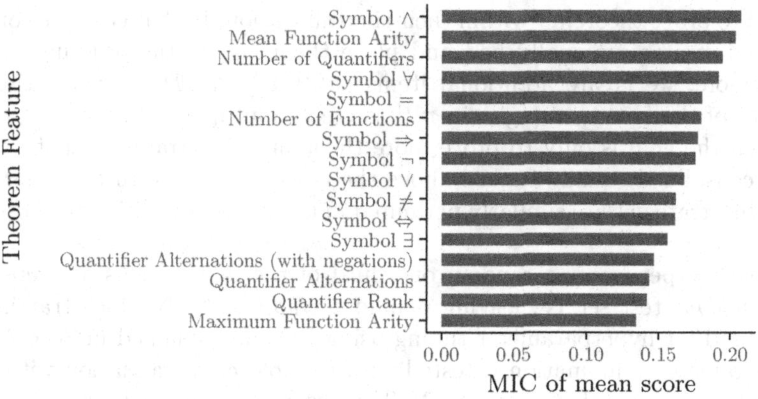

Fig. 3. MIC between selected features and scores.

We transform our extracted features as such using Zero-phase Component Analysis (ZCA), which ensures the transformed data is as close as possible to the original [6].

Coverage. As mentioned above, we run iLeanCoP on every first-order theorem in TPTP with each strategy allocated 1200 s. Although every theorem in intuitionistic logic also holds for classical logic, the converse does not hold. For that reason and because of the limitations of iLeanCoP, many theorems remain unproved by any strategy. We exclude these theorems from our experiments, leaving us with a data-set of 2240 theorems.

9 Experiments

We present two experiments in this work, as noted in Remark 9. In this section, we describe our experimental apparatus in detail.

As noted in Sect. 8, our data contains:

- $N = 2240$ theorems that are usable in our experiments;
- five strategies, of which $M = 4$ are used to build strategy schedules since one is a fallback strategy; and
- features $\boldsymbol{x}^{(i)}$ of theorems where $i \in [1, N]$ and features $\boldsymbol{d}^{(j)}$ of strategies where $j \in [1, M]$.

This data needs to be presented to our model for training in the form of $\mathcal{D} = \{(\boldsymbol{\pi}^{(i)}, \boldsymbol{x}^{(i)})\}_{i=1}^{N}$, as described in Sect. 3. Since the two experiments have slightly different goals, we specialize $\mathcal{D}$ according to each.

When aiming to predict schedules that minimize the time taken to prove theorems, a natural value for $\boldsymbol{\pi}^{(i)}$ is the index order that sorts strategies in increasing amounts of time taken to prove theorem i. However, some strategies may fail to prove theorem i within their time allocation. In that case, we consider the failed strategies equally bad and place them last in the ordering in $\boldsymbol{\pi}^{(i)}$. Furthermore, we create additional items $(\boldsymbol{\pi}'^{(i)}, \boldsymbol{x}^{(i)})$ in $\mathcal{D}$, by permuting the positions of the failed strategies in $\boldsymbol{\pi}^{(i)}$ to create multiple $\boldsymbol{\pi}'^{(i)}$.

When the goal is only to prove more theorems, the strategies that succeed are all considered equally ranked above the failed strategies. In this mode, the successful strategies are similarly permuted in the data, in addition to those that failed.

In each experiment, a random one-third of the N theorems are separated into a *holdout* test set $\dot{\boldsymbol{N}}$, leaving behind a training set $\ddot{\boldsymbol{N}}$. This training set is first used for hyperparameter tuning using BO. As explained in Sect. 7, each hyperparameter combination is tested with five-fold cross-validation within $\ddot{\boldsymbol{N}}$, to penalize instances that overfit to $\ddot{\boldsymbol{N}}$. This results in estimated optimum values for the hyperparameters. These are used to set the model, which is then trained on $\ddot{\boldsymbol{N}}$ and then finally evaluated on $\dot{\boldsymbol{N}}$. The whole process is repeated ten times with new random splits $\dot{\boldsymbol{N}}$ and $\ddot{\boldsymbol{N}}$ to create one set of ten results for that experiment.

10 Results

Each experiment, repeated ten times, is conducted in two phases: first, hyperparameter optimization; and second, model training and evaluation. The bounds on the search space in the first phase were always the same (see Appendix A). The holdout test set contained 747 theorems. A proof time of 600 s was emulated.

10.1 Experiment 1: Optimizing Proof Attempt Time

The results are shown in Fig. 4. The total prediction time for all 747 theorems, averaged across the trials, is 0.14 s.

The times across proof attempts are not normally distributed, for both the unmodified iLeanCoP schedule and the predicted ones, as confirmed by a Jarque-Bera test. Therefore, we used the right-tailed Wilcoxon signed-rank test for a pair-wise comparison of the times taken for each theorem by the original schedule in iLeanCoP versus the predicted schedules, resulting in a p-value of less

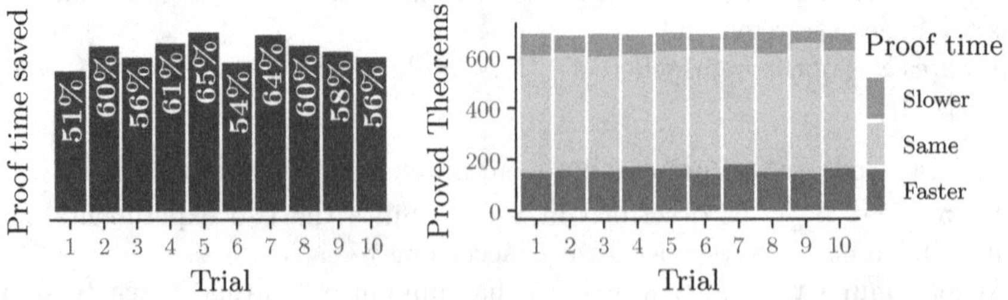

Fig. 4. Results of Experiment 1. Proof times are compared with precision 10^{-6}s.

than 10^{-6} in each trial, confirming the alternate hypothesis that the reduction in time taken to prove each theorem comes from a distribution with median greater than zero. This confirms that the time savings are statistically significant. Furthermore, we note from Fig. 4 a saving of more than 50% in the total proof-time in each trial.

10.2 Experiment 2: Proving More Theorems

We set our hyperparameter search to find time allocations for strategies. The resulting predicted schedules have gains and losses when compared to the original schedule, as shown in the four facets of Fig. 5. However, there is a consistent gain in the number of theorems proved and a gain of five theorems on average, evident from the mean values in (†) and (‡).

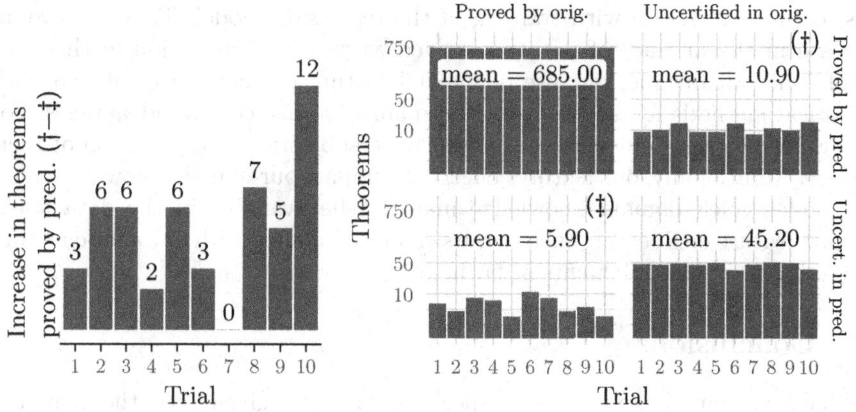

Fig. 5. Comparison of the proof attempts by the original (orig.) and predicted (pred.) schedules in Experiment 2. Theorems which are proved by pred. but could not be proved by orig. are counted in †, and the vice versa in ‡.

11 Related Work

Prior work on machine learning for *algorithm selection*, such as that introduced by Leyton-Brown et al. [13], is a precursor to our work. In that topic, the machine learning methods must perform the task of selecting a good algorithm from within a portfolio to solve the given problem instance. Typically, as was the case in the work by Leyton-Brown et al. [13], the learning methods predict the runtime of all algorithms, and then pick the fastest predicted one. This line of enquiry has been extended to select algorithms for SMT solvers—a recent example is MachSMT by Scott et al. [28]. The machine learning models in MachSMT are trained by considering all the portfolio members in pairs for each problem in the training set. This method is called *pairwise ranking*, which contrasts from our method, called *list-wise ranking*, in which we consider the full list of portfolio members all together.

In terms of the machine learning task, the work on scheduling solvers bears greater similarity to our presented work. In MedleySolver, for example, Pimpalkhare et al. [20] frame this task as a multi-armed bandit problem. They predict a sequence of solvers as well as the time allocation for each to generate schedules for the goal problems. MedleySolver is able to solve more problems than any individual solver would on its own.

With an approach that contrasts with ours, Hůla et al. [10] have made use of Graph Neural Networks (GNNs) for solver scheduling. They produce a regression model to predict, for the given problem, the runtime of all the solvers; which is used as the key to sort the solvers in increasing order of predicted runtime to build a schedule. This is an example of *point-wise ranking*. The authors use GNNs to automatically discover features for machine learning. They combine this feature extraction with training of the regression model. They achieve an increase in the number of problems solved as well as a reduction in the total proof time. Meanwhile, our use of manual feature engineering combined with statistical methods for selection and normalization has certain advantages. For one, we can analyse our features and derive a subjective interpretation of their efficacy. Additionally, our features effectively impart our domain knowledge onto the model. Such domain knowledge may not be available in the data itself. Manual feature engineering such as ours can be combined with automatic feature extraction to reap the benefits of both.

12 Conclusions

We have presented a method to specialize, for the given goal theorem, the sequence of strategies in the schedule used in each proof attempt. A Bayesian machine learning model is trained in this method using data generated by testing the prover of interest. When evaluated with the iLeanCoP prover using the TPTP library as a benchmark, our results show a significant reduction in the time taken to prove theorems. For theorems that are successfully proved, the average time saving is above 50%. The prediction time is on average low enough to have a negligible impact on the resources subtracted from the proof search itself.

We also extend this method to optimize time allocations to each strategy. In this setting, our results show a notable increase in the number of theorems proved.

This work shows, by example, that Bayesian machine learning models designed specifically to augment heuristics in theorem provers, with detailed consideration of the computational compromises required in this setting, can deliver substantial improvements.

Acknowledgments. Initial investigations for this work were co-supervised by Prof. Mateja Jamnik, Computer Laboratory, University of Cambridge, UK.

This work was supported by: the UK Engineering and Physical Sciences Research Council (EPSRC) through a Doctoral Training studentship, award reference 1788755; and the ERC Advanced Grant ALEXANDRIA (Project GA 742178). For the purpose of open access, the author has applied a Creative Commons Attribution (CC-BY-4.0) licence to any Author Accepted Manuscript version arising.

Computations for this work was performed using resources provided by the Cambridge Service for Data Driven Discovery (CSD3) operated by the University of Cambridge Research Computing Service, provided by Dell EMC and Intel using Tier-2 funding from the Engineering and Physical Sciences Research Council (capital grant EP/P020259/1), and DiRAC funding from the Science and Technology Facilities Council.

A Implementation, Code and Data

This work is implemented primarily in Matlab [36]. All experiments can be reproduced using the code, data and instructions available at [15]. The hyper-parameter search space in all experiments was restricted to $\varsigma \in [10, 300]$, $\sigma \in [0.01, 100.0]$ and $\eta \in [1, 100]$.

References

1. Balunovic, M., Bielik, P., Vechev, M.T.: Learning to solve SMT formulas. In: Annual Conference on Neural Information Processing Systems, pp. 10338–10349 (2018)
2. Barber, D.: Bayesian Reasoning and Machine Learning. Cambridge University Press, Cambridge (2012)
3. Bridge, J.P., Holden, S.B., Paulson, L.C.: Machine learning for first-order theorem proving. J. Autom. Reas. **53**(2), 141–172 (2014). https://doi.org/10.1007/s10817-014-9301-5
4. Cao, Z., Qin, T., Liu, T.Y., Tsai, M.F., Li, H.: Learning to rank: from pairwise approach to listwise approach. In: International Conference on Machine Learning, pp. 129–136. Association for Computing Machinery (2007)
5. Cheng, W., Dembczynski, K., Hüllermeier, E.: Label ranking methods based on the Plackett-Luce model. In: International Conference on Machine Learning, pp. 215–222. Omnipress (2010)
6. Duboue, P.: The Art of Feature Engineering: Essentials for Machine Learning. Cambridge University Press, Cambridge (2020)
7. Dummett, M.: Elements of Intuitionism, 2nd edn. Clarendon, Oxford (2000)
8. Guiver, J., Snelson, E.: Bayesian inference for Plackett-Luce ranking models. In: International Conference on Machine Learning, pp. 377–384. Association for Computing Machinery (2009)
9. Hales, T., et al.: A formal proof of the Kepler conjecture. In: Forum of Mathematics, Pi, vol. 5 (2017)
10. Hůla, J., Mojžíšek, D., Janota, M.: Graph neural networks for scheduling of SMT solvers. In: International Conference on Tools with Artificial Intelligence, pp. 447–451 (2021)
11. Johnson, S.R., Henderson, D.A., Boys, R.J.: On Bayesian inference for the Extended Plackett-Luce model (2020). arXiv:2002.05953

12. Kaliszyk, C., Urban, J., Vyskočil, J.: Machine learner for automated reasoning 0.4 and 0.5 (2014). arXiv:1402.2359
13. Leyton-Brown, K., Nudelman, E., Andrew, G., McFadden, J., Shoham, Y.: A portfolio approach to algorithm selection. In: International Joint Conference on Artificial Intelligence, pp. 1542–1543. Morgan Kaufmann Publishers Inc. (2003)
14. Luce, R.D.: Individual Choice Behavior: A Theoretical Analysis. Wiley, Hoboken (1959)
15. Mangla, C.: BRASS (2022). https://doi.org/10.5281/zenodo.6028568
16. Murphy, K.P.: Machine Learning: A Probabilistic Perspective. MIT press, Cambridge (2012)
17. Neal, R.M.: Bayesian Learning for Neural Networks. Springer, New York (1996). https://doi.org/10.1007/978-1-4612-0745-0
18. Otten, J.: Clausal connection-based theorem proving in intuitionistic first-order logic. In: Beckert, B. (ed.) TABLEAUX 2005. LNCS (LNAI), vol. 3702, pp. 245–261. Springer, Heidelberg (2005). https://doi.org/10.1007/11554554_19
19. Otten, J.: leanCoP 2.0 and ileanCoP 1.2: high performance lean theorem proving in classical and intuitionistic logic (system descriptions). In: Armando, A., Baumgartner, P., Dowek, G. (eds.) IJCAR 2008. LNCS (LNAI), vol. 5195, pp. 283–291. Springer, Heidelberg (2008). https://doi.org/10.1007/978-3-540-71070-7_23
20. Pimpalkhare, N., Mora, F., Polgreen, E., Seshia, S.A.: MedleySolver: online SMT algorithm selection. In: Li, C.-M., Manyà, F. (eds.) SAT 2021. LNCS, vol. 12831, pp. 453–470. Springer, Cham (2021). https://doi.org/10.1007/978-3-030-80223-3_31
21. Plackett, R.L.: The analysis of permutations. J. Roy. Stat. Soc. Ser. C (Appl. Stat.) **24**(2), 193–202 (1975)
22. Rasmussen, C.E., Williams, C.K.I.: Gaussian Processes for Machine Learning. Adaptive Computation and Machine Learning, MIT Press, Cambridge (2006)
23. Reshef, D.N., et al.: Detecting novel associations in large data sets. Science **334**(6062), 1518–1524 (2011)
24. Roberts, G.O., Rosenthal, J.S.: Optimal scaling for various Metropolis-Hastings algorithms. Stat. Sci. **16**(4), 351–367 (2001)
25. Rossi, P.E.: Bayesian Statistics and Marketing. Wiley, Hoboken (2006)
26. Schäfer, D., Hüllermeier, E.: Dyad ranking using Plackett-Luce models based on joint feature representations. Mach. Learn. **107**(5), 903–941 (2018)
27. Schulz, S.: E - a Brainiac Theorem Prover. AI Commun. **15**(23), 111–126 (2002)
28. Scott, J., Niemetz, A., Preiner, M., Nejati, S., Ganesh, V.: MachSMT: a machine learning-based algorithm selector for SMT solvers. In: TACAS 2021. LNCS, vol. 12652, pp. 303–325. Springer, Cham (2021). https://doi.org/10.1007/978-3-030-72013-1_16
29. Shahriari, B., Swersky, K., Wang, Z., Adams, R.P., De Freitas, N.: Taking the human out of the loop: a review of Bayesian optimization. Proc. IEEE **104**(1), 148–175 (2015)
30. Shawe-Taylor, J.: Kernel Methods for Pattern Analysis. Cambridge University Press, Cambridge (2004)
31. Silverthorn, B., Miikkulainen, R.: Latent class models for algorithm portfolio methods. In: AAAI Conference on Artificial Intelligence, pp. 167–172. AAAI Press (2010)
32. Sutcliffe, G.: The TPTP problem library and associated infrastructure: from CNF to TH0, TPTP v.6.4.0. J. Autom. Reason. **59**(4), 483–502 (2017)
33. Tammet, T.: Gandalf. J. Autom. Reason. **18**(2), 199–204 (1997)

34. Thornton, C.: Parity: the problem that won't go away. In: McCalla, G. (ed.) AI 1996. LNCS, vol. 1081, pp. 362–374. Springer, Heidelberg (1996). https://doi.org/10.1007/3-540-61291-2_65
35. Tipping, M.E.: Sparse Bayesian learning and the relevance vector machine. J. Mach. Learn. Res. **1**, 211–244 (2001)
36. Trauth, M.H.: MATLAB® Recipes for Earth Sciences. Springer, Heidelberg (2021). https://doi.org/10.1007/3-540-27984-9
37. Tsarkov, D., Horrocks, I.: FaCT++ description logic reasoner: system description. In: Furbach, U., Shankar, N. (eds.) IJCAR 2006. LNCS (LNAI), vol. 4130, pp. 292–297. Springer, Heidelberg (2006). https://doi.org/10.1007/11814771_26
38. Wasserman, L.: All of Statistics: A Concise Course in Statistical Inference. Springer, New York (2004). https://doi.org/10.1007/978-0-387-21736-9

A Framework for Approximate Generalization in Quantitative Theories

Temur Kutsia(✉) and Cleo Pau

RISC, Johannes Kepler University Linz, Linz, Austria
{kutsia,ipau}@risc.jku.at

Abstract. Anti-unification aims at computing generalizations for given terms, retaining their common structure and abstracting differences by variables. We study quantitative anti-unification where the notion of the common structure is relaxed into "proximal" up to the given degree with respect to the given fuzzy proximity relation. Proximal symbols may have different names and arities. We develop a generic set of rules for computing minimal complete sets of approximate generalizations and study their properties. Depending on the characterizations of proximities between symbols and the desired forms of solutions, these rules give rise to different versions of concrete algorithms.

Keywords: Generalization · Anti-unification · Quantiative theories · Fuzzy proximity relations

1 Introduction

Generalization problems play an important role in various areas of mathematics, computer science, and artificial intelligence. Anti-unification [12,14] is a logic-based method for computing generalizations. Being originally used for inductive and analogical reasoning, some recent applications include recursion scheme detection in functional programs [4], programming by examples in domain-specific languages [13], learning bug-fixing from software code repositories [3,15], automatic program repair [7], preventing bugs and misconfiguration in services [11], linguistic structure learning for chatbots [6], to name just a few.

In most of the existing theories where anti-unification is studied, the background knowledge is assumed to be precise. Therefore, those techniques are not suitable for reasoning with incomplete, imprecise information (which is very common in real-world communication), where the exact equality is replaced by its (quantitative) approximation. Fuzzy proximity and similarity relations are notable examples of such extensions. These kinds of quantitative theories have many useful applications, some most recent ones being related to artificial intelligence, program verification, probabilistic programming, or natural language processing. Many tasks arising in these areas require reasoning methods and computational tools that deal with quantitative information. For instance, approximate

J. Blanchette et al. (Eds.): IJCAR 2022, LNAI 13385, pp. 578–596, 2022.
https://doi.org/10.1007/978-3-031-10769-6_34

inductive reasoning, reasoning and programming by analogy, similarity detection in programming language statements or in natural language texts could benefit from solving approximate generalization constraints, which is a theoretically interesting and challenging task. Investigations in this direction have been started only recently. In [1], the authors proposed an anti-unification algorithm for fuzzy similarity (reflexive, symmetric, min-transitive) relations, where mismatches are allowed not only in symbol names, but also in their arities (fully fuzzy signatures). The algorithm from [9] is designed for fuzzy proximity (i.e., reflexive and symmetric) relations with mismatches only in symbol names.

In this paper, we study approximate anti-unification from a more general perspective. The considered relations are fuzzy proximity relations. Proximal symbols may have different names and arities. We consider four different variants of relating arguments between different proximal symbols: unrestricted relations/functions, and correspondence (i.e. left- and right-total) relations/functions. A generic set of rules for computing minimal complete sets of generalizations is introduced and its termination, soundness and completeness properties are proved. From these rules, we obtain concrete algorithms that deal with different kinds of argument relations. We also show how the existing approximate anti-unification algorithms and their generalizations fit into this framework.

Organization: In Sect. 2 we introduce the notation and definitions. Section 3 is devoted to a technical notion of term set consistency and to an algorithm for computing elements of consistent sets of terms. It is used later in the main set of anti-unification rules, which are introduced and characterized in Sect. 4. The concrete algorithms obtained from those rules are also described in this section. In Sect. 5, we discuss complexity. Section 6 offers a high-level picture of the studied problems and concludes.

An extended version of this work can be found in the technical report [8].

2 Preliminaries

Proximity Relations. Given a set S, a mapping $\mathcal{R}$ from $S \times S$ to the real interval $[0,1]$ is called a binary *fuzzy relation* on S. By fixing a number λ, $0 \leqslant \lambda \leqslant 1$, we can define the crisp (i.e., two-valued) counterpart of $\mathcal{R}$, named the λ-*cut* of $\mathcal{R}$, as $\mathcal{R}_\lambda := \{(s_1, s_2) \mid \mathcal{R}(s_1, s_2) \geqslant \lambda\}$. A fuzzy relation $\mathcal{R}$ on a set S is called a *proximity relation* if it is reflexive ($\mathcal{R}(s,s) = 1$ for all $s \in S$) and symmetric ($\mathcal{R}(s_1, s_2) = \mathcal{R}(s_2, s_1)$ for all $s_1, s_2 \in S$). A T-norm $\wedge$ is an associative, commutative, non-decreasing binary operation on $[0,1]$ with 1 as the unit element. We take minimum in the role of T-norm.

Terms and Substitutions. We consider a first-order alphabet consisting of a set of fixed arity function symbols $\mathcal{F}$ and a set of variables $\mathcal{V}$, which includes a special symbol _ (the anonymous variable). The set of *named* (i.e., non-anonymous) variables $\mathcal{V}\backslash\{\_\}$ is denoted by $\mathcal{V}^{\mathrm{N}}$. When the set of variables is not explicitly specified, we mean $\mathcal{V}$. The set of terms $\mathcal{T}(\mathcal{F}, \mathcal{V})$ over $\mathcal{F}$ and $\mathcal{V}$ is defined in the standard way: $t \in \mathcal{T}(\mathcal{F}, \mathcal{V})$ iff t is defined by the grammar $t := x \mid f(t_1, \ldots, t_n)$, where $x \in \mathcal{V}$ and $f \in \mathcal{F}$ is an n-ary symbol with $n \geqslant 0$. Terms over $\mathcal{T}(\mathcal{F}, \mathcal{V}^{\mathrm{N}})$ are defined similarly except that all variables are taken from $\mathcal{V}^{\mathrm{N}}$.

We denote arbitrary function symbols by f, g, h, constants by a, b, c, variables by x, y, z, v, and terms by s, t, r. The *head* of a term is defined as $\mathsf{head}(x) := x$ and $\mathsf{head}(f(t_1, \ldots, t_n)) := f$. For a term t, we denote with $\mathcal{V}(t)$ (resp. by $\mathcal{V}^{\mathrm{N}}(t)$) the set of all variables (resp. all named variables) appearing in t. A term is called *linear* if no named variable occurs in it more than once.

The deanonymization operation deanon replaces each occurrence of the anonymous variable in a term by a fresh variable. For instance, $\mathsf{deanon}(f(\_, x, g(\_))) = f(y', x, g(y'')))$, where y' and y'' are fresh. Hence, $\mathsf{deanon}(t) \in \mathcal{T}(\mathcal{F}, \mathcal{V}^{\mathrm{N}})$ is unique up to variable renaming for all $t \in \mathcal{T}(\mathcal{F}, \mathcal{V})$. $\mathsf{deanon}(t)$ is linear iff t is linear.

The notions of *term depth*, *term size* and a *position in a term* are defined in the standard way, see, e.g. [2]. By $t|_p$ we denote the subterm of t at position p and by $t[s]_p$ a term that is obtained from t by replacing the subterm at position p by the term s.

A *substitution* is a mapping from $\mathcal{V}^{\mathrm{N}}$ to $\mathcal{T}(\mathcal{F}, \mathcal{V}^{\mathrm{N}})$ (i.e., without anonymous variables), which is the identity almost everywhere. We use the Greek letters $\sigma, \vartheta, \varphi$ to denote substitutions, except for the identity substitution which is written as Id. We represent substitutions with the usual set notation. *Application* of a substitution σ to a term t, denoted by $t\sigma$, is defined as $\_\sigma := \_$, $x\sigma := \sigma(x)$, $f(t_1, \ldots, t_n)\sigma := f(t_1\sigma, \ldots, t_n\sigma)$. Substitution *composition* is defined as a composition of mappings. We write $\sigma\vartheta$ for the composition of σ with ϑ.

Argument Relations and Mappings. Given two sets $N = \{1, \ldots, n\}$ and $M = \{1, \ldots, m\}$, a binary *argument relation* over $N \times M$ is a (possibly empty) subset of $N \times M$. We denote argument relations by ρ. An argument relation $\rho \subseteq N \times M$ is (i) *left-total* if for all $i \in N$ there exists $j \in M$ such that $(i, j) \in \rho$; (ii) *right-total* if for all $j \in M$ there exists $i \in N$ such that $(i, j) \in \rho$. *Correspondence relations* are those that are both left- and right-total.

An *argument mapping* is an argument relation that is a partial injective function. In other words, an argument mapping π from $N = \{1, \ldots, n\}$ to $M = \{1, \ldots, m\}$ is a function $\pi : I_n \mapsto I_m$, where $I_n \subseteq N$, $I_m \subseteq M$ and $|I_n| = |I_m|$. Note that it can be also the empty mapping: $\pi : \emptyset \mapsto \emptyset$. The inverse of an argument mapping is again an argument mapping.

Given a proximity relation $\mathcal{R}$ over $\mathcal{F}$, we assume that for each pair of function symbols f and g with $\mathcal{R}(f, g) = \alpha > 0$, where f is n-ary and g is m-ary, there is also given an argument relation ρ over $\{1, \ldots, n\} \times \{1, \ldots, m\}$. We use the notation $f \sim^{\rho}_{\mathcal{R},\alpha} g$. These argument relations should satisfy the following conditions: ρ is the empty relation if f or g is a constant; ρ is the identity if $f = g$; $f \sim^{\rho}_{\mathcal{R},\alpha} g$ iff $g \sim^{\rho^{-1}}_{\mathcal{R},\alpha} f$, where ρ^{-1} is the inverse of ρ.

Example 1. Assume that we have four different versions of defining the notion of author (e.g., originated from four different knowledge bases) $author_1$(*first*-name, *middle-initial*, *last*-name), $author_2$(*first*-name, *last*-name), $author_3$(*last*-name, *first*-name, *middle-initial*), and $author_4$(*full*-name). One could define the argument relations/mappings between these function symbols e.g., as follows:

$$author_1 \sim^{\{(1,1),(3,2)\}}_{\mathcal{R},0.7} author_2, \quad author_1 \sim^{\{(3,1),(1,2),(2,3)\}}_{\mathcal{R},0.9} author_3,$$
$$author_1 \sim^{\{(1,1),(3,1)\}}_{\mathcal{R},0.5} author_4, \quad author_2 \sim^{\{(1,2),(2,1)\}}_{\mathcal{R},0.7} author_3,$$
$$author_2 \sim^{\{(1,1),(2,1)\}}_{\mathcal{R},0.5} author_4, \quad author_3 \sim^{\{(1,1),(2,1)\}}_{\mathcal{R},0.5} author_4.$$

Proximity Relations over Terms. Each proximity relation $\mathcal{R}$ in this paper is defined on $\mathcal{F} \cup \mathcal{V}$ such that $\mathcal{R}(f,x) = 0$ for all $f \in \mathcal{F}$ and $x \in \mathcal{V}$, and $\mathcal{R}(x,y) = 0$ for all $x \neq y$, $x, y \in \mathcal{V}$. We assume that $\mathcal{R}$ is *strict*: for all $w_1, w_2 \in \mathcal{F} \cup \mathcal{V}$, if $\mathcal{R}(w_1, w_2) = 1$, then $w_1 = w_2$. Yet another assumption is that for each $f \in \mathcal{F}$, its $(\mathcal{R}, \lambda)$-proximity class $\{g \mid \mathcal{R}(f,g) \geqslant \lambda\}$ is *finite* for any $\mathcal{R}$ and λ.

We extend such an $\mathcal{R}$ to terms from $\mathcal{T}(\mathcal{F}, \mathcal{V})$ as follows:

(a) $\mathcal{R}(t,s) := 0$ if $\mathcal{R}(\mathsf{head}(s), \mathsf{head}(t)) = 0$;
(b) $\mathcal{R}(t,s) := 1$ if $t = s$ and $t, s \in \mathcal{V}$;
(c) $\mathcal{R}(t,s) := \mathcal{R}(f,g) \wedge \mathcal{R}(t_{i_1}, s_{j_1}) \wedge \cdots \wedge \mathcal{R}(t_{i_k}, s_{j_k})$, if $t = f(t_1, \ldots, t_n)$, $s = g(s_1, \ldots, s_m)$, $f \sim^{\rho}_{\mathcal{R},\lambda} g$, and $\rho = \{(i_1, j_1), \ldots, (i_k, j_k)\}$.

If $\mathcal{R}(t,s) \geqslant \lambda$, we write $t \simeq_{\mathcal{R},\lambda} s$. When $\lambda = 1$, the relation $\simeq_{\mathcal{R},\lambda}$ does not depend on $\mathcal{R}$ due to strictness of the latter and is just the syntactic equality $=$.

The $(\mathcal{R}, \lambda)$-*proximity class* of a term t is $\mathbf{pc}_{\mathcal{R},\lambda}(t) := \{s \mid s \simeq_{\mathcal{R},\lambda} t\}$.

Generalizations. Given $\mathcal{R}$ and λ, a term r is an $(\mathcal{R}, \lambda)$-*generalization* of (alternatively, $(\mathcal{R}, \lambda)$-*more general than*) a term t, written as $r \precsim_{\mathcal{R},\lambda} t$, if there exists a substitution σ such that $\mathsf{deanon}(r)\sigma \simeq_{\mathcal{R},\lambda} \mathsf{deanon}(t)$. The strict part of $\precsim_{\mathcal{R},\lambda}$ is denoted by $\prec_{\mathcal{R},\lambda}$, i.e., $r \prec_{\mathcal{R},\lambda} t$ if $r \precsim_{\mathcal{R},\lambda} t$ and not $t \precsim_{\mathcal{R},\lambda} r$.

Example 2. Given a proximity relation $\mathcal{R}$, a cut value λ, constants $a \sim^{\varnothing}_{\mathcal{R},\alpha_1} b$ and $b \sim^{\varnothing}_{\mathcal{R},\alpha_2} c$, binary function symbols f and h, and a unary function symbol g such that $h \sim^{\{(1,1),(1,2)\}}_{\mathcal{R},\alpha_3} f$ and $h \sim^{\{(1,1)\}}_{\mathcal{R},\alpha_4} g$ with $\alpha_i \geqslant \lambda$, $1 \leqslant i \leqslant 4$, we have

- $h(x, \_) \precsim_{\mathcal{R},\lambda} h(a,x)$, because $h(x,x')\{x \mapsto a, x' \mapsto x\} = h(a,x) \simeq_{\mathcal{R},\lambda} h(a,x)$.
- $h(x, \_) \precsim_{\mathcal{R},\lambda} h(\_, x)$, because $h(x,x')\{x \mapsto y', x' \mapsto x\} = h(y',x) \simeq_{\mathcal{R},\lambda} h(y',x)$.
- $h(x,x) \not\precsim_{\mathcal{R},\lambda} h(\_, x)$, because $h(x,x) \not\precsim_{\mathcal{R},\lambda} h(y',x)$.
- $h(x, \_) \precsim_{\mathcal{R},\lambda} f(a,c)$, because $h(x,x')\{x \mapsto b\} = h(b,x') \simeq_{\mathcal{R},\lambda} f(a,c)$.
- $h(x, \_) \precsim_{\mathcal{R},\lambda} g(c)$, because $h(x,x')\{x \mapsto c\} = h(c,x') \simeq_{\mathcal{R},\lambda} g(c)$.

The notion of *syntactic generalization* of a term is a special case of $(\mathcal{R}, \lambda)$-generalization for $\lambda = 1$. We write $r \precsim t$ to indicate that r is a syntactic generalization of t. Its strict part is denoted by $\prec$.

Since $\mathcal{R}$ is strict, $r \precsim t$ is equivalent to $\mathsf{deanon}(r)\sigma = \mathsf{deanon}(t)$ for some σ (note the syntactic equality here).

Theorem 1. *If $r \lesssim t$ and $t \lesssim_{\mathcal{R},\lambda} s$, then $r \lesssim_{\mathcal{R},\lambda} s$.*

Proof. $r \lesssim t$ implies $\mathsf{deanon}(r)\sigma = \mathsf{deanon}(t)$ for some σ, while from $t \lesssim_{\mathcal{R},\lambda} s$ we have $\mathsf{deanon}(t)\vartheta \simeq_{\mathcal{R},\lambda} \mathsf{deanon}(s)$ for some ϑ. Then $\mathsf{deanon}(r)\sigma\vartheta \simeq_{\mathcal{R},\lambda} \mathsf{deanon}(s)$, which implies $r \lesssim_{\mathcal{R},\lambda} s$. □

Note that $r \lesssim_{\mathcal{R},\lambda} t$ and $t \lesssim_{\mathcal{R},\lambda} s$, in general, do not imply $r \lesssim_{\mathcal{R},\lambda} s$ due to non-transitivity of $\simeq_{\mathcal{R},\lambda}$.

Definition 1 (Minimal complete set of $(\mathcal{R},\lambda)$-generalizations). *Given $\mathcal{R}$, λ, t_1, and t_2, a set of terms T is a* complete set of $(\mathcal{R},\lambda)$-generalizations *of t_1 and t_2 if*

(a) every $r \in T$ is an $(\mathcal{R},\lambda)$-generalization of t_1 and t_2,
(b) if r' is an $(\mathcal{R},\lambda)$-generalization of t_1 and t_2, then there exists $r \in T$ such that $r' \lesssim r$ (note that we use syntactic generalization here).

In addition, T is minimal, if it satisfies the following property:

(c) if $r, r' \in T$, $r \neq r'$, then neither $r \prec_{\mathcal{R},\lambda} r'$ nor $r' \prec_{\mathcal{R},\lambda} r$.

A minimal complete set of $(\mathcal{R},\lambda)$-generalizations *($(\mathcal{R},\lambda)$-mcsg) of two terms is unique modulo variable renaming. The elements of the $(\mathcal{R},\lambda)$-mcsg of t_1 and t_2 are called least general $(\mathcal{R},\lambda)$-generalizations ($(\mathcal{R},\lambda)$-lggs) of t_1 and t_2.*

This definition directly extends to generalizations of finitely many terms.

The problem of computing an $(\mathcal{R},\lambda)$-generalization of terms t and s is called the $(\mathcal{R},\lambda)$-*anti-unification problem* of t and s. In anti-unification, the goal is to compute their least general $(\mathcal{R},\lambda)$-generalization.

The precise formulation of the anti-unification problem would be the following: Given $\mathcal{R}$, λ, t_1, t_2, find an $(\mathcal{R},\lambda)$-lgg r of t_1 and t_2, substitutions σ_1, σ_2, and the approximation degrees α_1, α_2 such that $\mathcal{R}(r\sigma_1, t_1) = \alpha_1$ and $\mathcal{R}(r\sigma_2, t_2) = \alpha_2$. A minimal complete algorithm to solve this problem would compute exactly the elements of $(\mathcal{R},\lambda)$-mcsg of t_1 and t_2 together with their approximation degrees. However, as we see below, it is problematic to solve the problem in this form. Therefore, we will consider a slightly modified variant, taking into account anonymous variables in generalizations and relaxing bounds on their degrees.

We assume that the terms to be generalized are ground. It is not a restriction because we can treat variables as constants that are close only to themselves.

Recall that the proximity class of any alphabet symbol is finite. Also, the symbols are related to each other by finitely many argument relations. One may think that it leads to finite proximity classes of terms, but this is not the case. Consider, e.g., $\mathcal{R}$ and λ, where $h \simeq_{\mathcal{R},\lambda}^{\{(1,1)\}} f$ with binary h and unary f. Then the $(\mathcal{R},\lambda)$-proximity class of $f(a)$ is infinite: $\{f(a)\} \cup \{h(a,t) \mid t \in \mathcal{T}(\mathcal{F},\mathcal{V})\}$. Also, the $(\mathcal{R},\lambda)$-mcsg for $f(a)$ and $f(b)$ is infinite: $\{f(x)\} \cup \{h(x,t) \mid t \in \mathcal{T}(\mathcal{F},\varnothing)\}$.

Definition 2. *Given the terms $t_1,\ldots,t_n$, $n \geqslant 1$, a position p in a term r is called* irrelevant for $(\mathcal{R},\lambda)$-generalizing *(resp.* for $(\mathcal{R},\lambda)$-proximity to*) $t_1,\ldots,t_n$ if $r[s]_p \lesssim_{\mathcal{R},\lambda} t_i$ (resp. $r[s]_p \simeq_{\mathcal{R},\lambda} t_i$) for all $1 \leqslant i \leqslant n$ and for all terms s.*

We say that r is a relevant $(\mathcal{R}, \lambda)$-generalization *(resp.* relevant $(\mathcal{R}, \lambda)$-proximal term*) of $t_1, \ldots, t_n$ if $r \lesssim_{\mathcal{R},\lambda} t_i$ (resp. $r \simeq_{\mathcal{R},\lambda} t_i$) for all $1 \leqslant i \leqslant n$ and $r|_p = \_$ for all positions p in r that is irrelevant for generalizing (resp. for proximity to) $t_1, \ldots, t_n$. The $(\mathcal{R}, \lambda)$*-relevant proximity class *of t is*

$$\mathbf{rpc}_{\mathcal{R},\lambda}(t) := \{s \mid s \text{ is a relevant } (\mathcal{R}, \lambda)\text{-proximal term of } t\}.$$

In the example above, position 2 in $h(x, t)$ is irrelevant for generalizing $f(a)$ and $f(b)$, and $h(x, \_)$ is one of their relevant generalizations. Note that $f(x)$ is also a relevant generalization of $f(a)$ and $f(b)$, since it contains no irrelevant positions. More general generalizations like, e.g., x, are relevant as well. Similarly, position 2 in $h(a, t)$ is irrelevant for proximity to $f(a)$ and $\mathbf{rpc}_{\mathcal{R},\lambda}(f(a)) = \{f(a), h(a, \_)\}$. Generally, $\mathbf{rpc}_{\mathcal{R},\lambda}(t)$ is finite for any t due to the finiteness of proximity classes of symbols and argument relations mentioned above.

Definition 3 (Minimal complete set of relevant $(\mathcal{R}, \lambda)$-generalizations). *Given $\mathcal{R}$, λ, t_1, and t_2, a set of terms T is a* complete set of relevant $(\mathcal{R}, \lambda)$-generalizations *of t_1 and t_2 if*

(a) every element of T is a relevant $(\mathcal{R}, \lambda)$-generalization of t_1 and t_2, and
(b) if r is a relevant $(\mathcal{R}, \lambda)$-generalization of t_1 and t_2, then there exists $r' \in T$ such that $r \lesssim r'$.

The minimality property is defined as in Definition 1.

This definition directly extends to relevant generalizations of finitely many terms. We use $(\mathcal{R}, \lambda)$-mcsrg as an abbreviation for minimal complete set of relevant $(\mathcal{R}, \lambda)$-generalization. Like relevant proximity classes, mcsrg's are also finite.

Lemma 1. *For given $\mathcal{R}$ and λ, if all argument relations are correspondence relations, then $(\mathcal{R}, \lambda)$-mcsg's and $(\mathcal{R}, \lambda)$-proximity classes for all terms are finite.*

Proof. Under correspondence relations no term contains an irrelevant position for generalization or for proximity. □

Hence, for correspondence relations the notions of mcsg and mcsrg coincide, as well as the notions of proximity class and relevant proximity class.

For a term r, we define its *linearized version* $\mathsf{lin}(r)$ as a term obtained from r by replacing each occurrence of a named variable in r by a fresh one. For instance, $\mathsf{lin}(f(x, \_, g(y, x, a), b)) = f(x', \_, g(y', x'', a), b)$, where x', x'', y' are fresh variables. Linearized versions of terms are unique modulo variable renaming.

Definition 4 (Generalization degree upper bound). *Given two terms r and t, a proximity relation $\mathcal{R}$, and a λ-cut, the $(\mathcal{R}, \lambda)$*-generalization degree upper bound of r and t, *denoted by $\mathsf{gdub}_{\mathcal{R},\lambda}(r, t)$, is defined as follows:*

Let $\alpha := \max\{\mathcal{R}(\mathsf{lin}(r)\sigma, t) \mid \sigma$ is a substitution$\}$. Then $\mathsf{gdub}_{\mathcal{R},\lambda}(r, t)$ is α if $\alpha \geqslant \lambda$, and 0 otherwise.

Intuitively, $\mathsf{gdub}_{\mathcal{R},\lambda}(r,t) = \alpha$ means that no instance of r can get closer than α to t in $\mathcal{R}$. From the definition it follows that if $r \lesssim_{\mathcal{R},\lambda} t$, then $0 < \lambda \leqslant \mathsf{gdub}_{\mathcal{R},\lambda}(r,t) \leq 1$ and if $r \not\lesssim_{\mathcal{R},\lambda} t$, then $\mathsf{gdub}_{\mathcal{R},\lambda}(r,t) = 0$.

The upper bound computed by gdub is more relaxed than it would be if the linearization function were not used, but this is what we will be able to compute in our algorithms later.

Example 3. Let $\mathcal{R}(a,b) = 0.6$, $\mathcal{R}(b,c) = 0.7$, and $\lambda = 0.5$. Then $\mathsf{gdub}_{\mathcal{R},\lambda}(f(x,b), f(a,c)) = 0.7$ and $\mathsf{gdub}_{\mathcal{R},\lambda}(f(x,x), f(a,c)) = \mathsf{gdub}_{\mathcal{R},\lambda}(f(x,y), f(a,c)) = 1$.

It is not difficult to see that if $r\sigma \simeq_{\mathcal{R},\lambda} t$, then $\mathcal{R}(r\sigma, t) \leqslant \mathsf{gdub}_{\mathcal{R},\lambda}(r,t)$. In Example 3, for $\sigma = \{x \mapsto b\}$ we have $\mathcal{R}(f(x,x)\sigma, f(a,c)) = \mathcal{R}(f(b,b), f(a,c)) = 0.6 < \mathsf{gdub}_{\mathcal{R},\lambda}(f(x,x), f(a,c)) = 1$.

We compute $\mathsf{gdub}_{\mathcal{R},\lambda}(r,t)$ as follows: If r is a variable, then $\mathsf{gdub}_{\mathcal{R},\lambda}(r,t) = 1$. Otherwise, if $\mathsf{head}(r) \sim^{\rho}_{\mathcal{R},\beta} \mathsf{head}(t)$, then $\mathsf{gdub}_{\mathcal{R},\lambda}(r,t) = \beta \wedge \bigwedge_{(i,j)\in\rho} \mathsf{gdub}_{\mathcal{R},\lambda}(r|_i, t|_j)$. Otherwise, $\mathsf{gdub}_{\mathcal{R},\lambda}(r,t) = 0$.

3 Term Set Consistency

The notion of term set consistency plays an important role in the computation of proximal generalizations. Intuitively, a set of terms is $(\mathcal{R},\lambda)$-consistent if all the terms in the set have a common $(\mathcal{R},\lambda)$-proximal term. In this section, we discuss this notion and the corresponding algorithms.

Definition 5 (Consistent set of terms). *A finite set of terms T is $(\mathcal{R},\lambda)$-*consistent *if there exists a term s such that $s \simeq_{\mathcal{R},\lambda} t$ for all $t \in T$.*

$(\mathcal{R},\lambda)$-consistency of a finite term set T is equivalent to $\bigcap_{t\in T} \mathbf{pc}_{\mathcal{R},\lambda}(t) \neq \emptyset$, but we cannot use this property to decide consistency, since proximity classes of terms can be infinite (when the argument relations are not restricted). For this reason, we introduce the operation $\sqcap$ on terms as follows: (i) $t \sqcap \_ = \_ \sqcap t = t$, (ii) $f(t_1,\ldots,t_n) \sqcap f(s_1,\ldots,s_n) = f(t_1 \sqcap s_1,\ldots,t_n \sqcap s_n)$, $n \geqslant 0$. Obviously, $\sqcap$ is associative (A), commutative (C), idempotent (I), and has $\_$ as its unit element (U). It can be extended to sets of terms: $T_1 \sqcap T_2 := \{t_1 \sqcap t_2 \mid t_1 \in T_1, t_2 \in T_2\}$. It is easy to see that $\sqcap$ on sets also satisfies the ACIU properties with the set $\{\_\}$ playing the role of the unit element.

Lemma 2. *A finite set of terms T is $(\mathcal{R},\lambda)$-consistent iff $\bigsqcap_{t\in T} \mathbf{rpc}_{\mathcal{R},\lambda}(t) \neq \emptyset$.*

Proof. ($\Rightarrow$) If $s \simeq_{\mathcal{R},\lambda} t$ for all $t \in T$, then $s_t \in \mathbf{rpc}_{\mathcal{R},\lambda}(t)$, where s_t is obtained from s by replacing all subterms that are irrelevant for its $(\mathcal{R},\lambda)$-proximity to t by $\_$. Assume $T = \{t_1,\ldots,t_n\}$. Then $s_{t_1} \sqcap \cdots \sqcap s_{t_n} \in \bigsqcap_{t\in T} \mathbf{rpc}_{\mathcal{R},\lambda}(t)$.

($\Leftarrow$) Obvious, since $s \simeq_{\mathcal{R},\lambda} t$ for $s \in \bigsqcap_{t\in T} \mathbf{rpc}_{\mathcal{R},\lambda}(t)$ and for all $t \in T$. □

Now we design an algorithm $\mathfrak{C}$ that computes $\bigsqcap_{t\in T} \mathbf{rpc}_{\mathcal{R},\lambda}(t)$ without actually computing $\mathbf{rpc}_{\mathcal{R},\lambda}(t)$ for each $t \in T$. A special version of the algorithm can be used to decide the $(\mathcal{R},\lambda)$-consistency of T.

The algorithm is rule-based. The rules work on states, that are pairs $\mathbf{I}; s$, where s is a term and $\mathbf{I}$ is a finite set of expressions of the form x in T, where T is a finite set of terms. $\mathcal{R}$ and λ are given. There are two rules ($\uplus$ stands for disjoint union):

Rem: **Removing the empty set**
$\{x \text{ in } \varnothing\} \uplus \mathbf{I}; s \Longrightarrow \mathbf{I}; s\{x \mapsto \_\}$.

Red: **Reduce a set to new sets**
$\{x \text{ in } \{t_1, \ldots, t_m\}\} \uplus \mathbf{I}; s \Longrightarrow \{y_1 \text{ in } T_1, \ldots, y_n \text{ in } T_n\} \cup \mathbf{I}; s\{x \mapsto h(y_1, \ldots, y_n)\}$, where $m \geqslant 1$, h is an n-ary function symbol such that $h \sim^{\rho_k}_{\mathcal{R},\gamma_k} \mathsf{head}(t_k)$ with $\gamma_k \geqslant \lambda$ for all $1 \leqslant k \leqslant m$, and $T_i := \{t_k|_j \mid (i,j) \in \rho_k, 1 \leqslant k \leqslant m\}$, $1 \leqslant i \leqslant n$, is the set of all those arguments of the terms $t_1, \ldots, t_m$ that are supposed to be $(\mathcal{R}, \lambda)$-proximal to the i's argument of h.

To compute $\bigsqcap_{t \in T} \mathbf{rpc}_{\mathcal{R},\lambda}(t)$, $\mathfrak{C}$ starts with $\{x \text{ in } T\}; x$ and applies the rules as long as possible. Red causes branching. A state of the form $\varnothing; s$ is called a success state. A failure state has the form $\mathbf{I}; s$, to which no rule applies and $\mathbf{I} \neq \varnothing$. In the full derivation tree, each leaf is a either success or a failure state.

Example 4. Assume a, b, c are constants, g, f, h are function symbols with the arities respectively 1, 2, and 3. Let λ be given and $\mathcal{R}$ be defined so that $\mathcal{R}(a,b) \geqslant \lambda$, $\mathcal{R}(b,c) \geqslant \lambda$, $h \sim^{\{(1,1),(1,2)\}}_{\mathcal{R},\beta} f$, $h \sim^{\{(2,1)\}}_{\mathcal{R},\gamma} g$ with $\beta \geqslant \lambda$ and $\gamma \geqslant \lambda$. Then

$$\mathbf{rpc}_{\mathcal{R},\lambda}(f(a,c)) = \{f(a,c),\ f(b,c),\ f(a,b),\ f(b,b),\ h(b, \_, \_)\},$$
$$\mathbf{rpc}_{\mathcal{R},\lambda}(g(a)) = \{g(a),\ g(b), h(\_, a, \_), h(\_, b, \_)\},$$

and $\mathbf{rpc}_{\mathcal{R},\lambda}(f(a,c)) \sqcap \mathbf{rpc}_{\mathcal{R},\lambda}(g(a)) = \{h(b,a,\_), h(b,b,\_)\}$. We show how to compute this set with $\mathfrak{C}$: $\{x \text{ in } \{f(a,c), g(a)\}\};\ x \Longrightarrow_{\mathsf{Red}} \{y_1 \text{ in } \{a,c\}, y_2 : \{a\}, y_3 \text{ in } \varnothing\};\ h(y_1,y_2,y_3) \Longrightarrow_{\mathsf{Rem}} \{y_1 \text{ in } \{a,c\},\ y_2 : \{a\}\};\ h(y_1,y_2,\_) \Longrightarrow_{\mathsf{Red}} \{y_2 \text{ in } \{a\}\}; h(b,y_2,\_)$. Here we have two ways to apply Red to the last state, leading to two elements of $\mathbf{rpc}_{\mathcal{R},\lambda}(f(a,c)) \sqcap \mathbf{rpc}_{\mathcal{R},\lambda}(g(a))$: $h(b,a,\_)$ and $h(b,b,\_)$.

Theorem 2. *Given a finite set of terms T, the algorithm $\mathfrak{C}$ always terminates starting from the state $\{x \text{ in } T\}; x$ (where x is a fresh variable). If S is the set of success states produced at the end, we have $\{s \mid \varnothing; s \in S\} = \bigsqcap_{t \in T} \mathbf{rpc}_{\mathcal{R},\lambda}(t)$.*

Proof. Termination: Associate to each state $\{x_1 \text{ in } T_1, \ldots x_n \text{ in } T_n\}; s$ the multiset $\{d_1, \ldots, d_n\}$, where d_i is the maximum depth of terms occurring in T_i. $d_i = 0$ if $T_i = \varnothing$. Compare these multisets by the Dershowitz-Manna ordering [5]. Each rule strictly reduces them, which implies termination.

By the definitions of $\mathbf{rpc}_{\mathcal{R},\lambda}$ and $\sqcap$, $h(s_1, \ldots, s_n) \in \bigsqcap_{t \in \{t_1,\ldots,t_m\}} \mathbf{rpc}_{\mathcal{R},\lambda}(t)$ iff $h \sim^{\rho_k}_{\mathcal{R},\gamma_k} \mathsf{head}(t_k)$ with $\gamma_k \geqslant \lambda$ for all $1 \leqslant k \leqslant m$ and $s_i \in \bigsqcap_{t \in T_i} \mathbf{rpc}_{\mathcal{R},\lambda}(t)$, where $T_i = \{t_k|_j \mid (i,j) \in \rho_k, 1 \leqslant k \leqslant m\}$, $1 \leqslant i \leqslant n$. Therefore, in the Rem rule, the instance of x (which is $h(y_1, \ldots, y_n)$) is in $\bigsqcap_{t \in \{t_1,\ldots,t_m\}} \mathbf{rpc}_{\mathcal{R},\lambda}(t)$ iff for each $1 \leqslant i \leqslant n$ we can find an instance of y_i in $\bigsqcap_{t \in T_i} \mathbf{rpc}_{\mathcal{R},\lambda}(t)$. If T_i is empty, it means that the i's argument of h is irrelevant for terms in $\{t_1, \ldots, t_m\}$ and can be replaced by _. (Rem does it in a subsequent step.) Hence, in each success branch of the derivation tree, the algorithm $\mathfrak{C}$ computes one element of $\bigsqcap_{t \in T} \mathbf{rpc}_{\mathcal{R},\lambda}(t)$. Branching at Red helps produce all elements of $\bigsqcap_{t \in T} \mathbf{rpc}_{\mathcal{R},\lambda}(t)$. □

It is easy to see how to use $\mathfrak{C}$ to decide the $(\mathcal{R}, \lambda)$-consistency of T: it is enough to find one successful branch in the $\mathfrak{C}$-derivation tree for $\{x \text{ in } T\}; x$. If there is no such branch, then T is not $(\mathcal{R}, \lambda)$-consistent. In fact, during the derivation we can even ignore the second component of the states.

4 Solving Generalization Problems

Now we can reformulate the anti-unification problem that will be solved in the remaining part of the paper. $\mathcal{R}$ is a proximity relation and λ is a cut value.

Given: $\mathcal{R}$, λ, and the ground terms $t_1, \ldots, t_n$, $n \geqslant 2$.

Find: a set S of tuples $(r, \sigma_1, \ldots, \sigma_n, \alpha_1, \ldots, \alpha_n)$ such that

- $\{r \mid (r, \ldots) \in \mathsf{S}\}$ is an $(\mathcal{R}, \lambda)$-mcsrg of $t_1, \ldots, t_n$,
- $r\sigma_i \simeq_{\mathcal{R},\lambda} t_i$ and $\alpha_i = \mathsf{gdub}_{\mathcal{R},\lambda}(r, t_i)$, $1 \leqslant i \leqslant n$, for each $(r, \sigma_1, \ldots, \sigma_n, \alpha_1, \ldots, \alpha_n) \in \mathsf{S}$.

(When $n = 1$, this is a problem of computing a relevant proximity class of a term.) Below we give a set of rules, from which one can obtain algorithms to solve the anti-unification problem for four versions of argument relations:

1. The most general (unrestricted) case; see algorithm $\mathfrak{A}_1$ below, the computed set of generalizations is an mcsrg;
2. Correspondence relations: using the same algorithm $\mathfrak{A}_1$, the computed set of generalizations is an mcsg;
3. Mappings: using a dedicated algorithm $\mathfrak{A}_2$, the computed set of generalizations is an mcsrg;
4. Correspondence mappings (bijections): using the same algorithm $\mathfrak{A}_2$, the computed set of generalizations is an mcsg.

Each of them has also the corresponding linear variant, computing minimal complete sets of (relevant) linear $(\mathcal{R}, \lambda)$-generalizations. They are denoted by adding the superscript lin to the corresponding algorithm name: $\mathfrak{A}_1^{\mathsf{lin}}$ and $\mathfrak{A}_2^{\mathsf{lin}}$.

For simplicity, we formulate the algorithms for the case $n = 2$. They can be extended for arbitrary n straightforwardly.

The main data structure in these algorithms is an anti-unification triple (AUT) $x : T_1 \triangleq T_2$, where T_1 and T_2 are finite *consistent* sets of ground terms. The idea is that x is a common generalization of all terms in $T_1 \cup T_2$. A configuration is a tuple $A; S; r; \alpha_1; \alpha_2$, where A is a set of AUTs to be solved, S is a set of solved AUTs (the store), r is the generalization computed so far, and the α's are the current approximations of generalization degree upper bounds of r for the input terms.

Before formulating the rules, we discuss one peculiarity of approximate generalizations:

Example 5. For a given $\mathcal{R}$ and λ, assume $\mathcal{R}(a, b) \geqslant \lambda$, $\mathcal{R}(b, c) \geqslant \lambda$, $h \sim_{\mathcal{R},\alpha}^{\{(1,1),(1,2)\}} f$ and $h \sim_{\mathcal{R},\beta}^{\{(1,1)\}} g$, where f is binary, g, h are unary, $\alpha \geqslant \lambda$ and $\beta \geqslant \lambda$. Then

- $h(b)$ is an $(\mathcal{R}, \lambda)$-generalization of $f(a, c)$ and $g(a)$.
- x is the only $(\mathcal{R}, \lambda)$-generalization of $f(a, d)$ and $g(a)$. One may be tempted to have h as the head of the generalization, e.g., $h(x)$, but x cannot be instantiated by any term that would be $(\mathcal{R}, \lambda)$-close to both a and d, since in the given $\mathcal{R}$, d is $(\mathcal{R}, \lambda)$-close only to itself. Hence, there would be no instance of $h(x)$ that is $(\mathcal{R}, \lambda)$-close to $f(a, d)$. Since there is no other alternative (except h) for the common neighbor of f and g, the generalization should be a fresh variable x.

This example shows that generalization algorithms should take into account not only the heads of the terms to be generalized, but also should look deeper, to make sure that the arguments grouped together by the given argument relation have a common neighbor. This justifies the requirement of consistency of a set of arguments, the notion introduced in the previous section and used in the decomposition rule below.

4.1 Anti-unification for Unrestricted Argument Relations

Algorithms $\mathfrak{A}_1^{\mathsf{lin}}$ and $\mathfrak{A}_1$ use the rules below to transform configurations into configurations. Given $\mathcal{R}$, λ, and the ground terms t_1 and t_2, we create the initial configuration $\{x : \{t_1\} \triangleq \{t_2\}\}; \varnothing; x; 1; 1$ and apply the rules as long as possible. Note that the rules preserve consistency of AUTs. The process generates a finite complete tree of derivations, whose terminal nodes have configurations with the first component empty. We will show how from these terminal configurations one collects the result as required in the anti-unification problem statement.

Tri: **Trivial**

$\{x : \varnothing \triangleq \varnothing\} \uplus A;\ S;\ r;\ \alpha_1; \alpha_2 \Longrightarrow A;\ S;\ r\{x \mapsto \_\};\ \alpha_1; \alpha_2.$

Dec: **Decomposition**

$\{x : T_1 \triangleq T_2\} \uplus A; S; r; \alpha_1; \alpha_2 \Longrightarrow$

$\quad \{y_i : Q_{i1} \triangleq Q_{i2} \mid 1 \leqslant i \leqslant n\} \cup A; S; r\{x \mapsto h(y_1, \ldots, y_n)\}; \alpha_1 \wedge \beta_1; \alpha_2 \wedge \beta_2,$

where $T_1 \cup T_2 \neq \varnothing$; h is n-ary with $n \geqslant 0$; $y_1, \ldots, y_n$ are fresh; and for $j = 1, 2$, if $T_j = \{t_1^j, \ldots, t_{m_j}^j\}$, then

- $h \sim_{\mathcal{R}, \gamma_k^j}^{\rho_k^j} \mathsf{head}(t_k^j)$ with $\gamma_k^j \geqslant \lambda$ for all $1 \leqslant k \leqslant m_j$ and $\beta_j = \gamma_1^j \wedge \cdots \wedge \gamma_{m_j}^j$ (note that $\beta_j = 1$ if $m_j = 0$),
- for all $1 \leqslant i \leqslant n$, $Q_{ij} = \cup_{k=1}^{m_j} \{t_k^j|_q \mid (i, q) \in \rho_k^j\}$ and is $(\mathcal{R}, \lambda)$-consistent.

Sol: **Solving**

$\{x : T_1 \triangleq T_2\} \uplus A;\ S;\ r;\ \alpha_1; \alpha_2 \Longrightarrow A;\ \{x : T_1 \triangleq T_2\} \cup S;\ r;\ \alpha_1; \alpha_2,$
if Tri and Dec rules are not applicable. (It means that at least one $T_i \neq \varnothing$ and either there is no h as it is required in the Dec rule, or at least one Q_{ij} from Dec is not $(\mathcal{R}, \lambda)$-consistent.)

Let expand be an *expansion operation* defined for sets of AUTs as

$$\mathsf{expand}(S) := \{x : \bigsqcap_{t \in T_1} \mathbf{rpc}_{\mathcal{R},\lambda}(t) \triangleq \bigsqcap_{t \in T_2} \mathbf{rpc}_{\mathcal{R},\lambda}(t) \mid x : T_1 \triangleq T_2 \in S\}.$$

Exhaustive application of the three rules above leads to configurations of the form $\emptyset; S; r; \alpha_1; \alpha_2$, where r is a linear term. These configurations are further postprocessed, replacing S by $\mathsf{expand}(S)$. We will use the letter E for expanded stores. Hence, terminal configurations obtained after the exhaustive rule application and expansion have the form $\emptyset; E; r; \alpha_1; \alpha_2$, where r is a linear term.[1] This is what Algorithm $\mathfrak{A}_1^{\mathsf{lin}}$ stops with.

To an expanded store $E = \{y_1 : Q_{11} \triangleq Q_{12}, \ldots, y_n : Q_{n1} \triangleq Q_{n2}\}$ we associate two sets of substitutions $\Sigma_L(E)$ and $\Sigma_R(E)$, defined as follows: $\sigma \in \Sigma_L(E)$ (resp. $\sigma \in \Sigma_R(E)$) iff $\mathsf{dom}(\sigma) = \{y_1, \ldots, y_n\}$ and $y_i\sigma \in Q_{i1}$ (resp. $y_i\sigma \in Q_{i2}$) for each $1 \leqslant i \leqslant n$. We call them the sets of *witness substitutions.*

Configurations containing expanded stores are called *expanded configurations.* From each expanded configuration $C = \emptyset; E; r; \alpha_1; \alpha_2$, we construct the set $\mathsf{S}(C) := \{(r, \sigma_1, \sigma_2, \alpha_1, \alpha_2) \mid \sigma_1 \in \Sigma_L(E), \sigma_2 \in \Sigma_R(E)\}$.

Given an anti-unification problem $\mathcal{R}$, λ, t_1 and t_2, the *answer computed by Algorithm* $\mathfrak{A}_1^{\mathsf{lin}}$ is the set $\mathsf{S} := \cup_{i=1}^m \mathsf{S}(C_i)$, where $C_1, \ldots, C_m$ are all of the final expanded configurations reached by $\mathfrak{A}_1^{\mathsf{lin}}$ for $\mathcal{R}$, λ, t_1, and t_2.[2]

Example 6. Assume a, b, c and d are constants with $b \sim_{\mathcal{R},0.5}^{\emptyset} c$, $c \sim_{\mathcal{R},0.6}^{\emptyset} d$, and f, g and h are respectively binary, ternary and quaternary function symbols with $h \sim_{\mathcal{R},0.7}^{\{(1,1),(3,2),(4,2)\}} f$ and $h \sim_{\mathcal{R},0.8}^{\{(1,1),(3,3)\}} g$. For the proximity relation $\mathcal{R}$ given in this way and $\lambda = 0.5$, Algorithm $\mathfrak{A}_1^{\mathsf{lin}}$ performs the following steps to anti-unify $f(a, b)$ and $g(a, c, d)$:

$$\{x : \{f(a, b)\} \triangleq \{g(a, c, d)\}\}; \emptyset; x; 1; 1 \Longrightarrow_{\mathsf{Dec}}$$
$$\{x_1 : \{a\} \triangleq \{a\},\ x_2 : \emptyset \triangleq \emptyset,\ x_3 : \{b\} \triangleq \{d\},$$
$$x_4 : \{b\} \triangleq \emptyset\}; \emptyset; h(x_1, x_2, x_3, x_4); 0.7; 0.8 \Longrightarrow_{\mathsf{Dec}}$$
$$\{x_2 : \emptyset \triangleq \emptyset,\ x_3 : \{b\} \triangleq \{d\},\ x_4 : \{b\} \triangleq \emptyset\}; \emptyset; h(a, x_2, x_3, x_4); 0.7; 0.8 \Longrightarrow_{\mathsf{Tri}}$$
$$\{x_3 : \{b\} \triangleq \{d\},\ x_4 : \{b\} \triangleq \emptyset\}; \emptyset; h(a, \_, x_3, x_4); 0.7; 0.8 \Longrightarrow_{\mathsf{Dec}}$$
$$\{x_4 : \{b\} \triangleq \emptyset\}; \emptyset; h(a, \_, c, x_4); 0.5; 0.6.$$

Here Dec applies in two different ways, with the substitutions $\{x_4 \mapsto b\}$ and $\{x_4 \mapsto c\}$, leading to two final configurations: $\emptyset; \emptyset; h(a, \_, c, b); 0.5; 0.6$ and $\emptyset; \emptyset; h(a, \_, c, c); 0.5; 0.6$. The witness substitutions are the identity substitutions. We have $\mathcal{R}(h(a, \_, c, b), f(a, b)) = 0.5$, $\mathcal{R}(h(a, \_, c, b), g(a, c, d)) = 0.6$, $\mathcal{R}(h(a, \_, c, c), f(a, b)) = 0.5$, and $\mathcal{R}(h(a, \_, c, c), g(a, c, d)) = 0.6$.

If we had $h \sim_{\mathcal{R},0.7}^{\{(1,1),(1,2),(4,2)\}} f$, then the algorithm would perform only the Sol step, because in the attempt to apply Dec to the initial configuration, the set

[1] Note that no side of the AUTs in E in those configurations is empty due to the condition at the **Decomposition** rule requiring the Q_{ij}'s to be $(\mathcal{R}, \lambda)$-consistent.

[2] If we are interested only in linear generalizations *without witness substitutions*, there is no need in computing expanded configurations in $\mathfrak{A}_1^{\mathsf{lin}}$.

$Q_{11} = \{a, b\}$ is inconsistent: $\mathbf{rpc}_{\mathcal{R},\lambda}(a) = \{a\}$, $\mathbf{rpc}_{\mathcal{R},\lambda}(b) = \{b, c\}$, and, hence, $\mathbf{rpc}_{\mathcal{R},\lambda}(a) \sqcap \mathbf{rpc}_{\mathcal{R},\lambda}(b) = \varnothing$.

Algorithm $\mathfrak{A}_1$ is obtained by further transforming the expanded configurations produced by $\mathfrak{A}_1^{\mathsf{lin}}$. This transformation is performed by applying the **Merge** rule below as long as possible. Intuitively, its purpose is to make the linear generalization obtained by $\mathfrak{A}_1^{\mathsf{lin}}$ less general by merging some variables.

Mer: **Merge**

$\varnothing; \{x_1 : R_{11} \triangleq R_{12},\, x_2 : R_{21} \triangleq R_{22}\} \uplus E;\, r;\, \alpha_1;\, \alpha_2 \Longrightarrow$
$\quad \varnothing;\, \{y : Q_1 \triangleq Q_2\} \cup E;\, r\sigma;\, \alpha_1;\, \alpha_2,$

where $Q_i = (R_{1i} \sqcap R_{2i}) \neq \varnothing$, $i = 1, 2$, y is fresh, and $\sigma = \{x_1 \mapsto y,\, x_2 \mapsto y\}$.

The answer computed by $\mathfrak{A}_1$ is defined similarly to the answer computed by $\mathfrak{A}_1^{\mathsf{lin}}$.

Example 7. Assume a, b are constants, f_1, f_2, g_1, and g_2 are unary function symbols, p is a binary function symbol, and h_1 and h_2 are ternary function symbols. Let λ be a cut value and $\mathcal{R}$ be defined as $f_i \sim_{\mathcal{R},\alpha_i}^{\{(1,1)\}} h_i$ and $g_i \sim_{\mathcal{R},\beta_i}^{\{(1,2)\}} h_i$ with $\alpha_i \geqslant \lambda$, $\beta_i \geqslant \lambda$, $i = 1, 2$. To generalize $p(f_1(a), g_1(b))$ and $p(f_2(a), g_2(b))$, we use $\mathfrak{A}_1$. The derivation starts as

$$\{x : \{p(f_1(a), g_1(b))\} \triangleq \{p(f_2(a), g_2(b))\}\};\, \varnothing;\, x;\, 1;\, 1 \Longrightarrow_{\mathsf{Dec}}$$
$$\{y_1 : \{f_1(a)\} \triangleq \{f_2(a)\},\, y_2 : \{g_1(b)\} \triangleq \{g_2(b)\}\};\, \varnothing;\, p(y_1, y_2);\, 1;\, 1 \Longrightarrow_{\mathsf{Sol}}^2$$
$$\varnothing;\, \{y_1 : \{f_1(a)\} \triangleq \{f_2(a)\},\, y_2 : \{g_1(b)\} \triangleq \{g_2(b)\}\};\, p(y_1, y_2);\, 1;\, 1.$$

At this stage, we expand the store, obtaining

$$\varnothing; \{y_1 : \{f_1(a), h_1(a, \_, \_)\} \triangleq \{f_2(a), h_2(a, \_, \_)\},$$
$$y_2 : \{g_1(b), h_1(\_, b, \_)\} \triangleq \{g_2(b), h_2(\_, b, \_)\}\};\, p(y_1, y_2);\, 1;\, 1.$$

If we had the standard intersection $\cap$ in the Mer rule, we would not be able to merge y_1 and y_2, because the obtained sets in the corresponding AUTs are disjoint. However, Mer uses $\sqcap$: we have $\{f_i(a), h_i(a, \_, \_)\} \sqcap \{g_i(b), h_i(\_, b, \_)\} = \{h_i(a, b, \_)\}$, $i = 1, 2$ and, therefore, can make the step

$$\varnothing; \{y_1 : \{f_1(a), h_1(a, \_, \_)\} \triangleq \{f_2(a), h_2(a, \_, \_)\},$$
$$y_2 : \{g_1(b), h_1(\_, b, \_)\} \triangleq \{g_2(b), h_2(\_, b, \_)\}\};\, p(y_1, y_2);\, 1;\, 1 \Longrightarrow_{\mathsf{Mer}}$$
$$\varnothing;\, \{z : \{h_1(a, b, \_)\} \triangleq \{h_2(a, b, \_)\}\};\, p(z, z);\, 1;\, 1.$$

Indeed, if we take the witness substitutions $\sigma_i = \{z \mapsto h_i(a, b, \_)\}$, $i = 1, 2$, and apply them to the obtained generalization, we get

$$p(z, z)\sigma_1 = p(h_1(a, b, \_), h_1(a, b, \_)) \simeq_{\mathcal{R},\lambda} p(f_1(a), g_1(b)),$$
$$p(z, z)\sigma_2 = p(h_2(a, b, \_), h_2(a, b, \_)) \simeq_{\mathcal{R},\lambda} p(f_2(a), g_2(b)).$$

Theorem 3. *Given $\mathcal{R}$, λ, and the ground terms t_1 and t_2, Algorithm $\mathfrak{A}_1$ terminates for $\{x : \{t_1\} \triangleq \{t_2\}\}; \varnothing; x; 1; 1$ and computes an answer set S such that*

1. *the set* $\{r \mid (r, \sigma_1, \sigma_2, \alpha_1, \alpha_2) \in \mathsf{S}\}$ *is an* $(\mathcal{R}, \lambda)$*-mcsrg of* t_1 *and* t_2,
2. *for each* $(r, \sigma_1, \sigma_2, \alpha_1, \alpha_2) \in \mathsf{S}$ *we have* $\mathcal{R}(r\sigma_i, t_i) \leqslant \alpha_i = \mathsf{gdub}_{\mathcal{R},\lambda}(r, t_i)$, $i = 1, 2$.

Proof. Termination: Define the depth of an AUT $x : \{t_1, \ldots, t_m\} \triangleq \{s_1, \ldots, s_n\}$ as the depth of the term $f(g(t_1, \ldots, t_m), h(s_1, \ldots, s_n))$. The rules Tri, Dec, and Sol strictly reduce the multiset of depths of AUTs in the first component of the configurations. Mer strictly reduces the number of distinct variables in generalizations. Hence, these rules cannot be applied infinitely often and $\mathfrak{A}_1$ terminates.

In order to prove (1), we need to verify three properties:

- Soundness: If $(r, \sigma_1, \sigma_2, \alpha_1, \alpha_2) \in \mathsf{S}$, then r is a relevant $(\mathcal{R}, \lambda)$-generalization of t_1 and t_2.
- Completeness: If r' is a relevant $(\mathcal{R}, \lambda)$-generalization of t_1 and t_2, then there exists $(r, \sigma_1, \sigma_2, \alpha_1, \alpha_2) \in \mathsf{S}$ such that $r' \lesssim r$.
- Minimality: If r and r' belong to two tuples from S such that $r \neq r'$, then neither $r \prec_{\mathcal{R},\lambda} r'$ nor $r' \prec_{\mathcal{R},\lambda} r$.

Soundness: We show that each rule transforms an $(\mathcal{R}, \lambda)$-generalization into an $(\mathcal{R}, \lambda)$-generalization. Since we start from a most general $(\mathcal{R}, \lambda)$-generalization of t_1 and t_2 (a fresh variable x), at the end of the algorithm we will get an $(\mathcal{R}, \lambda)$-generalization of t_1 and t_2. We also show that in this process all irrelevant positions are abstracted by anonymous variables, to guarantee that each computed generalization is relevant.

Dec: The computed h is $(\mathcal{R}, \lambda)$-close to the head of each term in $T_1 \cup T_2$. Q_{ij}'s correspond to argument relations between h and those heads, and each Q_{ij} is $(\mathcal{R}, \lambda)$-consistent, i.e., there exists a term that is $(\mathcal{R}, \lambda)$-close to each term in Q_{ij}. It implies that $x\sigma = h(y_1, \ldots, y_n)$ $(\mathcal{R}, \lambda)$-generalizes all the terms from $T_1 \cup T_2$. Note that at this stage, $h(y_1, \ldots, y_n)$ might not yet be a relevant $(\mathcal{R}, \lambda)$-generalization of T_1 and T_2: if there exists an irrelevant position $1 \leqslant i \leqslant n$ for the $(\mathcal{R}, \lambda)$-generalization of T_1 and T_2, then in the new configuration we will have an AUT $y_i : \varnothing \triangleq \varnothing$.

Tri: When Dec generates $y : \varnothing \triangleq \varnothing$, the Tri rule replaces y by _ in the computed generalization, making it relevant.

Sol does not change generalizations.

Mer merges AUTs whose terms have *nonempty* intersection of **rpc**'s. Hence, we can reuse the same variable in the corresponding positions in generalizations, i.e., Mer transforms a generalization computed so far into a less general one.

Completeness: We prove a slightly more general statement. Given two finite consistent sets of ground terms T_1 and T_2, if r' is a relevant $(\mathcal{R}, \lambda)$-generalization for all $t_1 \in T_1$ and $t_2 \in T_2$, then starting from $\{x : T_1 \triangleq T_2\}; \varnothing; x; 1; 1$, Algorithm $\mathfrak{A}_1$ computes a $(r, \sigma_1, \sigma_2, \alpha_1, \alpha_2)$ such that $r' \lesssim r$.

We may assume w.l.o.g. that r' is a relevant $(\mathcal{R}, \lambda)$-lgg. Due to the transitivity of $\lesssim$, completeness for such an r' will imply it for all terms more general than r'.

We proceed by structural induction on r'. If r' is a (named or anonymous) variable, the statement holds. Assume $r' = h(r'_1, \dots, r'_n)$, $T_1 = \{u_1, \dots, u_m\}$, and $T_2 = \{w_1, \dots, w_l\}$. Then h is such that $h \sim^{\rho_i}_{\mathcal{R},\beta_i} \mathsf{head}(u_i)$ for all $1 \leqslant i \leqslant m$ and $h \sim^{\mu_j}_{\mathcal{R},\gamma_j} \mathsf{head}(w_j)$ for all $1 \leqslant j \leqslant l$. Moreover, each r'_k is a relevant $(\mathcal{R}, \lambda)$-generalization of $Q_{k1} = \cup_{i=1}^{m}\{u_i|_q \mid (k,q) \in \rho_i\}$ and $Q_{k2} = \cup_{j=1}^{l}\{w_j|_q \mid (k,q) \in \mu_j\}$ and, hence, Q_{k1} and Q_{k2} are $(\mathcal{R}, \lambda)$-consistent. Therefore, we can perform a step by Dec, choosing $h(y_1, \dots, y_k)$ as the generalization term and $y_i : Q_{i1} \triangleq Q_{i2}$ as the new AUTs. By the induction hypothesis, for each $1 \leqslant i \leqslant n$ we can compute a relevant $(\mathcal{R}, \lambda)$-generalization r_i for Q_{i_1} and Q_{i2} such that $r'_i \lesssim r_i$.

If r' is linear, then the combination of the current Dec step with the derivations that lead to those r_i's computes a tuple $(r, \dots) \in \mathsf{S}$, where $r = h(r_1, \dots, r_n)$ and, hence, $r' \lesssim r$.

If r' is non-linear, assume without loss of generality that all occurrences of a shared variable z appear as the direct arguments of h: $z = r'_{k_1} = \dots = r'_{k_p}$ for $1 \leqslant k_1 < \dots < k_p \leqslant n$. Since r' is an lgg, Q_{k_i1} and Q_{k_i2} cannot be generalized by a non-variable term, thus, Tri and Dec are not applicable. Therefore, the AUTs $y_i : Q_{k_i1} \triangleq Q_{k_i2}$ would be transformed by Sol. Since all pairs Q_{k_i1} and Q_{k_i2}, $1 \leqslant i \leqslant p$, are generalized by the same variable, we have $\sqcap_{t \in Q_j} \mathbf{rpc}_{\mathcal{R},\lambda}(t) \neq \emptyset$, where $Q_j = \cup_{i=1}^{p} Q_{k_ij}$, $j = 1, 2$. Additionally, $r'_{k_1}, \dots, r'_{k_p}$ are all occurrences of z in r'. Hence, the condition of Mer is satisfied and we can extend our derivation with $p-1$-fold application of this rule, obtaining $r = h(r_1, \dots, r_n)$ with $z = r_{k_1} = \dots = r_{k_p}$, implying $r' \lesssim r$.

Minimality: Alternative generalizations are obtained by branching in Dec or Mer. If the current generalization r is transformed by Dec into two generalizations r_1 and r_2 on two branches, then $r_1 = h_1(y_1, \dots, y_m)$ and $r_2 = h_2(z_1, \dots, z_n)$ for some h's, and fresh y's and z's. It may happen that $r_1 \lesssim_{\mathcal{R},\lambda} r_2$ or vice versa (if h_1 and h_2 are $(\mathcal{R}, \lambda)$-close to each other), but neither $r_1 \prec_{\mathcal{R},\lambda} r_2$ nor $r_2 \prec_{\mathcal{R},\lambda} r_1$ holds. Hence, the set of generalizations computed before applying Mer is minimal. Mer groups AUTs together maximally, and different groupings are not comparable. Therefore, variables in generalizations are merged so that distinct generalizations are not $\prec_{\mathcal{R},\lambda}$-comparable. Hence, (1) is proven.

As for (2), for $i = 1, 2$, from the construction in Dec follows $\mathcal{R}(r\sigma_i, t_i) \leqslant \alpha_i$. Mer does not change α_i, thus, $\alpha_i = \mathsf{gdub}_{\mathcal{R},\lambda}(r, t_i)$ also holds, since the way how α_i is computed corresponds exactly to the computation of $\mathsf{gdub}_{\mathcal{R},\lambda}(r, t_i)$: $r \lesssim_{\mathcal{R},\lambda} t_i$ and only the decomposition changes the degree during the computation. □

The corollary below is proved similarly to Theorem 3:

Corollary 1. *Given $\mathcal{R}$, λ, and the ground terms t_1 and t_2, Algorithm $\mathfrak{A}_1^{\mathsf{lin}}$ terminates for $\{x : \{t_1\} \triangleq \{t_2\}\}; \emptyset; x; 1; 1$ and computes an answer set S such that*

1. *the set $\{r \mid (r, \sigma_1, \sigma_2, \alpha_1, \alpha_2) \in \mathsf{S}\}$ is a minimal complete set of relevant* linear *$(\mathcal{R}, \lambda)$-generalizations of t_1 and t_2,*
2. *for each $(r, \sigma_1, \sigma_2, \alpha_1, \alpha_2) \in \mathsf{S}$ we have $\mathcal{R}(r\sigma_i, t_i) \leqslant \alpha_i = \mathsf{gdub}_{\mathcal{R},\lambda}(r, t_i)$, $i = 1, 2$.*

4.2 Anti-unification with Correspondence Argument Relations

Correspondence relations make sure that for a pair of proximal symbols, no argument is irrelevant for proximity. Left- and right-totality of those relations guarantee that each argument of a term is close to at least one argument of its proximal term and the inverse relation remains a correspondence relation. Consequently, in the Dec rule of $\mathfrak{A}_1$, the sets Q_{ij} never get empty. Therefore, the Tri rule becomes obsolete and no anonymous variable appears in generalizations. As a result, the $(\mathcal{R}, \lambda)$-mcsrg and the $(\mathcal{R}, \lambda)$-mcsg coincide, and the algorithm computes a solution from which we get an $(\mathcal{R}, \lambda)$-mcsg for the given anti-unification problem. The linear version $\mathfrak{A}_1^{\mathsf{lin}}$ works analogously.

4.3 Anti-unification with Argument Mappings

When the argument relations are mappings, we are able to design a more constructive method for computing generalizations and their degree bounds (Recall that our mappings are partial injective functions, which guarantees that their inverses are also mappings.) We denote this algorithm by $\mathfrak{A}_2$. The configurations stay the same as in before, but the AUTs in A will contain only empty or singleton sets of terms. In the store, we may still get (after the expansion) AUTs with term sets containing more than one element. Only the Dec rule differs from its previous counterpart, having a simpler condition:

Dec: **Decomposition**

$$\{x : T_1 \triangleq T_2\} \uplus A; S; r; \alpha_1; \alpha_2 \Longrightarrow$$
$$\{y_i : Q_{i1} \triangleq Q_{i2} \mid 1 \leqslant i \leqslant n\} \cup A; S; r\{x \mapsto h(y_1, \ldots, y_n)\}; \alpha_1 \wedge \beta_1; \alpha_2 \wedge \beta_2,$$

where $T_1 \cup T_2 \neq \emptyset$; h is n-ary with $n \geqslant 0$; $y_1, \ldots, y_n$ are fresh; for $j = 1, 2$ and for all $1 \leqslant i \leqslant n$, if $T_j = \{t_j\}$ then $h \sim^{\pi_j}_{\mathcal{R},\beta_j} \mathsf{head}(t_j)$ and $Q_{ij} = \{t_j|_{\pi_j(i)}\}$, and if $T_j = \emptyset$ then $\beta_j = 1$ and $Q_{ij} = \emptyset$.

This Dec rule is equivalent to the special case of Dec for argument relations where $m_j \leqslant 1$. The new Q_{ij}'s contain at most one element (due to mappings) and, thus, are always $(\mathcal{R}, \lambda)$-consistent. Various choices of h in Dec and alternatives in grouping AUTs in Mer cause branching in the same way as in $\mathfrak{A}_1$. It is easy to see that the counterparts of Theorem 3 hold for $\mathfrak{A}_2$ and $\mathfrak{A}_2^{\mathsf{lin}}$ as well.

A special case of this fragment of anti-unification is anti-unification for similarity relations in fully fuzzy signatures from [1]. Similarity relations are min-transitive proximity relations. The position mappings in [1] can be modeled by our argument mappings, requiring them to be total for symbols of the smaller arity and to satisfy the similarity-specific consistency restrictions from [1].

4.4 Anti-unification with Correspondence Argument Mappings

Correspondence argument mappings are bijections between arguments of function symbols of the same arity. For such mappings, if $h \simeq^{\pi}_{\mathcal{R},\lambda} f$ and h is n-ary, then f is also n-ary and π is a permutation of $(1, \ldots, n)$. Hence, $\mathfrak{A}_2$ combines in this case the properties of $\mathfrak{A}_1$ for correspondence relations (Sect. 4.2) and of $\mathfrak{A}_2$ for argument mappings (Sect. 4.3): all generalizations are relevant, computed answer gives an mcsg of the input terms, and the algorithm works with term sets of cardinality at most 1.

5 Remarks About the Complexity

The proximity relation $\mathcal{R}$ can be naturally represented as an undirected graph, where the vertices are function symbols and an edge between them indicates that they are proximal. Graphs induced by proximity relations are usually sparse. Therefore we can represent them by (sorted) adjacency lists. In the adjacency lists, we can also accommodate the argument relations and proximity degrees.

In the rest of this section we use the following notation:

- n: the size of the input (number of symbols) of the corresponding algorithms,
- Δ: the maximum degree of $\mathcal{R}$ considered as a graph,
- $\mathfrak{a}$: the maximum arity of function symbols that occur in $\mathcal{R}$.
- $m^{\bullet n}$: a function defined on natural numbers m and n such that $1^{\bullet n} = n$ and $m^{\bullet n} = m^n$ for $m \neq 1$.

We assume that the given anti-unification problem is represented as a completely shared directed acyclic graph (dag). Each node of the dag has a pointer to the adjacency list (with respect to $\mathcal{R}$) of the symbol in the node.

Theorem 4. *Time complexities of $\mathfrak{C}$ and the linear versions of the generalization algorithms are as follows:*

- *$\mathfrak{C}$ for argument relations and $\mathfrak{A}_1^{\mathsf{lin}}$:* $O(n \cdot \Delta \cdot \Delta^{\bullet \mathfrak{a}^{\bullet n}})$,
- *$\mathfrak{C}$ for argument mappings and $\mathfrak{A}_2^{\mathsf{lin}}$:* $O(n \cdot \Delta \cdot \Delta^{\bullet n})$.

Proof (Sketch). In $\mathfrak{C}$, in the case of argument relations, an application of the Red rule to a state $\mathbf{I}; s$ replaces one element of $\mathbf{I}$ of size m by at most $\mathfrak{a}$ new elements, each of them of size $m-1$. Hence, one branch in the search tree for $\mathfrak{C}$, starting from a singleton set $\mathbf{I}$ of size n, will have the length at most $l = \sum_{i=0}^{n-1} \mathfrak{a}^i$. At each node on it there are at most Δ choices of applying Red with different h's, which gives the total size of the search tree to be at most $\sum_{i=0}^{l-1} \Delta^i$, i.e., the number of steps performed by $\mathfrak{C}$ in the worst case is $O(\Delta^{\bullet \mathfrak{a}^{\bullet n}})$. Those different h's are obtained by intersecting the proximity classes of the heads of terms $\{t_1, \ldots, t_m\}$ in the Red rule. In our graph representation of the proximity relation, proximity classes of symbols are exactly the adjacency lists of those symbols which we assume are sorted. Their maximal length is Δ. Hence, the work to be done at each node of the search tree of $\mathfrak{C}$ is to find the intersection of at most n sorted lists, each containing at most Δ elements. It needs $O(n \cdot \Delta)$ time. It gives the time complexity $O(n \cdot \Delta \cdot \Delta^{\bullet \mathfrak{a}^{\bullet n}})$ of $\mathfrak{C}$ for the relation case.

In the mapping case, an application of the Red rule to a state $\mathbf{I}; s$ replaces one element of $\mathbf{I}$ of size m by at most $\mathfrak{a}$ new elements of the *total* size $m-1$. Therefore, the maximal length of a branch is n, the branching factor is Δ, and the amount of work at each node, like above, is $O(n \cdot \Delta)$. Hence, the number of steps in the worst case is $O(\Delta^{\bullet n})$ and the time complexity of $\mathfrak{C}$ is $O(n \cdot \Delta \cdot \Delta^{\bullet n})$.

The fact that consistency check is incorporated in the Dec rule in $\mathfrak{A}_1^{\mathsf{lin}}$ can be used to guide the application of this rule, using the values memoized by the previous applications of Red. The very first time, the appropriate h in Dec is chosen arbitrarily. In any subsequent application of this rule, h is chosen according to the result of the Red rule that has already been applied to the arguments of the current AUT for their consistency check, as required by the condition of Dec. In this way, the applications of Dec and Sol will correspond to the applications of Red. There is a natural correspondence between the applications of Rem and Tri rules. Therefore, $\mathfrak{A}_1^{\mathsf{lin}}$ will have the search tree analogous to that of $\mathfrak{C}$. Hence the complexity of $\mathfrak{A}_1^{\mathsf{lin}}$ is $O(n \cdot \Delta \cdot \Delta^{\bullet\mathfrak{a}^{\bullet n}})$. $\mathfrak{A}_2^{\mathsf{lin}}$ does not call the consistency check, but does the same work as $\mathfrak{C}$ and, hence, has the same complexity $O(n \cdot \Delta \cdot \Delta^{\bullet n})$. □

6 Discussion and Conclusion

The diagram below illustrates the connections between different anti-unification problems based on argument relations:

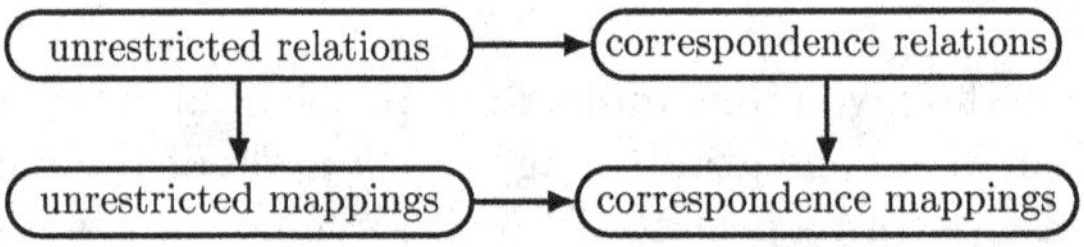

The arrows indicate the direction from more general problems to more specific ones. For the unrestricted cases (left column) we compute mcsrg's. For correspondence relations and correspondence mappings (right column), mcsg's are computed. (In fact, for them, the notions of mcsrg and mcsg coincide). The algorithms for relations (upper row) are more involved than those for mappings (lower row): Those for relations deal with AUTs containing arbitrary sets of terms, while for mappings, those sets have cardinality at most one, thus simplifying the conditions in the rules. Moreover, the two cases in the lower row generalize the existing anti-unification problems:

- the unrestricted mappings case generalizes the problem from [1] by extending similarity to proximity and relaxing the smaller-side-totality restriction;
- the correspondence mappings case generalizes the problem from [9] by allowing permutations between arguments of proximal function symbols.

All our algorithms can be easily turned into anti-unification algorithms for crisp tolerance relations[3] by taking lambda-cuts and ignoring the computation of the approximation degrees. Besides, they are modular and can be used to compute only linear generalizations by just skipping the merging rule. We provided complexity estimations for the algorithms that compute linear generalizations (that often are of practical interest).

[3] Tolerance: reflexive, symmetric, not necessarily transitive relation. According to Poincaré, a fundamental notion for mathematics applied to the physical world.

In this paper, we did not consider cases when the same pair of symbols is related to each other by more than one argument relation. Our results can be extended to them, that would open a way towards approximate anti-unification modulo background theories specified by shallow collapse-free axioms. Another interesting direction of future work would be extending our results to quantitative algebras [10] that also deal with quantitative extensions of equality.

Acknowledgments. Supported by the Austrian Science Fund, project P 35530.

References

1. Aït-Kaci, H., Pasi, G.: Fuzzy lattice operations on first-order terms over signatures with similar constructors: a constraint-based approach. Fuzzy Sets Syst. **391**, 1–46 (2020). https://doi.org/10.1016/j.fss.2019.03.019
2. Baader, F., Nipkow, T.: Term Rewriting and All That. Cambridge University Press, Cambridge (1998)
3. Bader, J., Scott, A., Pradel, M., Chandra, S.: Getafix: learning to fix bugs automatically. Proc. ACM Program. Lang. **3**(OOPSLA), 159:1–159:27 (2019). https://doi.org/10.1145/3360585
4. Barwell, A.D., Brown, C., Hammond, K.: Finding parallel functional pearls: Automatic parallel recursion scheme detection in Haskell functions via anti-unification. Future Gener. Comput. Syst. **79**, 669–686 (2018). https://doi.org/10.1016/j.future.2017.07.024
5. Dershowitz, N., Manna, Z.: Proving termination with multiset orderings. Commun. ACM **22**(8), 465–476 (1979). https://doi.org/10.1145/359138.359142
6. Galitsky, B.: Developing Enterprise Chatbots - Learning Linguistic Structures. Springer, Heidelberg (2019). https://doi.org/10.1007/978-3-030-04299-8
7. Kirbas, S., et al.: On the introduction of automatic program repair in Bloomberg. IEEE Softw. **38**(4), 43–51 (2021). https://doi.org/10.1109/MS.2021.3071086
8. Kutsia, T., Pau, C.: A framework for approximate generalization in quantitative theories. RISC Report Series 22-04, Research Institute for Symbolic Computation, Johannes Kepler University Linz (2022). https://doi.org/10.35011/risc.22-04
9. Kutsia, T., Pau, C.: Matching and generalization modulo proximity and tolerance relations. In: Özgün, A., Zinova, Y. (eds.) TbiLLC 2019. LNCS, vol. 13206, pp. 323–342. Springer, Cham (2019). https://doi.org/10.1007/978-3-030-98479-3_16
10. Mardare, R., Panangaden, P., Plotkin, G.D.: Quantitative algebraic reasoning. In: Grohe, M., Koskinen, E., Shankar, N. (eds.) Proceedings of the 31st Annual ACM/IEEE Symposium on Logic in Computer Science, LICS 2016, pp. 700–709. ACM (2016). https://doi.org/10.1145/2933575.2934518
11. Mehta, S., et al.: Rex: preventing bugs and misconfiguration in large services using correlated change analysis. In: Bhagwan, R., Porter, G. (eds.) 17th USENIX Symposium on Networked Systems Design and Implementation, NSDI 2020, Santa Clara, CA, USA, 25–27 February 2020, pp. 435–448. USENIX Association (2020). https://www.usenix.org/conference/nsdi20/presentation/mehta
12. Plotkin, G.D.: A note on inductive generalization. Mach. Intell. **5**(1), 153–163 (1970)
13. Raza, M., Gulwani, S., Milic-Frayling, N.: Programming by example using least general generalizations. In: Brodley, C.E., Stone, P. (eds.) Proceedings of the Twenty-Eighth AAAI Conference on Artificial Intelligence, 27–31 July 2014, Québec City, Québec, Canada, pp. 283–290. AAAI Press (2014)

14. Reynolds, J.C.: Transformational systems and the algebraic structure of atomic formulas. Mach. Intell. **5**(1), 135–151 (1970)
15. Rolim, R., Soares, G., Gheyi, R., D'Antoni, L.: Learning quick fixes from code repositories. CoRR abs/1803.03806 (2018). http://arxiv.org/abs/1803.03806

Guiding an Automated Theorem Prover with Neural Rewriting

Jelle Piepenbrock[1,2(✉)], Tom Heskes[2], Mikoláš Janota[1], and Josef Urban[1]

[1] Czech Technical University in Prague, Prague, Czech Republic
Jelle.Piepenbrock@cvut.cz
[2] Radboud University, Nijmegen, The Netherlands

Abstract. Automated theorem provers (ATPs) are today used to attack open problems in several areas of mathematics. An ongoing project by Kinyon and Veroff uses Prover9 to search for the proof of the Abelian Inner Mapping (AIM) Conjecture, one of the top open conjectures in quasigroup theory. In this work, we improve Prover9 on a benchmark of AIM problems by neural synthesis of useful alternative formulations of the goal. In particular, we design the 3SIL (stratified shortest solution imitation learning) method. 3SIL trains a neural predictor through a reinforcement learning (RL) loop to propose correct rewrites of the conjecture that guide the search.

3SIL is first developed on a simpler, Robinson arithmetic rewriting task for which the reward structure is similar to theorem proving. There we show that 3SIL outperforms other RL methods. Next we train 3SIL on the AIM benchmark and show that the final trained network, deciding what actions to take within the equational rewriting environment, proves 70.2% of problems, outperforming Waldmeister (65.5%). When we combine the rewrites suggested by the network with Prover9, we prove 8.3% more theorems than Prover9 in the same time, bringing the performance of the combined system to 90%.

Keywords: Automated theorem proving · Machine learning

1 Introduction

Machine learning (ML) has recently proven its worth in a number of fields, ranging from computer vision [17], to speech recognition [15], to playing games [28,40] with *reinforcement learning* (RL) [45]. It is also increasingly applied in automated and interactive theorem proving. Learned predictors have been used for premise selection [1] in hammers [6], to improve clause selection in saturation-based theorem provers [9], to synthesize functions in higher-order logic [12], and to guide connection-tableau provers [21] and interactive theorem provers [2,5,14].

Future growth of the knowledge base of mathematics and the complexity of mathematical proofs will increase the need for proof checking and its better computer support and automation. Simultaneously, the growing complexity of software will increase the need for formal verification to prevent failure modes [10].

J. Blanchette et al. (Eds.): IJCAR 2022, LNAI 13385, pp. 597–617, 2022.
https://doi.org/10.1007/978-3-031-10769-6_35

Automated theorem proving and mathematics will benefit from more advanced ML integration. One of the mathematical subfields that makes substantial use of automated theorem provers is the field of quasigroup and loop theory [32].

1.1 Contributions

In this paper, we propose to use a neural network to suggest lemmas to the Prover9 [25] ATP system by rewriting parts of the conjecture (Sect. 2). We test our method on a dataset of theorems collected in the work on the Abelian Inner Mapping (AIM) Conjecture [24] in loop theory. For this, we use the AIMLEAP proof system [7] as a reinforcement learning environment. This setup is described in Sect. 3. For development we used a simpler Robinson arithmetic rewriting task (Sect. 4). With the insights derived from this and a comparison with other methods, we describe our own 3SIL method in Sect. 5. We use a neural network to process the state of the proving attempt, for which the architecture is described in Sect. 6. The results on the Robinson arithmetic task are described in Sect. 7.1. We show our results on the AIMLEAP proving task, both using our predictor as a stand-alone prover and by suggesting lemmas to Prover9 in Sect. 7.2. Our contributions are:

1. We propose a training method for reinforcement learning in theorem proving settings: *stratified shortest solution imitation learning* (3SIL). This method is suited to the structure of theorem proving tasks. This method and the reasoning behind it is explained in Sect. 5.
2. We show that 3SIL outperforms other baseline RL methods on a simpler, Robinson arithmetic rewriting task for which the reward structure is similar to theorem proving (Sect. 7.1).
3. We show that a standalone neurally guided prover trained by the 3SIL method outperforms the hand-engineered Waldmeister prover on the AIMLEAP benchmark (Sect. 7.2).
4. We show that using a neural rewriting step that suggests rephrased versions of the conjecture to be added as lemmas improves the ATP performance on equational problems (Sects. 2 and 7.2).

2 ATP and Suggestion of Lemmas by Neural Rewriting

Saturation-based ATPs make use of the *given clause* [30] algorithm, which we briefly explain as background. A problem is expressed as a conjunction of many initial clauses (i.e., the clausified axioms and the negated goal which is always an equation in the AIM dataset). The algorithm starts with all the initial clauses in the *unprocessed set*. We then pick a clause from this set to be the given clause and move it to the *processed set* and do all inferences with the clauses in the processed set. The newly inferred clauses are added to the unprocessed set. This concludes one iteration of the algorithm, after which we pick a new given

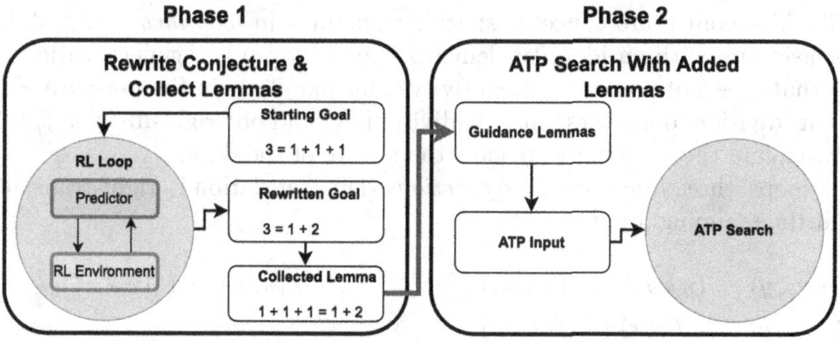

Fig. 1. Schematic representation of the proposed guidance method. In the first phase, we run a reinforcement learning loop to propose actions that rewrite a conjecture. This predictor is trained using the AIMLEAP proof environment. We collect the rewrites of the LHS and RHS of the conjecture. In the second phase, we add the rewrites to the ATP search input, to act as guidance. In this specific example, we only rewrote the conjecture for 1 step, but the added guidance lemmas are in reality the product of many steps in the RL loop.

clause and repeat [23]. Typically, this approach is designed to be *refutationally complete*, i.e., the algorithm is guaranteed to eventually find a contradiction if the original goal follows from the axioms.

This process can produce a lot of new clauses and the search space can become quite large. In this work, we modify the standard loop by adding useful lemmas to the initial clause set. These lemmas are proposed by a neural network that was trained *from zero knowledge* to rewrite the left- and right-hand sides of the initial goal to make them equal by using the axioms as the available rewrite actions. Even though the neural rewriting might not fully succeed, the rewrites produced by this process are likely to be useful as additional lemmas when added to the problem. This idea is schematically represented in Fig. 1.

3 AIM Conjecture and the AIMLEAP RL Environment

Automated theorem proving has been applied in the theory surrounding the Abelian Inner Mapping Conjecture, known as the AIM Conjecture. This is one of the top open conjectures in quasigroup theory. Work on the conjecture has been going on for more than a decade. Automated theorem provers use hundreds of thousands of inference steps when run on problems from this theory.

As a testbed for our machine learning and prover guidance methods we use a previously published dataset of problems generated by the AIM conjecture [7]. The dataset comes with a simple prover called AIMLEAP that can take machine learning advice.[1] We use this system as an RL environment. AIMLEAP keeps the state and carries out the cursor movements (the cursor determines the location of the rewrite) and rewrites that a neural predictor chooses.

[1] https://github.com/ai4reason/aimleap.

The AIM conjecture concerns specific structures in *loop theory* [24]. A loop is a quasigroup with an identity element. A quasigroup is a generalization of a group that does not preserve associativity. This manifests in the presence of two different 'division' operators, one left-division (\) and one right-division (/). We briefly explain the conjecture to show the nature of the data.

For loops, three *inner mapping functions* (left-translation L, right-translation R, and the mapping T) are:

$$L(u,x,y) := (y*x)\backslash(y*(x*u)) \qquad T(u,x) := x\backslash(u*x)$$
$$R(u,x,y) := ((u*x)*y)/(x*y)$$

These mappings can be seen as measures of the deviation from commutativity and associativity. The conjecture concerns the consequences of these three inner mapping functions forming an Abelian (commutative) group. There are two more notions, that of the *associator* function a and the *commutator* function K:

$$a(x,y,z) := (x*(y*z))\backslash((x*y)*z) \qquad K(x,y) := (y*x)/(x*y)$$

From these definitions, the conjecture can be stated. There are two parts to the conjecture. For both parts, the following equalities need to hold for all u, v, x, y, and z:

$$a(a(x,y,z),u,v) = 1 \qquad a(x,a(y,z,u),v) = 1 \qquad a(x,y,a(z,u,v)) = 1$$

where 1 is the identity element. These are necessary, but not sufficient for the two main parts of the conjecture. The first part of the conjecture asks whether a loop modulo its center is a group. In this context, the *center* is the set of all elements that commute with all other elements. This is the case if

$$K(a(x,y,z),u) = 1.$$

The second part of the conjecture asks whether a loop modulo its nucleus is an Abelian group. The *nucleus* is the set of elements that associate with all other elements. This is the case if

$$a(K(x,y),z,u) = 1 \qquad a(x,K(y,z),u) = 1 \qquad a(x,y,K(z,u)) = 1$$

3.1 The AIMLEAP RL Environment

Currently, work in this area is done using automated theorem provers such as *Prover9* [24,25]. This has led to some promising results, but the search space is enormous. The main strategy for proving the AIM conjecture thus far has been to prove weaker versions of the conjecture (using additional assumptions) and then import crucial proof steps into the stronger version of the proof. The *Prover9* theorem prover is especially suited to this approach because of its well-established *hints* mechanism [48]. The AIMLEAP dataset is derived from this

Prover9 approach and contains around 3468 theorems that can be proven with the supplied definitions and lemmas [7].

There are 177 possible actions in the AIMLEAP environment [7]. We handle the proof state as a tree, with the root node being an equality node. Three actions are cursor movements, where the cursor can be moved to an argument of the current position. The other actions all rewrite the current term at the cursor position with various axioms, definitions and lemmas that hold in the AIM context. As an example, this is one of the theorems in the dataset (\ and = are part of the language):

$$T(T(T(x, T(x, y)\backslash 1), T(x, y)\backslash 1), y) = T((T(x, y)\backslash 1)\backslash 1, T(x, y)\backslash 1)\,.$$

The task of the machine learning predictor is to process the proof state and recognize which actions are most likely to lead to a proof, meaning that the two sides of the starting equation are equal according to the AIMLEAP system. The only feedback that the environment gives is whether a proof has been found or not: there is no intermediate reward (i.e. rewards are *sparse*). The ramifications of this are further discussed in Sect. 5.1.

4 Rewriting in Robinson Arithmetic as an RL Task

To develop a machine learning method that can help solve equational theorem proving problems, we considered a simpler arithmetic task, which also has a tree-structured input and *a sparse reward structure*: the normalization of Robinson arithmetic expressions. The task is to normalize a mathematical expression to one specific form. This task has been implemented as a Python RL environment, which we make available.[2] The learning environment incorporates an existing dataset, constructed by Gauthier for RL experiments in the interactive theorem prover HOL4 [11]. Our RL setup for the task is also modeled after [11].

In more detail, the formalism that we use as an RL environment is Robinson arithmetic (RA). RA is a simple arithmetic theory. Its language contains the successor function S, addition $+$ and multiplication $*$ and one constant, the 0. The theory considers only non-negative numbers and we only use four axioms of RA. Numbers are represented by the constant 0 with the appropriate number of successor functions applied to it. The task for the agent is to rewrite an expression until there are only nodes of the successor or 0 types. Effectively, we are asking the agent to calculate the value of the expression. As an example, $S(S(0)) + S(0)$, representing $2 + 1$, needs to be rewritten to $S(S(S(0)))$.

The expressions are represented as a tree data structure. Within the environment, there are seven different rewrite actions available to the agent. The four axioms (equations) defining these actions are $x + 0 = x$, $x + S(y) = S(x + y)$, $x * 0 = 0$ and $x * S(y) = (x * y) + x$, where the agent can apply the equations in either direction. There is one exception: the multiplication by 0 cannot be applied from right to left, as this would require the agent to introduce a fresh

[2] https://github.com/learningeqtp/rewriteRL.

term which is out of scope for the current work. The place where the rewrite is applied is denoted by the location of the *cursor* in the expression tree.

In addition to the seven rewrite actions, the agent can move the cursor to one of the children of the current cursor node. This gives a total number of nine actions. Moving to a child of a node with only one child counts as moving to the left child. After a rewriting action, the cursor is reset to the root of the expression. More details on the actions are in the RewriteRL repository.

5 Reinforcement Learning Methods

This section describes the reinforcement learning methods, while Sect. 6 then further explains the particular neural architectures that are trained in the RL loops. We first briefly explain here the approaches that we used as reinforcement learning (RL) baselines, then we go into detail about the proposed 3SIL method.

5.1 Reinforcement Learning Baselines

General RL Setup. For comparison, we used implementations of four established reinforcement learning baseline methods. In reinforcement learning, we consider an *agent* that is acting within an *environment.* The agent can take actions a from the action-space $\mathcal{A}$ to change the state $s \in \mathcal{S}$ of the environment. The agent can be rewarded for certain actions taken in a certain states, with reward given by the *reward function* $\mathcal{R} : (\mathcal{S} \times \mathcal{A}) \rightarrow \mathbb{R}$. The behavior of the environment is given by the *state transition function* $\mathcal{P} : (\mathcal{S} \times \mathcal{A}) \rightarrow \mathcal{S}$. The history of the agent's actions and the environments states and rewards at each timestep t are collected in tuples (s_t, a_t, r_t). For a given history of a certain agent within an environment, we call the list of tuples (s_t, a_t, r_t) describing this history an *episode.* The *policy function* $\pi : \mathcal{S} \rightarrow \mathcal{A}$ allows the agent to decide which action to take. The agent's goal is to maximize the return R: the sum of discounted rewards $\sum_{t \geq 0} \gamma^t r_t$, where γ is a *discount factor* that allows control over how heavily rewards further in the future should be weighted. We will use R_t when we mean R, but calculated only from rewards from timestep t on. In the end, we are thus looking for a policy function π that maximizes the sum R of (discounted) expected rewards [45].

In our setting, every proof attempt (in the AIM setting) or normalization attempt (in the Robinson arithmetic setting) corresponds to an episode. The reward structure of theorem proving is such that there is only a reward of 1 at the end of a successful episode (i.e. a proof was found in AIM). Unsuccessful episodes get a reward of 0 at every timestep t.

A2C. The first method, *Advantage Actor-Critic*, or *A2C* [27] contains ideas on which the other three RL baseline methods build, so we will go into more detail for this method, while keeping the explanation for the other methods brief. For details we refer to the corresponding papers.

A2C attempts to find suitable parameters for an agent by minimizing a *loss function* consisting of two parts:

$$\mathcal{L} = \mathcal{L}^{\text{A2C}}_{\text{policy}} + \mathcal{L}^{\text{A2C}}_{\text{value}} .$$

In addition to the policy function π, the agent has access to a *value function* $\mathcal{V} : \mathcal{S} \rightarrow \mathbb{R}$, that predicts the sum of future rewards obtained when given a state. In practice, both the policy and the value function are computed by a neural network *predictor*. The parameters of the predictor are set by *stochastic gradient descent* to minimize $\mathcal{L}$. The set of parameters of the predictor that defines the policy function π is named θ, while the parameters that define the value function are named μ. The first part of the loss is the *policy loss*, which for one time step has the form

$$\mathcal{L}^{\text{A2C}}_{\text{policy}} = -\log \pi_\theta(a_t|s_t) A(s_t, a_t) ,$$

where $A(s, a)$ is the *advantage function*. The advantage function can be formulated in multiple ways, but the simplest is as $R_t - \mathcal{V}_\mu(s_t)$. That is to say: the advantage of an action in a certain state is the difference between the discounted rewards R_t after taking that action and the value estimate of the current state.

Minimizing $\mathcal{L}^{\text{A2C}}_{\text{policy}}$ amounts to maximizing the log probability of predicting actions that are judged by the advantage function to lead to high reward.

The value estimates $V_\mu(s)$ for computing the advantage function are supplied by the *value predictor* V_μ with parameters μ, which is trained using the loss:

$$\mathcal{L}^{\text{A2C}}_{\text{value}} = \frac{1}{2} \left(R_t - \mathcal{V}_\mu(s_t)\right)^2 ,$$

which minimizes the advantage function. The logic of this is that the value estimate at timestep t, $\mathcal{V}_\mu(s_t)$, will learn to incorporate the later rewards R_t, ensuring that when later seeing the same state, the possible future reward will be considered. Note that the sets of parameters θ and μ are not necessarily disjoint (see Sect. 6).

Note how the above equations are affected if there is no non-zero reward r_t obtained at any timestep. In that case, the value function $\mathcal{V}_\mu(s_t)$ will estimate (correctly) that any state will get 0 reward, which means that the advantage function $A(s, a)$ will also be 0 everywhere. This means that $\mathcal{L}^{\text{A2C}}_{\text{policy}}$ will be 0 in most cases, which will lead to no or little change in the parameters of the predictor: learning will be very slow. This is the difficult aspect of the structure of theorem proving: there is only reward at the end of a successful proof, and nowhere else. This implies a possible strategy is to imitate successful episodes, without a value function. In this case, we would only need to train a *policy function*, and no approximate *value function*. This an aspect we explore in the design of our own method 3SIL, which we will explain shortly.

Compared to two-player games, such as chess and go, for which many approaches have been tailored and successfully used [41], theorem-proving has the property that it is hard to collect useful examples to learn from, as only successful proofs are likely to contain useful knowledge. In chess or go, however, one player almost always wins and the other loses, which means that we can at least learn from the difference between the two strategies used by those players. As an example, we executed 2 million random proof attempts on the AIMLEAP environment, which led to 300 proofs to learn from, whereas in a two-player setting like chess, we would get 2 million games in which one player would likely win.

ACER. The second RL baseline method we tested in our experiments is ACER, *Actor-Critic with Experience Replay* [49]. This approach can make use of data from older episodes to train the current predictor. ACER applies corrections to the value estimates so that data from old episodes may be used to train the current policy. It also uses trust region policy optimization [35] to limit the size of the policy updates. This method is included as a baseline to check if using a larger replay buffer to update the parameters would be advantageous.

PPO. Our third RL baseline is the widely used *proximal policy optimization* (PPO) algorithm [36]. It restricts the size of the parameter update to avoid causing a large difference between the original predictor's behavior and the updated version's behavior. The method is related to the above trust region policy optimization method. In this way, PPO addresses the training instability of many reinforcement learning approaches. It has been used in various settings, for example complex video games [4]. With its versatility, the PPO algorithm is well-positioned. We use the PPO algorithm with clipped objective, as in [36].

SIL-PAAC. Our final RL baseline uses only the transitions with positive advantage to train on for a portion of the training procedure, to learn more from good episodes. This was proposed as *self-imitation learning* (SIL) [29]. To avoid confusion with the method that we are proposing, we extend the acronym to SIL-PAAC, for positive advantage actor-critic. This algorithm outperformed A2C on the sparse-reward task Montezuma's Revenge (a puzzle game). As theorem proving has a sparse reward structure, we included SIL-PAAC as a baseline. More information about the implementations for the baselines can be found in the Implementation Details section at the end of this work.

5.2 Stratified Shortest Solution Imitation Learning

We introduce stratified shortest solution imitation learning (**3SIL**) to tackle the equational theorem proving domain. It learns to explicitly imitate the actions taken during the shortest solutions found for each problem in the dataset. We do this by minimizing the cross-entropy $-\log p(a_{solution}|s_t)$ between the predictor output and the actions taken in the shortest solution. This is in contrast to the baseline methods, where value functions are used to judge the utility of decisions.

In our procedure this is not the case. Instead, we build upon the assumption for data selection that shorter proofs are better in the context of theorem proving

Algorithm 1. CollectEpisode

Input: problem p, policy π_θ, problem history H
Generate episode by following noisy version of π_θ on p
If solution, add list of tuples (s, a) to H[p]
Keep k shortest solutions in H[p]

Algorithm 2. 3SIL

```
Input: set of problems P, randomly initialized policy π_θ, batch size B, number of
batches NB, problem history H, number of warmup episodes m, number of episodes
f, max epochs ME
Output: trained policy π_θ, problem history H
for e = 0 to ME − 1 do
   if e = 0 then num = m else num = f
   for i = 0 to num − 1 do
      CollectEpisode(sample(P), π_θ, H) (Algorithm 1)
   end for
   for i = 0 to NB − 1 do
      Sample B tuples (s, a) with uniform probability for each problem from H
      Update θ to lower − Σ_{b=0}^{B} log π_θ(a_b|s_b) by gradient descent
   end for
end for
```

and expression normalization. In a sense, we value decisions from shorter proofs more and explicitly imitate those transitions. We keep a history H for each problem, where we store the current shortest solution (states seen and actions taken) found for that problem in the training dataset. We can also store multiple shortest solutions for each problem if there are multiple strategies for a proof (the number of solutions kept is governed by the parameter k).

During training, in the case $k = 1$, we sample state-action pairs from each problem's current shortest solution at an equal probability (if a solution was found). To be precise, we first randomly pick a theorem for which we have a solution, and then randomly sample one transition from the shortest encountered solution. This directly counters one of the phenomena that we had observed: the training examples for the baseline methods tend to be dominated by very long episodes (as they contribute more states and actions). This *stratified* sampling method ensures that problems with short proofs get represented equally in the training process.

The 3SIL algorithm is described in more detail in Algorithm 2. Sampling from a noisy version of policy π_θ means that actions are sampled from the predictor-defined distribution and in 5% of cases a random valid action is selected. This is also known as the ϵ-greedy policy (with ϵ at 0.05).

Related Methods. Our approach is similar to the imitation learning algorithm DAGGER (Dataset Aggregation), which was used for several games [34] and modified for branch-and-bound algorithms in [16]. The behavioral cloning (BC) technique used in robotics [47] also shares some elements. 3SIL significantly differs from DAGGER and BC because it does not use an outside expert to obtain useful data, because of the stratified sampling procedure, and because of the selection of the shortest solutions for each problem in the training dataset. We include as an additional baseline an implementation of behavioral cloning (BC), where we regard proofs already encountered as coming from an expert. We minimize cross-entropy between the actions in proofs we have found and the predictions to train the predictor. For BC, there is no stratified sampling or shortest solution selection, only the minimization of cross-entropy between actions taken from recent successful solutions and the predictor's output.

Extensions. For the AIM tasks, we introduce two other techniques, *biased sampling* and *episode pruning.* In biased sampling, problems without a solution in the history are sampled 5 times more during episode collection than solved problems to accelerate progress. This was determined by testing 1, 2, 5 and 10 as sampling proportions. For episode pruning, when the agent encountered the same state twice, we prune the episode to exclude the looping before storing the episode. This helps the predictor learn to avoid these loops.

6 Neural Architectures

The tree-structured states representing expressions occurring during the tasks will be processed by a neural network. The neural network takes the tree-structured state and predicts an action to take that will bring the expression closer to being normalized or the theorem closer to being proven.

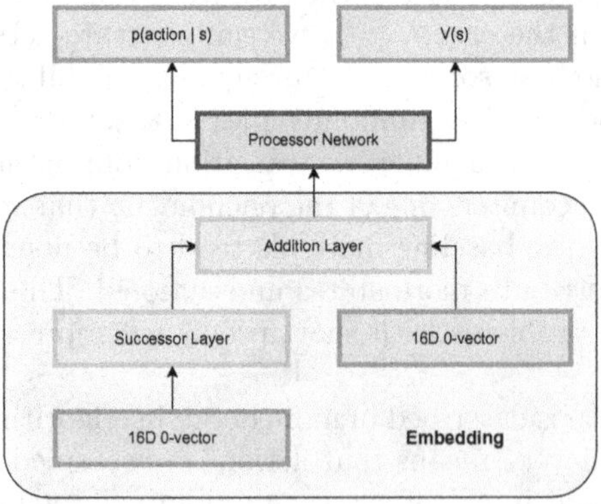

Fig. 2. Schematic representation of the creation of a representation of an expression (*an embedding*) using different neural network layers to represent different operations. The figure depicts the creation of a numerical representation for the Robinson arithmetic expression $(S(0) + 0)$. Note that the successor layer and the addition layer consist of trainable parameters, for which the values are set through gradient descent.

There are two main components to the neural network we use: an *embedding* tree neural network that outputs a numerical vector representing the tree-structured proof state and a second *processor* network that takes this vector representation of the state and outputs a distribution of the actions possible in the environment.[3]

Tree neural networks have been used in various settings, such as natural language processing [20] and also in Robinson arithmetic expression embedding [13]. These networks consist of smaller neural networks, each representing one of the possible functions that occur in the expressions. For example, there will be separate networks representing addition and multiplication. The cursor is a special unary operation node with its own network that we insert into the tree at the current location. For each unique constant, such as the constant 0 in RA or the identity element 1 for the AIM task, we generate a random vector (from a standard normal distribution) that will represent this leaf. In the case of the AIM task, these vectors are parameters that can be optimized during training.

At prediction time, the numerical representation of a tree is constructed by starting at the leaves of the tree, for which we can look up the generated vectors. These vectors act as input to the neural networks that represent the parent node's operation, yielding a new vector, which now represents the subtree of the parent node. The process repeats until there is a single vector for the entire tree after the root node is processed (see also Fig. 2).

The neural networks representing each operation consist of a linear transformation, a non-linearity in the form of a rectified linear unit (ReLU) and another linear transformation. In the case of binary operations, the first linear transformation will have an input dimension of $2n$ and an output dimension of n, where n is the dimension of the vectors representing leaves of the tree (the *internal representation size*). The weights representing these transformations are randomly initialized at the beginning of training.

When we have obtained a single vector embedding representing the entire tree data structure, this vector serves as the input to the *predictor* neural network, which consists of three linear layers, with non-linearities (Sigmoid/ReLU) in between these layers. The last layer has an output dimension equal to the number of possible actions in the environment. We obtain a probability distribution over the actions, e.g. by applying the softmax function to the output of this last layer. In the cases where we also need a value prediction, there is a parallel last layer that predicts the state's value (usually referred to as a *two-headed* network [41]). The internal representation size n for the Robinson arithmetic experiments is set to 16, for the AIM task this is 32. The number of neurons in each layer (except for the last one) of the predictor networks is 64.

In the AIM dataset task, an arbitrary number of variables can be introduced during the proof. These are represented by untrainable random vectors. We add a special neural network (with the same architecture as the networks representing unary operations, so from size n to n) that processes these vectors before they are

[3] In the reinforcement learning baselines that we use, this second *processor* network has the additional task of predicting the value of a state.

processed by the rest of the tree neural network embedding. The idea is that this neural network learns to project these new variable vectors into a subspace and that an arbitrary number of variables can be handled. The vectors are resampled at the start of each episode, so the agent cannot learn to recognize specific variables. This approach was partly inspired by the *prime* mechanism in [13], but we use separate vectors for all variables instead of building vectors sequentially. All our neural networks are implemented using the PyTorch library [31].

7 Experiments

We first describe our experiments on the Robinson arithmetic task, with which we designed the properties of our 3SIL approach with the help of comparisons with other algorithms. We then train a predictor using 3SIL on the AIMLEAP loop theory dataset, which we evaluate both as a standalone prover within the RL environment and as a neural guidance mechanism for the ATP Prover9.

7.1 Robinson Arithmetic Dataset

Dataset Details. The Robinson arithmetic dataset [11] is split into three distinct sets, based on the number of steps that it takes a fixed rewriting strategy to normalize the expression. This fixed strategy, LOPL, which stands for *left outermost proof length*, always rewrites the leftmost possible element. If it takes this strategy less than 90 steps to solve the problem, it is in the *low* difficulty category. Problems with a difficulty between 90 and 130 are in the *medium* category and a greater difficulty than 130 leads to the *high* category. The *high* dataset also contains problems the LOPL strategy could not solve within the time limit. The *low* dataset is split into a training and testing set. We train on the *low* difficulty problems, but after training we also test on problems with a higher difficulty. Because we have a difficulty measure for this dataset, we use a curriculum setup. We start by learning to normalize the expressions that a fixed strategy can normalize in a small amount of steps. This setup is similar to [11].

Training Setup. The 400 problems with the lowest difficulty are the starting point. Every time an agent reaches 95 percent success rate when evaluated on a sample of size 400 from these problems, we add 400 more difficult problems to set of training problems P. One iteration of the *collection* and *training* phase is called an *epoch*. Agents are evaluated after every epoch. The blocks of size 400 are called *levels*. The number of episodes m and f are set to 1000. For 3SIL and BC, the batch size BS is 32 and the number of batches NB is 250. The baselines are configured so that the number of episodes and training transitions is at least as many as the 3SIL/BC approaches. Episodes that take over 100 steps are stopped. ADAM [22] is used as an optimizer.

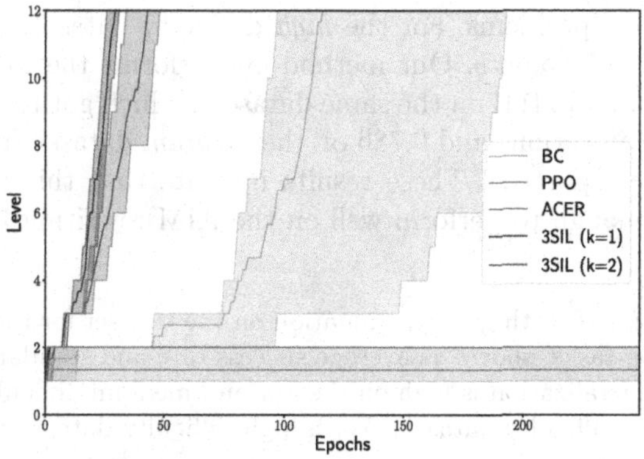

Fig. 3. The level in the curriculum reached by each method. Each method was run three times. The bold line shows the mean performance and the shaded region shows the minimum and maximum performance. K is the number of proofs stored per problem.

Results on RA Curriculum. In Fig. 3, we show the progression through the training curriculum for behavioral cloning (BC), the RL methods (PPO, ACER) and two configurations of 3SIL. Behavioral cloning simply imitates actions from successful episodes. Of the RL baselines, PPO reaches the second level in one run, while ACER steadily solves the first level and in the best run solves around 80% of the second level. Both methods do not learn enough solutions for the second level to advance to the third. A2C and SIL-PAAC do not reach the second level, so these are left out of the plot. However, they do learn to solve about 70–80% of the first 400 problems. From these results we can conclude that the RL baselines do not perform well on this task in our experiment. We attribute this to the difficulty of learning a good value function due to the sparse rewards (Sect. 5.1). Our hypothesis is that because this value estimate influences the policy updates, the RL methods do not learn well on this task. Note that the two methods with a trust region update mechanism, ACER and PPO, perform better than the methods without this mechanism. From these results, it is clear that 3SIL with 1 shortest proof stored, $k = 1$, is the best-performing configuration. It reaches the end of the training curriculum of about 5000 problems in 40 epochs. We experimented with $k = 3$ and $k = 4$, but these were both worse than $k = 2$.

Generalization. While our approach works well on the training set, we must check if the predictors generalize to unseen examples. Only the methods that reached the end of the curriculum are tested. In Table 1, we show the results of evaluating the performance of our predictors on the three different test sets: the unseen examples from the *low* dataset and the unseen examples from the *medium* and *high* datasets. Because we expect longer solutions, the episode limits are expanded from 100 steps to 200 and 250 for the *medium* and *high* datasets respectively. For the *low* and *medium* datasets, the second of which contains problems with more difficult solutions than the training data, the predictors solve almost all test problems. For the *high* difficulty dataset, the performance drops by at least 20% points. Our method outperforms the Monte Carlo Tree Search approach used in [11] on the same datasets, which got to 0.954 on the *low* dataset with 1600 iterations and 0.786 on the *medium* dataset (no results on the *high* dataset were reported). These results indicate that this training method might be strong enough to perform well on the AIM rewriting RL task.

Table 1. Generalization with greedy evaluation on the test set for the Robinson arithmetic normalization tasks, shown as average success rate and standard deviation from 3 training runs. Generalization is high on the low and medium difficulty (training data is similar to the low difficulty dataset). With high difficulty data, performance drops.

	Low	Medium	High
3SIL (κ = 1)	1.00 ± 0.01	0.98 ± 0.03	0.77 ± 0.10
3SIL (κ = 2)	0.99 ± 0.00	0.96 ± 0.01	0.66 ± 0.08
BC	0.98 ± 0.01	0.98 ± 0.01	0.56 ± 0.05

7.2 AIM Conjecture Dataset

Training Setup. Finally, we train and evaluate 3SIL on the AIM Conjecture dataset. We apply 3SIL ($k = 1$) to train predictors in the AIMLEAP environment. Ten percent of the AIM dataset is used as a hold-out test set, not seen during training. As there is no estimate for the difficulty of the problems in terms of the actions available to the predictor, we do not use a curriculum ordering for these experiments. The number m of episodes collected before training is set to 2,000,000. These random proof attempts result in about 300 proofs. The predictor learns from these proofs and afterwards the search for new proofs is also guided by its predictions. For the AIM experiments, episodes are stopped after 30 steps in the AIMLEAP environment. The predictors are trained for 100 epochs. The number of collected episodes per epoch f is 10,000. The successful proofs are stored, and the shortest proof for each theorem is kept. NB is 500 and BS is set to 32. The number of problems with a solution in the history after each epoch of the training run is shown in Fig. 4.

Results as a Standalone Prover. After 100 epochs, about 2500 of 3114 problems in the training dataset have a solution in their history. To test the generalization capability of the predictors, we inspect their performance on the holdout test set problems. In Table 2 we compare the success rate of the trained predictors on the holdout test set with three different automated theorem provers: E [37,38], Waldmeister [19] and Prover9. E is currently one of the best overall automated theorem provers [44], Waldmeister is a prover specialized in memory-efficient equational theorem proving [18] and Prover9 is the theorem prover that

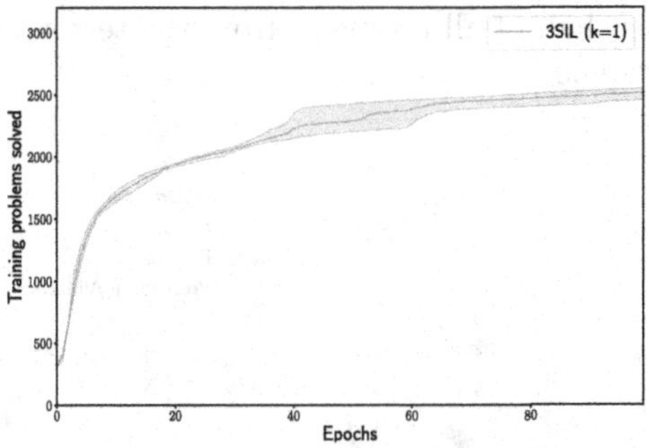

Fig. 4. The number of training problems for which a solution was encountered and stored (cumulative). At the start of the training, the models rapidly collect more solutions, but after 100 epochs, the process slows down and settles at about 2500 problems with known solutions. The minimum, maximum and mean of three runs are shown.

is used for AIM conjecture research and the prover that the dataset was generated by. Waldmeister and E are the best performing solvers in competitions for the relevant unit equality (UEQ) category [44].

Table 2. Theorem proving performance on the hold-out test set in fraction of problems solved. Means and standard deviations are the results of evaluations of 3 different predictors from 3 different training runs on the 354 unseen test set problems.

Method	Success Rate
Prover9 (60 s)	0.833
E (60 s)	0.802
Predictor + AIMLEAP(60 s)	0.702 ± 0.015
Waldmeister (60 s)	0.655
Predictor + AIMLEAP (1×)	0.586 ± 0.029

The results show that a single greedy evaluation of the predictor trying to solve the problem in the AIMLEAP environment is not as strong as the theorem proving software. However, the theorem provers got 60 s of execution time, and the execution of the predictor, including interaction with AIMLEAP, takes on average less than 1 s. We allowed the predictor setup to use 60 s, by running attempts in AIMLEAP until the time was up, sampling actions from the predictor's distribution with 5% noise, instead of using greedy execution. With this approach, the predictor setup outperforms Waldmeister.[4] Figure 5 shows the overlap between the problems solved by each prover. The diagram shows that each theorem prover found a few solutions that no other prover could find within

[4] After the initial experiments, we also evaluated Twee [42], which won the most recent UEQ track: it can prove most of the test problems in 60 s, only failing for 1 problem.

the time limit. Almost half of all problems from the test set that are solved are solved by all four systems.

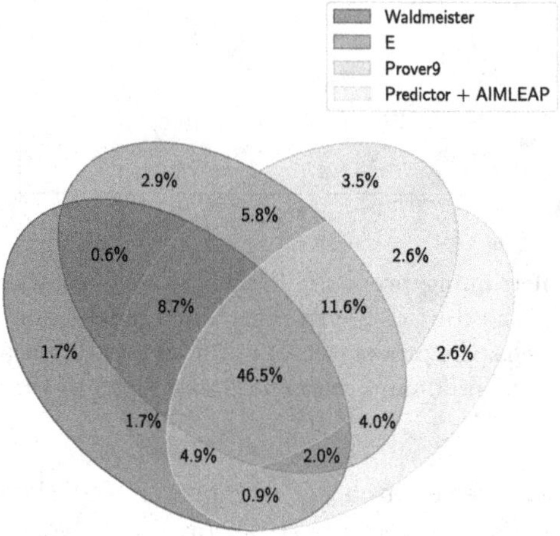

Fig. 5. Venn diagram of the test set problems solved by each solver with 60 s time limit.

Results of Neural Rewriting Combined with Prover9. We also combine the predictor with *Prover9*. In this setup, the predictor modifies the starting form of the goal, for a maximum of 1 s in the AIMLEAP environment. This produces new expressions on one or both sides of the equality. We then add, as lemmas, equalities between the left-hand side of the goal before the predictor's rewriting and after each rewriting (see Fig. 1). The same is done for the right-hand side. For each problem, this procedure yields new lemmas that are added to the problem specification file that is given to *Prover9*.

Table 3. Prover9 theorem proving performance on the hold-out test set when injecting lemmas suggested by the learned predictor. *Prover9*'s performance increases when using the suggested lemmas.

Method	Success Rate
Prover9 (1 s)	0.715
Prover9 (2 s)	0.746
Prover9 (60 s)	0.833
Rewriting (1 s) + Prover9 (1 s)	0.841 ± 0.019
Rewriting (1 s) + Prover9 (59 s)	**0.902** ± 0.016

In Table 3, it is shown that adding lemmas suggested by the rewriting actions of the trained predictor improves the performance of *Prover9*. Running *Prover9* for 2 s results in better performance than running it for 1 s, as expected. The combined (1 s + 1 s) system improved on *Prover9's* 2-s performance by 12.7% (= 0.841/0.746), indicating that the predictor suggests useful lemmas. Additionally, 1 s of neural rewriting combined with 59 s of Prover9 search proves almost 8.3% (= 0.902/0.833) more theorems than Prover9 with a 60 s time limit (Table 2).

7.3 Implementation Details

All experiments for the Robinson task were run on a 16 core Intel(R) Xeon(R) CPU E5-2670 0 @ 2.60 GHz. The AIM experiments were run on a 72 core Intel(R) Xeon(R) Gold 6140 CPU @ 2.30 GHz. All calculations were done on CPU. The PPO implementation was adapted from an existing implementation [3]. The model was updated every 2000 timesteps, the PPO clip coefficient was set to 0.2. The learning rate was 0.002 and the discount factor γ was set to 0.99. The ACER implementation was adapted from an available implementation [8]. The replay buffer size was 20,000. The truncation parameter was 10 and the model was updated every 100 steps. The replay ratio was set to 4. Trust region decay was set to 0.99 and the constraint was set to 1. The discount factor was set to 0.99 and the learning rate to 0.001. Off-policy minibatch size was set to 1. The A2C and SIL implementations were based on Pytorch actor-critic example code available at the PyTorch repository [33]. For the A2C algorithm, we experimented with two formulations of the advantage function: the 1-step lookahead estimate $(r_t + \gamma\mathcal{V}_\mu(s_{t+1})) - \mathcal{V}_\mu(s_t)$ and the $R_t - \mathcal{V}_\mu(s_t)$ formulation. However, we did not observe different performance, so we opted in the end for the 1-step estimate favored in the original A2C publication. For SIL-PAAC, we implemented the SIL loss on top of the A2C implementation. There is also a prioritized replay buffer with an exponent of 0.6, as in the original paper. Each epoch, 8000 (250 batches of size 32) transitions were taken from the prioritized replay buffer in the SIL step of the algorithm. The size of the prioritized replay buffer was 40,000. The critic loss weight was set to 0.01 as in the original paper. For the 3SIL and behavioral cloning implementations, we sample 8000 transitions (250 batches of size 32) from the replay buffer or history. For the behavioral cloning, we used a buffer of size 40,000. An example implementation of 3SIL can be found in the RewriteRL repository. On the Robinson arithmetic task, for 3SIL and BC, the evaluation is done greedily (always take the highest probability actions). For the other methods, we performed experiments with both greedy and non-greedy (sample from the predictor distribution and add 5% noise) evaluation and show the results the best-performing setting (which in most cases was the non-greedy evaluation, except for PPO). On the AIM task, we evaluate greedily with 3SIL.

AIMLEAP expects a distance estimate for each applicable action. This represents the estimated distance to a proof. This behavior was converted to a reinforcement learning setup by always setting the chosen action of the model to the minimum distance and all other actions to a distance larger than the maximum proof length. Only the chosen action is then carried out.

Versions of the automated theorem provers used: Version 2.5 of E [39], the Nov 2017 version of Prover9 [26] and the Feb 2018 version of Waldmeister [46] and version 2.4.1 of Twee [43].

8 Conclusion and Future Work

Our experiments show that a neural rewriter, trained with the 3SIL method that we designed, can learn to suggest useful lemmas that assist an ATP and improve its proving performance. With the same limit of 1 min, Prover9 managed to prove close to 8.3% more theorems. Furthermore, our 3SIL training method is powerful enough to train an equational prover from zero knowledge that can compete with hand-engineered provers, such as Waldmeister. Our system on its own proves 70.2% of the unseen test problems in 60s, while Waldmeister proved 65.5%.

In future work, we will apply our method to other equational reasoning tasks. An especially interesting research direction concerns selecting which proofs to learn from: some sub-proofs might be more general than other sub-proofs. The incorporation of graph neural networks instead of tree neural networks may improve the performance of the predictor, since in graph neural networks information not only propagates from the leaves to the root, but also through all other connections.

Acknowledgements. We would like to thank Chad Brown for his work with the AIMLEAP software. In addition, we thank Thibault Gauthier and Bartosz Piotrowski for their help with the Robinson arithmetic rewriting task and the AIM rewriting task respectively. We also thank the referees of the IJCAR conference for their useful comments.

This work was partially supported by the European Regional Development Fund under the Czech project AI&Reasoning no. CZ.02.1.01/0.0/0.0/15_003/ 0000466 (JP, JU), Amazon Research Awards (JP, JU) and by the Czech MEYS under the ERC CZ project *POSTMAN* no. LL1902 (JP, MJ).

This article is part of the RICAIP project that has received funding from the European Union's Horizon 2020 research and innovation programme under grant agreement No. 857306.

References

1. Alama, J., Heskes, T., Kühlwein, D., Tsivtsivadze, E., Urban, J.: Premise selection for mathematics by corpus analysis and kernel methods. J. Autom. Reason. **52**(2), 191–213 (2013). https://doi.org/10.1007/s10817-013-9286-5
2. Bansal, K., Loos, S., Rabe, M., Szegedy, C., Wilcox, S.: HOList: an environment for machine learning of higher order logic theorem proving. In: International Conference on Machine Learning, pp. 454–463 (2019)
3. Barhate, N.: Implementation of PPO algorithm. https://github.com/nikhilbarhate99

4. Berner, C., et al.: DOTA 2 with large scale deep reinforcement learning. arXiv preprint arXiv:1912.06680 (2019)
5. Blaauwbroek, L., Urban, J., Geuvers, H.: The Tactician. In: Benzmüller, C., Miller, B. (eds.) CICM 2020. LNCS (LNAI), vol. 12236, pp. 271–277. Springer, Cham (2020). https://doi.org/10.1007/978-3-030-53518-6_17
6. Blanchette, J.C., Kaliszyk, C., Paulson, L.C., Urban, J.: Hammering towards QED. J. Formalized Reason. **9**(1), 101–148 (2016). https://doi.org/10.6092/issn.1972-5787/4593
7. Brown, C.E., Piotrowski, B., Urban, J.: Learning to advise an equational prover. Artif. Intell. Theorem Proving, 1–13 (2020)
8. Chételat, D.: Implementation of ACER algorithm. https://github.com/dchetelat/acer
9. Chvalovský, K., Jakubův, J., Suda, M., Urban, J.: ENIGMA-NG: efficient neural and gradient-boosted inference guidance for E. In: Fontaine, P. (ed.) CADE 2019. LNCS (LNAI), vol. 11716, pp. 197–215. Springer, Cham (2019). https://doi.org/10.1007/978-3-030-29436-6_12
10. de Moura, L., Bjørner, N.: Z3: an efficient SMT solver. In: Ramakrishnan, C.R., Rehof, J. (eds.) TACAS 2008. LNCS, vol. 4963, pp. 337–340. Springer, Heidelberg (2008). https://doi.org/10.1007/978-3-540-78800-3_24
11. Gauthier, T.: Deep reinforcement learning in HOL4. arXiv preprint arXiv:1910.11797v1 (2019)
12. Gauthier, T.: Deep reinforcement learning for synthesizing functions in higher-order logic. In: International Conference on Logic for Programming, Artificial Intelligence and Reasoning (2020)
13. Gauthier, T.: Tree neural networks in HOL4. In: Benzmüller, C., Miller, B. (eds.) CICM 2020. LNCS (LNAI), vol. 12236, pp. 278–283. Springer, Cham (2020). https://doi.org/10.1007/978-3-030-53518-6_18
14. Gauthier, T., Kaliszyk, C., Urban, J., Kumar, R., Norrish, M.: TacticToe: learning to prove with tactics. J. Autom. Reason. **65**, 1–30 (2020)
15. Graves, A., Fernández, S., Gomez, F., Schmidhuber, J.: Connectionist temporal classification: labelling unsegmented sequence data with recurrent neural networks. In: Proceedings of the 23rd International Conference on Machine Learning, pp. 369–376 (2006)
16. He, H., Daume, H., III., Eisner, J.M.: Learning to search in branch and bound algorithms. Adv. Neural Inf. Process. Syst. **27**, 3293–3301 (2014)
17. He, K., Zhang, X., Ren, S., Sun, J.: Deep residual learning for image recognition. In: Proceedings of the IEEE Conference on Computer Vision and Pattern Recognition, pp. 770–778 (2016)
18. Hillenbrand, T., Buch, A., Vogt, R., Löchner, B.: WALDMEISTER - high-performance equational deduction. J. Autom. Reasoning **18**, 265–270 (2004)
19. Hillenbrand, T.: Citius altius fortius: lessons learned from the theorem prover Waldmeister. ENTCS **86**(1), 9–21 (2003)
20. Irsoy, O., Cardie, C.: Deep recursive neural networks for compositionality in language. Adv. Neural Inf. Process. Syst. **27**, 2096–2104 (2014)
21. Kaliszyk, C., Urban, J., Michalewski, H., Olšák, M.: Reinforcement learning of theorem proving. Adv. Neural Inf. Process. Syst. **31**, 8822–8833 (2018)
22. Kingma, D.P., Ba, J.: Adam: a method for stochastic optimization. arXiv preprint arXiv:1412.6980 (2014)
23. Kinyon, M.: Proof simplification and automated theorem proving. CoRR abs/1808.04251 (2018). http://arxiv.org/abs/1808.04251

24. Kinyon, M., Veroff, R., Vojtěchovský, P.: Loops with abelian inner mapping groups: an application of automated deduction. In: Bonacina, M.P., Stickel, M.E. (eds.) Automated Reasoning and Mathematics. LNCS (LNAI), vol. 7788, pp. 151–164. Springer, Heidelberg (2013). https://doi.org/10.1007/978-3-642-36675-8_8
25. McCune, W.: Prover9 and Mace (2010). http://www.cs.unm.edu/~mccune/prover9/
26. McCune, W.: Prover9. https://github.com/ai4reason/Prover9
27. Mnih, V., et al.: Asynchronous methods for deep reinforcement learning. In: International Conference on Machine Learning, pp. 1928–1937 (2016)
28. Mnih, V., et al.: Human-level control through deep reinforcement learning. Nature **518**(7540), 529–533 (2015)
29. Oh, J., Guo, Y., Singh, S., Lee, H.: Self-imitation learning. In: International Conference on Machine Learning, pp. 3878–3887 (2018)
30. Overbeek, R.A.: A new class of automated theorem-proving algorithms. J. ACM **21**(2), 191–200 (1974). https://doi.org/10.1145/321812.321814
31. Paszke, A., et al.: PyTorch: An imperative style, high-performance deep learning library. In: Wallach, H., Larochelle, H., Beygelzimer, A., d'Alché-Buc, F., Fox, E., Garnett, R. (eds.) Advances in Neural Information Processing Systems, vol. 32, pp. 8024–8035. Curran Associates, Inc. (2019). http://papers.neurips.cc/paper/9015-pytorch-an-imperative-style-high-performance-deep-learning-library.pdf
32. Phillips, J., Stanovský, D.: Automated theorem proving in quasigroup and loop theory. AI Commun. **23**(2–3), 267–283 (2010)
33. PyTorch: RL Examples. https://github.com/pytorch/examples/tree/main/reinforcement_learning
34. Ross, S., Gordon, G., Bagnell, D.: A reduction of imitation learning and structured prediction to no-regret online learning. In: Proceedings of the 14th International Conference on Artificial Intelligence and Statistics, pp. 627–635 (2011)
35. Schulman, J., Levine, S., Abbeel, P., Jordan, M., Moritz, P.: Trust region policy optimization. In: Bach, F., Blei, D. (eds.) Proceedings of the 32nd International Conference on Machine Learning. Proceedings of Machine Learning Research, vol. 37, pp. 1889–1897. PMLR, Lille (2015). https://proceedings.mlr.press/v37/schulman15.html
36. Schulman, J., Wolski, F., Dhariwal, P., Radford, A., Klimov, O.: Proximal policy optimization algorithms. arXiv preprint arXiv:1707.06347 (2017)
37. Schulz, S.: E - a brainiac theorem prover. AI Commun. **15**(2–3), 111–126 (2002)
38. Schulz, S., Cruanes, S., Vukmirović, P.: Faster, Higher, Stronger: E 2.3. In: Fontaine, P. (ed.) CADE 2019. LNCS (LNAI), vol. 11716, pp. 495–507. Springer, Cham (2019). https://doi.org/10.1007/978-3-030-29436-6_29
39. Schulz, S.: Eprover. https://wwwlehre.dhbw-stuttgart.de/~sschulz/E/E.html
40. Silver, D.: Mastering the game of go with deep neural networks and tree search. Nature **529**(7587), 484–489 (2016)
41. Silver, D., et al.: Mastering the game of go without human knowledge. Nature **550**(7676), 354–359 (2017)
42. Smallbone, N.: Twee: an equational theorem prover. In: Platzer, A., Sutcliffe, G. (eds.) CADE 2021. LNCS (LNAI), vol. 12699, pp. 602–613. Springer, Cham (2021). https://doi.org/10.1007/978-3-030-79876-5_35
43. Smallbone, N.: Twee 2.4.1. https://github.com/nick8325/twee/releases/download/2.4.1/twee-2.4.1-linux-amd64
44. Sutcliffe, G.: The CADE-27 automated theorem proving system competition - CASC-27. AI Commun. **32**(5–6), 373–389 (2020)

45. Sutton, R.S., Barto, A.G.: Reinforcement Learning: An Introduction. MIT press, Cambridge (2018)
46. Hillenbrand, T., Buch, A., Vogt, R., Löchner, B.: Waldmeister (2022). https://www.mpi-inf.mpg.de/departments/automation-of-logic/software/waldmeister/download
47. Torabi, F., Warnell, G., Stone, P.: Behavioral cloning from observation. In: Proceedings of the 27th International Joint Conference on Artificial Intelligence, IJCAI 2018, pp. 4950–4957. AAAI Press (2018)
48. Veroff, R.: Using hints to increase the effectiveness of an automated reasoning program: Case studies. J. Autom. Reason. **16**(3), 223–239 (1996). https://doi.org/10.1007/BF00252178
49. Wang, Z., et al.: Sample efficient actor-critic with experience replay. In: International Conference on Learning Representations (2016)

Rensets and Renaming-Based Recursion for Syntax with Bindings

Andrei Popescu(✉)

Department of Computer Science, University of Sheffield, Sheffield, UK
a.popescu@sheffield.ac.uk

Abstract. I introduce *renaming-enriched sets* (*rensets* for short), which are algebraic structures axiomatizing fundamental properties of renaming (also known as variable-for-variable substitution) on syntax with bindings. Rensets compare favorably in some respects with the well-known foundation based on nominal sets. In particular, renaming is a more fundamental operator than the nominal swapping operator and enjoys a simpler, equationally expressed relationship with the variable-freshness predicate. Together with some natural axioms matching properties of the syntactic constructors, rensets yield a truly minimalistic characterization of λ-calculus terms as an abstract datatype – one involving an infinite set of *unconditional equations*, referring only to the most fundamental term operators: the constructors and renaming. This characterization yields a recursion principle, which (similarly to the case of nominal sets) can be improved by incorporating Barendregt's variable convention. When interpreting syntax in semantic domains, my renaming-based recursor is easier to deploy than the nominal recursor. My results have been validated with the proof assistant Isabelle/HOL.

1 Introduction

Formal reasoning about syntax with bindings is necessary for the meta-theory of logics, calculi and programming languages, and is notoriously error-prone. A great deal of research has been put into formal frameworks that make the specification of, and the reasoning about bindings more manageable.

Researchers wishing to formalize work involving syntax with bindings must choose a paradigm for representing and manipulating syntax—typically a variant of one of the "big three": nameful (sometimes called "nominal" reflecting its best known incarnation, nominal logic [23,39]), nameless (De Bruijn) [4,13,49,51] and higher-order abstract syntax (HOAS) [19,20,28,34,35]. Each paradigm has distinct advantages and drawbacks compared with each of the others, some discussed at length, e.g., in [1,9] and [25, §8.5]. And there are also hybrid approaches, which combine some of the advantages [14,18,42,47].

A significant advantage of the nameful paradigm is that it stays close to the way one informally defines and manipulates syntax when describing systems in textbooks and research papers—where the binding variables are explicitly

J. Blanchette et al. (Eds.): IJCAR 2022, LNAI 13385, pp. 618–639, 2022.
https://doi.org/10.1007/978-3-031-10769-6_36

indicated. This can in principle ensure transparency of the formalization and allows the formalizer to focus on the high-level ideas. However, it only works if the technical challenge faced by the nameful paradigm is properly addressed: enabling the seamless definition and manipulation of concepts "up to alpha-equivalence", i.e., in such a way that the names of the bound variables are (present but nevertheless) inconsequential. This is particularly stringent in the case of recursion due to the binding constructors of terms not being free, hence not being *a priori* traversable recursively—in that simply writing some recursive clauses that traverse the constructors is not *a priori* guaranteed to produce a correct definition, but needs certain favorable conditions. The problem has been addressed by researchers in the form of tailored *nameful recursors* [23,33,39,43, 56,57], which are theorems that identify such favorable conditions and, based on them, guarantee the existence of functions that recurse over the non-free constructors.

In this paper, I make a contribution to the nameful paradigm in general, and to nameful recursion in particular. I introduce *rensets*, which are algebraic structures axiomatizing the properties of renaming, also known as variable-for-variable substitution, on terms with bindings (Sect. 3). Rensets differ from nominal sets (Sect. 2.2), which form the foundation of nominal logic, by their focus on (not necessarily injective) renaming rather than swapping (or permutation). Similarly to nominal sets, rensets are pervasive: Not only do the variables and terms form rensets, but so do any container-type combinations of rensets.

While lacking the pleasant symmetry of swapping, my axiomatization of renaming has its advantages. First, renaming is more fundamental than swapping because, at an abstract axiomatic level, renaming can define swapping but not vice versa (Sect. 4). The second advantage is about the ability to define another central operator: the variable freshness predicate. While the definability of freshness from swapping is a signature trait of nominal logic, my renaming-based alternative fares even better: In rensets freshness has a simple, first-order definition (Sect. 3). This contrasts the nominal logic definition, which involves a second-order statement about (co)finiteness of a set of variables. The third advantage is largely a consequence of the second: Rensets enriched with constructor-like operators facilitate an equational characterization of terms with bindings (using an infinite set of unconditional equations), which does not seem possible for swapping (Sect. 5.1). This produces a recursion principle (Sect. 5.2) which, like the nominal recursor, caters for Barendregt's variable convention, and in some cases is easier to apply than the nominal recursor—for example when interpreting syntax in semantic domains (Sect. 5.3).

In summary, I argue that my renaming-based axiomatization offers some benefits that strengthen the arsenal of the nameful paradigm: a simpler representation of freshness, a minimalistic equational characterization of terms, and a convenient recursion principle. My results are established with high confidence thanks to having been mechanized in Isabelle/HOL [32]. The mechanization is available [44] from Isabelle's Archive of Formal Proofs.

Here is the structure of the rest of this paper: Sect. 2 provides background on terms with bindings and on nominal logic. Section 3 introduces rensets and describes their basic properties. Section 4 establishes a formal connection to nominal sets. Section 5 discusses substitutive-set-based recursion. Section 6 discusses related work. A technical report [45] associated to this paper includes an appendix with more examples and results and more background on nominal sets.

2 Background

This section recalls the terms of λ-calculus and their basic operators (Sect. 2.1), and aspects of nominal logic including nominal sets and nominal recursion (Sect. 2.2).

2.1 Terms with Bindings

I work with the paradigmatic syntax of (untyped) λ-calculus. However, my results generalize routinely to syntaxes specified by arbitrary binding signatures such as the ones in [22, §2], [39,59] or [12].

Let Var be a countably infinite set of variables, ranged over by x, y, z etc. The set Trm of *λ-terms* (or *terms* for short), ranged over by t, t_1, t_2 etc., is defined by the grammar $t ::= \ \mathsf{Vr}\ x \mid \mathsf{Ap}\ t_1\ t_2 \mid \mathsf{Lm}\ x\ t$
with the proviso that terms are equated (identified) modulo alpha-equivalence (also known as naming equivalence). Thus, for example, if $x \neq z \neq y$ then $\mathsf{Lm}\ x\ (\mathsf{Ap}\ (\mathsf{Vr}\ x)\ (\mathsf{Vr}\ z))$ and $\mathsf{Lm}\ y\ (\mathsf{Ap}\ (\mathsf{Vr}\ y)\ (\mathsf{Vr}\ z))$ are considered to be the same term. I will often omit Vr when writing terms, as in, e.g., $\mathsf{Lm}\ x\ x$.

What the above specification means is (something equivalent to) the following: One first defines the set PTrm of *pre-terms* as freely generated by the grammar $p ::= \ \mathsf{PVr}\ x \mid \mathsf{PAp}\ p_1\ p_2 \mid \mathsf{PLm}\ x\ p$. Then one defines the alpha-equivalence relation $\equiv\ : \mathsf{PTrm} \to \mathsf{PTrm} \to \mathsf{Bool}$ inductively, proves that it is an equivalence, and defines Trm by quotienting PTrm to alpha-equivalence, i.e., $\mathsf{Trm} = \mathsf{PTrm}/\equiv$. Finally, one proves that the pre-term constructors are compatible with $\equiv$, and defines the term counterpart of these constructors: $\mathsf{Vr} : \mathsf{Var} \to \mathsf{Trm}$, $\mathsf{Ap} : \mathsf{Trm} \to \mathsf{Trm} \to \mathsf{Trm}$ and $\mathsf{Lm} : \mathsf{Var} \to \mathsf{Trm} \to \mathsf{Trm}$.

The above constructions are technical, but well-understood, and can be fully automated for an arbitrary syntax with bindings (not just that of λ-calculus); and tools such as the Isabelle/Nominal package [59,60] provide this automation, hiding pre-terms completely from the end user. In formal and informal presentations alike, one usually prefers to forget about pre-terms, and work with terms only. This has several advantages, including (1) being able to formalize concepts at the right abstraction level (since in most applications the naming of bound variables should be inconsequential) and (2) the renaming operator being well-behaved. However, there are some difficulties that need to be overcome when working with terms, and in this paper I focus on one of the major ones: providing recursion principles, i.e., mechanisms for defining functions by recursing over terms. This difficulty arises essentially because, unlike in the case of pre-term constructors, the binding constructor for terms is not free.

The main characters of my paper will be (generalizations of) some common operations and relations on Trm, namely:

- the constructors $\mathsf{Vr} : \mathsf{Var} \to \mathsf{Trm}$, $\mathsf{Ap} : \mathsf{Trm} \to \mathsf{Trm} \to \mathsf{Trm}$ and $\mathsf{Lm} : \mathsf{Var} \to \mathsf{Trm} \to \mathsf{Trm}$
- (capture-avoiding) renaming, also known as (capture-avoiding) substitution of variables for variables $\_[\_/\_] : \mathsf{Trm} \to \mathsf{Var} \to \mathsf{Var} \to \mathsf{Trm}$; e.g., we have $(\mathsf{Lm}\ x\ (\mathsf{Ap}\ x\ y))\ [x/y] = \mathsf{Lm}\ x'\ (\mathsf{Ap}\ x'\ x)$
- swapping $\_[\_\wedge\_] : \mathsf{Trm} \to \mathsf{Var} \to \mathsf{Var} \to \mathsf{Trm}$; e.g., we have $(\mathsf{Lm}\ x\ (\mathsf{Ap}\ x\ y))\ [x \wedge y] = \mathsf{Lm}\ y\ (\mathsf{Ap}\ y\ x)$
- the free-variable operator $\mathsf{FV} : \mathsf{Trm} \to \mathsf{Pow}(\mathsf{Var})$ (where $\mathsf{Pow}(\mathsf{Var})$ is the powerset of Var); e.g., we have $\mathsf{FV}(\mathsf{Lm}\ x\ (\mathsf{Ap}\ y\ x)) = \{y\}$
- freshness $\_\#\_ : \mathsf{Var} \to \mathsf{Trm} \to \mathsf{Bool}$; e.g., we have $x \# (\mathsf{Lm}\ x\ x)$; and assuming $x \neq y$, we have $\neg\, x \# (\mathsf{Lm}\ y\ x)$

The free-variable and freshness operators are of course related: A variable x is fresh for a term t (i.e., $x \# t$) if and only if it is not free in t (i.e., $x \notin \mathsf{FV}(t)$). The renaming operator $\_[\_/\_] : \mathsf{Trm} \to \mathsf{Var} \to \mathsf{Var} \to \mathsf{Trm}$ substitutes (in terms) *variables* for variables, not terms for variables. (But an algebraization of term-for-variable substitution is discussed in [45, Appendix D].)

2.2 Background on Nominal Logic

I will employ a formulation of nominal logic [38,39,57] that does not require any special logical foundation, e.g., axiomatic nominal set theory. For simplicity, I prefer the swapping-based formulation [38] to the equivalent permutation-based formulation—[45, Appendix C] gives details on these two alternatives.

A *pre-nominal set* is a pair $\mathcal{A} = (A, \_[\_\wedge\_])$ where A is a set and $\_[\_\wedge\_] : A \to \mathsf{Perm} \to A$ is a function called *the swapping operator of* $\mathcal{A}$ satisfying the following properties for all $a \in A$ and $x, x_1, x_2, y_1, y_2 \in \mathsf{Var}$:

$$\begin{aligned} \text{Identity:}\quad & a[x \wedge x] = a \\ \text{Involution:}\quad & a[x_1 \wedge x_2][x_1 \wedge x_2] = a \\ \text{Compositionality:}\quad & a[x_1 \wedge x_2][y_1 \wedge y_2] = a[y_1 \wedge y_2][(x_1[y_1 \wedge y_2]) \wedge (x_2[y_1 \wedge y_2])] \end{aligned}$$

Given a pre-nominal set $\mathcal{A} = (A, \_[\_\wedge\_])$, an element $a \in A$ and a set $X \subseteq \mathsf{Var}$, one says that *a is supported by X* if $a[x \wedge y] = a$ holds for all $x, y \in \mathsf{Var}$ such that $x, y \notin X$. An element $a \in A$ is called *finitely supported* if there exists a finite set $X \subseteq A$ such that a is supported by X. A *nominal set* is a pre-nominal set $\mathcal{A} = (A, \_[\_\wedge\_])$ such that every element of a is finitely supported. If $\mathcal{A} = (A, \_[\_\wedge\_])$ is a nominal set and $a \in A$, then the smallest set $X \subseteq A$ such that a is supported by X exists, and is denoted by $\mathsf{supp}^{\mathcal{A}}\, a$ and called the *support of* a. One calls a variable x *fresh for* a, written $x \# a$, if $x \notin \mathsf{supp}^{\mathcal{A}}\, a$.

An alternative, more direct definition of freshness (which is preferred, e.g., by Isabelle/Nominal [59,60]) is provided by the following proposition:

Proposition 1. For any nominal set $\mathcal{A} = (A, \_[\_\wedge\_])$ and any $x \in \mathsf{Var}$ and $a \in A$, it holds that $x \# a$ if and only if the set $\{y \mid a[y \wedge x] \neq a\}$ is finite.

Given two pre-nominal sets $\mathcal{A} = (A, \_[\_\wedge\_])$ and $\mathcal{B} = (B, \_[\_\wedge\_])$, the set $F = (A \to B)$ of functions from A to B becomes a pre-nominal set $\mathcal{F} = (F, \_[\_\wedge\_])$ by defining $f[x \wedge y]$ to send each $a \in A$ to $(f(a[x \wedge y]))[x \wedge y]$. $\mathcal{F}$ is not a nominal set because not all functions are finitely supported (though of course one obtains a nominal set by restricting to finitely supported functions).

The set of terms together with their swapping operator, $(\mathsf{Trm}, \_[\_\wedge\_])$, forms a nominal set, where the support of a term is precisely its set of free variables. However, the power of nominal logic resides in the fact that not only the set of terms, but also many other sets can be organized as nominal sets—including the target domains of many functions one may wish to define on terms. This gives rise to a convenient mechanism for defining functions recursively on terms:

Theorem 2 [39]. Let $\mathcal{A} = (A, \_[\_])$ be a nominal set and let $\mathsf{Vr}^{\mathcal{A}} : \mathsf{Var} \to A$, $\mathsf{Ap}^{\mathcal{A}} : A \to A \to A$ and $\mathsf{Lm}^{\mathcal{A}} : \mathsf{Var} \to A \to A$ be some functions, all supported by a finite set X of variables and with $\mathsf{Lm}^{\mathcal{A}}$ satisfying the following freshness condition for binders (FCB): There exists $x \in \mathsf{Var}$ such that $x \notin X$ and $x \# \mathsf{Lm}^{\mathcal{A}}\ x\ a$ for all $a \in A$.

Then there exists a unique function $f : \mathsf{Trm} \to A$ that is supported by X and such that the following hold for all $x \in \mathsf{Var}$ and $t_1, t_2, t \in \mathsf{Trm}$:

(i) $f\,(\mathsf{Vr}\ x) = \mathsf{Vr}^{\mathcal{A}}\ x$ (ii) $f\,(\mathsf{Ap}\ t_1\ t_2) = \mathsf{Ap}^{\mathcal{A}}\,(f\ t_1)\,(f\ t_2)$
(iii) $f\,(\mathsf{Lm}\ x\ t) = \mathsf{Lm}^{\mathcal{A}}\ x\ (f\ t)$ if $x \notin X$

A useful feature of nominal recursion is the support for Barendregt's famous *variable convention* [8, p. 26]: "If [the terms] $t_1, \ldots, t_n$ occur in a certain mathematical context (e.g. definition, proof), then in these terms all bound variables are chosen to be different from the free variables." The above recursion principle adheres to this convention by fixing a finite set X of variables meant to be free in the definition context and guaranteeing that the bound variables in the definitional clauses are distinct from them. Formally, the target domain operators $\mathsf{Vr}^{\mathcal{A}}$, $\mathsf{Ap}^{\mathcal{A}}$ and $\mathsf{Lm}^{\mathcal{A}}$ are supported by X, and the clause for λ-abstraction is conditioned by the binding variable x being outside of X. (The Barendregt convention is also present in nominal logic via induction principles [39,58–60].)

3 Rensets

This section introduces rensets, an alternative to nominal sets that axiomatize renaming rather than swapping or permutation.

A *renaming-enriched set* (*renset* for short) is a pair $\mathcal{A} = (A, \_[\_/\_])$ where A is a set and $\_[\_/\_] : A \to \mathsf{Var} \to \mathsf{Var} \to A$ is an operator such that the following hold for all $x, x_1, x_2, x_3, y, y_1, y_2 \in \mathsf{Var}$ and $a \in A$:

Identity: $a[x/x] = a$
Idempotence: If $x_1 \neq y$ then $a[x_1/y][x_2/y] = a[x_1/y]$
Chaining: If $y \neq x_2$ then $a[y/x_2][x_2/x_1][x_3/x_2] = a[y/x_2][x_3/x_1]$
Commutativity: If $x_2 \neq y_1 \neq x_1 \neq y_2$ then $a[x_2/x_1][y_2/y_1] = a[y_2/y_1][x_2/x_1]$

Let us call A the *carrier* of $\mathcal{A}$ and $\_[\_/\_]$ the *renaming operator* of $\mathcal{A}$. Similarly to the case of terms, we think of the elements $a \in A$ as some kind of variable-bearing entities and of $a[y/x]$ as the result of substituting x with y in a. With this intuition, the above properties are natural: Identity says that substituting a variable with itself has no effect. Idempotence acknowledges the fact that, after its renaming, a variable y is no longer there, so substituting it again has no effect. Chaining says that a chain of renamings $x_3/x_2/x_1$ has the same effect as the end-to-end renaming x_3/x_1 provided there is no interference from x_2, which is ensured by initially substituting x_2 with some other variable y. Finally, Commutativity allows the reordering of any two independent renamings.

Examples. $(\mathsf{Var}, \_[\_/\_])$ and $(\mathsf{Trm}, \_[\_/\_])$, the sets of variables and terms with the standard renaming operator on them, form rensets. Moreover, given any functor F on the category of sets and a renset $\mathcal{A} = (A, \_[\_/\_])$, let us define the renset $F\,\mathcal{A} = (F\,A, \_[\_/\_])$ as follows: for any $k \in F\,A$ and $x, y \in \mathsf{Var}$, $k[x/y] = F\,(\_[x/y])\,k$, where the last occurrence of F refers to the action of the functor on morphisms. This means that one can freely build new rensets from existing ones using container types (which are particular kinds of functors)—e.g., lists, sets, trees etc. Another way to put it: Rensets are closed under datatype and codatatype constructions [55].

In what follows, let us fix a renset $\mathcal{A} = (A, \_[\_/\_])$. One can define the notion of freshness of a variable for an element of a in the style of nominal logic. But the next proposition shows that simpler formulations are available.

Proposition 3. The following are equivalent:
(1) The set $\{y \in \mathsf{Var} \mid a[y/x] \neq a\}$ is finite.
(2) $a[y/x] = a$ for all $y \in \mathsf{Var}$. (3) $a[y/x] = a$ for some $y \in \mathsf{Var} \smallsetminus \{x\}$.

Let us define the predicate $\_\#\_ : \mathsf{Var} \to A \to \mathsf{Bool}$ as follows: $x \# a$, read x *is fresh for* a, if either of Proposition 3's equivalent properties holds.

Thus, points (1)–(3) above are three alternative formulations of $x \# a$, all referring to the lack of effect of substituting y for x, expressed as $a[y/x] = a$: namely that this phenomenon affects (1) all but a finite number of variables y, (2) all variables y, or (3) some variable $y \neq x$. The first formulation is the most complex of the three—it is the nominal definition, but using renaming instead of swapping. The other two formulations do not have counterparts in nominal logic, essentially because swapping is not as "efficient" as renaming at exposing freshness. In particular, (3) does not have a nominal counterpart because there is no single-swapping litmus test for freshness. The closest we can get to property (3) in a nominal set is the following: x is fresh for a if and only $a[y \wedge x] = a$ holds for some fresh y—but this needs freshness to explain freshness!

Examples (continued). For the rensets of variables and terms, freshness defined as above coincides with the expected operators: distinctness in the case of variables and standard freshness in the case of terms. And applying the definition of freshness to rensets obtained using finitary container types has similarly intuitive outcomes; for example, the freshness of a variable x for a list of items $[a_1, \ldots, a_n]$ means that x is fresh for each item a_i in the list.

Freshness satisfies some intuitive properties, which can be easily proved from its definition and the renset axioms. In particular, point (2) of the next proposition is the freshness-based version of the Chaining axiom.

Proposition 4. The following hold:
(1) If $x \# a$ then $a[y/x] = a$ (2) $x_2 \# a$ then $a[x_2/x_1][x_3/x_2] = a[x_3/x_1]$
(3) If $z \# a$ or $z = x$, and $x \# a$ or $z \neq y$, then $z \# a[y/x]$

4 Connection to Nominal Sets

So far I focused on consequences of the purely equational theory of rensets, without making any assumption about cardinality. But after additionally postulating a nominal-style finite support property, one can show that rensets give rise to nominal sets—which is what I will do in this section.

Let us say that a renset $\mathcal{A} = (A, \_[\_/\_])$ has the *Finite Support* property if, for all $a \in A$, the set $\{x \in \mathsf{Var} \mid \neg\, x \# a\}$ is finite.

Let $\mathcal{A} = (A, \_[\_/\_])$ be a renset satisfying Finite Support. Let us define the swapping operator $\_[\_ \wedge \_] : A \to \mathsf{Var} \to \mathsf{Var} \to A$ as follows: $a[x_1 \wedge x_2] = a[y/x_1][x_1/x_2][x_2/y]$, where y is a variable that is fresh for all the involved items, namely $y \notin \{x_1, x_2\}$ and $y \# a$. Indeed, this is how one would define swapping from renaming on terms: using a fresh auxiliary variable y, and exploiting that such a fresh y exists and that its choice is immaterial for the end result. The next lemma shows that this style of definition also works abstractly, i.e., all it needs are the renset axioms plus Finite Support.

Lemma 5. The following hold for all $x_1, x_2 \in \mathsf{Var}$ and $a \in A$:

(1) There exists $y \in \mathsf{Var}$ such that $y \notin \{x_1, x_2\}$ and $y \# a$.
(2) For all $y, y' \in \mathsf{Var}$ such that $y \notin \{x_1, x_2\}$, $y \# a$, $y' \notin \{x_1, x_2\}$ and $y' \# a$, $a[y/x_1][x_1/x_2][x_2/y] = a[y'/x_1][x_1/x_2][x_2/y']$.

And one indeed obtains an operator satisfying the nominal axioms:

Proposition 6. If $(A, \_[\_/\_])$ is a renset satisfying Finite Support, then $(A, \_[\_ \wedge \_])$ is a nominal set. Moreover, $(A, \_[\_/\_])$ and $(A, \_[\_ \wedge \_])$ have the same notion of freshness, in that the freshness operator defined from renaming coincides with that defined from swapping.

The above construction is functorial, as I detail next. Given two nominal sets $\mathcal{A} = (A, \_[\_ \wedge \_])$ and $\mathcal{B} = (B, \_[\_ \wedge \_])$, a *nominal morphism* $f : \mathcal{A} \to \mathcal{B}$ is a function $f : A \to B$ with the property that it commutes with swapping, in that $(f\ a)[x \wedge y] = f(a[x \wedge y])$ for all $a \in A$ and $x, y \in \mathsf{Var}$. Nominal sets and nominal morphisms form a category that I will denote by $\underline{Nom}$. Similarly, let us define a morphism $f : \mathcal{A} \to \mathcal{B}$ between two rensets $\mathcal{A} = (A, \_[\_/\_])$ and $\mathcal{B} = (B, \_[\_])$ to be a function $f : A \to B$ that commutes with renaming, yielding the category $\underline{Sbs}$ of rensets. Let us write $\underline{FSbs}$ for the full subcategory of $\underline{Sbs}$ given by rensets that satisfy Finite Support. Let us define $F : \underline{FSbs} \to \underline{Nom}$ to be an operator on objects and morphisms that sends each finite-support renset to the above described nominal set constructed from it, and sends each substitutive morphism to itself.

Theorem 7. F is a functor between $\underline{FSbs}$ and $\underline{Nom}$ which is injective on objects and full and faithful (i.e., bijective on morphisms).

One may ask whether it is also possible to make the trip back: from nominal to rensets. The answer is negative, at least if one wants to retain the same notion of freshness, i.e., have the freshness predicate defined in the nominal set be identical to the one defined in the resulting renset. This is because swapping preserves the cardinality of the support, whereas renaming must be allowed to change it since it might perform a non-injective renaming. The following example captures this idea:

Counterexample. Let $\mathcal{A} = (A, \_[\_\wedge\_])$ be a nominal set such that all elements of A have their support consisting of exactly two variables, x and y (with $x \neq y$). (For example, A can be the set of all terms with these free variables—this is indeed a nominal subset of the term nominal set because it is closed under swapping.) Assume for a contradiction that $\_[\_/\_]$ is an operation on A that makes $(A, \_[\_/\_])$ a renset with its induced freshness operator equal to that of $\mathcal{A}$. Then, by the definition of A, $a[y/x]$ needs to have exactly two non-fresh variables. But this is impossible, since by Proposition 4(3), all the variables different from y (including x) must be fresh for $a[y/x]$. In particular, $\mathcal{A}$ is not in the image of the functor $F : \underline{FSbs} \to \underline{Nom}$, which is therefore not surjective on objects.

Thus, at an abstract algebraic level renaming can define swapping, but not the other way around. This is not too surprising, since swapping is fundamentally bijective whereas renaming is not; but it further validates our axioms for renaming, highlighting their ability to define a well-behaved swapping.

5 Recursion Based on Rensets

Proposition 3 shows that, in rensets, renaming can define freshness using only equality and universal or existential quantification over variables—without needing any cardinality condition like in the case of swapping. As I am about to discuss, this forms the basis of a characterization of terms as the initial algebra of an equational theory (Sect. 5.1) and an expressive recursion principle (Sect. 5.2) that fares better than the nominal one for interpretations in semantic domains (Sect. 5.3).

5.1 Equational Characterization of the Term Datatype

Rensets contain elements that are "term-like" in as much as there is a renaming operator on them satisfying familiar properties of renaming on terms. This similarity with terms can be strengthened by enriching rensets with operators having arities that match those of the term constructors.

A *constructor-enriched renset* (*CE renset* for short) is a tuple $\mathcal{A} = (A, \_[\_/\_], \mathsf{Vr}^{\mathcal{A}}, \mathsf{Ap}^{\mathcal{A}}, \mathsf{Lm}^{\mathcal{A}})$ where:

- $(A, \_[\_/\_])$ is a renset
- $\mathsf{Vr}^{\mathcal{A}} : \mathsf{Var} \to A$, $\mathsf{Ap}^{\mathcal{A}} : A \to A \to A$ and $\mathsf{Lm}^{\mathcal{A}} : \mathsf{Var} \to A \to A$ are functions

such that the following hold for all $a, a_1, a_2 \in A$ and $x, y, z \in \mathsf{Var}$:

(S1) $(\mathsf{Vr}^{\mathcal{A}}\ x)[y/z] = \mathsf{Vr}^{\mathcal{A}}(x[y/z])$
(S2) $(\mathsf{Ap}^{\mathcal{A}}\ a_1\ a_2)[y/z] = \mathsf{Ap}^{\mathcal{A}}(a_1[y/z])\,(a_2[y/z])$
(S3) if $x \notin \{y, z\}$ then $(\mathsf{Lm}^{\mathcal{A}}\ x\ a)[y/z] = \mathsf{Lm}^{\mathcal{A}}\ x\,(a[y/z])$
(S4) $(\mathsf{Lm}^{\mathcal{A}}\ x\ a)[y/x] = \mathsf{Lm}^{\mathcal{A}}\ x\ a$
(S5) if $z \neq y$ then $\mathsf{Lm}^{\mathcal{A}}\ x\ (a[z/y]) = \mathsf{Lm}^{\mathcal{A}}\ y\ (a[z/y][y/x])$

Let us call $\mathsf{Vr}^{\mathcal{A}}, \mathsf{Ap}^{\mathcal{A}}, \mathsf{Lm}^{\mathcal{A}}$ the *constructors* of $\mathcal{A}$. (S1)–(S3) express the constructors' commutation with renaming (with capture-avoidance provisions in the case of (S3)), (S4) the lack of effect of substituting for a bound variable, and (S5) the possibility to rename a bound variable without changing the abstracted item (where the inner renaming of $z \neq y$ for y ensures the freshness of the "new name" y, hence its lack of interference with the other names in the "term-like" entity where the renaming takes place). All these are well-known to hold for terms:

Example. Terms with renaming and the constructors, namely $(\mathsf{Trm}, \_[\_/\_], \mathsf{Vr}, \mathsf{Ap}, \mathsf{Lm})$, form a CE renset which will be denoted by $\mathcal{T}rm$.

As it turns out, the CE renset axioms capture exactly the term structure $\mathcal{T}rm$, via initiality. The notion of *CE substitutive morphism* $f : \mathcal{A} \to \mathcal{B}$ between two CE rensets $\mathcal{A} = (A, \_[\_/\_], \mathsf{Vr}^{\mathcal{A}}, \mathsf{Ap}^{\mathcal{A}}, \mathsf{Lm}^{\mathcal{A}})$ and $\mathcal{B} = (B, \_[\_/\_], \mathsf{Vr}^{\mathcal{B}}, \mathsf{Ap}^{\mathcal{B}}, \mathsf{Lm}^{\mathcal{B}})$ is the expected one: a function $f : A \to B$ that is a substitutive morphism and also commutes with the constructors. Let us write $\underline{Sbs}_{\mathsf{CE}}$ for the category of CE rensets and morphisms.

Theorem 8. $\mathcal{T}rm$ is the initial CE renset, i.e., initial object in $\underline{Sbs}_{\mathsf{CE}}$.

Proof Idea. Let $\mathcal{A} = (A, \_[\_/\_], \mathsf{Vr}^{\mathcal{A}}, \mathsf{Ap}^{\mathcal{A}}, \mathsf{Lm}^{\mathcal{A}})$ be a CE renset. Instead of directly going after a function $f : \mathsf{Trm} \to A$, one first inductively defines a relation $R : \mathsf{Trm} \to A \to \mathsf{Bool}$, with inductive clauses reflecting the desired properties concerning the commutation with the constructors, e.g., $\frac{R\ t\ a}{R\ (\mathsf{Lm}\ x\ t)\ (\mathsf{Lm}^{\mathcal{A}}\ x\ a)}$. It suffices to prove that R is total and functional and preserves renaming, since that allows one to define a constructor- and renaming-preserving function (a morphism) f by taking $f\ t$ to be the unique a with $R\ t\ a$.

Proving that R is total is easy by standard induction on terms. Proving the other two properties, namely functionality and preservation of renaming, is more elaborate and requires their simultaneous proof together with a third property: that R preserves freshness. The simultaneous three-property proof follows by a form of "substitutive induction" on terms: Given a predicate $\phi : \mathsf{Trm} \to \mathsf{Bool}$, to show $\forall t \in \mathsf{Trm}.\ \phi\ t$ it suffices to show the following: (1) $\forall x \in \mathsf{Var}.\ \phi\ (\mathsf{Vr}\ x)$, (2) $\forall t_1, t_2 \in \mathsf{Trm}.\ \phi\ t_1 \,\&\, \phi\ t_2 \to \phi\,(\mathsf{Ap}\ t_1\ t_2)$, and (3) $\forall x \in \mathsf{Var}, t \in \mathsf{Trm}.\ (\forall s \in \mathsf{Trm}.\ \mathsf{Con}_{\_[\_/\_]}\ t\ s \to \phi\ s) \to \phi\,(\mathsf{Lm}\ x\ t)$, where $\mathsf{Con}_{\_[\_/\_]}\ t\ s$ means that t is connected to s by a chain of renamings.

Roughly speaking, R turns out to be functional because the λ-abstraction operator on the "term-like" inhabitants of A is, thanks to the axioms of CE renset, at least as non-injective as (i.e., identifies at least as many items as) the λ-abstraction operator on terms. □

Theorem 8 is the central result of this paper, from both practical and theoretical perspectives. Practically, it enables a useful form of recursion on terms (as I will discuss in the following sections). Theoretically, this is a characterization of terms as the initial algebra of an equational theory that only the most fundamental term operations, namely the constructors and renaming. The equational theory consists of the axioms of CE rensets (i.e., those of rensets plus (S1)–(S5)), which are an infinite set of unconditional equations—for example, axiom (S5) gives one equation for each pair of distinct variables y, z.

It is instructive to compare this characterization with the one offered by nominal logic, namely by Theorem 2. To do this, one first needs a lemma:

Lemma 9. Let $f : A \to B$ be a function between two nominal sets $\mathcal{A} = (A, \_[\_\wedge\_])$ and $\mathcal{B} = (B, \_[\_\wedge\_])$ and X a set of variables. Then f is supported by X if and only if $f(a[x \wedge y]) = (f\, a)[x \wedge y]$ for all $x, y \in \mathsf{Var} \smallsetminus X$.

Now Theorem 2 (with the variable avoidance set X taken to be $\emptyset$) can be rephrased as an initiality statement, as I describe below.

Let us define a *constructor-enriched nominal set* (*CE nominal set*) to be any tuple $\mathcal{A} = (A, \_[\_\wedge\_], \mathsf{Vr}^{\mathcal{A}}, \mathsf{Ap}^{\mathcal{A}}, \mathsf{Lm}^{\mathcal{A}})$ where $(A, \_[\_\wedge\_])$ is a nominal set and $\mathsf{Vr}^{\mathcal{A}} : \mathsf{Var} \to A$, $\mathsf{Ap}^{\mathcal{A}} : A \to A \to A$, $\mathsf{Lm}^{\mathcal{A}} : \mathsf{Var} \to A \to A$ are operators on A such that the following properties hold for all $a, a_1, a_2 \in A$ and $x, y, z \in \mathsf{Var}$:

(N1) $(\mathsf{Vr}^{\mathcal{A}}\, x)[y \wedge z] = \mathsf{Vr}^{\mathcal{A}}(x[y \wedge z])$
(N2) $(\mathsf{Ap}^{\mathcal{A}}\, a_1\, a_2)[y \wedge z] = \mathsf{Ap}^{\mathcal{A}}(a_1[y \wedge z])\,(a_2[y \wedge z])$
(N3) $(\mathsf{Lm}^{\mathcal{A}}\, x\, a)[y \wedge z] = \mathsf{Lm}^{\mathcal{A}}\,(x[y \wedge z])\,(a[y \wedge z])$
(N4) $x \,\#\, \mathsf{Lm}\, x\, a$, i.e., $\{y \in \mathsf{Var} \mid (\mathsf{Lm}\, x\, a)[y \wedge x] \neq \mathsf{Lm}\, x\, a\}$ is finite.

The notion of *CE nominal morphism* is defined as the expected extension of that of nominal morphism: a function that commutes with swapping and the constructors. Let $\underline{Nom}_{\mathsf{CE}}$ be the category of CE nominal sets morphisms.

Theorem 10 ([39], rephrased)**.** $(\mathsf{Trm}, \_[\_\wedge\_], \mathsf{Vr}, \mathsf{Ap}, \mathsf{Lm})$ is the initial CE nominal set, i.e., the initial object in $\underline{Nom}_{\mathsf{CE}}$.

The above theorem indeed corresponds exactly to Theorem 2 with $X = \emptyset$:

- the conditions (N1)–(N3) in the definition of CE nominal sets correspond (via Lemma 9) to the constructors being supported by $\emptyset$
- (N4) is the freshness condition for binders
- initiality, i.e., the existence of a unique morphism, is the same as the existence of the unique function $f : \mathsf{Trm} \to A$ stipulated in Theorem 2: commutation with the constructors is the Theorem 2 conditions (i)–(iii), and commutation with swapping means (via Lemma 9) f being supported by $\emptyset$.

Unlike the renaming-based characterization of terms (Theorem 8), the nominal logic characterization (Theorem 10) is not purely equational. This is due to a combination of two factors: (1) two of the axioms ((N4) and the Finite

Support condition) referring to freshness and (2) the impossibility of expressing freshness equationally from swapping. The problem seems fundamental, in that a nominal-style characterization does not seem to be expressible purely equationally. By contrast, while the freshness idea is implicit in the CE renset axioms, the freshness predicate itself is absent from Theorem 8.

5.2 Barendregt-Enhanced Recursion Principle

While Theorem 8 already gives a recursion principle, it is possible to improve it by incorporating Barendregt's variable convention (in the style of Theorem 2):

Theorem 11. Let X be a finite set, $(A, \_[\_/\_])$ a renset and $\mathsf{Vr}^{\mathcal{A}} : \mathsf{Var} \to A$, $\mathsf{Ap}^{\mathcal{A}} : A \to A \to A$ and $\mathsf{Lm}^{\mathcal{A}} : \mathsf{Var} \to A \to A$ some functions that satisfy the clauses (S1)–(S5) from the definition of CE renset, but only under the assumption that $x, y, z \notin X$. Then there exists a unique function $f : \mathsf{Trm} \to A$ such that th following hold:

(i) $f\,(\mathsf{Vr}\;x) = \mathsf{Vr}^{\mathcal{A}}\;x$ (ii) $f\,(\mathsf{Ap}\;t_1\;t_2) = \mathsf{Ap}^{\mathcal{A}}\,(f\;t_1)\,(f\;t_2)$
(iii) $f\,(\mathsf{Lm}\;x\;t) = \mathsf{Lm}^{\mathcal{A}}\;x\;(f\;t)$ if $x \notin X$ (iv) $f\,(t[y/z]) = (f\;t)[y/z]$ if $y, z \notin X$

Proof Idea. The constructions in the proof of Theorem 8 can be adapted to avoid clashing with the finite set of variables X. For example, the clause for λ-abstraction in the inductive definition of the relation R becomes $\frac{x \notin X \quad R\;t\;a}{R\;(\mathsf{Lm}\;x\;t)\;(\mathsf{Lm}^{\mathcal{A}}\;x\;a)}$ and preservation of renaming and freshness are also formulated to avoid X. Totality is still ensured thanks to the possibility of renaming bound variables—in terms and inhabitants of A alike (via the modified axiom (S5)). □

The above theorem says that if the structure $\mathcal{A}$ is assumed to be "almost" a CE set, save for additional restrictions involving the avoidance of X, then there exists a unique "almost"-morphism—satisfying the CE substitutive morphism conditions restricted so that the bound and renaming-participating variables avoid X. It is the renaming-based counterpart of the nominal Theorem 2.

In regards to the relative expressiveness of these two recursion principles (Theorems 11 and 2), it seems difficult to find an example that is definable by one but not by the other. In particular, my principle can seamlessly define standard nominal examples [39,40] such as the length of a term, the counting of λ-abstractions or of the free-variables occurrences, and term-for-variable substitution—[45, Appendix A] gives details. However, as I am about to discuss, I found an important class of examples where my renaming-based principle is significantly easier to deploy: that of interpreting syntax in semantic domains.

5.3 Extended Example: Semantic Interpretation

Semantic interpretations, also known as denotations (or denotational semantics), are pervasive in the meta-theory of logics and λ-calculi, for example when interpretating first-order logic (FOL) formulas in FOL models, or untyped or simply-typed λ-calculus or higher-order logic terms in specific models (such as full-frame or Henkin models). In what follows, I will focus on λ-terms and Henkin models, but the ideas discussed apply broadly to any kind of statically scoped interpretation of terms or formulas involving binders.

Let D be a set and $\mathsf{ap} : D \to D \to D$ and $\mathsf{lm} : (D \to D) \to D$ be operators modeling semantic notions of application and abstraction. An environment will be a function $\xi : \mathsf{Var} \to D$. Given $x, y \in \mathsf{Var}$ and $d, e \in D$, let us write $\xi\langle x := d\rangle$ for ξ updated with value d for x (i.e., acting like ξ on all variables except for x where it returns d); and let us write $\xi\langle x := d, y := e\rangle$ instead of $\xi\langle x := d\rangle\langle y := e\rangle$.

Say one wants to interpret terms in the semantic domain D in the context of environments, i.e., define the function $\mathsf{sem} : \mathsf{Trm} \to (\mathsf{Var} \to D) \to D$ that maps syntactic to semantic constructs; e.g., one would like to have:

- $\mathsf{sem}\,(\mathsf{Lm}\ x\ (\mathsf{Ap}\ x\ x))\ \xi = \mathsf{lm}(d \mapsto \mathsf{ap}\ d\ d)$ (regardless of ξ)
- $\mathsf{sem}\,(\mathsf{Lm}\ x\ (\mathsf{Ap}\ x\ y))\ \xi = \mathsf{lm}(d \mapsto \mathsf{ap}\ d\ (\xi\ y))$ (assuming $x \neq y$)

where I use $d \mapsto \ldots$ to describe functions in $D \to D$, e.g., $d \mapsto \mathsf{ap}\ d\ d$ is the function sending every $d \in D$ to $\mathsf{ap}\ d\ d$.

The definition should therefore naturally go recursively by the clauses:
(1) $\mathsf{sem}\,(\mathsf{Vr}\ x)\,\xi = \xi\ x$ (2) $\mathsf{sem}\,(\mathsf{Ap}\ t_1\ t_2)\,\xi = \mathsf{ap}\,(\mathsf{sem}\ t_1\ \xi)\,(\mathsf{sem}\ t_2\ \xi)$
(3) $\mathsf{sem}\,(\mathsf{Lm}\ x\ t)\,\xi = \mathsf{lm}\,(d \mapsto \mathsf{sem}\ t\,(\xi\langle x := d\rangle))$

Of course, since Trm is not a free datatype, these clauses do not work out of the box, i.e., do not form a definition (yet)—this is where binding-aware recursion principles such as Theorems 11 and 2 could step in. I will next try them both.

The three clauses above already determine constructor operations $\mathsf{Vr}^{\mathcal{I}}$, $\mathsf{Ap}^{\mathcal{I}}$ and $\mathsf{Lm}^{\mathcal{I}}$ on the set of interpretations, $I = (\mathsf{Var} \to D) \to D$, namely:

- $\mathsf{Vr}^{\mathcal{I}} : \mathsf{Var} \to I$ by $\mathsf{Vr}^{\mathcal{I}}\ x\ i\ \xi = \xi\ x$
- $\mathsf{Ap}^{\mathcal{I}} : I \to I \to I$ by $\mathsf{Ap}^{\mathcal{I}}\ i_1\, i_2\ \xi = \mathsf{ap}\,(i_1\, \xi)\,(i_2\, \xi)$
- $\mathsf{Lm}^{\mathcal{I}} : \mathsf{Var} \to I \to I$ by $\mathsf{Lm}^{\mathcal{I}}\ x\ i\ \xi = \mathsf{lm}\,(d \mapsto i\,(\xi\langle x := d\rangle))$

To apply the renaming-based recursion principle from Theorem 11, one must further define a renaming operator on I. Since the only chance to successfully apply this principle is if sem commutes with renaming, the definition should be inspired by the question: How can $\mathsf{sem}(t[y/x])$ be determined from $\mathsf{sem}\ t$, y and x? The answer is (4) $\mathsf{sem}\,(t[y/x])\,\xi = (\mathsf{sem}\ t)\ (\xi\langle x := \xi\ y\rangle)$, yielding an operator $[\_/\_]^{\mathcal{I}} : I \to \mathsf{Var} \to \mathsf{Var} \to I$ defined by $i\,[y/x]^{\mathcal{I}}\,\xi = i\,(\xi\langle x := \xi\ y\rangle)$.

It is not difficult to verify that $\mathcal{I} = (I, [\_/\_]^{\mathcal{I}}, \mathsf{Vr}^{\mathcal{I}}, \mathsf{Ap}^{\mathcal{I}}, \mathsf{Lm}^{\mathcal{I}})$ is a CE renset—for example, Isabelle's automatic methods discharge all the goals. This means Theorem 11 (or, since here one doesn't need Barendregt's variable convention, already Theorem 8) is applicable, and gives us a unique function sem that commutes with the constructors, i.e., satisfies clauses (1)–(3) (which are instances of the clauses (i)–(iii) from Theorem 11), and additionally commutes with renaming, i.e., satisfies clause (4) (which is an instances of the clause (iv) from Theorem 11).

On the other hand, to apply nominal recursion for defining sem, one must identify a swapping operator on I. Similarly to the case of renaming, this identification process is guided by the goal of determining $\mathsf{sem}(t[x \wedge y])$ from $\mathsf{sem}\ t$, x and

y, leading to (4') $\mathsf{sem}\,(t[x \wedge y])\,\xi = \mathsf{sem}\,t\,(\xi\langle x := \xi\,y, y := \xi\,x\rangle)$, which yields the definition of $[\_\wedge\_]^{\mathcal{I}}$ by $i\,[x \wedge y]^{\mathcal{I}}\,\xi = i\,(\xi\langle x := \xi\,y, y := \xi\,x\rangle)$. However, as pointed out by Pitts [39, §6.3] (in the slightly different context of interpreting simply-typed λ-calculus), the nominal recursor (Theorem 2) does *not* directly apply (hence neither does my reformulation based on CE nominal sets, Theorem 10). This is because, in my terminology, the structure $\mathcal{I} = (I, [\_\wedge\_]^{\mathcal{I}}, \mathsf{Vr}^{\mathcal{I}}, \mathsf{Ap}^{\mathcal{I}}, \mathsf{Lm}^{\mathcal{I}})$ is not a CE nominal set. The problematic condition is FCB (the freshness condition for binders), requiring that $x \#^{\mathcal{I}} (\mathsf{Lm}^{\mathcal{I}}\,x\,i)$ holds for all $i \in I$. Expanding the definition of $\#^{\mathcal{I}}$ (the nominal definition of freshness from swapping, recalled in Sect. 2.2) and the definitions of $[\_\wedge\_]^{\mathcal{I}}$ and $\mathsf{Lm}^{\mathcal{I}}$, one can see that $x \#^{\mathcal{I}} (\mathsf{Lm}^{\mathcal{I}}\,x\,i)$ means the following:
$\mathsf{lm}\ (d \mapsto i\,(\xi\langle x := \xi\,y, y := \xi\,x\rangle\langle x := d\rangle)) = \mathsf{lm}\ (d \mapsto i\,(\xi\langle x := d\rangle))$, i.e., $\mathsf{lm}\ (d \mapsto i\,(\xi\langle x := d, y := \xi\,x\rangle) = \mathsf{lm}\ (d \mapsto i\,(\xi\langle x := d\rangle))$, holds for all but a finite number of variables y.

The only chance for the above to be true is if i, when applied to an environment, ignores the value of y in that environment for all but a finite number of variables y; in other words, i only analyzes the value of a finite number of variables in that environment—but this is not guaranteed to hold for arbitrary elements $i \in I$. To repair this, Pitts engages in a form of induction-recursion [17], carving out from I a smaller domain that is still large enough to interpret all terms, then proving that both FCB and the other axioms hold for this restricted domain. It all works out in the end, but the technicalities are quite involved.

Although FCB is not required by the renaming-based principle, note incidentally that this condition would actually be true (and immediate to check) if working with freshness defined not from swapping but from renaming. Indeed, the renaming-based version of $x \#^{\mathcal{I}} (\mathsf{Lm}^{\mathcal{I}}\,x\ i)$ says that $\mathsf{lm}\ (d \mapsto i\,(\xi\langle x := \xi\,y\rangle\langle x := d\rangle)) = \mathsf{lm}\ (d \mapsto i\,(\xi\langle x := d\rangle))$ holds for all y (or at least for some $y \neq x$)—which is immediate since $\xi\langle x := \xi\,y\rangle\langle x := d\rangle = \xi\langle x := d\rangle$. This further illustrates the idea that semantic domains 'favor' renaming over swapping.

In conclusion, for interpreting syntax in semantic domains, my renaming-based recursor is trivial to apply, whereas the nominal recursor requires some fairly involved additional definitions and proofs.

6 Conclusion and Related Work

This paper introduced and studied rensets, contributing (1) theoretically, a minimalistic equational characterization of the datatype of terms with bindings and (2) practically, an addition to the formal arsenal for manipulating syntax with bindings. It is part of a longstanding line of work by myself and collaborators on exploring convenient definition and reasoning principles for bindings [25,27,43,46,47], and will be incorporated into the ongoing implementation of a new Isabelle definitional package for binding-aware datatypes [12].

Initial Model Characterizations of the Terms Datatype. My results provide a truly elementary characterization of terms with bindings, as an "ordinary" datatype specified by the fundamental operations only (the constructors plus

	Fiore et al. [22] Hofmann [29]	Pitts [39]	Urban et al. [57,56]	Norrish [33]	Popescu &Gunter [46]	Gheri& Popescu [25]	This paper
Paradigm	nameless	nameful	nameful	nameful	nameful	nameful	nameful
Barendregt?	n/a	yes	yes	yes	no	no	yes
Underlying category	$Set^{\mathbb{F}}$	*Set*	*Set*	*Set*	*Set*	*Set*	*Set*
Required operations/ relations	ctors, rename, free-vars	ctors, perm	ctors, perm	ctors, swap, free-vars	ctors, term/var subst, fresh	ctors, swap, fresh	ctors, rename
Required properties	functoriality, naturality	Horn clauses, fresh-def, fin-supp	Horn clauses, fresh-def	Horn clauses	Horn clauses	Horn clauses	equations

Fig. 1. Initial model characterizations of the datatype of terms with bindings "ctors" = "constructors", "perm" = "permutation", "fresh" = "the freshness predicate", "fresh-def" = "clause for defining the freshness predicate", "fin-supp" = "Finite Support"

variable-for-variable renaming) and some equations (those defining CE rensets). As far as specification simplicity goes, this is "the next best thing" after a completely free datatype such as those of natural numbers or lists.

Figure 1 shows previous characterizations from the literature, in which terms with bindings are identified as an initial model (or algebra) of some kind. For each of these, I indicate (1) the employed reasoning paradigm, (2) whether the initiality/recursion theorem features an extension with Barendregt's variable convention, (3) the underlying category (from where the carriers of the models are taken), (4) the operations and relations on terms to which the models must provide counterparts and (5) the properties required on the models.

While some of these results enjoy elegant mathematical properties of intrinsic value, my main interest is in the recursors they enable, specifically in the ease of deploying these recursors. That is, I am interested in how easy it is in principle to organize the target domain as a model of the requested type, hence obtain the desired morphism, i.e., get the recursive definition done. By this measure, elementary approaches relying on standard FOL-like models whose carriers are sets rather than pre-sheaves have an advantage. Also, it seems intuitive that a recursor is easier to apply if there are fewer operators, and fewer and structurally simpler properties required on its models—although empirical evidence of successfully deploying the recursor in practice should complement the simplicity assessment, to ensure that simplicity is not sponsored by lack of expressiveness.

The first column in Fig. 1's table contains an influential representative of the nameless paradigm: the result obtained independently by Fiore et al. [22] and Hofmann [29] characterizing terms as initial in the category of algebras over the pre-sheaf topos $Set^{\mathbb{F}}$, where $\mathbb{F}$ is the category of finite ordinals and functions between them. The operators required by algebras are the constructors, as well as the free-variable operator (implicitly as part of the separation on levels) and the injective renamings (as part of the functorial structure). The algebra's carrier is required to be a functor and the constructors to be natural transformations. There are several variations of this approach, e.g., [5,11,29], some implemented in proof assistants, e.g., [3,4,31].

The other columns refer to initiality results that are more closely related to mine. They take place within the nameful paradigm, and they all rely on elementary models (with set carriers). Pitts's already discussed nominal recursor [39] (based on previous work by Gabbay and Pitts [23]) employs the constructors and permutation (or swapping), and requires that its models satisfy some Horn clauses for constructors, permutation and freshness, together with the second-order properties that (1) define freshness from swapping and (2) express Finite Support. Urban et al.'s version [56,57] implemented in Isabelle/Nominal is an improvement of Pitts's in that it removes the Finite Support requirement from the models—which is practically significant because it enables non-finitely supported target domains for recursion. Norrish's result [33] is explicitly inspired by nominal logic, but renounces the definability of the free-variable operator from swapping—with the price of taking both swapping and free-variables as primitives. My previous work with Gunter and Gheri takes as primitives either term-for-variable substitution and freshness [46] or swapping and freshness [25], and requires properties expressed by different Horn clauses (and does not explore a Barendregt dimension, like Pitts, Urban et al. and Norrish do). My previous focus on term-for-variable substitution [46] (as opposed to renaming, i.e., variable-for-variable substitution) impairs expressiveness—for example, the depth of a term is not definable using a recursor based on term-for-variable substitution because we cannot say how term-for-variable substitution affects the depth of a term based on its depth and that of the substitutee alone. My current result based on rensets keeps freshness out of the primitive operators base (like nominal logic does), and provides an unconditionally equational characterization using only constructors and renaming. The key to achieving this minimality is the simple expression of freshness from renaming in my axiomatization of rensets. In future work, I plan a systematic formal comparison of the relative expressiveness of all these nameful recursors.

Recursors in Other Paradigms. Figure 1 focuses on nameful recursors, while only the Fiore et al./Hofmann recursor for the sake of a rough comparison with the nameless approach. I should stress that such a comparison is necessarily rough, since the nameless recursors do not give the same "payload" as the nameful ones. This is because of the handling of bound variables. In the nameless paradigm, the λ-constructor does not explicitly take a variable as an input, as in $\mathsf{Lm}\ x\ t$, i.e., does not have type $\mathsf{Var} \to \mathsf{Trm} \to \mathsf{Trm}$. Instead, the bindings are indicated through nameless pointers to positions in a term. So the nameless λ-constructor, let's call it NLm, takes only a term, as in $\mathsf{NLm}\ t$, i.e., has type $\mathsf{Trm} \to \mathsf{Trm}$ or a scope-safe (polymorphic or dependently-typed) variation of this, e.g., $\prod_{n \in \mathbb{F}} \mathsf{Trm}_n \to \mathsf{Trm}_{n+1}$ [22,29] or $\prod_{\alpha \in \mathsf{Type}} \mathsf{Trm}_\alpha \to \mathsf{Trm}_{\alpha+\mathsf{unit}}$ [5,11]. The λ-constructor is of course matched by operators in the considered models, which appears in the clauses of the functions f defined recursively on terms: Instead of a clause of the form $f\ (\mathsf{Lm}\ x\ t) = \langle$expression depending on x and $f\,t\rangle$ from the nameful paradigm, in the nameless paradigm one gets a clause of the form $f\ (\mathsf{NLm}\ t) = \langle$expression depending on $f\,t\rangle$. A nameless recursor is usually easier to prove correct and easier to apply because the nameless constructor NLm is free—whereas a nameful recursor must wrestle with the non-freeness of Lm, handled by verifying certain properties of the target models. However, once the definition is done, having nameful clauses pays off by allowing "textbook-style" proofs that stay close to the informal presentation of a calculus or logic, whereas with the nameless definition some additional index shifting bureaucracy is necessary. (See [9] for a detailed discussion, and [14] for a hybrid solution.)

A comparison of nameful recursion with HOAS recursion is also generally difficult, since major HOAS frameworks such as Abella [7], Beluga [37] or Twelf [36] are developed within non-standard logical foundations, allowing a λ-constructor of type $(\mathsf{Trm} \to \mathsf{Trm}) \to \mathsf{Trm}$, which is not amenable to typical well-foundedness based recursion but requires some custom solutions (e.g., [21,50]). However, the *weak HOAS* variant [16,27] employs a constructor of the form $\mathsf{WHLm} : (\mathsf{Var} \to \mathsf{Trm}) \to \mathsf{Trm}$ which *is* recursable, and in fact yields a free datatype, let us call it WHTrm—one generated by $\mathsf{WHVr} : \mathsf{Var} \to \mathsf{WHTrm}$, $\mathsf{WHAp} : \mathsf{WHTrm} \to \mathsf{WHTrm} \to \mathsf{WHTrm}$ and WHLm. WHTrm contains (natural encodings of) all terms but also additional entities referred to as "exotic terms". Partly because of the exotic terms, this free datatype by itself is not very helpful for recursively defining useful functions on terms. But the situation is dramatically improved if one employs a variant of weak HOAS called *parametric HOAS (PHOAS)* [15], i.e., takes Var not as a fixed type but as a type parameter (type variable) and works with $\prod_{\mathsf{Var} \in \mathsf{Type}} \mathsf{Trm}_{\mathsf{Var}}$; this enables many useful definitions by choosing a suitable type Var (usually large enough to make the necessary distinctions) and then performing standard recursion. The functions definable in the style of PHOAS seem to be exactly those definable via the semantic domain interpretation pattern (Sect. 5.3): Choosing the instantiation of Var to a type T corresponds to employing environments in $\mathsf{Var} \to T$. (I illustrate this at the end of [45, Appendix A] by showing the semantic-domain version of a PHOAS example.)

As a hybrid nameful/HOAS approach we can count Gordon and Melham's characterization of the datatype of terms [26], which employs the nameful constructors but formulates recursion treating Lm as if recursing in the weak-HOAS datatype WHTrm. Norrish's recursor [33] (a participant in Fig. 1) has been inferred from Gordon and Melham's one. Weak-HOAS recursion also has interesting connections with nameless recursion: In presheaf toposes such as those employed by Fiore et al. [22], Hofmann [29] and Ambler et al. [6], for any object T the function space $\mathsf{Var} \Rightarrow T$ is isomorphic to the De Bruijn level shifting transformation applied to T; this effectively equates the weak-HOAS and nameless recursors. A final cross-paradigm note: In themselves, nominal sets are not confined to the nameful paradigm; their category is equivalent [23] to the Schanuel topos [30], which is attractive for pursuing the nameless approach.

Axiomatizations of Renaming. In his study of name-passing process calculi, Staton [52] considers an enrichment of nominal sets with renaming (in addition to swapping) and axiomatizes renaming with the help of the nominal (swapping-defined) freshness predicate. He shows that the resulted category is equivalent to the non-injective renaming counterpart of the Schanuel topos (i.e., the subcategory of $\mathit{Set}^{\mathbb{F}}$ consisting of functors that preserve pullbacks of monos). Gabbay and Hofmann [24] provide an elementary characterization of the above category, in terms of *nominal renaming sets*, which are sets equipped with a multiple-variable-renaming action satisfying identity and composition laws, and a form of Finite Support (FS). Nominal renaming sets seem very related to rensets satisfying FS. Indeed, any nominal renaming set forms a FS-satisfying renset when restricted to single-variable renaming. Conversely, I conjecture that any FS-satisfying renset gives rise to a nominal renaming set. This correspondence seems similar to the one between the permutation-based and swapping-based alternative axiomatizations of nominal sets—in that the two express the same concept up to an isomorphism of categories. In their paper, Gabbay and Hofmann do not study renaming-based recursion, beyond noting the availability of a recursor stemming from the functor-category view (which, as I discussed above, enables nameless recursion with a weak-HOAS flavor). Pitts [41] introduces *nominal sets with* 01-*substitution structure*, which axiomatize substitution of one of two possible constants for variables on top of the nominal axiomatization, and proves that they form a category that is equivalent with that of cubical sets [10], hence relevant for the univalent foundations [54].

Other Work. Sun [53] develops universal algebra for first-order languages with bindings (generalizing work by Aczel [2]) and proves a completeness theorem. In joint work with Roşu [48], I develop first-order logic and prove completeness on top of a generic syntax with axiomatized free-variables and substitution.

Renaming Versus Swapping and Nominal Logic, Final Round. I believe that my work complements rather than competes with nominal logic. My results do not challenge the swapping-based approach to defining syntax (defining the alpha-equivalence on pre-terms and quotienting to obtain terms) recommended by nominal logic, which is more elegant than a renaming-based alternative; but my easier-to-apply recursor can be a useful addition even on top of the nominal substratum. Moreover, some of my constructions are explicitly inspired by the nominal ones. For example, I started by adapting the nominal idea of defining freshness from swapping before noticing that renaming enables a simpler formulation. My formal treatment of Barendregt's variable convention also originates from nominal logic—as it turns out, this idea works equally well in my setting. In fact, I came to believe that the possibility of a Barendregt enhancement is largely orthogonal to the particularities of a binding-aware recursor. In future work, I plan to investigate this, i.e., seek general conditions under which an initiality principle (such as Theorems 10 and 8) is amenable to a Barendregt enhancement (such as Theorems 2 and 11, respectively).

Acknowledgments. I am grateful to the IJCAR reviewers for their insightful comments and suggestions, and for pointing out related work.

References

1. Abel, A., et al.: Poplmark reloaded: mechanizing proofs by logical relations. J. Funct. Program. **29**, e19 (2019). https://doi.org/10.1017/S0956796819000170
2. Aczel, P.: Frege structures and notations in propositions, truth and set. In: The Kleene Symposium, pp. 31–59. North Holland (1980)
3. Allais, G., Atkey, R., Chapman, J., McBride, C., McKinna, J.: A type and scope safe universe of syntaxes with binding: their semantics and proofs. Proc. ACM Program. Lang. **2**(International Conference on Functional Programming (ICFP)), 90:1–90:30 (2018). https://doi.acm.org/10.1145/3236785
4. Allais, G., Chapman, J., McBride, C., McKinna, J.: Type-and-scope safe programs and their proofs. In: Bertot, Y., Vafeiadis, V. (eds.) Proceedings of the 6th ACM SIGPLAN Conference on Certified Programs and Proofs, CPP 2017, Paris, France, 16–17 January 2017, pp. 195–207. ACM (2017). https://doi.org/10.1145/3018610.3018613
5. Altenkirch, T., Reus, B.: Monadic presentations of lambda terms using generalized inductive types. In: Flum, J., Rodriguez-Artalejo, M. (eds.) CSL 1999. LNCS, vol. 1683, pp. 453–468. Springer, Heidelberg (1999). https://doi.org/10.1007/3-540-48168-0_32
6. Ambler, S.J., Crole, R.L., Momigliano, A.: A definitional approach to primitivexs recursion over higher order abstract syntax. In: Eighth ACM SIGPLAN International Conference on Functional Programming, Workshop on Mechanized Reasoning About Languages with Variable Binding, MERLIN 2003, Uppsala, Sweden, August 2003. ACM (2003). https://doi.org/10.1145/976571.976572
7. Baelde, D., et al.: Abella: a system for reasoning about relational specifications. J. Formaliz. Reason. **7**(2), 1–89 (2014). https://doi.org/10.6092/issn.1972-5787/4650
8. Barendregt, H.P.: The Lambda Calculus: Its Syntax and Semantics, Studies in Logic, vol. 40. Elsevier (1984)
9. Berghofer, S., Urban, C.: A head-to-head comparison of de Bruijn indices and names. Electr. Notes Theor. Comput. Sci. **174**(5), 53–67 (2007). https://doi.org/10.1016/j.entcs.2007.01.018
10. Bezem, M., Coquand, T., Huber, S.: A model of type theory in cubical sets. In: Matthes, R., Schubert, A. (eds.) 19th International Conference on Types for Proofs and Programs, TYPES 2013, 22–26 April 2013, Toulouse, France. LIPIcs, vol. 26, pp. 107–128. Schloss Dagstuhl - Leibniz-Zentrum für Informatik (2013). https://doi.org/10.4230/LIPIcs.TYPES.2013.107
11. Bird, R.S., Paterson, R.: De Bruijn notation as a nested datatype. J. Funct. Program. **9**(1), 77–91 (1999). https://doi.org/10.1017/S0956796899003366
12. Blanchette, J.C., Gheri, L., Popescu, A., Traytel, D.: Bindings as bounded natural functors. Proc. ACM Program. Lang. **3**(POPL), 22:1–22:34 (2019). https://doi.org/10.1145/3290335
13. de Bruijn, N.G.: Lambda calculus notation with nameless dummies, a tool for automatic formula manipulation, with application to the Church-Rosser theorem. Indag. Math. **75**(5), 381–392 (1972). https://doi.org/10.1016/1385-7258(72)90034-0
14. Charguéraud, A.: The locally nameless representation. J. Autom. Reason. **49**(3), 363–408 (2012). https://doi.org/10.1007/s10817-011-9225-2

15. Chlipala, A.: Parametric higher-order abstract syntax for mechanized semantics. In: Hook, J., Thiemann, P. (eds.) International Conference on Functional Programming (ICFP) 2008, pp. 143–156. ACM (2008). https://doi.org/10.1145/1411204.1411226
16. Despeyroux, J., Felty, A., Hirschowitz, A.: Higher-order abstract syntax in Coq. In: Dezani-Ciancaglini, M., Plotkin, G. (eds.) TLCA 1995. LNCS, vol. 902, pp. 124–138. Springer, Heidelberg (1995). https://doi.org/10.1007/BFb0014049
17. Dybjer, P.: A general formulation of simultaneous inductive-recursive definitions in type theory. J. Symb. Log. **65**(2), 525–549 (2000). https://doi.org/10.2307/2586554
18. Felty, A.P., Momigliano, A.: Hybrid: a definitional two-level approach to reasoning with higher-order abstract syntax. J. Autom. Reason. **48**(1), 43–105 (2012). https://doi.org/10.1007/s10817-010-9194-x
19. Felty, A.P., Momigliano, A., Pientka, B.: The next 700 challenge problems for reasoning with higher-order abstract syntax representations - part 2 - a survey. J. Autom. Reason. **55**(4), 307–372 (2015). https://doi.org/10.1007/s10817-015-9327-3
20. Felty, A.P., Momigliano, A., Pientka, B.: An open challenge problem repository for systems supporting binders. In: Cervesato, I., Chaudhuri, K. (eds.) Proceedings Tenth International Workshop on Logical Frameworks and Meta Languages: Theory and Practice, LFMTP 2015, Berlin, Germany, 1 August 2015. EPTCS, vol. 185, pp. 18–32 (2015). https://doi.org/10.4204/EPTCS.185.2
21. Ferreira, F., Pientka, B.: Programs using syntax with first-class binders. In: Yang, H. (ed.) ESOP 2017. LNCS, vol. 10201, pp. 504–529. Springer, Heidelberg (2017). https://doi.org/10.1007/978-3-662-54434-1_19
22. Fiore, M.P., Plotkin, G.D., Turi, D.: Abstract syntax and variable binding. In: Logic in Computer Science (LICS) 1999, pp. 193–202. IEEE Computer Society (1999). https://doi.org/10.1109/LICS.1999.782615
23. Gabbay, M., Pitts, A.M.: A new approach to abstract syntax involving binders. In: Logic in Computer Science (LICS) 1999, pp. 214–224. IEEE Computer Society (1999). https://doi.org/10.1109/LICS.1999.782617
24. Gabbay, M.J., Hofmann, M.: Nominal renaming sets. In: Cervesato, I., Veith, H., Voronkov, A. (eds.) LPAR 2008. LNCS (LNAI), vol. 5330, pp. 158–173. Springer, Heidelberg (2008). https://doi.org/10.1007/978-3-540-89439-1_11
25. Gheri, L., Popescu, A.: A formalized general theory of syntax with bindings: extended version. J. Autom. Reason. **64**(4), 641–675 (2020). https://doi.org/10.1007/s10817-019-09522-2
26. Gordon, A.D., Melham, T.: Five axioms of alpha-conversion. In: Goos, G., Hartmanis, J., van Leeuwen, J., von Wright, J., Grundy, J., Harrison, J. (eds.) TPHOLs 1996. LNCS, vol. 1125, pp. 173–190. Springer, Heidelberg (1996). https://doi.org/10.1007/BFb0105404
27. Gunter, E.L., Osborn, C.J., Popescu, A.: Theory support for weak higher order abstract syntax in Isabelle/HOL. In: Cheney, J., Felty, A.P. (eds.) Logical Frameworks and Meta-Languages: Theory and Practice (LFMTP) 2009, pp. 12–20. ACM (2009). https://doi.org/10.1145/1577824.1577827
28. Harper, R., Honsell, F., Plotkin, G.D.: A framework for defining logics. In: Logic in Computer Science (LICS) 1987, pp. 194–204. IEEE Computer Society (1987). https://doi.org/10.1145/138027.138060
29. Hofmann, M.: Semantical analysis of higher-order abstract syntax. In: Logic in Computer Science (LICS) 1999, pp. 204–213. IEEE Computer Society (1999). https://doi.org/10.1109/LICS.1999.782616

30. Johnstone, P.T.: Quotients of decidable objects in a topos. Math. Proc. Camb. Philos. Soc. **93**, 409–419 (1983). https://doi.org/10.1017/S0305004100060734
31. Kaiser, J., Schäfer, S., Stark, K.: Binder aware recursion over well-scoped de bruijn syntax. In: Andronick, J., Felty, A.P. (eds.) Proceedings of the 7th ACM SIGPLAN International Conference on Certified Programs and Proofs, CPP 2018, Los Angeles, CA, USA, 8–9 January 2018, pp. 293–306. ACM (2018). https://doi.org/10.1145/3167098
32. Nipkow, T., Wenzel, M., Paulson, L.C. (eds.): Isabelle/HOL—A Proof Assistant for Higher-Order Logic. LNCS, vol. 2283. Springer, Heidelberg (2002). https://doi.org/10.1007/3-540-45949-9
33. Norrish, M.: Recursive function definition for types with binders. In: Slind, K., Bunker, A., Gopalakrishnan, G. (eds.) TPHOLs 2004. LNCS, vol. 3223, pp. 241–256. Springer, Heidelberg (2004). https://doi.org/10.1007/978-3-540-30142-4_18
34. Paulson, L.C.: The foundation of a generic theorem prover. J. Autom. Reason. **5**(3), 363–397 (1989). https://doi.org/10.1007/BF00248324
35. Pfenning, F., Elliott, C.: Higher-order abstract syntax. In: Wexelblat, R.L. (ed.) Programming Language Design and Implementation (PLDI) 1988, pp. 199–208. ACM (1988). https://doi.org/10.1145/53990.54010
36. Pfenning, F., Schürmann, C.: System description: twelf — a meta-logical framework for deductive systems. In: Ganzinger, H. (ed.) CADE 1999. LNCS (LNAI), vol. 1632, pp. 202–206. Springer, Heidelberg (1999). https://doi.org/10.1007/3-540-48660-7_14
37. Pientka, B.: Beluga: programming with dependent types, contextual data, and contexts. In: Blume, M., Kobayashi, N., Vidal, G. (eds.) FLOPS 2010. LNCS, vol. 6009, pp. 1–12. Springer, Heidelberg (2010). https://doi.org/10.1007/978-3-642-12251-4_1
38. Pitts, A.M.: Nominal logic, a first order theory of names and binding. Inf. Comput. **186**(2), 165–193 (2003). https://doi.org/10.1016/S0890-5401(03)00138-X
39. Pitts, A.M.: Alpha-structural recursion and induction. J. ACM **53**(3), 459–506 (2006). https://doi.org/10.1145/1147954.1147961
40. Pitts, A.M.: Nominal Sets: Names and Symmetry in Computer Science. Cambridge Tracts in Theoretical Computer Science. Cambridge University Press, Cambridge (2013)
41. Pitts, A.M.: Nominal presentation of cubical sets models of type theory. In: Herbelin, H., Letouzey, P., Sozeau, M. (eds.) 20th International Conference on Types for Proofs and Programs (TYPES 2014). Leibniz International Proceedings in Informatics (LIPIcs), vol. 39, pp. 202–220. Schloss Dagstuhl-Leibniz-Zentrum fuer Informatik, Dagstuhl, Germany (2015). http://drops.dagstuhl.de/opus/volltexte/2015/5498
42. Pollack, R., Sato, M., Ricciotti, W.: A canonical locally named representation of binding. J. Autom. Reason. **49**(2), 185–207 (2012). https://doi.org/10.1007/s10817-011-9229-y
43. Popescu, A.: Contributions to the theory of syntax with bindings and to process algebra. Ph.D. thesis, University of Illinois at Urbana-Champaign (2010). https://www.andreipopescu.uk/pdf/thesisUIUC.pdf
44. Popescu, A.: Renaming-Enriched Sets. Arch. Formal Proofs 2022 (2022). https://www.isa-afp.org/entries/Renaming_Enriched_Sets.html
45. Popescu, A.: Rensets and renaming-based recursion for syntax with bindings. arXiv (2022). https://arxiv.org/abs/2205.09233

46. Popescu, A., Gunter, E.L.: Recursion principles for syntax with bindings and substitution. In: Chakravarty, M.M.T., Hu, Z., Danvy, O. (eds.) Proceeding of the 16th ACM SIGPLAN International Conference on Functional Programming, ICFP 2011, Tokyo, Japan, 19–21 September 2011, pp. 346–358. ACM (2011). https://doi.org/10.1145/2034773.2034819
47. Popescu, A., Gunter, E.L., Osborn, C.J.: Strong normalization for system F by HOAS on top of FOAS. In: Logic in Computer Science (LICS) 2010, pp. 31–40. IEEE Computer Society (2010). https://doi.org/10.1109/LICS.2010.48
48. Popescu, A., Roşu, G.: Term-generic logic. Theor. Comput. Sci. **577**, 1–24 (2015)
49. Schäfer, S., Tebbi, T., Smolka, G.: Autosubst: reasoning with de Bruijn terms and parallel substitutions. In: Urban, C., Zhang, X. (eds.) ITP 2015. LNCS, vol. 9236, pp. 359–374. Springer, Cham (2015). https://doi.org/10.1007/978-3-319-22102-1_24
50. Schürmann, C., Despeyroux, J., Pfenning, F.: Primitive recursion for higher-order abstract syntax. Theor. Comput. Sci. **266**(1–2), 1–57 (2001). https://doi.org/10.1016/S0304-3975(00)00418-7
51. Stark, K.: Mechanising syntax with binders in Coq. Ph.D. thesis, Saarland University, Saarbrücken, Germany (2020). https://publikationen.sulb.uni-saarland.de/handle/20.500.11880/28822
52. Staton, S.: Name-passing process calculi: operational models and structural operational semantics. Technical report, UCAM-CL-TR-688, University of Cambridge, Computer Laboratory (2007). https://www.cl.cam.ac.uk/techreports/UCAM-CL-TR-688.pdf
53. Sun, Y.: An algebraic generalization of Frege structures–binding algebras. Theor. Comput. Sci. **211**(1–2), 189–232 (1999)
54. The Univalent Foundations Program: Homotopy Type Theory. Univalent Foundations of Mathematics. Institute for Advanced Study (2013). https://homotopytypetheory.org/book
55. Traytel, D., Popescu, A., Blanchette, J.C.: Foundational, compositional (co)datatypes for higher-order logic: category theory applied to theorem proving. In: Logic in Computer Science (LICS) 2012, pp. 596–605. IEEE Computer Society (2012). https://doi.org/10.1109/LICS.2012.75
56. Urban, C.: Nominal techniques in Isabelle/HOL. J. Autom. Reason. **40**(4), 327–356 (2008). https://doi.org/10.1007/s10817-008-9097-2
57. Urban, C., Berghofer, S.: A recursion combinator for nominal datatypes implemented in Isabelle/HOL. In: Furbach, U., Shankar, N. (eds.) IJCAR 2006. LNCS (LNAI), vol. 4130, pp. 498–512. Springer, Heidelberg (2006). https://doi.org/10.1007/11814771_41
58. Urban, C., Berghofer, S., Norrish, M.: Barendregt's variable convention in rule inductions. In: Pfenning, F. (ed.) CADE 2007. LNCS (LNAI), vol. 4603, pp. 35–50. Springer, Heidelberg (2007). https://doi.org/10.1007/978-3-540-73595-3_4
59. Urban, C., Kaliszyk, C.: General bindings and alpha-equivalence in Nominal Isabelle. Log. Methods Comput. Sci. **8**(2) (2012). https://doi.org/10.2168/LMCS-8(2:14)2012
60. Urban, C., Tasson, C.: Nominal techniques in Isabelle/HOL. In: Nieuwenhuis, R. (ed.) CADE 2005. LNCS (LNAI), vol. 3632, pp. 38–53. Springer, Heidelberg (2005). https://doi.org/10.1007/11532231_4

Finite Two-Dimensional Proof Systems for Non-finitely Axiomatizable Logics

Vitor Greati[1,2(✉)] and João Marcos[1]

[1] Programa de Pós-graduação em Sistemas e Computação & DIMAp, Universidade Federal do Rio Grande do Norte, Natal, Brazil
vitor.greati.017@ufrn.edu.br, jmarcos@dimap.ufrn.br

[2] Bernoulli Institute, University of Groningen, Groningen, The Netherlands

Abstract. The characterizing properties of a proof-theoretical presentation of a given logic may hang on the choice of proof formalism, on the shape of the logical rules and of the sequents manipulated by a given proof system, on the underlying notion of consequence, and even on the expressiveness of its linguistic resources and on the logical framework into which it is embedded. Standard (one-dimensional) logics determined by (non-deterministic) logical matrices are known to be axiomatizable by analytic and possibly finite proof systems as soon as they turn out to satisfy a certain constraint of sufficient expressiveness. In this paper we introduce a recipe for cooking up a two-dimensional logical matrix (or B-matrix) by the combination of two (possibly partial) non-deterministic logical matrices. We will show that such a combination may result in B-matrices satisfying the property of sufficient expressiveness, even when the input matrices are not sufficiently expressive in isolation, and we will use this result to show that one-dimensional logics that are not finitely axiomatizable may inhabit finitely axiomatizable two-dimensional logics, becoming, thus, finitely axiomatizable by the addition of an extra dimension. We will illustrate the said construction using a well-known logic of formal inconsistency called **mCi**. We will first prove that this logic is not finitely axiomatizable by a one-dimensional (generalized) Hilbert-style system. Then, taking advantage of a known 5-valued non-deterministic logical matrix for this logic, we will combine it with another one, conveniently chosen so as to give rise to a B-matrix that is axiomatized by a two-dimensional Hilbert-style system that is both finite and analytic.

Keywords: Hilbert-style proof systems · finite axiomatizability · consequence relations · non-deterministic semantics · paraconsistency

1 Introduction

A logic is commonly defined nowadays as a relation that connects collections of formulas from a formal language and satisfies some closure properties. The

V. Greati acknowledges support from CAPES—Finance Code 001 and from the FWF project P 33548. J. Marcos acknowledges support from CNPq.

J. Blanchette et al. (Eds.): IJCAR 2022, LNAI 13385, pp. 640–658, 2022.
https://doi.org/10.1007/978-3-031-10769-6_37

established connections are called consecutions and each of them has two parts, an antecedent and a succedent, the latter often being said to 'follow from' (or to be a consequence of) the former. A logic may be manufactured in a number of ways, in particular as being induced by the set of derivations justified by the rules of inference of a given proof system. There are different kinds of proof systems, the differences between them residing mainly in the shapes of their rules of inference and on the way derivations are built. We will be interested here in Hilbert-style proof systems ('H-systems', for short), whose rules of inference have the same shape of the consecutions of the logic they canonically induce and whose associated derivations consist in expanding a given antecedent by applications of rules of inference until the desired succedent is produced. A remarkable property of an H-system is that the logic induced by it is the least logic containing the rules of inference of the system; in the words of [24], the system constitutes a 'logical basis' for the said logic.

Conventional H-systems, which we here dub 'SET-FMLA H-systems', do not allow for more than one formula in the succedents of the consecutions that they manipulate. Since [23], however, we have learned that the simple elimination of this restriction on H-systems —that is, allowing for sets of formulas rather than single formulas in the succedents— brings numerous advantages, among which we mention: *modularity* (correspondence between rules of inference and properties satisfied by a semantical structure), *analyticity* (control over the resources demanded to produce a derivation), and the automatic generation of analytic proof systems for a wide class of logics specified by sufficiently expressive non-deterministics semantics, with an associated straightforward proof-search procedure [13,18]. Such generalized systems, here dubbed 'SET-SET H-systems', induce logics whose consecutions involve succedents consisting in a collection of formulas, intuitively understood as 'alternative conclusions'.

An H-system $\mathcal{H}$ is said to be an *axiomatization* for a given logic $\mathcal{L}$ when the logic induced by $\mathcal{H}$ coincides with $\mathcal{L}$. A desirable property for an axiomatization is *finiteness*, namely the property of consisting on a finite collection of schematic axioms and rules of inference. A logic having a finite axiomatization is said to be 'finitely based'. In the literature, one may find examples of logics having a quite simple, finite semantic presentation, being, in contrast, not finitely based in terms of SET-FMLA H-systems [21]. These very logics, however, when seen as companions of logics with multiple formulas in the succedent, turn out to be finitely based in terms of SET-SET H-systems [18]. In other words, by updating the underlying proof-theoretical and the logical formalisms, we are able to obtain a finite axiomatization for logics which in a more restricted setting could not be said to be finitely based. We may compare the above mentioned movement to the common mathematical practice of adding dimensions in order to provide better insight on some phenomenon. A well-known example of that is given by the Fundamental Theorem of Algebra, which provides an elegant solution to the problem of determining the roots of polynomials over a single variable, demanding only that real coefficients should be replaced by complex coefficients. Another example, from Machine Learning, is the 'kernel trick' employed in support vector machines: by increasing the dimensionality of the input space, the transformed data points become more easily separable by hyperplanes, making it possible to achieve better results in classification tasks.

It is worth noting that there are logics that fail to be finitely based in terms of SET-SET H-systems. An example of a logic designed with the sole purpose of illustrating this possibility was provided in [18]. One of the goals of the present work is to show that an important logic from the literature of logics of formal inconsistency (LFIs) called **mCi** is also an example of this phenomenon. This logic results from adding infinitely-many axiom schemas to the logic **mbC**, a logic that is obtained by extending positive classical logic with two axiom schemas. Incidentally, along the proof of this result, we will show that **mCi** is the limit of a strictly increasing chain of LFIs extending **mbC** (comparable to the case of $\mathcal{C}_{\text{Lim}}$ in da Costa's hierarchy of increasingly weaker paraconsistent calculi [16]). A natural question, then, is whether we can enrich our technology, in the same vein, in order to provide finite axiomatizations for all these logics. We answer that in the affirmative by means of the two-dimensional frameworks developed in [11,17]. Logics, in this case, connect pairs of collections of formulas. A consecution, in this setting, may be read as involving formulas that are accepted and those that are not, as well as formulas that are rejected and those that are not. 'Acceptance' and 'rejection' are seen, thus, as two orthogonal dimensions that may interact, making it possible, thus, to express more complex consecutions than those expressible in one-dimensional logics. Two-dimensional H-systems, which we call 'SET^2-SET^2 H-systems', generalize SET-SET H-systems so as to manipulate pairs of collections of formulas, canonically inducing two-dimensional logics and constituting logical bases for them. Another goal of the present work is, therefore, to show how to obtain a two-dimensional logic inhabited by a (possibly not finitely based) one-dimensional logic of interest. More than that, the logic we obtain will be finitely axiomatizable in terms of a SET^2-SET^2 analytic H-system. The only requirements is that the one-dimensional logic of interest must have an associated semantics in terms of a finite non-deterministic logical matrix and that this matrix can be combined with another one through a novel procedure that we will introduce, resulting in a two-dimensional non-deterministic matrix (a B-matrix [9]) satisfying a certain condition of sufficient expressiveness [17]. An application of this approach will be provided here in order to produce the first finite and analytic axiomatization of **mCi**.

The paper is organized as follows: Sect. 2 introduces basic terminology and definitions regarding algebras and languages. Section 3 presents the notions of one-dimensional logics and SET-SET H-systems. Section 4 proves that **mCi** is not finitely axiomatizable by one-dimensional H-systems. Section 5 introduces two-dimensional logics and H-systems, and describes the approach to extending a logical matrix to a B-matrix with the goal of finding a finite two-dimensional axiomatization for the logic associated with the former. Section 6 presents a two-dimensional finite analytic H-system for **mCi**. In the final remarks, we highlight some byproducts of our present approach and some features of the resulting proof systems, in addition to pointing to some directions for further research.[1]

[1] Detailed proofs of some results may be found in https://arxiv.org/abs/2205.08920.

2 Preliminaries

A *propositional signature* is a family $\Sigma := \{\Sigma_k\}_{k\in\omega}$, where each Σ_k is a collection of k-ary *connectives.* We say that Σ *is finite* when its base set $\bigcup_{k\in\omega}\Sigma_k$ is finite. A *non-deterministic algebra over* Σ, or simply Σ*-nd-algebra*, is a structure $\mathbf{A} := \langle A, \cdot_{\mathbf{A}}\rangle$, such that A is a non-empty collection of values called the *carrier* of $\mathbf{A}$, and, for each $k \in \omega$ and $© \in \Sigma_k$, the multifunction $©_{\mathbf{A}} : A^k \to \wp(A)$ is the *interpretation of* © *in* $\mathbf{A}$. When Σ and A are finite, we say that $\mathbf{A}$ is *finite.* When the range of all interpretations of $\mathbf{A}$ contains only singletons, $\mathbf{A}$ is said to be a *deterministic algebra over* Σ, or simply a Σ*-algebra,* meeting the usual definition from Universal Algebra [12]. When $\varnothing$ is not in the range of each $©_{\mathbf{A}}$, $\mathbf{A}$ is said to be *total.* Given a Σ-algebra $\mathbf{A}$ and a $© \in \Sigma_1$, we let $©^0_{\mathbf{A}}(x) := x$ and $©^{i+1}_{\mathbf{A}}(x) := ©_{\mathbf{A}}(©^i_{\mathbf{A}}(x))$. A mapping $v : A \to B$ is a *homomorphism* from $\mathbf{A}$ to $\mathbf{B}$ when, for all $k \in \omega$, $© \in \Sigma_k$ and $x_1,\ldots,x_k \in A$, we have $f[©_{\mathbf{A}}(x_1,\ldots,x_k)] \subseteq ©_{\mathbf{B}}(f(x_1),\ldots,f(x_k))$. The set of all homomorphisms from $\mathbf{A}$ to $\mathbf{B}$ is denoted by $\mathsf{Hom}_{\Sigma}(\mathbf{A},\mathbf{B})$. When $\mathbf{B} = \mathbf{A}$, we write $\mathsf{End}_{\Sigma}(\mathbf{A})$, rather than $\mathsf{Hom}_{\Sigma}(\mathbf{A},\mathbf{A})$, for the set of *endomorphisms on* $\mathbf{A}$.

Let P be a denumerable collection of *propositional variables* and Σ be a propositional signature. The absolutely free Σ-algebra freely generated by P is denoted by $\mathbf{L}_{\Sigma}(P)$ and called the Σ*-language generated by* P. The elements of $L_{\Sigma}(P)$ are called Σ*-formulas,* and those among them that are not propositional variables are called Σ*-compounds.* Given $\Phi \subseteq L_{\Sigma}(P)$, we denote by Φ^{c} the set $L_{\Sigma}(P)\backslash\Phi$. The homomorphisms from $\mathbf{L}_{\Sigma}(P)$ to $\mathbf{A}$ are called *valuations on* $\mathbf{A}$, and we denote by $\mathsf{Val}_{\Sigma}(\mathbf{A})$ the collection thereof. Additionally, endomorphisms on $\mathbf{L}_{\Sigma}(P)$ are dubbed Σ*-substitutions,* and we let $\mathsf{Subs}^P_{\Sigma} := \mathsf{End}_{\Sigma}(\mathbf{L}_{\Sigma}(P))$; when there is no risk of confusion, we may omit the superscript from this notation.

Given $\varphi \in L_{\Sigma}(P)$, let $\mathsf{props}(\varphi)$ be the set of propositional variables occurring in φ. If $\mathsf{props}(\varphi) = \{p_1,\ldots,p_k\}$, we say that φ is k-ary (*unary,* for $k = 1$; *binary,* for $k = 2$) and let $\varphi_{\mathbf{A}} : A^k \to \wp(A)$ be *the* k*-ary multifunction on* $\mathbf{A}$ *induced by* φ, where, for all $x_1,\ldots,x_k \in A$, we have $\varphi_{\mathbf{A}}(x_1,\ldots,x_k) := \{v(\varphi) \mid v \in \mathsf{Val}_{\Sigma}(\mathbf{A})$ and $v(p_i) = x_i$, for $1 \le i \le k\}$. Moreover, given $\psi_1,\ldots,\psi_k \in L_{\Sigma}(P)$, we write $\varphi(\psi_1,\ldots,\psi_k)$ for the Σ-formula $\varphi_{\mathbf{L}_{\Sigma}(P)}(\psi_1,\ldots,\psi_k)$, and, where $\Phi \subseteq L_{\Sigma}(P)$ is a set of k-ary Σ-formulas, we let $\Phi(\psi_1,\ldots,\psi_k) := \{\varphi(\psi_1,\ldots,\psi_k) \mid \varphi \in \Phi\}$. Given $\varphi \in L_{\Sigma}(P)$, by $\mathsf{subf}(\varphi)$ we refer to the set of *subformulas of* φ. Where θ is a unary Σ-formula, we define the set $\mathsf{subf}^{\theta}(\varphi)$ as $\{\sigma(\theta) \mid \sigma : P \to \mathsf{subf}(\varphi)\}$. Given a set $\Theta \supseteq \{p\}$ of unary Σ-formulas, we set $\mathsf{subf}^{\Theta}(\varphi) := \bigcup_{\theta\in\Theta}\mathsf{subf}^{\theta}(\varphi)$. For example, if $\Theta = \{p, \neg p\}$, we will have $\mathsf{subf}^{\Theta}(\neg(q \vee r)) = \{q, r, q \vee r, \neg(q \vee r)\} \cup \{\neg q, \neg r, \neg(q \vee r), \neg\neg(q \vee r)\}$. Such generalized notion of subformulas will be used in the next section to provide a more generous proof-theoretical concept of *analyticity.*

3 One-Dimensional Consequence Relations

A Set-Set *statement* (or *sequent*) is a pair $(\Phi, \Psi) \in \wp(L_{\Sigma}(P)) \times \wp(L_{\Sigma}(P))$, where Φ is dubbed the *antecedent* and Ψ the *succedent.* A *one-dimensional con-*

sequence relation on $L_\Sigma(P)$ is a collection $\rhd$ of SET-SET statements satisfying, for all $\Phi, \Psi, \Phi', \Psi' \subseteq L_\Sigma(P)$,

(O) if $\Phi \cap \Psi \neq \varnothing$, then $\Phi \rhd \Psi$
(D) if $\Phi \rhd \Psi$, then $\Phi \cup \Phi' \rhd \Psi \cup \Psi'$
(C) if $\Pi \cup \Phi \rhd \Psi \cup \Pi^{\mathsf{c}}$ for all $\Pi \subseteq L_\Sigma(P)$, then $\Phi \rhd \Psi$

Properties **(O)**, **(D)** and **(C)** are called *overlap*, *dilution* and *cut*, respectively. The relation $\rhd$ is called *substitution-invariant* when it satisfies, for every $\sigma \in \mathsf{Subs}_\Sigma$,

(S) if $\Phi \rhd \Psi$, then $\sigma[\Phi] \rhd \sigma[\Psi]$

and it is called *finitary* when it satisfies

(F) if $\Phi \rhd \Psi$, then $\Phi^{\mathsf{f}} \rhd \Psi^{\mathsf{f}}$ for some finite $\Phi^{\mathsf{f}} \subseteq \Phi$ and $\Psi^{\mathsf{f}} \subseteq \Psi$

One-dimensional consequence relations will also be referred to as *one-dimensional logics.* Substitution-invariant finitary one-dimensional logics will be called *standard.* We will denote by $\blacktriangleright$ the complement of $\rhd$, called the *compatibility relation associated with* $\rhd$ [10].

A SET-FMLA *statement* is a sequent having a single formula as consequent. When we restrict standard consequence relations to collections of SET-FMLA statements, we define the so-called (substitution-invariant finitary) *Tarskian consequence relations.* Every one-dimensional consequence relation $\rhd$ determines a Tarskian consequence relation $\vdash_{\rhd} \subseteq \wp(L_\Sigma(P)) \times L_\Sigma(P)$, dubbed *the* SET-FMLA *Tarskian companion of* $\rhd$, such that, for all $\Phi \cup \{\psi\} \subseteq L_\Sigma(P)$, $\Phi \vdash_{\rhd} \psi$ if, and only if, $\Phi \rhd \{\psi\}$. It is well-known that the collection of all Tarskian consequence relations over a fixed language constitutes a complete lattice under set-theoretical inclusion [25]. Given a set C of such relations, we will denote by $\bigsqcup C$ its supremum in the latter lattice.

We present in what follows two ways of obtaining one-dimensional consequence relations: one semantical, via non-deterministic logical matrices [6], and the other proof-theoretical, via SET-SET Hilbert-style systems [18,23].

A *non-deterministic* Σ*-matrix*, or simply Σ*-nd-matrix*, is a structure $\mathbb{M} := \langle \mathbf{A}, D \rangle$, where $\mathbf{A}$ is a Σ-nd-algebra, whose carrier is the set of *truth-values*, and $D \subseteq A$ is the set of *designated truth-values.* Such structures are also known in the literature as 'PNmatrices' [7]; they generalize the so-called 'Nmatrices' [5], which are Σ-nd-matrices with the restriction that $\mathbf{A}$ must be total. From now on, whenever $X \subseteq A$, we denote $A \backslash X$ by $\overline{X}$. In case $\mathbf{A}$ is deterministic, we simply say that $\mathbb{M}$ is a Σ*-matrix.* Also, $\mathbb{M}$ is said to be *finite* when $\mathbf{A}$ is finite. Every Σ-nd-matrix $\mathbb{M}$ determines a substitution-invariant one-dimensional consequence relation over Σ, denoted by $\rhd_{\mathbb{M}}$, such that $\Phi \rhd_{\mathbb{M}} \Psi$ if, and only if, for all $v \in \mathsf{Val}_\Sigma(\mathbf{A})$, $v[\Phi] \cap \overline{D} \neq \varnothing$ or $v[\Psi] \cap D \neq \varnothing$. It is worth noting that $\rhd_{\mathbb{M}}$ is finitary whenever the carrier of $\mathbf{A}$ is finite (the proof runs very similar to that of the same result for Nmatrices [5, Theorem 3.15]).

A *strong homomorphism* between Σ-matrices $\mathbb{M}_1 := \langle \mathbf{A}_1, D_1 \rangle$ and $\mathbb{M}_2 := \langle \mathbf{A}_2, D_2 \rangle$ is a homomorphism h between $\mathbf{A}_1$ and $\mathbf{A}_2$ such that $x \in D_1$ if, and only if, $h(x) \in D_2$. When there is a surjective strong homomorphism between $\mathbb{M}_1$ and $\mathbb{M}_2$, we have that $\rhd_{\mathbb{M}_1} = \rhd_{\mathbb{M}_2}$.

Now, to the Hilbert-style systems. A (*schematic*) SET-SET *rule of inference* R_{s} is the collection of all substitution instances of the SET-SET statement s, called the *schema* of R_{s}. Each $r \in R_{\mathsf{s}}$ is called a *rule instance of* R_{s}. A (*schematic*) SET-SET *H-system* R is a collection of SET-SET rules of inference. When we constrain the rule instances of R to having only singletons as succedents, we obtain the conventional notion of Hilbert-style system, called here SET-FMLA *H-system.*

An R-*derivation* in a SET-SET H-system R is a rooted directed tree t such that every node is labelled with sets of formulas or with a discontinuation symbol $*$, and in which every non-leaf node (that is, a node with child nodes) n in t is an *expansion of* n *by a rule instance* r of R. This means that the antecedent of r is contained in the label of n and that n has exactly one child node for each formula ψ in the succedent of r. These child nodes are, in turn, labelled with the same formulas as those of n plus the respective formula ψ. In case r has an empty succedent, then n has a single child node labelled with $*$. Here we will consider only *finitary* SET-SET H-systems, in which each rule instance has finite antecedent and succedent. In such cases, we only need to consider finite derivations. Figure 1 illustrates how derivations using only finitary rules of inference may be graphically represented. We denote by $\ell^{\mathsf{t}}(\mathsf{n})$ the label of the node n in the tree t. It is worth observing that, for SET-FMLA H-systems, derivations are linear trees (as rule instances have a single formula in their succedents), or, in other words, just sequences of formulas built by applications of the rule instances, matching thus the conventional definition of Hilbert-style systems.

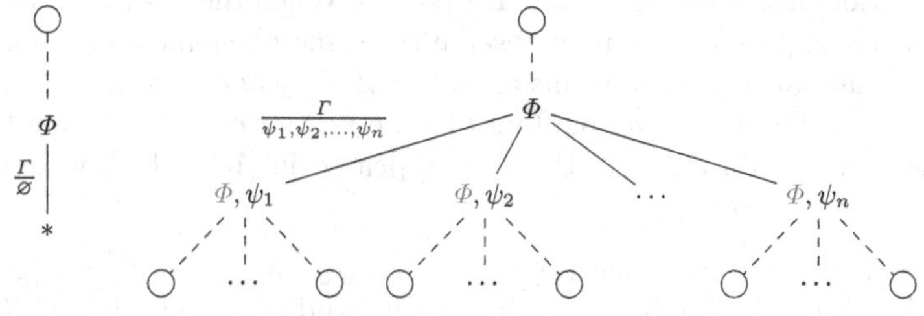

Fig. 1. Graphical representation of R-derivations, for R finitary. The dashed edges and blank circles represent other branches that may exist in the derivation. We usually omit the formulas inherited from the parent node, exhibiting only the ones introduced by the applied rule of inference. In both cases, we must have $\Gamma \subseteq \Phi$ to enable the application of the rule.

A node n of an R-derivation t is called Δ-*closed* in case it is a leaf node with $\ell^{\mathsf{t}}(\mathsf{n}) = *$ or $\ell^{\mathsf{t}}(\mathsf{n}) \cap \Delta \neq \varnothing$. A branch of t is Δ-closed when it ends in a Δ-closed node. When every branch in t is Δ-closed, we say that R is itself Δ-*closed.* An R-*proof* of a SET-SET statement (Φ, Ψ) is a Ψ-closed R-derivation t such that $\ell^{\mathsf{t}}(\mathsf{rt}(\mathsf{t})) \subseteq \Phi$.

Consider the binary relation $\rhd_{\mathsf{R}}$ on $\mathscr{P}(L_\Sigma(P))$ such that $\Phi \rhd_{\mathsf{R}} \Psi$ if, and only if, there is an R-proof of (Φ, Ψ). This relation is the smallest substitution-invariant one-dimensional consequence relation containing the rules of inference of R, and it is finitary when R is finitary. Since Set-Set (and Set-Fmla) H-systems canonically induce one-dimensional consequence relations, we may refer to them as *one-dimensional H-systems* or *one-dimensional axiomatizations.* In case there is a proof of (Φ, Ψ) whose nodes are labelled only with subsets of $\mathsf{subf}^\Theta[\Phi \cup \Psi]$, we write $\Phi \rhd_{\mathsf{R}}^{\Theta} \Psi$. In case $\rhd_{\mathsf{R}} = \rhd_{\mathsf{R}}^{\Theta}$, we say that R is Θ-*analytic.* Note that the ordinary notion of analyticity obtains when $\Theta = \{p\}$. From now on, whenever we use the word "analytic" we will mean this extended notion of Θ-analyticity, for some Θ implicit in the context. When the Θ happens to be important for us or we identify any risk of confusion, we will mention it explicitly.

In [13], based on the seminal results on axiomatizability via Set-Set H-systems by Shoesmith and Smiley [23], it was proved that any non-deterministic logical matrix $\mathbb{M}$ satisfying a criterion of sufficient expressiveness is axiomatizable by a Θ-analytic Set-Set Hilbert-style system, which is finite whenever $\mathbb{M}$ is finite, where Θ is the set of separators for the pairs of truth-values of $\mathbb{M}$. According to such criterion, an nd-matrix is *sufficiently expressive* when, for every pair (x, y) of distinct truth-values, there is a unary formula S, called a *separator for* (x, y), such that $\mathrm{S}_{\mathbf{A}}(x) \subseteq D$ and $\mathrm{S}_{\mathbf{A}}(y) \subseteq \overline{D}$, or vice-versa; in other words, when every pair of distinct truth-values is *separable in* $\mathbb{M}$.

We emphasize that it is essential for the above result the adoption of Set-Set H-systems, instead of the more restricted Set-Fmla H-systems. In fact, while two-valued matrices may always be finitely axiomatized by Set-Fmla H-systems [22], there are sufficiently expressive three-valued deterministic matrices [21] and even quite simple two-valued non-deterministic matrices [19] that fail to be finitely axiomatized by Set-Fmla H-systems. When the nd-matrix at hand is not sufficiently expressive, we may observe the same phenomenon of not having a finite axiomatization also in terms of Set-Set H-systems, even if the said nd-matrix is finite. The first example (and, to the best of our knowledge, the only one in the current literature) of this fact appeared in [13], which we reproduce here for later reference:

Example 1. Consider the signature $\Sigma := \{\Sigma_k\}_{k \in \omega}$ such that $\Sigma_1 := \{g, h\}$ and $\Sigma_k := \varnothing$ for all $k \neq 1$. Let $\mathbb{M} := \langle \mathbf{A}, \{\mathbf{a}\} \rangle$ be a Σ-nd-matrix, with $A := \{\mathbf{a}, \mathbf{b}, \mathbf{c}\}$ and

$$g_{\mathbf{A}}(x) = \begin{cases} \{\mathbf{a}\}, & \text{if } x = \mathbf{c} \\ A, & \text{otherwise} \end{cases} \qquad h_{\mathbf{A}}(x) = \begin{cases} \{\mathbf{b}\}, & \text{if } x = \mathbf{b} \\ A, & \text{otherwise} \end{cases}$$

This matrix is not sufficiently expressive because there is no separator for the pair $(\mathbf{b}, \mathbf{c})$, and [13] proved that it is not axiomatizable by a finite Set-Set H-system, even though an infinite Set-Set system that captures it has a quite simple description in terms of the following infinite collection of schemas:

$$\frac{h^i(p)}{p, g(p)}, \text{ for all } i \in \omega.$$

In the next section, we reveal another example of this same phenomenon, this time of the known LFI [14] called **mCi**. In the path of proving that this logic is not axiomatizable by a finite SET-SET H-system, we will show that there are infinitely many LFIs between **mbC** and **mCi**, organized in a strictly increasing chain whose limit is **mCi** itself.

Before continuing, it is worth emphasizing that any given non-sufficiently expressive nd-matrix may be conservatively extended to a sufficiently expressive nd-matrix provided new connectives are added to the language [18]. These new connectives have the sole purpose of separating the pairs of truth-values for which no separator is available in the original language. The SET-SET system produced from this extended nd-matrix can, then, be used to reason over the original logic, since the extension is conservative. However, these new connectives, which a priori have no meaning, are very likely to appear in derivations of consecutions of the original logic. This might not look like an attractive option to inferentialists who believe that purity of the schematic rules governing a given logical constant is essential for the meaning of the latter to be coherently fixed. In the subsequent sections, we will introduce and apply a potentially more expressive notion of logic in order to provide a *finite* and *analytic* H-system for logics that are not finitely axiomatizable in one dimension, while preserving their original languages.

4 The Logic mCi is Not Finitely Axiomatizable

A one-dimensional logic $\rhd$ over Σ is said to be $\neg$*-paraconsistent* when we have $p, \neg p \;\blacktriangleright\; q$, for $p, q \in P$. Moreover, $\rhd$ is $\neg$*-gently explosive* in case there is a collection $\bigcirc(p) \subseteq L_\Sigma(P)$ of unary formulas such that, for some $\varphi \in L_\Sigma(P)$, we have $\bigcirc(\varphi), \varphi \blacktriangleright \varnothing$; $\bigcirc(\varphi), \neg\varphi \blacktriangleright \varnothing$, and, for all $\varphi \in L_\Sigma(P)$, $\bigcirc(\varphi), \varphi, \neg\varphi \rhd \varnothing$. We say that $\rhd$ is a *logic of formal inconsistency (LFI)* in case it is $\neg$-paraconsistent yet $\neg$-gently explosive. In case $\bigcirc(p) = \{\circ p\}$, for $\circ$ a (primitive or composite) *consistency connective*, the logic is said also to be a ***C****-system.* In what follows, let Σ° be the propositional signature such that $\Sigma^\circ_1 := \{\neg, \circ\}$, $\Sigma^\circ_2 := \{\wedge, \vee, \supset\}$, and $\Sigma^\circ_k := \varnothing$ for all $k \notin \{1, 2\}$.

One of the simplest **C**-systems is the logic **mbC**, which was first presented in terms of a SET-FMLA H-system over Σ° obtained by extending any SET-FMLA H-system for positive classical logic (**CPL**$^+$) with the following pair of axiom schemas:

(em) $p \vee \neg p$
(bc1) $\circ p \supset (p \supset (\neg p \supset q))$

The logic **mCi**, in turn, is the **C**-system resulting from extending the H-system for **mbC** with the following (infinitely many) axiom schemas [20] (the resulting SET-FMLA H-system is denoted here by $\mathcal{H}_{\mathbf{mCi}}$):

(ci) $\neg\circ p \supset (p \wedge \neg p)$
(ci)$_j$ $\circ\neg^j\circ p$ (for all $0 \leq j < \omega$)

A unary connective © is said to constitute a *classical negation* in a one-dimensional logic $\rhd$ extending $\mathbf{CPL}^+$ in case, for all $\varphi, \psi \in L_\Sigma(P)$, $\varnothing \rhd \varphi \vee ©(\varphi)$ and $\varnothing \rhd \varphi \supset (©(\varphi) \supset \psi)$. One of the main differences between **mCi** and **mbC** is that an inconsistency connective $\bullet$ may be defined in the former using the paraconsistent negation, instead of a classical negation, by setting $\bullet\varphi := \neg\circ\varphi$ [20].

Both logics above were presented in [15] in ways other than H-systems: via tableau systems, via bivaluation semantics and via possible-translations semantics. In addition, while these logics are known not to be characterizable by a single finite deterministic matrix [20], a characteristic nd-matrix is available for **mbC** [1] and a 5-valued non-deterministic logical matrix is available for **mCi** [2], witnessing the importance of non-deterministic semantics in the study of non-classical logics. Such characterizations, moreover, allow for the extraction of sequent-style systems for these logics by the methodologies developed in [3,4]. Since **mCi**'s 5-valued nd-matrix will be useful for us in future sections, we recall it below for ease of reference.

Definition 1. *Let* $\mathcal{V}_5 := \{f, F, I, T, t\}$ *and* $\mathsf{Y}_5 := \{I, T, t\}$. *Define the* Σ°*-matrix* $\mathbb{M}_{\mathbf{mCi}} := \langle \mathbf{A}_5, \mathsf{Y}_5 \rangle$ *such that* $\mathbf{A}_5 := \langle \mathcal{V}_5, \cdot_{\mathbf{A}_5} \rangle$ *interprets the connectives of* Σ° *according to the following:*

$$\wedge_{\mathbf{A}_5}(x_1, x_2) := \begin{cases} \{f\} & \text{if either } x_1 \notin \mathsf{Y}_5 \text{ or } x_2 \notin \mathsf{Y}_5 \\ \{I, t\} & \text{otherwise} \end{cases}$$

$$\vee_{\mathbf{A}_5}(x_1, x_2) := \begin{cases} \{I, t\} & \text{if either } x_1 \in \mathsf{Y}_5 \text{ or } x_2 \in \mathsf{Y}_5 \\ \{f\} & \text{if } x_1, x_2 \notin \mathsf{Y}_5 \end{cases}$$

$$\supset_{\mathbf{A}_5}(x_1, x_2) := \begin{cases} \{I, t\} & \text{if either } x_1 \notin \mathsf{Y}_5 \text{ or } x_2 \in \mathsf{Y}_5 \\ \{f\} & \text{if } x_1 \in \mathsf{Y}_5 \text{ and } x_2 \notin \mathsf{Y}_5 \end{cases}$$

	f	F	I	T	t
$\neg_{\mathbf{A}_5}$	$\{I,t\}$	$\{T\}$	$\{I,t\}$	$\{F\}$	$\{f\}$

	f	F	I	T	t
$\circ_{\mathbf{A}_5}$	$\{T\}$	$\{T\}$	$\{F\}$	$\{T\}$	$\{T\}$

One might be tempted to apply the axiomatization algorithm of [13] to the finite non-deterministic logical matrix defined above to obtain a finite and analytic Set-Set system for **mCi**. However, it is not obvious, at first, whether this matrix is sufficiently expressive or not (we will, in fact, prove that it is not). In what follows, we will show now **mCi** is actually axiomatizable neither by a finite Set-Fmla H-system (first part), nor by a finite Set-Set H-system (second part); it so happens, thus, that it was not by chance that $\mathcal{H}_{\mathbf{mCi}}$ has been originally presented with infinitely many rule schemas. For the first part, we rely on the following general result:

Theorem 1 ([25], Theorem 2.2.8, adapted). *Let* $\vdash$ *be a standard Tarskian consequence relation. Then* $\vdash$ *is axiomatizable by a finite* Set-Fmla *H-system if, and only if, there is no strictly increasing sequence* $\vdash_0, \vdash_1, \ldots, \vdash_n, \ldots$ *of standard Tarskian consequence relations such that* $\vdash = \bigsqcup_{i \in \omega} \vdash_i$.

In order to apply the above theorem, we first present a family of finite SET-FMLA H-systems that, in the sequel, will be used to provide an increasing sequence of standard Tarskian consequence relations whose supremum is precisely **mCi**. Next, we show that this sequence is stricly increasing, by employing the matrix methodology traditionally used for showing the independence of axioms in a proof system.

Definition 2. *For each $k \in \omega$, let $\mathcal{H}^k_{\mathbf{mCi}}$ be a* SET-FMLA *H-system for positive classical logic together with the schemas* (em), (bc1), (ci) *and* $(\mathsf{ci})_j$*, for all* $0 \leq j \leq k$.

Since $\mathcal{H}^k_{\mathbf{mCi}}$ may be obtained from $\mathcal{H}_{\mathbf{mCi}}$ by deleting some (infinitely many) axioms, it is immediate that:

Proposition 1. *For every $k \in \omega$, $\vdash_{\mathcal{H}^k_{\mathbf{mCi}}} \subseteq \vdash_{\mathbf{mCi}}$.*

The way we define the promised increasing sequence of consequence relations in the next result is by taking the systems $\mathcal{H}^k_{\mathbf{mCi}}$ with odd superscripts, namely, we will be working with the sequence $\vdash_{\mathcal{H}^1_{\mathbf{mCi}}}, \vdash_{\mathcal{H}^3_{\mathbf{mCi}}}, \vdash_{\mathcal{H}^5_{\mathbf{mCi}}}, \ldots$ Excluding the cases where k is even will facilitate, in particular, the proof of Lemma 3.

Lemma 1. *For each $1 \leq k < \omega$, let $\vdash_k := \vdash_{\mathcal{H}^{2k-1}_{\mathbf{mCi}}}$. Then $\vdash_1 \subseteq \vdash_2 \subseteq \ldots$, and*

$$\vdash_{\mathbf{mCi}} = \bigsqcup\nolimits_{1 \leq k < \omega} \vdash_k .$$

Finally, we prove that the sequence outlined in the paragraph before Lemma 1 is strictly increasing. In order to achieve this, we define, for each $1 \leq k < \omega$, a Σ°-matrix $\mathbb{M}_k$ and prove that $\mathcal{H}^{2k-1}_{\mathbf{mCi}}$ is sound with respect to such matrix. Then, in the second part of the proof (the "independence part"), we show that, for each $1 \leq k < \omega$, $\mathbb{M}_k$ fails to validate the rule schema $(\mathsf{ci})_j$, for $j = 2k$, which is present in $\mathcal{H}^{2(k+1)-1}_{\mathbf{mCi}}$. In this way, by the contrapositive of the soundness result proved in the first part, we will have $(\mathsf{ci})_j$ provable in $\mathcal{H}^{2(k+1)-1}_{\mathbf{mCi}}$ while unprovable in $\mathcal{H}^{2k-1}_{\mathbf{mCi}}$. In what follows, for any $k \in \omega$, we use k^* to refer to the successor of k.

Definition 3. *Let $1 \leq k < \omega$. Define the $2k^*$-valued Σ°-matrix $\mathbb{M}_k := \langle \mathbf{A}_k, D_k \rangle$ such that $D_k := \{k^*+1, \ldots, 2k^*\}$ and $\mathbf{A}_k := \langle \{1, \ldots, 2k^*\}, \cdot_{\mathbf{A}_k} \rangle$, the interpretation of Σ° in $\mathbf{A}_k$ given by the following operations:*

$$x \vee_{\mathbf{A}_k} y := \begin{cases} 1 & \text{if } x, y \in \overline{D_k} \\ k^*+1 & \text{otherwise} \end{cases} \qquad x \wedge_{\mathbf{A}_k} y := \begin{cases} k^*+1 & \text{if } x, y \in D_k \\ 1 & \text{otherwise} \end{cases}$$

$$x \supset_{\mathbf{A}_k} y := \begin{cases} 1 & \text{if } x \in D_k \text{ and } y \notin \overline{D_k} \\ k^*+1 & \text{otherwise} \end{cases}$$

$$\circ_{\mathbf{A}_k} x := \begin{cases} 1 & \text{if } x = 2k^* \\ k^*+1 & \text{otherwise} \end{cases} \qquad \neg_{\mathbf{A}_k} x := \begin{cases} k^*+1 & \text{if } x \in \{1, 2k^*\} \\ x + k^* & \text{if } 2 \leq x \leq k^* \\ x - (k^*-1) & \text{if } k^*+1 \leq x \leq 2k^*-1 \end{cases}$$

Before continuing, we state results concerning this construction, which will be used in the remainder of the current line of argumentation. In what follows, when there is no risk of confusion, we omit the subscript '$\mathbf{A}_k$' from the interpretations to simplify the notation.

Lemma 2. *For all* $k \geq 1$ *and* $1 \leq m \leq 2k$,

$$\neg^m_{\mathbf{A}_k}(k^* + 1) = \begin{cases} (k^* + 1) + \frac{m}{2}, & \textit{if } m \textit{ is even} \\ 1 + \frac{m+1}{2}, & \textit{otherwise} \end{cases}$$

Lemma 3. *For all* $1 \leq k < \omega$, *we have* $\vdash_{\mathcal{H}^{2k^*-1}_{\mathbf{mCi}}} \circ\neg^{2k}\circ p$ *but* $\nvdash_{\mathcal{H}^{2k-1}_{\mathbf{mCi}}} \circ\neg^{2k}\circ p$.

Finally, Theorem 1, Lemma 1 and Lemma 3 give us the main result:

Theorem 2. **mCi** *is not axiomatizable by a finite* Set-Fmla *H-system.*

For the second part —namely, that no finite Set-Set H-system axiomatizes **mCi**—, we make use of the following result:

Theorem 3 ([23], Theorem 5.37, adapted). *Let* $\rhd$ *be a one-dimensional consequence relation over a propositional signature containing the binary connective* $\vee$. *Suppose that the* Set-Fmla *Tarskian companion of* $\rhd$, *denoted by* $\vdash_{\rhd}$, *satisfies the following property:*

$$\Phi, \varphi \vee \psi \vdash_{\rhd} \gamma \textit{ if, and only if, } \Phi, \varphi \vdash_{\rhd} \gamma \textit{ and } \Phi, \psi \vdash_{\rhd} \gamma \qquad \text{(Disj)}$$

If a Set-Set *H-system* R *axiomatizes* $\rhd$, *then* R *may be converted into a* Set-Fmla *H-system for* $\vdash_{\rhd}$ *that is finite whenever* R *is finite.*

It turns out that:

Lemma 4. **mCi** *satisfies* (Disj).

Proof. The non-deterministic semantics of **mCi** gives us that, for all $\varphi, \psi \in L_{\Sigma^\circ}(P)$, $\varphi \rhd_{\mathbb{M}_{\mathbf{mCi}}} \varphi \vee \psi$; $\psi \rhd_{\mathbb{M}_{\mathbf{mCi}}} \varphi \vee \psi$, and $\varphi \vee \psi \rhd_{\mathbb{M}_{\mathbf{mCi}}} \varphi, \psi$, and such facts easily imply (Disj).

Theorem 4. **mCi** *is not axiomatizable by a finite* Set-Set *H-system.*

Proof. If R were a finite Set-Set H-system for **mCi**, then, by Lemma 4 and Theorem 3, it could be turned into a finite Set-Fmla H-system for this very logic. This would contradict Theorem 2.

Finding a finite one-dimensional H-system for **mCi** (analytic or not) over the same language, then, proved to be impossible. The previous result also tells us that there is no sufficiently expressive non-deterministic matrix that characterizes **mCi** (for otherwise the recipe in [13] would deliver a finite analytic Set-Set H-system for it), and we may conclude, in particular, that:

Corollary 1. *The nd-matrix* $\mathbb{M}_{\mathbf{mCi}}$ *is not sufficiently expressive.*

The pairs of truth-values of $\mathbb{M}_{\mathbf{mCi}}$ that seem not to be separable (at least one of these pairs must not be, in view of the above corollary) are (t, T) and (f, F). The insufficiency of expressive power to take these specific pairs of values apart, however, would be circumvented if we had considered instead the matrix defined below, obtained from $\mathbb{M}_{\mathbf{mCi}}$ by changing its set of designated values:

Definition 4. *Let* $\mathbb{M}^{\mathsf{n}}_{\mathbf{mCi}} := \langle \mathbf{A}_5, \mathsf{N}_5 \rangle$, *where* $\mathsf{N}_5 := \{f, I, T\}$.

Note that, in $\mathbb{M}^{\mathsf{n}}_{\mathbf{mCi}}$, we have $t \notin \mathsf{N}_5$, while $T \in \mathsf{N}_5$, and we have that $f \in \mathsf{N}_5$, while $F \notin \mathsf{N}_5$. Therefore, the single propositional variable p separates in $\mathbb{M}^{\mathsf{n}}_{\mathbf{mCi}}$ the pairs (t, T) and (f, F). On the other hand, it is not clear now whether the pairs (t, F) and (f, T) are separable in this new matrix. Nonetheless, we will see, in the next section, how we can take advantage of the semantics of non-deterministic B-matrices in order to combine the expressiveness of $\mathbb{M}_{\mathbf{mCi}}$ and $\mathbb{M}^{\mathsf{n}}_{\mathbf{mCi}}$ in a very simple and intuitive manner, preserving the language and the algebra shared by these matrices. The notion of logic induced by the resulting structure will not be one-dimensional, as the one presented before, but rather two-dimensional, in a sense we shall detail in a moment. We identify two important aspects of this combination: first, the logics determined by the original matrices can be fully recovered from the combined logic; and, second, since the notions of H-systems and sufficient expressiveness, as well as the axiomatization algorithm of [13], were generalized in [17], the resulting two-dimensional logic may be algorithmically axiomatized by an *analytic* two-dimensional H-system that is *finite* if the combining matrices are finite, provided the criterion of sufficient expressiveness is satisfied after the combination. This will be the case, in particular, when we combine $\mathbb{M}_{\mathbf{mCi}}$ and $\mathbb{M}^{\mathsf{n}}_{\mathbf{mCi}}$. Consequently, this novel way of combining logics provides a quite general approach for producing finite and analytic axiomatizations for logics determined by non-deterministic logical matrices that fail to be finitely axiomatizable in one dimension; this includes the logics from Example 1, and also **mCi**.

5 Two-Dimensional Logics

From now on, we will employ the symbols Y, ⅄, N and И to informally refer to, respectively, the cognitive attitudes of *acceptance*, *non-acceptance*, *rejection* and *non-rejection*, collected in the set $\mathsf{Atts} := \{\mathsf{Y}, \text{⅄}, \mathsf{N}, \text{И}\}$. Given a set $\Phi \subseteq L_\Sigma(P)$, we will write Φ_α to intuitively mean that a given agent entertains the cognitive attitude $\alpha \in \mathsf{Atts}$ with respect to the formulas in Φ, that is: the formulas in Φ_{Y} will be understood as being accepted by the agent; the ones in $\Phi_{\text{⅄}}$, as non-accepted; the ones in Φ_{N}, as rejected; and the ones in $\Phi_{\text{И}}$, as non-rejected. Where $\alpha \in \mathsf{Atts}$, we let $\tilde{\alpha}$ be its flipped version, that is, $\tilde{\mathsf{Y}} := \text{⅄}$, $\tilde{\text{⅄}} := \mathsf{Y}$, $\tilde{\mathsf{N}} := \text{И}$ and $\tilde{\text{И}} := \mathsf{N}$.

We refer to each $\left(\begin{array}{c|c} \Phi_{\text{И}} & \Phi_{\text{⅄}} \\ \hline \Phi_{\mathsf{Y}} & \Phi_{\mathsf{N}} \end{array}\right) \in \mathscr{P}(L_\Sigma(P))^2 \times \mathscr{P}(L_\Sigma(P))^2$ as a B-*statement*, where $(\Phi_{\mathsf{Y}}, \Phi_{\mathsf{N}})$ is the *antecedent* and $(\Phi_{\text{⅄}}, \Phi_{\text{И}})$ is the *succedent*. The sets in the latter pairs are called *components.* A B-*consequence relation* is a collection $\mathrel{\vdots|\vdots}$ of B-statements satisfying:

(O2) if $\Phi_{\mathsf{Y}} \cap \Phi_{\text{⅄}} \neq \varnothing$ or $\Phi_{\mathsf{N}} \cap \Phi_{\text{И}} \neq \varnothing$, then $\frac{\Phi_{\text{И}}}{\Phi_{\mathsf{Y}}} \mid \frac{\Phi_{\text{⅄}}}{\Phi_{\mathsf{N}}}$
(D2) if $\frac{\Psi_{\text{И}}}{\Psi_{\mathsf{Y}}} \mid \frac{\Psi_{\text{⅄}}}{\Psi_{\mathsf{N}}}$ and $\Psi_\alpha \subseteq \Phi_\alpha$ for every $\alpha \in \mathsf{Atts}$, then $\frac{\Phi_{\text{И}}}{\Phi_{\mathsf{Y}}} \mid \frac{\Phi_{\text{⅄}}}{\Phi_{\mathsf{N}}}$
(C2) if $\frac{\Omega_{\mathsf{Z}}^{\mathsf{c}}}{\Omega_{\mathsf{S}}} \mid \frac{\Omega_{\mathsf{S}}^{\mathsf{c}}}{\Omega_{\mathsf{Z}}}$ for all $\Phi_{\mathsf{Y}} \subseteq \Omega_{\mathsf{S}} \subseteq \Phi_{\text{⅄}}^{\mathsf{c}}$ and $\Phi_{\mathsf{N}} \subseteq \Omega_{\mathsf{Z}} \subseteq \Phi_{\text{И}}^{\mathsf{c}}$, then $\frac{\Phi_{\text{И}}}{\Phi_{\mathsf{Y}}} \mid \frac{\Phi_{\text{⅄}}}{\Phi_{\mathsf{N}}}$

A B-consequence relation is called *substitution-invariant* if, in addition, $\frac{\Phi_{\text{И}}}{\Phi_{\mathsf{Y}}} \mid \frac{\Phi_{\text{⅄}}}{\Phi_{\mathsf{N}}}$ holds whenever, for every $\sigma \in \mathsf{Subs}_\Sigma$:

(S2) $\frac{\Psi_{\text{И}}}{\Psi_{\mathsf{Y}}} \mid \frac{\Psi_{\text{⅄}}}{\Psi_{\mathsf{N}}}$ and $\Phi_\alpha = \sigma(\Psi_\alpha)$ for every $\alpha \in \mathsf{Atts}$

Moreover, a B-consequence relation is called *finitary* when it enjoys the property

(F2) if $\frac{\Phi_{\text{И}}}{\Phi_{\mathsf{Y}}} \mid \frac{\Phi_{\text{⅄}}}{\Phi_{\mathsf{N}}}$, then $\frac{\Phi_{\text{И}}^{\mathsf{f}}}{\Phi_{\mathsf{Y}}^{\mathsf{f}}} \mid \frac{\Phi_{\text{⅄}}^{\mathsf{f}}}{\Phi_{\mathsf{N}}^{\mathsf{f}}}$, for some finite $\Phi_\alpha^{\mathsf{f}} \subseteq \Phi_\alpha$, and each $\alpha \in \mathsf{Atts}$

In what follows, B-consequence relations will also be referred to as *two-dimensional logics.* The complement of $\mathrel{\vdots|\vdots}$, sometimes called the *compatibility relation associated with* $\mathrel{\vdots|\vdots}$ [10], will be denoted by $\mathrel{\vdots\!\not|\!\vdots}$. Every B-consequence relation $\mathsf{C} := \mathrel{\vdots|\vdots}$ induces one-dimensional consequence relations $\rhd_{\mathsf{t}}^{\mathsf{C}}$ and $\rhd_{\mathsf{f}}^{\mathsf{C}}$, such that $\Phi_{\mathsf{Y}} \rhd_{\mathsf{t}}^{\mathsf{C}} \Phi_{\text{⅄}}$ iff $\frac{\varnothing}{\Phi_{\mathsf{Y}}} \mid \frac{\Phi_{\text{⅄}}}{\varnothing}$, and $\Phi_{\mathsf{N}} \rhd_{\mathsf{f}}^{\mathsf{C}} \Phi_{\text{И}}$ iff $\frac{\Phi_{\text{И}}}{\varnothing} \mid \frac{\varnothing}{\Phi_{\mathsf{N}}}$. Given a one-dimensional consequence relation $\rhd$, we say that it *inhabits the* t*-aspect of* C if $\rhd = \rhd_{\mathsf{t}}^{\mathsf{C}}$, and that it *inhabits the* f*-aspect of* C if $\rhd = \rhd_{\mathsf{f}}^{\mathsf{C}}$. B-consequence relations actually induce many other (even non-Tarskian) one-dimensional notions of logics; the reader is referred to [9,11] for a thorough presentation on this topic.

As we did for one-dimensional consequence relations, we present now realizations of B-consequence relations, first via the semantics of nd-B-matrices, then by means of two-dimensional H-systems.

A *non-deterministic* B-*matrix over* Σ, or simply Σ-*nd*-B-*matrix*, is a structure $\mathfrak{M} := \langle \mathbf{A}, \mathsf{Y}, \mathsf{N} \rangle$, where $\mathbf{A}$ is a Σ-nd-algebra, $\mathsf{Y} \subseteq A$ is the set of *designated values* and $\mathsf{N} \subseteq A$ is the set of *antidesignated values* of $\mathfrak{M}$. For convenience, we define $\text{⅄} := A \backslash \mathsf{Y}$ to be the set of *non-designated values*, and $\text{И} := A \backslash \mathsf{N}$ to be the set of *non-antidesignated values* of $\mathfrak{M}$. The elements of $\mathsf{Val}_\Sigma(\mathbf{A})$ are dubbed $\mathfrak{M}$*-valuations.* The B-*entailment relation determined by* $\mathfrak{M}$ is a collection $\mathrel{\vdots|\vdots}_{\mathfrak{M}}$ of B-statements such that

$$\textbf{(B-ent)} \quad \frac{\Phi_{\text{И}}}{\Phi_{\mathsf{Y}}} \Big| \frac{\Phi_{\text{⅄}}}{\Phi_{\mathsf{N}}} \mathfrak{M} \quad \text{iff} \quad \begin{array}{l}\text{there is no } \mathfrak{M}\text{-valuation } v \text{ such that} \\ v(\Phi_\alpha) \subseteq \alpha \text{ for each } \alpha \in \mathsf{Atts},\end{array}$$

for every $\Phi_{\mathsf{Y}}, \Phi_{\mathsf{N}}, \Phi_{\text{⅄}}, \Phi_{\text{И}} \subseteq L_\Sigma(P)$. Whenever $\frac{\Phi_{\text{И}}}{\Phi_{\mathsf{Y}}} \mid \frac{\Phi_{\text{⅄}}}{\Phi_{\mathsf{N}}} \mathfrak{M}$, we say that the B-statement $\left(\frac{\Phi_{\text{И}}}{\Phi_{\mathsf{Y}}} \middle| \frac{\Phi_{\text{⅄}}}{\Phi_{\mathsf{N}}}\right)$ *holds in* $\mathfrak{M}$ or *is valid in* $\mathfrak{M}$. An $\mathfrak{M}$-valuation that bears witness to $\frac{\Phi_{\text{И}}}{\Phi_{\mathsf{Y}}} \not| \frac{\Phi_{\text{⅄}}}{\Phi_{\mathsf{N}}} \mathfrak{M}$ is called a *countermodel for* $\left(\frac{\Phi_{\text{И}}}{\Phi_{\mathsf{Y}}} \middle| \frac{\Phi_{\text{⅄}}}{\Phi_{\mathsf{N}}}\right)$ *in* $\mathfrak{M}$. One may easily check that $\mathrel{\vdots|\vdots}_{\mathfrak{M}}$ is a substitution-invariant B-consequence relation, that is finitary when A is finite. Taking C as $\mathrel{\vdots|\vdots}_{\mathfrak{M}}$, we define $\rhd_{\mathsf{t}}^{\mathfrak{M}} := \rhd_{\mathsf{t}}^{\mathsf{C}}$ and $\rhd_{\mathsf{f}}^{\mathfrak{M}} := \rhd_{\mathsf{f}}^{\mathsf{C}}$.

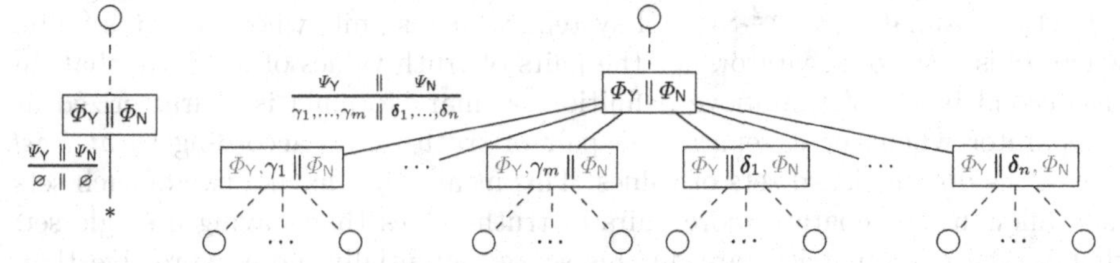

Fig. 2. Graphical representation of finite $\mathfrak{R}$-derivations. We emphasize that, in both cases, we must have $\Psi_{\mathsf{Y}} \subseteq \Phi_{\mathsf{Y}}$ and $\Psi_{\mathsf{N}} \subseteq \Phi_{\mathsf{N}}$ to enable the application of the rule.

We move now to two-dimensional, or $\textsc{Set}^2$-$\textsc{Set}^2$, H-systems, first introduced in [17]. A *(schematic)* $\textsc{Set}^2$-$\textsc{Set}^2$ *rule of inference* $R_{\mathfrak{s}}$ is the collection of all substitution instances of the $\textsc{Set}^2$-$\textsc{Set}^2$ statement $\mathfrak{s}$, called the *schema* of $R_{\mathfrak{s}}$. Each $r \in R_{\mathfrak{s}}$ is said to be a *rule instance of* $R_{\mathfrak{s}}$. In a proof-theoretic context, rather than writing the B-statement $\left(\frac{\Phi_{\text{И}} \,|\, \Phi_{\text{⅄}}}{\Phi_{\mathsf{Y}} \,|\, \Phi_{\mathsf{N}}}\right)$, we shall denote the corresponding rule by $\frac{\Phi_{\mathsf{Y}} \,\|\, \Phi_{\mathsf{N}}}{\Phi_{\text{⅄}} \,\|\, \Phi_{\text{И}}}$. A *(schematic)* $\textsc{Set}^2$-$\textsc{Set}^2$ *H-system* $\mathfrak{R}$ is a collection of $\textsc{Set}^2$-$\textsc{Set}^2$ rules of inference. $\textsc{Set}^2$-$\textsc{Set}^2$ *derivations* are as in the $\textsc{Set}$-$\textsc{Set}$ H-systems, but now the nodes are labelled with pairs of sets of formulas, instead of a single set. When applying a rule instance, each formula in the succedent produces a new branch as before, but now the formula goes to the same component in which it was found in the rule instance. See Fig. 2 for a general representation and compare it with Fig. 1.

Let $\mathfrak{t}$ be an $\mathfrak{R}$-derivation. A node $\mathfrak{n}$ of $\mathfrak{t}$ is $(\Psi_{\text{⅄}}, \Psi_{\text{И}})$*-closed* in case it is discontinued (namely, labelled with $*$) or it is a leaf node with $\ell^{\mathfrak{t}}(\mathfrak{n}) = (\Phi_{\mathsf{Y}}, \Phi_{\mathsf{N}})$ and either $\Phi_{\mathsf{Y}} \cap \Psi_{\text{⅄}} \neq \varnothing$ or $\Phi_{\mathsf{N}} \cap \Psi_{\text{И}} \neq \varnothing$. A branch of $\mathfrak{t}$ is $(\Psi_{\text{⅄}}, \Psi_{\text{И}})$*-closed* when it ends in a $(\Psi_{\text{⅄}}, \Psi_{\text{И}})$-closed node. An $\mathfrak{R}$-derivation $\mathfrak{t}$ is said to be $(\Psi_{\text{⅄}}, \Psi_{\text{И}})$-*closed* when all of its branches are $(\Psi_{\text{⅄}}, \Psi_{\text{И}})$-closed. An $\mathfrak{R}$*-proof* of $\left(\frac{\Phi_{\text{И}} \,|\, \Phi_{\text{⅄}}}{\Phi_{\mathsf{Y}} \,|\, \Phi_{\mathsf{N}}}\right)$ is a $(\Phi_{\text{⅄}}, \Phi_{\text{И}})$-closed $\mathfrak{R}$-derivation $\mathfrak{t}$ with $\ell^{\mathfrak{t}}(\mathsf{rt}(\mathfrak{t})) \subseteq (\Phi_{\mathsf{Y}}, \Phi_{\mathsf{N}})$. The definitions of the (finitary) substitution-invariant B-consequence relation $\div\!|\!\div_{\mathfrak{R}}$ induced by a (finitary) $\textsc{Set}^2$-$\textsc{Set}^2$ H-system $\mathfrak{R}$ and Θ-analyticity are obvious generalizations of the corresponding $\textsc{Set}$-$\textsc{Set}$ definitions.

In [17], the notion of sufficient expressiveness was generalized to nd-B-matrices. We reproduce here the main definitions for self-containment:

Definition 5. *Let* $\mathfrak{M} := \langle \mathbf{A}, \mathsf{Y}, \mathsf{N} \rangle$ *be a* Σ*-nd-*B*-matrix.*

- *Given* $X, Y \subseteq A$ *and* $\alpha \in \{\mathsf{Y}, \mathsf{N}\}$, *we say that* X *and* Y *are* α-separated, *denoted by* $X \#_\alpha Y$, *if* $X \subseteq \alpha$ *and* $Y \subseteq \tilde{\alpha}$, *or vice-versa.*
- *Given distinct truth-values* $x, y \in A$, *a unary formula* S *is a* separator *for* (x, y) *whenever* $\mathrm{S}_{\mathbf{A}}(x) \#_\alpha \mathrm{S}_{\mathbf{A}}(y)$ *for some* $\alpha \in \{\mathsf{Y}, \mathsf{N}\}$. *If there is a separator for each pair of distinct truth-values in* A, *then* $\mathfrak{M}$ *is said to be* sufficiently expressive.

In the same work [17], the axiomatization algorithm of [13] was also generalized, guaranteeing that every sufficiently expressive nd-B-matrix $\mathfrak{M}$ is axiomatizable by a Θ-analytic SET^2-SET^2 H-system, which is finite whenever $\mathfrak{M}$ is finite, where Θ is a set of separators for the pairs of truth-values of $\mathfrak{M}$. Note that, in the second bullet of the above definition, a unary formula is characterized as a separator whenever it separates a pair of truth-values according to *at least one* of the distinguished sets of values. This means that having two of such sets may allow us to separate more pairs of truth-values than having a single set, that is, the nd-B-matrices are, in this sense, potentially more expressive than the (one-dimensional) logical matrices.

Example 2. Let $\mathbf{A}$ be the Σ-nd-algebra from Example 1, and consider the nd-B-matrix $\mathfrak{M} := \langle \mathbf{A}, \{\mathbf{a}\}, \{\mathbf{b}\} \rangle$. As we know, in this matrix the pair $(\mathbf{b}, \mathbf{c})$ is not separable if we consider only the set of designated values $\{\mathbf{a}\}$. However, as we have now the set $\{\mathbf{b}\}$ of antidesignated truth-values, the separation becomes evident: the propositional variable p is a separator for this pair now, since $\mathbf{b} \in \{\mathbf{b}\}$ and $\mathbf{c} \notin \{\mathbf{b}\}$. The recipe from [17] produces the following SET^2-SET^2 axiomatization for $\mathfrak{M}$, with only three very simple schematic rules of inference:

$$\frac{p \parallel p}{\parallel} \qquad \frac{\parallel}{f(p), p \parallel p} \qquad \frac{\parallel \; p}{\parallel t(p)}$$

By construction, the one-dimensional logic determined by the nd-matrix of Example 1 inhabits the t-aspect of $\dot{\div}|\dot{\div}_{\mathfrak{M}}$, thus it can be seen as being axiomatized by this *finite* and *analytic* two-dimensional system (contrast with the *infinite* SET-SET axiomatization known for this logic provided in that same example).

We constructed above a Σ-nd-B-matrix from two Σ-nd-matrices in such a way that the one-dimensional logics determined by latter are fully recoverable from the former. We formalize this construction below:

Definition 6. *Let* $\mathbb{M} := \langle \mathbf{A}, D \rangle$ *and* $\mathbb{M}' := \langle \mathbf{A}, D' \rangle$ *be* Σ*-nd-matrices. The* B-product *between* $\mathbb{M}$ *and* $\mathbb{M}'$ *is the* Σ*-nd-*B*-matrix* $\mathbb{M} \odot \mathbb{M}' := \langle \mathbf{A}, D, D' \rangle$.

Note that $\Phi \rhd_{\mathbb{M}} \Psi$ iff ${}_{\overline{\Phi}}|^{\underline{\Psi}}{}_{\mathbb{M}\odot\mathbb{M}'}$ iff $\Phi \rhd_{\mathsf{t}}^{\mathbb{M}\odot\mathbb{M}'} \Psi$, and $\Phi \rhd_{\mathbb{M}'} \Psi$ iff ${}^{\underline{\Psi}}|_{\overline{\Phi}}{}_{\mathbb{M}\odot\mathbb{M}'}$ iff $\Phi \rhd_{\mathsf{f}}^{\mathbb{M}\odot\mathbb{M}'} \Psi$. Therefore, $\rhd_{\mathbb{M}}$ and $\rhd_{\mathbb{M}'}$ are easily recoverable from $\dot{\div}|\dot{\div}_{\mathbb{M}\odot\mathbb{M}'}$, since they inhabit, respectively, the t-aspect and the f-aspect of the latter. One of the applications of this novel way of putting two distinct logics together was illustrated in that same Example 2 to produce a two-dimensional analytic and finite axiomatization for a one-dimensional logic characterized by a Σ-nd-matrix. As we have shown, the latter one-dimensional logic does not need to be finitely axiomatizable by a SET-SET H-system. We present this application of B-products with more generality below:

Proposition 2. *Let* $\mathbb{M} := \langle \mathbf{A}, D \rangle$ *be a* Σ*-nd-matrix and suppose that* $U \subseteq A \times A$ *contains all and only the pairs of distinct truth-values that fail to be separable in* $\mathbb{M}$*. If, for some* $\mathbb{M}' := \langle \mathbf{A}, D' \rangle$*, the pairs in* U *are separable in* $\mathbb{M}'$*, then* $\mathbb{M}\odot\mathbb{M}'$ *is sufficiently expressive (thus, axiomatizable by an analytic* SET^2-SET^2 *H-system, that is finite whenever* $\mathbf{A}$ *is finite).*

6 A Finite and Analytic Proof System for mCi

In the spirit of Proposition 2, we define below a nd-B-matrix by combining the matrices $\mathbb{M}_{\mathbf{mCi}} := \langle \mathbf{A}_5, \mathsf{Y}_5 \rangle$ and $\mathbb{M}^{\mathsf{n}}_{\mathbf{mCi}} := \langle \mathbf{A}_5, \mathsf{N}_5 \rangle$ introduced in Sect. 4 (Definition 1 and Definition 4):

Definition 7. *Let* $\mathfrak{M}_{\mathbf{mCi}} := \mathbb{M}_{\mathbf{mCi}} \odot \mathbb{M}^{\mathsf{n}}_{\mathbf{mCi}} = \langle \mathbf{A}_5, \mathsf{Y}_5, \mathsf{N}_5 \rangle$*, with* $\mathsf{Y}_5 := \{I, T, t\}$ *and* $\mathsf{N}_5 := \{f, I, T\}$.

When we consider now both sets Y_5 and N_5 of designated and antidesignated truth-values, the separation of all truth-values of $\mathbf{A}_5$ becomes possible, that is, $\mathfrak{M}_{\mathbf{mCi}}$ is sufficiently expressive, as guaranteed by Proposition 2. Furthermore, notice that we have two alternatives for separating the pairs (I, t) and (I, T): either using the formula $\neg p$ or the formula $\circ p$. With this finite sufficiently expressive nd-B-matrix in hand, producing a *finite* $\{p, \circ p\}$-analytic two-dimensional H-system for it is immediate by [17, Theorem 2]. Since **mCi** inhabits the t-aspect of $\div|\div_{\mathfrak{M}_{\mathbf{mCi}}}$, we may then conclude that:

Theorem 5. **mCi** *is axiomatizable by a finite and analytic two-dimensional H-system.*

Our axiomatization recipe delivers an H-system with about 300 rule schemas. When we simplify it using the streamlining procedures indicated in that paper, we obtain a much more succinct and insightful presentation, with 28 rule schemas, which we call $\mathfrak{R}_{\mathbf{mCi}}$. The full presentation of this system is given below:

$$\frac{q \parallel}{p \supset q \parallel}{\supset}_1^{\mathbf{mCi}} \quad \frac{\parallel}{p, p \supset q \parallel}{\supset}_2^{\mathbf{mCi}} \quad \frac{p \supset q, p \parallel}{q \parallel}{\supset}_3^{\mathbf{mCi}} \quad \frac{p \parallel}{q \parallel p \supset q}{\supset}_4^{\mathbf{mCi}} \quad \frac{p \supset q, \circ(p \supset q) \parallel p \supset q}{\parallel}{\supset}_5^{\mathbf{mCi}}$$

$$\frac{p, q \parallel}{p \wedge q \parallel}{\wedge}_1^{\mathbf{mCi}} \quad \frac{p \wedge q \parallel}{p \parallel}{\wedge}_2^{\mathbf{mCi}} \quad \frac{p \wedge q \parallel}{q \parallel}{\wedge}_3^{\mathbf{mCi}} \quad \frac{\parallel}{p \wedge q \parallel p \wedge q}{\wedge}_4^{\mathbf{mCi}} \quad \frac{p \wedge q, \circ(p \wedge q) \parallel p \wedge q}{\parallel}{\wedge}_5^{\mathbf{mCi}}$$

$$\frac{p \parallel}{p \vee q \parallel}{\vee}_1^{\mathbf{mCi}} \quad \frac{q \parallel}{p \vee q \parallel}{\vee}_2^{\mathbf{mCi}} \quad \frac{p \vee q \parallel}{p, q \parallel}{\vee}_3^{\mathbf{mCi}} \quad \frac{\parallel}{p, q \parallel p \vee q}{\vee}_4^{\mathbf{mCi}} \quad \frac{p \vee q, \circ(p \vee q) \parallel p \vee q}{\parallel}{\vee}_5^{\mathbf{mCi}}$$

$$\frac{\circ p \parallel}{\parallel \circ p}{\circ}_1^{\mathbf{mCi}} \quad \frac{\parallel}{\circ\circ p \parallel}{\circ}_2^{\mathbf{mCi}} \quad \frac{\parallel \circ p}{\circ p \parallel}{\circ}_3^{\mathbf{mCi}} \quad \frac{\parallel}{\circ p \parallel p}{\circ}_4^{\mathbf{mCi}} \quad \frac{\parallel}{p \parallel \circ p}{\circ}_5^{\mathbf{mCi}}$$

$$\frac{\parallel}{\parallel \neg p, p}{\neg}_1^{\mathbf{mCi}} \quad \frac{\neg p, \circ p, p \parallel}{\parallel}{\neg}_2^{\mathbf{mCi}} \quad \frac{\neg p, p \parallel}{\parallel p}{\neg}_3^{\mathbf{mCi}} \quad \frac{\circ\neg p \parallel \neg p, p}{\parallel}{\neg}_4^{\mathbf{mCi}}$$

$$\frac{\parallel \neg p, p}{\neg p \parallel}{\neg}_5^{\mathbf{mCi}} \quad \frac{\parallel}{\neg p, \circ p \parallel}{\neg}_6^{\mathbf{mCi}} \quad \frac{\parallel}{\neg p, p \parallel}{\neg}_7^{\mathbf{mCi}} \quad \frac{\parallel}{\circ\neg p \parallel p}{\neg}_8^{\mathbf{mCi}}$$

Note that the set of rules $\{©_i^{\mathbf{mCi}} \mid © \in \{\wedge, \vee, \supset\}, i \in \{1, 2, 3\}\}$ makes it clear that the t-aspect of the induced B-consequence relation is inhabited by a logic extending positive classical logic, while the remaining rules for these connectives involve interactions between the two dimensions. Also, rule $\neg_2^{\mathbf{mCi}}$ indicates that $\circ$ satisfies one of the main conditions for being taken as a consistency connective in the logic inhabiting the t-aspect. In fact, all these observations are aligned with the fact that the logic inhabiting the t-aspect of $\div|\div_{\mathfrak{R}_{\mathbf{mCi}}}$ is precisely **mCi**. See, in Fig. 3, $\mathfrak{R}_{\mathbf{mCi}}$-derivations showing that, in **mCi**, $\neg\circ p$ and $p \wedge \neg p$ are logically equivalent and that $\circ\neg\circ p$ is a theorem.

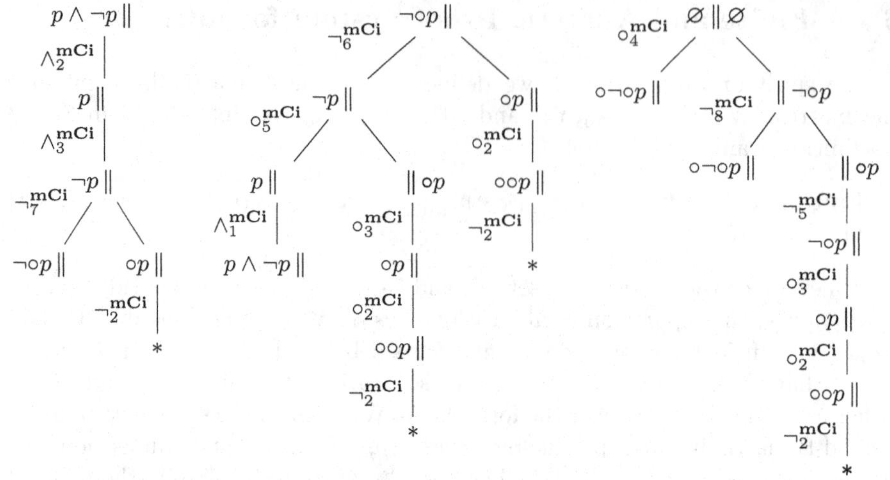

Fig. 3. $\mathfrak{R}_{\mathbf{mCi}}$-derivations showing, respectively, that $\frac{\varnothing}{p \wedge \neg p} \,|\, \frac{\neg \circ p}{\varnothing}\ \mathfrak{R}_{\mathbf{mCi}}$, $\frac{\varnothing}{\neg \circ p} \,|\, \frac{p \wedge \neg p}{\varnothing}\ \mathfrak{R}_{\mathbf{mCi}}$ and $\frac{\varnothing}{\varnothing} \,|\, \frac{\circ \neg \circ p}{\varnothing}\ \mathfrak{R}_{\mathbf{mCi}}$. Note that, for a cleaner presentation, we omit the formulas inherited from parent nodes.

7 Concluding Remarks

In this work, we introduced a mechanism for combining two non-deterministic logical matrices into a non-deterministic B-matrix, creating the possibility of producing finite and analytic two-dimensional axiomatizations for one-dimensional logics that may fail to be finitely axiomatizable in terms of one-dimensional Hilbert-style systems. It is worth mentioning that, as proved in [17], one may perform proof search and countermodel search over the resulting two-dimensional systems in time at most exponential on the size of the B-statement of interest through a straightforward proof-search algorithm.

We illustrated the above-mentioned combination mechanism with two examples, one of them corresponding to a well-known logic of formal inconsistency called **mCi**. We ended up proving not only that this logic is not finitely axiomatizable in one dimension, but also that it is the limit of a strictly increasing chain of LFIs extending the logic **mbC**. From the perspective of the study of B-consequence relations, these examples allow us to eliminate the suspicion that a two-dimensional H-system $\mathfrak{R}$ may always be converted into SET-SET H-systems for the logics inhabiting the one-dimensional aspects of $\frac{\cdot}{\cdot} \,|\, \frac{\cdot}{\cdot}\ \mathfrak{R}$ without losing any desirable property (in this case, finiteness of the presentation).

At first sight, the formalism of two-dimensional H-systems may be confused with the formalism of n-sided sequents [3,4], in which the objects manipulated by rules of inference (the so-called *n-sequents*) accommodate more than two sets of formulas in their structures. The reader interested in a comparison between these two different approaches is referred to the concluding remarks of [17].

We close with some observations regarding $\mathfrak{M}_{\mathbf{mCi}}$ and the two-dimensional H-system $\mathfrak{R}_{\mathbf{mCi}}$. A one-dimensional logic $\rhd$ is said to be $\neg$-*consistent* when

$\varphi, \neg\varphi \rhd \varnothing$ and $\neg$-*determined* when $\varnothing \rhd \varphi, \neg\varphi$ for all $\varphi \in L_{\Sigma}(P)$. A B-consequence relation $\frac{\cdot}{\cdot}|\frac{\cdot}{\cdot}$ is said to *allow for gappy reasoning* when $\frac{}{\varphi} \not\!\mid \frac{}{\varphi}$ and to *allow for glutty reasoning* when $\frac{\varphi}{} \not\!\mid \frac{\varphi}{}$, for some $\varphi \in L_{\Sigma}(P)$. Notice that $\neg$-determinedness in the logic inhabiting the t-aspect of a B-consequence relation by no means implies the disallowance of gappy reasoning in the two-dimensional setting: we still have $F \in \overline{\mathsf{Y}_5} \cap \overline{\mathsf{N}_5}$, so one may both non-accept and non-reject a formula φ in $\frac{\cdot}{\cdot}|\frac{\cdot}{\cdot}\,\mathfrak{R}_{\mathbf{mCi}}$, even though non-accepting both φ and its negation in **mCi** is not possible, in view of rule $\neg_7^{\mathbf{mCi}}$. Similarly, the recovery of $\neg$-consistency achieved via $\circ$ in such logic does not coincide with the gentle disallowance of glutty reasoning in $\frac{\cdot}{\cdot}|\frac{\cdot}{\cdot}\,\mathfrak{R}_{\mathbf{mCi}}$, that is, we do not have, in general, $\frac{}{p, \circ p}|\frac{}{p}\,\mathfrak{R}_{\mathbf{mCi}}$ or $\frac{}{p}|\frac{}{\circ p, p}\,\mathfrak{R}_{\mathbf{mCi}}$, even though for binary compounds both are derivable in view of rules $©_5^{\mathbf{mCi}}$, for $© \in \{\wedge, \vee, \supset\}$, and $\circ_1^{\mathbf{mCi}}$. With these observations we hope to call attention to the fact that B-consequence relations open the doors for further developments concerning the study of paraconsistency (and, dually, of paracompleteness), as well as the study of recovery operators [8].

References

1. Avron, A.: Non-deterministic matrices and modular semantics of rules. In: Beziau, J.Y. (ed.) Logica Universalis, pp. 149–167. Birkhäuser, Basel (2005). https://doi.org/10.1007/3-7643-7304-0_9
2. Avron, A.: 5-valued non-deterministic semantics for the basic paraconsistent logic mCi. Stud. Log. Grammar Rhetoric 127–136 (2008)
3. Avron, A., Ben-Naim, J., Konikowska, B.: Cut-free ordinary sequent calculi for logics having generalized finite-valued semantics. Log. Univers. **1**, 41–70 (2007). https://doi.org/10.1007/s11787-006-0003-6
4. Avron, A., Konikowska, B.: Multi-valued calculi for logics based on non-determinism. Log. J. IGPL **13**(4), 365–387 (2005). https://doi.org/10.1093/jigpal/jzi030
5. Avron, A., Lev, I.: Non-deterministic multiple-valued structures. J. Log. Comput. **15**(3), 241–261 (2005). https://doi.org/10.1093/logcom/exi001
6. Avron, A., Zamansky, A.: Non-deterministic semantics for logical systems. In: Gabbay, D.M., Guenthner, F. (eds.) Handbook of Philosophical Logic, vol. 16, pp. 227–304. Springer, Dordrecht (2011). https://doi.org/10.1007/978-94-007-0479-4_4
7. Baaz, M., Lahav, O., Zamansky, A.: Finite-valued semantics for canonical labelled calculi. J. Autom. Reason. **51**(4), 401–430 (2013). https://doi.org/10.1007/s10817-013-9273-x
8. Barrio, E.A., Carnielli, W.: Volume I: recovery operators in logics of formal inconsistency (special issue). Log. J. IGPL **28**(5), 615–623 (2019). https://doi.org/10.1093/jigpal/jzy053
9. Blasio, C.: Revisitando a lógica de Dunn-Belnap. Manuscrito **40**, 99–126 (2017). https://doi.org/10.1590/0100-6045.2017.v40n2.cb
10. Blasio, C., Caleiro, C., Marcos, J.: What is a logical theory? On theories containing assertions and denials. Synthese **198**(22), 5481–5504 (2019). https://doi.org/10.1007/s11229-019-02183-z
11. Blasio, C., Marcos, J., Wansing, H.: An inferentially many-valued two-dimensional notion of entailment. Bull. Sect. Log. **46**(3/4), 233–262 (2017). https://doi.org/10.18778/0138-0680.46.3.4.05

12. Burris, S., Sankappanavar, H.: A Course in Universal Algebra, vol. 91 (1981)
13. Caleiro, C., Marcelino, S.: Analytic calculi for monadic PNmatrices. In: Iemhoff, R., Moortgat, M., de Queiroz, R. (eds.) WoLLIC 2019. LNCS, vol. 11541, pp. 84–98. Springer, Heidelberg (2019). https://doi.org/10.1007/978-3-662-59533-6_6
14. Carnielli, W., Marcos, J.: A taxonomy of C-systems. In: Paraconsistency: The logical way to the inconsistent. Taylor and Francis (2002). https://doi.org/10.1201/9780203910139-3
15. Carnielli, W.A., Coniglio, M.E., Marcos, J.: Logics of formal inconsistency. In: Gabbay, D., Guenthner, F. (eds.) Handbook of Philosophical Logic, vol. 14, 2nd edn., pp. 1–93. Springer, Dordrecht (2007). https://doi.org/10.1007/978-1-4020-6324-4_1
16. Carnielli, W.A., Marcos, J.: Limits for paraconsistent calculi. Notre Dame J. Formal Log. **40**(3), 375–390 (1999). https://doi.org/10.1305/ndjfl/1022615617
17. Greati, V., Marcelino, S., Marcos, J.: Proof search on bilateralist judgments over non-deterministic semantics. In: Das, A., Negri, S. (eds.) TABLEAUX 2021. LNCS (LNAI), vol. 12842, pp. 129–146. Springer, Cham (2021). https://doi.org/10.1007/978-3-030-86059-2_8
18. Marcelino, S., Caleiro, C.: Axiomatizing non-deterministic many-valued generalized consequence relations. Synthese **198**(22), 5373–5390 (2019). https://doi.org/10.1007/s11229-019-02142-8
19. Marcelino, S.: An unexpected Boolean connective. Log. Univer. (2021). https://doi.org/10.1007/s11787-021-00280-7
20. Marcos, J.: Possible-translations semantics for some weak classically-based paraconsistent logics. J. Appl. Non-Classical Log. **18**(1), 7–28 (2008). https://doi.org/10.3166/jancl.18.7-28
21. Palasinska, K.: Deductive systems and finite axiomatization properties. Ph.D. thesis, Iowa State University (1994). https://doi.org/10.31274/rtd-180813-12680
22. Rautenberg, W.: 2-element matrices. Stud. Log. **40**(4), 315–353 (1981)
23. Shoesmith, D.J., Smiley, T.J.: Multiple-Conclusion Logic. Cambridge University Press, Cambridge (1978). https://doi.org/10.1017/CBO9780511565687
24. Wójcicki, R.: Some remarks on the consequence operation in sentential logics. Fundam. Math. **68**, 269–279 (1970)
25. Wójcicki, R.: Theory of Logical Calculi. Synthese Library, 1 edn., Springer, Dordrecht (1988). https://doi.org/10.1007/978-94-015-6942-2

Vampire Getting Noisy: Will Random Bits Help Conquer Chaos? (System Description)

Martin Suda(✉)

Czech Technical University in Prague, Prague, Czech Republic
martin.suda@cvut.cz

Abstract. Treating a saturation-based automatic theorem prover (ATP) as a Las Vegas randomized algorithm is a way to illuminate the chaotic nature of proof search and make it amenable to study by probabilistic tools. On a series of experiments with the ATP Vampire, the paper showcases some implications of this perspective for prover evaluation.

Keywords: Saturation-based proving · Evalutation · Randomization

1 Introduction

Saturation-based proof search is known to be fragile. Even seemingly insignificant changes in the search procedure, such as shuffling the order in which input formulas are presented to the prover, can have a huge impact on the prover's running time and thus on the ability to find a proof within a given time limit.

This *chaotic* aspect of the prover behaviour is relatively poorly understood, yet has obvious consequences for evaluation. A typical experimental evaluation of a new technique T compares the number of problems solved by a baseline run with a run enhanced by T (over an established benchmark and with a fixed timeout). While a higher number of problems solved by the run enhanced by T indicates a benefit of the new technique, it is hard to claim that a certain problem P is getting solved *thanks* to T. It might be that T just helps the prover get lucky on P by a complicated chain of cause and effect not related to the technique T—and the original idea behind it—in any reasonable sense.

We propose to expose and counter the effect of chaotic behaviours by deliberately *injecting randomness* into the prover and observing the results of many independently seeded runs. Although computationally more costly than standard evaluation, such an approach promises to bring new insights. We gain the ability to apply the tools of probability theory and statistics to analyze the results, assign confidences, and single out those problems that *robustly* benefit

This work was supported by the Czech Science Foundation project 20-06390Y and the project RICAIP no. 857306 under the EU-H2020 programme.

J. Blanchette et al. (Eds.): IJCAR 2022, LNAI 13385, pp. 659–667, 2022.
https://doi.org/10.1007/978-3-031-10769-6_38

from the evaluated technique. At the same time, by observing the changes in the corresponding runtime distributions we can even meaningfully establish the effect of the new technique on a single problem in isolation, something that is normally inconclusive due to the threat of chaotic fluctuations.

In this paper, we report on several experiments with a randomized version of the ATP Vampire [9]. After explaining the method in more detail (Sect. 2), we first demonstrate the extent in which the success of a typical Vampire proof search strategy can be ascribed to chance (Sect. 3). Next, we use the collected data to highlight the specifics of comparing two strategies probabilistically (Sect. 4). Finally, we focus on a single problem to see a chaotic behaviour smoothened into a distribution with a high variance (Sect. 5). The paper ends with an overview of related work (Sect. 6) and a discussion (Sect. 7).

2 Randomizing Out Chaos

Any developer of a saturation-based prover will confirm that the behaviour of a specific proving strategy on a specific problem is extremely hard to predict, that a typical experimental evaluation of a new technique (such as the one described earlier) invariably leads to both gains and losses in terms of the solved problems, and that a closer look at any of the "lost" problems often reveals just a complicated chain of cause and effect that steers the prover away from the original path (rather than a simple opportunity to improve the technique further).

These observations bring indirect evidence that the prover's behaviour is chaotic: A specific prover run can be likened to a single bead falling down through the pegs of the famous Galton board[1]. The bead follows a deterministic trajectory, but only because the code fixes every single detail of the execution, including many which the programmer did not care about and which were left as they are merely out of coincidence. We put forward here that any such fixed detail (which does not contribute to an officially implemented heuristic) represents a candidate location for randomization, since a different programmer could have fixed the detail differently and we would still call the code essentially the same.

Implementation: We implemented randomization on top of Vampire version 4.6.1; the code is available as a separate git branch[2]. We divided the randomization opportunities into three groups (governed by three new Vampire options).

Shuffling the input (`-si on`) randomly reorders the input formulas and, recursively, sub-formulas under commutative logical operations. This is done several times throughout the preprocessing pipeline, at the end of which a finished clause normal form is produced. Randomizing traversals (`-rtra on`) happens during saturation and consists of several randomized reorderings including: reordering literals in a newly generated clause and in each given clause before activation, and shuffling the order in which generated clauses are put into the

[1] https://en.wikipedia.org/wiki/Galton_board.

[2] https://github.com/vprover/vampire/tree/randire.

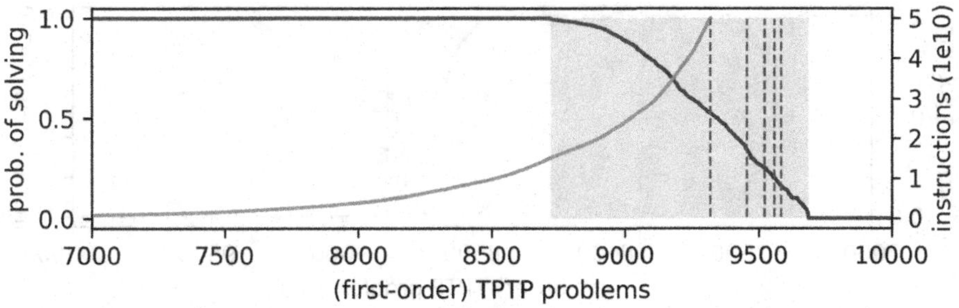

Fig. 1. Blue: first-order TPTP problems ordered by the decreasing probability of being solved by the `dis10` strategy within 50 billion instruction limit. Red: a cactus plot for the same strategy, showing the dependence between a given instruction budget (y-axis) and the number of problems on average solved within that budget (x-axis). (Color figure online)

passive set. It also (partially) randomizes term ids, which are used as tiebreakers in various term indexing operations and determine the default orientation of equational literals in the term sharing structure. Finally, "randomized age-weight ratio" (`-rawr on`) swaps the default, deterministic mechanism for choosing the next queue to select the given clause from [13] for a randomized one (which only respects the age-weight ratio probabilistically).

All the three options were active by default during our experiments.

3 Experiment 1: A Single-Strategy View

First, we set out to establish to what degree the performance of a Vampire strategy can be affected by randomization. We chose the default strategy of the prover except for the saturation algorithm, which we set to Discount, and the age-weight ratio, set to 1:10 (calling the strategy `dis10`). We ran our experiment on the first-order problems from the TPTP library [15] version 7.5.0[3].

To collect our data, we repeatedly (with different seeds) ran the prover on the problems, performing full randomization. We measured the executed instructions[4] needed to successfully solve a problem and used a limit of 50 billion instructions (which roughly corresponds to 15 s of running time on our machine[5]) after which a run was declared unsuccessful. We ran the prover 10 times on each problem and additionally as many times as required to observe the instruction count average (over both successful and unsuccessful runs) stabilize within 1% from any of its 10 previously recorded values[6].

A summary view of the experiment is given by Fig. 1. The most important to notice is the shaded region there, which spans 965 problems that were solved by

[3] Materials accompanying the experiments can be found at https://bit.ly/3JDCwea.
[4] As measured via the `perf_event_open` Linux performance monitoring feature.
[5] A server with Intel(R) Xeon(R) Gold 6140 CPUs @ 2.3 GHz and 500 GB RAM.
[6] Utilizing all the 72 cores of our machine, such data collection took roughly 12 h.

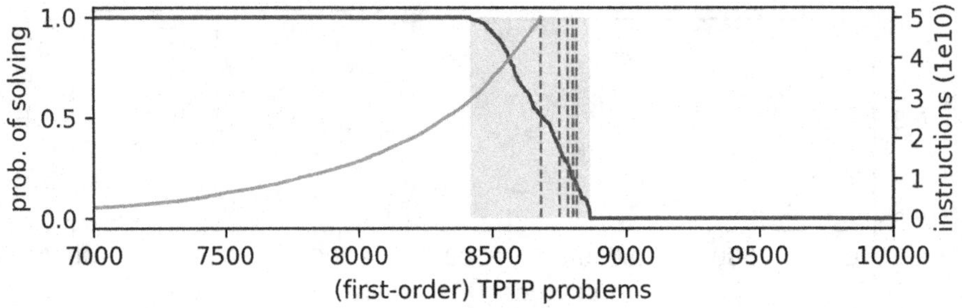

Fig. 2. The effect of turning AVATAR off in the dis10 strategy (cf. Figure 1).

dis10 at least once but not by every run. In other words, these problems have probability p of being solved between $0 < p < 1$. This is a relatively large number and can be compared to the 8720 "easy" problems solved by every run. The collected data implies that 9319.1 problems are being solved on average (marked by the left-most dashed line in Fig. 1) with a standard deviation $\sigma = 11.7$. The latter should be an interesting indicator for prover developers: beating a baseline by only 12 TPTP problems can easily be ascribed just to chance.

Figure 1 also contains the obligatory "cactus plot" (explained in the caption), which—thanks to the collected data—can be constructed with the "on average" qualifier. By definition, the plot reaches the left-most dashed line for the full instruction budged of 50 billion. The subsequent dashed lines mark the number of problems we would on average expect to solve by running the prover (independently) on each problem twice, three, four and five times. This is an information relevant for strategy scheduling: e.g., one can expect to solve whole additional 137 problems by running randomized dis10 for a second time.

Not every strategy exhibits the same degree of variability under randomization. Observe Fig. 2 with a plot analogous to Fig. 1, but for dis10 in which the AVATAR [16] has been turned off. The shaded area there is now much smaller (and only spans 448 problems). The powerful AVATAR architecture is getting convicted of making proof search more fragile and the prover less robust[7].

Remark. Randomization incurs a small but measurable computational overhead. On a single run of dis10 over the first-order TPTP (filtering out cases that took less than 1 s to finish, to prevent distortion by rounding errors) the observed median relative time spent randomizing on a single problem was 0.47%, the average 0.59%, and the worse[8] 13.86%. Without randomization, the dis10 strategy solved 9335 TPTP problems under the 50 billion instruction limit, i.e., 16 problems more than the average reported above. Such is the price we pay for turning our prover into a Las Vegas randomized algorithm.

[7] Another example of a strong but fragile heuristic is the lookahead literal selection [5], which selects literals in a clause based on the current content of the active set: dis10 enhanced with lookahead solves 9512.4 ($\pm$13.8) TPTP problems on average, 8672 problems with $p = 1$ and additional 1382 (!) problems with $0 < p < 1$.

[8] On the hard-to-parse, trivial-to-solve HWV094-1 with 361 199 clauses.

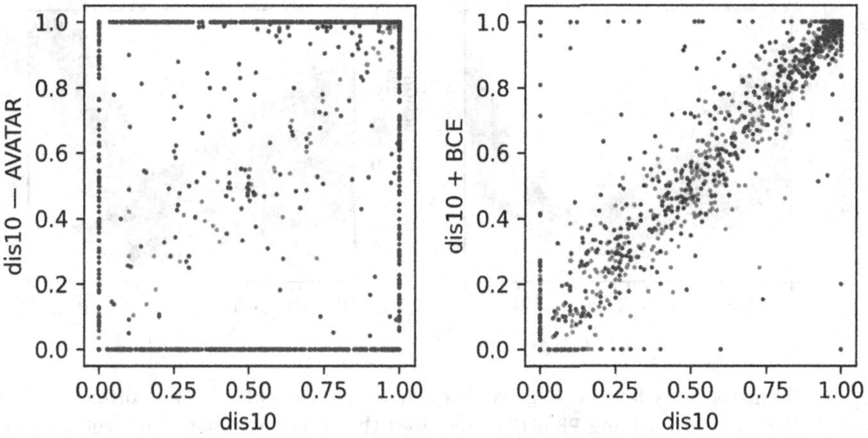

Fig. 3. Scatter plots comparing probabilities of solving a TPTP problem by the baseline `dis10` strategy and 1) `dis10` with AVATAR turned off (left), and 2) `dis10` with blocked clause elimination turned on (right). On problems marked red the respective technique could not be applied (no splittable clauses derived / no blocked clauses eliminated).

4 Experiment 2: Comparing Two Strategies

Once randomized performance profiles of multiple strategies are collected, it is interesting to look at two at a time. Figure 3 shows two very different scatter plots, each comparing our baseline `dis10` to its modified version in terms of the probabilities of solving individual problems.

On the left we see the effect of turning AVATAR off. The technique affects the proving landscape quite a lot and most problems have their mark along the edges of the plot where at least one of the two probabilities has the extreme value of either 0 or 1. What the plot does not show well, is how many marks end up at the extreme corners. These are: 7896 problems easy for both, 661 easy for AVATAR and hard without, 135 hard for AVATAR and easy without.

Such "purified", one-sided gains and losses constitute a new interesting indicator of the impact of a given technique. They should be the first to look at, e.g., during debugging, as they represent the most extreme but robust examples of how the new technique changes the capabilities of the prover.

The right plot is an analogous view, but now at the effect of turning on *blocked clause elimination* (BCE). This is a preprocessing technique coming from the context of propositional satisfiability [7] extended to first-order logic [8]. We see that here most of the visible problems show up as marks along the plot's main diagonal, suggesting a (mostly) negligible effect of the technique. The extreme corners hide: 8648 problems easy for both, 17 easy with BCE (11 satisfiable and 6 unsatisfiable), and 2 easy without BCE (1 satisfiable and 1 unsatisfiable).

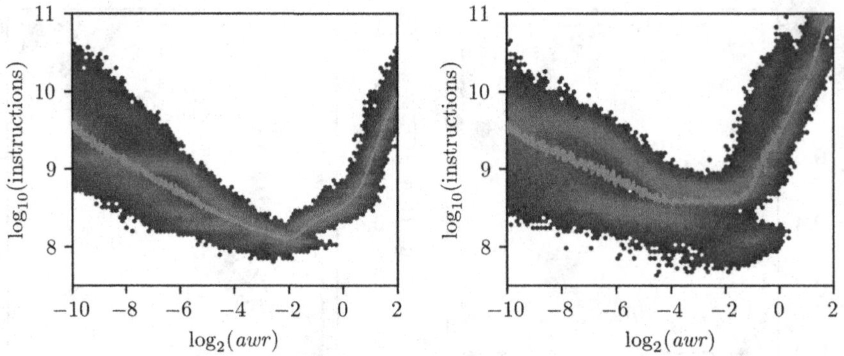

Fig. 4. 2D-histograms for the relative frequencies (color-scale) of how often, given a specific *awr* (x-axis), solving PRO017+2 required the shown number of instructions (y-axis). The curves in pink highlight the mean y-value for every x. The performance of dis10 (left) and the same strategy enhanced by a goal-directed heuristic (right). (Color figure online)

5 Experiment 3: Looking at One Problem at a Time

In their paper on age/weight shapes [13, Fig. 2], Rawson and Reger plot the number of given-clause loops required by Vampire to solve the TPTP problem PRO017+2 as a function of age/weight ratio (*awr*), a ratio specifying how often the prover selects the next clause to activate from its age-ordered and weight-ordered queues, respectively. The curve they obtain is quite "jiggly", indicating a fragile (discontinuous) dependence. Randomization allows us to smoothen the picture and reveal new, until now hidden, (probabilistic) patterns.

The 2D-histogram in Fig. 4 (left) was obtained from 100 independently seeded runs for each of 1200 distinct values of *awr* from between 1:1024 $= 2^{-10}$ and 4:1 $= 2^2$. We can confirm Rawson and Reger's observation of the best *awr* for PRO017+2 lying at around 1:2. However, we can now also attempt to explain the "jiggly-ness" of their curve: With a fragile proof search, even a slight change in *awr* effectively corresponds to an independent sample from the prover's execution resource[9] distribution, which—although changing continuously with *awr*—is of a high variance for our problem (note the log-scale of the y-axis)[10].

The distribution has another interesting property: At least for certain values of *awr* it is distinctly multi-modal. As if the prover can either find a proof quickly (after a lucky event?) or only after much harder effort later and almost nothing in between. Shedding more light on this phenomenon is left for further research.

It is also very interesting to observe the change of such a 2D-histogram when we modify the proof search strategy. Figure 4 (right) shows the effect of turning on SInE-level split queues [3], a goal directed clause selection heuristic

[9] Rawson and Reger [13] counted given-clause loops, we measure instructions.

[10] Even with 100 samples for each value of *awr*, the mean instruction count (rendered in pink in Fig. 4) looks jiggly towards the weight-heavy end of the plot.

(Vampire option `-slsq on`). We can see that the mean instruction count gets worse (for every tried *awr* value) and also the variance of the distribution distinctly increases. A curious effect of this is that we observe the shortest successful runs with `-slsq on`, while we still could not recommend (in the case of PRO017+2) this heuristic to the user. The probabilistic view makes us realize that there are competing criteria of prover performance for which one might want to optimize.

6 Related Work

The idea of randomizing a theorem prover is not new. Ertel [2] studied the speedup potential of running independently seeded instances of the connection prover SETHEO [10]. The dashed lines in our Figs. 1 and 2 capture an analogous notion in terms of "additional problems covered" for levels of parallelism 1–5. randoCoP [12] is a randomized version of another connection prover, leanCoP 2.0 [11]: especially in its incomplete setup, several restarts with different seeds helped randoCoP improve over leanCoP in terms of the number of solved problems.

Gomes et al. [4] notice that randomized complete backtracking algorithms for propositional satisfiability (SAT) lead to heavy-tailed runtime distributions on satisfiable instances. While we have not yet analyzed the runtime distributions coming from saturation-based first-order proof search in detail, we definitely observed high variance also for unsatisfiable problems. Also in the domain of SAT, Brglez et al. [1] proposed input shuffling as a way of turning solver's runtime into a random variable and studied the corresponding distributions.

An interesting view on the trade-offs between expected performance of a randomized solver and the risk associated with waiting for an especially long run to finish is given by Huberman et al. [6]. This is related to the last remark of the previous section.

Finally, in the satisfiability modulo theories (SMT) community, input shuffling, or scrambling, has been discussed as an obfuscation measure in competitions [17], where it should prevent the solvers to simply look up a precomputed answer upon recognising a previously seen problem. Notable is also the use of randomization in solver debugging via fuzz testing [14,18].

7 Discussion

As we have seen, the behaviour of a state-of-the-art saturation-based theorem prover is to a considerable degree chaotic and on many problems a mere perturbation of seemingly unimportant execution details decides about the success or the failure of the corresponding run. While this may be seen as a sign of our as-of-yet imperfect grasp of the technology, the author believes that an equally plausible view is that some form of chaos is inherent and originates from the complexity of the theorem proving task itself. (A higher-order logic proof search is expected to exhibit an even higher degree of fragility.)

This paper has proposed randomization as a key ingredient to a prover evaluation method that takes the chaotic nature of proof search into account. The extra cost required by the repeated runs, in itself not unreasonable to pay on contemporary parallel hardware, seems more than compensated by the new insights coming from the probabilistic picture that emerges. Moreover, other uses of randomization are easy to imagine, such as data augmentation for machine learning approaches or the construction of more robust strategy schedules. It feels that we only scratched the surface of the opened-up possibilities. More research will be needed to fully harness the potential of this perspective.

References

1. Brglez, F., Li, X.Y., Stallmann, M.F.M.: On SAT instance classes and a method for reliable performance experiments with SAT solvers. Ann. Math. Artif. Intell. **43**(1), 1–34 (2005). https://doi.org/10.1007/s10472-005-0417-5
2. Ertel, W.: OR-parallel theorem proving with random competition. In: Voronkov, A. (ed.) LPAR 1992. LNCS, vol. 624, pp. 226–237. Springer, Heidelberg (1992). https://doi.org/10.1007/BFb0013064
3. Gleiss, B., Suda, M.: Layered clause selection for saturation-based theorem proving. In: Fontaine, P., Korovin, K., Kotsireas, I.S., Rümmer, P., Tourret, S. (eds.) PAAR 7, Paris, France, June-July 2020. CEUR Workshop Proceedings, vol. 2752, pp. 34–52. CEUR-WS.org (2020). http://ceur-ws.org/Vol-2752/paper3.pdf
4. Gomes, C.P., Selman, B., Crato, N., Kautz, H.A.: Heavy-tailed phenomena in satisfiability and constraint satisfaction problems. J. Autom. Reason. **24**(1/2), 67–100 (2000). https://doi.org/10.1023/A:1006314320276
5. Hoder, K., Reger, G., Suda, M., Voronkov, A.: Selecting the selection. In: Olivetti, N., Tiwari, A. (eds.) IJCAR 2016. LNCS (LNAI), vol. 9706, pp. 313–329. Springer, Cham (2016). https://doi.org/10.1007/978-3-319-40229-1_22
6. Huberman, B., Lukose, R., Hogg, T.: An economics approach to hard computational problems. Science **275**, 51–4 (1997)
7. Järvisalo, M., Biere, A., Heule, M.: Blocked clause elimination. In: Esparza, J., Majumdar, R. (eds.) TACAS 2010. LNCS, vol. 6015, pp. 129–144. Springer, Heidelberg (2010). https://doi.org/10.1007/978-3-642-12002-2_10
8. Kiesl, B., Suda, M., Seidl, M., Tompits, H., Biere, A.: Blocked clauses in first-order logic. In: Eiter, T., Sands, D. (eds.) LPAR-21, Maun, Botswana, 7–12 May 2017. EPiC Series in Computing, vol. 46, pp. 31–48. EasyChair (2017)
9. Kovács, L., Voronkov, A.: First-order theorem proving and VAMPIRE. In: Sharygina, N., Veith, H. (eds.) CAV 2013. LNCS, vol. 8044, pp. 1–35. Springer, Heidelberg (2013). https://doi.org/10.1007/978-3-642-39799-8_1
10. Letz, R., Schumann, J., Bayerl, S., Bibel, W.: SETHEO: a high-performance theorem prover. J. Autom. Reason. **8**(2), 183–212 (1992). https://doi.org/10.1007/BF00244282
11. Otten, J.: leanCoP 2.0 and ileanCoP 1.2: high performance lean theorem proving in classical and intuitionistic logic (system descriptions). In: Armando, A., Baumgartner, P., Dowek, G. (eds.) IJCAR 2008. LNCS (LNAI), vol. 5195, pp. 283–291. Springer, Heidelberg (2008). https://doi.org/10.1007/978-3-540-71070-7_23
12. Raths, T., Otten, J.: randoCoP: randomizing the proof search order in the connection calculus. In: Konev, B., Schmidt, R.A., Schulz, S. (eds.) PAAR 1, Sydney, Australia, 10–11 August 2008. CEUR Workshop Proceedings, vol. 373. CEUR-WS.org (2008). http://ceur-ws.org/Vol-373/paper-08.pdf

13. Rawson, M., Reger, G.: Old or heavy? decaying gracefully with age/weight shapes. In: Fontaine, P. (ed.) CADE 2019. LNCS (LNAI), vol. 11716, pp. 462–476. Springer, Cham (2019). https://doi.org/10.1007/978-3-030-29436-6_27
14. Scott, J., Sudula, T., Rehman, H., Mora, F., Ganesh, V.: BanditFuzz: fuzzing SMT solvers with multi-agent reinforcement learning. In: Huisman, M., Păsăreanu, C., Zhan, N. (eds.) FM 2021. LNCS, vol. 13047, pp. 103–121. Springer, Cham (2021). https://doi.org/10.1007/978-3-030-90870-6_6
15. Sutcliffe, G.: The TPTP problem library and associated infrastructure. From CNF to TH0, TPTP v.6.4.0. J. Autom. Reason. **59**(4), 483–502 (2017). https://doi.org/10.1007/s10817-017-9407-7
16. Voronkov, A.: AVATAR: the architecture for first-order theorem provers. In: Biere, A., Bloem, R. (eds.) CAV 2014. LNCS, vol. 8559, pp. 696–710. Springer, Cham (2014). https://doi.org/10.1007/978-3-319-08867-9_46
17. Weber, T.: Scrambling and descrambling SMT-LIB benchmarks. In: King, T., Piskac, R. (eds.) SMT@IJCAR 2016, Coimbra, Portugal, 1–2 July 2016. CEUR Workshop Proceedings, vol. 1617, pp. 31–40. CEUR-WS.org (2016). http://ceur-ws.org/Vol-1617/paper3.pdf
18. Winterer, D., Zhang, C., Su, Z.: Validating SMT solvers via semantic fusion. In: Donaldson, A.F., Torlak, E. (eds.) PLDI 2020, London, UK, 15–20 June 2020, pp. 718–730. ACM (2020)

Evolution, Termination, and Decision Problems

On Eventual Non-negativity and Positivity for the Weighted Sum of Powers of Matrices

S. Akshay(✉), Supratik Chakraborty(✉), and Debtanu Pal

Indian Institute of Technology Bombay, Mumbai 400076, India
{akshayss,supratik,debtanu}@cse.iitb.ac.in

Abstract. The long run behaviour of linear dynamical systems is often studied by looking at eventual properties of matrices and recurrences that underlie the system. A basic problem in this setting is as follows: given a set of pairs of rational weights and matrices $\{(w_1, A_1), \ldots, (w_m, A_m)\}$, does there exist an integer N s.t for all $n \geq N$, $\sum_{i=1}^{m} w_i \cdot A_i^n \geq 0$ (resp. > 0). We study this problem, its applications and its connections to linear recurrence sequences. Our first result is that for $m \geq 2$, the problem is as hard as the ultimate positivity of linear recurrences, a long standing open question (known to be coNP-hard). Our second result is that for any $m \geq 1$, the problem reduces to ultimate positivity of linear recurrences. This yields upper bounds for several subclasses of matrices by exploiting known results on linear recurrence sequences. Our third result is a general reduction technique for a large class of problems (including the above) from diagonalizable case to the case where the matrices are simple (have non-repeated eigenvalues). This immediately gives a decision procedure for our problem for diagonalizable matrices.

Keywords: Eventual properties of matrices · Ultimate Positivity · linear recurrence sequences

1 Introduction

The study of eventual or asymptotic properties of discrete-time linear dynamical systems has long been of interest to both theoreticians and practitioners. Questions pertaining to (un)-decidability and/or computational complexity of predicting the long-term behaviour of such systems have been extensively studied over the last few decades. Despite significant advances, however, there remain simple-to-state questions that have eluded answers so far. In this work, we investigate one such problem, explore its significance and links with other known problems, and study its complexity and computability landscape.

This work was partly supported by DST/CEFIPRA/INRIA Project EQuaVE and DST/SERB Matrices Grant MTR/2018/000744.
Author names are in alphabetical order of last names.

J. Blanchette et al. (Eds.): IJCAR 2022, LNAI 13385, pp. 671–690, 2022.
https://doi.org/10.1007/978-3-031-10769-6_39

The time-evolution of linear dynamical systems is often modeled using linear recurrence sequences, or using sequences of powers of matrices. Asymptotic properties of powers of matrices are therefore of central interest in the study of linear differential systems, dynamic control theory, analysis of linear loop programs etc. (see e.g. [26,32,36,37]). The literature contains a rich body of work on the decidability and/or computational complexity of problems related to the long-term behaviour of such systems (see, e.g. [15,19,27,29,36,37]). A question of significant interest in this context is whether the powers of a given matrix of rational numbers eventually have only non-negative (resp. positive) entries. Such matrices, also called *eventually non-negative* (resp. *eventually positive*) matrices, enjoy beautiful algebraic properties ([13,16,25,38]), and have been studied by mathematicians, control theorists and computer scientists, among others. For example, the work of [26] investigates reachability and holdability of non-negative states for linear differential systems – a problem in which eventually non-negative matrices play a central role. Similarly, eventual non-negativity (or positivity) of a matrix modeling a linear dynamical system makes it possible to apply the elegant Perron-Frobenius theory [24,34] to analyze the long-term behaviour of the system beyond an initial number of time steps. Another level of complexity is added if the dynamics is controlled by a set of matrices rather than a single one. For instance, each matrix may model a mode of the linear dynamical system [23]. In a partial observation setting [22,39], we may not know which mode the system has been started in, and hence have to reason about eventual properties of this multi-modal system. This reduces to analyzing the sum of powers of the per-mode matrices, as we will see.

Motivated by the above considerations, we study the problem of determining whether a given matrix of rationals is eventually non-negative or eventually positive and also a generalized version of this problem, wherein we ask if the *weighted sum of powers of a given set of matrices of rationals* is eventually non-negative (resp. positive). Let us formalize the general problem statement. ***Given a set*** $\mathfrak{A} = \{(w_1, A_1), \ldots (w_m, A_m)\}$***, where each*** w_i ***is a rational number and each*** A_i ***is a*** $k \times k$ ***matrix of rationals, we wish to determine if*** $\sum_{i=1}^{m} w_i \cdot A_i^n$ ***has only non-negative (resp. positive) entries for all sufficiently large values of*** n***.*** We call this problem *Eventually Non-Negative (resp. Positive) Weighted Sum of Matrix Powers* problem, or $\mathsf{ENN}_{\mathsf{SoM}}$ (resp. $\mathsf{EP}_{\mathsf{SoM}}$) for short. The eventual non-negativity (resp. positivity) of powers of a single matrix is a special case of the above problem, where $\mathfrak{A} = \{(1, A)\}$. We call this special case the *Eventually Non-Negative (resp. Positive) Matrix* problem, or $\mathsf{ENN}_{\mathsf{Mat}}$ (resp. $\mathsf{EP}_{\mathsf{Mat}}$) for short.

Given the simplicity of the $\mathsf{ENN}_{\mathsf{SoM}}$ and $\mathsf{EP}_{\mathsf{SoM}}$ problem statements, one may be tempted to think that there ought to be simple algebraic characterizations that tell us whether $\sum_{i=1}^{m} w_i \cdot A_i^n$ is eventually non-negative or positive. But in fact, the landscape is significantly nuanced. On one hand, a solution to the general $\mathsf{ENN}_{\mathsf{SoM}}$ or $\mathsf{EP}_{\mathsf{SoM}}$ problem would resolve long-standing open questions in mathematics and computer science. On the other hand, efficient algorithms can indeed be obtained under certain well-motivated conditions. This paper is a study of both these aspects of the problem. Our primary contributions can be summarized as follows. Below, we use $\mathfrak{A} = \{(w_1, A_1), \ldots (w_m, A_m)\}$ to define an instance of $\mathsf{ENN_{SoM}}$ or $\mathsf{EP_{SoM}}$.

1. If $|\mathfrak{A}| \geq 2$, we show that both $\mathsf{ENN_{SoM}}$ and $\mathsf{EP_{SoM}}$ are as hard as the ultimate non-negativity problem for linear recurrence sequences ($\mathsf{UNN_{LRS}}$, for short). The decidability of $\mathsf{UNN_{LRS}}$ is closely related to Diophantine approximations, and remains unresolved despite extensive research (see e.g. [31]).
   Since $\mathsf{UNN_{LRS}}$ is coNP-hard (in fact, as hard as the decision problem for universal theory of reals), so is $\mathsf{ENN_{SoM}}$ and $\mathsf{EP_{SoM}}$, when $|\mathfrak{A}| \geq 2$. Thus, unless P = NP, we cannot hope for polynomial-time algorithms, and any algorithm would also resolve long-standing open problems.
2. On the other hand, regardless of $|\mathfrak{A}|$, we show a reduction in the other direction from $\mathsf{ENN_{SoM}}$ (resp. $\mathsf{EP_{SoM}}$) to $\mathsf{UNN_{LRS}}$ (resp. $\mathsf{UP_{LRS}}$, the strict version of $\mathsf{UNN_{LRS}}$). As a consequence, we get decidability and complexity bounds for special cases of $\mathsf{ENN_{SoM}}$ and $\mathsf{EP_{SoM}}$, by exploiting recent results on recurrence sequences [30,31,35]. For example, if each matrix A_i in $\mathfrak{A}$ is simple, i.e. has all distinct eigenvalues, we obtain PSPACE algorithms.
3. Finally, we consider the case where A_i is diagonalizable (also called non-defective or inhomogenous dilation map) for each $(w_i, A_i) \in \mathfrak{A}$. This is a practically useful class of matrices and strictly subsumes simple matrices. We present a novel reduction technique for a large family of problems (including eventual non-negativity/positivity, everywhere non-negativity/positivity etc.) over diagonalizable matrices to the corresponding problem over simple matrices. This yields effective decision procedures for $\mathsf{EP_{SoM}}$ and $\mathsf{ENN_{SoM}}$ for diagonalizable matrices. Our reduction makes use of a novel perturbation analysis that also has other interesting consequences.

As mentioned earlier, the eventual non-negativity and positivity problem for single rational matrices are well-motivated in the literature, and $\mathsf{EP_{Mat}}$ (or $\mathsf{EP_{SoM}}$ with $|\mathfrak{A}| = 1$) is known to be in PTIME [25]. But for $\mathsf{ENN_{Mat}}$, no decidability results are known to the best of our knowledge. From our work, we obtain two new results about $\mathsf{ENN_{Mat}}$: (i) in general $\mathsf{ENN_{Mat}}$ reduces to $\mathsf{UNN_{LRS}}$ and (ii) for diagonalizable matrices, we can decide $\mathsf{ENN_{Mat}}$. What is surprising (see Sect. 5) is that the latter decidability result goes via $\mathsf{ENN_{SoM}}$, i.e. the multiple matrices case. Thus, reasoning about sums of powers of matrices, viz. $\mathsf{ENN_{SoM}}$, is useful even when reasoning about powers of a single matrix, viz. $\mathsf{ENN_{Mat}}$.

Potential Applications of $\mathsf{ENN_{SoM}}$ *and* $\mathsf{EP_{SoM}}$. A prime motivation for defining the generalized problem statement $\mathsf{ENN_{SoM}}$ is that it is useful even when reasoning about the single matrix case $\mathsf{ENN_{Mat}}$. However and unsurprisingly, $\mathsf{ENN_{SoM}}$ and $\mathsf{EP_{SoM}}$ are also well-motivated independently. Indeed, for every application involving a linear dynamical system that reduces to $\mathsf{ENN_{Mat}}/\mathsf{EP_{Mat}}$, there is a naturally defined aggregated version of the application involving multiple independent linear dynamical systems that reduces to $\mathsf{ENN_{SoM}}/\mathsf{EP_{SoM}}$ (e.g., the *swarm of robots* example in [3]).

Beyond this, $\mathsf{ENN_{SoM}}/\mathsf{EP_{SoM}}$ arise naturally and directly when solving problems in different practical scenarios. Due to lack of space, we detail two applications here and describe more in the longer version of the paper [3].

Partially Observable Multi-modal Systems. Our first example comes from the domain of cyber-physical systems in a partially observable setting. Consider a system (e.g. a robot) with m modes of operation, where the ith mode dynamics is given by a linear transformation encoded as a $k \times k$ matrix of rationals, say A_i. Thus, if the system state at (discrete) time t is represented by a k-dimensional rational (row) vector $\mathbf{u_t}$, the state at time $t+1$, when operating in mode i, is given by $\mathbf{u_t} A_i$. Suppose the system chooses to operate in one of its various modes at time 0, and then sticks to this mode at all subsequent time. Further, the initial choice of mode is not observable, and we are only given a probability distribution over modes for the initial choice. This is natural, for instance, if our robot (multi-modal system) knows the terrain map and can make an initial choice of which path (mode) to take, but cannot change its path once it has chosen. If p_i is a rational number denoting the probability of choosing mode i initially, then the expected state at time n is given by $\sum_{i=1}^{m} p_i \cdot \mathbf{u_0} A_i^n = \mathbf{u_0}\big(\sum_{i=1}^{m} p_i \cdot A_i^n\big)$. A safety question in this context is whether starting from a state $\mathbf{u_0}$ with all non-negative (resp. positive) components, the system is expected to eventually stay locked in states that have all non-negative (resp. positive) components. In other words, does $\mathbf{u_0}\big(\sum_{i=1}^{m} p_i \cdot A_i^n\big)$ have all non-negative (resp. positive) entries for all sufficiently large n? Clearly, a sufficient condition for an affirmative answer to this question is to have $\sum_{i=1}^{n} p_i \cdot A_i^n$ eventually non-negative (resp. positive), which is an instance of $\mathsf{ENN_{SoM}}$ (resp. $\mathsf{EP_{SoM}}$).

Commodity Flow Networks. Consider a flow network where m different commodities $\{c_1, \ldots, c_m\}$ use the same flow infrastructure spanning k nodes, but have different loss/regeneration rates along different links. For every pair of nodes $i, j \in \{1, \ldots, k\}$ and for every commodity $c \in \{c_1, \ldots, c_m\}$, suppose $A_c[i,j]$ gives the fraction of the flow of commodity c starting from i that reaches j through the link connecting i and j (if it exists). In general, $A_c[i,j]$ is the product of the fraction of the flow of commodity c starting at i that is sent along the link to j, and the loss/regeneration rate of c as it flows in the link from i to j. Note that $A_c[i,j]$ can be 0 if commodity c is never sent directly from i to j, or the commodity is lost or destroyed in flowing along the link from i to j. It can be shown that $A_c^n[i,j]$ gives the fraction of the flow of c starting from i that reaches j after n hops through the network. If commodities keep circulating through the network ad-infinitum, we wish to find if the network gets *saturated*, i.e., for all sufficiently long enough hops through the network, there is a non-zero fraction of some commodity that flows from i to j for every pair i, j. This is equivalent to asking if there exists $N \in \mathbb{N}$ such that $\sum_{\ell=1}^{m} A_{c_\ell}^n > 0$. If different commodities have different weights (or costs) associated, with commodity c_i having the weight w_i, the above formulation asks if $\sum_{\ell=1}^{m} w_\ell . A_{c_\ell}^n$ is eventually positive, which is effectively the $\mathsf{EP_{SoM}}$ problem.

Other Related Work. Our problems of interest are different from other well-studied problems that arise if the system is allowed to choose its mode independently at each time step (e.g. as in Markov decision processes [5,21]). The crucial difference stems from the fact that we require that the mode be chosen once initially, and subsequently, the system must follow the same mode forever. Thus, our problems are prima facie different from those related to general probabilistic or weighted finite automata, where reachability of states and questions pertaining to long-run behaviour are either known to be undecidable or have remained open for long ([6,12,17]). Even in the case of unary probabilistic/weighted finite automata [1,4,8,11], reachability is known in general to be as hard as the Skolem problem on linear recurrences – a long-standing open problem, with decidability only known in very restricted cases. The difference sometimes manifests itself in the simplicity/hardness of solutions. For example, $\mathsf{EP}_{\mathsf{Mat}}$ (or $\mathsf{EP}_{\mathsf{SoM}}$ with $|\mathfrak{A}| = 1$) is known to be in PTIME [25] (not so for $\mathsf{ENN}_{\mathsf{Mat}}$ however), whereas it is still open whether the reachability problem for unary probabilistic/weighted automata is decidable. It is also worth remarking that instead of the sum of powers of matrices, if we considered the product of their powers, we would effectively be solving problems akin to the *mortality problem* [9,10] (which asks whether the all-0 matrix can be reached by multiplying with repetition from a set of matrices) – a notoriously difficult problem. The diagonalizable matrix restriction is a common feature in in the context of linear loop programs (see, e.g., [7,28]), where matrices are used for updates. Finally, logics to reason about temporal properties of linear loops have been studied, although decidability is known only in restricted settings, e.g. when each predicate defines a semi-algebraic set contained in some 3-dimensional subspace, or has intrinsic dimension 1 [20].

2 Preliminaries

The symbols $\mathbb{Q}, \mathbb{R}, \mathbb{A}$ and $\mathbb{C}$ denote the set of rational, real, algebraic and complex numbers respectively. Recall that an *algebraic number* is a root of a non-zero polynomial in one variable with rational coefficients. An algebraic number can be real or complex. We use $\mathbb{RA}$ to denote the set of real algebraic numbers (which includes all rationals). The sum, difference and product of two (real) algebraic numbers is again (real) algebraic. Furthermore, every root of a polynomial equation with (real) algebraic coefficients is again (real) algebraic. We call matrices with all rational (resp. real algebraic or real) entries *rational* (resp. *real algebraic* or *real*) *matrices*. We use $A \in \mathbb{Q}^{k \times l}$ (resp. $A \in \mathbb{R}^{k \times l}$ and $A \in \mathbb{RA}^{k \times l}$) to denote that A is a $k \times l$ rational (resp. real and real algebraic) matrix, with rows indexed 1 through k, and columns indexed 1 through l. The entry in the ith row and j^{th} column of a matrix A is denoted $A[i, j]$. If A is a column vector (i.e. $l = 1$), we often use boldface letters, viz. $\mathbf{A}$, to refer to it. In such cases, we use $\mathbf{A}[i]$ to denote the ith component of $\mathbf{A}$, i.e. $A[i, 1]$. The transpose of a $k \times l$ matrix A, denoted A^{T}, is the $l \times k$ matrix obtained by letting $A^{\mathsf{T}}[i, j] = A[j, i]$ for all $i \in \{1, \ldots l\}$ and $j \in \{1, \ldots k\}$. Matrix A is said to be *non-negative* (resp. *positive*) if all entries of A are non-negative (resp. positive) real numbers. Given a set $\mathfrak{A} = \{(w_1, A_1), \ldots (w_m, A_m)\}$ of (weight, matrix) pairs, where each $A_i \in \mathbb{Q}^{k \times k}$ (resp. $\in \mathbb{RA}^{k \times k}$) and each $w_i \in \mathbb{Q}$, we use $\sum \mathfrak{A}^n$ to denote the weighted matrix sum $\sum_{i=1}^{m} w_i \cdot A_i^n$, for every natural number $n > 0$. Note that $\sum \mathfrak{A}^n$ is itself a matrix in $\mathbb{Q}^{k \times k}$ (resp. $\mathbb{RA}^{k \times k}$).

Definition 1. *We say that $\mathfrak{A}$ is eventually non-negative (resp. positive) iff there is a positive integer N s.t., $\sum \mathfrak{A}^n$ is non-negative (resp. positive) for all $n \geq N$.*

The $\mathsf{ENN_{SoM}}$ (resp. $\mathsf{EP_{SoM}}$) problem, described in Sect. 1, can now be re-phrased as: *Given a set $\mathfrak{A}$ of pairs of rational weights and rational $k \times k$ matrices, is $\mathfrak{A}$ eventually non-negative (resp. positive)?* As mentioned in Sect. 1, if $\mathfrak{A} = \{(1, A)\}$, the $\mathsf{ENN_{SoM}}$ (resp. $\mathsf{EP_{SoM}}$) problem is also called $\mathsf{ENN_{Mat}}$ (resp. $\mathsf{EP_{Mat}}$). We note that the study of $\mathsf{ENN_{SoM}}$ and $\mathsf{EP_{SoM}}$ with $|\mathfrak{A}| = 1$ is effectively the study of $\mathsf{ENN_{Mat}}$ and $\mathsf{EP_{Mat}}$ i.e., wlog we can assume $w = 1$.

The *characteristic polynomial* of a matrix $A \in \mathbb{RA}^{k\times k}$ is given by $det(A-\lambda I)$, where I denotes the $k\times k$ identity matrix. Note that this is a degree k polynomial in λ. The roots of the characteristic polynomial are called the *eigenvalues* of A. The non-zero vector solution of the equation $A\mathbf{x} = \lambda_i \mathbf{x}$, where λ_i is an eigenvalue of A, is called an *eigenvector* of A. Although $A \in \mathbb{RA}^{k\times k}$, in general it can have eigenvalues $\lambda \in \mathbb{C}$ which are all algebraic numbers. An eigenvector is said to be positive (resp. non-negative) if each component of the eigenvector is a positive (resp. non-negative) rational number. A matrix is called *simple* if all its eigenvalues are distinct. Further, a matrix A is called *diagonalizable* if there exists an invertible matrix S and diagonal matrix D such that $SDS^{-1} = A$.

The study of weighted sum of powers of matrices is intimately related to the study of *linear recurrence sequences (LRS)*, as we shall see. We now present some definitions and useful properties of LRS. For more details on LRS, the reader is referred to the work of Everest et al. [14]. A sequence of rational numbers $\langle u\rangle = \langle u_n\rangle_{n=0}^{\infty}$ is called an LRS of *order* k (> 0) if the nth term of the sequence, for all $n \geq k$, can be expressed using the recurrence: $u_n = a_{k-1}u_{n-1} + \ldots + a_1 u_{n-k-1} + a_0 u_{n-k}$. Here, $a_0\ (\neq 0), a_1, \ldots, a_{k-1} \in \mathbb{Q}$ are called the *coefficients* of the LRS, and $u_0, u_1, \ldots, u_{k-1} \in \mathbb{Q}$ are called the *initial values* of the LRS. Given the coefficients and initial values, an LRS is uniquely defined. However, the same LRS may be defined by multiple sets of coefficients and corresponding initial values. An LRS $\langle u\rangle$ is said to be *periodic* with period ρ if it can be defined by the recurrence $u_n = u_{n-\rho}$ for all $n \geq \rho$. Given an LRS $\langle u\rangle$, its *characteristic polynomial* is $p_{\langle u\rangle}(x) = x^k - \sum_{i=0}^{k-1} a_i x^i$. We can factorize the characteristic polynomial as $p_{\langle u\rangle}(x) = \prod_{j=1}^{d}(x-\lambda_j)^{\rho_j}$, where λ_j is a root, called a *characteristic root* of *algebraic multiplicity* ρ_j. An LRS is called *simple* if $\rho_j = 1$ for all j, i.e. all characteristic roots are distinct. Let $\{\lambda_1, \lambda_2, \ldots, \lambda_d\}$ be distinct roots of $p_{\langle u\rangle}(x)$ with multiplicities $\rho_1, \rho_2, \ldots, \rho_d$ respectively. Then the nth term of the LRS, denoted u_n, can be expressed as $u_n = \sum_{j=1}^{d} q_j(n)\lambda_j^n$, where $q_j(x) \in \mathbb{C}(x)$ are univariate polynomials of degree at most $\rho_j - 1$ with complex coefficients such that $\sum_{j=1}^{d} \rho_j = k$. This representation of an LRS is known as the *exponential polynomial solution* representation. It is well known that scaling an LRS by a constant gives another LRS, and the sum and product of two LRSs is also an LRS (Theorem 4.1 in [14]). Given an LRS $\langle u\rangle$ defined by $u_n = a_{k-1}u_{n-1} + \ldots + a_1 u_{n-k-1} + a_0 u_{n-k}$, we define its *companion matrix* $M_{\langle u\rangle}$ to be the $k \times k$ matrix shown in Fig. 1.

$$M_{\langle u\rangle} = \begin{bmatrix} a_{k-1} & 1 \dots 0 & 0 \\ \vdots & \vdots \ddots \vdots & \vdots \\ a_2 & 0 \dots 1 & 0 \\ a_1 & 0 \dots 0 & 1 \\ a_0 & 0 \dots 0 & 0 \end{bmatrix}$$

Fig. 1. Companion matrix

When $\langle u\rangle$ is clear from the context, we often omit the subscript for clarity of notation, and use M for $M_{\langle u\rangle}$. Let $\mathbf{u} = (u_{k-1}, \dots, u_0)$ be a row vector containing the k initial values of the recurrence, and let $\mathbf{e_k} = (0, 0, \dots 1)^T$ be a column vector of k dimensions with the last element equal to 1 and the rest set to 0s. It is easy to see that for all $n \geq 1$, $\mathbf{u}M^n\mathbf{e_k}$ gives u_n. Note that the eigenvalues of the matrix M are exactly the roots of the characteristic polynomial of the LRS $\langle u\rangle$.

For $\mathbf{u} = (u_{k-1}, \dots, u_0)$, we call the matrix $G_{\langle u\rangle} = \begin{bmatrix} 0 & \mathbf{u} \\ \mathbf{0}^T & M_{\langle u\rangle} \end{bmatrix}$ the *generator matrix* of the LRS $\langle u\rangle$, where $\mathbf{0}$ is a k-dimensional vector of all 0s. We omit the subscript and use G instead of $G_{\langle u\rangle}$, when the LRS $\langle u\rangle$ is clear from the context. It is easy to show from the above that $u_n = G^{n+1}[1, k+1]$ for all $n \geq 0$.

We say that an LRS $\langle u\rangle$ is *ultimately non-negative* (resp. *ultimately positive*) iff there exists $N > 0$, such that $\forall n \geq N$, $u_n \geq 0$ (resp. $u_n > 0$)[1]. The problem of determining whether a given LRS is ultimately non-negative (resp. ultimately positive) is called the *Ultimate Non-negativity* (resp. *Ultimate Positivity*) problem for LRS. We use $\mathsf{UNN_{LRS}}$ (resp. $\mathsf{UP_{LRS}}$) to refer to this problem. It is known [19] that $\mathsf{UNN_{LRS}}$ and $\mathsf{UP_{LRS}}$ are polynomially inter-reducible, and these problems have been widely studied in the literature (e.g., [27,31,32]). A closely related problem is the *Skolem problem*, wherein we are given an LRS $\langle u\rangle$ and we are required to determine if there exists $n \geq 0$ such that $u_n = 0$. The relation between the Skolem problem and $\mathsf{UNN_{LRS}}$ (resp. $\mathsf{UP_{LRS}}$) has been extensively studied in the literature (e.g., [18,19,33]).

3 Hardness of Eventual Non-negativity and Positivity

In this section, we show that $\mathsf{UNN_{LRS}}$ (resp. $\mathsf{UP_{LRS}}$) polynomially reduces to $\mathsf{ENN_{SoM}}$ (resp. $\mathsf{EP_{SoM}}$) when $|\mathfrak{A}| \geq 2$. Since $\mathsf{UNN_{LRS}}$ and $\mathsf{UP_{LRS}}$ are known to be coNP-hard (in fact, as hard as the decision problem for the universal theory of reals Theorem 5.3 [31]), we conclude that $\mathsf{ENN_{SoM}}$ and $\mathsf{EP_{SoM}}$ are also coNP-hard and at least as hard as the decision problem for the universal theory of reals, when $|\mathfrak{A}| \geq 2$. Thus, unless $\mathsf{P} = \mathsf{NP}$, there is no hope of finding polynomial-time solutions to these problems.

Theorem 1. $\mathsf{UNN_{LRS}}$ *reduces to* $\mathsf{ENN_{SoM}}$ *with* $|\mathfrak{A}| \geq 2$ *in polynomial time.*

Proof. Given an LRS $\langle u\rangle$ of order k defined by the recurrence $u_n = a_{k-1}u_{n-1} + \dots + a_1 u_{n-k-1} + a_0 u_{n-k}$ and initial values $u_0, u_1, \dots, u_{k-1}$, construct two

[1] *Ultimately non-negative* (resp. *ultimately positive*) LRS, as defined by us, have also been called *ultimately positive* (resp. *strictly positive*) LRS elsewhere in the literature [31]. However, we choose to use terminology that is consistent across matrices and LRS, to avoid notational confusion.

matrices A_1 and A_2 such that $\langle u \rangle$ is ultimately non-negative iff $(A_1^n + A_2^n)$ is eventually non-negative. Consider $A_1 = \begin{bmatrix} 0 & \mathbf{u} \\ \mathbf{0}^T & M \end{bmatrix}$, the generator matrix of $\langle u \rangle$ and $A_2 = \begin{bmatrix} 0 & \mathbf{0} \\ \mathbf{0}^T & P \end{bmatrix}$, where $P \in \mathbb{Q}^{k \times k}$ is constructed such that : $P[i,j] \geq |M[i,j]|$. For example P can be constructed as: $P[i,j] = M[i,j]$ for all $j \in [2,k]$ and $i \in [1,k]$ and $P[i,j] = max(|a_0|, |a_1|, \dots, |a_{k-1}|) + 1$ for $j = 1$. Now consider the sequence of matrices defined by $A_1^n + A_2^n$, for all $n \geq 1$. By properties of the generator matrix, it is easily verified that $A_1^n = \begin{bmatrix} 0 & \mathbf{u}M^{n-1} \\ \mathbf{0}^T & M^n \end{bmatrix}$. Similarly, we get $A_2^n = \begin{bmatrix} 0 & \mathbf{0} \\ \mathbf{0}^T & P^n \end{bmatrix}$. Therefore, $A_1^n + A_2^n = \begin{bmatrix} 0 & \mathbf{u}M^{n-1} \\ \mathbf{0}^T & P^n + M^n \end{bmatrix}$, for all $n \geq 1$. Now, we can observe that $P^n + M^n$ is always non-negative, since $P[i,j] \geq |M[i,j]| \geq 0$ for all $i,j \in \{1, \dots k\}$ and hence $P^n[i,j] + M^n[i,j] \geq 0$ for all $i,j \in \{1, \dots k\}$ and $n \geq 1$. Thus we conclude that $A(n) = A_1^n + A_2^n \geq 0$ $(n \geq 1)$ iff $\langle u \rangle$ is ultimately non-negative, since the elements $A(n)[1,1] \dots, A(n)[1,k+1]$ consists of $(u_{n+k-2} \dots, u_n, u_{n-1})$ and the rest of the elements are non-negative. □

Observe that the same reduction technique works if we are required to use more than 2 matrices in $\mathsf{ENN}_{\mathsf{SoM}}$. Indeed, we can construct matrices $A_3, A_4, \dots, A_m$ similar to the construction of A_2 in the reduction above, by having the $k \times k$ matrix in the bottom right (see definition of A_2) to have positive values greater than the maximum absolute value of every element in the companion matrix.

A simple modification of the above proof setting $A_2 = \begin{bmatrix} 1 & \mathbf{0} \\ \mathbf{1}^T & P \end{bmatrix}$, where $\mathbf{1}$ denotes the k-dimensional vector of all 1's gives us the corresponding hardness result for $\mathsf{EP}_{\mathsf{SoM}}$ (see [3] for details).

Theorem 2. $\mathsf{UP}_{\mathsf{LRS}}$ *reduces to* $\mathsf{EP}_{\mathsf{SoM}}$ *with* $|\mathfrak{A}| \geq 2$ *in polynomial time.*

We remark that for the reduction technique used in Theorems 1 and 2 to work, we need at least two (weight, matrix) pairs in $\mathfrak{A}$. For explanation of why this reduction doesn't work when $|\mathfrak{A}| = 1$, we refer the reader to [3]. Having shown the hardness of $\mathsf{ENN}_{\mathsf{SoM}}$ and $\mathsf{EP}_{\mathsf{SoM}}$ when $|\mathfrak{A}| \geq 2$, we now proceed to establish upper bounds on the computational complexity of these problems.

4 Upper Bounds on Eventual Non-negativity and Positivity

In this section, we show that $\mathsf{ENN}_{\mathsf{SoM}}$ (resp. $\mathsf{EP}_{\mathsf{SoM}}$) is polynomially reducible to $\mathsf{UNN}_{\mathsf{LRS}}$ (resp. $\mathsf{UP}_{\mathsf{LRS}}$), regardless of $|\mathfrak{A}|$.

Theorem 3. $\mathsf{ENN}_{\mathsf{SoM}}$, *reduces to* $\mathsf{UNN}_{\mathsf{LRS}}$ *in polynomial time.*

The proof is in two parts. First, we show that for a single matrix A, we can construct a linear recurrence $\langle a \rangle$ such that A is eventually non-negative iff

$\langle a \rangle$ is ultimately non-negative. Then, we show that starting from such a linear recurrence for each matrix in $\mathfrak{A}$, we can construct a new LRS, say $\langle a^\star \rangle$, with the property that the weighted sum of powers of the matrices in $\mathfrak{A}$ is eventually non-negative iff $\langle a^\star \rangle$ is ultimately non-negative. Our proof makes crucial use of the following property of matrices.

Lemma 1 Adapted from Lemma 1.1 of [19]). *Let $A \in \mathbb{Q}^{k\times k}$ be a rational matrix with characteristic polynomial $p_A(\lambda) = det(A - \lambda I)$. Suppose we define the sequence $\langle a^{ij} \rangle$ for every $1 \leq i, j \leq k$ as follows: $a_n^{i,j} = A^{n+1}[i,j]$, for all $n \geq 0$. Then $\langle a^{i,j} \rangle$ is an LRS of order k with characteristic polynomial $p_A(x)$ and initial values given by $a_0^{ij} = A^1[i,j], \dots a_{k-1}^{ij} = A^k[i,j]$.*

This follows from the Cayley-Hamilton Theorem and the reader is referred to [19] for further details. From Lemma 1, it is easy to see that the LRS $\langle a^{i,j} \rangle$ for all $1 \leq i, j \leq k$ share the same order and characteristic polynomial (hence the defining recurrence) and differ only in their initial values. For notational convenience, we say that the LRS $\langle a^{i,j} \rangle$ is *generated by* $A[i,j]$.

Proposition 1. *A matrix $A \in \mathbb{Q}^{k\times k}$ is eventually non-negative iff all LRS $\langle a^{i,j} \rangle$ generated by $A[i,j]$ for all $1 \leq i, j \leq k$ are ultimately non-negative.*

The proof follows from the definition of eventually non-negative matrices and the definition of $\langle a^{ij} \rangle$. Next we define the notion of interleaving of LRS.

Definition 2. *Consider a set $S = \{\langle u^i \rangle : 0 \leq i < t\}$ of t LRSes, each having order k and the same characteristic polynomial. An LRS $\langle v \rangle$ is said to be the **LRS-interleaving** of S iff $v_{tn+s} = u_n^s$ for all $n \in \mathbb{N}$ and $0 \leq s < t$.*

Observe that, the order of $\langle v \rangle$ is tk and its initial values are given by the interleaving of the k initial values of the LRSes $\langle u^i \rangle$. Formally, the initial values are $v_{tj+i} = u_j^i$ for $0 \leq i < t$ and $0 \leq j < k$. The characteristic polynomial $p_{\langle v \rangle}(s)$ is equal to $p_{\langle u^i \rangle}(x^t)$.

Proposition 2. *The LRS-interleaving $\langle v \rangle$ of a set of LRSes $S = \{\langle u^i \rangle : 0 \leq i < t\}$ is ultimately non-negative iff each LRS $\langle u^i \rangle$ in S is ultimately non-negative.*

Now, from the definitions of LRSes $\langle a^{i,j} \rangle$, $\langle u^i \rangle$ and $\langle v \rangle$, and from Propositions 1 and 2, we obtain the following crucial lemma.

Lemma 2. *Given a matrix $A \in \mathbb{Q}^{k\times k}$, let $S = \{\langle u^i \rangle \mid u_n^i = a_n^{pq}$, where $p = \lfloor i/k \rfloor + 1$, $q = i \mod k + 1$, $0 \leq i < k^2\}$ be the set of k^2 LRSes mentioned in Lemma 1. The LRS $\langle v \rangle$ generated by LRS-interleaving of S satisfies the following:*

1. *A is eventually non-negative iff $\langle v \rangle$ is ultimately non-negative.*
2. *$p_{\langle v \rangle}(x) = \prod_{i=1}^{k}(x^{k^2} - \lambda_i)$, where $\lambda_1, \dots \lambda_k$ are the (possibly repeated) eigenvalues of A.*
3. *$v_{rk^2+sk+t} = u_r^{sk+t} = a_r^{s+1,t+1} = A^{r+1}[s+1, t+1]$ for all $r \in \mathbb{N}$, $0 \leq s, t < k$.*

We lift this argument from a single matrix to a weighted sum of matrices.

Lemma 3. *Given $\mathfrak{A} = \{(w_1, A_1), \ldots, (w_m, A_m)\}$, there exists a linear recurrence $\langle a^\star \rangle$, such that $\sum_{i=1}^m w_i A_i^n$ is eventually non-negative iff $\langle a^\star \rangle$ is ultimately non-negative.*

Proof. For each matrix A_i in $\mathfrak{A}$, let $\langle v^i \rangle$ be the interleaved LRS as constructed in Lemma 2. Let $w_i \langle v^i \rangle$ denote the scaled LRS whose n^{th} entry is $w_i v_n^i$ for all $n \geq 0$. The LRS $\langle a^\star \rangle$ is obtained by adding the scaled LRSes $w_1 \langle v^1 \rangle, w_2 \langle v^2 \rangle, \ldots w_m \langle v^m \rangle$. Clearly, $a_n^\star$ is non-negative iff $\sum_{i=1}^m w_i v_n^i$ is non-negative. From the definition of v^i (see Lemma 2), we also know that for all $n \geq 0$, $v_n^i = A_i^{r+1}[s + 1, t + 1]$, where $r = \lfloor n/k^2 \rfloor$, $s = \lfloor (n \mod k^2)/k \rfloor$ and $t = n \mod k$. Therefore, $a_n^\star$ is non-negative iff $\sum_{i=1}^m w_i A_i^{r+1}[s + 1, t + 1]$ is non-negative. It follows that $\langle a^\star \rangle$ is ultimately non-negative iff $\sum_{i=1}^m w_i A_i^n$ is eventually non-negative. □

From Lemma 3, we can conclude the main result of this section, i.e., proof of Theorem 3. The following corollary can be shown *mutatis mutandis*.

Corollary 1. $\mathsf{EP_{SoM}}$ *reduces to* $\mathsf{UP_{LRS}}$ *in polynomial time.*

We note that it is also possible to argue about the eventual non-negativity (positivity) of only certain indices of the matrix using a similar argument as above. By interleaving only the LRS's corresponding to certain indices of the matrices in $\mathfrak{A}$, we can show this problem's equivalence with $\mathsf{UNN_{LRS}}$ ($\mathsf{UP_{LRS}}$).

5 Decision Procedures for Special Cases

Since there are no known algorithms for solving $\mathsf{UNN_{LRS}}$ in general, the results of the previous section present a bleak picture for deciding $\mathsf{ENN_{SoM}}$ and $\mathsf{EP_{SoM}}$. We now show that these problems can be solved in some important special cases.

5.1 Simple Matrices and Matrices with Real Algebraic Eigenvalues

Our first positive result follows from known results for special classes of LRSes.

Theorem 4. $\mathsf{ENN_{SoM}}$ *and* $\mathsf{EP_{SoM}}$ *are decidable for* $\mathfrak{A} = \{(w_1, A_1), \ldots (w_m, A_m)\}$ *if one of the following conditions holds for all* $i \in \{1, \ldots m\}$.

1. *All A_i are simple. In this case,* $\mathsf{ENN_{SoM}}$ *and* $\mathsf{EP_{SoM}}$ *are in* PSPACE. *Additionally, if the rank k of all A_i is fixed,* $\mathsf{ENN_{SoM}}$ *and* $\mathsf{EP_{SoM}}$ *are in* PTIME.
2. *All eigenvalues of A_i are roots of real algebraic numbers. In this case,* $\mathsf{ENN_{SoM}}$ *and* $\mathsf{EP_{SoM}}$ *are in* $\mathsf{coNP^{PosSLP}}$ *(a complexity class in the Counting Hierarchy, contained in* PSPACE*).*

Proof. Suppose each $A_i \in \mathbb{Q}^{k \times k}$, and let $\lambda_{i,1}, \ldots \lambda_{i,k}$ be the (possibly repeated) eigenvalues of A_i. The characteristic polynomial of A_i is $p_{A_i}(x) = \prod_{j=1}^k (x - \lambda_{i,j})$. Denote the LRS obtained from A_i by LRS interleaving as in Lemma 2 as $\langle a^i \rangle$. By Lemma 2, we have (i) $a^i_{rk^2+sk+t} = A_i^{r+1}[s + 1, t + 1]$ for all $r \in \mathbb{N}$ and $0 \leq s, t < k$, and (ii) $p_{\langle a^i \rangle}(x) = \prod_{j=1}^k \left(x^{k^2} - \lambda_{i,j}\right)$. We now define the scaled LRS $\{\langle b^i \rangle$, where $\mid b^i_n = w_i\ a^i_n$ for all $n \in \mathbb{N}$. Since scaling does not change the characteristic polynomial of an LRS (refer [3] for a simple proof), we have $p_{\langle b^i \rangle}(x) = \prod_{j=1}^{k} \left(x^{k^2} - \lambda_{i,j}\right)$. Once the LRSes $\langle b^1 \rangle, \ldots \langle b^m \rangle$ are obtained as above, we sum them to obtain the LRS $\langle b^\star \rangle$. Thus, for all $n \in \mathbb{N}$, we have $b^\star_n = \sum_{i=1}^{m} b^i_n = \sum_{i=1}^{m} w_i\ a^i_n = \sum_{i=1}^{m} w_i\ A_i^r[s,t]$, where $n = rk^2 + sk + t$, $r \in \mathbb{N}$ and $0 \le s, t < k$. Hence, $\mathsf{ENN}_{\mathsf{SoM}}$ (resp. $\mathsf{EP}_{\mathsf{SoM}}$) for $\{(w_1, A_1), \ldots (w_m, A_m)\}$ polynomially reduces to $\mathsf{UNN}_{\mathsf{LRS}}$ (resp. $\mathsf{UP}_{\mathsf{LRS}}$) for $\langle b^\star \rangle$.

By [14], we know that the characteristic polynomial $p_{\langle b^\star \rangle}(x)$ is the LCM of the characteristic polynomials $p_{\langle b^i \rangle}(x)$ for $1 \le i \le m$. If A_i are simple, there are no repeated roots of $p_{\langle b^i \rangle}(x)$. If this holds for all $i \in \{1, \ldots m\}$, there are no repeated roots of the LCM of $p_{\langle b^1 \rangle}(x), \ldots p_{\langle b^m \rangle}(x)$ as well. Hence, $p_{\langle b^\star \rangle}(x)$ has no repeated roots. Similarly, if all eigenvalues of A_i are roots of real algebraic numbers, so are all roots of $p_{\langle b^i \rangle}(x)$. It follows that all roots of the LCM of $p_{\langle b^1 \rangle}(x), \ldots p_{\langle b^m \rangle}(x)$, i.e. $p_{\langle b^\star \rangle}(x)$, are also roots of real algebraic numbers.

The theorem now follows from the following two known results about LRS.

1. $\mathsf{UNN}_{\mathsf{LRS}}$ (resp. $\mathsf{UP}_{\mathsf{LRS}}$) for simple LRS is in PSPACE. Furthermore, if the LRS is of bounded order, $\mathsf{UNN}_{\mathsf{LRS}}$ (resp. $\mathsf{UP}_{\mathsf{LRS}}$) is in PTIME [31].
2. $\mathsf{UNN}_{\mathsf{LRS}}$ (resp. $\mathsf{UP}_{\mathsf{LRS}}$) for LRS in which all roots of characteristic polynomial are roots of real algebraic numbers is in $\mathsf{coNP}^{\mathsf{PosSLP}}$ [2]. □

Remark: The technique used in [31] to decide $\mathsf{UNN}_{\mathsf{LRS}}$ (resp. $\mathsf{UP}_{\mathsf{LRS}}$) for simple rational LRS also works for simple LRS with real algebraic coefficients and initial values. This allows us to generalize Theorem 4(1) to the case where all A_i's and w_i's are real algebraic matrices and weights respectively.

5.2 Diagonalizable Matrices

We now ask if $\mathsf{ENN}_{\mathsf{SoM}}$ and $\mathsf{EP}_{\mathsf{SoM}}$ can be decided if each matrix A_i is diagonalizable. Since diagonalizable matrices strictly generalize simple matrices, Theorem 4(1) cannot answer this question directly, unless one perhaps looks under the hood of the (highly non-trivial) proof of decidability of non-negativity/positivity of simple LRSes. The main contribution of this section is a reduction that allows us to decide $\mathsf{ENN}_{\mathsf{SoM}}$ and $\mathsf{EP}_{\mathsf{SoM}}$ for diagonalizable matrices using a black-box decision procedure (i.e. without knowing operational details of the procedure or details of its proof of correctness) for the corresponding problem for simple real-algebraic matrices.

Before we proceed further, let us consider an example of a non-simple matrix (i.e. one with repeated eigenvalues) that is diagonalizable.

$$A = \begin{bmatrix} 5 & 12 & -6 \\ -3 & -10 & 6 \\ -3 & -12 & 8 \end{bmatrix}$$

Fig. 2. Diagonalizable matrix

Specifically, matrix A in Fig. 2 has eigenvalues $2, 2$ and -1, and can be written as SDS^{-1}, where D is the 3×3 diagonal matrix with $D[1,1] = D[2,2] = 2$ and $D[3,3] = -1$, and S is the 3×3 matrix with columns $(-4,1,0)^{\mathsf{T}}$, $(2,0,1)^{\mathsf{T}}$ and $(-1,1,1)^{\mathsf{T}}$.

Interestingly, the reduction technique we develop applies to properties much more general than $\mathsf{ENN_{SoM}}$ and $\mathsf{EP_{SoM}}$. Formally, given a sequence of matrices B_n defined by $\sum_{i=1}^{m} w_i A_i^n$, we say that a property $\mathcal{P}$ of the sequence is *positive scaling invariant* if it stays unchanged even if we scale all A_is by the same positive real. Examples of such properties include $\mathsf{ENN_{SoM}}$, $\mathsf{EP_{SoM}}$, non-negativity and positivity of B_n (i.e. is $B_n[i,j] \geq 0$ or < 0, as the case may be, for all $n \geq 1$ and for all $1 \leq i, j \leq k$), existence of zero (i.e. is B_n equal to the all 0-matrix for some $n \geq 1$), existence of a zero element (i.e. is $B_n[i,j] = 0$ for some $n \geq 1$ and some $i, j \in \{1, \ldots k\}$), variants of the r-non-negativity (resp. r-positivity and r-zero) problem, i.e. does there exist at least/exactly/at most r non-negative (resp. positive/zero) elements in B_n for all $n \geq 1$, for a given $r \in [1, k]$) etc. The main result of this section is a reduction for deciding such properties, formalized in the following theorem.

Theorem 5. *The decision problem for every positive scaling invariant property on rational diagonalizable matrices effectively reduces to the decision problem for the property on real algebraic simple matrices.*

While we defer the proof of this theorem to later in the section, an immediate consequence of Theorem 5 and Theorem 4(1) (read with the note at the end of Sect. 5.1) is the following result.

Corollary 2. $\mathsf{ENN_{SoM}}$ *and* $\mathsf{EP_{SoM}}$ *are decidable for* $\mathfrak{A} = \{(w_1, A_1), \ldots (w_m, A_m)\}$ *if all* A_i*s are rational diagonalizable matrices and all* w_i*s are rational.*

It is important to note that Theorem 5 yields a decision procedure for checking any positive scaling invariant property of diagonalizable matrices from a corresponding decision procedure for real algebraic simple matrices *without making any assumptions* about the inner working of the latter decision procedure. Given *any* black-box decision procedure for checking *any* positive scaling property for a set of weighted simple matrices, our reduction tells us how a corresponding decision procedure for checking the same property for a set of weighted diagonalizable matrices can be constructed. Interestingly, since diagonalizable matrices have an exponential form solution with constant coefficients for exponential terms, we can use an algorithm that exploits this specific property of the exponential form (like Ouaknine and Worrell's algorithm [31], originally proposed for checking ultimate positivity of simple LRS) to deal with diagonalizable matrices. However, our reduction technique is neither specific to this algorithm nor does it rely on any special property the exponential form of the solution.

The proof of Theorem 5 crucially relies on the notion of perturbation of diagonalizable matrices, which we introduce first. Let A be a $k \times k$ real diagonalizable matrix. Then, there exists an invertible $k \times k$ matrix S and a diagonal $k \times k$ matrix D such that $A = SDS^{-1}$, where S and D may have complex entries. It follows from basic linear algebra that for every $i \in \{1, \ldots k\}$, $D[i,i]$ is an eigenvalue of A and if α is an eigenvalue of A with algebraic multiplicity ρ, then α appears exactly ρ times along the diagonal of D. Furthermore, for every $i \in \{1, \ldots k\}$, the ith column of S (resp. i^{th} row of S^{-1}) is an eigenvector of A (resp. of A^{T}) corresponding to the eigenvalue $D[i,i]$, and the columns of S

(resp. rows of S^{-1}) form a basis of the vector space $\mathbb{C}^k$. Let $\alpha_1, \ldots \alpha_m$ be the eigenvalues of A with algebraic multiplicities $\rho_1, \ldots \rho_m$ respectively. Wlog, we assume that $\rho_1 \geq \ldots \geq \rho_m$ and the diagonal of D is partitioned into segments as follows: the first ρ_1 entries along the diagonal are α_1, the next ρ_2 entries are α_2, and so on. We refer to these segments as the α_1-segment, α_2-segment and so on, of diagonal of D. Formally, if κ_i denotes $\sum_{j=1}^{i-1} \rho_j$, the α_i-segment of diagonal of D consists of entries $D[\kappa_i+1, \kappa_i+1], \ldots D[\kappa_i+\rho_i, \kappa_i+\rho_i]$, all of which are α_i.

Since A is a real matrix, its characteristic polynomial has all real coefficients and for every eigenvalue α of A (and hence of A^T), its complex conjugate, denoted $\overline{\alpha}$, is also an eigenvalue of A (and hence of A^T) with the same algebraic multiplicity. This allows us to define a bijection h_D from $\{1, \ldots, k\}$ to $\{1, \ldots k\}$ as follows. If $D[i,i]$ is real, then $h_D(i) = i$. Otherwise, let $D[i,i] = \alpha \in \mathbb{C}$ and let $D[i,i]$ be the lth element in the α-segment of the diagonal of D. Then $h_D(i) = j$, where $D[j,j]$ is the lth element in the $\overline{\alpha}$-segment of the diagonal of D. The matrix A being real also implies that for every real eigenvalue α of A (resp. of A^T), there exists a basis of *real eigenvectors* of the corresponding eigenspace. Additionally, for every non-real eigenvalue α and for every set of eigenvectors of A (resp. of A^T) that forms a basis of the eigenspace corresponding to α, the component-wise complex conjugates of these basis vectors serve as eigenvectors of A (resp. of A^T) and form a basis of the eigenspace corresponding to $\overline{\alpha}$.

Using the above notation, we choose matrix S^{-1} (and hence S) such that $A = SDS^{-1}$ as follows. Suppose α is an eigenvalue of A (and hence of A^T) with algebraic multiplicity ρ. Let $\{i+1, \ldots i+\rho\}$ be the set of indices j for which $D[j,j] = \alpha$. If α is real (resp. complex), the $i+1^{st}, \ldots i+\rho^{th}$ rows of S^{-1} are chosen to be real (resp. complex) eigenvectors of A^T that form a basis of the eigenspace corresponding to α. Moreover, if α is complex, the $h_D(i+s)^{th}$ row of S^{-1} is chosen to be the component-wise complex conjugate of the $i+s^{th}$ row of S^{-1}, for all $s \in \{1, \ldots \rho\}$.

Definition 3. *Let $A = SDS^{-1}$ be a $k \times k$ real diagonalizable matrix. We say that $\mathcal{E} = (\varepsilon_1, \ldots \varepsilon_k) \in \mathbb{R}^k$ is a* perturbation *w.r.t. D if $\varepsilon_i \neq 0$ and $\varepsilon_i = \varepsilon_{h_D(i)}$ for all $i \in \{1, \ldots k\}$. Further, the $\mathcal{E}$-*perturbed variant *of A is the matrix $A' = SD'S^{-1}$, where D' is the $k \times k$ diagonal matrix with $D'[i,i] = \varepsilon_i D[i,i]$ for all $i \in \{1, \ldots k\}$.*

In the following, we omit "w.r.t. D" and simply say "$\mathcal{E}$ is a perturbation", when D is clear from the context. Clearly, A' as defined above is a diagonalizable matrix and its eigenvalues are given by the diagonal elements of D'.

Recall that the diagonal of D is partitioned into α_i-segments, where each α_i is an eigenvalue of $A = SDS^{-1}$ with algebraic multiplicity ρ_i. We now use a similar idea to segment a perturbation $\mathcal{E}$ w.r.t. D. Specifically, the first ρ_1 elements of $\mathcal{E}$ constitute the α_1-segment of $\mathcal{E}$, the next ρ_2 elements of $\mathcal{E}$ constitute the α_2-segment of $\mathcal{E}$ and so on.

Definition 4. *A perturbation $\mathcal{E} = (\varepsilon_1, \ldots \varepsilon_k)$ is said to be* segmented *if the jth element (whenever present) in every segment of $\mathcal{E}$ has the same value, for all $1 \leq j \leq \rho_1$. Formally, if $i = \sum_{s=1}^{l-1} \rho_s + j$ and $1 \leq j \leq \rho_l \leq \rho_1$, then $\varepsilon_i = \varepsilon_j$.*

Clearly, the first ρ_1 elements of a segmented perturbation $\mathcal{E}$ define the whole of $\mathcal{E}$. As an example, suppose $(\alpha_1, \alpha_1, \alpha_1, \alpha_2, \alpha_2, \overline{\alpha_2}, \overline{\alpha_2}, \alpha_3)$ is the diagonal of D, where $\alpha_1, \alpha_2, \overline{\alpha_2}$ and α_3 are distinct eigenvalues of A. There are four segments of the diagonal of D (and of $\mathcal{E}$) of lengths $3, 2, 2$ and 1 respectively.

Example segmented perturbations in this case are $(\varepsilon_1, \varepsilon_2, \varepsilon_3, \varepsilon_1, \varepsilon_2, \varepsilon_1, \varepsilon_2, \varepsilon_1)$ and $(\varepsilon_3, \varepsilon_1, \varepsilon_2, \varepsilon_3, \varepsilon_1, \varepsilon_3, \varepsilon_1, \varepsilon_3)$. If $\varepsilon_1 \neq \varepsilon_2$ or $\varepsilon_2 \neq \varepsilon_3$, a perturbation that is *not segmented* is $\widetilde{\mathcal{E}} = (\varepsilon_1, \varepsilon_2, \varepsilon_3, \varepsilon_2, \varepsilon_3, \varepsilon_2, \varepsilon_3, \varepsilon_1)$.

Definition 5. *Given a segmented perturbation $\mathcal{E} = (\varepsilon_1, \ldots \varepsilon_k)$ w.r.t. D, a* rotation *of $\mathcal{E}$, denoted $\tau_D(\mathcal{E})$, is the segmented perturbation $\mathcal{E}' = (\varepsilon'_1, \ldots \varepsilon'_k)$ in which $\varepsilon'_{(i \bmod \rho_1)+1} = \varepsilon_i$ for $i \in \{1, \ldots \rho_1\}$, and all other ε'_is are as in Definition 4.*

Continuing with our example, if $\mathcal{E} = (\varepsilon_1, \varepsilon_2, \varepsilon_3, \varepsilon_1, \varepsilon_2, \varepsilon_1, \varepsilon_2, \varepsilon_1)$, then $\tau_D(\mathcal{E}) = (\varepsilon_3, \varepsilon_1, \varepsilon_2, \varepsilon_3, \varepsilon_1, \varepsilon_3, \varepsilon_1, \varepsilon_3)$, $\tau_D^2(\mathcal{E}) = (\varepsilon_2, \varepsilon_3, \varepsilon_1, \varepsilon_2, \varepsilon_3, \varepsilon_2, \varepsilon_3, \varepsilon_2)$ and $\tau_D^3(\mathcal{E}) = \mathcal{E}$.

Lemma 4. *Let $A = SDS^{-1}$ be a $k \times k$ real diagonalizable matrix with eigenvalues α_i of algebraic multiplicity ρ_i. Let $\mathcal{E} = (\varepsilon_1, \ldots \varepsilon_k)$ be a segmented perturbation w.r.t. D such that all ε_js have the same sign, and let A_u denote the $\tau_D^u(\mathcal{E})$-perturbed variant of A for $0 \leq u < \rho_1$, where $\tau^0(\mathcal{E}) = \mathcal{E}$. Then $A^n = \frac{1}{\left(\sum_{j=1}^{\rho_1} \varepsilon_j^n\right)} \sum_{u=0}^{\rho_1 - 1} A_u^n$, for all $n \geq 1$.*

Proof. Let $\mathcal{E}_u$ denote $\tau_D^u(\mathcal{E})$ for $0 \leq u < \rho_1$, and let $\mathcal{E}_u[i]$ denote the ith element of $\mathcal{E}_u$ for $1 \leq i \leq k$. It follows from Definitions 4 and 5 that for each $i, j \in \{1, \ldots \rho_1\}$, there is a unique $u \in \{0, \ldots \rho_1 - 1\}$ such that $\mathcal{E}_u[i] = \varepsilon_j$. Specifically, $u = i - j$ if $i \geq j$, and $u = (\rho_1 - j) + i$ if $i < j$. Furthermore, Definition 4 ensures that the above property holds not only for $i \in \{1, \ldots \rho_1\}$, but for all $i \in \{1, \ldots k\}$.

Let D_u denote the diagonal matrix with $D_u[i,i] = \mathcal{E}_u[i] D[i,i]$ for $0 \leq i < \rho_1$. Then D_u^n is the diagonal matrix with $D_u^n[i,i] = \left(\mathcal{E}_u[i] D[i,i]\right)^n$ for all $n \geq 1$. It follows from the definition of A_u that $A_u^n = S\ D_u^n\ S^{-1}$ for $0 \leq u < \rho$ and $n \geq 1$. Therefore, $\sum_{u=0}^{\rho_1-1} A_u^n = S\ \left(\sum_{u=0}^{\rho_1-1} D_u^n\right)\ S^{-1}$. Now, $\sum_{u=0}^{\rho_1-1} D_u^n$ is a diagonal matrix whose i^{th} element along the diagonal is $\sum_{u=0}^{\rho_1-1} \left(\mathcal{E}_u[i] D[i,i]\right)^n$ $= \left(\sum_{u=0}^{\rho_1-1} \mathcal{E}_u^n[i]\right) D^n[i,i]$. By virtue of the property mentioned in the previous paragraph, $\sum_{u=0}^{\rho_1-1} \mathcal{E}_u^n[i] = \sum_{j=1}^{\rho_1} \varepsilon_j^n$ for $1 \leq i \leq k$. Therefore, $\sum_{u=0}^{\rho_1-1} D_u^n = \left(\sum_{j=1}^{\rho_1} \varepsilon_j^n\right) D^n$, and hence, $\sum_{u=0}^{\rho_1-1} A_u^n = \left(\sum_{j=1}^{\rho_1} \varepsilon_j^n\right) S\ D^n\ S^{-1} = \left(\sum_{j=1}^{\rho_1} \varepsilon_j^n\right) A^n$. Since all ε_js have the same sign and are non-zero, $\left(\sum_{j=1}^{\rho_1} \varepsilon_j^n\right)$ is non-zero for all $n \geq 1$. It follows that $A^n = \frac{1}{\left(\sum_{j=1}^{\rho_1} \varepsilon_j^n\right)} \sum_{u=0}^{\rho_1-1} A_u^n$. □

We are now in a position to present the proof of the main result of this section, i.e. of Theorem 5. Our proof uses a variation of the idea used in the proof of Lemma 4 above.

Proof of Theorem 5. Consider a set $\{(w_1, A_1), \ldots (w_i, A_i)\}$ of (weight, matrix) pairs, where each matrix A_i is in $\mathbb{Q}^{k \times k}$ and each $w_i \in \mathbb{Q}$. Suppose further that each $A_i = S_i D_i S_i^{-1}$, where D_i is a diagonal matrix with segments along the diagonal arranged in descending order of algebraic multiplicities of the corresponding eigenvalues. Let ν_i be the number of distinct eigenvalues of A_i, and let these eigenvalues be $\alpha_{i,1}, \dots \alpha_{i,\nu_i}$. Let μ_i be the largest algebraic multiplicity among those of all eigenvalues of A_i, and let $\mu = lcm(\mu_1, \dots \mu_m)$. We now choose *positive* rationals $\varepsilon_1, \dots \varepsilon_\mu$ such that (i) all ε_js are distinct, and (ii) for every $i \in \{1, \dots m\}$, for every distinct $j, l \in \{1, \dots \nu_i\}$ and for every distinct $p, q \in \{1, \dots \mu\}$, we have $\frac{\varepsilon_p}{\varepsilon_q} \neq |\frac{\alpha_{i,j}}{\alpha i,l}|$. Since $\mathbb{Q}$ is a dense set, such a choice of $\varepsilon_1, \dots \varepsilon_\mu$ can always be made once all $|\frac{\alpha_{i,j}}{\alpha_{i,l}}|$s are known, even if within finite precision bounds.

For $1 \leq i \leq m$, let η_i denote μ/μ_i. We now define η_i distinct and segmented perturbations w.r.t. D_i as follows, and denote these as $\mathcal{E}_{i,1}, \dots \mathcal{E}_{i,\eta_i}$. For $1 \leq j \leq \eta_i$, the first μ_i elements (i.e. the first segment) of $\mathcal{E}_{i,j}$ are $\varepsilon_{(j-1)\mu_i+1}, \dots \varepsilon_{j\mu_i}$ (as chosen in the previous paragraph), and all other elements of $\mathcal{E}_{i,j}$ are defined as in Definition 4. For each $\mathcal{E}_{i,j}$ thus obtained, we also consider its rotations $\tau^u_{D_i}(\mathcal{E}_{i,j})$ for $0 \leq u < \mu_i$. For $1 \leq j \leq \eta_i$ and $0 \leq u < \mu_i$, let $A_{i,j,u} = S_i\ D_{i,j,u}\ S_i^{-1}$ denote the $\tau^u_{D_i}(\mathcal{E}_{i,j})$-perturbed variant of A_i. It follows from Definition 3 that if we consider the set of diagonal matrices $\{D_{i,j,u} \mid 1 \leq j \leq \eta_i, 0 \leq u < \mu_i\}$, then for every $p \in \{1, \dots k\}$ and for every $q \in \{1, \dots \mu\}$, there is a unique u and j such that $D_{i,j,u}[p,p] = \varepsilon_q$. Specifically, $j = \lfloor q/\mu_i \rfloor$. To find u, let $\mathcal{E}_{i,j}[p]$ be the $\widehat{p}^{th}$ element in a segment of $\mathcal{E}_{i,j}$, where $1 \leq \widehat{p} \leq \mu_i$, and let $\widehat{q}$ be $q \mod \mu_i$. Then, $u = (\widehat{p} - \widehat{q})$ if $\widehat{p} \geq \widehat{q}$ and $u = (\mu_i - \widehat{q}) + \widehat{p}$ otherwise. By our choice of ε_ts, we also know that for all $i \in \{1, \dots m\}$, for all $j, l \in \{1, \dots \nu_i\}$ and for all $p, q \in \{1, \dots \mu\}$, we have $\varepsilon_p \alpha_{i,l} \neq \varepsilon_q \alpha_{i,j}$ unless $p = q$ *and* $j = l$. This ensures that all $D_{i,j,u}$ matrices, and hence all $A_{i,j,u}$s matrices, are simple, i.e. have distinct eigenvalues.

Using the reasoning in Lemma 4, we can now show that $A_i^n = \frac{1}{\left(\sum_{j=1}^{\mu} \varepsilon_j^n\right)} \times \left(\sum_{j=1}^{\eta_i} \sum_{u=0}^{\mu_i-1} A_{i,j,u}^n\right)$ and so, $\sum_{i=1}^m w_i A_i^n = \frac{1}{\left(\sum_{j=1}^{\mu} \varepsilon_j^n\right)} \times \left(\sum_{i=1}^m \sum_{j=1}^{\eta_i} \sum_{u=0}^{\mu_i-1} w_i A_{i,j,u}^n\right)$. Since all ε_js are positive reals, $\sum_{j=1}^{\mu} \varepsilon_j^n$ is a positive real for all $n \geq 1$.

Hence, for each $p, q \in \{1, \dots k\}$, $\sum_{i=1}^m w_i A_i^n[p,q]$ is > 0, < 0 or $= 0$ if and only if $\left(\sum_{i=1}^m \sum_{j=1}^{\eta_i} \sum_{u=0}^{\mu_i-1} w_i A_{i,j,u}^n[p,q]\right)$ is > 0, < 0 or $= 0$, respectively. The only remaining helper result that is now needed to complete the proof of the theorem is that each $A_{i,j,u}$ is a real algebraic matrix. This is shown in Lemma 5, presented at the end of this section to minimally disturb the flow of arguments. □

The reduction in proof of Theorem 5 can be easily encoded as an algorithm, as shown in Algorithm 1. Further, in addition to Corollary 2, there are other consequences of our reduction. One such result (with proof in [3]) is below.

Corollary 3. *Given* $\mathfrak{A} = \{(w_1, A_1), \dots (w_m, A_m)\}$, *where each* $w_i \in \mathbb{Q}$ *and* $A_i \in \mathbb{Q}^{k\times k}$ *is diagonalizable, and a real value* $\varepsilon > 0$, *there exists* $\mathfrak{B} = \{(v_1, B_1), \dots (v_M, B_M)\}$, *where each* $v_i \in \mathbb{Q}$ *and each* $B_i \in \mathbb{RA}^{k\times k}$ *is simple, such that* $\left|\sum_{i=0}^m w_i A_i^n[p,q] - \sum_{j=0}^M v_j B_j^n[p,q]\right| < \varepsilon^n$ *for all* $p, q \in \{1, \dots k\}$ *and all* $n \geq 1$.

We end this section with the promised helper result used at the end of the proof of Theorem 5.

Algorithm 1. Reduction procedure for diagonalizable matrices

Input: $\mathfrak{A} = \{(w_i, A_i) \; : \; 1 \le i \le m,\; w_i \in \mathbb{Q},\; A_i \in \mathbb{Q}^{k\times k}$ and diagonalizable$\}$
Output: $\mathfrak{B} = \{(v_i, B_i) : \; 1 \le i \le t,\; v_i \in \mathbb{Q},\; B_i \in \mathbb{RA}^{k\times k}$ are simple$\}$
s.t. $\left(\sum_{i=1}^{m} w_i A_i^n\right) = f(n) \left(\sum_{i=1}^{t} v_i B_i^n\right)$, where $f(n) > 0$ for all $n \ge 0$?

$P \leftarrow \{1\}$; ▷ Initialize set of forbidden ratios of various ε_js
for i in 1 through m **do** ▷ For each matrix A_i
  $R_i \leftarrow \{(\alpha_{i,j}, \rho_{i,j}) \; : \; \alpha_{i,j}$ is eigenvalue of A_i with algebraic multiplicity $\rho_{i,j}\}$;
  $D_i \leftarrow$ Diagonal matrix of $\alpha_{i,j}$-segments ordered in decreasing order of $\rho_{i,j}$;
  $S_i \leftarrow$ Matrix of linearly independent eigenvectors of A_i s.t. $A_i = S_i D_i S_i^{-1}$;
  $P \leftarrow P \cup \left\{ |\alpha_{i,j}/\alpha_{i,l}| \; : \; \alpha_{i,j}, \alpha_{i,l} \text{ are eigenvalues in } R_i \right\}$; $\mu_i \leftarrow \max_j \rho_{i,j}$
$\mu = lcm(\mu_1, \ldots \mu_m)$; ▷ Count of ε_js needed
for j in 1 through μ **do** ▷ Generate all required ε_js
  Choose $\varepsilon_j \in \mathbb{Q}$ s.t. $\varepsilon_j > 0$ and $\varepsilon_j \notin \{\pi\varepsilon_p \; : \; 1 \le p < j,\; \pi \in P\}$;
$\mathfrak{B} \leftarrow \emptyset$; ▷ Initialize set of (weight, simple matrix) pairs
for i in 1 through m **do** ▷ For each matrix A_i
  $\nu_i \leftarrow \mu/\mu_i$; ▷ Count of segmented perturbations to be rotated for A_i
  for j in 0 through $\nu_i - 1$ **do** ▷ For each segmented perturbation
    $\mathcal{E}_{i,j} \leftarrow$ Seg. perturbn. w.r.t. D_i with first μ_i elements being $\varepsilon_{j\mu_i+1}, \ldots \varepsilon_{(j+1)\mu_i}$;
    for u in 0 through $\mu_i - 1$ **do** ▷ For each rotation of $\mathcal{E}_{i,j}$
      $A_{i,j,u} \leftarrow \tau_{D_i}^{u}(\mathcal{E}_{i,j})$-perturbed variant of A;
      $\mathfrak{B} \leftarrow \mathfrak{B} \cup \{(w_i, A_{i,j,u})\}$; ▷ Update $\mathfrak{A}'$
return $\mathfrak{B}$;

Lemma 5. *For every real (resp. real algebraic) diagonalizable matrix* $A = SDS^{-1}$ *and perturbation* $\mathcal{E} \in \mathbb{R}^k$ *(resp.* $\mathbb{RA}^k$*), the* $\mathcal{E}$*-perturbed variant of* A *is a real (resp. real algebraic) diagonalizable matrix.*

Proof. We first consider the case of $A \in \mathbb{R}^{k\times k}$ and $\mathcal{E} \in \mathbb{R}^k$. Given a perturbation $\mathcal{E}$ w.r.t. D, we first define k *simple* perturbations $\mathcal{E}_i$ $(1 \le i \le k)$ w.r.t. D as follows: $\mathcal{E}_i$ has all its components set to 1, except for the ith component, which is set to ε_i. Furthermore, if $D[i,i]$ is not real, then the $h_D(i)^{th}$ component of $\mathcal{E}_i$ is also set to ε_i. It is easy to see from Definition 3 that each $\mathcal{E}_i$ is a perturbation w.r.t. D. Moreover, if $j = h_D(i)$, then $\mathcal{E}_j = \mathcal{E}_i$.

Let $\widehat{\mathcal{E}} = \{\mathcal{E}_{i_1}, \ldots \mathcal{E}_{i_u}\}$ be the set of all *unique* perturbations w.r.t D among $\mathcal{E}_1, \ldots \mathcal{E}_k$. It follows once again from Definition 3 that the $\mathcal{E}$-perturbed variant of A can be obtained by a sequence of $\mathcal{E}_{i_j}$-perturbations, where $\mathcal{E}_{i_j} \in \widehat{\mathcal{E}}$. Specifically, let $A_{0,\widehat{\mathcal{E}}} = A$ and $A_{v,\widehat{\mathcal{E}}}$ be the $\mathcal{E}_{i_v}$-perturbed variant of $A_{v-1,\widehat{\mathcal{E}}}$ for all $v \in \{1, \ldots u\}$. Then, the $\mathcal{E}$-perturbed variant of A is identical to $A_{u,\widehat{\mathcal{E}}}$. This shows that it suffices to prove the lemma only for simple perturbations $\mathcal{E}_i$, as defined above. We focus on this special case below.

Let $A' = SD'S^{-1}$ be the $\mathcal{E}_i$-perturbed variant of A, and let $D[i,i] = \alpha$. For every $p \in \{1, \ldots k\}$, let $\mathbf{e_p}$ denote the p-dimensional unit vector whose p^{th} component is 1. Then, $A'\mathbf{e_p}$ gives the pth column of A'. We prove the first part of the lemma by showing that $A'\,\mathbf{e_p} = (S\,D'S^{-1})\,\mathbf{e_p} \in \mathbb{R}^{k\times 1}$ for all $p \in \{1, \ldots k\}$.

Let $\mathbf{T}$ denote $D' \; S^{-1} \; \mathbf{e_p}$. Then $\mathbf{T}$ is a column vector with $\mathbf{T}[r] = D'[r,r] \; S^{-1}[r,p]$ for all $r \in \{1,\ldots k\}$. Let $\mathbf{U}$ denote $S\mathbf{T}$. By definition, $\mathbf{U}$ is the pth column of the matrix A'. To compute $\mathbf{U}$, recall that the rows of S^{-1} form a basis of $\mathbb{C}^k$. Therefore, for every $q \in \{1,\ldots k\}$, $S^{-1} \; \mathbf{e_q}$ can be viewed as transforming the basis of the unit vector $\mathbf{e_q}$ to that given by the rows of S^{-1} (modulo possible scaling by real scalars denoting the lengths of the row vectors of S^{-1}). Similarly, computation of $\mathbf{U} = S\mathbf{T}$ can be viewed as applying the inverse basis transformation to $\mathbf{T}$. It follows that the components of $\mathbf{U}$ can be obtained by computing the dot product of $\mathbf{T}$ and the transformed unit vector $S^{-1} \; \mathbf{e_q}$, for each $q \in \{1,\ldots k\}$. In other words, $\mathbf{U}[q] = \mathbf{T} \cdot (S^{-1} \; \mathbf{e_q})$. We show below that each such $\mathbf{U}[q]$ is real.

By definition, $\mathbf{U}[q] = \sum_{r=1}^{k}(\mathbf{T}[r] \; S^{-1}[r,q]) = \sum_{r=1}^{k}(D'[r,r] \; S^{-1}[r,p] \; S^{-1}[r,q])$. We consider two cases below.

- If $D[i,i] = \alpha$ is real, recalling the definition of D', the expression for $\mathbf{U}[q]$ simplifies to $\sum_{r=1}^{k}(D[r,r] \; S^{-1}[r,p] \; S^{-1}[r,q]) + (\varepsilon_i - 1) \; \alpha \; S^{-1}[i,p] \; S^{-1}[i,q]$. Note that $\sum_{r=1}^{k}(D[r,r] \; S^{-1}[r,p] \; S^{-1}[r,q])$ is the qth component of the vector $(SDS^{-1}) \; \mathbf{e_p} = A \; \mathbf{e_p}$. Since A is real, so must be the qth component of $A \; \mathbf{e_p}$. Moreover, since α is real, by our choice of S^{-1}, both $S^{-1}[i,p]$ and $S^{-1}[i,q]$ are real. Since ε_i is also real, it follows that $(\varepsilon_i - 1) \; \alpha \; S^{-1}[i,p] \; S^{-1}[i,q]$ is real. Hence $\mathbf{U}[q]$ is real for all $q \in \{1,\ldots k\}$.
- If $D[i,i] = \alpha$ is not real, from Definition 3, we know that $D'[i,i] = \varepsilon_i \; \alpha$ and $D'[h_D(i), h_D(i)] = \varepsilon_i \; \overline{\alpha}$. The expression for $\mathbf{U}[q]$ then simplifies to $\sum_{r=1}^{k}\left(D[r,r] \; S^{-1}[r,p] \; S^{-1}[r,q]\right) + (\varepsilon_i - 1) \; (\beta + \gamma)$, where $\beta = \alpha \; S^{-1}[i,p] \; S^{-1}[i,q]$ and $\gamma = \overline{\alpha} \; S^{-1}[h_D(i),p] \; S^{-1}[h_D(i),q]$. By our choice of S^{-1}, we know that $S^{-1}[h_D(i),p] = \overline{S^{-1}[i,p]}$ and $S^{-1}[h_D(i),q] = \overline{S^{-1}[i,q]}$. Therefore, $\beta = \overline{\gamma}$ and hence $(\varepsilon_i - 1) \; (\beta + \gamma)$ is real. By a similar argument as in the previous case, it follows that $\mathbf{U}[q]$ is real for all $q \in \{1,\ldots k\}$.

The proof when $A \in \mathbb{RA}^{k\times k}$ and $\mathcal{E} \in \mathbb{Q}^k$ follows from a similar reasoning as above, and from the following facts about real algebraic matrices.

- If A is a real algebraic matrix, then every eigenvalue of A is either a real or complex algebraic number.
- If A is diagonalizable, then for every real (resp. complex) algebraic eigenvalue of A, there exists a set of real (resp. complex) algebraic eigenvectors that form a basis of the corresponding eigenspace. □

6 Conclusion

In this paper, we investigated eventual non-negativity and positivity for matrices and the weighted sum of powers of matrices ($\mathsf{ENN_{SoM}}/\mathsf{EP_{SoM}}$). First, we showed reductions from and to specific problems on linear recurrences, which allowed us give complexity lower and upper bounds. Second, we developed a new and generic perturbation-based reduction technique from simple matrices to diagonalizable matrices, which allowed us to transfer results between these settings.

Most of our results, that we showed in the rational setting, hold even with real-algebraic matrices by adapting the complexity notions and depending on corresponding results for ultimate positivity for linear recurrences and related problems over reals. As future work, we would like to extend our techniques for other problems of interest like the *existence* of a matrix power where all entries are non-negative or zero. Finally, the line of work started here could lead to effective algorithms and applications in varied areas ranging from control theory systems to cyber-physical systems, where eventual properties of matrices play a crucial role.

References

1. Akshay, S., Antonopoulos, T., Ouaknine, J., Worrell, J.: Reachability problems for Markov chains. Inf. Process. Lett. **115**(2), 155–158 (2015)
2. Akshay, S., Balaji, N., Murhekar, A., Varma, R., Vyas, N.: Near optimal complexity bounds for fragments of the Skolem problem. In: Paul, S., Bläser, M. (eds.) 37th International Symposium on Theoretical Aspects of Computer Science, STACS 2020, 10–13 March 2020, Montpellier, France, volume 154 of LIPIcs, pp. 37:1–37:18. Schloss Dagstuhl - Leibniz-Zentrum für Informatik (2020)
3. Akshay, S., Chakraborty, S., Pal, D.: On eventual non-negativity and positivity for the weighted sum of powers of matrices. arXiv preprint arXiv:2205.09190 (2022)
4. Akshay, S., Genest, B., Karelovic, B., Vyas, N.: On regularity of unary probabilistic automata. In: 33rd Symposium on Theoretical Aspects of Computer Science, STACS 2016, 17–20 February 2016, Orléans, France, volume 47 of LIPIcs, pp. 8:1–8:14. Schloss Dagstuhl - Leibniz-Zentrum für Informatik (2016)
5. S. Akshay, Blaise Genest, and Nikhil Vyas. Distribution based objectives for Markov decision processes. In: 33rd Symposium on Logic in Computer Science (LICS 2018), vol. IEEE, pp. 36–45 (2018)
6. Almagor, S., Boker, U., Kupferman, O.: What's decidable about weighted automata? Inf. Comput. **282**, 104651 (2020)
7. Almagor, S., Karimov, T., Kelmendi, E., Ouaknine, J., Worrell, J.: Deciding ω-regular properties on linear recurrence sequences. Proc. ACM Program. Lang. **5**(POPL), 1–24 (2021)
8. Barloy, C., Fijalkow, N., Lhote, N., Mazowiecki, F.: A robust class of linear recurrence sequences. In: Fernández, M., Muscholl, A. (eds.) 28th EACSL Annual Conference on Computer Science Logic, CSL 2020, 13–16 January 2020, Barcelona, Spain, volume 152 of LIPIcs, pp. 9:1–9:16. Schloss Dagstuhl - Leibniz-Zentrum für Informatik (2020)
9. Bell, P.C., Hirvensalo, M., Potapov, I.: Mortality for 2×2 matrices is NP-hard. In: Rovan, B., Sassone, V., Widmayer, P. (eds.) MFCS 2012. LNCS, vol. 7464, pp. 148–159. Springer, Heidelberg (2012). https://doi.org/10.1007/978-3-642-32589-2_16
10. Bell, P.C., Potapov, I., Semukhin, P.: On the mortality problem: from multiplicative matrix equations to linear recurrence sequences and beyond. Inf. Comput. **281**, 104736 (2021)
11. Bell, P.C., Semukhin, P.: Decision questions for probabilistic automata on small alphabets. arXiv preprint arXiv:2105.10293 (2021)
12. Blondel, V.D., Canterini, V.: Undecidable problems for probabilistic automata of fixed dimension. Theory Comput. Syst. **36**(3), 231–245 (2003). https://doi.org/10.1007/s00224-003-1061-2

13. Naqvi, S.C., McDonald, J.J.: Eventually nonnegative matrices are similar to seminonnegative matrices. Linear Algebra Appl. **381**, 245–258 (2004)
14. Everest, G., van der Poorten, A., Shparlinski, I., Ward, T.: Recurrence Sequences. Mathematical Surveys and Monographs, American Mathematical Society, United States (2003)
15. Fijalkow, N., Ouaknine, J., Pouly, A., Sousa-Pinto, J., Worrell, J.: On the decidability of reachability in linear time-invariant systems. In: Ozay, N., Prabhakar, P. (eds.) Proceedings of the 22nd ACM International Conference on Hybrid Systems: Computation and Control, HSCC 2019, Montreal, QC, Canada, 16–18 April 2019, pages 77–86. ACM (2019)
16. Friedland, S.: On an inverse problem for nonnegative and eventually nonnegative matrices. Isr. J. Math. **29**(1), 43–60 (1978). https://doi.org/10.1007/BF02760401
17. Gimbert, H., Oualhadj, Y.: Probabilistic automata on finite words: decidable and undecidable problems. In: Abramsky, S., Gavoille, C., Kirchner, C., Meyer auf der Heide, F., Spirakis, P.G. (eds.) ICALP 2010. LNCS, vol. 6199, pp. 527–538. Springer, . Probabilistic automata on finite words: Decidable and undecidable problems (2010). https://doi.org/10.1007/978-3-642-14162-1_44
18. Halava, V., Harju, T., Hirvensalo, M.: Positivity of second order linear recurrent sequences. Discrete Appl. Math. **154**(3), 447–451 (2006)
19. Halava, V., Harju, T., Hirvensalo, M., Karhumäki, J.: Skolem's Problem-on the Border Between Decidability and Undecidability. Technical report, Citeseer (2005)
20. Karimov, T., et al.: What's decidable about linear loops? Proc. ACM Program. Lang. **6**(POPL), 1–25 (2022)
21. Korthikanti, V.A., Viswanathan, M., Agha, G., Kwon, Y.: Reasoning about MDPs as transformers of probability distributions. In: QEST 2010, Seventh International Conference on the Quantitative Evaluation of Systems, Williamsburg, Virginia, USA, 15–18 September 2010, pp. 199–208. IEEE Computer Society (2010)
22. Lale, S., Azizzadenesheli, K., Hassibi, B., Anandkumar, A.: Logarithmic regret bound in partially observable linear dynamical systems. Adv. Neural Inf. Process. Syst. **33**, 20876–20888 (2020)
23. Lebacque, J.P., Ma, T.Y., Khoshyaran, M.M.: The cross-entropy field for multimodal dynamic assignment. In: Proceedings of Traffic and Granular Flow 2009 (2009)
24. MacCluer, C.R.: The many proofs and applications of Perron's theorem. Siam Rev. **42**(3), 487–498 (2000)
25. Noutsos, D.: On Perron-Frobenius property of matrices having some negative entries. Linear Algebra Appl. **412**(2), 132–153 (2006)
26. Noutsos, D., Tsatsomeros, M.J.: Reachability and holdability of nonnegative states. SIAM J. Matrix Anal. Appl. **30**(2), 700–712 (2008)
27. Ouaknine, J.: Decision problems for linear recurrence sequences. In: Gąsieniec, L., Wolter, F. (eds.) FCT 2013. LNCS, vol. 8070, pp. 2–2. Springer, Heidelberg (2013). https://doi.org/10.1007/978-3-642-40164-0_2
28. Ouaknine, J., Pinto, J.S., Worrell, J.: On termination of integer linear loops. In: Indyk, R. (ed.) Proceedings of the Twenty-Sixth Annual ACM-SIAM Symposium on Discrete Algorithms, SODA 2015, San Diego, CA, USA, 4–6 January 2015, pp. 957–969. SIAM (2015)
29. Ouaknine, J., Worrell, J.: Decision problems for linear recurrence sequences. In: Finkel, A., Leroux, J., Potapov, I. (eds.) RP 2012. LNCS, vol. 7550, pp. 21–28. Springer, Heidelberg (2012). https://doi.org/10.1007/978-3-642-33512-9_3

30. Ouaknine, J., Worrell, J.: Positivity problems for low-order linear recurrence sequences. In: Chekuri, C. (ed.) Proceedings of the Twenty-Fifth Annual ACM-SIAM Symposium on Discrete Algorithms, SODA 2014, Portland, Oregon, USA, 5–7 January 2014, pp. 366–379. SIAM (2014)
31. Ouaknine, J., Worrell, J.: Ultimate positivity is decidable for simple linear recurrence sequences. In: Esparza, J., Fraigniaud, P., Husfeldt, T., Koutsoupias, E. (eds.) ICALP 2014. LNCS, vol. 8573, pp. 330–341. Springer, Heidelberg (2014). https://doi.org/10.1007/978-3-662-43951-7_28
32. Ouaknine, J., Worrell, J.: On linear recurrence sequences and loop termination. ACM Siglog News **2**(2), 4–13 (2015)
33. Pan, V.Y., Chen, Z.Q.: The complexity of the matrix eigenproblem. In: Proceedings of the Thirty-First Annual ACM Symposium on Theory of Computing, STOC 2099, pp. 507–516, New York, NY, USA. Association for Computing Machinery (1999)
34. Rump, S.M.: Perron-Frobenius theory for complex matrices. Linear Algebra Appl. **363**, 251–273 (2003)
35. Akshay, S., Balaji, N., Vyas, N.: Complexity of Restricted Variants of Skolem and Related Problems. In Larsen, K.G., Bodlaender, H.L., Raskin, J.F. (eds.) 42nd International Symposium on Mathematical Foundations of Computer Science (MFCS 2017), volume 83 of Leibniz International Proceedings in Informatics (LIPIcs), pp. 78:1–78:14, Dagstuhl, Germany. Schloss Dagstuhl-Leibniz-Zentrum fuer Informatik (2017)
36. Tiwari, A.: Termination of linear programs. In: Alur, R., Peled, D.A. (eds.) CAV 2004. LNCS, vol. 3114, pp. 70–82. Springer, Heidelberg (2004). https://doi.org/10.1007/978-3-540-27813-9_6
37. Zaslavsky, B.G.: Eventually nonnegative realization of difference control systems. Dyn. Syst. Relat. Top. Adv. Ser. Dynam. Syst. **9**, 573–602 (1991)
38. Zaslavsky, B.G., McDonald, J.J.: Characterization of Jordan canonical forms which are similar to eventually nonnegative matrices with the properties of nonnegative matrices. Linear Algebra Appl. **372**, 253–285 (2003)
39. Zhang, A., et al.: Learning causal state representations of partially observable environments. arXiv preprint arXiv:1906.10437 (2019)

Decision Problems in a Logic for Reasoning About Reconfigurable Distributed Systems

Marius Bozga(✉), Lucas Bueri, and Radu Iosif

Univ. Grenoble Alpes, CNRS, Grenoble INP, VERIMAG, 38000 Saint-Martin-d'Hères, France
marius.bozga@univ-grenoble-alpes.fr

Abstract. We consider a logic used to describe sets of configurations of distributed systems, whose network topologies can be changed at runtime, by reconfiguration programs. The logic uses inductive definitions to describe networks with an unbounded number of components and interactions, written using a multiplicative conjunction, reminiscent of Bunched Implications [37] and Separation Logic [39]. We study the complexity of the satisfiability and entailment problems for the configuration logic under consideration. Additionally, we consider the robustness property of degree boundedness (is every component involved in a bounded number of interactions?), an ingredient for decidability of entailments.

1 Introduction

Distributed systems are increasingly used as critical parts of the infrastructure of our digital society, as in e.g., datacenters, e-banking and social networking. In order to address maintenance (e.g., replacement of faulty and obsolete network nodes by new ones) and data traffic issues (e.g., managing the traffic inside a datacenter [35]), the distributed systems community has recently put massive effort in designing algorithms for *reconfigurable systems*, whose network topologies change at runtime [23]. However, dynamic reconfiguration in the form of software or network upgrades has been recognized as one of the most important sources of cloud service outage [25].

This paper contributes to a logical framework that addresses the timely problems of formal *modeling* and *verification* of reconfigurable distributed systems. The basic building blocks of this framework are (i) a Hoare-style program proof calculus [1] used to write formal proofs of correctness of reconfiguration programs, and (ii) an invariant synthesis method [6] that proves the safety (i.e., absence of reachable error configurations) of the configurations defined by the assertions that annotate a reconfiguration program. These methods are combined to prove that an initially correct distributed system cannot reach an error state, following the execution of a given reconfiguration sequence.

The assertions of the proof calculus are written in a logic that defines infinite sets of configurations, consisting of *components* (i.e., processes running on different nodes of the network) connected by *interactions* (i.e., multi-party channels alongside which messages between components are transfered). Systems that share the same architectural style (e.g., pipeline, ring, star, tree, etc.) and differ by the number of components and interactions are described using inductively defined predicates. Such configurations can be modified either by (a) adding or removing components and interactions (reconfiguration), or (b) changing the local states of components, by firing interactions.

J. Blanchette et al. (Eds.): IJCAR 2022, LNAI 13385, pp. 691–711, 2022.
https://doi.org/10.1007/978-3-031-10769-6_40

The assertion logic views components and interactions as *resources*, that can be created or deleted, in the spirit of resource logics *à la* Bunched Implications [37], or Separation Logic [39]. The main advantage of using resource logics is their support for *local reasoning* [12]: reconfiguration actions are specified by pre- and postconditions mentioning only the resources involved, while framing out the rest of the configuration.

The price to pay for this expressive power is the difficulty of automating the reasoning in these logics. This paper makes several contributions in the direction of proof automation, by studying the complexity of the *satisfiability* and *entailment* problems, for the configuration logic under consideration. Additionally, we study the complexity of a robustness property [27], namely *degree boundedness* (is every component involved in a bounded number of interactions?). In particular, the latter problem is used as a prerequisite for defining a fragment with a decidable entailment problem. For space reasons, the proofs of the technical results are given in [5].

1.1 Motivating Example

The logic studied in this paper is motivated by the need for an assertion language that supports reasoning about dynamic reconfigurations in a distributed system. For instance, consider a distributed system consisting of a finite (but unknown) number of *components* (processes) placed in a ring, executing the same finite-state program and communicating via *interactions* that connect the *out* port of a component to the *in* port of its right neighbour, in a round-robin fashion, as in Fig. 1(a). The behavior of a component is a machine with two states, T and H, denoting whether the component has a token (T) or not (H). A component c_i without a token may receive one, by executing a transition $\mathsf{H} \xrightarrow{in} \mathsf{T}$, simultaneously with its left neighbour c_j, that executes the transition $\mathsf{T} \xrightarrow{out} \mathsf{H}$. Then, we say that the interaction (c_j, out, c_i, in) has fired, moving a token one position to the right in the ring. Note that there can be more than one token, moving independently in the system, as long as no token overtakes another token.

The token ring system is formally specified by the following inductive rules:

$$\begin{aligned}
\mathsf{ring}_{h,t}(x) &\leftarrow \exists y \exists z \,.\, [x]@q * \langle x.out, z.in\rangle * \mathsf{chain}_{h',t'}(z,y) * \langle y.out, x.in\rangle \\
\mathsf{chain}_{h,t}(x,y) &\leftarrow \exists z.\, [x]@q * \langle x.out, z.in\rangle * \mathsf{chain}_{h',t'}(z,y) \\
\mathsf{chain}_{0,1}(x,x) &\leftarrow [x]@\mathsf{T} \qquad \mathsf{chain}_{1,0}(x,x) \leftarrow [x]@\mathsf{H} \qquad \mathsf{chain}_{0,0}(x,x) \leftarrow [x]
\end{aligned}$$

$$\text{where } h' \stackrel{\text{def}}{=} \begin{cases} \max(h-1,0) & \text{, if } q = \mathsf{H} \\ h & \text{, if } q = \mathsf{T} \end{cases} \text{ and } t' \stackrel{\text{def}}{=} \begin{cases} \max(t-1,0) & \text{, if } q = \mathsf{T} \\ t & \text{, if } q = \mathsf{H} \end{cases}$$

The predicate $\mathsf{ring}_{h,t}(x)$ describes a ring with at least two components, such that at least h (resp. t) components are in state H (resp. T). The ring consists of a component x in state q, described by the formula $[x]@q$, an interaction from the *out* port of x to the *in* port of another component z, described as $\langle x.out, z.in\rangle$, a separate chain of components stretching from z to y ($\mathsf{chain}_{h',t'}(z,y)$), and an interaction connecting the *out* port of component y to the *in* port of component x ($\langle y.out, x.in\rangle$). Inductively, a chain consists of a component $[x]@q$, an interaction $\langle x.out, z.in\rangle$ and a separate $\mathsf{chain}_{h',t'}(z,y)$. Figure 1(b) depicts the unfolding of the inductive definition of the token ring, with the

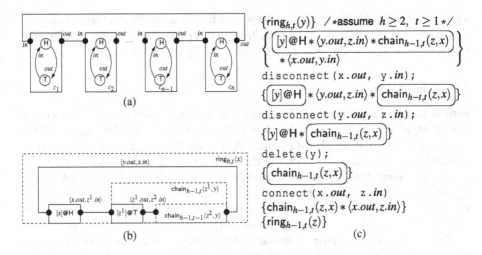

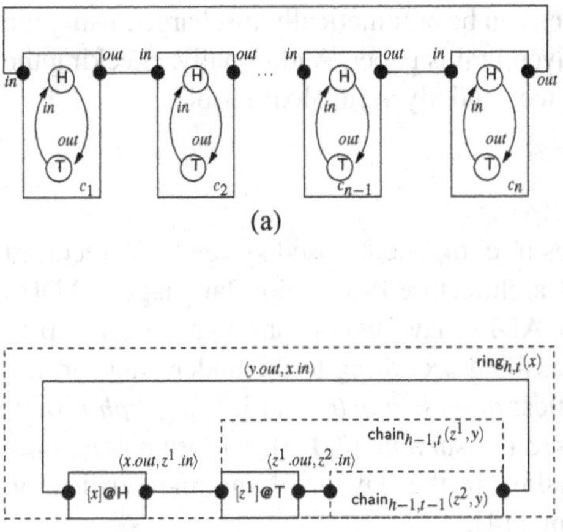

$\{\mathsf{ring}_{h,t}(y)\}$ /*assume $h \geq 2,\ t \geq 1$*/

$\left\{ \boxed{[y]@\mathsf{H} * \langle y.out, z.in\rangle * \mathsf{chain}_{h-1,t}(z,x)} * \langle x.out, y.in\rangle \right\}$

```
disconnect(x.out, y.in);
```

$\{\boxed{[y]@\mathsf{H}} * \langle y.out, z.in\rangle * \boxed{\mathsf{chain}_{h-t,t}(z,x)}\}$

```
disconnect(y.out, z.in);
```

$\{[y]@\mathsf{H} * \boxed{\mathsf{chain}_{h-1,t}(z,x)}\}$

```
delete(y);
```

$\{\boxed{\mathsf{chain}_{h-1,t}(z,x)}\}$

```
connect(x.out, z.in)
```

$\{\mathsf{chain}_{h-1,t}(z,x) * \langle x.out, z.in\rangle\}$

$\{\mathsf{ring}_{h-1,t}(z)\}$

(c)

Fig. 1. Inductive Specification and Reconfiguration of a Token Ring

existentially quantified variables z from the above rules α-renamed to $z^1, z^2, \ldots$ to avoid confusion.

A *reconfiguration program* takes as input a mapping of program variables to components and executes a sequence of *basic operations* i.e., component/interaction creation/deletion, involving the components and interactions denoted by these variables. For instance, the reconfiguration program in Fig. 1(c) takes as input three adjacent components, mapped to the variables x, y and z, respectively, removes the component y together with its left and right interactions and reconnects x directly with z. Programming reconfigurations is error-prone, because the interleaving between reconfiguration actions and interactions in a distributed system may lead to bugs that are hard to trace. For instance, if a reconfiguration program removes the last component in state T (resp. H) from the system, no token transfer interaction may fire and the system deadlocks.

We prove absence of such errors using a Hoare-style proof system [1], based on the logic introduced above as assertion language. For instance, the proof from Fig. 1(c) shows that the reconfiguration sequence applied to a component y in state H (i.e., $[y]@\mathsf{H}$) in a ring with at least $h \geq 2$ components in state H and at least $t \geq 1$ components in state T leads to a ring with at least $h-1$ components in state H and at least t in state T; note that the states of the components may change during the execution of the reconfiguration program, as tokens are moved by interactions.

The proof in Fig. 1(c) uses *local axioms* specifying, for each basic operation, only those components and interactions required to avoid faulting, with a *frame rule* $\{\phi\}\ \mathsf{P}\ \{\psi\} \Rightarrow \{\phi * \boxed{F}\}\ \mathsf{P}\ \{\psi * F\}$; for readability, the frame formulæ (from the preconditions of the conclusion of the frame rule applications) are enclosed in boxes.

The proof also uses the *consequence rule* $\{\phi\}\ \mathsf{P}\ \{\psi\} \Rightarrow \{\phi'\}\ \mathsf{P}\ \{\psi'\}$ that applies if ϕ' is stronger than ϕ and ψ' is weaker than ψ. The side conditions of the consequence rule require checking the validity of the entailments $\mathsf{ring}_{h,t}(y) \models \exists x \exists z\ .\ \langle x.out, y.in\rangle * [y]@\mathsf{H} * \langle y.out, z.in\rangle * \mathsf{chain}_{h-1,t}(z,x)$ and $\mathsf{chain}_{h-1,t}(z,x) * \langle x.out, z.in\rangle \models \mathsf{ring}_{h-1,t}(z)$, for all $h \geq 2$ and $t \geq 1$. These side conditions can be automatically discharged using the results on the decidability of entailments given in this paper. Additionally, checking the satisfiability of a precondition is used to detect trivially valid Hoare triples.

1.2 Related Work

Formal modeling coordinating architectures of component-based systems has received lots of attention, with the development of architecture description languages (ADL), such as BIP [3] or REO [2]. Many such ADLs have extensions that describe programmed reconfiguration, e.g., [19,30], classified according to the underlying formalism used to define their operational semantics: *process algebras* [13,33], *graph rewriting* [32,41,44], *chemical reactions* [43] (see the surveys [7,11]). Unfortunately, only few ADLs support formal verification, mainly in the flavour of runtime verification [10,17,20,31] or finite-state model checking [14].

Parameterized verification of unbounded networks of distributed processes uses mostly hard-coded coordinating architectures (see [4] for a survey). A first attempt at specifying architectures by logic is the *interaction logic* of Konnov et al. [29], a combination of Presburger arithmetic with monadic uninterpreted function symbols, that can describe cliques, stars and rings. More structured architectures (pipelines and trees) can be described using a second-order extension [34]. However, these interaction logics are undecidable and lack support for automated reasoning.

Specifying parameterized component-based systems by inductive definitions is not new. *Network grammars* [26,32,40] use context-free grammar rules to describe systems with linear (pipeline, token-ring) architectures obtained by composition of an unbounded number of processes. In contrast, we use predicates of unrestricted arities to describe architectural styles that are, in general, more complex than trees. Moreover, we write inductive definitions using a resource logic, suitable also for writing Hoare logic proofs of reconfiguration programs, based on local reasoning [12].

Local reasoning about concurrent programs has been traditionally the focus of Concurrent Separation Logic (CSL), based on a parallel composition rule [36], initially with a non-interfering (race-free) semantics [8] and later combining ideas of assume- and rely-guarantee [28,38] with local reasoning [22,42] and abstract notions of framing [15,16,21]. However, the body of work on CSL deals almost entirely with shared-memory multithreading programs, instead of distributed systems, which is the aim of our work. In contrast, we develop a resource logic in which the processes do not just share and own resources, but become mutable resources themselves.

The techniques developed in this paper are inspired by existing techniques for similar problems in the context of Separation Logic (SL) [39]. For instance, we use an abstract domain similar to the one defined by Brotherston et al. [9] for checking satisfiability of symbolic heaps in SL and reduce a fragment of the entailment problem in our logic to SL entailment [18]. In particular, the use of existing automated reasoning techniques for SL has pointed out several differences between the expressiveness of our logic and that of SL. First, the configuration logic describes hypergraph structures, in which edges are ℓ-tuples for $\ell \geq 2$, instead of directed graphs as in SL, where ℓ is a parameter of the problem: considering ℓ to be a constant strictly decreases the complexity of the problem. Second, the degree (number of hyperedges containing a given vertex) is unbounded, unlike in SL, where the degree of heaps is constant. Therefore, we dedicate an entire section (Sect. 4) to the problem of deciding the existence of a bound (and computing a cut-off) on the degree of the models of a formula, used as a prerequisite for the encoding of the entailment problems from the configuration logic as SL entailments.

2 Definitions

We denote by $\mathbb{N}$ the set of positive integers including zero. For a set A, we define $A^1 \stackrel{\text{def}}{=} A$, $A^{i+1} \stackrel{\text{def}}{=} A^i \times A$, for all $i \geq 1$, and $A^+ = \bigcup_{i \geq 1} A^i$, where $\times$ denotes the Cartesian product. We denote by $\text{pow}(A)$ the powerset of A and by $\text{mpow}(A)$ the power-multiset (set of multisets) of A. The cardinality of a finite set A is denoted as $||A||$. By writing $A \subseteq_{fin} B$ we mean that A is a finite subset of B. Given integers i and j, we write $[i, j]$ for the set $\{i, i+1, \ldots, j\}$, assumed to be empty if $i > j$. For a tuple $\mathbf{t} = \langle t_1, \ldots, t_n \rangle$, we define $|\mathbf{t}| \stackrel{\text{def}}{=} n$, $\langle \mathbf{t} \rangle_i \stackrel{\text{def}}{=} t_i$ and $\langle \mathbf{t} \rangle_{[i,j]} \stackrel{\text{def}}{=} \langle t_i, \ldots, t_j \rangle$. By writing $x = poly(y)$, for given $x, y \in \mathbb{N}$, we mean that there exists a polynomial function $f : \mathbb{N} \to \mathbb{N}$, such that $x \leq f(y)$.

2.1 Configurations

We model distributed systems as hypergraphs, whose vertices are *components* (i.e., the nodes of the network) and hyperedges are *interactions* (i.e., describing the way the components communicate with each other). The components are taken from a countably infinite set $\mathbb{C}$, called the *universe*. We consider that each component executes its own copy of the same *behavior*, represented as a finite-state machine $\mathbb{B} = (\mathcal{P}, \mathcal{Q}, \to)$, where $\mathcal{P}$ is a finite set of *ports*, $\mathcal{Q}$ is a finite set of *states* and $\to \subseteq \mathcal{Q} \times \mathcal{P} \times \mathcal{Q}$ is a transition relation. Intuitively, each transition $q \xrightarrow{p} q'$ of the behavior is triggerred by a visible event, represented by the port p. For instance, the behavior of the components of the token ring system from Fig. 1(a) is $\mathbb{B} = (\{in, out\}, \{\mathsf{H}, \mathsf{T}\}, \{\mathsf{H} \xrightarrow{in} \mathsf{T}, \mathsf{T} \xrightarrow{out} \mathsf{H}\})$. *The universe $\mathbb{C}$ and the behavior $\mathbb{B} = (\mathcal{P}, \mathcal{Q}, \to)$ are fixed in the rest of this paper.*

We introduce a logic for describing infinite sets of *configurations* of distributed systems with unboundedly many components and interactions. A configuration is a snapshot of the system, describing the topology of the network (i.e., the set of present components and interactions) together with the local state of each component:

Definition 1. *A* configuration *is a tuple $\gamma = (C, I, \rho)$, where:*

- *$C \subseteq_{fin} \mathbb{C}$ is a finite set of* components, *that are present in the configuration,*
- *$I \subseteq_{fin} (\mathbb{C} \times \mathcal{P})^+$ is a finite set of* interactions, *where each interaction is a sequence $(c_1, p_1, \ldots, c_n, p_n) \in (\mathbb{C} \times \mathcal{P})^n$ that binds together the ports $p_1, \ldots, p_n$ of the pairwise distinct components $c_1, \ldots, c_n$, respectively.*
- *$\rho : \mathbb{C} \to \mathcal{Q}$ is a* state map *associating each (possibly absent) component, a state of the behavior $\mathbb{B}$, such that the set $\{c \in \mathbb{C} \mid \rho(c) = q\}$ is infinite, for each $q \in \mathcal{Q}$.*

The last condition requires that there is an infinite pool of components in each state $q \in \mathcal{Q}$; since $\mathbb{C}$ is infinite and $\mathcal{Q}$ is finite, this condition is feasible. For example, the configurations of the token ring from Fig. 1(a) are $(\{c_1, \ldots, c_n\}, \{(c_i, out, c_{(i \bmod n)+1}, in) \mid$

$i \in [1,n]\}, \rho)$, where $\rho : \mathbb{C} \to \{\mathsf{H}, \mathsf{T}\}$ is a state map. The ring topology is described by the set of components $\{c_1, \ldots, c_n\}$ and interactions $\{(c_i, out, c_{(i \bmod n)+1}, in) \mid i \in [1,n]\}$.

Intuitively, an interaction $(c_1, p_1, \ldots, c_n, p_n)$ synchronizes transitions labeled by the ports $p_1, \ldots, p_n$ from the behaviors (i.e., replicas of the state machine $\mathbb{B}$) of $c_1, \ldots, c_n$, respectively. Note that the components c_i are not necessary part of the configuration. The interactions are classified according to their sequence of ports, called the *interaction type* and let $\mathsf{Inter} \stackrel{\mathsf{def}}{=} \mathcal{P}^+$ be the set of interaction types; an interaction type models, for instance, the passing of a certain kind of message (e.g., request, acknowledgement, etc.). From an operational point of view, two interactions that differ by a permutation of indices e.g., $(c_1, p_1, \ldots, c_n, p_n)$ and $(c_{i_1}, p_{i_1}, \ldots, c_{i_n}, p_{i_n})$ such that $\{i_1, \ldots, i_n\} = [1,n]$, are equivalent, since the set of transitions is the same; nevertheless, we chose to distinguish them in the following, exclusively for reasons of simplicity.

Below we define the composition of configurations, as the union of disjoint sets of components and interactions:

Definition 2. *The composition of two configurations $\gamma_i = (C_i, I_i, \rho)$, for $i = 1,2$, such that $C_1 \cap C_2 = \emptyset$ and $I_1 \cap I_2 = \emptyset$, is defined as $\gamma_1 \bullet \gamma_2 \stackrel{\mathsf{def}}{=} (C_1 \cup C_2, I_1 \cup I_2, \rho)$. The composition $\gamma_1 \bullet \gamma_2$ is undefined if $C_1 \cap C_2 \neq \emptyset$ or $I_1 \cap I_2 \neq \emptyset$.*

In analogy with graphs, the *degree* of a configuration is the maximum number of interactions from the configuration that involve a (possibly absent) component:

Definition 3. *The* degree *of a configuration $\gamma = (C, I, \rho)$ is defined as $\delta(\gamma) \stackrel{\mathsf{def}}{=} \max_{c \in \mathbb{C}} \delta_c(\gamma)$, where $\delta_c(\gamma) \stackrel{\mathsf{def}}{=} \|\{(c_1, p_1, \ldots, c_n, p_n) \in I \mid c = c_i,\ i \in [1,n]\}\|$.*

For instance, the configuration of the system from Fig. 1(a) has degree two.

2.2 Configuration Logic

Let $\mathbb{V}$ and $\mathbb{A}$ be countably infinite sets of *variables* and *predicates*, respectively. For each predicate $\mathsf{A} \in \mathbb{A}$, we denote its arity by $\#\mathsf{A}$. The formulæ of the *Configuration Logic* (CL) are described inductively by the following syntax:

$$\phi := \mathsf{emp} \mid [x] \mid \langle x_1.p_1, \ldots, x_n.p_n \rangle \mid x@q \mid x = y \mid x \neq y \mid \mathsf{A}(x_1, \ldots, x_{\#\mathsf{A}}) \mid \phi * \phi \mid \exists x \,.\, \phi$$

where $x, y, x_1, \ldots \in \mathbb{V}$, $q \in \mathcal{Q}$ and $\mathsf{A} \in \mathbb{A}$. A formula $[x]$, $\langle x_1.p_1, \ldots, x_n.p_n \rangle$, $x@q$ and $\mathsf{A}(x_1, \ldots, x_{\#\mathsf{A}})$ is called a *component*, *interaction*, *state* and *predicate* atom, respectively. These formulæ are also referred to as *atoms*. The connective $*$ is called the *separating conjunction*. We use the shorthand $[x]@q \stackrel{\mathsf{def}}{=} [x] * x@q$. For instance, the formula $[x]@q * [y]@q' * \langle x.out, y.in \rangle * \langle x.in, y.out \rangle$ describes a configuration consisting of two distinct components, denoted by the values of x and y, in states q and q', respectively, and two interactions binding the *out* port of one to the *in* port of the other component.

A formula is said to be *pure* if and only if it is a separating conjunction of state atoms, equalities and disequalities. A formula with no occurrences of predicate atoms (resp. existential quantifiers) is called *predicate-free* (resp. *quantifier-free*). A variable is *free* if it does not occur within the scope of an existential quantifier ; we note $\mathsf{fv}(\phi)$ the set of free variables of ϕ. A *sentence* is a formula with no free variables. A *substitution*

$\phi[x_1/y_1 \ldots x_n/y_n]$ replaces simultaneously every free occurrence of x_i by y_i in ϕ, for all $i \in [1,n]$. Before defining the semantics of CL formulæ, we introduce the set of inductive definitions that assigns meaning to predicates:

Definition 4. *A set of inductive definitions (SID) Δ consists of rules $\mathsf{A}(x_1,\ldots,x_{\#\mathsf{A}}) \leftarrow \phi$, where $x_1,\ldots,x_{\#\mathsf{A}}$ are pairwise distinct variables, called* parameters, *such that* $\mathrm{fv}(\phi) \subseteq \{x_1,\ldots,x_{\#\mathsf{A}}\}$. *The rule $\mathsf{A}(x_1,\ldots,x_{\#\mathsf{A}}) \leftarrow \phi$* defines *$\mathsf{A}$ and we denote by $\mathrm{def}_\Delta(\mathsf{A})$ the set of rules from Δ that define A.*

Note that having distinct parameters in a rule is without loss of generality, as e.g., a rule $\mathsf{A}(x_1,x_1) \leftarrow \phi$ can be equivalently written as $\mathsf{A}(x_1,x_2) \leftarrow x_1 = x_2 * \phi$. As a convention, we shall always use the names $x_1,\ldots,x_{\#\mathsf{A}}$ for the parameters of a rule that defines A.

The semantics of CL formulæ is defined by a satisfaction relation $\gamma \models^\nu_\Delta \phi$ between configurations and formulæ. This relation is parameterized by a *store* $\nu : \mathbb{V} \to \mathbb{C}$ mapping the free variables of a formula into components from the universe (possibly absent from γ) and an SID Δ. We write $\nu[x \leftarrow c]$ for the store that maps x into c and agrees with ν on all variables other than x. The definition of the satisfaction relation is by induction on the structure of formulæ, where $\gamma = (C, I, \rho)$ is a configuration (Definition 1):

$$\begin{array}{lll}
\gamma \models^\nu_\Delta \mathsf{emp} & \iff & C = \emptyset \text{ and } I = \emptyset \\
\gamma \models^\nu_\Delta [x] & \iff & C = \{\nu(x)\} \text{ and } I = \emptyset \\
\gamma \models^\nu_\Delta \langle x_1.p_1,\ldots,x_n.p_n\rangle & \iff & C = \emptyset \text{ and } I = \{(\nu(x_1),p_1,\ldots,\nu(x_n),p_n)\} \\
\gamma \models^\nu_\Delta x@q & \iff & \gamma \models^\nu_\Delta \mathsf{emp} \text{ and } \rho(\nu(x)) = q \\
\gamma \models^\nu_\Delta x \sim y & \iff & \gamma \models^\nu_\Delta \mathsf{emp} \text{ and } \nu(x) \sim \nu(y), \text{ for all } \sim \in \{=,\neq\} \\
\gamma \models^\nu_\Delta \mathsf{A}(y_1,\ldots,y_{\#\mathsf{A}}) & \iff & \gamma \models^\nu_\Delta \phi[x_1/y_1,\ldots,x_{\#\mathsf{A}}/y_{\#\mathsf{A}}], \text{ for some rule} \\
 & & \mathsf{A}(x_1,\ldots,x_{\#\mathsf{A}}) \leftarrow \phi \text{ from } \Delta \\
\gamma \models^\nu_\Delta \phi_1 * \phi_2 & \iff & \text{exist } \gamma_1,\gamma_2, \text{ such that } \gamma = \gamma_1 \bullet \gamma_2 \text{ and } \gamma_i \models^\nu_\Delta \phi_i, \text{ for } i = 1,2 \\
\gamma \models^\nu_\Delta \exists x \,.\, \phi & \iff & \gamma \models^{\nu[x \leftarrow c]}_\Delta \phi, \text{ for some } c \in \mathbb{C}
\end{array}$$

If ϕ is a sentence, the satisfaction relation $\gamma \models^\nu_\Delta \phi$ does not depend on the store, written $\gamma \models_\Delta \phi$, in which case we say that γ is a *model* of ϕ. If ϕ is a predicate-free formula, the satisfaction relation does not depend on the SID, written $\gamma \models^\nu \phi$. A formula ϕ is *satisfiable* if and only if the sentence $\exists x_1 \ldots \exists x_n \,.\, \phi$ has a model, where $\mathrm{fv}(\phi) = \{x_1,\ldots,x_n\}$. A formula ϕ *entails* a formula ψ, written $\phi \models_\Delta \psi$ if and only if, for any configuration γ and store ν, we have $\gamma \models^\nu_\Delta \phi$ only if $\gamma \models^\nu_\Delta \psi$.

2.3 Separation Logic

Separation Logic (SL) [39] will be used in the following to prove several technical results concerning the decidability and complexity of certain decision problems for CL. For self-containment reasons, we define SL below. The syntax of SL formulæ is described by the following grammar:

$$\phi := \mathsf{emp} \mid x_0 \mapsto (x_1,\ldots,x_{\mathfrak{K}}) \mid x = y \mid x \neq y \mid \mathsf{A}(x_1,\ldots,x_{\#\mathsf{A}}) \mid \phi * \phi \mid \exists x \,.\, \phi$$

where $x,y,x_0,x_1,\ldots \in \mathbb{V}$, $\mathsf{A} \in \mathbb{A}$ and $\mathfrak{K} \geq 1$ is an integer constant. Formulæ of SL are interpreted over finite partial functions $\mathsf{h} : \mathbb{C} \rightharpoonup_{fin} \mathbb{C}^{\mathfrak{K}}$, called *heaps*[1], by a satisfaction relation $\mathsf{h} \Vdash^\nu \phi$, defined inductively as follows:

[1] We use the universe $\mathbb{C}$ here for simplicity, the definition works with any countably infinite set.

$$\begin{array}{lcl}
\mathsf{h} \Vdash^{\nu}_{\Delta} \mathsf{emp} & \iff & \mathsf{h} = \emptyset \\
\mathsf{h} \Vdash^{\nu}_{\Delta} x_0 \mapsto (x_1, \ldots, x_{\mathfrak{K}}) & \iff & \mathrm{dom}(\mathsf{h}) = \{\nu(x_0)\} \text{ and } \mathsf{h}(\nu(x_0)) = \langle \nu(x_1), \ldots, \nu(x_{\mathfrak{K}}) \rangle \\
\mathsf{h} \Vdash^{\nu} \phi_1 * \phi_2 & \iff & \text{there exist } \mathsf{h}_1, \mathsf{h}_2 \text{ such that } \mathrm{dom}(\mathsf{h}_1) \cap \mathrm{dom}(\mathsf{h}_2) = \emptyset, \\
 & & \mathsf{h} = \mathsf{h}_1 \cup \mathsf{h}_2 \text{ and } \mathsf{h}_i \Vdash^{\nu}_{\Delta} \phi_i, \text{ for both } i = 1,2
\end{array}$$

where $\mathrm{dom}(\mathsf{h}) \stackrel{\mathrm{def}}{=} \{c \in \mathbb{C} \mid \mathsf{h}(c) \text{ is defined}\}$ is the domain of the heap and (dis-) equalities, predicate atoms and existential quantifiers are defined same as for CL.

2.4 Decision Problems

We define the decision problems that are the focus of the upcoming sections. As usual, a decision problem is a class of yes/no queries that differ only in their input. In our case, the input consists of an SID and one or two predicates, written between square brackets.

Definition 5. *We consider the following problems, for a SID Δ and predicates $\mathsf{A}, \mathsf{B} \in \mathbb{A}$:*

1. $\mathsf{Sat}[\Delta, \mathsf{A}]$*: is the sentence* $\exists x_1 \ldots \exists x_{\#\mathsf{A}} \,.\, \mathsf{A}(x_1, \ldots, x_{\#\mathsf{A}})$ *satisfiable for* Δ*?*
2. $\mathsf{Bnd}[\Delta, \mathsf{A}]$*: is the set* $\{\delta(\gamma) \mid \gamma \models_{\Delta} \exists x_1 \ldots \exists x_{\#\mathsf{A}} \,.\, \mathsf{A}(x_1, \ldots, x_{\#\mathsf{A}})\}$ *finite?*
3. $\mathsf{Entl}[\Delta, \mathsf{A}, \mathsf{B}]$*: does* $\mathsf{A}(x_1, \ldots, x_{\#\mathsf{A}}) \models_{\Delta} \exists x_{\#\mathsf{B}+1} \ldots \exists x_{\#\mathsf{A}} \,.\, \mathsf{B}(x_1, \ldots, x_{\#\mathsf{B}})$ *hold?*

The size of a formula ϕ is the total number of occurrences of symbols needed to write it down, denoted by $\mathrm{size}(\phi)$. The size of a SID Δ is $\mathrm{size}(\Delta) \stackrel{\mathrm{def}}{=} \sum_{\mathsf{A}(x_1,\ldots,x_{\#\mathsf{A}}) \leftarrow \phi \in \Delta} \mathrm{size}(\phi) + \#\mathsf{A} + 1$. Other parameters of a SID Δ are:

- $\mathrm{arity}(\Delta) \stackrel{\mathrm{def}}{=} \max\{\#\mathsf{A} \mid \mathsf{A}(x_1, \ldots, x_{\#\mathsf{A}}) \leftarrow \phi \in \Delta\}$,
- $\mathrm{width}(\Delta) \stackrel{\mathrm{def}}{=} \max\{\mathrm{size}(\phi) \mid \mathsf{A}(x_1, \ldots, x_{\#\mathsf{A}}) \leftarrow \phi \in \Delta\}$,
- $\mathrm{intersize}(\Delta) \stackrel{\mathrm{def}}{=} \max\{n \mid \langle x_1.p_1, \ldots, x_n.p_n \rangle \text{ occurs in } \phi, \mathsf{A}(x_1, \ldots, x_{\#\mathsf{A}}) \leftarrow \phi \in \Delta\}$.

For a decision problem $\mathsf{P}[\Delta, \mathsf{A}, \mathsf{B}]$, we consider its (k, ℓ)-bounded versions $\mathsf{P}^{(k,\ell)}[\Delta, \mathsf{A}, \mathsf{B}]$, obtained by restricting the predicates and interaction atoms occurring Δ to $\mathrm{arity}(\Delta) \leq k$ and $\mathrm{intersize}(\Delta) \leq \ell$, respectively, where k and ℓ are either positive integers or infinity. We consider, for each $\mathsf{P}[\Delta, \mathsf{A}, \mathsf{B}]$, the subproblems $\mathsf{P}^{(k,\ell)}[\Delta, \mathsf{A}, \mathsf{B}]$ corresponding to the three cases (1) $k < \infty$ and $\ell = \infty$, (2) $k = \infty$ and $\ell < \infty$, and (3) $k = \infty$ and $\ell = \infty$. As we explain next, this is because, for the decision problems considered (Definition 5), the complexity for the case $k < \infty, \ell < \infty$ matches the one for the case $k < \infty, \ell = \infty$.

Satisfiability (1) and entailment (3) arise naturally during verification of reconfiguration programs. For instance, $\mathsf{Sat}[\Delta, \phi]$ asks whether a specification ϕ of a set configurations (e.g., a pre-, post-condition, or a loop invariant) is empty or not (e.g., an empty precondition typically denotes a vacuous verification condition), whereas $\mathsf{Entl}[\Delta, \phi, \psi]$ is used as a side condition for the Hoare rule of consequence, as in e.g., the proof from Fig. 1(c). Moreover, entailments must be proved when checking inductiveness of a user-provided loop invariant.

The $\mathsf{Bnd}[\Delta, \phi]$ problem is used to check a necessary condition for the decidability of entailments i.e., $\mathsf{Entl}[\Delta, \phi, \psi]$. If $\mathsf{Bnd}[\Delta, \phi]$ has a positive answer, we can reduce the problem $\mathsf{Entl}[\Delta, \phi, \psi]$ to an entailment problem for SL, which is always interpreted over heaps of bounded degree [18]. Otherwise, the decidability status of the entailment problem is open, for configurations of unbounded degree, such as the one described by the example below.

Example 1. The following SID describes star topologies with a central controller connected to an unbounded number of workers stations:

$$Controller(x) \leftarrow [x] * Worker(x)$$
$$Worker(x) \leftarrow \exists y \,.\, \langle x.out, y.in\rangle * [y] * Worker(x) \qquad Worker(x) \leftarrow \mathsf{emp} \quad \blacksquare$$

3 Satisfiability

We show that the satisfiability problem (Definition 5, point 1) is decidable, using a method similar to the one pioneered by Brotherston et al. [9], for checking satisfiability of inductively defined symbolic heaps in SL. We recall that a formula π is *pure* if and only if it is a separating conjunction of equalities, disequalities and state atoms. In the following, the order of terms in (dis-)equalities is not important i.e., we consider $x = y$ (resp. $x \neq y$) and $y = x$ (resp. $y \neq x$) to be the same formula.

Definition 6. *The* closure $\mathrm{cl}(\pi)$ *of a pure formula* π *is the limit of the sequence* $\pi^0, \pi^1, \pi^2, \ldots$ *such that* $\pi^0 = \pi$ *and, for each* $i \geq 0$, π^{i+1} *is obtained by joining (with* $*$*) all of the following formulæ to* π^i*:*

- $x = z$*, where* x *and* z *are the same variable, or* $x = y$ *and* $y = z$ *both occur in* π^i*,*
- $x \neq z$*, where* $x = y$ *and* $y \neq z$ *both occur in* π^i*, or*
- $y@q$*, where* $x@q$ *and* $x = y$ *both occur in* π^i*.*

Because only finitely many such formulæ can be added, the sequence of pure formulæ from Definition 6 is bound to stabilize after polynomially many steps. A pure formula is satisfiable if and only if its closure does not contain contradictory literals i.e., $x = y$ and $x \neq y$, or $x@q$ and $x@q'$, for $q \neq q' \in Q$. We write $x \approx_\pi y$ (resp. $x \not\approx_\pi y$) if and only if $x = y$ (resp. $x \neq y$) occurs in $\mathrm{cl}(\pi)$ and $\mathrm{not}(x \approx_\pi y)$ (resp. $\mathrm{not}(x \not\approx_\pi y)$) whenever $x \approx_\pi y$ (resp. $x \not\approx_\pi y$) does not hold. Note that e.g., $\mathrm{not}(x \approx_\pi y)$ is not the same as $x \not\approx_\pi y$.

Base tuples constitute the abstract domain used by the algorithms for checking satisfiability (point 1 of Definition 5) and boundedness (point 2 of Definition 5), defined as follows:

Definition 7. *A* base tuple *is a triple* $\mathsf{t} = (C^\sharp, I^\sharp, \pi)$*, where:*

- $C^\sharp \in mpow\mathbb{V}$ *is a multiset of variables denoting present components,*
- $I^\sharp : \mathsf{Inter} \to mpow\mathbb{V}^+$ *maps each interaction type* $\tau \in \mathsf{Inter}$ *into a multiset of tuples of variables of length* $|\tau|$ *each, and*
- π *is a pure formula.*

A base tuple is called satisfiable *if and only if* π *is satisfiable and the following hold:*

1. *for all* $x, y \in C^\sharp$, $\mathrm{not}(x \approx_\pi y)$*,*
2. *for all* $\tau \in \mathsf{Inter}$, $\langle x_1, \ldots, x_{|\tau|}\rangle, \langle y_1, \ldots, y_{|\tau|}\rangle \in I^\sharp(\tau)$*, there exists* $i \in [1, |\tau|]$ *such that* $\mathrm{not}(x_i \approx_\pi y_i)$*,*
3. *for all* $\tau \in \mathsf{Inter}$, $\langle x_1, \ldots, x_{\#\tau}\rangle \in I^\sharp(\tau)$ *and* $1 \leq i < j \leq |\tau|$*, we have* $\mathrm{not}(x_i \approx_\pi x_j)$*.*

We denote by $\mathsf{SatBase}$ *the set of satisfiable base tuples.*

Intuitively, a base tuple is an abstract representation of a configuration, where components (resp. interactions) are represented by variables (resp. tuples of variables). Note that a base tuple $(C^\sharp, I^\sharp, \pi)$ is unsatisfiable if $C^\sharp$ $(I^\sharp)$ contains the same variable (tuple of variables) twice (for the same interaction type), hence the use of multisets in the definition of base tuples. It is easy to see that checking the satisfiability of a given base tuple $(C^\sharp, I^\sharp, \pi)$ can be done in time $poly(\|C^\sharp\| + \Sigma_{\tau \in \mathsf{Inter}} \|I^\sharp(\tau)\| + \mathrm{size}(\pi))$.

We define a partial *composition* operation on satisfiable base tuples, as follows:

$$(C_1^\sharp, I_1^\sharp, \pi_1) \otimes (C_2^\sharp, I_2^\sharp, \pi_2) \stackrel{\mathsf{def}}{=} (C_1^\sharp \cup C_2^\sharp, I_1^\sharp \cup I_2^\sharp, \pi_1 * \pi_2)$$

where the union of multisets is lifted to functions $\mathsf{Inter} \rightarrow \mathrm{mpow}(\mathbb{V}^+)$ in the usual way. The composition operation $\otimes$ is undefined if $(C_1^\sharp, I_1^\sharp, \pi_1) \otimes (C_2^\sharp, I_2^\sharp, \pi_2)$ is not satisfiable e.g., if $C_1^\sharp \cap C_2^\sharp \neq \emptyset$, $I_1^\sharp(\tau) \cap I_2^\sharp(\tau) \neq \emptyset$, for some $\tau \in \mathsf{Inter}$, or $\pi_1 * \pi_2$ is not satisfiable.

Given a pure formula π and a set of variables X, the projection $\pi\downarrow_X$ removes from π all atoms α, such that $\mathrm{fv}(\alpha) \not\subseteq X$. The *projection* of a base tuple $(C^\sharp, I^\sharp, \pi)$ on a variable set X is formally defined below:

$$(C^\sharp, I^\sharp, \pi)\downarrow_X \stackrel{\mathsf{def}}{=} \Big(C^\sharp \cap X, \lambda\tau \,.\, \{\langle x_1, \ldots, x_{|\tau|}\rangle \in I^\sharp(\tau) \mid x_1, \ldots, x_{|\tau|} \in X\}, \mathrm{cl}(\mathrm{dist}(I^\sharp) * \pi)\downarrow_X\Big)$$

$$\text{where } \mathrm{dist}(I^\sharp) \stackrel{\mathsf{def}}{=} \mathop{\ast}_{\tau \in \mathsf{Inter}} \mathop{\ast}_{\langle x_1, \ldots, x_{|\tau|}\rangle \in I^\sharp(\tau)} \mathop{\ast}_{1 \leq i < j \leq |\tau|} x_i \neq x_j$$

The *substitution* operation $(C^\sharp, I^\sharp, \pi)[x_1/y_1, \ldots, x_n/y_n]$ replaces simultaneously each x_i with y_i in $C^\sharp$, $I^\sharp$ and π, respectively. We lift the composition, projection and substitution operations to sets of satisfiable base tuples, as usual.

Next, we define the base tuple corresponding to a quantifier- and predicate-free formula $\phi = \psi * \pi$, where ψ consists of component and interaction atoms and π is pure. Since, moreover, we are interested in those components and interactions that are visible through a given indexed set of parameters $X = \{x_1, \ldots, x_n\}$, for a variable y, we denote by $\{\!\{y\}\!\}_\pi^X$ the parameter x_i with the least index, such that $y \approx_\pi x_i$, or y itself, if no such parameter exists. We define the following sets of formulæ:

$$\mathsf{Base}(\phi, X) \stackrel{\mathsf{def}}{=} \begin{cases} \{(C^\sharp, I^\sharp, \pi)\} & \text{, if } (C^\sharp, I^\sharp, \pi) \text{ is satisfiable} \\ \emptyset & \text{, otherwise} \end{cases}$$

$$\text{where } C^\sharp \stackrel{\mathsf{def}}{=} \{\{\!\{x\}\!\}_\pi^X \mid [x] \text{ occurs in } \psi\}$$

$$I^\sharp \stackrel{\mathsf{def}}{=} \lambda\langle p_1, \ldots, p_s\rangle . \{\langle \{\!\{y_1\}\!\}_\pi^X, \ldots, \{\!\{y_s\}\!\}_\pi^X\rangle \mid \langle y_1.p_1, \ldots, y_s.p_s\rangle \text{ occurs in } \psi\}$$

We consider a tuple of variables $\overrightarrow{\mathcal{X}}$, having a variable $\mathcal{X}(\mathsf{A})$ ranging over $\mathrm{pow}(\mathsf{SatBase})$, for each predicate A that occurs in Δ. With these definitions, each rule of Δ:

$$\mathsf{A}(x_1, \ldots, x_{\#\mathsf{A}}) \leftarrow \exists y_1 \ldots \exists y_m \,.\, \phi * \mathsf{B}_1(z_1^1, \ldots, z_{\#\mathsf{B}_1}^1) * \ldots * \mathsf{B}_h(z_1^h, \ldots, z_{\#\mathsf{B}_h}^h)$$

where ϕ is a quantifier- and predicate-free formula, induces the constraint:

$$\mathcal{X}(\mathsf{A}) \supseteq \Big(\mathsf{Base}(\phi, \{x_1, \ldots, x_{\#\mathsf{A}}\}) \otimes \bigotimes_{\ell=1}^{h} \mathcal{X}(\mathsf{B}_\ell)[x_1/z_1^\ell, \ldots, x_{\#\mathsf{B}_\ell}/z_{\#\mathsf{B}_\ell}^\ell]\Big)\downarrow_{x_1, \ldots, x_{\#\mathsf{A}}} \quad (1)$$

input: a SID Δ **output:** $\mu\vec{\mathcal{X}}.\Delta^\sharp$

initially $\mu\vec{\mathcal{X}}.\Delta^\sharp := \lambda \mathsf{A} . \emptyset$

for $\mathsf{A}(x_1,\ldots,x_{\#\mathsf{A}}) \leftarrow \exists y_1 \ldots \exists y_m . \phi \in \Delta$, with ϕ quantifier- and predicate-free **do**

$\quad \mu\vec{\mathcal{X}}.\Delta^\sharp(\mathsf{A}) := \mu\vec{\mathcal{X}}.\Delta^\sharp(\mathsf{A}) \cup \mathsf{Base}(\phi, \{x_1,\ldots,x_{\#\mathsf{A}}\})\downarrow_{x_1,\ldots,x_{\#\mathsf{A}}}$

while $\mu\vec{\mathcal{X}}.\Delta^\sharp$ still change **do**

$\quad$ **for** $\mathsf{r} : \mathsf{A}(x_1,\ldots,x_{\#\mathsf{A}}) \leftarrow \exists y_1 \ldots \exists y_m . \phi * \mathop{\Large *}_{\ell=1}^{h} \mathsf{B}_\ell(z_1^\ell,\ldots,z_{\#\mathsf{B}_\ell}^\ell) \in \Delta$ **do**

$\quad\quad$ **if there exist** $\mathsf{t}_1 \in \mu\vec{\mathcal{X}}.\Delta^\sharp(\mathsf{B}_1),\ldots,\mathsf{t}_h \in \mu\vec{\mathcal{X}}.\Delta^\sharp(\mathsf{B}_h)$ **then**

$\quad\quad\quad \mu\vec{\mathcal{X}}.\Delta^\sharp(\mathsf{A}) := \mu\vec{\mathcal{X}}.\Delta^\sharp(\mathsf{A}) \cup \left(\mathsf{Base}(\phi, \{x_1,\ldots,x_{\#\mathsf{A}}\}) \otimes \bigotimes_{\ell=1}^{h} \mathsf{t}_\ell[x_1/z_1^\ell,\ldots,x_{\#\mathsf{B}_\ell}/z_{\#\mathsf{B}_\ell}^\ell]\right)\downarrow_{x_1,\ldots,x_{\#\mathsf{A}}}$

Fig. 2. Algorithm for the Computation of the Least Solution

Let $\Delta^\sharp$ be the set of such constraints, corresponding to the rules in Δ and let $\mu\vec{\mathcal{X}}.\Delta^\sharp$ be the tuple of least solutions of the constraint system generated from Δ, indexed by the tuple of predicates that occur in Δ, such that $\mu\vec{\mathcal{X}}.\Delta^\sharp(\mathsf{A})$ denotes the entry of $\mu\vec{\mathcal{X}}.\Delta^\sharp$ correponding to A. Since the composition and projection are monotonic operations, such a least solution exists and is unique. Since $\mathsf{SatBase}$ is finite, the least solution can be attained in a finite number of steps, using a Kleene iteration (see Fig. 2).

We state below the main result leading to an elementary recursive algorithm for the satisfiability problem (Theorem 1). The intuition is that, if $\mu\vec{\mathcal{X}}.\Delta^\sharp(\mathsf{A})$ is not empty, then it contains only satisfiable base tuples, from which a model of $\mathsf{A}(x_1,\ldots,x_{\#\mathsf{A}})$ can be built.

Lemma 1. $\mathsf{Sat}[\Delta,\mathsf{A}]$ *has a positive answer if and only if* $\mu\vec{\mathcal{X}}.\Delta^\sharp(\mathsf{A}) \neq \emptyset$.

If the maximal arity of the predicates occurring in Δ is bound by a constant k, no satisfiable base tuple $(C^\sharp, I^\sharp, \pi)$ can have a tuple $\langle y_1,\ldots,y_{|\tau|}\rangle \in I^\sharp(\tau)$, for some $\tau \in \mathsf{Inter}$, such that $|\tau| > k$, since all variables $y_1,\ldots,y_{|\tau|}$ are parameters denoting distinct components (point 3 of Definition 7). Hence, the upper bound on the size of a satisfiable base tuple is constant, in both the $k < \infty, \ell < \infty$ and $k < \infty, \ell = \infty$ cases, which are, moreover indistinguishable complexity-wise (i.e., both are NP-complete). In contrast, in the cases $k = \infty, \ell < \infty$ and $k = \infty, \ell = \infty$, the upper bound on the size of satisfiable base tuples is polynomial and simply exponential in $\mathrm{size}(\Delta)$, incurring a complexity gap of one and two exponentials, respectively. The theorem below states the main result of this section:

Theorem 1. $\mathsf{Sat}^{(k,\infty)}[\Delta,\mathsf{A}]$ *is* NP*-complete for* $k \geq 4$, $\mathsf{Sat}^{(\infty,\ell)}[\Delta,\mathsf{A}]$ *is* EXP*-complete and* $\mathsf{Sat}[\Delta,\mathsf{A}]$ *is in* $2\mathsf{EXP}$.

The upper bounds are consequences of the fact that the size of a satisfiable base tuple is bounded by a simple exponential in the $\min(\mathrm{arity}(\Delta), \mathrm{intersize}(\Delta))$, hence the number of such tuples is doubly exponential in $\min(\mathrm{arity}(\Delta), \mathrm{intersize}(\Delta))$. The lower bounds are by a polynomial reduction from the satisfiability problem for SL [9].

Example 2. The doubly-exponential upper bound for the algorithm computing the least solution of a system of constraints of the form (1) is necessary, in general, as illustrated by the following worst-case example. Let n be a fixed parameter and consider the n-arity predicates $A_1,\ldots,A_n$ defined by the following SID:

$$A_i(x_1,\ldots,x_n) \leftarrow \mathop{\ast}_{j=0}^{n-i} A_{i+1}(x_1,\ldots,x_{i-1},[x_i,\ldots,x_n]^j), \quad \text{for all } i \in [1,n-1]$$
$$A_n(x_1,\ldots,x_n) \leftarrow \langle x_1.p,\ldots,x_n.p\rangle \quad A_n(x_1,\ldots,x_n) \leftarrow \mathsf{emp}$$

where, for a list of variables $x_i,\ldots,x_n$ and an integer $j \geq 0$, we write $[x_i,\ldots,x_n]^j$ for the list rotated to the left j times (e.g., $[x_1,x_2,x_3,x_4,x_5]^2 = x_3,x_4,x_5,x_1,x_2$). In this example, when starting with $A_1(x_1,\ldots,x_n)$ one eventually obtains predicate atoms $A_n(x_{i_1},\ldots,x_{i_n})$, for any permutation $x_{i_1},\ldots,x_{i_n}$ of $x_1,\ldots,x_n$. Since A_n may choose to create or not an interaction with that permutation of variables, the total number of base tuples generated for A_1 is $2^{n!}$. That is, the fixpoint iteration generates $2^{2^{O(n\log n)}}$ base tuples, whereas the size of the input of $\mathsf{Sat}[\Delta,\mathsf{A}]$ is *poly*(n). ∎

4 Degree Boundedness

The boundedness problem (Definition 5, point 2) asks for the existence of a bound on the degree (Definition 3) of the models of a sentence $\exists x_1 \ldots \exists x_{\#\mathsf{A}} \,.\, \mathsf{A}(x_1,\ldots,x_{\#\mathsf{A}})$. Intuitively, the $\mathsf{Bnd}[\Delta,\mathsf{A}]$ problem has a negative answer if and only if there are increasingly large unfoldings (i.e., expansions of a formula by replacement of a predicate atom with one of its definitions) of $\mathsf{A}(x_1,\ldots,x_{\#\mathsf{A}})$ repeating a rule that contains an interaction atom involving a parameter of the rule, which is always bound to the same component. We formalize the notion of unfolding below:

Definition 8. *Given a predicate* A *and a sequence* $(\mathsf{r}_1,i_1),\ldots,(\mathsf{r}_n,i_n) \in (\Delta\times\mathbb{N})^+$, *where* $\mathsf{r}_1 : \mathsf{A}(x_1,\ldots,x_{\#\mathsf{A}}) \leftarrow \phi \in \Delta$, *the* unfolding $\mathsf{A}(x_1,\ldots,x_{\#\mathsf{A}}) \xrightarrow{(\mathsf{r}_1,i_1)\ldots(\mathsf{r}_n,i_n)}_\Delta \psi$ *is inductively defined as (1)* $\psi = \phi$ *if* $n = 1$, *and (2)* ψ *is obtained from* ϕ *by replacing its* i_1*-th predicate atom* $\mathsf{B}(y_1,\ldots,y_{\#\mathsf{B}})$ *with* $\psi_1[x_1/y_1,\ldots,x_{\#\mathsf{B}}/y_{\#\mathsf{B}}]$, *where* $\mathsf{B}(x_1,\ldots,x_{\#\mathsf{B}}) \xrightarrow{(\mathsf{r}_2,i_2)\ldots(\mathsf{r}_n,i_n)}_\Delta \psi_1$ *is an unfolding, if* $n > 1$.

We show that the $\mathsf{Bnd}[\Delta,\mathsf{A}]$ problem can be reduced to the existence of increasingly large unfoldings or, equivalently, a cycle in a finite directed graph, built by a variant of the least fixpoint iteration algorithm used to solve the satisfiability problem (Fig. 3).

Definition 9. *Given satisfiable base pairs* $\mathsf{t},\mathsf{u} \in \mathsf{SatBase}$ *and a rule from* Δ*:*

$$\mathsf{r} : \mathsf{A}(x_1,\ldots,x_{\#\mathsf{A}}) \leftarrow \exists y_1 \ldots \exists y_m \,.\, \phi * \mathsf{B}_1(z_1^1,\ldots,z_{\#\mathsf{B}_1}^1) * \ldots * \mathsf{B}_h(z_1^h,\ldots,z_{\#\mathsf{B}_h}^h)$$

where ϕ *is a quantifier- and predicate-free formula, we write* $(\mathsf{A},\mathsf{t}) \overset{(\mathsf{r},i)}{\rightsquigarrow} (\mathsf{B},\mathsf{u})$ *if and only if* $\mathsf{B} = \mathsf{B}_i$ *and there exist satisfiable base tuples* $\mathsf{t}_1,\ldots,\mathsf{u} = \mathsf{t}_i,\ldots,\mathsf{t}_h \in \mathsf{SatBase}$, *such that* $\mathsf{t} \in (\mathsf{Base}(\phi,\{x_1,\ldots,x_{\#\mathsf{A}}\}) \otimes \bigotimes_{\ell=1}^h \mathsf{t}_\ell[x_1/z_1^\ell,\ldots,x_{\#\mathsf{B}_\ell}/z_{\#\mathsf{B}_\ell}^\ell]) \downarrow_{x_1,\ldots,x_{\#\mathsf{A}}}$. *We define the directed graph with edges labeled by pairs* $(\mathsf{r},i) \in \Delta\times\mathbb{N}$*:*

$$\mathcal{G}(\Delta) \stackrel{\mathsf{def}}{=} (\{\mathsf{def}(\Delta)\times\mathsf{SatBase}\}, \{\langle(\mathsf{A},\mathsf{t}),(\mathsf{r},i),(\mathsf{B},\mathsf{u})\rangle \mid (\mathsf{A},\mathsf{t}) \overset{(\mathsf{r},i)}{\rightsquigarrow} (\mathsf{B},\mathsf{u})\})$$

The graph $\mathcal{G}(\Delta)$ is built by the algorithm in Fig. 3, a slight variation of the classical Kleene iteration algorithm for the computation of the least solution of the constraints of the form (1). A path $(\mathsf{A}_1,\mathsf{t}_1) \overset{(\mathsf{r}_1,i_1)}{\rightsquigarrow} (\mathsf{A}_2,\mathsf{t}_2) \overset{(\mathsf{r}_2,i_2)}{\rightsquigarrow} \ldots \overset{(\mathsf{r}_n,i_n)}{\rightsquigarrow} (\mathsf{A}_n,\mathsf{t}_n)$ in $\mathcal{G}(\Delta)$ induces a unique

input: a SID Δ **output:** $\mathcal{G}(\Delta) = (V, E)$

initially $V := \emptyset, E := \emptyset$
for $\mathsf{A}(x_1, \ldots, x_{\#\mathsf{A}}) \leftarrow \exists y_1 \ldots \exists y_m \,.\, \phi \in \Delta$, with ϕ quantifier- and predicate-free **do**
    $V := V \cup (\{\mathsf{A}\} \times \mathsf{Base}(\phi, \{x_1, \ldots, x_{\#\mathsf{A}}\}){\downarrow}_{x_1,\ldots,x_{\#\mathsf{A}}})$
while V or E still change **do**
    for $\mathsf{r} : \mathsf{A}(x_1, \ldots, x_{\#\mathsf{A}}) \leftarrow \exists y_1 \ldots \exists y_m \,.\, \phi * \mathop{\ast}_{\ell=1}^{h} \mathsf{B}_\ell(z_1^\ell, \ldots, z_{\#\mathsf{B}_\ell}^\ell) \in \Delta$ **do**
        if there exist $(\mathsf{B}_1, \mathsf{t}_1), \ldots, (\mathsf{B}_h, \mathsf{t}_h) \in V$ **then**
            $\mathcal{X} := \left(\mathsf{Base}(\phi, \{x_1, \ldots, x_{\#\mathsf{A}}\}) \otimes \bigotimes_{\ell=1}^{h} \mathsf{t}_\ell[x_1/z_1^\ell, \ldots, x_{\#\mathsf{B}_\ell}/z_{\#\mathsf{B}_\ell}^\ell]\right){\downarrow}_{x_1,\ldots,x_{\#\mathsf{A}}}$
            $V := V \cup (\{\mathsf{A}\} \times \mathcal{X})$
            $E := E \cup \{\langle(\mathsf{A}, \mathsf{t}), (\mathsf{r}, \ell), (\mathsf{B}_\ell, \mathsf{t}_\ell)\rangle \mid \mathsf{t} \in \mathcal{X}, \ell \in [1, h]\}$

Fig. 3. Algorithm for the Construction of $\mathcal{G}(\Delta)$

unfolding $\mathsf{A}_1(x_1, \ldots, x_{\#\mathsf{A}_1}) \xrightarrow{(\mathsf{r}_1,i_1)\ldots(\mathsf{r}_n,i_n)}_\Delta \phi$ (Definition 8). Since the vertices of $\mathcal{G}(\Delta)$ are pairs (A, t), where t is a satisfiable base tuple and the edges of $\mathcal{G}(\Delta)$ reflect the construction of the base tuples from the least solution of the constraints (1), the outcome ϕ of this unfolding is always a satisfiable formula.

An *elementary cycle* of $\mathcal{G}(\Delta)$ is a path from some vertex (B, u) back to itself, such that (B, u) does not occur on the path, except at its endpoints. The cycle is, moreover, *reachable* from (A, t) if and only if there exists a path $(\mathsf{A}, \mathsf{t}) \overset{(\mathsf{r}_1,i_1)}{\rightsquigarrow} \ldots \overset{(\mathsf{r}_n,i_n)}{\rightsquigarrow} (\mathsf{B}, \mathsf{u})$ in $\mathcal{G}(\Delta)$. We reduce the complement of the $\mathsf{Bnd}[\Delta, \mathsf{A}]$ problem, namely the existence of an infinite set of models of $\exists x_1 \ldots \exists x_{\#\mathsf{A}} \,.\, \mathsf{A}(x_1, \ldots, x_{\#\mathsf{A}})$ of unbounded degree, to the existence of a reachable elementary cycle in $\mathcal{G}(\Delta')$, where Δ' is obtained from Δ, as described in the following.

First, we consider, for each predicate $\mathsf{B} \in \mathrm{def}(\Delta)$, a predicate B', of arity $\#\mathsf{B} + 1$, not in $\mathrm{def}(\Delta)$ i.e., the set of predicates for which there exists a rule in Δ. Second, for each rule $\mathsf{B}_0(x_1, \ldots, x_{\#\mathsf{B}_0}) \leftarrow \exists y_1 \ldots \exists y_m \,.\, \phi * \mathop{\ast}_{\ell=2}^{h} \mathsf{B}_\ell(z_1^\ell, \ldots, z_{\#\mathsf{B}_\ell}^\ell) \in \Delta$, where ϕ is a quantifier- and predicate-free formula and $\mathrm{iv}(\phi) \subseteq \mathrm{fv}(\phi)$ denotes the subset of variables occurring in interaction atoms in ϕ, the SID Δ' has the following rules:

$$\mathsf{B}_0'(x_1, \ldots, x_{\#\mathsf{B}_0}, x_{\#\mathsf{B}_0+1}) \leftarrow \exists y_1 \ldots \exists y_m \,.\, \phi * \mathop{\ast}_{\xi \in \mathrm{iv}(\phi)} x_{\#\mathsf{B}_0+1} \neq \xi * \\ \mathop{\ast}_{\ell=2}^{h} \mathsf{B}_\ell'(z_1^\ell, \ldots, z_{\#\mathsf{B}_\ell}^\ell, x_{\#\mathsf{B}_0+1}) \tag{2}$$

$$\mathsf{B}_0'(x_1, \ldots, x_{\#\mathsf{B}_0}, x_{\#\mathsf{B}_0+1}) \leftarrow \exists y_1 \ldots \exists y_m \,.\, \phi * x_{\#\mathsf{B}_0+1} = \xi * \\ \mathop{\ast}_{\ell=2}^{h} \mathsf{B}_\ell'(z_1^\ell, \ldots, z_{\#\mathsf{B}_\ell}^\ell, x_{\#\mathsf{B}_0+1}) \tag{3}$$

for each variable $\xi \in \mathrm{iv}(\phi)$, that occurs in an interaction atom in ϕ.

There exists a family of models (with respect to Δ) of $\exists x_1 \ldots \exists x_{\#\mathsf{A}} \,.\, \mathsf{A}(x_1, \ldots, x_{\#\mathsf{A}})$ of unbounded degree if and only if these are models of $\exists x_1 \ldots \exists x_{\#\mathsf{A}+1} \,.\, \mathsf{A}'(x_1, \ldots, x_{\#\mathsf{A}+1})$ (with respect to Δ') and the last parameter of each predicate $\mathsf{B}' \in \mathrm{def}(\Delta')$ can be mapped, in each of the these models, to a component that occurs in unboundedly many interactions. The latter condition is equivalent to the existence of an elementary cycle, containing a rule of the form (3), that it, moreover, reachable from some vertex $(\mathsf{A}', \mathsf{t})$ of $\mathcal{G}(\Delta')$, for some $\mathsf{t} \in \mathsf{SatBase}$. This reduction is formalized below:

Lemma 2. *There exists an infinite sequence of configurations* $\gamma_1, \gamma_2, \ldots$ *such that* $\gamma_i \models_\Delta \exists x_1 \ldots \exists x_{\#\mathsf{A}} . \mathsf{A}(x_1, \ldots, x_{\#\mathsf{A}})$ *and* $\delta(\gamma_i) < \delta(\gamma_{i+1})$, *for all* $i \geq 1$ *if and only if* $\mathcal{G}(\Delta')$ *has an elementary cycle containing a rule (3), reachable from a node* $(\mathsf{A}', \mathfrak{t})$, *for* $\mathfrak{t} \in \mathsf{SatBase}$.

The complexity result below uses a similar argument on the maximal size of (hence the number of) base tuples as in Theorem 1, leading to similar complexity gaps:

Theorem 2. $\mathsf{Bnd}^{(k,\infty)}[\Delta, \mathsf{A}]$ *is in* co-NP, $\mathsf{Bnd}^{(\infty,\ell)}[\Delta, \mathsf{A}]$ *is in* EXP, $\mathsf{Bnd}[\Delta, \mathsf{A}]$ *is in* 2EXP.

Moreover, the construction of $\mathcal{G}(\Delta')$ allows to prove the following cut-off result:

Proposition 1. *Let* γ *be a configuration and* ν *be a store, such that* $\gamma \models^\nu_\Delta \mathsf{A}(x_1, \ldots, x_{\#\mathsf{A}})$. *If* $\mathsf{Bnd}^{(k,\ell)}[\Delta, \mathsf{A}]$ *then (1)* $\delta(\gamma) = poly(\mathrm{size}(\Delta))$ *if* $k < \infty$, $\ell = \infty$, *(2)* $\delta(\gamma) = 2^{poly(\mathrm{size}(\Delta))}$ *if* $k = \infty$, $\ell < \infty$ *and (3)* $\delta(\gamma) = 2^{2^{poly(\mathrm{size}(\Delta))}}$ *if* $k = \infty$, $\ell = \infty$.

5 Entailment

This section is concerned with the entailment problem $\mathsf{Entl}[\Delta, \mathsf{A}, \mathsf{B}]$, that asks whether $\gamma \models^\nu_\Delta \exists x_{\#\mathsf{A}+1} \ldots \exists x_{\#\mathsf{B}} . \mathsf{B}(x_1, \ldots, x_{\#\mathsf{B}})$, for every configuration γ and store ν, such that $\gamma \models^\nu_\Delta \mathsf{A}(x_1, \ldots, x_{\#\mathsf{A}})$. For instance, the proof from Fig. 1(c) relies on the following entailments, that occur as the side conditions of the Hoare logic rule of consequence:

$$\mathsf{ring}_{h,t}(y) \models_\Delta \exists x \exists z . [y]@\mathsf{H} * \langle y.out, z.in \rangle * \mathsf{chain}_{h-1,t}(z,x) * \langle x.out, y.in \rangle$$
$$[z]@\mathsf{H} * \langle z.out, x.in \rangle * \mathsf{chain}_{h-1,t}(x,y) * \langle y.out, z.in \rangle \models_\Delta \mathsf{ring}_{h,t}(z)$$

By introducing two fresh predicates A_1 and A_2, defined by the rules:

$$\mathsf{A}_1(x_1) \leftarrow \exists y \exists z . [x_1]@\mathsf{H} * \langle x_1.out, z.in \rangle * \mathsf{chain}_{h-1,t}(z,y) * \langle y.out, x_1.in \rangle \quad (4)$$
$$\mathsf{A}_2(x_1,x_2) \leftarrow \exists z . [x_1]@\mathsf{H} * \langle x_1.out, z.in \rangle * \mathsf{chain}_{h-1,t}(z,x_2) * \langle x_2.out, x_1.in \rangle \quad (5)$$

the above entailments are equivalent to $\mathsf{Entl}[\Delta, \mathsf{ring}_{h,t}, \mathsf{A}_1]$ and $\mathsf{Entl}[\Delta, \mathsf{A}_2, \mathsf{ring}_{h,t}]$, respectively, where Δ consists of the rules (4) and (5), together with the rules that define the $\mathsf{ring}_{h,t}$ and $\mathsf{chain}_{h,t}$ predicates (Sect. 1.1).

We show that the entailment problem is undecidable, in general (Thm. 3), and recover a decidable fragment, by means of three syntactic conditions, typically met in our examples. These conditions use the following notion of *profile*:

Definition 10. *The* profile *of a SID* Δ *is the pointwise greatest function* $\lambda_\Delta : \mathbb{A} \rightarrow pow(\mathbb{N})$, *mapping each predicate* A *into a subset of* $[1, \#\mathsf{A}]$, *such that, for each rule* $\mathsf{A}(x_1, \ldots, x_{\#\mathsf{A}}) \leftarrow \phi$ *from* Δ, *each atom* $\mathsf{B}(y_1, \ldots, y_{\#\mathsf{B}})$ *from* ϕ *and each* $i \in \lambda_\Delta(\mathsf{B})$, *there exists* $j \in \lambda_\Delta(\mathsf{A})$, *such that* x_j *and* y_i *are the same variable.*

The profile identifies the parameters of a predicate that are always replaced by a variable $x_1, \ldots, x_{\#\mathsf{A}}$ in each unfolding of $\mathsf{A}(x_1, \ldots, x_{\#\mathsf{A}})$, according to the rules in Δ; it is computed by a greatest fixpoint iteration, in time $poly(\mathrm{size}(\Delta))$.

Definition 11. *A rule* $\mathsf{A}(x_1, \ldots, x_{\#\mathsf{A}}) \leftarrow \exists y_1 \ldots \exists y_m . \phi * \mathop{\Large *}_{\ell=1}^{h} \mathsf{B}_\ell(z^\ell_1, \ldots, z^\ell_{\#\mathsf{B}_\ell})$, *where* ϕ *is a quantifier- and predicate-free formula, is said to be:*

1. progressing *if and only if* $\phi = [x_1] * \psi$, *where* ψ *consists of interaction atoms involving* x_1 *and (dis-)equalities, such that* $\bigcup_{\ell=1}^{h}\{z_1^\ell,\ldots,z_{\#\mathsf{B}_\ell}^\ell\} = \{x_2,\ldots,x_{\#\mathsf{A}}\} \cup \{y_1,\ldots,y_m\}$,
2. connected *if and only if, for each* $\ell \in [1,h]$ *there exists an interaction atom in* ψ *that contains both* z_1^ℓ *and a variable from* $\{x_1\} \cup \{x_i \mid i \in \lambda_\Delta(\mathsf{A})\}$,
3. equationally-restricted (e-restricted) *if and only if, for every disequation* $x \neq y$ *from* ϕ, *we have* $\{x,y\} \cap \{x_i \mid i \in \lambda_\Delta(\mathsf{A})\} \neq \emptyset$.

A SID Δ *is* progressing, connected *and* e-restricted *if and only if each rule in* Δ *is* progressing, connected *and* e-restricted, *respectively.*

For example, the SID consisting of the rules from Sect. 1.1, together with rules (4) and (5) is progressing, connected and e-restricted.

We recall that $\mathrm{def}_\Delta(\mathsf{A})$ is the set of rules from Δ that define A and denote by $\mathrm{def}_\Delta^*(\mathsf{A})$ the least superset of $\mathrm{def}_\Delta(\mathsf{A})$ containing the rules that define a predicate from a rule in $\mathrm{def}_\Delta^*(\mathsf{A})$. The following result shows that the entailment problem becomes undecidable as soon as the connectivity condition is even slightly lifted:

Theorem 3. $\mathsf{Entl}[\Delta,\mathsf{A},\mathsf{B}]$ *is undecidable, even when* Δ *is progressing and e-restricted, and only the rules in* $\mathrm{def}_\Delta^*(\mathsf{A})$ *are connected (the rules in* $\mathrm{def}_\Delta^*(\mathsf{B})$ *may be disconnected).*

On the positive side, we prove that $\mathsf{Entl}[\Delta,\mathsf{A},\mathsf{B}]$ is decidable, if Δ is progressing, connected and e-restricted, assuming further that $\mathsf{Bnd}[\Delta,\mathsf{A}]$ has a positive answer. In this case, the bound on the degree of the models of $\mathsf{A}(x_1,\ldots,x_{\#\mathsf{A}})$ is effectively computable, using the algorithm from Fig. 3 (see Proposition 1 for a cut-off result) and denote by $\mathfrak{B}$ this bound, throughout this section.

The proof uses a reduction of $\mathsf{Entl}[\Delta,\mathsf{A},\mathsf{B}]$ to a similar problem for SL, showed to be decidable [18]. We recall the definition of SL, interpreted over heaps $\mathsf{h} : \mathbb{C} \rightharpoonup_{fin} \mathbb{C}^{\mathfrak{K}}$, introduced in Sect. 2.3. SL rules are denoted as $\overline{\mathsf{A}}(x_1,\ldots,x_{\#(\overline{\mathsf{A}})}) \leftarrow \phi$, where ϕ is a SL formula, such that $\mathrm{fv}(\phi) \subseteq \{x_1,\ldots,x_{\#(\overline{\mathsf{A}})}\}$ and SL SIDs are denoted as $\overline{\Delta}$. The profile $\lambda_{\overline{\Delta}}$ is defined for SL same as for CL (Definition 10).

Definition 12. *A* SL *rule* $\overline{\mathsf{A}}(x_1,\ldots,x_{\#(\overline{\mathsf{A}})}) \leftarrow \phi$ *from a SID* $\overline{\Delta}$ *is said to be:*

1. progressing *if and only if* $\phi = \exists t_1 \ldots \exists t_m \,.\, x_1 \mapsto (y_1,\ldots,y_{\mathfrak{K}}) * \psi$, *where* ψ *contains only predicate and equality atoms,*
2. connected *if and only if* $z_1 \in \{x_i \mid i \in \lambda_{\overline{\Delta}}(\overline{\mathsf{A}})\} \cup \{y_1,\ldots,y_{\mathfrak{K}}\}$, *for every predicate atom* $\overline{\mathsf{B}}(z_1,\ldots,z_{\#(\overline{\mathsf{B}})})$ *from* ϕ.

Note that the definitions of progressing and connected rules are different for SL, compared to CL (Definition 11); in the rest of this section, we rely on the context to distinguish progressing (connected) SL rules from progressing (connected) CL rules. Moreover, e-restricted rules are defined in the same way for CL and SL (point 3 of Definition 11). A tight upper bound on the complexity of the entailment problem between SL formulæ, interpreted by progressing, connected and e-restricted SIDs, is given below:

Theorem 4 ([18]). *The* SL *entailment problem is in* $2^{2^{poly(\mathrm{width}(\overline{\Delta})\cdot\log\mathrm{size}(\overline{\Delta}))}}$, *for progressing, connected and e-restricted SIDs.*

The reduction of $\mathsf{Entl}[\Delta, \mathsf{A}, \mathsf{B}]$ to SL entailments is based on the idea of viewing a configuration as a logical structure (hypergraph), represented by a undirected *Gaifman graph*, in which every tuple from a relation (hyperedge) becomes a clique [24]. In a similar vein, we encode a configuration, of degree at most $\mathfrak{B}$, by a heap of degree $\mathfrak{K}$ (Definition 13), such that $\mathfrak{K}$ is defined using the following integer function:

$$\mathrm{pos}(i,j,k) \stackrel{\mathsf{def}}{=} 1 + \mathfrak{B} \cdot \sum_{\ell=1}^{j-1} |\tau_\ell| + i \cdot |\tau_j| + k$$

where $\mathsf{Inter} \stackrel{\mathsf{def}}{=} \{\tau_1, \ldots, \tau_M\}$ is the set of interaction types and $Q \stackrel{\mathsf{def}}{=} \{q_1, \ldots, q_N\}$ is the set of states of the behavior $\mathbb{B} = (\mathcal{P}, Q, \rightarrow)$ (Sect. 2). Here $i \in [0, \mathfrak{B}-1]$ denotes an interaction of type $j \in [1, M]$ and $k \in [0, N-1]$ denotes a state. We use M and N throughout the rest of this section, to denote the number of interaction types and states, respectively.

For a set I of interactions, let $\mathsf{Tuples}_I^j(c) \stackrel{\mathsf{def}}{=} \{\langle c_1, \ldots, c_n \rangle \mid (c_1, p_1, \ldots, c_n, p_n) \in I,\ \tau_j = \langle p_1, \ldots, p_n \rangle,\ c \in \{c_1, \ldots, c_n\}\}$ be the tuples of components from an interaction of type τ_j from I, that contain a given component c.

Definition 13. *Given a configuration* $\gamma = (C, I, \rho)$, *such that* $\delta(\gamma) \leq \mathfrak{B}$, *a* Gaifman heap *for* γ *is a heap* $\mathsf{h} : \mathbb{C} \rightharpoonup_{fin} \mathbb{C}^{\mathfrak{K}}$, *where* $\mathfrak{K} \stackrel{\mathsf{def}}{=} \mathrm{pos}(0, M+1, N)$, $\mathrm{dom}(\mathsf{h}) = \mathrm{nodes}(\gamma)$ *and, for all* $c_0 \in \mathrm{dom}(\mathsf{h})$, *such that* $\mathsf{h}(c_0) = \langle c_1, \ldots, c_{\mathfrak{K}} \rangle$, *the following hold:*

1. $c_1 = c_0$ *if and only if* $c_0 \in C$,
2. *for all* $j \in [1, M]$, $\mathsf{Tuples}_I^j(c) = \{\mathbf{c}_1, \ldots, \mathbf{c}_s\}$ *if and only if there exist integers* $0 \leq k_1 < \ldots < k_s < \mathfrak{B}$, *such that* $\langle \mathsf{h}(c_0) \rangle_{\mathrm{inter}(k_i, j)} = \mathbf{c}_i$, *for all* $i \in [1, s]$, *where* $\mathrm{inter}(i, j) \stackrel{\mathsf{def}}{=} [\mathrm{pos}(i-1, j, 0), \mathrm{pos}(i, j, 0)]$ *are the entries of the i-th interaction of type* τ_j *in* $\mathsf{h}(c_0)$,
3. *for all* $k \in [1, N]$, *we have* $\langle \mathsf{h}(c_0) \rangle_{\mathrm{state}(k)} = c_0$ *if and only if* $\rho(c_0) = q_k$, *where the entry* $\mathrm{state}(k) \stackrel{\mathsf{def}}{=} \mathrm{pos}(0, M+1, k-1)$ *in* $\mathsf{h}(c_0)$ *corresponds to the state* $q_k \in Q$.

We denote by $\mathbb{G}(\gamma)$ *the set of Gaifman heaps for* γ.

Intuitively, if h is a Gaifman heap for γ and $c_0 \in \mathrm{dom}(\mathsf{h})$, then the first entry of $\mathsf{h}(c_0)$ indicates whether c_0 is present (condition 1 of Definition 13), the next $\mathfrak{B} \cdot \sum_{j=1}^{M} |\tau_j|$ entries are used to encode the interactions of each type τ_j (condition 2 of Definition 13), whereas the last N entries are used to represent the state of the component (condition 3 of Definition 13). Note that the encoding of configurations by Gaifman heaps is not unique: two Gaifman heaps for the same configuration may differ in the order of the tuples from the encoding of an interaction type and the choice of the unconstrained entries from $\mathsf{h}(c_0)$, for each $c_0 \in \mathrm{dom}(\mathsf{h})$. On the other hand, if two configurations have the same Gaifman heap encoding, they must be the same configuration.

Example 3. Figure 4(b) shows a Gaifman heap for the configuration in Fig. 4(a), where each component belongs to at most 2 interactions of type $\langle out, in \rangle$. ■

We build a SL SID $\overline{\Delta}$ that generates the Gaifman heaps of the models of the predicate atoms occurring in a progressing CL SID Δ. The construction associates to each variable x, that occurs free or bound in a rule from Δ, a unique $\mathfrak{K}$-tuple of variables $\eta(x) \in \mathbb{V}^{\mathfrak{K}}$,

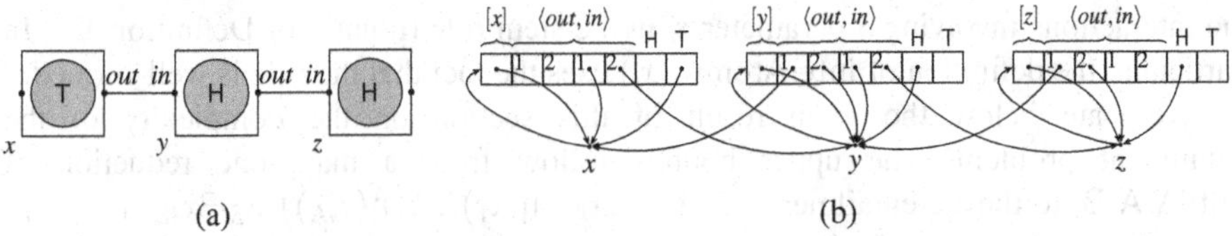

Fig. 4. Gaifman Heap for a Chain Configuration

that represents the image of the store value $\nu(x)$ in a Gaifman heap h i.e., $\mathsf{h}(\nu(x)) = \nu(\eta(x))$. Moreover, we consider, for each predicate symbol $\mathsf{A} \in \mathrm{def}(\Delta)$, an annotated predicate symbol $\overline{\mathsf{A}}_\iota$ of arity $\#\overline{\mathsf{A}}_\iota = (\mathfrak{K}+1)\cdot\#\mathsf{A}$, where $\iota : [1,\#\mathsf{A}] \times [1,M] \rightarrow 2^{[0,\mathfrak{B}-1]}$ is a map associating each parameter $i \in [1,\#\mathsf{A}]$ and each interaction type τ_j, for $j \in [1,M]$, a set of integers $\iota(i,j)$ denoting the positions of the encodings of the interactions of type τ_j, involving the value of x_i, in the models of $\overline{\mathsf{A}}_\iota(x_1,\ldots,x_{\#\mathsf{A}},\eta(x_1),\ldots,\eta(x_{\#\mathsf{A}}))$ (point 2 of Definition 13). Then $\overline{\Delta}$ contains rules of the form:

$$\overline{\mathsf{A}}_\iota(x_1,\ldots,x_{\#(\mathsf{A})},\eta(x_1),\ldots,\eta(x_{\#(\mathsf{A})})) \leftarrow \tag{6}$$
$$\exists y_1\ldots\exists y_m \exists\eta(y_1)\ldots\exists\eta(y_m)\,.\,\overline{\psi} * \pi * \mathop{\Large *}_{\ell=1}^{h} \overline{\mathsf{B}}^{\ell}_{\iota\ell}(z^\ell_1,\ldots,z^\ell_{\#(\mathsf{B}^\ell)},\eta(z^\ell_1),\ldots,\eta(z^\ell_{\#(\mathsf{B}^\ell)}))$$

for which Δ has a *stem rule* $\mathsf{A}(x_1,\ldots,x_{\#(\mathsf{A})}) \leftarrow \exists y_1\ldots\exists y_m\,.\,\psi*\pi*\mathop{\Large *}_{\ell=1}^{h}\mathsf{B}^\ell(z^\ell_1,\ldots,z^\ell_{\#\mathsf{B}^\ell})$, where $\psi*\pi$ is a quantifier- and predicate-free formula and π is the conjunction of equalities and disequalities from $\psi*\pi$. However, not all rules (6) are considered in $\overline{\Delta}$, but only the ones meeting the following condition:

Definition 14. *A rule of the form (6) is* well-formed *if and only if, for each $i \in [1,\#\mathsf{A}]$ and each $j \in [1,M]$, there exists a set of integers $Y_{i,j} \subseteq [0,\mathfrak{B}-1]$, such that:*

- $\|Y_{i,j}\| = \|I^j_{\psi,\pi}(x_i)\|$, *where $I^j_{\psi,\pi}(x)$ is the set of interaction atoms $\langle z_1.p_1,\ldots,z_n.p_n\rangle$ from ψ of type $\tau_j = \langle p_1,\ldots,p_n\rangle$, such that $z_s \approx_\pi x$, for some $s \in [1,n]$,*
- $Y_{i,j} \subseteq \iota(i,j)$ *and* $\iota(i,j)\backslash Y_{i,j} = Z_j(x_i)$, *where* $Z_j(x) \stackrel{\text{def}}{=} \bigcup_{\ell=1}^{h}\bigcup_{k=1}^{\#\mathsf{B}^\ell}\{\iota^\ell(k,j) \mid x \approx_\pi z^\ell_k\}$ *is the set of positions used to encode the interactions of type τ_j involving the store value of the parameter x, in the sub-configuration corresponding to an atom $\mathsf{B}_\ell(z^\ell_1,\ldots,z^\ell_{\#(\mathsf{B}^\ell)})$, for some $\ell \in [1,h]$.*

We denote by $\overline{\Delta}$ the set of well-formed rules (6), such that, moreover:

$$\overline{\psi} \stackrel{\text{def}}{=} x_1 \mapsto \eta(x_1) * \mathop{\Large *}_{x\in\mathrm{fv}(\psi)} \mathsf{CompStates}_\psi(x) * \mathop{\Large *}_{i=1}^{\#\mathsf{A}} \mathsf{InterAtoms}_\psi(x_i), \text{ where:}$$
$$\mathsf{CompStates}_\psi(x) \stackrel{\text{def}}{=} \mathop{\Large *}_{[x]\text{ occurs in }\psi} \langle\eta(x)\rangle_1 = x * \mathop{\Large *}_{x@q_k\text{ occurs in }\psi} \langle\eta(x)\rangle_{\mathrm{state}(k)} = x$$
$$\mathsf{InterAtoms}_\psi(x_i) \stackrel{\text{def}}{=} \mathop{\Large *}_{j=1}^{M}\mathop{\Large *}_{p=1}^{r_j} \langle\eta(x_i)\rangle_{\mathrm{inter}(j,k^j_p)} = \mathbf{x}^j_p \text{ and } \{k^j_1,\ldots,k^j_{r_j}\} \stackrel{\text{def}}{=} \iota(i,j)\backslash Z_j(x_i)$$

Here for two tuples of variables $\mathbf{x} = \langle x_1,\ldots,x_k\rangle$ and $\mathbf{y} = \langle y_1,\ldots,y_k\rangle$, we denote by $\mathbf{x} = \mathbf{y}$ the formula $\mathop{\Large *}_{i=1}^{k} x_i = y_i$. Intuitively, the SL formula $\mathsf{CompStates}_\psi(x)$ realizes the encoding of the component and state atoms from ψ, in the sense of points (1) and (3) from Definition 13, whereas the formula $\mathsf{InterAtoms}_\psi(x_i)$ realizes the encodings of the interactions involving a parameter x_i in the stem rule (point 2 of Definition 13). In particular, the definition of $\mathsf{InterAtoms}_\psi(x_i)$ uses the fact that the rule is well-formed.

We state below the main result of this section on the complexity of the entailment problem. The upper bounds follow from a many-one reduction of $\mathsf{Entl}[\Delta, \mathsf{A}, \mathsf{B}]$ to the SL entailment $\overline{\mathsf{A}}_\iota(x_1, \ldots, x_{\#\mathsf{A}}, \eta(x_1), \ldots, \eta(x_{\#\mathsf{A}})) \Vdash_{\overline{\Delta}} \exists x_{\#\mathsf{B}+1} \ldots \exists x_{\#\mathsf{B}} \exists \eta(x_{\#\mathsf{B}+1}) \ldots \exists \eta(x_{\#\mathsf{B}})$. $\overline{\mathsf{B}}_{\iota'}(x_1, \ldots, x_{\#\mathsf{B}}, \eta(x_1), \ldots, \eta(x_{\#\mathsf{B}}))$, in combination with the upper bound provided by Theorem 4, for SL entailments. If $k < \infty$, the complexity is tight for CL, whereas gaps occur for $k = \infty, \ell < \infty$ and $k = \infty, \ell = \infty$, due to the cut-off on the degree bound (Proposition 1), which impacts the size of $\overline{\Delta}$ and time needed to generate it from Δ.

Theorem 5. *If Δ is progressing, connected and e-restricted and, moreover,* $\mathsf{Bnd}[\Delta, \mathsf{A}]$ *has a positive answer,* $\mathsf{Entl}^{k,\ell}[\Delta, \mathsf{A}, \mathsf{B}]$ *is in* 2EXP, $\mathsf{Entl}^{\infty,\ell}[\Delta, \mathsf{A}, \mathsf{B}]$ *is in* 3EXP $\cap$ 2EXP-*hard, and* $\mathsf{Entl}[\Delta, \mathsf{A}, \mathsf{B}]$ *is in* 4EXP $\cap$ 2EXP-*hard.*

6 Conclusions and Future Work

We study the satisfiability and entailment problems in a logic used to write proofs of correctness for dynamically reconfigurable distributed systems. The logic views the components and interactions from the network as resources and reasons also about the local states of the components. We reuse existing techniques for Separation Logic [39], showing that our configuration logic is more expressive than SL, fact which is confirmed by a number of complexity gaps. Closing up these gaps and finding tight complexity classes in the more general cases is considered for future work. In particular, we aim at lifting the boundedness assumption on the degree of the configurations that must be considered to check the validity of entailments.

References

1. Ahrens, E., Bozga, M., Iosif, R., Katoen, J.: Local reasoning about parameterized reconfigurable distributed systems. CoRR, abs/2107.05253 (2021)
2. Arbab, F.: Reo: a channel-based coordination model for component composition. Math. Struct. Comput. Sci. **14**(3), 329–366 (2004)
3. Basu, A., Bozga, M., Sifakis, J.: Modeling heterogeneous real-time components in BIP. In: Fourth IEEE International Conference on Software Engineering and Formal Methods (SEFM 2006), pp. 3–12. IEEE Computer Society (2006)
4. Bloem, R., et al.: Decidability of Parameterized Verification. Synthesis Lectures on Distributed Computing Theory. Morgan & Claypool Publishers (2015)
5. Bozga, M., Bueri, L., Iosif, R.: Decision problems in a logic for reasoning about reconfigurable distributed systems. CoRR, abs/2202.09637 (2022)
6. Bozga, M., Iosif, R., Sifakis, J.: Verification of component-based systems with recursive architectures. CoRR, abs/2112.08292 (2021)
7. Bradbury, J., Cordy, J., Dingel, J., Wermelinger, M.: A survey of self-management in dynamic software architecture specifications. In: Proceedings of the 1st ACM SIGSOFT workshop on Self-managed systems, pp. 28–33. ACM (2004)
8. Brookes, S., O'Hearn, P.W.: Concurrent separation logic. ACM SIGLOG News **3**(3), 47–65 (2016)

9. Brotherston, J., Fuhs, C., Pérez, J.A.N., Gorogiannis, N.: A decision procedure for satisfiability in separation logic with inductive predicates. In: CSL-LICS, pp. 25:1–25:10. ACM (2014)
10. Bucchiarone, A., Galeotti, J.P.: Dynamic software architectures verification using dynalloy. Electron. Commun. Eur. Assoc. Softw. Sci. Technol. **10** (2008). https://doi.org/10.14279/tuj.eceasst.10.145
11. Butting, A., Heim, R., Kautz, O., Ringert, J.O., Rumpe, B., Wortmann, A.: A classification of dynamic reconfiguration in component and connector architecture description. In: Proceedings of MODELS 2017 Satellite Event: Workshops (ModComp). CEUR Workshop Proceedings, vol. 2019, pp. 10–16. CEUR-WS.org (2017)
12. Calcagno, C., O'Hearn, P.W., Yang, H.: Local action and abstract separation logic. In: 22nd IEEE Symposium on Logic in Computer Science (LICS 2007), 10–12 July 2007, Wroclaw, Poland, Proceedings, pp. 366–378. IEEE Computer Society (2007)
13. Cavalcante, E., Batista, T.V., Oquendo, F.: Supporting dynamic software architectures: from architectural description to implementation. In: Bass, L., Lago, P., Kruchten, P. (eds.) 12th Working IEEE/IFIP Conference on Software Architecture, WICSA 2015, pp. 31–40. IEEE Computer Society (2015)
14. Clarke, D.: A basic logic for reasoning about connector reconfiguration. Fundam. Inf. **82**(4), 361–390 (2008)
15. Dinsdale-Young, T., Birkedal, L., Gardner, P., Parkinson, M., Yang, H.: Views: compositional reasoning for concurrent programs. SIGPLAN Not. **48**(1), 287–300 (2013)
16. Dinsdale-Young, T., Dodds, M., Gardner, P., Parkinson, M.J., Vafeiadis, V.: Concurrent abstract predicates. In: D'Hondt, T. (ed.) ECOOP 2010. LNCS, vol. 6183, pp. 504–528. Springer, Heidelberg (2010). https://doi.org/10.1007/978-3-642-14107-2_24
17. Dormoy, J., Kouchnarenko, O., Lanoix, A.: Using temporal logic for dynamic reconfigurations of components. In: Barbosa, L.S., Lumpe, M. (eds.) FACS 2010. LNCS, vol. 6921, pp. 200–217. Springer, Heidelberg (2012). https://doi.org/10.1007/978-3-642-27269-1_12
18. Echenim, M., Iosif, R., Peltier, N.: Unifying decidable entailments in separation logic with inductive definitions. In: Platzer, A., Sutcliffe, G. (eds.) CADE 2021. LNCS (LNAI), vol. 12699, pp. 183–199. Springer, Cham (2021). https://doi.org/10.1007/978-3-030-79876-5_11
19. El-Ballouli, R., Bensalem, S., Bozga, M., Sifakis, J.: Programming dynamic reconfigurable systems. Int. J. Softw. Tools Technol. Transf. **23**, 701–719 (2021)
20. El-Hokayem, A., Bozga, M., Sifakis, J.: A temporal configuration logic for dynamic reconfigurable systems. In: Hung, C., Hong, J., Bechini, A., Song, E. (eds.) SAC 2021: The 36th ACM/SIGAPP Symposium on Applied Computing, Virtual Event, Republic of Korea, 22–26 March 2021, pp. 1419–1428. ACM (2021)
21. Farka, F., Nanevski, A., Banerjee, A., Delbianco, G.A., Fábregas, I.: On algebraic abstractions for concurrent separation logics. Proc. ACM Program. Lang. **5**(POPL), 1–32 (2021)
22. Feng, X., Ferreira, R., Shao, Z.: On the relationship between concurrent separation logic and assume-guarantee reasoning. In: De Nicola, R. (ed.) ESOP 2007. LNCS, vol. 4421, pp. 173–188. Springer, Heidelberg (2007). https://doi.org/10.1007/978-3-540-71316-6_13
23. Foerster, K., Schmid, S.: Survey of reconfigurable data center networks: enablers, algorithms, complexity. SIGACT News **50**(2), 62–79 (2019)
24. Gaifman, H.: On local and non-local properties. Stud. Log. Found. Math. **107**, 105–135 (1982)
25. Gunawi, H.S., et al.: Why does the cloud stop computing? Lessons from hundreds of service outages. In: Proceedings of the Seventh ACM Symposium on Cloud Computing, SoCC 2016, pp. 1–16. Association for Computing Machinery, New York (2016)
26. Hirsch, D., Inverardi, P., Montanari, U.: Graph grammars and constraint solving for software architecture styles. In: Proceedings of the Third International Workshop on Software Architecture, ISAW 1998, pp. 69–72. Association for Computing Machinery, New York (1998)

27. Jansen, C., Katelaan, J., Matheja, C., Noll, T., Zuleger, F.: Unified reasoning about robustness properties of symbolic-heap separation logic. In: Yang, H. (ed.) ESOP 2017. LNCS, vol. 10201, pp. 611–638. Springer, Heidelberg (2017). https://doi.org/10.1007/978-3-662-54434-1_23
28. Jones, C.B.: Developing methods for computer programs including a notion of interference. Ph.D. thesis, University of Oxford, UK (1981)
29. Konnov, I.V., Kotek, T., Wang, Q., Veith, H., Bliudze, S., Sifakis, J.: Parameterized systems in BIP: design and model checking. In: 27th International Conference on Concurrency Theory, CONCUR 2016, volume 59 of LIPIcs, pp. 30:1–30:16. Schloss Dagstuhl - Leibniz-Zentrum für Informatik (2016)
30. Krause, C., Maraikar, Z., Lazovik, A., Arbab, F.: Modeling dynamic reconfigurations in Reo using high-level replacement systems. Sci. Comput. Program. **76**, 23–36 (2011)
31. Lanoix, A., Dormoy, J., Kouchnarenko, O.: Combining proof and model-checking to validate reconfigurable architectures. Electron. Notes Theor. Comput. Sci. **279**(2), 43–57 (2011)
32. Le Metayer, D.: Describing software architecture styles using graph grammars. IEEE Trans. Softw. Eng. **24**(7), 521–533 (1998)
33. Magee, J., Kramer, J.: Dynamic structure in software architectures. In: ACM SIGSOFT Software Engineering Notes, vol. 21, no. 6, pp. 3–14. ACM (1996)
34. Mavridou, A., Baranov, E., Bliudze, S., Sifakis, J.: Configuration logics: modeling architecture styles. J. Log. Algebr. Meth. Program. **86**(1), 2–29 (2017)
35. Noormohammadpour, M., Raghavendra, C.S.: Datacenter traffic control: understanding techniques and tradeoffs. IEEE Commun. Surv. Tutor. **20**(2), 1492–1525 (2018)
36. O'Hearn, P.W.: Resources, concurrency, and local reasoning. Theor. Comput. Sci. **375**(1–3), 271–307 (2007)
37. O'Hearn, P.W., Pym, D.J.: The logic of bunched implications. Bull. Symb. Log. **5**(2), 215–244 (1999)
38. Owicki, S., Gries, D.: An axiomatic proof technique for parallel programs. In: Gries, D. (ed.) Programming Methodology. Texts and Monographs in Computer Science, pp. 130–152. Springer, New York (1978). https://doi.org/10.1007/978-1-4612-6315-9_12
39. Reynolds, J.C.: Separation logic: a logic for shared mutable data structures. In: Proceedings of 17th IEEE Symposium on Logic in Computer Science (LICS 2002), 22–25 July 2002, Copenhagen, Denmark, pp. 55–74. IEEE Computer Society (2002)
40. Shtadler, Z., Grumberg, O.: Network grammars, communication behaviors and automatic verification. In: Sifakis, J. (ed.) CAV 1989. LNCS, vol. 407, pp. 151–165. Springer, Heidelberg (1990). https://doi.org/10.1007/3-540-52148-8_13
41. Taentzer, G., Goedicke, M., Meyer, T.: Dynamic change management by distributed graph transformation: towards configurable distributed systems. In: Ehrig, H., Engels, G., Kreowski, H.-J., Rozenberg, G. (eds.) TAGT 1998. LNCS, vol. 1764, pp. 179–193. Springer, Heidelberg (2000). https://doi.org/10.1007/978-3-540-46464-8_13
42. Vafeiadis, V., Parkinson, M.: A marriage of rely/guarantee and separation logic. In: Caires, L., Vasconcelos, V.T. (eds.) CONCUR 2007. LNCS, vol. 4703, pp. 256–271. Springer, Heidelberg (2007). https://doi.org/10.1007/978-3-540-74407-8_18
43. Wermelinger, M.: Towards a chemical model for software architecture reconfiguration. IEE Proc.-Softw. **145**(5), 130–136 (1998)
44. Wermelinger, M., Fiadeiro, J.L.: A graph transformation approach to software architecture reconfiguration. Sci. Comput. Program. **44**(2), 133–155 (2002)

Proving Non-Termination and Lower Runtime Bounds with LoAT (System Description)

Florian Frohn(✉) and Jürgen Giesl(✉)

LuFG Informatik 2, RWTH Aachen University, Aachen, Germany
florian.frohn@cs.rwth-aachen.de, giesl@informatik.rwth-aachen.de

Abstract. We present the *Loop Acceleration Tool* (LoAT), a powerful tool for proving non-termination and worst-case lower bounds for programs operating on integers. It is based on the novel calculus from [10,11] for *loop acceleration*, i.e., transforming loops into non-deterministic straight-line code, and for finding non-terminating configurations. To implement it efficiently, LoAT uses a new approach based on unsat cores. We evaluate LoAT's power and performance by extensive experiments.

1 Introduction

Efficiency is one of the most important properties of software. Consequently, *automated complexity analysis* is of high interest to the software verification community. Most research in this area has focused on deducing *upper* bounds on the worst-case complexity of programs. In contrast, the *Loop Acceleration Tool* LoAT aims to find performance bugs by deducing *lower* bounds on the worst-case complexity of programs operating on integers. Since non-termination implies the lower bound ∞, LoAT is also equipped with non-termination techniques.

LoAT is based on *loop acceleration* [4,5,9–11,15], which replaces loops by non-deterministic code: The resulting program chooses a value n, representing the number of loop iterations in the original program. To be sound, suitable constraints on n are synthesized to ensure that the original loop allows for at least n iterations. Moreover, the transformed program updates the program variables to the same values as n iterations of the original loop, but it does so in a single step. To achieve that, the loop body is transformed into a *closed form*, which is parameterized in n. In this way, LoAT is able to compute *symbolic under-approximations* of programs, i.e., every execution path in the resulting transformed program corresponds to a path in the original program, but not necessarily vice versa. In contrast to many other techniques for computing under-approximations, the symbolic approximations of LoAT cover *infinitely many runs* of *arbitrary length*.

Funded by the Deutsche Forschungsgemeinschaft (DFG, German Research Foundation) - 235950644 (Project GI 274/6-2).

J. Blanchette et al. (Eds.): IJCAR 2022, LNAI 13385, pp. 712–722, 2022.
https://doi.org/10.1007/978-3-031-10769-6_41

Contributions: The main new feature of the novel version of LoAT presented in this paper is the *integration* of the *loop acceleration calculus* from [10,11], which combines different loop acceleration techniques in a modular way, into LoAT's framework. This enables LoAT to use the loop acceleration calculus for the analysis of full integer programs, whereas the standalone implementation of the calculus from [10,11] was only applicable to single loops without branching in the body. To control the application of the calculus, we use a new technique based on unsat cores (see Sect. 5). The new version of LoAT is evaluated in extensive experiments. See [14] for all proofs.

2 Preliminaries

Let $\mathcal{L} \supseteq \{main\}$ be a finite set of *locations*, where *main* is the *canonical start location* (i.e., the entry point of the program), and let $\vec{x} := [x_1, \ldots, x_d]$ be the vector of *program variables*. Furthermore, let $\mathcal{TV}$ be a countably infinite set of *temporary variables*, which are used to model non-determinism, and let $\sup \mathbb{Z} := \infty$. We call an arithmetic expression e an *integer expression* if it evaluates to an integer when all variables in e are instantiated by integers. LoAT analyzes tail-recursive programs operating on integers, represented as *integer transition systems* (ITSs), i.e., sets of *transitions* $f(\vec{x}) \xrightarrow{p} g(\vec{a})\,[\varphi]$ where $f, g \in \mathcal{L}$, the *update* $\vec{a}$ is a vector of d integer expressions over $\mathcal{TV} \cup \vec{x}$, the *cost* p is either an arithmetic expression over $\mathcal{TV} \cup \vec{x}$ or ∞, and the *guard* φ is a conjunction of inequations over integer expressions with variables from $\mathcal{TV} \cup \vec{x}$.[1] For example, consider the loop on the left and the corresponding transition t_{loop} on the right.

$$\textbf{while } x > 0 \textbf{ do } x \leftarrow x - 1 \qquad f(x) \xrightarrow{1} f(x-1)\,[x > 0] \quad (t_{loop})$$

Here, the cost 1 instructs LoAT to use the number of loop iterations as cost measure. LoAT allows for arbitrary *user defined* cost measures, since the user can choose any polynomials over the program variables as costs. LoAT synthesizes transitions with cost ∞ to represent non-terminating runs, i.e., such transitions are not allowed in the input.

A *configuration* is of the form $f(\vec{c})$ with $f \in \mathcal{L}$ and $\vec{c} \in \mathbb{Z}^d$. For any entity $s \notin \mathcal{L}$ and any arithmetic expressions $\vec{b} = [b_1, \ldots, b_d]$, let $s(\vec{b})$ denote the result of replacing each variable x_i in s by b_i, for all $1 \leq i \leq d$. Moreover, $\mathcal{V}ars(s)$ denotes the program variables and $\mathcal{TV}(s)$ denotes the temporary variables occurring in s. For an integer transition system $\mathcal{T}$, a configuration $f(\vec{c})$ *evaluates to* $g(\vec{c}\,')$ *with cost* $k \in \mathbb{Z} \cup \{\infty\}$, written $f(\vec{c}) \xrightarrow{k}_{\mathcal{T}} g(\vec{c}\,')$, if there exist a transition $f(\vec{x}) \xrightarrow{p} g(\vec{a})\,[\varphi] \in \mathcal{T}$ and an instantiation of its temporary variables with integers such that the following holds:

$$\varphi(\vec{c}) \land \vec{c}\,' = \vec{a}(\vec{c}) \land k = p(\vec{c}).$$

[1] LoAT can also analyze the complexity of certain non-tail-recursive programs, see [9]. For simplicity, we restrict ourselves to tail-recursive programs in the current paper.

As usual, we write $f(\vec{c}) \xrightarrow{k}^*_{\mathcal{T}} g(\vec{c}')$ if $f(\vec{c})$ evaluates to $g(\vec{c}')$ in arbitrarily many steps, and the sum of the costs of all steps is k. We omit the costs if they are irrelevant. The *derivation height* of $f(\vec{c})$ is

$$dh_{\mathcal{T}}(f(\vec{c})) := \sup\{k \mid \exists g(\vec{c}').\, f(\vec{c}) \xrightarrow{k}^*_{\mathcal{T}} g(\vec{c}')\}$$

and the *runtime complexity* of $\mathcal{T}$ is

$$rc_{\mathcal{T}}(n) := \sup\{dh_{\mathcal{T}}(main(c_1, \ldots, c_d)) \mid |c_1| + \ldots + |c_d| \leq n\}.$$

$\mathcal{T}$ terminates if no configuration $main(\vec{c})$ admits an infinite $\to_{\mathcal{T}}$-sequence and $\mathcal{T}$ is *finitary* if no configuration $main(\vec{c})$ admits a $\to_{\mathcal{T}}$-sequence with cost ∞. Otherwise, $\vec{c}$ is a *witness of non-termination* or a *witness of infinitism*, respectively. Note that termination implies finitism for ITSs where no transition has cost ∞. However, our approach may transform non-terminating ITSs into terminating, infinitary ITSs, as it replaces non-terminating loops by transitions with cost ∞.

3 Overview of LoAT

The goal of LoAT is to compute a lower bound on $rc_{\mathcal{T}}$ or even prove non-termination of $\mathcal{T}$. To this end, it repeatedly applies program simplifications, so-called *processors*. When applying them with a suitable strategy (see [8,9]), one eventually obtains *simplified transitions* of the form $main(\vec{x}) \xrightarrow{p} f(\vec{a})\,[\varphi]$ where $f \neq main$. As LoAT's processors are *sound for lower bounds* (i.e., if they transform $\mathcal{T}$ to $\mathcal{T}'$, then $dh_{\mathcal{T}} \geq dh_{\mathcal{T}'}$), such a simplified transition gives rise to the lower bound $I_\varphi \cdot p$ on $dh_{\mathcal{T}}(main(\vec{x}))$ (where I_φ denotes the indicator function of φ, which is 1 for values where φ holds and 0 otherwise). This bound can be lifted to $rc_{\mathcal{T}}$ by solving a so-called *limit problem*, see [9].

LoAT's processors are also *sound for non-termination*, as they preserve finitism. So if $p = \infty$, then it suffices to prove satisfiability of φ to prove infinitism, which implies non-termination of the original ITS, where transitions with cost ∞ are forbidden (see Sect. 2). LoAT's most important processors are:

Loop Acceleration (Sect. 4) transforms a *simple loop*, i.e., a single transition $f(\vec{x}) \xrightarrow{p} f(\vec{a})\,[\varphi]$, into a non-deterministic transition that can simulate several loop iterations in one step. For example, loop acceleration transforms t_{loop} to

$$f(x) \xrightarrow{n} f(x-n)\,[x \geq n \wedge n > 0], \qquad (t_{loop^n})$$

where $n \in \mathcal{TV}$, i.e., the value of n can be chosen non-deterministically.

Instantiation [9, Theorem 3.12] replaces temporary variables by integer expressions. For example, it could instantiate n with x in t_{loop^n}, resulting in

$$f(x) \xrightarrow{x} f(0)\,[x > 0]. \qquad (t_{loop^x})$$

Chaining [9, Theorem 3.18] combines two subsequent transitions into one transition. For example, chaining combines the transitions

$$\begin{aligned} & main(x) \xrightarrow{1} f(x) \\ \text{and } t_{loop^x} \text{ to } \quad & main(x) \xrightarrow{x+1} f(0)\,[x>0]\,. \end{aligned}$$

Nonterm (Sect. 6) searches for witnesses of non-termination, characterized by a formula ψ. So it turns, e.g.,

$$f(x_1,x_2) \xrightarrow{1} f(x_1-x_2,x_2)\,[x_1>0] \qquad (t_{nonterm})$$

$$\text{into } \; f(x_1,x_2) \xrightarrow{\infty} sink(x_1,x_2)\,[x_1>0 \wedge x_2 \le 0]$$

(where $sink \in \mathcal{L}$ is fresh), as each $\vec{c} \in \mathbb{Z}^2$ with $c_1 > 0 \wedge c_2 \le 0$ witnesses non-termination of $t_{nonterm}$, i.e., here ψ is $x_1 > 0 \wedge x_2 \le 0$.

Intuitively, LoAT uses **Chaining** to transform non-simple loops into simple loops. **Instantiation** resolves non-determinism heuristically and thus reduces the number of temporary variables, which is crucial for scalability. In addition to these processors, LoAT removes transitions after processing them, as explained in [9]. See [8,9] for heuristics and a suitable strategy to apply LoAT's processors.

4 Modular Loop Acceleration

For **Loop Acceleration**, LoAT uses *conditional acceleration techniques* [10]. Given two formulas ξ and $\check{\varphi}$, and a loop with update $\vec{a}$, a conditional acceleration technique yields a formula $accel(\xi,\check{\varphi},\vec{a})$ which implies that ξ holds throughout n loop iterations (i.e., ξ is an *n-invariant*), provided that $\check{\varphi}$ is an n-invariant, too. In the following, let $\vec{a}^0(\vec{x}) := \vec{x}$ and $\vec{a}^{m+1}(\vec{x}) := \vec{a}(\vec{a}^m(\vec{x})) = \vec{a}[\vec{x}/\vec{a}^m(\vec{x})]$.

Definition 1 (Conditional Acceleration Technique). *A function accel is a* conditional acceleration technique *if the following implication holds for all formulas ξ and $\check{\varphi}$ with variables from $\mathcal{TV} \cup \vec{x}$, all updates $\vec{a}$, all $n > 0$, and all instantiations of the variables with integers:*

$$(accel(\xi,\check{\varphi},\vec{a}) \wedge \forall i \in [0,n).\ \check{\varphi}(\vec{a}^i(\vec{x}))) \implies \forall i \in [0,n).\ \xi(\vec{a}^i(\vec{x})).$$

The prerequisite $\forall i \in [0,n).\ \check{\varphi}(\vec{a}^i(\vec{x}))$ is ensured by previous acceleration steps, i.e., $\check{\varphi}$ is initially $\top$ (*true*), and it is refined by conjoining a part ξ of the loop guard in each acceleration step. When formalizing acceleration techniques, we only specify the result of *accel* for certain arguments ξ, $\check{\varphi}$, and $\vec{a}$, and assume $accel(\xi,\check{\varphi},\vec{a}) = \bot$ (*false*) otherwise.

Definition 2 (LoAT's Conditional Acceleration Techniques [10,11]).

Increase $accel_{inc}(\xi, \breve{\varphi}, \vec{a}) := \xi \quad \text{if} \models \xi \wedge \breve{\varphi} \implies \xi(\vec{a})$

Decrease $accel_{dec}(\xi, \breve{\varphi}, \vec{a}) := \xi(\vec{a}^{n-1}(\vec{x})) \text{ if} \models \xi(\vec{a}) \wedge \breve{\varphi} \implies \xi$

Eventual Decrease $accel_{ev\text{-}dec}(t > 0, \breve{\varphi}, \vec{a}) := t > 0 \wedge t(\vec{a}^{n-1}(\vec{x})) > 0$
$\text{if} \models (t \geq t(\vec{a}) \wedge \breve{\varphi}) \implies t(\vec{a}) \geq t(\vec{a}^2(\vec{x}))$

Eventual Increase $accel_{ev\text{-}inc}(t > 0, \breve{\varphi}, \vec{a}) := t > 0 \wedge t \leq t(\vec{a})$
$\text{if} \models (t \leq t(\vec{a}) \wedge \breve{\varphi}) \implies t(\vec{a}) \leq t(\vec{a}^2(\vec{x}))$

Fixpoint $accel_{fp}(t > 0, \breve{\varphi}, \vec{a}) := t > 0 \wedge \bigwedge_{x \in closure_{\vec{a}}(t)} x = x(\vec{a})$
$\text{where } closure_{\vec{a}}(t) := \bigcup_{i \in \mathbb{N}} \mathcal{V}ars(t(\vec{a}^i(\vec{x})))$

The above five techniques are taken from [10,11], where only deterministic loops are considered (i.e., there are no temporary variables). Lifting them to non-deterministic loops in a way that allows for *exact* conditional acceleration techniques (which capture all possible program runs) is non-trivial and beyond the scope of this paper. Thus, we sacrifice exactness and treat temporary variables like additional constant program variables whose update is the identity, resulting in a sound under-approximation (that captures a subset of all possible runs).

So essentially, **Increase** and **Decrease** handle inequations $t > 0$ in the loop guard where t increases or decreases (weakly) monotonically when applying the loop's update. The canonical examples where **Increase** or **Decrease** applies are

$$f(x, \ldots) \to f(x+1, \ldots)\,[x > 0 \wedge \ldots] \quad \text{or} \quad f(x, \ldots) \to f(x-1, \ldots)\,[x > 0 \wedge \ldots],$$

respectively. **Eventual Decrease** applies if t never increases again once it starts to decrease. The canonical example is $f(x, y, \ldots) \to f(x + y, y - 1, \ldots)\,[x > 0 \wedge \ldots]$. Similarly, **Eventual Increase** applies if t never decreases again once it starts to increase. **Fixpoint** can be used for inequations $t > 0$ that do not behave (eventually) monotonically. It should only be used if $accel_{fp}(t > 0, \breve{\varphi}, \vec{a})$ is satisfiable.

LoAT uses the *acceleration calculus* of [10]. It operates on *acceleration problems* $[\![\psi \mid \breve{\varphi} \mid \widehat{\varphi}]\!]_{\vec{a}}$, where ψ (which is initially $\top$) is repeatedly refined. When it stops, ψ is used as the guard of the resulting accelerated transition. The formulas $\breve{\varphi}$ and $\widehat{\varphi}$ are the parts of the loop guard that have already or have not yet been handled, respectively. So $\breve{\varphi}$ is initially $\top$, and $\widehat{\varphi}$ and $\vec{a}$ are initialized with the guard φ and the update of the loop $f(\vec{x}) \xrightarrow{p} f(\vec{a})\,[\varphi]$ under consideration, i.e., the initial acceleration problem is $[\![\top \mid \top \mid \varphi]\!]_{\vec{a}}$. Once $\widehat{\varphi}$ is $\top$, the loop is accelerated to $f(\vec{x}) \xrightarrow{q} f(\vec{a}^n(\vec{x}))\,[\psi \wedge n > 0]$, where the cost q and a closed form for $\vec{a}^n(\vec{x})$ are computed by the recurrence solver PURRS [2].

Definition 3 (Acceleration Calculus for Conjunctive Loops). *The relation* $\rightsquigarrow$ *on acceleration problems is defined as*

$$\frac{accel(\xi, \breve{\varphi}, \vec{a}) = \psi_2}{[\![\psi_1 \mid \breve{\varphi} \mid \xi \wedge \widehat{\varphi}]\!]_{\vec{a}} \rightsquigarrow [\![\psi_1 \wedge \psi_2 \mid \breve{\varphi} \wedge \xi \mid \widehat{\varphi}]\!]_{\vec{a}}} \quad \textit{accel is a conditional acceleration technique}$$

So to accelerate a loop, one picks a not yet handled part ξ of the guard in each step. When accelerating $f(\vec{x}) \to f(\vec{a})\,[\xi]$ using a conditional acceleration technique *accel*, one may assume $\forall i \in [0,n).\ \breve{\varphi}(\vec{a}^i(\vec{x}))$. The result of *accel* is conjoined to the result ψ_1 computed so far, and ξ is moved from the third to the second component of the problem, i.e., to the already handled part of the guard.

Example 4 (Acceleration Calculus). We show how to accelerate the loop

$$f(x,y) \xrightarrow{x} f(x-y,y)\,[x>0 \land y \geq 0] \qquad \text{to}$$

$$f(x,y) \xrightarrow{(x+\frac{y}{2})\cdot n-\frac{y}{2}\cdot n^2} f(x-n\cdot y,y)\,[y\geq 0 \land x-(n-1)\cdot y>0 \land n>0].$$

The closed form $\vec{a}^n(x) = (x - n\cdot y, y)$ can be computed via recurrence solving. Similarly, the cost $(x+\frac{y}{2})\cdot n - \frac{y}{2}\cdot n^2$ of n loop iterations is obtained by solving the following recurrence relation (where $c^{(n)}$ and $x^{(n)}$ denote the cost and the value of x after n applications of the transition, respectively).

$$c^{(n)} = c^{(n-1)} + x^{(n-1)} = c^{(n-1)} + x - (n-1)\cdot y \qquad \text{and} \qquad c^{(1)} = x.$$

The guard is computed as follows:

$$\begin{aligned}\llbracket \top \mid \top \mid x>0 \land y\geq 0 \rrbracket_{\vec{a}} &\rightsquigarrow \llbracket y\geq 0 \mid y \geq 0 \mid x>0\rrbracket_{\vec{a}}\\ &\rightsquigarrow \llbracket y\geq 0 \land x-(n-1)\cdot y>0 \mid y\geq 0 \land x>0 \mid \top\rrbracket_{\vec{a}}.\end{aligned}$$

In the 1st step, we have $\xi = (y \geq 0)$ and $accel_{inc}(y\geq 0, \top, \vec{a}) = (y \geq 0)$. In the 2nd step, we have $\xi = (x>0)$ and $accel_{dec}(x>0, y\geq 0, \vec{a}) = (x-(n-1)\cdot y>0)$. So the inequation $x-(n-1)\cdot y>0$ ensures n-invariance of $x>0$.

5 Efficient Loop Acceleration Using Unsat Cores

Each attempt to apply a conditional acceleration technique other than **Fixpoint** requires proving an implication, which is implemented via SMT solving by proving unsatisfiability of its negation. For **Fixpoint**, satisfiability of $accel_{fp}(t>0, \breve{\varphi}, \vec{a})$ is checked via SMT. So even though LoAT restricts ξ to atoms, up to $\Theta(m^2)$ attempts to apply a conditional acceleration technique are required to accelerate a loop whose guard contains m inequations using a naive strategy ($5\cdot m$ attempts for the 1st $\rightsquigarrow$-step, $5\cdot(m-1)$ attempts for the 2nd step, ...).

To improve efficiency, LoAT uses a novel encoding that requires just $5\cdot m$ attempts. For any $\alpha \in AT_{imp} = \{inc, dec, ev\text{-}dec, ev\text{-}inc\}$, let $encode_\alpha(\xi, \breve{\varphi}, \vec{a})$ be the implication that has to be valid in order to apply $accel_\alpha$, whose premise is of the form $\ldots \land \breve{\varphi}$. Instead of repeatedly refining $\breve{\varphi}$, LoAT tries to prove validity[2] of $encode_{\alpha,\xi} := encode_\alpha(\xi, \varphi\setminus\{\xi\}, \vec{a})$ for each $\alpha \in AT_{imp}$ and each $\xi \in \varphi$, where φ is the (conjunctive) guard of the transition that should be accelerated. Again,

[2] Here and in the following, we unify conjunctions of atoms with sets of atoms.

proving validity of an implication is equivalent to proving unsatisfiability of its negation. So if validity of $encode_{\alpha,\xi}$ can be shown, then SMT solvers can also provide an *unsat core* for $\neg encode_{\alpha,\xi}$.

Definition 5 (Unsat Core). *Given a conjunction ψ, we call each unsatisfiable subset of ψ an* unsat core *of ψ.*

Theorem 6 shows that when handling an inequation ξ, one only has to require n-invariance for the elements of $\varphi \setminus \{\xi\}$ that occur in an unsat core of $\neg encode_{\alpha,\xi}$. Thus, an unsat core of $\neg encode_{\alpha,\xi}$ can be used to determine which prerequisites $\check{\varphi}$ are needed for the inequation ξ. This information can then be used to find a suitable order for handling the inequations of the guard. Thus, in this way one only has to check (un)satisfiability of the $4 \cdot m$ formulas $\neg encode_{\alpha,\xi}$. If no such order is found, then LoAT either fails to accelerate the loop under consideration, or it resorts to using **Fixpoint**, as discussed below.

Theorem 6 (Unsat Core Induces $\rightsquigarrow$-Step). *Let $deps_{\alpha,\xi}$ be the intersection of $\varphi \setminus \{\xi\}$ and an unsat core of $\neg encode_{\alpha,\xi}$. If $\check{\varphi}$ implies $deps_{\alpha,\xi}$, then $accel_\alpha(\xi, \check{\varphi}, \vec{a}) = accel_\alpha(\xi, \varphi \setminus \{\xi\}, \vec{a})$.*

Example 7 (Controlling Acceleration Steps via Unsat Cores). Reconsider Example 4. Here, LoAT would try to prove, among others, the following implications:

$$encode_{dec,x>0} = (x - y > 0 \land y > 0) \implies x > 0 \quad (1)$$

$$encode_{inc,y>0} = (y > 0 \land x > 0) \implies y > 0 \quad (2)$$

To do so, it would try to prove unsatisfiability of $\neg encode_{\alpha,\xi}$ via SMT. For (1), we get $\neg encode_{dec,x>0} = (x - y > 0 \land y > 0 \land x \leq 0)$, whose only unsat core is $\neg encode_{dec,x>0}$, and its intersection with $\varphi \setminus \{x > 0\} = \{y > 0\}$ is $\{y > 0\}$.

For (2), we get $\neg encode_{inc,y>0} = (y > 0 \land x > 0 \land y \leq 0)$, whose minimal unsat core is $y > 0 \land y \leq 0$, and its intersection with $\varphi \setminus \{y > 0\} = \{x > 0\}$ is empty. So by Theorem 6, we have $accel_{inc}(y > 0, \top, \vec{a}) = accel_{inc}(y > 0, x > 0, \vec{a})$.

In this way, validity of $encode_{\alpha_1,x>0}$ and $encode_{\alpha_2,y>0}$ is proven for all $\alpha_1 \in AT_{imp} \setminus \{inc\}$ and all $\alpha_2 \in AT_{imp}$. However, the premise $x \leq x - y \land y > 0$ of $encode_{ev\text{-}inc,x>0}$ is unsatisfiable and thus a corresponding acceleration step would yield a transition with unsatisfiable guard. To prevent that, LoAT only uses a technique $\alpha \in AT_{imp}$ for ξ if the premise of $encode_{\alpha,\xi}$ is satisfiable.

So for each inequation ξ from φ, LoAT synthesizes up to 4 potential $\rightsquigarrow$-steps corresponding to $accel_\alpha(\xi, deps_{\alpha,\xi}, \vec{a})$, where $\alpha \in AT_{imp}$. If validity of $encode_{\alpha,\xi}$ cannot be shown for any $\alpha \in AT_{imp}$, then LoAT tries to prove satisfiability of $accel_{fp}(\xi, \top, \vec{a})$ to see if **Fixpoint** should be applied. Note that the 2nd argument of $accel_{fp}$ is irrelevant, i.e., **Fixpoint** does not benefit from previous acceleration steps and thus $\rightsquigarrow$-steps that use it do not have any dependencies.

It remains to find a suitably ordered subset S of m $\rightsquigarrow$-steps that constitutes a successful $\rightsquigarrow$-sequence. In the following, we define $AT := AT_{imp} \cup \{fp\}$ and we extend the definition of $deps_{\alpha,\xi}$ to the case $\alpha = fp$ by defining $deps_{fp,\xi} := \varnothing$.

Lemma 8. *Let $C \subseteq AT \times \varphi$ be the smallest set such that $(\alpha, \xi) \in C$ implies*

(a) if $\alpha \in AT_{imp}$, then $encode_{\alpha,\xi}$ is valid and its premise is satisfiable,
(b) if $\alpha = fp$, then $accel_{fp}(\xi, \top, \vec{a})$ is satisfiable, and
(c) $deps_{\alpha,\xi} \subseteq \{\xi' \mid (\alpha', \xi') \in C$ for some $\alpha' \in AT\}$.

Let $S := \{(\alpha, \xi) \in C \mid \alpha \geq_{AT} \alpha'$ for all $(\alpha', \xi) \in C\}$ where $>_{AT}$ is the total order inc $>_{AT}$ dec $>_{AT}$ ev-dec $>_{AT}$ ev-inc $>_{AT}$ fp. We define $(\alpha', \xi') \prec (\alpha, \xi)$ if $\xi' \in deps_{\alpha,\xi}$. Then $\prec$ is a strict (and hence, well-founded) order on S.

The order $>_{AT}$ in Lemma 8 corresponds to the order proposed in [10]. Note that the set C can be computed without further (potentially expensive) SMT queries by a straightforward fixpoint iteration, and well-foundedness of $\prec$ follows from minimality of C. For Example 7, we get

$$C = \{(dec, x > 0), (ev\text{-}dec, x > 0)\} \cup \{(\alpha, y > 0) \mid \alpha \in AT\} \qquad \text{and}$$
$$S = \{(dec, x > 0), (inc, y > 0)\} \text{ with } (inc, y > 0) \prec (dec, x > 0).$$

Finally, we can construct a valid $\rightsquigarrow$-sequence via the following theorem.

Theorem 9. (Finding $\rightsquigarrow$-Sequences). *Let S be defined as in Lemma 8 and assume that for each $\xi \in \varphi$, there is an $\alpha \in AT$ such that $(\alpha, \xi) \in S$. W.l.o.g., let $\varphi = \bigwedge_{i=1}^{m} \xi_i$ where $(\alpha_1, \xi_1) \prec' \ldots \prec' (\alpha_m, \xi_m)$ for some strict total order $\prec'$ containing $\prec$, and let $\check{\varphi}_j := \bigwedge_{i=1}^{j} \xi_i$. Then for all $j \in [0, m)$, we have:*

$$\left[\!\left[\bigwedge_{i=1}^{j} accel_{\alpha_i}(\xi_i, \check{\varphi}_{i-1}, \vec{a}) \;\middle|\; \check{\varphi}_j \;\middle|\; \bigwedge_{i=j+1}^{m} \xi_i \right]\!\right]_{\vec{a}} \rightsquigarrow \left[\!\left[\bigwedge_{i=1}^{j+1} accel_{\alpha_i}(\xi_i, \check{\varphi}_{i-1}, \vec{a}) \;\middle|\; \check{\varphi}_{j+1} \;\middle|\; \bigwedge_{i=j+2}^{m} \xi_i \right]\!\right]_{\vec{a}}$$

In our example, we have $\prec' = \prec$ as $\prec$ is total. Thus, we obtain a $\rightsquigarrow$-sequence by first processing $y > 0$ with **Increase** and then processing $x > 0$ with **Decrease**.

6 Proving Non-Termination of Simple Loops

To prove non-termination, LoAT uses a variation of the calculus from Sect. 4, see [11]. To adapt it for proving non-termination, further restrictions have to be imposed on the conditional acceleration techniques, resulting in the notion of *conditional non-termination techniques*, see [11, Def. 10]. We denote a $\rightsquigarrow$-step that uses a conditional non-termination technique with $\rightsquigarrow_{nt}$.

Theorem 10. (Proving Non-Termination via $\rightsquigarrow_{nt}$). *Let $f(\vec{x}) \to f(\vec{a})\,[\varphi] \in \mathcal{T}$. If $[\![\top \mid \top \mid \varphi]\!]_{\vec{a}} \rightsquigarrow_{nt}^{*} [\![\psi \mid \varphi \mid \top]\!]_{\vec{a}}$, then for every $\vec{c} \in \mathbb{Z}^d$ where $\psi(\vec{c})$ is satisfiable, the configuration $f(\vec{c})$ admits an infinite $\to_{\mathcal{T}}$-sequence.*

The conditional non-termination techniques used by LoAT are **Increase**, **Eventual Increase**, and **Fixpoint**. So non-termination proofs can be synthesized while trying to accelerate a loop with very little overhead. After successfully accelerating a loop as explained in Sect. 5, LoAT tries to find a second suitably ordered $\rightsquigarrow$-sequence, where it only considers the conditional non-termination techniques mentioned above. If LoAT succeeds, then it has found a $\rightsquigarrow_{nt}$-sequence which gives rise to a proof of non-termination via Theorem 10.

7 Implementation, Experiments, and Conclusion

Our implementation in LoAT can parse three widely used formats for ITSs (see [13]), and it is configurable via a minimalistic set of command-line options:

`--timeout` to set a timeout in seconds
`--proof-level` to set the verbosity of the proof output
`--plain` to switch from colored to monochrome proof-output
`--limit-strategy` to choose a strategy for solving limit problems, see [9]
`--mode` to choose an analysis mode for LoAT (`complexity` or `non_termination`)

We evaluate three versions of LoAT: LoAT '19 uses templates to find invariants that facilitate loop acceleration for proving non-termination [8]; LoAT '20 deduces worst-case lower bounds based on loop acceleration via *metering functions* [9]; and LoAT '22 applies the calculus from [10,11] as described in Sect. 5 and 6. We also include three other state-of-the-art termination tools in our evaluation: T2 [6], VeryMax [16], and iRankFinder [3,7]. Regarding complexity, the only other tool for worst-case lower bounds of ITSs is LOBER [1]. However, we do not compare with LOBER, as it only analyses (multi-path) loops instead of full ITSs.

We use the examples from the categories *Termination* (1222 examples) and *Complexity of ITSs* (781 examples), respectively, of the *Termination Problems Data Base* [19]. All benchmarks have been performed on *StarExec* [18] (Intel Xeon E5-2609, 2.40GHz, 264GB RAM [17]) with a wall clock timeout of 300 s.

	No	Yes	Avg. Rt	Median Rt	Std. Dev. Rt
LoAT '22	493	0	9.4	0.2	41.5
LoAT '19	459	0	22.6	1.5	67.5
T2	438	610	22.6	1.2	66.7
VeryMax	419	628	29.9	1.0	66.7
iRankFinder	399	634	44.1	4.9	89.1

		LoAT '22					
	$rc_{\mathcal{T}}(n)$	$\Omega(1)$	$\Omega(n)$	$\Omega(n^2)$	$\Omega(n^{>2})$	EXP	$\Omega(\omega)$
LoAT '20	$\Omega(1)$	180	63	1	–	–	12
	$\Omega(n)$	6	218	3	–	–	–
	$\Omega(n^2)$	–	1	69	–	–	–
	$\Omega(n^{>2})$	–	–	–	7	–	–
	EXP	1	–	–	–	4	–
	$\Omega(\omega)$	–	–	–	–	–	216

By the table on the left, LoAT '22 is the most powerful tool for non-termination. The improvement over LoAT '19 demonstrates that the calculus from [10,11] is more powerful and efficient than the approach from [8]. The last three columns show the average, the median, and the standard deviation of the wall clock runtime, including examples where the timeout was reached.

The table on the right shows the results for complexity. The diagonal corresponds to examples where LoAT '20 and LoAT '22 yield the same result. The entries above or below the diagonal correspond to examples where LoAT '22 or LoAT '20 is better, respectively. There are 8 regressions and 79 improvements, so the calculus from [10,11] used by LoAT '22 is also beneficial for lower bounds.

LoAT is open source and its source code is available on GitHub [12]. See [13,14] for details on our evaluation, related work, all proofs, and a pre-compiled binary.

References

1. Albert, E., Genaim, S., Martin-Martin, E., Merayo, A., Rubio, A.: Lower-bound synthesis using loop specialization and Max-SMT. In: Silva, A., Leino, K.R.M. (eds.) CAV 2021. LNCS, vol. 12760, pp. 863–886. Springer, Cham (2021). https://doi.org/10.1007/978-3-030-81688-9_40
2. Bagnara, R., Pescetti, A., Zaccagnini, A., Zaffanella, E.: PURRS: towards computer algebra support for fully automatic worst-case complexity analysis. CoRR abs/cs/0512056 (2005). https://arxiv.org/abs/cs/0512056
3. Ben-Amram, A.M., Doménech, J.J., Genaim, S.: Multiphase-linear ranking functions and their relation to recurrent sets. In: Chang, B.-Y.E. (ed.) SAS 2019. LNCS, vol. 11822, pp. 459–480. Springer, Cham (2019). https://doi.org/10.1007/978-3-030-32304-2_22
4. Bozga, M., Gîrlea, C., Iosif, R.: Iterating octagons. In: Kowalewski, S., Philippou, A. (eds.) TACAS 2009. LNCS, vol. 5505, pp. 337–351. Springer, Heidelberg (2009). https://doi.org/10.1007/978-3-642-00768-2_29
5. Bozga, M., Iosif, R., Konečný, F.: Fast acceleration of ultimately periodic relations. In: Touili, T., Cook, B., Jackson, P. (eds.) CAV 2010. LNCS, vol. 6174, pp. 227–242. Springer, Heidelberg (2010). https://doi.org/10.1007/978-3-642-14295-6_23
6. Brockschmidt, M., Cook, B., Ishtiaq, S., Khlaaf, H., Piterman, N.: T2: temporal property verification. In: Chechik, M., Raskin, J.-F. (eds.) TACAS 2016. LNCS, vol. 9636, pp. 387–393. Springer, Heidelberg (2016). https://doi.org/10.1007/978-3-662-49674-9_22
7. Doménech, J.J., Genaim, S.: iRankFinder. In: Lucas, S. (ed.) WST 2018, p. 83 (2018). http://wst2018.webs.upv.es/wst2018proceedings.pdf
8. Frohn, F., Giesl, J.: Proving non-termination via loop acceleration. In: Barrett, C.W., Yang, J. (eds.) FMCAD 2019, pp. 221–230 (2019). https://doi.org/10.23919/FMCAD.2019.8894271
9. Frohn, F., Naaf, M., Brockschmidt, M., Giesl, J.: Inferring lower runtime bounds for integer programs. ACM TOPLAS **42**(3), 13:1–13:50 (2020). https://doi.org/10.1145/3410331. Revised and extended version of a paper which appeared in IJCAR 2016, pp. 550–567. LNCS, vol. 9706 (2016)
10. Frohn, F.: A calculus for modular loop acceleration. In: Biere, A., Parker, D. (eds.) TACAS 2020. LNCS, vol. 12078, pp. 58–76. Springer, Cham (2020). https://doi.org/10.1007/978-3-030-45190-5_4
11. Frohn, F., Fuhs, C.: A calculus for modular loop acceleration and non-termination proofs. CoRR abs/2111.13952 (2021). https://arxiv.org/abs/2111.13952, to appear in STTT
12. Frohn, F.: LoAT on GitHub. https://github.com/aprove-developers/LoAT
13. Frohn, F., Giesl, J.: Empirical evaluation of: proving non-termination and lower runtime bounds with LoAT. https://ffrohn.github.io/loat-tool-paper-evaluation
14. Frohn, F., Giesl, J.: Proving non-termination and lower runtime bounds with LoAT (System Description). CoRR abs/2202.04546 (2022). https://arxiv.org/abs/2202.04546
15. Kroening, D., Lewis, M., Weissenbacher, G.: Under-approximating loops in C programs for fast counterexample detection. Formal Methods Syst. Des. **47**(1), 75–92 (2015). https://doi.org/10.1007/s10703-015-0228-1
16. Larraz, D., Nimkar, K., Oliveras, A., Rodríguez-Carbonell, E., Rubio, A.: Proving non-termination using Max-SMT. In: Biere, A., Bloem, R. (eds.) CAV 2014. LNCS, vol. 8559, pp. 779–796. Springer, Cham (2014). https://doi.org/10.1007/978-3-319-08867-9_52

17. StarExec hardware specifications. https://www.starexec.org/starexec/public/machine-specs.txt
18. Stump, A., Sutcliffe, G., Tinelli, C.: StarExec: a cross-community infrastructure for logic solving. In: Demri, S., Kapur, D., Weidenbach, C. (eds.) IJCAR 2014. LNCS (LNAI), vol. 8562, pp. 367–373. Springer, Cham (2014). https://doi.org/10.1007/978-3-319-08587-6_28
19. Termination Problems Data Base (TPDB, Git SHA 755775). https://github.com/TermCOMP/TPDB

Implicit Definitions with Differential Equations for KeYmaera X (System Description)

James Gallicchio(✉), Yong Kiam Tan(✉), Stefan Mitsch(✉), and André Platzer(✉)

Computer Science Department, Carnegie Mellon University, Pittsburgh, USA
jgallicc@andrew.cmu.edu, {yongkiat,smitsch,aplatzer}@cs.cmu.edu

Abstract. Definition packages in theorem provers provide users with means of defining and organizing concepts of interest. This system description presents a new definition package for the hybrid systems theorem prover KeYmaera X based on differential dynamic logic (dL). The package adds KeYmaera X support for user-defined smooth functions whose graphs can be implicitly characterized by dL formulas. Notably, this makes it possible to implicitly characterize functions, such as the exponential and trigonometric functions, as solutions of differential equations and then prove properties of those functions using dL's differential equation reasoning principles. Trustworthiness of the package is achieved by minimally extending KeYmaera X's soundness-critical kernel with a single axiom scheme that expands function occurrences with their implicit characterization. Users are provided with a high-level interface for defining functions and non-soundness-critical tactics that automate low-level reasoning over implicit characterizations in hybrid system proofs.

Keywords: Definitions · Differential dynamic logic · Verification of hybrid systems · Theorem proving

1 Introduction

KeYmaera X [7] is a theorem prover implementing differential dynamic logic dL [17,19–21] for specifying and verifying properties of hybrid systems mixing discrete dynamics and differential equations. Definitions enable users to express complex theorem statements in concise terms, e.g., by modularizing hybrid system models and their proofs [14]. Prior to this work, KeYmaera X had only one mechanism for definition, namely, non-recursive abbreviations via uniform substitution [14,20]. This restriction meant that common and useful functions, e.g., the trigonometric and exponential functions, could not be directly used in KeYmaera X, even though they can be uniquely characterized by dL formulas [17].

This system description introduces a new KeYmaera X definitional mechanism where functions are *implicitly defined* in dL as solutions of ordinary differential equations (ODEs). Although definition packages are available in most

J. Blanchette et al. (Eds.): IJCAR 2022, LNAI 13385, pp. 723–733, 2022.
https://doi.org/10.1007/978-3-031-10769-6_42

general-purpose proof assistants, our package is novel in tackling the question of how best to support user-defined functions in the *domain-specific* setting for hybrid systems. In contrast to tools with builtin support for *some* fixed subsets of special functions [1,9,23]; or higher-order logics that can work with functions via their infinitary series expansions [4], e.g., $\exp(t) = \sum_{i=0}^{\infty} \frac{t^i}{i!}$; our package strikes a balance between practicality and generality by allowing users to define and reason about *any* function characterizable in dL as the solution of an ODE (Sect. 2), e.g., $\exp(t)$ solves the ODE $e' = e$ with initial value $e(0) = 1$.

Theoretically, implicit definitions strictly expand the class of ODE invariants amenable to dL's complete ODE invariance proof principles [22]; such invariants play a key role in ODE safety proofs [21] (see Proposition 3). In practice, arithmetical identities and other specifications involving user-defined functions are proved by automatically unfolding their implicit ODE characterizations and reusing existing KeYmaera X support for ODE reasoning (Sect. 3). The package is designed to provide seamless integration of implicit definitions in KeYmaera X and its usability is demonstrated on several hybrid system verification examples drawn from the literature that involve special functions (Sect. 4).

All proofs are in the supplement [8]. The definitions package is part of KeYmaera X with a usage guide at: http://keymaeraX.org/keymaeraXfunc/.

2 Interpreted Functions in Differential Dynamic Logic

This section briefly recalls differential dynamic logic (dL) [17,18,20,21] and explains how its term language is extended to support implicit function definitions.

Syntax. Terms $e, \tilde{e}$ and formulas ϕ, ψ in dL are generated by the following grammar, with variable x, rational constant c, k-ary function symbols h (for any $k \in \mathbb{N}$), comparison operator $\sim \in \{=, \neq, \geq, >, \leq, <\}$, and hybrid program α:

$$e, \tilde{e} ::= x \mid c \mid e + \tilde{e} \mid e \cdot \tilde{e} \mid h(e_1, \ldots, e_k) \tag{1}$$

$$\phi, \psi ::= e \sim \tilde{e} \mid \phi \land \psi \mid \phi \lor \psi \mid \neg\phi \mid \forall x \phi \mid \exists x \phi \mid [\alpha]\,\phi \mid \langle\alpha\rangle\,\phi \tag{2}$$

The terms and formulas above extend the first-order language of real arithmetic ($\mathrm{FOL}_{\mathbb{R}}$) with the box ($[\alpha]\,\phi$) and diamond ($\langle\alpha\rangle\,\phi$) modality formulas which express that *all* or *some* runs of hybrid program α satisfy postcondition ϕ, respectively. Table 1 gives an intuitive overview of dL's hybrid programs language for modeling systems featuring discrete and continuous dynamics and their interactions thereof. In dL's uniform substitution calculus, function symbols h are *uninterpreted*, i.e., they semantically correspond to an arbitrary (smooth) function. Such uninterpreted function symbols (along with uninterpreted predicate and program symbols) are crucially used to give a parsimonious axiomatization of dL based on uniform substitution [20] which, in turn, enables a trustworthy microkernel implementation of the logic in the theorem prover KeYmaera X [7,16].

Table 1. Syntax and informal semantics of hybrid programs

Program	Behavior
$?\phi$	Stay in the current state if ϕ is true, otherwise abort and discard run
$x := e$	Store the value of term e in variable x
$x := *$	Store an arbitrary real value in variable x
$x' = f(x) \,\&\, Q$	Continuously follow ODE $x' = f(x)$ in domain Q for any duration ≥ 0
$\mathtt{if}(\phi)\,\alpha$	Run program α if ϕ is true, otherwise skip. Definable by $?\phi; \alpha \cup ?\neg\phi$
$\alpha; \beta$	Run program α, then run program β in any resulting state(s)
$\alpha \cup \beta$	Nondeterministically run either program α or program β
α^*	Nondeterministically repeat program α for n iterations, for any $n \in \mathbb{N}$
$\{\alpha\}$	For readability, braces are used to group and delimit hybrid programs

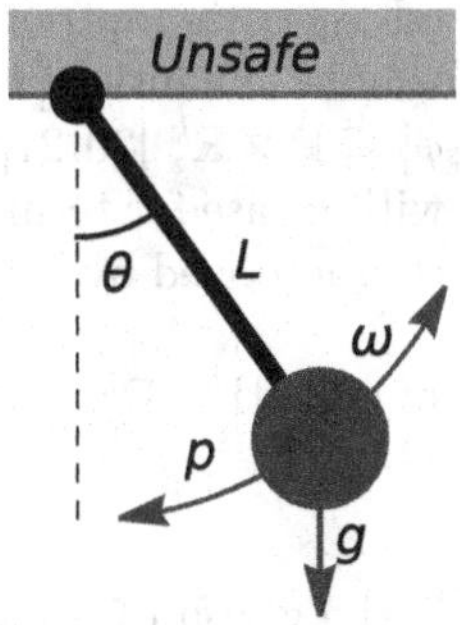

Hybrid program model (auxiliary variables s, c):

$$\hat{\alpha}_s \equiv \begin{pmatrix} s := *; c := *; ?\phi_{\sin}(s,\theta); ?\phi_{\cos}(c,\theta); \\ p := *; \mathtt{if}\left(\frac{1}{2}(\omega - p)^2 < \frac{g}{L}c\right) \{\omega := \omega - p\}; \\ \{\theta' = \omega, \omega' = -\frac{g}{L}s - k\omega, s' = \omega c, c' = -\omega s\} \end{pmatrix}^*$$

Hybrid program model (trigonometric functions):

$$\alpha_s \equiv \begin{pmatrix} p := *; \mathtt{if}\left(\frac{1}{2}(\omega - p)^2 < \frac{g}{L}\cos(\theta)\right) \{\omega := \omega - p\}; \\ \{\theta' = \omega, \omega' = -\frac{g}{L}\sin(\theta) - k\omega\} \end{pmatrix}^*$$

dL safety specification:

$$\phi_s \equiv g > 0 \wedge L > 0 \wedge k > 0 \wedge \theta = 0 \wedge \omega = 0 \rightarrow [\alpha_s]\, |\theta| < \frac{\pi}{2}$$

Fig. 1. Running example of a swinging pendulum driven by an external force (left), its hybrid program models and dL safety specification (right). Program α_s uses trigonometric functions directly, while program $\hat{\alpha}_s$ uses variables s, c to implicitly track the values of $\sin(\theta)$ and $\cos(\theta)$, respectively (additions in red). The implicit characterizations $\phi_{\sin}(s,\theta), \phi_{\cos}(c,\theta)$ are defined in (4), (5) and are not repeated here for brevity. (Color figure online)

Running Example. Adequate modeling of hybrid systems often requires the use of *interpreted* function symbols that denote specific functions of interest. As a running example, consider the swinging pendulum shown in Fig. 1. The ODEs describing its continuous motion are $\theta' = \omega, \omega' = -\frac{g}{L}\sin(\theta) - k\omega$, where θ is the swing angle, ω is the angular velocity, and g, k, L are the gravitational constant, coefficient of friction, and length of the rigid rod suspending the pendulum, respectively. The hybrid program α_s models an external force that repeatedly pushes the pendulum and changes its angular velocity by a nondeterministically chosen value p; the guard `if(...)` condition is designed to ensure that the push does not cause the pendulum to swing above the horizontal as specified by ϕ_s. Importantly, the function symbols sin, cos must denote the usual real trigonometric functions in α_s. Program $\hat{\alpha}_s$ shows the same pendulum modeled in dL *without* the use of interpreted symbols, but instead using auxiliary variables s, c. Note that $\hat{\alpha}_s$ is cumbersome and subtle to get right: the implicit characterizations $\phi_{\sin}(s, \theta), \phi_{\cos}(c, \theta)$ from (4), (5) are lengthy and the differential equations $s' = \omega c, c' = -\omega s$ must be manually calculated and added to ensure that s, c correctly track the trigonometric functions as θ evolves continuously [18,22].

Interpreted Functions. To enable extensible use of interpreted functions in dL, the term grammar (1) is enriched with k-ary function symbols h that carry an *interpretation* annotation [5,27], $h_{\ll\phi\gg}$, where $\phi \equiv \phi(x_0, y_1, \ldots, y_k)$ is a dL formula with free variables in $x_0, y_1, \ldots, y_k$ and no uninterpreted symbols. Intuitively, ϕ is a formula that characterizes the graph of the intended interpretation for h, where $y_1, \ldots, y_k$ are inputs to the function and x_0 is the output. Since ϕ depends only on the values of its free variables, its formula semantics $[\![\phi]\!]$ can be equivalently viewed as a subset of Euclidean space $[\![\phi]\!] \subseteq \mathbb{R} \times \mathbb{R}^k$ [20,21]. The dL term semantics $\nu[\![e]\!]$ [20,21] in a state ν is extended with a case for terms $h_{\ll\phi\gg}(e_1, \ldots, e_k)$ by evaluation of the smooth C^∞ function characterized by $[\![\phi]\!]$:

$$\nu[\![h_{\ll\phi\gg}(e_1, \ldots, e_k)]\!] = \begin{cases} \hat{h}(\nu[\![e_1]\!], \ldots, \nu[\![e_k]\!]) & \text{if } [\![\phi]\!] \text{ graph of smooth } \hat{h}{:}\mathbb{R}^k \to \mathbb{R} \\ 0 & \text{otherwise} \end{cases}$$

This semantics says that, if the relation $[\![\phi]\!] \subseteq \mathbb{R} \times \mathbb{R}^k$ is the graph of some smooth C^∞ function $\hat{h} : \mathbb{R}^k \to \mathbb{R}$, then the annotated syntactic symbol $h_{\ll\phi\gg}$ is interpreted semantically as $\hat{h}$. Note that the graph relation uniquely defines $\hat{h}$ (if it exists). Otherwise, $h_{\ll\phi\gg}$ is interpreted as the constant zero function which ensures that the term semantics remain well-defined for all terms. An alternative is to leave the semantics of some terms (possibly) undefined, but this would require more extensive changes to the semantics of dL and extra case distinctions during proofs [2].

Axiomatics and Differentially-Defined Functions. To support reasoning for implicit definitions, annotated interpretations are reified to characterization axioms for expanding interpreted functions in the following lemma.

Lemma 1. (Function interpretation). *The FI axiom (below) for* dL *is sound where h is a k-ary function symbol and the formula semantics $[\![\phi]\!]$ is the graph of a smooth C^∞ function $\hat{h} : \mathbb{R}^k \to \mathbb{R}$.*

$$\text{FI} \quad e_0 = h_{\ll\phi\gg}(e_1, \ldots, e_k) \leftrightarrow \phi(e_0, e_1, \ldots, e_k)$$

Axiom FI enables reasoning for terms $h_{\ll\phi\gg}(e_1, \ldots, e_k)$ through their implicit interpretation ϕ, but Lemma 1 does not directly yield an implementation because it has a soundness-critical side condition that interpretation ϕ characterizes the graph of a smooth C^∞ function. It is possible to syntactically characterize this side condition [2], e.g., the formula $\forall y_1, \ldots, y_k \exists x_0 \phi(x_0, y_1, \ldots, y_k)$ expresses that the graph represented by ϕ has at least one output value x_0 for each input value $y_1, \ldots, y_k$, but this burdens users with the task of proving this side condition in dL before working with their desired function. The KeYmaera X definition package opts for a middle ground between generality and ease-of-use by implementing FI for univariate, *differentially-defined* functions, i.e., the interpretation ϕ has the following shape, where $x = (x_0, x_1, \ldots, x_n)$ abbreviates a vector of variables, there is one input $t = y_1$, and $X = (X_0, X_1, \ldots, X_n)$, T are dL terms that do not mention any free variables, e.g., are rational constants, which have constant value in any dL state:

$$\phi(x_0, t) \equiv \left\langle x_1, \ldots, x_n := *; \left\{ \begin{matrix} x' = -f(x,t), t' = -1 \cup \\ x' = f(x,t), t' = 1 \end{matrix} \right\} \right\rangle \left(\begin{matrix} x = X \land \\ t = T \end{matrix} \right) \tag{3}$$

Formula (3) says from point x_0, there exists a choice of the remaining coordinates $x_1, \ldots, x_n$ such that it is possible to follow the defining ODE either forward $x' = f(x,t), t' = 1$ or backward $x' = -f(x,t), t' = -1$ in time to reach the initial values $x = X$ at time $t = T$. In other words, the implicitly defined function $h_{\ll \phi(x_0,t) \gg}$ is the x_0-coordinate projected solution of the ODE starting from initial values X at initial time T. For example, the trigonometric functions used in Fig. 1 are differentially-definable as respective projections:

$$\phi_{\sin}(s,t) \equiv \left\langle c := *; \left\{ \begin{matrix} s' = -c, c' = s, t' = -1 \cup \\ s' = c, c' = -s, t' = 1 \end{matrix} \right\} \right\rangle \left(\begin{matrix} s = 0 \land c = 1 \land \\ t = 0 \end{matrix} \right) \tag{4}$$

$$\phi_{\cos}(c,t) \equiv \left\langle s := *; \left\{ \begin{matrix} s' = -c, c' = s, t' = -1 \cup \\ s' = c, c' = -s, t' = 1 \end{matrix} \right\} \right\rangle \left(\begin{matrix} s = 0 \land c = 1 \land \\ t = 0 \end{matrix} \right) \tag{5}$$

By Picard-Lindelöf [21, Thm. 2.2], the ODE $x' = f(x,t)$ has a unique solution $\Phi : (a,b) \to \mathbb{R}^{n+1}$ on an open interval (a,b) for some $-\infty \leq a < b \leq \infty$. Moreover, $\Phi(t)$ is C^∞ smooth in t because the ODE right-hand sides are dL terms with smooth interpretations [20]. Therefore, the side condition for Lemma 1 reduces to showing that Φ exists globally, i.e., it is defined on $t \in (-\infty, \infty)$.

Lemma 2. (Smooth interpretation). *If formula $\exists x_0 \phi(x_0, t)$ is valid, $\phi(x_0, t)$ from* (3) *characterizes a smooth C^∞ function and axiom FI is sound for $\phi(x_0, t)$.*

Lemma 2 enables an implementation of axiom FI in KeYmaera X that combines a syntactic check (the interpretation has the shape of formula (3)) and a side condition check (requiring users to prove existence for their interpretations).

The addition of differentially-defined functions to dL strictly increases the deductive power of ODE invariants, a key tool in deductive ODE safety reasoning [21]. Intuitively, the added functions allow direct, syntactic descriptions of invariants, e.g., the exponential or trigonometric functions, that have effective invariance proofs using dL's complete ODE invariance reasoning principles [22].

Proposition 3. (Invariant expressivity). *There are valid polynomial* dL *differential equation safety properties which are provable using differentially-defined function invariants but are not provable using polynomial invariants.*

3 KeYmaera X Implementation

The implicit definition package adds interpretation annotations and axiom FI based on Lemma 2 in ≈170 lines of code extensions to KeYmaera X's soundness-critical core [7, 16]. This section focuses on non-soundness-critical usability features provided by the package that build on those core changes.

3.1 Core-Adjacent Changes

KeYmaera X has a browser-based user interface with concrete, ASCII-based dL syntax [14]. The package extends KeYmaera X's parsers and pretty printers with support for interpretation annotations `h«...»(...)` and users can simultaneously define a family of functions as respective coordinate projections of the solution of an n-dimensional ODE (given initial conditions) with sugared syntax:

```
implicit Real h1(Real t), ..., hn(Real t) = {{initcond};{ODE}}
```

For example, the implicit definitions (4), (5) can be written with the following sugared syntax; KeYmaera X automatically inserts the associated interpretation annotations for the trigonometric function symbols, see the supplement [8] for a KeYmaera X snippet of formula ϕ_s from Fig. 1 using this sugared definition.

```
implicit Real sin(Real t), cos(Real t)
  = {{sin:=0; cos:=1;}; {sin'=cos, cos'=-sin}}
```

In fact, the functions sin, cos, exp are so ubiquitous in hybrid system models that the package builds their definitions in automatically without requiring users to write them explicitly. In addition, although arithmetic involving those functions is undecidable [11, 24], KeYmaera X can export those functions whenever its external arithmetic tools have partial arithmetic support for those functions.

3.2 Intermediate and User-Level Proof Automation

The package automatically proves three important lemmas about user-defined functions that can be transparently re-used in all subsequent proofs:

1. It proves the side condition of axiom FI using KeYmaera X's automation for proving sufficient duration existence of solutions for ODEs [26] which automatically shows global existence of solutions for all affine ODEs and some univariate nonlinear ODEs. As an example of the latter, the hyperbolic tanh function is differentially-defined as the solution of ODE $x' = 1 - x^2$ with initial value $x = 0$ at $t = 0$ whose global existence is proved automatically.

2. It proves that the functions have initial values as specified by their interpretation, e.g., $\sin(0) = 0$, $\cos(0) = 1$, and $\tanh(0) = 0$.
3. It proves the *differential axiom* [20] for each function that is used to enable syntactic derivative calculations in dL, e.g., the differential axioms for sin, cos are $(\sin(e))' = \cos(e)(e)'$ and $(\cos(e))' = -\sin(e)(e)'$, respectively. Briefly, these axioms are automatically derived in a correct-by-construction manner using dL's syntactic version of the chain rule for differentials [20, Fig. 3], so the rate of change of $\sin(e)$ is the rate of change of $\sin(\cdot)$ with respect to its argument e, multiplied by the rate of change of its argument $(e)'$.

These lemmas enable the use of differentially-defined functions with all existing ODE automation in KeYmaera X [22,26]. In particular, since differentially-defined functions are univariate Noetherian functions, they admit complete ODE invariance reasoning principles in dL [22] as implemented in KeYmaera X.

The package also adds specialized support for arithmetical reasoning over differential definitions to supplement external arithmetic tools in proofs. First, it allows users to manually prove identities and bounds using KeYmaera X's ODE reasoning. For example, the bound $\tanh(\lambda x)^2 < 1$ used in the example α_n from Sect. 4 is proved by *differential unfolding* as follows (see supplement [8]):

$$\frac{\vdash \tanh(0)^2 < 1 \quad \tanh(\lambda v)^2 < 1 \vdash [\{v' = 1 \,\&\, v \leq x\} \cup \{v' = -1 \,\&\, v \geq x\}]\tanh(\lambda v)^2 < 1}{\vdash \tanh(\lambda x)^2 < 1}$$

This deduction step says that, to show the conclusion (below rule bar), it suffices to prove the premises (above rule bar), i.e., the bound is true at $v = 0$ (left premise) and it is preserved as v is evolved forward $v' = 1$ or backward $v' = -1$ along the real line until it reaches x (right premise). The left premise is proved using the initial value lemma for tanh while the right premise is proved by ODE invariance reasoning with the differential axiom for tanh [22].

Second, the package uses KeYmaera X's uniform substitution mechanism [20] to implement (untrusted) abstraction of functions with fresh variables when solving arithmetic subgoals, e.g., the following arithmetic bound for example α_n is proved by abstraction after adding the bounds $\tanh(\lambda x)^2 < 1, \tanh(\lambda y)^2 < 1$.

$$\textbf{Bound:} \quad x(\tanh(\lambda x) - \tanh(\lambda y)) + y(\tanh(\lambda x) + \tanh(\lambda y)) \leq 2\sqrt{x^2 + y^2}$$

$$\textbf{Abstracted:} \quad t_x^2 < 1 \land t_y^2 < 1 \rightarrow x(t_x - t_y) + y(t_x + t_y) \leq 2\sqrt{x^2 + y^2}$$

4 Examples

The definition package enables users to work with differentially-defined functions in KeYmaera X, including modeling and expressing their design intuitions in proofs. This section applies the package to verify various continuous and hybrid system examples from the literature featuring such functions.

Discretely Driven Pendulum. The specification ϕ_s from Fig. 1 contains a discrete loop whose safety property is proved by a loop invariant, i.e., a formula that is preserved by the discrete and continuous dynamics in each loop iteration [21].

The key invariant is $Inv \equiv \frac{g}{L}(1 - \cos\theta) + \frac{1}{2}\omega^2 < \frac{g}{L}$, which expresses that the total energy of the system (sum of potential and kinetic energy on the LHS) is less than the energy needed to cross the horizontal (RHS). The main steps are as follows (proofs for these steps are automated by KeYmaera X):

1. $Inv \rightarrow \left[\texttt{if}\left(\frac{1}{2}(\omega - p)^2 < \frac{g}{L}\cos(\theta)\right) \{\omega := \omega - p\}\right] Inv$, which shows that the discrete guard only allows push p if it preserves the energy invariant, and
2. $Inv \rightarrow \left[\{\theta' = \omega, \omega' = -\frac{g}{L}\sin(\theta) - k\omega\}\right] Inv$, which shows that Inv is an energy invariant of the pendulum's ODE.

Neuron Interaction. The ODE α_n models the interaction between a pair of neurons [12]; its specification ϕ_n nests dL's diamond and box modalities to express that the system norm ($\sqrt{x^2+y^2}$) is asymptotically bounded by 2τ.

$$\alpha_n \equiv x' = -\frac{x}{\tau} + \tanh(\lambda x) - \tanh(\lambda y), y' = -\frac{y}{\tau} + \tanh(\lambda x) + \tanh(\lambda y)$$

$$\phi_n \equiv \tau > 0 \rightarrow \forall \varepsilon{>}0 \langle \alpha_n \rangle [\alpha_n] \sqrt{x^2+y^2} \leq 2\tau + \varepsilon$$

The verification of ϕ_n uses differentially-defined functions in concert with KeYmaera X's symbolic ODE safety and liveness reasoning [26]. The proof uses a decaying exponential bound $\sqrt{x^2+y^2} \leq \exp(-\frac{t}{\tau})\sqrt{x_0^2+y_0^2} + 2\tau(1 - \exp(-\frac{t}{\tau}))$, where the constants x_0, y_0 are symbolic initial values for x, y at initial time $t = 0$, respectively. Notably, the arithmetic subgoals from this example are all proved using abstraction and differential unfolding (Sect. 3) without relying on external arithmetic solver support for tanh.

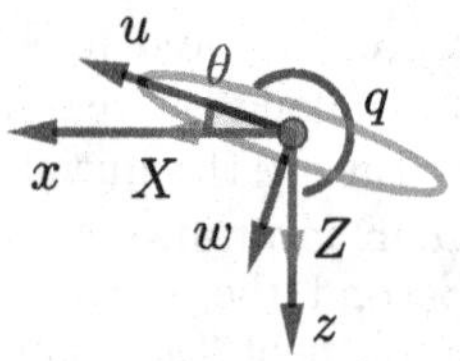

Longitudinal Flight Dynamics. The differential equations α_a below describe the 6th order longitudinal motion of an airplane while climbing or descending [10,25]. The airplane adjusts its pitch angle θ with pitch rate q, which determines its axial velocity u and vertical velocity w, and, in turn, range x and altitude z (illustrated on the right). The physical parameters are: gravity g, mass m, aerodynamic thrust and moment M along the lateral axis, aerodynamic and thrust forces X, Z along x and z, respectively, and the moment of inertia I_{yy}, see [10, Sect. 6.2].

$$\alpha_a \equiv u' = \frac{X}{m} - g\sin(\theta) - qw, \qquad w' = \frac{Z}{m} + g\cos(\theta) + qu, \qquad q' = \frac{M}{I_{yy}},$$
$$x' = \cos(\theta)u + \sin(\theta)w, \qquad z' = -\sin(\theta)u + \cos(\theta)w, \qquad \theta' = q$$

The verification of specification $J \rightarrow [\alpha_a]J$ shows that the safety envelope $J \equiv J_1 \wedge J_2 \wedge J_3$ is invariant along the flow of α_a with algebraic invariants J_i:

$$J_1 \equiv \frac{Mz}{I_{yy}} + g\theta + \left(\frac{X}{m} - qw\right)\cos(\theta) + \left(\frac{Z}{m} + qu\right)\sin(\theta) = 0$$

$$J_2 \equiv \frac{Mz}{I_{yy}} - \left(\frac{Z}{m} + qu\right)\cos(\theta) + \left(\frac{X}{m} - qw\right)\sin(\theta) = 0 \quad J_3 \equiv -q^2 + \frac{2M\theta}{I_{yy}} = 0$$

Additional examples are available in the supplement [8], including: a bouncing ball on a sinusoidal surface [6,13] and a robot collision avoidance model [15].

5 Conclusion

This work presents a convenient mechanism for extending the dL term language with differentially-defined functions, thereby furthering the class of real-world systems amenable to modeling and formalization in KeYmaera X. Minimal soundness-critical changes are made to the KeYmaera X kernel, which maintains its trustworthiness while allowing the use of newly defined functions in concert with all existing dL hybrid systems reasoning principles implemented in KeYmaera X. Future work could formally verify these kernel changes by extending the existing formalization of dL [3]. Further integration of external arithmetic tools [1,9,23] will also help to broaden the classes of arithmetic sub-problems that can be solved effectively in hybrid systems proofs.

Acknowledgments. We thank the anonymous reviewers for their helpful feedback on this paper. This material is based upon work supported by the National Science Foundation under Grant No. CNS-1739629. This research was sponsored by the AFOSR under grant number FA9550-16-1-0288.

References

1. Akbarpour, B., Paulson, L.C.: MetiTarski: an automatic theorem prover for real-valued special functions. J. Autom. Reason. **44**(3), 175–205 (2010). https://doi.org/10.1007/s10817-009-9149-2
2. Bohrer, B., Fernández, M., Platzer, A.: dL_ι: definite descriptions in differential dynamic logic. In: Fontaine, P. (ed.) CADE 2019. LNCS (LNAI), vol. 11716, pp. 94–110. Springer, Cham (2019). https://doi.org/10.1007/978-3-030-29436-6_6
3. Bohrer, R., Rahli, V., Vukotic, I., Völp, M., Platzer, A.: Formally verified differential dynamic logic. In: Bertot, Y., Vafeiadis, V. (eds.) CPP, pp. 208–221. ACM (2017). https://doi.org/10.1145/3018610.3018616
4. Boldo, S., Lelay, C., Melquiond, G.: Formalization of real analysis: a survey of proof assistants and libraries. Math. Struct. Comput. Sci. **26**(7), 1196–1233 (2016). https://doi.org/10.1017/S0960129514000437
5. Bonichon, R., Delahaye, D., Doligez, D.: Zenon: an extensible automated theorem prover producing checkable proofs. In: Dershowitz, N., Voronkov, A. (eds.) LPAR 2007. LNCS (LNAI), vol. 4790, pp. 151–165. Springer, Heidelberg (2007). https://doi.org/10.1007/978-3-540-75560-9_13
6. Denman, W.: Automated verification of continuous and hybrid dynamical systems. Ph.D. thesis, University of Cambridge, UK (2015)
7. Fulton, N., Mitsch, S., Quesel, J.-D., Völp, M., Platzer, A.: KeYmaera X: an axiomatic tactical theorem prover for hybrid systems. In: Felty, A.P., Middeldorp, A. (eds.) CADE 2015. LNCS (LNAI), vol. 9195, pp. 527–538. Springer, Cham (2015). https://doi.org/10.1007/978-3-319-21401-6_36
8. Gallicchio, J., Tan, Y.K., Mitsch, S., Platzer, A.: Implicit definitions with differential equations for KeYmaera X (system description). CoRR abs/2203.01272 (2022). http://arxiv.org/abs/2203.01272
9. Gao, S., Kong, S., Clarke, E.M.: dReal: an SMT solver for nonlinear theories over the reals. In: Bonacina, M.P. (ed.) CADE 2013. LNCS (LNAI), vol. 7898, pp. 208–214. Springer, Heidelberg (2013). https://doi.org/10.1007/978-3-642-38574-2_14

10. Ghorbal, K., Platzer, A.: Characterizing algebraic invariants by differential radical invariants. In: Ábrahám, E., Havelund, K. (eds.) TACAS 2014. LNCS, vol. 8413, pp. 279–294. Springer, Heidelberg (2014). https://doi.org/10.1007/978-3-642-54862-8_19
11. Gödel, K.: Über formal unentscheidbare Sätze der Principia Mathematica und verwandter Systeme I. Monatshefte für Mathematik und Physik **38**(1), 173–198 (1931). https://doi.org/10.1007/BF01700692
12. Khalil, H.K.: Nonlinear Systems. Macmillan, New York (1992)
13. Liu, J., Zhan, N., Zhao, H., Zou, L.: Abstraction of elementary hybrid systems by variable transformation. In: Bjørner, N., de Boer, F. (eds.) FM 2015. LNCS, vol. 9109, pp. 360–377. Springer, Cham (2015). https://doi.org/10.1007/978-3-319-19249-9_23
14. Mitsch, S.: Implicit and explicit proof management in KeYmaera X. In: Proença, J., Paskevich, A. (eds.) F-IDE. EPTCS, vol. 338, pp. 53–67 (2021). https://doi.org/10.4204/EPTCS.338.8
15. Mitsch, S., Ghorbal, K., Vogelbacher, D., Platzer, A.: Formal verification of obstacle avoidance and navigation of ground robots. Int. J. Robot. Res. **36**(12), 1312–1340 (2017). https://doi.org/10.1177/0278364917733549
16. Mitsch, S., Platzer, A.: A retrospective on developing hybrid system provers in the KeYmaera family. In: Ahrendt, W., Beckert, B., Bubel, R., Hähnle, R., Ulbrich, M. (eds.) Deductive Software Verification: Future Perspectives. LNCS, vol. 12345, pp. 21–64. Springer, Cham (2020). https://doi.org/10.1007/978-3-030-64354-6_2
17. Platzer, A.: Differential dynamic logic for hybrid systems. J. Autom. Reason. **41**(2), 143–189 (2008). https://doi.org/10.1007/s10817-008-9103-8
18. Platzer, A.: Logical Analysis of Hybrid Systems: Proving Theorems for Complex Dynamics. Springer, Heidelberg (2010). https://doi.org/10.1007/978-3-642-14509-4
19. Platzer, A.: The complete proof theory of hybrid systems. In: LICS, pp. 541–550. IEEE Computer Society (2012). https://doi.org/10.1109/LICS.2012.64
20. Platzer, A.: A complete uniform substitution calculus for differential dynamic logic. J. Autom. Reason. **59**(2), 219–265 (2016). https://doi.org/10.1007/s10817-016-9385-1
21. Platzer, A.: Logical foundations of cyber-physical systems. Springer, Cham (2018). https://doi.org/10.1007/978-3-319-63588-0
22. Platzer, A., Tan, Y.K.: Differential equation invariance axiomatization. J. ACM **67**(1) (2020). https://doi.org/10.1145/3380825
23. Ratschan, S., She, Z.: Safety verification of hybrid systems by constraint propagation-based abstraction refinement. ACM Trans. Embed. Comput. Syst. **6**(1), 8 (2007). https://doi.org/10.1145/1210268.1210276
24. Richardson, D.: Some undecidable problems involving elementary functions of a real variable. J. Symb. Log. **33**(4), 514–520 (1968). https://doi.org/10.2307/2271358
25. Stengel, R.F.: Flight Dynamics. Princeton University Press (2004)
26. Tan, Y.K., Platzer, A.: An axiomatic approach to existence and liveness for differential equations. Form. Asp. Comput. **33**(4), 461–518 (2021). https://doi.org/10.1007/s00165-020-00525-0
27. Wiedijk, F.: Stateless HOL. In: Hirschowitz, T. (ed.) TYPES. EPTCS, vol. 53, pp. 47–61 (2009). https://doi.org/10.4204/EPTCS.53.4

Automatic Complexity Analysis of Integer Programs via Triangular Weakly Non-Linear Loops

Nils Lommen(✉), Fabian Meyer, and Jürgen Giesl(✉)

LuFG Informatik 2, RWTH Aachen University, Aachen, Germany
lommen@cs.rwth-aachen.de, giesl@informatik.rwth-aachen.de

Abstract. There exist several results on deciding termination and computing runtime bounds for *triangular weakly non-linear loops* (twn-loops). We show how to use results on such subclasses of programs where complexity bounds are computable within incomplete approaches for complexity analysis of full integer programs. To this end, we present a novel modular approach which computes local runtime bounds for subprograms which can be transformed into twn-loops. These local runtime bounds are then lifted to global runtime bounds for the whole program. The power of our approach is shown by our implementation in the tool KoAT which analyzes complexity of programs where all other state-of-the-art tools fail.

1 Introduction

Most approaches for automated complexity analysis of programs are based on incomplete techniques like ranking functions (see, e.g., [1–4,6,11,12,18,20,21,31]). However, there also exist numerous results on subclasses of programs where questions concerning termination or complexity are *decidable*, e.g., [5,14,15,19,22,24,25,32,34]. In this work we consider the subclass of *triangular weakly non-linear loops* (twn-loops), where there exist *complete* techniques for analyzing termination and runtime complexity (we discuss the "completeness" and decidability of these techniques below). An example for a twn-loop is:

$$\textbf{while } (x_1^2+x_3^5 < x_2 \wedge x_1 \neq 0) \textbf{ do } (x_1, x_2, x_3) \leftarrow (-2 \cdot x_1,\ 3 \cdot x_2 - 2 \cdot x_3^3,\ x_3) \qquad (1)$$

Its guard is a propositional formula over (possibly *non-linear*) polynomial inequations. The update is *weakly non-linear*, i.e., no variable x_i occurs non-linear in its own update. Furthermore, it is *triangular*, i.e., we can order the variables such that the update of any x_i does not depend on the variables $x_1, \ldots, x_{i-1}$ with smaller indices. Then, by handling one variable after the other one can compute a *closed form* which corresponds to applying the loop's update n times. Using

Funded by the Deutsche Forschungsgemeinschaft (DFG, German Research Foundation) - 235950644 (Project GI 274/6-2) and DFG Research Training Group 2236 UnRAVeL.

J. Blanchette et al. (Eds.): IJCAR 2022, LNAI 13385, pp. 734–754, 2022.
https://doi.org/10.1007/978-3-031-10769-6_43

these closed forms, termination can be reduced to an existential formula over $\mathbb{Z}$ [15] (whose validity is decidable for linear arithmetic and where SMT solvers often also prove (in)validity in the non-linear case). In this way, one can show that non-termination of twn-loops over $\mathbb{Z}$ is semi-decidable (and it is decidable over the real numbers).

While termination of twn-loops over $\mathbb{Z}$ is not decidable, by using the closed forms, [19] presented a "*complete*" complexity analysis technique. More precisely, for every twn-loop over $\mathbb{Z}$, it infers a polynomial which is an upper bound on the runtime for all those inputs where the loop terminates. So for all (possibly non-linear) terminating twn-loops over $\mathbb{Z}$, the technique of [19] *always* computes polynomial runtime bounds. In contrast, existing tools based on incomplete techniques for complexity analysis often fail for programs with non-linear arithmetic.

In [6,18] we presented such an incomplete modular technique for complexity analysis which uses individual ranking functions for different subprograms. Based on this, we now introduce a novel approach to automatically infer runtime bounds for programs possibly consisting of multiple consecutive or nested loops by handling some subprograms as twn-loops and by using ranking functions for others. In order to compute runtime bounds, we analyze subprograms in topological order, i.e., in case of multiple consecutive loops, we start with the first loop and propagate knowledge about the resulting values of variables to subsequent loops. By inferring runtime bounds for one subprogram after the other, in the end we obtain a bound on the runtime complexity of the whole program. We first try to compute runtime bounds for subprograms by so-called multiphase linear ranking functions (MΦRFs, see [3,4,18,20]). If MΦRFs do not yield a finite runtime bound for the respective subprogram, then we use our novel twn-technique on the unsolved parts of the subprogram. So for the first time, "complete" complexity analysis techniques like [19] for subclasses of programs with *non-linear* arithmetic are combined with incomplete techniques based on (linear) ranking functions like [6,18]. Based on our approach, in future work one could integrate "complete" techniques for further subclasses (e.g., for *solvable loops* [24,25,30,34] which can be transformed into twn-loops by suitable automorphisms [15]).

Structure: After introducing preliminaries in Sect. 2, in Sect. 3 we show how to lift a (local) runtime bound which is only sound for a subprogram to an overall global runtime bound. In contrast to previous techniques [6,18], our lifting approach works for any method of bound computation (not only for ranking functions). In Sect. 4, we improve the existing results on complexity analysis of twn- loops [14,15,19] such that they yield concrete polynomial bounds, we refine these bounds by considering invariants, and we show how to apply these results to full programs which contain twn-loops as subprograms. Section 5 extends this technique to larger subprograms which can be transformed into twn-loops. In Sect. 6 we evaluate the implementation of our approach in the complexity analysis tool KoAT and show that one can now also successfully analyze the runtime of programs containing non-linear arithmetic. We refer to [26] for all proofs.

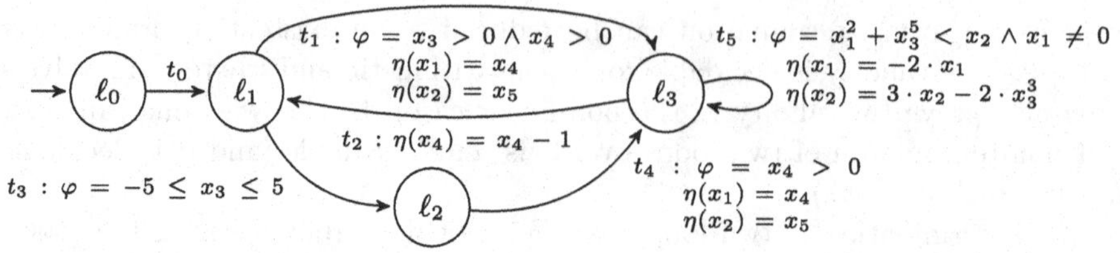

Fig. 1. An Integer Program with a Nested Self-Loop

2 Preliminaries

This section recapitulates preliminaries for complexity analysis from [6,18].

Definition 1 (Atoms and Formulas). *We fix a set $\mathcal{V}$ of variables. The set of* atoms *$\mathcal{A}(\mathcal{V})$ consists of all inequations $p_1 < p_2$ for polynomials $p_1, p_2 \in \mathbb{Z}[\mathcal{V}]$. $\mathcal{F}(\mathcal{V})$ is the set of all propositional* formulas *built from atoms $\mathcal{A}(\mathcal{V})$, $\land$, and $\lor$.*

In addition to "<", we also use "≥", "=", "≠", etc., and negations "¬" which can be simulated by formulas (e.g., $p_1 \geq p_2$ is equivalent to $p_2 < p_1 + 1$ for integers).

For integer programs, we use a formalism based on transitions, which also allows us to represent **while**-programs like (1) easily. Our programs may have *non-deterministic branching*, i.e., the guards of several applicable transitions can be satisfied. Moreover, *non-deterministic sampling* is modeled by *temporary variables* whose values are updated arbitrarily in each evaluation step.

Definition 2 (Integer Program). *$(\mathcal{PV}, \mathcal{L}, \ell_0, \mathcal{T})$ is an* integer program *where*

- *$\mathcal{PV} \subseteq \mathcal{V}$ is a finite set of* program variables, *$\mathcal{V} \setminus \mathcal{PV}$ are* temporary variables
- *$\mathcal{L}$ is a finite set of* locations *with an* initial location *$\ell_0 \in \mathcal{L}$*
- *$\mathcal{T}$ is a finite set of* transitions. *A transition is a 4-tuple $(\ell, \varphi, \eta, \ell')$ with a* start location *$\ell \in \mathcal{L}$,* target location *$\ell' \in \mathcal{L} \setminus \{\ell_0\}$,* guard *$\varphi \in \mathcal{F}(\mathcal{V})$, and* update function *$\eta : \mathcal{PV} \to \mathbb{Z}[\mathcal{V}]$ mapping program variables to update polynomials.*

Transitions $(\ell_0, \_, \_, \_)$ are called *initial.* Note that ℓ_0 has no incoming transitions.

Example 3. Consider the program in Fig. 1 with $\mathcal{PV} = \{x_i \mid 1 \leq i \leq 5\}$, $\mathcal{L} = \{\ell_i \mid 0 \leq i \leq 3\}$, and $\mathcal{T} = \{t_i \mid 0 \leq i \leq 5\}$, where t_5 has non-linear arithmetic in its guard and update. We omitted trivial guards, i.e., $\varphi = \texttt{true}$, and identity updates of the form $\eta(v) = v$. Thus, t_5 corresponds to the **while**-program (1).

A *state* is a mapping $\sigma : \mathcal{V} \to \mathbb{Z}$, Σ denotes the set of all states, and $\mathcal{L} \times \Sigma$ is the set of *configurations.* We also apply states to arithmetic expressions p or formulas φ, where the number $\sigma(p)$ resp. the Boolean value $\sigma(\varphi)$ results from replacing each variable v by $\sigma(v)$. So for a state with $\sigma(x_1) = -8$, $\sigma(x_2) = 55$, and $\sigma(x_3) = 1$, the expression $x_1^2 + x_3^5$ evaluates to $\sigma(x_1^2 + x_3^5) = 65$ and the formula $\varphi = (x_1^2 + x_3^5 < x_2)$ evaluates to $\sigma(\varphi) = (65 < 55) = \texttt{false}$. From now on, we fix a program $(\mathcal{PV}, \mathcal{L}, \ell_0, \mathcal{T})$.

Definition 4 (Evaluation of Programs). *For configurations* (ℓ, σ), (ℓ', σ') *and* $t = (\ell_t, \varphi, \eta, \ell'_t) \in \mathcal{T}$, $(\ell, \sigma) \to_t (\ell', \sigma')$ *is an* evaluation step *if* $\ell = \ell_t$, $\ell' = \ell'_t$, $\sigma(\varphi) = \mathtt{true}$, *and* $\sigma(\eta(v)) = \sigma'(v)$ *for all* $v \in \mathcal{PV}$. *Let* $\to_{\mathcal{T}} = \bigcup_{t \in \mathcal{T}} \to_t$, *where we also write* $\to$ *instead of* $\to_t$ *or* $\to_{\mathcal{T}}$. *Let* $(\ell_0, \sigma_0) \to^k (\ell_k, \sigma_k)$ *abbreviate* $(\ell_0, \sigma_0) \to \ldots \to (\ell_k, \sigma_k)$ *and let* $(\ell, \sigma) \to^* (\ell', \sigma')$ *if* $(\ell, \sigma) \to^k (\ell', \sigma')$ *for some* $k \geq 0$.

So when denoting states σ as tuples $(\sigma(x_1), \ldots, \sigma(x_5)) \in \mathbb{Z}^5$ for the program in Fig. 1, we have $(\ell_0, (1,5,7,1,3)) \to_{t_0} (\ell_1, (1,5,7,1,3)) \to_{t_1} (\ell_3, (1,1,3,1,3)) \to_{t_5}^3 (\ell_3, (1,-8,55,1,3)) \to_{t_2} \ldots$. The runtime complexity $\mathrm{rc}(\sigma_0)$ of a program corresponds to the length of the longest evaluation starting in the initial state σ_0.

Definition 5 (Runtime Complexity). *The* runtime complexity *is* $\mathrm{rc}: \Sigma \to \overline{\mathbb{N}}$ *with* $\overline{\mathbb{N}} = \mathbb{N} \cup \{\omega\}$ *and* $\mathrm{rc}(\sigma_0) = \sup\{k \in \mathbb{N} \mid \exists(\ell', \sigma').\, (\ell_0, \sigma_0) \to^k (\ell', \sigma')\}$.

3 Computing Global Runtime Bounds

We now introduce our general approach for computing (upper) runtime bounds. We use weakly monotonically increasing functions as bounds, since they can easily be "composed" (i.e., if f and g increase monotonically, then so does $f \circ g$).

Definition 6 (Bounds [6,18]**).** *The set of* bounds $\mathcal{B}$ *is the smallest set with* $\overline{\mathbb{N}} \subseteq \mathcal{B}$, $\mathcal{PV} \subseteq \mathcal{B}$, *and* $\{b_1 + b_2,\, b_1 \cdot b_2,\, k^{b_1}\} \subseteq \mathcal{B}$ *for all* $k \in \mathbb{N}$ *and* $b_1, b_2 \in \mathcal{B}$.

A bound constructed from $\mathbb{N}$, $\mathcal{PV}$, $+$, and $\cdot$ is *polynomial*. So for $\mathcal{PV} = \{x, y\}$, we have ω, x^2, $x + y$, $2^{x+y} \in \mathcal{B}$. Here, x^2 and $x + y$ are polynomial bounds.

We measure the size of variables by their absolute values. For any $\sigma \in \Sigma$, $|\sigma|$ is the state with $|\sigma|(v) = |\sigma(v)|$ for all $v \in \mathcal{V}$. So if σ_0 denotes the initial state, then $|\sigma_0|$ maps every variable to its initial "size", i.e., its initial absolute value. $\mathcal{RB}_{\mathrm{glo}} : \mathcal{T} \to \mathcal{B}$ is a *global runtime bound* if for each transition t and initial state $\sigma_0 \in \Sigma$, $\mathcal{RB}_{\mathrm{glo}}(t)$ evaluated in the state $|\sigma_0|$ over-approximates the number of evaluations of t in any run starting in the configuration (ℓ_0, σ_0). Let $\to_{\mathcal{T}}^* \circ \to_t$ denote the relation where arbitrary many evaluation steps are followed by a step with t.

Definition 7 (Global Runtime Bound [6,18]**).** *The function* $\mathcal{RB}_{glo} : \mathcal{T} \to \mathcal{B}$ *is a* global runtime bound *if for all* $t \in \mathcal{T}$ *and all states* $\sigma_0 \in \Sigma$ *we have* $|\sigma_0|(\mathcal{RB}_{glo}(t)) \geq \sup\{k \in \mathbb{N} \mid \exists\, (\ell', \sigma').\, (\ell_0, \sigma_0)\, (\to_{\mathcal{T}}^* \circ \to_t)^k\, (\ell', \sigma')\}$.

For the program in Fig. 1, in Example 12 we will infer $\mathcal{RB}_{\mathrm{glo}}(t_0) = 1$, $\mathcal{RB}_{\mathrm{glo}}(t_i) = x_4$ for $1 \leq i \leq 4$, and $\mathcal{RB}_{\mathrm{glo}}(t_5) = 8 \cdot x_4 \cdot x_5 + 13006 \cdot x_4$. By adding the bounds for all transitions, a global runtime bound $\mathcal{RB}_{\mathrm{glo}}$ yields an upper bound on the program's runtime complexity. So for all $\sigma_0 \in \Sigma$ we have $|\sigma_0|(\sum_{t \in \mathcal{T}} \mathcal{RB}_{\mathrm{glo}}(t)) \geq \mathrm{rc}(\sigma_0)$.

For *local runtime bounds*, we consider the *entry transitions* of subsets $\mathcal{T}' \subseteq \mathcal{T}$.

Definition 8 (Entry Transitions [6,18]**).** *Let* $\varnothing \neq \mathcal{T}' \subseteq \mathcal{T}$. *Its* entry transitions *are* $\mathcal{E}_{\mathcal{T}'} = \{t \mid t = (\ell, \varphi, \eta, \ell') \in \mathcal{T} \setminus \mathcal{T}' \wedge$ *there is a transition* $(\ell', \_, \_, \_) \in \mathcal{T}'\}$.

So in Fig. 1, we have $\mathcal{E}_{\mathcal{T}\setminus\{t_0\}} = \{t_0\}$ and $\mathcal{E}_{\{t_5\}} = \{t_1, t_4\}$.

In contrast to global runtime bounds, a *local* runtime bound $\mathcal{RB}_{\text{loc}} : \mathcal{E}_{\mathcal{T}'} \to \mathcal{B}$ only takes a subset $\mathcal{T}'$ into account. A *local run* is started by an entry transition $r \in \mathcal{E}_{\mathcal{T}'}$ followed by transitions from $\mathcal{T}'$. A *local runtime bound* considers a subset $\mathcal{T}'_> \subseteq \mathcal{T}'$ and over-approximates the number of evaluations of any transition from $\mathcal{T}'_>$ in an arbitrary local run of the subprogram with the transitions $\mathcal{T}'$. More precisely, for every $t \in \mathcal{T}'_>$, $\mathcal{RB}_{\text{loc}}(r)$ over-approximates the number of applications of t in any run of $\mathcal{T}'$, if $\mathcal{T}'$ is entered via $r \in \mathcal{E}_{\mathcal{T}'}$. However, local runtime bounds do not consider how often an entry transition from $\mathcal{E}_{\mathcal{T}'}$ is evaluated or how large a variable is when we evaluate an entry transition. To illustrate that $\mathcal{RB}_{\text{loc}}(r)$ is a bound on the number of evaluations of transitions from $\mathcal{T}'_>$ after evaluating r, we often write $\mathcal{RB}_{\text{loc}}(\to_r \mathcal{T}'_>)$ instead of $\mathcal{RB}_{\text{loc}}(r)$.

Definition 9 (Local Runtime Bound). *Let* $\varnothing \neq \mathcal{T}'_> \subseteq \mathcal{T}' \subseteq \mathcal{T}$. *The function* $\mathcal{RB}_{loc} : \mathcal{E}_{\mathcal{T}'} \to \mathcal{B}$ *is a* local runtime bound *for* $\mathcal{T}'_>$ *w.r.t.* $\mathcal{T}'$ *if for all* $t \in \mathcal{T}'_>$, *all* $r \in \mathcal{E}_{\mathcal{T}'}$ *with* $r = (\ell, \_, \_, \_)$, *and all* $\sigma \in \Sigma$ *we have* $|\sigma|(\mathcal{RB}_{loc}(\to_r \mathcal{T}'_>)) \geq \sup\{k \in \mathbb{N} \mid \exists \sigma_0, (\ell', \sigma').\ (\ell_0, \sigma_0) \to^*_{\mathcal{T}} \circ \to_r (\ell, \sigma)\ (\to^*_{\mathcal{T}'} \circ \to_t)^k\ (\ell', \sigma')\}$.

Our approach is *modular* since it computes local bounds for program parts separately. To lift local to global runtime bounds, we use *size bounds* $\mathcal{SB}(t, v)$ to over-approximate the size (i.e., absolute value) of the variable v after evaluating t in any run of the program. See [6] for the automatic computation of size bounds.

Definition 10 (Size Bound [6,18]**).** *The function* $\mathcal{SB} : (\mathcal{T} \times \mathcal{PV}) \to \mathcal{B}$ *is a* size bound *if for all* $(t, v) \in \mathcal{T} \times \mathcal{PV}$ *and all states* $\sigma_0 \in \Sigma$ *we have* $|\sigma_0|(\mathcal{SB}(t, v)) \geq \sup\{|\sigma'(v)| \mid \exists (\ell', \sigma').\ (\ell_0, \sigma_0)\ (\to^* \circ \to_t)\ (\ell', \sigma')\}$.

To compute global from local runtime bounds $\mathcal{RB}_{\text{loc}}(\to_r \mathcal{T}'_>)$ and size bounds $\mathcal{SB}(r, v)$, Theorem 11 generalizes the approach of [6,18]. Each local run is started by an entry transition r. Hence, we use an already computed global runtime bound $\mathcal{RB}_{\text{glo}}(r)$ to over-approximate the number of times that such a local run is started. To over-approximate the size of each variable v when entering the local run, we instantiate it by the size bound $\mathcal{SB}(r, v)$. So size bounds on previous transitions are needed to compute runtime bounds, and similarly, runtime bounds are needed to compute size bounds in [6]. For any bound b, "$b\ [v/\mathcal{SB}(r, v) \mid v \in \mathcal{PV}]$" results from b by replacing every program variable v by $\mathcal{SB}(r, v)$. Here, weak monotonic increase of b ensures that the over-approximation of the variables v in b by $\mathcal{SB}(r, v)$ indeed also leads to an over-approximation of b. The analysis starts with an *initial* runtime bound $\mathcal{RB}_{\text{glo}}$ and an *initial* size bound $\mathcal{SB}$ which map all transitions resp. all pairs from $\mathcal{T} \times \mathcal{PV}$ to ω, except for the transitions t which do not occur in cycles of $\mathcal{T}$, where $\mathcal{RB}_{\text{glo}}(t) = 1$. Afterwards, $\mathcal{RB}_{\text{glo}}$ and $\mathcal{SB}$ are refined repeatedly, where we alternate between computing runtime and size bounds.

Theorem 11 (Computing Global Runtime Bounds). *Let $\mathcal{RB}_{glo}$ be a global runtime bound, $\mathcal{SB}$ be a size bound, and $\varnothing \neq \mathcal{T}'_{>} \subseteq \mathcal{T}' \subseteq \mathcal{T}$ such that $\mathcal{T}'$ contains no initial transitions. Moreover, let $\mathcal{RB}_{loc}$ be a local runtime bound for $\mathcal{T}'_{>}$ w.r.t. $\mathcal{T}'$. Then $\mathcal{RB}'_{glo}$ is also a global runtime bound, where for all $t \in \mathcal{T}$ we define:*

$$\mathcal{RB}'_{glo}(t) = \begin{cases} \mathcal{RB}_{glo}(t), & \text{if } t \in \mathcal{T} \setminus \mathcal{T}'_{>} \\ \sum_{r \in \mathcal{E}_{\mathcal{T}'}} \mathcal{RB}_{glo}(r) \cdot (\mathcal{RB}_{loc}(\rightarrow_r \mathcal{T}'_{>})\, [v/\mathcal{SB}(r,v) \mid v \in \mathcal{PV}]), & \text{if } t \in \mathcal{T}'_{>} \end{cases}$$

Example 12. For the example in Fig. 1, we first use $\mathcal{T}'_{>} = \{t_2\}$ and $\mathcal{T}' = \mathcal{T} \setminus \{t_0\}$. With the ranking function x_4 one obtains $\mathcal{RB}_{\text{loc}}(\rightarrow_{t_0} \mathcal{T}'_{>}) = x_4$, since t_2 decreases the value of x_4 and no transition increases it. Then we can infer the global runtime bound $\mathcal{RB}_{\text{glo}}(t_2) = \mathcal{RB}_{\text{glo}}(t_0) \cdot (x_4\, [v/\mathcal{SB}(t_0, v) \mid v \in \mathcal{PV}]) = x_4$ as $\mathcal{RB}_{\text{glo}}(t_0) = 1$ (since t_0 is evaluated at most once) and $\mathcal{SB}(t_0, x_4) = x_4$ (since t_0 does not change any variables). Similarly, we can infer $\mathcal{RB}_{\text{glo}}(t_1) = \mathcal{RB}_{\text{glo}}(t_3) = \mathcal{RB}_{\text{glo}}(t_4) = x_4$.

For $\mathcal{T}'_{>} = \mathcal{T}' = \{t_5\}$, our twn-approach in Sect. 4 will infer the local runtime bound $\mathcal{RB}_{\text{loc}} : \mathcal{E}_{\{t_5\}} \to \mathcal{B}$ with $\mathcal{RB}_{\text{loc}}(\rightarrow_{t_1} \{t_5\}) = 4 \cdot x_2 + 3$ and $\mathcal{RB}_{\text{loc}}(\rightarrow_{t_4} \{t_5\}) = 4 \cdot x_2 + 4 \cdot x_3^3 + 4 \cdot x_3^5 + 3$ in Example 30. By Theorem 11 we obtain the global bound

$$\begin{aligned} \mathcal{RB}_{\text{glo}}(t_5) &= \mathcal{RB}_{\text{glo}}(t_1) \cdot (\mathcal{RB}_{\text{loc}}(\rightarrow_{t_1} \{t_5\})[v/\mathcal{SB}(t_1, v) \mid v \in \mathcal{PV}]) + \\ &\quad\; \mathcal{RB}_{\text{glo}}(t_4) \cdot (\mathcal{RB}_{\text{loc}}(\rightarrow_{t_4} \{t_5\})[v/\mathcal{SB}(t_4, v) \mid v \in \mathcal{PV}]) \\ &= x_4 \cdot (4 \cdot x_5 + 3) + x_4 \cdot (4 \cdot x_5 + 4 \cdot 5^3 + 4 \cdot 5^5 + 3) \\ &\qquad\qquad (\text{as } \mathcal{SB}(t_1, x_2) = \mathcal{SB}(t_4, x_2) = x_5 \text{ and } \mathcal{SB}(t_4, x_3) = 5) \\ &= 8 \cdot x_4 \cdot x_5 + 13006 \cdot x_4. \end{aligned}$$

Thus, $\text{rc}(\sigma_0) \in \mathcal{O}(n^2)$ where n is the largest initial absolute value of all program variables. While the approach of [6,18] was limited to local bounds resulting from ranking functions, here we need our Theorem 11. It allows us to use both local bounds resulting from twn-loops (for the non-linear transition t_5 where tools based on ranking functions cannot infer a bound, see Sect. 6) and local bounds resulting from ranking functions (for $t_1, \ldots, t_4$, since our twn-approach of Sect. 4 and 5 is limited to so-called simple cycles and cannot handle the full program).

In contrast to [6,18], we allow different local bounds for different entry transitions in Definition 9 and Theorem 11. Our example demonstrates that this can indeed lead to a smaller asymptotic bound for the whole program: By distinguishing the cases where t_5 is reached via t_1 or t_4, we end up with a quadratic bound, because the local bound $\mathcal{RB}_{\text{loc}}(\rightarrow_{t_1} \{t_5\})$ is linear and while x_3 occurs with degrees 5 and 3 in $\mathcal{RB}_{\text{loc}}(\rightarrow_{t_4} \{t_5\})$, the size bound for x_3 is constant after t_3 and t_4.

To improve size and runtime bounds repeatedly, we treat the strongly connected components (SCCs)[1] of the program in topological order such that

[1] As usual, a graph is *strongly connected* if there is a path from every node to every other node. A *strongly connected component* is a maximal strongly connected subgraph.

improved bounds for previous transitions are already available when handling the next SCC. We first try to infer local runtime bounds by multiphase-linear ranking functions (see [18] which also contains a heuristic for choosing $\mathcal{T}'_>$ and $\mathcal{T}'$ when using ranking functions). If ranking functions do not yield finite local bounds for all transitions of the SCC, then we apply the twn-technique from Sect. 4 and 5 on the remaining unbounded transitions (see Sect. 5 for choosing $\mathcal{T}'_>$ and $\mathcal{T}'$ in that case). Afterwards, the global runtime bound is updated according to Theorem 11.

4 Local Runtime Bounds for Twn-Self-Loops

In Sect. 4.1 we recapitulate twn-loops and their termination in our setting. Then in Sect. 4.2 we present a (complete) algorithm to infer polynomial runtime bounds for all terminating twn-loops. Compared to [19], we increased its precision considerably by computing bounds that take the different roles of the variables into account and by using over-approximations to remove monomials. Moreover, we show how our algorithm can be used to infer local runtime bounds for twn-loops occurring in integer programs. Section 5 will show that our algorithm can also be applied to infer runtime bounds for larger cycles in programs instead of just self-loops.

4.1 Termination of Twn-Loops

Definition 13 extends the definition of twn-loops in [15,19] by an initial transition and an update-invariant. Here, ψ is an *update-invariant* if $\models \psi \to \eta(\psi)$ where η is the update of the transition (i.e., invariance must hold independent of the guard).

Definition 13. (Twn-Loop). *An integer program* $(\mathcal{PV}, \mathcal{L}, \ell_0, \mathcal{T})$ *is a* triangular weakly non-linear loop (twn-loop) *if* $\mathcal{PV} = \{x_1, \ldots, x_d\}$ *for some* $d \geq 1$, $\mathcal{L} = \{\ell_0, \ell\}$, *and* $\mathcal{T} = \{t_0, t\}$ *with* $t_0 = (\ell_0, \psi, \mathrm{id}, \ell)$ *and* $t = (\ell, \varphi, \eta, \ell)$ *for some* $\psi, \varphi \in \mathcal{F}(\mathcal{PV})$ *with* $\models \psi \to \eta(\psi)$, *where* $\mathrm{id}(v) = v$ *for all* $v \in \mathcal{PV}$, *and for all* $1 \leq i \leq d$ *we have* $\eta(x_i) = c_i \cdot x_i + p_i$ *for some* $c_i \in \mathbb{Z}$ *and some polynomial* $p_i \in \mathbb{Z}[x_{i+1}, \ldots, x_d]$. *We often denote the loop by* (ψ, φ, η) *and refer to* ψ, φ, η *as its (update-) invariant, guard, and update, respectively. If* $c_i \geq 0$ *holds for all* $1 \leq i \leq d$, *then the program is a* non-negative *triangular weakly non-linear loop* (tnn-loop).

Example 14. The program consisting of the initial transition $(\ell_0, \texttt{true}, \mathrm{id}, \ell_3)$ and the self-loop t_5 in Fig. 1 is a twn-loop (corresponding to the **while**-loop (1)). This loop terminates as every iteration increases x_1^2 by a factor of 4 whereas x_2 is only tripled. Thus, $x_1^2 + x_3^5$ eventually outgrows the value of x_2.

To transform programs into twn- or tnn-form, one can combine subsequent transitions by *chaining*. Here, similar to states σ, we also apply the update η to polynomials and formulas by replacing each program variable v by $\eta(v)$.

Definition 15 (Chaining). *Let $t_1, \ldots, t_n$ be a sequence of transitions without temporary variables where $t_i = (\ell_i, \varphi_i, \eta_i, \ell_{i+1})$ for all $1 \leq i \leq n-1$, i.e., the target location of t_i is the start location of t_{i+1}. We may have $t_i = t_j$ for $i \neq j$, i.e., a transition may occur several times in the sequence. Then the transition $t_1 \star \ldots \star t_n = (\ell_1, \varphi, \eta, \ell_{n+1})$ results from* chaining *$t_1, \ldots, t_n$ where*

$$\varphi = \varphi_1 \wedge \eta_1(\varphi_2) \wedge \eta_2(\eta_1(\varphi_3)) \wedge \ldots \wedge \eta_{n-1}(\ldots \eta_1(\varphi_n) \ldots)$$
$$\eta(v) = \eta_n(\ldots \eta_1(v) \ldots) \text{ for all } v \in \mathcal{PV}, \text{ i.e., } \eta = \eta_n \circ \ldots \circ \eta_1.$$

Similar to [15,19], we can restrict ourselves to tnn-loops, since chaining transforms any twn-loop L into a tnn-loop $L \star L$. Chaining preserves the termination behavior, and a bound on $L \star L$'s runtime can be transformed into a bound for L.

Lemma 16 (Chaining Preserves Asymptotic Runtime, see [19, Lemma 18]**).** *For the twn-loop $L = (\psi, \varphi, \eta)$ with the transitions $t_0 = (\ell_0, \psi, \mathrm{id}, \ell)$, $t = (\ell, \varphi, \eta, \ell)$, and runtime complexity rc_L, the program $L \star L$ with the transitions t_0 and $t \star t = (\psi, \varphi \wedge \eta(\varphi), \eta \circ \eta)$ is a tnn-loop. For its runtime complexity $\mathrm{rc}_{L \star L}$, we have $2 \cdot \mathrm{rc}_{L \star L}(\sigma) \leq \mathrm{rc}_L(\sigma) \leq 2 \cdot \mathrm{rc}_{L \star L}(\sigma) + 1$ for all $\sigma \in \Sigma$.*

Example 17. The program of Example 14 is only a twn-loop and not a tnn-loop as x_1 occurs with a negative coefficient -2 in its own update. Hence, we chain the loop and consider $t_5 \star t_5$. The update of $t_5 \star t_5$ is $(\eta \circ \eta)(x_1) = 4 \cdot x_1$, $(\eta \circ \eta)(x_2) = 9 \cdot x_2 - 8 \cdot x_3^3$, and $(\eta \circ \eta)(x_3) = x_3$. To ease the presentation, in this example we will keep the guard φ instead of using $\varphi \wedge \eta(\varphi)$ (ignoring $\eta(\varphi)$ in the conjunction of the guard does not decrease the runtime complexity).

Our algorithm starts with computing a closed form for the loop update, which describes the values of the program variables after n iterations of the loop. Formally, a tuple of arithmetic expressions $\mathtt{cl}_{\boldsymbol{x}}^n = (\mathtt{cl}_{x_1}^n, \ldots, \mathtt{cl}_{x_d}^n)$ over the variables $\boldsymbol{x} = (x_1, \ldots, x_d)$ and the distinguished variable n is a *(normalized) closed form* for the update η with *start value* $n_0 \geq 0$ if for all $1 \leq i \leq d$ and all $\sigma : \{x_1, \ldots, x_d, n\} \to \mathbb{Z}$ with $\sigma(n) \geq n_0$, we have $\sigma(\mathtt{cl}_{x_i}^n) = \sigma(\eta^n(x_i))$. As shown in [14,15,19], for tnn-loops such a normalized closed form and the start value n_0 can be computed by handling one variable after the other, and these normalized closed forms can be represented as so-called *normalized poly-exponential expressions*. Here, $\mathbb{N}_{\geq m}$ stands for $\{x \in \mathbb{N} \mid x \geq m\}$.

Definition 18. (Normalized Poly-Exponential Expression [14,15,19]**).** *Let $\mathcal{PV} = \{x_1, \ldots, x_d\}$. Then we define the set of all* normalized poly-exponential expressions *by* $\mathbb{NPE} = \{\sum_{j=1}^{\ell} p_j \cdot n^{a_j} \cdot b_j^n \mid \ell, a_j \in \mathbb{N},\ p_j \in \mathbb{Q}[\mathcal{PV}],\ b_j \in \mathbb{N}_{\geq 1}\}$.

Example 19. A normalized closed form (with start value $n_0 = 0$) for the tnn-loop in Example 17 is $\mathtt{cl}_{x_1}^n = x_1 \cdot 4^n$, $\mathtt{cl}_{x_2}^n = (x_2 - x_3^3) \cdot 9^n + x_3^3$, and $\mathtt{cl}_{x_3}^n = x_3$.

Using the normalized closed form, similar to [15] one can represent non-termination of a tnn-loop (ψ, φ, η) by the formula

$$\exists \boldsymbol{x} \in \mathbb{Z}^d,\ m \in \mathbb{N}.\ \forall n \in \mathbb{N}_{\geq m}.\ \psi \wedge \varphi[\boldsymbol{x}/\mathtt{cl}_{\boldsymbol{x}}^n]. \tag{2}$$

Here, $\varphi[\boldsymbol{x}/\mathtt{cl}_{\boldsymbol{x}}^n]$ means that each variable x_i in φ is replaced by $\mathtt{cl}_{x_i}^n$. Since ψ is an update-invariant, if ψ holds, then $\psi[\boldsymbol{x}/\mathtt{cl}_{\boldsymbol{x}}^n]$ holds as well for all $n \geq n_0$. Hence, whenever $\forall n \in \mathbb{N}_{\geq m}.\ \psi \wedge \varphi[\boldsymbol{x}/\mathtt{cl}_{\boldsymbol{x}}^n]$ holds, then $\mathtt{cl}_{\boldsymbol{x}}^{\max\{n_0,m\}}$ witnesses non-termination. Thus, invalidity of (2) is equivalent to termination of the loop.

Normalized poly-exponential expressions have the advantage that it is always clear which addend determines their asymptotic growth when increasing n. So as in [15], (2) can be transformed into an existential formula and we use an SMT solver to prove its invalidity in order to prove termination of the loop. As shown in [15, Theorem 42], non-termination of twn-loops over $\mathbb{Z}$ is semi-decidable and deciding termination is Co-NP-complete if the loop is linear and the eigenvalues of the update matrix are rational.

4.2 Runtime Bounds for Twn-Loops via Stabilization Thresholds

As observed in [19], since the closed forms for tnn-loops are poly-exponential expressions that are weakly monotonic in n, every tnn-loop (ψ, φ, η) *stabilizes* for each input $\boldsymbol{e} \in \mathbb{Z}^d$. So there is a number of loop iterations (a *stabilization threshold* $\mathrm{sth}_{(\psi,\varphi,\eta)}(\boldsymbol{e})$), such that the truth value of the loop guard φ does not change anymore when performing further loop iterations. Hence, the runtime of every terminating tnn-loop is bounded by its stabilization threshold.

Definition 20 (Stabilization Threshold). *Let (ψ, φ, η) be a tnn-loop with $\mathcal{PV} = \{x_1, \ldots, x_d\}$. For each $\boldsymbol{e} = (e_1, \ldots, e_d) \in \mathbb{Z}^d$, let $\sigma_{\boldsymbol{e}} \in \Sigma$ with $\sigma_{\boldsymbol{e}}(x_i) = e_i$ for all $1 \leq i \leq d$. Let $\Psi \subseteq \mathbb{Z}^d$ such that $\boldsymbol{e} \in \Psi$ iff $\sigma_{\boldsymbol{e}}(\psi)$ holds. Then $\mathrm{sth}_{(\psi,\varphi,\eta)} : \mathbb{Z}^d \to \mathbb{N}$ is the* stabilization threshold *of (ψ, φ, η) if for all $\boldsymbol{e} \in \Psi$, $\mathrm{sth}_{(\psi,\varphi,\eta)}(\boldsymbol{e})$ is the smallest number such that $\sigma_{\boldsymbol{e}}\left(\, \eta^n(\varphi) \leftrightarrow \eta^{\mathrm{sth}_{(\psi,\varphi,\eta)}(\boldsymbol{e})}(\varphi)\,\right)$ holds for all $n \geq \mathrm{sth}_{(\psi,\varphi,\eta)}(\boldsymbol{e})$.*

For the tnn-loop from Example 17, it will turn out that $2 \cdot x_2 + 2 \cdot x_3^3 + 2 \cdot x_3^5 + 1$ is an upper bound on its stabilization threshold, see Example 28.

To compute such upper bounds on a tnn-loop's stabilization threshold (i.e., upper bounds on its runtime if the loop is terminating), we now present a construction based on *monotonicity thresholds*, which are computable [19, Lemma 12].

Definition 21 (Monotonicity Threshold [19]). *Let $(b_1, a_1), (b_2, a_2) \in \mathbb{N}^2$ such that $(b_1, a_1) >_{\mathrm{lex}} (b_2, a_2)$ (i.e., $b_1 > b_2$ or both $b_1 = b_2$ and $a_1 > a_2$). For any $k \in \mathbb{N}_{\geq 1}$, the k-*monotonicity threshold *of (b_1, a_1) and (b_2, a_2) is the smallest $n_0 \in \mathbb{N}$ such that for all $n \geq n_0$ we have $n^{a_1} \cdot b_1^n > k \cdot n^{a_2} \cdot b_2^n$.*

For example, the 1-monotonicity threshold of $(4, 0)$ and $(3, 1)$ is 7 as the largest root of $f(n) = 4^n - n \cdot 3^n$ is approximately 6.5139.

Our procedure again instantiates the variables of the loop guard φ by the normalized closed form $\mathtt{cl}_{\boldsymbol{x}}^n$ of the loop's update. However, in the poly-exponential expressions $\sum_{j=1}^{\ell} p_j \cdot n^{a_j} \cdot b_j^n$ resulting from $\varphi[\boldsymbol{x}/\mathtt{cl}_{\boldsymbol{x}}^n]$, the corresponding technique of [19, Lemma 21] over-approximated the polynomials p_j by a polynomial that did not distinguish the effects of the different variables $x_1, \ldots, x_d$. Such an over-approximation is only useful for a direct asymptotic bound on the runtime of the twn-loop, but it is too coarse for a useful *local* runtime bound within the complexity analysis of a larger program. For instance, in Example 12 it is crucial to obtain local bounds like $4 \cdot x_2 + 4 \cdot x_3^3 + 4 \cdot x_3^5 + 3$ which indicate that only the variable x_3 may influence the runtime with an exponent of 3 or 5. Thus, if the size of x_3 is bound by a constant, then the resulting global bound becomes linear.

So we now improve precision and over-approximate the polynomials p_j by the polynomial $\sqcup\{p_1, \ldots, p_\ell\}$ which contains every monomial $x_1^{e_1} \cdot \ldots \cdot x_d^{e_d}$ of $\{p_1, \ldots, p_\ell\}$, using the absolute value of the largest coefficient with which the monomial occurs in $\{p_1, \ldots, p_\ell\}$. Thus, $\sqcup\{x_3^3 - x_3^5, x_2 - x_3^3\} = x_2 + x_3^3 + x_3^5$. In the following let $\boldsymbol{x} = (x_1, \ldots, x_d)$, and for $\boldsymbol{e} = (e_1, \ldots, e_d) \in \mathbb{N}^d$, $\boldsymbol{x}^{\boldsymbol{e}}$ denotes $x_1^{e_1} \cdot \ldots \cdot x_d^{e_d}$.

Definition 22 (Over-Approximation of Polynomials). *Let $p_1, \ldots, p_\ell \in \mathbb{Z}[\boldsymbol{x}]$, and for all $1 \leq j \leq \ell$, let $\mathcal{I}_j \subseteq (\mathbb{Z} \setminus \{0\}) \times \mathbb{N}^d$ be the* index set *of the polynomial p_j where $p_j = \sum_{(c,\boldsymbol{e}) \in \mathcal{I}_j} c \cdot \boldsymbol{x}^{\boldsymbol{e}}$ and there are no $c \neq c'$ with $(c, \boldsymbol{e}), (c', \boldsymbol{e}) \in \mathcal{I}_j$. For all $\boldsymbol{e} \in \mathbb{N}^d$ we define $c_{\boldsymbol{e}} \in \mathbb{N}$ with $c_{\boldsymbol{e}} = \max\{|c| \mid (c, \boldsymbol{e}) \in \mathcal{I}_1 \cup \ldots \cup \mathcal{I}_\ell\}$, where $\max \varnothing = 0$. Then the* over-approximation *of $p_1, \ldots, p_\ell$ is $\sqcup\{p_1, \ldots, p_\ell\} = \sum_{\boldsymbol{e} \in \mathbb{N}^d} c_{\boldsymbol{e}} \cdot \boldsymbol{x}^{\boldsymbol{e}}$.*

Clearly, $\sqcup\{p_1, \ldots, p_\ell\}$ indeed over-approximates the absolute value of each p_j.

Corollary 23 (Soundness of $\sqcup\{p_1, \ldots, p_\ell\}$). *For all $\sigma : \{x_1, \ldots, x_d\} \to \mathbb{Z}$ and all $1 \leq j \leq \ell$, we have $|\sigma|(\sqcup\{p_1, \ldots, p_\ell\}) \geq |\sigma(p_j)|$.*

A drawback is that $\sqcup\{p_1, \ldots, p_\ell\}$ considers all monomials and to obtain weakly monotonically increasing bounds from $\mathcal{B}$, it uses the absolute values of their coefficients. This can lead to polynomials of unnecessarily high degree. To improve the precision of the resulting bounds, we now allow to over-approximate the poly-exponential expressions $\sum_{j=1}^{\ell} p_j \cdot n^{a_j} \cdot b_j^n$ which result from instantiating the variables of the loop guard by the closed form. For this over-approximation, we take the invariant ψ of the tnn-loop into account. So while (2) showed that update-invariants ψ can restrict the sets of possible witnesses for non-termination and thus simplify the termination proofs of twn-loops, we now show that pre-conditions ψ can also be useful to improve the bounds on twn-loops.

More precisely, Definition 24 allows us to replace addends $p \cdot n^a \cdot b^n$ by $p \cdot n^i \cdot j^n$ where $(j, i) >_{\text{lex}} (b, a)$ if the monomial p is always positive (when the precondition ψ is fulfilled) and where $(b, a) >_{\text{lex}} (i, j)$ if p is always non-positive.

Definition 24 (Over-Approximation of Poly-Exponential Expressions). *Let $\psi \in \mathcal{F}(\mathcal{PV})$ and let $npe = \sum_{(p,a,b) \in \Lambda} p \cdot n^a \cdot b^n \in \mathbb{NPE}$ where Λ is a set of tuples (p, a, b) containing a monomial[2] p and two numbers $a, b \in \mathbb{N}$. Here, we*

[2] Here, we consider monomials of the form $p = c \cdot x_1^{e_1} \cdot \ldots \cdot x_d^{e_d}$ with coefficients $c \in \mathbb{Q}$.

may have $(p, a, b), (p', a, b) \in \Lambda$ *for* $p \neq p'$. *Let* $\Delta, \Gamma \subseteq \Lambda$ *such that* $\models \psi \to (p > 0)$ *holds for all* $(p, a, b) \in \Delta$ *and* $\models \psi \to (p \leq 0)$ *holds for all* $(p, a, b) \in \Gamma$.[3] *Then*

$$\lceil npe \rceil^{\psi}_{\Delta,\Gamma} = \sum\nolimits_{(p,a,b)\in\Delta\uplus\Gamma} p \cdot n^{i_{(p,a,b)}} \cdot j^n_{(p,a,b)} + \sum\nolimits_{(p,a,b)\in\Lambda\setminus(\Delta\uplus\Gamma)} p \cdot n^a \cdot b^n$$

is an over-approximation *of npe if* $i_{(p,a,b)}, j_{(p,a,b)} \in \mathbb{N}$ *are numbers such that* $(j_{(p,a,b)}, i_{(p,a,b)}) >_{\text{lex}} (b, a)$ *holds if* $(p, a, b) \in \Delta$ *and* $(b, a) >_{\text{lex}} (j_{(p,a,b)}, i_{(p,a,b)})$ *holds if* $(p, a, b) \in \Gamma$. *Note that* $i_{(p,a,b)}$ *or* $j_{(p,a,b)}$ *can also be* 0.

Example 25. Let $npe = q_3 \cdot 16^n + q_2 \cdot 9^n + q_1 = q_3 \cdot 16^n + q_2' \cdot 9^n + q_2'' \cdot 9^n + q_1' + q_1''$, where $q_3 = -x_1^2$, $q_2 = q_2' + q_2''$, $q_2' = x_2$, $q_2'' = -x_3^3$, $q_1 = q_1' + q_1''$, $q_1' = x_3^3$, $q_1'' = -x_3^5$, and $\psi = (x_3 > 0)$. We can choose $\Delta = \{(x_3^3, 0, 1)\}$ since $\models \psi \to (x_3^3 > 0)$ and $\Gamma = \{(-x_3^5, 0, 1)\}$ since $\models \psi \to (-x_3^5 \leq 0)$. Moreover, we choose $j_{(x_3^3,0,1)} = 9$, $i_{(x_3^3,0,1)} = 0$, which is possible since $(9, 0) >_{\text{lex}} (1, 0)$. Similarly, we choose $j_{(-x_3^5,0,1)} = 0$, $i_{(-x_3^5,0,1)} = 0$, since $(1, 0) >_{\text{lex}} (0, 0)$. Thus, we replace x_3^3 and $-x_3^5$ by the larger addends $x_3^3 \cdot 9^n$ and 0. The motivation for the latter is that this removes all addends with exponent 5 from *npe*. The motivation for the former is that then, we have both the addends $-x_3^3 \cdot 9^n$ and $x_3^3 \cdot 9^n$ in the expression which cancel out, i.e., this removes all addends with exponent 3. Hence, we obtain $\lceil npe \rceil^{\psi}_{\Delta,\Gamma} = p_2 \cdot 16^n + p_1 \cdot 9^n$ with $p_2 = -x_1^2$ and $p_1 = x_2$. To find a suitable over-approximation which removes addends with high exponents, our implementation uses a heuristic for the choice of Δ, Γ, $i_{(p,a,b)}$, and $j_{(p,a,b)}$.

The following lemma shows the soundness of the over-approximation $\lceil npe \rceil^{\psi}_{\Delta,\Gamma}$.

Lemma 26 (Soundness of $\lceil npe \rceil^{\psi}_{\Delta,\Gamma}$**).** *Let* ψ, *npe*, Δ, Γ, $i_{(p,a,b)}$, $j_{(p,a,b)}$, *and* $\lceil npe \rceil^{\psi}_{\Delta,\Gamma}$ *be as in Definition 24, and let* $D_{\lceil npe \rceil^{\psi}_{\Delta,\Gamma}} =$

$$\max(\{1\text{-}monotonicity\ threshold\ of\ (j_{(p,a,b)}, i_{(p,a,b)})\ and\ (b, a) \mid (p, a, b) \in \Delta\} \cup \{1\text{-}monotonicity\ threshold\ of\ (b, a)\ and\ (j_{(p,a,b)}, i_{(p,a,b)}) \mid (p, a, b) \in \Gamma\}).$$

Then for all $\mathbf{e} \in \Psi$ *and all* $n \geq D_{\lceil npe \rceil^{\psi}_{\Delta,\Gamma}}$, *we have* $\sigma_{\mathbf{e}}(\lceil npe \rceil^{\psi}_{\Delta,\Gamma}) \geq \sigma_{\mathbf{e}}(npe)$.

For any terminating tnn-loop (ψ, φ, η), Theorem 27 now uses the new concepts of Definition 22 and 24 to compute a polynomial $\text{sth}^{\sqcup}$ which is an upper bound on the loop's stabilization threshold (and hence, on its runtime). For any atom $\alpha = (s_1 < s_2)$ (resp. $s_2 - s_1 > 0$) in the loop guard φ, let $npe_\alpha \in \mathbb{NPE}$ be a poly-exponential expression which results from multiplying $(s_2 - s_1)[\boldsymbol{x}/\mathtt{cl}^n_{\boldsymbol{x}}]$ with the least common multiple of all denominators occurring in $(s_2 - s_1)[\boldsymbol{x}/\mathtt{cl}^n_{\boldsymbol{x}}]$. Since the loop is terminating, for some of these atoms this expression will become non-positive for large enough n and our goal is to compute bounds on their corresponding stabilization thresholds. First, one can replace npe_α by an over-approximation $\lceil npe_\alpha \rceil^{\psi'}_{\Delta,\Gamma}$ where $\psi' = (\psi \wedge \varphi)$ considers both the invariant ψ

[3] Δ and Γ do not have to contain *all* such tuples, but can be (possibly empty) subsets.

and the guard φ. Let $\Psi' \subseteq \mathbb{Z}^d$ such that $\boldsymbol{e} \in \Psi'$ iff $\sigma_{\boldsymbol{e}}(\psi')$ holds. By Lemma 26 (i.e., $\sigma_{\boldsymbol{e}}(\lceil npe_\alpha \rceil_{\Delta,\Gamma}^{\psi'}) \geq \sigma_{\boldsymbol{e}}(npe_\alpha)$ for all $\boldsymbol{e} \in \Psi'$), it suffices to compute a bound on the stabilization threshold of $\lceil npe_\alpha \rceil_{\Delta,\Gamma}^{\psi'}$ if it is always non-positive for large enough n, because if $\lceil npe_\alpha \rceil_{\Delta,\Gamma}^{\psi'}$ is non-positive, then so is npe_α. We say that an over-approximation $\lceil npe_\alpha \rceil_{\Delta,\Gamma}^{\psi'}$ is *eventually non-positive* iff whenever $\lceil npe_\alpha \rceil_{\Delta,\Gamma}^{\psi'} \neq npe_\alpha$, then one can show that for all $\boldsymbol{e} \in \Psi'$, $\sigma_{\boldsymbol{e}}(\lceil npe_\alpha \rceil_{\Delta,\Gamma}^{\psi'})$ is always non-positive for large enough n.[4] Using over-approximations $\lceil npe_\alpha \rceil_{\Delta,\Gamma}^{\psi'}$ can be advantageous because $\lceil npe_\alpha \rceil_{\Delta,\Gamma}^{\psi'}$ may contain less monomials than npe_α and thus, the construction $\sqcup$ from Definition 22 can yield a polynomial of lower degree. So although npe_α's stabilization threshold might be smaller than the one of $\lceil npe_\alpha \rceil_{\Delta,\Gamma}^{\psi'}$, our technique might compute a smaller bound on the stabilization threshold when considering $\lceil npe_\alpha \rceil_{\Delta,\Gamma}^{\psi'}$ instead of *npe*.

Theorem 27 (Bound on Stabilization Threshold). *Let $L = (\psi, \varphi, \eta)$ be a terminating tnn-loop, let $\psi' = (\psi \wedge \varphi)$, and let $\mathtt{cl}_{\boldsymbol{x}}^n$ be a normalized closed form for η with start value n_0. For every atom $\alpha = (s_1 < s_2)$ in φ, let $\lceil npe_\alpha \rceil_{\Delta,\Gamma}^{\psi'}$ be an eventually non-positive over-approximation of npe_α and let $D_\alpha = D_{\lceil npe_\alpha \rceil_{\Delta,\Gamma}^{\psi'}}$.*

If $\lceil npe_\alpha \rceil_{\Delta,\Gamma}^{\psi'} = \sum_{j=1}^{\ell} p_j \cdot n^{a_j} \cdot b_j^n$ with $p_j \neq 0$ for all $1 \leq j \leq \ell$ and $(b_\ell, a_\ell) >_{\mathrm{lex}} \ldots >_{\mathrm{lex}} (b_1, a_1)$, then let $C_\alpha = \max\{1, N_2, M_2, \ldots, N_\ell, M_\ell\}$, where we have:

$$M_j = \begin{cases} 0, & \text{if } b_j = b_{j-1} \\ \text{1-monotonicity threshold of} & \\ \quad (b_j, a_j) \text{ and } (b_{j-1}, a_{j-1}+1), & \text{if } b_j > b_{j-1} \end{cases} \qquad N_j = \begin{cases} 1, & \text{if } j = 2 \\ mt', & \text{if } j = 3 \\ \max\{mt, mt'\}, & \text{if } j > 3 \end{cases}$$

Here, mt' is the $(j-2)$-monotonicity threshold of (b_{j-1}, a_{j-1}) and (b_{j-2}, a_{j-2}) and $mt = \max\{$1-monotonicity threshold of (b_{j-2}, a_{j-2}) and $(b_i, a_i) \mid 1 \leq i \leq j-3\}$. Let $Pol_\alpha = \{p_1, \ldots, p_{\ell-1}\}$, $Pol = \bigcup_{\text{atom } \alpha \text{ occurs in } \varphi} Pol_\alpha$, $C = \max\{C_\alpha \mid$ atom α occurs in $\varphi\}$, $D = \max\{D_\alpha \mid$ atom α occurs in $\varphi\}$, and $\mathrm{sth}^\sqcup \in \mathbb{Z}[\boldsymbol{x}]$ with $\mathrm{sth}^\sqcup = 2 \cdot \sqcup Pol + \max\{n_0, C, D\}$. Then for all $\boldsymbol{e} \in \Psi'$, we have $|\sigma_{\boldsymbol{e}}|(\mathrm{sth}^\sqcup) \geq \mathrm{sth}_{(\psi,\varphi,\eta)}(\boldsymbol{e})$. If the tnn-loop has the initial transition t_0 and looping transition t, then $\mathcal{RB}_{glo}(t_0) = 1$ and $\mathcal{RB}_{glo}(t) = \mathrm{sth}^\sqcup$ is a global runtime bound for L.

Example 28. The guard φ of the tnn-loop in Example 17 has the atoms $\alpha = (x_1^2 + x_3^5 < x_2)$, $\alpha' = (0 < x_1)$, and $\alpha'' = (0 < -x_1)$ (since $x_1 \neq 0$ is transformed into $\alpha' \vee \alpha''$). When instantiating the variables by the closed forms of Example 19 with start value $n_0 = 0$, Theorem 27 computes the bound 1 on the stabilization thresholds for α' and α''. So the only interesting atom is $\alpha = (0 < s_2 - s_1)$ for $s_1 = x_1^2 + x_3^5$ and $s_2 = x_2$. We get $npe_\alpha = (s_2 - s_1)[\boldsymbol{x}/\mathtt{cl}_{\boldsymbol{x}}^n] = q_3 \cdot 16^n + q_2 \cdot 9^n + q_1$, with q_j as in Example 25.

---

[4] This can be shown similar to the proof of (2) for (non-)termination of the loop. Thus, we transform $\exists\, \boldsymbol{x} \in \mathbb{Z}^d,\ m \in \mathbb{N}.\ \forall n \in \mathbb{N}_{\geq m}.\ \psi' \wedge \lceil npe_\alpha \rceil_{\Delta,\Gamma}^{\psi'} > 0$ into an existential formula as in [15] and try to prove its invalidity by an SMT solver.

In the program of Fig. 1, the corresponding self-loop t_5 has two entry transitions t_4 and t_1 which result in two tnn-loops with the update-invariants $\psi_1 = \mathtt{true}$ resulting from transition t_4 and $\psi_2 = (x_3 > 0)$ from t_1. So ψ_2 is an update-invariant of t_5 which always holds when reaching t_5 via transition t_1.

For $\psi_1 = \mathtt{true}$, we choose $\Delta = \Gamma = \varnothing$, i.e., $\lceil npe_\alpha \rceil_{\Delta,\Gamma}^{\psi_1'} = npe_\alpha$. So we have $b_3 = 16$, $b_2 = 9$, $b_1 = 1$, and $a_j = 0$ for all $1 \le j \le 3$. We obtain

$$M_2 = 0, \text{ as } 0 \text{ is the 1-monotonicity threshold of } (9,0) \text{ and } (1,1)$$
$$M_3 = 0, \text{ as } 0 \text{ is the 1-monotonicity threshold of } (16,0) \text{ and } (9,1)$$
$$N_2 = 1 \text{ and } N_3 = 1, \text{ as } 1 \text{ is the 1-monotonicity threshold of } (9,0) \text{ and } (1,0).$$

Hence, we get $C = C_\alpha = \max\{1, N_2, M_2, N_3, M_3\} = 1$. So we obtain the runtime bound $\mathrm{sth}_{\psi_1}^{\sqcup} = 2 \cdot \sqcup\{q_1, q_2\} + \max\{n_0, C_\alpha\} = 2 \cdot x_2 + 2 \cdot x_3^3 + 2 \cdot x_3^5 + 1$ for the loop $t_5 \star t_5$ w.r.t. ψ_1. By Lemma 16, this means that $2 \cdot \mathrm{sth}_{\psi_1}^{\sqcup} + 1 = 4 \cdot x_2 + 4 \cdot x_3^3 + 4 \cdot x_3^5 + 3$ is a runtime bound for the loop at transition t_5.

For the update-invariant $\psi_2 = (x_3 > 0)$, we use the over-approximation $\lceil npe_\alpha \rceil_{\Delta,\Gamma}^{\psi_2'} = p_2 \cdot 16^n + p_1 \cdot 9^n$ with $p_2 = -x_1^2$ and $p_1 = x_2$ from Example 25, where $\psi_2' = (\psi_2 \wedge \varphi)$ implies that it is always non-positive for large enough n. Now we obtain $M_2 = 0$ (the 1-monotonicity threshold of $(16,0)$ and $(9,1)$) and $N_2 = 1$, where $C = C_\alpha = \max\{1, N_2, M_2\} = 1$. Moreover, we have $D_\alpha = \max\{1, 0\} = 1$, since

$$1 \text{ is the 1-monotonicity threshold of } (9,0) \text{ and } (1,0), \text{ and}$$
$$0 \text{ is the 1-monotonicity threshold of } (1,0) \text{ and } (0,0).$$

We now get the tighter bound $\mathrm{sth}_{\psi_2}^{\sqcup} = 2 \cdot \sqcup\{p_1\} + \max\{n_0, C_\alpha, D_\alpha\} = 2 \cdot x_2 + 1$ for $t_5 \star t_5$. So t_5's runtime bound is $2 \cdot \mathrm{sth}_{\psi_2}^{\sqcup} + 1 = 4 \cdot x_2 + 3$ when using invariant ψ_2.

Theorem 29 shows how the technique of Lemma 16 and Theorem 27 can be used to compute local runtime bounds for twn-loops whenever such loops occur within an integer program. To this end, one needs the new Theorem 11 where in contrast to [6,18] these local bounds do not have to result from ranking functions.

To turn a self-loop t and $r \in \mathcal{E}_{\{t\}}$ from a larger program $\mathcal{P}$ into a twn-loop (ψ, φ, η), we use t's guard φ and update η. To obtain an update-invariant ψ, our implementation uses the Apron library [23] for computing invariants on a version of the full program where we remove all entry transitions $\mathcal{E}_{\{t\}}$ except r.[5] From the invariants computed for t, we take those that are also update-invariants of t.

Theorem 29 (Local Bounds for Twn-Loops). *Let $\mathcal{P} = (\mathcal{PV}, \mathcal{L}, \ell_0, \mathcal{T})$ be an integer program with $\mathcal{PV}' = \{x_1, \ldots, x_d\} \subseteq \mathcal{PV}$. Let $t = (\ell, \varphi, \eta, \ell) \in \mathcal{T}$ with $\varphi \in \mathcal{F}(\mathcal{PV}')$, $\eta(v) \in \mathbb{Z}[\mathcal{PV}']$ for all $v \in \mathcal{PV}'$, and $\eta(v) = v$ for all $v \in \mathcal{PV} \setminus \mathcal{PV}'$. For any entry transition $r \in \mathcal{E}_{\{t\}}$, let $\psi \in \mathcal{F}(\mathcal{PV}')$ such that $\models \psi \to \eta(\psi)$ and*

[5] Regarding invariants for the full program in the computation of local bounds for t is possible since in contrast to [6,18] our definition of local bounds from Definition 9 is restricted to states that are reachable from an initial configuration (ℓ_0, σ_0).

such that $\sigma(\psi)$ *holds whenever there is a* $\sigma_0 \in \Sigma$ *with* $(\ell_0, \sigma_0) \to_{\mathcal{T}}^* \circ \to_r (\ell, \sigma)$. *If* $L = (\psi, \varphi, \eta)$ *is a terminating tnn-loop, then let* $\mathcal{RB}_{loc}(\to_r \{t\}) = \mathrm{sth}^{\sqcup}$, *where* $\mathrm{sth}^{\sqcup}$ *is defined as in Theorem 27. If* L *is a terminating twn-loop but no tnn-loop, let* $\mathcal{RB}_{loc}(\to_r \{t\}) = 2 \cdot \mathrm{sth}^{\sqcup} + 1$, *where* $\mathrm{sth}^{\sqcup}$ *is the bound of Theorem 27 computed for* $L \star L$. *Otherwise, let* $\mathcal{RB}_{loc}(\to_r \{t\}) = \omega$. *Then* $\mathcal{RB}_{loc}$ *is a local runtime bound for* $\{t\} = \mathcal{T}'_{>} = \mathcal{T}'$ *in the program* $\mathcal{P}$.

Example 30. In Fig. 1, we consider the self-loop t_5 with $\mathcal{E}_{\{t_5\}} = \{t_4, t_1\}$ and the update-invariants $\psi_1 = \texttt{true}$ resp. $\psi_2 = (x_3 > 0)$. For t_5's guard φ and update η, both (ψ_i, φ, η) are terminating twn-loops (see Example 14), i.e., (2) is invalid.

By Theorem 29 and Example 28, $\mathcal{RB}_{\mathrm{loc}}$ with $\mathcal{RB}_{\mathrm{loc}}(\to_{t_4} \{t_5\}) = 4 \cdot x_2 + 4 \cdot x_3^3 + 4 \cdot x_3^5 + 3$ and $\mathcal{RB}_{\mathrm{loc}}(\to_{t_1} \{t_5\}) = 4 \cdot x_2 + 3$ is a local runtime bound for $\{t_5\} = \mathcal{T}'_{>} = \mathcal{T}'$ in the program of Fig. 1. As shown in Example 12, Theorem 11 then yields the global runtime bound $\mathcal{RB}_{\mathrm{glo}}(t_5) = 8 \cdot x_4 \cdot x_5 + 13006 \cdot x_4$.

5 Local Runtime Bounds for Twn-Cycles

Section 4 introduced a technique to determine local runtime bounds for twn-self-loops in a program. To increase its applicability, we now extend it to larger cycles. For every entry transition of the cycle, we *chain* the transitions of the cycle, starting with the transition which follows the entry transition. In this way, we obtain loops consisting of a single transition. If the chained loop is a twn-loop, we can apply Theorem 29 to compute a local runtime bound. Any local bound on the chained transition is also a bound on each of the original transitions.[6]

By Theorem 29, we obtain a bound on the number of evaluations of the *complete cycle*. However, we also have to consider a *partial execution* which stops before traversing the full cycle. Therefore, we increase every local runtime bound by 1.

Note that this replacement of a cycle by a self-loop which results from chaining its transitions is only sound for *simple* cycles. A cycle is simple if each iteration through the cycle can only be done in a unique way. So the cycle must not have any subcycles and there also must not be any indeterminisms concerning the next transition to be taken. Formally, $\mathcal{C} = \{t_1, \ldots, t_n\} \subset \mathcal{T}$ is a simple cycle if $\mathcal{C}$ does not contain temporary variables and there are pairwise different locations $\ell_1, \ldots, \ell_n$ such that $t_i = (\ell_i, \_, \_, \ell_{i+1})$ for $1 \leq i \leq n-1$ and $t_n = (\ell_n, \_, \_, \ell_1)$. This ensures that if there is an evaluation with $\to_{t_i} \circ \to_{\mathcal{C}\setminus\{t_i\}}^* \circ \to_{t_i}$, then the steps with $\to_{\mathcal{C}\setminus\{t_i\}}^*$ have the form $\to_{t_{i+1}} \circ \ldots \circ \to_{t_n} \circ \to_{t_1} \circ \ldots \circ \to_{t_{i-1}}$.

Algorithm 1 describes how to compute a local runtime bound for a simple cycle $\mathcal{C} = \{t_1, \ldots, t_n\}$ as above. In the loop of Line 2, we iterate over all entry transitions r of $\mathcal{C}$. If r reaches the transition t_i, then in Line 3 and 4 we chain $t_i \star \ldots \star t_n \star t_1 \star \ldots \star t_{i-1}$ which corresponds to one iteration of the cycle starting

[6] This is sufficient for our improved definition of local bounds in Definition 9 where in contrast to [6, 18] we do not require a bound on the *sum* but only on *each* transition in the considered set $\mathcal{T}'$. Moreover, here we again benefit from our extension to compute individual local bounds for different entry transitions.

Algorithm 1. Algorithm to Compute Local Runtime Bounds for Cycles

input : A program $(\mathcal{PV}, \mathcal{L}, \ell_0, \mathcal{T})$ and a simple cycle $\mathcal{C} = \{t_1, \dots, t_n\} \subset \mathcal{T}$
output : A local runtime bound $\mathcal{RB}_{\mathrm{loc}}$ for $\mathcal{C} = \mathcal{T}'_> = \mathcal{T}'$

1 Initialize $\mathcal{RB}_{\mathrm{loc}}$: $\mathcal{RB}_{\mathrm{loc}}(\to_r \mathcal{C}) = \omega$ for all $r \in \mathcal{E}_\mathcal{C}$.
2 **forall** $r \in \mathcal{E}_\mathcal{C}$ **do**
3 Let $i \in \{1, \dots, n\}$ such that r's target location is the start location ℓ_i of t_i.
4 Let $t = t_i \star \dots \star t_n \star t_1 \star \dots \star t_{i-1}$.
5 **if** *there exists a renaming* π *of* $\mathcal{PV}$ *such that* $\pi(t)$ *results in a twn-loop* **then**
6 Set $\mathcal{RB}_{\mathrm{loc}}(\to_r \mathcal{C}) \leftarrow \pi^{-1}(1+$ result of Theorem 29 on $\pi(t)$ and $\pi(r))$.
7 **return** local runtime bound $\mathcal{RB}_{\mathrm{loc}}$.

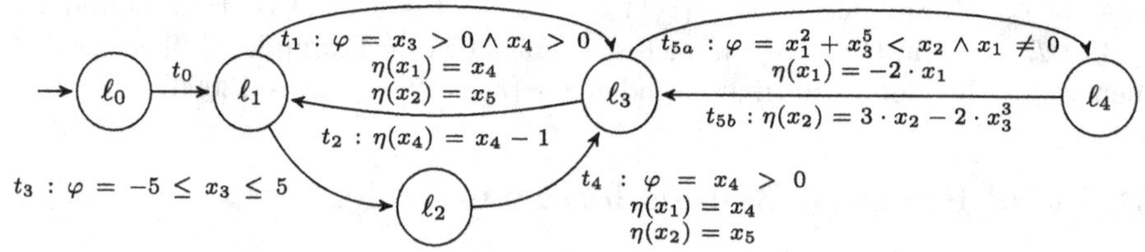

Fig. 2. An Integer Program with a Nested Non-Self-Loop

in t_i. If a suitable renaming (and thus also reordering) of the variables turns the chained transition into a twn-loop, then we use Theorem 29 to compute a local runtime bound $\mathcal{RB}_{\mathrm{loc}}(\to_r \mathcal{C})$ in Lines 5 and 6. If the chained transition does not give rise to a twn-loop, then $\mathcal{RB}_{\mathrm{loc}}(\to_r \mathcal{C})$ is ω (Line 1). In practice, to use the twn-technique for a transition t in a program, our tool KoAT searches for those simple cycles that contain t and where the chained cycle is a twn-loop. Among those cycles it chooses the one with the smallest runtime bounds for its entry transitions.

Theorem 31 (Correctness of Algorithm 1). *Let* $\mathcal{P} = (\mathcal{PV}, \mathcal{L}, \ell_0, \mathcal{T})$ *be an integer program and let* $\mathcal{C} \subset \mathcal{T}$ *be a simple cycle in* $\mathcal{P}$. *Then the result* $\mathcal{RB}_{loc} : \mathcal{E}_\mathcal{C} \to \mathcal{B}$ *of Algorithm 1 is a local runtime bound for* $\mathcal{C} = \mathcal{T}'_> = \mathcal{T}'$.

Example 32. We apply Algorithm 1 on the cycle $\mathcal{C} = \{t_{5a}, t_{5b}\}$ of the program in Fig. 2. $\mathcal{C}$'s entry transitions t_1 and t_4 both end in ℓ_3. Chaining t_{5a} and t_{5b} yields the transition t_5 of Fig. 1, i.e., $t_5 = t_{5a} \star t_{5b}$. Thus, Algorithm 1 essentially transforms the program of Fig. 2 into Fig. 1. As in Example 28 and 30, we obtain $\mathcal{RB}_{\mathrm{loc}}(\to_{t_4} \mathcal{C}) = 1 + (2 \cdot \mathrm{sth}^{\sqcup}_{\mathtt{true}} + 1) = 4 \cdot x_2 + 4 \cdot x_3^3 + 4 \cdot x_3^5 + 4$ and $\mathcal{RB}_{\mathrm{loc}}(\to_{t_1} \mathcal{C}) = 1 + (2 \cdot \mathrm{sth}^{\sqcup}_{x_3 > 0} + 1) = 4 \cdot x_2 + 4$, resulting in the global runtime bound $\mathcal{RB}_{\mathrm{glo}}(t_{5a}) = \mathcal{RB}_{\mathrm{glo}}(t_{5b}) = 8 \cdot x_4 \cdot x_5 + 13008 \cdot x_4$, which again yields $\mathrm{rc}(\sigma_0) \in \mathcal{O}(n^2)$.

6 Conclusion and Evaluation

We showed that results on subclasses of programs with computable complexity bounds like [19] are not only theoretically interesting, but they have an important practical value. To our knowledge, our paper is the first to integrate such results into an incomplete approach for automated complexity analysis like [6,18]. For this integration, we developed several novel contributions which extend and improve the previous approaches in [6,18,19] substantially:

(a) We extended the concept of local runtime bounds such that they can now depend on entry transitions (Definition 9).
(b) We generalized the computation of global runtime bounds such that one can now lift arbitrary local bounds to global bounds (Theorem 11). In particular, the local bounds might be due to either ranking functions or twn-loops.
(c) We improved the technique for the computation of bounds on twn-loops such that these bounds now take the roles of the different variables into account (Definition 22, Corollary 23, and Theorem 27).
(d) We extended the notion of twn-loops by update-invariants and developed a new over-approximation of their closed forms which takes invariants into account (Definition 13 and 24, Lemma 26, and Theorem 27).
(e) We extended the handling of twn-loops to twn-cycles (Theorem 31).

The need for these improvements is demonstrated by our leading example in Fig. 1 (where the contributions (a)–(d) are needed to infer quadratic runtime complexity) and by the example in Fig. 2 (which illustrates (e)). In this way, the power of automated complexity analysis is increased substantially, because now one can also infer runtime bounds for programs containing non-linear arithmetic.

To demonstrate the power of our approach, we evaluated the integration of our new technique to infer local runtime bounds for twn-cycles in our re-implementation of the tool KoAT (written in OCaml) and compare the results to other state-of-the-art tools. To distinguish our re-implementation of KoAT from the original version of the tool from [6], let KoAT1 refer to the tool from [6] and let KoAT2 refer to our new re-implementation. KoAT2 applies a local control-flow refinement technique [18] (using the tool iRankFinder [8]) and preprocesses the program in the beginning, e.g., by extending the guards of transitions by invariants inferred using the Apron library [23]. For all occurring SMT problems, KoAT2 uses Z3 [28]. We tested the following configurations of KoAT2, which differ in the techniques used for the computation of local runtime bounds:

- KoAT2+RF only uses linear ranking functions to compute local runtime bounds
- KoAT2+MΦRF5 uses multiphase-linear ranking functions of depth ≤ 5
- KoAT2+TWN only uses twn-cycles to compute local runtime bounds (Algorithm 1)
- KoAT2+TWN+RF uses Algorithm 1 for twn-cycles and linear ranking functions
- KoAT2+TWN+MΦRF5 uses Algorithm 1 for twn-cycles and MΦRFs of depth ≤ 5

Existing approaches for automated complexity analysis are already very powerful on programs that only use linear arithmetic in their guards and updates.

	$\mathcal{O}(1)$	$\mathcal{O}(n)$	$\mathcal{O}(n^2)$	$\mathcal{O}(n^{>2})$	$\mathcal{O}(EXP)$	$< \infty$	AVG$^+$(s)	AVG(s)
KoAT2 + TWN + MΦRF5	26	231 (5)	73 (5)	13 (4)	1 (1)	344 (15)	8.72	23.93
KoAT2 + TWN + RF	27	227 (5)	73 (5)	13 (4)	1 (1)	341 (15)	8.11	19.77
KoAT2 + MΦRF5	24	226 (1)	68	10	0	328 (1)	8.23	21.63
KoAT2 + RF	25	214 (1)	68	10	1	318 (1)	8.49	16.56
MaxCore	23	216 (2)	66	7	0	312 (2)	2.02	5.31
CoFloCo	22	196 (1)	66	5	0	289 (1)	0.62	2.66
KoAT1	25	169 (1)	74	12	6	286 (1)	1.77	2.77
Loopus	17	170 (1)	49	5 (1)	0	241 (2)	0.42	0.43
KoAT2 + TWN	20 (1)	111 (4)	3 (2)	2 (2)	0	136 (9)	2.54	26.59

Fig. 3. Evaluation on the Collection CINT$^+$

The corresponding benchmarks for *Complexity of Integer Transitions Systems* (CITS) and *Complexity of* C *Integer Programs* (CINT) from the *Termination Problems Data Base* [33] which is used in the annual *Termination and Complexity Competition (TermComp)* [17] contain almost only examples with linear arithmetic. Here, the existing tools already infer finite runtimes for more than 89% of those examples in the collections CITS and CINT where this *might*[7] be possible.

The main benefit of our new integration of the twn-technique is that in this way one can also infer finite runtime bounds for programs that contain non-linear guards or updates. To demonstrate this, we extended both collections CITS and CINT by 20 examples that represent typical such programs, including several benchmarks from the literature [3,14,15,18,20,34], as well as our programs from Fig. 1 and 2. See [27] for a detailed list and description of these examples.

Figure 3 presents our evaluation on the collection CINT$^+$, consisting of the 484 examples from CINT and our 20 additional examples for non-linear arithmetic. We refer to [27] for the (similar) results on the corresponding collection CITS$^+$.

In the C programs of CINT$^+$, all variables are interpreted as integers over $\mathbb{Z}$ (i.e., without overflows). For KoAT2 and KoAT1, we used Clang [7] and llvm2kittel [10] to transform C programs into integer transitions systems as in Definition 2. We compare KoAT2 with KoAT1 [6] and the tools CoFloCo [11,12], MaxCore [2] with CoFloCo in the backend, and Loopus [31]. We do not compare with RaML [21], as it does not support programs whose complexity depends on (possibly negative) integers (see [29]). We also do not compare with PUBS [1], because as stated in [9] by one of its authors, CoFloCo is stronger than PUBS. For the same reason, we only consider MaxCore with the backend CoFloCo instead of PUBS.

All tools were run inside an Ubuntu Docker container on a machine with an AMD Ryzen 7 3700X octa-core CPU and 48 GB of RAM. As in *TermComp*, we applied a timeout of 5 min for every program.

In Fig. 3, the first entry in every cell denotes the number of benchmarks from CINT$^+$ where the respective tool inferred the corresponding bound. The number

[7] The tool LoAT [13,16] proves unbounded runtime for 217 of the 781 examples from CITS and iRankFinder [4,8] proves non-termination for 118 of 484 programs of CINT.

in brackets is the corresponding number of benchmarks when only regarding our 20 new examples for non-linear arithmetic. The runtime bounds computed by the tools are compared asymptotically as functions which depend on the largest initial absolute value n of all program variables. So for instance, there are $26 + 231 = 257$ programs in CINT^+ (and 5 of them come from our new examples) where KoAT2+TWN+MΦRF5 can show that $\mathrm{rc}(\sigma_0) \in \mathcal{O}(n)$ holds for all initial states σ_0 where $|\sigma_0(v)| \leq n$ for all $v \in \mathcal{PV}$. For 26 of these programs, KoAT2+TWN+MΦRF5 can even show that $\mathrm{rc}(\sigma_0) \in \mathcal{O}(1)$, i.e., their runtime complexity is constant. Overall, this configuration succeeds on 344 examples, i.e., "$< \infty$" is the number of examples where a finite bound on the runtime complexity could be computed by the respective tool within the time limit. "AVG$^+$(s)" is the average runtime of the tool on successful runs in seconds, i.e., where the tool inferred a finite time bound before reaching the timeout, whereas "AVG(s)" is the average runtime of the tool on all runs including timeouts.

On the original benchmarks CINT where very few examples contain non-linear arithmetic, integrating TWN into a configuration that already uses multiphase-linear ranking functions does not increase power much: KoAT2+TWN+MΦRF5 succeeds on $344 - 15 = 329$ such programs and KoAT2+MΦRF5 solves $328 - 1 = 327$ examples. On the other hand, if one only has linear ranking functions, then an improvement via our twn-technique has similar effects as an improvement with multiphase-linear ranking functions (here, the success rate of KoAT2+MΦRF5 is similar to KoAT2+TWN+RF which solves $341 - 15 = 326$ such programs).

But the main benefit of our technique is that it also allows to successfully handle examples with non-linear arithmetic. Here, our new technique is significantly more powerful than previous ones. Other tools and configurations without TWN in Fig. 3 solve at most 2 of the 20 new examples. In contrast, KoAT2+TWN+RF and KoAT2+TWN+MΦRF5 both succeed on 15 of them.[8] In particular, our running examples from Fig. 1 and 2 and even isolated twn-loops like t_5 or $t_5 \star t_5$ from Example 14 and 17 can *only* be solved by KoAT2 with our twn-technique.

To summarize, our evaluations show that KoAT2 with the added twn-technique outperforms all other configurations and tools for automated complexity analysis on all considered benchmark sets (i.e., CINT^+, CINT, CITS^+, and CITS) and it is the only tool which is also powerful on examples with non-linear arithmetic.

KoAT's source code, a binary, and a Docker image are available at https://aprove-developers.github.io/KoAT_TWN/. The website also has details on our experiments and *web interfaces* to run KoAT's configurations directly online.

Acknowledgments. We are indebted to M. Hark for many fruitful discussions about complexity, twn-loops, and KoAT. We are grateful to S. Genaim and J. J. Doménech for a suitable version of iRankFinder which we could use for control-flow refinement in KoAT's backend. Moreover, we thank A. Rubio and E. Martín-Martín for a static binary of MaxCore, A. Flores-Montoya and F. Zuleger for help in running CoFloCo and Loopus, F. Frohn for help and advice, and the reviewers for their feedback to improve the paper.

[8] One is the non-terminating leading example of [15], so at most 19 *might* terminate.

References

1. Albert, E., Arenas, P., Genaim, S., Puebla, G.: Automatic inference of upper bounds for recurrence relations in cost analysis. In: Alpuente, M., Vidal, G. (eds.) SAS 2008. LNCS, vol. 5079, pp. 221–237. Springer, Heidelberg (2008). https://doi.org/10.1007/978-3-540-69166-2_15
2. Albert, E., Bofill, M., Borralleras, C., Martín-Martín, E., Rubio, A.: Resource analysis driven by (conditional) termination proofs. Theory Pract. Logic Program. **19**, 722–739 (2019). https://doi.org/10.1017/S1471068419000152
3. Ben-Amram, A.M., Genaim, S.: On multiphase-linear ranking functions. In: Majumdar, R., Kunčak, V. (eds.) CAV 2017. LNCS, vol. 10427, pp. 601–620. Springer, Cham (2017). https://doi.org/10.1007/978-3-319-63390-9_32
4. Ben-Amram, A.M., Doménech, J.J., Genaim, S.: Multiphase-linear ranking functions and their relation to recurrent sets. In: Chang, B.-Y.E. (ed.) SAS 2019. LNCS, vol. 11822, pp. 459–480. Springer, Cham (2019). https://doi.org/10.1007/978-3-030-32304-2_22
5. Braverman, M.: Termination of integer linear programs. In: Ball, T., Jones, R.B. (eds.) CAV 2006. LNCS, vol. 4144, pp. 372–385. Springer, Heidelberg (2006). https://doi.org/10.1007/11817963_34
6. Brockschmidt, M., Emmes, F., Falke, S., Fuhs, C., Giesl, J.: Analyzing runtime and size complexity of integer programs. ACM Trans. Program. Lang. Syst. **38**, 1–50 (2016). https://doi.org/10.1145/2866575
7. Clang Compiler. https://clang.llvm.org/
8. Doménech, J.J., Genaim, S.: iRankFinder. In: Lucas, S. (ed.) WST 2018, p. 83 (2018). http://wst2018.webs.upv.es/wst2018proceedings.pdf
9. Doménech, J.J., Gallagher, J.P., Genaim, S.: Control-flow refinement by partial evaluation, and its application to termination and cost analysis. Theory Pract. Logic Program. **19**, 990–1005 (2019). https://doi.org/10.1017/S1471068419000310
10. Falke, S., Kapur, D., Sinz, C.: Termination analysis of C programs using compiler intermediate languages. In: Schmidt-Schauß, M. (ed.) RTA 2011. LIPIcs, vol. 10, pp. 41–50 (2011). https://doi.org/10.4230/LIPIcs.RTA.2011.41
11. Flores-Montoya, A., Hähnle, R.: Resource analysis of complex programs with cost equations. In: Garrigue, J. (ed.) APLAS 2014. LNCS, vol. 8858, pp. 275–295. Springer, Cham (2014). https://doi.org/10.1007/978-3-319-12736-1_15
12. Flores-Montoya, A.: Upper and lower amortized cost bounds of programs expressed as cost relations. In: Fitzgerald, J., Heitmeyer, C., Gnesi, S., Philippou, A. (eds.) FM 2016. LNCS, vol. 9995, pp. 254–273. Springer, Cham (2016). https://doi.org/10.1007/978-3-319-48989-6_16
13. Frohn, F., Giesl, J.: Proving non-termination via loop acceleration. In: Barrett, C.W., Yang, J. (eds.) FMCAD 2019, pp. 221–230 (2019). https://doi.org/10.23919/FMCAD.2019.8894271
14. Frohn, F., Giesl, J.: Termination of triangular integer loops is decidable. In: Dillig, I., Tasiran, S. (eds.) CAV 2019. LNCS, vol. 11562, pp. 426–444. Springer, Cham (2019). https://doi.org/10.1007/978-3-030-25543-5_24
15. Frohn, F., Hark, M., Giesl, J.: Termination of polynomial loops. In: Pichardie, D., Sighireanu, M. (eds.) SAS 2020. LNCS, vol. 12389, pp. 89–112. Springer, Cham (2020). https://doi.org/10.1007/978-3-030-65474-0_5, https://arxiv.org/abs/1910.11588
16. Frohn, F., Naaf, M., Brockschmidt, M., Giesl, J.: Inferring lower runtime bounds for integer programs. ACM Trans. Program. Lang. Syst. **42**, 1–50 (2020). https://doi.org/10.1145/3410331

17. Giesl, J., Rubio, A., Sternagel, C., Waldmann, J., Yamada, A.: The termination and complexity competition. In: Beyer, D., Huisman, M., Kordon, F., Steffen, B. (eds.) TACAS 2019. LNCS, vol. 11429, pp. 156–166. Springer, Cham (2019). https://doi.org/10.1007/978-3-030-17502-3_10
18. Giesl, J., Lommen, N., Hark, M., Meyer, F.: Improving automatic complexity analysis of integer programs. In: The Logic of Software: A Tasting Menu of Formal Methods. LNCS, vol. 13360 (to appear). Also appeared in CoRR, abs/2202.01769. https://arxiv.org/abs/2202.01769
19. Hark, M., Frohn, F., Giesl, J.: Polynomial loops: beyond termination. In: Albert, E., Kovács, L. (eds.) LPAR 2020, EPiC, vol. 73, pp. 279–297 (2020). https://doi.org/10.29007/nxv1
20. Heizmann, M., Leike, J.: Ranking templates for linear loops. Log. Methods Comput. Sci. **11**(1), 16 (2015). https://doi.org/10.2168/LMCS-11(1:16)2015
21. Hoffmann, J., Das, A., Weng, S.-C.: Towards automatic resource bound analysis for OCaml. In: Castagna, G., Gordon, A.D. (eds.) POPL 2017, pp. 359–373 (2017). https://doi.org/10.1145/3009837.3009842
22. Hosseini, M., Ouaknine, J., Worrell, J.: Termination of linear loops over the integers. In: Baier, C., Chatzigiannakis, I., Flocchini, P., Leonardi, S. (eds.) ICALP 2019, LIPIcs, vol. 132 (2019). https://doi.org/10.4230/LIPIcs.ICALP.2019.118
23. Jeannet, B., Miné, A.: Apron: a library of numerical abstract domains for static analysis. In: Bouajjani, A., Maler, O. (eds.) CAV 2009. LNCS, vol. 5643, pp. 661–667. Springer, Heidelberg (2009). https://doi.org/10.1007/978-3-642-02658-4_52
24. Kincaid, Z., Breck, J., Cyphert, J., Reps, T.W.: Closed forms for numerical loops. Proc. ACM Program. Lang. **3**(POPL), 1–29 (2019). https://doi.org/10.1145/3290368
25. Kovács, L.: Reasoning algebraically about p-solvable loops. In: Ramakrishnan, C.R., Rehof, J. (eds.) TACAS 2008. LNCS, vol. 4963, pp. 249–264. Springer, Heidelberg (2008). https://doi.org/10.1007/978-3-540-78800-3_18
26. Lommen, N., Meyer, F., Giesl, J.: Automatic complexity analysis of integer programs via triangular weakly non-linear loops. CoRR abs/2205.08869 (2022). https://arxiv.org/abs/2205.08869
27. Lommen, N., Meyer, F., Giesl, J.: Empirical evaluation of: "Automatic complexity analysis of integer programs via triangular weakly non-linear loops". https://aprove-developers.github.io/KoAT_TWN/
28. de Moura, L., Bjørner, N.: Z3: an efficient SMT solver. In: Ramakrishnan, C.R., Rehof, J. (eds.) TACAS 2008. LNCS, vol. 4963, pp. 337–340. Springer, Heidelberg (2008). https://doi.org/10.1007/978-3-540-78800-3_24
29. RaML (Resource Aware ML). https://www.raml.co/interface/
30. Rodríguez-Carbonell, E., Kapur, D.: Automatic generation of polynomial loop invariants: algebraic foundation. In: Gutierrez, J. (ed.) ISSAC 2004, pp. 266–273 (2004). https://doi.org/10.1145/1005285.1005324
31. Sinn, M., Zuleger, F., Veith, H.: Complexity and resource bound analysis of imperative programs using difference constraints. J. Autom. Reason. **59**, 3–45 (2017). https://doi.org/10.1007/s10817-016-9402-4
32. Tiwari, A.: Termination of linear programs. In: Alur, R., Peled, D.A. (eds.) CAV 2004. LNCS, vol. 3114, pp. 70–82. Springer, Heidelberg (2004). https://doi.org/10.1007/978-3-540-27813-9_6
33. TPDB (Termination Problems Data Base). https://github.com/TermCOMP/TPDB
34. Xu, M., Li, Z.-B.: Symbolic termination analysis of solvable loops. J. Symb. Comput. **50**, 28–49 (2013). https://doi.org/10.1016/j.jsc.2012.05.005

Author Index

Akshay, S. 671
Albert, Elvira 3
Alrabbaa, Christian 271

Baader, Franz 271
Barbosa, Haniel 15
Barrett, Clark 15, 95, 125
Berg, Jeremias 75
Bernreiter, Michael 331
Bidoit, Nicole 310
Bílková, Marta 429
Blaisdell, Eben 449
Borgwardt, Stefan 271
Bouziane, Hinde Lilia 359
Bozga, Marius 691
Bromberger, Martin 147
Brown, Chad E. 350
Bryant, Randal E. 106
Bueri, Lucas 691

Cailler, Julie 359
Cauli, Claudia 281
Chakraborty, Supratik 671
Chlebowski, Szymon 407

Dachselt, Raimund 271
Das, Anupam 509
Delahaye, David 359
Dill, David 125
Dixon, Clare 486
Dowek, Gilles 8
Draheim, Dirk 300
Duarte, André 169
Durán, Francisco 529

Eker, Steven 529
Escobar, Santiago 529

Felli, Paolo 36
Ferrari, Mauro 57
Fiorentini, Camillo 57
Frittella, Sabine 429
Frohn, Florian 712
Fujita, Tomohiro 388

Gallicchio, James 723
Giesl, Jürgen 712, 734
Girlando, Marianna 509
Gordillo, Pablo 3
Greati, Vitor 640
Grieskamp, Wolfgang 125

Haifani, Fajar 188, 208
Hernández-Cerezo, Alejandro 3
Heskes, Tom 597
Heule, Marijn J. H. 106
Holden, Sean B. 559
Holub, Štěpán 369
Hustadt, Ullrich 486

Ihalainen, Hannes 75
Indrzejczak, Andrzej 541
Iosif, Radu 691

Janota, Mikoláš 597
Järv, Priit 300
Järvisalo, Matti 75
Jukiewicz, Marcin 407

Kaliszyk, Cezary 350
Kanovich, Max 449
Koopmann, Patrick 188, 271
Korovin, Konstantin 169
Kozhemiachenko, Daniil 429
Kremer, Gereon 15, 95
Kutsia, Temur 578
Kuznetsov, Stepan L. 449

Lachnitt, Hanna 15
Lahav, Ori 468
Leidinger, Hendrik 228
Leszczyńska-Jasion, Dorota 407
Leutgeb, Lorenz 147
Lolic, Anela 331
Lommen, Nils 734

Ma, Yue 310
Maly, Jan 331
Mangla, Chaitanya 559
Marcos, João 640
Martí-Oliet, Narciso 529

Matsuzaki, Takuya 388
Méndez, Julián 271
Meseguer, José 529
Meyer, Fabian 734
Mitsch, Stefan 723
Montali, Marco 36

Nalon, Cláudia 486
Niemetz, Aina 15
Nötzli, Andres 15, 125

Ortiz, Magdalena 281
Ozdemir, Alex 15

Pal, Debtanu 671
Papacchini, Fabio 486
Park, Junkil 125
Pau, Cleo 578
Paulson, Lawrence C. 559
Piepenbrock, Jelle 597
Pimentel, Elaine 449
Piterman, Nir 281
Platzer, André 723
Popescu, Andrei 618
Preiner, Mathias 15

Qadeer, Shaz 125

Raška, Martin 369
Reeves, Joseph E. 106
Reynolds, Andrew 15, 95, 125
Robillard, Simon 359
Rodríguez-Núñez, Clara 3
Rosain, Johann 359
Rubio, Albert 3
Rubio, Rubén 529

Scedrov, Andre 449
Sheng, Ying 125
Sochański, Michał 407
Starosta, Štěpán 369
Suda, Martin 659

Talcott, Carolyn 529
Tammet, Tanel 300
Tan, Yong Kiam 723
Tinelli, Cesare 15, 95, 125
Tomczyk, Agata 407
Tourret, Sophie 188

Urban, Josef 597

Viswanathan, Arjun 15
Viteri, Scott 15

Weidenbach, Christoph 147, 188, 208, 228
Winkler, Sarah 36
Woltran, Stefan 331

Yamada, Akihisa 248
Yang, Hui 310

Zohar, Yoni 15, 125, 468

Printed in the United States
by Baker & Taylor Publisher Services